Differential Equations with Boundary-Value Problems

SIXTH EDITION

DIFFERENTIAL EQUATIONS WITH BOUNDARY-VALUE PROBLEMS

DENNIS G. ZILL
Loyola Marymount University

MICHAEL R. CULLEN
Late of Loyola Marymount University

THOMSON

BROOKS/COLE

Australia • Canada • Mexico • Singapore • Spain
United Kingdom • United States

THOMSON
BROOKS/COLE

Executive Publisher: Curt Hinrichs
Acquisitions Editor: John-Paul Ramin
Development Editor: Leslie Lahr
Assistant Editor: Katherine Brayton and Lisa Chow
Editorial Assistant: Darlene Amidon-Brent
Technology Project Manager: Earl Perry
Marketing Manager: Tom Ziolkowski
Marketing Assistant: Erin Mitchell
Advertising Project Manager: Bryan Vann
Project Manager, Editorial Production: Cheryll Linthicum
Print/Media Buyer: Doreen Suruki
Permissions Editor: Kiely Sexton

Production Service: Hearthside Publishing Services/Anne Seitz
Text Designer: Kim Rokusek/Rokusek Design
Art Director: Vernon T. Boes
Photo Researcher: Gretchen Miller
Copy Editor: Barbara Willette
Illustrator: Hearthside Publishing Services/Jade Myers
Cover Designer: Larry Didona
Cover Image: Imtek Imagineering, Inc./Corbis
Cover Printer: Quebecor World—Taunton
Compositor: Progressive Information Technologies
Printer: Quebecor World—Taunton

Printed in Canada
5   6   7   07

For more information about our products, contact us at:
Thomson Learning Academic Resource Center
1-800-423-0563
For permission to use material from this text, contact us by:
Phone: 1-800-730-2214
Fax: 1-800-730-2215
Web: http://www.thomsonrights.com

Library of Congress Control Number: 2004104624

Student Edition
ISBN-13: 978-0-534-41887-8
ISBN-10: 0-534-41887-2

International Student Edition
ISBN-13: 978-0-534-42074-1
ISBN-10: 0-534-42074-5

Brooks/Cole—Thomson Learning
10 Davis Drive
Belmont, CA 94002
USA

Asia
Thomson Learning
5 Shenton Way #01-01
UIC Building
Singapore 068808

Australia/New Zealand
Thomson Learning
102 Dodds Street
Southbank, Victoria 3006
Australia

Canada
Nelson
1120 Birchmount Road
Toronto, Ontario M1K 5G4
Canada

Europe/Middle East/Africa
Thomson Learning
High Holborn House
50/51 Bedford Row
London WC1R 4LR
United Kingdom

Latin America
Thomson Learning
Seneca, 53
Colonia Polanco
11560 Mexico D.F.
Mexico

Spain/Portugal
Paraninfo
Calle/Magallanes, 25
28015 Madrid, Spain

CONTENTS

CHAPTER 5

MODELING WITH HIGHER-ORDER DIFFERENTIAL EQUATIONS   194

CHAPTER 6

SERIES SOLUTIONS OF LINEAR EQUATIONS   237

CHAPTER 7

THE LAPLACE TRANSFORM   277

CHAPTER 8

SYSTEMS OF LINEAR FIRST-ORDER DIFFERENTIAL EQUATIONS   329

CHAPTER 9

NUMERICAL SOLUTIONS OF ORDINARY DIFFERENTIAL EQUATIONS   367

CHAPTER 10

PLANE AUTONOMOUS SYSTEMS   394

CHAPTER 11

ORTHOGONAL FUNCTIONS AND FOURIER SERIES   429

CHAPTER 12

BOUNDARY-VALUE PROBLEMS IN RECTANGULAR COORDINATES   464

CHAPTER 13

BOUNDARY-VALUE PROBLEMS IN OTHER COORDINATE SYSTEMS   503

CHAPTER 14

INTEGRAL TRANSFORM METHOD   520

CHAPTER 15

NUMERICAL SOLUTIONS OF PARTIAL DIFFERENTIAL EQUATIONS   543

APPENDICES

PREFACE

TO THE STUDENT

Authors of books live with the hope that someone actually *reads* them. Contrary to what you might believe, almost everything in a typical college-level mathematics text is written for you and not the instructor. True, the topics that are covered in the text are chosen to appeal to instructors because instructors make the decision on whether to use the book in their classes, but everything written in it is aimed directly at you, the student. So I want to encourage you—no, actually I want to *tell* you—to read this textbook! But do not read this text as you would a novel; you should not read it fast and you should not skip anything. Think of it as a *work*book. By this, I mean that mathematics should always be read with pencil and paper at the ready because, most likely, you will have to *work* your way through the examples and the discussion. Read—oops, *work*—all the examples in a section before attempting any of the exercises; the examples are constructed to illustrate what I consider the most important aspects of the section, and therefore, reflect the procedures necessary to work most of the problems in the exercise sets. I tell my students, "When reading an example, cover up the solution, try working it first, compare your work against the solution given, and then resolve any differences." I have tried to include most of the important steps in each example, but if something is not clear, you should always try—and here is where the pencil and paper come in again—to fill in the details or missing steps. This might not be always easy, but that is part of the learning process. The accumulation of facts followed by the slow assimilation of understanding simply cannot be achieved without a struggle.

Specifically for you, I have revamped the old *Student Solutions Manual* supplement and have renamed it the *Student Resource and Solutions Manual* (herein referred to by the initials *SRSM*). Now in addition to containing solutions of selected problems from the exercises sets, the *SRSM* has hints for solving problems, extra examples, and a review of those areas of algebra and calculus that I feel are particularly important to the successful study of differential equations. Bear in mind that you do not have to purchase the *SRSM;* by following my pointers given at the beginning of most sections, you can review the appropriate mathematics from your old precalculus or calculus texts.

In conclusion, I wish you good luck and success. I hope you enjoy the text and the course you are about to embark on. When I was an undergraduate math major, it was one of my favorites because I especially liked mathematics that connected with the physical world. If you have any comments, if you find any errors as you read and work your way through the text, or if you come up with a good idea for improving either it or the *SRSM,* please feel free to contact either me or my editor at Brooks/Cole Publishing Company:

John-Paul.Ramin@thomson.com

TO THE INSTRUCTOR

WHAT HAS CHANGED IN THIS EDITION?

First, let me say what has *not* changed. The chapter lineup by topics, the number and order of sections within a chapter, and the basic underlying philosophy remain the same as in the previous editions.

In case you are examining this text for the first time, *Differential Equations with Boundary-Value Problems, 6th Edition,* can be used for either a one-semester course in ordinary differential equations, or a two-semester course covering ordinary and partial differential equations. The shorter version of the text, *A First Course in Differential Equations with Modeling Applications, 8th Edition,* ends with Chapter 9. For a one-semester course, I assume that the students have successfully completed at least two semesters of calculus. Since you are reading this, undoubtedly you have already examined the table of contents for the topics that are covered. You will not find a "suggested syllabus" in this preface; I will not pretend to be so wise as to tell other teachers what to teach. I feel that there is plenty of material here to pick from and to form a course to your liking. The text strikes a reasonable balance between the analytical, qualitative, and quantitative approaches to the study of differential equations. As far as my "underlying philosophy" goes, it is this: An undergraduate text should be written with the student's understanding kept firmly in mind, which means to me that the material should be presented in a straightforward, readable, and helpful manner, while keeping the level of theory consistent with the notion of a "first course."

For those who are familiar with the previous editions, I would like to mention some of the improvements made in this edition.

WHAT IS NEW IN THIS EDITION?

- *Design* It should be obvious that the look of the text has changed considerably. The new trim size, larger pages, and new colorful design should make using this text easier and a more pleasant experience for both students and instructors.
- *Projects* In the fifth edition, there were three project modules; these modules written by Gil Lewis proved to be so popular that it was decided to include additional modeling projects in this edition. Each chapter begins with a table of contents, and remarks on the chapter content, and each of the first nine chapters ends with a new modeling project.
- *CD-ROM* The inclusion of a CD-ROM is another feature that is new to this edition. This CD contains a variety of demonstration tools; these are either *Project Tools* or *Text Tools.* The nine *Project Tools* were designed to illustrate aspects of the new modeling projects, whereas the twelve *Text Tools* tie in with the discussions in the various chapters. As students progress through the text, they can use these tools, either on their own or guided by the instructor, to explore the analytic, qualitative, and numerical aspects of differential equations. The *Text Tools* can also be used by the instructor for classroom demonstrations with the purposes of motivating the discussion of certain topics. *Text Tools* illustrate the interval of definition of a solution of an ordinary differential equation, direction fields, phase lines, Euler's method, growth and decay, *LR*-series circuits, mixtures, predator-prey interactions, spring/mass systems, a linear double pendulum, phase portraits of linear systems, and numerical methods.
- *Section Introduction and Review Material* Each section begins with an *Introduction,* and in most sections, that heading is followed by another called *Review Material.* The *Introduction* lays out the topics and goals of that section. Additionally, in *Review Material,* the student is told what topics in mathematics are used in that section and what sections in the text, previously covered, will be relevant to the current discussion. For example, since the traditional third semester of multivariable calculus may not be required for a one-semester course in differential equations, the recommendation for Section 2.4 ("Exact Equations") states that the student should review—or perhaps learn for the first time—the concepts of partial differentiation, the differential of a function of two variables, and partial

integration. Recommendations such as this include pointers to suitable texts or indicate whether that material is reviewed in the *SRSM*.

- **Chapter 2: First-Order Differential Equations** The method of isoclines has been restored to the discussion of direction fields in Section 2.1.

- **Chapter 5: Modeling with Higher-Order Differential Equations** I have (finally) been persuaded to simplify the notation for eigenvalues throughout the text. The confusing mixture of λ and λ^2 and $\sqrt{-\lambda}$ in the solution of two-point boundary-value problems has been done away with in favor of the consistent use of the symbol λ. In all examples that illustrate how to find eigenvalues and eigenfunctions of a boundary-value problem (such as Example 2 in Section 5.2), the three cases $\lambda = -\alpha^2 < 0$, $\lambda = 0$, and $\lambda = \alpha^2 > 0$ are emphasized repeatedly. The impact of this change occurs principally in Chapters 11, 12, and 13, where λ is used as the separation constant in the solution of a linear partial differential equation by separation of variables.

- **Chapter 6: Series Solutions of Linear Equations** The title of Section 6.3 has been changed to "Special Functions." The discussion of differential equations solvable in terms of Bessel functions has been expanded. In Section 6.3, the modified Bessel equation of order ν and the modified Bessel functions $I_\nu(x)$ and $K_\nu(x)$ are introduced for the first time.

- **Chapter 7: The Laplace Transform** For the convenience of those who wish to go a bit more slowly through the introduction of the use of the Laplace transform in solving linear initial-value problems, I have divided Section 7.2 into two subsections. Similarly, because of its length, Section 7.4 has been divided into three subsections. To facilitate the assignment of homework, Exercises 7.2 and Exercises 7.4 have also been divided into two and three subsections, respectively.

- **Chapter 8: Systems of Linear First-Order Differential Equations** After a leave of absence from the last two editions, the method of undetermined coefficients has been restored to this chapter. Section 8.3 now covers both undetermined coefficients and variation of parameters. Although this section is unified by the single theme of finding a particular solution of a nonhomogeneous linear system, it might be too long to cover conveniently in one lecture. Here again, I have divided the section and the corresponding exercises into two subsections. This division will also it make it easier for instructors who wish to skip one of the topics.

- **Chapter 12:** The discussion on solving nonhomogeneous boundary-value problems has been expanded. See Section 12.6.

- **Chapters 13 and 14:** Computer renditions of solution surfaces defined by Fourier series and Fourier integral representations have been added.

- **Miscellaneous Changes and Additions** Many exercises sets have been rethought and in some cases rewritten. New problems, most of them conceptual, have been added, and some old problems have been retired. For ease in assigning homework, the exercise sets corresponding to sections devoted to applications are partitioned by boldface headings that reflect the same order in which the these applications were considered in that section. (For example, see Exercises 1.3.) When mathematical models appear as problems in an exercise set corresponding to a section whose main thrust is not those applications, the problems are called out by a boldface heading, either **Mathematical Models** or **Miscellaneous Mathematics Models,** and each problem is given a boldface title to make it easier to select homework problems. (For example, see Exercises 1.3 and Exercises 2.1.) In an effort to improve the student's visualization of solutions, more graphs have been added throughout. In addition, many figures have been redrawn to improve their clarity. All figures in the text now have captions. Many of the end-of-section *Remarks* have been expanded. Also, *Remarks* have been added to some sections where previously there were none. Much of the command syntax for computer algebra systems that was found under the heading *Use of Computers* in the fifth edition has been moved to the *SRSM*.

SUPPLEMENTS FOR STUDENTS

- *Student Resource and Solutions Manual*, by Dennis G. Zill and Warren S. Wright (0-534-41879-1, 0-534-41888-0) provides reviews of relevant material from algebra and calculus, the solution of every third problem in each exercise set (with the exception of the Discussion/Project Problems and Computer Lab Assignments), command syntax for the computer algebra systems *Mathematica* and *Maple*, lists of important concepts, as well as helpful hints on how to start certain problems.
- *DE Tools* (CD), by Hubert Hohn, provides visual illustration of important topics from Chapters 1-9 and from the projects at the end of these chapters.
- *iLrn Tutorial Student Version* Free access to this text-specific, interactive, Web-based tutorial system is included with the text. *iLrn Tutorial Student Version* is browser based, making it easy to use even for students who have little technical proficiency. Because *iLrn Tutorial* allows students to work with real mathematical notation in real time it gives instant analysis and feedback. The algorithmic regeneration of practice problems provides students unlimited practice based on problems from their text. When stuck on a particular problem or concept, a student need only log on to vMentor, accessed through *iLrn Tutorial*, where they can talk (using their own computer microphones) to tutors, who will guide them through the problem using an interactive whiteboard for illustration.

SUPPLEMENTS FOR INSTRUCTORS

- *Instructor's Complete Solutions Manual*, by Warren S. Wright (0-534-41889-9), gives worked-out solutions to all problems in the text.
- *Supplemental Exercises* by Kevin TeBeest (0-534-41893-7), includes multiple-choice and free-response questions for each section of the text.
- *iLrn Instructor Version* is made up of two components, *iLrn Testing* and *iLrn Tutorial*, and offers full algorithmic generation of problems and the ability to assign free-response mathematics problems online. *iLrn Testing* is an Internet-ready, text-specific testing suite that allows instructors to create customized exams and track student progress in an accessible, browser-based format. *iLrn Tutorial* is a text specific, interactive tutorial software program that is delivered via the Web (at http://ilrn.com) and is offered in both student and instructor versions. Like *iLrn Testing*, it is browser based, making it an intuitive mathematical guide for students. Because *iLrn Tutorial* allows students to work in real time, it provides instant analysis and feedback. The tracking program built into the instructor version of the software enables instructors to monitor student progress, with results flowing automatically to the instructor's gradebook.

ACKNOWLEDGMENTS

Compiling a mathematics textbook such as this and making sure that its thousands of symbols and hundreds of equations are (mostly) accurate is an enormous task, but since I am called "the author" that is my job and responsibility. But many people besides myself have expended large amounts of time and energy in working towards its eventual publication. So I would like to take this opportunity to express my sincerest appreciation to everyone–most of them unknown to me–at

Hubert F. Hohn is the Director of Computer Arts at Massachusetts College of Art, Boston, Massachusetts, and is currently working under a grant at MIT to write mathematics software for its online course initiative. Hu has been the software designer for many of the CDs that appear in contemporary mathematics texts, most notably, *Interactive Differential Equations*, which Hohn designed in consultation with Professors John Cantwell, Jean McDill, Steve Strogatz, and Beverly West, published by Addison-Wesley. Hu previously served as professor of photography and artistic director of the photography department at the Banff Centre for the Arts in Alberta, Canada.

Kevin Cooper, Ph.D., Colorado State University, is the Computing Coordinator for Mathematics at Washington State University, Pullman, Washington. His main interest is numerical analysis, and he has written papers and one text in that area. Kevin also devotes considerable time to creating mathematical software components, such as *DynaSys,* a program to analyze dynamical systems numerically.

Thomas P. LoFaro, Ph.D., Boston University, is an Associate Professor in the Department of Mathematics and Computer Science at Gustavus Adolphus College, St. Peter, Minnesota. His interests are in applications of differential equations and dynamical systems. Tom has written papers on neuroscience, population genetics, and the mathematical foundations of ranking algorithms, and was a contributor to *ODE ARCHITECT,* published by John Wiley & Sons, Inc., Kevin and Tom were co-creators of the WSU *IDEA* website and collaborators on "Differential Equations on the Internet," in *REVOLUTIONS IN DIFFERENTIAL EQUATIONS,* published by the MAA.

Brooks/Cole Publishing Company, at Thomson Learning, and at Hearthside Publication Services who were involved in the publication of this sixth edition. Through their combined efforts, I feel that this edition is the best yet. I would, however, like to single out a few individuals for special recognition: At Brooks/Cole/Thomson, Cheryll Linthicum, Production Project Manager, for her willingness to at least listen to an author's ideas; Larry Didona for the excellent cover designs; Kim Rokusek for the interior design; Vernon Boes for supervising all the art and design; John-Paul Ramin, sponsoring editor, for his cooperation and readiness to try new things; Lisa Chow, assistant editor, for coordinating all the supplements; Leslie Lahr, developmental editor, for her suggestions, support, boundless patience, and at times, her sympathetic ear; and at Hearthside Production Services, Anne Seitz, production editor, for putting all the pieces of the puzzle together. I also extend my heartfelt appreciation to: the reviewers of this revision,

William Atherton, *Cleveland State University*
Dominic P. Clemence, *North Carolina Agricultural and Technical State University*
Richard A. DiDio, *La Salle University*
James M. Edmondson, *Santa Barbara City College*
Steve B. Khlief, *Tennessee Technological University* (retired)
Gerald Mueller, *Columbus State Community College*
Jacek Polewczak, *California State University Northridge*
Seenith Sivasundaram, *Embry-Riddle Aeronautical University*
Don E. Soash, *Hillsborough Community College*
Gregory Stein, *The Cooper Union*
Patrick Ward, *Illinois Central College*
Jianping Zhu, *University of Akron*
Jan Zijlstra, *Middle Tennessee State University*
Jay Zimmerman, *Towson University*

for their many thoughtful suggestions and criticisms; my long-time colleague and friend Michael Berg for rendering several new cartoons for this edition; and Gilbert Lewis, Michigan Technological University, for allowing us to use two of his project modules from the previous edition as the basis for new chapter projects.

Finally, my hat is off to Hubert Hohn, Massachusetts College of Art, Kevin Cooper, Washington State University, and Tom LoFaro, Gustavus Adolphus College, for agreeing to take time out of their very demanding schedules to participate in this revision. Hu crafted the excellent tools on the enclosed CD and gave the entire manuscript an unexpected, though welcome, final critical reading. The nine new end-of-chapter projects were written jointly by Kevin and Tom. I confess to having tweaked a few of their words here and there, but I left untouched the vein of humor, so uniquely theirs, that runs throughout all these projects. I am indebted to Kevin and Tom for good-naturedly putting up with these changes, my many other demands, and my not-too-subtle reminders of deadlines. Because Kevin and Tom have worked, together and separately, for a long time to improve student understanding of differential equations as a subject by involving them in conceptual discussions, mathematical modeling, and in the use of software and websites, they not only welcome but are eager for feedback. Student—and, of course, instructor—comments about the chapter projects can be communicated directly to the authors via email:

kcooper@math.wsu.edu
tlofaro@gustavus.edu

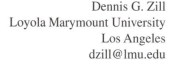

Dennis G. Zill
Loyola Marymount University
Los Angeles
dzill@lmu.edu

DIFFERENTIAL EQUATIONS WITH BOUNDARY-VALUE PROBLEMS

1

The differential equation

$$\frac{dA}{dt} = kA$$

can be used to estimate the age of a fossil. See page 21.

© Dave Harman, Gallo Images/CORBIS

The differential equation

$$\frac{dT}{dt} = k(T - T_m)$$

is a model of how fast an object cools. See page 21.

Royalty-Free/CORBIS

The differential equation

$$\frac{dx}{dt} = kx(n + 1 - x)$$

can be used to predict the number of people infected with a virus. See page 22.

Gloria-Leigh Logan, www.acclaimstockphotography.com

A differential equation such as

$$m\frac{d^2x}{dt^2} + F(x) = g(t)$$

can help us to understand why a physical system fails. See *Project 5* on page 233.

University of Washington Libraries, Special Collections, UW21422

The words *differential* and *equations* certainly suggest solving some kind of equation that contains derivatives. Indeed, one of your tasks in this course will be to solve equations such as $y'' + 2y' + y = 0$ for an unknown function $y = \phi(x)$. You will also see, as the course unfolds, that there is more to the study of differential equations than just mastering methods someone has devised to solve them.

But first things first. In order to read, study, and be conversant in a specialized subject, you have to learn the terminology of that discipline. This is the thrust of the first two sections of this chapter. In the last section we briefly examine the link between differential equations and the real world. Practical questions such as *How fast does a disease spread*? *How fast does a population change*? involve rates of change or derivatives. And so the mathematical formulation—or the mathematical model—of experiments, observations, or theories may be a differential equation.

1.1   DEFINITIONS AND TERMINOLOGY

INTRODUCTION: The derivative dy/dx of a function $y = \phi(x)$ is itself another function $\phi'(x)$ found by an appropriate rule. The function $y = e^{0.1x^2}$ is differentiable on the interval $(-\infty, \infty)$, and by the Chain Rule its derivative is $dy/dx = 0.2xe^{0.1x^2}$. If we replace $e^{0.1x^2}$ on the right-hand side of the last equation by the symbol y, the derivative becomes

$$\frac{dy}{dx} = 0.2xy. \tag{1}$$

Now imagine that a friend of yours simply hands you equation (1)—you have no idea how it was constructed—and asks, "What is the function represented by the symbol y?" You are now face-to-face with one of the basic problems in this course: How do you solve such an equation for the unknown function $y = \phi(x)$?

REVIEW MATERIAL: Since we are concerned primarily with functions that are differentiable, we recommend that you review the following concepts: the definition of the derivative, the derivative as slope, conditions under which a derivative exists, conditions under which a derivative does not exist, the rules of differentiation, the derivative as a rate of change, the relationship between the first derivative and the notions of increasing and decreasing functions, and the connection between the second derivative and the concavity of a graph.

CD: The **Interval of Definition Tool** on the *DE Tools* CD can be used in conjunction with the discussion of that topic on pages 5 and 6.

A DEFINITION   The equation that we made up in (1) is called a **differential equation.** Before proceeding any further, let us consider a more precise definition of this concept.

DEFINITION 1.1   **Differential Equation**

An equation containing the derivatives of one or more dependent variables, with respect to one or more independent variables, is said to be a **differential equation (DE).**

To talk about them, we shall classify differential equations by **type, order,** and **linearity.**

CLASSIFICATION BY TYPE   If an equation contains only ordinary derivatives of one or more dependent variables with respect to a single independent variable it is said to be an **ordinary differential equation (ODE).** For example,

A DE can contain more
than one dependent variable
↓   ↓

$$\frac{dy}{dx} + 5y = e^x, \quad \frac{d^2y}{dx^2} - \frac{dy}{dx} + 6y = 0, \quad \text{and} \quad \frac{dx}{dt} + \frac{dy}{dt} = 2x + y \tag{2}$$

are ordinary differential equations. An equation involving partial derivatives of one or more dependent variables of two or more independent variables is called a **partial differential equation (PDE).** For example,

$$\frac{\partial^2 u}{\partial x^2} + \frac{\partial^2 u}{\partial y^2} = 0, \quad \frac{\partial^2 u}{\partial x^2} = \frac{\partial^2 u}{\partial t^2} - 2\frac{\partial u}{\partial t}, \quad \text{and} \quad \frac{\partial u}{\partial y} = -\frac{\partial v}{\partial x} \tag{3}$$

are partial differential equations.*

Throughout this text ordinary derivatives will be written using either the **Leibniz notation** dy/dx, d^2y/dx^2, d^3y/dx^3, ... or the **prime notation** y', y'', y''', ... By using the latter notation, the first two differential equations in (2) can be written a little more compactly as $y' + 5y = e^x$ and $y'' - y' + 6y = 0$. Actually, the prime notation is used to denote only the first three derivatives; the fourth derivative is written $y^{(4)}$ instead of y''''. In general, the nth derivative of y is written $d^n y/dx^n$ or $y^{(n)}$. Although less convenient to write and to typeset, the Leibniz notation has an advantage over the prime notation in that it clearly displays both the dependent and independent variables. For example, in the equation

$$\overset{\displaystyle \text{unknown function}}{\underset{\text{independent variable}}{\frac{d^2x}{dt^2} + 16x = 0}}$$

it is immediately seen that the symbol x now represents a dependent variable whereas the independent variable is t. You should also be aware that in physical sciences and engineering Newton's **dot notation** (derogatively referred to by some as the "flyspeck" notation) is sometimes used to denote derivatives with respect to time t. Thus the differential equation $d^2s/dt^2 = -32$ becomes $\ddot{s} = -32$. Partial derivatives are often denoted by a **subscript notation** indicating the independent variables. For example, with the subscript notation the second equation in (3) becomes $u_{xx} = u_{tt} - 2u_t$.

CLASSIFICATION BY ORDER The **order of a differential equation** (either ODE or PDE) is the order of the highest derivative in the equation. For example,

$$\overset{\displaystyle \text{second order}}{\frac{d^2y}{dx^2}} + 5\overset{\displaystyle \text{first order}}{\left(\frac{dy}{dx}\right)^3} - 4y = e^x$$

is a second-order ordinary differential equation. First-order ordinary differential equations are occasionally written in differential form $M(x, y)dx + N(x, y)dy = 0$. For example, if we assume that y denotes the dependent variable in $(y - x)dx + 4x\,dy = 0$ then $y' = dy/dx$, and so by dividing by the differential dx, we get the alternative form $4xy' + y = x$. See the *Remarks* at the end of this section.

In symbols we can express an nth-order ordinary differential equation in one dependent variable by the general form

$$F(x, y, y', \ldots, y^{(n)}) = 0, \tag{4}$$

where F is a real-valued function of $n + 2$ variables: x, y, y', ..., $y^{(n)}$. For both practical and theoretical reasons we shall also make the assumption hereafter that it is possible to solve an ordinary differential equation in the form (4) uniquely for the

---

*Except for this introductory section, only ordinary differential equations are considered in *A First Course in Differential Equations with Modeling Applications*, Eighth Edition. In that text, the word *equation* and the abbreviation DE refer only to ODEs. Partial differential equations or PDEs are considered in the expanded volume *Differential Equations with Boundary-Value Problems*, Sixth Edition.

highest derivative $y^{(n)}$ in terms of the remaining $n + 1$ variables. The differential equation

$$\frac{d^n y}{dx^n} = f(x, y, y', \ldots, y^{(n-1)}), \tag{5}$$

where f is a real-valued continuous function, is referred to as the **normal form** of (4). Thus, when it suits our purposes, we shall use the normal forms

$$\frac{dy}{dx} = f(x, y) \quad \text{and} \quad \frac{d^2 y}{dx^2} = f(x, y, y')$$

to represent general first- and second-order ordinary differential equations. For example, the normal form of the first-order equation $4xy' + y = x$ is $y' = (x - y)/4x$; the normal form of the second-order equation $y'' - y' + 6y = 0$ is $y'' = y' - 6y$. See the Remarks.

CLASSIFICATION BY LINEARITY   An nth-order ordinary differential equation (4) is said to be **linear** if F is linear in $y, y', \ldots, y^{(n)}$. This means that an nth-order ODE is linear when (4) is $a_n(x)y^{(n)} + a_{n-1}(x)y^{(n-1)} + \cdots + a_1(x)y' + a_0(x)y - g(x) = 0$ or

$$a_n(x)\frac{d^n y}{dx^n} + a_{n-1}(x)\frac{d^{n-1}y}{dx^{n-1}} + \cdots + a_1(x)\frac{dy}{dx} + a_0(x)y = g(x). \tag{6}$$

Two important special cases of (6) are linear first-order ($n = 1$) and linear second-order ($n = 2$) DEs:

$$a_1(x)\frac{dy}{dx} + a_0(x)y = g(x) \quad \text{and} \quad a_2(x)\frac{d^2 y}{dx^2} + a_1(x)\frac{dy}{dx} + a_0(x)y = g(x). \tag{7}$$

In the additive combination on the left-hand side of equation (6) we see that the characteristic two properties of a linear ODE are as follows:

- The dependent variable y and all its derivatives $y', y'', \ldots, y^{(n)}$ are of the first degree, that is, the power of each term involving y is 1.
- The coefficients $a_0, a_1, \ldots, a_n$ of $y, y', \ldots, y^{(n)}$ depend at most on the independent variable x.

The equations

$$(y - x)dx + 4x\, dy = 0, \quad y'' - 2y' + y = 0, \quad \text{and} \quad \frac{d^3 y}{dx^3} + x\frac{dy}{dx} - 5y = e^x$$

are, in turn, linear first-, second-, and third-order ordinary differential equations. We have just demonstrated that the first equation is linear in the variable y by writing it in the alternative form $4xy' + y = x$. A **nonlinear** ordinary differential equation is simply one that is not linear. Nonlinear functions of the dependent variable or its derivatives, such as $\sin y$ or $e^{y'}$, cannot appear in a linear equation. Therefore,

nonlinear term: nonlinear term: nonlinear term:
coefficient depends on y nonlinear function of y power not 1
$\downarrow$ $\downarrow$ $\downarrow$

$$(1 - y)y' + 2y = e^x, \quad \frac{d^2 y}{dx^2} + \sin y = 0, \quad \text{and} \quad \frac{d^4 y}{dx^4} + y^2 = 0$$

are examples of nonlinear first-, second-, and fourth-order ordinary differential equations, respectively.

SOLUTIONS   As was stated before, one of the goals in this course is to solve, or find solutions of, differential equations. In the next definition we consider the concept of a solution of an ordinary differential equation.

> **DEFINITION 1.2**   **Solution of an ODE**
>
> Any function ϕ, defined on an interval I and possessing at least n derivatives that are continuous on I, which when substituted into an nth-order ordinary differential equation reduces the equation to an identity, is said to be a **solution** of the equation on the interval.

In other words, a solution of an nth-order ordinary differential equation (4) is a function ϕ that possesses at least n derivatives and for which

$$F(x, \phi(x), \phi'(x), \ldots, \phi^{(n)}(x)) = 0 \quad \text{for all } x \text{ in } I.$$

We say that ϕ *satisfies* the differential equation on I. For our purposes we shall also assume that a solution ϕ is a real-valued function. In our introductory discussion we saw that $y = e^{0.1x^2}$ is a solution of $dy/dx = 0.2xy$ on the interval $(-\infty, \infty)$.

Occasionally, it will be convenient to denote a solution by the alternative symbol $y(x)$.

INTERVAL OF DEFINITION   You cannot think *solution* of an ordinary differential equation without simultaneously thinking *interval*. The interval I in Definition 1.2 is variously called the **interval of definition,** the **interval of existence,** the **interval of validity,** or the **domain of the solution** and can be an open interval (a, b), a closed interval $[a, b]$, an infinite interval (a, ∞), and so on.

EXAMPLE 1   Verification of a Solution

Verify that the indicated function is a solution of the given differential equation on the interval $(-\infty, \infty)$.

(a) $dy/dx = xy^{1/2}; \quad y = \frac{1}{16}x^4$ **(b)** $y'' - 2y' + y = 0; \quad y = xe^x$

SOLUTION   One way of verifying that the given function is a solution is to see, after substituting, whether each side of the equation is the same for every x in the interval.

(a) From

left-hand side: $\quad \dfrac{dy}{dx} = \dfrac{1}{16}(4 \cdot x^3) = \dfrac{1}{4}x^3,$

right-hand side: $\quad xy^{1/2} = x \cdot \left(\dfrac{1}{16}x^4\right)^{1/2} = x \cdot \left(\dfrac{1}{4}x^2\right) = \dfrac{1}{4}x^3,$

we see that each side of the equation is the same for every real number x. Note that $y^{1/2} = \frac{1}{4}x^2$ is, by definition, the nonnegative square root of $\frac{1}{16}x^4$.

(b) From the derivatives $y' = xe^x + e^x$ and $y'' = xe^x + 2e^x$ we have, for every real number x,

left-hand side: $\quad y'' - 2y' + y = (xe^x + 2e^x) - 2(xe^x + e^x) + xe^x = 0,$

right-hand side: $\quad 0.$

Note, too, that in Example 1 each differential equation possesses the constant solution $y = 0$, $-\infty < x < \infty$. A solution of a differential equation that is identically zero on an interval I is said to be a **trivial solution.**

SOLUTION CURVE   The graph of a solution ϕ of an ODE is called a **solution curve.** Since ϕ is a differentiable function, it is continuous on its interval I of definition. Thus there may be a difference between the graph of the *function* ϕ and the graph of the *solution* ϕ. Put another way, the domain of the function ϕ need not be the same as the interval I of definition (or domain) of the solution ϕ. Example 2 illustrates the difference.

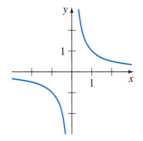

(a) function $y = 1/x$, $x \neq 0$

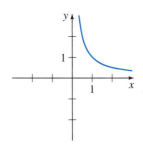

(b) solution $y = 1/x$, $(0, \infty)$

FIGURE 1.1   The function $y = 1/x$ is not the same as the solution $y = 1/x$

EXAMPLE 2   Function versus Solution

The domain of $y = 1/x$, considered simply as a *function*, is the set of all real numbers x except 0. When we graph $y = 1/x$, we plot points in the xy-plane corresponding to a judicious sampling of numbers taken from its domain. The rational function $y = 1/x$ is discontinuous at 0, and its graph, in a neighborhood of the origin, is given in Figure 1.1(a). The function $y = 1/x$ is not differentiable at $x = 0$, since the y-axis (whose equation is $x = 0$) is a vertical asymptote of the graph.

Now $y = 1/x$ is also a solution of the linear first-order differential equation $xy' + y = 0$. (Verify.) But when we say $y = 1/x$ is a *solution* of this DE, we mean it is a function defined on an interval I on which it is differentiable and satisfies the equation. In other words, $y = 1/x$ is a solution of the DE on *any* interval not containing 0, such as $(-3, -1)$, $\left(\frac{1}{2}, 10\right)$, $(-\infty, 0)$, or $(0, \infty)$. Because the solution curves defined by $y = 1/x$ on the intervals $-3 < x < -1$ and $\frac{1}{2} < x < 10$ are simply segments, or pieces, of the solution curves defined by $y = 1/x$ on $-\infty < x < 0$ and $0 < x < \infty$, respectively, it makes sense to take the interval I to be as large as possible. Thus we take I to be either $(-\infty, 0)$ or $(0, \infty)$. The solution curve on $(0, \infty)$ is shown in Figure 1.1(b).

EXPLICIT AND IMPLICIT SOLUTIONS   You should be familiar with the terms *explicit functions* and *implicit functions* from your study of calculus. A solution in which the dependent variable is expressed solely in terms of the independent variable and constants is said to be an **explicit solution.** For our purposes, let us think of an explicit solution as an explicit formula $y = \phi(x)$ that we can manipulate, evaluate, and differentiate using the standard rules. We have just seen in the last two examples that $y = \frac{1}{16}x^4$, $y = xe^x$, and $y = 1/x$ are, in turn, explicit solutions of $dy/dx = xy^{1/2}$, $y'' - 2y' + y = 0$, and $xy' + y = 0$. Moreover, the trivial solution $y = 0$ is an explicit solution of all three equations. When we get down to the business of actually solving some ordinary differential equations, you will see that methods of solution do not always lead directly to an explicit solution $y = \phi(x)$. This is particularly true when we attempt to solve nonlinear first-order differential equations. Often we have to be content with a relation or expression $G(x, y) = 0$ that defines a solution ϕ implicitly.

DEFINITION 1.3   Implicit Solution of an ODE

A relation $G(x, y) = 0$ is said to be an **implicit solution** of an ordinary differential equation (4) on an interval I, provided there exists at least one function ϕ that satisfies the relation as well as the differential equation on I.

It is beyond the scope of this course to investigate the conditions under which a relation $G(x, y) = 0$ defines a differentiable function ϕ. So we shall assume that if the formal implementation of a method of solution leads to a relation $G(x, y) = 0$, then there exists at least one function ϕ that satisfies both the relation (that is, $G(x, \phi(x)) = 0$) and the differential equation on an interval I. If the implicit solution $G(x, y) = 0$ is fairly simple, we may be able to solve for y in terms of x and obtain one or more explicit solutions. See the Remarks.

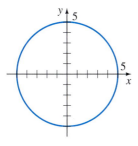

(a) implicit solution

$$x^2 + y^2 = 25$$

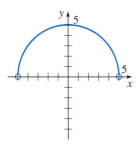

(b) explicit solution

$$y_1 = \sqrt{25 - x^2}, \; -5 < x < 5$$

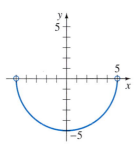

(c) explicit solution

$$y_2 = -\sqrt{25 - x^2}, \; -5 < x < 5$$

FIGURE 1.2 An implicit and two explicit solutions of $y' = -x/y$

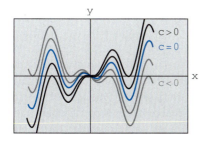

FIGURE 1.3 Some solutions of $xy' - y = x^2 \sin x$

| **EXAMPLE 3** | **Verification of an Implicit Solution** |

The relation $x^2 + y^2 = 25$ is an implicit solution of the differential equation

$$\frac{dy}{dx} = -\frac{x}{y} \tag{8}$$

on the interval $-5 < x < 5$. By implicit differentiation we obtain

$$\frac{d}{dx}x^2 + \frac{d}{dx}y^2 = \frac{d}{dx}25 \quad \text{or} \quad 2x + 2y\frac{dy}{dx} = 0.$$

Solving the last equation for the symbol dy/dx gives (8). Moreover, solving $x^2 + y^2 = 25$ for y in terms of x yields $y = \pm\sqrt{25 - x^2}$. The two functions $y = \phi_1(x) = \sqrt{25 - x^2}$ and $y = \phi_2(x) = -\sqrt{25 - x^2}$ satisfy the relation (that is, $x^2 + \phi_1^2 = 25$ and $x^2 + \phi_2^2 = 25$) and are explicit solutions defined on the interval $-5 < x < 5$. The solution curves given in Figures 1.2(b) and 1.2(c) are segments of the graph of the implicit solution in Figure 1.2(a).

Any relation of the form $x^2 + y^2 - c = 0$ *formally* satisfies (8) for any constant c. However, it is understood that the relation should always make sense in the real number system; thus, for example, if $c = -25$ we cannot say that $x^2 + y^2 + 25 = 0$ is an implicit solution of the equation. (Why not?)

Because the distinction between an explicit solution and an implicit solution should be intuitively clear, we will not belabor the issue by always saying "Here is an explicit (implicit) solution."

FAMILIES OF SOLUTIONS The study of differential equations is similar to that of integral calculus. In some texts, a solution ϕ is sometimes referred to as an **integral** of the equation, and its graph is called an **integral curve.** When evaluating an antiderivative or indefinite integral in calculus, we use a single constant c of integration. Analogously, when solving a first-order differential equation $F(x, y, y') = 0$, we *usually* obtain a solution containing a single arbitrary constant or parameter c. A solution containing an arbitrary constant represents a set $G(x, y, c) = 0$ of solutions called a **one-parameter family of solutions.** When solving an nth-order differential equation $F(x, y, y', \ldots, y^{(n)}) = 0$, we seek an **$n$-parameter family of solutions** $G(x, y, c_1, c_2, \ldots, c_n) = 0$. This means that *a single differential equation can possess an infinite number of solutions* corresponding to the unlimited number of choices for the parameter(s). A solution of a differential equation that is free of arbitrary parameters is called a **particular solution.** For example, the one-parameter family $y = cx - x \cos x$ is an explicit solution of the linear first-order equation $xy' - y = x^2 \sin x$ on the interval $(-\infty, \infty)$. (Verify.) Figure 1.3, obtained using graphing software, shows the graphs of some of the solutions in this family. The solution $y = -x \cos x$, the colored curve in the figure, is a particular solution corresponding to $c = 0$. Similarly, on the interval $(-\infty, \infty)$, $y = c_1e^x + c_2xe^x$ is a two-parameter family of solutions of the linear second-order equation $y'' - 2y' + y = 0$ in Example 1. (Verify.) Some particular solutions of the equation are the trivial solution $y = 0$ ($c_1 = c_2 = 0$), $y = xe^x$ ($c_1 = 0, c_2 = 1$), $y = 5e^x - 2xe^x$ ($c_1 = 5, c_2 = -2$), and so on.

Sometimes a differential equation possesses a solution that is not a member of a family of solutions of the equation—that is, a solution that cannot be obtained by specializing *any* of the parameters in the family of solutions. Such an extra solution is called a **singular solution.** For example, we have seen that $y = \frac{1}{16}x^4$ and $y = 0$ are solutions of the differential equation $dy/dx = xy^{1/2}$ on $(-\infty, \infty)$. In Section 2.2 we shall demonstrate, by actually solving it, that the differential equation $dy/dx = xy^{1/2}$ possesses the one-parameter family of solutions $y = \left(\frac{1}{4}x^2 + c\right)^2$. When $c = 0$, the resulting particular solution is $y = \frac{1}{16}x^4$. But notice that the trivial solution $y = 0$ is a singular solution since it is not a member of the family $y = \left(\frac{1}{4}x^2 + c\right)^2$; there is no way of assigning a value to the constant c to obtain $y = 0$.

In all the preceding examples we have used x and y to denote the independent and dependent variables, respectively. But you should become accustomed to seeing and working with other symbols to denote these variables. For example, we could denote the independent variable by t and the dependent variable by x.

EXAMPLE 4   Using Different Symbols

The functions $x = c_1 \cos 4t$ and $x = c_2 \sin 4t$, where c_1 and c_2 are arbitrary constants or parameters, are both solutions of the linear differential equation

$$x'' + 16x = 0.$$

For $x = c_1 \cos 4t$ the first two derivatives with respect to t are $x' = -4c_1 \sin 4t$ and $x'' = -16c_1 \cos 4t$. Substituting x'' and x then gives

$$x'' + 16x = -16c_1 \cos 4t + 16(c_1 \cos 4t) = 0.$$

In like manner, for $x = c_2 \sin 4t$ we have $x'' = -16c_2 \sin 4t$, and so

$$x'' + 16x = -16c_2 \sin 4t + 16(c_2 \sin 4t) = 0.$$

Finally, it is straightforward to verify that the linear combination of solutions, or the two-parameter family $x = c_1 \cos 4t + c_2 \sin 4t$, is also a solution of the differential equation. ∎

The next example shows that a solution of a differential equation can be a piecewise-defined function.

EXAMPLE 5   A Piecewise-Defined Solution

You should verify that the one-parameter family $y = cx^4$ is a one-parameter family of solutions of the differential equation $xy' - 4y = 0$ on the inverval $(-\infty, \infty)$. See Figure 1.4(a). The piecewise-defined differentiable function

$$y = \begin{cases} -x^4, & x < 0 \\ x^4, & x \geq 0 \end{cases}$$

is a particular solution of the equation but cannot be obtained from the family $y = cx^4$ by a single choice of c; the solution is constructed from the family by choosing $c = -1$ for $x < 0$ and $c = 1$ for $x \geq 0$. See Figure 1.4(b). ∎

SYSTEMS OF DIFFERENTIAL EQUATIONS   Up to this point we have been discussing single differential equations containing one unknown function. But often in theory, as well as in many applications, we must deal with systems of differential equations. A **system of ordinary differential equations** is two or more equations involving the derivatives of two or more unknown functions of a single independent variable. For example, if x and y denote dependent variables and t denotes the independent variable, then a system of two first-order differential equations is given by

$$\frac{dx}{dt} = f(t, x, y)$$

$$\frac{dy}{dt} = g(t, x, y).$$ (9)

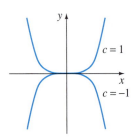

(a) two explicit solutions

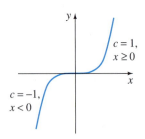

(b) piecewise-defined solution

FIGURE 1.4   Some solutions of $xy' - 4y = 0$

A **solution** of a system such as (9) is a pair of differentiable functions $x = \phi_1(t)$, $y = \phi_2(t)$, defined on a common interval I, that satisfy each equation of the system on this interval.

REMARKS

(*i*) A few last words about implicit solutions of differential equations are in order. In Example 3 we were able to solve the relation $x^2 + y^2 = 25$ for y in terms of x to get two explicit solutions, $\phi_1(x) = \sqrt{25 - x^2}$ and $\phi_2(x) = -\sqrt{25 - x^2}$, of the differential equation (8). But don't read too much into this one example. Unless it is easy or important or you are instructed to, there is usually no need to try to solve an implicit solution $G(x, y) = 0$ for y explicitly in terms of x. Also do not misinterpret the second sentence following Definition 1.3. An implicit solution $G(x, y) = 0$ can define a perfectly good differentiable function ϕ that is a solution of a DE, but yet we might not be able to solve $G(x, y) = 0$ using analytical methods such as algebra. The solution curve of ϕ may be a segment or piece of the graph of $G(x, y) = 0$. See Problems 41 and 42 in Exercises 1.1. Also, read the discussion following Example 4 in Section 2.2.

(*ii*) Although the concept of a solution has been emphasized in this section, you should also be aware that a DE does not necessarily have to possess a solution. See Problem 35 in Exercises 1.1. The question of whether a solution exists will be touched on in the next section.

(*iii*) It might not be apparent whether a first-order ODE written in differential form $M(x, y)dx + N(x, y)dy = 0$ is linear or nonlinear because there is nothing in this form that tells us which symbol denotes the dependent variable. See Problems 9 and 10 in Exercises 1.1.

(*iv*) It might not seem like a big deal to assume that $F(x, y, y', \ldots, y^{(n)}) = 0$ can be solved for $y^{(n)}$, but one should be a little bit careful here. There are exceptions, and there certainly are some problems connected with this assumption. See Problems 48 and 49 in Exercises 1.1.

(*v*) You may run across the term *closed form solutions* in DE texts or in lectures in courses in differential equations. Translated, this phrase usually refers to explicit solutions that are expressible in terms of *elementary* (or familiar) *functions:* finite combinations of integer powers of x, roots, exponential and logarithmic functions, and trigonometric and inverse trigonometric functions.

(*vi*) If *every* solution of an nth-order ODE $F(x, y, y', \ldots, y^{(n)}) = 0$ on an interval I can be obtained from an n-parameter family $G(x, y, c_1, c_2, \ldots, c_n) = 0$ by appropriate choices of the parameters c_i, $i = 1, 2, \ldots, n$, we then say that the family is the **general solution** of the DE. In solving linear ODEs, we shall impose relatively simple restrictions on the coefficients of the equation; with these restrictions one can be assured that not only does a solution exist on an interval but also that a family of solutions yields all possible solutions. Nonlinear ODEs, with the exception of some first-order equations, are usually difficult or impossible to solve in terms of elementary functions. Furthermore, if we happen to obtain a family of solutions for a nonlinear equation, it is not obvious whether this family contains all solutions. On a practical level, then, the designation "general solution" is applied only to linear ODEs. Don't be concerned about this concept at this point, but store the words "general solution" in the back of your mind—we will come back to this notion in Section 2.3 and again in Chapter 4.

EXERCISES 1.1

Answers to selected odd-numbered problems begin on page ANS-1.

In Problems 1–8 state the order of the given ordinary differential equation. Determine whether the equation is linear or nonlinear by matching it with (6).

1. $(1 - x)y'' - 4xy' + 5y = \cos x$

2. $x\dfrac{d^3y}{dx^3} - \left(\dfrac{dy}{dx}\right)^4 + y = 0$

3. $t^5 y^{(4)} - t^3 y'' + 6y = 0$

4. $\dfrac{d^2u}{dr^2} + \dfrac{du}{dr} + u = \cos(r + u)$

5. $\dfrac{d^2y}{dx^2} = \sqrt{1 + \left(\dfrac{dy}{dx}\right)^2}$

6. $\dfrac{d^2R}{dt^2} = -\dfrac{k}{R^2}$

7. $(\sin \theta)y''' - (\cos \theta)y' = 2$

8. $\ddot{x} - \left(1 - \dfrac{\dot{x}^2}{3}\right)\dot{x} + x = 0$

In Problems 9 and 10 determine whether the given first-order differential equation is linear in the indicated dependent variable by matching it with the first differential equation given in (7).

9. $(y^2 - 1)dx + x\,dy = 0$; in y; in x

10. $u\,dv + (v + uv - ue^u)du = 0$; in v; in u

In Problems 11–14 verify that the indicated function is an explicit solution of the given differential equation. Assume an appropriate interval I of definition for each solution.

11. $2y' + y = 0$; $y = e^{-x/2}$

12. $\dfrac{dy}{dt} + 20y = 24$; $y = \dfrac{6}{5} - \dfrac{6}{5}e^{-20t}$

13. $y'' - 6y' + 13y = 0$; $y = e^{3x}\cos 2x$

14. $y'' + y = \tan x$; $y = -(\cos x)\ln(\sec x + \tan x)$

In Problems 15–18 verify that the indicated function $y = \phi(x)$ is an explicit solution of the given first-order differential equation. Proceed as in Example 2, by considering ϕ simply as a *function*, give its domain. Then by considering ϕ as a *solution* of the differential equation, give at least one interval I of definition.

15. $(y - x)y' = y - x + 8$; $y = x + 4\sqrt{x + 2}$.

16. $y' = 25 + y^2$; $y = 5\tan 5x$

17. $y' = 2xy^2$; $y = 1/(4 - x^2)$

18. $2y' = y^3 \cos x$; $y = (1 - \sin x)^{-1/2}$

In Problems 19 and 20 verify that the indicated expression is an implicit solution of the given first-order differential equation. Find at least one explicit solution $y = \phi(x)$ in each case. Use a graphing utility to obtain the graph of an explicit solution. Give an interval I of definition of each solution ϕ.

19. $\dfrac{dX}{dt} = (X - 1)(1 - 2X)$; $\ln\left(\dfrac{2X - 1}{X - 1}\right) = t$

20. $2xy\,dx + (x^2 - y)dy = 0$; $-2x^2y + y^2 = 1$

In Problems 21–24 verify that the indicated family of functions is a solution of the given differential equation. Assume an appropriate interval I of definition for each solution.

21. $\dfrac{dP}{dt} = P(1 - P)$; $P = \dfrac{c_1 e^t}{1 + c_1 e^t}$

22. $\dfrac{dy}{dx} + 2xy = 1$; $y = e^{-x^2}\displaystyle\int_0^x e^{t^2}\,dt + c_1 e^{-x^2}$

23. $\dfrac{d^2y}{dx^2} - 4\dfrac{dy}{dx} + 4y = 0$; $y = c_1 e^{2x} + c_2 x e^{2x}$

24. $x^3\dfrac{d^3y}{dx^3} + 2x^2\dfrac{d^2y}{dx^2} - x\dfrac{dy}{dx} + y = 12x^2$;
$y = c_1 x^{-1} + c_2 x + c_3 x \ln x + 4x^2$

25. Verify that the piecewise-defined function

$$y = \begin{cases} -x^2, & x < 0 \\ x^2, & x \geq 0 \end{cases}$$

is a solution of the differential equation $xy' - 2y = 0$ on $(-\infty, \infty)$.

26. In Example 3 we saw that $y = \phi_1(x) = \sqrt{25 - x^2}$ and $y = \phi_2(x) = -\sqrt{25 - x^2}$ are solutions of $dy/dx = -x/y$ on the interval $(-5, 5)$. Explain why the piecewise-defined function

$$y = \begin{cases} \sqrt{25 - x^2}, & -5 < x < 0 \\ -\sqrt{25 - x^2}, & 0 \leq x < 5 \end{cases}$$

is not a solution of the differential equation on the interval $(-5, 5)$.

27. Find values of m so that the function $y = e^{mx}$ is a solution of the given differential equation. Explain your reasoning.
   (a) $y' + 2y = 0$ (b) $y'' - 5y' + 6y = 0$

28. Find values of m so that the function $y = x^m$ is a solution of the given differential equation. Explain your reasoning.

    (a) $xy'' + 2y' = 0$ **(b)** $x^2y'' - 7xy' + 15y = 0$

In Problems 29–32 use the concept that $y = c$, $-\infty < x < \infty$, is a constant function if and only if $y' = 0$ to determine whether the given differential equation possesses constant solutions.

29. $3xy' + 5y = 10$ **30.** $y' = y^2 + 2y - 3$

31. $(y - 1)y' = 1$ **32.** $y'' + 4y' + 6y = 10$

In Problems 33 and 34 verify that the indicated pair of functions is a solution of the given system of differential equations on the interval $(-\infty, \infty)$.

33. $\dfrac{dx}{dt} = x + 3y$ **34.** $\dfrac{d^2x}{dt^2} = 4y + e^t$

    $\dfrac{dy}{dt} = 5x + 3y;$ $\dfrac{d^2y}{dt^2} = 4x - e^t;$

    $x = e^{-2t} + 3e^{6t},$ $x = \cos 2t + \sin 2t + \tfrac{1}{5}e^t,$

    $y = -e^{-2t} + 5e^{6t}$ $y = -\cos 2t - \sin 2t - \tfrac{1}{5}e^t$

DISCUSSION/PROJECT PROBLEMS

35. Make up a differential equation that does not possess any real solutions.

36. Make up a differential equation that you feel confident possesses only the trivial solution $y = 0$. Explain your reasoning.

37. What function do you know from calculus is such that its first derivative is itself? Its first derivative is a constant multiple k of itself? Write each answer in the form of a first-order differential equation with a solution.

38. What function (or functions) do you know from calculus is such that its second derivative is itself? Its second derivative is the negative of itself? Write each answer in the form of a second-order differential equation with a solution.

39. Given that $y = \sin x$ is an explicit solution of the first-order differential equation $\dfrac{dy}{dx} = \sqrt{1 - y^2}$. Find an interval I of definition. [*Hint: I* is *not* the interval $-\infty < x < \infty$.]

40. Discuss why it makes intuitive sense to presume that the linear differential equation $y'' + 2y' + 4y = 5 \sin t$ has a solution of the form $y = A \sin t + B \cos t$, where A and B are constants. Then find specific constants A and B so that $y = A \sin t + B \cos t$ is a particular solution of the DE.

In Problems 41 and 42 the given figure represents the graph of an implicit solution $G(x, y) = 0$ of a differential equation $dy/dx = f(x, y)$. In each case the relation $G(x, y) = 0$ implicitly defines several solutions of the DE. Carefully reproduce each figure on a piece of paper. Use different colored pencils to mark off segments, or pieces, on each graph that correspond to graphs of solutions. Keep in mind that a solution ϕ must be a function and differentiable. Use the solution curve to estimate an interval I of definition of each solution ϕ.

41.

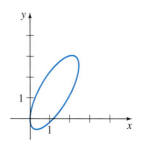

FIGURE 1.5   Graph for Problem 41

42.

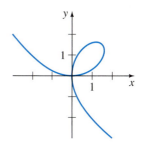

FIGURE 1.6   Graph for Problem 42

43. The graphs of members of the one-parameter family $x^3 + y^3 = 3cxy$ are called **folia of Descartes.** Verify that this family is an implicit solution of the first-order differential equation

$$\frac{dy}{dx} = \frac{y(y^3 - 2x^3)}{x(2y^3 - x^3)}.$$

44. The graph in Figure 1.6 is the member of the family of folia in Problem 43 corresponding to $c = 1$. Discuss: How can the DE in Problem 43 help in finding points on the graph of $x^3 + y^3 = 3xy$ where the tangent line is vertical? How does knowing where a tangent line is vertical help in determining an interval I of definition of a solution ϕ of the DE? Carry out your ideas and compare with your estimates of the intervals in Problem 42.

45. In Example 3 the largest interval I over which the explicit solutions $y = \phi_1(x)$ and $y = \phi_2(x)$ are defined is the open interval $(-5, 5)$. Why can't the interval I of definition be the closed interval $[-5, 5]$?

46. In Problem 21 a one-parameter family of solutions of the DE $P' = P(1 - P)$ is given. Does any solution curve pass through the point $(0, 3)$? Through the point $(0, 1)$?

47. Discuss, and illustrate with examples, how to solve differential equations of the forms $dy/dx = f(x)$ and $d^2y/dx^2 = f(x)$.

48. The differential equation $x(y')^2 - 4y' - 12x^3 = 0$ has the form given in (4). Determine whether the equation can be put into the normal form $dy/dx = f(x, y)$.

49. The normal form (5) of an nth-order differential equation is equivalent to (4) whenever both forms have exactly the same solutions. Make up a first-order differential equation for which $F(x, y, y') = 0$ is not equivalent to the normal form $dy/dx = f(x, y)$.

50. Find a linear second-order differential equation $F(x, y, y', y'') = 0$ for which $y = c_1x + c_2x^2$ is a two-parameter family of solutions. Make sure that your equation is free of the arbitrary parameters c_1 and c_2.

Qualitative information about a solution $y = \phi(x)$ of a differential equation can often be obtained from the equation itself. Before working Problems 51–54, recall the geometric significance of the derivatives dy/dx and d^2y/dx^2.

51. Consider the differential equation $dy/dx = e^{-x^2}$.
   (a) Explain why a solution of the DE must be an increasing function on any interval of the x-axis.
   (b) What are $\lim_{x \to -\infty} dy/dx$ and $\lim_{x \to \infty} dy/dx$? What does this suggest about a solution curve as $x \to \pm\infty$?
   (c) Determine an interval over which a solution curve is concave down and an interval over which the curve is concave up.
   (d) Sketch the graph of a solution $y = \phi(x)$ of the differential equation whose shape is suggested by parts (a)–(c).

52. Consider the differential equation $dy/dx = 5 - y$.
   (a) Either by inspection or by the method suggested in Problems 29–32, find a constant solution of the DE.
   (b) Using only the differential equation, find intervals on the y-axis on which a nonconstant solution $y = \phi(x)$

is increasing. Find intervals on the y-axis on which $y = \phi(x)$ is decreasing.

53. Consider the differential equation $dy/dx = y(a - by)$, where a and b are positive constants.
   (a) Either by inspection or by the method suggested in Problems 29–32, find two constant solutions of the DE.
   (b) Using only the differential equation, find intervals on the y-axis on which a nonconstant solution $y = \phi(x)$ is increasing. Find intervals on which $y = \phi(x)$ is decreasing.
   (c) Using only the differential equation, explain why $y = a/2b$ is the y-coordinate of a point of inflection of the graph of a nonconstant solution $y = \phi(x)$.
   (d) On the same coordinate axes, sketch the graphs of the two constant solutions found in part (a). These constant solutions partition the xy-plane into three regions. In each region, sketch the graph of a nonconstant solution $y = \phi(x)$ whose shape is suggested by the results in parts (b) and (c).

54. Consider the differential equation $y' = y^2 + 4$.
   (a) Explain why there exist no constant solutions of the DE.
   (b) Describe the graph of a solution $y = \phi(x)$. For example, can a solution curve have any relative extrema?
   (c) Explain why $y = 0$ is the y-coordinate of a point of inflection of a solution curve.
   (d) Sketch the graph of a solution $y = \phi(x)$ of the differential equation whose shape is suggested by parts (a)–(c).

COMPUTER LAB ASSIGNMENTS

In Problems 55 and 56 use a CAS to compute all derivatives and to carry out the simplifications needed to verify that the indicated function is a particular solution of the given differential equation.

55. $y^{(4)} - 20y''' + 158y'' - 580y' + 841y = 0$; $y = xe^{5x}\cos 2x$

56. $x^3y''' + 2x^2y'' + 20xy' - 78y = 0$;
$$y = 20\frac{\cos(5\ln x)}{x} - 3\frac{\sin(5\ln x)}{x}$$

1.2   INITIAL-VALUE PROBLEMS

INTRODUCTION: We are often interested in problems in which we seek a solution $y(x)$ of a differential equation so that $y(x)$ satisfies prescribed side conditions—that is, conditions imposed on the unknown $y(x)$ or its derivatives. On some interval I containing x_0, the problem

$$\textit{Solve:} \qquad \frac{d^n y}{dx^n} = f(x, y, y', \ldots, y^{(n-1)})$$

$$\textit{Subject to:} \qquad y(x_0) = y_0, y'(x_0) = y_1, \ldots, y^{(n-1)}(x_0) = y_{n-1},$$

(1)

where $y_0, y_1, \ldots, y_{n-1}$ are arbitrarily specified real constants, is called an **initial-value problem** **(IVP).** The values of $y(x)$ and its first $n - 1$ derivatives at a single point x_0: $y(x_0) = y_0$, $y'(x_0) = y_1, \ldots, y^{(n-1)}(x_0) = y_{n-1}$ are called **initial conditions.**

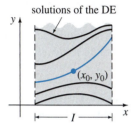

solutions of the DE

FIGURE 1.7   Solution of first-order IVP

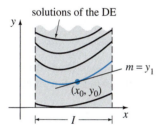

solutions of the DE

FIGURE 1.8   Solution of second-order IVP

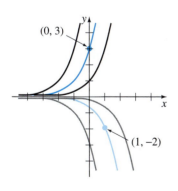

FIGURE 1.9   Solutions of two IVPs

FIRST- AND SECOND-ORDER IVPS   The problem given in (1) is also called an ***nth*-order initial-value problem.** For example,

$$\textit{Solve:} \qquad \frac{dy}{dx} = f(x, y)$$

$$\textit{Subject to:} \qquad y(x_0) = y_0$$

(2)

and

$$\textit{Solve:} \qquad \frac{d^2 y}{dx^2} = f(x, y, y')$$

$$\textit{Subject to:} \qquad y(x_0) = y_0, y'(x_0) = y_1$$

(3)

are **first-** and **second-order** initial-value problems, respectively. These two problems are easy to interpret in geometric terms. For (2) we are seeking a solution $y(x)$ of the differential equation $y' = f(x, y)$ on an interval I containing x_0 so that its graph passes through the specified point (x_0, y_0). A solution curve is shown in color in Figure 1.7. For (3) we want to find a solution $y(x)$ of the differential equation $y'' = f(x, y, y')$ on an interval I containing x_0 so that its graph not only passes through (x_0, y_0) but the slope of the curve at this point is the number y_1. A solution curve is shown in color in Figure 1.8. The words *initial conditions* derive from physical systems where the independent variable is time t and where $y(t_0) = y_0$ and $y'(t_0) = y_1$ represent the position and velocity, respectively, of an object at some beginning, or initial, time t_0.

Solving an *n*th-order initial-value problem such as (1) frequently entails first finding an *n*-parameter family of solutions of the given differential equation and then using the *n* initial conditions at x_0 to determine numerical values of the *n* constants in the family. The resulting particular solution is defined on some interval I containing the initial point x_0.

EXAMPLE 1   Two First-Order IVPs

In Problem 37 in Exercises 1.1 you were asked to deduce that $y = ce^x$ is a one-parameter family of solutions of the simple first-order equation $y' = y$. All the solutions in this family are defined on the interval $(-\infty, \infty)$. If we impose an initial condition, say, $y(0) = 3$, then substituting $x = 0$, $y = 3$ in the family determines the constant $3 = ce^0 = c$. Thus $y = 3e^x$ is a solution of the IVP

$$y' = y, \quad y(0) = 3.$$

Now if we demand that a solution curve pass through the point $(1, -2)$ rather than $(0, 3)$, then $y(1) = -2$ will yield $-2 = ce$ or $c = -2e^{-1}$. In this case $y = -2e^{x-1}$ is a solution of the IVP

$$y' = y, \quad y(1) = -2.$$

The two solution curves are shown in color in Figure 1.9.

The next example illustrates another first-order initial-value problem. In this example notice how the interval I of definition of the solution $y(x)$ depends on the initial condition $y(x_0) = y_0$.

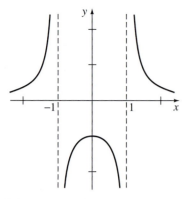

(a) function defined for all x except $x = \pm 1$

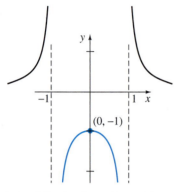

(b) solution defined on interval containing $x = 0$

FIGURE 1.10   Graphs of function and solution of IVP in Example 2

EXAMPLE 2   Interval I of Definition of a Solution

In Problem 6 of Exercises 2.2 you will be asked to show that a one-parameter family of solutions of the first-order differential equation $y' + 2xy^2 = 0$ is $y = 1/(x^2 + c)$. If we impose the initial condition $y(0) = -1$, then substituting $x = 0$ and $y = -1$ into the family of solutions gives $-1 = 1/c$ or $c = -1$. Thus, $y = 1/(x^2 - 1)$. We now emphasize the following three distinctions:

- Considered as a *function,* the domain of $y = 1/(x^2 - 1)$ is the set of real numbers x for which $y(x)$ is defined; this is the set of all real numbers except $x = -1$ and $x = 1$. See Figure 1.10(a).
- Considered as a *solution of the differential equation* $y' + 2xy^2 = 0$, the interval I of definition of $y = 1/(x^2 - 1)$ could be taken to be any interval over which $y(x)$ is defined and differentiable. As can be seen in Figure 1.10(a), the largest intervals on which $y = 1/(x^2 - 1)$ is a solution are $-\infty < x < -1$, $-1 < x < 1$, and $1 < x < \infty$.
- Considered as a *solution of the initial-value problem* $y' + 2xy^2 = 0$, $y(0) = -1$, the interval I of definition of $y = 1/(x^2 - 1)$ could be taken to be any interval over which $y(x)$ is defined, differentiable, *and* contains the initial point $x = 0$; the largest interval for which this is true is $-1 < x < 1$. See Figure 1.10(b).

See Problems 3–6 in Exercises 1.2 for a continuation of Example 2.

EXAMPLE 3   Second-Order IVP

In Example 4 of Section 1.1 we saw that $x = c_1 \cos 4t + c_2 \sin 4t$ is a two-parameter family of solutions of $x'' + 16x = 0$. Find a solution of the initial-value problem

$$x'' + 16x = 0, \quad x\left(\frac{\pi}{2}\right) = -2, \quad x'\left(\frac{\pi}{2}\right) = 1. \tag{4}$$

SOLUTION   We first apply $x(\pi/2) = -2$ to the given family of solutions: $c_1 \cos 2\pi + c_2 \sin 2\pi = -2$. Since $\cos 2\pi = 1$ and $\sin 2\pi = 0$, we find that $c_1 = -2$. We next apply $x'(\pi/2) = 1$ to the one-parameter family $x(t) = -2 \cos 4t + c_2 \sin 4t$. Differentiating and then setting $t = \pi/2$ and $x' = 1$ gives $8 \sin 2\pi + 4c_2 \cos 2\pi = 1$, from which we see that $c_2 = \frac{1}{4}$. Hence $x = -2 \cos 4t + \frac{1}{4} \sin 4t$ is a solution of (4).

EXISTENCE AND UNIQUENESS   Two fundamental questions arise in considering an initial-value problem:

> *Does a solution of the problem exist?*
> *If a solution exists, is it unique?*

For the first-order initial-value problem (2), we ask:

Existence $\begin{cases} \textit{Does the differential equation } dy/dx = f(x, y) \textit{ possess solutions?} \\ \textit{Do any of the solution curves pass through the point } (x_0, y_0)? \end{cases}$

Uniqueness $\begin{cases} \textit{When can we be certain that there is precisely one solution} \\ \textit{curve passing through the point } (x_0, y_0)? \end{cases}$

Note that in Examples 1 and 3 the phrase "*a* solution" is used rather than "*the* solution" of the problem. The indefinite article "a" is used deliberately to suggest the possibility that other solutions may exist. At this point it has not been demonstrated that there is a single solution of each problem. The next example illustrates an initial-value problem with two solutions.

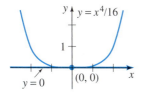

FIGURE 1.11   Two solutions of the same IVP

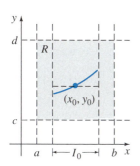

FIGURE 1.12   Rectangular region R

EXAMPLE 4   **An IVP Can Have Several Solutions**

Each of the functions $y = 0$ and $y = \frac{1}{16}x^4$ satisfies the differential equation $dy/dx = xy^{1/2}$ and the initial condition $y(0) = 0$, and so the initial-value problem

$$\frac{dy}{dx} = xy^{1/2}, \quad y(0) = 0$$

has at least two solutions. As illustrated in Figure 1.11, the graphs of both functions pass through the same point $(0, 0)$.

Within the safe confines of a formal course in differential equations one can be fairly confident that *most* differential equations will have solutions and that solutions of initial-value problems will *probably* be unique. Real life, however, is not so idyllic. Therefore it is desirable to know in advance of trying to solve an initial-value problem whether a solution exists and, when it does, whether it is the only solution of the problem. Since we are going to consider first-order differential equations in the next two chapters, we state here without proof a straightforward theorem that gives conditions that are sufficient to guarantee the existence and uniqueness of a solution of a first-order initial-value problem of the form given in (2). We shall wait until Chapter 4 to address the question of existence and uniqueness of a second-order initial-value problem.

THEOREM 1.1   **Existence of a Unique Solution**

Let R be a rectangular region in the xy-plane defined by $a \le x \le b$, $c \le y \le d$ that contains the point (x_0, y_0) in its interior. If $f(x, y)$ and $\partial f/\partial y$ are continuous on R, then there exists some interval I_0: $x_0 - h < x < x_0 + h$, $h > 0$, contained in $a \le x \le b$, and a unique function $y(x)$, defined on I_0, that is a solution of the initial-value problem (2).

The foregoing result is one of the most popular existence and uniqueness theorems for first-order differential equations because the criteria of continuity of $f(x, y)$ and $\partial f/\partial y$ are relatively easy to check. The geometry of Theorem 1.1 is illustrated in Figure 1.12.

EXAMPLE 5   **Example 4 Revisited**

We saw in Example 4 that the differential equation $dy/dx = xy^{1/2}$ possesses at least two solutions whose graphs pass through $(0, 0)$. Inspection of the functions

$$f(x, y) = xy^{1/2} \quad \text{and} \quad \frac{\partial f}{\partial y} = \frac{x}{2y^{1/2}}$$

shows that they are continuous in the upper half-plane defined by $y > 0$. Hence Theorem 1.1 enables us to conclude that through any point (x_0, y_0), $y_0 > 0$ in the upper half-plane there is some interval centered at x_0 on which the given differential equation has a unique solution. Thus, for example, even without solving it, we know that there exists some interval centered at 2 on which the initial-value problem $dy/dx = xy^{1/2}$, $y(2) = 1$ has a unique solution.

In Example 1, Theorem 1.1 guarantees that there are no other solutions of the initial-value problems $y' = y$, $y(0) = 3$ and $y' = y$, $y(1) = -2$ other than $y = 3e^x$ and $y = -2e^{x-1}$, respectively. This follows from the fact that $f(x, y) = y$ and $\partial f/\partial y = 1$ are continuous throughout the entire xy-plane. It can be further shown that the interval I on which each solution is defined is $(-\infty, \infty)$.

INTERVAL OF EXISTENCE/UNIQUENESS Suppose $y(x)$ represents a solution of the initial-value problem (2). The following three sets on the real x-axis may not be the same: the domain of the function $y(x)$, the interval I over which the solution $y(x)$ is defined or exists, and the interval I_0 of existence *and* uniqueness. Example 2 of Section 1.1 illustrated the difference between the domain of a function and the interval I of definition. Now suppose (x_0, y_0) is a point in the interior of the rectangular region R in Theorem 1.1. It turns out that the continuity of the function $f(x, y)$ on R by itself is sufficient to guarantee the existence of at least one solution of $dy/dx = f(x, y)$, $y(x_0) = y_0$, defined on some interval I. The interval I of definition for this initial-value problem is usually taken to be the largest interval containing x_0 over which the solution $y(x)$ is defined and differentiable. The interval I depends on both $f(x, y)$ and the initial condition $y(x_0) = y_0$. See Problems 31–34 in Exercises 1.2. The extra condition of continuity of the first partial derivative $\partial f/\partial y$ on R enables us to say that not only does a solution exist on some interval I_0 containing x_0, but it is the *only* solution satisfying $y(x_0) = y_0$. However, Theorem 1.1 does not give any indication of the sizes of intervals I and I_0; *the interval I of definition need not be as wide as the region R, and the interval I_0 of existence and uniqueness may not be as large as I.* The number $h > 0$ that defines the interval I_0: $x_0 - h < x < x_0 + h$ could be very small, and so it is best to think that the solution $y(x)$ is *unique in a local sense*—that is, a solution defined near the point (x_0, y_0). See Problem 44 in Exercises 1.2.

REMARKS

(*i*) The conditions in Theorem 1.1 are sufficient but not necessary. This means that when $f(x, y)$ and $\partial f/\partial y$ are continuous on a rectangular region R, it must always follow that a solution of (2) exists and is unique whenever (x_0, y_0) is a point interior to R. However, if the conditions stated in the hypothesis of Theorem 1.1 do not hold, then anything could happen: Problem (2) *may* still have a solution and this solution *may* be unique, or (2) may have several solutions, or it may have no solution at all. A rereading of Example 5 reveals that the hypotheses of Theorem 1.1 do not hold on the line $y = 0$ for the differential equation $dy/dx = xy^{1/2}$, and so it is not surprising, as we saw in Example 4 of this section, that there are two solutions defined on a common interval $-h < x < h$ satisfying $y(0) = 0$. On the other hand, the hypotheses of Theorem 1.1 do not hold on the line $y = 1$ for the differential equation $dy/dx = |y - 1|$. Nevertheless it can be proved that the solution of the initial-value problem $dy/dx = |y - 1|$, $y(0) = 1$, is unique. Can you guess this solution?

(*ii*) You are encouraged to read, think about, work, and then keep in mind Problem 43 in Exercises 1.2.

EXERCISES 1.2

Answers to selected odd-numbered problems begin on page ANS-1.

In Problems 1 and 2, $y = 1/(1 + c_1 e^{-x})$ is a one-parameter family of solutions of the first-order DE $y' = y - y^2$. Find a solution of the first-order IVP consisting of this differential equation and the given initial condition.

1. $y(0) = -\frac{1}{3}$ **2.** $y(-1) = 2$

In Problems 3–6, $y = 1/(x^2 + c)$ is a one-parameter family of solutions of the first-order DE $y' + 2xy^2 = 0$. Find a solution of the first-order IVP consisting of this differential equation and the given initial condition. Give the largest interval I over which the solution is defined.

3. $y(2) = \frac{1}{3}$ **4.** $y(-2) = \frac{1}{2}$

5. $y(0) = 1$ **6.** $y\left(\frac{1}{2}\right) = -4$

In Problems 7–10, $x = c_1 \cos t + c_2 \sin t$ is a two-parameter family of solutions of the second-order DE $x'' + x = 0$. Find a solution of the second-order IVP consisting of this differential equation and the given initial conditions.

7. $x(0) = -1, \quad x'(0) = 8$

8. $x(\pi/2) = 0, \quad x'(\pi/2) = 1$

9. $x(\pi/6) = \frac{1}{2}, \quad x'(\pi/6) = 0$

10. $x(\pi/4) = \sqrt{2}, \quad x'(\pi/4) = 2\sqrt{2}$

In Problems 11–14, $y = c_1 e^x + c_2 e^{-x}$ is a two-parameter family of solutions of the second-order DE $y'' - y = 0$. Find a solution of the second-order IVP consisting of this differential equation and the given initial conditions.

11. $y(0) = 1, \quad y'(0) = 2$

12. $y(1) = 0, \quad y'(1) = e$

13. $y(-1) = 5, \quad y'(-1) = -5$

14. $y(0) = 0, \quad y'(0) = 0$

In Problems 15 and 16 determine by inspection at least two solutions of the given first-order IVP.

15. $y' = 3y^{2/3}, \quad y(0) = 0$

16. $xy' = 2y, \quad y(0) = 0$

In Problems 17–24 determine a region of the xy-plane for which the given differential equation would have a unique solution whose graph passes through a point (x_0, y_0) in the region.

17. $\dfrac{dy}{dx} = y^{2/3}$

18. $\dfrac{dy}{dx} = \sqrt{xy}$

19. $x\dfrac{dy}{dx} = y$

20. $\dfrac{dy}{dx} - y = x$

21. $(4 - y^2)y' = x^2$

22. $(1 + y^3)y' = x^2$

23. $(x^2 + y^2)y' = y^2$

24. $(y - x)y' = y + x$

In Problems 25–28 determine whether Theorem 1.1 guarantees that the differential equation $y' = \sqrt{y^2 - 9}$ possesses a unique solution through the given point.

25. $(1, 4)$

26. $(5, 3)$

27. $(2, -3)$

28. $(-1, 1)$

29. (a) By inspection find a one-parameter family of solutions of the differential equation $xy' = y$. Verify that each member of the family is a solution of the initial-value problem $xy' = y$, $y(0) = 0$.

(b) Explain part (a) by determining a region R in the xy-plane for which the differential equation $xy' = y$ would have a unique solution through a point (x_0, y_0) in R.

(c) Verify that the piecewise-defined function
$$y = \begin{cases} 0, & x < 0 \\ x, & x \geq 0 \end{cases}$$
satisfies the condition $y(0) = 0$. Determine whether this function is also a solution of the initial-value problem in part (a).

30. (a) Verify that $y = \tan(x + c)$ is a one-parameter family of solutions of the differential equation $y' = 1 + y^2$.

(b) Since $f(x, y) = 1 + y^2$ and $\partial f/\partial y = 2y$ are continuous everywhere, the region R in Theorem 1.1 can be taken to be the entire xy-plane. Use the family of solutions in part (a) to find an explicit solution of the first-order initial-value problem $y' = 1 + y^2$, $y(0) = 0$. Even though $x_0 = 0$ is in the interval $-2 < x < 2$, explain why the solution is not defined on this interval.

(c) Determine the largest interval I of definition for the solution of the initial-value problem in part (b).

31. (a) Verify that $y = -1/(x + c)$ is a one-parameter family of solutions of the differential equation $y' = y^2$.

(b) Since $f(x, y) = y^2$ and $\partial f/\partial y = 2y$ are continuous everywhere, the region R in Theorem 1.1 can be taken to be the entire xy-plane. Find a solution from the family in part (a) that satisfies $y(0) = 1$. Find a solution from the family in part (a) that satisfies $y(0) = -1$. Determine the largest interval I of definition for the solution of each initial-value problem.

32. (a) Find a solution from the family in part (a) of Problem 31 that satisfies $y' = y^2$, $y(0) = y_0$, where $y_0 \neq 0$. Explain why the largest interval I of definition for this solution is either $-\infty < x < 1/y_0$ or $1/y_0 < x < \infty$.

(b) Determine the largest interval I of definition for the solution of the first-order initial-value problem $y' = y^2, y(0) = 0$.

33. (a) Verify that $3x^2 - y^2 = c$ is a one-parameter family of solutions of the differential equation $y\, dy/dx = 3x$.

(b) By hand, sketch the graph of the implicit solution $3x^2 - y^2 = 3$. Find all explicit solutions $y = \phi(x)$ of the DE in part (a) defined by this relation. Give the interval I of definition of each explicit solution.

(c) The point $(-2, 3)$ is on the graph of $3x^2 - y^2 = 3$, but which of the explicit solutions in part (b) satisfies $y(-2) = 3$?

34. (a) Use the family of solutions in part (a) of Problem 33 to find an implicit solution of the initial-value problem $y\, dy/dx = 3x$, $y(2) = -4$. Then, by hand, sketch the graph of the explicit solution of this problem and give its interval I of definition.

(b) Are there any explicit solutions of $y\, dy/dx = 3x$ that pass through the origin?

In Problems 35–38 the graph of a member of a family of solutions of a second-order differential equation $d^2y/dx^2 = f(x, y, y')$ is given. Match the solution curve with at least one pair of the following initial conditions.

(a) $y(1) = 1$, $y'(1) = -2$
(b) $y(-1) = 0$, $y'(-1) = -4$
(c) $y(1) = 1$, $y'(1) = 2$
(d) $y(0) = -1$, $y'(0) = 2$
(e) $y(0) = -1$, $y'(0) = 0$
(f) $y(0) = -4$, $y'(0) = -2$

35.

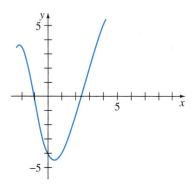

FIGURE 1.13 Graph for Problem 35

36.

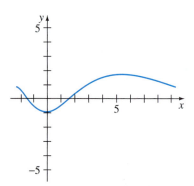

FIGURE 1.14 Graph for Problem 36

37.

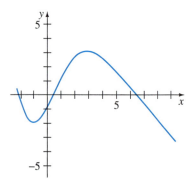

FIGURE 1.15 Graph for Problem 37

38.

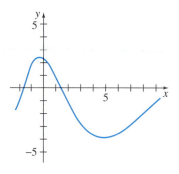

FIGURE 1.16 Graph for Problem 38

DISCUSSION/PROJECT PROBLEMS

In Problems 39 and 40 use Problem 47 in Exercises 1.1 and (2) and (3) of this section.

39. Find a function $y = f(x)$ whose graph at each point (x, y) has the slope given by $8e^{2x} + 6x$ and has the y-intercept $(0, 9)$.

40. Find a function $y = f(x)$ whose second derivative is $y'' = 12x - 2$ at each point (x, y) on its graph and $y = -x + 5$ is tangent to the graph at the point corresponding to $x = 1$.

41. Consider the initial-value problem $y' = x - 2y$, $y(0) = \frac{1}{2}$. Determine which of the two curves shown in Figure 1.17 is the only plausible solution curve. Explain your reasoning.

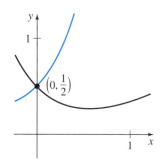

FIGURE 1.17 Graphs for Problem 41

42. Determine a plausible value of x_0 for which the graph of the solution of the initial-value problem $y' + 2y = 3x - 6$, $y(x_0) = 0$ is tangent to the x-axis at $(x_0, 0)$. Explain your reasoning.

43. Suppose that the first-order differential equation $dy/dx = f(x, y)$ possesses a one-parameter family of solutions and that $f(x, y)$ satisfies the hypotheses of Theorem 1.1 in some rectangular region R of the xy-plane. Explain why two different solution curves cannot intersect or be tangent to each other at a point (x_0, y_0) in R.

44. The functions $y(x) = \frac{1}{16}x^4$, $-\infty < x < \infty$ and

$$y(x) = \begin{cases} 0, & x < 0 \\ \frac{1}{16}x^4, & x \geq 0 \end{cases}$$

have the same domain but are clearly different. See Figures 1.18(a) and 1.18(b), respectively. Show that both functions are solutions of the initial-value problem $dy/dx = xy^{1/2}$, $y(2) = 1$ on the interval $(-\infty, \infty)$. Resolve the apparent contradiction between this fact and the last sentence in Example 5.

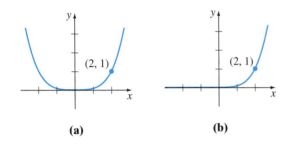

(a)   (b)

FIGURE 1.18   Two solutions of the IVP in Problem 44

MATHEMATICAL MODEL

45. Population Growth Beginning in the next section we will see that differential equations can be used to describe or *model* many different physical systems. In this problem suppose that a model of the growing population of a small community is given by the initial-value problem

$$\frac{dP}{dt} = 0.15P(t) + 20, \quad P(0) = 100,$$

where P is the number of individuals in the community and time t is measured in years. How fast, that is, at what *rate,* is the population increasing at $t = 0$? How fast is the population increasing when the population is 500?

1.3   DIFFERENTIAL EQUATIONS AS MATHEMATICAL MODELS

INTRODUCTION: In this section we introduce the notion of a differential equation as a *mathematical model* and discuss some specific models in biology, chemistry and physics. Once we have studied some methods for solving DEs in Chapters 2 and 4 we return to, and solve, some of these models in Chapters 3 and 5.

REVIEW MATERIAL: If you have taken a course in elementary physics, you might review the following concepts and laws: units of measurement, weight, mass, density, Archimedes' principle, Hooke's law, Kirchhoff's laws, and Newton's laws of motion (especially his second law of motion).

MATHEMATICAL MODELS   It is often desirable to describe the behavior of some real-life system or phenomenon, whether physical, sociological, or even economic, in mathematical terms. The mathematical description of a system or a phenomenon is called a **mathematical model** and is constructed with certain goals in mind. For example, we may wish to understand the mechanisms of a certain ecosystem by studying the growth of animal populations in that system, or we may wish to date fossils by analyzing the decay of a radioactive substance either in the fossil or in the stratum in which it was discovered.

Construction of a mathematical model of a system starts with

(*i*)   *identification of the variables that are responsible for changing the system. We may choose not to incorporate all these variables into the model at first. In this step we are specifying the **level of resolution** of the model.*

Next

> (*ii*) *we make a set of reasonable assumptions, or hypotheses, about the system we are trying to describe. These assumptions will also include any empirical laws that may be applicable to the system.*

For some purposes it may be perfectly within reason to be content with low-resolution models. For example, you may already be aware that in beginning physics courses, the retarding force of air friction is sometimes ignored in modeling the motion of a body falling near the surface of the earth, but if you are a scientist whose job it is to accurately predict the flight path of a long-range projectile, you have to take into account air resistance and other factors such as the curvature of the earth.

Since the assumptions made about a system frequently involve *a rate of change* of one or more of the variables, the mathematical depiction of all these assumptions may be one or more equations involving *derivatives*. In other words, the mathematical model may be a differential equation or a system of differential equations.

Once we have formulated a mathematical model that is either a differential equation or a system of differential equations, we are faced with the not insignificant problem of trying to solve it. *If* we can solve it, then we deem the model to be reasonable if its solution is consistent with either experimental data or known facts about the behavior of the system. But if the predictions produced by the solution are poor, we can either increase the level of resolution of the model or make alternative assumptions about the mechanisms for change in the system. The steps of the modeling process are then repeated, as shown in the following diagram:

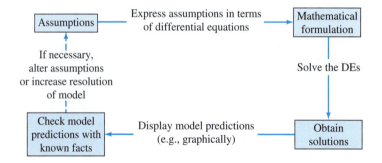

Of course, by increasing the resolution, we add to the complexity of the mathematical model and increase the likelihood that we cannot obtain an explicit solution.

A mathematical model of a physical system will often involve the variable time *t*. A solution of the model then gives the **state of the system;** in other words, the values of the dependent variable (or variables) for appropriate values of *t* describe the system in the past, present, and future.

POPULATION DYNAMICS   One of the earliest attempts to model human **population growth** by means of mathematics was by the English economist Thomas Malthus in 1798. Basically, the idea behind the Malthusian model is the assumption that the rate at which the population of a country grows at a certain time is proportional* to the total population of the country at that time. In other words, the more people there are at time *t*, the more there are going to be in the future. In mathematical terms, if $P(t)$ denotes the total population at time *t*, then this assumption can be expressed as

---

*If two quantities *u* and *v* are proportional, we write $u \propto v$. This means one quantity is a constant multiple of the other: $u = kv$.

$$\frac{dP}{dt} \propto P \quad \text{or} \quad \frac{dP}{dt} = kP, \tag{1}$$

where k is a constant of proportionality. This simple model, which fails to take into account many factors that can influence human populations to either grow or decline (immigration and emigration, for example), nevertheless turned out to be fairly accurate in predicting the population of the United States during the years 1790–1860. Populations that grow at a rate described by (1) are rare; nevertheless, (1) is still used to model *growth of small populations over short intervals of time* (bacteria growing in a petri dish, for example).

RADIOACTIVE DECAY The nucleus of an atom consists of combinations of protons and neutrons. Many of these combinations of protons and neutrons are unstable—that is, the atoms decay or transmute into atoms of another substance. Such nuclei are said to be radioactive. For example, over time the highly radioactive radium, Ra-226, transmutes into the radioactive gas radon, Rn-222. To model the phenomenon of **radioactive decay,** it is assumed that the rate dA/dt at which the nuclei of a substance decay is proportional to the amount (more precisely, the number of nuclei) $A(t)$ of the substance remaining at time t:

$$\frac{dA}{dt} \propto A \quad \text{or} \quad \frac{dA}{dt} = kA. \tag{2}$$

Of course, equations (1) and (2) are exactly the same; the difference is only in the interpretation of the symbols and the constants of proportionality. For growth, as we expect in (1), $k > 0$, and for decay, as in (2), $k < 0$.

The model (1) for growth can also be seen as the equation $dS/dt = rS$, which describes the growth of capital S when an annual rate of interest r is compounded continuously. The model (2) for decay also occurs in biological applications such as determining the half-life of a drug—the time that it takes for 50% of a drug to be eliminated from a body by excretion or metabolism. In chemistry the decay model (2) appears in the mathematical description of a first-order chemical reaction. The point is this:

A single differential equation can serve as a mathematical model for many different phenomena.

Mathematical models are often accompanied by certain side conditions. For example, in (1) and (2) we would expect to know, in turn, the initial population P_0 and the initial amount of radioactive substance A_0 on hand. If the initial point in time is taken to be $t = 0$, then we know that $P(0) = P_0$ and $A(0) = A_0$. In other words, a mathematical model can consist of either an initial-value problem or, as we shall see later on in Section 5.2, a boundary-value problem.

NEWTON'S LAW OF COOLING/WARMING According to Newton's empirical law of cooling/warming, the rate at which the temperature of a body changes is proportional to the difference between the temperature of the body and the temperature of the surrounding medium, the so-called ambient temperature. If $T(t)$ represents the temperature of a body at time t, T_m the temperature of the surrounding medium, and dT/dt the rate at which the temperature of the body changes, then Newton's law of cooling/warming translates into the mathematical statement

$$\frac{dT}{dt} \propto T - T_m \quad \text{or} \quad \frac{dT}{dt} = k(T - T_m), \tag{3}$$

where k is a constant of proportionality. In either case, cooling or warming, if T_m is a constant it stands to reason that $k < 0$.

SPREAD OF A DISEASE   A contagious disease—for example, a flu virus—is spread throughout a community by people coming into contact with other people. Let $x(t)$ denote the number of people who have contracted the disease and $y(t)$ denote the number of people who have not yet been exposed. It seems reasonable to assume that the rate dx/dt at which the disease spreads is proportional to the number of encounters, or *interactions,* between these two groups of people. If we assume that the number of interactions is jointly proportional to $x(t)$ and $y(t)$—that is, proportional to the product xy—then

$$\frac{dx}{dt} = kxy, \tag{4}$$

where k is the usual constant of proportionality. Suppose a small community has a fixed population of n people. If one infected person is introduced into this community, then it could be argued that $x(t)$ and $y(t)$ are related by $x + y = n + 1$. Using this last equation to eliminate y in (4) gives us the model

$$\frac{dx}{dt} = kx(n + 1 - x). \tag{5}$$

An obvious initial condition accompanying equation (5) is $x(0) = 1$.

CHEMICAL REACTIONS   The disintegration of a radioactive substance, governed by the differential equation (1), is said to be a **first-order reaction.** In chemistry a few reactions follow this same empirical law: If the molecules of substance A decompose into smaller molecules, it is a natural assumption that the rate at which this decomposition takes place is proportional to the amount of the first substance that has not undergone conversion; that is, if $X(t)$ is the amount of substance A remaining at any time, then $dX/dt = kX,$ where k is a negative constant since X is decreasing. An example of a first-order chemical reaction is the conversion of t-butyl chloride, $(CH_3)_3CCl$, into t-butyl alcohol, $(CH_3)_3COH$:

$$(CH_3)_3CCl + NaOH \rightarrow (CH_3)_3COH + NaCl.$$

Only the concentration of the t-butyl chloride controls the rate of reaction. But in the reaction

$$CH_3Cl + NaOH \rightarrow CH_3OH + NaCl$$

one molecule of sodium hydroxide, $NaOH$, is consumed for every molecule of methyl chloride, CH_3Cl, thus forming one molecule of methyl alcohol, CH_3OH, and one molecule of sodium chloride, $NaCl$. In this case the rate at which the reaction proceeds is proportional to the product of the remaining concentrations of CH_3Cl and $NaOH$. To describe this second reaction in general, let us suppose *one* molecule of a substance A combines with *one* molecule of a substance B to form *one* molecule of a substance C. If X denotes the amount of chemical C formed at time t and if α and β are, in turn, the amounts of the two chemicals A and B at $t = 0$ (the initial amounts), then the instantaneous amounts of A and B not converted to chemical C are $\alpha - X$ and $\beta - X$, respectively. Hence the rate of formation of C is given by

$$\frac{dX}{dt} = k(\alpha - X)(\beta - X), \tag{6}$$

where k is a constant of proportionality. A reaction whose model is equation (6) is said to be a **second-order reaction.**

MIXTURES   The mixing of two salt solutions of differing concentrations gives rise to a first-order differential equation for the amount of salt contained in the

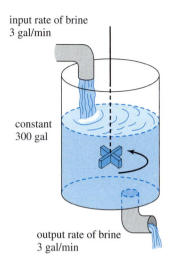

input rate of brine
3 gal/min

constant
300 gal

output rate of brine
3 gal/min

FIGURE 1.19   Mixing tank

mixture. Let us suppose that a large mixing tank initially holds 300 gallons of brine (that is, water in which a certain number of pounds of salt has been dissolved). Another brine solution is pumped into the large tank at a rate of 3 gallons per minute; the concentration of the salt in this inflow is 2 pounds per gallon. When the solution in the tank is well stirred, it is pumped out at the same rate as the entering solution. See Figure 1.19. If $A(t)$ denotes the amount of salt (measured in pounds) in the tank at time t, then the rate at which $A(t)$ changes is a net rate:

$$\frac{dA}{dt} = \left(\begin{array}{c} input\ rate \\ of\ salt \end{array}\right) - \left(\begin{array}{c} output\ rate \\ of\ salt \end{array}\right) = R_{in} - R_{out}. \tag{7}$$

The input rate R_{in} at which salt enters the tank is the product of the inflow concentration of salt and the inflow rate of fluid. Note that R_{in} is measured in pounds per minute:

$$R_{in} = \underset{\substack{\uparrow \\ \text{concentration} \\ \text{of salt} \\ \text{in inflow}}}{(2\ \text{lb/gal})} \cdot \underset{\substack{\uparrow \\ \text{input rate} \\ \text{of brine}}}{(3\ \text{gal/min})} = \underset{\substack{\uparrow \\ \text{input rate} \\ \text{of salt}}}{(6\ \text{lb/min})}.$$

Now, since the solution is being pumped out of the tank at the same rate that it is pumped in, the number of gallons of brine in the tank at time t is a constant 300 gallons. Hence the concentration of the salt in the tank as well as in the outflow is $c(t) = A(t)/300$ lb/gal, and so the output rate R_{out} of salt is

$$R_{out} = \underset{\substack{\uparrow \\ \text{concentration} \\ \text{of salt} \\ \text{in outflow}}}{\left(\frac{A(t)}{300}\ \text{lb/gal}\right)} \cdot \underset{\substack{\uparrow \\ \text{output rate} \\ \text{of brine}}}{(3\ \text{gal/min})} = \underset{\substack{\uparrow \\ \text{output rate} \\ \text{of salt}}}{\frac{A(t)}{100}\ \text{lb/min}}.$$

The net rate (7) then becomes

$$\frac{dA}{dt} = 6 - \frac{A}{100} \quad \text{or} \quad \frac{dA}{dt} + \frac{1}{100}A = 6. \tag{8}$$

If r_{in} and r_{out} denote general input and output rates of the brine solutions,* then there are three possibilities: $r_{in} = r_{out}$, $r_{in} > r_{out}$, and $r_{in} < r_{out}$. In the analysis leading to (8) we have assumed that $r_{in} = r_{out}$. In the latter two cases, the number of gallons of brine in the tank is either increasing ($r_{in} > r_{out}$) or decreasing ($r_{in} < r_{out}$) at the net rate $r_{in} - r_{out}$. See Problems 10–12 in Exercises 1.3.

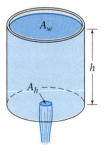

A_w

h

A_h

FIGURE 1.20   Draining tank

DRAINING A TANK   In hydrodynamics, **Torricelli's law** states that the speed v of efflux of water though a sharp-edged hole at the bottom of a tank filled to a depth h is the same as the speed that a body (in this case a drop of water) would acquire in falling freely from a height h—that is, $v = \sqrt{2gh}$, where g is the acceleration due to gravity. This last expression comes from equating the kinetic energy $\frac{1}{2}mv^2$ with the potential energy mgh and solving for v. Suppose a tank filled with water is allowed to drain through a hole under the influence of gravity. We would like to find the depth h of water remaining in the tank at time t. Consider the tank shown in Figure 1.20. If the area of the hole is A_h (in ft^2) and the speed of the water leaving the tank is $v = \sqrt{2gh}$ (in ft/s), then the volume of water leaving the tank per

*Don't confuse these symbols with R_{in} and R_{out}, which are input and output rates of *salt*.

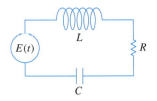

(a) *LRC*-series circuit

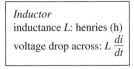

Inductor
inductance L: henries (h)
voltage drop across: $L\dfrac{di}{dt}$

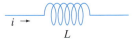

$i \rightarrow$ L

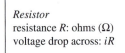

Resistor
resistance R: ohms (Ω)
voltage drop across: iR

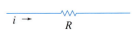

$i \rightarrow$ R

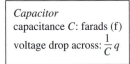

Capacitor
capacitance C: farads (f)
voltage drop across: $\dfrac{1}{C}q$

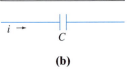

$i \rightarrow$ C

(b)

FIGURE 1.21 Symbols, units, and voltages. Current $i(t)$ and charge $q(t)$ are measured in amperes (A) and coulombs (C), respectively

second is $A_h\sqrt{2gh}$ (in ft³/s). Thus if $V(t)$ denotes the volume of water in the tank at time t, then

$$\frac{dV}{dt} = -A_h\sqrt{2gh}, \tag{9}$$

where the minus sign indicates that V is decreasing. Note here that we are ignoring the possibility of friction at the hole that might cause a reduction of the rate of flow there. Now if the tank is such that the volume of water in it at time t can be written $V(t) = A_w h$, where A_w (in ft²) is the *constant* area of the upper surface of the water (see Figure 1.20), then $dV/dt = A_w\, dh/dt$. Substituting this last expression into (9) gives us the desired differential equation for the height of the water at time t:

$$\frac{dh}{dt} = -\frac{A_h}{A_w}\sqrt{2gh}. \tag{10}$$

It is interesting to note that (10) remains valid even when A_w is not constant. In this case we must express the upper surface area of the water as a function of h—that is, $A_w = A(h)$. See Problem 14 in Exercises 1.3.

SERIES CIRCUITS Consider the single-loop series circuit shown in Figure 1.21(a), containing an inductor, resistor, and capacitor. The current in a circuit after a switch is closed is denoted by $i(t)$; the charge on a capacitor at time t is denoted by $q(t)$. The letters L, R, and C are known as inductance, resistance, and capacitance, respectively, and are generally constants. Now according to **Kirchhoff's second law,** the impressed voltage $E(t)$ on a closed loop must equal the sum of the voltage drops in the loop. Figure 1.21(b) shows the symbols and the formulas for the respective voltage drops across an inductor, a capacitor, and a resistor. Since current $i(t)$ is related to charge $q(t)$ on the capacitor by $i = dq/dt$, adding the three voltages

$$\overset{\text{inductor}}{L\frac{di}{dt} = L\frac{d^2q}{dt^2}}, \quad \overset{\text{resistor}}{iR = R\frac{dq}{dt}}, \quad \text{and} \quad \overset{\text{capacitor}}{\frac{1}{C}q}$$

and equating the sum to the impressed voltage yields a second-order differential equation

$$L\frac{d^2q}{dt^2} + R\frac{dq}{dt} + \frac{1}{C}q = E(t). \tag{11}$$

We will examine a differential equation analogous to (11) in great detail in Section 5.1.

FALLING BODIES To construct a mathematical model of the motion of a body moving in a force field, one often starts with Newton's second law of motion. Recall from elementary physics that Newton's **first law of motion** states that a body either will remain at rest or will continue to move with a constant velocity unless acted on by an external force. In each case this is equivalent to saying that when the sum of the forces $F = \sum F_k$—that is, the *net* or resultant force—acting on the body is zero, then the acceleration a of the body is zero. **Newton's second law of motion** indicates that when the net force acting on a body is not zero, then the net force is proportional to its acceleration a or, more precisely, $F = ma$, where m is the mass of the body.

Now suppose a rock is tossed upward from the roof of a building as illustrated in Figure 1.22. What is the position $s(t)$ of the rock relative to the ground at time t? The acceleration of the rock is the second derivative d^2s/dt^2. If we assume that the

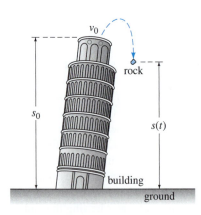

FIGURE 1.22 Position of rock measured from ground level

upward direction is positive and that no force acts on the rock other than the force of gravity, then Newton's second law gives

$$m\frac{d^2s}{dt^2} = -mg \quad \text{or} \quad \frac{d^2s}{dt^2} = -g. \tag{12}$$

In other words, the net force is simply the weight $F = F_1 = -W$ of the rock near the surface of the earth. Recall that the magnitude of the weight is $W = mg$, where m is the mass of the body and g is the acceleration due to gravity. The minus sign in (12) is used because the weight of the rock is a force directed downward, which is opposite to the positive direction. If the height of the building is s_0 and the initial velocity of the rock is v_0, then s is determined from the second-order initial-value problem

$$\frac{d^2s}{dt^2} = -g, \quad s(0) = s_0, \quad s'(0) = v_0. \tag{13}$$

Although we have not been stressing solutions of the equations we have constructed, note that (13) can be solved by integrating the constant $-g$ twice with respect to t. The initial conditions determine the two constants of integration. From elementary physics you might recognize the solution of (13) as the formula $s(t) = -\frac{1}{2}gt^2 + v_0t + s_0$.

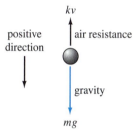

FIGURE 1.23 Falling body of mass m

FALLING BODIES AND AIR RESISTANCE Before Galileo's famous experiment from the leaning tower of Pisa, it was generally believed that heavier objects in free fall, such as a cannonball, fell with a greater acceleration than lighter objects, such as a feather. Obviously a cannonball and a feather when dropped simultaneously from the same height *do* fall at different rates, but it is not because a cannonball is heavier. The difference in rates is due to air resistance. The resistive force of air was ignored in the model given in (13). Under some circumstances a falling body of mass m, such as a feather with low density and irregular shape, encounters air resistance proportional to its instantaneous velocity v. If we take, in this circumstance, the positive direction to be oriented downward, then the net force acting on the mass is given by $F = F_1 + F_2 = mg - kv$, where the weight $F_1 = mg$ of the body is force acting in the positive direction and air resistance $F_2 = -kv$ is a force, called **viscous damping,** acting in the opposite or upward direction. See Figure 1.23. Now since v is related to acceleration a by $a = dv/dt$, Newton's second law becomes $F = ma = m \, dv/dt$. By equating the net force to this form of Newton's second law, we obtain a first-order differential equation for the velocity $v(t)$ of the body at time t,

$$m\frac{dv}{dt} = mg - kv. \tag{14}$$

Here k is a positive constant of proportionality. If $s(t)$ is the distance the body falls in time t from its initial point of release, then $v = ds/dt$ and $a = dv/dt = d^2s/dt^2$. In terms of s, (14) is a second-order differential equation

$$m\frac{d^2s}{dt^2} = mg - k\frac{ds}{dt} \quad \text{or} \quad m\frac{d^2s}{dt^2} + k\frac{ds}{dt} = mg. \tag{15}$$

A SLIPPING CHAIN Suppose a uniform chain of length L feet is draped over a metal peg anchored into a wall high above ground level. Let us assume that the peg is frictionless and that the chain weighs ρ lb/ft. Figure 1.24(a) illustrates the position of the chain where it hangs in equilibrium; if displaced a little to either the right or the left, the chain would slip off the peg. Suppose the positive direction is taken to be downward, and suppose $x(t)$ denotes the distance that the right end of the chain would fall in time t. The equilibrium position corresponds to $x = 0$. In Figure 1.24(b) the chain is displaced an amount x_0 feet and is held on the peg until

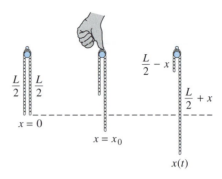

(a) equilibrium **(b)** chain held **(c)** motion
                     until $t = 0$ for $t > 0$

FIGURE 1.24   Chain slipping from frictionless peg

it is released at an initial time that is designated as $t = 0$. For the chain in motion, as shown in Figure 1.24(c), we have the following quantities:

weight of the chain: $W = (L \text{ ft}) \cdot (\rho \text{ lb/ft}) = L\rho,$

mass of the chain: $m = W/g = L\rho/32,$

net force: $F = \left(\dfrac{L}{2} + x\right)\rho - \left(\dfrac{L}{2} - x\right)\rho = 2x\rho.$

Since $a = d^2x/dt^2$, $ma = F$ becomes

$$\frac{L\rho}{32}\frac{d^2x}{dt^2} = 2\rho x \quad \text{or} \quad \frac{d^2x}{dt^2} - \frac{64}{L}x = 0. \tag{16}$$

(a) suspension bridge cable

(b) telephone wires

FIGURE 1.25   Cables suspended between vertical supports

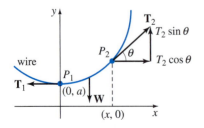

FIGURE 1.26   Element of cable

SUSPENDED CABLES   Suppose a flexible cable, wire, or heavy rope is suspended between two vertical supports. Physical examples of this could be one of the two cables supporting the roadbed of a suspension bridge as shown in Figure 1.25(a) or a long telephone wire strung between two posts as shown in Figure 1.25(b). Our goal is to construct a mathematical model that describes the shape that such a cable assumes.

To begin, let's agree to examine only a portion or element of the cable between its lowest point P_1 and any arbitrary point P_2. As drawn in color in Figure 1.26, this element of the cable is the curve in a rectangular coordinate system with y-axis chosen to pass through the lowest point P_1 on the curve and the x-axis chosen a units below P_1. Three forces are acting on the cable: the tensions $\mathbf{T}_1$ and $\mathbf{T}_2$ in the cable that are tangent to the cable at P_1 and P_2, respectively, and the portion $\mathbf{W}$ of the total vertical load between the points P_1 and P_2. Let $T_1 = |\mathbf{T}_1|$, $T_2 = |\mathbf{T}_2|$, and $W = |\mathbf{W}|$ denote the magnitudes of these vectors. Now the tension $\mathbf{T}_2$ resolves into horizontal and vertical components (scalar quantities) $T_2 \cos\theta$ and $T_2 \sin\theta$. Because of static equilibrium we can write

$$T_1 = T_2 \cos\theta \quad \text{and} \quad W = T_2 \sin\theta.$$

By dividing the last equation by the first, we eliminate T_2 and get $\tan\theta = W/T_1$. But because $dy/dx = \tan\theta$, we arrive at

$$\frac{dy}{dx} = \frac{W}{T_1}. \tag{17}$$

This simple first-order differential equation serves as a model for both the shape of a flexible wire such as a telephone wire hanging under its own weight and the shape of the cables that support the roadbed of a suspension bridge. We will come back to equation (17) in Exercises 2.2 and Section 5.3.

REMARKS

Each example in this section has described a dynamical system—a system that changes or evolves with the flow of time t. Since the study of dynamical systems is a branch of mathematics currently in vogue, we shall occasionally relate the terminology of that field to the discussion at hand.

In more precise terms, a **dynamical system** consists of a set of time-dependent variables, called **state variables,** together with a rule that enables us to determine (without ambiguity) the state of the system (this may be a past, present, or future state) in terms of a state prescribed at some time t_0. Dynamical systems are classified as either discrete-time systems or continuous-time systems. In this course we shall be concerned only with continuous-time systems—systems in which *all* variables are defined over a continuous range of time. The rule, or mathematical model, in a continuous-time dynamical system is a differential equation or a system of differential equations. The **state of the system** at a time t is the value of the state variables at that time; the specified state of the system at a time t_0 is simply the initial conditions that accompany the mathematical model. The solution of the initial-value problem is referred to as the **response of the system.** For example, in the case of radioactive decay, the rule is $dA/dt = kA$. Now if the quantity of a radioactive substance at some time t_0 is known, say $A(t_0) = A_0$, then by solving the rule we find that the response of the system for $t \geq t_0$ is $A(t) = A_0 e^{(t-t_0)}$ (see Section 3.1). The response $A(t)$ is the single state variable for this system. In the case of the rock tossed from the roof of a building, the response of the system—the solution of the differential equation $d^2s/dt^2 = -g$, subject to the initial state $s(0) = s_0$, $s'(0) = v_0$—is the function $s(t) = -\frac{1}{2}gt^2 + v_0 t + s_0, 0 \leq t \leq T$, where T represents the time when the rock hits the ground. The state variables are $s(t)$ and $s'(t)$, which are, respectively, the vertical position of the rock above ground and its velocity at time t. The acceleration $s''(t)$ is *not* a state variable, since we have to know only any initial position and initial velocity at a time t_0 to uniquely determine the rock's position $s(t)$ and velocity $s'(t) = v(t)$ for any time in the interval $t_0 \leq t \leq T$. The acceleration $s''(t) = a(t)$ is, of course, given by the differential equation $s''(t) = -g, 0 < t < T$.

One last point: Not every system studied in this text is a dynamical system. We shall also examine some static systems in which the model is a differential equation.

WHAT LIES AHEAD Throughout this text you will see three different types of approaches to, or analyses of, differential equations. Over the centuries, differential equations would often spring from the efforts of a scientist/engineer to describe some physical phenomenon or to translate an empirical or experimental law into mathematical terms. As a consequence, a scientist/engineer/mathematician would often spend many years of his or her life trying to find the solutions of a DE. With a solution in hand the study of its properties then followed. This quest for solutions is called by some the *analytical approach* to differential equations. Once they realized that explicit solutions are at best difficult to obtain and at worst impossible to obtain, mathematicians learned that a differential equation itself could be a font of valuable information. It is possible, in some instances, to glean directly from the differential equation answers to questions such as, Does the DE actually have solutions? If a solution of the DE exists and satisfies an initial condition, is it the only such solution? What are some of the properties of the unknown solutions? What can we say about the geometry of the solution curves? Such an approach is *qualitative analysis*. Finally, if a differential equation cannot be solved by analytical methods, yet we can prove that a solution exists, the next logical query is, Can we somehow approximate the values of an unknown solution? Here we enter the realm of *numerical analysis*. An affirmative answer to the last question stems from the fact that a differential equation can be used as a cornerstone for constructing very accurate approximation algorithms. In Chapter 2 we start with qualitative considerations of first-order ODEs, then examine analytical stratagems for solving some special first-order equations, and conclude with an introduction to an elementary numerical method. See Figure 1.27.

(a) analytical   (b) qualitative   (c) numerical

FIGURE 1.27   Different approaches to the study of differential equations

EXERCISES 1.3

Answers to selected odd-numbered problems begin on page ANS-1.

POPULATION DYNAMICS

1. Under the same assumptions underlying the model in (1), determine a differential equation for the population $P(t)$ of a country when individuals are allowed to immigrate into the country at a constant rate $r > 0$. What is the differential equation for the population $P(t)$ of the country when individuals are allowed to emigrate from the country at a constant rate $r > 0$?

2. The population model given in (1) fails to take death into consideration; the growth rate equals the birth rate. In another model of a changing population of a community, it is assumed that the rate at which the population changes is a *net* rate—that is, the difference between the rate of births and the rate of deaths in the community. Determine a model for the population $P(t)$ if both the birth rate and the death rate are proportional to the population present at time t.

3. Using the concept of net rate introduced in Problem 2, determine a model for a population $P(t)$ if the birth rate is proportional to the population present at time t but the death rate is proportional to the square of the population present at time t.

4. Modify the model in Problem 3 for net rate at which the population $P(t)$ of a certain kind of fish changes, by also assuming that the fish are harvested at a constant rate $h > 0$.

NEWTON'S LAW OF COOLING/WARMING

5. A cup of coffee cools according to Newton's law of cooling (3). Use data from the graph of the temperature $T(t)$ in Figure 1.28 to estimate the constants T_m, T_0, and k in a model of the form of a first-order initial-value problem: $dT/dt = k(T - T_m)$, $T(0) = T_0$.

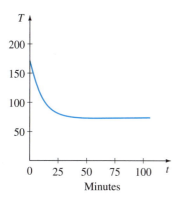

FIGURE 1.28 Cooling curve in Problem 5

6. The ambient temperature T_m in (3) could be a function of time t. Suppose that in an artificially controlled environment, $T_m(t)$ is periodic with a 24-hour period, as illustrated in Figure 1.29. Devise a mathematical model for the temperature $T(t)$ of a body within this environment.

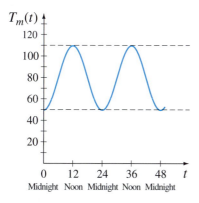

FIGURE 1.29 Ambient temperature in Problem 6

SPREAD OF A DISEASE/TECHNOLOGY

7. Suppose a student carrying a flu virus returns to an isolated college campus of 1000 students. Determine a differential equation for the number of people $x(t)$ who have contracted the flu if the rate at which the disease spreads is proportional to the number of interactions between the number of students who have the flu and the number of students who have not yet been exposed to it.

8. At a time denoted as $t = 0$ a technological innovation is introduced into a community that has a fixed population of n people. Determine a differential equation for the number of people $x(t)$ who have adopted the innovation at time t if it is assumed that the rate at which the innovations spread through the community is jointly propor-

tional to the number of people who have adopted it and the number of people who have not adopted it.

MIXTURES

9. Suppose that a large mixing tank initially holds 300 gallons of water in which 50 pounds of salt have been dissolved. Pure water is pumped into the tank at a rate of 3 gal/min, and when the solution is well stirred, it is then pumped out at the same rate. Determine a differential equation for the amount of salt $A(t)$ in the tank at time t. What is $A(0)$?

10. Suppose that a large mixing tank initially holds 300 gallons of water in which 50 pounds of salt have been dissolved. Another brine solution is pumped into the tank at a rate of 3 gal/min, and when the solution is well stirred, it is then pumped out at a *slower* rate of 2 gal/min. If the concentration of the solution entering is 2 lb/gal, determine a differential equation for the amount of salt $A(t)$ in the tank at time t.

11. What is the differential equation in Problem 10, if the well-stirred solution is pumped out at a *faster* rate of 3.5 gal/min?

12. Generalize the model given in (8) on page 23 by assuming that the large tank initially contains N_0 number of gallons of brine, r_{in} and r_{out} are the input and output rates of the brine, respectively (measured in gallons per minute), c_{in} is the concentration of the salt in the inflow, $c(t)$ the concentration of the salt in the tank as well as in the outflow at time t (measured in pounds of salt per gallon), and $A(t)$ is the amount of salt in the tank at time t.

DRAINING A TANK

13. Suppose water is leaking from a tank through a circular hole of area A_h at its bottom. When water leaks through a hole, friction and contraction of the stream near the hole reduce the volume of water leaving the tank per second to $cA_h\sqrt{2gh}$, where c $(0 < c < 1)$ is an empirical constant. Determine a differential equation for the height h of water at time t for the cubical tank shown in Figure 1.30. The radius of the hole is 2 in., and $g = 32$ ft/s^2.

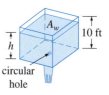

FIGURE 1.30 Cubical tank in Problem 13

14. The right-circular conical tank shown in Figure 1.31 loses water out of a circular hole at its bottom. Determine a differential equation for the height of the water h at time t. The radius of the hole is 2 in., $g = 32$ ft/s^2, and the friction/contraction factor introduced in Problem 13 is $c = 0.6$.

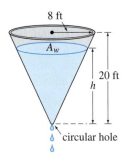

FIGURE 1.31 Conical tank in Problem 14

SERIES CIRCUITS

15. A series circuit contains a resistor and an inductor as shown in Figure 1.32. Determine a differential equation for the current $i(t)$ if the resistance is R, the inductance is L, and the impressed voltage is $E(t)$.

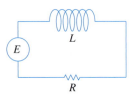

FIGURE 1.32 LR-series circuit in Problem 15

16. A series circuit contains a resistor and a capacitor as shown in Figure 1.33. Determine a differential equation for the charge $q(t)$ on the capacitor if the resistance is R, the capacitance is C, and the impressed voltage is $E(t)$.

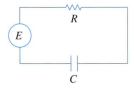

FIGURE 1.33 RC-series circuit in Problem 16

FALLING BODIES AND AIR RESISTANCE

17. For high-speed motion through the air—such as the skydiver shown in Figure 1.34, falling before the para-

chute is opened—air resistance is closer to a power of the instantaneous velocity $v(t)$. Determine a differential equation for the velocity $v(t)$ of a falling body of mass m if air resistance is proportional to the square of the instantaneous velocity.

FIGURE 1.34 Air resistance proportional to square of velocity in Problem 17

NEWTON'S SECOND LAW AND ARCHIMEDES' PRINCIPLE

18. A cylindrical barrel s feet in diameter of weight w lb is floating in water as shown in Figure 1.35(a). After an initial depression the barrel exhibits an up-and-down bobbing motion along a vertical line. Using Figure 1.35(b), determine a differential equation for the vertical displacement $y(t)$ if the origin is taken to be on the vertical axis at the surface of the water when the barrel is at rest. Use **Archimedes' principle:** Buoyancy, or upward force of the water on the barrel, is equal to the weight of the water displaced. Assume that the downward direction is positive, that the weight density of water is 62.4 lb/ft^3, and that there is no resistance between the barrel and the water.

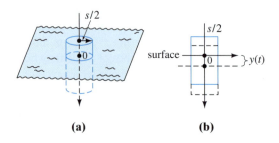

FIGURE 1.35 Bobbing motion of floating barrel in Problem 18

NEWTON'S SECOND LAW AND HOOKE'S LAW

19. After a mass m is attached to a spring it stretches it s units and then hangs at rest in the equilibrium position as shown in Figure 1.36(b). After the spring/mass system has been set in motion, let $x(t)$ denote the directed distance of the mass beyond the equilibrium position. As indicated in Figure 1.36(c), assume that the downward direction is positive, that the motion takes place in a vertical straight line through the center of gravity of the mass, and that the only forces acting on the system are the weight of the mass and the restoring force of the stretched spring. Use **Hooke's law:** The restoring force of a spring is proportional to its total elongation. Determine a differential equation for the displacement $x(t)$ at time t.

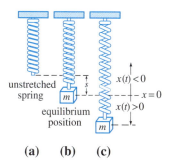

FIGURE 1.36 Spring/mass system in Problem 19

20. In Problem 19, what is a differential equation for the displacement $x(t)$ if the motion takes place in a medium that imparts a damping force on the spring/mass system that is proportional to the instantaneous velocity of the mass and acts in a direction opposite to that of motion?

NEWTON'S SECOND LAW AND VARIABLE MASS

When the mass m of a body moving through a force field is variable, Newton's second law takes on the form: If the net force acting on a body is not zero, then the net force F is equal to the time rate of change of momentum of the body—that is,

$$F = \frac{d}{dt}(mv),* \qquad (18)$$

where mv is momentum. Use this formulation of Newton's second law in Problems 21 and 22.

---

*Note that when m is constant, (18) is the same as $F = m \, dv/dt = ma$ where $a = dv/dt$ is acceleration.

21. A uniform 10-foot-long chain is coiled loosely on the ground. As shown in Figure 1.37, one end of the chain is pulled vertically upward by means of a constant force of 5 lb. The chain weighs 1 lb/ft. Determine a differential equation for the height $x(t)$ of the end above ground level at time t. Assume that the positive direction is upward.

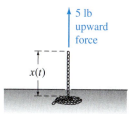

FIGURE 1.37 Chain pulled upward in Problem 21

22. A uniform chain of length L, measured in feet, is held vertically so that the lower end just touches the floor as shown in Figure 1.38. The chain weighs 2 lb/ft. The upper end that is held is released from rest at $t = 0$, and the chain falls straight down. Ignore air resistance, assume that the positive direction is downward, and let $x(t)$ denote the length of the chain on the floor at time t. Use the fact that the net force F in (18) acting on the chain at time $t \geq 0$ is the constant $2L$ to show that a differential equation for $x(t)$ is

$$(L - x)\frac{d^2x}{dt^2} - \left(\frac{dx}{dt}\right)^2 = Lg.$$

FIGURE 1.38 Vertically held chain in Problem 22

NEWTON'S SECOND LAW AND THE LAW OF UNIVERSAL GRAVITATION

23. By **Newton's universal law of gravitation** the free-fall acceleration a of a body, such as the satellite shown in Figure 1.39, falling a great distance to the surface is *not* the constant g. Rather, the acceleration a is inversely proportional to the square of the distance from the center of the earth, $a = k/r^2$, where k is the constant of proportionality. Use the fact that at the surface of the earth $r = R$ and

$a = g$ to determine k. If the positive direction is upward, use Newton's second law and his universal law of gravitation to find a differential equation for the distance r.

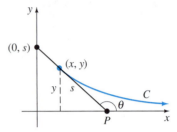

satellite of mass m

surface

r

R

earth of mass M

FIGURE 1.39 Satellite in Problem 23

24. Suppose a hole is drilled through the center of the earth and a bowling ball of mass m is dropped into the hole, as shown in Figure 1.40. Construct a mathematical model that describes the motion of the ball. At time t, let r denote the distance from the center of the earth to the mass m, M denote the mass of the earth, M_r denote the mass of that portion of the earth within a sphere of radius r, and δ denote the constant density of the earth.

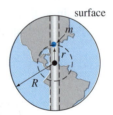

surface

m

r

R

FIGURE 1.40 Hole through earth in Problem 24

MISCELLANEOUS MATHEMATICAL MODELS

25. Learning Theory In the theory of learning, the rate at which a subject is memorized is assumed to be proportional to the amount that is left to be memorized. Suppose M denotes the total amount of a subject to be memorized and $A(t)$ is the amount memorized in time t. Determine a differential equation for the amount $A(t)$.

26. Forgetfulness In Problem 25 assume that the rate at which material is *forgotten* is proportional to the amount memorized in time t. Determine a differential equation for the amount $A(t)$ when forgetfulness is taken into account.

27. Infusion of a Drug A drug is infused into a patient's bloodstream at a constant rate of r grams per second. Simultaneously, the drug is removed at a rate proportional to the amount $x(t)$ of the drug present at time t. Determine a differential equation for the amount $x(t)$.

28. Tractrix A person P, starting at the origin, moves in the direction of the positive x-axis, pulling a weight along the curve C, called a **tractrix,** as shown in Figure 1.41. The weight, initially located on the y-axis at $(0, s)$ is pulled by a rope of constant length s, which is kept taut throughout the motion. Determine a differential equation for the path C of motion. Assume that the rope is always tangent to C.

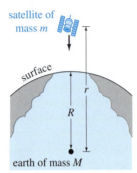

y

$(0, s)$

(x, y)

y | s

θ

P

C

x

FIGURE 1.41 Tractrix curve in Problem 28

29. Reflecting Surface Assume that when the plane curve C shown in Figure 1.42 is revolved about the x-axis it generates a surface of revolution with the property that all light rays L parallel to the x-axis striking the surface are reflected to a single point O (the origin). Use the fact that the angle of incidence is equal to the angle of reflection to determine a differential equation that describes the shape of the curve C. Such a curve C is important in applications ranging from construction of telescopes to satellite antennas, automobile headlights, and solar collectors. [*Hint:* Inspection of the figure shows that we can write $\phi = 2\theta$. Why? Now use an appropriate trigonometric identity.]

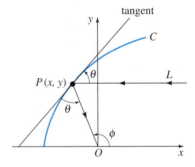

tangent

y

C

$P(x, y)$

θ

L

θ

ϕ

O

x

FIGURE 1.42 Reflecting surface in Problem 29

DISCUSSION/PROJECT PROBLEMS

30. Reread Problem 37 in Exercises 1.1 and then give an explicit solution $P(t)$ for equation (1). Find a one-parameter family of solutions of (1).

31. Reread the sentence following equation (3) and assume that T_m is a positive constant. Discuss why we would expect $k < 0$ in (3) in both cases of cooling and warming. You might start by interpreting, say, $T(t) > T_m$ in a graphical manner.

32. Reread the discussion leading up to equation (8). If we assume that initially the tank holds, say, 50 lbs of salt, it stands to reason that because salt is being added to the tank continuously for $t > 0$, $A(t)$ should be an increasing function. Discuss how you might determine from the DE, without actually solving it, the number of pounds of salt in the tank after a long period of time.

33. Population Model The differential equation $\dfrac{dP}{dt} = (k \cos t)P$, where k is a positive constant, is a model of human population $P(t)$ of a certain community. Discuss an interpretation for the solution of this equation, in other words, what kind of population do you think the differential equation describes?

34. Rotating Fluid As shown in Figure 1.43(a), a right circular cylinder partially filled with fluid is rotated with a constant angular velocity ω about a vertical y-axis through its center. The rotating fluid forms a surface of revolution S. To identify S, we first establish a coordinate system consisting of a vertical plane determined by the y-axis and an x-axis drawn perpendicular to the y-axis such that the point of intersection of the axes (the origin) is located at the lowest point on the surface S.

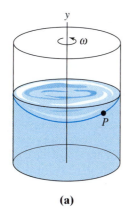

(a)

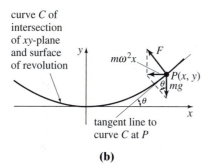

(b)

FIGURE 1.43 Rotating fluid in Problem 34

We then seek a function $y = f(x)$ that represents the curve C of intersection of the surface S and the vertical coordinate plane. Let the point $P(x, y)$ denote the position of a particle of the rotating fluid of mass m in the coordinate plane. See Figure 1.43(b).

(a) At P there is a reaction force of magnitude F due to the other particles of the fluid which is normal to the surface S. By Newton's second law the magnitude of the net force acting on the particle is $m\omega^2 x$. What is this force? Use Figure 1.43(b) to discuss the nature and origin of the equations

$$F \cos \theta = mg, \quad F \sin \theta = m\omega^2 x.$$

(b) Use part (a) to find a first-order differential equation that defines the function $y = f(x)$.

35. Falling Body In Problem 23, suppose $r = R + s$, where s is the distance from the surface of the earth to the falling body. What does the differential equation obtained in Problem 23 become when s is very small in comparison to R?

36. Raindrops Keep Falling In meteorology the term *virga* refers to falling raindrops or ice particles that evaporate before they reach the ground. Assume that a typical raindrop is spherical in shape. Starting at some time, which we can designate as $t = 0$, the raindrop of radius r_0 falls from rest from a cloud and begins to evaporate.

(a) If it is assumed that a raindrop evaporates in such a manner that its shape remains spherical, then it also makes sense to assume that the rate at which the raindrop evaporates, that is, the rate at which it loses mass, is proportional to its surface area. Show that this latter assumption implies that the rate at which the radius r of the raindrop decreases is a constant. Find $r(t)$. [*Hint:* See Problem 47 in Exercises 1.1.]

(b) If the positive direction is downward, construct a mathematical model for the velocity v of the falling raindrop at time t. Ignore air resistance. [*Hint:* See the introduction to Problems 21 and 22.]

37. Let It Snow The "snowplow problem" is a classic and appears in many differential equations texts but was probably made famous by Ralph Palmer Agnew:

> *"One day it started snowing at a heavy and steady rate. A snowplow started out at noon, going 2 miles the first hour and 1 mile the second hour. What time did it start snowing?"*

If possible, find the text *Differential Equations*, Ralph Palmer Agnew, McGraw-Hill Book Co., and then discuss the construction and solution of the mathematical model.

38. Reread this section and classify each mathematical model as linear or nonlinear.

39. Population Dynamics Suppose that $P'(t) = 0.15P(t)$ represents a mathematical model for the growth of a certain cell culture, where $P(t)$ is the size of the culture (measured in millions of cells) at time t (measured in hours). How fast is the culture growing at the time t when the size of the culture reaches 2 million cells?

40. Radioactive Decay Suppose that $A'(t) = -0.0004332\, A(t)$ represents a mathematical model for the decay of radium-226, where $A(t)$ is the amount of radium (measured in grams) remaining at time t (measured in years). How much of the radium sample remains at the time t when the sample is decaying at a rate of 0.002 gram per year?

CHAPTER 1 IN REVIEW

Answers to selected odd-numbered problems begin on page ANS-1.

In Problems 1 and 2 fill in the blank and then write this result as a linear first-order differential equation that is free of the symbol c_1 and has the form $dy/dx = f(x, y)$. The symbols c_1 and k represent constants.

1. $\dfrac{d}{dx} c_1 e^{kx} = $ _____

2. $\dfrac{d}{dx} (5 + c_1 e^{-2x}) = $ _____

In Problems 3 and 4 fill in the blank and then write this result as a linear second-order differential equation that is free of the symbols c_1 and c_2 and has the form $F(y, y'') = 0$. The symbols c_1, c_2, and k represent constants.

3. $\dfrac{d^2}{dx^2} (c_1 \cos kx + c_2 \sin kx) = $ _____

4. $\dfrac{d^2}{dx^2} (c_1 \cosh kx + c_2 \sinh kx) = $ _____

In Problems 5 and 6 compute y' and y'' and then combine these derivatives with y as a linear second-order differential equation that is free of the symbols c_1 and c_2 and has the form $F(y, y', y'') = 0$. The symbols c_1 and c_2 represent constants.

5. $y = c_1 e^x + c_2 x e^x$ **6.** $y = c_1 e^x \cos x + c_2 e^x \sin x$

In Problems 7–12 match each of the given differential equations with one or more of these solutions:
(a) $y = 0$,   **(b)** $y = 2$,   **(c)** $y = 2x$,   **(d)** $y = 2x^2$.

7. $xy' = 2y$ **8.** $y' = 2$

9. $y' = 2y - 4$ **10.** $xy' = y$

11. $y'' + 9y = 18$ **12.** $xy'' - y' = 0$

In Problems 13 and 14 determine by inspection at least one solution of the given differential equation.

13. $y'' = y'$ **14.** $y' = y(y - 3)$

In Problems 15 and 16 interpret each statement as a differential equation.

15. On the graph of $y = \phi(x)$, the slope of the tangent line at a point $P(x, y)$ is the square of the distance from $P(x, y)$ to the origin.

16. On the graph of $y = \phi(x)$, the rate at which the slope changes with respect to x at a point $P(x, y)$ is the negative of the slope of the tangent line at $P(x, y)$.

17. (a) Give the domain of the function $y = x^{2/3}$.
    (b) Give the largest interval I of definition over which $y = x^{2/3}$ is a solution of the differential equation $3xy' - 2y = 0$.

18. (a) Verify that the one-parameter family $y^2 - 2y = x^2 - x + c$ is an implicit solution of the differential equation $(2y - 2)y' = 2x - 1$.
    (b) Find a member of the one-parameter family in part (a) that satisfies the initial condition $y(0) = 1$.
    (c) Use your result in part (b) to find an explicit *function* $y = \phi(x)$ that satisfies $y(0) = 1$. Give the domain of the function ϕ. Is $y = \phi(x)$ a *solution* of the initial-value problem? If so, give its interval I of definition; if not, explain.

19. Given that $y = -\dfrac{2}{x} + x$ is a solution of the DE $xy' + y = 2x$. Find x_0 and the largest interval I for which $y(x)$ is a solution of the IVP $xy' + y = 2x$, $y(x_0) = 1$.

20. Suppose that $y(x)$ denotes a solution of the initial-value problem $y' = x^2 + y^2$, $y(1) = -1$ and that $y(x)$ possesses at least a second derivative at $x = 1$. In some neighborhood of $x = 1$, use the DE to determine whether $y(x)$ is increasing or decreasing, and whether the graph $y(x)$ is concave up or concave down.

21. A differential equation may possess more than one family of solutions.
    (a) Plot different members of the families $y = \phi_1(x) = x^2 + c_1$ and $y = \phi_2(x) = -x^2 + c_2$.
    (b) Verify that $y = \phi_1(x)$ and $y = \phi_2(x)$ are two solutions of the nonlinear first-order differential equation $(y')^2 = 4x^2$.

(c) Construct a piecewise-defined function that is a solution of the nonlinear DE in part (b) but is not a member of either family of solutions in part (a).

22. What is the slope of the tangent line to the graph of a solution of $y' = 6\sqrt{y} + 5x^3$ that passes through $(-1, 4)$?

In Problems 23–26 verify that the indicated function is a particular solution of the given differential equation. Give an interval of definition I for each solution.

23. $y'' + y = 2\cos x - 2\sin x;$ $y = x\sin x + x\cos x$

24. $y'' + y = \sec x;$ $y = x\sin x + (\cos x)\ln(\cos x)$

25. $x^2 y'' + xy' + y = 0;$ $y = \sin(\ln x)$

26. $x^2 y'' + xy' + y = \sec(\ln x);$
$y = \cos(\ln x)\ln(\cos(\ln x)) + (\ln x)\sin(\ln x)$

27. The graph of a solution of a second-order initial-value problem $d^2y/dx^2 = f(x, y, y')$, $y(2) = y_0$, $y'(2) = y_1$, is given in Figure 1.44. Use the graph to estimate the values of y_0 and y_1.

28. A tank in the form of a right circular cylinder of radius 2 feet and height 10 feet is standing on end. If the tank is initially full of water, and water leaks from a circular hole of radius $\frac{1}{2}$ inch at its bottom, determine a differential equation for the height h of the water at time t. Ignore friction and contraction of water at the hole.

29. The number of field mice in a certain pasture is given by the function $200 - 10t$, where time t is measured in years. Determine a differential equation governing a population of owls that feed on the mice if the rate at which the owl population grows is proportional to the difference between the number of owls at time t and number of field mice at time t.

30. Suppose a bug advances along a circular path of radius r at a constant angular velocity $\omega > 0$. See Figure 1.45. Show that the x- and y-coordinates of the bug's position both satisfy a differential equation of the form

$$\frac{d^2u}{dt^2} + \omega^2 u = 0$$

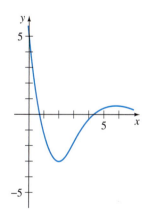

FIGURE 1.44 Graph for Problem 27

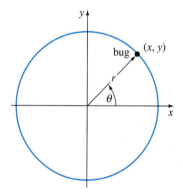

FIGURE 1.45 Circular path in Problem 30

DIVING DECEPTION PASS

You drag yourself, exhausted and gasping, out of the water onto an unfamiliar rocky shore. You see your dive partner, who got out sooner, staggering to her feet 100 feet up the strand. When you are finally close enough to talk, the question on both of your lips is "What was that?" You entered the water a half mile away, west of the Deception Pass Bridge,* on what was to be an easy and entertaining dive. Instead, you found yourself dragged willy-nilly—first along the bottom, then on the surface—through the narrows, past the islands in the tidal stream, in and out of the whirlpool, finally to land here, well east of the bridge. You thought you were diving at slack water. Now, as your dive partner glares at you venomously, you resort to the passive voice: "Mistakes were made."

Deception Pass is a narrow channel between two large islands near Seattle, Washington. There is a bridge across the channel that connects the two islands. See Figure 1. When the tide comes in, the entire Pacific Ocean pushes water through that channel to get to the inner part of Puget Sound. You knew that before the dive and tried to compensate by diving at slack water—the time at either high tide or low tide when there is no flow through the channel. When you consult the tide tables again, you realize that you made two mistakes. First, you overlooked the fact that the times were given in Pacific Daylight Time. You could have just looked at your watch, but it was unthinkable to you that the federal government would have made anything that easy, so you assumed that the annual predictions were made uniformly in Standard Time. In addition, you underestimated the force of the current away from the narrowest part of the pass. Thus you were diving an hour *after* slack water, and the current a quarter mile from the narrows was *stronger* than you guessed. You recognize that betting your life and that of your friend on a guess was not the best way to promote the relationship. You would like to try the dive again some time, but it seems desirable to avoid any further acquaintance with the whirlpool, so you want to understand the current through the pass better.

FIGURE 1 Deception Pass Bridge. Note the white water below

Tide tables give the predicted velocity of the current under the bridge. The question is, how is that data related to the velocity of the current farther away? After some thought you realize that the velocity of the current is inversely proportional to the size of the channel through which the water must flow. You write this relationship as the differential equation

$$x' = \frac{v_0}{S(x)}, \tag{1}$$

where $S(x)$ represents the relative size of the channel a distance x from the bridge. Time is measured in seconds, and x is measured in feet from the bridge. Note also that x increases in the eastward direction and that v_0 represents the velocity of the current under the bridge. You may think of $x = x(t)$ as the position of a diver t seconds after entering the channel. Equation (1) lets you relate the single number you know—the velocity v_0 of the water under the bridge—to the velocity x' of the water (and any diver in that water) at any other point in the channel.

---

*The bridge is located in Deception Pass State Park north of Oak Harbor, Washington.

You notice in the schematic of Deception Pass in Figure 2 that the size of the channel is somewhat like an hourglass—wide at either end but narrow in the middle. So your first approximation to the size $S(x)$ is a quadratic function—a function whose values are small in the middle and increase away from the middle. The narrowest or smallest part of the channel occurs about 100 feet east of the bridge. After some measurements and calculations you conclude that the size function

$$S(x) = 1 + \frac{1}{200000}(x - 100)^2 \tag{2}$$

is reasonable from 600 feet to the west of the bridge to 1000 feet east of the bridge.

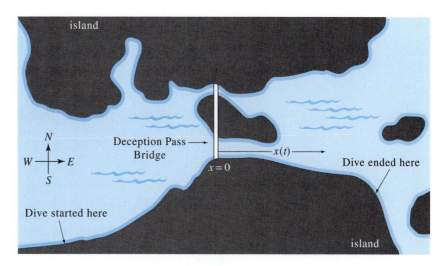

FIGURE 2   Schematic of Deception Pass

PROBLEM 1.   Suppose that the velocity v_0 of the current under the bridge is 7 mph. Convert v_0 to feet per second, and then find the velocity of the current 600 feet to the west of the bridge. Likewise find the velocity of the current 200 feet east of the bridge, and 800 feet east of the bridge.

PROBLEM 2.   Under the assumptions of Problem 1, where is the current fastest?

PROBLEM 3 (CD).   You will learn how to solve differential equations such as (1) in Chapter 2. For the moment, use the **Deception Pass Tool** on the *DE Tools* CD to see how your motion changes depending on where you start. In words, describe your motion if you enter the channel 1000 feet west of the bridge. Describe your motion if you enter 1000 feet east of the bridge.

The results you obtained in Problems 1–3 seem somewhat realistic, but they describe the current only when the tide is at its maximum flow. Now you wonder what the velocity actually was when you were in the channel. How big an error was it to miss slack water by an hour? You want to put the time variation of the tide into your analysis. On the June day that you had your adventure, there was a 13-hour (or $13(60)(60) = 46{,}800$ seconds) interval between successive high tides. Thus you model the tidal current using the sine function $\sin(\pi t/23400)$ because its period is $\dfrac{2\pi}{\pi/23400} = 46{,}800$. Your differential equation becomes

$$x' = \frac{v_0 \sin(\pi t/23400)}{S(x)}. \tag{3}$$

By consulting tide data from NOAA, you find that the maximum velocity of the current under the bridge on the day in question was 8 knots. That's about 9.2 mph, or $v_0 = 13.5$ ft/s. That's fast! Still, you were within an hour of slack water, and you didn't start in the fastest part of the channel.

> **PROBLEM 4.** Suppose that the time $t = 0$ corresponds to slack water. Compute the velocity of the current near your entry point 1000 feet west of the bridge, that is, with $x = -1000$, when you started your dive one hour after slack water. Remember that time is given in seconds.
>
> **PROBLEM 5.** You want to compute your likely position relative to the bridge from the time you entered the water (exactly 1 hour after slack water, at a point 1000 feet west of the bridge). What initial-value problem would you solve to do so?

It is evident that you should have done your homework before you went diving. Next time you go out, you will be armed with better information. Regrettably, you will also be armed with a new dive partner, since it is too late to retrieve your reputation with the other.

STILL CURIOUS?

The results you obtained above do not completely agree with your experience. When you entered the water to the west of the bridge, you were pulled along quickly by the current almost immediately. As you approached the narrows, however, the current became more insistent, and that's what you see in the plot of the solution. On looking at a depth chart of the channel, you notice that the passage to the west of the bridge looks wide but is shallow. Thus the size of the channel that the tide must pass through is deceptively small in that place (no, that is not why it is called Deception Pass). You decide to construct a different function $S(x)$ that better describes the size of the channel west of the bridge. You eventually find that

$$S(x) = 3 - \tfrac{1}{300}|x + 600| - \tfrac{1}{1000}x \tag{4}$$

seems to be a physically reasonable approximation to the channel size over a different range than the quadratic function, namely, from $x = -1200$ to $x = 0$ (the bridge). Substituting (4) into (3), you then find from the new model

$$x' = \frac{v_0 \sin(\pi t/23400)}{3 - \tfrac{1}{300}|x + 600| - \tfrac{1}{1000}x}, \tag{5}$$

that your velocity when you entered the water was about 3.45 ft/s. It seems that you swept along at a nice walking pace almost from the moment you entered the water. By the time you realized what was happening, the tide was stronger, and you were nearer to the narrows. A close encounter with the whirlpool was inevitable.

> **PROBLEM 6.** Show that the model (5) gives opposite signs for the velocity of the current at $x = -300$ and $x = 300$. What does this mean physically? Discuss the physical validity of this DE in view of the sign of the channel-size function $S(x)$.

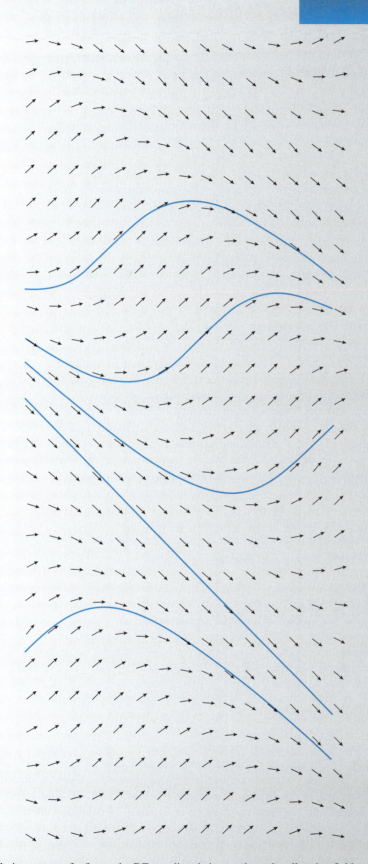

2 FIRST-ORDER DIFFERENTIAL EQUATIONS

The history of mathematics is rife with stories of people who devoted a lot of time to solving equations—algebraic equations at first and then eventually differential equations. We will study some of the more important methods for solving first-order DEs in Sections 2.2–2.5. However, a DE can possess a solution yet there might not exist any analytical method for finding it. In Sections 2.1 and 2.6 we do not solve any DEs but show how to glean information directly from the equation itself. In Section 2.1 we see how the DE yields qualitative information about graphs that enables us to sketch renditions of solution curves. In Section 2.6 we use the differential equation to construct a numerical procedure for approximating solutions.

Solution curves of a first-order DE wending their way through a direction field. See pages 40-41.

2.1   SOLUTION CURVES WITHOUT A SOLUTION

INTRODUCTION: Let us imagine for the moment that we have in front of us a first-order differential equation $dy/dx = f(x, y)$, and let us further imagine that we can neither find nor invent a method for solving it analytically. This is not as bad a predicament as one might think, since the differential equation itself can sometimes "tell" us specifics about how its solutions "behave."

We begin our study of first-order differential equations with two ways of analyzing a DE qualitatively. Both these ways enable us to determine, in an approximate sense, what a solution curve must look like without actually solving the equation.

REVIEW MATERIAL: In the first part of this section we will use the fact that the function f in the normal form of a first-order differential equation $dy/dx = f(x, y)$ is the **slope function** for the unknown solutions. In the second part we will use the property that the algebraic sign of the first derivative dy/dx indicates whether the differentiable function $y(x)$ is increasing or decreasing on an interval.

CD: The **Direction Field Tool** and the **Phase Line Tool** on the *DE Tools* CD can be used in conjunction with the discussion of those topics on pages 41 and 43.

2.1.1   DIRECTION FIELDS

SOME FUNDAMENTAL QUESTIONS   We have seen in Section 1.2 that whenever $f(x, y)$ and $\partial f/\partial y$ satisfy certain continuity conditions, qualitative questions about existence and uniqueness of solutions can be answered. In this section we shall see that other qualitative questions about properties of solutions—How does a solution behave near a certain point? How does a solution behave as $x \to \infty$?—can often be answered when the function f depends solely on the variable y. We begin, however, with a simple concept from calculus: A derivative dy/dx of a differentiable function $y = y(x)$ gives slopes of tangent lines at points on its graph.

SLOPE   Because a solution $y = y(x)$ of a first-order differential equation $dy/dx = f(x, y)$ is necessarily a differentiable function on its interval I of definition, it must also be continuous on I. Thus the corresponding solution curve on I must have no breaks and must possess a tangent line at each point $(x, y(x))$. The slope of the tangent line at $(x, y(x))$ on a solution curve is the value of the first derivative dy/dx at this point, and this we know from the differential equation: $f(x, y(x))$. Now suppose that (x, y) represents any point in a region of the xy-plane over which the function f is defined. The value $f(x, y)$ that the function f assigns to the point represents the slope of a line, or, as we shall envision it, a line segment called a **lineal element.** For example, consider the equation $dy/dx = 0.2xy$, where $f(x, y) = 0.2xy$. At, say, the point $(2, 3)$, the slope of a lineal element is $f(2, 3) = 0.2(2)(3) = 1.2$. Figure 2.1(a) shows a line segment with slope 1.2 passing though $(2, 3)$. As shown in Figure 2.1(b), *if* a solution curve also passes through the point $(2, 3)$, it does so tangent to this line segment; in other words, the lineal element is a miniature tangent line at that point.

DIRECTION FIELD   If we systematically evaluate f over a rectangular grid of points in the xy-plane and draw a lineal element at each point (x, y) of the grid with slope $f(x, y)$, then the collection of all these lineal elements is called a **direction field** or a **slope field** of the differential equation $dy/dx = f(x, y)$. Visually, the direction field suggests the appearance or shape of a family of solution curves of the differential equation, and consequently, it may be possible to see at a glance certain qualitative aspects of the solutions—regions in the plane, for example, in which a solution exhibits an unusual behavior. A single solution curve that passes through a direction field must follow the flow pattern of the field; it is tangent to a lineal element when it intersects a point in the grid.

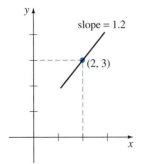

(a) lineal element at a point

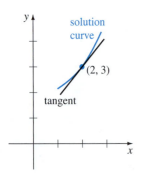

(b) lineal element is tangent to solution curve that passes through the point

FIGURE 2.1   A solution curve is tangent to lineal element at $(2, 3)$.

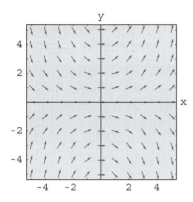

(a) direction field for
$dy/dx = 0.2xy$

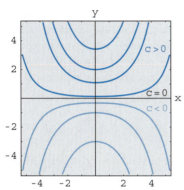

(b) some solution curves in the
family $y = ce^{0.1x^2}$

FIGURE 2.2 Direction field and
solution curves

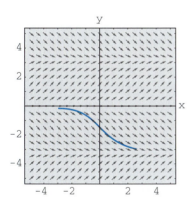

FIGURE 2.3 Direction field for
$dy/dx = \sin y$

EXAMPLE 1   Direction Field

The direction field for the differential equation $dy/dx = 0.2xy$ shown in Figure 2.2(a) was obtained by using computer software in which a 5×5 grid of points (mh, nh), m and n integers, was defined by letting $-5 \le m \le 5$, $-5 \le n \le 5$, and $h = 1$. Notice in Figure 2.2(a) that at any point along the x-axis ($y = 0$) and the y-axis ($x = 0$), the slopes are $f(x, 0) = 0$ and $f(0, y) = 0$, respectively, so the lineal elements are horizontal. Moreover, observe in the first quadrant that for a fixed value of x the values of $f(x, y) = 0.2xy$ increase as y increases; similarly, for a fixed y the values of $f(x, y) = 0.2xy$ increase as x increases. This means that as both x and y increase, the lineal elements almost become vertical and have positive slope ($f(x, y) = 0.2xy > 0$ for $x > 0$, $y > 0$). In the second quadrant, $|f(x, y)|$ increases as $|x|$ and y increase, and so the lineal elements again become almost vertical but this time have negative slope ($f(x, y) = 0.2xy < 0$ for $x < 0$, $y > 0$). Reading from left to right, imagine a solution curve that starts at a point in the second quadrant, moves steeply downward, becomes flat as it passes through the y-axis, and then, as it enters the first quadrant, moves steeply upward—in other words, its shape would be concave upward and similar to a horseshoe. From this it could be surmised that $y \to \infty$ as $x \to \pm\infty$. Now in the third and fourth quadrants, since $f(x, y) = 0.2xy > 0$ and $f(x, y) = 0.2xy < 0$, respectively, the situation is reversed; a solution curve increases and then decreases as we move from left to right. We saw in (1) of Section 1.1 that $y = e^{0.1x^2}$ is an explicit solution of the differential equation $dy/dx = 0.2xy$; you should verify that a one-parameter family of solutions of the same equation is given by $y = ce^{0.1x^2}$. For purposes of comparison with Figure 2.2(a) some representative graphs of members of this family are shown in Figure 2.2(b). ∎

EXAMPLE 2   Direction Field

Use a direction field to sketch an approximate solution curve for the initial-value problem $dy/dx = \sin y$, $y(0) = -\frac{3}{2}$.

SOLUTION   Before proceeding, recall that from the continuity of $f(x, y) = \sin y$ and $\partial f/\partial y = \cos y$, Theorem 1.1 guarantees the existence of a unique solution curve passing through any specified point (x_0, y_0) in the plane. Now we set our computer software again for a 5×5 rectangular region and specify (because of the initial condition) points in that region with vertical and horizontal separation of $\frac{1}{2}$ unit—that is, at points (mh, nh), $h = \frac{1}{2}$, m and n integers such that $-10 \le m \le 10$, $-10 \le n \le 10$. The result is shown in Figure 2.3. Because the right-hand side of $dy/dx = \sin y$ is 0 at $y = 0$ and at $y = -\pi$, the lineal elements are horizontal at all points whose second coordinates are $y = 0$ or $y = -\pi$. It makes sense then that a solution curve passing through the initial point $(0, -\frac{3}{2})$ has the shape shown in color in the figure. ∎

INCREASING/DECREASING   Interpretation of the derivative dy/dx as a function that gives slope plays the key role in the construction of a direction field. Another telling property of the first derivative will be used next, namely, if $dy/dx > 0$ (or $dy/dx < 0$) for all x in an interval I, then a differentiable function $y = y(x)$ is increasing (or decreasing) on I.

REMARKS

Sketching a direction field by hand is straightforward but time consuming; it is probably one of those tasks about which an argument can be made for doing it once or twice in a lifetime, but it is overall most efficiently carried out by means of computer software. Before calculators, PCs, and software the **method of isoclines** was used to facilitate sketching a direction field by hand. For the DE $dy/dx = f(x, y)$, any member of the family of curves $f(x, y) = c$, c a constant, is called an **isocline.** Lineal elements drawn through points on a specific isocline, say, $f(x, y) = c_1$ all have the same slope c_1. In Problem 15 in Exercises 2.1 you have your two opportunities to sketch a direction field by hand.

2.1.2 AUTONOMOUS FIRST-ORDER DEs

AUTONOMOUS FIRST-ORDER DEs In Section 1.1 we divided the class of ordinary differential equations into two types: linear and nonlinear. We now consider briefly another kind of classification of ordinary differential equations, a classification that is of particular importance in the qualitative investigation of differential equations. An ordinary differential equation in which the independent variable does not appear explicitly is said to be **autonomous.** If the symbol x denotes the independent variable, then an autonomous first-order differential equation can be written as $f(y, y') = 0$ or in normal form as

$$\frac{dy}{dx} = f(y). \tag{1}$$

We shall assume throughout that the function f in (1) and its derivative f' are continuous functions of y on some interval I. The first-order equations

$$\overset{\overset{\displaystyle f(y)}{\downarrow}}{\frac{dy}{dx} = 1 + y^2} \quad \text{and} \quad \overset{\overset{\displaystyle f(x, y)}{\downarrow}}{\frac{dy}{dx} = 0.2xy}$$

are autonomous and nonautonomous, respectively.

Many differential equations encountered in applications or equations that are models of physical laws that do not change over time are autonomous. As we have already seen in Section 1.3, in an applied context, symbols other than y and x are routinely used to represent the dependent and independent variables. For example, if t represents time then inspection of

$$\frac{dA}{dt} = kA, \quad \frac{dx}{dt} = kx(n + 1 - x), \quad \frac{dT}{dt} = k(T - T_m), \quad \frac{dA}{dt} = 6 - \frac{1}{100}A,$$

where k, n, and T_m are constants, shows that each equation is time independent. Indeed, *all* of the first-order differential equations introduced in Section 1.3 are time independent and so are autonomous.

CRITICAL POINTS The zeros of the function f in (1) are of special importance. We say that a real number c is a **critical point** of the autonomous differential equation (1) if it is a zero of f—that is, $f(c) = 0$. A critical point is also called an **equilibrium point** or **stationary point.** Now observe that if we substitute the constant function $y(x) = c$ into (1), then both sides of the equation are zero. This means:

> If c is a critical point of (1), then $y(x) = c$ is a constant solution of the autonomous differential equation.

A constant solution $y(x) = c$ of (1) is called an **equilibrium solution;** equilibria are the *only* constant solutions of (1).

As was already mentioned, we can tell when a nonconstant solution $y = y(x)$ of (1) is increasing or decreasing by determining the algebraic sign of the derivative *dy/dx;* in the case of (1) we do this by identifying intervals on the *y*-axis over which the function $f(y)$ is positive or negative.

EXAMPLE 3   An Autonomous DE

The differential equation

$$\frac{dP}{dt} = P(a - bP),$$

where *a* and *b* are positive constants, has the normal form $dP/dt = f(P)$, which is (1) with *t* and *P* playing the parts of *x* and *y*, respectively, and hence is autonomous. From $f(P) = P(a - bP) = 0$ we see that 0 and *a/b* are critical points of the equation, and so the equilibrium solutions are $P(t) = 0$ and $P(t) = a/b$. By putting the critical points on a vertical line, we divide the line into three intervals: $-\infty < P < 0, 0 < P < a/b$, $a/b < P < \infty$. The arrows on the line shown in Figure 2.4 indicate the algebraic sign of $f(P) = P(a - bP)$ on these intervals and whether a nonconstant solution $P(t)$ is increasing or decreasing on an interval. The following table explains the figure.

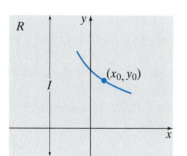

FIGURE 2.4 Phase portrait of *dP/dt* = P(a − bP)

Interval	Sign of $f(P)$	$P(t)$	Arrow
$(-\infty, 0)$	minus	decreasing	points down
$(0, a/b)$	plus	increasing	points up
$(a/b, \infty)$	minus	decreasing	points down

Figure 2.4 is called a **one-dimensional phase portrait,** or simply **phase portrait,** of the differential equation $dP/dt = P(a - bP)$. The vertical line is called a **phase line.**

SOLUTION CURVES   Without solving an autonomous differential equation, we can usually say a great deal about its solution curves. Since the function *f* in (1) is independent of the variable *x*, we may consider *f* defined for $-\infty < x < \infty$ or for $0 \leq x < \infty$. Also, since *f* and its derivative *f'* are continuous functions of *y* on some interval *I* of the *y*-axis, the fundamental results of Theorem 1.1 hold in some horizontal strip or region *R* in the *xy*-plane corresponding to *I*, and so through any point (x_0, y_0) in *R* there passes only one solution curve of (1). See Figure 2.5(a). For the sake of discussion, let us suppose that (1) possesses exactly two critical points c_1 and c_2 and that $c_1 < c_2$. The graphs of the equilibrium solutions $y(x) = c_1$ and $y(x) = c_2$ are horizontal lines, and these lines partition the region *R* into three subregions R_1, R_2, and R_3, as illustrated in Figure 2.5(b). Without proof, here are some conclusions that we can draw about a nonconstant solution $y(x)$ of (1):

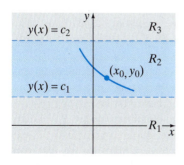

(a) region *R*

(b) subregions R_1, R_2, and R_3 of *R*

FIGURE 2.5 Lines $y(x) = c_1$ and $y(x) = c_2$ partition *R* into three horizontal subregions

- If (x_0, y_0) is in a subregion R_i, *i* = 1, 2, 3, and $y(x)$ is a solution whose graph passes through this point, then $y(x)$ remains in the subregion R_i for all *x*. As illustrated in Figure 2.5(b), the solution $y(x)$ in R_2 is bounded below by c_1 and above by c_2, that is, $c_1 < y(x) < c_2$ for all *x*. The solution curve stays within R_2 for all *x* because the graph of a nonconstant solution of (1) cannot cross the graph of either equilibrium solution $y(x) = c_1$ or $y(x) = c_2$. See Problem 33 in Exercises 2.1.

- By continuity of *f* we must then have either $f(y) > 0$ or $f(y) < 0$ for all *x* in a subregion R_i, *i* = 1, 2, 3. In other words, $f(y)$ cannot change signs in a subregion. See Problem 33 in Exercises 2.1.

- Since $dy/dx = f(y(x))$ is either positive or negative in a subregion R_i, *i* = 1, 2, 3, a solution $y(x)$ is strictly monotonic — that is, $y(x)$ is either increasing or decreasing in the subregion R_i. Therefore $y(x)$ cannot be oscillatory, nor can it have a relative extremum (maximum or minimum). See Problem 33 in Exercises 2.1.

- If $y(x)$ is *bounded above* by a critical point c_1 (as in subregion R_1 where $y(x) < c_1$ for all x), then the graph of $y(x)$ must approach the graph of the equilibrium solution $y(x) = c_1$ either as $x \to \infty$ or as $x \to -\infty$. If $y(x)$ is *bounded*—that is, bounded above and below by two consecutive critical points (as in subregion R_2 where $c_1 < y(x) < c_2$ for all x)—then the graph of $y(x)$ must approach the graphs of the equilibrium solutions $y(x) = c_1$ and $y(x) = c_2$, one as $x \to \infty$ and the other as $x \to -\infty$. If $y(x)$ is *bounded below* by a critical point (as in subregion R_3 where $c_2 < y(x)$ for all x), then the graph of $y(x)$ must approach the graph of the equilibrium solution $y(x) = c_2$ either as $x \to \infty$ or as $x \to -\infty$. See Problem 34 in Exercises 2.1.

With the foregoing facts in mind, let us reexamine the differential equation in Example 3.

EXAMPLE 4   Example 3 Revisited

The three intervals determined on the P-axis or phase line by the critical points $P = 0$ and $P = a/b$ now correspond in the tP-plane to three subregions:

$$R_1: -\infty < P < 0, \quad R_2: 0 < P < a/b, \quad \text{and} \quad R_3: a/b < P < \infty,$$

where $-\infty < t < \infty$. The phase portrait in Figure 2.4 tells us that $P(t)$ is decreasing in R_1, increasing in R_2, and decreasing in R_3. If $P(0) = P_0$ is an initial value, then in R_1, R_2, and R_3 we have, respectively, the following:

(*i*)   For $P_0 < 0$, $P(t)$ is bounded above. Since $P(t)$ is decreasing, $P(t)$ decreases without bound for increasing t, and so $P(t) \to 0$ as $t \to -\infty$. This means that the negative t-axis, the graph of the equilibrium solution $P(t) = 0$, is a horizontal asymptote for a solution curve.

(*ii*)   For $0 < P_0 < a/b$, $P(t)$ is bounded. Since $P(t)$ is increasing, $P(t) \to a/b$ as $t \to \infty$ and $P(t) \to 0$ as $t \to -\infty$. The graphs of the two equilibrium solutions, $P(t) = 0$ and $P(t) = a/b$, are horizontal lines that are horizontal asymptotes for any solution curve starting in this subregion.

(*iii*)   For $P_0 > a/b$, $P(t)$ is bounded below. Since $P(t)$ is decreasing, $P(t) \to a/b$ as $t \to \infty$. The graph of the equilibrium solution $P(t) = a/b$ is a horizontal asymptote for a solution curve.

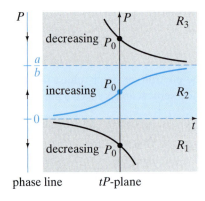

FIGURE 2.6 Phase portrait and solution curves in each of the three subregions

In Figure 2.6 the phase line is the P-axis in the tP-plane. For clarity the original phase line from Figure 2.4 is reproduced to the left of the plane in which the subregions R_1, R_2, and R_3 are shaded. The graphs of the equilibrium solutions $P(t) = a/b$ and $P(t) = 0$ (the t-axis) are shown in the figure as colored dashed lines; the solid graphs represent typical graphs of $P(t)$ illustrating the three cases just discussed.

In a subregion such as R_1 in Example 4, where $P(t)$ is decreasing and unbounded below, we must necessarily have $P(t) \to -\infty$. Do *not* interpret this last statement to mean $P(t) \to -\infty$ as $t \to \infty$; we could have $P(t) \to -\infty$ as $t \to T$, where $T > 0$ is a finite number that depends on the initial condition $P(t_0) = P_0$. Thinking in dynamic terms, $P(t)$ could "blow up" in finite time; thinking graphically, $P(t)$ could have a vertical asymptote at $t = T > 0$. A similar remark holds for the subregion R_3.

The differential equation $dy/dx = \sin y$ in Example 2 is autonomous and has an infinite number of critical points, since $\sin y = 0$ at $y = n\pi$, n an integer. Moreover, we now know that because the solution $y(x)$ that passes through $\left(0, -\frac{3}{2}\right)$ is bounded above and below by two consecutive critical points ($-\pi < y(x) < 0$) and is decreasing ($\sin y < 0$ for $-\pi < y < 0$), the graph of $y(x)$ must approach the graphs of the equilibrium solutions as horizontal asymptotes: $y(x) \rightarrow -\pi$ as $x \rightarrow \infty$ and $y(x) \rightarrow 0$ as $x \rightarrow -\infty$.

EXAMPLE 5   Solution Curves of an Autonomous DE

The autonomous equation $dy/dx = (y - 1)^2$ possesses the single critical point 1. From the phase portrait in Figure 2.7(a) we conclude that a solution $y(x)$ is an increasing function in the subregions $-\infty < y < 1$ and $1 < y < \infty$, where $-\infty < x < \infty$. For an initial condition $y(0) = y_0 < 1$, a solution $y(x)$ is increasing and bounded above by 1, and so $y(x) \rightarrow 1$ as $x \rightarrow \infty$; for $y(0) = y_0 > 1$ a solution $y(x)$ is increasing and unbounded.

Now $y(x) = 1 - 1/(x + c)$ is a one-parameter family of solutions of the differential equation. (See Problem 4 in Exercises 2.2) A given initial condition determines a value for c. For the initial conditions, say, $y(0) = -1 < 1$ and $y(0) = 2 > 1$, we find, in turn, that $y(x) = 1 - 1/\left(x + \frac{1}{2}\right)$, and so $y(x) = 1 - 1/(x - 1)$. As shown in Figures 2.7(b) and 2.7(c), the graph of each of these rational functions possesses a

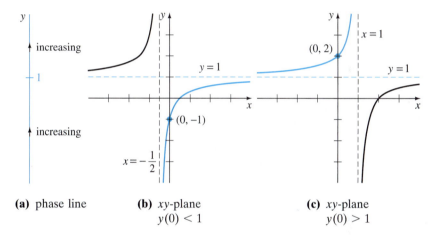

(a) phase line

(b) xy-plane $y(0) < 1$

(c) xy-plane $y(0) > 1$

FIGURE 2.7   Behavior of solutions near $y = 1$

vertical asymptote. But bear in mind that the solutions of the IVPs

$$\frac{dy}{dx} = (y - 1)^2, \quad y(0) = -1 \quad \text{and} \quad \frac{dy}{dx} = (y - 1)^2, \quad y(0) = 2$$

are defined on special intervals. They are, respectively,

$$y(x) = 1 - \frac{1}{x + \frac{1}{2}}, \quad -\frac{1}{2} < x < \infty \quad \text{and} \quad y(x) = 1 - \frac{1}{x - 1}, \quad -\infty < x < 1.$$

The solution curves are the portions of the graphs in Figures 2.7(b) and 2.7(c) shown in solid color. As predicted by the phase portrait, for the solution curve in Figure 2.7(b), $y(x) \rightarrow 1$ as $x \rightarrow \infty$; for the solution curve in Figure 2.7(c), $y(x) \rightarrow \infty$ as $x \rightarrow 1$ from the left.

ATTRACTORS AND REPELLERS   Suppose that $y(x)$ is a nonconstant solution of the autonomous differential equation given in (1) and that c is a critical point of the DE. There are basically three types of behavior $y(x)$ can exhibit near c. In Figure 2.8 we have placed c on four vertical phase lines. When both arrowheads on either side of the dot labeled c point *toward c,* as in Figure 2.8(a), all solutions $y(x)$ of (1) that start from an initial point (x_0, y_0) sufficiently near c exhibit the asymptotic behavior $\lim_{x\to\infty} y(x) = c$. For this reason the critical point c is said to be **asymptotically stable.** Using a physical analogy, a solution that starts near c is like a charged particle

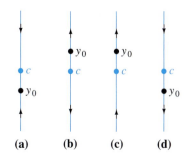

(a)   (b)   (c)   (d)

FIGURE 2.8   Critical point c is: an attractor in (a), a repeller in (b), and semi-stable in (c) and (d).

slopes of lineal
elements on a horizontal
line are all the same

slopes of lineal
elements on a
vertical line vary

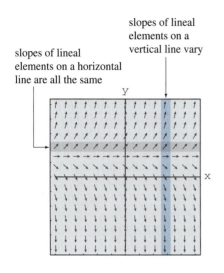

FIGURE 2.9   Direction field for an autonomous DE

that, over time, is drawn to a particle of opposite charge, and so c is also referred to as an **attractor.** When both arrowheads on either side of the dot labeled c point *away* from c, as in Figure 2.8(b), all solutions $y(x)$ of (1) that start from an initial point (x_0, y_0) move away from c as x increases. In this case the critical point c is said to be **unstable.** An unstable critical point is also called a **repeller,** for obvious reasons. The critical point c illustrated in Figures 2.8(c) and 2.8(d) is neither an attractor nor a repeller. But since c exhibits characteristics of both an attractor and a repeller—that is, a solution starting from an initial point (x_0, y_0) sufficiently near c is attracted to c from one side and repelled from the other side—we say that the critical point c is **semi-stable.** In Example 3 the critical point a/b is asymptotically stable (an attractor) and the critical point 0 is unstable (a repeller). The critical point 1 in Example 5 is semi-stable.

AUTONOMOUS DES AND DIRECTION FIELDS   If a first-order differential equation is autonomous, then we see from the right-hand side of its normal form $dy/dx = f(y)$ that slopes of lineal elements through points in the rectangular grid used to construct a direction field for the DE depend solely on the y-coordinate of the points. Put another way, lineal elements passing through points on any *horizontal* line must all have the same slope; slopes of lineal elements along any *vertical* line will, of course, vary. These facts are apparent from inspection of the horizontal gray strip and vertical colored strip in Figure 2.9. The figure exhibits a direction field for the autonomous equation $dy/dx = 2y - 2$. With these facts in mind, reexamine Figure 2.3.

EXERCISES 2.1

Answers to selected odd-numbered problems begin on page ANS-1.

2.1.1   DIRECTION FIELDS

In Problems 1–4 reproduce the given computer-generated direction field. Then sketch, by hand, an approximate solution curve that passes through each of the indicated points. Use different colored pencils for each solution curve.

1. $\dfrac{dy}{dx} = x^2 - y^2$

   (a) $y(-2) = 1$ (b) $y(3) = 0$
   (c) $y(0) = 2$ (d) $y(0) = 0$

2. $\dfrac{dy}{dx} = e^{-0.01xy^2}$

   (a) $y(-6) = 0$ (b) $y(0) = 1$
   (c) $y(0) = -4$ (d) $y(8) = -4$

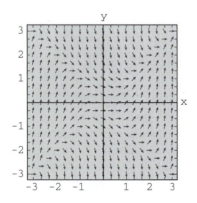

FIGURE 2.10   Direction field for Problem 1

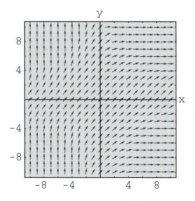

FIGURE 2.11   Direction field for Problem 2

3. $\dfrac{dy}{dx} = 1 - xy$

   (a) $y(0) = 0$ **(b)** $y(-1) = 0$

   (c) $y(2) = 2$ **(d)** $y(0) = -4$

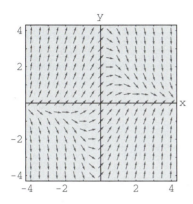

FIGURE 2.12 Direction field for Problem 3

4. $\dfrac{dy}{dx} = (\sin x) \cos y$

   (a) $y(0) = 1$ **(b)** $y(1) = 0$

   (c) $y(3) = 3$ **(d)** $y(0) = -\frac{5}{2}$

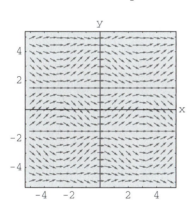

FIGURE 2.13 Direction field for Problem 4

In Problems 5–12 use computer software to obtain a direction field for the given differential equation. By hand, sketch an approximate solution curve passing through each of the given points.

5. $y' = x$

   (a) $y(0) = 0$

   (b) $y(0) = -3$

6. $y' = x + y$

   (a) $y(-2) = 2$

   (b) $y(1) = -3$

7. $y\dfrac{dy}{dx} = -x$

   (a) $y(1) = 1$

   (b) $y(0) = 4$

8. $\dfrac{dy}{dx} = \dfrac{1}{y}$

   (a) $y(0) = 1$

   (b) $y(-2) = -1$

9. $\dfrac{dy}{dx} = 0.2x^2 + y$

   (a) $y(0) = \frac{1}{2}$

   (b) $y(2) = -1$

10. $\dfrac{dy}{dx} = xe^y$

   (a) $y(0) = -2$

   (b) $y(1) = 2.5$

11. $y' = y - \cos\dfrac{\pi}{2}x$

   (a) $y(2) = 2$

   (b) $y(-1) = 0$

12. $\dfrac{dy}{dx} = 1 - \dfrac{y}{x}$

   (a) $y\left(-\frac{1}{2}\right) = 2$

   (b) $y\left(\frac{3}{2}\right) = 0$

In Problems 13 and 14 the given figure represents the graph of $f(y)$ and $f(x)$, respectively. By hand, sketch a direction field over an appropriate grid for $dy/dx = f(y)$ (Problem 13) and then for $dy/dx = f(x)$ (Problem 14).

13.

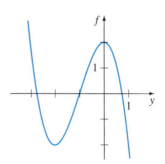

FIGURE 2.14 Graph for Problem 13

14.

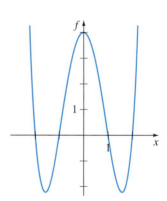

FIGURE 2.15 Graph for Problem 14

15. In parts (a) and (b) sketch **isoclines** $f(x, y) = c$ (see the Remarks on page 42) for the given differential equation using the indicated values of c. Construct a direction field over a grid by carefully drawing lineal elements with the appropriate slope at chosen points on each isocline. In each case, use this rough direction field to sketch an approximate solution curve for the IVP consisting of the DE and the initial condition $y(0) = 1$.

   (a) $dy/dx = x + y$; c an integer satisfying $-5 \le c \le 5$

   (b) $dy/dx = x^2 + y^2$; $c = \frac{1}{4}, c = 1, c = \frac{9}{4}, c = 4$

DISCUSSION/PROJECT PROBLEMS

16. (a) Consider the direction field of the differential equation $dy/dx = x(y - 4)^2 - 2$, but do not use technology to obtain it. Describe the slopes of the lineal elements on the lines $x = 0$, $y = 3$, $y = 4$, and $y = 5$.

(b) Consider the IVP $dy/dx = x(y - 4)^2 - 2$, $y(0) = y_0$, where $y_0 < 4$. Can a solution $y(x) \to \infty$ as $x \to \infty$? Based on the information in part (a), discuss.

17. For a first-order DE $dy/dx = f(x, y)$ a curve in the plane defined by $f(x, y) = 0$ is called a **nullcline** of the equation, since a lineal element at a point on the curve has zero slope. Use computer software to obtain a direction field over a rectangular grid of points for $dy/dx = x^2 - 2y$, and then superimpose the graph of the nullcline $y = \frac{1}{2}x^2$ over the direction field. Discuss the behavior of solution curves in regions of the plane defined by $y < \frac{1}{2}x^2$ and by $y > \frac{1}{2}x^2$. Sketch some approximate solution curves. Try to generalize your observations.

18. **(a)** Identify the nullclines (see Problem 17) in Problems 1, 3, and 4. With a colored pencil, circle any lineal elements in the Figures 2.10, 2.12, and 2.13 that you think may be a lineal element at a point on a nullcline.

(b) What are the nullclines of an autonomous first-order DE?

2.1.2   AUTONOMOUS FIRST-ORDER DEs

19. Consider the autonomous first-order differential equation $dy/dx = y - y^3$ and the initial condition $y(0) = y_0$. By hand, sketch the graph of a typical solution $y(x)$ when y_0 has the given values.

(a) $y_0 > 1$ **(b)** $0 < y_0 < 1$
(c) $-1 < y_0 < 0$ **(d)** $y_0 < -1$

20. Consider the autonomous first-order differential equation $dy/dx = y^2 - y^4$ and the initial condition $y(0) = y_0$. By hand, sketch the graph of a typical solution $y(x)$ when y_0 has the given values.

(a) $y_0 > 1$ **(b)** $0 < y_0 < 1$
(c) $-1 < y_0 < 0$ **(d)** $y_0 < -1$

In Problems 21–28 find the critical points and phase portrait of the given autonomous first-order differential equation. Classify each critical point as asymptotically stable, unstable, or semi-stable. By hand, sketch typical solution curves in the regions in the xy-plane determined by the graphs of the equilibrium solutions.

21. $\dfrac{dy}{dx} = y^2 - 3y$ **22.** $\dfrac{dy}{dx} = y^2 - y^3$

23. $\dfrac{dy}{dx} = (y - 2)^4$ **24.** $\dfrac{dy}{dx} = 10 + 3y - y^2$

25. $\dfrac{dy}{dx} = y^2(4 - y^2)$ **26.** $\dfrac{dy}{dx} = y(2 - y)(4 - y)$

27. $\dfrac{dy}{dx} = y \ln(y + 2)$ **28.** $\dfrac{dy}{dx} = \dfrac{ye^y - 9y}{e^y}$

In Problems 29 and 30 consider the autonomous differential equation $dy/dx = f(y)$, where the graph of f is given. Use the graph to locate the critical points of each differential equation. Sketch a phase portrait of each differential equation. By hand, sketch typical solution curves in the subregions in the xy-plane determined by the graphs of the equilibrium solutions.

29.

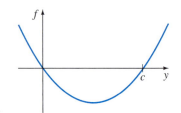

FIGURE 2.16   Graph for Problem 29

30.

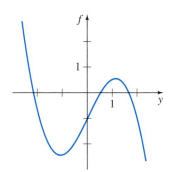

FIGURE 2.17   Graph for Problem 30

DISCUSSION/PROJECT PROBLEMS

31. Consider the autonomous DE $dy/dx = (2/\pi)y - \sin y$. Determine the critical points of the equation. Discuss a way of obtaining a phase portrait of the equation. Classify the critical points as asymptotically stable, unstable, or semi-stable.

32. A critical point c of an autonomous first-order DE is said to be **isolated** if there exists some open interval that contains c but no other critical point. Can there exist an autonomous DE of the form given in (1) for which *every* critical point is nonisolated? Discuss; do not think profound thoughts.

33. Suppose that $y(x)$ is a nonconstant solution of the autonomous equation $dy/dx = f(y)$ and that c is a critical point of the DE. Discuss. Why can't the graph of $y(x)$ cross the graph of the equilibrium solution $y = c$? Why can't $f(y)$ change signs in one of the subregions discussed on page 43? Why can't $y(x)$ be oscillatory or have a relative extremum (maximum or minimum)?

34. Suppose that $y(x)$ is a solution of the autonomous equation $dy/dx = f(y)$ and is bounded above and below by two consecutive critical points $c_1 < c_2$, as in subregion R_2 of Figure 2.5(b). If $f(y) > 0$ in the region, then $\lim_{x \to \infty} y(x) = c_2$. Discuss why there cannot exist a number $L < c_2$ such that $\lim_{x \to \infty} y(x) = L$. As part of your discussion, consider what happens to $y'(x)$ as $x \to \infty$.

35. Using the autonomous equation (1), discuss how it is possible to obtain information about the location of points of inflection of a solution curve.

36. Consider the autonomous DE $dy/dx = y^2 - y - 6$. Use your ideas from Problem 35 to find intervals on the y-axis for which solution curves are concave up and intervals for which solution curves are concave down. Discuss why *each* solution curve of an initial-value problem of the form $dy/dx = y^2 - y - 6$, $y(0) = y_0$, where $-2 < y_0 < 3$, has a point of inflection with the same y-coordinate. What is that y-coordinate? Carefully sketch the solution curve for which $y(0) = -1$. Repeat for $y(2) = 2$.

37. Suppose the autonomous DE in (1) has no critical points. Discuss the behavior of the solutions.

MATHEMATICAL MODELS

38. Population Model The differential equation in Example 3 is a well-known population model. Suppose the DE is changed to

$$\frac{dP}{dt} = P(aP - b),$$

where a and b are positive constants. Discuss what happens to the population P as time t increases.

39. Terminal Velocity The autonomous differential equation

$$m\frac{dv}{dt} = mg - kv,$$

where k is a positive constant of proportionality and g is the acceleration due to gravity, is a model for the velocity v of a body of mass m that is falling under the influence of gravity. Because the term $-kv$ represents air resistance, the velocity of a body falling from a great height does not increase without bound as time t increases.

(a) Use a phase portrait of the differential equation to find the limiting, or terminal, velocity of the body. Explain your reasoning.

(b) Find the terminal velocity of the body if air resistance is proportional to v^2. See pages 25 and 30.

40. Chemical Reactions When certain kinds of chemicals are combined, the rate at which the new compound is formed is modeled by the autonomous differential equation

$$\frac{dX}{dt} = k(\alpha - X)(\beta - X),$$

where $k > 0$ is a constant of proportionality and $\beta > \alpha > 0$. Here $X(t)$ denotes the number of grams of the new compound formed in time t. See page 22.

(a) Use a phase portrait of the differential equation to predict the behavior of X as $t \to \infty$.

(b) Consider the case when $\alpha = \beta$. Use a phase portrait of the differential equation to predict the behavior of X as $t \to \infty$ when $X(0) < \alpha$. When $X(0) > \alpha$.

(c) Verify that an explicit solution of the DE in the case when $k = 1$ and $\alpha = \beta$ is $X(t) = \alpha - 1/(t + c)$. Find a solution satisfying $X(0) = \alpha/2$. Find a solution satisfying $X(0) = 2\alpha$. Graph these two solutions. Does the behavior of the solutions as $t \to \infty$ agree with your answers to part (b)?

2.2 SEPARABLE VARIABLES

INTRODUCTION: We begin our study of how to solve differential equations with the simplest of all differential equations: first-order equations with separable variables.

REVIEW MATERIAL: The method studied in this section as well as many of the techniques for solving differential equations involve integration. So it may be worth your time to review basic integration formulas (such as $\int u^n \, du$ and $\int du/u$) and techniques of integration (such as integration by parts and partial fraction decomposition) from a calculus text or, if you use a CAS, the command syntax for integration. A brief review of some of these topics can also be found in the *Student Resource and Solutions Manual*.

SOLUTION BY INTEGRATION   Consider the first-order differential equation $dy/dx = f(x, y)$. When f does not depend on the variable y, that is, $f(x, y) = g(x)$, the differential equation

$$\frac{dy}{dx} = g(x) \tag{1}$$

can be solved by integration. If $g(x)$ is a continuous function, then integrating both sides of (1) gives $y = \int g(x)\, dx = G(x) + c$, where $G(x)$ is an antiderivative (indefinite integral) of $g(x)$. For example, if $dy/dx = 1 + e^{2x}$, then its solution is $y = \int (1 + e^{2x})dx$ or $y = x + \frac{1}{2}e^{2x} + c$.

A DEFINITION   Equation (1), as well as its method of solution, is just a special case when the function f in the normal form $dy/dx = f(x, y)$ can be factored into a function of x times a function of y.

DEFINITION 2.1   **Separable Equation**

A first-order differential equation of the form

$$\frac{dy}{dx} = g(x)h(y)$$

is said to be **separable** or to have **separable variables.**

For example, the equations

$$\frac{dy}{dx} = y^2 x e^{3x+4y} \quad \text{and} \quad \frac{dy}{dx} = y + \sin x$$

are separable and nonseparable, respectively. In the first equation we can factor $f(x, y) = y^2 x e^{3x+4y}$ as

$$f(x, y) \;=\; y^2 x e^{3x+4y} \;=\; \underset{\underset{g(x)}{\downarrow}}{(x e^{3x})}\underset{\underset{h(y)}{\downarrow}}{(y^2 e^{4y})},$$

but in the second equation there is no way of expressing $y + \sin x$ as a product of a function of x times a function of y.

Observe that by dividing by the function $h(y)$, we can write a separable equation $dy/dx = g(x)h(y)$ as

$$p(y)\frac{dy}{dx} = g(x), \tag{2}$$

where, for convenience, we have denoted $1/h(y)$ by $p(y)$. From this last form we can see immediately that (2) reduces to (1) when $h(y) = 1$.

Now if $y = \phi(x)$ represents a solution of (2), we must have $p(\phi(x))\phi'(x) = g(x)$, and therefore

$$\int p(\phi(x))\phi'(x)\, dx = \int g(x)\, dx. \tag{3}$$

But $dy = \phi'(x)\, dx$, and so (3) is the same as

$$\int p(y)\, dy = \int g(x)\, dx \quad \text{or} \quad H(y) = G(x) + c, \tag{4}$$

where $H(y)$ and $G(x)$ are antiderivatives of $p(y) = 1/h(y)$ and $g(x)$, respectively.

METHOD OF SOLUTION   Equation (4) indicates the procedure for solving separable equations. A one-parameter family of solutions, usually given implicitly, is obtained by integrating both sides of $p(y)\, dy = g(x)\, dx$.

NOTE There is no need to use two constants in the integration of a separable equation, because if we write $H(y) + c_1 = G(x) + c_2$, then the difference $c_2 - c_1$ can be replaced by a single constant c, as in (4). In many instances throughout the chapters that follow, we will relabel constants in a manner convenient to a given equation. For example, multiples of constants or combinations of constants can sometimes be replaced by a single constant.

EXAMPLE 1   Solving a Separable DE

Solve $(1 + x)\,dy - y\,dx = 0$.

SOLUTION Dividing by $(1 + x)y$, we can write $dy/y = dx/(1 + x)$, from which it follows that

$$\int \frac{dy}{y} = \int \frac{dx}{1 + x}$$

$$\ln|y| = \ln|1 + x| + c_1$$

$$y = e^{\ln|1+x|+c_1} = e^{\ln|1+x|} \cdot e^{c_1} \qquad \leftarrow \text{laws of exponents}$$

$$= |1 + x|\, e^{c_1}$$

$$= \pm e^{c_1}(1 + x). \qquad \leftarrow \begin{cases} |1 + x| = 1 + x, & x \geq -1 \\ |1 + x| = -(1 + x), & x < -1 \end{cases}$$

Relabeling $\pm e^{c_1}$ as c then gives $y = c(1 + x)$.

ALTERNATIVE SOLUTION Because each integral results in a logarithm, a judicious choice for the constant of integration is $\ln|c|$ rather than c. Rewriting the second line of the solution as $\ln|y| = \ln|1 + x| + \ln|c|$ enables us to combine the terms on the right-hand side by the properties of logarithms. From $\ln|y| = \ln|c(1 + x)|$ we immediately get $y = c(1 + x)$. Even if the indefinite integrals are not *all* logarithms, it may still be advantageous to use $\ln|c|$. However, no firm rule can be given.

In Section 1.1 we saw that a solution curve may be only a segment or an arc of the graph of an implicit solution $G(x, y) = 0$.

EXAMPLE 2   Solution Curve

Solve the initial-value problem $\dfrac{dy}{dx} = -\dfrac{x}{y}, \quad y(4) = -3$.

SOLUTION Rewriting the equation as $y\,dy = -x\,dx$, we get

$$\int y\,dy = -\int x\,dx \quad \text{and} \quad \frac{y^2}{2} = -\frac{x^2}{2} + c_1.$$

We can write the result of the integration as $x^2 + y^2 = c^2$ by replacing the constant $2c_1$ by c^2. This solution of the differential equation represents a family of concentric circles centered at the origin.

Now when $x = 4$, $y = -3$, so $16 + 9 = 25 = c^2$. Thus the initial-value problem determines the circle $x^2 + y^2 = 25$ with radius 5. Because of its simplicity we can solve this implicit solution for an explicit solution that satisfies the initial condition. We have seen this solution as $y = \phi_2(x)$ or $y = -\sqrt{25 - x^2}, -5 < x < 5$ in Example 3 of Section 1.1. A solution curve is the graph of a differentiable function. In this case the solution curve is the lower semicircle, shown in color in Figure 2.18, containing the point $(4, -3)$.

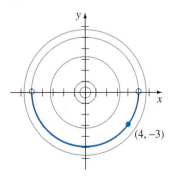

FIGURE 2.18 Solution curve for the IVP in Example 2

LOSING A SOLUTION   Some care should be exercised in separating variables, since the variable divisors could be zero at a point. Specifically, if r is a zero of the function $h(y)$, then substituting $y = r$ into $dy/dx = g(x)h(y)$ makes both sides zero; in other words, $y = r$ is a constant solution of the differential equation. But after variables are separated, the left-hand side of $\dfrac{dy}{h(y)} = g(x)\,dx$ is undefined at r. As a consequence, $y = r$ might not show up in the family of solutions obtained after integration and simplification. Recall that such a solution is called a singular solution.

EXAMPLE 3   Losing a Solution

Solve $\dfrac{dy}{dx} = y^2 - 4$.

SOLUTION   We put the equation in the form

$$\frac{dy}{y^2 - 4} = dx \quad \text{or} \quad \left[\frac{\frac{1}{4}}{y - 2} - \frac{\frac{1}{4}}{y + 2} \right] dy = dx. \tag{5}$$

The second equation in (5) is the result of using partial fractions on the left-hand side of the first equation. Integrating and using the laws of logarithms gives

$$\frac{1}{4} \ln|y - 2| - \frac{1}{4} \ln|y + 2| = x + c_1$$

$$\text{or} \quad \ln\left| \frac{y - 2}{y + 2} \right| = 4x + c_2 \quad \text{or} \quad \frac{y - 2}{y + 2} = \pm e^{4x + c_2}.$$

Here we have replaced $4c_1$ by c_2. Finally, after replacing $\pm e^{c_2}$ by c and solving the last equation for y, we get the one-parameter family of solutions

$$y = 2\,\frac{1 + ce^{4x}}{1 - ce^{4x}}. \tag{6}$$

Now if we factor the right-hand side of the differential equation as $dy/dx = (y - 2)(y + 2)$, we know from the discussion of critical points in Section 2.1 that $y = 2$ and $y = -2$ are two constant (equilibrium) solutions. The solution $y = 2$ is a member of the family of solutions defined by (6) corresponding to the value $c = 0$. However, $y = -2$ is a singular solution; it cannot be obtained from (6) for any choice of the parameter c. This latter solution was lost early on in the solution process. Inspection of (5) clearly indicates that we must preclude $y = \pm 2$ in these steps. ▪

EXAMPLE 4   An Initial-Value Problem

Solve $(e^{2y} - y) \cos x\, \dfrac{dy}{dx} = e^y \sin 2x, \quad y(0) = 0$.

SOLUTION   Dividing the equation by $e^y \cos x$ gives

$$\frac{e^{2y} - y}{e^y}\, dy = \frac{\sin 2x}{\cos x}\, dx.$$

Before integrating, we use termwise division on the left-hand side and the trigonometric identity $\sin 2x = 2 \sin x \cos x$ on the right-hand side. Then

integration by parts → $\displaystyle \int (e^y - ye^{-y})\, dy = 2 \int \sin x\, dx$

yields $e^y + ye^{-y} + e^{-y} = -2 \cos x + c. \tag{7}$

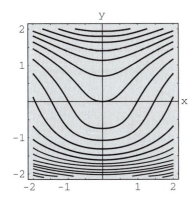

FIGURE 2.19 Level curves $G(x, y) = c$, where $G(x, y) = e^y + ye^{-y} + e^{-y} + 2 \cos x$

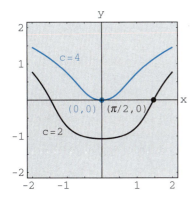

FIGURE 2.20 Level curves $c = 2$ and $c = 4$

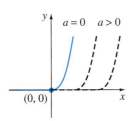

FIGURE 2.21 Piecewise-defined solutions of (9)

The initial condition $y = 0$ when $x = 0$ implies $c = 4$. Thus a solution of the initial-value problem is

$$e^y + ye^{-y} + e^{-y} = 4 - 2 \cos x. \qquad (8) \qquad \blacksquare$$

USE OF COMPUTERS The Remarks at the end of Section 1.1 mentioned that it may be difficult to use an implicit solution $G(x, y) = 0$ to find an explicit solution $y = \phi(x)$. Equation (8) shows that the task of solving for y in terms of x may present more problems than just the drudgery of symbol pushing—sometimes it simply cannot be done! Implicit solutions such as (8) are somewhat frustrating; neither the graph of the equation nor an interval over which a solution satisfying $y(0) = 0$ is defined is apparent. The problem of "seeing" what an implicit solution looks like can be overcome in some cases by means of technology. One way* of proceeding is to use the contour plot application of a CAS. Recall from multivariate calculus that for a function of two variables $z = G(x, y)$ the *two-dimensional* curves defined by $G(x, y) = c$, where c is constant, are called the *level curves* of the function. With the aid of a CAS, some of the level curves of the function $G(x, y) = e^y + ye^{-y} + e^{-y} + 2 \cos x$ have been reproduced in Figure 2.19. The family of solutions defined by (7) is the level curves $G(x, y) = c$. Figure 2.20 illustrates the level curve $G(x, y) = 4$, which is the particular solution (8), in solid color. The other curve in Figure 2.20 is the level curve $G(x, y) = 2$, which is the member of the family $G(x, y) = c$ that satisfies $y(\pi/2) = 0$.

If an initial condition leads to a particular solution by yielding a specific value of the parameter c in a family of solutions for a first-order differential equation, there is a natural inclination for most students (and instructors) to relax and be content. However, a solution of an initial-value problem may not be unique. We saw in Example 4 of Section 1.2 that the initial-value problem

$$\frac{dy}{dx} = xy^{1/2}, \quad y(0) = 0 \qquad (9)$$

has at least two solutions, $y = 0$ and $y = \frac{1}{16}x^4$. We are now in a position to solve the equation. Separating variables and integrating $y^{-1/2}\, dy = x\, dx$ gives

$$2y^{1/2} = \frac{x^2}{2} + c_1 \quad \text{or} \quad y = \left(\frac{x^2}{4} + c\right)^2.$$

When $x = 0$, then $y = 0$, and so necessarily, $c = 0$. Therefore $y = \frac{1}{16}x^4$. The trivial solution $y = 0$ was lost by dividing by $y^{1/2}$. In addition, the initial-value problem (9) possesses infinitely many more solutions, since for any choice of the parameter $a \geq 0$ the piecewise-defined function

$$y = \begin{cases} 0, & x < a \\ \frac{1}{16}(x^2 - a^2)^2, & x \geq a \end{cases}$$

satisfies both the differential equation and the initial condition. See Figure 2.21.

REMARKS

(*i*) If g is a function continuous on an interval I containing x_0, then from the fundamental theorem of calculus we have

$$\frac{d}{dx} \int_{x_0}^{x} g(t)\, dt = g(x).$$

---

*In Section 2.6 we will discuss several other ways of proceeding that are based on the concept of a numerical solver.

In other words, $\int_{x_0}^{x} g(t)\, dt$ is an antiderivative of the function g. There are times when this form is convenient. For example, if g is continuous on an interval I containing x_0, then a solution of the simple initial-value problem $dy/dx = g(x)$, $y(x_0) = y_0$ that is defined on I is given by

$$y(x) = y_0 + \int_{x_0}^{x} g(t)\, dt.$$

You should verify this. Since an antiderivative of a continuous function cannot always be expressed in terms of elementary functions, this might be the best we can do in obtaining an explicit solution of an IVP. For example, e^{-x^2} has no elementary antiderivative, and so a solution of the initial-value problem $dy/dx = e^{-x^2}$, $y(3) = 5$, is $y(x) = 5 + \int_{3}^{x} e^{-t^2}\, dt$.

(*ii*) In some of the preceding examples we saw that the constant in the one-parameter family of solutions for a first-order differential equation can be relabeled when convenient. Also, it can easily happen that two individuals solving the same equation correctly, arrive at dissimilar expressions for their answers. For example, by separation of variables we can show that one-parameter families of solutions for the DE $(1 + y^2)dx + (1 + x^2)dy = 0$ are

$$\arctan x + \arctan y = c \quad \text{or} \quad \frac{x + y}{1 - xy} = c.$$

As you work your way through the next several sections, keep in mind that families of solutions may be equivalent in the sense that one family may be obtained from another by either relabeling the constant or applying algebra and trigonometry. See Problems 27 and 28 in Exercises 2.2.

EXERCISES 2.2

Answers to selected odd-numbered problems begin on page ANS-1.

In Problems 1–22 solve the given differential equation by separation of variables.

1. $\dfrac{dy}{dx} = \sin 5x$

2. $\dfrac{dy}{dx} = (x + 1)^2$

3. $dx + e^{3x}dy = 0$

4. $dy - (y - 1)^2 dx = 0$

5. $x\dfrac{dy}{dx} = 4y$

6. $\dfrac{dy}{dx} + 2xy^2 = 0$

7. $\dfrac{dy}{dx} = e^{3x+2y}$

8. $e^x y \dfrac{dy}{dx} = e^{-y} + e^{-2x-y}$

9. $y \ln x \dfrac{dx}{dy} = \left(\dfrac{y + 1}{x}\right)^2$

10. $\dfrac{dy}{dx} = \left(\dfrac{2y + 3}{4x + 5}\right)^2$

11. $\csc y\, dx + \sec^2 x\, dy = 0$

12. $\sin 3x\, dx + 2y \cos^3 3x\, dy = 0$

13. $(e^y + 1)^2 e^{-y}\, dx + (e^x + 1)^3 e^{-x}\, dy = 0$

14. $x(1 + y^2)^{1/2}\, dx = y(1 + x^2)^{1/2}\, dy$

15. $\dfrac{dS}{dr} = kS$

16. $\dfrac{dQ}{dt} = k(Q - 70)$

17. $\dfrac{dP}{dt} = P - P^2$

18. $\dfrac{dN}{dt} + N = Nte^{t+2}$

19. $\dfrac{dy}{dx} = \dfrac{xy + 3x - y - 3}{xy - 2x + 4y - 8}$

20. $\dfrac{dy}{dx} = \dfrac{xy + 2y - x - 2}{xy - 3y + x - 3}$

21. $\dfrac{dy}{dx} = x\sqrt{1 - y^2}$

22. $(e^x + e^{-x})\dfrac{dy}{dx} = y^2$

In Problems 23–28 find an implicit and an explicit solution of the given initial-value problem.

23. $\dfrac{dx}{dt} = 4(x^2 + 1)$, $x(\pi/4) = 1$

24. $\dfrac{dy}{dx} = \dfrac{y^2 - 1}{x^2 - 1}, \quad y(2) = 2$

25. $x^2 \dfrac{dy}{dx} = y - xy, \quad y(-1) = -1$

26. $\dfrac{dy}{dt} + 2y = 1, \quad y(0) = \frac{5}{2}$

27. $\sqrt{1 - y^2}\, dx - \sqrt{1 - x^2}\, dy = 0, \quad y(0) = \dfrac{\sqrt{3}}{2}$

28. $(1 + x^4)dy + x(1 + 4y^2)dx = 0, \quad y(1) = 0$

29. (a) Find a solution of the initial-value problem consisting of the differential equation in Example 3 and the initial conditions $y(0) = 2$, $y(0) = -2$, and $y\!\left(\frac{1}{4}\right) = 1$.

(b) Find the solution of the differential equation in Example 4 when $\ln c_1$ is used as the constant of integration on the *left-hand* side in the solution and $4 \ln c_1$ is replaced by $\ln c$. Then solve the same initial-value problems in part (a).

30. Find a solution of $x\dfrac{dy}{dx} = y^2 - y$ that passes through the indicated points.

(a) $(0, 1)$ **(b)** $(0, 0)$ **(c)** $\left(\frac{1}{2}, \frac{1}{2}\right)$ **(d)** $\left(2, \frac{1}{4}\right)$

31. Find a singular solution of Problem 21. Of Problem 22.

32. Show that an implicit solution of

$$2x \sin^2 y\, dx - (x^2 + 10)\cos y\, dy = 0$$

is given by $\ln(x^2 + 10) + \csc y = c$. Find the constant solutions, if any, that were lost in the solution of the differential equation.

Often a radical change in the form of the solution of a differential equation corresponds to a very small change in either the initial condition or the equation itself. In Problems 33–36 find an explicit solution of the given initial-value problem. Use a graphing utility to plot the graph of each solution. Compare each solution curve in a neighborhood of $(0, 1)$.

33. $\dfrac{dy}{dx} = (y - 1)^2, \quad y(0) = 1$

34. $\dfrac{dy}{dx} = (y - 1)^2, \quad y(0) = 1.01$

35. $\dfrac{dy}{dx} = (y - 1)^2 + 0.01, \quad y(0) = 1$

36. $\dfrac{dy}{dx} = (y - 1)^2 - 0.01, \quad y(0) = 1$

37. Every autonomous first-order equation $dy/dx = f(y)$ is separable. Find explicit solutions $y_1(x)$, $y_2(x)$, $y_3(x)$, and $y_4(x)$ of the differential equation $dy/dx = y - y^3$ that satisfy, in turn, the initial conditions $y_1(0) = 2$, $y_2(0) = \frac{1}{2}$, $y_3(0) = -\frac{1}{2}$, and $y_4(0) = -2$. Use a graphing

utility to plot the graphs of each solution. Compare these graphs with those predicted in Problem 19 of Exercises 2.1. Give the exact interval of definition for each solution.

38. (a) The autonomous first-order differential equation $dy/dx = 1/(y - 3)$ has no critical points. Nevertheless, place 3 on the phase line and obtain a phase portrait of the equation. Compute d^2y/dx^2 to determine where solution curves are concave up and where they are concave down (see Problems 35 and 36 in Exercises 2.1). Use the phase portrait and concavity to sketch, by hand, some typical solution curves.

(b) Find explicit solutions $y_1(x)$, $y_2(x)$, $y_3(x)$, and $y_4(x)$ of the differential equation in part (a) that satisfy, in turn, the initial conditions $y_1(0) = 4$, $y_2(0) = 2$, $y_3(1) = 2$, and $y_4(-1) = 4$. Graph each solution and compare with your sketches in part (a). Give the exact interval of definition for each solution.

39. (a) Find an explicit solution of the initial-value problem

$$\dfrac{dy}{dx} = \dfrac{2x + 1}{2y}, \quad y(-2) = -1.$$

(b) Use a graphing utility to plot the graph of the solution in part (a). Use the graph to estimate the interval I of definition of the solution.

(c) Determine the exact interval I of definition by analytical methods.

40. Repeat parts (a)–(c) of Problem 39 for the IVP consisting of the differential equation in Problem 7 and the initial condition $y(0) = 0$.

DISCUSSION/PROJECT PROBLEMS

41. (a) Explain why the interval of definition of the explicit solution $y = \phi_2(x)$ of the initial-value problem in Example 2 is the *open* interval $-5 < x < 5$.

(b) Can any solution of the differential equation cross the x-axis? Do you think that $x^2 + y^2 = 1$ is an implicit solution of the initial-value problem $dy/dx = -x/y$, $y(1) = 0$?

42. (a) If $a > 0$, discuss the differences, if any, between the solutions of the initial-value problems consisting of the differential equation $dy/dx = x/y$ and each of the initial conditions $y(a) = a$, $y(a) = -a$, $y(-a) = a$, and $y(-a) = -a$.

(b) Does the initial-value problem $dy/dx = x/y$, $y(0) = 0$ have a solution?

(c) Solve $dy/dx = x/y$, $y(1) = 2$ and give the exact interval I of definition of its solution.

43. In Problems 37 and 38 we saw that every autonomous first-order differential equation $dy/dx = f(y)$ is separable. Does this fact help in the solution of the initial-value problem $\dfrac{dy}{dx} = \sqrt{1 + y^2}\,\sin^2 y, \quad y(0) = \tfrac{1}{2}$? Discuss. Sketch, by hand, a plausible solution curve of the problem.

44. Without the use of technology, how would you solve

$$\left(\sqrt{x} + x\right)\dfrac{dy}{dx} = \sqrt{y} + y?$$

Carry out your ideas.

45. Find a function whose square plus the square of its derivative is 1.

46. (a) The differential equation in Problem 27 is equivalent to the normal form

$$\dfrac{dy}{dx} = \sqrt{\dfrac{1 - y^2}{1 - x^2}}$$

in the square region in the xy-plane defined by $|x| < 1, |y| < 1$. But the quantity under the radical is nonnegative also in the regions defined by $|x| > 1$, $|y| > 1$. Sketch all regions in the xy-plane for which this differential equation possesses real solutions.

(b) Solve the DE in part (a) in the regions defined by $|x| > 1, |y| > 1$. Then find an implicit and an explicit solution of the differential equation subject to $y(2) = 2$.

MATHEMATICAL MODEL

47. Suspension Bridge   In (17) of Section 1.3 we saw that a mathematical model for the shape of a flexible cable strung between two vertical supports is

$$\dfrac{dy}{dx} = \dfrac{W}{T_1}, \tag{10}$$

where W denotes the portion of the total vertical load between the points P_1 and P_2 shown in Figure 1.26. The DE (10) is separable under the following conditions that describe a suspension bridge.

Let us assume that the x- and y-axes are as shown in Figure 2.22—that is, the x-axis runs along the horizontal roadbed, and the y-axis passes through $(0, a)$, which is the lowest point on one cable over the span of the bridge, coinciding with the interval $[-L/2, L/2]$. In the case of a suspension bridge, the usual assumption is that the vertical load in (10) is only a uniform roadbed distributed along the horizontal axis. In other words, it is assumed that the weight of all cables is negligible in comparison to the weight of the roadbed and that the weight per unit length of the roadbed (say, pounds per horizontal foot) is a constant ρ. Use this information to

set up and solve an appropriate initial-value problem from which the shape (a curve with equation $y = \phi(x)$) of each of the two cables in a suspension bridge is determined. Express your solution of the IVP in terms of the sag h and span L. See Figure 2.22.

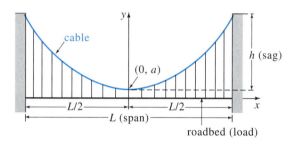

FIGURE 2.22   Shape of a cable in Problem 47

COMPUTER LAB ASSIGNMENTS

48. (a) Use a CAS and the concept of level curves to plot representative graphs of members of the family of solutions of the differential equation $\dfrac{dy}{dx} = -\dfrac{8x + 5}{3y^2 + 1}$. Experiment with different numbers of level curves as well as various rectangular regions defined by $a \leq x \leq b, \ c \leq y \leq d$.

(b) On separate coordinate axes plot the graphs of the particular solutions corresponding to the initial conditions: $y(0) = -1; \quad y(0) = 2; \quad y(-1) = 4;$ $y(-1) = -3$.

49. (a) Find an implicit solution of the IVP

$$(2y + 2)dy - (4x^3 + 6x)dx = 0, \quad y(0) = -3.$$

(b) Use part (a) to find an explicit solution $y = \phi(x)$ of the IVP.

(c) Consider your answer to part (b) as a *function* only. Use a graphing utility or a CAS to graph this function, and then use the graph to estimate its domain.

(d) With the aid of a root-finding application of a CAS, determine the approximate largest interval I of definition of the *solution* $y = \phi(x)$ in part (b). Use a graphing utility or a CAS to graph the solution curve for the IVP on this interval.

50. (a) Use a CAS and the concept of level curves to plot representative graphs of members of the family of solutions of the differential equation $\dfrac{dy}{dx} = \dfrac{x(1 - x)}{y(-2 + y)}$. Experiment with different numbers of level curves as well as various rectangular regions in the xy-plane until your result resembles Figure 2.23.

(b) On separate coordinate axes, plot the graph of the implicit solution corresponding to the initial condition $y(0) = \frac{3}{2}$. Use a colored pencil to mark off that segment of the graph that corresponds to the solution curve of a solution ϕ that satisfies the initial condition. With the aid of a root-finding application of a CAS, determine the approximate largest interval I of definition of the solution ϕ. (Hint: First find the points on the curve in part (a) where the tangent is vertical.)

(c) Repeat part (b) for the initial condition $y(0) = -2$.

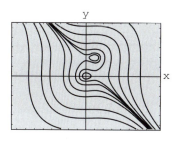

FIGURE 2.23   Level curves in Problem 50

2.3   LINEAR EQUATIONS

INTRODUCTION: We continue our quest for solutions of first-order DEs by next examining linear equations. Linear differential equations are an especially "friendly" family of differential equations in that, given a linear equation, whether first-order or a higher-order kin, there is always a good possibility that we can find some sort of solution of the equation that we can look at.

REVIEW MATERIAL: The general form of a linear nth-order differential equation was given in (6) of Section 1.1. If x and y denote independent and dependent variables, respectively, then the characteristics of a linear equation are as follows: y and all its derivatives are of the first degree, and the coefficients either are constants or depend on x but not on y.

A DEFINITION   The form of a linear first-order DE was given in (7) of Section 1.1. This form, the case when $n = 1$ in (6) of that section, is reproduced here for convenience.

DEFINITION 2.2   **Linear Equation**

A first-order differential equation of the form

$$a_1(x)\frac{dy}{dx} + a_0(x)y = g(x) \qquad (1)$$

is said to be a **linear equation** in the dependent variable y.

When $g(x) = 0$, the linear equation (1) is said to be **homogeneous;** otherwise, it is **nonhomogeneous.**

STANDARD FORM   By dividing both sides of (1) by the lead coefficient $a_1(x)$, we obtain a more useful form, the **standard form,** of a linear equation:

$$\frac{dy}{dx} + P(x)y = f(x). \qquad (2)$$

We seek a solution of (2) on an interval I for which both coefficient functions P and f are continuous.

In the discussion that follows we illustrate a property and a procedure and end up with a formula representing the form that every solution of (2) must have. But more than the formula, the property and the procedure are important, because these two concepts carry over to linear equations of higher order.

THE PROPERTY   The differential equation (2) has the property that its solution is the **sum** of the two solutions: $y = y_c + y_p$, where y_c is a solution of the associated homogeneous equation

$$\frac{dy}{dx} + P(x)y = 0 \tag{3}$$

and y_p is a particular solution of the nonhomogeneous equation (2). To see this, observe that

$$\frac{d}{dx}\,[y_c + y_p] + P(x)[y_c + y_p] = \underbrace{\left[\frac{dy_c}{dx} + P(x)y_c\right]}_{0} + \underbrace{\left[\frac{dy_p}{dx} + P(x)y_p\right]}_{f(x)} = f(x).$$

Now the homogeneous equation (3) is also separable. This fact enables us to find y_c by writing (3) as

$$\frac{dy}{y} + P(x)\,dx = 0$$

and integrating. Solving for y gives $y_c = ce^{-\int P(x)dx}$. For convenience let us write $y_c = cy_1(x)$, where $y_1 = e^{-\int P(x)dx}$. The fact that $dy_1/dx + P(x)y_1 = 0$ will be used next to determine y_p.

THE PROCEDURE   We can now find a particular solution of equation (2) by a procedure known as **variation of parameters.** The basic idea here is to find a function u so that $y_p = u(x)y_1(x) = u(x)e^{-\int P(x)dx}$ is a solution of (2). In other words, our assumption for y_p is the same as $y_c = cy_1(x)$ except that c is replaced by the "variable parameter" u. Substituting $y_p = uy_1$ into (2) gives

$$\overset{\text{Product Rule}}{\downarrow}$$

$$u\frac{dy_1}{dx} + y_1\frac{du}{dx} + P(x)uy_1 = f(x) \quad \text{or} \quad u\overset{\text{zero}}{\underset{\downarrow}{\left[\frac{dy_1}{dx} + P(x)y_1\right]}} + y_1\frac{du}{dx} = f(x)$$

so

$$y_1\frac{du}{dx} = f(x).$$

Separating variables and integrating then gives

$$du = \frac{f(x)}{y_1(x)}\,dx \quad \text{and} \quad u = \int \frac{f(x)}{y_1(x)}\,dx.$$

Since $y_1(x) = e^{-\int P(x)dx}$, we see that $1/y_1(x) = e^{\int P(x)dx}$. Therefore

$$y_p = uy_1 = \left(\int \frac{f(x)}{y_1(x)}\,dx\right)e^{-\int P(x)dx} = e^{-\int P(x)dx}\int e^{\int P(x)dx}f(x)\,dx,$$

and

$$y = \underbrace{ce^{-\int P(x)dx}}_{y_c} + \underbrace{e^{-\int P(x)dx}\int e^{\int P(x)dx}f(x)\,dx}_{y_p}. \tag{4}$$

Hence if (2) has a solution, it must be of form (4). Conversely, it is a straightforward exercise in differentiation to verify that (4) constitutes a one-parameter family of solutions of equation (2).

You should not memorize the formula given in (4). *However,* you should remember the special term

$$e^{\int P(x)dx} \tag{5}$$

because it is used in an equivalent but easier way of solving (2). If equation (4) is multiplied by (5),

$$e^{\int P(x)dx}y = c + \int e^{\int P(x)dx}f(x)\,dx, \tag{6}$$

and then (6) is differentiated,

$$\frac{d}{dx}[e^{\int P(x)dx}y] = e^{\int P(x)dx}f(x), \tag{7}$$

we get

$$e^{\int P(x)dx}\frac{dy}{dx} + P(x)e^{\int P(x)dx}y = e^{\int P(x)dx}f(x). \tag{8}$$

Dividing the last result by $e^{\int P(x)dx}$ gives (2).

METHOD OF SOLUTION   The recommended method of solving (2) actually consists of (6)–(8) worked in reverse order. In other words, if (2) is multiplied by (5), we get (8). The left-hand side of (8) is recognized as the derivative of the product of $e^{\int P(x)dx}$ and y. This gets us to (7). We then integrate both sides of (7) to get the solution (6). Because we can solve (2) by integration after multiplication by $e^{\int P(x)dx}$, we call this function an **integrating factor** for the differential equation.* For convenience we summarize these results. We again emphasize that you should not memorize formula (4) but work through the following procedure each time.

Solving a Linear First-Order Equation

(*i*)   Put a linear equation of form (1) into the standard form (2).

(*ii*)   From the standard form identify $P(x)$ and then find the integrating factor $e^{\int P(x)dx}$.

(*iii*)   Multiply the standard form of the equation by the integrating factor. The left-hand side of the resulting equation is automatically the derivative of the integrating factor and y:

$$\frac{d}{dx}[e^{\int P(x)dx}y] = e^{\int P(x)dx}f(x).$$

(*iv*)   Integrate both sides of this last equation.

EXAMPLE 1   Solving a Homogeneous Linear DE

Solve $\dfrac{dy}{dx} - 3y = 0$.

SOLUTION   This linear equation can be solved by separation of variables. Alternatively, since the equation is already in the standard form (2), we see that $P(x) = -3$, and so the integrating factor is $e^{\int(-3)dx} = e^{-3x}$. We multiply the equation by this factor and recognize that

$$e^{-3x}\frac{dy}{dx} - 3e^{-3x}y = 0 \quad \text{is the same as} \quad \frac{d}{dx}[e^{-3x}y] = 0.$$

*This integrating factor can be derived by an alternative procedure discussed in Section 2.4.

Integrating both sides of the last equation gives $e^{-3x}y = c$. Solving for y gives us the explicit solution $y = ce^{3x}$, $-\infty < x < \infty$.

EXAMPLE 2   Solving a Nonhomogeneous Linear DE

Solve $\dfrac{dy}{dx} - 3y = 6$.

SOLUTION   The associated homogeneous equation for this DE was solved in Example 1. Again the equation is already in the standard form (2), and the integrating factor is still $e^{\int(-3)dx} = e^{-3x}$. This time multiplying the given equation by this factor gives

$$e^{-3x}\frac{dy}{dx} - 3e^{-3x}y = 6e^{-3x}, \quad \text{which is the same as} \quad \frac{d}{dx}[e^{-3x}y] = 6e^{-3x}.$$

Integrating both sides of the last equation gives $e^{-3x}y = -2e^{-3x} + c$ or $y = -2 + ce^{3x}$, $-\infty < x < \infty$.

The final solution in Example 2 is the sum of two solutions: $y = y_c + y_p$, where $y_c = ce^{3x}$ is the solution of the homogeneous equation in Example 1 and $y_p = -2$ is a particular solution of the nonhomogeneous equation $y' - 3y = 6$. You need not be concerned about whether a linear first-order equation is homogeneous or nonhomogeneous; when you follow the solution procedure outlined above, a solution of a nonhomogeneous equation necessarily turns out to be $y = y_c + y_p$. However, the distinction between solving a homogeneous DE and solving a nonhomogeneous DE becomes more important in Chapter 4, where we solve linear higher-order equations.

When a_1, a_0, and g in (1) are constants, the differential equation is autonomous. In Example 2 you can verify from the normal form $dy/dx = 3(y + 2)$ that -2 is a critical point and that it is unstable (a repeller). Thus a solution curve with an initial point either above or below the graph of the equilibrium solution $y = -2$ pushes away from this horizontal line as x increases. Figure 2.24, obtained with the aid of a graphing utility, shows the graph of $y = -2$ along with some additional solution curves.

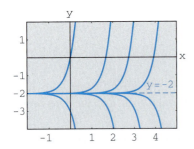

FIGURE 2.24   Some solutions of $y' - 3y = 6$

CONSTANT OF INTEGRATION   Notice that in the general discussion and in Examples 1 and 2 we disregarded a constant of integration in the evaluation of the indefinite integral in the exponent of $e^{\int P(x)dx}$. If you think about the laws of exponents and the fact that the integrating factor multiplies both sides of the differential equation, you should be able to explain why writing $\int P(x)\,dx + c$ is unnecessary. See Problem 44 in Exercises 2.3.

GENERAL SOLUTION   Suppose again that the functions P and f in (2) are continuous on a common interval I. In the steps leading to (4) we showed that *if* (2) has a solution on I, then it must be of the form given in (4). Conversely, it is a straightforward exercise in differentiation to verify that any function of the form given in (4) is a solution of the differential equation (2) on I. In other words, (4) is a one-parameter family of solutions of equation (2) and *every solution of (2) defined on I is a member of this family*. Therefore we call (4) the **general solution** of the differential equation on the interval I. (See the Remarks at the end of Section 1.1.) Now by writing (2) in the normal form $y' = F(x, y)$, we can identify $F(x, y) = -P(x)y + f(x)$ and $\partial F/\partial y = -P(x)$. From the continuity of P and f on the interval I we see that F and $\partial F/\partial y$ are also continuous on I. With Theorem 1.1 as our justification, we conclude that there exists one and only one solution of the initial-value problem

$$\frac{dy}{dx} + P(x)y = f(x), \quad y(x_0) = y_0 \tag{9}$$

defined on *some* interval I_0 containing x_0. But when x_0 is in I, finding a solution of (9) is just a matter of finding an appropriate value of c in (4)—that is, to each x_0 in I there corresponds a distinct c. In other words, the interval I_0 of existence and uniqueness in Theorem 1.1 for the initial-value problem (9) is the entire interval I.

EXAMPLE 3   General Solution

Solve $x\dfrac{dy}{dx} - 4y = x^6 e^x$.

SOLUTION   Dividing by x, we get the standard form

$$\frac{dy}{dx} - \frac{4}{x}y = x^5 e^x. \tag{10}$$

From this form we identify $P(x) = -4/x$ and $f(x) = x^5 e^x$ and further observe that P and f are continuous on $(0, \infty)$. Hence the integrating factor is

we can use ln x instead of ln |x| since x > 0
↓

$$e^{-4\int dx/x} = e^{-4\ln x} = e^{\ln x^{-4}} = x^{-4}.$$

Here we have used the basic identity $b^{\log_b N} = N, N > 0$. Now we multiply (10) by x^{-4} and rewrite

$$x^{-4}\frac{dy}{dx} - 4x^{-5}y = xe^x \quad \text{as} \quad \frac{d}{dx}[x^{-4}y] = xe^x.$$

It follows from integration by parts that the general solution defined on the interval $(0, \infty)$ is $x^{-4}y = xe^x - e^x + c$ or $y = x^5 e^x - x^4 e^x + cx^4$. ∎

Except in the case when the lead coefficient is 1, the recasting of equation (1) into the standard form (2) requires division by $a_1(x)$. Values of x for which $a_1(x) = 0$ are called **singular points** of the equation. Singular points are potentially troublesome. Specifically, in (2), if $P(x)$ (formed by dividing $a_0(x)$ by $a_1(x)$) is discontinuous at a point, the discontinuity may carry over to solutions of the differential equation.

EXAMPLE 4   General Solution

Find the general solution of $(x^2 - 9)\dfrac{dy}{dx} + xy = 0$.

SOLUTION   We write the differential equation in standard form

$$\frac{dy}{dx} + \frac{x}{x^2 - 9}y = 0 \tag{11}$$

and identify $P(x) = x/(x^2 - 9)$. Although P is continuous on $(-\infty, -3)$, $(-3, 3)$, and $(3, \infty)$, we shall solve the equation on the first and third intervals. On these intervals the integrating factor is

$$e^{\int x\,dx/(x^2-9)} = e^{\frac{1}{2}\int 2x\,dx/(x^2-9)} = e^{\frac{1}{2}\ln|x^2-9|} = \sqrt{x^2 - 9}.$$

After multiplying the standard form (11) by this factor, we get

$$\frac{d}{dx}\left[\sqrt{x^2 - 9}\, y\right] = 0.$$

Integrating both sides of the last equation gives $\sqrt{x^2 - 9}\, y = c$. Thus for either $x > 3$ or $x < -3$ the general solution of the equation is $y = \dfrac{c}{\sqrt{x^2 - 9}}$.

Notice in the preceding example that $x = 3$ and $x = -3$ are singular points of the equation and that every function in the general solution $y = c/\sqrt{x^2 - 9}$ is discontinuous at these points. On the other hand, $x = 0$ is a singular point of the differential equation in Example 3, but the general solution $y = x^5 e^x - x^4 e^x + cx^4$ is noteworthy in that every function in this one-parameter family is continuous at $x = 0$ and is defined on the interval $(-\infty, \infty)$ and not just on $(0, \infty)$, as stated in the solution. However, the family $y = x^5 e^x - x^4 e^x + cx^4$ defined on $(-\infty, \infty)$ cannot be considered the general solution of the DE, since the singular point $x = 0$ still causes a problem. See Problem 39 in Exercises 2.3.

EXAMPLE 5   An Initial-Value Problem

Solve $\dfrac{dy}{dx} + y = x, \quad y(0) = 4$.

SOLUTION The equation is in standard form, and $P(x) = 1$ and $f(x) = x$ are continuous on $(-\infty, \infty)$. The integrating factor is $e^{\int dx} = e^x$, and so integrating

$$\frac{d}{dx}[e^x y] = xe^x$$

gives $e^x y = xe^x - e^x + c$. Solving this last equation for y yields the general solution $y = x - 1 + ce^{-x}$. But from the initial condition we know that $y = 4$ when $x = 0$. Substituting these values into the general solution implies $c = 5$. Hence the solution of the problem is

$$y = x - 1 + 5e^{-x}, \quad -\infty < x < \infty. \tag{12}$$

Figure 2.25, obtained with the aid of a graphing utility, shows the graph of (12) in solid color, along with the graphs of other representative solutions in the one-parameter family $y = x - 1 + ce^{-x}$. In this general solution we identify $y_c = ce^{-x}$ and $y_p = x - 1$. It is interesting to observe that as x increases, the graphs of *all* members of the family are close to the graph of the particular solution $y_p = x - 1$, which is shown in solid black in Figure 2.25. This is because the contribution of $y_c = ce^{-x}$ to the values of a solution becomes negligible for increasing values of x. We say that $y_c = ce^{-x}$ is a **transient term**, since $y_c \to 0$ as $x \to \infty$. While this behavior is not a characteristic of all general solutions of linear equations (see Example 2), the notion of a transient is often important in applied problems.

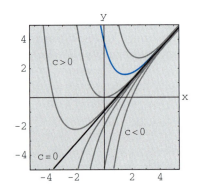

FIGURE 2.25 Some solutions of $y' + y = x$

DISCONTINUOUS COEFFICIENTS In applications the coefficients $P(x)$ and $f(x)$ in (2) may be piecewise continuous. In the next example $f(x)$ is piecewise continuous on $[0, \infty)$ with a single discontinuity, namely, a (finite) jump discontinuity at $x = 1$. We solve the problem in two parts corresponding to the two intervals over which f is defined. It is then possible to piece together the two solutions at $x = 1$ so that $y(x)$ is continuous on $[0, \infty)$.

EXAMPLE 6   An Initial-Value Problem

Solve $\dfrac{dy}{dx} + y = f(x)$, $y(0) = 0$ where $f(x) = \begin{cases} 1, & 0 \le x \le 1, \\ 0, & x > 1. \end{cases}$

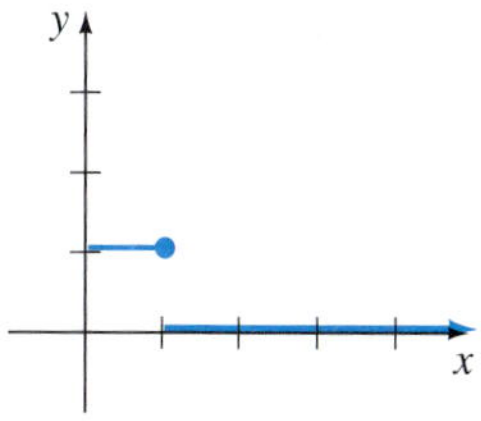

SOLUTION   The graph of the discontinuous function f is shown in Figure 2.26. We solve the DE for $y(x)$ first on the interval $0 \le x \le 1$ and then on the interval $1 < x < \infty$. For $0 \le x \le 1$ we have

$$\frac{dy}{dx} + y = 1 \quad \text{or, equivalently,} \quad \frac{d}{dx}[e^x y] = e^x.$$

Integrating this last equation and solving for y gives $y = 1 + c_1 e^{-x}$. Since $y(0) = 0$, we must have $c_1 = -1$, and therefore $y = 1 - e^{-x}$, $0 \le x \le 1$. Then for $x > 1$, the equation

$$\frac{dy}{dx} + y = 0$$

leads to $y = c_2 e^{-x}$. Hence we can write

$$y = \begin{cases} 1 - e^{-x}, & 0 \le x \le 1, \\ c_2 e^{-x}, & x > 1. \end{cases}$$

By appealing to the definition of continuity at a point it is possible to determine c_2 so that the foregoing function is continuous at $x = 1$. The requirement that $\lim_{x \to 1^+} y(x) = y(1)$ implies that $c_2 e^{-1} = 1 - e^{-1}$ or $c_2 = e - 1$. As seen in Figure 2.27, the function

$$y = \begin{cases} 1 - e^{-x}, & 0 \le x \le 1, \\ (e - 1)e^{-x}, & x > 1. \end{cases} \tag{13}$$

is continuous on $(0, \infty)$.

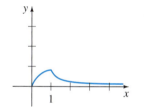

FIGURE 2.26   Discontinuous $f(x)$

FIGURE 2.27   Graph of function in (13)

It is worthwhile to think about (13) and Figure 2.27 a little bit; you are urged to read and answer Problem 42 in Exercises 2.3.

FUNCTIONS DEFINED BY INTEGRALS   Some simple functions do not possess antiderivatives that are elementary functions, and integrals of these kinds of functions are called **nonelementary.** For example, you may have seen in calculus that $\int e^{x^2} dx$ and $\int \sin x^2 dx$ are nonelementary integrals. In applied mathematics some important functions are *defined* in terms of nonelementary integrals. Two such **special functions** are the **error function** and the **complementary error function:**

$$\text{erf}(x) = \frac{2}{\sqrt{\pi}} \int_0^x e^{-t^2} dt \quad \text{and} \quad \text{erfc}(x) = \frac{2}{\sqrt{\pi}} \int_x^\infty e^{-t^2} dt. \tag{14}$$

Since $\left(2/\sqrt{\pi}\right) \int_0^\infty e^{-t^2} dt = 1$, it is seen from (14) that the error function $\text{erf}(x)$ and the complementary error function $\text{erfc}(x)$ are related by $\text{erf}(x) + \text{erfc}(x) = 1$. Because of its importance in areas such as probability and statistics, the error function has been extensively tabulated. Note that $\text{erf}(0) = 0$ is one obvious functional value. Values of $\text{erf}(x)$ can also be found using a CAS. Before working through the next example, you are urged to reread (*i*) of the *Remarks* at the end of Section 2.2.

EXAMPLE 7   **The Error Function**

Solve the initial-value problem $\dfrac{dy}{dx} - 2xy = 2$, $y(0) = 1$.

SOLUTION   Since the equation is already in standard form, we see that the integrating factor is $e^{-x^2}\,dx$, and so from

$$\frac{d}{dx}[e^{-x^2}y] = 2e^{-x^2} \quad \text{we get} \quad y = 2e^{x^2}\int_0^x e^{-t^2}\,dt + ce^{x^2}. \tag{15}$$

Applying $y(0) = 1$ to the last expression then gives $c = 1$. Hence the solution of the problem is

$$y = 2e^{x^2}\int_0^x e^{-t^2}\,dt + e^{x^2} \quad \text{or} \quad y = e^{x^2}\left[1 + \sqrt{\pi}\,\text{erf}(x)\right].$$

The graph of this solution on the interval $(-\infty, \infty)$, shown in color in Figure 2.28 among other members of the family defined in (15), was obtained with the aid of a computer algebra system.

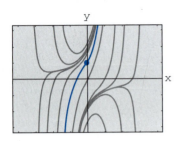

FIGURE 2.28   Some solutions of $y' - 2xy = 2$

USE OF COMPUTERS   The computer algebra systems *Mathematica* and *Maple* are capable of producing implicit or explicit solutions for some kinds of differential equations using their *dsolve* commands.*

REMARKS

(*i*) In general, a linear DE of any order is said to be homogeneous when $g(x) = 0$ in (6) of Section 1.1. For example, the linear second-order DE $y'' - 2y' + 6y = 0$ is homogeneous. As can be seen in this example and in the special case (3) of this section, the trivial solution $y = 0$ is always a solution of a homogeneous linear DE.

(*ii*) Occasionally, a first-order differential equation is not linear in one variable but is linear in the other variable. For example, the differential equation

$$\frac{dy}{dx} = \frac{1}{x + y^2}$$

is not linear in the variable y. But its reciprocal

$$\frac{dx}{dy} = x + y^2 \quad \text{or} \quad \frac{dx}{dy} - x = y^2$$

is recognized as linear in the variable x. You should verify that the integrating factor $e^{\int(-1)dy} = e^{-y}$ and integration by parts yield the explicit solution $x = -y^2 - 2y - 2 + ce^y$ for the second equation. This expression is, then, an implicit solution of the first equation.

(*iii*) Mathematicians have "adopted" as their own certain words from engineering, which they found appropriately descriptive. The word *transient*, used earlier, is one of these terms. In future discussions the words *input* and

---

*Certain commands have the same spelling, but in *Mathematica* commands begin with a capital letter (**Dsolve**), whereas in *Maple* the same command begins with a lower case letter (**dsolve**). When discussing such common syntax, we compromise and write, for example, *dsolve*. See the *Student Resource and Solutions Manual* for the complete input commands used to solve a linear first-order DE.

$F_1 \rightarrow 8$

$P(x) = -\frac{1}{x}$

$Q(x) = \dfrac{(x^2 \sin x)}{x}$

output will occasionally pop up. The function *f* in (2) is called the **input** or **driving function;** a solution *y(x)* of the differential equation for a given input is called the **output** or **response.**

(*iv*) The term **special functions** mentioned in conjunction with the error function also applies to the **sine integral function** and the **Fresnel sine integral** introduced in Problems 49 and 50 in Exercises 2.3. "Special Functions" is actually a well-defined branch of mathematics. More special functions are studied in Section 6.3.

EXERCISES 2.3

Answers to selected odd-numbered problems begin on page ANS-2.

In Problems 1–24 find the general solution of the given differential equation. Give the largest interval *I* over which the general solution is defined. Determine whether there are any transient terms in the general solution.

1. $\dfrac{dy}{dx} = 5y$ **2.** $\dfrac{dy}{dx} + 2y = 0$

3. $\dfrac{dy}{dx} + y = e^{3x}$ **4.** $3\dfrac{dy}{dx} + 12y = 4$

5. $y' + 3x^2 y = x^2$ **6.** $y' + 2xy = x^3$

7. $x^2 y' + xy = 1$ **8.** $y' = 2y + x^2 + 5$

9. $x\dfrac{dy}{dx} - y = x^2 \sin x$ **10.** $x\dfrac{dy}{dx} + 2y = 3$

11. $x\dfrac{dy}{dx} + 4y = x^3 - x$ **12.** $(1 + x)\dfrac{dy}{dx} - xy = x + x^2$

13. $x^2 y' + x(x + 2)y = e^x$

14. $xy' + (1 + x)y = e^{-x}\sin 2x$

15. $y\,dx - 4(x + y^6)\,dy = 0$

16. $y\,dx = (ye^y - 2x)\,dy$

17. $\cos x\dfrac{dy}{dx} + (\sin x)y = 1$

18. $\cos^2 x \sin x\dfrac{dy}{dx} + (\cos^3 x)y = 1$

19. $(x + 1)\dfrac{dy}{dx} + (x + 2)y = 2xe^{-x}$

20. $(x + 2)^2\dfrac{dy}{dx} = 5 - 8y - 4xy$

21. $\dfrac{dr}{d\theta} + r\sec\theta = \cos\theta$

22. $\dfrac{dP}{dt} + 2tP = P + 4t - 2$

23. $x\dfrac{dy}{dx} + (3x + 1)y = e^{-3x}$

24. $(x^2 - 1)\dfrac{dy}{dx} + 2y = (x + 1)^2$

In Problems 25–30 solve the given initial-value problem. Give the largest interval *I* over which the solution is defined.

25. $xy' + y = e^x$, $y(1) = 2$

26. $y\dfrac{dx}{dy} - x = 2y^2$, $y(1) = 5$

27. $L\dfrac{di}{dt} + Ri = E$, $i(0) = i_0$,

L, R, E, and i_0 constants

28. $\dfrac{dT}{dt} = k(T - T_m)$; $T(0) = T_0$,

k, T_m, and T_0 constants

29. $(x + 1)\dfrac{dy}{dx} + y = \ln x$, $y(1) = 10$

30. $y' + (\tan x)y = \cos^2 x$, $y(0) = -1$

In Problems 31–34 proceed as in Example 6 to solve the given initial-value problem. Use a graphing utility to graph the continuous function *y(x)*.

31. $\dfrac{dy}{dx} + 2y = f(x)$, $y(0) = 0$, where

$$f(x) = \begin{cases} 1, & 0 \le x \le 3 \\ 0, & x > 3 \end{cases}$$

32. $\dfrac{dy}{dx} + y = f(x)$, $y(0) = 1$, where

$$f(x) = \begin{cases} 1, & 0 \le x \le 1 \\ -1, & x > 1 \end{cases}$$

33. $\dfrac{dy}{dx} + 2xy = f(x)$, $y(0) = 2$, where

$$f(x) = \begin{cases} x, & 0 \le x < 1 \\ 0, & x \ge 1 \end{cases}$$

34. $(1 + x^2)\dfrac{dy}{dx} + 2xy = f(x)$, $y(0) = 0$, where

$$f(x) = \begin{cases} x, & 0 \le x < 1 \\ -x, & x \ge 1 \end{cases}$$

35. Proceed in a manner analogous to Example 6 to solve the initial-value problem $y' + P(x)y = 4x$, $y(0) = 3$, where

$$P(x) = \begin{cases} 2, & 0 \le x \le 1, \\ -2/x, & x > 1. \end{cases}$$

Use a graphing utility to graph the continuous function $y(x)$.

36. Consider the initial-value problem $y' + e^x y = f(x)$, $y(0) = 1$. Express the solution of the IVP for $x > 0$ as a nonelementary integral when $f(x) = 1$. What is the solution when $f(x) = 0$? When $f(x) = e^x$?

37. Express the solution of the initial-value problem $y' - 2xy = 1$, $y(1) = 1$, in terms of erf(x).

DISCUSSION/PROJECT PROBLEMS

38. Reread the discussion following Example 2. Construct a linear first-order differential equation for which all nonconstant solutions approach the horizontal asymptote $y = 4$ as $x \to \infty$.

39. Reread Example 3 and then discuss, with reference to Theorem 1.1, the existence and uniqueness of a solution of the initial-value problem consisting of $xy' - 4y = x^6 e^x$ and the given initial condition.
(a) $y(0) = 0$ (b) $y(0) = y_0$, $y_0 > 0$
(c) $y(x_0) = y_0$, $x_0 > 0$, $y_0 > 0$

40. Reread Example 4 and then find the general solution of the differential equation on the interval $(-3, 3)$.

41. Reread the discussion following Example 5. Construct a linear first-order differential equation for which all solutions are asymptotic to the line $y = 3x - 5$ as $x \to \infty$.

42. Reread Example 6 and then discuss why it is technically incorrect to say that the function in (13) is a solution of the IVP on the interval $[0, \infty)$.

43. (a) Construct a linear first-order differential equation of the form $xy' + a_0(x)y = g(x)$ for which $y_c = c/x^3$ and $y_p = x^3$. Give an interval on which $y = x^3 + c/x^3$ is the general solution of the DE.
(b) Give an initial condition $y(x_0) = y_0$ for the DE found in part (a) so that the solution of the IVP is $y = x^3 - 1/x^3$. Repeat if the solution is

$y = x^3 + 2/x^3$. Give an interval I of definition of each of these solutions. Graph the solution curves. Is there an initial-value problem whose solution is defined on $(-\infty, \infty)$?
(c) Is each IVP found in part (b) unique? That is, can there be more than one IVP for which, say, $y = x^3 - 1/x^3$, x in some interval I, is the solution?

44. In determining the integrating factor (5), we did not use a constant of integration in the evaluation of $\int P(x)\, dx$. Explain why using $\int P(x)\, dx + c$ has no effect on the solution of (2).

45. Suppose $P(x)$ is continuous on some interval I and a is a number in I. What can be said about the solution of the initial-value problem $y' + P(x)y = 0$, $y(a) = 0$?

MATHEMATICAL MODELS

46. Radioactive Decay Series The following system of differential equations is encountered in the study of the decay of a special type of radioactive series of elements:

$$\frac{dx}{dt} = -\lambda_1 x,$$

$$\frac{dy}{dt} = \lambda_1 x - \lambda_2 y,$$

where λ_1 and λ_2 are constants. Discuss how to solve this system subject to $x(0) = x_0$, $y(0) = y_0$. Carry out your ideas.

47. Heart Pacemaker A heart pacemaker consists of a switch, a battery of constant voltage E_0, a capacitor with constant capacitance C, and the heart as a resistor with constant resistance R. When the switch is closed, the capacitor charges; when the switch is open, the capacitor discharges, sending an electrical stimulus to the heart. During the time the heart is being stimulated, the voltage E across the heart satisfies the linear differential equation

$$\frac{dE}{dt} = -\frac{1}{RC}E.$$

Solve the DE subject to $E(4) = E_0$.

COMPUTER LAB ASSIGNMENTS

48. (a) Express the solution of the initial-value problem $y' - 2xy = -1$, $y(0) = \sqrt{\pi}/2$, in terms of erfc(x).
(b) Use tables or a CAS to find the value of $y(2)$. Use a CAS to graph the solution curve for the IVP on $(-\infty, \infty)$.

49. (a) The **sine integral function** is defined by $\text{Si}(x) = \displaystyle\int_0^x \frac{\sin t}{t}\, dt$, where the integrand is defined to be 1 at $t = 0$. Express the solution $y(x)$ of the initial-value problem $x^3y' + 2x^2y = 10 \sin x$, $y(1) = 0$ in terms of Si(x).

(b) Use a CAS to graph the solution curve for the IVP for $x > 0$.

(c) Use a CAS to find the value of the absolute maximum of the solution $y(x)$ for $x > 0$.

50. (a) The **Fresnel sine integral** is defined by $S(x) = \int_0^x \sin\left(\frac{\pi}{2}t^2\right) dt$. Express the solution $y(x)$

of the initial-value problem $y' - (\sin x^2)y = 0$, $y(0) = 5$, in terms of $S(x)$.

(b) Use a CAS to graph the solution curve for the IVP on $(-\infty, \infty)$.

(c) It is known that $S(x) \to \frac{1}{2}$ as $x \to \infty$ and $S(x) \to -\frac{1}{2}$ as $x \to -\infty$. What does the solution $y(x)$ approach as $x \to \infty$? As $x \to -\infty$?

(d) Use a CAS to find the values of the absolute maximum and the absolute minimum of the solution $y(x)$.

2.4 EXACT EQUATIONS

INTRODUCTION: Although the simple first-order equation $y\,dx + x\,dy = 0$ is separable, we can solve the equation in an alternative manner by recognizing that the expression on the left-hand side of the equality is the differential of the function $f(x, y) = xy$; that is, $d(xy) = y\,dx + x\,dy$. In this section we examine first-order equations in differential form $M(x, y)\,dx + N(x, y)\,dy = 0$. By applying a simple test to M and N, we can determine whether $M(x, y)\,dx + N(x, y)\,dy$ is a differential of a function $f(x, y)$. If the answer is yes, we can construct f by partial integration.

REVIEW MATERIAL: In the discussion that follows we assume knowledge of partial differentiation and partial integration. A brief review of these topics can be found in the *Student Resource and Solutions Manual*. For a more extensive coverage of multivariate calculus please consult any calculus text.

DIFFERENTIAL OF A FUNCTION OF TWO VARIABLES If $z = f(x, y)$ is a function of two variables with continuous first partial derivatives in a region R of the xy-plane, then its differential is

$$dz = \frac{\partial f}{\partial x}dx + \frac{\partial f}{\partial y}dy. \tag{1}$$

In the special case when $f(x, y) = c$, where c is a constant, then (1) implies

$$\frac{\partial f}{\partial x}dx + \frac{\partial f}{\partial y}dy = 0. \tag{2}$$

In other words, given a one-parameter family of functions $f(x, y) = c$, we can generate a first-order differential equation by computing the differential of both sides of the equality. For example, if $x^2 - 5xy + y^3 = c$, then (2) gives the first-order DE

$$(2x - 5y)\,dx + (-5x + 3y^2)\,dy = 0. \tag{3}$$

A DEFINITION Of course, not every first-order DE written in differential form $M(x, y)\,dx + N(x, y)\,dy = 0$ corresponds to a differential of $f(x, y) = c$. So for our purposes it is more important to turn the foregoing example around; namely, if we are given a first-order DE such as (3), is there some way we can recognize that the differential expression $(2x - 5y)\,dx + (-5x + 3y^2)\,dy$ is the differential $d(x^2 - 5xy + y^3)$? If there is, then an implicit solution of (3) is $x^2 - 5xy + y^3 = c$. We answer this question after the next definition.

DEFINITION 2.3   Exact Equation

A differential expression $M(x, y)\, dx + N(x, y)\, dy$ is an **exact differential** in a region R of the xy-plane if it corresponds to the differential of some function $f(x, y)$ defined in R. A first-order differential equation of the form

$$M(x, y)\, dx \; + \; N(x, y)\, dy \; = \; 0$$

is said to be an **exact equation** if the expression on the left-hand side is an exact differential.

For example, $x^2 y^3\, dx + x^3 y^2\, dy = 0$ is an exact equation, because its left-hand side is an exact differential:

$$d\left(\frac{1}{3}x^3 y^3\right) = x^2 y^3\, dx + x^3 y^2\, dy.$$

Notice that if we make the identifications $M(x, y) = x^2 y^3$ and $N(x, y) = x^3 y^2$, then $\partial M/\partial y = 3x^2 y^2 = \partial N/\partial x$. Theorem 2.1, given next, shows that the equality of the partial derivatives $\partial M/\partial y$ and $\partial N/\partial x$ is no coincidence.

THEOREM 2.1   Criterion for an Exact Differential

Let $M(x, y)$ and $N(x, y)$ be continuous and have continuous first partial derivatives in a rectangular region R defined by $a < x < b$, $c < y < d$. Then a necessary and sufficient condition that $M(x, y)\, dx + N(x, y)\, dy$ be an exact differential is

$$\frac{\partial M}{\partial y} = \frac{\partial N}{\partial x}. \tag{4}$$

PROOF OF THE NECESSITY   For simplicity let us assume that $M(x, y)$ and $N(x, y)$ have continuous first partial derivatives for all (x, y). Now if the expression $M(x, y)\, dx + N(x, y)\, dy$ is exact, there exists some function f such that for all x in R,

$$M(x, y)\, dx + N(x, y)\, dy = \frac{\partial f}{\partial x}\, dx + \frac{\partial f}{\partial y}\, dy.$$

Therefore

$$M(x, y) = \frac{\partial f}{\partial x}, \quad N(x, y) = \frac{\partial f}{\partial y},$$

and

$$\frac{\partial M}{\partial y} = \frac{\partial}{\partial y}\left(\frac{\partial f}{\partial x}\right) = \frac{\partial^2 f}{\partial y\, \partial x} = \frac{\partial}{\partial x}\left(\frac{\partial f}{\partial y}\right) = \frac{\partial N}{\partial x}.$$

The equality of the mixed partials is a consequence of the continuity of the first partial derivatives of $M(x, y)$ and $N(x, y)$. ∎

The sufficiency part of Theorem 2.1 consists of showing that there exists a function f for which $\partial f/\partial x = M(x, y)$ and $\partial f/\partial y = N(x, y)$ whenever (4) holds. The construction of the function f actually reflects a basic procedure for solving exact equations.

METHOD OF SOLUTION   Given an equation in the differential form $M(x, y)\, dx + N(x, y)\, dy = 0$, determine whether the equality in (4) holds. If it does, then there exists a function f for which

$$\frac{\partial f}{\partial x} = M(x, y).$$

We can find f by integrating $M(x, y)$ with respect to x while holding y constant:

$$f(x, y) = \int M(x, y) \, dx + g(y), \qquad (5)$$

where the arbitrary function $g(y)$ is the "constant" of integration. Now differentiate (5) with respect to y and assume $\partial f / \partial y = N(x, y)$:

$$\frac{\partial f}{\partial y} = \frac{\partial}{\partial y} \int M(x, y) \, dx + g'(y) = N(x, y).$$

This gives

$$g'(y) = N(x, y) - \frac{\partial}{\partial y} \int M(x, y) \, dx. \qquad (6)$$

Finally, integrate (6) with respect to y and substitute the result in (5). The implicit solution of the equation is $f(x, y) = c$.

Some observations are in order. First, it is important to realize that the expression $N(x, y) - (\partial / \partial y) \int M(x, y) \, dx$ in (6) is independent of x, because

$$\frac{\partial}{\partial x} \left[N(x, y) - \frac{\partial}{\partial y} \int M(x, y) \, dx \right] = \frac{\partial N}{\partial x} - \frac{\partial}{\partial y} \left(\frac{\partial}{\partial x} \int M(x, y) \, dx \right) = \frac{\partial N}{\partial x} - \frac{\partial M}{\partial y} = 0.$$

Second, we could just as well start the foregoing procedure with the assumption that $\partial f / \partial y = N(x, y)$. After integrating N with respect to y and then differentiating that result, we would find the analogues of (5) and (6) to be, respectively,

$$f(x, y) = \int N(x, y) \, dy + h(x) \quad \text{and} \quad h'(x) = M(x, y) - \frac{\partial}{\partial x} \int N(x, y) \, dy.$$

In either case *none of these formulas should be memorized.*

EXAMPLE 1   Solving an Exact DE

Solve $2xy \, dx + (x^2 - 1) \, dy = 0$.

SOLUTION   With $M(x, y) = 2xy$ and $N(x, y) = x^2 - 1$ we have

$$\frac{\partial M}{\partial y} = 2x = \frac{\partial N}{\partial x}.$$

Thus the equation is exact, and so, by Theorem 2.1, there exists a function $f(x, y)$ such that

$$\frac{\partial f}{\partial x} = 2xy \quad \text{and} \quad \frac{\partial f}{\partial y} = x^2 - 1.$$

From the first of these equations we obtain, after integrating,

$$f(x, y) = x^2 y + g(y).$$

Taking the partial derivative of the last expression with respect to y and setting the result equal to $N(x, y)$ gives

$$\frac{\partial f}{\partial y} = x^2 + g'(y) = x^2 - 1. \qquad \leftarrow N(x, y)$$

It follows that $g'(y) = -1$ and $g(y) = -y$. Hence $f(x, y) = x^2 y - y$, and so the solution of the equation in implicit form is $x^2 y - y = c$. The explicit form of the solution is easily seen to be $y = c/(1 - x^2)$ and is defined on any interval not containing either $x = 1$ or $x = -1$. ▪

NOTE The solution of the DE in Example 1 is *not* $f(x, y) = x^2y - y$. Rather, it is $f(x, y) = c$; if a constant is used in the integration of $g'(y)$, we can then write the solution as $f(x, y) = 0$. Note, too, that the equation could be solved by separation of variables.

EXAMPLE 2   **Solving an Exact DE**

Solve $(e^{2y} - y \cos xy) \, dx + (2xe^{2y} - x \cos xy + 2y) \, dy = 0$.

SOLUTION   The equation is exact because

$$\frac{\partial M}{\partial y} = 2e^{2y} + xy \sin xy - \cos xy = \frac{\partial N}{\partial x}.$$

Hence a function $f(x, y)$ exists for which

$$M(x, y) = \frac{\partial f}{\partial x} \quad \text{and} \quad N(x, y) = \frac{\partial f}{\partial y}.$$

Now for variety we shall start with the assumption that $\partial f/\partial y = N(x, y)$; that is,

$$\frac{\partial f}{\partial y} = 2xe^{2y} - x \cos xy + 2y$$

$$f(x, y) = 2x \int e^{2y} \, dy - x \int \cos xy \, dy + 2 \int y \, dy.$$

Remember, the reason x can come out in front of the symbol $\int$ is that in the integration with respect to y, x is treated as an ordinary constant. It follows that

$$f(x, y) = xe^{2y} - \sin xy + y^2 + h(x)$$

$$\frac{\partial f}{\partial x} = e^{2y} - y \cos xy + h'(x) = e^{2y} - y \cos xy, \quad \leftarrow M(x, y)$$

and so $h'(x) = 0$ or $h(x) = c$. Hence a family of solutions is

$$xe^{2y} - \sin xy + y^2 + c = 0.$$

EXAMPLE 3   **An Initial-Value Problem**

Solve $\dfrac{dy}{dx} = \dfrac{xy^2 - \cos x \sin x}{y(1 - x^2)}, \quad y(0) = 2$.

SOLUTION   By writing the differential equation in the form

$$(\cos x \sin x - xy^2) \, dx + y(1 - x^2) \, dy = 0,$$

we recognize that the equation is exact because

$$\frac{\partial M}{\partial y} = -2xy = \frac{\partial N}{\partial x}.$$

Now

$$\frac{\partial f}{\partial y} = y(1 - x^2)$$

$$f(x, y) = \frac{y^2}{2}(1 - x^2) + h(x)$$

$$\frac{\partial f}{\partial x} = -xy^2 + h'(x) = \cos x \sin x - xy^2.$$

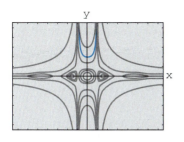

y

x

FIGURE 2.29 Some graphs of members of the family $y^2(1 - x^2) - \cos^2 x = c$

The last equation implies that $h'(x) = \cos x \sin x$. Integrating gives

$$h(x) = -\int (\cos x)(-\sin x \, dx) = -\frac{1}{2}\cos^2 x.$$

Thus $\dfrac{y^2}{2}(1 - x^2) - \dfrac{1}{2}\cos^2 x = c_1$ or $y^2(1 - x^2) - \cos^2 x = c,$ (7)

where $2c_1$ has been replaced by c. The initial condition $y = 2$ when $x = 0$ demands that $4(1) - \cos^2(0) = c$, and so $c = 3$. An implicit solution of the problem is then $y^2(1 - x^2) - \cos^2 x = 3$.

The solution curve of the IVP is the curve drawn in color in Figure 2.29; it is part of an interesting family of curves. The graphs of the members of the one-parameter family of solutions given in (7) can be obtained in several ways, two of which are using software to graph level curves (as discussed in Section 2.2) and using a graphing utility to carefully graph the explicit functions obtained for various values of c by solving $y^2 = (c + \cos^2 x)/(1 - x^2)$ for y.

INTEGRATING FACTORS   Recall from the last section that the left-hand side of the linear equation $y' + P(x)y = f(x)$ can be transformed into a derivative when we multiply the equation by an integrating factor. The same basic idea sometimes works for a nonexact differential equation $M(x, y) \, dx + N(x, y) \, dy = 0$. That is, it is sometimes possible to find an **integrating factor** $\mu(x, y)$ so that, after multiplying, the left-hand side of

$$\mu(x, y)M(x, y) \, dx + \mu(x, y)N(x, y) \, dy = 0$$ (8)

is an exact differential. In an attempt to find μ, we turn to the criterion (4) for exactness. Equation (8) is exact if and only if $(\mu M)_y = (\mu N)_x$, where the subscripts denote partial derivatives. By the Product Rule of differentiation the last equation is the same as $\mu M_y + \mu_y M = \mu N_x + \mu_x N$ or

$$\mu_x N - \mu_y M = (M_y - N_x)\mu.$$ (9)

Although M, N, M_y, and N_x are known functions of x and y, the difficulty here in determining the unknown $\mu(x, y)$ from (9) is that we must solve a partial differential equation. Since we are not prepared to do that, we make a simplifying assumption. Suppose μ is a function of one variable; for example, say that μ depends only on x. In this case, $\mu_x = d\mu/dx$ and $\mu_y = 0$, so (9) can be written as

$$\frac{d\mu}{dx} = \frac{M_y - N_x}{N}\mu.$$ (10)

We are still at an impasse if the quotient $(M_y - N_x)/N$ depends on both x and y. However, if after all obvious algebraic simplifications are made, the quotient $(M_y - N_x)/N$ turns out to depend solely on the variable x, then (10) is a first-order ordinary differential equation. We can finally determine μ because (10) is *separable* as well as *linear*. It follows from either Section 2.2 or Section 2.3 that $\mu(x) = e^{\int ((M_y - N_x)/N) dx}$. In like manner, it follows from (9) that if μ depends only on the variable y, then

$$\frac{d\mu}{dy} = \frac{N_x - M_y}{M}\mu.$$ (11)

In this case, if $(N_x - M_y)/M$ is a function of y only, then we can solve (11) for μ.

We summarize the results for the differential equation

$$M(x, y)\,dx + N(x, y)\,dy = 0. \tag{12}$$

- If $(M_y - N_x)/N$ is a function of x alone, then an integrating factor for (12) is

$$\mu(x) = e^{\int \frac{M_y - N_x}{N}\,dx}. \tag{13}$$

- If $(N_x - M_y)/M$ is a function of y alone, then an integrating factor for (12) is

$$\mu(y) = e^{\int \frac{N_x - M_y}{M}\,dy}. \tag{14}$$

EXAMPLE 4   A Nonexact DE Made Exact

The nonlinear first-order differential equation

$$xy\,dx + (2x^2 + 3y^2 - 20)\,dy = 0$$

is not exact. With the identifications $M = xy$, $N = 2x^2 + 3y^2 - 20$, we find the partial derivatives $M_y = x$ and $N_x = 4x$. The first quotient from (13) gets us nowhere, since

$$\frac{M_y - N_x}{N} = \frac{x - 4x}{2x^2 + 3y^2 - 20} = \frac{-3x}{2x^2 + 3y^2 - 20}$$

depends on x and y. However, (14) yields a quotient that depends only on y:

$$\frac{N_x - M_y}{M} = \frac{4x - x}{xy} = \frac{3x}{xy} = \frac{3}{y}.$$

The integrating factor is then $e^{\int 3dy/y} = e^{3\ln y} = e^{\ln y^3} = y^3$. After we multiply the given DE by $\mu(y) = y^3$, the resulting equation is

$$xy^4\,dx + (2x^2y^3 + 3y^5 - 20y^3)\,dy = 0.$$

You should verify that the last equation is now exact as well as show, using the method of this section, that a family of solutions is $\frac{1}{2}x^2y^4 + \frac{1}{2}y^6 - 5y^4 = c$.

REMARKS

(*i*) When testing an equation for exactness, make sure it is of the precise form $M(x, y)\,dx + N(x, y)\,dy = 0$. Sometimes a differential equation is written $G(x, y)\,dx = H(x, y)\,dy$. In this case, first rewrite it as $G(x, y)\,dx - H(x, y)\,dy = 0$ and then identify $M(x, y) = G(x, y)$ and $N(x, y) = -H(x, y)$ before using (4).

(*ii*) In some texts on differential equations the study of exact equations precedes that of linear DEs. Then the method for finding integrating factors just discussed can be used to derive an integrating factor for $y' + P(x)y = f(x)$. By rewriting the last equation in the differential form $(P(x)y - f(x))\,dx + dy = 0$, we see that

$$\frac{M_y - N_x}{N} = P(x).$$

From (13) we arrive at the already familiar integrating factor $e^{\int P(x)dx}$, used in Section 2.3.

EXERCISES 2.4

Answers to selected odd-numbered problems begin on page ANS-2.

In Problems 1–20 determine whether the given differential equation is exact. If it is exact, solve it.

1. $(2x - 1)\,dx + (3y + 7)\,dy = 0$

2. $(2x + y)\,dx - (x + 6y)\,dy = 0$

3. $(5x + 4y)\,dx + (4x - 8y^3)\,dy = 0$

4. $(\sin y - y \sin x)\,dx + (\cos x + x \cos y - y)\,dy = 0$

5. $(2xy^2 - 3)\,dx + (2x^2y + 4)\,dy = 0$

6. $\left(2y - \dfrac{1}{x} + \cos 3x\right)\dfrac{dy}{dx} + \dfrac{y}{x^2} - 4x^3 + 3y \sin 3x = 0$

7. $(x^2 - y^2)\,dx + (x^2 - 2xy)\,dy = 0$

8. $\left(1 + \ln x + \dfrac{y}{x}\right)dx = (1 - \ln x)\,dy$

9. $(x - y^3 + y^2 \sin x)\,dx = (3xy^2 + 2y \cos x)\,dy$

10. $(x^3 + y^3)\,dx + 3xy^2\,dy = 0$

11. $(y \ln y - e^{-xy})\,dx + \left(\dfrac{1}{y} + x \ln y\right)dy = 0$

12. $(3x^2y + e^y)\,dx + (x^3 + xe^y - 2y)\,dy = 0$

13. $x\dfrac{dy}{dx} = 2xe^x - y + 6x^2$

14. $\left(1 - \dfrac{3}{y} + x\right)\dfrac{dy}{dx} + y = \dfrac{3}{x} - 1$

15. $\left(x^2y^3 - \dfrac{1}{1 + 9x^2}\right)\dfrac{dx}{dy} + x^3y^2 = 0$

16. $(5y - 2x)y' - 2y = 0$

17. $(\tan x - \sin x \sin y)\,dx + \cos x \cos y\,dy = 0$

18. $(2y \sin x \cos x - y + 2y^2 e^{xy^2})\,dx = (x - \sin^2 x - 4xye^{xy^2})\,dy$

19. $(4t^3y - 15t^2 - y)\,dt + (t^4 + 3y^2 - t)\,dy = 0$

20. $\left(\dfrac{1}{t} + \dfrac{1}{t^2} - \dfrac{y}{t^2 + y^2}\right)dt + \left(ye^y + \dfrac{t}{t^2 + y^2}\right)dy = 0$

In Problems 21–26 solve the given initial-value problem.

21. $(x + y)^2\,dx + (2xy + x^2 - 1)\,dy = 0, \quad y(1) = 1$

22. $(e^x + y)\,dx + (2 + x + ye^y)\,dy = 0, \quad y(0) = 1$

23. $(4y + 2t - 5)\,dt + (6y + 4t - 1)\,dy = 0, \quad y(-1) = 2$

24. $\left(\dfrac{3y^2 - t^2}{y^5}\right)\dfrac{dy}{dt} + \dfrac{t}{2y^4} = 0, \quad y(1) = 1$

25. $(y^2 \cos x - 3x^2y - 2x)\,dx + (2y \sin x - x^3 + \ln y)\,dy = 0, \quad y(0) = e$

26. $\left(\dfrac{1}{1 + y^2} + \cos x - 2xy\right)\dfrac{dy}{dx} = y(y + \sin x), \quad y(0) = 1$

In Problems 27 and 28 find the value of k so that the given differential equation is exact.

27. $(y^3 + kxy^4 - 2x)\,dx + (3xy^2 + 20x^2y^3)\,dy = 0$

28. $(6xy^3 + \cos y)\,dx + (2kx^2y^2 - x \sin y)\,dy = 0$

In Problems 29 and 30 verify that the given differential equation is not exact. Multiply the given differential equation by the indicated integrating factor $\mu(x, y)$ and verify that the new equation is exact. Solve.

29. $(-xy \sin x + 2y \cos x)\,dx + 2x \cos x\,dy = 0; \quad \mu(x, y) = xy$

30. $(x^2 + 2xy - y^2)\,dx + (y^2 + 2xy - x^2)\,dy = 0; \quad \mu(x, y) = (x + y)^{-2}$

In Problems 31–36 solve the given differential equation by finding, as in Example 4, an appropriate integrating factor.

31. $(2y^2 + 3x)\,dx + 2xy\,dy = 0$

32. $y(x + y + 1)\,dx + (x + 2y)\,dy = 0$

33. $6xy\,dx + (4y + 9x^2)\,dy = 0$

34. $\cos x\,dx + \left(1 + \dfrac{2}{y}\right)\sin x\,dy = 0$

35. $(10 - 6y + e^{-3x})\,dx - 2\,dy = 0$

36. $(y^2 + xy^3)\,dx + (5y^2 - xy + y^3 \sin y)\,dy = 0$

In Problems 37 and 38 solve the given initial-value problem by finding, as in Example 4, an appropriate integrating factor.

37. $x\,dx + (x^2y + 4y)\,dy = 0, \quad y(4) = 0$

38. $(x^2 + y^2 - 5)\,dx = (y + xy)\,dy, \quad y(0) = 1$

39. (a) Show that a one-parameter family of solutions of the equation

$$(4xy + 3x^2)\, dx + (2y + 2x^2)\, dy = 0$$

is $x^3 + 2x^2 y + y^2 = c$.

(b) Show that the initial conditions $y(0) = -2$ and $y(1) = 1$ determine the same implicit solution.

(c) Find explicit solutions $y_1(x)$ and $y_2(x)$ of the differential equation in part (a) such that $y_1(0) = -2$ and $y_2(1) = 1$. Use a graphing utility to graph $y_1(x)$ and $y_2(x)$.

DISCUSSION/PROJECT PROBLEMS

40. Consider the concept of an integrating factor used in Problems 29–38. Are the two equations $M\, dx + N\, dy = 0$ and $\mu M\, dx + \mu N\, dy = 0$ necessarily equivalent in the sense that a solution of one is also a solution of the other? Discuss.

41. Reread Example 3 and then discuss why we can conclude that the interval of definition of the explicit solution of the IVP (the colored curve in Figure 2.29) is $(-1, 1)$.

42. Discuss how the functions $M(x, y)$ and $N(x, y)$ can be found so that each differential equation is exact. Carry out your ideas.

(a) $M(x, y)\, dx + \left(xe^{xy} + 2xy + \dfrac{1}{x} \right) dy = 0$

(b) $\left(x^{-1/2} y^{1/2} + \dfrac{x}{x^2 + y} \right) dx + N(x, y)\, dy = 0$

43. Differential equations are sometimes solved by having a clever idea. Here is a little exercise in cleverness: Although the differential equation $(x - \sqrt{x^2 + y^2})\, dx + y\, dy = 0$ is not exact, show how the rearrangement $(x\, dx + y\, dy)/\sqrt{x^2 + y^2} = dx$ and the observation $\frac{1}{2} d(x^2 + y^2) = x\, dx + y\, dy$ can lead to a solution.

44. True or False: Every separable first-order equation $dy/dx = g(x)h(y)$ is exact.

MATHEMATICAL MODEL

45. Falling Chain A portion of a uniform chain of length 8 ft is loosely coiled around a peg at the edge of a high horizontal platform, and the remaining portion of the chain hangs at rest over the edge of the platform. See Figure 2.30. Suppose the length of the overhanging chain is 3 ft, that the chain weighs 2 lb/ft, and that the

positive direction is downward. Starting at $t = 0$ seconds, the weight of the overhanging portion causes the chain on the table to uncoil smoothly and to fall to the floor. If $x(t)$ denotes the length of the chain overhanging the table at time $t > 0$, then $v = dx/dt$ is its velocity. When all resistive forces are ignored, it can be shown that a mathematical model relating v to x is given by

$$xv\frac{dv}{dx} + v^2 = 32x.$$

(a) Rewrite this model in differential form. Proceed as in Problems 31–36 and solve the DE for v in terms of x by finding an appropriate integrating factor. Find an explicit solution $v(x)$.

(b) Determine the velocity with which the chain leaves the platform.

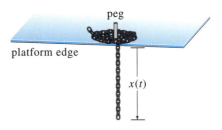

FIGURE 2.30   Uncoiling chain in Problem 45

COMPUTER LAB ASSIGNMENTS

46. Streamlines (a) The solution of the differential equation

$$\frac{2xy}{(x^2 + y^2)^2}\, dx + \left[1 + \frac{y^2 - x^2}{(x^2 + y^2)^2} \right] dy = 0$$

is a family of curves that can be interpreted as streamlines of a fluid flow around a circular object whose boundary is described by the equation $x^2 + y^2 = 1$. Solve this DE and note the solution $f(x, y) = c$ for $c = 0$.

(b) Use a CAS to plot the streamlines for $c = 0$, ± 0.2, ± 0.4, ± 0.6, and ± 0.8 in three different ways. First, use the *contourplot* of a CAS. Second, solve for x in terms of the variable y. Plot the resulting two functions of y for the given values of c, and then combine the graphs. Third, use the CAS to solve a cubic equation for y in terms of x.

2.5   SOLUTIONS BY SUBSTITUTIONS

INTRODUCTION: We usually solve a differential equation by recognizing it as a certain *kind* of equation (say, separable or linear) and then carry out a procedure, consisting of equation-specific mathematical steps, that yields a solution of the equation. But it is not uncommon to be stumped by a differential equation because it does not fall into one of the classes of equations that we know how to solve. The procedure discussed in this section may be helpful in this situation.

SUBSTITUTIONS   Often the first step in solving a differential equation consists of transforming it into another differential equation by means of a **substitution.** For example, suppose we wish to transform the first-order differential equation $dy/dx = f(x, y)$ by the substitution $y = g(x, u)$, where u is regarded as a function of the variable x. If g possesses first-partial derivatives, then the Chain Rule

$$\frac{dy}{dx} = \frac{\partial g}{\partial x}\frac{dx}{dx} + \frac{\partial g}{\partial u}\frac{du}{dx} \quad \text{gives} \quad \frac{dy}{dx} = g_x(x, u) + g_u(x, u)\frac{du}{dx}.$$

If we replace dy/dx by the foregoing derivative and replace y in $f(x, y)$ by $g(x, u)$, then the DE $dy/dx = f(x, y)$ becomes $g_x(x, u) + g_u(x, u)\dfrac{du}{dx} = f(x, g(x, u))$, which, solved for du/dx, has the form $\dfrac{du}{dx} = F(x, u)$. If we can determine a solution $u = \phi(x)$ of this last equation, then a solution of the original differential equation is $y = g(x, \phi(x))$.

In the discussion that follows we examine three different kinds of first-order differential equations that are solvable by means of a substitution.

HOMOGENEOUS EQUATIONS   If a function f possesses the property $f(tx, ty) = t^\alpha f(x, y)$ for some real number α, then f is said to be a **homogeneous function** of degree α. For example, $f(x, y) = x^3 + y^3$ is a homogeneous function of degree 3, since

$$f(tx, ty) = (tx)^3 + (ty)^3 = t^3(x^3 + y^3) = t^3 f(x, y),$$

whereas $f(x, y) = x^3 + y^3 + 1$ is not homogeneous. A first-order DE in differential form

$$M(x, y)\, dx + N(x, y)\, dy = 0 \tag{1}$$

is said to be **homogeneous*** if both coefficient functions M and N are homogeneous equations of the *same* degree. In other words, (1) is homogeneous if

$$M(tx, ty) = t^\alpha M(x, y) \quad \text{and} \quad N(tx, ty) = t^\alpha N(x, y).$$

In addition, if M and N are homogeneous functions of degree α, we can also write

$$M(x, y) = x^\alpha M(1, u) \quad \text{and} \quad N(x, y) = x^\alpha N(1, u), \quad \text{where } u = y/x, \tag{2}$$

and

$$M(x, y) = y^\alpha M(v, 1) \quad \text{and} \quad N(x, y) = y^\alpha N(v, 1), \quad \text{where } v = x/y. \tag{3}$$

See Problem 31 in Exercises 2.5. Properties (2) and (3) suggest the substitutions that can be used to solve a homogeneous differential equation. Specifically, *either* of the substitutions $y = ux$ or $x = vy$, where u and v are new dependent variables, will reduce a homogeneous equation to a *separable* first-order differential equation.

---

*Here the word *homogeneous* does not mean the same as it did in Section 2.3. Recall that a linear first-order equation $a_1(x)y' + a_0(x)y = g(x)$ is homogeneous when $g(x) = 0$.

To show this, observe that as a consequence of (2) a homogeneous equation $M(x, y)\, dx + N(x, y)\, dy = 0$ can be rewritten as

$$x^\alpha M(1, u)\, dx + x^\alpha N(1, u)\, dy = 0 \quad \text{or} \quad M(1, u)\, dx + N(1, u)\, dy = 0,$$

where $u = y/x$ or $y = ux$. By substituting the differential $dy = u\, dx + x\, du$ into the last equation and gathering terms, we obtain a separable DE in the variables u and x:

$$M(1, u)\, dx + N(1, u)[u\, dx + x\, du] = 0$$

$$[M(1, u) + uN(1, u)]\, dx + xN(1, u)\, du = 0$$

or

$$\frac{dx}{x} + \frac{N(1, u)\, du}{M(1, u) + uN(1, u)} = 0.$$

At this point we offer the same advice as in the preceding sections: Do not memorize anything here (especially the last formula); rather, *work through the procedure each time*. The proof that the substitutions $x = vy$ and $dx = v\, dy + y\, dv$ also lead to a separable equation follows in an analogous manner from (3).

EXAMPLE 1   Solving a Homogeneous DE

Solve $(x^2 + y^2)\, dx + (x^2 - xy)\, dy = 0$.

SOLUTION   Inspection of $M(x, y) = x^2 + y^2$ and $N(x, y) = x^2 - xy$ shows that these coefficients are homogeneous functions of degree 2. If we let $y = ux$, then $dy = u\, dx + x\, du$, so, after substituting, the given equation becomes

$$(x^2 + u^2 x^2)\, dx + (x^2 - ux^2)[u\, dx + x\, du] = 0$$

$$x^2(1 + u)\, dx + x^3(1 - u)\, du = 0$$

$$\frac{1 - u}{1 + u}\, du + \frac{dx}{x} = 0$$

$$\left[-1 + \frac{2}{1 + u}\right] du + \frac{dx}{x} = 0. \quad \leftarrow \text{long division}$$

After integration the last line gives

$$-u + 2 \ln|1 + u| + \ln|x| = \ln|c|$$

$$-\frac{y}{x} + 2 \ln\left|1 + \frac{y}{x}\right| + \ln|x| = \ln|c|. \quad \leftarrow \text{resubstituting } u = y/x$$

Using the properties of logarithms, we can write the preceding solution as

$$\ln\left|\frac{(x + y)^2}{cx}\right| = \frac{y}{x} \quad \text{or} \quad (x + y)^2 = cxe^{y/x}.$$

Although either of the indicated substitutions can be used for every homogeneous differential equation, in practice we try $x = vy$ whenever the function $M(x, y)$ is simpler than $N(x, y)$. Also it could happen that after using one substitution, we may encounter integrals that are difficult or impossible to evaluate in closed form; switching substitutions may result in an easier problem.

BERNOULLI'S EQUATION   The differential equation

$$\frac{dy}{dx} + P(x)y = f(x)y^n, \tag{4}$$

where n is any real number, is called **Bernoulli's equation.** Note that for $n = 0$ and $n = 1$ equation (4) is linear. For $n \neq 0$ and $n \neq 1$ the substitution $u = y^{1-n}$ reduces any equation of form (4) to a linear equation.

EXAMPLE 2   **Solving a Bernoulli DE**

Solve $x\dfrac{dy}{dx} + y = x^2y^2$.

SOLUTION   We first rewrite the equation as

$$\frac{dy}{dx} + \frac{1}{x}y = xy^2$$

by dividing by x. With $n = 2$ we have $u = y^{-1}$ or $y = u^{-1}$. We then substitute

$$\frac{dy}{dx} = \frac{dy}{du}\frac{du}{dx} = -u^{-2}\frac{du}{dx} \qquad \leftarrow \text{Chain Rule}$$

into the given equation and simplify. The result is

$$\frac{du}{dx} - \frac{1}{x}u = -x.$$

The integrating factor for this linear equation on, say, $(0, \infty)$ is

$$e^{-\int dx/x} = e^{-\ln x} = e^{\ln x^{-1}} = x^{-1}.$$

Integrating $\dfrac{d}{dx}[x^{-1}u] = -1$

gives $x^{-1}u = -x + c$ or $u = -x^2 + cx$. Since $u = y^{-1}$, we have $y = 1/u$, and so a solution of the given equation is $y = 1/(-x^2 + cx)$.   ∎

Note that we have not obtained the general solution of the original nonlinear differential equation in Example 2, since $y = 0$ is a singular solution of the equation.

REDUCTION TO SEPARATION OF VARIABLES   A differential equation of the form

$$\frac{dy}{dx} = f(Ax + By + C) \tag{5}$$

can always be reduced to an equation with separable variables by means of the substitution $u = Ax + By + C$, $B \neq 0$. Example 3 illustrates the technique.

EXAMPLE 3   **An Initial-Value Problem**

Solve $\dfrac{dy}{dx} = (-2x + y)^2 - 7$, $y(0) = 0$.

SOLUTION   If we let $u = -2x + y$, then $du/dx = -2 + dy/dx$, and so the differential equation is transformed into

$$\frac{du}{dx} + 2 = u^2 - 7 \quad \text{or} \quad \frac{du}{dx} = u^2 - 9.$$

The last equation is separable. Using partial fractions

$$\frac{du}{(u-3)(u+3)} = dx \quad \text{or} \quad \frac{1}{6}\left[\frac{1}{u-3} - \frac{1}{u+3}\right]du = dx$$

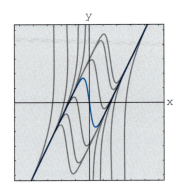

FIGURE 2.31   Some solutions of
$y' = (-2x + y)^2 - 7$

and then integrating yields

$$\frac{1}{6}\ln\left|\frac{u-3}{u+3}\right| = x + c_1 \quad \text{or} \quad \frac{u-3}{u+3} = e^{6x+6c_1} = ce^{6x}. \qquad \leftarrow \text{replace } e^{6c_1} \text{ by } c$$

Solving the last equation for u and then resubstituting gives the solution

$$u = \frac{3(1 + ce^{6x})}{1 - ce^{6x}} \quad \text{or} \quad y = 2x + \frac{3(1 + ce^{6x})}{1 - ce^{6x}}. \qquad (6)$$

Finally, applying the initial condition $y(0) = 0$ to the last equation in (6) gives $c = -1$. Figure 2.31, obtained with the aid of a graphing utility, shows the graph of the particular solution $y = 2x + \dfrac{3(1 - e^{6x})}{1 + e^{6x}}$ in color, along with the graphs of some other members of the family of solutions (6).

EXERCISES 2.5

Answers to selected odd-numbered problems begin on page ANS-2.

Each DE in Problems 1–14 is homogeneous.

In Problems 1–10 solve the given differential equation by using an appropriate substitution.

1. $(x - y)\, dx + x\, dy = 0$

2. $(x + y)\, dx + x\, dy = 0$

3. $x\, dx + (y - 2x)\, dy = 0$

4. $y\, dx = 2(x + y)\, dy$

5. $(y^2 + yx)\, dx - x^2\, dy = 0$

6. $(y^2 + yx)\, dx + x^2\, dy = 0$

7. $\dfrac{dy}{dx} = \dfrac{y - x}{y + x}$

8. $\dfrac{dy}{dx} = \dfrac{x + 3y}{3x + y}$

9. $-y\, dx + \left(x + \sqrt{xy}\right) dy = 0$

10. $x\dfrac{dy}{dx} = y + \sqrt{x^2 - y^2}, \quad x > 0$

In Problems 11–14 solve the given initial-value problem.

11. $xy^2\dfrac{dy}{dx} = y^3 - x^3, \quad y(1) = 2$

12. $(x^2 + 2y^2)\dfrac{dx}{dy} = xy, \quad y(-1) = 1$

13. $(x + ye^{y/x})\, dx - xe^{y/x}\, dy = 0, \quad y(1) = 0$

14. $y\, dx + x(\ln x - \ln y - 1)\, dy = 0, \quad y(1) = e$

Each DE in Problems 15–22 is a Bernoulli equation.

In Problems 15–20 solve the given differential equation by using an appropriate substitution.

15. $x\dfrac{dy}{dx} + y = \dfrac{1}{y^2}$

16. $\dfrac{dy}{dx} - y = e^x y^2$

17. $\dfrac{dy}{dx} = y(xy^3 - 1)$

18. $x\dfrac{dy}{dx} - (1 + x)y = xy^2$

19. $t^2\dfrac{dy}{dt} + y^2 = ty$

20. $3(1 + t^2)\dfrac{dy}{dt} = 2ty(y^3 - 1)$

In Problems 21 and 22 solve the given initial-value problem.

21. $x^2\dfrac{dy}{dx} - 2xy = 3y^4, \quad y(1) = \frac{1}{2}$

22. $y^{1/2}\dfrac{dy}{dx} + y^{3/2} = 1, \quad y(0) = 4$

Each DE in Problems 23–30 is of the form given in (5).

In Problems 23–28 solve the given differential equation by using an appropriate substitution.

23. $\dfrac{dy}{dx} = (x + y + 1)^2$

24. $\dfrac{dy}{dx} = \dfrac{1 - x - y}{x + y}$

25. $\dfrac{dy}{dx} = \tan^2(x + y)$

26. $\dfrac{dy}{dx} = \sin(x + y)$

27. $\dfrac{dy}{dx} = 2 + \sqrt{y - 2x + 3}$

28. $\dfrac{dy}{dx} = 1 + e^{y-x+5}$

In Problems 29 and 30 solve the given initial-value problem.

29. $\dfrac{dy}{dx} = \cos(x + y), \quad y(0) = \pi/4$

30. $\dfrac{dy}{dx} = \dfrac{3x + 2y}{3x + 2y + 2}, \quad y(-1) = -1$

DISCUSSION/PROJECT PROBLEMS

31. Explain why it is always possible to express any homogeneous differential equation $M(x, y)\, dx + N(x, y)\, dy = 0$ in the form

$$\frac{dy}{dx} = F\left(\frac{y}{x}\right).$$

You might start by proving that

$$M(x, y) = x^\alpha M(1, y/x) \quad \text{and} \quad N(x, y) = x^\alpha N(1, y/x).$$

32. Put the homogeneous differential equation

$$(5x^2 - 2y^2)\, dx - xy\, dy = 0$$

into the form given in Problem 31.

33. **(a)** Determine two singular solutions of the DE in Problem 10.
  (b) If the initial condition $y(5) = 0$ is as prescribed in Problem 10, then what is the largest interval I over which the solution is defined? Use a graphing utility to graph the solution curve for the IVP.

34. In Example 3 the solution $y(x)$ becomes unbounded as $x \to \pm\infty$. Nevertheless, $y(x)$ is asymptotic to a curve as $x \to -\infty$ and to a different curve as $x \to \infty$. What are the equations of these curves?

35. The differential equation $dy/dx = P(x) + Q(x)y + R(x)y^2$ is known as **Riccati's equation.**
  (a) A Riccati equation can be solved by a succession of two substitutions *provided* that we know a particular solution y_1 of the equation. Show that the substitution $y = y_1 + u$ reduces Riccati's equation to a Bernoulli equation (4) with $n = 2$. The Bernoulli equation can then be reduced to a linear equation by the substitution $w = u^{-1}$.

(b) Find a one-parameter family of solutions for the differential equation

$$\frac{dy}{dx} = -\frac{4}{x^2} - \frac{1}{x}y + y^2$$

where $y_1 = 2/x$ is a known solution of the equation.

36. Determine an appropriate substitution to solve

$$xy' = y \ln(xy).$$

MATHEMATICAL MODELS

37. **Falling Chain** In Problem 45 in Exercises 2.4 we saw that a mathematical model for the velocity v of a chain slipping off the edge of a high horizontal platform is

$$xv\frac{dv}{dx} + v^2 = 32x.$$

In that problem you were asked to solve the DE by converting it into an exact equation using an integrating factor. This time solve the DE using the fact that it is a Bernoulli equation.

38. **Population Growth** In the study of population dynamics one of the most famous models for a growing but bounded population is the **logistic equation**

$$\frac{dP}{dt} = P(a - bP),$$

where a and b are positive constants. Although we will come back to this equation and solve it by an alternative method in Section 3.2, solve the DE this first time using the fact that it is a Bernoulli equation.

2.6 A NUMERICAL METHOD

INTRODUCTION: A first-order differential equation $dy/dx = f(x, y)$ is a source of information. We started this chapter by observing that we could garner *qualitative* information from a first-order DE about its solutions even before we attempted to solve the equation. Then in Sections 2.2–2.5 we examined first-order DEs *analytically*—that is, we developed some procedures for obtaining explicit and implicit solutions. But a differential equation can possess a solution, yet we might not be able to obtain it analytically. So to round out the picture of the different types of analyses of differential equations, we conclude this chapter with a method by which we can "solve" the differential equation *numerically*—this means that the DE is used as the cornerstone of an algorithm for approximating the unknown solution.

In this section we are going to develop only the simplest of numerical methods—a method that utilizes the idea that a tangent line can be used to approximate the values of a function in a small neighborhood of the point of tangency. A more extensive treatment of numerical methods is given in Chapter 9.

CD: The **Euler Method Tool** on the *DE Tools* CD can be used in conjunction with the discussion of that topic on pages 80 and 81.

USING THE TANGENT LINE Let us assume that the first-order initial-value problem

$$y' = f(x, y), \quad y(x_0) = y_0 \tag{1}$$

possesses a solution. One way of approximating this solution is to use tangent lines. For example, let $y(x)$ denote the unknown solution of the first-order initial-value problem $y' = 0.1\sqrt{y} + 0.4x^2$, $y(2) = 4$. The nonlinear differential equation in this IVP cannot be solved directly by any of the methods considered in Sections 2.2, 2.4, and 2.5, nevertheless we can still find approximate numerical values of the unknown $y(x)$. Specifically, suppose we wish to know the value of $y(2.5)$. The IVP has a solution, and, as the flow of the direction field of the DE in Figure 2.32(a) suggests, a solution curve must have a shape similar to the curve shown in color.

The direction field in Figure 2.32(a) was generated with lineal elements passing through points in a grid with integer coordinates. As the solution curve passes through the initial point $(2, 4)$, the lineal element at this point is a tangent line with slope given by $f(2, 4) = 0.1\sqrt{4} + 0.4(2)^2 = 1.8$. As is apparent in Figure 2.32(a) and the "zoom in" in Figure 2.32(b), when x is close to 2, the points on the solution curve are close to the points on the tangent line (the lineal element). Using the point $(2, 4)$, the slope $f(2, 4) = 1.8$, and the point-slope form of a line, we find that an equation of the tangent line is $y = L(x)$, where $L(x) = 1.8x + 0.4$. This last equation, called a **linearization** of $y(x)$ at $x = 2$, can be used to approximate values of $y(x)$ within a small neighborhood of $x = 2$. If $y_1 = L(x_1)$ denotes the y-coordinate on the tangent line and $y(x_1)$ is the y-coordinate on the solution curve corresponding to an x-coordinate x_1 that is close to $x = 2$, then $y(x_1) \approx y_1$. If we choose, say, $x_1 = 2.1$, then $y_1 = L(2.1) = 1.8(2.1) + 0.4 = 4.18$, and so $y(2.1) \approx 4.18$.

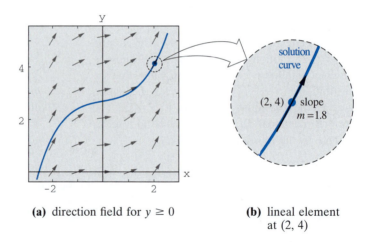

(a) direction field for $y \geq 0$ **(b)** lineal element at (2, 4)

FIGURE 2.32 Magnification of a neighborhood about the point (2,4)

EULER'S METHOD To generalize the procedure just illustrated, we use the linearization of the unknown solution $y(x)$ of (1) at $x = x_0$:

$$L(x) = y_0 + f(x_0, y_0)(x - x_0). \tag{2}$$

The graph of this linearization is a straight line tangent to the graph of $y = y(x)$ at the point (x_0, y_0). We now let h be a positive increment of the x-axis, as shown in Figure 2.33. Then by replacing x by $x_1 = x_0 + h$ in (2), we get

$$L(x_1) = y_0 + f(x_0, y_0)(x_0 + h - x_0) \quad \text{or} \quad y_1 = y_0 + hf(x_1, y_1),$$

where $y_1 = L(x_1)$. The point (x_1, y_1) on the tangent line is an approximation to the point $(x_1, y(x_1))$ on the solution curve. Of course, the accuracy of the approximation $L(x_1) \approx y(x_1)$ or $y_1 \approx y(x_1)$ depends heavily on the size of the increment h.

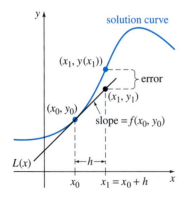

FIGURE 2.33 Approximating $y(x_1)$ using a tangent line

Usually, we must choose this **step size** to be "reasonably small." We now repeat the process using a second "tangent line" at (x_1, y_1).* By identifying the new starting point as (x_1, y_1) with (x_0, y_0) in the above discussion, we obtain an approximation $y_2 \approx y(x_2)$ corresponding to two steps of length h from x_0, that is, $x_2 = x_1 + h = x_0 + 2h$, and

$$y(x_2) = y(x_0 + 2h) = y(x_1 + h) \approx y_2 = y_1 + hf(x_1, y_1).$$

Continuing in this manner, we see that $y_1, y_2, y_3, \ldots$, can be defined recursively by the general formula

$$y_{n+1} = y_n + hf(x_n, y_n), \tag{3}$$

where $x_n = x_0 + nh$, $n = 0, 1, 2, \ldots$. This procedure of using successive "tangent lines" is called **Euler's method.**

TABLE 2.1 $h = 0.1$

x_n	y_n
2.00	4.0000
2.10	4.1800
2.20	4.3768
2.30	4.5914
2.40	4.8244
2.50	5.0768

TABLE 2.2 $h = 0.05$

x_n	y_n
2.00	4.0000
2.05	4.0900
2.10	4.1842
2.15	4.2826
2.20	4.3854
2.25	4.4927
2.30	4.6045
2.35	4.7210
2.40	4.8423
2.45	4.9686
2.50	5.0997

EXAMPLE 1 Euler's Method

Consider the initial-value problem $y' = 0.1\sqrt{y} + 0.4x^2$, $y(2) = 4$. Use Euler's method to obtain an approximation of $y(2.5)$ using first $h = 0.1$ and then $h = 0.05$.

SOLUTION With the identification $f(x, y) = 0.1\sqrt{y} + 0.4x^2$, (3) becomes

$$y_{n+1} = y_n + h(0.1\sqrt{y_n} + 0.4x_n^2).$$

Then for $h = 0.1$, $x_0 = 2$, $y_0 = 4$, and $n = 0$, we find

$$y_1 = y_0 + h(0.1\sqrt{y_0} + 0.4x_0^2) = 4 + 0.1(0.1\sqrt{4} + 0.4(2)^2) = 4.18,$$

which, as we have already seen, is an estimate to the value of $y(2.1)$. However, if we use the smaller step size $h = 0.05$, it takes two steps to reach $x = 2.1$. From

$$y_1 = 4 + 0.05(0.1\sqrt{4} + 0.4(2)^2) = 4.09$$
$$y_2 = 4.09 + 0.05(0.1\sqrt{4.09} + 0.4(2.05)^2) = 4.18416187$$

we have $y_1 \approx y(2.05)$ and $y_2 \approx y(2.1)$. The remainder of the calculations were carried out using software. The results are summarized in Tables 2.1 and 2.2, where each entry has been rounded to four decimal places. We see in Tables 2.1 and 2.2 that it takes five steps with $h = 0.1$ and 10 steps with $h = 0.05$, respectively, to get to $x = 2.5$. Intuitively, we would expect that $y_{10} = 5.0997$ corresponding to $h = 0.05$ is the better approximation of $y(2.5)$ than the value $y_5 = 5.0768$ corresponding to $h = 0.1$.

In Example 2 we apply Euler's method to a differential equation for which we have already found a solution. We do this to compare the values of the approximations y_n at each step with the true or actual values of the solution $y(x_n)$ of the initial-value problem.

EXAMPLE 2 Comparison of Approximate and Actual Values

Consider the initial-value problem $y' = 0.2xy$, $y(1) = 1$. Use Euler's method to obtain an approximation of $y(1.5)$ using first $h = 0.1$ and then $h = 0.05$.

SOLUTION With the identification $f(x, y) = 0.2xy$, (3) becomes

$$y_{n+1} = y_n + h(0.2x_n y_n)$$

*This is not an actual tangent line since (x_1, y_1) lies on the first tangent and not on the solution curve.

where $x_0 = 1$ and $y_0 = 1$. Again with the aid of computer software we obtain the values in Tables 2.3 and 2.4.

TABLE 2.3 $h = 0.1$

x_n	y_n	Actual value	Abs. error	% Rel. error
1.00	1.0000	1.0000	0.0000	0.00
1.10	1.0200	1.0212	0.0012	0.12
1.20	1.0424	1.0450	0.0025	0.24
1.30	1.0675	1.0714	0.0040	0.37
1.40	1.0952	1.1008	0.0055	0.50
1.50	1.1259	1.1331	0.0073	0.64

TABLE 2.4 $h = 0.05$

x_n	y_n	Actual value	Abs. error	% Rel. error
1.00	1.0000	1.0000	0.0000	0.00
1.05	1.0100	1.0103	0.0003	0.03
1.10	1.0206	1.0212	0.0006	0.06
1.15	1.0318	1.0328	0.0009	0.09
1.20	1.0437	1.0450	0.0013	0.12
1.25	1.0562	1.0579	0.0016	0.16
1.30	1.0694	1.0714	0.0020	0.19
1.35	1.0833	1.0857	0.0024	0.22
1.40	1.0980	1.1008	0.0028	0.25
1.45	1.1133	1.1166	0.0032	0.29
1.50	1.1295	1.1331	0.0037	0.32

In Example 1 the true or actual values were calculated from the known solution $y = e^{0.1(x^2-1)}$. (Verify.) The **absolute error** is defined to be

$$|\, actual\ value - approximation\,|.$$

The **relative error** and **percentage relative error** are, in turn,

$$\frac{absolute\ error}{|\,actual\ value\,|} \quad \text{and} \quad \frac{absolute\ error}{|\,actual\ value\,|} \times 100.$$

It is apparent from Tables 2.3 and 2.4 that the accuracy of the approximations improves as the step size h decreases. Also, we see that even though the percentage relative error is growing with each step, it does not appear to be that bad. But you should not be deceived by one example. If we simply change the coefficient of the right side of the DE in Example 2 from 0.2 to 2, then at $x_n = 1.5$ the percentage relative errors increase dramatically. See Problem 4 in Exercises 2.6.

A CAVEAT   Euler's method is just one of many different ways a solution of a differential equation can be approximated. Although attractive for its simplicity, *Euler's method is seldom used in serious calculations.* It was introduced here simply to give you a first taste of numerical methods. We will go into greater detail in discussing numerical methods that give significantly greater accuracy, notably the **fourth order Runge-Kutta method,** referred to as the **RK4 method,** in Chapter 9.

NUMERICAL SOLVERS   Regardless of whether we can actually find an explicit or implicit solution, if a solution of a differential equation exists, it represents a smooth curve in the Cartesian plane. The basic idea behind *any* numerical method for first-order ordinary differential equations is to somehow approximate the *y*-values of a solution for preselected values of *x*. We start at a specified initial point (x_0, y_0) on a solution curve and proceed to calculate in a step-by-step fashion a sequence of points $(x_1, y_1), (x_2, y_2), \ldots, (x_n, y_n)$ whose *y*-coordinates y_i approximate the *y*-coordinates $y(x_i)$ of points $(x_1, y(x_1)), (x_2, y(x_2)), \ldots, (x_n, y(x_n))$ that lie on the graph of the usually unknown solution $y(x)$. By taking the *x*-coordinates close together (that is, for small values of *h*) and by joining the points (x_1, y_1), $(x_2, y_2), \ldots, (x_n, y_n)$ with short line segments, we obtain a polygonal curve whose qualitative characteristics we hope are close to those of an actual solution curve. Drawing curves is something well suited to a computer. A computer program written to either implement a numerical method or render a visual representation of an approximate solution curve fitting the numerical data produced by this method is referred to as a **numerical solver.** There are many different numerical solvers commercially available, either embedded in a larger software package, such as a computer algebra system, or provided as a stand-alone package. Some software packages simply plot the generated numerical approximations, whereas others generate hard numerical data as well as the corresponding approximate or **numerical solution curves.** By way of illustration of the connect-the-dots nature of the graphs produced by a numerical solver, the two black polygonal graphs in Figure 2.34 are the numerical solution curves for the initial-value problem $y' = 0.2xy$, $y(0) = 1$ on the interval $[0, 4]$ obtained from Euler's method and the RK4 method using the step size $h = 1$. The colored smooth curve is the graph of the exact solution $y = e^{0.1x^2}$ of the IVP. Notice in Figure 2.34 that, even with the ridiculously large step size of $h = 1$, the RK4 method produces the more believable "solution curve." The numerical solution curve obtained from the RK4 method is indistinguishable from the actual solution curve on the interval $[0, 4]$ when a more typical step size of $h = 0.1$ is used.

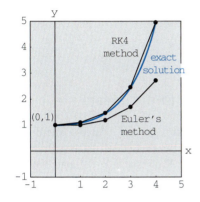

FIGURE 2.34   Comparison of the Runge-Kutta (RK4) and Euler methods

USING A NUMERICAL SOLVER   Knowledge of the various numerical methods is not necessary in order to use a numerical solver. A solver usually requires that the differential equation be expressed in normal form $dy/dx = f(x, y)$. Numerical solvers that generate only curves usually require that you supply $f(x, y)$ and the initial data x_0 and y_0 and specify the desired numerical method. If the idea is to approximate the numerical value of $y(a)$, then a solver may additionally require that you state a value for *h* or, equivalently, give the number of steps that you want to take to get from $x = x_0$ to $x = a$. For example, if we wanted to approximate $y(4)$ for the IVP illustrated in Figure 2.34, then, starting at $x = 0$ it would take four steps to reach $x = 4$ with a step size of $h = 1$; 40 steps is equivalent to a step size of $h = 0.1$. Although we will not delve here into the many problems that one can encounter when attempting to approximate mathematical quantities, you should at least be aware of the fact that a numerical solver may break down near certain points or give an incomplete or misleading picture when applied to some first-order differential equations in the normal form. Figure 2.35 illustrates the graph obtained by applying Euler's method to a certain first-order initial-value problem $dy/dx = f(x, y)$, $y(0) = 1$. Equivalent results were obtained using three different commercial numerical solvers, yet the graph is hardly a plausible solution curve. (Why?) There are several avenues of recourse when a numerical solver has difficulties; three of the more obvious are decrease the step size, use another numerical method, and try a different numerical solver.

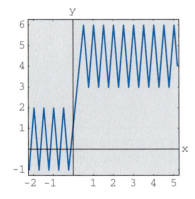

FIGURE 2.35   A not very helpful numerical solution curve

EXERCISES 2.6

Answers to selected odd-numbered problems begin on page ANS-2.

In Problems 1 and 2 use Euler's method to obtain a four-decimal approximation of the indicated value. Carry out the recursion of (3) by hand, first using $h = 0.1$ and then using $h = 0.05$.

1. $y' = 2x - 3y + 1$, $y(1) = 5$; $y(1.2)$

2. $y' = x + y^2$, $y(0) = 0$; $y(0.2)$

In Problems 3 and 4 use Euler's method to obtain a four-decimal approximation of the indicated value. First use $h = 0.1$ and then use $h = 0.05$. Find an explicit solution for each initial-value problem and then construct tables similar to Tables 2.3 and 2.4.

3. $y' = y$, $y(0) = 1$; $y(1.0)$

4. $y' = 2xy$, $y(1) = 1$; $y(1.5)$

In Problems 5–10 use a numerical solver and Euler's method to obtain a four-decimal approximation of the indicated value. First use $h = 0.1$ and then use $h = 0.05$.

5. $y' = e^{-y}$, $y(0) = 0$; $y(0.5)$

6. $y' = x^2 + y^2$, $y(0) = 1$; $y(0.5)$

7. $y' = (x - y)^2$, $y(0) = 0.5$; $y(0.5)$

8. $y' = xy + \sqrt{y}$, $y(0) = 1$; $y(0.5)$

9. $y' = xy^2 - \dfrac{y}{x}$, $y(1) = 1$; $y(1.5)$

10. $y' = y - y^2$, $y(0) = 0.5$; $y(0.5)$

In Problems 11 and 12 use a numerical solver to obtain a numerical solution curve for the given initial-value problem. First use Euler's method and then the RK4 method. Use $h = 0.25$ in each case. Superimpose both solution curves on the same coordinate axes. If possible, use a different color for each curve. Repeat, using $h = 0.1$ and $h = 0.05$.

11. $y' = 2(\cos x)y$, $y(0) = 1$

12. $y' = y(10 - 2y)$, $y(0) = 1$

DISCUSSION/PROJECT PROBLEMS

13. Use a numerical solver and Euler's method to approximate $y(1.0)$, where $y(x)$ is the solution to $y' = 2xy^2$, $y(0) = 1$. First use $h = 0.1$ and then use $h = 0.05$. Repeat, using the RK4 method. Discuss what might cause the approximations to $y(1.0)$ to differ so greatly.

COMPUTER LAB ASSIGNMENTS

14. (a) Use a numerical solver and the RK4 method to graph the solution of the initial-value problem $y' = -2xy + 1$, $y(0) = 0$.
   (b) Solve the initial-value problem by one of the analytic procedures developed earlier in this chapter.
   (c) Use the analytic solution $y(x)$ found in part (b) and a CAS to find the coordinates of all relative extrema.

CHAPTER 2 IN REVIEW

Answers to selected odd-numbered problems begin on page ANS-2.

In Problems 1 and 2 fill in the blanks.

1. The DE $y' - ky = A$, where k and A are constants, is autonomous. The critical point _____ of the equation is a(n) _____ (attractor or repeller) for $k > 0$ and a(n) _____(attractor or repeller) for $k < 0$.

2. The initial-value problem $x\dfrac{dy}{dx} - 4y = 0$, $y(0) = k$ has an infinite number of solutions for $k = $ _____ and no solution for $k = $ _____.

In Problems 3 and 4 construct an autonomous first-order differential equation $dy/dx = f(y)$ whose phase portrait is consistent with the given figure.

3.

FIGURE 2.36   Graph for Problem 3

4.

FIGURE 2.37   Graph for Problem 4

5. The number 0 is a critical point of the autonomous differential equation $dx/dt = x^n$, where n is a positive integer. For what values of n is 0 asymptotically stable? Semi-stable? Unstable? Repeat for the differential equation $dx/dt = -x^n$.

6. Consider the differential equation

$$\frac{dP}{dt} = f(P), \quad \text{where} \quad f(P) = -0.5P^3 - 1.7P + 3.4.$$

The function $f(P)$ has one real zero, as shown in Figure 2.38. Without attempting to solve the differential equation, estimate the value of $\lim_{t\to\infty} P(t)$.

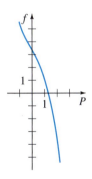

FIGURE 2.38   Graph for Problem 6

7. Figure 2.39 is a portion of a direction field of a differential equation $dy/dx = f(x, y)$. By hand, sketch two different solution curves—one that is tangent to the lineal element shown in black and one that is tangent to the lineal element shown in color.

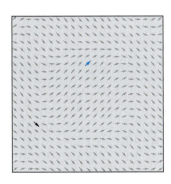

FIGURE 2.39   Portion of a direction field for Problem 7

8. Classify each differential equation as separable, exact, linear, homogeneous, or Bernoulli. Some equations may be more than one kind. Do not solve.

(a) $\dfrac{dy}{dx} = \dfrac{x - y}{x}$ (b) $\dfrac{dy}{dx} = \dfrac{1}{y - x}$

(c) $(x + 1)\dfrac{dy}{dx} = -y + 10$ (d) $\dfrac{dy}{dx} = \dfrac{1}{x(x - y)}$

(e) $\dfrac{dy}{dx} = \dfrac{y^2 + y}{x^2 + x}$ (f) $\dfrac{dy}{dx} = 5y + y^2$

(g) $y\,dx = (y - xy^2)\,dy$ (h) $x\dfrac{dy}{dx} = ye^{x/y} - x$

(i) $xy\,y' + y^2 = 2x$ (j) $2xy\,y' + y^2 = 2x^2$

(k) $y\,dx + x\,dy = 0$

(l) $\left(x^2 + \dfrac{2y}{x}\right)dx = (3 - \ln x^2)\,dy$

(m) $\dfrac{dy}{dx} = \dfrac{x}{y} + \dfrac{y}{x} + 1$

(n) $\dfrac{y}{x^2}\dfrac{dy}{dx} + e^{2x^3 + y^2} = 0$

In Problems 9–16 solve the given differential equation.

9. $(y^2 + 1)\,dx = y\sec^2 x\,dy$

10. $y(\ln x - \ln y)\,dx = (x\ln x - x\ln y - y)\,dy$

11. $(6x + 1)y^2\dfrac{dy}{dx} + 3x^2 + 2y^3 = 0$

12. $\dfrac{dx}{dy} = -\dfrac{4y^2 + 6xy}{3y^2 + 2x}$

13. $t\dfrac{dQ}{dt} + Q = t^4\ln t$

14. $(2x + y + 1)y' = 1$

15. $(x^2 + 4)\,dy = (2x - 8xy)\,dx$

16. $(2r^2\cos\theta\sin\theta + r\cos\theta)\,d\theta$
$+ (4r + \sin\theta - 2r\cos^2\theta)\,dr = 0$

In Problems 17 and 18 solve the given initial-value problem and give the largest interval I on which the solution is defined.

17. $\sin x\dfrac{dy}{dx} + (\cos x)y = 0,\quad y\left(\dfrac{7\pi}{6}\right) = -2$

18. $\dfrac{dy}{dt} + 2(t + 1)y^2 = 0,\quad y(0) = -\dfrac{1}{8}$

19. (a) Without solving, explain why the initial-value problem

$$\frac{dy}{dx} = \sqrt{y}, \quad y(x_0) = y_0$$

has no solution for $y_0 < 0$.

(b) Solve the initial-value problem in part (a) for $y_0 > 0$ and find the largest interval I on which the solution is defined.

20. (a) Find an implicit solution of the initial-value problem

$$\frac{dy}{dx} = \frac{y^2 - x^2}{xy}, \quad y(1) = -\sqrt{2}.$$

(b) Find an explicit solution of the problem in part (a) and give the largest interval I over which the solution is defined. A graphing utility may be helpful here.

21. Graphs of some members of a family of solutions for a first-order differential equation $dy/dx = f(x, y)$ are shown in Figure 2.40. The graphs of two implicit solutions, one that passes through the point $(1, -1)$ and one that passes through $(-1, 3)$, are shown in black. Reproduce the figure on a piece of paper. With colored pencils trace out the solution curves for the solutions $y = y_1(x)$ and $y = y_2(x)$ defined by the implicit solutions such that $y_1(1) = -1$ and $y_2(-1) = 3$, respectively. Estimate the intervals on which the solutions $y = y_1(x)$ and $y = y_2(x)$ are defined.

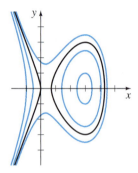

FIGURE 2.40   Graph for Problem 21

22. Use Euler's method with step size $h = 0.1$ to approximate $y(1.2)$, where $y(x)$ is a solution of the initial-value problem $y' = 1 + x\sqrt{y}$, $y(1) = 9$.

In Problems 23 and 24 each figure represents a portion of a direction field of an autonomous first-order differential equation $dy/dx = f(y)$. Reproduce the figure on a separate piece of paper and then complete the direction field over the grid. The points of the grid are (mh, nh), where $h = \frac{1}{2}$, m and n integers, $-7 \le m \le 7$, $-7 \le n \le 7$. In each direction field, sketch by hand an approximate solution curve that passes through each of the solid points shown in color. Discuss: Does it appear that the DE possesses critical points in the interval $-3.5 \le y \le 3.5$? If so, classify the critical points as asymptotically stable, unstable, or semi-stable.

23.

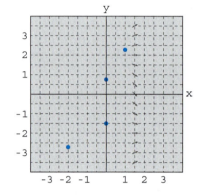

FIGURE 2.41   Portion of a direction field for Problem 23

24.

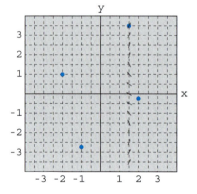

FIGURE 2.42   Portion of a direction field for Problem 24

HARVESTING NATURAL RESOURCES

You envy these salmon farmers. They don't have to rope or tie their product, they don't have to brand it, and best of all they don't have to smell its wastes. Moreover, they always know how many salmon they will be able to sell. All they have to do is feed their product and apply makeup to it on the way to the market. If only you could have that same kind of security.

You are a fisheries manager for the state of Alaska, and it is your job to help establish regulations for commercial salmon fishing in territorial waters near Glacier Bay, Alaska.* Your problem is that you never know how many fish are out there or which species will dominate in a given year. People who fish salmon for a living depend on the limits you set to put food on their tables and cars in their garages. On the other hand, you have a fishery that could collapse if you set limits that are too high.

Commercial fishing exploits a natural and wild resource. In contrast to, say, a cattle farmer who can breed new calves to replenish the herd continually, fishermen taking a wild resource cannot control their supply except by crude harvest limitations. Thus commercial and recreational fishing are regulated in an attempt to limit the harvesting and preserve the resource. The effects of harvesting a renewable natural resource such as fish can be modeled using a modification of the **logistic equation:**

$$\frac{dP}{dt} = rP\left(1 - \frac{P}{K}\right) - H(P). \tag{1}$$

FIGURE 1   Fish trawler near Glacier Bay, Alaska

National Oceanic and Atmospheric Administration/Department of Commerce

The first term on the right-hand side of the DE in (1) is a model of the population growth with a growth rate $r > 0$ and a carrying capacity (maximum sustainable population) of $K > 0$. See Section 3.2 for a more detailed discussion of this population model. The second term $H(P)$ is the *harvesting* term. You want to look at two different forms of $H(P)$. Each form will correspond to a possible regulatory strategy, and it is your task to understand how these strategies affect the long-term fish population.

You begin by considering what happens in the logistic model without harvesting—that is, when $H(P) = 0$. Let P denote the salmon population in thousands and define

$$\ell(P) = rP\left(1 - \frac{P}{K}\right).$$

PROBLEM 1.   Graph $y = \ell(P)$ and use this graph to sketch approximate solution curves of (1) in the tP-plane when $H(P) = 0$. What are the critical points of this autonomous DE? Use phase line analysis—that is, a one-dimensional phase portrait—to give the stability classification of each critical point.

PROBLEM 2.   Interpret the solution curves sketched in Problem 1 in terms of the long-time behavior of the salmon population. Begin with the equilibrium solutions of the DE and then interpret the solution curves corresponding to various possible initial conditions.

---

*Glacier Bay is located in Glacier Bay National Park in the Panhandle of southeast Alaska.

Now that you understand what happens without human intervention, you begin to consider what happens if we start harvesting the salmon. You begin by assuming that $r = 1$ and that the carrying capacity is $K - 1000$. Your objective is to determine a harvest rate for the fishing industry. Thus instead of assigning a number to the harvest rate, you simply assume that the harvesting occurs at a constant rate of h thousand salmon per year. With $H(P) = h$, (1) becomes

$$\frac{dP}{dt} = rP\left(1 - \frac{P}{K}\right) - h, \quad h > 0. \tag{2}$$

The DE in (2) is called the **constant-harvest model.** Your goal is to understand what happens to the salmon population as h increases.

PROBLEM 3 (CD). In (2), let $f_h(P) = rP(1 - P/1000) - h$. Use the **Logistic Harvest Tool** on the *DE Tools* CD to examine $f_h(P)$ for $0 < h < 300$. Describe what happens to the critical points of the DE and the corresponding phase lines. Relate this behavior to the population model. In particular, for each harvesting level h, what initial population levels ultimately lead to extinction, and how does this change with h?

PROBLEM 4. For what value of h is there only one critical point? What does the phase line look like for values of h slightly smaller than and slightly larger than this value? Again, interpret this in terms of the model for the salmon population.

One thing that you need to care about is a dependence on specific numbers in the model. You don't actually know the ocean's carrying capacity K or a species growth rate r. Studies and experiments suggest values for these parameters, but even then they are only estimates. It is very important for you to understand the behavior of the model as a whole. Otherwise, you could make a mistake in parameter value and wipe out the fishery.

PROBLEM 5. Repeat Problem 3 without assigning numerical values to r and K. Use calculus techniques rather than the CD to sketch a graph of $f_h(P)$.

The constant-harvest model (2) that you have examined corresponds to a simple approach to licensing in which fishermen are allowed a constant take, regardless of the time required for that. Another approach is to assume that harvesting is proportional to the population present. In other words, instead of allowing the same number of fish to be harvested each year, you allow only a fraction of the present population to be caught. In this scenario we write $H(P) = \alpha P, 0 \le \alpha < 1$, and (1) becomes

$$\frac{dP}{dt} = rP\left(1 - \frac{P}{K}\right) - \alpha P. \tag{3}$$

The differential equation in (3) is called the **proportional harvesting model.**

PROBLEM 6. Compute the critical points of the proportional harvesting model (3) and use phase line analysis to classify their stability. Interpret your results in terms of the salmon population.

PROBLEM 7. Now it's time to prepare your report and suggest a policy to your supervisor. Compare and contrast these two harvesting strategies. Be sure to make note of the strengths and limitations of each model. Brainstorm with your colleagues on other possible harvesting strategies and model these as well. Write a report making a suggestion for a regulatory strategy based on the work you've done.

STILL CURIOUS?

The problems above dealing with constant harvesting provide an example of a **bifurcation.** A bifurcation is essentially a dramatic change in the qualitative structure of the phase line, such as the appearance or disappearance of a critical point. In what follows we look more carefully at bifurcations in an attempt to determine when that might occur.

We say that a critical point of the autonomous DE $x' = f(x)$ is **hyperbolic** if small perturbations of $f(x)$ *do not* change the qualitative structure of the differential equation. Critical points are allowed to move slightly but not vanish, appear out of the blue, or change their stability. If arbitrary small changes *do* change the nature of the system, we say that the critical point is **nonhyperbolic.**

Consider a family of autonomous differential equations $x' = f_a(x)$, where a is a parameter. A bifurcation diagram for a family of DEs is simply a graph that shows the location and stability of the critical points for each parameter a. Consider the example above with $r = 1$, $K = 1000$, and constant harvesting. The critical points satisfy the quadratic equation

$$P\left(1 - \frac{P}{1000}\right) - h = 0. \qquad (4)$$

The equations of the top and bottom curves of the parabola shown in Figure 2 are thus obtained from (4) using the quadratic formula. Their stability can be easily determined as well: The top curve is the asymptotically stable critical point, and the bottom curve is the unstable critical point.

PROBLEM 8 (CAS). Find the equations defined by (4). Then use a computer algebra system to plot the graph shown in Figure 2.

Note that for $h < 250$ there are two critical points, and a small change in h does not change either the number or the stability of these critical points. These critical

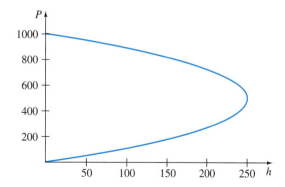

FIGURE 2 Bifurcation diagram for (1) when $H(P)$ is a constant

points are hyperbolic. However, when $h = 250$, there is only one critical point which is nonhyperbolic, since an infinitesimally small change of h leads to a dramatic change in the qualitative structure. In other words, if you move h any distance at all from 250, then the number of critical points changes from one to two or from one to zero.

You see that qualitative changes in the dynamical structure of the family of differential equations only occur at a nonhyperbolic critical point. This change is called a **bifurcation,** and the parameter value at which the bifurcation occurs is called a **bifurcation point.**

Your supervisor does not care about hyperbolic critical points, and she does *not* want to hear about bifurcations. However, she does care about the health of the fishery. For that reason you want to choose limits on your harvest that cause the equilibria to be well to the left of the bifurcation point.

PROBLEM 9.   Why do you want to choose a value of h that is to the left of the bifurcation point on the graph?

3

MODELING WITH FIRST-ORDER DIFFERENTIAL EQUATIONS

3.1 Linear Models
3.2 Nonlinear Models
3.3 Modeling with Systems of Differential Equations
CHAPTER 3 IN REVIEW

In Section 1.3 we saw how a first-order differential equation could be used as a mathematical model in the study of population growth, radioactive decay, continuous compound interest, cooling of bodies, mixtures, chemical reactions, fluid draining through a hole in a tank, velocity of a falling body, rate of memorization, and current in a series circuit. With the methods developed in Chapter 2, we are now able to solve some of the linear DEs (Section 3.1) and nonlinear DEs (Section 3.2) that commonly appear in applications. The chapter concludes with the natural next step—in Section 3.3 we examine how systems of simultaneous first-order DEs can arise as mathematical models in coupled physical systems (for example, a population of predators such as foxes interacting with a population of prey such as rabbits).

Malthusian Model

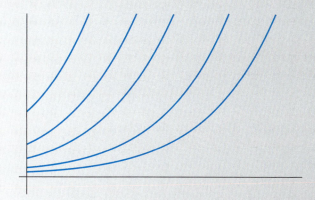

Logistic Model

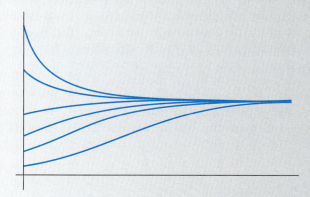

Predator-Prey Model

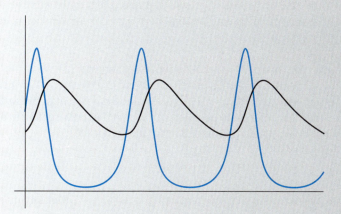

Families of solution curves for three famous population models. See pages 92, 104, and 115.

3.1 LINEAR MODELS

INTRODUCTION: In this section we solve some of the linear first-order models that were introduced in Section 1.3.

REVIEW MATERIAL: It is recommended that you take another look at the concept of a differential equation as a mathematical model in Section 1.3, especially equations (1), (2), (3), (7), (8), and (14) of that section. Also review steps (i)–(iv) in the summary "Solving a Linear First-Order Equation" on page 59 of Section 2.3.

CD: The **Growth and Decay**, the **Mixture**, and the *LR* **Circuit Tools** on the *DE tools* CD can be used in conjunction with the discussion of those topics on pages 93, 95, and 96.

GROWTH AND DECAY The initial-value problem

$$\frac{dx}{dt} = kx, \quad x(t_0) = x_0, \tag{1}$$

where k is a constant of proportionality, serves as a model for diverse phenomena involving either growth or decay. We saw in Section 1.3 that, in biological applications, the rate of growth of certain populations (bacteria, small animals) over short periods of time is proportional to the population present at time t. Knowing the population at some arbitrary initial time t_0, we can then use the solution of (1) to predict the population in the future—that is, at times $t > t_0$. The constant of proportionality k in (1) can be determined from the solution of the initial-value problem, using a subsequent measurement of x at a time $t_1 > t_0$. In physics and chemistry, (1) is seen in the form of a *first-order reaction*—that is, a reaction whose rate, or velocity, dx/dt is directly proportional to the amount x of a substance that is unconverted or remaining at time t. The decomposition, or decay, of U-238 (uranium) by radioactivity into Th-234 (thorium) is a first-order reaction.

EXAMPLE 1 Bacterial Growth

A culture initially has P_0 number of bacteria. At $t = 1$ h the number of bacteria is measured to be $\frac{3}{2}P_0$. If the rate of growth is proportional to the number of bacteria $P(t)$ present at time t, determine the time necessary for the number of bacteria to triple.

SOLUTION We first solve the differential equation in (1), with the symbol x replaced by P. With $t_0 = 0$, the initial condition is $P(0) = P_0$. We then use the empirical observation that $P(1) = \frac{3}{2}P_0$ to determine the constant of proportionality k.

Notice that the differential equation $dP/dt = kP$ is both separable and linear. When it is put in the standard form of a linear first-order DE,

$$\frac{dP}{dt} - kP = 0,$$

we can see by inspection that the integrating factor is e^{-kt}. Multiplying both sides of the equation by this term and integrating gives, in turn,

$$\frac{d}{dt}[e^{-kt}P] = 0 \quad \text{and} \quad e^{-kt}P = c.$$

Therefore $P(t) = ce^{kt}$. At $t = 0$ it follows that $P_0 = ce^0 = c$, and so $P(t) = P_0e^{kt}$. At $t = 1$ we have $\frac{3}{2}P_0 = P_0e^k$ or $e^k = \frac{3}{2}$. From the last equation we get

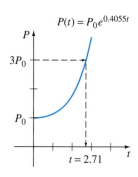

FIGURE 3.1 Time in which population triples

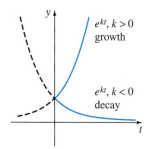

FIGURE 3.2 Growth ($k > 0$) and decay ($k < 0$)

$k = \ln \frac{3}{2} = 0.4055$, and so $P(t) = P_0 e^{0.4055t}$. To find the time at which the number of bacteria has tripled, we solve $3P_0 = P_0 e^{0.4055t}$ for t. It follows that $0.4055t = \ln 3$, or

$$t = \frac{\ln 3}{0.4055} \approx 2.71 \text{ h.}$$

See Figure 3.1.

Notice in Example 1 that the actual number P_0 of bacteria present at time $t = 0$ played no part in determining the time required for the number in the culture to triple. The time necessary for an initial population of, say, 100 or 1,000,000 bacteria to triple is still approximately 2.71 hours.

As shown in Figure 3.2, the exponential function e^{kt} increases as t increases for $k > 0$ and decreases as t increases for $k < 0$. Thus problems describing growth (whether of populations, bacteria, or even capital) are characterized by a positive value of k, whereas problems involving decay (as in radioactive disintegration) yield a negative k value. Accordingly, we say that k is either a **growth constant** ($k > 0$) or a **decay constant** ($k < 0$).

HALF-LIFE In physics the **half-life** is a measure of the stability of a radioactive substance. The half-life is simply the time it takes for one-half of the atoms in an initial amount A_0 to disintegrate, or transmute, into the atoms of another element. The longer the half-life of a substance, the more stable it is. For example, the half-life of highly radioactive radium, Ra-226, is about 1700 years. In 1700 years one-half of a given quantity of Ra-226 is transmuted into radon, Rn-222. The most commonly occurring uranium isotope, U-238, has a half-life of approximately 4,500,000,000 years. In about 4.5 billion years, one-half of a quantity of U-238 is transmuted into lead, Pb-206.

EXAMPLE 2   **Half-Life of Plutonium**

A breeder reactor converts relatively stable uranium 238 into the isotope plutonium 239. After 15 years it is determined that 0.043% of the initial amount A_0 of plutonium has disintegrated. Find the half-life of this isotope if the rate of disintegration is proportional to the amount remaining.

SOLUTION   Let $A(t)$ denote the amount of plutonium remaining at time t. As in Example 1, the solution of the initial-value problem

$$\frac{dA}{dt} = kA, \quad A(0) = A_0$$

is $A(t) = A_0 e^{kt}$. If 0.043% of the atoms of A_0 have disintegrated, then 99.957% of the substance remains. To find the decay constant k, we use $0.99957A_0 = A(15)$ — that is, $0.99957A_0 = A_0 e^{15k}$. Solving for k then gives $k = \frac{1}{15} \ln 0.99957 = -0.00002867$. Hence $A(t) = A_0 e^{-0.00002867t}$. Now the half-life is the corresponding value of time at which $A(t) = \frac{1}{2}A_0$. Solving for t gives $\frac{1}{2}A_0 = A_0 e^{-0.00002867t}$ or $\frac{1}{2} = e^{-0.00002867t}$. The last equation yields

$$t = \frac{\ln 2}{0.00002867} \approx 24{,}180 \text{ yr.}$$

CARBON DATING About 1950 the chemist Willard Libby devised a method of using radioactive carbon as a means of determining the approximate ages of fossils. The theory of **carbon dating** is based on the fact that the isotope carbon 14 is produced in the atmosphere by the action of cosmic radiation on nitrogen. The ratio of the amount of C-14 to ordinary carbon in the atmosphere appears to be a constant, and as a consequence the proportionate amount of the isotope present in all living organisms is the same as that in the atmosphere. When an organism dies, the absorption of C-14, by either breathing or eating, ceases. Thus by comparing the proportionate amount of C-14 present, say, in a fossil with the constant ratio found in the atmosphere, it is possible to obtain a reasonable estimation of the fossil's age. The method is based on the knowledge that the half-life of radioactive C-14 is approximately 5600 years. For his work Libby won the Nobel Prize for chemistry in 1960. Libby's method has been used to date wooden furniture in Egyptian tombs, the woven flax wrappings of the Dead Sea scrolls, and the cloth of the enigmatic shroud of Turin.

EXAMPLE 3   Age of a Fossil

A fossilized bone is found to contain one-thousandth of the C-14 level found in living matter. Estimate the age of the fossil.

SOLUTION   The starting point is again $A(t) = A_0 e^{kt}$. To determine the value of the decay constant k, we use the fact that $\frac{1}{2}A_0 = A(5600)$ or $\frac{1}{2}A_0 = A_0 e^{5600k}$. From $5600k = \ln \frac{1}{2} = -\ln 2$ we then get $k = -(\ln 2)/5600 = -0.00012378$. Therefore $A(t) = A_0 e^{-0.00012378t}$. With $A(t) = \frac{1}{1000}A_0$ we have $\frac{1}{1000}A_0 = A_0 e^{-0.00012378t}$, so $-0.00012378t = \ln \frac{1}{1000} = -\ln 1000$. Thus the age of the fossil is about

$$t = \frac{\ln 1000}{0.00012378} \approx 55{,}800 \text{ yr.}$$

The age found in Example 3 is really at the border of accuracy for this method. The usual carbon-14 technique is limited to about 9 half-lives of the isotope, or about 50,000 years. One reason for this limitation is that the chemical analysis needed to obtain an accurate measurement of the remaining C-14 becomes somewhat formidable around the point of $\frac{1}{1000}A_0$. Also, this analysis demands the destruction of a rather large sample of the specimen. If this measurement is accomplished indirectly, based on the actual radioactivity of the specimen, then it is very difficult to distinguish between the radiation from the fossil and the normal background radiation.* But recently, the use of a particle accelerator has enabled scientists to separate C-14 from stable C-12 directly. When the precise value of the ratio of C-14 to C-12 is computed, the accuracy of this method can be extended to 70,000–100,000 years. Other isotopic techniques such as using potassium 40 and argon 40 can give ages of several million years.† Nonisotopic methods based on the use of amino acids are also sometimes possible.

NEWTON'S LAW OF COOLING/WARMING   In equation (3) of Section 1.3 we saw that the mathematical formulation of Newton's empirical law of cooling/warming of an object is given by the linear first-order differential equation

$$\frac{dT}{dt} = k(T - T_m), \tag{2}$$

where k is a constant of proportionality, $T(t)$ is the temperature of the object for $t > 0$, and T_m is the ambient temperature—that is, the temperature of the medium around the object. In Example 4 we assume that T_m is constant.

---

*The number of disintegrations per minute per gram of carbon is recorded by using a Geiger counter. The lower level of detectability is about 0.1 disintegrations per minute per gram.

†Potassium-argon dating is used in dating terrestrial materials such as minerals, rocks, and lava and extraterrestrial materials such as meteorites and lunar rocks. The age of a fossil can be estimated by determining the age of the rock strata in which it was found.

EXAMPLE 4   Cooling of a Cake

When a cake is removed from an oven, its temperature is measured at 300° F. Three minutes later its temperature is 200° F. How long will it take for the cake to cool off to a room temperature of 70° F?

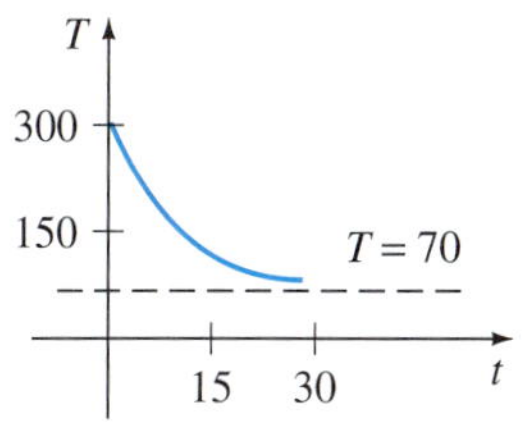

SOLUTION   In (2) we make the identification $T_m = 70$. We must then solve the initial-value problem

$$\frac{dT}{dt} = k(T - 70), \quad T(0) = 300 \tag{3}$$

and determine the value of k so that $T(3) = 200$.

Equation (3) is both linear and separable. If we separate variables,

$$\frac{dT}{T - 70} = k \, dt,$$

yields $\ln|T - 70| = kt + c_1$, and so $T = 70 + c_2 e^{kt}$. When $t = 0$, $T = 300$, so $300 = 70 + c_2$ gives $c_2 = 230$; therefore $T = 70 + 230e^{kt}$. Finally, the measurement $T(3) = 200$ leads to $e^{3k} = \frac{13}{23}$ or $k = \frac{1}{3}\ln\frac{13}{23} = -0.19018$. Thus

$$T(t) = 70 + 230e^{-0.19018t}. \tag{4}$$

We note that (4) furnishes no finite solution to $T(t) = 70$, since $\lim_{t\to\infty} T(t) = 70$. Yet we intuitively expect the cake to reach room temperature after a reasonably long period of time. How long is "long"? Of course, we should not be disturbed by the fact that the model (3) does not quite live up to our physical intuition. Parts (a) and (b) of Figure 3.3 clearly show that the cake will be approximately at room temperature in about one-half hour.

(a)

$T(t)$	t (min)
75°	20.1
74°	21.3
73°	22.8
72°	24.9
71°	28.6
70.5°	32.3

(b)

FIGURE 3.3 Temperature of cooling cake approaches room temperature

The ambient temperature in (2) need not be a constant but could be a function $T_m(t)$ of time t. See Problem 18 in Exercises 3.1 and *Project 7*.

MIXTURES   The mixing of two fluids sometimes gives rise to a linear first-order differential equation. When we discussed the mixing of two brine solutions in Section 1.3, we assumed that the rate $A'(t)$ at which the amount of salt in the mixing tank changes was a net rate:

$$\frac{dA}{dt} = (\text{input rate of salt}) - (\text{output rate of salt}) = R_{in} - R_{out}. \tag{5}$$

In Example 5 we solve equation (8) of Section 1.3.

EXAMPLE 5   Mixture of Two Salt Solutions

Recall that the large tank considered in Section 1.3 held 300 gallons of a brine solution. Salt was entering and leaving the tank; a brine solution was being pumped into the tank at the rate of 3 gal/min, it mixed with the solution there, and then the mixture was pumped out at the rate of 3 gal/min. The concentration of the salt in the inflow, or solution entering, was 2 lb/gal, and so salt was entering the tank at the rate $R_{in} = (2 \text{ lb/gal}) \cdot (3 \text{ gal/min}) = 6$ lb/min and leaving the tank at the rate $R_{out} = (A/300 \text{ lb/gal}) \cdot (3 \text{ gal/min}) = A/100$ lb/min. From this data and (5) we get equation (8) of Section 1.3. Let us pose the question: If 50 pounds of salt were dissolved initially in the 300 gallons, how much salt is in the tank after a long time?

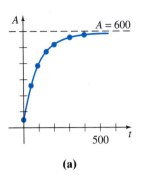

(a)

t (min)	A (lb)
50	266.41
100	397.67
150	477.27
200	525.57
300	572.62
400	589.93

(b)

FIGURE 3.4 Pounds of salt in tank as a function of time t

SOLUTION To find the amount of salt $A(t)$ in the tank at time t, we solve the initial-value problem

$$\frac{dA}{dt} + \frac{1}{100}A = 6, \quad A(0) = 50.$$

Note here that the side condition is the initial amount of salt $A(0) = 50$ in the tank and *not* the initial amount of liquid in the tank. Now since the integrating factor of the linear differential equation is $e^{t/100}$, we can write the equation as

$$\frac{d}{dt}[e^{t/100}A] = 6e^{t/100}.$$

Integrating the last equation and solving for A gives the general solution $A(t) = 600 + ce^{-t/100}$. When $t = 0$, $A = 50$, so we find that $c = -550$. Thus the amount of salt in the tank at time t is given by

$$A(t) = 600 - 550e^{-t/100}. \tag{6}$$

The solution (6) was used to construct the table in Figure 3.4(b). Also, it can be seen from (6) and Figure 3.4(a) that $A(t) \rightarrow 600$ as $t \rightarrow \infty$. Of course, this is what we would intuitively expect; over a long time the number of pounds of salt in the solution must be (300 gal)(2 lb/gal) = 600 lb.

In Example 5 we assumed that the rate at which the solution was pumped in was the same as the rate at which the solution was pumped out. However, this need not be the case; the mixed brine solution could be pumped out at a rate r_{out} faster or slower than the rate r_{in} at which the other brine solution is pumped in. For example, if the well-stirred solution in Example 5 is pumped out at a slower rate of, say, $r_{out} = 2$ gal/min, then liquid will accumulate in the tank at the rate of $r_{in} - r_{out} = (3 - 2)$ gal/min = 1 gal/min. After t minutes, (1 gal/min) $\cdot$ (t min) = t gal will accumulate, and so the tank will contain $300 + t$ gallons of brine. The concentration of the outflow is then $c(t) = A/(300 + t)$, and the output rate of salt is $R_{out} = c(t) \cdot r_{out}$, or

$$R_{out} = \left(\frac{A}{300 + t} \text{ lb/gal}\right) \cdot (2 \text{ gal/min}) = \frac{2A}{300 + t} \text{ lb/min}.$$

Hence equation (5) becomes

$$\frac{dA}{dt} = 6 - \frac{2A}{300 + t} \quad \text{or} \quad \frac{dA}{dt} + \frac{2}{300 + t}A = 6.$$

You should verify that the solution of the last equation, subject to $A(0) = 50$, is $A(t) = 600 + 2t - (4.95 \times 10^7)(300 + t)^{-2}$. See the discussion following (8) of Section 1.3, Problem 12 in Exercises 1.3, and Problems 23–26 in Exercises 3.1.

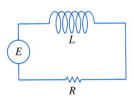

FIGURE 3.5 *LR* series circuit

SERIES CIRCUITS For a series circuit containing only a resistor and an inductor, Kirchhoff's second law states that the sum of the voltage drop across the inductor ($L(di/dt)$) and the voltage drop across the resistor (iR) is the same as the impressed voltage ($E(t)$) on the circuit. See Figure 3.5.

Thus we obtain the linear differential equation for the current $i(t)$,

$$L\frac{di}{dt} + Ri = E(t), \tag{7}$$

where L and R are constants known as the inductance and the resistance, respectively. The current $i(t)$ is also called the **response** of the system.

The voltage drop across a capacitor with capacitance C is given by $q(t)/C$, where q is the charge on the capacitor. Hence, for the series circuit shown in Figure 3.6, Kirchhoff's second law gives

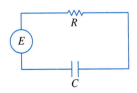

FIGURE 3.6 *RC* series circuit

$$Ri + \frac{1}{C}q = E(t). \tag{8}$$

But current i and charge q are related by $i = dq/dt$, so (8) becomes the linear differential equation

$$R\frac{dq}{dt} + \frac{1}{C}q = E(t). \tag{9}$$

EXAMPLE 6   Series Circuit

A 12-volt battery is connected to a series circuit in which the inductance is $\frac{1}{2}$ henry and the resistance is 10 ohms. Determine the current i if the initial current is zero.

SOLUTION   From (7) we see that we must solve

$$\frac{1}{2}\frac{di}{dt} + 10i = 12,$$

subject to $i(0) = 0$. First, we multiply the differential equation by 2 and read off the integrating factor e^{20t}. We then obtain

$$\frac{d}{dt}[e^{20t}i] = 24e^{20t}.$$

Integrating each side of the last equation and solving for i gives $i(t) = \frac{6}{5} + ce^{-20t}$. Now $i(0) = 0$ implies $0 = \frac{6}{5} + c$ or $c = -\frac{6}{5}$. Therefore the response is $i(t) = \frac{6}{5} - \frac{6}{5}e^{-20t}$.

From (4) of Section 2.3 we can write a general solution of (7):

$$i(t) = \frac{e^{-(R/L)t}}{L}\int e^{(R/L)t}E(t)\,dt + ce^{-(R/L)t}. \tag{10}$$

In particular, when $E(t) = E_0$ is a constant, (10) becomes

$$i(t) = \frac{E_0}{R} + ce^{-(R/L)t}. \tag{11}$$

Note that as $t \to \infty$, the second term in equation (11) approaches zero. Such a term is usually called a **transient term;** any remaining terms are called the **steady-state** part of the solution. In this case E_0/R is also called the **steady-state current;** for large values of time it appears that the current in the circuit is simply governed by Ohm's law ($E = iR$).

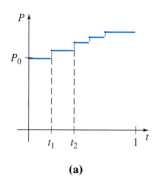

(a)

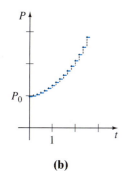

(b)

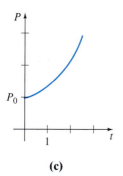

(c)

FIGURE 3.7   Population growth is a discrete process

REMARKS

The solution $P(t) = P_0 e^{0.4055t}$ of the initial-value problem in Example 1 described the population of a colony of bacteria at any time $t > 0$. Of course, $P(t)$ is a continuous function that takes on *all* real numbers in the interval $P_0 \leq P < \infty$. But since we are talking about a population, common sense dictates that P can take on only positive integer values. Moreover, we would not expect the population to grow continuously—that is, every second, every microsecond, and so on—as predicted by our solution; there may be intervals of time $[t_1, t_2]$ over which there is no growth at all. Perhaps, then, the graph shown in Figure 3.7(a) is a more realistic description of P than is the graph of an exponential function. Using a continuous function to describe a discrete phenomenon is often more a matter of convenience than of accuracy. However, for some purposes we may be satisfied if our model describes the system fairly closely when viewed macroscopically in time, as in Figures 3.7(b) and 3.7(c), rather than microscopically, as in Figure 3.7(a).

EXERCISES 3.1

Answers to selected odd-numbered problems begin on page ANS-2.

GROWTH AND DECAY

1. The population of a community is known to increase at a rate proportional to the number of people present at time t. If an initial population P_0 has doubled in 5 years, how long will it take to triple? To quadruple?

2. Suppose it is known that the population of the community in Problem 1 is 10,000 after 3 years. What was the initial population P_0? What will be the population in 10 years? How fast is the population growing at $t = 10$?

3. The population of a town grows at a rate proportional to the population present at time t. The initial population of 500 increases by 15% in 10 years. What will be the population in 30 years? How fast is the population growing at $t = 30$?

4. The population of bacteria in a culture grows at a rate proportional to the number of bacteria present at time t. After 3 hours it is observed that 400 bacteria are present. After 10 hours 2000 bacteria are present. What was the initial number of bacteria?

5. The radioactive isotope of lead, Pb-209, decays at a rate proportional to the amount present at time t and has a half-life of 3.3 hours. If 1 gram of this isotope is present initially, how long will it take for 90% of the lead to decay?

6. Initially 100 milligrams of a radioactive substance was present. After 6 hours the mass had decreased by 3%. If the rate of decay is proportional to the amount of the substance present at time t, find the amount remaining after 24 hours.

7. Determine the half-life of the radioactive substance described in Problem 6.

8. (a) Consider the initial-value problem $dA/dt = kA$, $A(0) = A_0$ as the model for the decay of a radioactive substance. Show that, in general, the half-life T of the substance is $T = -(\ln 2)/k$.
   (b) Show that the solution of the initial-value problem in part (a) can be written $A(t) = A_0 2^{-t/T}$.

(c) If a radioactive substance has the half-life T given in part (a), how long will it take an initial amount A_0 of the substance to decay to $\frac{1}{8}A_0$?

9. When a vertical beam of light passes through a transparent medium, the rate at which its intensity I decreases is proportional to $I(t)$, where t represents the thickness of the medium (in feet). In clear seawater, the intensity 3 feet below the surface is 25% of the initial intensity I_0 of the incident beam. What is the intensity of the beam 15 feet below the surface?

10. When interest is compounded continuously, the amount of money increases at a rate proportional to the amount S present at time t, that is, $dS/dt = rS$, where r is the annual rate of interest.
   (a) Find the amount of money accrued at the end of 5 years when $5000 is deposited in a savings account drawing $5\frac{3}{4}$% annual interest compounded continuously.
   (b) In how many years will the initial sum deposited have doubled?
   (c) Use a calculator to compare the amount obtained in part (a) with the amount $S = 5000(1 + \frac{1}{4}(0.0575))^{5(4)}$ that is accrued when interest is compounded quarterly.

CARBON DATING

11. Archaeologists used pieces of burned wood, or charcoal, found at the site to date prehistoric paintings and drawings on walls and ceilings of a cave in Lascaux, France. See Figure 3.8. Use the information on page 94 to determine the approximate age of a piece of burned wood, if it was found that 85.5% of the C-14 found in living trees of the same type had decayed.

FIGURE 3.8 Cave wall painting in Problem 11

12. The shroud of Turin, which shows the negative image of the body of a man who appears to have been crucified, is believed by many to be the burial shroud of Jesus of Nazareth. See Figure 3.9. In 1988 the Vatican granted permission to have the shroud carbon dated. Three independent scientific laboratories analyzed the cloth and concluded that the shroud was approximately 660 years old,* an age consistent with its historical appearance. Using this age, determine what percentage of the original amount of C-14 remained in the cloth as of 1988.

© Bettmann/CORBIS

FIGURE 3.9 Shroud image in Problem 12

NEWTON'S LAW OF COOLING/WARMING

13. A thermometer is removed from a room where the temperature is $70°$ F and is taken outside, where the air temperature is $10°$ F. After one-half minute the thermometer reads $50°$ F. What is the reading of the thermometer at $t = 1$ min? How long will it take for the thermometer to reach $15°$ F?

14. A thermometer is taken from an inside room to the outside, where the air temperature is $5°$ F. After 1 minute the thermometer reads $55°$ F, and after 5 minutes it reads $30°$ F. What is the initial temperature of the inside room?

15. A small metal bar, whose initial temperature was $20°$ C, is dropped into a large container of boiling water. How long will it take the bar to reach $90°$ C if it is known that its temperature increases $2°$ in 1 second? How long will it take the bar to reach $98°$ C?

*Some scholars have disagreed with this finding. For more information on this fascinating mystery see the Shroud of Turin home page at http://www.shroud.com/.

16. Two large containers A and B of the same size are filled with different fluids. The fluids in containers A and B are maintained at $0°$ C and $100°$ C, respectively. A small metal bar, whose initial temperature is $100°$ C, is lowered into container A. After 1 minute the temperature of the bar is $90°$ C. After 2 minutes the bar is removed and instantly transferred to the other container. After 1 minute in container B the temperature of the bar rises $10°$. How long, measured from the start of the entire process, will it take the bar to reach $99.9°$ C?

17. A thermometer reading $70°$ F is placed in an oven preheated to a constant temperature. Through a glass window in the oven door, an observer records that the thermometer reads $110°$ F after $\frac{1}{2}$ minute and $145°$ F after 1 minute. How hot is the oven?

18. At $t = 0$ a sealed test tube containing a chemical is immersed in a liquid bath. The initial temperature of the chemical in the test tube is $80°$ F. The liquid bath has a controlled temperature (measured in degrees Fahrenheit) given by $T_m(t) = 100 - 40e^{-0.1t}$, $t \geq 0$, where t is measured in minutes.

(a) Assume that $k = -0.1$ in (2). Before solving the IVP, describe in words, what you expect the temperature $T(t)$ of the chemical to be like in the short term. In the long term.

(b) Solve the initial-value problem. Use a graphing utility to plot the graph of $T(t)$ on time intervals of various lengths. Do the graphs agree with your predictions in part (a)?

MIXTURES

19. A tank contains 200 liters of fluid in which 30 grams of salt is dissolved. Brine containing 1 gram of salt per liter is then pumped into the tank at a rate of 4 L/min; the well-mixed solution is pumped out at the same rate. Find the number $A(t)$ of grams of salt in the tank at time t.

20. Solve Problem 19 assuming that pure water is pumped into the tank.

21. A large tank is filled to capacity with 500 gallons of pure water. Brine containing 2 pounds of salt per gallon is pumped into the tank at a rate of 5 gal/min. The well-mixed solution is pumped out at the same rate. Find the number $A(t)$ of pounds of salt in the tank at time t.

22. In Problem 21, what is the concentration $c(t)$ of the salt in the tank at time t? At $t = 5$ min? What is the concentration of the salt in the tank after a long time, that is, as $t \to \infty$? At what time is the concentration of the salt in the tank equal to one-half this limiting value?

23. Solve Problem 21 under the assumption that the solution is pumped out at a faster rate of 10 gal/min. When is the tank empty?

24. Determine the amount of salt in the tank at time t in Example 5 if the concentration of salt in the inflow is variable and given by $c_{in}(t) = 2 + \sin(t/4)$ lb/gal. Without actually graphing, conjecture what the solution curve of the IVP should look like. Then use a graphing utility to plot the graph of the solution on the interval [0, 300]. Repeat for the interval [0, 600] and compare your graph with that in Figure 3.4(a).

25. A large tank is partially filled with 100 gallons of fluid in which 10 pounds of salt is dissolved. Brine containing $\frac{1}{2}$ pound of salt per gallon is pumped into the tank at a rate of 6 gal/min. The well-mixed solution is then pumped out at a slower rate of 4 gal/min. Find the number of pounds of salt in the tank after 30 minutes.

26. In Example 5 the size of the tank containing the salt mixture was not given. Suppose, as in the discussion following Example 5, that the rate at which brine is pumped into the tank is 3 gal/min but that the well-stirred solution is pumped out at a rate of 2 gal/min. It stands to reason that since brine is accumulating in the tank at the rate of 1 gal/min, any finite tank must eventually overflow. Now suppose that the tank has an open top and has a total capacity of 400 gallons.

(a) When will the tank overflow?

(b) What will be the number of pounds of salt in the tank at the instant it overflows?

(c) Assume that although the tank is overflowing, brine solution continues to be pumped in at a rate of 3 gal/min and the well-stirred solution continues to be pumped out at a rate of 2 gal/min. Devise a method for determining the number of pounds of salt in the tank at $t = 150$ minutes.

(d) Determine the number of pounds of salt in the tank as $t \to \infty$. Does your answer agree with your intuition?

(e) Use a graphing utility to plot the graph of $A(t)$ on the interval [0, 500].

SERIES CIRCUITS

27. A 30-volt electromotive force is applied to an LR series circuit in which the inductance is 0.1 henry and the resistance is 50 ohms. Find the current $i(t)$ if $i(0) = 0$. Determine the current as $t \to \infty$.

28. Solve equation (7) under the assumption that $E(t) = E_0 \sin \omega t$ and $i(0) = i_0$.

29. A 100-volt electromotive force is applied to an RC series circuit in which the resistance is 200 ohms and the capacitance is 10^{-4} farad. Find the charge $q(t)$ on the capacitor if $q(0) = 0$. Find the current $i(t)$.

30. A 200-volt electromotive force is applied to an RC series circuit in which the resistance is 1000 ohms and the capacitance is 5×10^{-6} farad. Find the charge $q(t)$ on

the capacitor if $i(0) = 0.4$. Determine the charge and current at $t = 0.005$ s. Determine the charge as $t \to \infty$.

31. An electromotive force

$$E(t) = \begin{cases} 120, & 0 \le t \le 20 \\ 0, & t > 20 \end{cases}$$

is applied to an LR series circuit in which the inductance is 20 henries and the resistance is 2 ohms. Find the current $i(t)$ if $i(0) = 0$.

32. Suppose an RC series circuit has a variable resistor. If the resistance at time t is given by $R = k_1 + k_2 t$, where k_1 and k_2 are known positive constants, then (9) becomes

$$(k_1 + k_2 t)\frac{dq}{dt} + \frac{1}{C}q = E(t).$$

If $E(t) = E_0$ and $q(0) = q_0$, where E_0 and q_0 are constants, show that

$$q(t) = E_0 C + (q_0 - E_0 C)\left(\frac{k_1}{k_1 + k_2 t}\right)^{1/Ck_2}.$$

MISCELLANEOUS MATHEMATICAL MODELS

33. Air Resistance In (14) of Section 1.3 we saw that a differential equation describing the velocity v of a falling mass subject to air resistance proportional to the instantaneous velocity is

$$m\frac{dv}{dt} = mg - kv,$$

where $k > 0$ is a constant of proportionality. The positive direction is downward.

(a) Solve the equation subject to the initial condition $v(0) = v_0$.

(b) Use the solution in part (a) to determine the limiting, or terminal, velocity of the mass. We saw how to determine the terminal velocity without solving the DE in Problem 39 in Exercises 2.1.

(c) If the distance s, measured from the point where the mass was released above ground, is related to velocity v by $ds/dt = v(t)$, find an explicit expression for $s(t)$ if $s(0) = 0$.

34. How High?—No Air Resistance Suppose a small cannonball weighing 16 pounds is shot vertically upward, as shown in Figure 3.10, with an initial velocity $v_0 = 300$ ft/s. The answer to the question "How high does the cannonball go?" depends on whether we take air resistance into account.

(a) Suppose air resistance is ignored. If the positive direction is upward, then a model for the state of the cannonball is given by $d^2s/dt^2 = -g$ (equation (12) of Section 1.3). Since $ds/dt = v(t)$ the last differential equation is the same as $dv/dt = -g$, where we take $g = 32$ ft/s². Find the velocity $v(t)$ of the cannonball at time t.

(b) Use the result obtained in part (a) to determine the height $s(t)$ of the cannonball measured from ground level. Find the maximum height attained by the cannonball.

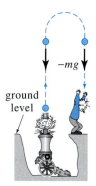

FIGURE 3.10 Find the maximum height of the cannonball in Problem 34

35. How High?—Linear Air Resistance Repeat Problem 34, but this time assume that air resistance is proportional to instantaneous velocity. It stands to reason that the maximum height attained by the cannonball must be *less* than that in part (b) of Problem 34. Show this by supposing that the constant of proportionality is $k = 0.0025$. (*Hint:* Slightly modify the DE in Problem 33.)

36. Skydiving A skydiver weighs 125 pounds, and her parachute and equipment combined weigh another 35 pounds. After exiting from a plane at an altitude of 15,000 feet, she waits 15 seconds and opens her parachute. Assume that the constant of proportionality in the model in Problem 33 has the value $k = 0.5$ during free fall and $k = 10$ after the parachute is opened. Assume that her initial velocity on leaving the plane is zero. What is her velocity and how far has she traveled 20 seconds after leaving the plane? See Figure 3.11. How does her velocity at 20 seconds compare with her terminal velocity? How long does it take her to reach the ground? (*Hint:* Think in terms of two distinct IVPs.)

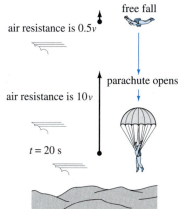

FIGURE 3.11
Find the time to reach the ground in Problem 36

37. Evaporating Raindrop As a raindrop falls, it evaporates while retaining its spherical shape. If we make the further assumptions that the rate at which the raindrop evaporates is proportional to its surface area and that air resistance is negligible, then a model for the velocity $v(t)$ of the raindrop is

$$\frac{dv}{dt} + \frac{3(k/\rho)}{(k/\rho)t + r_0}v = g.$$

Here ρ is the density of water, r_0 is the radius of the raindrop at $t = 0$, $k < 0$ is the constant of proportionality, and the downward direction is taken to be the positive direction.

(a) Solve for $v(t)$ if the raindrop falls from rest.

(b) Reread Problem 36 of Exercises 1.3 and then show that the radius of the raindrop at time t is $r(t) = (k/\rho)t + r_0$.

(c) If $r_0 = 0.01$ ft and $r = 0.007$ ft 10 seconds after the raindrop falls from a cloud, determine the time at which the raindrop has evaporated completely.

38. Fluctuating Population The differential equation $dP/dt = (k\cos t)P$, where k is a positive constant, is a mathematical model for a population $P(t)$ that undergoes yearly seasonal fluctuations. Solve the equation subject to $P(0) = P_0$. Use a graphing utility to graph the solution for different choices of P_0.

39. Population Model In one model of the changing population $P(t)$ of a community, it is assumed that

$$\frac{dP}{dt} = \frac{dB}{dt} - \frac{dD}{dt},$$

where dB/dt and dD/dt are the birth and death rates, respectively.

(a) Solve for $P(t)$ if $dB/dt = k_1 P$ and $dD/dt = k_2 P$.

(b) Analyze the cases $k_1 > k_2$, $k_1 = k_2$, and $k_1 < k_2$.

40. Memorization When forgetfulness is taken into account, the rate of memorization of a subject is given by

$$\frac{dA}{dt} = k_1(M - A) - k_2 A,$$

where $k_1 > 0$, $k_2 > 0$, $A(t)$ is the amount to be memorized in time t, M is the total amount to be memorized, and $M - A$ is the amount remaining to be memorized. (See Problems 25 and 26 in Exercises 1.3.)

(a) Since the DE is autonomous, use the phase portrait concept of Section 2.1 to find the limiting value of $A(t)$ as $t \to \infty$. Interpret the result.

(b) Solve for $A(t)$ subject to $A(0) = 0$. Sketch the graph of $A(t)$ and verify your prediction in part (a).

41. Drug Dissemination A mathematical model for the rate at which a drug disseminates into the bloodstream is given by $dx/dt = r - kx$, where r and k are positive constants. The function $x(t)$ describes the concentration of the drug in the bloodstream at time t.

(a) Since the DE is autonomous, use the phase portrait concept of Section 2.1 to find the limiting value of $x(t)$ as $t \to \infty$.

(b) Solve the DE subject to $x(0) = 0$. Sketch the graph of $x(t)$ and verify your prediction in part (a). At what time is the concentration one-half this limiting value?

DISCUSSION/PROJECT PROBLEMS

42. **Cooling and Warming** A small metal bar is removed from an oven whose temperature is a constant 300° F into a room whose temperature is a constant 70° F. Simultaneously, an identical metal bar is removed from the room and placed into the oven. Assume that time t is measured in minutes. Discuss: Why is there a future value of time, call it $t^* > 0$, at which the temperature of each bar is the same?

43. **Heart Pacemaker** A heart pacemaker, shown in Figure 3.12, consists of a switch, a battery, a capacitor, and the heart as a resistor. When the switch S is at P, the capacitor charges; when S is at Q, the capacitor discharges, sending an electrical stimulus to the heart. In Problem 47 in Exercises 2.3 we saw that during this time the electrical stimulus is being applied to the heart, the voltage E across the heart satisfies the linear DE

$$\frac{dE}{dt} = -\frac{1}{RC} E.$$

(a) Let us assume that over the time interval of length t_1, $0 < t < t_1$, the switch S is at position P shown in Figure 3.12 and the capacitor is being charged. When the switch is moved to position Q at time t_1 the capacitor discharges, sending an impulse to the heart over the time interval of length t_2: $t_1 \le t < t_1 + t_2$. Thus over the initial charging/discharging interval $0 < t < t_1 + t_2$ the voltage to the heart is actually modeled by the piecewise defined differential equation

$$\frac{dE}{dt} = \begin{cases} 0, & 0 \le t < t_1 \\ -\dfrac{1}{RC} E, & t_1 \le t < t_1 + t_2. \end{cases}$$

By moving S between P and Q, the charging and discharging over time intervals of lengths t_1 and t_2 is repeated indefinitely. Suppose $t_1 = 4$ s, $t_2 = 2$ s, $E_0 = 12$ V, and $E(0) = 0$, $E(4) = 12$, $E(6) = 0$, $E(10) = 12$, $E(12) = 0$, and so on. Solve for $E(t)$ on the interval $0 \le t \le 24$.

(b) Suppose for the sake of illustration that $R = C = 1$. Use a graphing utility to graph the solution for the IVP in part (a) on the interval $0 \le t \le 24$.

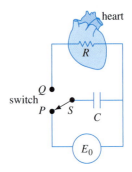

FIGURE 3.12   Model of a pacemaker in Problem 43

44. **Sliding Box (a)** A box of mass m slides down an inclined plane that makes an angle θ with the horizontal as shown in Figure 3.13. Find a differential equation for the velocity $v(t)$ of the box at time t in each of the following three cases:

    (i)   No sliding friction and no air resistance
    (ii) With sliding friction and no air resistance
    (iii) With sliding friction and air resistance

In cases *(ii)* and *(iii)*, use the fact that the force of friction opposing the motion of the box is μN, where μ is the coefficient of sliding friction and N is the normal component of the weight of the box. In case *(iii)* assume that air resistance is proportional to the instantaneous velocity.

(b) In part (a), suppose that the box weighs 96 pounds, that the angle of inclination of the plane is $\theta = 30°$, that the coefficient of sliding friction is $\mu = \sqrt{3}/4$, and that the additional retarding force due to air resistance is numerically equal to $\frac{1}{4}v$. Solve the differential equation in each of the three cases, assuming that the box starts from rest from the highest point 50 ft above ground.

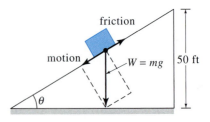

FIGURE 3.13   Box sliding down inclined plane in Problem 44

COMPUTER LAB ASSIGNMENTS

45. Sliding Box—Continued (a) In Problem 44, let $s(t)$ be the distance measured down the inclined plane from the highest point. Use $ds/dt = v(t)$ and the solution for each of the three cases in part (b) of Problem 44 to find the time that it takes the box to slide completely down the inclined plane. A root-finding application of a CAS may be useful here.

(b) In the case in which there is friction ($\mu \neq 0$) but no air resistance, explain why the box will not slide down the plane starting from *rest* from the highest point above ground when the inclination angle θ satisfies $\tan \theta \leq \mu$.

(c) The box *will* slide downward on the plane when $\tan \theta \leq \mu$ if it is given an initial velocity $v(0) = v_0 > 0$. Suppose that $\mu = \sqrt{3}/4$ and $\theta = 23°$. Verify that $\tan \theta \leq \mu$. How far will the box slide down the plane if $v_0 = 1$ ft/s?

(d) Using the values $\mu = \sqrt{3}/4$ and $\theta = 23°$, approximate the smallest initial velocity v_0 that can be given to the box so that, starting at the highest point 50 ft above ground, it will slide completely down the inclined plane. Then find the corresponding time it takes to slide down the plane.

46. What Goes Up ... (a) It is well known that the model in which air resistance is ignored, part (a) of Problem 34, predicts that the time t_a it takes the cannonball to attain its maximum height is the same as the time t_d it takes the cannonball to fall from the maximum height to the ground. Moreover, the magnitude of the impact velocity v_i will be the same as the initial velocity v_0 of the cannonball. Verify both of these results.

(b) Then, using the model in Problem 35 that takes air resistance into account, compare the value of t_a with t_d and the value of the magnitude of v_i with v_0. A root-finding application of a CAS (or graphic calculator) may be useful here.

3.2   NONLINEAR MODELS

INTRODUCTION: We finish our study of single first-order differential equations with an examination of some nonlinear models.

REVIEW MATERIAL: Most—but not all—of the differential equations in this section can be solved by separation of variables. Thus we recommend that, if necessary, you review again algebraic and analytical techniques involved in the art of integration. Another quick review of Section 1.3 is recommended; this time reexamine equations (5), (6), and (10). Also, check to see whether you worked Problems 7, 8, 13, 14, and 17 in Exercises 1.3. Similar problems appear in Exercises 3.2.

POPULATION DYNAMICS If $P(t)$ denotes the size of a population at time t, the model for exponential growth begins with the assumption that $dP/dt = kP$ for some $k > 0$. In this model, the **relative**, or **specific, growth rate** defined by

$$\frac{dP/dt}{P} \tag{1}$$

is a constant k. True cases of exponential growth over long periods of time are hard to find because the limited resources of the environment will at some time exert restrictions on the growth of a population. Thus for other models, (1) can be expected to decrease as the population P increases in size.

The assumption that the rate at which a population grows (or declines) is dependent only on the number present and not on any time-dependent mechanisms such as seasonal phenomena (see Problem 33 in Exercises 1.3) can be stated as

$$\frac{dP/dt}{P} = f(P) \quad \text{or} \quad \frac{dP}{dt} = Pf(P). \tag{2}$$

The differential equation in (2), which is widely assumed in models of animal populations, is called the **density-dependent hypothesis.**

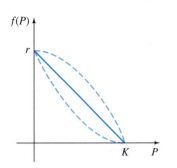

FIGURE 3.14 Simplest assumption for $f(P)$ is a straight line (solid color)

LOGISTIC EQUATION   Suppose an environment is capable of sustaining no more than a fixed number K of individuals in its population. The quantity K is called the **carrying capacity** of the environment. Hence for the function f in (2) we have $f(K) = 0$, and we simply let $f(0) = r$. Figure 3.14 shows three functions f that satisfy these two conditions. The simplest assumption that we can make is that $f(P)$ is linear—that is, $f(P) = c_1 P + c_2$. If we use the conditions $f(0) = r$ and $f(K) = 0$, we find, in turn, $c_2 = r$ and $c_1 = -r/K$, and so f takes on the form $f(P) = r - (r/K)P$. Equation (2) becomes

$$\frac{dP}{dt} = P\left(r - \frac{r}{K}P\right). \tag{3}$$

With constants relabeled, the nonlinear equation (3) is the same as

$$\frac{dP}{dt} = P(a - bP). \tag{4}$$

Around 1840 the Belgian mathematician-biologist P. F. Verhulst was concerned with mathematical models for predicting the human populations of various countries. One of the equations he studied was (4), where $a > 0$ and $b > 0$. Equation (4) came to be known as the **logistic equation,** and its solution is called the **logistic function.** The graph of a logistic function is called a **logistic curve.**

The linear differential equation $dP/dt = kP$ does not provide a very accurate model for population when the population itself is very large. Overcrowded conditions, with the resulting detrimental effects on the environment such as pollution and excessive and competitive demands for food and fuel, can have an inhibiting effect on population growth. As we shall now see, the solution of (4) is bounded as $t \to \infty$. If we rewrite (4) as $dP/dt = aP - bP^2$, the nonlinear term $-bP^2$, $b > 0$, can be interpreted as an "inhibition" or "competition" term. Also, in most applications the positive constant a is much larger than the constant b.

Logistic curves have proved to be quite accurate in predicting the growth patterns, in a limited space, of certain types of bacteria, protozoa, water fleas (*Daphnia*), and fruit flies (*Drosophila*).

SOLUTION OF THE LOGISTIC EQUATION   One method of solving (4) is separation of variables. Decomposing the left side of $dP/P(a - bP) = dt$ into partial fractions and integrating gives

$$\left(\frac{1/a}{P} + \frac{b/a}{a - bP}\right)dP = dt$$

$$\frac{1}{a}\ln|P| - \frac{1}{a}\ln|a - bP| = t + c$$

$$\ln\left|\frac{P}{a - bP}\right| = at + ac$$

$$\frac{P}{a - bP} = c_1 e^{at}.$$

It follows from the last equation that

$$P(t) = \frac{ac_1 e^{at}}{1 + bc_1 e^{at}} = \frac{ac_1}{bc_1 + e^{-at}}.$$

If $P(0) = P_0$, $P_0 \neq a/b$, we find $c_1 = P_0/(a - bP_0)$, and so, after substituting and simplifying, the solution becomes

$$P(t) = \frac{aP_0}{bP_0 + (a - bP_0)e^{-at}}. \tag{5}$$

GRAPHS OF $P(T)$ The basic shape of the graph of the logistic function $P(t)$ can be obtained without too much effort. Although the variable t usually represents time and we are seldom concerned with applications in which $t < 0$, it is nonetheless of some interest to include this interval when displaying the various graphs of P. From (5) we see that

$$P(t) \rightarrow \frac{aP_0}{bP_0} = \frac{a}{b} \quad \text{as} \quad t \rightarrow \infty \quad \text{and} \quad P(t) \rightarrow 0 \quad \text{as} \quad t \rightarrow -\infty.$$

The dashed line $P = a/2b$ shown in Figure 3.15 corresponds to the ordinate of a point of inflection of the logistic curve. To show this, we differentiate (4) by the product rule:

$$\frac{d^2P}{dt^2} = P\left(-b\frac{dP}{dt}\right) + (a - bP)\frac{dP}{dt} = \frac{dP}{dt}(a - 2bP)$$

$$= P(a - bP)(a - 2bP)$$

$$= 2b^2 P\left(P - \frac{a}{b}\right)\left(P - \frac{a}{2b}\right).$$

From calculus recall that the points where $d^2P/dt^2 = 0$ are possible points of inflection, but $P = 0$ and $P = a/b$ can obviously be ruled out. Hence $P = a/2b$ is the only possible ordinate value at which the concavity of the graph can change. For $0 < P < a/2b$ it follows that $P'' > 0$, and $a/2b < P < a/b$ implies that $P'' < 0$. Thus, as we read from left to right, the graph changes from concave up to concave down at the point corresponding to $P = a/2b$. When the initial value satisfies $0 < P_0 < a/2b$, the graph of $P(t)$ assumes the shape of an S, as we see in Figure 3.15(a). For $a/2b < P_0 < a/b$ the graph is still S-shaped, but the point of inflection occurs at a negative value of t, as shown in Figure 3.15(b).

We have already seen equation (4) in (5) of Section 1.3 in the form $dx/dt = kx(n + 1 - x)$, $k > 0$. This differential equation provides a reasonable model for describing the spread of an epidemic brought about initially by introducing an infected individual into a static population. The solution $x(t)$ represents the number of individuals infected with the disease at time t.

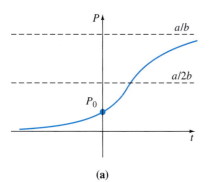

(a)

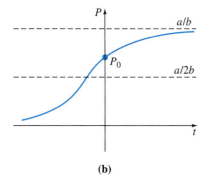

(b)

FIGURE 3.15   Logistic curves for different initial conditions

| **EXAMPLE 1** | **Logistic Growth** |

Suppose a student carrying a flu virus returns to an isolated college campus of 1000 students. If it is assumed that the rate at which the virus spreads is proportional not only to the number x of infected students but also to the number of students not infected, determine the number of infected students after 6 days if it is further observed that after 4 days $x(4) = 50$.

SOLUTION   Assuming that no one leaves the campus throughout the duration of the disease, we must solve the initial-value problem

$$\frac{dx}{dt} = kx(1000 - x), \quad x(0) = 1.$$

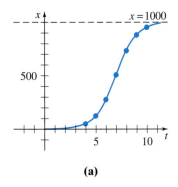

(a)

t (days)	x (number infected)
4	50 (observed)
5	124
6	276
7	507
8	735
9	882
10	953

(b)

FIGURE 3.16 Number of infected students $x(t)$ approaches 1000 as time t increases

By making the identification $a = 1000k$ and $b = k$, we have immediately from (5) that

$$x(t) = \frac{1000k}{k + 999ke^{-1000kt}} = \frac{1000}{1 + 999e^{-1000kt}}.$$

Now, using the information $x(4) = 50$, we determine k from

$$50 = \frac{1000}{1 + 999e^{-4000k}}.$$

We find $-1000k = \frac{1}{4}\ln\frac{19}{999} = -0.9906$. Thus

$$x(t) = \frac{1000}{1 + 999e^{-0.9906t}}.$$

Finally, $\qquad x(6) = \dfrac{1000}{1 + 999e^{-5.9436}} = 276$ students.

Additional calculated values of $x(t)$ are given in the table in Figure 3.16(b).

MODIFICATIONS OF THE LOGISTIC EQUATION   There are many variations of the logistic equation. For example, the differential equations

$$\frac{dP}{dt} = P(a - bP) - h \quad \text{and} \quad \frac{dP}{dt} = P(a - bP) + h \qquad (6)$$

could serve, in turn, as models for the population in a fishery where fish are **harvested** or are **restocked** at rate h. When $h > 0$ is a constant, the DEs in (6) can be readily analyzed qualitatively or solved analytically by separation of variables. The equations in (6) could also serve as models of the human population decreased by emigration or increased by immigration, respectively. The rate h in (6) could be a function of time t or could be population dependent; for example, harvesting might be done periodically over time or might be done at a rate proportional to the population P at time t. In the latter instance, the model would look like $P' = P(a - bP) - cP$, $c > 0$. The human population of a community might change because of immigration in such a manner that the contribution due to immigration was large when the population P of the community was itself small but small when P was large; a reasonable model for the population of the community would then be $P' = P(a - bP) + ce^{-kP}$, $c > 0$, $k > 0$. See Problem 22 in Exercises 3.2. Another equation of the form given in (2),

$$\frac{dP}{dt} = P(a - b\ln P), \qquad (7)$$

is a modification of the logistic equation known as the **Gompertz differential equation.** This DE is sometimes used as a model in the study of the growth or decline of populations, the growth of solid tumors, and certain kinds of actuarial predictions. See Problem 8 in Exercises 3.2.

CHEMICAL REACTIONS   Suppose that a grams of chemical A are combined with b grams of chemical B. If there are M parts of A and N parts of B formed in the compound and $X(t)$ is the number of grams of chemical C formed, then the number of grams of chemical A and the number of grams of chemical B remaining at time t are, respectively,

$$a - \frac{M}{M + N}X \quad \text{and} \quad b - \frac{N}{M + N}X.$$

The law of mass action states that when no temperature change is involved, the rate at which the two substances react is proportional to the product of the amounts of A and B that are untransformed (remaining) at time t:

$$\frac{dX}{dt} \propto \left(a - \frac{M}{M+N}X\right)\left(b - \frac{N}{M+N}X\right). \qquad (8)$$

If we factor out $M/(M+N)$ from the first factor and $N/(M+N)$ from the second and introduce a constant of proportionality $k > 0$, (8) has the form

$$\frac{dX}{dt} = k(\alpha - X)(\beta - X), \qquad (9)$$

where $\alpha = a(M+N)/M$ and $\beta = b(M+N)/N$. Recall from (6) of Section 1.3 that a chemical reaction governed by the nonlinear differential equation (9) is said to be a **second-order reaction.**

EXAMPLE 2   Second-Order Chemical Reaction

A compound C is formed when two chemicals A and B are combined. The resulting reaction between the two chemicals is such that for each gram of A, 4 grams of B is used. It is observed that 30 grams of the compound C is formed in 10 minutes. Determine the amount of C at time t if the rate of the reaction is proportional to the amounts of A and B remaining and if initially there are 50 grams of A and 32 grams of B. How much of the compound C is present at 15 minutes? Interpret the solution as $t \to \infty$.

SOLUTION   Let $X(t)$ denote the number of grams of the compound C present at time t. Clearly, $X(0) = 0$ g and $X(10) = 30$ g.

If, for example, 2 grams of compound C is present, we must have used, say, a grams of A and b grams of B, so $a + b = 2$ and $b = 4a$. Thus we must use $a = \frac{2}{5} = 2\left(\frac{1}{5}\right)$ g of chemical A and $b = \frac{8}{5} = 2\left(\frac{4}{5}\right)$ g of B. In general, for X grams of C we must use

$$\frac{1}{5}X \text{ grams of } A \quad \text{and} \quad \frac{4}{5}X \text{ grams of } B.$$

The amounts of A and B remaining at time t are then

$$50 - \frac{1}{5}X \quad \text{and} \quad 32 - \frac{4}{5}X,$$

respectively.

Now we know that the rate at which compound C is formed satisfies

$$\frac{dX}{dt} \propto \left(50 - \frac{1}{5}X\right)\left(32 - \frac{4}{5}X\right).$$

To simplify the subsequent algebra, we factor $\frac{1}{5}$ from the first term and $\frac{4}{5}$ from the second and then introduce the constant of proportionality:

$$\frac{dX}{dt} = k(250 - X)(40 - X).$$

By separation of variables and partial fractions, we can write

$$-\frac{1/210}{250 - X}\,dX + \frac{1/210}{40 - X}\,dX = k\,dt.$$

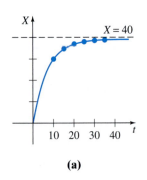

(a)

t (min)	X (g)
10	30 (measured)
15	34.78
20	37.25
25	38.54
30	39.22
35	39.59

(b)

FIGURE 3.17 $X(t)$ starts at 0 and approaches 40 as t increases

Integrating gives

$$\ln \frac{250 - X}{40 - X} = 210kt + c_1 \quad \text{or} \quad \frac{250 - X}{40 - X} = c_2 e^{210kt}. \tag{10}$$

When $t = 0$, $X = 0$, so it follows at this point that $c_2 = \frac{25}{4}$. Using $X = 30$ g at $t = 10$, we find $210k = \frac{1}{10} \ln \frac{88}{25} = 0.1258$. With this information we solve the last equation in (10) for X:

$$X(t) = 1000 \frac{1 - e^{-0.1258t}}{25 - 4e^{-0.1258t}}. \tag{11}$$

The behavior of X as a function of time is displayed in Figure 3.17. It is clear from the accompanying table and (11) that $X \to 40$ as $t \to \infty$. This means that 40 grams of compound C is formed, leaving

$$50 - \frac{1}{5}(40) = 42 \text{ g of } A \quad \text{and} \quad 32 - \frac{4}{5}(40) = 0 \text{ g of } B.$$

REMARKS

The indefinite integral $\int du/(a^2 - u^2)$ can be evaluated in terms of logarithms, the inverse hyperbolic tangent, or the inverse hyperbolic cotangent. For example, of the two results

$$\int \frac{du}{a^2 - u^2} = \frac{1}{a} \tanh^{-1} \frac{u}{a} + c, \quad |u| < a \tag{12}$$

$$\int \frac{du}{a^2 - u^2} = \frac{1}{2a} \ln \left| \frac{a + u}{a - u} \right| + c, \quad |u| \neq a, \tag{13}$$

(12) may be convenient in Problems 15 and 24 in Exercises 3.2, whereas (13) may be preferable in Problem 25.

EXERCISES 3.2

Answers to selected odd-numbered problems begin on page ANS-3.

LOGISTIC EQUATION

1. The number $N(t)$ of supermarkets throughout the country that are using a computerized check out system is described by the initial-value problem

$$\frac{dN}{dt} = N(1 - 0.0005N), \quad N(0) = 1.$$

(a) Use the phase portrait concept of Section 2.1 to predict how many supermarkets are expected to adopt the new procedure over a long period of time. By hand, sketch a solution curve of the given initial-value problem.

(b) Solve the initial-value problem and then use a graphing utility to verify the solution curve in part

(a). How many companies are expected to adopt the new technology when $t = 10$?

2. The number $N(t)$ of people in a community who are exposed to a particular advertisement is governed by the logistic equation. Initially, $N(0) = 500$, and it is observed that $N(1) = 1000$. Solve for $N(t)$ if it is predicted that the limiting number of people in the community who will see the advertisement is 50,000.

3. A model for the population $P(t)$ in a suburb of a large city is given by the initial-value problem

$$\frac{dP}{dt} = P(10^{-1} - 10^{-7}P), \quad P(0) = 5000,$$

where t is measured in months. What is the limiting value of the population? At what time will the population be equal to one-half of this limiting value?

4. (a) Census data for the United States between 1790 and 1950 is given in Table 3.1. Construct a logistic population model using the data from 1790, 1850, and 1910.

(b) Construct a table comparing actual census population with the population predicted by the model in part (a). Compute the error and the percentage error for each entry pair.

TABLE 3.1

Year	Population (in millions)
1790	3.929
1800	5.308
1810	7.240
1820	9.638
1830	12.866
1840	17.069
1850	23.192
1860	31.433
1870	38.558
1880	50.156
1890	62.948
1900	75.996
1910	91.972
1920	105.711
1930	122.775
1940	131.669
1950	150.697

MODIFICATIONS OF THE LOGISTIC MODEL

5. (a) If a constant number h of fish are harvested from a fishery per unit time, then a model for the population $P(t)$ of the fishery at time t is given by

$$\frac{dP}{dt} = P(a - bP) - h, \quad P(0) = P_0,$$

where a, b, h, and P_0 are positive constants. Suppose $a = 5$, $b = 1$, and $h = 4$. Since the DE is autonomous, use the phase portrait concept of Section 2.1 to sketch representative solution curves corresponding to the cases $P_0 > 4$, $1 < P_0 < 4$, and $0 < P_0 < 1$. Determine the long-term behavior of the population in each case.

(b) Solve the IVP in part (a). Verify the results of your phase portrait in part (a) by using a graphing utility to plot the graph of $P(t)$ with an initial condition taken from each of the three intervals given.

(c) Use the information in parts (a) and (b) to determine whether the fishery population becomes extinct in finite time. If so, find that time.

6. Investigate the harvesting model in Problem 5 both qualitatively and analytically in the case $a = 5$, $b = 1$, $h = \frac{25}{4}$. Determine whether the population becomes extinct in finite time. If so, find that time.

7. Repeat Problem 6 in the case $a = 5$, $b = 1$, $h = 7$.

8. (a) Suppose $a = b = 1$ in the Gompertz differential equation (7). Since the DE is autonomous, use the phase portrait concept of Section 2.1 to sketch representative solution curves corresponding to the cases $P_0 > e$ and $0 < P_0 < e$.

(b) Suppose $a = 1$, $b = -1$ in (7). Use a new phase portrait to sketch representative solution curves corresponding to the cases $P_0 > e^{-1}$ and $0 < P_0 < e^{-1}$.

(c) Find an explicit solution of (7) subject to $P(0) = P_0$.

CHEMICAL REACTIONS

9. Two chemicals A and B are combined to form a chemical C. The rate, or velocity, of the reaction is proportional to the product of the instantaneous amounts of A and B not converted to chemical C. Initially, there are 40 grams of A and 50 grams of B, and for each gram of B, 2 grams of A is used. It is observed that 10 grams of C is formed in 5 minutes. How much is formed in 20 minutes? What is the limiting amount of C after a long time? How much of chemicals A and B remains after a long time?

10. Solve Problem 9 if 100 grams of chemical A is present initially. At what time is chemical C half-formed?

MISCELLANEOUS MATHEMATICAL MODELS

11. Leaking Cylindrical Tank A tank in the form of a right circular cylinder standing on end is leaking water through a circular hole in its bottom. As we saw in (10) of Section 1.3, when friction and contraction of water at the hole are ignored, the height h of water in the tank is described by

$$\frac{dh}{dt} = -\frac{A_h}{A_w}\sqrt{2gh},$$

where A_w and A_h are the cross-sectional areas of the water and the hole, respectively.

(a) Solve for $h(t)$ if the initial height of the water is H. By hand, sketch the graph of $h(t)$ and give its interval I of definition in terms of the symbols A_w, A_h, and H. Use $g = 32$ ft/s^2.

(b) Suppose the tank is 10 feet high and has radius 2 feet and the circular hole has radius $\frac{1}{2}$ inch. If the tank is initially full, how long will it take to empty?

12. Leaking Cylindrical Tank-Continued When friction and contraction of the water at the hole are taken into account, the model in Problem 11 becomes

$$\frac{dh}{dt} = -c\frac{A_h}{A_w}\sqrt{2gh},$$

where $0 < c < 1$. How long will it take the tank in Problem 11(b) to empty if $c = 0.6$? See Problem 13 in Exercises 1.3.

13. Leaking Conical Tank A tank in the form of a right-circular cone standing on end, vertex down, is leaking water through a circular hole in its bottom.

   (a) Suppose the tank is 20 feet high and has radius 8 feet and the circular hole has radius 2 inches. In Problem 14 in Exercises 1.3 you were asked to show that the differential equation governing the height h of water leaking from a tank is

$$\frac{dh}{dt} = -\frac{5}{6h^{3/2}}.$$

   In this model, friction and contraction of the water at the hole were taken into account with $c = 0.6$, and g was taken to be 32 ft/s^2. See Figure 1.31. If the tank is initially full, how long will it take the tank to empty?

   (b) Suppose the tank has a vertex angle of 60° and the circular hole has radius 2 inches. Determine the differential equation governing the height h of water. Use $c = 0.6$ and $g = 32$ ft/s^2. If the height of the water is initially 9 feet, how long will it take the tank to empty?

14. Inverted Conical Tank Suppose that the conical tank in Problem 13(a) is inverted, as shown in Figure 3.18 and that water leaks out a circular hole of radius 2 inches in the center of its circular base. Is the time it takes to empty a full tank the same as for the tank with vertex down in Problem 13? Take the friction/contraction coefficient to be $c = 0.6$ and $g = 32$ ft/s^2.

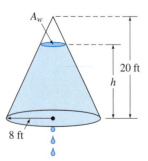

FIGURE 3.18   Inverted conical tank in Problem 14

15. Air Resistance A differential equation for the velocity v of a falling mass m subjected to air resistance

proportional to the square of the instantaneous velocity is

$$m\frac{dv}{dt} = mg - kv^2,$$

where $k > 0$ is a constant of proportionality. The positive direction is downward.

   (a) Solve the equation subject to the initial condition $v(0) = v_0$.

   (b) Use the solution in part (a) to determine the limiting, or terminal, velocity of the mass. We saw how to determine the terminal velocity without solving the DE in Problem 39 in Exercises 2.1.

   (c) If the distance s, measured from the point where the mass was released above ground, is related to velocity v by $ds/dt = v(t)$, find an explicit expression for $s(t)$ if $s(0) = 0$.

16. How High?—Nonlinear Air Resistance Consider the 16-pound cannonball shot vertically upward in Problems 34 and 35 in Exercises 3.1 with an initial velocity $v_0 = 300$ ft/s. Determine the maximum height attained by the cannonball if air resistance is assumed to be proportional to the square of the instantaneous velocity. Assume that the positive direction is upward and take $k = 0.0003$. (*Hint:* Slightly modify the DE in Problem 15.)

17. That Sinking Feeling **(a)** Determine a differential equation for the velocity $v(t)$ of a mass m sinking in water that imparts a resistance proportional to the square of the instantaneous velocity and also exerts an upward buoyant force whose magnitude is given by Archimedes' principle. See Problem 18 in Exercises 1.3. Assume that the positive direction is downward.

   (b) Solve the differential equation in part (a).

   (c) Determine the limiting, or terminal, velocity of the sinking mass.

18. Solar Collector The differential equation

$$\frac{dy}{dx} = \frac{-x + \sqrt{x^2 + y^2}}{y}$$

describes the shape of a plane curve C that will reflect all incoming light beams to the same point and could be a model for the mirror of a reflecting telescope, a satellite antenna, or a solar collector. See Problem 29 in Exercises 1.3. There are several ways of solving this DE.

   (a) Verify that the differential equation is homogeneous (see Section 2.5). Show that the substitution $y = ux$ yields

$$\frac{u\,du}{\sqrt{1 + u^2}\left(1 - \sqrt{1 + u^2}\right)} = \frac{dx}{x}.$$

   Use a CAS (or another judicious substitution) to integrate the left-hand side of the equation. Show that the curve C must be a parabola with focus at the origin and is symmetric with respect to the x-axis.

(b) Show that the first differential equation can also be solved by means of the substitution $u = x^2 + y^2$.

19. **Tsunami** (a) A simple model for the shape of a tsunami, or tidal wave, is given by

$$\frac{dW}{dx} = W\sqrt{4 - 2W},$$

where $W(x) > 0$ is the height of the wave expressed as a function of its position relative to a point offshore. By inspection, find all constant solutions of the DE.

(b) Solve the differential equation in part (a). A CAS may be useful for integration.

(c) Use a graphing utility to obtain the graphs of all solutions that satisfy the initial condition $W(0) = 2$.

20. **Evaporation** An outdoor decorative pond in the shape of a hemispherical tank is to be filled with water pumped into the tank through an inlet in its bottom. Suppose that the radius of the tank is $R = 10$ ft, that water is pumped in at a rate of π ft^3/min, and that the tank is initially empty. See Figure 3.19. As the tank fills, it loses water through evaporation. Assume that the rate of evaporation is proportional to the area A of the surface of the water and that the constant of proportionality is $k = 0.01$.

(a) The rate of change dV/dt of the volume of the water at time t is a net rate. Use this net rate to determine a differential equation for the height h of the water at time t. The volume of the water shown in the figure is $V = \pi Rh^2 - \frac{1}{3}\pi h^3$, where $R = 10$. Express the area of the surface of the water $A = \pi r^2$ in terms of h.

(b) Solve the differential equation in part (a). Graph the solution.

(c) If there were no evaporation, how long would it take the tank to fill?

(d) With evaporation, what is the depth of the water at the time found in part (c)? Will the tank ever be filled? Prove your assertion.

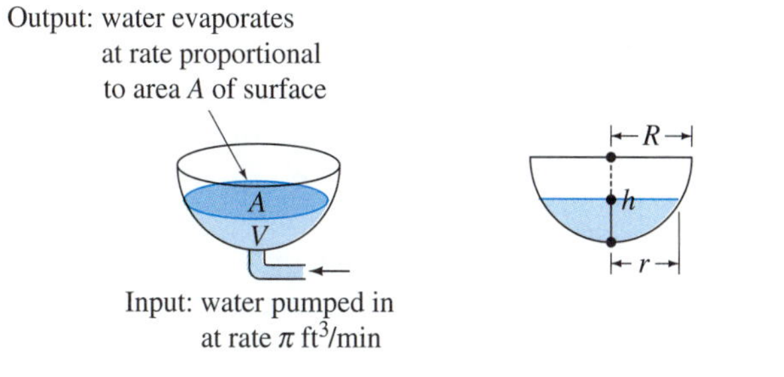

Output: water evaporates
at rate proportional
to area A of surface

Input: water pumped in
at rate π ft^3/min

(a) hemispherical tank   (b) cross-section of tank

FIGURE 3.19   Decorative pond in Problem 20

Computer Lab Assignments

21. **Regression Line** Read the documentation for your CAS on *scatter plots* (or *scatter diagrams*) and *least-squares linear fit*. The straight line that best fits a set of data points is called a **regression line** or a **least squares line.** Your task is to construct a logistic model for the population of the United States, defining $f(P)$ in (2) as an equation of a regression line based on the population data in the table in Problem 4. One way of doing this is to approximate the left-hand side $\frac{1}{P}\frac{dP}{dt}$ of the first equation in (2), using the forward difference quotient in place of dP/dt:

$$Q(t) = \frac{1}{P(t)}\frac{P(t+h) - P(t)}{h}.$$

(a) Make a table of the values t, $P(t)$, and $Q(t)$ using $t = 0, 10, 20, \ldots, 160$ and $h = 10$. For example, the first line of the table should contain $t = 0$, $P(0)$, and $Q(0)$. With $P(0) = 3.929$ and $P(10) = 5.308$,

$$Q(0) = \frac{1}{P(0)}\frac{P(10) - P(0)}{10} = 0.035.$$

Note that $Q(160)$ depends on the 1960 census population $P(170)$. Look up this value.

(b) Use a CAS to obtain a scatter plot of the data $(P(t), Q(t))$ computed in part (a). Also use a CAS to find an equation of the regression line and to superimpose its graph on the scatter plot.

(c) Construct a logistic model $dP/dt = Pf(P)$, where $f(P)$ is the equation of the regression line found in part (b).

(d) Solve the model in part (c) using the initial condition $P(0) = 3.929$.

(e) Use a CAS to obtain another scatter plot, this time of the ordered pairs $(t, P(t))$ from your table in part (a). Use your CAS to superimpose the graph of the solution in part (d) on the scatter plot.

(f) Look up the U.S. census data for 1970, 1980, and 1990. What population does the logistic model in part (c) predict for these years? What does the model predict for the U.S. population $P(t)$ as $t \to \infty$?

22. **Immigration Model** (a) In Examples 3 and 4 of Section 2.1, we saw that any solution $P(t)$ of (4) possesses the asymptotic behavior $P(t) \to a/b$ as $t \to \infty$ for $P_0 > a/b$ and for $0 < P_0 < a/b$; as a consequence, the equilibrium solution $P = a/b$ is called an attractor. Use a root-finding application of a CAS (or a graphic calculator) to approximate the equilibrium solution of the immigration model

$$\frac{dP}{dt} = P(1 - P) + 0.3e^{-P}.$$

(b) Use a graphing utility to graph the function $F(P) = P(1 - P) + 0.3e^{-P}$. Explain how this graph can be used to determine whether the number found in part (a) is an attractor.

(c) Use a numerical solver to compare the solution curves for the IVPs

$$\frac{dP}{dt} = P(1 - P), \quad P(0) = P_0$$

for $P_0 = 0.2$ and $P_0 = 1.2$ with the solution curves for the IVPs

$$\frac{dP}{dt} = P(1 - P) + 0.3e^{-P}, \quad P(0) = P_0$$

for $P_0 = 0.2$ and $P_0 = 1.2$. Superimpose all curves on the same coordinate axes but, if possible, use a different color for the curves of the second initial-value problem. Over a long period of time, what percentage increase does the immigration model predict in the population compared to the logistic model?

23. What Goes Up ... In Problem 16 let t_a be the time it takes the cannonball to attain its maximum height and let t_d be the time it takes the cannonball to fall from the maximum height to the ground. Compare the value of t_a with the value of t_d and compare the magnitude of the impact velocity v_i with the initial velocity v_0. See Problem 46 in Exercises 3.1. A root-finding application of a CAS might be useful here. (*Hint:* Use the model in Problem 15 when the cannonball is falling.)

24. Skydiving A skydiver is equipped with a stopwatch and an altimeter. As shown in Figure 3.20, he opens his parachute 25 seconds after exiting a plane flying at an altitude of 20,000 feet and observes that his altitude is 14,800 feet. Assume that air resistance is proportional to the square of the instantaneous velocity, his initial velocity on leaving the plane is zero, and $g = 32$ ft/s^2.

(a) Find the distance $s(t)$, measured from the plane, the skydiver has traveled during freefall in time t.

(*Hint:* The constant of proportionality k in the model given in Problem 15 is not specified. Use the expression for terminal velocity v_t obtained in part (b) of Problem 15 to eliminate k from the IVP. Then eventually solve for v_t.)

(b) How far does the skydiver fall and what is his velocity at $t = 15$ s?

FIGURE 3.20   Skydiver in Problem 24

25. Hitting Bottom A helicopter hovers 500 feet above a large open tank full of liquid (not water). A dense compact object weighing 160 pounds is dropped (released from rest) from the helicopter into the liquid. Assume that air resistance is proportional to instantaneous velocity v while the object is in the air and that viscous damping is proportional to v^2 after the object has entered the liquid. For air, take $k = \frac{1}{4}$, and for the liquid, $k = 0.1$. Assume that the positive direction is downward. If the tank is 75 feet high, determine the time and the impact velocity when the object hits the bottom of the tank. (*Hint:* Think in terms of two distinct IVPs. If you use (13), be careful in removing the absolute value sign. You might compare the velocity when the object hits the liquid—the initial velocity for the second problem—with the terminal velocity v_t of the object falling through the liquid.)

3.3   MODELING WITH SYSTEMS OF DIFFERENTIAL EQUATIONS

INTRODUCTION: This section is similar to Section 1.3 in that we are just going to discuss certain mathematical models, but instead of a single differential equation the models will be systems of first-order differential equations. Although some of the models will be based on topics explored in the preceding two sections, we are not going to develop any general methods for solving these systems. There are reasons for this: First, we do not possess the necessary mathematical tools for solving systems at this point. Second, some of the systems that we discuss—notably the systems of *nonlinear* first-order DEs—simply cannot be solved analytically. We shall examine solution methods for systems of *linear* DEs in Chapters 4, 7, and 8.

CD: The **Predator-Prey Tool** on the *DE Tools* CD can be used in conjunction with the discussion of that topic on pages 115, and 116.

LINEAR/NONLINEAR SYSTEMS We have seen that a single differential equation can serve as a mathematical model for a single population in an environment. But if there are, say, two interacting and perhaps competing species living in the same environment (for example, rabbits and foxes), then a model for their populations $x(t)$ and $y(t)$ might be a system of two first-order differential equations such as

$$\frac{dx}{dt} = g_1(t, x, y)$$

$$\frac{dy}{dt} = g_2(t, x, y). \tag{1}$$

When g_1 and g_2 are linear in the variables x and y—that is, g_1 and g_2 have the forms

$$g_1(t, x, y) = c_1 x + c_2 y + f_1(t) \quad \text{and} \quad g_2(t, x, y) = c_3 x + c_4 y + f_2(t),$$

where the coefficients c_i could depend on t—then (1) is said to be a **linear system.** A system of differential equations that is not linear is said to be **nonlinear.**

RADIOACTIVE SERIES In the discussion of radioactive decay in Sections 1.3 and 3.1 we assumed that the rate of decay was proportional to the number $A(t)$ of nuclei of the substance present at time t. When a substance decays by radioactivity, it usually doesn't just transmute in one step into a stable substance; rather, the first substance decays into another radioactive substance, which in turn decays into a third substance, and so on. This process, called a **radioactive decay series,** continues until a stable element is reached. For example, the uranium decay series is U-238 $\rightarrow$ Th-234 $\rightarrow \cdots \rightarrow$ Pb-206, where Pb-206 is a stable isotope of lead. The half-lives of the various elements in a radioactive series can range from billions of years (4.5×10^9 years for U-238) to a fraction of a second. Suppose a radioactive series is described schematically by $X \xrightarrow{-\lambda_1} Y \xrightarrow{-\lambda_2} Z$, where $k_1 = -\lambda_1 < 0$ and $k_2 = -\lambda_2 < 0$ are the decay constants for substances X and Y, respectively, and Z is a stable element. Suppose, too, that $x(t)$, $y(t)$, and $z(t)$ denote amounts of substances X, Y, and Z, respectively, remaining at time t. The decay of element X is described by

$$\frac{dx}{dt} = -\lambda_1 x,$$

whereas the rate at which the second element Y decays is the net rate

$$\frac{dy}{dt} = \lambda_1 x - \lambda_2 y,$$

since Y is *gaining* atoms from the decay of X and at the same time *losing* atoms because of its own decay. Since Z is a stable element, it is simply gaining atoms from the decay of element Y:

$$\frac{dz}{dt} = \lambda_2 y.$$

In other words, a model of the radioactive decay series for three elements is the linear system of three first-order differential equations

$$\frac{dx}{dt} = -\lambda_1 x$$

$$\frac{dy}{dt} = \lambda_1 x - \lambda_2 y \tag{2}$$

$$\frac{dz}{dt} = \lambda_2 y.$$

MIXTURES   Consider the two tanks shown in Figure 3.21. Let us suppose for the sake of discussion that tank A contains 50 gallons of water in which 25 pounds of salt is dissolved. Suppose tank B contains 50 gallons of pure water. Liquid is pumped in and out of the tanks as indicated in the figure; the mixture exchanged between the two tanks and the liquid pumped out of tank B are assumed to be well stirred. We wish to construct a mathematical model that describes the number of pounds $x_1(t)$ and $x_2(t)$ of salt in tanks A and B, respectively, at time t.

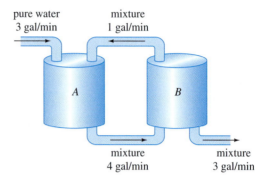

pure water
3 gal/min

mixture
1 gal/min

A B

mixture
4 gal/min

mixture
3 gal/min

FIGURE 3.21   Connected mixing tanks

By an analysis similar to that on page 23 in Section 1.3 and Example 5 of Section 3.1 we see that the net rate of change of $x_1(t)$ for tank A is

$$\frac{dx_1}{dt} = \overbrace{(3 \text{ gal/min}) \cdot (0 \text{ lb/gal}) + (1 \text{ gal/min}) \cdot \left(\frac{x_2}{50} \text{ lb/gal}\right)}^{\substack{\text{input rate} \\ \text{of salt}}} - \overbrace{(4 \text{ gal/min}) \cdot \left(\frac{x_1}{50} \text{ lb/gal}\right)}^{\substack{\text{output rate} \\ \text{of salt}}}$$

$$= -\frac{2}{25}x_1 + \frac{1}{50}x_2.$$

Similarly, for tank B the net rate of change of $x_2(t)$ is

$$\frac{dx_2}{dt} = 4 \cdot \frac{x_1}{50} - 3 \cdot \frac{x_2}{50} - 1 \cdot \frac{x_2}{50}$$

$$= \frac{2}{25}x_1 - \frac{2}{25}x_2.$$

Thus we obtain the linear system

$$\frac{dx_1}{dt} = -\frac{2}{25}x_1 + \frac{1}{50}x_2$$

$$\frac{dx_2}{dt} = \frac{2}{25}x_1 - \frac{2}{25}x_2. \tag{3}$$

Observe that the foregoing system is accompanied by the initial conditions $x_1(0) = 25$, $x_2(0) = 0$.

A PREDATOR-PREY MODEL   Suppose that two different species of animals interact within the same environment or ecosystem, and suppose further that the first species eats only vegetation and the second eats only the first species. In other words, one species is a predator and the other is a prey. For example, wolves hunt grass-eating caribou, sharks devour little fish, and the snowy owl pursues an arctic rodent called the lemming. For the sake of discussion, let us imagine that the predators are foxes and the prey are rabbits.

Let $x(t)$ and $y(t)$ denote the fox and rabbit populations, respectively, at time t. If there were no rabbits, then one might expect that the foxes, lacking an adequate food supply, would decline in number according to

$$\frac{dx}{dt} = -ax, \quad a > 0. \tag{4}$$

When rabbits are present in the environment, however, it seems reasonable that the number of encounters or interactions between these two species per unit time is jointly proportional to their populations x and y—that is, proportional to the product xy. Thus when rabbits are present there is a supply of food, and so foxes are added to the system at a rate bxy, $b > 0$. Adding this last rate to (4) gives a model for the fox population:

$$\frac{dx}{dt} = -ax + bxy. \tag{5}$$

On the other hand, if there were no foxes, then the rabbits would, with an added assumption of unlimited food supply, grow at a rate that is proportional to the number of rabbits present at time t:

$$\frac{dy}{dt} = dy, \quad d > 0. \tag{6}$$

But when foxes are present, a model for the rabbit population is (6) decreased by cxy, $c > 0$—that is, decreased by the rate at which the rabbits are eaten during their encounters with the foxes:

$$\frac{dy}{dt} = dy - cxy. \tag{7}$$

Equations (5) and (7) constitute a system of nonlinear differential equations

$$\frac{dx}{dt} = -ax + bxy = x(-a + by)$$
$$\frac{dy}{dt} = dy - cxy = y(d - cx), \tag{8}$$

where a, b, c, and d are positive constants. This famous system of equations is known as the **Lotka-Volterra predator-prey model.**

Except for two constant solutions, $x(t) = 0$, $y(t) = 0$ and $x(t) = d/c$, $y(t) = a/b$, the nonlinear system (8) cannot be solved in terms of elementary functions. However, we can analyze such systems quantitatively and qualitatively. See Chapter 9, "Numerical Solutions of Differential Equations," and Chapter 10, "Plane Autonomous Systems and Stability."*

EXAMPLE 1   Predator-Prey Model

Suppose

$$\frac{dx}{dt} = -0.16x + 0.08xy$$

$$\frac{dy}{dt} = 4.5y - 0.9xy$$

---

*Chapters 10–15 are in the expanded version of this text, *Differential Equations with Boundary-Value Problems.*

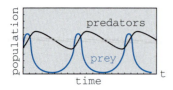

FIGURE 3.22 Populations of predators (black) and prey (color) appear to be periodic

represents a predator-prey model. Because we are dealing with populations, we have $x(t) \geq 0$, $y(t) \geq 0$. Figure 3.22, obtained with the aid of a numerical solver, shows typical population curves of the predators and prey for this model superimposed on the same coordinate axes. The initial conditions used were $x(0) = 4$, $y(0) = 4$. The curve in black represents the population $x(t)$ of the predators (foxes), and the colored curve is the population $y(t)$ of the prey (rabbits). Observe that the model seems to predict that both populations $x(t)$ and $y(t)$ are periodic in time. This makes intuitive sense because as the number of prey decreases, the predator population eventually decreases because of a diminished food supply; but attendant to a decrease in the number of predators is an increase in the number of prey; this in turn gives rise to an increased number of predators, which ultimately brings about another decrease in the number of prey.

COMPETITION MODELS Now suppose two different species of animals occupy the same ecosystem, not as predator and prey but rather as competitors for the same resources (such as food and living space) in the system. In the absence of the other, let us assume that the rate at which each population grows is given by

$$\frac{dx}{dt} = ax \quad \text{and} \quad \frac{dy}{dt} = cy, \tag{9}$$

respectively.

Since the two species compete, another assumption might be that each of these rates is diminished simply by the influence, or existence, of the other population. Thus a model for the two populations is given by the linear system

$$\frac{dx}{dt} = ax - by$$

$$\frac{dy}{dt} = cy - dx, \tag{10}$$

where a, b, c, and d are positive constants.

On the other hand, we might assume, as we did in (5), that each growth rate in (9) should be reduced by a rate proportional to the number of interactions between the two species:

$$\frac{dx}{dt} = ax - bxy$$

$$\frac{dy}{dt} = cy - dxy. \tag{11}$$

Inspection shows that this nonlinear system is similar to the Lotka-Volterra predator-prey model. Finally, it might be more realistic to replace the rates in (9), which indicate that the population of each species in isolation grows exponentially, with rates indicating that each population grows logistically (that is, over a long time the population is bounded):

$$\frac{dx}{dt} = a_1 x - b_1 x^2 \quad \text{and} \quad \frac{dy}{dt} = a_2 y - b_2 y^2. \tag{12}$$

When these new rates are decreased by rates proportional to the number of interactions, we obtain another nonlinear model

$$\frac{dx}{dt} = a_1 x - b_1 x^2 - c_1 xy = x(a_1 - b_1 x - c_1 y)$$

$$\frac{dy}{dt} = a_2 y - b_2 y^2 - c_2 xy = y(a_2 - b_2 y - c_2 x), \tag{13}$$

where all coefficients are positive. The linear system (10) and the nonlinear systems (11) and (13) are, of course, called **competition models.**

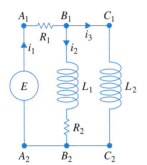

NETWORKS   An electrical network having more than one loop also gives rise to simultaneous differential equations. As shown in Figure 3.23, the current $i_1(t)$ splits in the directions shown at point B_1, called a *branch point* of the network. By **Kirchhoff's first law** we can write

$$i_1(t) = i_2(t) + i_3(t). \qquad (14)$$

In addition, we can also apply **Kirchhoff's second law** to each loop. For loop $A_1B_1B_2A_2A_1$, summing the voltage drops across each part of the loop gives

$$E(t) = i_1R_1 + L_1\frac{di_2}{dt} + i_2R_2. \qquad (15)$$

Similarly, for loop $A_1B_1C_1C_2B_2A_2A_1$ we find

$$E(t) = i_1R_1 + L_2\frac{di_3}{dt}. \qquad (16)$$

FIGURE 3.23   Network whose model is given in (17)

Using (14) to eliminate i_1 in (15) and (16) yields two linear first-order equations for the currents $i_2(t)$ and $i_3(t)$:

$$L_1\frac{di_2}{dt} + (R_1 + R_2)i_2 + R_1i_3 = E(t)$$
$$L_2\frac{di_3}{dt} + \qquad R_1i_2 + R_1i_3 = E(t). \qquad (17)$$

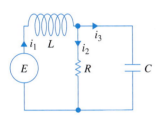

We leave it as an exercise (see Problem 14) to show that the system of differential equations describing the currents $i_1(t)$ and $i_2(t)$ in the network containing a resistor, an inductor, and a capacitor shown in Figure 3.24 is

$$L\frac{di_1}{dt} + Ri_2 \qquad = E(t)$$
$$RC\frac{di_2}{dt} + \quad i_2 - i_1 = 0. \qquad (18)$$

FIGURE 3.24   Network whose model is given in (18)

Answers to selected odd-numbered problems begin on page ANS-3.

RADIOACTIVE SERIES

1. We have not discussed methods by which systems of first-order differential equations can be solved. Nevertheless, systems such as (2) can be solved with no knowledge other than how to solve a single linear first-order equation. Find a solution of (2) subject to the initial conditions $x(0) = x_0$, $y(0) = 0$, $z(0) = 0$.

2. In Problem 1, suppose that time is measured in days, that the decay constants are $k_1 = -0.138629$ and $k_2 = -0.004951$, and that $x_0 = 20$. Use a graphing utility to obtain the graphs of the solutions $x(t)$, $y(t)$, and

$z(t)$ on the same set of coordinate axes. Use the graphs to approximate the half-lives of substances X and Y.

3. Use the graphs in Problem 2 to approximate the times when the amounts $x(t)$ and $y(t)$ are the same, the times when the amounts $x(t)$ and $z(t)$ are the same, and the times when the amounts $y(t)$ and $z(t)$ are the same. Why does the time determined when the amounts $y(t)$ and $z(t)$ are the same make intuitive sense?

4. Construct a mathematical model for a radioactive series of four elements W, X, Y, and Z, where Z is a stable element.

MIXTURES

5. Consider two tanks A and B, with liquid being pumped in and out at the same rates, as described by the system

of equations (3). What is the system of differential equations if, instead of pure water, a brine solution containing 2 pounds of salt per gallon is pumped into tank A?

6. Use the information given in Figure 3.25 to construct a mathematical model for the number of pounds of salt $x_1(t)$, $x_2(t)$ and $x_3(t)$ at time t in tanks A, B, and C, respectively.

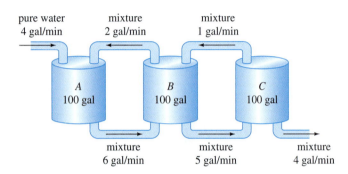

FIGURE 3.25 Mixing tanks in Problem 6

7. Two very large tanks A and B are each partially filled with 100 gallons of brine. Initially, 100 pounds of salt is dissolved in the solution in tank A and 50 pounds of salt is dissolved in the solution in tank B. The system is closed in that the well-stirred liquid is pumped only between the tanks, as shown in Figure 3.26.

(a) Use the information given in the figure to construct a mathematical model for the number of pounds of salt $x_1(t)$ and $x_2(t)$ at time t in tanks A and B, respectively.

(b) Find a relationship between the variables $x_1(t)$ and $x_2(t)$ that holds at time t. Explain why this relationship makes intuitive sense. Use this relationship to help find the amount of salt in tank B at $t = 30$ min.

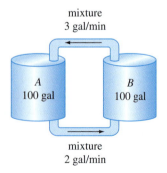

FIGURE 3.26 Mixing tanks in Problem 7

8. Three large tanks contain brine, as shown in Figure 3.27. Use the information in the figure to construct a

mathematical model for the number of pounds of salt $x_1(t)$, $x_2(t)$, and $x_3(t)$ at time t in tanks A, B, and C, respectively. Without solving the system, predict limiting values of $x_1(t)$, $x_2(t)$, and $x_3(t)$ as $t \to \infty$.

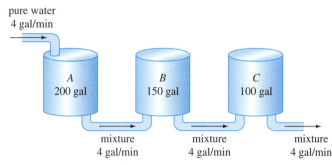

FIGURE 3.27 Mixing tanks in Problem 8

PREDATOR-PREY MODELS

9. Consider the Lotka-Volterra predator-prey model defined by

$$\frac{dx}{dt} = -0.1x + 0.02xy$$

$$\frac{dy}{dt} = 0.2y - 0.025xy,$$

where the populations $x(t)$ (predators) and $y(t)$ (prey) are measured in thousands. Suppose $x(0) = 6$ and $y(0) = 6$. Use a numerical solver to graph $x(t)$ and $y(t)$. Use the graphs to approximate the time $t > 0$ when the two populations are first equal. Use the graphs to approximate the period of each population.

COMPETITION MODELS

10. Consider the competition model defined by

$$\frac{dx}{dt} = x(2 - 0.4x - 0.3y)$$

$$\frac{dy}{dt} = y(1 - 0.1y - 0.3x),$$

where the populations $x(t)$ and $y(t)$ are measured in thousands and t in years. Use a numerical solver to analyze the populations over a long period of time for each of the following cases:
(a) $x(0) = 1.5$, $y(0) = 3.5$
(b) $x(0) = 1$, $y(0) = 1$
(c) $x(0) = 2$, $y(0) = 7$
(d) $x(0) = 4.5$, $y(0) = 0.5$

11. Consider the competition model defined by

$$\frac{dx}{dt} = x(1 - 0.1x - 0.05y)$$

$$\frac{dy}{dt} = y(1.7 - 0.1y - 0.15x),$$

where the populations $x(t)$ and $y(t)$ are measured in thousands and t in years. Use a numerical solver to analyze the populations over a long period of time for each of the following cases:
(a) $x(0) = 1,\quad y(0) = 1$
(b) $x(0) = 4,\quad y(0) = 10$
(c) $x(0) = 9,\quad y(0) = 4$
(d) $x(0) = 5.5,\quad y(0) = 3.5$

NETWORKS

12. Show that a system of differential equations that describes the currents $i_2(t)$ and $i_3(t)$ in the electrical network shown in Figure 3.28 is

$$L\frac{di_2}{dt} + L\frac{di_3}{dt} + R_1 i_2 = E(t)$$

$$-R_1\frac{di_2}{dt} + R_2\frac{di_3}{dt} + \frac{1}{C}i_3 = 0.$$

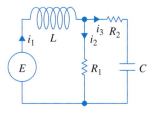

FIGURE 3.28   Network in Problem 12

13. Determine a system of first-order differential equations that describes the currents $i_2(t)$ and $i_3(t)$ in the electrical network shown in Figure 3.29.

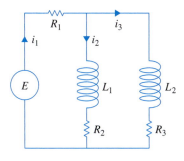

FIGURE 3.29   Network in Problem 13

14. Show that the linear system given in (18) describes the currents $i_1(t)$ and $i_2(t)$ in the network shown in Figure 3.24. (*Hint: dq/dt = i_3.*)

MISCELLANEOUS MATHEMATICAL MODELS

15. SIR Model A communicable disease is spread throughout a small community, with a fixed population of n people, by contact between infected individuals and people who are susceptible to the disease. Suppose initially that everyone is susceptible to the disease and that no one leaves the community while the epidemic is spreading. At time t, let $s(t)$, $i(t)$, and $r(t)$ denote, in turn, the number of people in the community (measured in hundreds) who are *susceptible* to the disease but not yet infected with it, the number of people who are *infected* with the disease, and the number of people who have *recovered* from the disease. Explain why the system of differential equations

$$\frac{ds}{dt} = -k_1 si$$

$$\frac{di}{dt} = -k_2 i + k_1 si$$

$$\frac{dr}{dt} = k_2 i,$$

where k_1 (called the *infection rate*) and k_2 (called the *removal rate*) are positive constants, is a reasonable mathematical model, commonly called a **SIR model,** for the spread of the epidemic throughout the community. Give plausible initial conditions associated with this system of equations.

16. (a) In Problem 15, explain why it is sufficient to analyze only

$$\frac{ds}{dt} = -k_1 si$$

$$\frac{di}{dt} = -k_2 i + k_1 si.$$

(b) Suppose $k_1 = 0.2$, $k_2 = 0.7$, and $n = 10$. Choose various values of $i(0) = i_0$, $0 < i_0 < 10$. Use a numerical solver to determine what the model predicts about the epidemic in the two cases $s_0 > k_2/k_1$ and $s_0 \leq k_2/k_1$. In the case of an epidemic, estimate the number of people who are eventually infected.

DISCUSSION/PROJECT PROBLEMS

17. Concentration of a Nutrient Suppose compartments A and B shown in Figure 3.30 are filled with fluids and are separated by a permeable membrane. The figure is a compartmental representation of the exterior and interior of a cell. Suppose, too, that a nutrient necessary for cell growth passes through the membrane. A model for the concentrations $x(t)$ and $y(t)$ of the nutrient in compartments A and B, respectively, at time t is given by the linear system of differential equations

$$\frac{dx}{dt} = \frac{\kappa}{V_A}(y - x)$$

$$\frac{dy}{dt} = \frac{\kappa}{V_B}(x - y),$$

where V_A and V_B are the volumes of the compartments, and $\kappa > 0$ is a permeability factor. Let $x(0) = x_0$ and $y(0) = y_0$ denote the initial concentrations of the nutrient. Solely on the basis of the equations in the system and the assumption $x_0 > y_0 > 0$, sketch, on the same set of coordinate axes, possible solution curves of the system. Explain your reasoning. Discuss the behavior of the solutions over a long period of time.

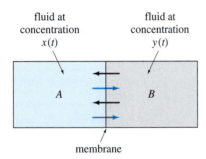

fluid at concentration $x(t)$

fluid at concentration $y(t)$

A B

membrane

FIGURE 3.30 Nutrient flow through a membrane in Problem 17

18. The system in Problem 17, like the system in (2), can be solved with no advanced knowledge. Solve for $x(t)$ and $y(t)$ and compare their graphs with your sketches in Problem 17. Determine the limiting values of $x(t)$

and $y(t)$ as $t \to \infty$. Explain why the answer to the last question makes intuitive sense.

19. Solely on the basis of the physical description of the mixture problem on page 114 and in Figure 3.21, discuss the nature of the functions $x_1(t)$ and $x_2(t)$. What is the behavior of each function over a long period of time? Sketch possible graphs of $x_1(t)$ and $x_2(t)$. Check your conjectures by using a numerical solver to obtain numerical solution curves of (3) subject to the initial conditions $x_1(0) = 25$, $x_2(0) = 0$.

20. Newton's Law of Cooling/Warming As shown in Figure 3.31, a small metal bar is placed inside container A, and container A then is placed within a much larger container B. As the metal bar cools, the ambient temperature $T_A(t)$ of the medium within container A changes according to Newton's law of cooling. As container A cools, the temperature of the medium inside container B does not change significantly and can be considered to be a constant T_B. Construct a mathematical model for the temperatures $T(t)$ and $T_A(t)$, where $T(t)$ is the temperature of the metal bar inside container A. As in Problems 1 and 18, this model can be solved by using prior knowledge. Find a solution of the system subject to the initial conditions $T(0) = T_0$, $T_A(0) = T_1$.

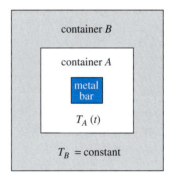

container B

container A

metal bar

$T_A(t)$

T_B = constant

FIGURE 3.31 Container within a container in Problem 20

CHAPTER 3 IN REVIEW

Answers to selected odd-numbered problems begin on page ANS-3.

Answer Problems 1 and 2 without referring back to the text. Fill in the blank or answer true or false.

1. If $P(t) = P_0 e^{0.15t}$ gives the population in an environment at time t, then a differential equation satisfied by $P(t)$ is _____.

2. If the rate of decay of a radioactive substance is proportional to the amount $A(t)$ remaining at time t, then the half-life of the substance is necessarily

$T = -(\ln 2)/k$. The rate of decay of the substance at time $t = T$ is one-half the rate of decay at $t = 0$.__

3. In March 1976 the world population reached 4 billion. A popular news magazine predicted that with an average yearly growth rate of 1.8%, the world population would be 8 billion in 45 years. How does this value compare with that predicted by the model that assumes the rate of increase in population is proportional to the population present at time t?

4. Air containing 0.06% carbon dioxide is pumped into a room whose volume is 8000 ft^3. The air is pumped in at a rate of 2000 ft^3/min, and the circulated air is then pumped out at the same rate. If there is an initial concentration of 0.2% carbon dioxide in the room, determine the subsequent amount in the room at time t. What is the concentration at 10 minutes? What is the steady-state, or equilibrium, concentration of carbon dioxide?

5. Solve the differential equation

$$\frac{dy}{dx} = -\frac{y}{\sqrt{s^2 - y^2}}$$

of the tractrix. See Problem 28 in Exercises 1.3. Assume that the initial point on the y-axis in (0, 10) and that the length of the rope is $x = 10$ ft.

6. Suppose a cell is suspended in a solution containing a solute of constant concentration C_s. Suppose further that the cell has constant volume V and that the area of its permeable membrane is the constant A. By **Fick's law** the rate of change of its mass m is directly proportional to the area A and the difference $C_s - C(t)$, where $C(t)$ is the concentration of the solute inside the cell at time t. Find $C(t)$ if $m = V \cdot C(t)$ and $C(0) = C_0$. See Figure 3.32.

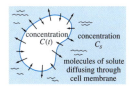

FIGURE 3.32 Cell in Problem 6

7. Suppose that as a body cools, the temperature of the surrounding medium increases because it completely absorbs the heat being lost by the body. Let $T(t)$ and $T_m(t)$ be the temperatures of the body and the medium at time t, respectively. If the initial temperature of the body is T_1 and the initial temperature of the medium is T_2, then it can be shown in this case that Newton's law of cooling is $dT/dt = k(T - T_m)$, $k < 0$, where $T_m = T_2 + B(T_1 - T)$, $B > 0$ is a constant.

(a) The foregoing DE is autonomous. Use the phase portrait concept of Section 2.1 to determine the limiting value of the temperature $T(t)$ as $t \to \infty$. What is the limiting value of $T_m(t)$ as $t \to \infty$?

(b) Verify your answers in part (a) by actually solving the differential equation.

(c) Discuss a physical interpretation of your answers in part (a).

8. According to **Stefan's law of radiation,** the absolute temperature T of a body cooling in a medium at constant absolute temperature T_m is given by

$$\frac{dT}{dt} = k(T^4 - T_m^4),$$

where k is a constant. Stefan's law can be used over a greater temperature range than Newton's law of cooling.

(a) Solve the differential equation.

(b) Show that when $T - T_m$ is small compared to T_m then Newton's law of cooling approximates Stefan's law. (*Hint:* Think binomial series of the right-hand side of the DE.)

9. An *LR* series circuit has a variable inductor with the inductance defined by

$$L(t) = \begin{cases} 1 - \dfrac{1}{10}t, & 0 \le t < 10 \\ 0, & t \ge 10. \end{cases}$$

Find the current $i(t)$ if the resistance is 0.2 ohm, the impressed voltage is $E(t) = 4$, and $i(0) = 0$. Graph $i(t)$.

10. A classical problem in the calculus of variations is to find the shape of a curve $\mathscr{C}$ such that a bead, under the influence of gravity, will slide from point $A(0, 0)$ to point $B(x_1, y_1)$ in the least time. See Figure 3.33. It can be shown that a nonlinear differential for the shape $y(x)$ of the path is $y[1 + (y')^2] = k$, where k is a constant. First solve for dx in terms of y and dy, and then use the substitution $y = k \sin^2\theta$ to obtain a parametric form of the solution. The curve $\mathscr{C}$ turns out to be a cycloid.

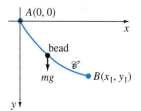

FIGURE 3.33 Sliding bead in Problem 10

The **clepsydra,** or water clock, was a device used by the ancient Egyptians, Greeks, Romans, and Chinese to measure the passage of time by observing the change in the height of water that was permitted to flow out of a small hole in the bottom of a container or tank. In Problems 11–14 use the differential equation (see Problems 11–14 in Exercises 3.2),

$$\frac{dh}{dt} = -c\frac{A_h}{A_w}\sqrt{2gh}$$

as a model for the height h of water in a tank at time t. Assume in each of these problems that $h(0) = 2$ ft corresponds to water filled to the top of the tank, the hole in the bottom is circular with radius $\frac{1}{32}$ in., $g = 32$ ft/s^2, and that $c = 0.6$.

11. Suppose that a tank is made of glass and has the shape of a right circular cylinder of radius 1 ft. Find the height $h(t)$ of the water.

12. For the tank in Problem 11, how far up from its bottom should a mark be made on its side, as shown in Figure 3.34, that corresponds to the passage of 1 hour? Continue and determine where to place the marks corresponding to the passage of 2 h, 3 h, . . . , 12 h. Explain why these marks are not evenly spaced.

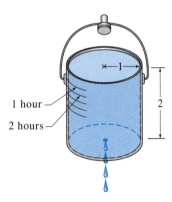

FIGURE 3.34   Clepsydra in Problem 12

13. Suppose that the glass tank has the shape of a cone with circular cross sections as shown in Figure 3.35. Can this water clock measure 12 time intervals of length equal to 1 hour? Explain using sound mathematics.

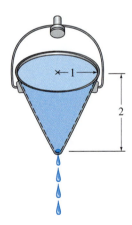

FIGURE 3.35   Clepsydra in Problem 13

14. Suppose that $r = f(h)$ defines the shape of a water clock for which the time marks are equally spaced. Use the above differential equation to find $f(h)$ and sketch a typical graph of h as a function of r. Assume that the cross-sectional area A_h of the hole is constant. (*Hint:* In this situation, $dh/dt = -a$, where $a > 0$ is a constant.)

15. A model for the populations of two interacting species of animals is

$$\frac{dx}{dt} = k_1 x(\alpha - x)$$

$$\frac{dy}{dt} = k_2 xy.$$

Solve for x and y in terms of t.

16. Initially, two large tanks A and B each hold 100 gallons of brine. The well-stirred liquid is pumped between the tanks as shown in Figure 3.36. Use the information given in the figure to construct a mathematical model for the number of pounds of salt $x_1(t)$ and $x_2(t)$ at time t in tanks A and B, respectively.

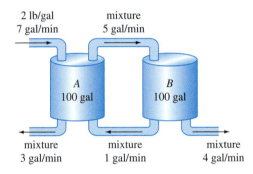

FIGURE 3.36   Mixing tanks in Problem 16

When all the curves in a family $G(x, y, c_1) = 0$ intersect orthogonally all the curves in another family $H(x, y, c_2) = 0$, the families are said to be **orthogonal trajectories** of each other. See Figure 3.37. If $dy/dx = f(x, y)$ is the differential equation of one family, then the differential equation for the orthogonal trajectories of this family is $dy/dx = -1/f(x, y)$. In Problems 17 and 18, find the differential equation of the given family. Find the orthogonal trajectories of this family. Use a graphing utility to graph both families on the same set of coordinate axes.

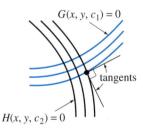

FIGURE 3.37   Orthogonal trajectories

17. $y = -x - 1 + c_1 e^x$ 18. $y = \dfrac{1}{x + c_1}$

SWIMMING THE SALMON RIVER

FIGURE 1   Rafting on the Salmon River

It's been a hot day on the Salmon River. See Figure 1.* Your rafting party has stopped for lunch on a pleasant stretch of sand, and you are basking in the warmth of the sun on the beach. After a few minutes, you notice two members of the opposite sex pull their kayaks out onto the beach directly across from you. Suddenly, the beach on the other side of the river seems more inviting than the one on your side. Your rafting companions don't share that interest. You ask yourself, "How fast must I swim to get across and not be swept downstream into Fiddle Creek Rapids?" It would completely ruin the panache you show by swimming the river if you died in the attempt.

Let's derive a system of differential equations that describes your velocity at any point in the river. The first step is to introduce a coordinate system. Let the river flow in the positive y-direction or northward, locate the kayakers at the point $(0, 0)$ on the west beach, and locate your position on the east beach at $(w, 0)$. See Figure 2(a). Suppose that the river is w feet across and is flowing at a constant rate of v_r ft/s.

Suppose further that you swim at a rate of v_s ft/s relative to the river and that your *velocity is always directed toward the kayak party*. See Figure 2(b). We want to know the answer to the question "What does v_s have to be in order for there to be a solution from $(w, 0)$ to $(0, 0)$?"

If you are at the point $(x(t), y(t))$ at time t, then your velocity vector **v** at this point is the sum of two vectors $\mathbf{v} = \mathbf{v}_s + \mathbf{v}_r$ representing, in turn, your speed and direction of swimming ($\mathbf{v}_s$) and the speed and direction of the river ($\mathbf{v}_r$). Because your swimming direction is always directed toward the kayaks at $(0, 0)$, $\mathbf{v}_s$ has components in both the x- and y-directions, while

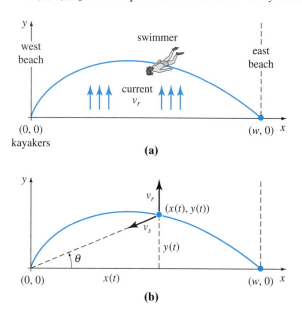

FIGURE 2   Path of swimmer

*The Lower Main Salmon River runs through Idaho.

the river's velocity $\mathbf{v}_r$ has only a component in the y-direction. Using the fact that dx/dt is the component of $\mathbf{v}$ in the x-direction and dy/dt is the component of $\mathbf{v}$ in the y-direction we see from Figure 2(b) that

$$\mathbf{v} = \left(\frac{dx}{dt}, \frac{dy}{dt}\right) = \mathbf{v}_s + \mathbf{v}_r = (-v_s \cos\theta, -v_s \sin\theta) + (0, v_r) \tag{1}$$

$$= (-v_s \cos\theta, v_r - v_s \sin\theta),$$

where the magnitudes $|\mathbf{v}_r| = v_r$ and $|\mathbf{v}_s| = v_s$ are speeds. By using the fact that the corresponding components in (1) are equal and that $\cos\theta = x/\sqrt{x^2 + y^2}$, $\sin\theta = y/\sqrt{x^2 + y^2}$ we've constructed the system of first-order differential equations

$$\frac{dx}{dt} = -v_s \frac{x}{\sqrt{x^2 + y^2}}$$

$$\frac{dy}{dt} = v_r - v_s \frac{y}{\sqrt{x^2 + y^2}}. \tag{2}$$

Because $\dfrac{dy}{dx} = \dfrac{dy/dt}{dx/dt}$, the solution curves of this system of differential equations satisfy the single first-order equation

$$\frac{dy}{dx} = \frac{y - (v_r/v_s)\sqrt{x^2 + y^2}}{x}. \tag{3}$$

Your goal is to now solve equation (3) and determine what values of v_r/v_s allow you to reach the kayakers and avoid the embarrassment of being dashed on the rocks below.

PROBLEM 1. Because w is the initial distance between you and the kayakers, the initial condition associated with (3) is $y(w) = 0$. Solve this initial-value problem using one of the methods discussed in Section 2.5.

PROBLEM 2 (CAS). Without loss of generality we can let $w = 1$ so that your position on the east beach is $(1, 0)$. Use a CAS to graph the solution in Problem 1 in the case $v_r = 1$ for various values of v_s—that is, for $0 < v_s < 1$, $v_s = 1$, and $v_s > 1$. Determine under what conditions you either make it to the kayaks or don't. Repeat this problem for other values of v_r. Make a conjecture on the relationship between v_r and v_s necessary to ensure that you safely make it to the kayaks. Also see the **Swimming Tool** on the *DE Tools* CD.

PROBLEM 3. Prove that if you swim at a speed v_s greater that the current speed of the river—that is, $v_s > v_r$, then you will reach the kayaks. Explain from a physical perspective why this condition is necessary even though you are not swimming directly into the current.

Another approach to the problem is to use polar coordinates. This is reasonable since we assume that your swimming velocity is always directed toward the kayaks at $(0, 0)$. In other words, the angular component of this velocity is zero. Let's begin by expressing the system in (2) in polar coordinates. Recall that $x = r\cos\theta$, $y = r\sin\theta$, $r^2 = x^2 + y^2$, and that $\tan\theta = y/x$. Differentiating the last two equations implicitly with respect to t gives

$$\frac{dr}{dt} = \frac{1}{r}\left[x\frac{dx}{dt} + y\frac{dy}{dt}\right]$$

$$\frac{d\theta}{dt} = \frac{1}{r^2}\left[x\frac{dy}{dt} - y\frac{dx}{dt}\right]. \tag{4}$$

PROBLEM 4. Use the differential equations in (4) to show that

$$\frac{dr}{dt} = v_r \sin\theta - v_s$$

$$\frac{d\theta}{dt} = \frac{v_r}{r}\cos\theta,$$

$$(5)$$

and thus

$$\frac{dr}{d\theta} - \left(\frac{v_r\sin\theta - v_s}{v_r\cos\theta}\right)r = 0.$$

$$(6)$$

PROBLEM 5. Use the methods of either Section 2.2 or Section 2.3 to solve the initial-value problem consisting of equation (6) and the initial condition $r(0) = w$.

PROBLEM 6. Repeat Problem 3 using the solution of equation (6). In particular, at what angle do you approach the kayakers? Explain the physical interpretation of this condition and relate it to your answer in Problem 3.

STILL CURIOUS?

Your friend Bubba decides to follow you and swim to the west beach to meet the kayakers. Bubba does not understand why you swam directly at the kayakers. He thinks that he must be able to reach the opposite beach by just swimming directly west (relative to the river) at constant rate v_s. He is confident that he can swim fast enough to avoid being swept into Fiddle Creek Rapids 3 miles downstream from the point (1, 0). He plans to simply walk to the kayakers' position when he hits the beach.

PROBLEM 7. Show that the model for the path Bubba takes in the river is

$$\frac{dy}{dx} = -\frac{v_r}{v_s}.$$

PROBLEM 8. The current speed v_r of a river is usually not a constant. Rather, an approximation to the current speed (measured in miles per hour) could be a function such as $v_r = 30x(1 - x)$, $0 \leq x \leq 1$, the values of which are smallest at the shores (in this case, $x = 0$ and $x = 1$) and largest in the middle of the river. Assume that Bubba starts from (1, 0) and that $v_s = 2$ mph. Solve the DE in Problem 7 with v_r as given. Will he make it across the river, or will he be swept into the rapids? If he makes it across the river, how far will he have to walk to reach the kayakers?

You are also encouraged to explore the **Bungee Jumping Tool** on the *DE Tools* CD; it may provide you additional insights into this project.

4

HIGHER-ORDER DIFFERENTIAL EQUATIONS

We turn now to the solution of ordinary differential equations of order two or higher. In the first seven sections of this chapter we examine the underlying theory and the methods for solving certain kinds of *linear* DEs. The elimination method for solving system of linear DEs is introduced in Section 4.8 because this method simply uncouples a system into individual linear equations in each dependent variable. The chapter concludes with a brief examination of *nonlinear* higher-order equations.

Solution Curves: $\Delta > 0$

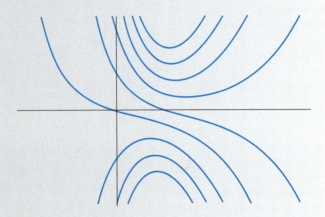

Solution Curves: $\Delta = 0$

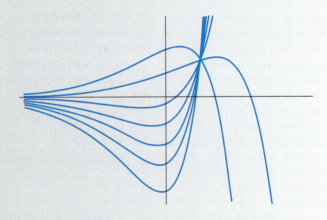

Solution Curves: $\Delta < 0$

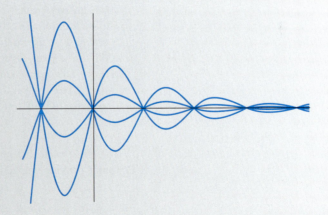

The general solution of a linear constant-coefficient second-order DE $ay'' + by' + cy = 0$ depends on the value of $\Delta = b^2 - 4ac$. See pages 143 and 144.

4.1 LINEAR DIFFERENTIAL EQUATIONS: BASIC THEORY

INTRODUCTION: In Chapter 2 we saw that we could solve a few first-order differential equations by recognizing them as separable, linear, exact, homogeneous, or perhaps Bernoulli equations. Even though the solutions of these equations were in the form of a one-parameter family, this family, with one exception, did not represent the general solution of the differential equation. Only in the case of *linear* first-order DEs were we able to obtain general solutions, by paying attention to certain continuity conditions. Recall that a **general solution** is a family of solutions defined on some interval *I* that contains all solutions of the DE that are defined on *I*. Because our primary goal in the present chapter is to find general solutions of linear higher-order DEs, we first need to examine some theory of linear equations.

REVIEW MATERIAL: Reread the *Remarks* at the end of Section 1.1 along with pages 58–62 of Section 2.3. Also, if you have taken a course in linear algebra, we suggest that you review the concept of *linear independence*.

4.1.1 INITIAL-VALUE AND BOUNDARY-VALUE PROBLEMS

INITIAL-VALUE PROBLEM In Section 1.2 we defined an initial-value problem for a general *n*th-order differential equation. For a linear differential equation, an **nth-order initial-value problem** is

Solve:
$$a_n(x)\frac{d^ny}{dx^n} + a_{n-1}(x)\frac{d^{n-1}y}{dx^{n-1}} + \cdots + a_1(x)\frac{dy}{dx} + a_0(x)y = g(x) \tag{1}$$

Subject to: $\quad y(x_0) = y_0, \quad y'(x_0) = y_1, \ldots, \quad y^{(n-1)}(x_0) = y_{n-1}.$

Recall that for a problem such as this one we seek a function defined on some interval *I*, containing x_0, that satisfies the differential equation and the *n* initial conditions specified at x_0: $y(x_0) = y_0, y'(x_0) = y_1, \ldots, y^{(n-1)}(x_0) = y_{n-1}$. We have already seen that in the case of a second-order initial-value problem, a solution curve must pass through the point (x_0, y_0) and have slope y_1 at this point.

EXISTENCE AND UNIQUENESS In Section 1.2 we stated a theorem that gave conditions under which the existence and uniqueness of a solution of a first-order initial-value problem were guaranteed. The theorem that follows gives sufficient conditions for the existence of a unique solution of the problem in (1).

THEOREM 4.1	**Existence of a Unique Solution**

Let $a_n(x), a_{n-1}(x), \ldots, a_1(x), a_0(x)$ and $g(x)$ be continuous on an interval *I*, and let $a_n(x) \neq 0$ for every *x* in this interval. If $x = x_0$ is any point in this interval, then a solution $y(x)$ of the initial-value problem (1) exists on the interval and is unique.

EXAMPLE 1   **Unique Solution of an IVP**

The initial-value problem

$$3y''' + 5y'' - y' + 7y = 0, \quad y(1) = 0, \quad y'(1) = 0, \quad y''(1) = 0$$

possesses the trivial solution $y = 0$. Because the third-order equation is linear with constant coefficients, it follows that all the conditions of Theorem 4.1 are fulfilled. Hence $y = 0$ is the *only* solution on any interval containing $x = 1$.

EXAMPLE 2   Unique Solution of an IVP

You should verify that the function $y = 3e^{2x} + e^{-2x} - 3x$ is a solution of the initial-value problem

$$y'' - 4y = 12x, \quad y(0) = 4, \quad y'(0) = 1.$$

Now the differential equation is linear, the coefficients as well as $g(x) = 12x$ are continuous, and $a_2(x) = 1 \neq 0$ on any interval I containing $x = 0$. We conclude from Theorem 4.1 that the given function is the unique solution on I.

The requirements in Theorem 4.1 that $a_i(x)$, $i = 0, 1, 2, \ldots, n$ be continuous and $a_n(x) \neq 0$ for every x in I are both important. Specifically, if $a_n(x) = 0$ for some x in the interval, then the solution of a linear initial-value problem may not be unique or even exist. For example, you should verify that the function $y = cx^2 + x + 3$ is a solution of the initial-value problem

$$x^2 y'' - 2xy' + 2y = 6, \quad y(0) = 3, \quad y'(0) = 1$$

on the interval $(-\infty, \infty)$ for any choice of the parameter c. In other words, there is no unique solution of the problem. Although most of the conditions of Theorem 4.1 are satisfied, the obvious difficulties are that $a_2(x) = x^2$ is zero at $x = 0$ and that the initial conditions are also imposed at $x = 0$.

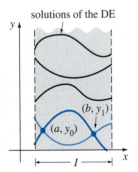

solutions of the DE

FIGURE 4.1   Solution curves of a BVP that pass through two points

BOUNDARY-VALUE PROBLEM   Another type of problem consists of solving a linear differential equation of order two or greater in which the dependent variable y or its derivatives are specified at *different points*. A problem such as

Solve: $$a_2(x)\frac{d^2 y}{dx^2} + a_1(x)\frac{dy}{dx} + a_0(x)y = g(x)$$

Subject to: $$y(a) = y_0, \quad y(b) = y_1$$

is called a **boundary-value problem (BVP).** The prescribed values $y(a) = y_0$ and $y(b) = y_1$ are called **boundary conditions.** A solution of the foregoing problem is a function satisfying the differential equation on some interval I, containing a and b, whose graph passes through the two points (a, y_0) and (b, y_1). See Figure 4.1.

For a second-order differential equation, other pairs of boundary conditions could be

$$y'(a) = y_0, \quad y(b) = y_1$$
$$y(a) = y_0, \quad y'(b) = y_1$$
$$y'(a) = y_0, \quad y'(b) = y_1,$$

where y_0 and y_1 denote arbitrary constants. These three pairs of conditions are just special cases of the general boundary conditions

$$\alpha_1 y(a) + \beta_1 y'(a) = \gamma_1$$
$$\alpha_2 y(b) + \beta_2 y'(b) = \gamma_2.$$

The next example shows that even when the conditions of Theorem 4.1 are fulfilled, a boundary-value problem may have several solutions (as suggested in Figure 4.1), a unique solution, or no solution at all.

EXAMPLE 3   **A BVP Can Have Many, One, or No Solutions**

In Example 4 of Section 1.1 we saw that the two-parameter family of solutions of the differential equation $x'' + 16x = 0$ is

$$x = c_1 \cos 4t + c_2 \sin 4t. \tag{2}$$

(a) Suppose we now wish to determine that solution of the equation that further satisfies the boundary conditions $x(0) = 0$, $x(\pi/2) = 0$. Observe that the first condition $0 = c_1 \cos 0 + c_2 \sin 0$ implies $c_1 = 0$, so $x = c_2 \sin 4t$. But when $t = \pi/2$, $0 = c_2 \sin 2\pi$ is satisfied for any choice of c_2 since $\sin 2\pi = 0$. Hence the boundary-value problem

$$x'' + 16x = 0, \quad x(0) = 0, \quad x\left(\frac{\pi}{2}\right) = 0 \tag{3}$$

has infinitely many solutions. Figure 4.2 shows the graphs of some of the members of the one-parameter family $x = c_2 \sin 4t$ that pass through the two points $(0, 0)$ and $(\pi/2, 0)$.

(b) If the boundary-value problem in (3) is changed to

$$x'' + 16x = 0, \quad x(0) = 0, \quad x\left(\frac{\pi}{8}\right) = 0 \tag{4}$$

then $x(0) = 0$ still requires $c_1 = 0$ in the solution (2). But applying $x(\pi/8) = 0$ to $x = c_2 \sin 4t$ demands that $0 = c_2 \sin (\pi/2) = c_2 \cdot 1$. Hence $x = 0$ is a solution of this new boundary-value problem. Indeed, it can be proved that $x = 0$ is the *only* solution of (4).

(c) Finally, if we change the problem to

$$x'' + 16x = 0, \quad x(0) = 0, \quad x\left(\frac{\pi}{2}\right) = 1 \tag{5}$$

we find again from $x(0) = 0$ that $c_1 = 0$, but applying $x(\pi/2) = 1$ to $x = c_2 \sin 4t$ leads to the contradiction $1 = c_2 \sin 2\pi = c_2 \cdot 0 = 0$. Hence the boundary-value problem (5) has no solution.

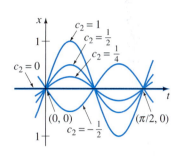

FIGURE 4.2   Some solution curves of (3)

4.1.2   HOMOGENEOUS EQUATIONS

A linear nth-order differential equation of the form

$$a_n(x) \frac{d^n y}{dx^n} + a_{n-1}(x) \frac{d^{n-1} y}{dx^{n-1}} + \cdots + a_1(x) \frac{dy}{dx} + a_0(x)y = 0 \tag{6}$$

is said to be **homogeneous,** whereas an equation

$$a_n(x) \frac{d^n y}{dx^n} + a_{n-1}(x) \frac{d^{n-1} y}{dx^{n-1}} + \cdots + a_1(x) \frac{dy}{dx} + a_0(x)y = g(x), \tag{7}$$

with $g(x)$ not identically zero, is said to be **nonhomogeneous.** For example, $2y'' + 3y' - 5y = 0$ is a homogeneous linear second-order differential equation, whereas $x^3 y''' + 6y' + 10y = e^x$ is a nonhomogeneous linear third-order differential equation. The word *homogeneous* in this context does not refer to coefficients that are homogeneous functions, as in Section 2.5.

We shall see that to solve a nonhomogeneous linear equation (7), we must first be able to solve the **associated homogeneous equation** (6).

To avoid needless repetition throughout the remainder of this text, we shall, as a matter of course, make the following important assumptions when

Please remember these two assumptions.

stating definitions and theorems about linear equations (1). On some common interval I,

- the coefficient functions $a_i(x)$, $i = 0, 1, 2, \ldots, n$ and $g(x)$ are continuous;
- $a_n(x) \neq 0$ for every x in the interval.

DIFFERENTIAL OPERATORS   In calculus, differentiation is often denoted by the capital letter D—that is, $dy/dx = Dy$. The symbol D is called a **differential operator** because it transforms a differentiable function into another function. For example, $D(\cos 4x) = -4 \sin 4x$ and $D(5x^3 - 6x^2) = 15x^2 - 12x$. Higher-order derivatives can be expressed in terms of D in a natural manner:

$$\frac{d}{dx}\left(\frac{dy}{dx}\right) = \frac{d^2y}{dx^2} = D(Dy) = D^2y \quad \text{and, in general,} \quad \frac{d^ny}{dx^n} = D^ny,$$

where y represents a sufficiently differentiable function. Polynomial expressions involving D, such as $D + 3$, $D^2 + 3D - 4$, and $5x^3D^3 - 6x^2D^2 + 4xD + 9$, are also differential operators. In general, we define an ***n*th-order differential operator** or **polynomial operator** to be

$$L = a_n(x)D^n + a_{n-1}(x)D^{n-1} + \cdots + a_1(x)D + a_0(x). \tag{8}$$

As a consequence of two basic properties of differentiation, $D(cf(x)) = cDf(x)$, c is a constant, and $D\{f(x) + g(x)\} = Df(x) + Dg(x)$, the differential operator L possesses a linearity property; that is, L operating on a linear combination of two differentiable functions is the same as the linear combination of L operating on the individual functions. In symbols this means that

$$L\{\alpha f(x) + \beta g(x)\} = \alpha L(f(x)) + \beta L(g(x)), \tag{9}$$

where α and β are constants. Because of (9) we say that the nth-order differential operator L is a **linear operator.**

DIFFERENTIAL EQUATIONS   Any linear differential equation can be expressed in terms of the D notation. For example, the differential equation $y'' + 5y' + 6y = 5x - 3$ can be written as $D^2y + 5Dy + 6y = 5x - 3$ or $(D^2 + 5D + 6)y = 5x - 3$. Using (8), we can write the linear nth-order differential equations (6) and (7) compactly as

$$L(y) = 0 \quad \text{and} \quad L(y) = g(x),$$

respectively.

SUPERPOSITION PRINCIPLE   In the next theorem we see that the sum, or **superposition,** of two or more solutions of a homogeneous linear differential equation is also a solution.

THEOREM 4.2   **Superposition Principle—Homogeneous Equations**

Let $y_1, y_2, \ldots, y_k$ be solutions of the homogeneous nth-order differential equation (6) on an interval I. Then the linear combination

$$y = c_1 y_1(x) + c_2 y_2(x) + \cdots + c_k y_k(x),$$

where the c_i, $i = 1, 2, \ldots, k$ are arbitrary constants, is also a solution on the interval.

PROOF   We prove the case $k = 2$. Let L be the differential operator defined in (8), and let $y_1(x)$ and $y_2(x)$ be solutions of the homogeneous equation $L(y) = 0$. If we define $y = c_1 y_1(x) + c_2 y_2(x)$, then by linearity of L we have

$$L(y) = L\{c_1 y_1(x) + c_2 y_2(x)\} = c_1 L(y_1) + c_2 L(y_2) = c_1 \cdot 0 + c_2 \cdot 0 = 0. \blacksquare$$

Corollaries to Theorem 4.2

(A) A constant multiple $y = c_1 y_1(x)$ of a solution $y_1(x)$ of a homogeneous linear differential equation is also a solution.

(B) A homogeneous linear differential equation always possesses the trivial solution $y = 0$.

EXAMPLE 4   Superposition—Homogeneous DE

The functions $y_1 = x^2$ and $y_2 = x^2 \ln x$ are both solutions of the homogeneous linear equation $x^3 y''' - 2xy' + 4y = 0$ on the interval $(0, \infty)$. By the superposition principle the linear combination

$$y = c_1 x^2 + c_2 x^2 \ln x$$

is also a solution of the equation on the interval.

The function $y = e^{7x}$ is a solution of $y'' - 9y' + 14y = 0$. Because the differential equation is linear and homogeneous, the constant multiple $y = ce^{7x}$ is also a solution. For various values of c we see that $y = 9e^{7x}$, $y = 0$, $y = -\sqrt{5}e^{7x}$, ... are all solutions of the equation.

LINEAR DEPENDENCE AND LINEAR INDEPENDENCE   The next two concepts are basic to the study of linear differential equations.

DEFINITION 4.1   Linear Dependence/Independence

A set of functions $f_1(x), f_2(x), \ldots, f_n(x)$ is said to be **linearly dependent** on an interval I if there exist constants $c_1, c_2, \ldots, c_n$, not all zero, such that

$$c_1 f_1(x) + c_2 f_2(x) + \cdots + c_n f_n(x) = 0$$

for every x in the interval. If the set of functions is not linearly dependent on the interval, it is said to be **linearly independent.**

In other words, a set of functions is linearly independent on an interval I if the only constants for which

$$c_1 f_1(x) + c_2 f_2(x) + \cdots + c_n f_n(x) = 0$$

for every x in the interval are $c_1 = c_2 = \cdots = c_n = 0$.

It is easy to understand these definitions for a set consisting of two functions $f_1(x)$ and $f_2(x)$. If the set of functions is linearly dependent on an interval, then there exist constants c_1 and c_2 that are not both zero such that, for every x in the interval, $c_1 f_1(x) + c_2 f_2(x) = 0$. Therefore if we assume that $c_1 \neq 0$, it follows that $f_1(x) = (-c_2/c_1)f_2(x)$; that is, *if a set of two functions is linearly dependent, then one function is simply a constant multiple of the other.* Conversely, if $f_1(x) = c_2 f_2(x)$ for some constant c_2, then $(-1) \cdot f_1(x) + c_2 f_2(x) = 0$ for every x in the interval. Hence the set of functions is linearly dependent because at least one of the constants (namely, $c_1 = -1$) is not zero. We conclude that *a set of two functions $f_1(x)$ and $f_2(x)$ is linearly independent when neither function is a constant multiple of the other* on the interval. For example, the set of functions $f_1(x) = \sin 2x, f_2(x) = \sin x \cos x$ is linearly dependent on $(-\infty, \infty)$ because $f_1(x)$ is a constant multiple of $f_2(x)$. Recall from the double-angle formula for the sine that $\sin 2x = 2 \sin x \cos x$. On the other hand, the set of functions $f_1(x) = x, f_2(x) = |x|$ is linearly independent on $(-\infty, \infty)$. Inspection of Figure 4.3 should convince you that neither function is a constant multiple of the other on the interval.

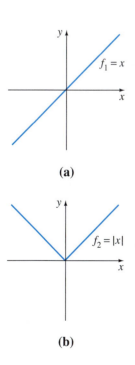

FIGURE 4.3   Set consisting of f_1 and f_2 is linearly independent on $(-\infty, \infty)$

It follows from the preceding discussion that the quotient $f_2(x)/f_1(x)$ is not a constant on an interval on which the set $f_1(x)$, $f_2(x)$ is linearly independent. This little fact will be used in the next section.

EXAMPLE 5   **Linearly Dependent Set of Functions**

The set of functions $f_1(x) = \cos^2 x$, $f_2(x) = \sin^2 x$, $f_3(x) = \sec^2 x$, $f_4(x) = \tan^2 x$ is linearly dependent on the interval $(-\pi/2, \pi/2)$ because

$$c_1 \cos^2 x + c_2 \sin^2 x + c_3 \sec^2 x + c_4 \tan^2 x = 0$$

when $c_1 = c_2 = 1$, $c_3 = -1$, $c_4 = 1$. We used here $\cos^2 x + \sin^2 x = 1$ and $1 + \tan^2 x = \sec^2 x$. ∎

A set of functions $f_1(x)$, $f_2(x)$, . . . , $f_n(x)$ is linearly dependent on an interval if at least one function can be expressed as a linear combination of the remaining functions.

EXAMPLE 6   **Linearly Dependent Set of Functions**

The set of functions $f_1(x) = \sqrt{x} + 5, f_2(x) = \sqrt{x} + 5x, f_3(x) = x - 1, f_4(x) = x^2$ is linearly dependent on the interval $(0, \infty)$ because f_2 can be written as a linear combination of f_1, f_3, and f_4. Observe that

$$f_2(x) = 1 \cdot f_1(x) + 5 \cdot f_3(x) + 0 \cdot f_4(x)$$

for every x in the interval $(0, \infty)$. ∎

SOLUTIONS OF DIFFERENTIAL EQUATIONS   We are primarily interested in linearly independent functions or, more to the point, linearly independent solutions of a linear differential equation. Although we could always appeal directly to Definition 4.1, it turns out that the question of whether the set of n solutions $y_1, y_2, \ldots, y_n$ of a homogeneous linear nth-order differential equation (6) is linearly independent can be settled somewhat mechanically by using a determinant.

DEFINITION 4.2   **Wronskian**

Suppose each of the functions $f_1(x), f_2(x), \ldots, f_n(x)$ possesses at least $n - 1$ derivatives. The determinant

$$W(f_1, f_2, \ldots, f_n) = \begin{vmatrix} f_1 & f_2 & \cdots & f_n \\ f_1' & f_2' & \cdots & f_n' \\ \vdots & \vdots & & \vdots \\ f_1^{(n-1)} & f_2^{(n-1)} & \cdots & f_n^{(n-1)} \end{vmatrix},$$

where the primes denote derivatives, is called the **Wronskian** of the functions.

THEOREM 4.3   **Criterion for Linearly Independent Solutions**

Let $y_1, y_2, \ldots, y_n$ be n solutions of the homogeneous linear nth-order differential equation (6) on an interval I. Then the set of solutions is **linearly independent** on I if and only if $W(y_1, y_2, \ldots, y_n) \neq 0$ for every x in the interval.

It follows from Theorem 4.3 that when $y_1, y_2, \ldots, y_n$ are n solutions of (6) on an interval I, the Wronskian $W(y_1, y_2, \ldots, y_n)$ is either identically zero or never zero on the interval.

A set of n linearly independent solutions of a homogeneous linear nth-order differential equation is given a special name.

DEFINITION 4.3   **Fundamental Set of Solutions**

Any set $y_1, y_2, \ldots, y_n$ of n linearly independent solutions of the homogeneous linear nth-order differential equation (6) on an interval I is said to be a **fundamental set of solutions** on the interval.

The basic question of whether a fundamental set of solutions exists for a linear equation is answered in the next theorem.

THEOREM 4.4   **Existence of a Fundamental Set**

There exists a fundamental set of solutions for the homogeneous linear nth-order differential equation (6) on an interval I.

Analogous to the fact that any vector in three dimensions can be expressed as a linear combination of the *linearly independent* vectors **i, j, k,** any solution of an nth-order homogeneous linear differential equation on an interval I can be expressed as a linear combination of n linearly independent solutions on I. In other words, n linearly independent solutions $y_1, y_2, \ldots, y_n$ are the basic building blocks for the general solution of the equation.

THEOREM 4.5   **General Solution—Homogeneous Equations**

Let $y_1, y_2, \ldots, y_n$ be a fundamental set of solutions of the homogeneous linear nth-order differential equation (6) on an interval I. Then the **general solution** of the equation on the interval is

$$y = c_1 y_1(x) + c_2 y_2(x) + \cdots + c_n y_n(x),$$

where c_i, $i = 1, 2, \ldots, n$ are arbitrary constants.

Theorem 4.5 states that if $Y(x)$ is any solution of (6) on the interval, then constants $C_1, C_2, \ldots, C_n$ can always be found so that

$$Y(x) = C_1 y_1(x) + C_2 y_2(x) + \cdots + C_n y_n(x).$$

We will prove the case when $n = 2$.

PROOF   Let Y be a solution and y_1 and y_2 be linearly independent solutions of $a_2 y'' + a_1 y' + a_0 y = 0$ on an interval I. Suppose $x = t$ is a point in I for which $W(y_1(t), y_2(t)) \neq 0$. Suppose also that $Y(t) = k_1$ and $Y'(t) = k_2$. If we now examine the equations

$$C_1 y_1(t) + C_2 y_2(t) = k_1$$

$$C_1 y_1'(t) + C_2 y_2'(t) = k_2,$$

it follows that we can determine C_1 and C_2 uniquely, provided that the determinant of the coefficients satisfies

$$\begin{vmatrix} y_1(t) & y_2(t) \\ y_1'(t) & y_2'(t) \end{vmatrix} \neq 0.$$

But this determinant is simply the Wronskian evaluated at $x = t$, and, by assumption, $W \neq 0$. If we define $G(x) = C_1 y_1(x) + C_2 y_2(x)$, we observe that $G(x)$ satisfies the differential equation since it is a superposition of two known solutions; $G(x)$ satisfies the initial conditions

$$G(t) = C_1 y_1(t) + C_2 y_2(t) = k_1 \quad \text{and} \quad G'(t) = C_1 y_1'(t) + C_2 y_2'(t) = k_2;$$

and $Y(x)$ satisfies the *same* linear equation and the *same* initial conditions. Because the solution of this linear initial-value problem is unique (Theorem 4.1), we have $Y(x) = G(x)$ or $Y(x) = C_1 y_1(x) + C_2 y_2(x)$. ∎

EXAMPLE 7   **General Solution of a Homogeneous DE**

The functions $y_1 = e^{3x}$ and $y_2 = e^{-3x}$ are both solutions of the homogeneous linear equation $y'' - 9y = 0$ on the interval $(-\infty, \infty)$. By inspection the solutions are linearly independent on the x-axis. This fact can be corroborated by observing that the Wronskian

$$W(e^{3x}, e^{-3x}) = \begin{vmatrix} e^{3x} & e^{-3x} \\ 3e^{3x} & -3e^{-3x} \end{vmatrix} = -6 \neq 0$$

for every x. We conclude that y_1 and y_2 form a fundamental set of solutions and, consequently $y = c_1 e^{3x} + c_2 e^{-3x}$ is the general solution of the equation on the interval. ∎

EXAMPLE 8   **A Solution Obtained from a General Solution**

The function $y = 4 \sinh 3x - 5e^{3x}$ is a solution of the differential equation in Example 7. (Verify this.) In view of Theorem 4.5, we must be able to obtain this solution from the general solution $y = c_1 e^{3x} + c_2 e^{-3x}$. Observe that if we choose $c_1 = 2$ and $c_2 = -7$, then $y = 2e^{3x} - 7e^{-3x}$ can be rewritten as

$$y = 2e^{3x} - 2e^{-3x} - 5e^{-3x} = 4\left(\frac{e^{3x} - e^{-3x}}{2}\right) - 5e^{-3x}.$$

The last expression is recognized as $y = 4 \sinh 3x - 5e^{-3x}$. ∎

EXAMPLE 9   **General Solution of a Homogeneous DE**

The functions $y_1 = e^x$, $y_2 = e^{2x}$, and $y_3 = e^{3x}$ satisfy the third-order equation $y''' - 6y'' + 11y' - 6y = 0$. Since

$$W(e^x, e^{2x}, e^{3x}) = \begin{vmatrix} e^x & e^{2x} & e^{3x} \\ e^x & 2e^{2x} & 3e^{3x} \\ e^x & 4e^{2x} & 9e^{3x} \end{vmatrix} = 2e^{6x} \neq 0$$

for every real value of x, the functions y_1, y_2, and y_3 form a fundamental set of solutions on $(-\infty, \infty)$. We conclude that $y = c_1 e^x + c_2 e^{2x} + c_3 e^{3x}$ is the general solution of the differential equation on the interval. ∎

4.1.3 NONHOMOGENEOUS EQUATIONS

Any function y_p, free of arbitrary parameters, that satisfies (7) is said to be a **particular solution** or **particular integral** of the equation. For example, it is a straightforward task to show that the constant function $y_p = 3$ is a particular solution of the nonhomogeneous equation $y'' + 9y = 27$.

Now if $y_1, y_2, \ldots, y_k$ are solutions of (6) on an interval I and y_p is any particular solution of (7) on I, then the linear combination

$$y = c_1 y_1(x) + c_2 y_2(x) + \cdots + c_k y_k(x) + y_p \qquad (10)$$

is also a solution of the nonhomogeneous equation (7). If you think about it, this makes sense, because the linear combination $c_1 y_1(x) + c_2 y_2(x) + \cdots + c_k y_k(x)$ is transformed into 0 by the operator $L = a_n D^n + a_{n-1} D^{n-1} + \cdots + a_1 D + a_0$, whereas y_p is transformed into $g(x)$. If we use $k = n$ linearly independent solutions of the nth-order equation (6), then the expression in (10) becomes the general solution of (7).

THEOREM 4.6   **General Solution – Nonhomogeneous Equations**

Let y_p be any particular solution of the nonhomogeneous linear nth-order differential equation (7) on an interval I, and let $y_1, y_2, \ldots, y_n$ be a fundamental set of solutions of the associated homogeneous differential equation (6) on I. Then the **general solution** of the equation on the interval is

$$y = c_1 y_1(x) + c_2 y_2(x) + \cdots + c_n y_n(x) + y_p,$$

where the $c_i, i = 1, 2, \ldots, n$ are arbitrary constants.

PROOF   Let L be the differential operator defined in (8), and let $Y(x)$ and $y_p(x)$ be particular solutions of the nonhomogeneous equation $L(y) = g(x)$. If we define $u(x) = Y(x) - y_p(x)$, then by linearity of L we have

$$L(u) = L\{Y(x) - y_p(x)\} = L(Y(x)) - L(y_p(x)) = g(x) - g(x) = 0.$$

This shows that $u(x)$ is a solution of the homogeneous equation $L(y) = 0$. Hence, by Theorem 4.5, $u(x) = c_1 y_1(x) + c_2 y_2(x) + \cdots + c_n y_n(x)$, and so

$$Y(x) - y_p(x) = c_1 y_1(x) + c_2 y_2(x) + \cdots + c_n y_n(x)$$

or $$Y(x) = c_1 y_1(x) + c_2 y_2(x) + \cdots + c_n y_n(x) + y_p(x). \qquad ∎$$

COMPLEMENTARY FUNCTION   We see in Theorem 4.6 that the general solution of a nonhomogeneous linear equation consists of the sum of two functions:

$$y = c_1 y_1(x) + c_2 y_2(x) + \cdots + c_n y_n(x) + y_p(x) = y_c(x) + y_p(x).$$

The linear combination $y_c(x) = c_1 y_1(x) + c_2 y_2(x) + \cdots + c_n y_n(x)$, which is the general solution of (6), is called the **complementary function** for equation (7). In other words, to solve a nonhomogeneous linear differential equation, we first solve the associated homogeneous equation and then find any particular solution of the nonhomogeneous equation. The general solution of the nonhomogeneous equation is then

$$y = complementary\ function + any\ particular\ solution$$
$$= y_c + y_p.$$

EXAMPLE 10   **General Solution of a Nonhomogeneous DE**

By substitution the function $y_p = -\frac{11}{12} - \frac{1}{2}x$ is readily shown to be a particular solution of the nonhomogeneous equation

$$y''' - 6y'' + 11y' - 6y = 3x. \qquad (11)$$

To write the general solution of (11), we must also be able to solve the associated homogeneous equation

$$y''' - 6y'' + 11y' - 6y = 0.$$

But in Example 9 we saw that the general solution of this latter equation on the interval $(-\infty, \infty)$ was $y_c = c_1 e^x + c_2 e^{2x} + c_3 e^{3x}$. Hence the general solution of (11) on the interval is

$$y = y_c + y_p = c_1 e^x + c_2 e^{2x} + c_3 e^{3x} - \frac{11}{12} - \frac{1}{2}x.$$

ANOTHER SUPERPOSITION PRINCIPLE The last theorem of this discussion will be useful in Section 4.4 when we consider a method for finding particular solutions of nonhomogeneous equations.

THEOREM 4.7   **Superposition Principle—Nonhomogeneous Equations**

Let $y_{p_1}, y_{p_2}, \ldots, y_{p_k}$ be k particular solutions of the nonhomogeneous linear nth-order differential equation (7) on an interval I corresponding, in turn, to k distinct functions $g_1, g_2, \ldots, g_k$. That is, suppose y_{p_i} denotes a particular solution of the corresponding differential equation

$$a_n(x)y^{(n)} + a_{n-1}(x)y^{(n-1)} + \cdots + a_1(x)y' + a_0(x)y = g_i(x), \quad (12)$$

where $i = 1, 2, \ldots, k$. Then

$$y_p = y_{p_1}(x) + y_{p_2}(x) + \cdots + y_{p_k}(x) \quad (13)$$

is a particular solution of

$$a_n(x)y^{(n)} + a_{n-1}(x)y^{(n-1)} + \cdots + a_1(x)y' + a_0(x)y$$
$$= g_1(x) + g_2(x) + \cdots + g_k(x). \quad (14)$$

PROOF   We prove the case $k = 2$. Let L be the differential operator defined in (8), and let $y_{p_1}(x)$ and $y_{p_2}(x)$ be particular solutions of the nonhomogeneous equations $L(y) = g_1(x)$ and $L(y) = g_2(x)$, respectively. If we define $y_p = y_{p_1}(x) + y_{p_2}(x)$, we want to show that y_p is a particular solution of $L(y) = g_1(x) + g_2(x)$. The result follows again by the linearity of the operator L:

$$L(y_p) = L\{y_{p_1}(x) + y_{p_2}(x)\} = L(y_{p_1}(x)) + L(y_{p_2}(x)) = g_1(x) + g_2(x). \quad\blacksquare$$

EXAMPLE 11   **Superposition—Nonhomogeneous DE**

You should verify that

$y_{p_1} = -4x^2$ is a particular solution of $y'' - 3y' + 4y = -16x^2 + 24x - 8$,

$y_{p_2} = e^{2x}$ is a particular solution of $y'' - 3y' + 4y = 2e^{2x}$,

$y_{p_3} = xe^x$ is a particular solution of $y'' - 3y' + 4y = 2xe^x - e^x$.

It follows from (13) of Theorem 4.7 that the superposition of y_{p_1}, y_{p_2}, and y_{p_3},

$$y = y_{p_1} + y_{p_2} + y_{p_3} = -4x^2 + e^{2x} + xe^x,$$

is a solution of

$$y'' - 3y' + 4y = \underbrace{-16x^2 + 24x - 8}_{g_1(x)} + \underbrace{2e^{2x}}_{g_2(x)} + \underbrace{2xe^x - e^x}_{g_3(x)}.$$

NOTE   If the y_{pi} are particular solutions of (12) for $i = 1, 2, \ldots, k$, then the linear combination

$$y_p = c_1 y_{p_1} + c_2 y_{p_2} + \cdots + c_k y_{p_k},$$

where the c_i are constants, is also a particular solution of (14) when the right-hand member of the equation is the linear combination

$$c_1 g_1(x) + c_2 g_2(x) + \cdots + c_k g_k(x).$$

Before we actually start solving homogeneous and nonhomogeneous linear differential equations, we need one additional bit of theory, which is presented in the next section.

REMARKS

This remark is a continuation of the brief discussion of dynamical systems given at the end of Section 1.3.

A dynamical system whose rule or mathematical model is a linear nth-order differential equation

$$a_n(t)y^{(n)} + a_{n-1}(t)y^{(n-1)} + \cdots + a_1(t)y' + a_0(t)y = g(t)$$

is said to be an nth-order **linear system.** The n time-dependent functions $y(t)$, $y'(t), \ldots, y^{(n-1)}(t)$ are the **state variables** of the system. Recall that their values at some time t give the **state of the system.** The function g is variously called the **input function, forcing function,** or **excitation function.** A solution $y(t)$ of the differential equation is said to be the **output** or **response of the system.** Under the conditions stated in Theorem 4.1, the output or response $y(t)$ is uniquely determined by the input and the state of the system prescribed at a time t_0 — that is, by the initial conditions $y(t_0), y'(t_0), \ldots, y^{(n-1)}(t_0)$.

For a dynamical system to be a linear system, it is necessary that the superposition principle (Theorem 4.7) holds in the system; that is, the response of the system to a superposition of inputs is a superposition of outputs. We have already examined some simple linear systems in Section 3.1 (linear first-order equations); in Section 5.1 we examine linear systems in which the mathematical models are second-order differential equations.

EXERCISES 4.1

Answers to selected odd-numbered problems begin on page ANS-3.

4.1.1   INITIAL-VALUE AND BOUNDARY-VALUE PROBLEMS

In Problems 1–4 the given family of functions is the general solution of the differential equation on the indicated interval. Find a member of the family that is a solution of the initial-value problem.

1. $y = c_1 e^x + c_2 e^{-x}, (-\infty, \infty)$;
$y'' - y = 0, y(0) = 0, y'(0) = 1$

2. $y = c_1 e^{4x} + c_2 e^{-x}, (-\infty, \infty)$;
$y'' - 3y' - 4y = 0, y(0) = 1, y'(0) = 2$

3. $y = c_1 x + c_2 x \ln x, (0, \infty)$;
$x^2 y'' - xy' + y = 0, y(1) = 3, y'(1) = -1$

4. $y = c_1 + c_2 \cos x + c_3 \sin x, (-\infty, \infty)$;
$y''' + y' = 0, y(\pi) = 0, y'(\pi) = 2, y''(\pi) = -1$

5. Given that $y = c_1 + c_2 x^2$ is a two-parameter family of solutions of $xy'' - y' = 0$ on the interval $(-\infty, \infty)$, show that constants c_1 and c_2 cannot be found so that a member of the family satisfies the initial conditions $y(0) = 0$, $y'(0) = 1$. Explain why this does not violate Theorem 4.1.

6. Find two members of the family of solutions in Problem 5 that satisfy the initial conditions $y(0) = 0$, $y'(0) = 0$.

7. Given that $x(t) = c_1 \cos \omega t + c_2 \sin \omega t$ is the general solution of $x'' + \omega^2 x = 0$ on the interval $(-\infty, \infty)$, show that a solution satisfying the initial conditions $x(0) = x_0$, $x'(0) = x_1$ is given by

$$x(t) = x_0 \cos \omega t + \frac{x_1}{\omega} \sin \omega t.$$

8. Use the general solution of $x'' + \omega^2 x = 0$ given in Problem 7 to show that a solution satisfying the initial conditions $x(t_0) = x_0$, $x'(t_0) = x_1$ is the solution given in Problem 7 shifted by an amount t_0:

$$x(t) = x_0 \cos \omega(t - t_0) + \frac{x_1}{\omega} \sin \omega(t - t_0).$$

In Problems 9 and 10 find an interval centered about $x = 0$ for which the given initial-value problem has a unique solution.

9. $(x - 2)y'' + 3y = x$, $y(0) = 0, y'(0) = 1$

10. $y'' + (\tan x)y = e^x$, $y(0) = 1, y'(0) = 0$

11. (a) Use the family in Problem 1 to find a solution of $y'' - y = 0$ that satisfies the boundary conditions $y(0) = 0, y(1) = 1$.
 (b) The DE in part (a) has the alternative general solution $y = c_3 \cosh x + c_4 \sinh x$ on $(-\infty, \infty)$. Use this family to find a solution that satisfies the boundary conditions in part (a).
 (c) Show that the solutions in parts (a) and (b) are equivalent

12. Use the family in Problem 5 to find a solution of $xy'' - y' = 0$ that satisfies the boundary conditions $y(0) = 1, y'(1) = 6$.

In Problems 13 and 14 the given two-parameter family is a solution of the indicated differential equation on the interval $(-\infty, \infty)$. Determine whether a member of the family can be found that satisfies the boundary conditions.

13. $y = c_1 e^x \cos x + c_2 e^x \sin x$; $y'' - 2y' + 2y = 0$
 (a) $y(0) = 1, y'(\pi) = 0$ **(b)** $y(0) = 1, y(\pi) = -1$
 (c) $y(0) = 1, y\left(\dfrac{\pi}{2}\right) = 1$ **(d)** $y(0) = 0, y(\pi) = 0$.

14. $y = c_1 x^2 + c_2 x^4 + 3$; $x^2 y'' - 5xy' + 8y = 24$
 (a) $y(-1) = 0, y(1) = 4$ **(b)** $y(0) = 1, y(1) = 2$
 (c) $y(0) = 3, y(1) = 0$ **(d)** $y(1) = 3, y(2) = 15$

4.1.2 HOMOGENEOUS EQUATIONS

In Problems 15–22 determine whether the given set of functions is linearly independent on the interval $(-\infty, \infty)$.

15. $f_1(x) = x$, $f_2(x) = x^2$, $f_3(x) = 4x - 3x^2$

16. $f_1(x) = 0$, $f_2(x) = x$, $f_3(x) = e^x$

17. $f_1(x) = 5$, $f_2(x) = \cos^2 x$, $f_3(x) = \sin^2 x$

18. $f_1(x) = \cos 2x$, $f_2(x) = 1$, $f_3(x) = \cos^2 x$

19. $f_1(x) = x$, $f_2(x) = x - 1$, $f_3(x) = x + 3$

20. $f_1(x) = 2 + x$, $f_2(x) = 2 + |x|$

21. $f_1(x) = 1 + x$, $f_2(x) = x$, $f_3(x) = x^2$

22. $f_1(x) = e^x$, $f_2(x) = e^{-x}$, $f_3(x) = \sinh x$

In Problems 23–30 verify that the given functions form a fundamental set of solutions of the differential equation on the indicated interval. Form the general solution.

23. $y'' - y' - 12y = 0$; $e^{-3x}, e^{4x}, (-\infty, \infty)$

24. $y'' - 4y = 0$; $\cosh 2x, \sinh 2x, (-\infty, \infty)$

25. $y'' - 2y' + 5y = 0$; $e^x \cos 2x, e^x \sin 2x, (-\infty, \infty)$

26. $4y'' - 4y' + y = 0$; $e^{x/2}, xe^{x/2}, (-\infty, \infty)$

27. $x^2 y'' - 6xy' + 12y = 0$; $x^3, x^4, (0, \infty)$

28. $x^2 y'' + xy' + y = 0$; $\cos(\ln x), \sin(\ln x), (0, \infty)$

29. $x^3 y''' + 6x^2 y'' + 4xy' - 4y = 0$; $x, x^{-2}, x^{-2} \ln x, (0, \infty)$

30. $y^{(4)} + y'' = 0$; $1, x, \cos x, \sin x, (-\infty, \infty)$

4.1.3 NONHOMOGENEOUS EQUATIONS

In Problems 31–34 verify that the given two-parameter family of functions is the general solution of the nonhomogeneous differential equation on the indicated interval.

31. $y'' - 7y' + 10y = 24e^x$;
 $y = c_1 e^{2x} + c_2 e^{5x} + 6e^x, (-\infty, \infty)$

32. $y'' + y = \sec x$;
 $y = c_1 \cos x + c_2 \sin x + x \sin x + (\cos x) \ln(\cos x)$,
 $(-\pi/2, \pi/2)$

33. $y'' - 4y' + 4y = 2e^{2x} + 4x - 12$;
 $y = c_1 e^{2x} + c_2 xe^{2x} + x^2 e^{2x} + x - 2, (-\infty, \infty)$

34. $2x^2 y'' + 5xy' + y = x^2 - x$;
 $y = c_1 x^{-1/2} + c_2 x^{-1} + \frac{1}{15}x^2 - \frac{1}{6}x, (0, \infty)$

35. (a) Verify that $y_{p_1} = 3e^{2x}$ and $y_{p_2} = x^2 + 3x$ are, respectively, particular solutions of

$$y'' - 6y' + 5y = -9e^{2x}$$

and $y'' - 6y' + 5y = 5x^2 + 3x - 16.$

(b) Use part (a) to find particular solutions of

$$y'' - 6y' + 5y = 5x^2 + 3x - 16 - 9e^{2x}$$

and $y'' - 6y' + 5y = -10x^2 - 6x + 32 + e^{2x}$.

36. (a) By inspection find a particular solution of

$$y'' + 2y = 10.$$

(b) By inspection find a particular solution of

$$y'' + 2y = -4x.$$

(c) Find a particular solution of $y'' + 2y = -4x + 10$.
(d) Find a particular solution of $y'' + 2y = 8x + 5$.

DISCUSSION/PROJECT PROBLEMS

37. Let $n = 1, 2, 3, \ldots$. Discuss how the observations $D^n x^{n-1} = 0$ and $D^n x^n = n!$ can be used to find the general solutions of the given differential equations.
(a) $y'' = 0$ **(b)** $y''' = 0$
(c) $y^{(4)} = 0$ **(d)** $y'' = 2$
(e) $y''' = 6$ **(f)** $y^{(4)} = 24$

38. Suppose that $y_1 = e^x$ and $y_2 = e^{-x}$ are two solutions of a homogeneous linear differential equation. Explain why $y_3 = \cosh x$ and $y_4 = \sinh x$ are also solutions of the equation.

39. (a) Verify that $y_1 = x^3$ and $y_2 = |x|^3$ are linearly independent solutions of the differential equation $x^2 y'' - 4xy' + 6y = 0$ on the interval $(-\infty, \infty)$.

(b) Show that $W(y_1, y_2) = 0$ for every real number x. Does this result violate Theorem 4.3? Explain.
(c) Verify that $Y_1 = x^3$ and $Y_2 = x^2$ are also linearly independent solutions of the differential equation in part (a) on the interval $(-\infty, \infty)$.
(d) Find a solution of the differential equation satisfying $y(0) = 0$, $y'(0) = 0$.
(e) By the superposition principle, Theorem 4.2, both linear combinations $y = c_1 y_1 + c_2 y_2$ and $Y = c_1 Y_1 + c_2 Y_2$ are solutions of the differential equation. Discuss whether one, both, or neither of the linear combinations is a general solution of the differential equation on the interval $(-\infty, \infty)$.

40. Is the set of functions $f_1(x) = e^{x+2}$, $f_2(x) = e^{x-3}$ linearly dependent or linearly independent on $(-\infty, \infty)$? Discuss.

41. Suppose $y_1, y_2, \ldots, y_k$ are k linearly independent solutions on $(-\infty, \infty)$ of a homogeneous linear nth-order differential equation with constant coefficients. By Theorem 4.2 it follows that $y_{k+1} = 0$ is also a solution of the differential equation. Is the set of solutions $y_1, y_2, \ldots, y_k, y_{k+1}$ linearly dependent or linearly independent on $(-\infty, \infty)$? Discuss.

42. Suppose that $y_1, y_2, \ldots, y_k$ are k nontrivial solutions of a homogeneous linear nth-order differential equation with constant coefficients and that $k = n + 1$. Is the set of solutions $y_1, y_2, \ldots, y_k$ linearly dependent or linearly independent on $(-\infty, \infty)$? Discuss.

4.2   REDUCTION OF ORDER

INTRODUCTION: In the preceding section we saw that the general solution of a homogeneous linear second-order differential equation

$$a_2(x)y'' + a_1(x)y' + a_0(x)y = 0 \tag{1}$$

was a linear combination $y = c_1 y_1 + c_2 y_2$, where y_1 and y_2 are solutions that constitute a linearly independent set on some interval I. Beginning in the next section, we examine a method for determining these solutions when the coefficients of the DE in (1) are constants. This method, which is a straightforward exercise in algebra, breaks down in a few cases and yields only a single solution y_1 of the DE. It turns out that we can construct a second solution y_2 of a homogeneous equation (1) (even when the coefficients in (1) are variable) provided that we know a nontrivial solution y_1 of the DE. The basic idea described in this section is that *equation (1) can be reduced to a linear first-order DE by means of a substitution* involving the known solution y_1. A second solution y_2 of (1) is apparent after this first-order DE is solved.

REDUCTION OF ORDER   Suppose that y_1 denotes a nontrivial solution of (1) and that y_1 is defined on an interval I. We seek a second solution, y_2, so that y_1, y_2 is a linearly independent set on I. Recall from Section 4.1 that if y_1 and y_2 are linearly independent, then their quotient y_2/y_1 is nonconstant on I—that is, $y_2(x)/y_1(x) = u(x)$ or $y_2(x) = u(x)y_1(x)$. The function $u(x)$ can be found by substituting $y_2(x) = u(x)y_1(x)$ into the given differential equation. This method is called **reduction of order** because we must solve a linear first-order differential equation to find u.

EXAMPLE 1   A Second Solution by Reduction of Order

Given that $y_1 = e^x$ is a solution of $y'' - y = 0$ on the interval $(-\infty, \infty)$, use reduction of order to find a second solution y_2.

SOLUTION   If $y = u(x)y_1(x) = u(x)e^x$, then the product rule gives

$$y' = ue^x + e^x u', \quad y'' = ue^x + 2e^x u' + e^x u'',$$

and so

$$y'' - y = e^x(u'' + 2u') = 0.$$

Since $e^x \neq 0$, the last equation requires $u'' + 2u' = 0$. If we make the substitution $w = u'$, this linear second-order equation in u becomes $w' + 2w = 0$, which is a linear first-order equation in w. Using the integrating factor e^{2x}, we can write $\frac{d}{dx}[e^{2x}w] = 0$. After integrating, we get $w = c_1 e^{-2x}$ or $u' = c_1 e^{-2x}$. Integrating again then yields $u = -\frac{1}{2}c_1 e^{-2x} + c_2$. Thus

$$y = u(x)e^x = -\frac{c_1}{2}e^{-x} + c_2 e^x. \tag{2}$$

By picking $c_2 = 0$ and $c_1 = -2$, we obtain the desired second solution, $y_2 = e^{-x}$. Because $W(e^x, e^{-x}) \neq 0$ for every x, the solutions are linearly independent on $(-\infty, \infty)$.

Since we have shown that $y_1 = e^x$ and $y_2 = e^{-x}$ are linearly independent solutions of a linear second-order equation, the expression in (2) is actually the general solution of $y'' - y = 0$ on $(-\infty, \infty)$.

GENERAL CASE   Suppose we divide by $a_2(x)$ in order to put equation (1) in the **standard form**

$$y'' + P(x)y' + Q(x)y = 0, \tag{3}$$

where $P(x)$ and $Q(x)$ are continuous on some interval I. Let us suppose further that $y_1(x)$ is a known solution of (3) on I and that $y_1(x) \neq 0$ for every x in the interval. If we define $y = u(x)y_1(x)$, it follows that

$$y' = uy_1' + y_1 u', \quad y'' = uy_1'' + 2y_1' u' + y_1 u''$$

$$y'' + Py' + Qy = u[\underbrace{y_1'' + Py_1' + Qy_1}_{\text{zero}}] + y_1 u'' + (2y_1' + Py_1)u' = 0.$$

This implies that we must have

$$y_1 u'' + (2y_1' + Py_1)u' = 0 \quad \text{or} \quad y_1 w' + (2y_1' + Py_1)w = 0, \tag{4}$$

where we have let $w = u'$. Observe that the last equation in (4) is both linear and separable. Separating variables and integrating, we obtain

$$\frac{dw}{w} + 2\frac{y_1'}{y_1}dx + P\,dx = 0$$

$$\ln|wy_1^2| = -\int P\,dx + c \quad \text{or} \quad wy_1^2 = c_1 e^{-\int P\,dx}.$$

We solve the last equation for w, use $w = u'$, and integrate again:

$$u = c_1 \int \frac{e^{-\int P\,dx}}{y_1^2}\,dx + c_2.$$

By choosing $c_1 = 1$ and $c_2 = 0$, we find from $y = u(x)y_1(x)$ that a second solution of equation (3) is

$$y_2 = y_1(x) \int \frac{e^{-\int P(x)\,dx}}{y_1^2(x)}\,dx. \qquad (5)$$

It makes a good review of differentiation to verify that the function $y_2(x)$ defined in (5) satisfies equation (3) and that y_1 and y_2 are linearly independent on any interval on which $y_1(x)$ is not zero.

| **EXAMPLE 2** | **A Second Solution by Formula (5)** |

The function $y_1 = x^2$ is a solution of $x^2 y'' - 3xy' + 4y = 0$. Find the general solution of the differential equation on the interval $(0, \infty)$.

SOLUTION   From the standard form of the equation,

$$y'' - \frac{3}{x}y' + \frac{4}{x^2}y = 0,$$

we find from (5)

$$y_2 = x^2 \int \frac{e^{3\int dx/x}}{x^4}\,dx \qquad \leftarrow e^{3\int dx/x} = e^{\ln x^3} = x^3$$

$$= x^2 \int \frac{dx}{x} = x^2 \ln x$$

The general solution on the interval $(0, \infty)$ is given by $y = c_1 y_1 + c_2 y_2$; that is, $y = c_1 x^2 + c_2 x^2 \ln x$.

REMARKS

(*i*) The derivation and use of formula (5) have been illustrated here because this formula appears again in the next section and in Sections 4.7 and 6.2. We use (5) simply to save time in obtaining a desired result. Your instructor will tell you whether you should memorize (5) or whether you should know the first principles of reduction of order.

(*ii*) Reduction of order can be used to find the general solution of a nonhomogeneous equation $a_2(x)y'' + a_1(x)y' + a_0(x)y = g(x)$ whenever a solution y_1 of the associated homogeneous equation is known. See Problems 17–20 in Exercises 4.2.

EXERCISES 4.2

Answers to selected odd-numbered problems begin on page ANS-4.

In Problems 1–16 the indicated function $y_1(x)$ is a solution of the given differential equation. Use reduction of order or formula (5), as instructed, to find a second solution $y_2(x)$.

1. $y'' - 4y' + 4y = 0$; $y_1 = e^{2x}$

2. $y'' + 2y' + y = 0$; $y_1 = xe^{-x}$

3. $y'' + 16y = 0;$ $y_1 = \cos 4x$

4. $y'' + 9y = 0;$ $y_1 = \sin 3x$

5. $y'' - y = 0;$ $y_1 = \cosh x$

6. $y'' - 25y = 0;$ $y_1 = e^{5x}$

7. $9y'' - 12y' + 4y = 0;$ $y_1 = e^{2x/3}$

8. $6y'' + y' - y = 0;$ $y_1 = e^{x/3}$

9. $x^2 y'' - 7xy' + 16y = 0;$ $y_1 = x^4$

10. $x^2 y'' + 2xy' - 6y = 0;$ $y_1 = x^2$

11. $xy'' + y' = 0;$ $y_1 = \ln x$

12. $4x^2 y'' + y = 0;$ $y_1 = x^{1/2} \ln x$

13. $x^2 y'' - xy' + 2y = 0;$ $y_1 = x \sin(\ln x)$

14. $x^2 y'' - 3xy' + 5y = 0;$ $y_1 = x^2 \cos(\ln x)$

15. $(1 - 2x - x^2)y'' + 2(1 + x)y' - 2y = 0;$ $y_1 = x + 1$

16. $(1 - x^2)y'' + 2xy' = 0;$ $y_1 = 1$

In Problems 17–20 the indicated function $y_1(x)$ is a solution of the associated homogeneous equation. Use the method of reduction of order to find a second solution $y_2(x)$ of the homogeneous equation and a particular solution of the given nonhomogeneous equation.

17. $y'' - 4y = 2;$ $y_1 = e^{-2x}$

18. $y'' + y' = 1;$ $y_1 = 1$

19. $y'' - 3y' + 2y = 5e^{3x};$ $y_1 = e^x$

20. $y'' - 4y' + 3y = x;$ $y_1 = e^x$

DISCUSSION/PROJECT PROBLEMS

21. (a) Give a convincing demonstration that the second-order equation $ay'' + by' + cy = 0$, a, b, and c constants, always possesses at least one solution of the form $y_1 = e^{m_1 x}$, m_1 a constant.

(b) Explain why the differential equation in part (a) must then have a second solution either of the form $y_2 = e^{m_2 x}$ or of the form $y_2 = xe^{m_1 x}$, m_1 and m_2 constants.

(c) Reexamine Problems 1–8. Can you explain why the statements in parts (a) and (b) above are not contradicted by the answers to Problems 3–5?

22. Verify that $y_1(x) = x$ is a solution of $xy'' - xy' + y = 0$. Use reduction of order to find a second solution $y_2(x)$ in the form of an infinite series. Conjecture an interval of definition for $y_2(x)$.

COMPUTER LAB ASSIGNMENTS

23. (a) Verify that $y_1(x) = e^x$ is a solution of

$$xy'' - (x + 10)y' + 10y = 0.$$

(b) Use (5) to find a second solution $y_2(x)$. Use a CAS to carry out the required integration.

(c) Explain, using Corollary (A) of Theorem 4.2, why the second solution can be written compactly as

$$y_2(x) = \sum_{n=0}^{10} \frac{1}{n!} x^n.$$

| 4.3 | **HOMOGENEOUS LINEAR EQUATIONS WITH CONSTANT COEFFICIENTS** |

INTRODUCTION: As a means of motivating the discussion in this section, let us return to first-order differential equations, more specifically, to *homogeneous* linear equations $ay' + by = 0$, where the coefficients $a \neq 0$ and b are constants. This type of equation can be solved either by separation of variables or with the aid of an integrating factor, *but* there is another solution method, one that uses only algebra. Before illustrating this alternative method, we make one observation: Solving $ay' + by = 0$ for y' yields $y' = ky$, where k is a constant. This observation reveals the nature of the unknown solution y; the only nontrivial elementary function whose derivative is a constant multiple of itself is an exponential function e^{mx}.

Now the new solution method: If we substitute $y = e^{mx}$ and $y' = me^{mx}$ into $ay' + by = 0$, we get

$$ame^{mx} + be^{mx} = 0 \quad \text{or} \quad e^{mx}(am + b) = 0.$$

Because e^{mx} is never zero for real values of x, the last equation is satisfied only when m is a solution or root of the first-degree polynomial equation $am + b = 0$. For this single value of m, $y = e^{mx}$ is a solution of the DE. To illustrate, consider the constant-coefficient equation $2y' + 5y = 0$. It is not necessary to go through the differentiation and substitution of $y = e^{mx}$ into the DE; we merely have to form the equation $2m + 5 = 0$ and solve it for m. From $m = -\frac{5}{2}$ we conclude that $y = e^{-\frac{5}{2}x}$ is a solution of $2y' + 5y = 0$, and its general solution on the interval $(-\infty, \infty)$ is $y = c_1 e^{-\frac{5}{2}x}$.

In this section we shall see that the foregoing procedure can produce exponential solutions for homogeneous linear higher-order DEs,

$$a_n y^{(n)} + a_{n-1} y^{(n-1)} + \cdots + a_2 y'' + a_1 y' + a_0 y = 0 \tag{1}$$

where the coefficients a_i, $i = 0, 1, \ldots, n$ are real constants and $a_n \neq 0$.

REVIEW MATERIAL: Review Problem 27 in Exercises 1.1, Section 2.3, and Theorem 4.5. To solve higher-order DEs such as (1), you will be expected to solve higher-degree polynomial equations. A brief review of finding rational roots of polynomial equations is given in the *Student Resource and Solutions Manual*.

AUXILIARY EQUATION We begin by considering the special case of the second-order equation

$$ay'' + by' + cy = 0, \tag{2}$$

where a, b, and c are constants. If we try to find a solution of the form $y = e^{mx}$, then after substituting $y' = me^{mx}$ and $y'' = m^2 e^{mx}$, equation (2) becomes

$$am^2 e^{mx} + bme^{mx} + ce^{mx} = 0 \quad \text{or} \quad e^{mx}(am^2 + bm + c) = 0.$$

As in the introduction, we argue that because $e^{mx} \neq 0$ for all x, it is apparent that the only way $y = e^{mx}$ can satisfy the differential equation (2) is when m is chosen as a root of the quadratic equation

$$am^2 + bm + c = 0. \tag{3}$$

This last equation is called the **auxiliary equation** of the differential equation (2). Since the two roots of (3) are $m_1 = \left(-b + \sqrt{b^2 - 4ac}\right)/2a$ and $m_2 = \left(-b - \sqrt{b^2 - 4ac}\right)/2a$, there will be three forms of the general solution of (2) corresponding to the three cases:

- m_1 and m_2 real and distinct ($b^2 - 4ac > 0$),
- m_1 and m_2 real and equal ($b^2 - 4ac = 0$), and
- m_1 and m_2 conjugate complex numbers ($b^2 - 4ac < 0$).

We discuss each of these cases in turn.

CASE I: DISTINCT REAL ROOTS Under the assumption that the auxiliary equation (3) has two unequal real roots m_1 and m_2, we find two solutions, $y_1 = e^{m_1 x}$ and $y_2 = e^{m_2 x}$. We see that these functions are linearly independent on $(-\infty, \infty)$ and hence form a fundamental set. It follows that the general solution of (2) on this interval is

$$y = c_1 e^{m_1 x} + c_2 e^{m_2 x}. \tag{4}$$

CASE II: REPEATED REAL ROOTS When $m_1 = m_2$, we necessarily obtain only one exponential solution, $y_1 = e^{m_1 x}$. From the quadratic formula we find that $m_1 = -b/2a$ since the only way to have $m_1 = m_2$ is to have $b^2 - 4ac = 0$. It follows from (5) in Section 4.2 that a second solution of the equation is

$$y_2 = e^{m_1 x} \int \frac{e^{2m_1 x}}{e^{2m_1 x}} \, dx = e^{m_1 x} \int dx = x e^{m_1 x}. \tag{5}$$

In (5) we have used the fact that $-b/a = 2m_1$. The general solution is then

$$y = c_1 e^{m_1 x} + c_2 x e^{m_1 x}. \tag{6}$$

CASE III: CONJUGATE COMPLEX ROOTS   If m_1 and m_2 are complex, then we can write $m_1 = \alpha + i\beta$ and $m_2 = \alpha - i\beta$, where α and $\beta > 0$ are real and $i^2 = -1$. Formally, there is no difference between this case and Case I, and hence

$$y = C_1 e^{(\alpha + i\beta)x} + C_2 e^{(\alpha - i\beta)x}.$$

However, in practice we prefer to work with real functions instead of complex exponentials. To this end we use Euler's formula:

$$e^{i\theta} = \cos\theta + i\sin\theta,$$

where θ is any real number.* It follows from this formula that

$$e^{i\beta x} = \cos\beta x + i\sin\beta x \quad \text{and} \quad e^{-i\beta x} = \cos\beta x - i\sin\beta x, \qquad (7)$$

where we have used $\cos(-\beta x) = \cos\beta x$ and $\sin(-\beta x) = -\sin\beta x$. Note that by first adding and then subtracting the two equations in (7), we obtain, respectively,

$$e^{i\beta x} + e^{-i\beta x} = 2\cos\beta x \quad \text{and} \quad e^{i\beta x} - e^{-i\beta x} = 2i\sin\beta x.$$

Since $y = C_1 e^{(\alpha + i\beta)x} + C_2 e^{(\alpha - i\beta)x}$ is a solution of (2) for any choice of the constants C_1 and C_2, the choices $C_1 = C_2 = 1$ and $C_1 = 1, C_2 = -1$ give, in turn, two solutions:

$$y_1 = e^{(\alpha + i\beta)x} + e^{(\alpha - i\beta)x} \quad \text{and} \quad y_2 = e^{(\alpha + i\beta)x} - e^{(\alpha - i\beta)x}.$$

But

$$y_1 = e^{\alpha x}(e^{i\beta x} + e^{-i\beta x}) = 2e^{\alpha x}\cos\beta x$$

and

$$y_2 = e^{\alpha x}(e^{i\beta x} - e^{-i\beta x}) = 2ie^{\alpha x}\sin\beta x.$$

Hence from Corollary (A) of Theorem 4.2 the last two results show that $e^{\alpha x}\cos\beta x$ and $e^{\alpha x}\sin\beta x$ are *real* solutions of (2). Moreover, these solutions form a fundamental set on $(-\infty, \infty)$. Consequently, the general solution is

$$y = c_1 e^{\alpha x}\cos\beta x + c_2 e^{\alpha x}\sin\beta x = e^{\alpha x}(c_1\cos\beta x + c_2\sin\beta x). \qquad (8)$$

EXAMPLE 1   Second-Order DEs

Solve the following differential equations.

(a) $2y'' - 5y' - 3y = 0$ **(b)** $y'' - 10y' + 25y = 0$ **(c)** $y'' + 4y' + 7y = 0$

SOLUTION   We give the auxiliary equations, the roots, and the corresponding general solutions.

(a) $2m^2 - 5m - 3 = (2m + 1)(m - 3) = 0, m_1 = -\frac{1}{2}, m_2 = 3$

From (4), $y = c_1 e^{-x/2} + c_2 e^{3x}$.

(b) $m^2 - 10m + 25 = (m - 5)^2 = 0, m_1 = m_2 = 5$

From (6), $y = c_1 e^{5x} + c_2 x e^{5x}$.

(c) $m^2 + 4m + 7 = 0, m_1 = -2 + \sqrt{3}i, m_2 = -2 - \sqrt{3}i$

From (8) with $\alpha = -2, \beta = \sqrt{3}, y = e^{-2x}\left(c_1\cos\sqrt{3}x + c_2\sin\sqrt{3}x\right)$.   ∎

---

*A formal derivation of Euler's formula can be obtained from the Maclaurin series $e^x = \sum\limits_{n=0}^{\infty} \dfrac{x^n}{n!}$ by substituting $x = i\theta$, using $i^2 = -1, i^3 = -i, \ldots$, and then separating the series into real and imaginary parts. The plausibility thus established, we can adopt $\cos\theta + i\sin\theta$ as the *definition* of $e^{i\theta}$.

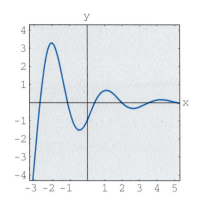

FIGURE 4.4   Solution curve of IVP in Example 2

EXAMPLE 2   **An Initial-Value Problem**

Solve $4y'' + 4y' + 17y = 0$, $y(0) = -1$, $y'(0) = 2$.

SOLUTION   By the quadratic formula we find that the roots of the auxiliary equation $4m^2 + 4m + 17 = 0$ are $m_1 = -\frac{1}{2} + 2i$ and $m_2 = -\frac{1}{2} - 2i$. Thus from (8) we have $y = e^{-x/2}(c_1 \cos 2x + c_2 \sin 2x)$. Applying the condition $y(0) = -1$, we see from $e^0(c_1 \cos 0 + c_2 \sin 0) = -1$ that $c_1 = -1$. Differentiating $y = e^{-x/2}(-\cos 2x + c_2 \sin 2x)$ and then using $y'(0) = 2$ gives $2c_2 + \frac{1}{2} = 2$ or $c_2 = \frac{3}{4}$. Hence the solution of the IVP is $y = e^{-x/2}(-\cos 2x + \frac{3}{4} \sin 2x)$. In Figure 4.4 we see that the solution is oscillatory, but $y \to 0$ as $x \to \infty$ and $|y| \to \infty$ as $x \to -\infty$.

TWO EQUATIONS WORTH KNOWING   The two differential equations

$$y'' + k^2y = 0 \quad \text{and} \quad y'' - k^2y = 0,$$

k real, are important in applied mathematics. For $y'' + k^2y = 0$, the auxiliary equation $m^2 + k^2 = 0$ has imaginary roots $m_1 = ki$ and $m_2 = -ki$. With $\alpha = 0$ and $\beta = k$ in (8), the general solution of the DE is seen to be

$$y = c_1 \cos kx + c_2 \sin kx. \tag{9}$$

On the other hand, the auxiliary equation $m^2 - k^2 = 0$ for $y'' - k^2y = 0$ has distinct real roots $m_1 = k$ and $m_2 = -k$, and so by (4) the general solution of the DE is

$$y = c_1 e^{kx} + c_2 e^{-kx}. \tag{10}$$

Notice that if we choose $c_1 = c_2 = \frac{1}{2}$ and $c_1 = \frac{1}{2}$, $c_2 = -\frac{1}{2}$ in (10), we get the particular solutions $y = \frac{1}{2}(e^{kx} + e^{-kx}) = \cosh kx$ and $y = \frac{1}{2}(e^{kx} - e^{-kx}) = \sinh kx$. Since $\cosh kx$ and $\sinh kx$ are linearly independent on any interval of the x-axis, an alternative form for the general solution of $y'' - k^2y = 0$ is

$$y = c_1 \cosh kx + c_2 \sinh kx. \tag{11}$$

See Problems 41, 42, and 56 in Exercises 4.3.

HIGHER-ORDER EQUATIONS   In general, to solve an nth-order differential equation (1) where the a_i, $i = 0, 1, \ldots, n$ are real constants, we must solve an nth-degree polynomial equation

$$a_n m^n + a_{n-1} m^{n-1} + \cdots + a_2 m^2 + a_1 m + a_0 = 0. \tag{12}$$

If all the roots of (12) are real and distinct, then the general solution of (1) is

$$y = c_1 e^{m_1 x} + c_2 e^{m_2 x} + \cdots + c_n e^{m_n x}.$$

It is somewhat harder to summarize the analogues of Cases II and III because the roots of an auxiliary equation of degree greater than two can occur in many combinations. For example, a fifth-degree equation could have five distinct real roots, or three distinct real and two complex roots, or one real and four complex roots, or five real but equal roots, or five real roots but two of them equal, and so on. When m_1 is a root of multiplicity k of an nth-degree auxiliary equation (that is, k roots are equal to m_1), it can be shown that the linearly independent solutions are

$$e^{m_1 x}, \quad x e^{m_1 x}, \quad x^2 e^{m_1 x}, \ldots, \quad x^{k-1} e^{m_1 x}$$

and the general solution must contain the linear combination

$$c_1 e^{m_1 x} + c_2 x e^{m_1 x} + c_3 x^2 e^{m_1 x} + \cdots + c_k x^{k-1} e^{m_1 x}.$$

Finally, it should be remembered that when the coefficients are real, complex roots of an auxiliary equation always appear in conjugate pairs. Thus, for example, a cubic polynomial equation can have at most two complex roots.

EXAMPLE 3	Third-Order DE

Solve $y''' + 3y'' - 4y = 0$.

SOLUTION It should be apparent from inspection of $m^3 + 3m^2 - 4 = 0$ that one root is $m_1 = 1$ and so $m - 1$ is a factor of $m^3 + 3m^2 - 4$. By division we find

$$m^3 + 3m^2 - 4 = (m - 1)(m^2 + 4m + 4) = (m - 1)(m + 2)^2,$$

and so the other roots are $m_2 = m_3 = -2$. Thus the general solution of the DE is $y = c_1e^x + c_2e^{-2x} + c_3xe^{-2x}$.

EXAMPLE 4	Fourth-Order DE

Solve $\dfrac{d^4y}{dx^4} + 2\dfrac{d^2y}{dx^2} + y = 0$.

SOLUTION The auxiliary equation $m^4 + 2m^2 + 1 = (m^2 + 1)^2 = 0$ has roots $m_1 = m_3 = i$ and $m_2 = m_4 = -i$. Thus from Case II the solution is

$$y = C_1e^{ix} + C_2e^{-ix} + C_3xe^{ix} + C_4xe^{-ix}.$$

By Euler's formula the grouping $C_1e^{ix} + C_2e^{-ix}$ can be rewritten as

$$c_1 \cos x + c_2 \sin x$$

after a relabeling of constants. Similarly, $x(C_3e^{ix} + C_4e^{-ix})$ can be expressed as $x(c_3 \cos x + c_4 \sin x)$. Hence the general solution is

$$y = c_1 \cos x + c_2 \sin x + c_3x \cos x + c_4x \sin x.$$

Example 4 illustrates a special case when the auxiliary equation has repeated complex roots. In general, if $m_1 = \alpha + i\beta$, $\beta > 0$ is a complex root of multiplicity k of an auxiliary equation with real coefficients, then its conjugate $m_2 = \alpha - i\beta$ is also a root of multiplicity k. From the $2k$ complex-valued solutions

$$e^{(\alpha+i\beta)x}, \quad xe^{(\alpha+i\beta)x}, \quad x^2e^{(\alpha+i\beta)x}, \ldots, \quad x^{k-1}e^{(\alpha+i\beta)x},$$

$$e^{(\alpha-i\beta)x}, \quad xe^{(\alpha-i\beta)x}, \quad x^2e^{(\alpha-i\beta)x}, \ldots, \quad x^{k-1}e^{(\alpha-i\beta)x},$$

we conclude, with the aid of Euler's formula, that the general solution of the corresponding differential equation must then contain a linear combination of the $2k$ real linearly independent solutions

$$e^{\alpha x} \cos \beta x, \quad xe^{\alpha x} \cos \beta x, \quad x^2e^{\alpha x} \cos \beta x, \ldots, \quad x^{k-1}e^{\alpha x} \cos \beta x,$$

$$e^{\alpha x} \sin \beta x, \quad xe^{\alpha x} \sin \beta x, \quad x^2e^{\alpha x} \sin \beta x, \ldots, \quad x^{k-1}e^{\alpha x} \sin \beta x.$$

In Example 4 we identify $k = 2$, $\alpha = 0$, and $\beta = 1$.

Of course the most difficult aspect of solving constant-coefficient differential equations is finding roots of auxiliary equations of degree greater than two. For example, to solve $3y''' + 5y'' + 10y' - 4y = 0$, we must solve $3m^3 + 5m^2 + 10m - 4 = 0$. Something we can try is to test the auxiliary equation for rational roots. Recall that if $m_1 = p/q$ is a rational root (expressed in lowest terms) of an auxiliary equation $a_nm^n + \cdots + a_1m + a_0 = 0$ with integer coefficients, then p is a factor of a_0 and q is a factor of a_n. For our specific cubic auxiliary equation, all the factors of $a_0 = -4$ and $a_n = 3$ are p: $\pm 1, \pm 2, \pm 4$ and q: $\pm 1, \pm 3$, so the possible rational roots are p/q: $\pm 1, \pm 2, \pm 4, \pm\frac{1}{3}, \pm\frac{2}{3}, \pm\frac{4}{3}$. Each of these numbers can then be tested—say, by synthetic division. In this way we discover both the root $m_1 = \frac{1}{3}$ and the factorization

$$3m^3 + 5m^2 + 10m - 4 = \left(m - \tfrac{1}{3}\right)(3m^2 + 6m + 12).$$

The quadratic formula then yields the remaining roots $m_2 = -1 + \sqrt{3}i$ and $m_3 = -1 - \sqrt{3}i$. Therefore the general solution of $3y''' + 5y'' + 10y' - 4y = 0$ is $y = c_1 e^{x/3} + e^{-x}\left(c_2 \cos \sqrt{3}x + c_3 \sin \sqrt{3}x\right)$.

There is more on this in the **SRSM**.

USE OF COMPUTERS Finding roots or approximation of roots of auxiliary equations is a routine problem with an appropriate calculator or computer software. Polynomial equations (in one variable) of degree less than five can be solved by means of algebraic formulas using the *solve* commands in *Mathematica* and *Maple*. For auxiliary equations of degree five or greater it might be necessary to resort to numerical commands such as **NSolve** and **FindRoot** in *Mathematica*. Because of their capability of solving polynomial equations, it is not surprising that these computer algebra systems are also able, by means of their *dsolve* commands, to provide explicit solutions of homogeneous linear constant-coefficient differential equations.

In his classic text *Differential Equations,* by Ralph Palmer Agnew* (used by the author as a student), the following statement is made:

> *It is not reasonable to expect students in this course to have computing skill and equipment necessary for efficient solving of equations such as*

$$4.317 \frac{d^4 y}{dx^4} + 2.179 \frac{d^3 y}{dx^3} + 1.416 \frac{d^2 y}{dx^2} + 1.295 \frac{dy}{dx} + 3.169y = 0. \qquad (13)$$

Although it is debatable whether computing skills have improved in the intervening years, it is a certainty that technology has. If one has access to a computer algebra system, equation (13) could now be considered reasonable. After simplification and some relabeling of output, *Mathematica* yields the (approximate) general solution

$$y = c_1 e^{-0.728852x} \cos(0.618605x) + c_2 e^{-0.728852x} \sin(0.618605x)$$
$$+ c_3 e^{0.476478x} \cos(0.759081x) + c_4 e^{0.476478x} \sin(0.759081x).$$

Finally, if we are faced with an initial-value problem consisting of, say, a fourth-order equation, then to fit the general solution of the DE to the four initial conditions, we must solve four linear equations in four unknowns (the c_1, c_2, c_3, c_4 in the general solution). Using a CAS to solve the system can save lots of time. See Problems 61 and 62 in Exercises 4.3 and Problem 35 in Chapter 4 in Review.

*McGraw-Hill, New York, 1960.

EXERCISES 4.3

Answers to selected odd-numbered problems begin on page ANS-4.

In Problems 1–14 find the general solution of the given second-order differential equation.

1. $4y'' + y' = 0$

2. $y'' - 36y = 0$

3. $y'' - y' - 6y = 0$

4. $y'' - 3y' + 2y = 0$

5. $y'' + 8y' + 16y = 0$

6. $y'' - 10y' + 25y = 0$

7. $12y'' - 5y' - 2y = 0$

8. $y'' + 4y' - y = 0$

9. $y'' + 9y = 0$

10. $3y'' + y = 0$

11. $y'' - 4y' + 5y = 0$

12. $2y'' + 2y' + y = 0$

13. $3y'' + 2y' + y = 0$

14. $2y'' - 3y' + 4y = 0$

In Problems 15–28 find the general solution of the given higher-order differential equation.

15. $y''' - 4y'' - 5y' = 0$

16. $y''' - y = 0$

17. $y''' - 5y'' + 3y' + 9y = 0$

18. $y''' + 3y'' - 4y' - 12y = 0$

19. $\dfrac{d^3u}{dt^3} + \dfrac{d^2u}{dt^2} - 2u = 0$

20. $\dfrac{d^3x}{dt^3} - \dfrac{d^2x}{dt^2} - 4x = 0$

21. $y''' + 3y'' + 3y' + y = 0$

22. $y''' - 6y'' + 12y' - 8y = 0$

23. $y^{(4)} + y''' + y'' = 0$

24. $y^{(4)} - 2y'' + y = 0$

25. $16\dfrac{d^4y}{dx^4} + 24\dfrac{d^2y}{dx^2} + 9y = 0$

26. $\dfrac{d^4y}{dx^4} - 7\dfrac{d^2y}{dx^2} - 18y = 0$

27. $\dfrac{d^5u}{dr^5} + 5\dfrac{d^4u}{dr^4} - 2\dfrac{d^3u}{dr^3} - 10\dfrac{d^2u}{dr^2} + \dfrac{du}{dr} + 5u = 0$

28. $2\dfrac{d^5x}{ds^5} - 7\dfrac{d^4x}{ds^4} + 12\dfrac{d^3x}{ds^3} + 8\dfrac{d^2x}{ds^2} = 0$

In Problems 29–36 solve the given initial-value problem.

29. $y'' + 16y = 0, \quad y(0) = 2, y'(0) = -2$

30. $\dfrac{d^2y}{d\theta^2} + y = 0, \quad y\left(\dfrac{\pi}{3}\right) = 0, y'\left(\dfrac{\pi}{3}\right) = 2$

31. $\dfrac{d^2y}{dt^2} - 4\dfrac{dy}{dt} - 5y = 0, \quad y(1) = 0, y'(1) = 2$

32. $4y'' - 4y' - 3y = 0, \quad y(0) = 1, y'(0) = 5$

33. $y'' + y' + 2y = 0, \quad y(0) = y'(0) = 0$

34. $y'' - 2y' + y = 0, \quad y(0) = 5, y'(0) = 10$

35. $y''' + 12y'' + 36y' = 0, \quad y(0) = 0, y'(0) = 1, y''(0) = -7$

36. $y''' + 2y'' - 5y' - 6y = 0, \quad y(0) = y'(0) = 0, y''(0) = 1$

In Problems 37–40 solve the given boundary-value problem.

37. $y'' - 10y' + 25y = 0, \quad y(0) = 1, y(1) = 0$

38. $y'' + 4y = 0, \quad y(0) = 0, y(\pi) = 0$

39. $y'' + y = 0, \quad y'(0) = 0, y'\left(\dfrac{\pi}{2}\right) = 0$

40. $y'' - 2y' + 2y = 0, \quad y(0) = 1, y(\pi) = 1$

In Problems 41 and 42 solve the given problem first using the form of the general solution given in (10). Solve again, this time using the form given in (11).

41. $y'' - 3y = 0, \quad y(0) = 1, y'(0) = 5$

42. $y'' - y = 0, \quad y(0) = 1, y'(1) = 0.$

In Problems 43–48 each figure represents the graph of a particular solution of one of the following differential equations:

    (a) $y'' - 3y' - 4y = 0$ **(b)** $y'' + 4y = 0$

    (c) $y'' + 2y' + y = 0$ **(d)** $y'' + y = 0$

    (e) $y'' + 2y' + 2y = 0$ **(f)** $y'' - 3y' + 2y = 0$

Match a solution curve with one of the differential equations. Explain your reasoning.

43.

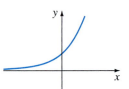

FIGURE 4.5   Graph for Problem 43

44.

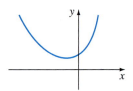

FIGURE 4.6   Graph for Problem 44

45.

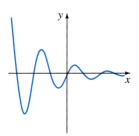

FIGURE 4.7   Graph for Problem 45

46.

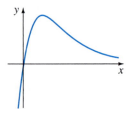

FIGURE 4.8   Graph for Problem 46

47.

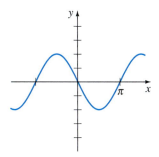

FIGURE 4.9 Graph for Problem 47

48.

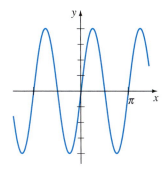

FIGURE 4.10 Graph for Problem 48

DISCUSSION/PROJECT PROBLEMS

49. The roots of a cubic auxiliary equation are $m_1 = 4$ and $m_2 = m_3 = -5$. What is the corresponding homogeneous linear differential equation? Discuss: Is your answer unique?

50. Two roots of a cubic auxiliary equation with real coefficients are $m_1 = -\frac{1}{2}$ and $m_2 = 3 + i$. What is the corresponding homogeneous linear differential equation?

51. Find the general solution of $y''' + 6y'' + y' - 34y = 0$ if it is known that $y_1 = e^{-4x} \cos x$ is one solution.

52. To solve $y^{(4)} + y = 0$, we must find the roots of $m^4 + 1 = 0$. This is a trivial problem using a CAS but can also be done by hand working with complex numbers. Observe that $m^4 + 1 = (m^2 + 1)^2 - 2m^2$. How does this help? Solve the differential equation.

53. Verify that $y = \sinh x - 2\cos\left(x + \dfrac{\pi}{6}\right)$ is a particular solution of $y^{(4)} - y = 0$. Reconcile this particular solution with the general solution of the DE.

54. Consider the boundary-value problem $y'' + \lambda y = 0$, $y(0) = 0$, $y(\pi/2) = 0$. Discuss: Is it possible to determine values of λ so that the problem possesses **(a)** trivial solutions? **(b)** nontrivial solutions?

55. In the study of techniques of integration in calculus, certain indefinite integrals of the form $\int e^{ax} f(x)\, dx$

could be evaluated by applying integration by parts twice, recovering the original integral on the right-hand side, solving for the original integral, and obtaining a constant multiple $k \int e^{ax} f(x)\, dx$ on the left-hand side. Then the value of the integral is found by dividing by k. Discuss: For what kind of functions f does the described procedure work? Your solution should lead to a differential equation. Carefully analyze this equation and solve for f.

MATHEMATICAL MODEL

56. Slipping Chain Reread the discussion on the slipping chain in Section 1.3 and illustrated in Figure 1.24 on page 26.

(a) Use the form of the solution given in (11) of this section to find the general solution of equation (16) of Section 1.3,

$$\frac{d^2x}{dt^2} - \frac{64}{L} x = 0.$$

(b) Find a particular solution that satisfies the initial conditions stated in the discussion on pages 25–26.

(c) Suppose that the total length of the chain is $L = 20$ ft and that $x_0 = 1$. Find the velocity at which the slipping chain will leave the supporting peg.

COMPUTER LAB ASSIGNMENTS

In Problems 57–60 use a computer either as an aid in solving the auxiliary equation or as a means of directly obtaining the general solution of the given differential equation. If you use a CAS to obtain the general solution, simplify the output and, if necessary, write the solution in terms of real functions.

57. $y''' - 6y'' + 2y' + y = 0$

58. $6.11y''' + 8.59y'' + 7.93y' + 0.778y = 0$

59. $3.15y^{(4)} - 5.34y'' + 6.33y' - 2.03y = 0$

60. $y^{(4)} + 2y'' - y' + 2y = 0$

In Problems 61 and 62 use a CAS as an aid in solving the auxiliary equation. Form the general solution of the differential equation. Then use a CAS as an aid in solving the system of equations for the coefficients c_i, $i = 1, 2, 3, 4$ that results when the initial conditions are applied to the general solution.

61. $2y^{(4)} + 3y''' - 16y'' + 15y' - 4y = 0$,
$y(0) = -2, y'(0) = 6, y''(0) = 3, y'''(0) = \frac{1}{2}$

62. $y^{(4)} - 3y''' + 3y'' - y' = 0$,
$y(0) = y'(0) = 0, y''(0) = y'''(0) = 1$

4.4 UNDETERMINED COEFFICIENTS—SUPERPOSITION APPROACH*

INTRODUCTION: To solve a nonhomogeneous linear differential equation

$$a_n y^{(n)} + a_{n-1} y^{(n-1)} + \cdots + a_1 y' + a_0 y = g(x), \tag{1}$$

we must do two things: (*i*) find the complementary function y_c, and (*ii*) find *any* particular solution y_p of the nonhomogeneous equation (1). Then, as was discussed in Section 4.1, the general solution of (1) is $y = y_c + y_p$.

The complementary function y_c is the general solution of the associated homogeneous DE of (1), that is, $a_n y^{(n)} + a_{n-1} y^{(n-1)} + \cdots + a_1 y' + a_0 y = 0$. In the preceding section we saw how to solve these kinds of equations when the coefficients were constants. Our goal then in the present section is to develop a method for obtaining particular solutions.

REVIEW MATERIAL: Review Subsection 4.1.3 of Section 4.1.

METHOD OF UNDETERMINED COEFFICIENTS The first of two ways we shall consider for obtaining a particular solution y_p for a nonhomogeneous linear DE is called the **method of undetermined coefficients.** The underlying idea behind this method is a conjecture about the form of y_p, an educated guess really, that is motivated by the kinds of functions that make up the input function $g(x)$. The general method is limited to linear DEs such as (1) where

- the coefficients a_i, $i = 0, 1, \ldots, n$ are constants and
- $g(x)$ is a constant k, a polynomial function, an exponential function $e^{\alpha x}$, a sine or cosine function $\sin \beta x$ or $\cos \beta x$, or finite sums and products of these functions.

NOTE Strictly speaking, $g(x) = k$ (constant) is a polynomial function. Since a constant function is probably not the first thing that comes to mind when you think of polynomial functions, for emphasis we shall continue to use the redundancy "constant functions, polynomials, ..."

The following functions are some examples of the types of inputs $g(x)$ that are appropriate for this discussion:

$$g(x) = 10, \quad g(x) = x^2 - 5x, \quad g(x) = 15x - 6 + 8e^{-x},$$

$$g(x) = \sin 3x - 5x \cos 2x, \qquad g(x) = xe^x \sin x + (3x^2 - 1)e^{-4x}.$$

That is, $g(x)$ is a linear combination of functions of the type

$$P(x) = a_n x^n + a_{n-1} x^{n-1} + \cdots + a_1 x + a_0, \quad P(x)\, e^{\alpha x}, \quad P(x)\, e^{\alpha x} \sin \beta x, \quad \text{and} \quad P(x)\, e^{\alpha x} \cos \beta x,$$

where n is a nonnegative integer and α and β are real numbers. The method of undetermined coefficients is not applicable to equations of form (1) when

---

*__Note to the Instructor:__ In this section the method of undetermined coefficients is developed from the viewpoint of the superposition principle for nonhomogeneous equations (Theorem 4.7). In Section 4.5 an entirely different approach will be presented, one utilizing the concept of differential annihilator operators. Take your pick.

$$g(x) = \ln x, \quad g(x) = \frac{1}{x}, \quad g(x) = \tan x, \quad g(x) = \sin^{-1}x,$$

and so on. Differential equations in which the input $g(x)$ is a function of this last kind will be considered in Section 4.6.

The set of functions that consists of constants, polynomials, exponentials $e^{\alpha x}$, sines, and cosines has the remarkable property that derivatives of their sums and products are again sums and products of constants, polynomials, exponentials $e^{\alpha x}$, sines, and cosines. Because the linear combination of derivatives $a_n y_p^{(n)} + a_{n-1} y_p^{(n-1)} + \cdots + a_1 y_p' + a_0 y_p$ must be identical to $g(x)$, it seems reasonable to assume that y_p *has the same form as* $g(x)$.

The next two examples illustrate the basic method.

EXAMPLE 1   General Solution Using Undetermined Coefficients

Solve $y'' + 4y' - 2y = 2x^2 - 3x + 6.$ (2)

SOLUTION

Step 1.   We first solve the associated homogeneous equation $y'' + 4y' - 2y = 0$. From the quadratic formula we find that the roots of the auxiliary equation $m^2 + 4m - 2 = 0$ are $m_1 = -2 - \sqrt{6}$ and $m_2 = -2 + \sqrt{6}$. Hence the complementary function is

$$y_c = c_1 e^{-(2+\sqrt{6})x} + c_2 e^{(-2+\sqrt{6})x}.$$

Step 2.   Now, because the function $g(x)$ is a quadratic polynomial, let us assume a particular solution that is also in the form of a quadratic polynomial:

$$y_p = Ax^2 + Bx + C.$$

We seek to determine *specific* coefficients A, B, and C for which y_p is a solution of (2). Substituting y_p and the derivatives

$$y_p' = 2Ax + B \quad \text{and} \quad y_p'' = 2A$$

into the given differential equation (2), we get

$$y_p'' + 4y_p' - 2y_p = 2A + 8Ax + 4B - 2Ax^2 - 2Bx - 2C = 2x^2 - 3x + 6.$$

Because the last equation is supposed to be an identity, the coefficients of like powers of x must be equal:

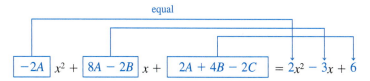

That is, $-2A = 2, \quad 8A - 2B = -3, \quad 2A + 4B - 2C = 6.$

Solving this system of equations leads to the values $A = -1$, $B = -\frac{5}{2}$, and $C = -9$. Thus a particular solution is

$$y_p = -x^2 - \frac{5}{2}x - 9.$$

Step 3.   The general solution of the given equation is

$$y = y_c + y_p = c_1 e^{-(2+\sqrt{6})x} + c_1 e^{(-2+\sqrt{6})x} - x^2 - \frac{5}{2}x - 9.$$

EXAMPLE 2   Particular Solution Using Undetermined Coefficients

Find a particular solution of $y'' - y' + y = 2 \sin 3x$.

SOLUTION   A natural first guess for a particular solution would be $A \sin 3x$. But because successive differentiations of $\sin 3x$ produce $\sin 3x$ *and* $\cos 3x$, we are prompted instead to assume a particular solution that includes both of these terms:

$$y_p = A \cos 3x + B \sin 3x.$$

Differentiating y_p and substituting the results into the differential equation gives, after regrouping,

$$y_p'' - y_p' + y_p = (-8A - 3B) \cos 3x + (3A - 8B) \sin 3x = 2 \sin 3x$$

or

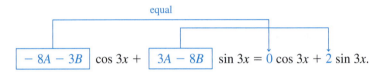

$$\boxed{-8A - 3B}\; \cos 3x + \boxed{3A - 8B}\; \sin 3x = 0 \cos 3x + 2 \sin 3x.$$

From the resulting system of equations,

$$-8A - 3B = 0, \quad 3A - 8B = 2,$$

we get $A = \frac{6}{73}$ and $B = -\frac{16}{73}$. A particular solution of the equation is

$$y_p = \frac{6}{73} \cos 3x - \frac{16}{73} \sin 3x.$$

As we mentioned, the form that we assume for the particular solution y_p is an educated guess; it is not a blind guess. This educated guess must take into consideration not only the types of functions that make up $g(x)$ but also, as we shall see in Example 4, the functions that make up the complementary function y_c.

EXAMPLE 3   Forming y_p by Superposition

Solve $y'' - 2y' - 3y = 4x - 5 + 6xe^{2x}$.   (3)

SOLUTION

Step 1. First, the solution of the associated homogeneous equation $y'' - 2y' - 3y = 0$ is found to be $y_c = c_1 e^{-x} + c_2 e^{3x}$.

Step 2. Next, the presence of $4x - 5$ in $g(x)$ suggests that the particular solution includes a linear polynomial. Furthermore, because the derivative of the product xe^{2x} produces $2xe^{2x}$ and e^{2x}, we also assume that the particular solution includes both xe^{2x} and e^{2x}. In other words, g is the sum of two basic kinds of functions:

$$g(x) = g_1(x) + g_2(x) = polynomial + exponentials.$$

Correspondingly, the superposition principle for nonhomogeneous equations (Theorem 4.7) suggests that we seek a particular solution

$$y_p = y_{p_1} + y_{p_2},$$

where $y_{p_1} = Ax + B$ and $y_{p_2} = Cxe^{2x} + Ee^{2x}$. Substituting

$$y_p = Ax + B + Cxe^{2x} + Ee^{2x}$$

into the given equation (3) and grouping like terms gives

$$y_p'' - 2y_p' - 3y_p = -3Ax - 2A - 3B - 3Cxe^{2x} + (2C - 3E)e^{2x} = 4x - 5 + 6xe^{2x}. \quad (4)$$

From this identity we obtain the four equations

$$-3A = 4, \quad -2A - 3B = -5, \quad -3C = 6, \quad 2C - 3E = 0.$$

The last equation in this system results from the interpretation that the coefficient of e^{2x} in the right member of (4) is zero. Solving, we find $A = -\frac{4}{3}$, $B = \frac{23}{9}$, $C = -2$, and $E = -\frac{4}{3}$. Consequently,

$$y_p = -\frac{4}{3}x + \frac{23}{9} - 2xe^{2x} - \frac{4}{3}e^{2x}.$$

Step 3. The general solution of the equation is

$$y = c_1 e^{-x} + c_2 e^{3x} - \frac{4}{3}x + \frac{23}{9} - \left(2x + \frac{4}{3}\right)e^{2x}. \qquad \blacksquare$$

In light of the superposition principle (Theorem 4.7) we can also approach Example 3 from the viewpoint of solving two simpler problems. You should verify that substituting

$$y_{p_1} = Ax + B \qquad \text{into} \quad y'' - 2y' - 3y = 4x - 5$$

and

$$y_{p_2} = Cxe^{2x} + Ee^{2x} \quad \text{into} \quad y'' - 2y' - 3y = 6xe^{2x}$$

yields, in turn, $y_{p_1} = -\frac{4}{3}x + \frac{23}{9}$ and $y_{p_2} = -\left(2x + \frac{4}{3}\right)e^{2x}$. A particular solution of (3) is then $y_p = y_{p_1} + y_{p_2}$.

The next example illustrates that sometimes the "obvious" assumption for the form of y_p is not a correct assumption.

EXAMPLE 4   A Glitch in the Method

Find a particular solution of $y'' - 5y' + 4y = 8e^x$.

SOLUTION   Differentiation of e^x produces no new functions. Thus, proceeding as we did in the earlier examples, we can reasonably assume a particular solution of the form $y_p = Ae^x$. But substitution of this expression into the differential equation yields the contradictory statement $0 = 8e^x$ so we have clearly made the wrong guess for y_p.

The difficulty here is apparent on examining the complementary function $y_c = c_1 e^x + c_2 e^{4x}$. Observe that our assumption Ae^x is already present in y_c. This means that e^x is a solution of the associated homogeneous differential equation, and a constant multiple Ae^x when substituted into the differential equation necessarily produces zero.

What then should be the form of y_p? Inspired by Case II of Section 4.3, let's see whether we can find a particular solution of the form

$$y_p = Axe^x.$$

Substituting $y_p' = Axe^x + Ae^x$ and $y_p'' = Axe^x + 2Ae^x$ into the differential equation and simplifying gives

$$y_p'' - 5y_p' + 4y_p = -3Ae^x = 8e^x.$$

From the last equality we see that the value of A is now determined as $A = -\frac{8}{3}$. Therefore a particular solution of the given equation is $y_p = -\frac{8}{3}xe^x$. $\blacksquare$

The difference in the procedures used in Examples 1–3 and in Example 4 suggests that we consider two cases. The first case reflects the situation in Examples 1–3.

CASE I   No function in the assumed particular solution is a solution of the associated homogeneous differential equation.

In Table 4.1 we illustrate some specific examples of $g(x)$ in (1) along with the corresponding form of the particular solution. We are, of course, taking for granted that no function in the assumed particular solution y_p is duplicated by a function in the complementary function y_c.

TABLE 4.1   Trial Particular Solutions

$g(x)$	Form of y_p
1. 1 (any constant)	A
2. $5x + 7$	$Ax + B$
3. $3x^2 - 2$	$Ax^2 + Bx + C$
4. $x^3 - x + 1$	$Ax^3 + Bx^2 + Cx + E$
5. $\sin 4x$	$A \cos 4x + B \sin 4x$
6. $\cos 4x$	$A \cos 4x + B \sin 4x$
7. e^{5x}	Ae^{5x}
8. $(9x - 2)e^{5x}$	$(Ax + B)e^{5x}$
9. $x^2 e^{5x}$	$(Ax^2 + Bx + C)e^{5x}$
10. $e^{3x} \sin 4x$	$Ae^{3x} \cos 4x + Be^{3x} \sin 4x$
11. $5x^2 \sin 4x$	$(Ax^2 + Bx + C) \cos 4x + (Ex^2 + Fx + G) \sin 4x$
12. $xe^{3x} \cos 4x$	$(Ax + B)e^{3x} \cos 4x + (Cx + E)e^{3x} \sin 4x$

EXAMPLE 5   Forms of Particular Solutions — Case I

Determine the form of a particular solution of

(a) $y'' - 8y' + 25y = 5x^3 e^{-x} - 7e^{-x}$ **(b)** $y'' + 4y = x \cos x$

SOLUTION   **(a)** We can write $g(x) = (5x^3 - 7)e^{-x}$. Using entry 9 in Table 4.1 as a model, we assume a particular solution of the form

$$y_p = (Ax^3 + Bx^2 + Cx + E)e^{-x}.$$

Note that there is no duplication between the terms in y_p and the terms in the complementary function $y_c = e^{4x}(c_1 \cos 3x + c_2 \sin 3x)$.

(b) The function $g(x) = x \cos x$ is similar to entry 11 in Table 4.1 except, of course, that we use a linear rather than a quadratic polynomial and $\cos x$ and $\sin x$ instead of $\cos 4x$ and $\sin 4x$ in the form of y_p:

$$y_p = (Ax + B) \cos x + (Cx + E) \sin x.$$

Again observe that there is no duplication of terms between y_p and $y_c = c_1 \cos 2x + c_2 \sin 2x$.   ∎

If $g(x)$ consists of a sum of, say, m terms of the kind listed in the table, then (as in Example 3) the assumption for a particular solution y_p consists of the sum of the trial forms $y_{p_1}, y_{p_2}, \ldots, y_{p_m}$ corresponding to these terms:

$$y_p = y_{p_1} + y_{p_2} + \cdots + y_{p_m}.$$

The foregoing sentence can be put another way.

Form Rule for Case I   *The form of y_p is a linear combination of all linearly independent functions that are generated by repeated differentiations of $g(x)$.*

EXAMPLE 6   Forming y_p by Superposition—Case I

Determine the form of a particular solution of

$$y'' - 9y' + 14y = 3x^2 - 5 \sin 2x + 7xe^{6x}.$$

SOLUTION

Corresponding to $3x^2$ we assume $y_{p_1} = Ax^2 + Bx + C.$

Corresponding to $-5 \sin 2x$ we assume $y_{p_2} = E \cos 2x + F \sin 2x.$

Corresponding to $7xe^{6x}$ we assume $y_{p_3} = (Gx + H)e^{6x}.$

The assumption for the particular solution is then

$$y_p = y_{p_1} + y_{p_2} + y_{p_3} = Ax^2 + Bx + C + E \cos 2x + F \sin 2x + (Gx + H)e^{6x}.$$

No term in this assumption duplicates a term in $y_c = c_1 e^{2x} + c_2 e^{7x}.$

CASE II   A function in the assumed particular solution is also a solution of the associated homogeneous differential equation.

The next example is similar to Example 4.

EXAMPLE 7   Particular Solution—Case II

Find a particular solution of $y'' - 2y' + y = e^x.$

SOLUTION   The complementary function is $y_c = c_1 e^x + c_2 x e^x.$ As in Example 4, the assumption $y_p = Ae^x$ will fail since it is apparent from y_c that e^x is a solution of the associated homogeneous equation $y'' - 2y' + y = 0.$ Moreover, we will not be able to find a particular solution of the form $y_p = Axe^x$ since the term xe^x is also duplicated in y_c. We next try

$$y_p = Ax^2 e^x.$$

Substituting into the given differential equation yields $2Ae^x = e^x$, and so $A = \frac{1}{2}$. Thus a particular solution is $y_p = \frac{1}{2} x^2 e^x.$

Suppose again that $g(x)$ consists of m terms of the kind given in Table 4.1, and suppose further that the usual assumption for a particular solution is

$$y_p = y_{p_1} + y_{p_2} + \cdots + y_{p_m},$$

where the y_{p_i}, $i = 1, 2, \ldots, m$ are the trial particular solution forms corresponding to these terms. Under the circumstances described in Case II, we can make up the following general rule.

> ***Multiplication Rule for Case II***   *If any y_{p_i} contains terms that duplicate terms in y_c, then that y_{p_i} must be multiplied by x^n, where n is the smallest positive integer that eliminates that duplication.*

EXAMPLE 8   An Initial-Value Problem

Solve $y'' + y = 4x + 10 \sin x$, $y(\pi) = 0$, $y'(\pi) = 2.$

SOLUTION   The solution of the associated homogeneous equation $y'' + y = 0$ is $y_c = c_1 \cos x + c_2 \sin x$. Because $g(x) = 4x + 10 \sin x$ is the sum of a linear polynomial and a sine function, our normal assumption for y_p, from entries 2 and 5 of Table 4.1, would be the sum of $y_{p_1} = Ax + B$ and $y_{p_2} = C \cos x + E \sin x$:

$$y_p = Ax + B + C \cos x + E \sin x. \tag{5}$$

But there is an obvious duplication of the terms $\cos x$ and $\sin x$ in this assumed form and two terms in the complementary function. This duplication can be eliminated by simply multiplying y_{p_2} by x. Instead of (5) we now use

$$y_p = Ax + B + Cx \cos x + Ex \sin x. \tag{6}$$

Differentiating this expression and substituting the results into the differential equation gives

$$y_p'' + y_p = Ax + B - 2C \sin x + 2E \cos x = 4x + 10 \sin x,$$

and so $A = 4$, $B = 0$, $-2C = 10$, and $2E = 0$. The solutions of the system are immediate: $A = 4$, $B = 0$, $C = -5$, and $E = 0$. Therefore from (6) we obtain $y_p = 4x - 5x \cos x$. The general solution of the given equation is

$$y = y_c + y_p = c_1 \cos x + c_2 \sin x + 4x - 5x \cos x.$$

We now apply the prescribed initial conditions to the general solution of the equation. First, $y(\pi) = c_1 \cos \pi + c_2 \sin \pi + 4\pi - 5\pi \cos \pi = 0$ yields $c_1 = 9\pi$ since $\cos \pi = -1$ and $\sin \pi = 0$. Next, from the derivative

$$y' = -9\pi \sin x + c_2 \cos x + 4 + 5x \sin x - 5 \cos x$$

and $\qquad y'(\pi) = -9\pi \sin \pi + c_2 \cos \pi + 4 + 5\pi \sin \pi - 5 \cos \pi = 2$

we find $c_2 = 7$. The solution of the initial-value is then

$$y = 9\pi \cos x + 7 \sin x + 4x - 5x \cos x.$$

EXAMPLE 9 Using the Multiplication Rule

Solve $y'' - 6y' + 9y = 6x^2 + 2 - 12e^{3x}$.

SOLUTION The complementary function is $y_c = c_1 e^{3x} + c_2 x e^{3x}$. And so, based on entries 3 and 7 of Table 4.1, the usual assumption for a particular solution would be

$$y_p = \underbrace{Ax^2 + Bx + C}_{y_{p_1}} + \underbrace{Ee^{3x}}_{y_{p_2}}.$$

Inspection of these functions shows that the one term in y_{p_2} is duplicated in y_c. If we multiply y_{p_2} by x, we note that the term xe^{3x} is still part of y_c. But multiplying y_{p_2} by x^2 eliminates all duplications. Thus the operative form of a particular solution is

$$y_p = Ax^2 + Bx + C + Ex^2 e^{3x}.$$

Differentiating this last form, substituting into the differential equation, and collecting like terms gives

$$y_p'' - 6y_p' + 9y_p = 9Ax^2 + (-12A + 9B)x + 2A - 6B + 9C + 2Ee^{3x} = 6x^2 + 2 - 12e^{3x}.$$

It follows from this identity that $A = \frac{2}{3}$, $B = \frac{8}{9}$, $C = \frac{2}{3}$, and $E = -6$. Hence the general solution $y = y_c + y_p$ is $y = c_1 e^{3x} + c_2 x e^{3x} + \frac{2}{3}x^2 + \frac{8}{9}x + \frac{2}{3} - 6x^2 e^{3x}$.

EXAMPLE 10   **Third-Order DE—Case I**

Solve $y''' + y'' = e^x \cos x$.

SOLUTION   From the characteristic equation $m^3 + m^2 = 0$ we find $m_1 = m_2 = 0$ and $m_3 = -1$. Hence the complementary function of the equation is $y_c = c_1 + c_2 x + c_3 e^{-x}$. With $g(x) = e^x \cos x$, we see from entry 10 of Table 4.1 that we should assume

$$y_p = A e^x \cos x + B e^x \sin x.$$

Because there are no functions in y_p that duplicate functions in the complementary solution, we proceed in the usual manner. From

$$y_p''' + y_p'' = (-2A + 4B)e^x \cos x + (-4A - 2B)e^x \sin x = e^x \cos x$$

we get $-2A + 4B = 1$ and $-4A - 2B = 0$. This system gives $A = -\frac{1}{10}$ and $B = \frac{1}{5}$, so a particular solution is $y_p = -\frac{1}{10}e^x \cos x + \frac{1}{5}e^x \sin x$. The general solution of the equation is

$$y = y_c + y_p = c_1 + c_2 x + c_3 e^{-x} - \frac{1}{10}e^x \cos x + \frac{1}{5}e^x \sin x.$$ ∎

EXAMPLE 11   **Fourth-Order DE—Case II**

Determine the form of a particular solution of $y^{(4)} + y''' = 1 - x^2 e^{-x}$.

SOLUTION   Comparing $y_c = c_1 + c_2 x + c_3 x^2 + c_4 e^{-x}$ with our normal assumption for a particular solution

$$y_p = \underbrace{A}_{y_{p_1}} + \underbrace{Bx^2 e^{-x} + Cxe^{-x} + Ee^{-x}}_{y_{p_2}},$$

we see that the duplications between y_c and y_p are eliminated when y_{p_1} is multiplied by x^3 and y_{p_2} is multiplied by x. Thus the correct assumption for a particular solution is $y_p = Ax^3 + Bx^3 e^{-x} + Cx^2 e^{-x} + Exe^{-x}$.   ∎

REMARKS

(*i*) In Problems 27–36 in Exercises 4.4 you are asked to solve initial-value problems, and in Problems 37–40 you are asked to solve boundary-value problems. As illustrated in Example 8, be sure to apply the initial conditions or the boundary conditions to the general solution $y = y_c + y_p$. Students often make the mistake of applying these conditions only to the complementary function y_c because it is that part of the solution that contains the constants $c_1, c_2, \ldots, c_n$.

(*ii*) From the "Form Rule for Case I" on page 154 of this section you see why the method of undetermined coefficients is not well suited to nonhomogeneous linear DEs when the input function $g(x)$ is something other than one of the four basic types highlighted in color on page 150. For example, if $P(x)$ is a polynomial, then continued differentiation of $P(x)e^{\alpha x} \sin \beta x$ will generate an independent set containing only a *finite* number of functions—all of the same type, namely, a polynomial times $e^{\alpha x} \sin \beta x$ or a polynomial times $e^{\alpha x} \cos \beta x$. On the other hand, repeated differentiation of input functions such as $g(x) = \ln x$ or $g(x) = \tan^{-1}x$ generates an independent set containing an *infinite* number of functions:

$$\text{derivatives of } \ln x: \quad \frac{1}{x}, \frac{-1}{x^2}, \frac{2}{x^3}, \dots,$$

$$\text{derivatives of } \tan^{-1}x: \quad \frac{1}{1+x^2}, \frac{-2x}{(1+x^2)^2}, \frac{-2+6x^2}{(1+x^2)^3}, \dots.$$

EXERCISES 4.4

Answers to selected odd-numbered problems begin on page ANS-4.

In Problems 1–26 solve the given differential equation by undetermined coefficients.

1. $y'' + 3y' + 2y = 6$

2. $4y'' + 9y = 15$

3. $y'' - 10y' + 25y = 30x + 3$

4. $y'' + y' - 6y = 2x$

5. $\frac{1}{4}y'' + y' + y = x^2 - 2x$

6. $y'' - 8y' + 20y = 100x^2 - 26xe^x$

7. $y'' + 3y = -48x^2e^{3x}$

8. $4y'' - 4y' - 3y = \cos 2x$

9. $y'' - y' = -3$

10. $y'' + 2y' = 2x + 5 - e^{-2x}$

11. $y'' - y' + \frac{1}{4}y = 3 + e^{x/2}$

12. $y'' - 16y = 2e^{4x}$

13. $y'' + 4y = 3\sin 2x$

14. $y'' - 4y = (x^2 - 3)\sin 2x$

15. $y'' + y = 2x\sin x$

16. $y'' - 5y' = 2x^3 - 4x^2 - x + 6$

17. $y'' - 2y' + 5y = e^x\cos 2x$

18. $y'' - 2y' + 2y = e^{2x}(\cos x - 3\sin x)$

19. $y'' + 2y' + y = \sin x + 3\cos 2x$

20. $y'' + 2y' - 24y = 16 - (x + 2)e^{4x}$

21. $y''' - 6y'' = 3 - \cos x$

22. $y''' - 2y'' - 4y' + 8y = 6xe^{2x}$

23. $y''' - 3y'' + 3y' - y = x - 4e^x$

24. $y''' - y'' - 4y' + 4y = 5 - e^x + e^{2x}$

25. $y^{(4)} + 2y'' + y = (x - 1)^2$

26. $y^{(4)} - y'' = 4x + 2xe^{-x}$

In Problems 27–36 solve the given initial-value problem.

27. $y'' + 4y = -2, \quad y\left(\frac{\pi}{8}\right) = \frac{1}{2}, y'\left(\frac{\pi}{8}\right) = 2$

28. $2y'' + 3y' - 2y = 14x^2 - 4x - 11, \quad y(0) = 0, y'(0) = 0$

29. $5y'' + y' = -6x, \quad y(0) = 0, y'(0) = -10$

30. $y'' + 4y' + 4y = (3 + x)e^{-2x}, \quad y(0) = 2, y'(0) = 5$

31. $y'' + 4y' + 5y = 35e^{-4x}, \quad y(0) = -3, y'(0) = 1$

32. $y'' - y = \cosh x, \quad y(0) = 2, y'(0) = 12$

33. $\frac{d^2x}{dt^2} + \omega^2x = F_0\sin\omega t, \quad x(0) = 0, x'(0) = 0$

34. $\frac{d^2x}{dt^2} + \omega^2x = F_0\cos\gamma t, \quad x(0) = 0, x'(0) = 0$

35. $y''' - 2y'' + y' = 2 - 24e^x + 40e^{5x}, \quad y(0) = \frac{1}{2}, y'(0) = \frac{5}{2}, y''(0) = -\frac{9}{2}$

36. $y''' + 8y = 2x - 5 + 8e^{-2x}, \quad y(0) = -5, y'(0) = 3, y''(0) = -4$

In Problems 37–40 solve the given boundary-value problem.

37. $y'' + y = x^2 + 1, \quad y(0) = 5, y(1) = 0$

38. $y'' - 2y' + 2y = 2x - 2, \quad y(0) = 0, y(\pi) = \pi$

39. $y'' + 3y = 6x$, $\quad y(0) = 0$, $y(1) + y'(1) = 0$

40. $y'' + 3y = 6x$, $\quad y(0) + y'(0) = 0$, $y(1) = 0$

In Problems 41 and 42 solve the given initial-value problem in which the input function $g(x)$ is discontinuous. (*Hint:* Solve each problem on two intervals, and then find a solution so that y and y' are continuous at $x = \pi/2$ (Problem 41) and at $x = \pi$ (Problem 42).)

41. $y'' + 4y = g(x)$, $\quad y(0) = 1$, $y'(0) = 2$, where

$$g(x) = \begin{cases} \sin x, & 0 \le x \le \dfrac{\pi}{2} \\[2mm] 0, & x > \dfrac{\pi}{2} \end{cases}$$

42. $y'' - 2y' + 10y = g(x)$, $\quad y(0) = 0$, $\ y'(0) = 0$, where

$$g(x) = \begin{cases} 20, & 0 \le x \le \pi \\ 0, & x > \pi \end{cases}$$

DISCUSSION/PROJECT PROBLEMS

43. Consider the differential equation $ay'' + by' + cy = e^{kx}$, where a, b, c, and k are constants. The auxiliary equation of the associated homogeneous equation is $am^2 + bm + c = 0$.
   (a) If k is not a root of the auxiliary equation, show that we can find a particular solution of the form $y_p = Ae^{kx}$, where $A = 1/(ak^2 + bk + c)$.
   (b) If k is a root of the auxiliary equation of multiplicity one, show that we can find a particular solution of the form $y_p = Axe^{kx}$, where $A = 1/(2ak + b)$. Explain how we know that $k \ne -b/2a$.
   (c) If k is a root of the auxiliary equation of multiplicity two, show that we can find a particular solution of the form $y = Ax^2 e^{kx}$, where $A = 1/(2a)$.

44. Discuss how the method of this section can be used to find a particular solution of $y'' + y = \sin x \cos 2x$. Carry out your idea.

45. Without solving, match a solution curve of $y'' + y = f(x)$ shown in the figure with one of the following functions:
   (*i*) $f(x) = 1$,   (*ii*) $f(x) = e^{-x}$,
   (*iii*) $f(x) = e^x$,   (*iv*) $f(x) = \sin 2x$,
   (*v*) $f(x) = e^x \sin x$,   (*vi*) $f(x) = \sin x$.
   Briefly discuss your reasoning.

(a)

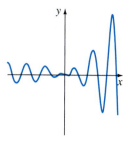

FIGURE 4.11   Solution curve

(b)

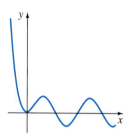

FIGURE 4.12   Solution curve

(c)

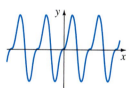

FIGURE 4.13   Solution curve

(d)

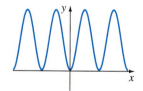

FIGURE 4.14   Solution curve

COMPUTER LAB ASSIGNMENTS

In Problems 46 and 47 find a particular solution of the given differential equation. Use a CAS as an aid in carrying out differentiations, simplifications, and algebra.

46. $y'' - 4y' + 8y = (2x^2 - 3x)e^{2x} \cos 2x + (10x^2 - x - 1)e^{2x} \sin 2x$

47. $y^{(4)} + 2y'' + y = 2 \cos x - 3x \sin x$

4.5 UNDETERMINED COEFFICIENTS—ANNIHILATOR APPROACH

INTRODUCTION: We saw in Section 4.1 that an nth-order differential equation can be written

$$a_n D^n y + a_{n-1} D^{n-1} y + \cdots + a_1 D y + a_0 y = g(x), \tag{1}$$

where $D^k y = d^k y/dy^k$, $k = 0, 1, \ldots, n$. When it suits our purpose, (1) is also written as $L(y) = g(x)$, where L denotes the linear nth-order differential, or polynomial, operator

$$a_n D^n + a_{n-1} D^{n-1} + \cdots + a_1 D + a_0. \tag{2}$$

Not only is the operator notation a helpful shorthand, but on a very practical level the application of differential operators enables us to justify the somewhat mind-numbing rules for determining the form of particular solution y_p presented in the preceding section. In this section there are no special rules; the form of y_p follows almost automatically once we have found an appropriate linear differential operator that *annihilates* $g(x)$ in (1). Before investigating how this is done, we need to examine two concepts.

FACTORING OPERATORS When the coefficients a_i, $i = 0, 1, \ldots, n$ are real constants, a linear differential operator (1) can be factored whenever the characteristic polynomial $a_n m^n + a_{n-1} m^{n-1} + \cdots + a_1 m + a_0$ factors. In other words, if r_1 is a root of the auxiliary equation

$$a_n m^n + a_{n-1} m^{n-1} + \cdots + a_1 m + a_0 = 0,$$

then $L = (D - r_1) P(D)$, where the polynomial expression $P(D)$ is a linear differential operator of order $n - 1$. For example, if we treat D as an algebraic quantity, then the operator $D^2 + 5D + 6$ can be factored as $(D + 2)(D + 3)$ or as $(D + 3)(D + 2)$. Thus if a function $y = f(x)$ possesses a second derivative, then

$$(D^2 + 5D + 6)y = (D + 2)(D + 3)y = (D + 3)(D + 2)y.$$

This illustrates a general property:

Factors of a linear differential operator with constant coefficients commute.

A differential equation such as $y'' + 4y' + 4y = 0$ can be written as

$$(D^2 + 4D + 4)y = 0 \quad \text{or} \quad (D + 2)(D + 2)y = 0 \quad \text{or} \quad (D + 2)^2 y = 0.$$

ANNIHILATOR OPERATOR If L is a linear differential operator with constant coefficients and f is a sufficiently differentiable function such that

$$L(f(x)) = 0,$$

then L is said to be an **annihilator** of the function. For example, a constant function $y = k$ is annihilated by D since $Dk = 0$. The function $y = x$ is annihilated by the differential operator D^2 since the first and second derivatives of x are 1 and 0, respectively. Similarly, $D^3 x^2 = 0$, and so on.

The differential operator D^n annihilates each of the functions

$$1, \quad x, \quad x^2, \quad \ldots, \quad x^{n-1}. \tag{3}$$

As an immediate consequence of (3) and the fact that differentiation can be done term by term, a polynomial

$$c_0 + c_1 x + c_2 x^2 + \cdots + c_{n-1} x^{n-1} \qquad (4)$$

can be annihilated by finding an operator that annihilates the highest power of x.

The functions that are annihilated by a linear nth-order differential operator L are simply those functions that can be obtained from the general solution of the homogeneous differential equation $L(y) = 0$.

The differential operator $(D - \alpha)^n$ annihilates each of the functions

$$e^{\alpha x}, \quad x e^{\alpha x}, \quad x^2 e^{\alpha x}, \quad \ldots, \quad x^{n-1} e^{\alpha x}. \qquad (5)$$

To see this, note that the auxiliary equation of the homogeneous equation $(D - \alpha)^n y = 0$ is $(m - \alpha)^n = 0$. Since α is a root of multiplicity n, the general solution is

$$y = c_1 e^{\alpha x} + c_2 x e^{\alpha x} + \cdots + c_n x^{n-1} e^{\alpha x}. \qquad (6)$$

EXAMPLE 1   Annihilator Operators

Find a differential operator that annihilates the given function.

(a) $1 - 5x^2 + 8x^3$ **(b)** e^{-3x} **(c)** $4e^{2x} - 10xe^{2x}$

SOLUTION   **(a)** From (3) we know that $D^4 x^3 = 0$, and so it follows from (4) that

$$D^4(1 - 5x^2 + 8x^3) = 0.$$

(b) From (5), with $\alpha = -3$ and $n = 1$, we see that

$$(D + 3)e^{-3x} = 0.$$

(c) From (5) and (6), with $\alpha = 2$ and $n = 2$, we have

$$(D - 2)^2(4e^{2x} - 10xe^{2x}) = 0. \qquad \blacksquare$$

When α and β, $\beta > 0$ are real numbers, the quadratic formula reveals that $[m^2 - 2\alpha m + (\alpha^2 + \beta^2)]^n = 0$ has complex roots $\alpha + i\beta$, $\alpha - i\beta$, both of multiplicity n. From the discussion at the end of Section 4.3 we have the next result.

The differential operator $[D^2 - 2\alpha D + (\alpha^2 + \beta^2)]^n$ annihilates each of the functions

$$
\begin{aligned}
&e^{\alpha x} \cos \beta x, \quad x e^{\alpha x} \cos \beta x, \quad x^2 e^{\alpha x} \cos \beta x, \ldots, \quad x^{n-1} e^{\alpha x} \cos \beta x, \\
&e^{\alpha x} \sin \beta x, \quad x e^{\alpha x} \sin \beta x, \quad x^2 e^{\alpha x} \sin \beta x, \ldots, \quad x^{n-1} e^{\alpha x} \sin \beta x.
\end{aligned}
\qquad (7)
$$

EXAMPLE 2   Annihilator Operator

Find a differential operator that annihilates $5e^{-x} \cos 2x - 9e^{-x} \sin 2x$.

SOLUTION   Inspection of the functions $e^{-x} \cos 2x$ and $e^{-x} \sin 2x$ shows that $\alpha = -1$ and $\beta = 2$. Hence from (7) we conclude that $D^2 + 2D + 5$ will annihilate each function. Since $D^2 + 2D + 5$ is a linear operator, it will annihilate *any* linear combination of these functions such as $5e^{-x} \cos 2x - 9e^{-x} \sin 2x$. $\qquad \blacksquare$

When $\alpha = 0$ and $n = 1$, a special case of (7) is

$$(D^2 + \beta^2) \begin{cases} \cos \beta x \\ \sin \beta x \end{cases} = 0. \tag{8}$$

For example, $D^2 + 16$ will annihilate any linear combination of $\sin 4x$ and $\cos 4x$.

We are often interested in annihilating the sum of two or more functions. As just seen in Examples 1 and 2, if L is a linear differential operator such that $L(y_1) = 0$ and $L(y_2) = 0$, then L will annihilate the linear combination $c_1y_1(x) + c_2y_2(x)$. This is a direct consequence of Theorem 4.2. Let us now suppose that L_1 and L_2 are linear differential operators with constant coefficients such that L_1 annihilates $y_1(x)$ and L_2 annihilates $y_2(x)$, but $L_1(y_2) \neq 0$ and $L_2(y_1) \neq 0$. Then the *product* of differential operators L_1L_2 annihilates the sum $c_1y_1(x) + c_2y_2(x)$. We can easily demonstrate this, using linearity and the fact that $L_1L_2 = L_2L_1$:

$$L_1L_2(y_1 + y_2) = L_1L_2(y_1) + L_1L_2(y_2)$$
$$= L_2L_1(y_1) + L_1L_2(y_2)$$
$$= L_2[\underbrace{L_1(y_1)}_{\text{zero}}] + L_1[\underbrace{L_2(y_2)}_{\text{zero}}] = 0.$$

For example, we know from (3) that D^2 annihilates $7 - x$ and from (8) that $D^2 + 16$ annihilates $\sin 4x$. Therefore the product of operators $D^2(D^2 + 16)$ will annihilate the linear combination $7 - x + 6 \sin 4x$.

NOTE The differential operator that annihilates a function is not unique. We saw in part (b) of Example 1 that $D + 3$ will annihilate e^{-3x}, but so will differential operators of higher order as long as $D + 3$ is one of the factors of the operator. For example, $(D + 3)(D + 1)$, $(D + 3)^2$, and $D^3(D + 3)$ all annihilate e^{-3x}. (Verify this.) As a matter of course, when we seek a differential annihilator for a function $y = f(x)$, we want the operator of *lowest possible order* that does the job.

UNDETERMINED COEFFICIENTS This brings us to the point of the preceding discussion. Suppose that $L(y) = g(x)$ is a linear differential equation with constant coefficients and that the input $g(x)$ consists of finite sums and products of the functions listed in (3), (5), and (7)—that is, $g(x)$ is a linear combination of functions of the form

$$k \text{ (constant)}, \quad x^m, \quad x^m e^{\alpha x}, \quad x^m e^{\alpha x} \cos \beta x, \quad \text{and} \quad x^m e^{\alpha x} \sin \beta x,$$

where m is a nonnegative integer and α and β are real numbers. We now know that such a function $g(x)$ can be annihilated by a differential operator L_1 of lowest order, consisting of a product of the operators D^n, $(D - \alpha)^n$, and $(D^2 - 2\alpha D + \alpha^2 + \beta^2)^n$. Applying L_1 to both sides of the equation $L(y) = g(x)$ yields $L_1L(y) = L_1(g(x)) = 0$. By solving the *homogeneous higher-order* equation $L_1L(y) = 0$, we can discover the *form* of a particular solution y_p for the original *nonhomogeneous* equation $L(y) = g(x)$. We then substitute this assumed form into $L(y) = g(x)$ to find an explicit particular solution. This procedure for determining y_p, called the **method of undetermined coefficients,** is illustrated in the next several examples.

Before proceeding, recall that the general solution of a nonhomogeneous linear differential equation $L(y) = g(x)$ is $y = y_c + y_p$, where y_c is the complementary function—that is, the general solution of the associated homogeneous equation $L(y) = 0$. The general solution of each equation $L(y) = g(x)$ is defined on the interval $(-\infty, \infty)$.

EXAMPLE 3   **General Solution Using Undetermined Coefficients**

Solve $y'' + 3y' + 2y = 4x^2$.   (9)

SOLUTION

Step 1.   First, we solve the homogeneous equation $y'' + 3y' + 2y = 0$. Then, from the auxiliary equation $m^2 + 3m + 2 = (m + 1)(m + 2) = 0$ we find $m_1 = -1$ and $m_2 = -2$, and so the complementary function is

$$y_c = c_1 e^{-x} + c_2 e^{-2x}.$$

Step 2.   Now, since $4x^2$ is annihilated by the differential operator D^3, we see that $D^3(D^2 + 3D + 2)y = 4D^3 x^2$ is the same as

$$D^3(D^2 + 3D + 2)y = 0.$$ (10)

The auxiliary equation of the fifth-order equation in (10),

$$m^3(m^2 + 3m + 2) = 0 \quad \text{or} \quad m^3(m + 1)(m + 2) = 0,$$

has roots $m_1 = m_2 = m_3 = 0$, $m_4 = -1$, and $m_5 = -2$. Thus its general solution must be

$$y = c_1 + c_2 x + c_3 x^2 + \boxed{c_4 e^{-x} + c_5 e^{-2x}}.$$ (11)

The terms in the shaded box in (11) constitute the complementary function of the original equation (9). We can then argue that a particular solution y_p of (9) should also satisfy equation (10). This means that the terms remaining in (11) must be the basic form of y_p:

$$y_p = A + Bx + Cx^2,$$ (12)

where, for convenience, we have replaced c_1, c_2, and c_3 by A, B, and C, respectively. For (12) to be a particular solution of (9), it is necessary to find *specific* coefficients A, B, and C. Differentiating (12), we have

$$y_p' = B + 2Cx, \quad y_p'' = 2C,$$

and substitution into (9) then gives

$$y_p'' + 3y_p' + 2y_p = 2C + 3B + 6Cx + 2A + 2Bx + 2Cx^2 = 4x^2.$$

Because the last equation is supposed to be an identity, the coefficients of like powers of x must be equal:

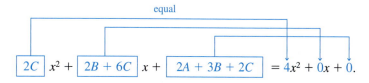

$$\boxed{2C} \; x^2 + \boxed{2B + 6C} \; x + \boxed{2A + 3B + 2C} = 4x^2 + 0x + 0.$$

That is $2C = 4, \quad 2B + 6C = 0, \quad 2A + 3B + 2C = 0.$ (13)

Solving the equations in (13) gives $A = 7$, $B = -6$, and $C = 2$. Thus $y_p = 7 - 6x + 2x^2$.

Step 3.   The general solution of the equation in (9) is $y = y_c + y_p$ or

$$y = c_1 e^{-x} + c_2 e^{-2x} + 7 - 6x + 2x^2.$$

EXAMPLE 4 General Solution Using Undetermined Coefficients

Solve $y'' - 3y' = 8e^{3x} + 4 \sin x$. $\hspace{2cm}$ (14)

SOLUTION

Step 1. The auxiliary equation for the associated homogeneous equation $y'' - 3y' = 0$ is $m^2 - 3m = m(m - 3) = 0$, and so $y_c = c_1 + c_2 e^{3x}$.

Step 2. Now, since $(D - 3)e^{3x} = 0$ and $(D^2 + 1) \sin x = 0$, we apply the differential operator $(D - 3)(D^2 + 1)$ to both sides of (14):

$$(D - 3)(D^2 + 1)(D^2 - 3D)y = 0. \hspace{1cm} (15)$$

The auxiliary equation of (15) is

$$(m - 3)(m^2 + 1)(m^2 - 3m) = 0 \quad \text{or} \quad m(m - 3)^2(m^2 + 1) = 0.$$

Thus $\hspace{2cm} y = \boxed{c_1 + c_2 e^{3x}} + c_3 x e^{3x} + c_4 \cos x + c_5 \sin x.$

After excluding the linear combination of terms in the box that corresponds to y_c, we arrive at the form of y_p:

$$y_p = Axe^{3x} + B \cos x + C \sin x.$$

Substituting y_p in (14) and simplifying yield

$$y_p'' - 3y_p' = 3Ae^{3x} + (-B - 3C) \cos x + (3B - C) \sin x = 8e^{3x} + 4 \sin x.$$

Equating coefficients gives $3A = 8$, $-B - 3C = 0$, and $3B - C = 4$. We find $A = \frac{8}{3}$, $B = \frac{6}{5}$, and $C = -\frac{2}{5}$, and consequently,

$$y_p = \frac{8}{3} xe^{3x} + \frac{6}{5} \cos x - \frac{2}{5} \sin x.$$

Step 3. The general solution of (14) is then

$$y = c_1 + c_2 e^{3x} + \frac{8}{3} xe^{3x} + \frac{6}{5} \cos x - \frac{2}{5} \sin x.$$

EXAMPLE 5 General Solution Using Undetermined Coefficients

Solve $y'' + y = x \cos x - \cos x$. $\hspace{2cm}$ (16)

SOLUTION The complementary function is $y_c = c_1 \cos x + c_2 \sin x$. Now by comparing $\cos x$ and $x \cos x$ with the functions in the first row of (7), we see that $\alpha = 0$ and $n = 1$, and so $(D^2 + 1)^2$ is an annihilator for the right-hand member of the equation in (16). Applying this operator to the differential equation gives

$$(D^2 + 1)^2 (D^2 + 1)y = 0 \quad \text{or} \quad (D^2 + 1)^3 y = 0.$$

Since i and $-i$ are both complex roots of multiplicity 3 of the auxiliary equation of the last differential equation, we conclude that

$$y = \boxed{c_1 \cos x + c_2 \sin x} + c_3 x \cos x + c_4 x \sin x + c_5 x^2 \cos x + c_6 x^2 \sin x.$$

We substitute

$$y_p = Ax \cos x + Bx \sin x + Cx^2 \cos x + Ex^2 \sin x$$

into (16) and simplify:

$$y_p'' + y_p = 4 Ex \cos x - 4 Cx \sin x + (2B + 2C) \cos x + (-2A + 2E) \sin x$$
$$= x \cos x - \cos x.$$

Equating coefficients gives the equations $4E = 1$, $-4C = 0$, $2B + 2C = -1$, and $-2A + 2E = 0$, from which we find $A = \frac{1}{4}$, $B = -\frac{1}{2}$, $C = 0$, and $E = \frac{1}{4}$. Hence the general solution of (16) is

$$y = c_1 \cos x + c_2 \sin x + \frac{1}{4}x \cos x - \frac{1}{2}x \sin x + \frac{1}{4}x^2 \sin x.$$

EXAMPLE 6 Form of a Particular Solution

Determine the form of a particular solution for

$$y'' - 2y' + y = 10e^{-2x} \cos x. \tag{17}$$

SOLUTION The complementary function for the given equation is $y_c = c_1 e^x + c_2 x e^x$.

Now from (7), with $\alpha = -2$, $\beta = 1$, and $n = 1$, we know that

$$(D^2 + 4D + 5)e^{-2x} \cos x = 0.$$

Applying the operator $D^2 + 4D + 5$ to (17) gives

$$(D^2 + 4D + 5)(D^2 - 2D + 1)y = 0. \tag{18}$$

Since the roots of the auxiliary equation of (18) are $-2 - i$, $-2 + i$, 1, and 1, we see from

$$y = c_1 e^x + c_2 x e^x + c_3 e^{-2x} \cos x + c_4 e^{-2x} \sin x$$

that a particular solution of (17) can be found with the form

$$y_p = Ae^{-2x} \cos x + Be^{-2x} \sin x.$$

EXAMPLE 7 Form of a Particular Solution

Determine the form of a particular solution for

$$y''' - 4y'' + 4y' = 5x^2 - 6x + 4x^2 e^{2x} + 3e^{5x}. \tag{19}$$

SOLUTION Observe that

$$D^3(5x^2 - 6x) = 0, \quad (D - 2)^3 x^2 e^{2x} = 0, \quad \text{and} \quad (D - 5)e^{5x} = 0.$$

Therefore $D^3(D - 2)^3(D - 5)$ applied to (19) gives

$$D^3(D - 2)^3(D - 5)(D^3 - 4D^2 + 4D)y = 0$$

or $$D^4(D - 2)^5(D - 5)y = 0.$$

The roots of the auxiliary equation for the last differential equation are easily seen to be 0, 0, 0, 0, 2, 2, 2, 2, 2, and 5. Hence

$$y = c_1 + c_2 x + c_3 x^2 + c_4 x^3 + c_5 e^{2x} + c_6 x e^{2x} + c_7 x^2 e^{2x} + c_8 x^3 e^{2x} + c_9 x^4 e^{2x} + c_{10} e^{5x}. \tag{20}$$

Because the linear combination $c_1 + c_5 e^{2x} + c_6 x e^{2x}$ corresponds to the complementary function of (19), the remaining terms in (20) give the form of a particular solution of the differential equation:

$$y_p = Ax + Bx^2 + Cx^3 + Ex^2 e^{2x} + Fx^3 e^{2x} + Gx^4 e^{2x} + He^{5x}.$$

SUMMARY OF THE METHOD For your convenience the method of undetermined coefficients is summarized as follows.

Undetermined Coefficients—Annihilator Approach

The differential equation $L(y) = g(x)$ has constant coefficients, and the function $g(x)$ consists of finite sums and products of constants, polynomials, exponential functions $e^{\alpha x}$, sines, and cosines.

(*i*)   Find the complementary solution y_c for the homogeneous equation $L(y) = 0$.

(*ii*)   Operate on both sides of the nonhomogeneous equation $L(y) = g(x)$ with a differential operator L_1 that annihilates the function $g(x)$.

(*iii*)   Find the general solution of the higher-order homogeneous differential equation $L_1 L(y) = 0$.

(*iv*)   Delete from the solution in step (*iii*) all those terms that are duplicated in the complementary solution y_c found in step (*i*). Form a linear combination y_p of the terms that remain. This is the form of a particular solution of $L(y) = g(x)$.

(*v*)   Substitute y_p found in step (*iv*) into $L(y) = g(x)$. Match coefficients of the various functions on each side of the equality, and solve the resulting system of equations for the unknown coefficients in y_p.

(*vi*)   With the particular solution found in step (*v*), form the general solution $y = y_c + y_p$ of the given differential equation.

REMARKS

The method of undetermined coefficients is not applicable to linear differential equations with variable coefficients nor is it applicable to linear equations with constant coefficients when $g(x)$ is a function such as

$$g(x) = \ln x, \quad g(x) = \frac{1}{x}, \quad g(x) = \tan x, \quad g(x) = \sin^{-1} x,$$

and so on. Differential equations in which the input $g(x)$ is a function of this last kind will be considered in the next section.

EXERCISES 4.5

Answers to selected odd-numbered problems begin on page ANS-4.

In Problems 1–10 write the given differential equation in the form $L(y) = g(x)$, where L is a linear differential operator with constant coefficients. If possible, factor L.

1. $9y'' - 4y = \sin x$

2. $y'' - 5y = x^2 - 2x$

3. $y'' - 4y' - 12y = x - 6$

4. $2y'' - 3y' - 2y = 1$

5. $y''' + 10y'' + 25y' = e^x$

6. $y''' + 4y' = e^x \cos 2x$

7. $y''' + 2y'' - 13y' + 10y = xe^{-x}$

8. $y''' + 4y'' + 3y' = x^2 \cos x - 3x$

9. $y^{(4)} + 8y' = 4$

10. $y^{(4)} - 8y'' + 16y = (x^3 - 2x)e^{4x}$

In Problems 11–14 verify that the given differential operator annihilates the indicated functions.

11. D^4; $y = 10x^3 - 2x$

12. $2D - 1$; $y = 4e^{x/2}$

13. $(D - 2)(D + 5)$; $y = e^{2x} + 3e^{-5x}$

14. $D^2 + 64$; $y = 2\cos 8x - 5\sin 8x$

In Problems 15–26 find a linear differential operator that annihilates the given function.

15. $1 + 6x - 2x^3$

16. $x^3(1 - 5x)$

17. $1 + 7e^{2x}$

18. $x + 3xe^{6x}$

19. $\cos 2x$

20. $1 + \sin x$

21. $13x + 9x^2 - \sin 4x$ **22.** $8x - \sin x + 10 \cos 5x$

23. $e^{-x} + 2xe^x - x^2e^x$ **24.** $(2 - e^x)^2$

25. $3 + e^x \cos 2x$ **26.** $e^{-x} \sin x - e^{2x} \cos x$

In Problems 27–34 find linearly independent functions that are annihilated by the given differential operator.

27. D^5 **28.** $D^2 + 4D$

29. $(D - 6)(2D + 3)$ **30.** $D^2 - 9D - 36$

31. $D^2 + 5$ **32.** $D^2 - 6D + 10$

33. $D^3 - 10D^2 + 25D$ **34.** $D^2(D - 5)(D - 7)$

In Problems 35–64 solve the given differential equation by undetermined coefficients.

35. $y'' - 9y = 54$ **36.** $2y'' - 7y' + 5y = -29$

37. $y'' + y' = 3$ **38.** $y''' + 2y'' + y' = 10$

39. $y'' + 4y' + 4y = 2x + 6$

40. $y'' + 3y' = 4x - 5$

41. $y''' + y'' = 8x^2$ **42.** $y'' - 2y' + y = x^3 + 4x$

43. $y'' - y' - 12y = e^{4x}$ **44.** $y'' + 2y' + 2y = 5e^{6x}$

45. $y'' - 2y' - 3y = 4e^x - 9$

46. $y'' + 6y' + 8y = 3e^{-2x} + 2x$

47. $y'' + 25y = 6 \sin x$

48. $y'' + 4y = 4 \cos x + 3 \sin x - 8$

49. $y'' + 6y' + 9y = -xe^{4x}$

50. $y'' + 3y' - 10y = x(e^x + 1)$

51. $y'' - y = x^2e^x + 5$

52. $y'' + 2y' + y = x^2e^{-x}$

53. $y'' - 2y' + 5y = e^x \sin x$

54. $y'' + y' + \dfrac{1}{4}y = e^x(\sin 3x - \cos 3x)$

55. $y'' + 25y = 20 \sin 5x$ **56.** $y'' + y = 4 \cos x - \sin x$

57. $y'' + y' + y = x \sin x$ **58.** $y'' + 4y = \cos^2 x$

59. $y''' + 8y'' = -6x^2 + 9x + 2$

60. $y''' - y'' + y' - y = xe^x - e^{-x} + 7$

61. $y''' - 3y'' + 3y' - y = e^x - x + 16$

62. $2y''' - 3y'' - 3y' + 2y = (e^x + e^{-x})^2$

63. $y^{(4)} - 2y''' + y'' = e^x + 1$

64. $y^{(4)} - 4y'' = 5x^2 - e^{2x}$

In Problems 65–72 solve the given initial-value problem.

65. $y'' - 64y = 16$, $y(0) = 1, y'(0) = 0$

66. $y'' + y' = x$, $y(0) = 1, y'(0) = 0$

67. $y'' - 5y' = x - 2$, $y(0) = 0, y'(0) = 2$

68. $y'' + 5y' - 6y = 10e^{2x}$, $y(0) = 1, y'(0) = 1$

69. $y'' + y = 8 \cos 2x - 4 \sin x$, $y\left(\dfrac{\pi}{2}\right) = -1, y'\left(\dfrac{\pi}{2}\right) = 0$

70. $y''' - 2y'' + y' = xe^x + 5$, $y(0) = 2, y'(0) = 2$, $y''(0) = -1$

71. $y'' - 4y' + 8y = x^3$, $y(0) = 2, y'(0) = 4$

72. $y^{(4)} - y''' = x + e^x$, $y(0) = 0, y'(0) = 0, y''(0) = 0$, $y'''(0) = 0$

DISCUSSION/PROJECT PROBLEMS

73. Suppose L is a linear differential operator that factors but has variable coefficients. Do the factors of L commute? Defend your answer.

4.6 VARIATION OF PARAMETERS

INTRODUCTION: The procedure that we used to find a particular solution y_p of a linear first-order differential equation on an interval is applicable to linear higher-order DEs as well. To adapt the method of **variation of parameters** to a linear second-order differential equation

$$a_2(x)y'' + a_1(x)y' + a_0(x)y = g(x), \tag{1}$$

we begin by putting the equation into the standard form

$$y'' + P(x)y' + Q(x)y = f(x). \tag{2}$$

by dividing through by the lead coefficient $a_2(x)$. Equation (2) is the second-order analogue of the standard form of a linear first-order equation: $dy/dx + P(x)y = f(x)$. In (2) we suppose that $P(x)$, $Q(x)$, and $f(x)$ are continuous on some common interval I. As we have already seen in Section 4.3, there is no difficulty in obtaining the complementary function y_c, the general solution of the associated homogeneous equation of (2), when the coefficients are constant.

REVIEW MATERIAL: Variation of parameters was introduced in Section 2.3 and used again in Section 4.2. A review of those sections is recommended. Also, you might have forgotten some of the indefinite integrals encountered in this section. A brief review of these forms is given in the *Student Resource and Solutions Manual*.

ASSUMPTIONS   Corresponding to the assumption $y_p = u_1(x)y_1(x)$ that we used in Section 2.3 to find a particular solution y_p of $dy/dx + P(x)y = f(x)$, for the linear second-order equation (2) we seek a solution of the form

$$y_p = u_1(x)y_1(x) + u_2(x)y_2(x), \tag{3}$$

where y_1 and y_2 form a fundamental set of solutions on I of the associated homogeneous form of (1). Using the Product Rule to differentiate y_p twice, we get

$$y_p' = u_1 y_1' + y_1 u_1' + u_2 y_2' + y_2 u_2'$$

$$y_p'' = u_1 y_1'' + y_1' u_1' + y_1 u_1'' + u_1' y_1' + u_2 y_2'' + y_2' u_2' + y_2 u_2'' + u_2' y_2'.$$

Substituting (3) and the foregoing derivatives into (2) and grouping terms yields

$$y_p'' + P(x)y_p' + Q(x)y_p = u_1[\overset{\text{zero}}{\overbrace{y_1'' + Py_1' + Qy_1}}] + u_2[\overset{\text{zero}}{\overbrace{y_2'' + Py_2' + Qy_2}}] + y_1 u_1'' + u_1' y_1'$$

$$+ y_2 u_2'' + u_2' y_2' + P[y_1 u_1' + y_2 u_2'] + y_1' u_1' + y_2' u_2'$$

$$= \frac{d}{dx}[y_1 u_1'] + \frac{d}{dx}[y_2 u_2'] + P[y_1 u_1' + y_2 u_2'] + y_1' u_1' + y_2' u_2'$$

$$= \frac{d}{dx}[y_1 u_1' + y_2 u_2'] + P[y_1 u_1' + y_2 u_2'] + y_1' u_1' + y_2' u_2' = f(x). \tag{4}$$

Because we seek to determine two unknown functions u_1 and u_2, reason dictates that we need two equations. We can obtain these equations by making the further assumption that the functions u_1 and u_2 satisfy $y_1 u_1' + y_2 u_2' = 0$. This assumption does not come out of the blue but is prompted by the first two terms in (4) since if we demand that $y_1 u_1' + y_2 u_2' = 0$, then (4) reduces to $y_1' u_1' + y_2' u_2' = f(x)$. We now have our desired two equations, albeit two equations for determining the derivatives u_1' and u_2'. By Cramer's rule, the solution of the system

$$y_1 u_1' + y_2 u_2' = 0$$

$$y_1' u_1' + y_2' u_2' = f(x)$$

can be expressed in terms of determinants:

$$u_1' = \frac{W_1}{W} = -\frac{y_2 f(x)}{W} \quad \text{and} \quad u_2' = \frac{W_2}{W} = \frac{y_1 f(x)}{W}, \tag{5}$$

where $\quad W = \begin{vmatrix} y_1 & y_2 \\ y_1' & y_2' \end{vmatrix}, \quad W_1 = \begin{vmatrix} 0 & y_2 \\ f(x) & y_2' \end{vmatrix}, \quad W_2 = \begin{vmatrix} y_1 & 0 \\ y_1' & f(x) \end{vmatrix}. \tag{6}$

The functions u_1 and u_2 are found by integrating the results in (5). The determinant W is recognized as the Wronskian of y_1 and y_2. By linear independence of y_1 and y_2 on I, we know that $W(y_1(x), y_2(x)) \neq 0$ for every x in the interval.

SUMMARY OF THE METHOD   Usually, it is not a good idea to memorize formulas in lieu of understanding a procedure. However, the foregoing procedure is too long and complicated to use each time we wish to solve a differential equation. In this case it is more efficient to simply use the formulas in (5). Thus to solve $a_2 y'' + a_1 y' + a_0 y = g(x)$, first find the complementary function $y_c = c_1 y_1 + c_2 y_2$ and then compute the Wronskian $W(y_1(x), y_2(x))$. By dividing by a_2, we put the equation into the standard form $y'' + Py' + Qy = f(x)$ to determine $f(x)$. We find u_1 and u_2 by integrating $u_1' = W_1/W$ and $u_2' = W_2/W$, where W_1 and W_2 are defined as in (6). A particular solution is $y_p = u_1 y_1 + u_2 y_2$. The general solution of the equation is then $y = y_c + y_p$.

EXAMPLE 1   **General Solution Using Variation of Parameters**

Solve $y'' - 4y' + 4y = (x + 1)e^{2x}$.

SOLUTION   From the auxiliary equation $m^2 - 4m + 4 = (m - 2)^2 = 0$ we have $y_c = c_1 e^{2x} + c_2 x e^{2x}$. With the identifications $y_1 = e^{2x}$ and $y_2 = x e^{2x}$, we next compute the Wronskian:

$$W(e^{2x}, xe^{2x}) = \begin{vmatrix} e^{2x} & xe^{2x} \\ 2e^{2x} & 2xe^{2x} + e^{2x} \end{vmatrix} = e^{4x}.$$

Since the given differential equation is already in form (2) (that is, the coefficient of y'' is 1), we identify $f(x) = (x + 1)e^{2x}$. From (6) we obtain

$$W_1 = \begin{vmatrix} 0 & xe^{2x} \\ (x + 1)e^{2x} & 2xe^{2x} + e^{2x} \end{vmatrix} = -(x + 1)xe^{4x}, \quad W_2 = \begin{vmatrix} e^{2x} & 0 \\ 2e^{2x} & (x + 1)e^{2x} \end{vmatrix} = (x + 1)e^{4x},$$

and so from (5)

$$u_1' = -\frac{(x + 1)xe^{4x}}{e^{4x}} = -x^2 - x, \quad u_2' = \frac{(x + 1)e^{4x}}{e^{4x}} = x + 1.$$

It follows that $u_1 = -\frac{1}{3}x^3 - \frac{1}{2}x^2$ and $u_2 = \frac{1}{2}x^2 + x$. Hence

$$y_p = \left(-\frac{1}{3}x^3 - \frac{1}{2}x^2\right)e^{2x} + \left(\frac{1}{2}x^2 + x\right)xe^{2x} = \frac{1}{6}x^3 e^{2x} + \frac{1}{2}x^2 e^{2x}$$

and

$$y = y_c + y_p = c_1 e^{2x} + c_2 x e^{2x} + \frac{1}{6}x^3 e^{2x} + \frac{1}{2}x^2 e^{2x}.$$

EXAMPLE 2   **General Solution Using Variation of Parameters**

Solve $4y'' + 36y = \csc 3x$.

SOLUTION   We first put the equation in the standard form (2) by dividing by 4:

$$y'' + 9y = \frac{1}{4}\csc 3x.$$

Because the roots of the auxiliary equation $m^2 + 9 = 0$ are $m_1 = 3i$ and $m_2 = -3i$, the complementary function is $y_c = c_1 \cos 3x + c_2 \sin 3x$. Using $y_1 = \cos 3x$, $y_2 = \sin 3x$, and $f(x) = \frac{1}{4}\csc 3x$, we obtain

$$W(\cos 3x, \sin 3x) = \begin{vmatrix} \cos 3x & \sin 3x \\ -3\sin 3x & 3\cos 3x \end{vmatrix} = 3,$$

$$W_1 = \begin{vmatrix} 0 & \sin 3x \\ \frac{1}{4}\csc 3x & 3\cos 3x \end{vmatrix} = -\frac{1}{4}, \quad W_2 = \begin{vmatrix} \cos 3x & 0 \\ -3\sin 3x & \frac{1}{4}\csc 3x \end{vmatrix} = \frac{1}{4}\frac{\cos 3x}{\sin 3x}.$$

Integrating $\quad u_1' = \dfrac{W_1}{W} = -\dfrac{1}{12} \quad$ and $\quad u_2' = \dfrac{W_2}{W} = \dfrac{1}{12}\dfrac{\cos 3x}{\sin 3x}$

gives $u_1 = -\frac{1}{12}x$ and $u_2 = \frac{1}{36}\ln|\sin 3x|$. Thus a particular solution is

$$y_p = -\frac{1}{12}x\cos 3x + \frac{1}{36}(\sin 3x)\ln|\sin 3x|.$$

The general solution of the equation is

$$y = y_c + y_p = c_1\cos 3x + c_2\sin 3x - \frac{1}{12}x\cos 3x + \frac{1}{36}(\sin 3x)\ln|\sin 3x|. \quad (7)$$

Equation (7) represents the general solution of the differential equation on, say, the interval $(0, \pi/6)$.

CONSTANTS OF INTEGRATION   When computing the indefinite integrals of u_1' and u_2', we need not introduce any constants. This is because

$$y = y_c + y_p = c_1 y_1 + c_2 y_2 + (u_1 + a_1)y_1 + (u_2 + b_1)y_2$$
$$= (c_1 + a_1)y_1 + (c_2 + b_1)y_2 + u_1 y_1 + u_2 y_2$$
$$= C_1 y_1 + C_2 y_2 + u_1 y_1 + u_2 y_2.$$

EXAMPLE 3   **General Solution Using Variation of Parameters**

Solve $y'' - y = \dfrac{1}{x}$.

SOLUTION   The auxiliary equation $m^2 - 1 = 0$ yields $m_1 = -1$ and $m_2 = 1$. Therefore $y_c = c_1 e^x + c_2 e^{-x}$. Now $W(e^x, e^{-x}) = -2$, and

$$u_1' = -\frac{e^{-x}(1/x)}{-2}, \quad u_1 = \frac{1}{2}\int_{x_0}^x \frac{e^{-t}}{t}\,dt,$$

$$u_2' = \frac{e^x(1/x)}{-2}, \quad u_2 = -\frac{1}{2}\int_{x_0}^x \frac{e^t}{t}\,dt.$$

Since the foregoing integrals are nonelementary, we are forced to write

$$y_p = \frac{1}{2}e^x\int_{x_0}^x \frac{e^{-t}}{t}\,dt - \frac{1}{2}e^{-x}\int_{x_0}^x \frac{e^t}{t}\,dt,$$

and so $\quad y = y_c + y_p = c_1 e^x + c_2 e^{-x} + \frac{1}{2}e^x\int_{x_0}^x \frac{e^{-t}}{t}\,dt - \frac{1}{2}e^{-x}\int_{x_0}^x \frac{e^t}{t}\,dt.$

In Example 3 we can integrate on any interval $x_0 \le t \le x$ not containing the origin.

HIGHER-ORDER EQUATIONS   The method we have just examined for nonhomogeneous second-order differential equations can be generalized to linear nth-order equations that have been put into the standard form

$$y^{(n)} + P_{n-1}(x)y^{(n-1)} + \cdots + P_1(x)y' + P_0(x)y = f(x). \tag{8}$$

If $y_c = c_1y_1 + c_2y_2 + \cdots + c_ny_n$ is the complementary function for (8), then a particular solution is

$$y_p = u_1(x)y_1(x) + u_2(x)y_2(x) + \cdots + u_n(x)y_n(x),$$

where the u'_k, $k = 1, 2, \ldots, n$ are determined by the n equations

$$
\begin{aligned}
y_1u'_1 + y_2u'_2 + \cdots + y_nu'_n &= 0 \\
y'_1u'_1 + y'_2u'_2 + \cdots + y'_nu'_n &= 0 \\
&\vdots \\
y_1^{(n-1)}u'_1 + y_2^{(n-1)}u'_2 + \cdots + y_n^{(n-1)}u'_n &= f(x).
\end{aligned} \tag{9}
$$

The first $n-1$ equations in this system, like $y_1u'_1 + y_2u'_2 = 0$ in (4), are assumptions made to simplify the resulting equation after $y_p = u_1(x)y_1(x) + \cdots + u_n(x)y_n(x)$ is substituted in (8). In this case Cramer's rule gives

$$u'_k = \frac{W_k}{W}, \quad k = 1, 2, \ldots, n,$$

where W is the Wronskian of $y_1, y_2, \ldots, y_n$ and W_k is the determinant obtained by replacing the kth column of the Wronskian by the column consisting of the right-hand side of (9)—that is, the column consisting of $(0, 0, \ldots, f(x))$. When $n = 2$, we get (5). When $n = 3$, the particular solution is $y_p = u_1y_1 + u_2y_2 + u_3y_3$, where y_1, y_2, and y_3 constitute a linearly independent set of solutions of the associated homogeneous DE and u_1, u_2, u_3 are determined from

$$u'_1 = \frac{W_1}{W}, \quad u'_2 = \frac{W_2}{W}, \quad u'_3 = \frac{W_3}{W}, \tag{10}$$

$$W_1 = \begin{vmatrix} 0 & y_2 & y_3 \\ 0 & y'_2 & y'_3 \\ f(x) & y''_2 & y''_3 \end{vmatrix}, \quad W_2 = \begin{vmatrix} y_1 & 0 & y_3 \\ y'_1 & 0 & y'_3 \\ y''_1 & f(x) & y''_3 \end{vmatrix}, \quad W_3 = \begin{vmatrix} y_1 & y_2 & 0 \\ y'_1 & y'_2 & 0 \\ y''_1 & y''_2 & f(x) \end{vmatrix}, \quad \text{and} \quad W = \begin{vmatrix} y_1 & y_2 & y_3 \\ y'_1 & y'_2 & y'_3 \\ y''_1 & y''_2 & y''_3 \end{vmatrix}.$$

See Problems 25 and 26 in Exercises 4.6.

REMARKS

(i) Variation of parameters has a distinct advantage over the method of undetermined coefficients in that it will *always* yield a particular solution y_p provided that the associated homogeneous equation can be solved. The present method is not limited to a function $f(x)$ that is a combination of the four types listed on page 150. As we shall see in the next section, variation of parameters, unlike undetermined coefficients, is applicable to linear DEs with variable coefficients.

(ii) In the problems that follow, do not hesitate to simplify the form of y_p. Depending on how the antiderivatives of u'_1 and u'_2 are found, you might not obtain the same y_p as given in the answer section. For example, in Problem 3 in Exercises 4.6 both $y_p = \frac{1}{2}\sin x - \frac{1}{2}x\cos x$ and $y_p = \frac{1}{4}\sin x - \frac{1}{2}x\cos x$ are valid answers. In either case the general solution $y = y_c + y_p$ simplifies to $y = c_1\cos x + c_2\sin x - \frac{1}{2}x\cos x$. Why?

EXERCISES 4.6

Answers to selected odd-numbered problems begin on page ANS-5.

In Problems 1–18 solve each differential equation by variation of parameters.

1. $y'' + y = \sec x$

2. $y'' + y = \tan x$

3. $y'' + y = \sin x$

4. $y'' + y = \sec \theta \tan \theta$

5. $y'' + y = \cos^2 x$

6. $y'' + y = \sec^2 x$

7. $y'' - y = \cosh x$

8. $y'' - y = \sinh 2x$

9. $y'' - 4y = \dfrac{e^{2x}}{x}$

10. $y'' - 9y = \dfrac{9x}{e^{3x}}$

11. $y'' + 3y' + 2y = \dfrac{1}{1 + e^x}$

12. $y'' - 2y' + y = \dfrac{e^x}{1 + x^2}$

13. $y'' + 3y' + 2y = \sin e^x$

14. $y'' - 2y' + y = e^t \arctan t$

15. $y'' + 2y' + y = e^{-t} \ln t$ **16.** $2y'' + 2y' + y = 4\sqrt{x}$

17. $3y'' - 6y' + 6y = e^x \sec x$

18. $4y'' - 4y' + y = e^{x/2}\sqrt{1 - x^2}$

In Problems 19–22 solve each differential equation by variation of parameters, subject to the initial conditions $y(0) = 1$, $y'(0) = 0$.

19. $4y'' - y = xe^{x/2}$ **20.** $2y'' + y' - y = x + 1$

21. $y'' + 2y' - 8y = 2e^{-2x} - e^{-x}$

22. $y'' - 4y' + 4y = (12x^2 - 6x)e^{2x}$

In Problems 23 and 24 the indicated functions are known linearly independent solutions of the associated homogeneous differential equation on $(0, \infty)$. Find the general solution of the given nonhomogeneous equation.

23. $x^2 y'' + xy' + \left(x^2 - \frac{1}{4}\right)y = x^{3/2}$;
$y_1 = x^{-1/2}\cos x$, $y_2 = x^{-1/2}\sin x$

24. $x^2 y'' + xy' + y = \sec(\ln x)$;
$y_1 = \cos(\ln x)$, $y_2 = \sin(\ln x)$

In Problems 25 and 26 solve the given third-order differential equation by variation of parameters.

25. $y''' + y' = \tan x$

26. $y''' + 4y' = \sec 2x$

DISCUSSION/PROJECT PROBLEMS

In Problems 27 and 28 discuss how the methods of undetermined coefficients and variation of parameters can be combined to solve the given differential equation. Carry out your ideas.

27. $3y'' - 6y' + 30y = 15 \sin x + e^x \tan 3x$

28. $y'' - 2y' + y = 4x^2 - 3 + x^{-1}e^x$

29. What are the intervals of definition of the general solutions in Problems 1, 7, 9, and 18? Discuss why the interval of definition of the general solution in Problem 24 is *not* $(0, \infty)$.

30. Find the general solution of $x^4 y'' + x^3 y' - 4x^2 y = 1$ given that $y_1 = x^2$ is a solution of the associated homogeneous equation.

4.7   CAUCHY-EULER EQUATION

INTRODUCTION: The same relative ease with which we were able to find explicit solutions of higher-order linear differential equations with constant coefficients in the preceding sections does not, in general, carry over to linear equations with variable coefficients. We shall see in Chapter 6 that when a linear DE has variable coefficients, the best that we can *usually* expect is to find a solution in the form of an infinite series. However, the type of DE considered in this section is an exception to this rule; it is a linear equation with variable coefficients whose general solution can always be expressed in terms of powers of x, sines, cosines, and logarithmic functions. Moreover, its method of solution is quite similar to that for constant-coefficient equations in that an auxiliary equation must be solved.

CAUCHY-EULER EQUATION   A linear differential equation of the form

$$a_n x^n \frac{d^n y}{dx^n} + a_{n-1} x^{n-1} \frac{d^{n-1} y}{dx^{n-1}} + \cdots + a_1 x \frac{dy}{dx} + a_0 y = g(x),$$

where the coefficients $a_n, a_{n-1}, \ldots, a_0$ are constants, is known as a **Cauchy-Euler equation.** The observable characteristic of this type of equation is that the degree $k = n, n - 1, \ldots, 1, 0$ of the monomial coefficients x^k matches the order k of differentiation $d^k y/dx^k$:

$$\underset{\uparrow}{\overset{\text{same}}{\downarrow}} \qquad \underset{\uparrow}{\overset{\text{same}}{\downarrow}}$$

$$a_n x^n \frac{d^n y}{dx^n} + a_{n-1} x^{n-1} \frac{d^{n-1} y}{dx^{n-1}} + \cdots .$$

As in Section 4.3, we start the discussion with a detailed examination of the forms of the general solutions of the homogeneous second-order equation

$$ax^2 \frac{d^2 y}{dx^2} + bx \frac{dy}{dx} + cy = 0.$$

The solution of higher-order equations follows analogously. Also, we can solve the nonhomogeneous equation $ax^2 y'' + bxy' + cy = g(x)$ by variation of parameters, once we have determined the complementary function y_c.

NOTE   The coefficient ax^2 of y'' is zero at $x = 0$. Hence to guarantee that the fundamental results of Theorem 4.1 are applicable to the Cauchy-Euler equation, we confine our attention to finding the general solutions defined on the interval $(0, \infty)$. Solutions on the interval $(-\infty, 0)$ can be obtained by substituting $t = -x$ into the differential equation. See Problems 37 and 38 in Exercises 4.7.

METHOD OF SOLUTION   We try a solution of the form $y = x^m$, where m is to be determined. Analogous to what happened when we substituted e^{mx} into a linear equation with constant coefficients, when we substitute x^m, each term of a Cauchy-Euler equation becomes a polynomial in m times x^m, since

$$a_k x^k \frac{d^k y}{dx^k} = a_k x^k m(m - 1)(m - 2) \cdots (m - k + 1)x^{m-k} = a_k m(m - 1)(m - 2) \cdots (m - k + 1)x^m.$$

For example, when we substitute $y = x^m$ the second-order equation becomes

$$ax^2 \frac{d^2 y}{dx^2} + bx \frac{dy}{dx} + cy = am(m - 1)x^m + bmx^m + cx^m = (am(m - 1) + bm + c)x^m.$$

Thus $y = x^m$ is a solution of the differential equation whenever m is a solution of the **auxiliary equation**

$$am(m - 1) + bm + c = 0 \quad \text{or} \quad am^2 + (b - a)m + c = 0. \tag{1}$$

There are three different cases to be considered, depending on whether the roots of this quadratic equation are real and distinct, real and equal, or complex. In the last case the roots appear as a conjugate pair.

CASE I: DISTINCT REAL ROOTS   Let m_1 and m_2 denote the real roots of (1) such that $m_1 \neq m_2$. Then $y_1 = x^{m_1}$ and $y_2 = x^{m_2}$ form a fundamental set of solutions. Hence the general solution is

$$y = c_1 x^{m_1} + c_2 x^{m_2}. \tag{2}$$

EXAMPLE 1 Distinct Roots

Solve $x^2 \dfrac{d^2y}{dx^2} - 2x \dfrac{dy}{dx} - 4y = 0$.

SOLUTION Rather than just memorizing equation (1), it is preferable to assume $y = x^m$ as the solution a few times to understand the origin and the difference between this new form of the auxiliary equation and that obtained in Section 4.3. Differentiate twice,

$$\frac{dy}{dx} = mx^{m-1}, \quad \frac{d^2y}{dx^2} = m(m-1)x^{m-2},$$

and substitute back into the differential equation:

$$x^2 \frac{d^2y}{dx^2} - 2x \frac{dy}{dx} - 4y = x^2 \cdot m(m-1)x^{m-2} - 2x \cdot mx^{m-1} - 4x^m$$

$$= x^m(m(m-1) - 2m - 4) = x^m(m^2 - 3m - 4) = 0$$

if $m^2 - 3m - 4 = 0$. Now $(m + 1)(m - 4) = 0$ implies $m_1 = -1$, $m_2 = 4$, so $y = c_1 x^{-1} + c_2 x^4$.

CASE II: REPEATED REAL ROOTS If the roots of (1) are repeated (that is, $m_1 = m_2$), then we obtain only one solution—namely, $y = x^{m_1}$. When the roots of the quadratic equation $am^2 + (b - a)m + c = 0$ are equal, the discriminant of the coefficients is necessarily zero. It follows from the quadratic formula that the root must be $m_1 = -(b - a)/2a$.

Now we can construct a second solution y_2, using (5) of Section 4.2. We first write the Cauchy-Euler equation in the standard form

$$\frac{d^2y}{dx^2} + \frac{b}{ax} \frac{dy}{dx} + \frac{c}{ax^2} y = 0$$

and make the identifications $P(x) = b/ax$ and $\int (b/ax)\, dx = (b/a) \ln x$. Thus

$$y_2 = x^{m_1} \int \frac{e^{-(b/a)\ln x}}{x^{2m_1}}\, dx$$

$$= x^{m_1} \int x^{-b/a} \cdot x^{-2m_1}\, dx \qquad \leftarrow e^{-(b/a)\ln x} = e^{\ln x^{-b/a}} = x^{-b/a}$$

$$= x^{m_1} \int x^{-b/a} \cdot x^{(b-a)/a}\, dx \qquad \leftarrow -2m_1 = (b-a)/a$$

$$= x^{m_1} \int \frac{dx}{x} = x^{m_1} \ln x.$$

The general solution is then

$$y = c_1 x^{m_1} + c_2 x^{m_1} \ln x. \tag{3}$$

EXAMPLE 2 Repeated Roots

Solve $4x^2 \dfrac{d^2y}{dx^2} + 8x \dfrac{dy}{dx} + y = 0$.

SOLUTION The substitution $y = x^m$ yields

$$4x^2 \frac{d^2y}{dx^2} + 8x \frac{dy}{dx} + y = x^m(4m(m-1) + 8m + 1) = x^m(4m^2 + 4m + 1) = 0$$

when $4m^2 + 4m + 1 = 0$ or $(2m + 1)^2 = 0$. Since $m_1 = -\frac{1}{2}$, the general solution is $y = c_1 x^{-1/2} + c_2 x^{-1/2} \ln x$. ◼

For higher-order equations, if m_1 is a root of multiplicity k, then it can be shown that

$$x^{m_1}, \quad x^{m_1} \ln x, \quad x^{m_1} (\ln x)^2, \ldots, \quad x^{m_1} (\ln x)^{k-1}$$

are k linearly independent solutions. Correspondingly, the general solution of the differential equation must then contain a linear combination of these k solutions.

CASE III: CONJUGATE COMPLEX ROOTS   If the roots of (1) are the conjugate pair $m_1 = \alpha + i\beta$, $m_2 = \alpha - i\beta$, where α and $\beta > 0$ are real, then a solution is

$$y = C_1 x^{\alpha + i\beta} + C_2 x^{\alpha - i\beta}.$$

But when the roots of the auxiliary equation are complex, as in the case of equations with constant coefficients, we wish to write the solution in terms of real functions only. We note the identity

$$x^{i\beta} = (e^{\ln x})^{i\beta} = e^{i\beta \ln x},$$

which, by Euler's formula, is the same as

$$x^{i\beta} = \cos(\beta \ln x) + i \sin(\beta \ln x).$$

Similarly, $$x^{-i\beta} = \cos(\beta \ln x) - i \sin(\beta \ln x).$$

Adding and subtracting the last two results yields

$$x^{i\beta} + x^{-i\beta} = 2 \cos(\beta \ln x) \quad \text{and} \quad x^{i\beta} - x^{-i\beta} = 2i \sin(\beta \ln x),$$

respectively. From the fact that $y = C_1 x^{\alpha + i\beta} + C_2 x^{\alpha - i\beta}$ is a solution for any values of the constants, we see, in turn, for $C_1 = C_2 = 1$ and $C_1 = 1$, $C_2 = -1$ that

$$y_1 = x^{\alpha}(x^{i\beta} + x^{-i\beta}) \quad \text{and} \quad y_2 = x^{\alpha}(x^{i\beta} - x^{-i\beta})$$

or $$y_1 = 2x^{\alpha} \cos(\beta \ln x) \quad \text{and} \quad y_2 = 2ix^{\alpha} \sin(\beta \ln x)$$

are also solutions. Since $W(x^{\alpha} \cos(\beta \ln x), x^{\alpha} \sin(\beta \ln x)) = \beta x^{2\alpha - 1} \neq 0$, $\beta > 0$ on the interval $(0, \infty)$, we conclude that

$$y_1 = x^{\alpha} \cos(\beta \ln x) \quad \text{and} \quad y_2 = x^{\alpha} \sin(\beta \ln x)$$

constitute a fundamental set of real solutions of the differential equation. Hence the general solution is

$$y = x^{\alpha}[c_1 \cos(\beta \ln x) + c_2 \sin(\beta \ln x)]. \tag{4}$$

EXAMPLE 3   **An Initial-Value Problem**

Solve $4x^2 y'' + 17y = 0$, $y(1) = -1$, $y'(1) = -\frac{1}{2}$.

SOLUTION   The y' term is missing in the given Cauchy-Euler equation; nevertheless, the substitution $y = x^m$ yields

$$4x^2 y'' + 17y = x^m (4m(m - 1) + 17) = x^m (4m^2 - 4m + 17) = 0$$

when $4m^2 - 4m + 17 = 0$. From the quadratic formula we find that the roots are $m_1 = \frac{1}{2} + 2i$ and $m_2 = \frac{1}{2} - 2i$. With the identifications $\alpha = \frac{1}{2}$ and $\beta = 2$, we see from (4) that the general solution of the differential equation is

$$y = x^{1/2}[c_1 \cos(2 \ln x) + c_2 \sin(2 \ln x)].$$

By applying the initial conditions $y(1) = -1$, $y'(1) = -\frac{1}{2}$ to the foregoing solution and using $\ln 1 = 0$, we then find, in turn, that $c_1 = -1$ and $c_2 = 0$. Hence the solution

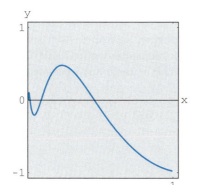

(a) solution on $0 < x \leq 1$

(b) solution on $0 < x \leq 100$

FIGURE 4.15   Solution curve of IVP in Example 3

of the initial-value problem is $y = -x^{1/2} \cos(2 \ln x)$. The graph of this function, obtained with the aid of computer software, is given in Figure 4.15. The particular solution is seen to be oscillatory and unbounded as $x \rightarrow \infty$. ∎

The next example illustrates the solution of a third-order Cauchy-Euler equation.

EXAMPLE 4   Third-Order Equation

Solve $x^3 \dfrac{d^3y}{dx^3} + 5x^2 \dfrac{d^2y}{dx^2} + 7x \dfrac{dy}{dx} + 8y = 0$.

SOLUTION   The first three derivatives of $y = x^m$ are

$$\frac{dy}{dx} = mx^{m-1}, \quad \frac{d^2y}{dx^2} = m(m-1)x^{m-2}, \quad \frac{d^3y}{dx^3} = m(m-1)(m-2)x^{m-3},$$

so the given differential equation becomes

$$x^3 \frac{d^3y}{dx^3} + 5x^2 \frac{d^2y}{dx^2} + 7x \frac{dy}{dx} + 8y = x^3 m(m-1)(m-2)x^{m-3} + 5x^2 m(m-1)x^{m-2} + 7xmx^{m-1} + 8x^m$$

$$= x^m(m(m-1)(m-2) + 5m(m-1) + 7m + 8)$$

$$= x^m(m^3 + 2m^2 + 4m + 8) = x^m(m+2)(m^2 + 4) = 0.$$

In this case we see that $y = x^m$ will be a solution of the differential equation for $m_1 = -2$, $m_2 = 2i$, and $m_3 = -2i$. Hence the general solution is $y = c_1 x^{-2} + c_2 \cos(2 \ln x) + c_3 \sin(2 \ln x)$. ∎

The method of undetermined coefficients described in Sections 4.5 and 4.6 does not carry over, *in general,* to linear differential equations with variable coefficients. Consequently, in our next example the method of variation of parameters is employed.

EXAMPLE 5   Variation of Parameters

Solve $x^2 y'' - 3xy' + 3y = 2x^4 e^x$.

SOLUTION   Since the equation is nonhomogeneous, we first solve the associated homogeneous equation. From the auxiliary equation $(m-1)(m-3) = 0$ we find $y_c = c_1 x + c_2 x^3$. Now before using variation of parameters to find a particular solution $y_p = u_1 y_1 + u_2 y_2$, recall that the formulas $u_1' = W_1/W$ and $u_2' = W_2/W$, where W_1, W_2, and W are the determinants defined on page 168, were derived under the assumption that the differential equation has been put into the standard form $y'' + P(x)y' + Q(x)y = f(x)$. Therefore we divide the given equation by x^2, and from

$$y'' - \frac{3}{x} y' + \frac{3}{x^2} y = 2x^2 e^x$$

we make the identification $f(x) = 2x^2 e^x$. Now with $y_1 = x$, $y_2 = x^3$ and

$$W = \begin{vmatrix} x & x^3 \\ 1 & 3x^2 \end{vmatrix} = 2x^3, \quad W_1 = \begin{vmatrix} 0 & x^3 \\ 2x^2 e^x & 3x^2 \end{vmatrix} = -2x^5 e^x, \quad W_2 = \begin{vmatrix} x & 0 \\ 1 & 2x^2 e^x \end{vmatrix} = 2x^3 e^x,$$

we find $\quad u_1' = -\dfrac{2x^5 e^x}{2x^3} = -x^2 e^x \quad$ and $\quad u_2' = \dfrac{2x^3 e^x}{2x^3} = e^x$.

The integral of the last function is immediate, but in the case of u_1' we integrate by parts twice. The results are $u_1 = -x^2 e^x + 2xe^x - 2e^x$ and $u_2 = e^x$. Hence $y_p = u_1 y_1 + u_2 y_2$ is

$$y_p = (-x^2 e^x + 2xe^x - 2e^x)x + e^x x^3 = 2x^2 e^x - 2xe^x.$$

Finally $\qquad\qquad y = y_c + y_p = c_1 x + c_2 x^3 + 2x^2 e^x - 2xe^x.$

REDUCTION TO CONSTANT COEFFICIENTS   The similarities between the forms of solutions of Cauchy-Euler equations and solutions of linear equations with constant coefficients are not just a coincidence. For example, when the roots of the auxiliary equations for $ay'' + by' + cy = 0$ and $ax^2 y'' + bxy' + cy = 0$ are distinct and real, the respective general solutions are

$$y = c_1 e^{m_1 x} + c_2 e^{m_2 x} \quad \text{and} \quad y = c_1 x^{m_1} + c_2 x^{m_2}, \quad x > 0. \qquad (5)$$

In view of the identity $e^{\ln x} = x$, $x > 0$, the second solution given in (5) can be expressed in the same form as the first solution:

$$y = c_1 e^{m_1 \ln x} + c_2 e^{m_2 \ln x} = c_1 e^{m_1 t} + c_2 e^{m_2 t},$$

where $t = \ln x$. This last result illustrates the fact that any Cauchy-Euler equation can *always* be rewritten as a linear differential equation with constant coefficients by means of the substitution $x = e^t$. The idea is to solve the new differential equation in terms of the variable t, using the methods of the previous sections, and once the general solution is obtained, resubstitute $t = \ln x$. This method, illustrated in the last example, requires the use of the Chain Rule of differentiation.

EXAMPLE 6   **Changing to Constant Coefficients**

Solve $x^2 y'' - xy' + y = \ln x$.

SOLUTION   With the substitution $x = e^t$ or $t = \ln x$, it follows that

$$\frac{dy}{dx} = \frac{dy}{dt}\frac{dt}{dx} = \frac{1}{x}\frac{dy}{dt} \qquad \leftarrow \text{Chain Rule}$$

$$\frac{d^2 y}{dx^2} = \frac{1}{x}\frac{d}{dx}\left(\frac{dy}{dt}\right) + \frac{dy}{dt}\left(-\frac{1}{x^2}\right) \qquad \leftarrow \text{Product Rule and Chain Rule}$$

$$= \frac{1}{x}\left(\frac{d^2 y}{dt^2}\frac{1}{x}\right) + \frac{dy}{dt}\left(-\frac{1}{x^2}\right) = \frac{1}{x^2}\left(\frac{d^2 y}{dt^2} - \frac{dy}{dt}\right).$$

Substituting in the given differential equation and simplifying yields

$$\frac{d^2 y}{dt^2} - 2\frac{dy}{dt} + y = t.$$

Since this last equation has constant coefficients, its auxiliary equation is $m^2 - 2m + 1 = 0$, or $(m-1)^2 = 0$. Thus we obtain $y_c = c_1 e^t + c_2 t e^t$.

By undetermined coefficients we try a particular solution of the form $y_p = A + Bt$. This assumption leads to $-2B + A + Bt = t$, so $A = 2$ and $B = 1$. Using $y = y_c + y_p$, we get

$$y = c_1 e^t + c_2 t e^t + 2 + t,$$

and so the general solution of the original differential equation on the interval $(0, \infty)$ is $y = c_1 x + c_2 x \ln x + 2 + \ln x.$

EXERCISES 4.7

Answers to selected odd-numbered problems begin on page ANS-5.

In Problems 1–18 solve the given differential equation.

1. $x^2y'' - 2y = 0$ **2.** $4x^2y'' + y = 0$

3. $xy'' + y' = 0$ **4.** $xy'' - 3y' = 0$

5. $x^2y'' + xy' + 4y = 0$ **6.** $x^2y'' + 5xy' + 3y = 0$

7. $x^2y'' - 3xy' - 2y = 0$ **8.** $x^2y'' + 3xy' - 4y = 0$

9. $25x^2y'' + 25xy' + y = 0$ **10.** $4x^2y'' + 4xy' - y = 0$

11. $x^2y'' + 5xy' + 4y = 0$ **12.** $x^2y'' + 8xy' + 6y = 0$

13. $3x^2y'' + 6xy' + y = 0$ **14.** $x^2y'' - 7xy' + 41y = 0$

15. $x^3y''' - 6y = 0$ **16.** $x^3y''' + xy' - y = 0$

17. $xy^{(4)} + 6y''' = 0$

18. $x^4y^{(4)} + 6x^3y''' + 9x^2y'' + 3xy' + y = 0$

In Problems 19–24 solve the given differential equation by variation of parameters.

19. $xy'' - 4y' = x^4$

20. $2x^2y'' + 5xy' + y = x^2 - x$

21. $x^2y'' - xy' + y = 2x$ **22.** $x^2y'' - 2xy' + 2y = x^4e^x$

23. $x^2y'' + xy' - y = \ln x$ **24.** $x^2y'' + xy' - y = \dfrac{1}{x+1}$

In Problems 25–30 solve the given initial-value problem. Use a graphing utility to graph the solution curve.

25. $x^2y'' + 3xy' = 0,\quad y(1) = 0, y'(1) = 4$

26. $x^2y'' - 5xy' + 8y = 0,\quad y(2) = 32, y'(2) = 0$

27. $x^2y'' + xy' + y = 0,\quad y(1) = 1, y'(1) = 2$

28. $x^2y'' - 3xy' + 4y = 0,\quad y(1) = 5, y'(1) = 3$

29. $xy'' + y' = x,\quad y(1) = 1, y'(1) = -\frac{1}{2}$

30. $x^2y'' - 5xy' + 8y = 8x^6,\quad y(\frac{1}{2}) = 0, y'(\frac{1}{2}) = 0$

In Problems 31–36 use the substitution $x = e^t$ to transform the given Cauchy-Euler equation to a differential equation with constant coefficients. Solve the original equation by solving the new equation using the procedures in Sections 4.3–4.5.

31. $x^2y'' + 9xy' - 20y = 0$

32. $x^2y'' - 9xy' + 25y = 0$

33. $x^2y'' + 10xy' + 8y = x^2$

34. $x^2y'' - 4xy' + 6y = \ln x^2$

35. $x^2y'' - 3xy' + 13y = 4 + 3x$

36. $x^3y''' - 3x^2y'' + 6xy' - 6y = 3 + \ln x^3$

In Problems 37 and 38 solve the given initial-value problem on the interval $(-\infty, 0)$.

37. $4x^2y'' + y = 0,\quad y(-1) = 2, y'(-1) = 4$

38. $x^2y'' - 4xy' + 6y = 0,\quad y(-2) = 8, y'(-2) = 0$

DISCUSSION/PROJECT PROBLEMS

39. How would you use the method of this section to solve

$$(x + 2)^2y'' + (x + 2)y' + y = 0?$$

Carry out your ideas. State an interval over which the solution is defined.

40. Can a Cauchy-Euler differential equation of lowest order with real coefficients be found if it is known that 2 and $1 - i$ are roots of its auxiliary equation? Carry out your ideas.

41. The initial-conditions $y(0) = y_0$, $y'(0) = y_1$ apply to each of the following differential equations:

$$x^2y'' = 0,$$
$$x^2y'' - 2xy' + 2y = 0,$$
$$x^2y'' - 4xy' + 6y = 0.$$

For what values of y_0 and y_1 does each initial-value problem have a solution?

42. What are the x-intercepts of the solution curve shown in Figure 4.15? How many x-intercepts are there in the interval $0 < x < \frac{1}{2}$?

COMPUTER LAB ASSIGNMENTS

In Problems 43–46 solve the given differential equation by using a CAS to find the (approximate) roots of the auxiliary equation.

43. $2x^3y''' - 10.98x^2y'' + 8.5xy' + 1.3y = 0$

44. $x^3y''' + 4x^2y'' + 5xy' - 9y = 0$

45. $x^4y^{(4)} + 6x^3y''' + 3x^2y'' - 3xy' + 4y = 0$

46. $x^4y^{(4)} - 6x^3y''' + 33x^2y'' - 105xy' + 169y = 0$

47. Solve $x^3y''' - x^2y'' - 2xy' + 6y = x^2$ by variation of parameters. Use a CAS as an aid in computing roots of the auxiliary equation and the determinants given in (10) of Section 4.6.

4.8 SOLVING SYSTEMS OF LINEAR EQUATIONS BY ELIMINATION

INTRODUCTION: Simultaneous ordinary differential equations involve two or more equations that contain derivatives of two or more dependent variables—the unknown functions—with respect to a single independent variable. The method of **systematic elimination** for solving systems of differential equations with constant coefficients is based on the algebraic principle of elimination of variables. We shall see that the analogue of *multiplying* an algebraic equation by a constant is *operating* on an ODE with some combination of derivatives.

REVIEW MATERIAL: Because the method of systematic elimination simply uncouples a system into distinct linear ODEs in each dependent variable, this section gives you an opportunity to practice what you learned in Sections 4.3, 4.4 (or 4.5), and 4.6.

SYSTEMATIC ELIMINATION   The elimination of an unknown in a system of linear differential equations is expedited by rewriting each equation in the system in differential operator notation. Recall from Section 4.1 that a single linear equation

$$a_n y^{(n)} + a_{n-1} y^{(n-1)} + \cdots + a_1 y' + a_0 y = g(t),$$

where the a_i, $i = 0, 1, \ldots, n$ are constants, can be written as

$$(a_n D^n + a_{n-1} D^{(n-1)} + \cdots + a_1 D + a_0)y = g(t).$$

If the nth-order differential operator $a_n D^n + a_{n-1} D^{(n-1)} + \cdots + a_1 D + a_0$ factors into differential operators of lower order, then the factors commute. Now, for example, to rewrite the system

$$x'' + 2x' + y'' = x + 3y + \sin t$$

$$x' + y' = -4x + 2y + e^{-t}$$

in terms of the operator D, we first bring all terms involving the dependent variables to one side and group the same variables:

$$\begin{matrix} x'' + 2x' - x + y'' - 3y = \sin t \\ x' - 4x + y' - 2y = e^{-t} \end{matrix} \quad \text{is the same as} \quad \begin{matrix} (D^2 + 2D - 1)x + (D^2 - 3)y = \sin t \\ (D - 4)x + (D - 2)y = e^{-t}. \end{matrix}$$

SOLUTION OF A SYSTEM   A **solution** of a system of differential equations is a set of sufficiently differentiable functions $x = \phi_1(t)$, $y = \phi_2(t)$, $z = \phi_3(t)$, and so on, that satisfies each equation in the system on some common interval I.

METHOD OF SOLUTION   Consider the simple system of linear first-order equations

$$\begin{aligned} \frac{dx}{dt} &= 3y \\ \frac{dy}{dt} &= 2x \end{aligned} \quad \text{or, equivalently,} \quad \begin{aligned} Dx - 3y &= 0 \\ 2x - Dy &= 0. \end{aligned} \tag{1}$$

Operating on the first equation in (1) by D while multiplying the second by -3 and then adding eliminates y from the system and gives $D^2 x - 6x = 0$. Since the roots of the auxiliary equation of the last DE are $m_1 = \sqrt{6}$ and $m_2 = -\sqrt{6}$, we obtain

$$x(t) = c_1 e^{-\sqrt{6}t} + c_2 e^{\sqrt{6}t}. \tag{2}$$

Multiplying the first equation in (1) by 2 while operating on the second by D and then subtracting gives the differential equation for y, $D^2y - 6y = 0$. It follows immediately that

$$y(t) = c_3 e^{-\sqrt{6}t} + c_4 e^{\sqrt{6}t}. \tag{3}$$

Now (2) and (3) do not satisfy the system (1) for every choice of c_1, c_2, c_3, and c_4 because the system itself puts a constraint on the number of parameters in a solution that can be chosen arbitrarily. To see this, observe that substituting $x(t)$ and $y(t)$ into the first equation of the original system (1) gives, after simplification,

$$(-\sqrt{6}c_1 - 3c_3)e^{-\sqrt{6}t} + (\sqrt{6}c_2 - 3c_4)e^{\sqrt{6}t} = 0.$$

Since the latter expression is to be zero for all values of t, we must have $-\sqrt{6}c_1 - 3c_3 = 0$ and $\sqrt{6}c_2 - 3c_4 = 0$. These two equations enable us to write c_3 as a multiple of c_1 and c_4 as a multiple of c_2:

$$c_3 = -\frac{\sqrt{6}}{3}c_1 \quad \text{and} \quad c_4 = \frac{\sqrt{6}}{3}c_2. \tag{4}$$

Hence we conclude that a solution of the system must be

$$x(t) = c_1 e^{-\sqrt{6}t} + c_2 e^{\sqrt{6}t}, \quad y(t) = -\frac{\sqrt{6}}{3}c_1 e^{-\sqrt{6}t} + \frac{\sqrt{6}}{3}c_2 e^{\sqrt{6}t}.$$

You are urged to substitute (2) and (3) into the second equation of (1) and verify that the same relationship (4) holds between the constants.

EXAMPLE 1   Solution by Elimination

Solve
$$Dx + (D + 2)y = 0$$
$$(D - 3)x - 2y = 0. \tag{5}$$

SOLUTION Operating on the first equation by $D - 3$ and on the second by D and then subtracting eliminates x from the system. It follows that the differential equation for y is

$$[(D - 3)(D + 2) + 2D]y = 0 \quad \text{or} \quad (D^2 + D - 6)y = 0.$$

Since the characteristic equation of this last differential equation is $m^2 + m - 6 = (m - 2)(m + 3) = 0$, we obtain the solution

$$y(t) = c_1 e^{2t} + c_2 e^{-3t}. \tag{6}$$

Eliminating y in a similar manner yields $(D^2 + D - 6)x = 0$, from which we find

$$x(t) = c_3 e^{2t} + c_4 e^{-3t}. \tag{7}$$

As we noted in the foregoing discussion, a solution of (5) does not contain four independent constants. Substituting (6) and (7) into the first equation of (5) gives

$$(4c_1 + 2c_3)e^{2t} + (-c_2 - 3c_4)e^{-3t} = 0.$$

From $4c_1 + 2c_3 = 0$ and $-c_2 - 3c_4 = 0$ we get $c_3 = -2c_1$ and $c_4 = -\frac{1}{3}c_2$. Accordingly, a solution of the system is

$$x(t) = -2c_1 e^{2t} - \frac{1}{3}c_2 e^{-3t}, \quad y(t) = c_1 e^{2t} + c_2 e^{-3t}.$$

Because we could just as easily solve for c_3 and c_4 in terms of c_1 and c_2, the solution in Example 1 can be written in the alternative form

$$x(t) = c_3 e^{2t} + c_4 e^{-3t}, \quad y(t) = -\frac{1}{2}c_3 e^{2t} - 3c_4 e^{-3t}.$$

It sometimes pays to keep one's eyes open when solving systems. Had we solved for x first in Example 1, then y could be found, along with the relationship between the constants, using the last equation in the system (5). You should verify that substituting $x(t)$ into $y = \frac{1}{2}(Dx - 3x)$ yields $y = -\frac{1}{2}c_3 e^{2t} - 3c_4 e^{-3t}$. Also note in the initial discussion that the relationship given in (4) and the solution $y(t)$ of (1) could also have been obtained by using $x(t)$ in (2) and the first equation of (1) in the form

$$y = \frac{1}{3}Dx = -\frac{1}{3}\sqrt{6}c_1 e^{-\sqrt{6}t} + \frac{1}{3}\sqrt{6}c_2 e^{\sqrt{6}t}.$$

EXAMPLE 2   Solution by Elimination

Solve
$$\begin{aligned} x' - 4x + y'' &= t^2 \\ x' + x + y' &= 0. \end{aligned} \tag{8}$$

SOLUTION   First we write the system in differential operator notation:

$$\begin{aligned} (D - 4)x + D^2 y &= t^2 \\ (D + 1)x + Dy &= 0. \end{aligned} \tag{9}$$

Then, by eliminating x, we obtain

$$[(D + 1)D^2 - (D - 4)D]y = (D + 1)t^2 - (D - 4)0$$

or
$$(D^3 + 4D)y = t^2 + 2t.$$

Since the roots of the auxiliary equation $m(m^2 + 4) = 0$ are $m_1 = 0$, $m_2 = 2i$, and $m_3 = -2i$, the complementary function is $y_c = c_1 + c_2 \cos 2t + c_3 \sin 2t$. To determine the particular solution y_p, we use undetermined coefficients by assuming that $y_p = At^3 + Bt^2 + Ct$. Therefore $y_p' = 3At^2 + 2Bt + C$, $y_p'' = 6At + 2B$, $y_p''' = 6A$,

$$y_p''' + 4y_p' = 12At^2 + 8Bt + 6A + 4C = t^2 + 2t.$$

The last equality implies that $12A = 1$, $8B = 2$, and $6A + 4C = 0$; hence $A = \frac{1}{12}$, $B = \frac{1}{4}$, and $C = -\frac{1}{8}$. Thus

$$y = y_c + y_p = c_1 + c_2 \cos 2t + c_3 \sin 2t + \frac{1}{12}t^3 + \frac{1}{4}t^2 - \frac{1}{8}t. \tag{10}$$

Eliminating y from the system (9) leads to

$$[(D - 4) - D(D + 1)]x = t^2 \quad \text{or} \quad (D^2 + 4)x = -t^2.$$

It should be obvious that $x_c = c_4 \cos 2t + c_5 \sin 2t$ and that undetermined coefficients can be applied to obtain a particular solution of the form $x_p = At^2 + Bt + C$. In this case the usual differentiations and algebra yield $x_p = -\frac{1}{4}t^2 + \frac{1}{8}$, and so

$$x = x_c + x_p = c_4 \cos 2t + c_5 \sin 2t - \frac{1}{4}t^2 + \frac{1}{8}. \tag{11}$$

Now c_4 and c_5 can be expressed in terms of c_2 and c_3 by substituting (10) and (11) into either equation of (8). By using the second equation, we find, after combining terms,

$$(c_5 - 2c_4 - 2c_2) \sin 2t + (2c_5 + c_4 + 2c_3) \cos 2t = 0,$$

so $c_5 - 2c_4 - 2c_2 = 0$ and $2c_5 + c_4 + 2c_3 = 0$. Solving for c_4 and c_5 in terms of c_2 and c_3 gives $c_4 = -\frac{1}{5}(4c_2 + 2c_3)$ and $c_5 = \frac{1}{5}(2c_2 - 4c_3)$. Finally, a solution of (8) is found to be

$$x(t) = -\frac{1}{5}(4c_2 + 2c_3) \cos 2t + \frac{1}{5}(2c_2 - 4c_3) \sin 2t - \frac{1}{4}t^2 + \frac{1}{8},$$

$$y(t) = c_1 + c_2 \cos 2t + c_3 \sin 2t + \frac{1}{12}t^3 + \frac{1}{4}t^2 - \frac{1}{8}t.$$

EXAMPLE 3

In (3) of Section 3.3 we saw that the system of linear first-order differential equations

$$\frac{dx_1}{dt} = -\frac{2}{25}x_1 + \frac{1}{50}x_2$$

$$\frac{dx_2}{dt} = \frac{2}{25}x_1 - \frac{2}{25}x_2$$

is a model for the number of pounds of salt $x_1(t)$ and $x_2(t)$ in brine mixtures in tanks A and B, respectively, shown in Figure 3.21. At that time we were not able to solve the system. But now, in terms of differential operators, the foregoing system can be written as

$$\left(D + \frac{2}{25}\right)x_1 - \frac{1}{50}x_2 = 0$$

$$-\frac{2}{25}x_1 + \left(D + \frac{2}{25}\right)x_2 = 0.$$

Operating on the first equation by $D + \frac{2}{25}$, multiplying the second equation by $\frac{1}{50}$, adding, and then simplifying gives $(625D^2 + 100D + 3)x_1 = 0$. From the auxiliary equation

$$625m^2 + 100m + 3 = (25m + 1)(25m + 3) = 0$$

we see immediately that $x_1(t) = c_1 e^{-t/25} + c_2 e^{-3t/25}$. We can now obtain $x_2(t)$ by using the first DE of the system in the form $x_2 = 50\left(D + \frac{2}{25}\right)x_1$. In this manner we

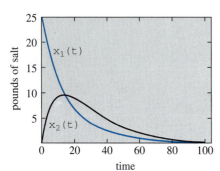

find the solution of the system to be

$$x_1(t) = c_1 e^{-t/25} + c_2 e^{-3t/25}, \quad x_2(t) = 2c_1 e^{-t/25} - 2c_2 e^{-3t/25}.$$

In the original discussion on page 114 we assumed that the initial conditions were $x_1(0) = 25$ and $x_2(0) = 0$. Applying these conditions to the solution yields $c_1 + c_2 = 25$ and $2c_1 - 2c_2 = 0$. Solving these equations simultaneously gives $c_1 = c_2 = \frac{25}{2}$. Finally, a solution of the initial-value problem is

$$x_1(t) = \frac{25}{2} e^{-t/25} + \frac{25}{2} e^{-3t/25}, \quad x_2(t) = 25e^{-t/25} - 25e^{-3t/25}.$$

FIGURE 4.16 Pounds of salt in tanks A and B

The graphs of both of these equations are given in Figure 4.16. Consistent with the fact that pure water is being pumped into tank A we see in the figure that $x_1(t) \to 0$ and $x_2(t) \to 0$ as $t \to \infty$.

EXERCISES 4.8

Answers to selected odd-numbered problems begin on page ANS-5.

In Problems 1–20 solve the given system of differential equations by systematic elimination.

1. $\dfrac{dx}{dt} = 2x - y$

$\dfrac{dy}{dt} = x$

2. $\dfrac{dx}{dt} = 4x + 7y$

$\dfrac{dy}{dt} = x - 2y$

3. $\dfrac{dx}{dt} = -y + t$

$\dfrac{dy}{dt} = x - t$

4. $\dfrac{dx}{dt} - 4y = 1$

$\dfrac{dy}{dt} + x = 2$

5. $(D^2 + 5)x - 2y = 0$

$-2x + (D^2 + 2)y = 0$

6. $(D + 1)x + (D - 1)y = 2$

$3x + (D + 2)y = -1$

7. $\dfrac{d^2x}{dt^2} = 4y + e^t$

$\dfrac{d^2y}{dt^2} = 4x - e^t$

8. $\dfrac{d^2x}{dt^2} + \dfrac{dy}{dt} = -5x$

$\dfrac{dx}{dt} + \dfrac{dy}{dt} = -x + 4y$

9. $Dx + D^2y = e^{3t}$

$(D + 1)x + (D - 1)y = 4e^{3t}$

10. $D^2x - Dy = t$

$(D + 3)x + (D + 3)y = 2$

11. $(D^2 - 1)x - y = 0$

$(D - 1)x + Dy = 0$

12. $(2D^2 - D - 1)x - (2D + 1)y = 1$

$(D - 1)x + Dy = -1$

13. $2\dfrac{dx}{dt} - 5x + \dfrac{dy}{dt} = e^t$

$\dfrac{dx}{dt} - x + \dfrac{dy}{dt} = 5e^t$

14. $\dfrac{dx}{dt} + \dfrac{dy}{dt} = e^t$

$-\dfrac{d^2x}{dt^2} + \dfrac{dx}{dt} + x + y = 0$

15. $(D - 1)x + (D^2 + 1)y = 1$

$(D^2 - 1)x + (D + 1)y = 2$

16. $D^2x - 2(D^2 + D)y = \sin t$

$x + Dy = 0$

17. $Dx = y$

$Dy = z$

$Dz = x$

18. $Dx + z = e^t$

$(D - 1)x + Dy + Dz = 0$

$x + 2y + Dz = e^t$

19. $\dfrac{dx}{dt} = 6y$

$\dfrac{dy}{dt} = x + z$

$\dfrac{dz}{dt} = x + y$

20. $\dfrac{dx}{dt} = -x + z$

$\dfrac{dy}{dt} = -y + z$

$\dfrac{dz}{dt} = -x + y$

In Problems 21 and 22 solve the given initial-value problem.

21. $\dfrac{dx}{dt} = -5x - y$

$\dfrac{dy}{dt} = 4x - y$

$x(1) = 0,\ y(1) = 1$

22. $\dfrac{dx}{dt} = y - 1$

$\dfrac{dy}{dt} = -3x + 2y$

$x(0) = 0,\ y(0) = 0$

MATHEMATICAL MODELS

23. Projectile Motion A projectile shot from a gun has weight $w = mg$ and velocity $\mathbf{v}$ tangent to its path of motion. Ignoring air resistance and all other forces acting on the projectile except its weight, determine a system of differential equations that describes its path of motion. See Figure 4.17. Solve the system. (*Hint:* Use Newton's second law of motion in the x and y directions.)

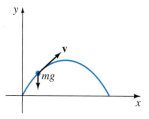

FIGURE 4.17 Path of projectile in Problem 23

24. Projectile Motion with Air Resistance Determine a system of differential equations that describes the path of motion in Problem 23 if air resistance is a retarding force $\mathbf{k}$ (of magnitude k) acting tangent to the path of the projectile but opposite to its motion. See Figure 4.18. Solve the system. (*Hint:* $\mathbf{k}$ is a multiple of velocity, say $c\mathbf{v}$.)

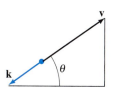

FIGURE 4.18 Forces in Problem 24

DISCUSSION/PROJECT PROBLEMS

25. Examine and discuss the following system:

$$Dx - 2Dy = t^2$$
$$(D + 1)x - 2(D + 1)y = 1.$$

COMPUTER LAB ASSIGNMENTS

26. Reexamine Figure 4.16 in Example 3. Then use a root-finding application to determine when tank B contains more salt than tank A.

27. (a) Reread Problem 8 of Exercises 3.3. In that problem you were asked to show that the system of differential equations

$$\dfrac{dx_1}{dt} = -\dfrac{1}{50}x_1$$
$$\dfrac{dx_2}{dt} = \dfrac{1}{50}x_1 - \dfrac{2}{75}x_2$$
$$\dfrac{dx_3}{dt} = \dfrac{2}{75}x_2 - \dfrac{1}{25}x_3$$

is a model for the amounts of salt in the connected mixing tanks *A, B,* and *C* shown in Figure 3.27. Solve the system subject to $x_1(0) = 15$, $x_2(t) = 10$, $x_3(t) = 5$.

(b) Use a CAS to graph $x_1(t)$, $x_2(t)$, and $x_3(t)$ in the same coordinate plane (as in Figure 4.16) on the interval [0, 200].

(c) Because only pure water is pumped into Tank *A,* it stands to reason that the salt will eventually be flushed out of all three tanks. Use a root-finding application of a CAS to determine the time when the amount of salt in each tank is less than or equal to 0.5 pound. When will the amounts of salt $x_1(t)$, $x_2(t)$, and $x_3(t)$ be simultaneously less than or equal to 0.5 pound?

4.9 NONLINEAR DIFFERENTIAL EQUATIONS

INTRODUCTION: The difficulties that surround higher-order *nonlinear* DEs and the few methods that yield analytic solutions are examined next.

REVIEW MATERIAL: Two of the solution methods considered in this section employ a change of variable to reduce a second-order DE to a first-order DE. In that sense these methods are analogous to the material in Section 4.2. Reviews of Sections 2.2 and 2.5 and the Taylor series concept from calculus are also recommended.

SOME DIFFERENCES There are several significant differences between linear and nonlinear differential equations. We saw in Section 4.1 that homogeneous linear equations of order two or higher have the property that a linear combination of solutions is also a solution (Theorem 4.2). Nonlinear equations do not possess this property of superposability. See Problems 1 and 18 in Exercises 4.9. We can find general solutions of linear first-order DEs and higher-order equations with constant coefficients. Even when we can solve a nonlinear first-order differential equation in the form of a one-parameter family, this family does not, as a rule, represent a general solution. Stated another way, nonlinear first-order DEs can possess singular solutions, whereas linear equations cannot. But the major difference between linear and nonlinear equations of order two or higher lies in the realm of solvability. Given a linear equation, there is a chance that we can find some form of a solution that we can look at—an explicit solution or perhaps a solution in the form of an infinite series (see Chapter 6). On the other hand, nonlinear higher-order differential equations virtually defy solution by analytical methods. Although this might sound disheartening, there are still things that can be done. As was pointed out at the end of Section 1.3, we can always analyze a nonlinear DE qualitatively and numerically.

Let us make it clear at the outset that nonlinear higher-order differential equations are important—dare we say even more important than linear equations?—because as we fine-tune the mathematical model of, say, a physical system, we also increase the likelihood that this higher-resolution model will be nonlinear.

We begin by illustrating an analytical method that *occasionally* enables us to find explicit/implicit solutions of special kinds of nonlinear second-order differential equations.

REDUCTION OF ORDER Nonlinear second-order differential equations $F(x, y', y'') = 0$, where the dependent variable y is missing, and $F(y, y', y'') = 0$, where the independent variable x is missing, can sometimes be solved by using first-order methods. Each equation can be reduced to a first-order equation by means of the substitution $u = y'$.

The next example illustrates the substitution technique for an equation of the form $F(x, y', y'') = 0$. If $u = y'$, then the differential equation becomes $F(x, u, u') = 0$. If we can solve this last equation for u, we can find y by integration. Note that since we are solving a second-order equation, its solution will contain two arbitrary constants.

EXAMPLE 1   Dependent Variable y Is Missing

Solve $y'' = 2x(y')^2$.

SOLUTION If we let $u = y'$, then $du/dx = y''$. After substituting, the second-order equation reduces to a first-order equation with separable variables; the independent variable is x and the dependent variable is u:

$$\frac{du}{dx} = 2xu^2 \quad \text{or} \quad \frac{du}{u^2} = 2x\,dx$$

$$\int u^{-2}\,du = \int 2x\,dx$$

$$-u^{-1} = x^2 + c_1^2.$$

The constant of integration is written as c_1^2 for convenience. The reason should be obvious in the next few steps. Because $u^{-1} = 1/y'$, it follows that

$$\frac{dy}{dx} = -\frac{1}{x^2 + c_1^2},$$

and so $\qquad y = -\int \frac{dx}{x^2 + c_1^2} \quad \text{or} \quad y = -\frac{1}{c_1}\tan^{-1}\frac{x}{c_1} + c_2.$

Next we show how to solve an equation that has the form $F(y, y', y'') = 0$. Once more we let $u = y'$, but because the independent variable x is missing we use this substitution to transform the differential equation into one in which the independent variable is y and the dependent variable is u. To this end we use the Chain Rule to compute the second derivative of y:

$$y'' = \frac{du}{dx} = \frac{du}{dy}\frac{dy}{dx} = u\frac{du}{dy}.$$

In this case the first-order equation that we must now solve is

$$F\left(y, u, u\frac{du}{dy}\right) = 0.$$

EXAMPLE 2   Independent Variable x Is Missing

Solve $yy'' = (y')^2$.

SOLUTION With the aid of $u = y'$, the Chain Rule shown above, and separation of variables, the given differential equation becomes

$$y\left(u\frac{du}{dy}\right) = u^2 \quad \text{or} \quad \frac{du}{u} = \frac{dy}{y}.$$

Integrating the last equation then yields $\ln|u| = \ln|y| + c_1$, which, in turn, gives $u = c_2 y$, where the constant $\pm e^{c_1}$ has been relabeled as c_2. We now resubstitute $u = dy/dx$, separate variables once again, integrate, and relabel constants a second time:

$$\int \frac{dy}{y} = c_2 \int dx \quad \text{or} \quad \ln|y| = c_2 x + c_3 \quad \text{or} \quad y = c_4 e^{c_2 x}. \qquad \blacksquare$$

USE OF TAYLOR SERIES In some instances a solution of a nonlinear initial-value problem, in which the initial conditions are specified at x_0, can be approximated by a Taylor series centered at x_0.

EXAMPLE 3   **Taylor Series Solution of an IVP**

Let us assume that a solution of the initial-value problem

$$y'' = x + y - y^2, \quad y(0) = -1, \quad y'(0) = 1 \qquad (1)$$

exists. If we further assume that the solution $y(x)$ of the problem is analytic at 0, then $y(x)$ possesses a Taylor series expansion centered at 0:

$$y(x) = y(0) + \frac{y'(0)}{1!}x + \frac{y''(0)}{2!}x^2 + \frac{y'''(0)}{3!}x^3 + \frac{y^{(4)}(0)}{4!}x^4 + \frac{y^{(5)}(0)}{5!}x^5 + \cdots. \qquad (2)$$

Note that the values of the first and second terms in the series (2) are known since those values are the specified initial conditions $y(0) = -1$, $y'(0) = 1$. Moreover, the differential equation itself defines the value of the second derivative at 0: $y''(0) = 0 + y(0) - y(0)^2 = 0 + (-1) - (-1)^2 = -2$. We can then find expressions for the higher derivatives y''', $y^{(4)}$, ... by calculating the successive derivatives of the differential equation:

$$y'''(x) = \frac{d}{dx}(x + y - y^2) = 1 + y' - 2yy' \qquad (3)$$

$$y^{(4)}(x) = \frac{d}{dx}(1 + y' - 2yy') = y'' - 2yy'' - 2(y')^2 \qquad (4)$$

$$y^{(5)}(x) = \frac{d}{dx}(y'' - 2yy'' - 2(y')^2) = y''' - 2yy''' - 6y'y'', \qquad (5)$$

and so on. Now using $y(0) = -1$ and $y'(0) = 1$, we find from (3) that $y'''(0) = 4$. From the values $y(0) = -1$, $y'(0) = 1$, and $y''(0) = -2$ we find $y^{(4)}(0) = -8$ from (4). With the additional information that $y'''(0) = 4$, we then see from (5) that $y^{(5)}(0) = 24$. Hence from (2) the first six terms of a series solution of the initial-value problem (1) are

$$y(x) = -1 + x - x^2 + \frac{2}{3}x^3 - \frac{1}{3}x^4 + \frac{1}{5}x^5 + \cdots. \qquad \blacksquare$$

USE OF A NUMERICAL SOLVER Numerical methods, such as Euler's method or the Runge-Kutta method, are developed solely for first-order differential equations and then are extended to systems of first-order equations. To analyze an nth-order initial-value problem numerically, we express the nth-order ODE as a system of n first-order equations. In brief, here is how it is done for a second-order initial-value problem: First, solve for y''—that is, put the DE into normal form $y'' = f(x, y, y')$—and then let $y' = u$. For example, if we substitute $y' = u$ in

$$\frac{d^2 y}{dx^2} = f(x, y, y'), \quad y(x_0) = y_0, \quad y'(x_0) = u_0, \qquad (6)$$

then $y'' = u'$ and $y'(x_0) = u(x_0)$, so the initial-value problem (6) becomes

Solve:
$$\begin{cases} y' = u \\ u' = f(x, y, u) \end{cases}$$

Subject to: $y(x_0) = y_0, u(x_0) = u_0$.

However, it should be noted that a commercial numerical solver *may not* require*
that you supply the system.

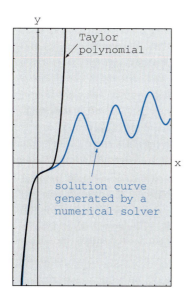

FIGURE 4.19   Comparison of two
approximate solutions

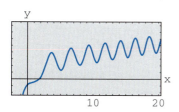

FIGURE 4.20   Numerical solution
curve for the IVP in (1)

EXAMPLE 4   **Graphical Analysis of Example 3**

Following the foregoing procedure, we find that the second-order initial-value problem in Example 3 is equivalent to

$$\frac{dy}{dx} = u$$

$$\frac{du}{dx} = x + y - y^2$$

with initial conditions $y(0) = -1$, $u(0) = 1$. With the aid of a numerical solver, we get the solution curve shown in color in Figure 4.19. For comparison the graph of the fifth-degree Taylor polynomial $T_5(x) = -1 + x - x^2 + \frac{2}{3}x^3 - \frac{1}{3}x^4 + \frac{1}{5}x^5$ is shown in black. Although we do not know the interval of convergence of the Taylor series obtained in Example 3, the closeness of the two curves in a neighborhood of the origin suggests that the power series may converge on the interval $(-1, 1)$. ∎

QUALITATIVE QUESTIONS   The colored graph in Figure 4.19 raises some questions of a qualitative nature: Is the solution of the original initial-value problem oscillatory as $x \to \infty$? The graph generated by a numerical solver on the larger interval shown in Figure 4.20 would seem to *suggest* that the answer is yes. But this single example—or even an assortment of examples—does not answer the basic question as to whether *all* solutions of the differential equation $y'' = x + y - y^2$ are oscillatory in nature. Also, what is happening to the solution curve in Figure 4.20 when x is near -1? What is the behavior of solutions of the differential equation as $x \to -\infty$? Are solutions bounded as $x \to \infty$? Questions such as these are not easily answered, in general, for nonlinear second-order differential equations. But certain kinds of second-order equations lend themselves to a systematic qualitative analysis, and these, like their first-order relatives encountered in Section 2.1, are the kind that have no explicit dependence on the independent variable. Second-order ODEs of the form

$$F(y, y', y'') = 0 \quad \text{or} \quad \frac{d^2y}{dx^2} = f(y, y'),$$

equations free of the independent variable *x,* are called **autonomous.** The differential equation in Example 2 is autonomous, and because of the presence of the *x* term on its right-hand side, the equation in Example 3 is nonautonomous. For an in-depth treatment of the topic of stability of autonomous second-order differential equations and autonomous systems of differential equations refer to Chapter 10 in *Differential Equations with Boundary-Value Problems.*

---

*Some numerical solvers only require that a second-order differential equation be expressed in normal form $y'' = f(x, y, y')$. The translation of the single equation into a system of two equations is then built into the computer program, since the first equation of the system is always $y' = u$ and the second equation is $u' = f(x, y, u)$.

EXERCISES 4.9

Answers to selected odd-numbered problems begin on page ANS-5.

In Problems 1 and 2 verify that y_1 and y_2 are solutions of the given differential equation but that $y = c_1 y_1 + c_2 y_2$ is, in general, not a solution.

1. $(y'')^2 = y^2;$ $y_1 = e^x, y_2 = \cos x$

2. $yy'' = \dfrac{1}{2}(y')^2;$ $y_1 = 1, y_2 = x^2$

In Problems 3–8 solve the given differential equation by using the substitution $u = y'$.

3. $y'' + (y')^2 + 1 = 0$ **4.** $y'' = 1 + (y')^2$

5. $x^2 y'' + (y')^2 = 0$ **6.** $(y + 1)y'' = (y')^2$

7. $y'' + 2y(y')^3 = 0$ **8.** $y^2 y'' = y'$

9. Consider the initial-value problem

$$y'' + yy' = 0, \quad y(0) = 1, y'(0) = -1.$$

  (a) Use the DE and a numerical solver to graph the solution curve.

  (b) Find an explicit solution of the IVP. Use a graphing utility to graph this solution.

  (c) Find an interval of definition for the solution in part (b).

10. Find two solutions of the initial-value problem

$$(y'')^2 + (y')^2 = 1, \quad y\left(\frac{\pi}{2}\right) = \frac{1}{2}, \quad y'\left(\frac{\pi}{2}\right) = \frac{\sqrt{3}}{2}.$$

Use a numerical solver to graph the solution curves.

In Problems 11 and 12 show that the substitution $u = y'$ leads to a Bernoulli equation. Solve this equation (see Section 2.5).

11. $xy'' = y' + (y')^3$

12. $xy'' = y' + x(y')^2$

In Problems 13–16 proceed as in Example 3 and obtain the first six nonzero terms of a Taylor series solution, centered at 0, of the given initial-value problem. Use a numerical solver and a graphing utility to compare the solution curve with the graph of the Taylor polynomial.

13. $y'' = x + y^2,$ $y(0) = 1, y'(0) = 1$

14. $y'' + y^2 = 1,$ $y(0) = 2, y'(0) = 3$

15. $y'' = x^2 + y^2 - 2y',$ $y(0) = 1, y'(0) = 1$

16. $y'' = e^y,$ $y(0) = 0, y'(0) = -1$

17. In calculus the curvature of a curve that is defined by a function $y = f(x)$ is defined as

$$\kappa = \frac{y''}{[1 + (y')^2]^{3/2}}.$$

Find $y = f(x)$ for which $\kappa = 1$. (*Hint:* For simplicity, ignore constants of integration.)

DISCUSSION/PROJECT PROBLEMS

18. In Problem 1 we saw that $\cos x$ and e^x were solutions of the nonlinear equation $(y'')^2 - y^2 = 0$. Verify that $\sin x$ and e^{-x} are also solutions. Without attempting to solve the differential equation, discuss how these explicit solutions can be found by using knowledge about linear equations. Without attempting to verify, discuss why the linear combinations $y = c_1 e^x + c_2 e^{-x} + c_3 \cos x + c_4 \sin x$ and $y = c_2 e^{-x} + c_4 \sin x$ are not, in general, solutions, but the two special linear combinations $y = c_1 e^x + c_2 e^{-x}$ and $y = c_3 \cos x + c_4 \sin x$ *must* satisfy the differential equation.

19. Discuss how the method of reduction of order considered in this section can be applied to the third-order differential equation $y''' = \sqrt{1 + (y'')^2}$. Carry out your ideas and solve the equation.

20. Discuss how to find an alternative two-parameter family of solutions for the nonlinear differential equation $y'' = 2x(y')^2$ in Example 1. (*Hint:* Suppose that $-c_1^2$ is used as the constant of integration instead of $+c_1^2$.)

MATHEMATICAL MODELS

21. Motion in a Force Field A mathematical model for the position $x(t)$ of a body moving rectilinearly on the x-axis in an inverse-square force field is given by

$$\frac{d^2 x}{dt^2} = -\frac{k^2}{x^2}.$$

Suppose that at $t = 0$ the body starts from rest from the position $x = x_0$, $x_0 > 0$. Show that the velocity of the body at time t is given by $v^2 = 2k^2(1/x - 1/x_0)$. Use the last expression and a CAS to carry out the integration to express time t in terms of x.

22. A mathematical model for the position $x(t)$ of a moving object is

$$\frac{d^2 x}{dt^2} + \sin x = 0.$$

Use a numerical solver to graphically investigate the solutions of the equation subject to $x(0) = 0$, $x'(0) = x_1$, $x_1 \geq 0$. Discuss the motion of the object for $t \geq 0$ and for various choices of x_1. Investigate the equation

$$\frac{d^2x}{dt^2} + \frac{dx}{dt} + \sin x = 0$$

in the same manner. Give a possible physical interpretation of the dx/dt term.

CHAPTER 4 IN REVIEW

Answers to selected odd-numbered problems begin on page ANS-5.

Answer Problems 1–4 without referring back to the text. Fill in the blank or answer true or false.

1. The only solution of the initial-value problem $y'' + x^2 y = 0$, $y(0) = 0$, $y'(0) = 0$ is _____.

2. For the method of undetermined coefficients, the assumed form of the particular solution y_p for $y'' - y = 1 + e^x$ is _____.

3. A constant multiple of a solution of a linear differential equation is also a solution. _____

4. If the set consisting of two functions f_1 and f_2 is linearly independent on an interval I, then the Wronskian $W(f_1, f_2) \neq 0$ for all x in I. _____

5. Give an interval over which the set of two functions $f_1(x) = x^2$ and $f_2(x) = x|x|$ is linearly independent. Then give an interval over which the set consisting of f_1 and f_2 is linearly dependent.

6. Without the aid of the Wronskian, determine whether the given set of functions is linearly independent or linearly dependent on the indicated interval.
   (a) $f_1(x) = \ln x, f_2(x) = \ln x^2, (0, \infty)$
   (b) $f_1(x) = x^n, f_2(x) = x^{n+1}, n = 1, 2, \ldots, (-\infty, \infty)$
   (c) $f_1(x) = x, f_2(x) = x + 1, (-\infty, \infty)$
   (d) $f_1(x) = \cos\left(x + \frac{\pi}{2}\right), f_2(x) = \sin x, (-\infty, \infty)$
   (e) $f_1(x) = 0, f_2(x) = x, (-5, 5)$
   (f) $f_1(x) = 2, f_2(x) = 2x, (-\infty, \infty)$
   (g) $f_1(x) = x^2, f_2(x) = 1 - x^2, f_3(x) = 2 + x^2, (-\infty, \infty)$
   (h) $f_1(x) = xe^{x+1}, f_2(x) = (4x - 5)e^x,$
      $f_3(x) = xe^x, (-\infty, \infty)$

7. Suppose $m_1 = 3$, $m_2 = -5$, and $m_3 = 1$ are roots of multiplicity one, two, and three, respectively, of an auxiliary equation. Write down the general solution of the corresponding homogeneous linear DE if it is
   (a) an equation with constant coefficients,
   (b) a Cauchy-Euler equation.

8. Consider the differential equation $ay'' + by' + cy = g(x)$, where a, b, and c are constants. Choose the input functions $g(x)$ for which the method of undetermined coeffi-

cients is applicable and the input functions for which the method of variation of parameters is applicable.
   (a) $g(x) = e^x \ln x$
   (b) $g(x) = x^3 \cos x$
   (c) $g(x) = \dfrac{\sin x}{e^x}$
   (d) $g(x) = 2x^{-2}e^x$
   (e) $g(x) = \sin^2 x$
   (f) $g(x) = \dfrac{e^x}{\sin x}$

In Problems 9–24 use the procedures developed in this chapter to find the general solution of each differential equation.

9. $y'' - 2y' - 2y = 0$

10. $2y'' + 2y' + 3y = 0$

11. $y''' + 10y'' + 25y' = 0$

12. $2y''' + 9y'' + 12y' + 5y = 0$

13. $3y''' + 10y'' + 15y' + 4y = 0$

14. $2y^{(4)} + 3y''' + 2y'' + 6y' - 4y = 0$

15. $y'' - 3y' + 5y = 4x^3 - 2x$

16. $y'' - 2y' + y = x^2 e^x$

17. $y''' - 5y'' + 6y' = 8 + 2\sin x$

18. $y''' - y'' = 6$

19. $y'' - 2y' + 2y = e^x \tan x$

20. $y'' - y = \dfrac{2e^x}{e^x + e^{-x}}$

21. $6x^2 y'' + 5xy' - y = 0$

22. $2x^3 y''' + 19x^2 y'' + 39xy' + 9y = 0$

23. $x^2 y'' - 4xy' + 6y = 2x^4 + x^2$

24. $x^2 y'' - xy' + y = x^3$

25. Write down the form of the general solution $y = y_c + y_p$ of the given differential equation in the two cases $\omega \neq \alpha$ and $\omega = \alpha$. Do not determine the coefficients in y_p.
   (a) $y'' + \omega^2 y = \sin \alpha x$
   (b) $y'' - \omega^2 y = e^{\alpha x}$

26. (a) Given that $y = \sin x$ is a solution of

$$y^{(4)} + 2y''' + 11y'' + 2y' + 10y = 0,$$

   find the general solution of the DE *without the aid of a calculator or a computer.*

(b) Find a linear second-order differential equation with constant coefficients for which $y_1 = 1$ and $y_2 = e^{-x}$ are solutions of the associated homogeneous equation and $y_p = \frac{1}{2}x^2 - x$ is a particular solution of the nonhomogeneous equation.

27. (a) Write the general solution of the fourth-order DE $y^{(4)} - 2y'' + y = 0$ entirely in terms of hyperbolic functions.

(b) Write down the form of a particular solution of $y^{(4)} - 2y'' + y = \sinh x$.

28. Consider the differential equation

$$x^2y'' - (x^2 + 2x)y' + (x + 2)y = x^3.$$

Verify that $y_1 = x$ is one solution of the associated homogeneous equation. Then show that the method of reduction of order discussed in Section 4.2 leads to a second solution y_2 of the homogeneous equation as well as a particular solution y_p of the nonhomogeneous equation. Form the general solution of the DE on the interval $(0, \infty)$.

In Problems 29–34 solve the given differential equation subject to the indicated conditions.

29. $y'' - 2y' + 2y = 0$, $\quad y\left(\frac{\pi}{2}\right) = 0, y(\pi) = -1$

30. $y'' + 2y' + y = 0$, $\quad y(-1) = 0, y'(0) = 0$

31. $y'' - y = x + \sin x$, $\quad y(0) = 2, y'(0) = 3$

32. $y'' + y = \sec^3 x$, $\quad y(0) = 1, y'(0) = \frac{1}{2}$

33. $y'y'' = 4x$, $\quad y(1) = 5, y'(1) = 2$

34. $2y'' = 3y^2$, $\quad y(0) = 1, y'(0) = 1$

35. (a) Use a CAS as an aid in finding the roots of the auxiliary equation for

$$12y^{(4)} + 64y''' + 59y'' - 23y' - 12y = 0.$$

Give the general solution of the equation.

(b) Solve the DE in part (a) subject to the initial conditions $y(0) = -1$, $y'(0) = 2$, $y''(0) = 5$, $y'''(0) = 0$. Use a CAS as an aid in solving the resulting systems of four equations in four unknowns.

36. Find a member of the family of solutions of $xy'' + y' + \sqrt{x} = 0$ whose graph is tangent to the x-axis at $x = 1$. Use a graphing utility to graph the solution curve.

In Problems 37–40 use systematic elimination to solve the given system.

37. $\dfrac{dx}{dt} + \dfrac{dy}{dt} = 2x + 2y + 1$

$\dfrac{dx}{dt} + 2\dfrac{dy}{dt} = \quad y + 3$

38. $\dfrac{dx}{dt} = 2x + y + t - 2$

$\dfrac{dy}{dt} = 3x + 4y - 4t$

39. $(D - 2)x \quad\quad - y = -e^t$

$-3x + (D - 4)y = -7e^t$

40. $(D + 2)x + (D + 1)y = \sin 2t$

$5x + (D + 3)y = \cos 2t$

BUNGEE JUMPING

To the Instructor: You might want to wait until Chapter 5 is covered to assign this project.

Suppose that you have no sense. Suppose that you are standing on a bridge above the Malad River Canyon* and that you plan to jump off that bridge. See Figure 1. You have no suicide wish. Instead, you plan to attach a bungee cord to your feet, to dive gracefully into the void, and to be pulled back gently by the cord before you hit the river that is 174 feet below. You have brought a number of different cords to affix to your feet, including several standard bungee cords, a climbing rope, and a steel cable. You need to choose the stiffness and length of a cord to avoid the unpleasantness associated with an unexpected water landing. You are undaunted by this task, because you know math!

Each of the cords you have brought will be tied off so as to be 100 feet long when hanging from the bridge. Call the position at the bottom of the cord 0, and measure the position of your feet below that "natural length" as $x(t)$, where x increases as you go down and is a function of time t. See Figure 2. Then at the time you jump, $x(0) = -100$, and if your 6-foot frame hits the water head first, then at that time $x(t) = 174 - 100 - 6 = 68$.

You know that the acceleration due to gravity is a constant g, so the force pulling downwards on your body is mg. You know that when you leap from the bridge, air resistance will increase proportionally to your speed, providing a force in the opposite direction to your motion of about βv, where β is a constant and v is your velocity. Finally, you know that Hooke's law describing the action of springs says that the bungee cord will eventually exert a force on you proportional to your distance past the natural length of the cord. Thus you know that the force of the cord pulling you back from destruction can be expressed as

$$b(x) = \begin{cases} 0, & x \le 0 \\ -kx, & x > 0. \end{cases} \tag{1}$$

The number $k > 0$ in (1) is called the *spring constant* and is where the stiffness of the cord you use influences the equation. For example, if you used the steel cable, then k would be very large, giving a tremendous stopping force very suddenly as you passed the natural length of the cable. This could lead to discomfort, injury, or even a Darwin award. You want to choose the cord with a value of k large enough to stop you above or just touching the water but not too suddenly. Consequently, you are interested in finding the distance you fall below the natural length of the cord as a function of the spring constant. To do that, you must solve the differential equation that we have derived in words above: The net force mx'' on your body is given by

$$mx'' = mg + b(x) - \beta x'. \tag{2}$$

Here mg is your weight, 160 pounds, and x' is the rate of change of your position below the equilibrium with respect to time—that is, your velocity. The constant β for air resistance depends on a number of things, including whether you wear your

FIGURE 1   The falls of the Malad River. Note the pedestrian bridge

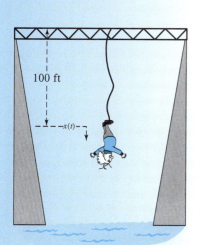

100 ft

$x(t)$

FIGURE 2   The fall from the bridge. As depicted here, $x(t) < 0$

*The Malad River Canyon is located in the Malad Gorge State Park near Hagerman, Idaho.

skin-tight pink Spandex or your skater shorts and XXL T-shirt, but you know that the value today is about 1.

Differential equation (2) is nonlinear, but inside it are two linear equations struggling to get out. You know how to solve such equations from your work in Chapters 4 and 5. When $x < 0$, the equation is $mx'' = mg - \beta x'$, while after you pass the natural length of the cord it is $mx'' = mg - kx - \beta x'$. You solve each equation separately and then piece the solutions together when $x(t) = 0$.

PROBLEM 1. Solve the equation $mx'' + \beta x' = mg$ for $x(t)$, given that you *step* off the bridge—that is, *no jumping, no diving!* "Stepping off" means that the initial conditions are $x(0) = -100$, $x'(0) = 0$. Use $mg = 160$, $\beta = 1$, and $g = 32$.

PROBLEM 2. Use the solution from Problem 1 to compute the length of time you free-fall (that is, the time it takes to go the natural length of the cord: 100 feet).

PROBLEM 3. Compute the derivative of the solution you found in Problem 1 and evaluate it at the time you found in Problem 2. You have found your downward speed when you pass the point where the cord starts to pull.

Problem 1 has given you an expression for your position t seconds after you step off the bridge, before the bungee cord starts to pull you back. Notice that it does not depend on the value of k. When you pass the natural length of the bungee cord, it does start to pull back, so the differential equation changes. Let t_1 denote the time you computed in Problem 2, and v_1 denote the speed you calculated in Problem 3.

PROBLEM 4. Solve the initial-value problem

$$mx'' + \beta x' + kx = mg, \quad x(t_1) = 0, \quad x'(t_1) = v_1.$$

For now you may use the value $k = 14$, but eventually you will need to replace this number with the values of k for the cords you brought. The solution $x(t)$ represents your position below the natural length of the cord after it starts to pull back.

Now you have an expression for your position as the cord pulls on your body. All you have to do is find the time t_2 at which you stop going down. When you stop going down, your velocity is zero—that is, $x'(t_2) = 0$.

PROBLEM 5. Compute the derivative of the expression you found in Problem 4 and solve for the value of t where the derivative is zero. Denote this time as t_2. Be careful that the time you compute is greater than t_1—there are several times when your motion stops at the top and bottom of your bounces! After you find t_2, substitute it back into the solution you found in Problem 4 to find your lowest position.

PROBLEM 6 (CAS). You have brought a soft bungee cord with $k = 8.5$, a stiffer cord with $k = 10.7$, and a climbing rope for which $k = 16.4$. Which, if any, of these cords can you use safely under the given conditions?

As you see, knowing a little bit of math is a dangerous thing. The assumption that the drag due to air resistance is linear applies only for low speeds. By the time you swoop past the natural length of the cord, that approximation is only wishful thinking, so your actual mileage may vary. Moreover, springs behave nonlinearly in large oscillations, so Hooke's law is only an approximation. Do not trust your life to an approximation made by a man who has been dead for two hundred years. Leave bungee jumping to the professionals.

STILL CURIOUS?

PROBLEM 7.   You have a bungee cord for which you have not determined the spring constant k. To do so, you suspend a weight of 10 pounds from the end of the 100-foot cord, causing it to stretch 1.2 feet. What is the value of k for this cord?

PROBLEM 8 (CAS).   What would happen if your 220-pound friend uses the bungee cord whose spring constant is $k = 10.7$?

PROBLEM 9 (CAS).   If your heavy friend wants to jump anyway, then how short should you make the cord so that he does not get wet?

PROBLEM 10 (CAS).   Graph the solutions you found in Problems 1 and 2 on the same coordinate axes. Explain the differences.

5

MODELING WITH HIGHER-ORDER DIFFERENTIAL EQUATIONS

A single differential equation can serve as a model for diverse physical systems. For this reason we focus on one application in Section 5.1. Except for terminology and physical interpretations of the four terms in the linear equation $ay'' + by' + cy = g(t)$, the mathematical description of a mechanical spring/mass system is identical to, say, that of an electrical series circuit.

    In Section 5.1 we deal exclusively with initial-value problems, whereas in Section 5.2 we examine applications described by boundary-value problems. In Section 5.3 we show how the simple pendulum and a suspended wire lead to nonlinear mathematical models.

Driven Motion: $\gamma > \omega$

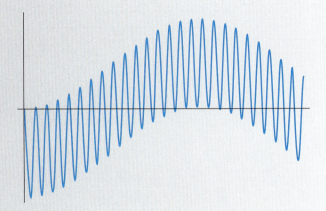

Driven Motion: $\gamma \approx \omega$ (Beats)

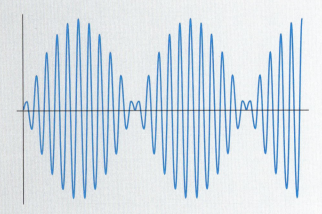

Driven Motion: $\gamma = \omega$ (Pure Resonance)

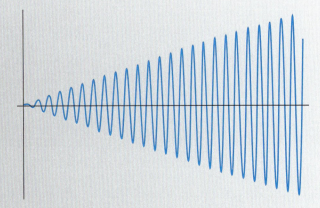

The IVP $x'' + \omega^2 x = F_0 \cos \gamma t$, $x(0) = 0$, $x'(0) = 0$ is a model of an undamped but sinusoidally driven spring/mass system. See pages 202–204, 210.

5.1 LINEAR MODELS: INITIAL-VALUE PROBLEMS

INTRODUCTION: In this section we are going to consider several linear dynamical systems in which each mathematical model is a second-order differential equation with constant coefficients along with initial conditions specified at a time that we shall take to be $t = 0$:

$$a\frac{d^2y}{dt^2} + b\frac{dy}{dt} + cy = g(t), \quad y(0) = y_0, \quad y'(0) = y_1.$$

Recall that the function g is the **input, driving function,** or **forcing function** of the system. A solution $y(t)$ of the differential equation on an interval I containing $t = 0$ that satisfies the initial conditions is called the **output** or **response** of the system.

REVIEW MATERIAL: A review of Sections 4.1, 4.3, and 4.4, along with Problems 29–36 in Exercises 4.3 and Problems 27–36 in Exercises 4.4 is recommended.

CD: The **Spring/Mass Tool** on the *DE Tools* CD can be used in conjunction with the discussion of sinusoidally driven spring/mass systems on page 202.

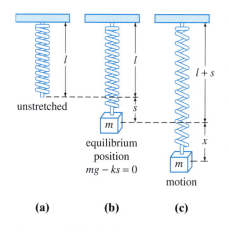

unstretched

equilibrium position
$mg - ks = 0$

$l + s$

x

motion

(a)   **(b)**   **(c)**

FIGURE 5.1 Spring/mass system

5.1.1 SPRING/MASS SYSTEMS: FREE UNDAMPED MOTION

HOOKE'S LAW Suppose that a flexible spring is suspended vertically from a rigid support and then a mass m is attached to its free end. The amount of stretch, or elongation, of the spring will of course depend on the mass; masses with different weights stretch the spring by differing amounts. By Hooke's law, the spring itself exerts a restoring force F opposite to the direction of elongation and proportional to the amount of elongation s. Simply stated, $F = ks$, where k is a constant of proportionality called the **spring constant.** The spring is essentially characterized by the number k. For example, if a mass weighing 10 pounds stretches a spring $\frac{1}{2}$ foot, then $10 = k\left(\frac{1}{2}\right)$ implies $k = 20$ lb/ft. Necessarily then, a mass weighing, say, 8 pounds stretches the same spring only $\frac{2}{5}$ foot.

NEWTON'S SECOND LAW After a mass m is attached to a spring, it stretches the spring by an amount s and attains a position of equilibrium at which its weight W is balanced by the restoring force ks. Recall that weight is defined by $W = mg$, where mass is measured in slugs, kilograms, or grams and $g = 32$ ft/s², 9.8 m/s², or 980 cm/s², respectively. As indicated in Figure 5.1(b), the condition of equilibrium is $mg = ks$ or $mg - ks = 0$. If the mass is displaced by an amount x from its equilibrium position, the restoring force of the spring is then $k(x + s)$. Assuming that there are no retarding forces acting on the system and assuming that the mass vibrates free of other external forces—**free motion**—we can equate Newton's second law with the net, or resultant, force of the restoring force and the weight:

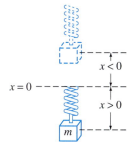

$x < 0$

$x = 0$

$x > 0$

m

FIGURE 5.2 Direction below the equilibrium position is positive.

$$m\frac{d^2x}{dt^2} = -k(s + x) + mg = \underbrace{-kx + mg - ks}_{\text{zero}} = -kx. \tag{1}$$

The negative sign in (1) indicates that the restoring force of the spring acts opposite to the direction of motion. Furthermore, we adopt the convention that displacements measured *below* the equilibrium position are positive. See Figure 5.2.

DE OF FREE UNDAMPED MOTION By dividing (1) by the mass m, we obtain the second-order differential equation $d^2x/dt^2 + (k/m)x = 0$, or

$$\frac{d^2x}{dt^2} + \omega^2 x = 0, \tag{2}$$

where $\omega^2 = k/m$. Equation (2) is said to describe **simple harmonic motion** or **free undamped motion.** Two obvious initial conditions associated with (2) are $x(0) = x_0$ and $x'(0) = x_1$, the initial displacement and initial velocity of the mass, respectively. For example, if $x_0 > 0$, $x_1 < 0$, the mass starts from a point *below* the equilibrium position with an imparted *upward* velocity. When $x'(0) = 0$, the mass is said to be released from rest. For example, if $x_0 < 0$, $x_1 = 0$, the mass is released from *rest* from a point $|x_0|$ units *above* the equilibrium position.

EQUATION OF MOTION To solve equation (2), we note that the solutions of its auxiliary equation $m^2 + \omega^2 = 0$ are the complex numbers $m_1 = \omega i$, $m_2 = -\omega i$. Thus from (8) of Section 4.3 we find the general solution of (2) to be

$$x(t) = c_1 \cos \omega t + c_2 \sin \omega t. \tag{3}$$

The **period** of motion described by (3) is $T = 2\pi/\omega$. The number T represents the time (measured in seconds) it takes the mass to execute one cycle of motion. A cycle is one complete oscillation of the mass, that is, the mass m moving from, say, the lowest point below the equilibrium position to the point highest above the equilibrium position and then back to the lowest point. From a graphical viewpoint $T = 2\pi/\omega$ seconds is the length of the time interval between two successive maxima (or minima) of $x(t)$. Keep in mind that a maximum of $x(t)$ is a positive displacement corresponding to the mass attaining its greatest distance below the equilibrium position, whereas a minimum of $x(t)$ is negative displacement corresponding to the mass attaining its greatest height above the equilibrium position. We refer to either case as an **extreme displacement** of the mass. The **frequency** of motion is $f = 1/T = \omega/2\pi$ and is the number of cycles completed each second. For example, if $x(t) = 2 \cos 3\pi t - 4 \sin 3\pi t$, then the period is $T = 2\pi/3\pi = 2/3$ s, and the frequency is $f = 3/2$ cycles/s. From a graphical viewpoint the graph of $x(t)$ repeats every $\frac{2}{3}$ second, that is, $x\left(t + \frac{2}{3}\right) = x(t)$, and $\frac{3}{2}$ cycles of the graph are completed each second (or, equivalently, three cycles of the graph are completed every 2 seconds). The number $\omega = \sqrt{k/m}$ (measured in radians per second) is called the **circular frequency** of the system. Depending on which text you read, both $f = \omega/2\pi$ and ω are also referred to as the **natural frequency** of the system. Finally, when the initial conditions are used to determine the constants c_1 and c_2 in (3), we say that the resulting particular solution or response is the **equation of motion.**

| **EXAMPLE 1** | **Free Undamped Motion** |

A mass weighing 2 pounds stretches a spring 6 inches. At $t = 0$ the mass is released from a point 8 inches below the equilibrium position with an upward velocity of $\frac{4}{3}$ ft/s. Determine the equation of motion.

SOLUTION Because we are using the engineering system of units, the measurements given in terms of inches must be converted into feet: 6 in. $= \frac{1}{2}$ ft; 8 in. $= \frac{2}{3}$ ft. In addition, we must convert the units of weight given in pounds into units of mass. From $m = W/g$ we have $m = \frac{2}{32} = \frac{1}{16}$ slug. Also, from Hooke's law, $2 = k\left(\frac{1}{2}\right)$ implies that the spring constant is $k = 4$ lb/ft. Hence (1) gives

$$\frac{1}{16}\frac{d^2x}{dt^2} = -4x \quad \text{or} \quad \frac{d^2x}{dt^2} + 64x = 0.$$

The initial displacement and initial velocity are $x(0) = \frac{2}{3}$, $x'(0) = -\frac{4}{3}$, where the negative sign in the last condition is a consequence of the fact that the mass is given an initial velocity in the negative, or upward, direction.

Now $\omega^2 = 64$ or $\omega = 8$, so the general solution of the differential equation is

$$x(t) = c_1 \cos 8t + c_2 \sin 8t. \tag{4}$$

Applying the initial conditions to $x(t)$ and $x'(t)$ gives $c_1 = \frac{2}{3}$ and $c_2 = -\frac{1}{6}$. Thus the equation of motion is

$$x(t) = \frac{2}{3} \cos 8t - \frac{1}{6} \sin 8t. \tag{5}$$

ALTERNATIVE FORM OF $x(t)$ When $c_1 \neq 0$ and $c_2 \neq 0$, the actual **amplitude** A of free vibrations is not obvious from inspection of equation (3). For example, although the mass in Example 1 is initially displaced $\frac{2}{3}$ foot beyond the equilibrium position, the amplitude of vibrations is a number larger than $\frac{2}{3}$. Hence it is often convenient to convert a solution of form (3) to the simpler form

$$x(t) = A \sin(\omega t + \phi), \tag{6}$$

where $A = \sqrt{c_1^2 + c_2^2}$ and ϕ is a **phase angle** defined by

$$\left. \begin{array}{l} \sin \phi = \dfrac{c_1}{A} \\[2ex] \cos \phi = \dfrac{c_2}{A} \end{array} \right\} \tan \phi = \dfrac{c_1}{c_2}. \tag{7}$$

To verify this, we expand (6) by the addition formula for the sine function:

$$A \sin \omega t \cos \phi + A \cos \omega t \sin \phi = (A \sin \phi)\cos \omega t + (A \cos \phi)\sin \omega t. \tag{8}$$

It follows from Figure 5.3 that if ϕ is defined by

$$\sin \phi = \frac{c_1}{\sqrt{c_1^2 + c_2^2}} = \frac{c_1}{A}, \quad \cos \phi = \frac{c_2}{\sqrt{c_1^2 + c_2^2}} = \frac{c_2}{A},$$

then (8) becomes

$$A\frac{c_1}{A} \cos \omega t + A\frac{c_2}{A} \sin \omega t = c_1 \cos \omega t + c_2 \sin \omega t = x(t).$$

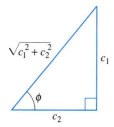

FIGURE 5.3   A relationship between $c_1 > 0$, $c_2 > 0$ and phase angle ϕ

EXAMPLE 2   **Alternative Form of Solution (5)**

In view of the foregoing discussion we can write solution (5) in the alternative form $x(t) = A \sin(8t + \phi)$. Computation of the amplitude is straightforward, $A = \sqrt{(\frac{2}{3})^2 + (-\frac{1}{6})^2} = \sqrt{\frac{17}{36}} \approx 0.69$ ft, but some care should be exercised when computing the phase angle ϕ defined by (7). With $c_1 = \frac{2}{3}$ and $c_2 = -\frac{1}{6}$ we find $\tan \phi = -4$, and a calculator then gives $\tan^{-1}(-4) = -1.326$ rad. This is *not* the phase angle, since $\tan^{-1}(-4)$ is located in the *fourth quadrant* and therefore contradicts the fact that $\sin \phi > 0$ and $\cos \phi < 0$ because $c_1 > 0$ and $c_2 < 0$. Hence we must take ϕ to be the *second-quadrant* angle $\phi = \pi + (-1.326) = 1.816$ rad. Thus (5) is the same as

$$x(t) = \frac{\sqrt{17}}{6} \sin(8t + 1.816). \tag{9}$$

The period of this function is $T = 2\pi/8 = \pi/4$ s.

Figure 5.4(a) illustrates the mass in Example 2 going through approximately two complete cycles of motion. Reading from left to right, the first five positions (marked with black dots) correspond to the initial position of the mass below the equilibrium position $\left(x = \frac{2}{3}\right)$, the mass passing through the equilibrium position for the first time heading upward ($x = 0$), the mass at its extreme displacement above

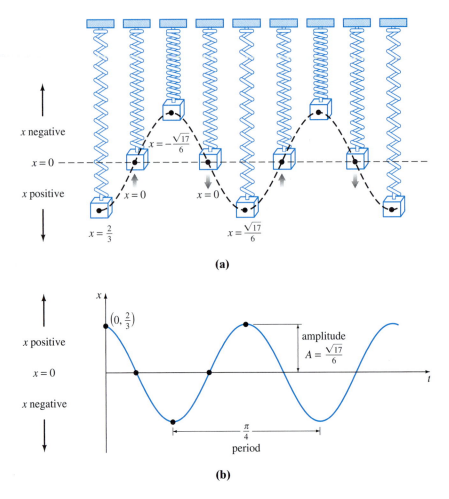

FIGURE 5.4 Simple harmonic motion

the equilibrium position $\left(x = -\sqrt{17}/6\right)$, the mass at the equilibrium position for the second time heading downward $(x = 0)$, and the mass at its extreme displacement below the equilibrium position $\left(x = \sqrt{17}/6\right)$. The black dots on the graph of (9), given in Figure 5.4(b), also agree with the five positions just given. Note, however, that in Figure 5.4(b) the positive direction in the tx-plane is the usual upward direction and so is opposite to the positive direction indicated in Figure 5.4(a). Hence the solid color graph representing the motion of the mass in Figure 5.4(b) is the reflection through the t-axis of the black dashed curve in Figure 5.4(a).

Form (6) is very useful because it is easy to find values of time for which the graph of $x(t)$ crosses the positive t-axis (the line $x = 0$). We observe that $\sin(\omega t + \phi) = 0$ when $\omega t + \phi = n\pi$, where n is a nonnegative integer.

SYSTEMS WITH VARIABLE SPRING CONSTANTS In the model discussed above we assumed an ideal world—a world in which the physical characteristics of the spring do not change over time. In the nonideal world, however, it seems reasonable to expect that when a spring/mass system is in motion for a long period, the spring will weaken; in other words, the "spring constant" will vary—or, more specifically, decay—with time. In one model for the **aging spring** the spring constant k in (1) is replaced by the decreasing function $K(t) = ke^{-\alpha t}$, $k > 0$, $\alpha > 0$. The linear differential equation $mx'' + ke^{-\alpha t}x = 0$ cannot be solved by the methods considered in Chapter 4. Nevertheless, we can obtain two linearly independent solutions using the methods in Chapter 6. See Problem 15 in Exercises 5.1, Example 4 in Section 6.3, and Problems 33 and 39 in Exercises 6.3.

When a spring/mass system is subjected to an environment in which the temperature is rapidly decreasing, it might make sense to replace the constant k with $K(t) = kt$, $k > 0$, a function that increases with time. The resulting model, $mx'' + ktx = 0$, is a form of **Airy's differential equation.** Like the equation for an aging spring, Airy's equation can be solved by the methods of Chapter 6. See Problem 16 in Exercises 5.1, Example 3 in Section 6.1, and Problems 34, 35, and 40 in Exercises 6.3.

(a)

(b)

FIGURE 5.5 Damping devices

5.1.2 SPRING/MASS SYSTEMS: FREE DAMPED MOTION

The concept of free harmonic motion is somewhat unrealistic, since the motion described by equation (1) assumes that there are no retarding forces acting on the moving mass. Unless the mass is suspended in a perfect vacuum, there will be at least a resisting force due to the surrounding medium. As Figure 5.5 shows, the mass could be suspended in a viscous medium or connected to a dashpot damping device.

DE OF FREE DAMPED MOTION In the study of mechanics, damping forces acting on a body are considered to be proportional to a power of the instantaneous velocity. In particular, we shall assume throughout the subsequent discussion that this force is given by a constant multiple of dx/dt. When no other external forces are impressed on the system, it follows from Newton's second law that

$$m\frac{d^2x}{dt^2} = -kx - \beta\frac{dx}{dt}, \tag{10}$$

where β is a positive *damping constant* and the negative sign is a consequence of the fact that the damping force acts in a direction opposite to the motion.

Dividing (10) by the mass m, we find that the differential equation of **free damped motion** is $d^2x/dt^2 + (\beta/m)dx/dt + (k/m)x = 0$ or

$$\frac{d^2x}{dt^2} + 2\lambda\frac{dx}{dt} + \omega^2 x = 0, \tag{11}$$

where

$$2\lambda = \frac{\beta}{m}, \qquad \omega^2 = \frac{k}{m}. \tag{12}$$

The symbol 2λ is used only for algebraic convenience because the auxiliary equation is $m^2 + 2\lambda m + \omega^2 = 0$ and the corresponding roots are then

$$m_1 = -\lambda + \sqrt{\lambda^2 - \omega^2}, \qquad m_2 = -\lambda - \sqrt{\lambda^2 - \omega^2}.$$

We can now distinguish three possible cases depending on the algebraic sign of $\lambda^2 - \omega^2$. Since each solution contains the *damping factor* $e^{-\lambda t}$, $\lambda > 0$, the displacements of the mass become negligible as time t increases.

CASE I: $\lambda^2 - \omega^2 > 0$ In this situation the system is said to be **overdamped** because the damping coefficient β is large when compared to the spring constant k. The corresponding solution of (11) is $x(t) = c_1 e^{m_1 t} + c_2 e^{m_2 t}$ or

$$x(t) = e^{-\lambda t}\left(c_1 e^{\sqrt{\lambda^2 - \omega^2}\,t} + c_2 e^{-\sqrt{\lambda^2 - \omega^2}\,t}\right). \tag{13}$$

This equation represents a smooth and nonoscillatory motion. Figure 5.6 shows two possible graphs of $x(t)$.

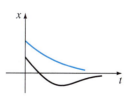

FIGURE 5.6 Motion of an overdamped system

CASE II: $\lambda^2 - \omega^2 = 0$ The system is said to be **critically damped** because any slight decrease in the damping force would result in oscillatory motion. The general solution of (11) is $x(t) = c_1 e^{m_1 t} + c_2 t e^{m_1 t}$ or

$$x(t) = e^{-\lambda t}(c_1 + c_2 t). \tag{14}$$

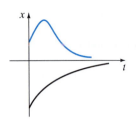

FIGURE 5.7   Motion of a critically damped system

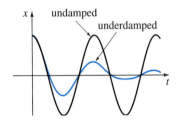

FIGURE 5.8   Motion of an underdamped system

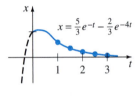

(a)

t	$x(t)$
1	0.601
1.5	0.370
2	0.225
2.5	0.137
3	0.083

(b)

FIGURE 5.9   Overdamped system

Some graphs of typical motion are given in Figure 5.7. Notice that the motion is quite similar to that of an overdamped system. It is also apparent from (14) that the mass can pass through the equilibrium position at most one time.

CASE III: $\lambda^2 - \omega^2 < 0$ In this case the system is said to be **underdamped** since the damping coefficient is small compared to the spring constant. The roots m_1 and m_2 are now complex:

$$m_1 = -\lambda + \sqrt{\omega^2 - \lambda^2}\,i, \quad m_2 = -\lambda - \sqrt{\omega^2 - \lambda^2}\,i.$$

Thus the general solution of equation (11) is

$$x(t) = e^{-\lambda t}\left(c_1 \cos\sqrt{\omega^2 - \lambda^2}\,t + c_2 \sin\sqrt{\omega^2 - \lambda^2}\,t\right). \tag{15}$$

As indicated in Figure 5.8, the motion described by (15) is oscillatory; but because of the coefficient $e^{-\lambda t}$, the amplitudes of vibration $\to 0$ as $t \to \infty$.

EXAMPLE 3   Overdamped Motion

It is readily verified that the solution of the initial-value problem

$$\frac{d^2x}{dt^2} + 5\frac{dx}{dt} + 4x = 0, \quad x(0) = 1, \quad x'(0) = 1$$

is

$$x(t) = \frac{5}{3}e^{-t} - \frac{2}{3}e^{-4t}. \tag{16}$$

The problem can be interpreted as representing the overdamped motion of a mass on a spring. The mass is initially released from a position 1 unit *below* the equilibrium position with a *downward* velocity of 1 ft/s.

To graph $x(t)$, we find the value of t for which the function has an extremum—that is, the value of time for which the first derivative (velocity) is zero. Differentiating (16) gives $x'(t) = -\frac{5}{3}e^{-t} + \frac{8}{3}e^{-4t}$, so that $x'(t) = 0$ implies $e^{3t} = \frac{8}{5}$ or $t = \frac{1}{3}\ln\frac{8}{5} = 0.157$. It follows from the first derivative test, as well as our physical intuition, that $x(0.157) = 1.069$ ft is actually a maximum. In other words, the mass attains an extreme displacement of 1.069 feet below the equilibrium position.

We should also check to see whether the graph crosses the t-axis—that is, whether the mass passes through the equilibrium position. This cannot happen in this instance because the equation $x(t) = 0$, or $e^{3t} = \frac{2}{5}$, has the physically irrelevant solution $t = \frac{1}{3}\ln\frac{2}{5} = -0.305$.

The graph of $x(t)$, along with some other pertinent data, is given in Figure 5.9.

EXAMPLE 4   Critically Damped Motion

A mass weighing 8 pounds stretches a spring 2 feet. Assuming that a damping force numerically equal to 2 times the instantaneous velocity acts on the system, determine the equation of motion if the mass is initially released from the equilibrium position with an upward velocity of 3 ft/s.

SOLUTION   From Hooke's law we see that $8 = k(2)$ gives $k = 4$ lb/ft and that $W = mg$ gives $m = \frac{8}{32} = \frac{1}{4}$ slug. The differential equation of motion is then

$$\frac{1}{4}\frac{d^2x}{dt^2} = -4x - 2\frac{dx}{dt} \quad \text{or} \quad \frac{d^2x}{dt^2} + 8\frac{dx}{dt} + 16x = 0. \tag{17}$$

The auxiliary equation for (17) is $m^2 + 8m + 16 = (m + 4)^2 = 0$, so $m_1 = m_2 = -4$. Hence the system is critically damped, and

$$x(t) = c_1 e^{-4t} + c_2 t e^{-4t}. \qquad (18)$$

Applying the initial conditions $x(0) = 0$ and $x'(0) = -3$, we find, in turn, that $c_1 = 0$ and $c_2 = -3$. Thus the equation of motion is

$$x(t) = -3t e^{-4t}. \qquad (19)$$

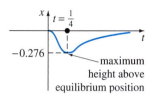

FIGURE 5.10 Critically damped system

To graph $x(t)$, we proceed as in Example 3. From $x'(t) = -3e^{-4t}(1 - 4t)$ we see that $x'(t) = 0$ when $t = \frac{1}{4}$. The corresponding extreme displacement is $x\left(\frac{1}{4}\right) = -3\left(\frac{1}{4}\right)e^{-1} = -0.276$ ft. As shown in Figure 5.10, we interpret this value to mean that the mass reaches a maximum height of 0.276 foot above the equilibrium position.

EXAMPLE 5   Underdamped Motion

A mass weighing 16 pounds is attached to a 5-foot-long spring. At equilibrium the spring measures 8.2 feet. If the mass is initially released from rest at a point 2 feet above the equilibrium position, find the displacements $x(t)$ if it is further known that the surrounding medium offers a resistance numerically equal to the instantaneous velocity.

SOLUTION The elongation of the spring after the mass is attached is $8.2 - 5 = 3.2$ ft, so it follows from Hooke's law that $16 = k(3.2)$ or $k = 5$ lb/ft. In addition, $m = \frac{16}{32} = \frac{1}{2}$ slug, so the differential equation is given by

$$\frac{1}{2}\frac{d^2x}{dt^2} = -5x - \frac{dx}{dt} \quad \text{or} \quad \frac{d^2x}{dt^2} + 2\frac{dx}{dt} + 10x = 0. \qquad (20)$$

Proceeding, we find that the roots of $m^2 + 2m + 10 = 0$ are $m_1 = -1 + 3i$ and $m_2 = -1 - 3i$, which then implies that the system is underdamped and

$$x(t) = e^{-t}(c_1 \cos 3t + c_2 \sin 3t). \qquad (21)$$

Finally, the initial conditions $x(0) = -2$ and $x'(0) = 0$ yield $c_1 = -2$ and $c_2 = -\frac{2}{3}$, so the equation of motion is

$$x(t) = e^{-t}\left(-2\cos 3t - \frac{2}{3}\sin 3t\right). \qquad (22)$$

ALTERNATIVE FORM OF $x(t)$ In a manner identical to the procedure used on page 197, we can write any solution

$$x(t) = e^{-\lambda t}\left(c_1 \cos \sqrt{\omega^2 - \lambda^2}\, t + c_2 \sin \sqrt{\omega^2 - \lambda^2}\, t\right)$$

in the alternative form

$$x(t) = A e^{-\lambda t} \sin\left(\sqrt{\omega^2 - \lambda^2}\, t + \phi\right), \qquad (23)$$

where $A = \sqrt{c_1^2 + c_2^2}$ and the phase angle ϕ is determined from the equations

$$\sin \phi = \frac{c_1}{A}, \quad \cos \phi = \frac{c_2}{A}, \quad \tan \phi = \frac{c_1}{c_2}.$$

The coefficient $A e^{-\lambda t}$ is sometimes called the **damped amplitude** of vibrations. Because (23) is not a periodic function, the number $2\pi/\sqrt{\omega^2 - \lambda^2}$ is called the **quasi period** and $\sqrt{\omega^2 - \lambda^2}/2\pi$ is the **quasi frequency.** The quasi period is the time interval between two successive maxima of $x(t)$. You should verify, for the equation of motion in Example 5, that $A = 2\sqrt{10}/3$ and $\phi = 4.391$. Therefore an equivalent form of (22) is

$$x(t) = \frac{2\sqrt{10}}{3} e^{-t} \sin(3t + 4.391).$$

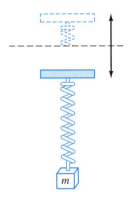

FIGURE 5.11   Oscillatory vertical motion of the support

5.1.3   SPRING/MASS SYSTEMS: DRIVEN MOTION

DE OF DRIVEN MOTION WITH DAMPING   Suppose we now take into consideration an external force $f(t)$ acting on a vibrating mass on a spring. For example, $f(t)$ could represent a driving force causing an oscillatory vertical motion of the support of the spring. See Figure 5.11. The inclusion of $f(t)$ in the formulation of Newton's second law gives the differential equation of **driven** or **forced motion**:

$$m\frac{d^2x}{dt^2} = -kx - \beta\frac{dx}{dt} + f(t). \tag{24}$$

Dividing (24) by m gives

$$\frac{d^2x}{dt^2} + 2\lambda\frac{dx}{dt} + \omega^2 x = F(t), \tag{25}$$

where $F(t) = f(t)/m$ and, as in the preceding section, $2\lambda = \beta/m$, $\omega^2 = k/m$. To solve the latter nonhomogeneous equation, we can use either the method of undetermined coefficients or variation of parameters.

EXAMPLE 6   Interpretation of an Initial-Value Problem

Interpret and solve the initial-value problem

$$\frac{1}{5}\frac{d^2x}{dt^2} + 1.2\frac{dx}{dt} + 2x = 5\cos 4t, \quad x(0) = \frac{1}{2}, \quad x'(0) = 0. \tag{26}$$

SOLUTION   We can interpret the problem to represent a vibrational system consisting of a mass ($m = \frac{1}{5}$ slug or kilogram) attached to a spring ($k = 2$ lb/ft or N/m). The mass is initially released from rest $\frac{1}{2}$ unit (foot or meter) below the equilibrium position. The motion is damped ($\beta = 1.2$) and is being driven by an external periodic ($T = \pi/2$ s) force beginning at $t = 0$. Intuitively, we would expect that even with damping the system would remain in motion until such time as the forcing function was "turned off," in which case the amplitudes would diminish. However, as the problem is given, $f(t) = 5\cos 4t$ will remain "on" forever.

We first multiply the differential equation in (26) by 5 and solve

$$\frac{dx^2}{dt^2} + 6\frac{dx}{dt} + 10x = 0$$

by the usual methods. Because $m_1 = -3 + i$, $m_2 = -3 - i$, it follows that $x_c(t) = e^{-3t}(c_1 \cos t + c_2 \sin t)$. Using the method of undetermined coefficients, we assume a particular solution of the form $x_p(t) = A\cos 4t + B\sin 4t$. Differentiating $x_p(t)$ and substituting into the DE gives

$$x_p'' + 6x_p' + 10x_p = (-6A + 24B)\cos 4t + (-24A - 6B)\sin 4t = 25\cos 4t.$$

The resulting system of equations

$$-6A + 24B = 25, \quad -24A - 6B = 0$$

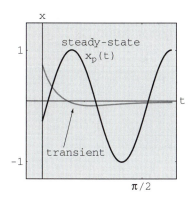

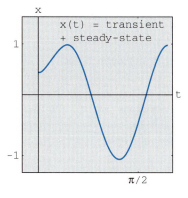

(a)

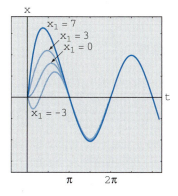

(b)

FIGURE 5.12 Graph of solution given in (28)

yields $A = -\frac{25}{102}$ and $B = \frac{50}{51}$. It follows that

$$x(t) = e^{-3t}(c_1 \cos t + c_2 \sin t) - \frac{25}{102} \cos 4t + \frac{50}{51} \sin 4t. \qquad (27)$$

When we set $t = 0$ in the above equation, we obtain $c_1 = \frac{38}{51}$. By differentiating the expression and then setting $t = 0$, we also find that $c_2 = -\frac{86}{51}$. Therefore the equation of motion is

$$x(t) = e^{-3t}\left(\frac{38}{51} \cos t - \frac{86}{51} \sin t\right) - \frac{25}{102} \cos 4t + \frac{50}{51} \sin 4t. \quad (28)$$

TRANSIENT AND STEADY-STATE TERMS When F is a periodic function, such as $F(t) = F_0 \sin \gamma t$ or $F(t) = F_0 \cos \gamma t$, the general solution of (25) for $\lambda > 0$ is the sum of a nonperiodic function $x_c(t)$ and a periodic function $x_p(t)$. Moreover, $x_c(t)$ dies off as time increases—that is, $\lim_{t \to \infty} x_c(t) = 0$. Thus for large values of time, the displacements of the mass are closely approximated by the particular solution $x_p(t)$. The complementary function $x_c(t)$ is said to be a **transient term** or **transient solution,** and the function $x_p(t)$, the part of the solution that remains after an interval of time, is called a **steady-state term** or **steady-state solution.** Note therefore that the effect of the initial conditions on a spring/mass system driven by F is transient. In the particular solution (28), $e^{-3t}\left(\frac{38}{51} \cos t - \frac{86}{51} \sin t\right)$ is a transient term and $x_p(t) = -\frac{25}{102} \cos 4t + \frac{50}{51} \sin 4t$ is a steady-state term. The graphs of these two terms and the solution (28) are given in Figures 5.12(a) and 5.12(b), respectively.

EXAMPLE 7 Transient/Steady-State Solutions

The solution of the initial-value problem

$$\frac{d^2x}{dt^2} + 2\frac{dx}{dt} + 2x = 4 \cos t + 2 \sin t, \quad x(0) = 0, \quad x'(0) = x_1,$$

where x_1 is constant, is given by

$$x(t) = \underbrace{(x_1 - 2)\, e^{-t} \sin t}_{\text{transient}} + \underbrace{2 \sin t}_{\text{steady-state}}.$$

Solution curves for selected values of the initial velocity x_1 are shown in Figure 5.13. The graphs show that the influence of the transient term is negligible for about $t > 3\pi/2$.

DE OF DRIVEN MOTION WITHOUT DAMPING With a periodic impressed force and no damping force, there is no transient term in the solution of a problem. Also, we shall see that a periodic impressed force with a frequency near or the same as the frequency of free undamped vibrations can cause a severe problem in any oscillatory mechanical system.

EXAMPLE 8 Undamped Forced Motion

Solve the initial-value problem

$$\frac{d^2x}{dt^2} + \omega^2 x = F_0 \sin \gamma t, \quad x(0) = 0, \quad x'(0) = 0, \qquad (29)$$

where F_0 is a constant and $\gamma \neq \omega$.

FIGURE 5.13 Graph of solution in Example 7 for various x_1

SOLUTION   The complementary function is $x_c(t) = c_1 \cos \omega t + c_2 \sin \omega t$. To obtain a particular solution we assume $x_p(t) = A \cos \gamma t + B \sin \gamma t$ so that

$$x_p'' + \omega^2 x_p = A(\omega^2 - \gamma^2) \cos \gamma t + B(\omega^2 - \gamma^2) \sin \gamma t = F_0 \sin \gamma t.$$

Equating coefficients immediately gives $A = 0$ and $B = F_0/(\omega^2 - \gamma^2)$. Therefore

$$x_p(t) = \frac{F_0}{\omega^2 - \gamma^2} \sin \gamma t.$$

Applying the given initial conditions to the general solution

$$x(t) = c_1 \cos \omega t + c_2 \sin \omega t + \frac{F_0}{\omega^2 - \gamma^2} \sin \gamma t$$

yields $c_1 = 0$ and $c_2 = -\gamma F_0/\omega(\omega^2 - \gamma^2)$. Thus the solution is

$$x(t) = \frac{F_0}{\omega(\omega^2 - \gamma^2)} (-\gamma \sin \omega t + \omega \sin \gamma t), \quad \gamma \neq \omega. \qquad (30)$$

PURE RESONANCE   Although equation (30) is not defined for $\gamma = \omega$, it is interesting to observe that its limiting value as $\gamma \to \omega$ can be obtained by applying L'Hôpital's Rule. This limiting process is analogous to "tuning in" the frequency of the driving force ($\gamma/2\pi$) to the frequency of free vibrations ($\omega/2\pi$). Intuitively, we expect that over a length of time we should be able to substantially increase the amplitudes of vibration. For $\gamma = \omega$ we define the solution to be

$$x(t) = \lim_{\gamma \to \omega} F_0 \frac{-\gamma \sin \omega t + \omega \sin \gamma t}{\omega(\omega^2 - \gamma^2)} = F_0 \lim_{\gamma \to \omega} \frac{\dfrac{d}{d\gamma}(-\gamma \sin \omega t + \omega \sin \gamma t)}{\dfrac{d}{d\gamma}(\omega^3 - \omega\gamma^2)}$$

$$= F_0 \lim_{\gamma \to \omega} \frac{-\sin \omega t + \omega t \cos \gamma t}{-2\omega\gamma} \qquad (31)$$

$$= F_0 \frac{-\sin \omega t + \omega t \cos \omega t}{-2\omega^2}$$

$$= \frac{F_0}{2\omega^2} \sin \omega t - \frac{F_0}{2\omega} t \cos \omega t.$$

As suspected, when $t \to \infty$ the displacements become large; in fact, $|x(t_n)| \to \infty$ when $t_n = n\pi/\omega$, $n = 1, 2, \ldots$. The phenomenon we have just described is known as **pure resonance.** The graph given in Figure 5.14 shows typical motion in this case.

In conclusion it should be noted that there is no actual need to use a limiting process on (30) to obtain the solution for $\gamma = \omega$. Alternatively, equation (31) follows by solving the initial-value problem

$$\frac{d^2 x}{dt^2} + \omega^2 x = F_0 \sin \omega t, \quad x(0) = 0, \quad x'(0) = 0$$

directly by conventional methods.

If the displacements of a spring/mass system were actually described by a function such as (31), the system would necessarily fail. Large oscillations of the mass would eventually force the spring beyond its elastic limit. One might argue too that the resonating model presented in Figure 5.14 is completely unrealistic because it ignores the retarding effects of ever-present damping forces. Although it is true that pure resonance cannot occur when the smallest amount of damping is taken into consideration, large and equally destructive amplitudes of vibration (although bounded as $t \to \infty$) can occur. See Problem 43 in Exercises 5.1.

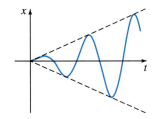

FIGURE 5.14   Pure resonance

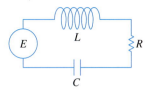

FIGURE 5.15 *LRC* series circuit

5.1.4   SERIES CIRCUIT ANALOGUE

LRC **SERIES CIRCUITS**   As was mentioned in the introduction to this chapter, many different physical systems can be described by a linear second-order differential equation similar to the differential equation of forced motion with damping:

$$m\frac{d^2x}{dt^2} + \beta\frac{dx}{dt} + kx = f(t). \tag{32}$$

If $i(t)$ denotes current in the *LRC* **series electrical circuit** shown in Figure 5.15, then the voltage drops across the inductor, resistor, and capacitor are as shown in Figure 1.21. By Kirchhoff's second law the sum of these voltages equals the voltage $E(t)$ impressed on the circuit; that is,

$$L\frac{di}{dt} + Ri + \frac{1}{C}q = E(t). \tag{33}$$

But the charge $q(t)$ on the capacitor is related to the current $i(t)$ by $i = dq/dt$, so (33) becomes the linear second-order differential equation

$$L\frac{d^2q}{dt^2} + R\frac{dq}{dt} + \frac{1}{C}q = E(t). \tag{34}$$

The nomenclature used in the analysis of circuits is similar to that used to describe spring-mass systems.

If $E(t) = 0$, the **electrical vibrations** of the circuit are said to be **free.** Because the auxiliary equation for (34) is $Lm^2 + Rm + 1/C = 0$, there will be three forms of the solution with $R \neq 0$, depending on the value of the discriminant $R^2 - 4L/C$. We say that the circuit is

$$\text{**overdamped** if} \qquad R^2 - 4L/C > 0,$$

$$\text{**critically damped** if} \quad R^2 - 4L/C = 0,$$

and $\qquad$ **underdamped** if $\qquad R^2 - 4L/C < 0.$

In each of these three cases the general solution of (34) contains the factor $e^{-Rt/2L}$, so $q(t) \rightarrow 0$ as $t \rightarrow \infty$. In the underdamped case when $q(0) = q_0$, the charge on the capacitor oscillates as it decays; in other words, the capacitor is charging and discharging as $t \rightarrow \infty$. When $E(t) = 0$ and $R = 0$, the circuit is said to be undamped, and the electrical vibrations do not approach zero as t increases without bound; the response of the circuit is **simple harmonic.**

EXAMPLE 9   **Underdamped Series Circuit**

Find the charge $q(t)$ on the capacitor in an *LRC* series circuit when $L = 0.25$ henry (h), $R = 10$ ohms (Ω), $C = 0.001$ farad (f), $E(t) = 0$, $q(0) = q_0$ coulombs (C), and $i(0) = 0$.

SOLUTION   Since $1/C = 1000$, equation (34) becomes

$$\frac{1}{4}q'' + 10q' + 1000q = 0 \quad \text{or} \quad q'' + 40q' + 4000q = 0.$$

Solving this homogeneous equation in the usual manner, we find that the circuit is underdamped and $q(t) = e^{-20t}(c_1 \cos 60t + c_2 \sin 60t)$. Applying the initial conditions, we find $c_1 = q_0$ and $c_2 = \frac{1}{3}q_0$. Thus

$$q(t) = q_0 e^{-20t}\left(\cos 60t + \frac{1}{3}\sin 60t\right).$$

Using (23), we can write the foregoing solution as

$$q(t) = \frac{q_0\sqrt{10}}{3}\, e^{-20t}\, \sin(60t + 1.249).$$

When there is an impressed voltage $E(t)$ on the circuit, the electrical vibrations are said to be **forced.** In the case when $R \neq 0$, the complementary function $q_c(t)$ of (34) is called a **transient solution.** If $E(t)$ is periodic or a constant, then the particular solution $q_p(t)$ of (34) is a **steady-state solution.**

EXAMPLE 10   Steady-State Current

Find the steady-state solution $q_p(t)$ and the **steady-state current** in an *LRC* series circuit when the impressed voltage is $E(t) = E_0 \sin \gamma t$.

SOLUTION   The steady-state solution $q_p(t)$ is a particular solution of the differential equation

$$L\frac{d^2q}{dt^2} + R\frac{dq}{dt} + \frac{1}{C}q = E_0 \sin \gamma t.$$

Using the method of undetermined coefficients, we assume a particular solution of the form $q_p(t) = A \sin \gamma t + B \cos \gamma t$. Substituting this expression into the differential equation, simplifying, and equating coefficients gives

$$A = \frac{E_0\left(L\gamma - \dfrac{1}{C\gamma}\right)}{-\gamma\left(L^2\gamma^2 - \dfrac{2L}{C} + \dfrac{1}{C^2\gamma^2} + R^2\right)}, \quad B = \frac{E_0 R}{-\gamma\left(L^2\gamma^2 - \dfrac{2L}{C} + \dfrac{1}{C^2\gamma^2} + R^2\right)}.$$

It is convenient to express A and B in terms of some new symbols.

If $\qquad X = L\gamma - \dfrac{1}{C\gamma}$, then $\quad X^2 = L^2\gamma^2 - \dfrac{2L}{C} + \dfrac{1}{C^2\gamma^2}$.

If $\qquad Z = \sqrt{X^2 + R^2}$, then $\quad Z^2 = L^2\gamma^2 - \dfrac{2L}{C} + \dfrac{1}{C^2\gamma^2} + R^2$.

Therefore $A = E_0 X/(-\gamma Z^2)$ and $B = E_0 R/(-\gamma Z^2)$, so the steady-state charge is

$$q_p(t) = -\frac{E_0 X}{\gamma Z^2}\sin \gamma t - \frac{E_0 R}{\gamma Z^2}\cos \gamma t.$$

Now the steady-state current is given by $i_p(t) = q_p'(t)$:

$$i_p(t) = \frac{E_0}{Z}\left(\frac{R}{Z}\sin \gamma t - \frac{X}{Z}\cos \gamma t\right). \tag{35}$$

The quantities $X = L\gamma - 1/C\gamma$ and $Z = \sqrt{X^2 + R^2}$ defined in Example 11 are called the **reactance** and **impedance,** respectively, of the circuit. Both the reactance and the impedance are measured in ohms.

EXERCISES 5.1

Answers to selected odd-numbered problems begin on page ANS-6.

5.1.1 SPRING/MASS SYSTEMS: FREE UNDAMPED MOTION

1. A mass weighing 4 pounds is attached to a spring whose spring constant is 16 lb/ft. What is the period of simple harmonic motion?

2. A 20-kilogram mass is attached to a spring. If the frequency of simple harmonic motion is $2/\pi$ cycles/s, what is the spring constant k? What is the frequency of simple harmonic motion if the original mass is replaced with an 80-kilogram mass?

3. A mass weighing 24 pounds, attached to the end of a spring, stretches it 4 inches. Initially, the mass is released from rest from a point 3 inches above the equilibrium position. Find the equation of motion.

4. Determine the equation of motion if the mass in Problem 3 is initially released from the equilibrium position with a downward velocity of 2 ft/s.

5. A mass weighing 20 pounds stretches a spring 6 inches. The mass is initially released from rest from a point 6 inches below the equilibrium position.
   (a) Find the position of the mass at the times $t = \pi/12$, $\pi/8$, $\pi/6$, $\pi/4$, and $9\pi/32$ s.
   (b) What is the velocity of the mass when $t = 3\pi/16$ s? In which direction is the mass heading at this instant?
   (c) At what times does the mass pass through the equilibrium position?

6. A force of 400 newtons stretches a spring 2 meters. A mass of 50 kilograms is attached to the end of the spring and is initially released from the equilibrium position with an upward velocity of 10 m/s. Find the equation of motion.

7. Another spring whose constant is 20 N/m is suspended from the same rigid support but parallel to the spring/mass system in Problem 6. A mass of 20 kilograms is attached to the second spring, and both masses are initially released from the equilibrium position with an upward velocity of 10 m/s.
   (a) Which mass exhibits the greater amplitude of motion?
   (b) Which mass is moving faster at $t = \pi/4$ s? At $\pi/2$ s?
   (c) At what times are the two masses in the same position? Where are the masses at these times? In which directions are the masses moving?

8. A mass weighing 32 pounds stretches a spring 2 feet. Determine the amplitude and period of motion if the mass is initially released from a point 1 foot above the equilibrium position with an upward velocity of 2 ft/s. How many complete cycles will the mass have completed at the end of 4π seconds?

9. A mass weighing 8 pounds is attached to a spring. When set in motion, the spring/mass system exhibits simple harmonic motion. Determine the equation of motion if the spring constant is 1 lb/ft and the mass is initially released from a point 6 inches below the equilibrium position with a downward velocity of $\frac{3}{2}$ ft/s. Express the equation of motion in the form given in (6).

10. A mass weighing 10 pounds stretches a spring $\frac{1}{4}$ foot. This mass is removed and replaced with a mass of 1.6 slugs, which is initially released from a point $\frac{1}{3}$ foot above the equilibrium position with a downward velocity of $\frac{5}{4}$ ft/s. Express the equation of motion in the form given in (6). At what times does the mass attain a displacement below the equilibrium position numerically equal to $\frac{1}{2}$ the amplitude?

11. A mass weighing 64 pounds stretches a spring 0.32 foot. The mass is initially released from a point 8 inches above the equilibrium position with a downward velocity of 5 ft/s.
    (a) Find the equation of motion.
    (b) What are the amplitude and period of motion?
    (c) How many complete cycles will the mass have completed at the end of 3π seconds?
    (d) At what time does the mass pass through the equilibrium position heading downward for the second time?
    (e) At what times does the mass attain its extreme displacements on either side of the equilibrium position?
    (f) What is the position of the mass at $t = 3$ s?
    (g) What is the instantaneous velocity at $t = 3$ s?
    (h) What is the acceleration at at $t = 3$ s?
    (i) What is the instantaneous velocity at the times when the mass passes through the equilibrium position?
    (j) At what times is the mass 5 inches below the equilibrium position?
    (k) At what times is the mass 5 inches below the equilibrium position heading in the upward direction?

12. A mass of 1 slug is suspended from a spring whose spring constant is 9 lb/ft. The mass is initially released from a point 1 foot above the equilibrium position with an upward velocity of $\sqrt{3}$ ft/s. Find the times at which the mass is heading downward at a velocity of 3 ft/s.

13. Under some circumstances when two parallel springs, with constants k_1 and k_2, support a single mass, the **effective spring constant** of the system is given by $k = 4k_1k_2/(k_1 + k_2)$. A mass weighing 20 pounds stretches one spring 6 inches and another spring 2 inches. The springs are attached to a common rigid support and then to a metal plate. As shown in Figure 5.16, the mass is attached to the center of the plate in the double-spring arrangement. Determine the effective spring constant of this system. Find the equation of motion if the mass is initially released from the equilibrium position with a downward velocity of 2 ft/s.

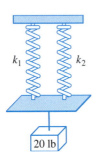

FIGURE 5.16   Double-spring system in Problem 13

14. A certain mass stretches one spring $\frac{1}{3}$ foot and another spring $\frac{1}{2}$ foot. The two springs are attached to a common rigid support in the manner described in Problem 13 and Figure 5.16. The first mass is set aside, a mass weighing 8 pounds is attached to the double-spring arrangement, and the system is set in motion. If the period of motion is $\pi/15$ second, determine how much the first mass weighs.

15. A model of a spring/mass system is $4x'' + e^{-0.1t}x = 0$. By inspection of the differential equation only, discuss the behavior of the system over a long period of time.

16. A model of a spring/mass system is $4x'' + tx = 0$. By inspection of the differential equation only, discuss the behavior of the system over a long period of time.

5.1.2 SPRING/MASS SYSTEMS: FREE DAMPED MOTION

In Problems 17–20 the given figure represents the graph of an equation of motion for a damped spring/mass system. Use the graph to determine
(a) whether the initial displacement is above or below the equilibrium position and
(b) whether the mass is initially released from rest, heading downward, or heading upward.

17.

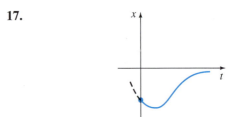

FIGURE 5.17   Graph for Problem 17

18.

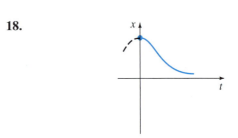

FIGURE 5.18   Graph for Problem 18

19.

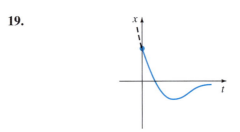

FIGURE 5.19   Graph for Problem 19

20.

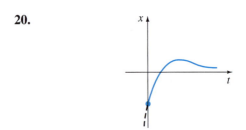

FIGURE 5.20   Graph for Problem 20

21. A mass weighing 4 pounds is attached to a spring whose constant is 2 lb/ft. The medium offers a damping force that is numerically equal to the instantaneous velocity. The mass is initially released from a point 1 foot above the equilibrium position with a downward velocity of 8 ft/s. Determine the time at which the mass passes through the equilibrium position. Find the time at which the mass attains its extreme displacement from the equilibrium position. What is the position of the mass at this instant?

22. A 4-foot spring measures 8 feet long after a mass weighing 8 pounds is attached to it. The medium through which the mass moves offers a damping force numerically equal to $\sqrt{2}$ times the instantaneous velocity. Find the equation of motion if the mass is initially released from the equilibrium position with a downward velocity of 5 ft/s. Find the time at which the mass attains its extreme displacement from the equilibrium position. What is the position of the mass at this instant?

23. A 1-kilogram mass is attached to a spring whose constant is 16 N/m, and the entire system is then submerged in a liquid that imparts a damping force numerically equal to 10 times the instantaneous velocity. Determine the equations of motion if
   (a) the mass is initially released from rest from a point 1 meter below the equilibrium position, and then
   (b) the mass is initially released from a point 1 meter below the equilibrium position with an upward velocity of 12 m/s.

24. In parts (a) and (b) of Problem 23 determine whether the mass passes through the equilibrium position. In each case find the time at which the mass attains its extreme displacement from the equilibrium position. What is the position of the mass at this instant?

25. A force of 2 pounds stretches a spring 1 foot. A mass weighing 3.2 pounds is attached to the spring, and the system is then immersed in a medium that offers a damping force numerically equal to 0.4 times the instantaneous velocity.
   (a) Find the equation of motion if the mass is initially released from rest from a point 1 foot above the equilibrium position.
   (b) Express the equation of motion in the form given in (23).
   (c) Find the first time at which the mass passes through the equilibrium position heading upward.

26. After a mass weighing 10 pounds is attached to a 5-foot spring, the spring measures 7 feet. This mass is removed and replaced with another mass that weighs 8 pounds. The entire system is placed in a medium that offers a damping force numerically equal to the instantaneous velocity.
   (a) Find the equation of motion if the mass is initially released from a point $\frac{1}{2}$ foot below the equilibrium position with a downward velocity of 1 ft/s.
   (b) Express the equation of motion in the form given in (23).
   (c) Find the times at which the mass passes through the equilibrium position heading downward.
   (d) Graph the equation of motion.

27. A mass weighing 10 pounds stretches a spring 2 feet. The mass is attached to a dashpot device that offers a damping force numerically equal to β ($\beta > 0$) times the instanta-

neous velocity. Determine the values of the damping constant β so that the subsequent motion is **(a)** overdamped, **(b)** critically damped, and **(c)** underdamped.

28. A mass weighing 24 pounds stretches a spring 4 feet. The subsequent motion takes place in medium that offers a damping force numerically equal to β ($\beta > 0$) times the instantaneous velocity. If the mass is initially released from the equilibrium position with an upward velocity of 2 ft/s, show that when $\beta > 3\sqrt{2}$ the equation of motion is

$$x(t) = \frac{-3}{\sqrt{\beta^2 - 18}} e^{-2\beta t/3} \sinh\frac{2}{3}\sqrt{\beta^2 - 18}\,t.$$

5.1.3 SPRING/MASS SYSTEMS: DRIVEN MOTION

29. A mass weighing 16 pounds stretches a spring $\frac{8}{3}$ feet. The mass is initially released from rest from a point 2 feet below the equilibrium position, and the subsequent motion takes place in a medium that offers a damping force numerically equal to $\frac{1}{2}$ the instantaneous velocity. Find the equation of motion if the mass is driven by an external force equal to $f(t) = 10\cos 3t$.

30. A mass of 1 slug is attached to a spring whose constant is 5 lb/ft. Initially, the mass is released 1 foot below the equilibrium position with a downward velocity of 5 ft/s, and the subsequent motion takes place in a medium that offers a damping force numerically equal to 2 times the instantaneous velocity.
   (a) Find the equation of motion if the mass is driven by an external force equal to $f(t) = 12\cos 2t + 3\sin 2t$.
   (b) Graph the transient and steady-state solutions on the same coordinate axes.
   (c) Graph the equation of motion.

31. A mass of 1 slug, when attached to a spring, stretches it 2 feet and then comes to rest in the equilibrium position. Starting at $t = 0$, an external force equal to $f(t) = 8\sin 4t$ is applied to the system. Find the equation of motion if the surrounding medium offers a damping force numerically equal to 8 times the instantaneous velocity.

32. In Problem 31 determine the equation of motion if the external force is $f(t) = e^{-t}\sin 4t$. Analyze the displacements for $t \to \infty$.

33. When a mass of 2 kilograms is attached to a spring whose constant is 32 N/m, it comes to rest in the equilibrium position. Starting at $t = 0$, a force equal to $f(t) = 68e^{-2t}\cos 4t$ is applied to the system. Find the equation of motion in the absence of damping.

34. In Problem 33 write the equation of motion in the form $x(t) = A\sin(\omega t + \phi) + Be^{-2t}\sin(4t + \theta)$. What is the amplitude of vibrations after a very long time?

35. A mass m is attached to the end of a spring whose constant is k. After the mass reaches equilibrium, its support begins to oscillate vertically about a horizontal line L according to a formula $h(t)$. The value of h represents the distance in feet measured from L. See Figure 5.21.

 (a) Determine the differential equation of motion if the entire system moves through a medium offering a damping force numerically equal to $\beta(dx/dt)$.

 (b) Solve the differential equation in part (a) if the spring is stretched 4 feet by a mass weighing 16 pounds and $\beta = 2$, $h(t) = 5 \cos t$, $x(0) = x'(0) = 0$.

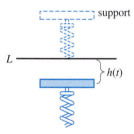

FIGURE 5.21   Oscillating support in Problem 35

36. A mass of 100 grams is attached to a spring whose constant is 1600 dynes/cm. After the mass reaches equilibrium, its support oscillates according to the formula $h(t) = \sin 8t$, where h represents displacement from its original position. See Problem 35 and Figure 5.21.

 (a) In the absence of damping, determine the equation of motion if the mass starts from rest from the equilibrium position.

 (b) At what times does the mass pass through the equilibrium position?

 (c) At what times does the mass attain its extreme displacements?

 (d) What are the maximum and minimum displacements?

 (e) Graph the equation of motion.

In Problems 37 and 38 solve the given initial-value problem.

37. $\dfrac{d^2x}{dt^2} + 4x = -5 \sin 2t + 3 \cos 2t,$

 $x(0) = -1, \quad x'(0) = 1$

38. $\dfrac{d^2x}{dt^2} + 9x = 5 \sin 3t, \quad x(0) = 2, \quad x'(0) = 0$

39. (a) Show that the solution of the initial-value problem

 $$\dfrac{d^2x}{dt^2} + \omega^2 x = F_0 \cos \gamma t, \quad x(0) = 0, \quad x'(0) = 0$$

is $x(t) = \dfrac{F_0}{\omega^2 - \gamma^2} (\cos \gamma t - \cos \omega t).$

 (b) Evaluate $\displaystyle\lim_{\gamma \to \omega} \dfrac{F_0}{\omega^2 - \gamma^2} (\cos \gamma t - \cos \omega t).$

40. Compare the result obtained in part (b) of Problem 39 with the solution obtained using variation of parameters when the external force is $F_0 \cos \omega t$.

41. (a) Show that $x(t)$ given in part (a) of Problem 39 can be written in the form

 $$x(t) = \dfrac{-2F_0}{\omega^2 - \gamma^2} \sin \dfrac{1}{2} (\gamma - \omega)t \sin \dfrac{1}{2} (\gamma + \omega)t.$$

 (b) If we define $\varepsilon = \frac{1}{2}(\gamma - \omega)$, show that when ε is small an approximate solution is

 $$x(t) = \dfrac{F_0}{2\varepsilon\gamma} \sin \varepsilon t \sin \gamma t.$$

 When ε is small, the frequency $\gamma/2\pi$ of the impressed force is close to the frequency $\omega/2\pi$ of free vibrations. When this occurs, the motion is as indicated in Figure 5.22. Oscillations of this kind are called **beats** and are due to the fact that the frequency of $\sin \varepsilon t$ is quite small in comparison to the frequency of $\sin \gamma t$. The dashed curves, or envelope of the graph of $x(t)$, are obtained from the graphs of $\pm(F_0/2\varepsilon\gamma) \sin \varepsilon t$. Use a graphing utility with various values of F_0, ε, and γ to verify the graph in Figure 5.22.

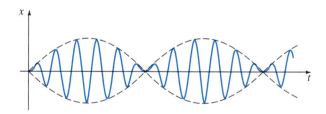

FIGURE 5.22   Beats phenomenon in Problem 41

COMPUTER LAB ASSIGNMENTS

42. Can there be beats when a damping force is added to the model in part (a) of Problem 39? Defend your position with graphs obtained either from the explicit solution of the problem

 $$\dfrac{d^2x}{dt^2} + 2\lambda\dfrac{dx}{dt} + \omega^2 x = F_0 \cos \gamma t, \quad x(0) = 0, \quad x'(0) = 0$$

or from solution curves obtained using a numerical solver.

43. (a) Show that the general solution of

$$\frac{d^2x}{dt^2} + 2\lambda\frac{dx}{dt} + \omega^2 x = F_0\sin\gamma t$$

is

$$x(t) = Ae^{-\lambda t}\sin\left(\sqrt{\omega^2 - \lambda^2}t + \phi\right)$$
$$+ \frac{F_0}{\sqrt{(\omega^2 - \gamma^2)^2 + 4\lambda^2\gamma^2}}\sin(\gamma t + \theta),$$

where $A = \sqrt{c_1^2 + c_2^2}$ and the phase angles ϕ and θ are, respectively, defined by $\sin\phi = c_1/A$, $\cos\phi = c_2/A$ and

$$\sin\theta = \frac{-2\lambda\gamma}{\sqrt{(\omega^2 - \gamma^2)^2 + 4\lambda^2\gamma^2}},$$

$$\cos\theta = \frac{\omega^2 - \gamma^2}{\sqrt{(\omega^2 - \gamma^2)^2 + 4\lambda^2\gamma^2}}.$$

(b) The solution in part (a) has the form $x(t) = x_c(t) + x_p(t)$. Inspection shows that $x_c(t)$ is transient, and hence for large values of time, the solution is approximated by $x_p(t) = g(\gamma)\sin(\gamma t + \theta)$, where

$$g(\gamma) = \frac{F_0}{\sqrt{(\omega^2 - \gamma^2)^2 + 4\lambda^2\gamma^2}}.$$

Although the amplitude $g(\gamma)$ of $x_p(t)$ is bounded as $t\to\infty$, show that the maximum oscillations will occur at the value $\gamma_1 = \sqrt{\omega^2 - 2\lambda^2}$. What is the maximum value of g? The number $\sqrt{\omega^2 - 2\lambda^2}/2\pi$ is said to be the **resonance frequency** of the system.

(c) When $F_0 = 2$, $m = 1$, and $k = 4$, g becomes

$$g(\gamma) = \frac{2}{\sqrt{(4 - \gamma^2)^2 + \beta^2\gamma^2}}.$$

Construct a table of the values of γ_1 and $g(\gamma_1)$ corresponding to the damping coefficients $\beta = 2$, $\beta = 1$, $\beta = \frac{3}{4}$, $\beta = \frac{1}{2}$, and $\beta = \frac{1}{4}$. Use a graphing utility to obtain the graphs of g corresponding to these damping coefficients. Use the same coordinate axes. This family of graphs is called the **resonance curve** or **frequency response curve** of the system. What is γ_1 approaching as $\beta\to 0$? What is happening to the resonance curve as $\beta\to 0$?

44. Consider a driven undamped spring/mass system described by the initial-value problem

$$\frac{d^2x}{dt^2} + \omega^2 x = F_0\sin^n\gamma t, \quad x(0) = 0, \quad x'(0) = 0.$$

(a) For $n = 2$, discuss why there is a single frequency $\gamma_1/2\pi$ at which the system is in pure resonance.

(b) For $n = 3$, discuss why there are two frequencies $\gamma_1/2\pi$ and $\gamma_2/2\pi$ at which the system is in pure resonance.

(c) Suppose $\omega = 1$ and $F_0 = 1$. Use a numerical solver to obtain the graph of the solution of the initial-value problem for $n = 2$ and $\gamma = \gamma_1$ in part (a). Obtain the graph of the solution of the initial-value problem for $n = 3$ corresponding, in turn, to $\gamma = \gamma_1$ and $\gamma = \gamma_2$ in part (b).

5.1.4 SERIES CIRCUIT ANALOGUE

45. Find the charge on the capacitor in an *LRC* series circuit at $t = 0.01$ s when $L = 0.05$ h, $R = 2\ \Omega$, $C = 0.01$ f, $E(t) = 0$ V, $q(0) = 5$ C, and $i(0) = 0$ A. Determine the first time at which the charge on the capacitor is equal to zero.

46. Find the charge on the capacitor in an *LRC* series circuit when $L = \frac{1}{4}$ h, $R = 20\ \Omega$, $C = \frac{1}{300}$ f, $E(t) = 0$ V, $q(0) = 4$ C, and $i(0) = 0$ A. Is the charge on the capacitor ever equal to zero?

In Problems 47 and 48 find the charge on the capacitor and the current in the given *LRC* series circuit. Find the maximum charge on the capacitor.

47. $L = \frac{5}{3}$ h, $R = 10\ \Omega$, $C = \frac{1}{30}$ f, $E(t) = 300$ V, $q(0) = 0$ C, $i(0) = 0$ A

48. $L = 1$ h, $R = 100\ \Omega$, $C = 0.0004$ f, $E(t) = 30$ V, $q(0) = 0$ C, $i(0) = 2$ A

49. Find the steady-state charge and the steady-state current in an *LRC* series circuit when $L = 1$ h, $R = 2\ \Omega$, $C = 0.25$ f, and $E(t) = 50\cos t$ V.

50. Show that the amplitude of the steady-state current in the *LRC* series circuit in Example 10 is given by E_0/Z, where Z is the impedance of the circuit.

51. Use Problem 50 to show that the steady-state current in an *LRC* series circuit when $L = \frac{1}{2}$ h, $R = 20\ \Omega$, $C = 0.001$ f, and $E(t) = 100\sin 60t$ V, is given by $i_p(t) = 4.160\sin(60t - 0.588)$.

52. Find the steady-state current in an *LRC* series circuit when $L = \frac{1}{2}$ h, $R = 20\ \Omega$, $C = 0.001$ f, and $E(t) = 100\sin 60t + 200\cos 40t$ V.

53. Find the charge on the capacitor in an *LRC* series circuit when $L = \frac{1}{2}$ h, $R = 10\ \Omega$, $C = 0.01$ f, $E(t) = 150$ V, $q(0) = 1$ C, and $i(0) = 0$ A. What is the charge on the capacitor after a long time?

54. Show that if L, R, C, and E_0 are constant, then the amplitude of the steady-state current in Example 10 is a maximum when $\gamma = 1/\sqrt{LC}$. What is the maximum amplitude?

55. Show that if L, R, E_0, and γ are constant, then the amplitude of the steady-state current in

Example 10 is a maximum when the capacitance is $C = 1/L\gamma^2$.

56. Find the charge on the capacitor and the current in an LC circuit when $L = 0.1$ h, $C = 0.1$ f, $E(t) = 100 \sin \gamma t$ V, $q(0) = 0$ C, and $i(0) = 0$ A.

57. Find the charge on the capacitor and the current in an LC circuit when $E(t) = E_0 \cos \gamma t$ V, $q(0) = q_0$ C, and $i(0) = i_0$ A.

58. In Problem 57 find the current when the circuit is in resonance.

5.2   LINEAR MODELS: BOUNDARY-VALUE PROBLEMS

INTRODUCTION: The preceding section was devoted to systems in which a second-order mathematical model was accompanied by initial conditions—that is, side conditions that are specified on the unknown function and its first derivative at a single point. But often the mathematical description of a physical system demands that we solve a homogeneous linear differential equation subject to boundary conditions—that is, conditions specified on the unknown function, or on one of its derivatives, or even on a linear combination of the unknown function and one of its derivatives, at two (or more) different points.

REVIEW MATERIAL: Review Problems 37–40 in Exercises 4.3 and Problems 37–40 in Exercises 4.4.

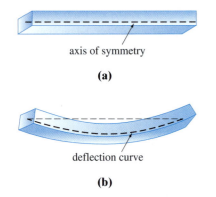

FIGURE 5.23   Deflection of a homogeneous beam

DEFLECTION OF A BEAM   Many structures are constructed by using girders or beams, and these beams deflect or distort under their own weight or under the influence of some external force. As we shall now see, this deflection $y(x)$ is governed by a relatively simple linear fourth-order differential equation.

To begin, let us assume that a beam of length L is homogeneous and has uniform cross sections along its length. In the absence of any load on the beam (including its weight), a curve joining the centroids of all its cross sections is a straight line called the **axis of symmetry.** See Figure 5.23(a). If a load is applied to the beam in a vertical plane containing the axis of symmetry, the beam, as shown in Figure 5.23(b), undergoes a distortion, and the curve connecting the centroids of all cross sections is called the **deflection curve** or **elastic curve.** The deflection curve approximates the shape of the beam. Now suppose that the x-axis coincides with the axis of symmetry and that the deflection $y(x)$, measured from this axis, is positive if downward. In the theory of elasticity it is shown that the bending moment $M(x)$ at a point x along the beam is related to the load per unit length $w(x)$ by the equation

$$\frac{d^2M}{dx^2} = w(x). \qquad (1)$$

In addition, the bending moment $M(x)$ is proportional to the curvature κ of the elastic curve

$$M(x) = EI\kappa, \qquad (2)$$

where E and I are constants; E is Young's modulus of elasticity of the material of the beam, and I is the moment of inertia of a cross section of the beam (about an axis known as the neutral axis). The product EI is called the **flexural rigidity** of the beam.

Now, from calculus, curvature is given by $\kappa = y''/[1 + (y')^2]^{3/2}$. When the deflection $y(x)$ is small, the slope $y' \approx 0$, and so $[1 + (y')^2]^{3/2} \approx 1$. If we let $\kappa \approx y''$, equation (2) becomes $M = EI\,y''$. The second derivative of this last expression is

$$\frac{d^2M}{dx^2} = EI\frac{d^2}{dx^2}\,y'' = EI\frac{d^4y}{dx^4}. \tag{3}$$

Using the given result in (1) to replace d^2M/dx^2 in (3), we see that the deflection $y(x)$ satisfies the fourth-order differential equation

$$EI\frac{d^4y}{dx^4} = w(x). \tag{4}$$

Boundary conditions associated with equation (4) depend on how the ends of the beam are supported. A cantilever beam is **embedded** or **clamped** at one end and **free** at the other. A diving board, an outstretched arm, an airplane wing, and a balcony are common examples of such beams, but even trees, flagpoles, skyscrapers, and the George Washington Monument can act as cantilever beams because they are embedded at one end and are subject to the bending force of the wind. For a cantilever beam the deflection $y(x)$ must satisfy the following two conditions at the embedded end $x = 0$:

- $y(0) = 0$ because there is no deflection, and
- $y'(0) = 0$ because the deflection curve is tangent to the x-axis (in other words, the slope of the deflection curve is zero at this point).

At $x = L$ the free-end conditions are

- $y''(L) = 0$ because the bending moment is zero, and
- $y'''(L) = 0$ because the shear force is zero.

The function $F(x) = dM/dx = EI\,d^3y/dx^3$ is called the shear force. If an end of a beam is **simply supported** or **hinged** (also called **pin supported** and **fulcrum supported**) then we must have $y = 0$ and $y'' = 0$ at that end. Table 5.1 summarizes the boundary conditions that are associated with (4). See Figure 5.24.

(a) embedded at both ends

$x = 0$ $x = L$

$x = 0$ $x = L$

(b) cantilever beam: embedded at the left end, free at the right end

$x = 0$ $x = L$

(c) simply supported at both ends

FIGURE 5.24 Beams with various end conditions

TABLE 5.1

Ends of the Beam	Boundary Conditions
embedded	$y = 0, \quad y' = 0$
free	$y'' = 0, \quad y''' = 0$
simply supported or hinged	$y = 0, \quad y'' = 0$

EXAMPLE 1   **An Embedded Beam**

A beam of length L is embedded at both ends. Find the deflection of the beam if a constant load w_0 is uniformly distributed along its length—that is, $w(x) = w_0$, $0 < x < L$.

SOLUTION   From (4) we see that the deflection $y(x)$ satisfies

$$EI\frac{d^4y}{dx^4} = w_0.$$

Because the beam is embedded at both its left end ($x = 0$) and its right end ($x = L$), there is no vertical deflection and the line of deflection is horizontal at these points. Thus the boundary conditions are

$$y(0) = 0, \quad y'(0) = 0, \quad y(L) = 0, \quad y'(L) = 0.$$

We can solve the nonhomogeneous differential equation in the usual manner (find y_c by observing that $m = 0$ is root of multiplicity four of the auxiliary equation $m^4 = 0$ and then find a particular solution y_p by undetermined coefficients) or we can simply integrate the equation $d^4y/dx^4 = w_0/EI$ four times in succession. Either way, we find the general solution of the equation $y = y_c + y_p$ to be

$$y(x) = c_1 + c_2 x + c_3 x^2 + c_4 x^3 + \frac{w_0}{24EI} x^4.$$

Now the conditions $y(0) = 0$ and $y'(0) = 0$ give, in turn, $c_1 = 0$ and $c_2 = 0$, whereas the remaining conditions $y(L) = 0$ and $y'(L) = 0$ applied to $y(x) = c_3 x^2 + c_4 x^3 + \frac{w_0}{24EI} x^4$ yield the simultaneous equations

$$c_3 L^2 + c_4 L^3 + \frac{w_0}{24EI} L^4 = 0$$

$$2c_3 L + 3c_4 L^2 + \frac{w_0}{6EI} L^3 = 0.$$

Solving this system gives $c_3 = w_0 L^2/24EI$ and $c_4 = -w_0 L/12EI$. Thus the deflection is

$$y(x) = \frac{w_0 L^2}{24EI} x^2 - \frac{w_0 L}{12EI} x^3 + \frac{w_0}{24EI} x^4$$

or $y(x) = \frac{w_0}{24EI} x^2 (x - L)^2$. By choosing $w_0 = 24EI$, and $L = 1$, we obtain the deflection curve in Figure 5.25.

FIGURE 5.25   Deflection curve for Example 1

EIGENVALUES AND EIGENFUNCTIONS   Many applied problems demand that we solve a two-point boundary-value problem (BVP) involving a linear differential equation that contains a parameter λ. We seek the values of λ for which the boundary-value problem has *nontrivial,* that is, *nonzero,* solutions.

EXAMPLE 2   Nontrivial Solutions of a BVP

Solve the boundary-value problem

$$y'' + \lambda y = 0, \quad y(0) = 0, \quad y(L) = 0.$$

SOLUTION   We shall consider three cases: $\lambda = 0$, $\lambda < 0$, and $\lambda > 0$.

CASE I: For $\lambda = 0$ the solution of $y'' = 0$ is $y = c_1 x + c_2$. The conditions $y(0) = 0$ and $y(L) = 0$ applied to this solution imply, in turn, $c_2 = 0$ and $c_1 = 0$. Hence for $\lambda = 0$ the only solution of the boundary-value problem is the trivial solution $y = 0$.

CASE II: For $\lambda < 0$ it is convenient to write $\lambda = -\alpha^2$, where α denotes a positive number. With this notation the roots of the auxiliary equation $m^2 - \alpha^2 = 0$ are $m_1 = \alpha$ and $m_2 = -\alpha$. Since the interval on which we are working is finite, we choose to write the general solution of $y'' - \alpha^2 y = 0$ as $y = c_1 \cosh \alpha x + c_2 \sinh \alpha x$. Now $y(0)$ is

$$y(0) = c_1 \cosh 0 + c_2 \sinh 0 = c_1 \cdot 1 + c_2 \cdot 0 = c_1,$$

and so $y(0) = 0$ implies that $c_1 = 0$. Thus $y = c_2 \sinh \alpha x$. The second condition $y(L) = 0$ demands that $c_2 \sinh \alpha L = 0$. For $\alpha \neq 0$, $\sinh \alpha L \neq 0$; consequently, we are forced to choose $c_2 = 0$. Again the only solution of the BVP is the trivial solution $y = 0$.

Note that we use hyperbolic functions here. Reread "Two Equations Worth Knowing" on page 145

CASE III: For $\lambda > 0$ we write $\lambda = \alpha^2$, where α is a positive number. Because the auxiliary equation $m^2 + \alpha^2 = 0$ has complex roots $m_1 = i\alpha$ and $m_2 = -i\alpha$, the general solution of $y'' + \alpha^2 y = 0$ is $y = c_1 \cos \alpha x + c_2 \sin \alpha x$. As before, $y(0) = 0$ yields $c_1 = 0$, and so $y = c_2 \sin \alpha x$. Now the last condition $y(L) = 0$, or

$$c_2 \sin \alpha L = 0,$$

is satisfied by choosing $c_2 = 0$. But this means that $y = 0$. If we require $c_2 \neq 0$, then $\sin \alpha L = 0$ is satisfied whenever αL is an integer multiple of π.

$$\alpha L = n\pi \quad \text{or} \quad \alpha = \frac{n\pi}{L} \quad \text{or} \quad \lambda_n = \alpha_n^2 = \left(\frac{n\pi}{L}\right)^2, \quad n = 1, 2, 3, \ldots.$$

Therefore for any real nonzero c_2, $y = c_2 \sin(n\pi x/L)$ is a solution of the problem for each n. Because the differential equation is homogeneous, any constant multiple of a solution is also a solution, so we may, if desired, simply take $c_2 = 1$. In other words, for each number in the sequence

$$\lambda_1 = \frac{\pi^2}{L^2}, \quad \lambda_2 = \frac{4\pi^2}{L^2}, \quad \lambda_3 = \frac{9\pi^2}{L^2}, \cdots,$$

the *corresponding* function in the sequence

$$y_1 = \sin\frac{\pi}{L}x, \quad y_2 = \sin\frac{2\pi}{L}x, \quad y_3 = \sin\frac{3\pi}{L}x, \cdots,$$

is a nontrivial solution of the original problem.

The numbers $\lambda_n = n^2\pi^2/L^2$, $n = 1, 2, 3, \ldots$ for which the boundary-value problem in Example 2 possesses nontrivial solutions are known as **eigenvalues.** The nontrivial solutions that depend on these values of λ_n, $y_n = c_2 \sin(n\pi x/L)$ or simply $y_n = \sin(n\pi x/L)$, are called **eigenfunctions.**

BUCKLING OF A THIN VERTICAL COLUMN In the eighteenth century Leonhard Euler was one of the first mathematicians to study an eigenvalue problem in analyzing how a thin elastic column buckles under a compressive axial force.

Consider a long slender vertical column of uniform cross-section and length L. Let $y(x)$ denote the deflection of the column when a constant vertical compressive force, or load, P is applied to its top, as shown in Figure 5.26. By comparing bending moments at any point along the column, we obtain

$$EI \frac{d^2 y}{dx^2} = -Py \quad \text{or} \quad EI \frac{d^2 y}{dx^2} + Py = 0, \tag{5}$$

where E is Young's modulus of elasticity and I is the moment of inertia of a cross-section about a vertical line through its centroid.

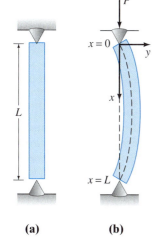

(a)   **(b)**

FIGURE 5.26 Elastic column buckling under a compressive force

EXAMPLE 3   **The Euler Load**

Find the deflection of a thin vertical homogeneous column of length L subjected to a constant axial load P if the column is hinged at both ends.

SOLUTION The boundary-value problem to be solved is

$$EI \frac{d^2 y}{dx^2} + Py = 0, \quad y(0) = 0, \quad y(L) = 0.$$

First note that $y = 0$ is a perfectly good solution of this problem. This solution has a simple intuitive interpretation: If the load P is not great enough, there is no deflection. The question then is this: For what values of P will the column bend? In mathematical terms: For what values of P does the given boundary-value problem possess nontrivial solutions?

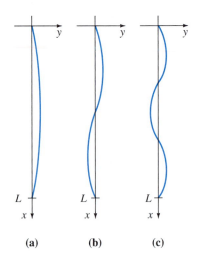

FIGURE 5.27 Deflection curves corresponding to compressive forces P_1, P_2, P_3

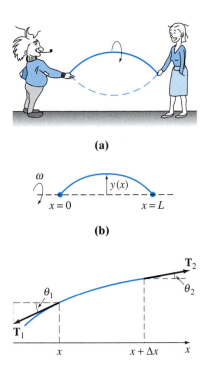

FIGURE 5.28 Rotating string and forces acting on it

By writing $\lambda = P/EI$, we see that

$$y'' + \lambda y = 0, \quad y(0) = 0, \quad y(L) = 0$$

is identical to the problem in Example 2. From Case III of that discussion we see that the deflections are $y_n(x) = c_2 \sin(n\pi x/L)$ corresponding to the eigenvalues $\lambda_n = P_n/EI = n^2\pi^2/L^2$, $n = 1, 2, 3, \ldots$. Physically, this means that the column will buckle or deflect only when the compressive force is one of the values $P_n = n^2\pi^2 EI/L^2$, $n = 1, 2, 3, \ldots$. These different forces are called **critical loads**. The deflection corresponding to the smallest critical load $P_1 = \pi^2 EI/L^2$, called the **Euler load**, is $y_1(x) = c_2 \sin(\pi x/L)$ and is known as the **first buckling mode**.

The deflection curves in Example 3 corresponding to $n = 1$, $n = 2$, and $n = 3$ are shown in Figure 5.27. Note that if the original column has some sort of physical restraint put on it at $x = L/2$, then the smallest critical load will be $P_2 = 4\pi^2 EI/L^2$, and the deflection curve will be as shown in Figure 5.27(b). If restraints are put on the column at $x = L/3$ and at $x = 2L/3$, then the column will not buckle until the critical load $P_3 = 9\pi^2 EI/L^2$ is applied, and the deflection curve will be as shown in Figure 5.27(c). See Problem 23 in Exercises 5.2.

ROTATING STRING   The simple linear second-order differential equation

$$y'' + \lambda y = 0 \tag{6}$$

occurs again and again as a mathematical model. In Section 5.1 we saw (6) in the forms $d^2x/dt^2 + (k/m)x = 0$ and $d^2q/dt^2 + (1/LC)q = 0$ as models for, respectively, the simple harmonic motion of a spring/mass system and the simple harmonic response of a series circuit. It is apparent when the model for the deflection of a thin column in (5) is written as $d^2y/dx^2 + (P/EI)y = 0$ that it is the same as (6). We encounter the basic equation (6) one more time in this section: as a model that defines the deflection curve or the shape $y(x)$ assumed by a rotating string. The physical situation is analogous to when two persons hold a jump rope and twirl it in a synchronous manner. See Figure 5.28(a) and 5.28(b).

Suppose a string of length L with constant linear density ρ (mass per unit length) is stretched along the x-axis and fixed at $x = 0$ and $x = L$. Suppose the string is then rotated about that axis at a constant angular speed ω. Consider a portion of the string on the interval $[x, x + \Delta x]$, where Δx is small. If the magnitude T of the tension $\mathbf{T}$, acting tangential to the string, is constant along the string, then the desired differential equation can be obtained by equating two different formulations of the net force acting on the string on the interval $[x, x + \Delta x]$. First, we see from Figure 5.28(c) that the net vertical force is

$$F = T \sin \theta_2 - T \sin \theta_1. \tag{7}$$

When angles θ_1 and θ_2 (measured in radians) are small, we have $\sin \theta_2 \approx \tan \theta_2$ and $\sin \theta_1 \approx \tan \theta_1$. Moreover, since $\tan \theta_2$ and $\tan \theta_1$ are, in turn, slopes of the lines containing the vectors $\mathbf{T}_2$ and $\mathbf{T}_1$, we can also write

$$\tan \theta_2 = y'(x + \Delta x) \quad \text{and} \quad \tan \theta_1 = y'(x).$$

Thus (7) becomes

$$F \approx T[y'(x + \Delta x) - y'(x)]. \tag{8}$$

Second, we can obtain a different form of this same net force using Newton's second law, $F = ma$. Here the mass of the string on the interval is $m = \rho \, \Delta x$; the centripetal acceleration of a body rotating with angular speed ω in a circle of radius r is $a = r\omega^2$. With Δx small we take $r = y$. Thus the net vertical force is also approximated by

$$F \approx -(\rho \, \Delta x)y\omega^2, \tag{9}$$

where the minus sign comes from the fact that the acceleration points in the direction opposite to the positive y-direction. Now by equating (8) and (9), we have

$$T[y'(x + \Delta x) - y'(x)] = -(\rho\Delta x)y\omega^2 \quad \text{or} \quad T\underbrace{\frac{y'(x + \Delta x) - y'(x)}{\Delta x}}_{\text{difference quotient}} + \rho\omega^2 y = 0. \quad (10)$$

For Δx close to zero the difference quotient in (10) is approximately the second derivative d^2y/dx^2. Finally, we arrive at the model

$$T\frac{d^2y}{dx^2} + \rho\omega^2 y = 0. \quad (11)$$

Since the string is anchored at its ends $x = 0$ and $x = L$, we expect that the solution $y(x)$ of equation (11) should also satisfy the boundary conditions $y(0) = 0$ and $y(L) = 0$.

REMARKS

(*i*) Eigenvalues are not always easily found, as they were in Example 2; you might have to approximate roots of equations such as $\tan x = -x$ or $\cos x \cosh x = 1$. See Problems 34–38 in Exercises 5.2.

(*ii*) Boundary conditions applied to a general solution of a linear differential equation can lead to a homogeneous algebraic system of linear equations in which the unknowns are the coefficients c_i in the general solution. A homogeneous algebraic system of linear equations is always consistent because it possesses at least a trivial solution. But a homogeneous system of n linear equations in n unknowns has a nontrivial solution if and only if the determinant of the coefficients equals zero. You might need to use this last fact in Problems 19 and 20 in Exercises 5.2.

EXERCISES 5.2

Answers to selected odd-numbered problems begin on page ANS-6.

DEFLECTION OF A BEAM

In Problems 1–5 solve equation (4) subject to the appropriate boundary conditions. The beam is of length L, and w_0 is a constant.

1. (a) The beam is embedded at its left end and free at its right end, and $w(x) = w_0, 0 < x < L$.
   (b) Use a graphing utility to graph the deflection curve when $w_0 = 24EI$ and $L = 1$.

2. (a) The beam is simply supported at both ends, and $w(x) = w_0, 0 < x < L$.
   (b) Use a graphing utility to graph the deflection curve when $w_0 = 24EI$ and $L = 1$.

3. (a) The beam is embedded at its left end and simply supported at its right end, and $w(x) = w_0, 0 < x < L$.
   (b) Use a graphing utility to graph the deflection curve when $w_0 = 48EI$ and $L = 1$.

4. (a) The beam is embedded at its left end and simply supported at its right end, and $w(x) = w_0 \sin(\pi x/L)$, $0 < x < L$.
   (b) Use a graphing utility to graph the deflection curve when $w_0 = 2\pi^3 EI$ and $L = 1$.
   (c) Use a root-finding application of a CAS (or a graphic calculator) to approximate the point in the graph in part (b) at which the maximum deflection occurs. What is the maximum deflection?

5. (a) The beam is simply supported at both ends, and $w(x) = w_0 x, 0 < x < L$.
   (b) Use a graphing utility to graph the deflection curve when $w_0 = 36EI$ and $L = 1$.

(c) Use a root-finding application of a CAS (or a graphic calculator) to approximate the point in the graph in part (b) at which the maximum deflection occurs. What is the maximum deflection?

6. (a) Find the maximum deflection of the cantilever beam in Problem 1.

(b) How does the maximum deflection of a beam that is half as long compare with the value in part (a)?

(c) Find the maximum deflection of the simply supported beam in Problem 2.

(d) How does the maximum deflection of the simply supported beam in part (c) compare with the value of maximum deflection of the embedded beam in Example 1?

7. A cantilever beam of length L is embedded at its right end, and a horizontal tensile force of P pounds is applied to its free left end. When the origin is taken at its free end, as shown in Figure 5.29, the deflection $y(x)$ of the beam can be shown to satisfy the differential equation

$$EIy'' = Py - w(x)\frac{x}{2}.$$

Find the deflection of the cantilever beam if $w(x) = w_0x,\ 0 < x < L$, and $y(0) = 0, y'(L) = 0$.

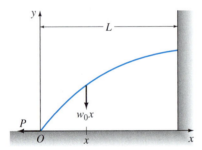

FIGURE 5.29   Deflection of cantilever beam in Problem 7

8. When a compressive instead of a tensile force is applied at the free end of the beam in Problem 7, the differential equation of the deflection is

$$EIy'' = -Py - w(x)\frac{x}{2}.$$

Solve this equation if $w(x) = w_0x,\ 0 < x < L$, and $y(0) = 0, y'(L) = 0$.

EIGENVALUES AND EIGENFUNCTIONS

In Problems 9–18 find the eigenvalues and eigenfunctions for the given boundary-value problem.

9. $y'' + \lambda y = 0,\quad y(0) = 0,\quad y(\pi) = 0$

10. $y'' + \lambda y = 0,\quad y(0) = 0,\quad y(\pi/4) = 0$

11. $y'' + \lambda y = 0,\quad y'(0) = 0,\quad y(L) = 0$

12. $y'' + \lambda y = 0,\quad y(0) = 0,\quad y'(\pi/2) = 0$

13. $y'' + \lambda y = 0,\quad y'(0) = 0,\quad y'(\pi) = 0$

14. $y'' + \lambda y = 0,\quad y(-\pi) = 0,\quad y(\pi) = 0$

15. $y'' + 2y' + (\lambda + 1)y = 0,\quad y(0) = 0,\quad y(5) = 0$

16. $y'' + (\lambda + 1)y = 0,\quad y'(0) = 0,\quad y'(1) = 0$

17. $x^2y'' + xy' + \lambda y = 0,\quad y(1) = 0,\quad y(e^\pi) = 0$

18. $x^2y'' + xy' + \lambda y = 0,\quad y'(e^{-1}) = 0,\quad y(1) = 0$

In Problems 19 and 20 find the eigenvalues and eigenfunctions for the given boundary-value problem. Consider only the case $\lambda = \alpha^4, \alpha > 0$.

19. $y^{(4)} - \lambda y = 0,\quad y(0) = 0,\quad y''(0) = 0,\quad y(1) = 0,$ $y''(1) = 0$

20. $y^{(4)} - \lambda y = 0,\quad y'(0) = 0,\quad y'''(0) = 0,\quad y(\pi) = 0,$ $y''(\pi) = 0$

BUCKLING OF A THIN COLUMN

21. Consider Figure 5.27. Where should physical restraints be placed on the column if we want the critical load to be P_4? Sketch the deflection curve corresponding to this load.

22. The critical loads of thin columns depend on the end conditions of the column. The value of the Euler load P_1 in Example 3 was derived under the assumption that the column was hinged at both ends. Suppose that a thin vertical homogeneous column is embedded at its base ($x = 0$) and free at its top ($x = L$) and that a constant axial load P is applied to its free end. This load either causes a small deflection δ as shown in Figure 5.30 or does not cause such a deflection.

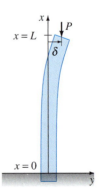

FIGURE 5.30   Deflection of vertical column in Problem 22

In either case the differential equation for the deflection $y(x)$ is

$$EI\frac{d^2y}{dx^2} + Py = P\delta.$$

(a) What is the predicted deflection when $\delta = 0$?

(b) When $\delta \neq 0$, show that the Euler load for this column is one-fourth of the Euler load for the hinged column in Example 3.

23. As was mentioned in Problem 22, the differential equation (5) that governs the deflection $y(x)$ of a thin elastic column subject to a constant compressive axial force P is valid only when the ends of the column are hinged. In general, the differential equation governing the deflection of the column is given by

$$\frac{d^2}{dx^2}\left(EI\frac{d^2y}{dx^2}\right) + P\frac{d^2y}{dx^2} = 0.$$

Assume that the column is uniform (EI is a constant) and that the ends of the column are hinged. Show that the solution of this fourth-order differential equation subject to the boundary conditions $y(0) = 0$, $y''(0) = 0$, $y(L) = 0$, $y''(L) = 0$ is equivalent to the analysis in Example 3.

24. Suppose that a uniform thin elastic column is hinged at the end $x = 0$ and embedded at the end $x = L$.

(a) Use the fourth-order differential equation given in Problem 23 to find the eigenvalues λ_n, the critical loads P_n, the Euler load P_1, and the deflections $y_n(x)$.

(b) Use a graphing utility to graph the first buckling mode.

ROTATING STRING

25. Consider the boundary-value problem introduced in the construction of the mathematical model for the shape of a rotating string:

$$T\frac{d^2y}{dx^2} + \rho\omega^2 y = 0, \quad y(0) = 0, \quad y(L) = 0.$$

For constant T and ρ, define the critical speeds of angular rotation ω_n as the values of ω for which the boundary-value problem has nontrivial solutions. Find the critical speeds ω_n and the corresponding deflections $y_n(x)$.

26. When the magnitude of tension T is not constant, then a model for the deflection curve or shape $y(x)$ assumed by a rotating string is given by

$$\frac{d}{dx}\left[T(x)\frac{dy}{dx}\right] + \rho\omega^2 y = 0.$$

Suppose that $1 < x < e$ and that $T(x) = x^2$.

(a) If $y(1) = 0$, $y(e) = 0$, and $\rho\omega^2 > 0.25$, show that the critical speeds of angular rotation are $\omega_n = \frac{1}{2}\sqrt{(4n^2\pi^2 + 1)/\rho}$ and the corresponding deflections are

$$y_n(x) = c_2 x^{-1/2} \sin(n\pi \ln x), \quad n = 1, 2, 3, \dots.$$

(b) Use a graphing utility to graph the deflection curves on the interval $[1, e]$ for $n = 1, 2, 3$. Choose $c_2 = 1$.

MISCELLANEOUS BOUNDARY-VALUE PROBLEMS

27. Temperature in a Sphere   Consider two concentric spheres of radius $r = a$ and $r = b$, $a < b$. See Figure 5.31. The temperature $u(r)$ in the region between the spheres is determined from the boundary-value problem

$$r\frac{d^2u}{dr^2} + 2\frac{du}{dr} = 0, \quad u(a) = u_0, \quad u(b) = u_1,$$

where u_0 and u_1 are constants. Solve for $u(r)$.

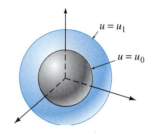

FIGURE 5.31   Concentric spheres in Problem 27

28. Temperature in a Ring   The temperature $u(r)$ in the circular ring shown in Figure 5.32 is determined from the boundary-value problem

$$r\frac{d^2u}{dr^2} + \frac{du}{dr} = 0, \quad u(a) = u_0, \quad u(b) = u_1,$$

where u_0 and u_1 are constants. Show that

$$u(r) = \frac{u_0 \ln(r/b) - u_1 \ln(r/a)}{\ln(a/b)}.$$

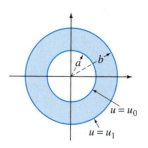

FIGURE 5.32   Circular ring in Problem 28

DISCUSSION/PROJECT PROBLEMS

29. **Simple Harmonic Motion** The model $mx'' + kx = 0$ for simple harmonic motion, discussed in Section 5.1, can be related to Example 2 of this section.

    Consider a free undamped spring/mass system for which the spring constant is, say, $k = 10$ lb/ft. Determine those masses m_n that can be attached to the spring so that when each mass is released at the equilibrium position at $t = 0$ with a nonzero velocity v_0, it will then pass through the equilibrium position at $t = 1$ second. How many times will each mass m_n pass through the equilibrium position in the time interval $0 < t < 1$?

30. **Damped Motion** Assume that the model for the spring/mass system in Problem 29 is replaced by $mx'' + 2x' + kx = 0$. In other words, the system is free but is subjected to damping numerically equal to 2 times the instantaneous velocity. With the same initial conditions and spring constant as in Problem 29, investigate whether a mass m can be found that will pass through the equilibrium position at $t = 1$ second.

In Problems 31 and 32 determine whether it is possible to find values y_0 and y_1 (Problem 31) and values of $L > 0$ (Problem 32) so that the given boundary-value problem has **(a)** precisely one nontrivial solution, **(b)** more than one solution, **(c)** no solution, **(d)** the trivial solution.

31. $y'' + 16y = 0, \quad y(0) = y_0, y(\pi/2) = y_1$

32. $y'' + 16y = 0, \quad y(0) = 1, y(L) = 1$

33. Consider the boundary-value problem

    $$y'' + \lambda y = 0, \quad y(-\pi) = y(\pi), \quad y'(-\pi) = y'(\pi).$$

    (a) The type of boundary conditions specified are called **periodic boundary conditions.** Give a geometric interpretation of these conditions.

    (b) Find the eigenvalues and eigenfunctions of the problem.

    (c) Use a graphing utility to graph some of the eigenfunctions. Verify your geometric interpretation of the boundary conditions given in part (a).

34. Show that the eigenvalues and eigenfunctions of the boundary-value problem

    $$y'' + \lambda y = 0, \quad y(0) = 0, \quad y(1) + y'(1) = 0$$

    are $\lambda_n = \alpha_n^2$ and $y_n = \sin \alpha_n x$, respectively, where α_n, $n = 1, 2, 3, \ldots$ are the consecutive positive roots of the equation $\tan \alpha = -\alpha$.

COMPUTER LAB ASSIGNMENTS

35. Use a CAS to plot graphs to convince yourself that the equation $\tan \alpha = -\alpha$ in Problem 34 has an infinite number of roots. Explain why the negative roots of the equation can be ignored. Explain why $\lambda = 0$ is not an eigenvalue even though $\alpha = 0$ is an obvious solution of the equation $\tan \alpha = -\alpha$.

36. Use a root-finding application of a CAS to approximate the first four eigenvalues $\lambda_1, \lambda_2, \lambda_3,$ and λ_4 for the BVP in Problem 34.

In Problems 37 and 38 find the eigenvalues and eigenfunctions of the given boundary-value problem. Use a CAS to approximate the first four eigenvalues $\lambda_1, \lambda_2, \lambda_3,$ and λ_4.

37. $y'' + \lambda y = 0, \quad y(0) = 0, \quad y(1) - \frac{1}{2} y'(1) = 0$

38. $y^{(4)} - \lambda y = 0, \quad y(0) = 0, y'(0) = 0, y(1) = 0, y'(1) = 0$
    (*Hint:* Consider only $\lambda = \alpha^4, \alpha > 0$.)

5.3 NONLINEAR MODELS

INTRODUCTION: In this section we examine some nonlinear higher-order mathematical models. We are able to solve some of these models using the substitution method (leading to reduction of the order of the DE) introduced on page 184. In some cases where the model cannot be solved, we show how a nonlinear DE can be replaced by a linear DE through a process called *linearization*.

REVIEW MATERIAL: A review of Section 4.9 is recommended.

NONLINEAR SPRINGS The mathematical model in (1) of Section 5.1 has the form

$$m \frac{d^2x}{dt^2} + F(x) = 0, \tag{1}$$

where $F(x) = kx$. Because x denotes the displacement of the mass from its equilibrium position, $F(x) = kx$ is Hooke's law—that is, the force exerted by the spring that tends to restore the mass to the equilibrium position. A spring acting under a linear restoring force $F(x) = kx$ is naturally referred to as a **linear spring.** But springs are seldom perfectly linear. Depending on how it is constructed and the material used, a spring can range from "mushy," or soft, to "stiff," or hard, so its restorative force may vary from something below to something above that given by the linear law. In the case of free motion, if we assume that a nonaging spring has some nonlinear characteristics, then it might be reasonable to assume that the restorative force of a spring—that is, $F(x)$ in (1)—is proportional to, say, the cube of the displacement x of the mass beyond its equilibrium position or that $F(x)$ is a linear combination of powers of the displacement such as that given by the nonlinear function $F(x) = kx + k_1 x^3$. A spring whose mathematical model incorporates a nonlinear restorative force, such as

$$m \frac{d^2x}{dt^2} + kx^3 = 0 \quad \text{or} \quad m \frac{d^2x}{dt^2} + kx + k_1 x^3 = 0, \tag{2}$$

is called a **nonlinear spring.** In addition, we examined mathematical models in which damping imparted to the motion was proportional to the instantaneous velocity dx/dt and the restoring force of a spring was given by the linear function $F(x) = kx$. But these were simply assumptions; in more realistic situations damping could be proportional to some power of the instantaneous velocity dx/dt. The nonlinear differential equation

$$m \frac{d^2x}{dt^2} + \beta \left| \frac{dx}{dt} \right| \frac{dx}{dt} + kx = 0 \tag{3}$$

is one model of a free spring/mass system in which the damping force is proportional to the square of the velocity. One can then envision other kinds of models: linear damping and nonlinear restoring force, nonlinear damping and nonlinear restoring force, and so on. The point is that nonlinear characteristics of a physical system lead to a mathematical model that is nonlinear.

Notice in (2) that both $F(x) = kx^3$ and $F(x) = kx + k_1 x^3$ are odd functions of x. To see why a polynomial function containing only odd powers of x provides a reasonable model for the restoring force, let us express F as a power series centered at the equilibrium position $x = 0$:

$$F(x) = c_0 + c_1 x + c_2 x^2 + c_3 x^3 + \cdots.$$

When the displacements x are small, the values of x^n are negligible for n sufficiently large. If we truncate the power series with, say, the fourth term, then $F(x) = c_0 + c_1 x + c_2 x^2 + c_3 x^3$. For the force at $x > 0$,

$$F(x) = c_0 + c_1 x + c_2 x^2 + c_3 x^3,$$

and for the force at $-x < 0$,

$$F(-x) = c_0 - c_1 x + c_2 x^2 - c_3 x^3$$

to have the same magnitude but act in the opposite direction, we must have $F(-x) = -F(x)$. Because this means that F is an odd function, we must have $c_0 = 0$ and $c_2 = 0$, and so $F(x) = c_1 x + c_3 x^3$. Had we used only the first two terms in the series, the same argument yields the linear function $F(x) = c_1 x$. A restoring force with mixed powers, such as $F(x) = c_1 x + c_2 x^2$, and the corresponding vibrations, are said to be unsymmetrical. In the next discussion we shall write $c_1 = k$ and $c_3 = k_1$.

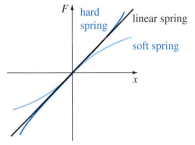

FIGURE 5.33   Hard and soft springs

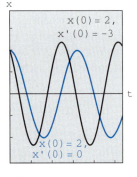

(a) hard spring

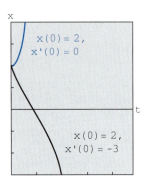

(b) soft spring

FIGURE 5.34   Numerical solution curves

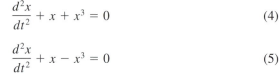

FIGURE 5.35   Simple pendulum

HARD AND SOFT SPRINGS   Let us take a closer look at the equation in (1) in the case in which the restoring force is given by $F(x) = kx + k_1x^3$, $k > 0$. The spring is said to be **hard** if $k_1 > 0$ and **soft** if $k_1 < 0$. Graphs of three types of restoring forces are illustrated in Figure 5.33. The next example illustrates these two special cases of the differential equation $m\ d^2x/dt^2 + kx + k_1x^3 = 0$, $m > 0$, $k > 0$.

EXAMPLE 1   **Comparison of Hard and Soft Springs**

The differential equations

$$\frac{d^2x}{dt^2} + x + x^3 = 0 \tag{4}$$

and

$$\frac{d^2x}{dt^2} + x - x^3 = 0 \tag{5}$$

are special cases of the second equation in (2) and are models of a hard spring and a soft spring, respectively. Figure 5.34(a) shows two solutions of (4) and Figure 5.34(b) shows two solutions of (5) obtained from a numerical solver. The curves shown in black are solutions that satisfy the initial conditions $x(0) = 2$, $x'(0) = -3$; the two curves in color are solutions that satisfy $x(0) = 2$, $x'(0) = 0$. These solution curves certainly suggest that the motion of a mass on the hard spring is oscillatory, whereas motion of a mass on the soft spring appears to be nonoscillatory. But we must be careful about drawing conclusions based on a couple of numerical solution curves. A more complete picture of the nature of the solutions of both of these equations can be obtained from the qualitative analysis discussed in Chapter 10. ▣

NONLINEAR PENDULUM   Any object that swings back and forth is called a **physical pendulum**. The **simple pendulum** is a special case of the physical pendulum and consists of a rod of length l to which a mass m is attached at one end. In describing the motion of a simple pendulum in a vertical plane, we make the simplifying assumptions that the mass of the rod is negligible and that no external damping or driving forces act on the system. The displacement angle θ of the pendulum, measured from the vertical as shown in Figure 5.35, is considered positive when measured to the right of OP and negative to the left of OP. Now recall the arc s of a circle of radius l is related to the central angle θ by the formula $s = l\theta$. Hence angular acceleration is

$$a = \frac{d^2s}{dt^2} = l\frac{d^2\theta}{dt^2}.$$

From Newton's second law we then have

$$F = ma = ml\frac{d^2\theta}{dt^2}.$$

From Figure 5.35 we see that the magnitude of the tangential component of the force due to the weight W is $mg \sin\theta$. In direction this force is $-mg \sin\theta$ because it points to the left for $\theta > 0$ and to the right for $\theta < 0$. We equate the two different versions of the tangential force to obtain $ml\ d^2\theta/dt^2 = -mg \sin\theta$, or

$$\frac{d^2\theta}{dt^2} + \frac{g}{l}\sin\theta = 0. \tag{6}$$

LINEARIZATION   Because of the presence of $\sin\theta$, the model in (6) is nonlinear. In an attempt to understand the behavior of the solutions of nonlinear higher-order differential equations, one sometimes tries to simplify the problem by replacing nonlinear terms by certain approximations. For example, the Maclaurin series for $\sin \theta$ is given by

$$\sin \theta = \theta - \frac{\theta^3}{3!} + \frac{\theta^5}{5!} - \cdots$$

so if we use the approximation $\sin \theta \approx \theta - \theta^3/6$, equation (6) becomes $d^2\theta/dt^2 + (g/l)\theta - (g/6l)\theta^3 = 0$. Observe that this last equation is the same as the second nonlinear equation in (2) with $m = 1$, $k = g/l$, and $k_1 = -g/6l$. However, if we assume that the displacements θ are small enough to justify using the replacement $\sin \theta \approx \theta$, then (6) becomes

$$\frac{d^2\theta}{dt^2} + \frac{g}{l}\theta = 0. \tag{7}$$

See Problem 22 in Exercises 5.3. If we set $\omega^2 = g/l$, we recognize (7) as the differential equation (2) of Section 5.1 that is a model for the free undamped vibrations of a linear spring/mass system. In other words, (7) is again the basic linear equation $y'' + \lambda y = 0$ discussed on page 214 of Section 5.2. As a consequence we say that equation (7) is a **linearization** of equation (6). Because the general solution of (7) is $\theta(t) = c_1 \cos \omega t + c_2 \sin \omega t$, this linearization suggests that for initial conditions amenable to small oscillations the motion of the pendulum described by (6) will be periodic.

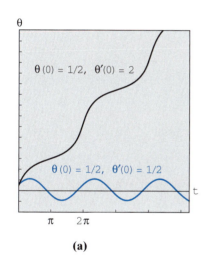

θ

$\theta(0) = 1/2, \quad \theta'(0) = 2$

$\theta(0) = 1/2, \quad \theta'(0) = 1/2$

t

$\pi \quad 2\pi$

(a)

(b) $\theta(0) = \frac{1}{2},$
$\theta'(0) = \frac{1}{2}$

(c) $\theta(0) = \frac{1}{2},$
$\theta'(0) = 2$

FIGURE 5.36 Oscillating pendulum in (b); whirling pendulum in (c)

| | EXAMPLE 2 | Two Initial-Value Problems |

The graphs in Figure 5.36(a) were obtained with the aid of a numerical solver and represent solution curves of equation (6) when $\omega^2 = 1$. The colored curve depicts the solution of (6) that satisfies the initial conditions $\theta(0) = \frac{1}{2}$, $\theta'(0) = \frac{1}{2}$, whereas the black curve is the solution of (6) that satisfies $\theta(0) = \frac{1}{2}$, $\theta'(0) = 2$. The colored curve represents a periodic solution—the pendulum oscillating back and forth as shown in Figure 5.36(b) with an apparent amplitude $A \leq 1$. The black curve shows that θ increases without bound as time increases—the pendulum, starting from the same initial displacement, is given an initial velocity of magnitude great enough to send it over the top; in other words, the pendulum is whirling about its pivot as shown in Figure 5.36(c). In the absence of damping, the motion in each case is continued indefinitely.

TELEPHONE WIRES The first-order differential equation $dy/dx = W/T_1$ is equation (17) of Section 1.3. This differential equation, established with the aid of Figure 1.26 on page 26 serves as a mathematical model for the shape of a flexible cable suspended between two vertical supports when the cable is carrying a vertical load. In Section 2.2 we solved this simple DE under the assumption that the vertical load carried by the cables of a suspension bridge was the weight of a horizontal roadbed distributed evenly along the x-axis. With $W = \rho x$, ρ the weight per unit length of the roadbed, the shape of each cable between the vertical supports turned out to be parabolic. We are now in a position to determine the shape of a uniform flexible cable hanging only under its own weight, such as a wire strung between two telephone posts. The vertical load is now the wire itself, and so if ρ is the linear density of the wire (measured, say, in pounds per feet) and s is the length of the segment P_1P_2 in Figure 1.26 then $W = \rho s$. Hence

$$\frac{dy}{dx} = \frac{\rho s}{T_1}. \tag{8}$$

Since the arc length between points P_1 and P_2 is given by

$$s = \int_0^x \sqrt{1 + \left(\frac{dy}{dx}\right)^2} \, dx, \tag{9}$$

it follows from the fundamental theorem of calculus that the derivative of (9) is

$$\frac{ds}{dx} = \sqrt{1 + \left(\frac{dy}{dx}\right)^2}.\tag{10}$$

Differentiating (8) with respect to x and using (10) lead to the second-order equation

$$\frac{d^2y}{dx^2} = \frac{\rho}{T_1}\frac{ds}{dx} \quad \text{or} \quad \frac{d^2y}{dx^2} = \frac{\rho}{T_1}\sqrt{1 + \left(\frac{dy}{dx}\right)^2}.\tag{11}$$

In the example that follows we solve (11) and show that the curve assumed by the suspended cable is a **catenary.** Before proceeding, observe that the nonlinear second-order differential equation (11) is one of those equations having the form $F(x, y', y'') = 0$ discussed in Section 4.9. Recall that we have a chance of solving an equation of this type by reducing the order of the equation by means of the substitution $u = y'$.

EXAMPLE 3   An Initial-Value Problem

From the position of the y-axis in Figure 1.26 it is apparent that initial conditions associated with the second differential equation in (11) are $y(0) = a$ and $y'(0) = 0$. If we substitute $u = y'$, then the equation in (11) becomes $\dfrac{du}{dx} = \dfrac{\rho}{T_1}\sqrt{1 + u^2}$. Separating variables, we find that

$$\int \frac{du}{\sqrt{1 + u^2}} = \frac{\rho}{T_1}\int dx \quad \text{gives} \quad \sinh^{-1}u = \frac{\rho}{T_1}x + c_1.$$

Now, $y'(0) = 0$ is equivalent to $u(0) = 0$. Since $\sinh^{-1} 0 = 0$, $c_1 = 0$, and so $u = \sinh(\rho x/T_1)$. Finally, by integrating both sides of

$$\frac{dy}{dx} = \sinh\frac{\rho}{T_1}x, \quad \text{we get} \quad y = \frac{T_1}{\rho}\cosh\frac{\rho}{T_1}x + c_2.$$

Using $y(0) = a$, $\cosh 0 = 1$, the last equation implies that $c_2 = a - T_1/\rho$. Thus we see that the shape of the hanging wire is given by $y = (T_1/\rho)\cosh(\rho x/T_1) + a - T_1/\rho$.

In Example 3, had we been clever enough at the start to choose $a = T_1/\rho$, then the solution of the problem would have been simply the hyperbolic cosine $y = (T_1/\rho)\cosh(\rho x/T_1)$.

ROCKET MOTION   In Section 1.3 we saw that the differential equation of a free-falling body of mass m near the surface of the earth is given by

$$m\frac{d^2s}{dt^2} = -mg, \quad \text{or simply} \quad \frac{d^2s}{dt^2} = -g,$$

where s represents the distance from the surface of the earth to the object and the positive direction is considered to be upward. In other words, the underlying assumption here is that the distance s to the object is small when compared with the radius R of the earth; put yet another way, the distance y from the center of the earth to the object is approximately the same as R. If, on the other hand, the distance y to the object, such as a rocket or a space probe, is large when compared to R, then we combine Newton's second law of motion and his universal law of gravitation to derive a differential equation in the variable y.

Suppose a rocket is launched vertically upward from the ground as shown in Figure 5.37. If the positive direction is upward and air resistance is ignored, then the differential equation of motion after fuel burnout is

$$m\frac{d^2y}{dt^2} = -k\frac{Mm}{y^2} \quad \text{or} \quad \frac{d^2y}{dt^2} = -k\frac{M}{y^2}, \tag{12}$$

where k is a constant of proportionality, y is the distance from the center of the earth to the rocket, M is the mass of the earth, and m is the mass of the rocket. To determine the constant k, we use the fact that when $y = R$, $kMm/R^2 = mg$ or $k = gR^2/M$. Thus the last equation in (12) becomes

$$\frac{d^2y}{dt^2} = -g\frac{R^2}{y^2}. \tag{13}$$

See Problem 14 in Exercises 5.3.

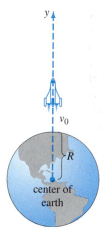

FIGURE 5.37 Distance to rocket is large compared to R.

VARIABLE MASS Notice in the preceding discussion that we described the motion of the rocket after it has burned all its fuel, when presumably its mass m is constant. Of course, during its powered ascent the total mass of the rocket varies as its fuel is being expended. The second law of motion, as originally advanced by Newton, states that when a body of mass m moves through a force field with velocity v, the time rate of change of the momentum mv of the body is equal to applied or net force F acting on the body:

$$F = \frac{d}{dt}(mv). \tag{14}$$

If m is constant, then (14) yields the more familiar form $F = m\, dv/dt = ma$, where a is acceleration. We use the form of Newton's second law given in (14) in the next example, in which the mass m of the body is variable.

EXAMPLE 4   **Chain Pulled Upward by a Constant Force**

A uniform 10-foot-long chain is coiled loosely on the ground. One end of the chain is pulled vertically upward by means of constant force of 5 pounds. The chain weighs 1 pound per foot. Determine the height of the end above ground level at time t. See Figure 1.37 and Problem 21 in Exercises 1.3.

SOLUTION   Let us suppose that $x = x(t)$ denotes the height of the end of the chain in the air at time t, $v = dx/dt$, and the positive direction is upward. For the portion of the chain that is in the air at time t we have the following variable quantities:

weight:	$W = (x \text{ ft}) \cdot (1 \text{ lb/ft}) = x,$
mass:	$m = W/g = x/32,$
net force:	$F = 5 - W = 5 - x.$

Thus from (14) we have

Product Rule

$$\frac{d}{dt}\left(\frac{x}{32}v\right) = 5 - x \quad \text{or} \quad x\frac{dv}{dt} + v\frac{dx}{dt} = 160 - 32x. \tag{15}$$

Because $v = dx/dt$, the last equation becomes

$$x\frac{d^2x}{dt^2} + \left(\frac{dx}{dt}\right)^2 + 32x = 160. \tag{16}$$

The nonlinear second-order differential equation (16) has the form $F(x, x', x'') = 0$, which is the second of the two forms considered in Section 4.9 that can possibly be solved by reduction of order. To solve (16), we revert back to (15) and use $v = x'$ along with the Chain Rule. From $\dfrac{dv}{dt} = \dfrac{dv}{dx}\dfrac{dx}{dt} = v\dfrac{dv}{dx}$ the second equation in (15) can be rewritten as

$$xv\frac{dv}{dx} + v^2 = 160 - 32x. \tag{17}$$

On inspection (17) might appear intractable, since it cannot be characterized as any of the first-order equations that were solved in Chapter 2. However, by rewriting (17) in differential form $M(x,v)dx + N(x,v)dv = 0$, we observe that, although the equation

$$(v^2 + 32x - 160)dx + xv\, dv = 0 \tag{18}$$

is not exact, it can be transformed into an exact equation by multiplying it by an integrating factor. From $(M_v - N_x)/N = 1/x$ we see from (13) of Section 2.4 that an integrating factor is $e^{\int dx/x} = e^{\ln x} = x$. When (18) is multiplied by $\mu(x) = x$, the resulting equation is exact (verify). By identifying $\partial f/\partial x = xv^2 + 32x^2 - 160x$, $\partial f/\partial v = x^2 v$ and then proceeding as in Section 2.4, we obtain

$$\frac{1}{2}x^2 v^2 + \frac{32}{3}x^3 - 80x^2 = c_1. \tag{19}$$

Since we have assumed that all of the chain is on the floor initially, we have $x(0) = 0$. This last condition applied to (19) yields $c_1 = 0$. By solving the algebraic equation $\frac{1}{2}x^2 v^2 + \frac{32}{3}x^3 - 80x^2 = 0$ for $v = dx/dt > 0$, we get another first-order differential equation,

$$\frac{dx}{dt} = \sqrt{160 - \frac{64}{3}x}.$$

The last equation can be solved by separation of variables. You should verify that

$$-\frac{3}{32}\left(160 - \frac{64}{3}x\right)^{1/2} = t + c_2. \tag{20}$$

This time the initial condition $x(0) = 0$ implies that $c_2 = -3\sqrt{10}/8$. Finally, by squaring both sides of (20) and solving for x, we arrive at the desired result,

$$x(t) = \frac{15}{2} - \frac{15}{2}\left(1 - \frac{4\sqrt{10}}{15}t\right)^2. \tag{21}$$

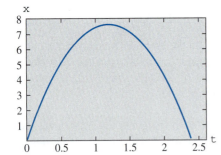

FIGURE 5.38   Graph of (21) for $x(t) \geq 0$

The graph of (21) given in Figure 5.38 should not, on physical grounds, be taken at face value. See Problem 15 in Exercises 5.3.

EXERCISES 5.3

Answers to selected odd-numbered problems begin on page ANS-7.

To the Instructor   In addition to Problems 24 and 25, all or portions of Problems 1–6, 8–13, 15, 17, 22, and 23 could serve as Computer Lab Assignments.

NONLINEAR SPRINGS

In Problems 1–4, the given differential equation is model of an undamped spring/mass system in which the restoring force $F(x)$ in (1) is nonlinear. For each equation use a numerical solver to plot the solution curves that satisfy the given

initial conditions. If the solutions appear to be periodic use
the solution curve to estimate the period T of oscillations.

1. $\dfrac{d^2x}{dt^2} + x^3 = 0$,

   $x(0) = 1, x'(0) = 1; \quad x(0) = \frac{1}{2}, x'(0) = -1$

2. $\dfrac{d^2x}{dt^2} + 4x - 16x^3 = 0$,

   $x(0) = 1, x'(0) = 1; \quad x(0) = -2, x'(0) = 2$

3. $\dfrac{d^2x}{dt^2} + 2x - x^2 = 0$,

   $x(0) = 1, x'(0) = 1; \quad x(0) = \frac{3}{2}, x'(0) = -1$

4. $\dfrac{d^2x}{dt^2} + xe^{0.01x} = 0$,

   $x(0) = 1, x'(0) = 1; \quad x(0) = 3, x'(0) = -1$

5. In Problem 3, suppose the mass is released from
   the initial position $x(0) = 1$ with an initial velocity
   $x'(0) = x_1$. Use a numerical solver to estimate the
   smallest value of $|x_1|$ at which the motion of the mass
   is nonperiodic.

6. In Problem 3, suppose the mass is released from an initial
   position $x(0) = x_0$ with the initial velocity $x'(0) = 1$. Use
   a numerical solver to estimate an interval $a \le x_0 \le b$ for
   which the motion is oscillatory.

7. Find a linearization of the differential equation in
   Problem 4.

8. Consider the model of an undamped nonlinear
   spring/mass system given by $x'' + 8x - 6x^3 + x^5 = 0$.
   Use a numerical solver to discuss the nature of the
   oscillations of the system corresponding to the initial
   conditions:

   $x(0) = 1, x'(0) = 1; \quad x(0) = -2, x'(0) = \frac{1}{2};$

   $x(0) = \sqrt{2}, x'(0) = 1; \quad x(0) = 2, x'(0) = \frac{1}{2};$

   $x(0) = 2, x'(0) = 0 \quad x(0) = -\sqrt{2}, x'(0) = -1.$

In Problems 9 and 10 the given differential equation is a
model of a damped nonlinear spring/mass system. Predict
the behavior of each system as $t \to \infty$. For each equation
use a numerical solver to obtain the solution curves satisfying the given initial conditions.

9. $\dfrac{d^2x}{dt^2} + \dfrac{dx}{dt} + x + x^3 = 0$,

   $x(0) = -3, x'(0) = 4; \quad x(0) = 0, x'(0) = -8$

10. $\dfrac{d^2x}{dt^2} + \dfrac{dx}{dt} + x - x^3 = 0$,

    $x(0) = 0, x'(0) = \frac{3}{2}; \quad x(0) = -1, x'(0) = 1$

11. The model $mx'' + kx + k_1x^3 = F_0\cos\omega t$ of an
    undamped periodically driven spring/mass system is

called **Duffing's differential equation.** Consider the
initial-value problem $x'' + x + k_1x^3 = 5\cos t$, $x(0) = 1$,
$x'(0) = 0$. Use a numerical solver to investigate the
behavior of the system for values of $k_1 > 0$ ranging from
$k_1 = 0.01$ to $k_1 = 100$. State your conclusions.

12. (a) Find values of $k_1 < 0$ for which the system in
       Problem 11 is oscillatory.
    (b) Consider the initial-value problem

    $$x'' + x + k_1x^3 = \cos\tfrac{3}{2}t, \ x(0) = 0, \ x'(0) = 0.$$

    Find values for $k_1 < 0$ for which the system is
    oscillatory.

NONLINEAR PENDULUM

13. Consider the model of the free damped nonlinear pen-
    dulum given by

    $$\dfrac{d^2\theta}{dt^2} + 2\lambda\dfrac{d\theta}{dt} + \omega^2\sin\theta = 0.$$

    Use a numerical solver to investigate whether the motion
    in the two cases $\lambda^2 - \omega^2 > 0$ and $\lambda^2 - \omega^2 < 0$ corre-
    sponds, respectively, to the overdamped and under-
    damped cases discussed in Section 5.1 for spring/mass
    systems. Choose appropriate initial conditions and values
    of λ and ω.

ROCKET MOTION

14. (a) Use the substitution $v = dy/dt$ to solve (13) for v in
       terms of y. Assuming that the velocity of the rocket
       at burnout is $v = v_0$ and $y \approx R$ at that instant, show
       that the approximate value of the constant c of
       integration is $c = -gR + \frac{1}{2}v_0^2$.
    (b) Use the solution for v in part (a) to show that the
       escape velocity of the rocket is given by $v_0 = \sqrt{2gR}$.
       (*Hint:* Take $y \to \infty$ and assume $v > 0$ for all time t.)
    (c) The result in part (b) holds for any body in the
       solar system. Use the values $g = 32$ ft/s² and
       $R = 4000$ mi to show that the escape velocity from
       the earth is (approximately) $v_0 = 25,000$ mi/h.
    (d) Find the escape velocity from the moon if the
       acceleration of gravity is $0.165g$ and $R = 1080$ mi.

VARIABLE MASS

15. (a) In Example 4, how much of the chain would you
       intuitively expect the constant 5-pound force to be
       able to lift?
    (b) What is the initial velocity of the chain?
    (c) Why is the time interval corresponding to $x(t) \ge 0$
       given in Figure 5.38, not the interval I of definition

of the solution (21)? Determine the interval I. How much chain is actually lifted? Explain any difference between this answer and your prediction in part (a).

16. A uniform chain of length L, measured in feet, is held vertically so that the lower end just touches the floor. The chain weighs 2 lb/ft. The upper end that is held is released from rest at $t = 0$ and the chain falls straight down. See Figure 1.38. As we saw in Problem 22 in Exercises 1.3, if $x(t)$ denotes the length of the chain on the floor at time t, air resistance is ignored, and the positive direction is taken to be downward, then

$$(L - x)\frac{d^2x}{dt^2} - \left(\frac{dx}{dt}\right)^2 = Lg.$$

(a) Solve for v in terms of x. Solve for x in terms of t. Express v in terms of t.
(b) Determine how long it takes for the chain to fall completely to the ground.
(c) What velocity does the model in part (a) predict for the upper end of the chain as it hits the ground?

17. A portion of a uniform chain of length 8 feet is loosely coiled around a peg at the edge of a high horizontal platform, and the remaining portion of the chain hangs at rest over the edge of the platform. Suppose that the length of the overhang is 3 feet and that the chain weighs 2 lb/ft. Starting at $t = 0$ the weight of the overhanging portion causes the chain on the platform to uncoil smoothly and fall to the floor.

(a) Ignore any resistive forces and assume that the positive direction is downward. If $x(t)$ denotes the length of the chain overhanging the platform at time $t > 0$ and $v = dx/dt$, find a differential equation that relates v to x.
(b) Proceed as in Example 4 and solve for v in terms of x by finding an appropriate integrating factor.
(c) Express time t in terms of x. Use a CAS as an aid in determining the time it takes for a 7-foot segment of chain to uncoil completely—that is, fall from the platform.

18. A portion of a uniform chain of length 8 feet lies stretched out on a high horizontal platform, and the remaining portion of the chain hangs over the edge of the platform as shown in Figure 5.39. Suppose the length of the overhang is 3 feet and that the chain weighs 2 lb/ft. The end of the chain on the platform is held until at $t = 0$ it is released from rest, and the chain begins to slide off the platform because of the weight of the overhanging portion.

(a) Ignore any resistive forces and assume that the positive direction is downward. If $x(t)$ denotes the length of the chain overhanging the platform at time $t > 0$ and $v = dx/dt$, show that v is related to x by the differential equation $v\dfrac{dv}{dx} = 4x$.

(b) Solve for v in terms of x. Solve for x in terms of t. Express v in terms of t.
(c) Approximate the time it takes for the rest of the chain to slide off the platform. Find the velocity at which the end of the chain leaves the edge of the platform.
(d) Suppose the chain is L feet long and weighs a total of W pounds. If the overhang at $t = 0$ is x_0 feet, show that the velocity at which the end of the chain leaves the edge of the platform is $v(L) = \sqrt{\dfrac{g}{L}(L^2 - x_0^2)}$.

FIGURE 5.39 Sliding chain in Problem 18

MISCELLANEOUS MATHEMATICAL MODELS

19. **Pursuit Curve** In a naval exercise a ship S_1 is pursued by a submarine S_2 as shown in Figure 5.40. Ship S_1 departs point $(0, 0)$ at $t = 0$ and proceeds along a straight-line course (the y-axis) at a constant speed v_1. The submarine S_2 keeps ship S_1 in visual contact, indicated by the straight dashed line L in the figure, while traveling at a constant speed v_2 along a curve C. Assume that ship S_2 starts at the point $(a, 0)$, $a > 0$, at $t = 0$ and that L is tangent to C. Determine a mathematical model that describes the curve C. Find an explicit solution of the differential equation. For convenience define $r = v_1/v_2$. Determine whether the paths of S_1 and S_2 will ever intersect by considering the cases $r > 1$, $r < 1$, and $r = 1$. (*Hint:* $\dfrac{dt}{dx} = \dfrac{dt}{ds}\dfrac{ds}{dx}$, where s is arc length measured along C.)

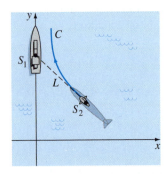

FIGURE 5.40 Pursuit curve in Problem 19

20. Pursuit Curve In another naval exercise a destroyer S_1 pursues a submerged submarine S_2. Suppose that S_1 at $(9, 0)$ on the x-axis detects S_2 at $(0, 0)$ and that S_2 simultaneously detects S_1. The captain of the destroyer S_1 assumes that the submarine will take immediate evasive action and conjectures that its likely new course is the straight line indicated in Figure 5.41. When S_1 is at $(3, 0)$, it changes from its straight-line course toward the origin to a pursuit curve C. Assume that the speed of the destroyer is, at all times, a constant 30 mi/h and that the submarine's speed is a constant 15 mi/h.

(a) Explain why the captain waits until S_1 reaches $(3, 0)$ before ordering a course change to C.

(b) Using polar coordinates, find an equation $r = f(\theta)$ for the curve C.

(c) Let T denote the time, measured from the initial detection, at which the destroyer intercepts the submarine. Find an upper bound for T.

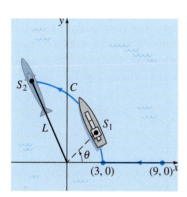

FIGURE 5.41 Pursuit curve in Problem 20

DISCUSSION/PROJECT PROBLEMS

21. Discuss why the damping term in equation (3) is written as

$$\beta \left| \frac{dx}{dt} \right| \frac{dx}{dt} \quad \text{instead of} \quad \beta \left(\frac{dx}{dt} \right)^2.$$

22. (a) Experiment with a calculator to find an interval $0 \le \theta < \theta_1$, where θ is measured in radians, for which you think $\sin \theta \approx \theta$ is a fairly good estimate. Then use a graphing utility to plot the graphs of $y = x$ and $y = \sin x$ on the same coordinate axes for $0 \le x \le \pi/2$. Do the graphs confirm your observations with the calculator?

(b) Use a numerical solver to plot the solution curves of the initial-value problems

$$\frac{d^2\theta}{dt^2} + \sin\theta = 0, \ \theta(0) = \theta_0, \ \theta'(0) = 0$$

and

$$\frac{d^2\theta}{dt^2} + \theta = 0, \ \theta(0) = \theta_0, \ \theta'(0) = 0$$

for several values of θ_0 in the interval $0 \le \theta < \theta_1$ found in part (a). Then plot solution curves of the initial-value problems for several values of θ_0 for which $\theta_0 > \theta_1$.

23. (a) Consider the nonlinear pendulum whose oscillations are defined by (6). Use a numerical solver as an aid to determine whether a pendulum of length l will oscillate faster on the earth or on the moon. Use the same initial conditions, but choose these initial conditions so that the pendulum oscillates back and forth.

(b) For which location in part (a) does the pendulum have greater amplitude?

(c) Are the conclusions in parts (a) and (b) the same when the linear model (7) is used?

COMPUTER LAB ASSIGNMENTS

24. Consider the initial-value problem

$$\frac{d^2\theta}{dt^2} + \sin\theta = 0, \quad \theta(0) = \frac{\pi}{12}, \quad \theta'(0) = -\frac{1}{3}$$

for a nonlinear pendulum. Since we cannot solve the differential equation, we can find no explicit solution of this problem. But suppose we wish to determine the first time $t_1 > 0$ for which the pendulum in Figure 5.35, starting from its initial position to the right, reaches the position OP—that is, the first positive root of $\theta(t) = 0$. In this problem and the next, we examine several ways to proceed.

(a) Approximate t_1 by solving the linear problem $d^2\theta/dt^2 + \theta = 0, \theta(0) = \pi/12, \theta'(0) = -\frac{1}{3}$.

(b) Use the method illustrated in Example 3 of Section 4.9 to find the first four nonzero terms of a Taylor series solution $\theta(t)$ centered at 0 for the nonlinear initial-value problem. Give the exact values of all coefficients.

(c) Use the first two terms of the Taylor series in part (b) to approximate t_1.

(d) Use the first three terms of the Taylor series in part (b) to approximate t_1.

(e) Use a root-finding application of a CAS (or a graphic calculator) and the first four terms of the Taylor series in part (b) to approximate t_1.

(f) In this part of the problem you are led through the commands in *Mathematica* that enable you to approximate the root t_1. The procedure is easily modified so that any root of $\theta(t) = 0$ can be approximated. (*If you do not have Mathematica, adapt the given procedure by finding the corresponding syntax for the CAS you have on hand.*) Precisely reproduce and then, in turn, execute each line in the given sequence of commands.

```
sol = NDSolve[{y"[t] + Sin[y[t]] == 0,
            y[0] == Pi/12, y'[0] == -1/3},
          y, {t, 0, 5}]//Flatten
solution = y[t]/.sol
Clear[y]
y[t_]: = Evaluate[solution]
y[t]
gr1 = Plot[y[t], {t, 0, 5}]
root = FindRoot[y[t] == 0, {t, 1}]
```

(g) Appropriately modify the syntax in part (f) and find the next two positive roots of $\theta(t) = 0$.

25. Consider a pendulum that is released from rest from an initial displacement of θ_0 radians. Solving the linear model (7) subject to the initial conditions $\theta(0) = \theta_0$, $\theta'(0) = 0$ gives $\theta(t) = \theta_0 \cos \sqrt{g/l}\, t$. The period of oscillations predicted by this model is given by the familiar formula $T = 2\pi/\sqrt{g/l} = 2\pi \sqrt{l/g}$. The interesting thing about this formula for T is that it does not depend on the magnitude of the initial displacement θ_0. In other words, the linear model predicts that the time it would take the pendulum to swing from an initial displacement of, say, $\theta_0 = \pi/2 \ (= 90°)$ to $-\pi/2$ and back again would be exactly the same as the time it would take to cycle from, say, $\theta_0 = \pi/360 \ (= 0.5°)$ to $-\pi/360$. This is intuitively unreasonable; the actual period must depend on θ_0.

If we assume that $g = 32$ ft/s^2 and $l = 32$ ft, then the period of oscillation of the linear model is $T = 2\pi$ s. Let us compare this last number with the period predicted by the nonlinear model when $\theta_0 = \pi/4$. Using a numerical solver that is capable of generating hard data, approximate the solution of

$$\frac{d^2\theta}{dt^2} + \sin\theta = 0, \quad \theta(0) = \frac{\pi}{4}, \quad \theta'(0) = 0$$

on the interval $0 \le t \le 2$. As in Problem 24, if t_1 denotes the first time the pendulum reaches the position OP in Figure 5.35, then the period of the nonlinear pendulum is $4t_1$. Here is another way of solving the equation $\theta(t) = 0$. Experiment with small step sizes and advance the time, starting at $t = 0$ and ending at $t = 2$. From your hard data, observe the time t_1 when $\theta(t)$ changes, for the first time, from positive to negative. Use the value t_1 to determine the true value of the period of the nonlinear pendulum. Compute the percentage relative error in the period estimated by $T = 2\pi$.

CHAPTER 5 IN REVIEW

Answers to selected odd-numbered problems begin on page ANS-7.

Answer Problems 1–8 without referring back to the text. Fill in the blank or answer true/false.

1. If a mass weighing 10 pounds stretches a spring 2.5 feet, a mass weighing 32 pounds will stretch it _____ feet.

2. The period of simple harmonic motion of mass weighing 8 pounds attached to a spring whose constant is 6.25 lb/ft is _____ seconds.

3. The differential equation of a spring/mass system is $x'' + 16x = 0$. If the mass is initially released from a point 1 meter above the equilibrium position with a downward velocity of 3 m/s, the amplitude of vibrations is _____ meters.

4. Pure resonance cannot take place in the presence of a damping force. _____

5. In the presence of a damping force, the displacements of a mass on a spring will always approach zero as $t \to \infty$. _____

6. A mass on a spring whose motion is critically damped can possibly pass through the equilibrium position twice. _____

7. At critical damping any increase in damping will result in an _____ system.

8. If simple harmonic motion is described by $x = (\sqrt{2}/2) \sin(2t + \phi)$, the phase angle ϕ is _____ when the initial conditions are $x(0) = -\frac{1}{2}$ and $x'(0) = 1$.

In Problems 9 and 10 the eigenvalues and eigenfunctions of the boundary-value problem $y'' + \lambda y = 0$, $y'(0) = 0$, $y'(\pi) = 0$ are $\lambda_n = n^2$, $n = 0, 1, 2, \ldots$, and $y = \cos nx$, respectively. Fill in the blanks.

9. A solution of the BVP when $\lambda = 8$ is $y = $ _____ because _____.

10. A solution of the BVP when $\lambda = 36$ is $y = $ _____ because _____.

11. A free undamped spring/mass system oscillates with a period of 3 seconds. When 8 pounds are removed from the spring, the system has a period of 2 seconds. What was the weight of the original mass on the spring?

12. A mass weighing 12 pounds stretches a spring 2 feet. The mass is initially released from a point 1 foot below the equilibrium position with an upward velocity of 4 ft/s.
   (a) Find the equation of motion.
   (b) What are the amplitude, period, and frequency of the simple harmonic motion?

(c) At what times does the mass return to the point 1 foot below the equilibrium position?

(d) At what times does the mass pass through the equilibrium position moving upward? Moving downward?

(e) What is the velocity of the mass at $t = 3\pi/16$ s?

(f) At what times is the velocity zero?

13. A force of 2 pounds stretches a spring 1 foot. With one end held fixed, a mass weighing 8 pounds is attached to the other end. The system lies on a table that imparts a frictional force numerically equal to $\frac{3}{2}$ times the instantaneous velocity. Initially, the mass is displaced 4 inches above the equilibrium position and released from rest. Find the equation of motion if the motion takes place along a horizontal straight line that is taken as the x-axis.

14. A mass weighing 32 pounds stretches a spring 6 inches. The mass moves through a medium offering a damping force that is numerically equal to β times the instantaneous velocity. Determine the values of $\beta > 0$ for which the spring/mass system will exhibit oscillatory motion.

15. A spring with constant $k = 2$ is suspended in a liquid that offers a damping force numerically equal to 4 times the instantaneous velocity. If a mass m is suspended from the spring, determine the values of m for which the subsequent free motion is nonoscillatory.

16. The vertical motion of a mass attached to a spring is described by the IVP $\frac{1}{4}x'' + x' + x = 0$, $x(0) = 4, x'(0) = 2$. Determine the maximum vertical displacement of the mass.

17. A mass weighing 4 pounds stretches a spring 18 inches. A periodic force equal to $f(t) = \cos \gamma t + \sin \gamma t$ is impressed on the system starting at $t = 0$. In the absence of a damping force, for what value of γ will the system be in a state of pure resonance?

18. Find a particular solution for $x'' + 2\lambda x' + \omega^2 x = A$, where A is a constant force.

19. A mass weighing 4 pounds is suspended from a spring whose constant is 3 lb/ft. The entire system is immersed in a fluid offering a damping force numerically equal to the instantaneous velocity. Beginning at $t = 0$, an external force equal to $f(t) = e^{-t}$ is impressed on the system. Determine the equation of motion if the mass is initially released from rest at a point 2 feet below the equilibrium position.

20. (a) Two springs are attached in series as shown in Figure 5.42. If the mass of each spring is ignored, show that the effective spring constant k of the system is defined by $1/k = 1/k_1 + 1/k_2$.

(b) A mass weighing W pounds stretches a spring $\frac{1}{2}$ foot and stretches a different spring $\frac{1}{4}$ foot. The two springs are attached, and the mass is then attached to the double spring as shown in Figure 5.42. Assume that the motion is free and that there is no damping force present. Determine the equation of motion if the mass is initially released at a point 1 foot below the equilibrium position with a downward velocity of $\frac{2}{3}$ ft/s.

(c) Show that the maximum speed of the mass is $\frac{2}{3}\sqrt{3g + 1}$.

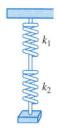

FIGURE 5.42 Attached springs in Problem 20

21. A series circuit contains an inductance of $L = 1$ h, a capacitance of $C = 10^{-4}$ f, and an electromotive force of $E(t) = 100 \sin 50t$ V. Initially, the charge q and current i are zero.

(a) Determine the charge $q(t)$.

(b) Determine the current $i(t)$.

(c) Find the times for which the charge on the capacitor is zero.

22. (a) Show that the current $i(t)$ in an LRC series circuit satisfies $L\dfrac{d^2 i}{dt^2} + R\dfrac{di}{dt} + \dfrac{1}{C}i = E'(t)$, where $E'(t)$ denotes the derivative of $E(t)$.

(b) Two initial conditions $i(0)$ and $i'(0)$ can be specified for the DE in part (a). If $i(0) = i_0$ and $q(0) = q_0$, what is $i'(0)$?

23. Consider the boundary-value problem

$$y'' + \lambda y = 0, \quad y(0) = y(2\pi), \quad y'(0) = y'(2\pi).$$

Show that except for the case $\lambda = 0$, there are two independent eigenfunctions corresponding to each eigenvalue.

24. A bead is constrained to slide along a frictionless rod of length L. The rod is rotating in a vertical plane with a constant angular velocity ω about a pivot P fixed at the midpoint of the rod, but the design of the pivot allows the bead to move along the entire length of the rod. Let $r(t)$ denote the position of the bead relative to this rotating coordinate system as shown in Figure 5.43. To apply Newton's second law of motion to this rotating frame of reference, it is necessary to use the fact that the net force acting on the bead is the sum of the real forces (in this case, the force due to gravity) and the inertial forces (coriolis,

transverse, and centrifugal). The mathematics is a little complicated, so we just give the resulting differential equation for r:

$$m\frac{d^2r}{dt^2} = m\omega^2 r - mg\sin\omega t.$$

(a) Solve the foregoing DE subject to the initial conditions $r(0) = r_0$, $r'(0) = v_0$.
(b) Determine the initial conditions for which the bead exhibits simple harmonic motion. What is the minimum length L of the rod for which it can accommodate simple harmonic motion of the bead?
(c) For initial conditions other than those obtained in part (b), the bead must eventually fly off the rod. Explain using the solution $r(t)$ in part (a).
(d) Suppose $\omega = 1$ rad/s. Use a graphing utility to graph the solution $r(t)$ for the initial conditions $r(0) = 0$, $r'(0) = v_0$, where v_0 is 0, 10, 15, 16, 16.1, and 17.
(e) Suppose the length of the rod is $L = 40$ ft. For each pair of initial conditions in part (d), use a root-finding application to find the total time that the bead stays on the rod.

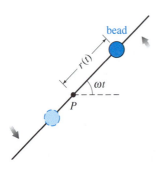

FIGURE 5.43   Rotating rod in Problem 24

25. Suppose a mass m lying on a flat, dry, frictionless surface is attached to the free end of a spring whose constant is k. In Figure 5.44(a) the mass is shown at the equilibrium position $x = 0$, that is, the spring is neither stretched nor compressed. As shown in Figure 5.44(b), the displacement $x(t)$ of the mass to the right of the equilibrium position is positive and negative to the left. Derive a differential equation for the free horizontal (sliding) motion of the mass. Discuss the difference between the derivation of this DE and the analysis leading to (1) of Section 5.1.

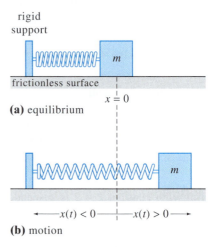

FIGURE 5.44   Sliding spring/mass system in Problem 25

26. What is the differential equation of motion in Problem 25 if kinetic friction (but no other damping forces) acts on the sliding mass? (*Hint:* Assume that the magnitude of the force of kinetic friction is $f_k = \mu mg$, where mg is the weight of the mass and the constant $\mu > 0$ is the coefficient of kinetic friction. Then consider two cases, $x' > 0$ and $x' < 0$. Interpret these cases physically.)

Smithsonian Institution, National Museum of American History

Tacoma Public Library, Richards Studio Collection TPL-562

Tacoma Public Library, Richards Studio Collection TPL-561

FIGURE 1   The Tacoma Narrows Bridge — before and after the collapse

You've probably seen the movies. The bridge begins to wobble from side to side. The oscillations get larger and larger. Leonard Coatsworth leaves his car with his dog Tubby inside and crawls on his hand and knees off the bridge to safety. Suddenly, the bridge collapses. The date is November 7, 1940, only four months after its grand opening. The collapse came as no surprise because the Tacoma Narrows Bridge,* or "Galloping Gertie," as it was fondly called by local residents, was notorious—even before it opened for traffic—for a swaying and vertical undulating motion of its roadway caused by the peculiar wind currents that pass through the narrows. See Figure 1.

After working through the problems in Chapter 5, you might suspect that resonance was the culprit. Somehow the forcing of the wind and the natural frequency of the suspension cables coincide, and thus the amplitude of the forced system (in this case the bridge) grew without bound, eventually causing the bridge to fall. But recent work by Lazer and McKenna suggests that the phenomenon that caused the bridge to fail was more complex than resonance.[†] Most of the models presented in their work are beyond the scope of this project. The **Still Curious?** section at the end of this project provides an introduction to one of their more elementary models.

In the previous edition of this text, Gilbert Lewis proposed a new model based on the work of Lazer and McKenna. This model illustrates one nonlinear mechanism that could have led to the demise of Galloping Gertie.

Imagine one cable of the suspension bridge hanging vertically. In some ways the cable is much like a very stiff spring. It has a natural length, the end of which we will place at $x = 0$. When a cable is stretched beyond this length ($x > 0$), it exerts a force in the upward direction; when compressed ($x < 0$), it exerts a force in the downward direction. But a cable is not a spring, and it is relatively easy to convince yourself that the upward restitution force that results from stretching the cable is greater than the downward force that comes from compressing it. In other words, we want to model the cable like a spring using Hooke's law with different spring constants depending on whether $x < 0$ or $x > 0$. Let a denote the

---

*The original Tacoma Narrows Bridge connected the city of Tacoma, Washington, and Gig Harbor, Washington.

[†]A.C. Lazer and P.J. McKenna, Large Amplitude Periodic Oscillations in Suspension Bridges: Some New Connections with Nonlinear Analysis, SIAM Review 32 (December 1990): 537–578.

spring constant for compression and b the spring constant for stretching so that $0 < a < b$. Define

$$F(x) = \begin{cases} ax, & x < 0 \\ bx, & x \geq 0. \end{cases} \tag{1}$$

In the absence of damping, our differential equation then becomes

$$mx'' + F(x) = g(t), \tag{2}$$

where the function $g(t)$ represents the forcing due to the wind.

PROBLEM 1 (CD).   To illustrate the core issue underlying this phenomenon, suppose in (2) that $m = 1$, $a = 1$, $b = 4$, $g(t) = \sin 4t$, and initial conditions $x(0) = 0$, $x'(0) = \alpha$. Use the **Tacoma Bridge Tool** on the *DE Tools* CD to plot solutions of (2) on the interval $0 \leq t \leq 100$ for a variety of values of α. What do you observe and what might these observations say about the fate of the bridge?

As was mentioned above, the model given in (2) is nonlinear. But the nonlinearity arises because $F(x)$ is piecewise linear. Thus we can find partial solutions that are defined over time intervals where the solution $x(t)$ does not change sign.

PROBLEM 2.   Use the parameters and initial conditions given in Problem 1 but assume that $\alpha > 0$. If $F(x) = 4x$ for t small, show that

$$x(t) = \frac{1}{6} \sin 2t \, [3\alpha + 1 - \cos 2t]$$

is a solution of (2) on the interval $0 \leq t \leq \pi/2$. Note that in addition to showing that $x(t)$ satisfies the differential equation and initial conditions you must also show that $x(\pi/2) = 0$. Plot this function on the given time interval. Find $x'(\pi/2)$ and show that it is negative.

For the next time interval we use $F(x) = x$.

PROBLEM 3.   Use the values of $x(\pi/2)$ and $x'(\pi/2)$ obtained in Problem 2 and show that

$$x(t) = \cos t \left[\left(\alpha + \frac{2}{5} \right) - \frac{4}{15} \sin t \cos 2t \right]$$

is a solution of (2) on the interval $\pi/2 \leq t \leq 3\pi/2$.

We refer to each solution piece as a **cycle**. Thus each cycle corresponds to either a positive or a negative displacement of the bridge.

PROBLEM 4 (CAS).   Use your solutions in Problems 2 and 3 to plot $x(t)$ on the interval $0 \leq t \leq 3\pi/2$. How does the velocity at $t = 3\pi/2$ compare with the velocity at $t = 0$?

Clearly, we could continue in this manner, successively solving the appropriate linear differential equation using the initial conditions $x = 0$ and a new value

x' calculated from the final x' of the previous solution. A careful analysis of this reveals that the amplitude of each successive cycle increases by a constant rate proportional to $\frac{2}{15}$. Thus as $t \to \infty$, the amplitudes of the oscillations grow without bound. This is what you should have observed in Problem 1.

It is instructive to see what happens for some other values of a and b in (1).

> **PROBLEM 5 (CD).**   Consider the three cases of (1) where $b = 1$, $a = 4$; $b = 64$, $a = 4$; and $b = 36$, $a = 25$. Notice that in the first case the condition $0 < a < b$ is not satisfied. Again using $g(t) = \sin 4t$, $m = 1$, and initial conditions $x(0) = 0$, $x'(0) = 1$, plot the solution of (2) on the interval $0 \le t \le 100$ in each of the three cases. Describe the long-term behavior of $x(t)$ in each of the three cases.

STILL CURIOUS?

The paper by Lazer and McKenna referred to above presents a somewhat different model. In this section we briefly discuss that model. Consider the differential equation

$$x'' + \beta x' + F(x) = -g + \lambda \sin \omega t \qquad (3)$$

where β is the damping constant, g is the acceleration due to gravity, and the function F is defined in (1). The new features in this model are the damping term $\beta x'$ and the forcing term $\lambda \sin \omega t$ due to the wind. In the next two problems we take $a = 17$, $b = 13$, $\beta = 0.01$, and $g = 10$.

> **PROBLEM 6 (CD).**   Let's begin by seeing what happens when $\lambda = 0$. This represents the state of the bridge when the wind is not blowing. Plot the solution of (3) with initial conditions $x(0) = x_0 < 0$, $x'(0) = 0$ on the interval $0 \le t \le 50$ and describe what happens to the bridge.

> **PROBLEM 7 (CD).**   Now let's see what might happen when the wind blows. Let $\omega = 4$ and $\lambda = 0.04$. Plot solutions of (3) with initial conditions $x(0) = -\frac{10}{17}$, $x'(0) = 0$ on the interval $0 \le t \le 50$ and describe what happens to the bridge. Does $x(t)$ ever cross the t-axis? What part of the piecewise defined function $F(x)$ is relevant in this case? What are the physical implications of this property?

In Problem 7 you probably noticed that $x(t) < 0$ for all t. Thus only one "piece" of $F(x)$ is relevant. In other words, we are really solving the differential equation

$$x'' + 0.01x' + 17x = -10 + 0.04 \sin 4t. \qquad (4)$$

> **PROBLEM 8 (CAS).**   Solve the differential equation in (4) with the initial conditions $x(0) = x_0 < 0$, $x'(0) = 0$. Indicate the transient and steady-state terms as $t \to \infty$.

In Problem 7 the small value of λ means that the wind is not blowing very hard. Let's see what happens as this parameter grows and the wind begins to howl.

PROBLEM 9 (CD). Let $\lambda = 0.2$ and $x'(0) = 0$. For each of the initial displacements $x(0) = -0.5, -0.4, \ldots, 0.4, 0.5$, plot solution curves on the interval $0 \le t \le 50$. Describe what happens to the oscillations as the values of $x(0)$ increase. Interpret this in terms of the bridge system being modeled.

THE REST OF THE STORY

The bridge over the Tacoma Narrows was eventually rebuilt. It opened in October 1950, 10 years after the collapse. Notice in Figure 2 that the original bridge, whose fatal flaw was its light and graceful design, was replaced by one whose span was greatly stiffened by truss work. But the "new" 1950 bridge is outdated by current highway standards; a new suspension bridge is currently under construction parallel to the existing bridge. See Figure 3.

FIGURE 2 Tacoma Narrows Bridge rebuilt in 1950

FIGURE 3 Design of parallel bridges; the addition is scheduled to open in 2007

6

SERIES SOLUTIONS OF LINEAR EQUATIONS

Up to now we have primarily solved linear DEs with constant coefficients. The only exception was the Cauchy-Euler equation. In applications, higher-order DEs with variable coefficients are just as important as, if not more important than, DEs with constant coefficients. As was mentioned in Section 4.7, even a simple linear second-order DE with variable coefficients such as $y'' + xy = 0$ does not possess elementary solutions. Nevertheless, we *can* find two linearly independent solutions of $y'' + xy = 0$, but as we shall see in this chapter, the solutions of this equation are defined by infinite series.

Bessel Functions of the First Kind

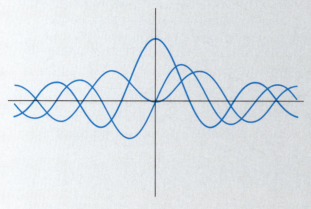

Bessel Functions of the Second Kind

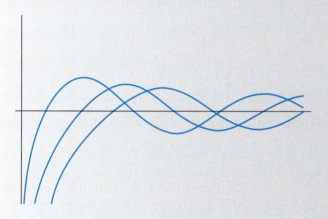

Legendre Polynomials

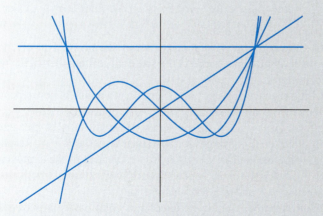

Special functions often derive from the solution of linear DEs with variable coefficients. Illustrated are families of solutions of two important DEs considered in Section 6.3. See pages 260, 261, and 267.

6.1 SOLUTIONS ABOUT ORDINARY POINTS

INTRODUCTION: In Section 4.3 we saw that solving a homogeneous linear DE with *constant coefficients* was essentially a problem in algebra. By finding the roots of the auxiliary equation we could write a general solution of the DE as a linear combination of the elementary functions x^k, $x^k e^{\alpha x}$, $x^k e^{\alpha x} \cos \beta x$, and $x^k e^{\alpha x} \sin \beta x$, k a nonnegative integer. But as pointed out in the introduction to Section 4.7, *most* linear higher-order DEs with *variable coefficients* cannot be solved in terms of elementary functions. A usual course of action for equations of this sort is to assume a solution in the form of infinite series and proceed in a manner similar to the method of undetermined coefficients (Section 4.4). In this section we consider linear second-order DEs with variable coefficients which possess solutions in the form of *power series.*

REVIEW MATERIAL: Although we begin with a brief review of power series, you should consult a standard calculus text for a more in-depth coverage of fundamental notions such as convergence, divergence, absolute convergence of an infinite series, and the interval and radius of convergence of a power series.

6.1.1 REVIEW OF POWER SERIES

Recall from calculus that a power series in $x - a$ is an infinite series of the form

$$\sum_{n=0}^{\infty} c_n(x-a)^n = c_0 + c_1(x-a) + c_2(x-a)^2 + \cdots.$$

Such a series is also said to be a **power series centered at a.** For example, the power series $\sum_{n=0}^{\infty}(x+1)^n$ is centered at $a = -1$. In this section we are concerned mainly with power series in x, in other words, power series such as $\sum_{n=1}^{\infty} 2^{n-1} x^n = x + 2x^2 + 4x^3 + \cdots$ that are centered at $a = 0$. The following list summarizes some important facts about power series.

- **Convergence** A power series $\sum_{n=0}^{\infty} c_n(x-a)^n$ is convergent at a specified value of x if its sequence of partial sums $\{S_N(x)\}$ converges—that is, $\lim_{N \to \infty} S_N(x) = \lim_{N \to \infty} \sum_{n=0}^{N} c_n(x-a)^n$ exists. If the limit does not exist at x, then the series is said to be divergent.

- **Interval of Convergence** Every power series has an interval of convergence. The interval of convergence is the set of all real numbers x for which the series converges.

- **Radius of Convergence** Every power series has a radius of convergence R. If $R > 0$, then the power series $\sum_{n=0}^{\infty} c_n(x-a)^n$ converges for $|x - a| < R$ and diverges for $|x - a| > R$. If the series converges only at its center a, then $R = 0$. If the series converges for all x, then we write $R = \infty$. Recall that the absolute-value inequality $|x - a| < R$ is equivalent to the simultaneous inequality $a - R < x < a + R$. A power series might or might not converge at the endpoints $a - R$ and $a + R$ of this interval.

- **Absolute Convergence** Within its interval of convergence a power series converges absolutely. In other words, if x is a number in the interval of convergence and is not an endpoint of the interval, then the series of absolute values $\sum_{n=0}^{\infty} |c_n(x-a)^n|$ converges. See Figure 6.1.

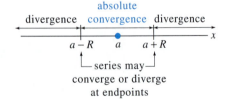

FIGURE 6.1 Absolute convergence within the interval of convergence and divergence outside of this interval

- **Ratio Test** Convergence of a power series can often be determined by the ratio test. Suppose that $c_n \neq 0$ for all n and that

$$\lim_{n \to \infty} \left| \frac{c_{n+1}(x-a)^{n+1}}{c_n(x-a)^n} \right| = |x-a| \lim_{n \to \infty} \left| \frac{c_{n+1}}{c_n} \right| = L.$$

If $L < 1$, the series converges absolutely; if $L > 1$, the series diverges; and if $L = 1$, the test is inconclusive. For example, for the power series $\sum_{n=1}^{\infty} (x-3)^n / 2^n n$ the ratio test gives

$$\lim_{n \to \infty} \left| \frac{\dfrac{(x-3)^{n+1}}{2^{n+1}(n+1)}}{\dfrac{(x-3)^n}{2^n n}} \right| = |x-3| \lim_{n \to \infty} \frac{n}{2(n+1)} = \frac{1}{2}|x-3|;$$

the series converges absolutely for $\frac{1}{2}|x-3| < 1$ or $|x-3| < 2$ or $1 < x < 5$. This last interval is referred to as the *open* interval of convergence. The series diverges for $|x-3| > 2$—that is, for $x > 5$ or $x < 1$. At the left endpoint $x = 1$ of the open interval of convergence, the series of constants $\sum_{n=1}^{\infty} ((-1)^n/n)$ is convergent by the alternating series test. At the right endpoint $x = 5$, the series $\sum_{n=1}^{\infty} (1/n)$ is the divergent harmonic series. The interval of convergence of the series is $[1, 5)$, and the radius of convergence is $R = 2$.

- **A Power Series Defines a Function** A power series defines a function $f(x) = \sum_{n=0}^{\infty} c_n(x-a)^n$ whose domain is the interval of convergence of the series. If the radius of convergence is $R > 0$, then f is continuous, differentiable, and integrable on the interval $(a - R, a + R)$. Moreover, $f'(x)$ and $\int f(x)\,dx$ can be found by term-by-term differentiation and integration. Convergence at an endpoint may be either lost by differentiation or gained through integration. If $y = \sum_{n=0}^{\infty} c_n x^n$ is a power series in x, then the first two derivatives are $y' = \sum_{n=0}^{\infty} n x^{n-1}$ and $y'' = \sum_{n=0}^{\infty} n(n-1)x^{n-2}$. Notice that the first term in the first derivative and the first two terms in the second derivative are zero. We omit these zero terms and write

$$y' = \sum_{n=1}^{\infty} c_n n x^{n-1} \quad \text{and} \quad y'' = \sum_{n=2}^{\infty} c_n n(n-1)x^{n-2}. \tag{1}$$

These results are important and will be used shortly.

- **Identity Property** If $\sum_{n=0}^{\infty} c_n(x-a)^n = 0$, $R > 0$ for all numbers x in the interval of convergence, then $c_n = 0$ for all n.

- **Analytic at a Point** A function f is analytic at a point a if it can be represented by a power series in $x - a$ with a positive or infinite radius of convergence. In calculus it is seen that functions such as e^x, $\cos x$, $\sin x$, $\ln(1 - x)$, and so on can be represented by Taylor series. Recall, for example, that

$$e^x = 1 + \frac{x}{1!} + \frac{x^2}{2!} + \cdots, \quad \sin x = x - \frac{x^3}{3!} + \frac{x^5}{5!} - \cdots, \quad \cos x = 1 - \frac{x^2}{2!} + \frac{x^4}{4!} - \frac{x^6}{6!} + \cdots \tag{2}$$

for $|x| < \infty$. These Taylor series centered at 0, called Maclaurin series, show that e^x, $\sin x$, and $\cos x$ are analytic at $x = 0$.

- **Arithmetic of Power Series** Power series can be combined through the operations of addition, multiplication, and division. The procedures for power series are similar to those by which two polynomials are added, multiplied, and divided—that is, we add coefficients of like powers of x, use the distributive law and collect like terms, and perform long division. For example, using the series in (2), we have

$$e^x \sin x = \left(1 + x + \frac{x^2}{2} + \frac{x^3}{6} + \frac{x^4}{24} + \cdots\right)\left(x - \frac{x^3}{6} + \frac{x^5}{120} - \frac{x^7}{5040} + \cdots\right)$$

$$= (1)x + (1)x^2 + \left(-\frac{1}{6} + \frac{1}{2}\right)x^3 + \left(-\frac{1}{6} + \frac{1}{6}\right)x^4 + \left(\frac{1}{120} - \frac{1}{12} + \frac{1}{24}\right)x^5 + \cdots$$

$$= x + x^2 + \frac{x^3}{3} - \frac{x^5}{30} - \cdots.$$

Since the power series for e^x and $\sin x$ converge for $|x| < \infty$, the product series converges on the same interval. Problems involving multiplication or division of power series can be done with minimal fuss by using a computer algebra system.

SHIFTING THE SUMMATION INDEX   For the remainder of this section, as well as this chapter, it is important that you become adept at simplifying the sum of two or more power series, each expressed in summation (sigma) notation, to an expression with a single Σ. As the next example illustrates, combining two or more summations as a single summation often requires a reindexing—that is, a shift in the index of summation.

EXAMPLE 1   **Adding Two Power Series**

Write $\sum_{n=2}^{\infty} n(n-1)c_n x^{n-2} + \sum_{n=0}^{\infty} c_n x^{n+1}$ as a single power series whose general term involves x^k.

SOLUTION   To add the two series, it is necessary that both summation indices start with the same number and the powers of x in each series be "in phase"; that is, if one series starts with a multiple of, say, x to the first power, then we want the other series to start with the same power. Note that in the given problem the first series starts with x^0 whereas the second series starts with x^1. By writing the first term of the first series outside the summation notation,

series starts
with x
for $n = 3$ ↓   series starts
with x
for $n = 0$ ↓

$$\sum_{n=2}^{\infty} n(n-1)c_n x^{n-2} + \sum_{n=0}^{\infty} c_n x^{n+1} = 2 \cdot 1 c_2 x^0 + \sum_{n=3}^{\infty} n(n-1)c_n x^{n-2} + \sum_{n=0}^{\infty} c_n x^{n+1},$$

we see that both series on the right-hand side start with the same power of x—namely, x^1. Now to get the same summation index, we are inspired by the exponents of x; we let $k = n - 2$ in the first series and at the same time let $k = n + 1$ in the second series. The right-hand side becomes

same

$$2c_2 + \sum_{k=1}^{\infty} (k+2)(k+1)c_{k+2} x^k + \sum_{k=1}^{\infty} c_{k-1} x^k. \tag{3}$$

same

Remember that the summation index is a "dummy" variable; the fact that $k = n - 1$ in one case and $k = n + 1$ in the other should cause no confusion if you keep in mind that it is the *value* of the summation index that is important. In both cases k takes on the same successive values $k = 1, 2, 3, \ldots$ when n takes on the values $n = 2, 3, 4, \ldots$ for $k = n - 1$ and $n = 0, 1, 2, \ldots$ for $k = n + 1$. We are now in a position to add the series in (3) term by term:

$$\sum_{n=2}^{\infty} n(n-1)c_n x^{n-2} + \sum_{n=0}^{\infty} c_n x^{n+1} = 2c_2 + \sum_{k=1}^{\infty} [(k+2)(k+1)c_{k+2} + c_{k-1}]x^k. \tag{4}$$

If you are not convinced of the result in (4), then write out a few terms on both sides of the equality.

6.1.2 POWER SERIES SOLUTIONS

A DEFINITION Suppose the linear second-order differential equation

$$a_2(x)y'' + a_1(x)y' + a_0(x)y = 0 \qquad (5)$$

is put into standard form

$$y'' + P(x)y' + Q(x)y = 0 \qquad (6)$$

by dividing by the leading coefficient $a_2(x)$. We have the following definition.

> **DEFINITION 6.1**   **Ordinary and Singular Points**
>
> A point x_0 is said to be an **ordinary point** of the differential equation (5) if both $P(x)$ and $Q(x)$ in the standard form (6) are analytic at x_0. A point that is not an ordinary point is said to be a **singular point** of the equation.

Every finite value of x is an ordinary point of the differential equation $y'' + (e^x)y' + (\sin x)y = 0$. In particular, $x = 0$ is an ordinary point because, as we have already seen in (2), both e^x and $\sin x$ are analytic at this point. The negation in the second sentence of Definition 6.1 stipulates that if at least one of the functions $P(x)$ and $Q(x)$ in (6) fails to be analytic at x_0, then x_0 is a singular point. Note that $x = 0$ is a singular point of the differential equation $y'' + (e^x)y' + (\ln x)y = 0$ because $Q(x) = \ln x$ is discontinuous at $x = 0$ and so cannot be represented by a power series in x.

POLYNOMIAL COEFFICIENTS We shall be interested primarily in the case when (5) has polynomial coefficients. A polynomial is analytic at any value x, and a rational function is analytic *except* at points where its denominator is zero. Thus if $a_2(x)$, $a_1(x)$, and $a_0(x)$ are polynomials with no common factors, then both rational functions $P(x) = a_1(x)/a_2(x)$ and $Q(x) = a_0(x)/a_2(x)$ are analytic except where $a_2(x) = 0$. It follows, then, that $x = x_0$ is an ordinary point of (5) if $a_2(x_0) \neq 0$ whereas $x = x_0$ is a singular point of (5) if $a_2(x_0) = 0$. For example, the only singular points of the equation $(x^2 - 1)y'' + 2xy' + 6y = 0$ are solutions of $x^2 - 1 = 0$ or $x = \pm 1$. All other finite values* of x are ordinary points. Inspection of the Cauchy-Euler equation $ax^2y'' + bxy' + cy = 0$ shows that it has a singular point at $x = 0$. Singular points need not be real numbers. The equation $(x^2 + 1)y'' + xy' - y = 0$ has singular points at the solutions of $x^2 + 1 = 0$—namely, $x = \pm i$. All other values of x, real or complex, are ordinary points.

We state the following theorem about the existence of power series solutions without proof.

> **THEOREM 6.1**   **Existence of Power Series Solutions**
>
> If $x = x_0$ is an ordinary point of the differential equation (5), we can always find two linearly independent solutions in the form of a power series centered at x_0—that is, $y = \sum_{n=0}^{\infty} c_n(x - x_0)^n$. A series solution converges at least on some interval defined by $|x - x_0| < R$, where R is the distance from x_0 to the closest singular point.

A solution of the form $y = \sum_{n=0}^{\infty} c_n(x - x_0)^n$ is said to be a **solution about the ordinary point x_0.** The distance R in Theorem 6.1 is the *minimum value* or the

*For our purposes, ordinary points and singular points will always be finite points. It is possible for an ODE to have, say, a singular point at infinity.

lower bound for the radius of convergence of series solutions of the differential equation about x_0.

In the next example we use the fact that in the complex plane, the distance between two complex numbers $a + bi$ and $c + di$ is just the distance between the two points (a, b) and (c, d).

EXAMPLE 2   Lower Bound for Radius of Convergence

The complex numbers $1 \pm 2i$ are singular points of the differential equation $(x^2 - 2x + 5)y'' + xy' - y = 0$. Because $x = 0$ is an ordinary point of the equation, Theorem 6.1 guarantees that we can find two power series solutions about 0, that is, solutions that look like $y = \sum_{n=0}^{\infty} c_n x^n$. Without actually finding these solutions, we know that *each* series must converge *at least* for $|x| < \sqrt{5}$ because $R = \sqrt{5}$ is the distance in the complex plane from 0 (the point $(0, 0)$) to either of the numbers $1 + 2i$ (the point $(1, 2)$) or $1 - 2i$ (the point $(1, -2)$). However, one of these two solutions is valid on an interval much larger than $-\sqrt{5} < x < \sqrt{5}$; in actual fact this solution is valid on $(-\infty, \infty)$ because it can be shown that one of the two power series solutions about 0 reduces to a polynomial. Therefore we also say that $\sqrt{5}$ is the lower bound for the radius of convergence of series solutions of the differential equation about 0.

If we seek solutions of the given DE about a different ordinary point, say, $x = -1$, then each series $y = \sum_{n=0}^{\infty} c_n(x + 1)^n$ converges at least for $|x| < 2\sqrt{2}$ because the distance from -1 to either $1 + 2i$ or $1 - 2i$ is $R = \sqrt{8} = 2\sqrt{2}$.   ∎

NOTE   In the examples that follow as well as in Exercises 6.1 we shall, for the sake of simplicity, find power series solutions only about the ordinary point $x = 0$. If it is necessary to find a power series solution of a linear DE about an ordinary point $x_0 \neq 0$, we can simply make the change of variable $t = x - x_0$ in the equation (this translates $x = x_0$ to $t = 0$), find solutions of the new equation of the form $y = \sum_{n=0}^{\infty} c_n t^n$, and then resubstitute $t = x - x_0$.

FINDING A POWER SERIES SOLUTION   The actual determination of a power series solution of a homogeneous linear second-order DE is quite analogous to what we did in Section 4.4 in finding particular solutions of nonhomogeneous DEs by the method of undetermined coefficients. Indeed, the power-series method of solving a linear DE with variable coefficients is often described as "the method of undetermined *series* coefficients." In brief, here is the idea: We substitute $y = \sum_{n=0}^{\infty} c_n x^n$ into the differential equation, combine series as we did in Example 1, and then equate all coefficients to the right-hand side of the equation to determine the coefficients c_n. But because the right-hand side is zero, the last step requires, by the identity property in the preceding bulleted list, that all coefficients of x must be equated to zero. No, this does *not* mean that all coefficients *are* zero; this would not make sense—after all, Theorem 6.1 guarantees that we can find two solutions. Example 3 illustrates how the single assumption that $y = \sum_{n=0}^{\infty} c_n x^n = c_0 + c_1 x + c_2 x^2 + \cdots$ leads to two sets of coefficients, so we have two distinct power series $y_1(x)$ and $y_2(x)$, both expanded about the ordinary point $x = 0$. The general solution of the differential equation is $y = C_1 y_1(x) + C_2 y_2(x)$; indeed, it can be shown that $C_1 = c_0$ and $C_2 = c_1$.

EXAMPLE 3   Power Series Solutions

Solve $y'' + xy = 0$.

SOLUTION   Since there are no finite singular points, Theorem 6.1 guarantees two power series solutions centered at 0, convergent for $|x| < \infty$. Substituting

$y = \sum_{n=0}^{\infty} c_n x^n$ and the second derivative $y'' = \sum_{n=2}^{\infty} n(n-1)c_n x^{n-2}$ (see (1)) into the differential equation gives

$$y'' + xy = \sum_{n=2}^{\infty} c_n n(n-1)x^{n-2} + x\sum_{n=0}^{\infty} c_n x^n = \sum_{n=2}^{\infty} c_n n(n-1)x^{n-2} + \sum_{n=0}^{\infty} c_n x^{n+1}. \quad (7)$$

In Example 1 we already added the last two series on the right-hand side of the equality in (7) by shifting the summation index. From the result given in (4),

$$y'' + xy = 2c_2 + \sum_{k=1}^{\infty} [(k+1)(k+2)c_{k+2} + c_{k-1}]x^k = 0. \quad (8)$$

At this point we invoke the identity property. Since (8) is identically zero, it is necessary that the coefficient of each power of x be set equal to zero—that is, $2c_2 = 0$ (it is the coefficient of x^0), and

$$(k+1)(k+2)c_{k+2} + c_{k-1} = 0, \quad k = 1, 2, 3, \dots. \quad (9)$$

Now $2c_2 = 0$ obviously dictates that $c_2 = 0$. But the expression in (9), called a **recurrence relation,** determines the c_k in such a manner that we can choose a certain subset of the set of coefficients to be *nonzero*. Since $(k+1)(k+2) \neq 0$ for all values of k, we can solve (9) for c_{k+2} in terms of c_{k-1}:

$$c_{k+2} = -\frac{c_{k-1}}{(k+1)(k+2)}, \quad k = 1, 2, 3, \dots. \quad (10)$$

This relation generates consecutive coefficients of the assumed solution one at a time as we let k take on the successive integers indicated in (10):

$$k = 1, \quad c_3 = -\frac{c_0}{2 \cdot 3}$$

$$k = 2, \quad c_4 = -\frac{c_1}{3 \cdot 4}$$

$$k = 3, \quad c_5 = -\frac{c_2}{4 \cdot 5} = 0 \qquad \leftarrow c_2 \text{ is zero}$$

$$k = 4, \quad c_6 = -\frac{c_3}{5 \cdot 6} = \frac{1}{2 \cdot 3 \cdot 5 \cdot 6} c_0$$

$$k = 5, \quad c_7 = -\frac{c_4}{6 \cdot 7} = \frac{1}{3 \cdot 4 \cdot 6 \cdot 7} c_1$$

$$k = 6, \quad c_8 = -\frac{c_5}{7 \cdot 8} = 0 \qquad \leftarrow c_5 \text{ is zero}$$

$$k = 7, \quad c_9 = -\frac{c_6}{8 \cdot 9} = \frac{1}{2 \cdot 3 \cdot 5 \cdot 6 \cdot 8 \cdot 9} c_0$$

$$k = 8, \quad c_{10} = -\frac{c_7}{9 \cdot 10} = \frac{1}{3 \cdot 4 \cdot 6 \cdot 7 \cdot 9 \cdot 10} c_1$$

$$k = 9, \quad c_{11} = -\frac{c_8}{10 \cdot 11} = 0 \qquad \leftarrow c_8 \text{ is zero}$$

and so on. Now substituting the coefficients just obtained into the original assumption

$$y = c_0 + c_1 x + c_2 x^2 + c_3 x^3 + c_4 x^4 + c_5 x^5 + c_6 x^6 + c_7 x^7 + c_8 x^8 + c_9 x^9 + c_{10} x^{10} + c_{11} x^{11} + \cdots,$$

we get

$$y = c_0 + c_1 x + 0 - \frac{c_0}{2 \cdot 3}x^3 - \frac{c_1}{3 \cdot 4}x^4 + 0 + \frac{c_0}{2 \cdot 3 \cdot 5 \cdot 6}x^6$$

$$+ \frac{c_1}{3 \cdot 4 \cdot 6 \cdot 7}x^7 + 0 - \frac{c_0}{2 \cdot 3 \cdot 5 \cdot 6 \cdot 8 \cdot 9}x^9 - \frac{c_1}{3 \cdot 4 \cdot 6 \cdot 7 \cdot 9 \cdot 10}x^{10} + 0 + \cdots.$$

After grouping the terms containing c_0 and the terms containing c_1, we obtain $y = c_0 y_1(x) + c_1 y_2(x)$, where

$$y_1(x) = 1 - \frac{1}{2 \cdot 3}x^3 + \frac{1}{2 \cdot 3 \cdot 5 \cdot 6}x^6 - \frac{1}{2 \cdot 3 \cdot 5 \cdot 6 \cdot 8 \cdot 9}x^9 + \cdots = 1 + \sum_{k=1}^{\infty} \frac{(-1)^k}{2 \cdot 3 \cdots (3k-1)(3k)}x^{3k}$$

$$y_2(x) = x - \frac{1}{3 \cdot 4}x^4 + \frac{1}{3 \cdot 4 \cdot 6 \cdot 7}x^7 - \frac{1}{3 \cdot 4 \cdot 6 \cdot 7 \cdot 9 \cdot 10}x^{10} + \cdots = x + \sum_{k=1}^{\infty} \frac{(-1)^k}{3 \cdot 4 \cdots (3k)(3k+1)}x^{3k+1}.$$

Because the recursive use of (10) leaves c_0 and c_1 completely undetermined, they can be chosen arbitrarily. As was mentioned prior to this example, the linear combination $y = c_0 y_1(x) + c_1 y_2(x)$ actually represents the general solution of the differential equation. Although we know from Theorem 6.1 that each series solution converges for $|x| < \infty$, this fact can also be verified by the ratio test. ∎

The differential equation in Example 3 is called **Airy's equation** and is encountered in the study of diffraction of light, diffraction of radio waves around the surface of the earth, aerodynamics, and the deflection of a uniform thin vertical column that bends under its own weight. Other common forms of Airy's equation are $y'' - xy = 0$ and $y'' + \alpha^2 xy = 0$. See Problem 41 in Exercises 6.3 for an application of the last equation.

EXAMPLE 4   Power Series Solution

Solve $(x^2 + 1)y'' + xy' - y = 0$.

SOLUTION As we have already seen on page 241 the given differential equation has singular points at $x = \pm i$, and so a power series solution centered at 0 will converge at least for $|x| < 1$, where 1 is the distance in the complex plane from 0 to either i or $-i$. The assumption $y = \sum_{n=0}^{\infty} c_n x^n$ and its first two derivatives (see (1)) lead to

$$(x^2 + 1) \sum_{n=2}^{\infty} n(n-1)c_n x^{n-2} + x \sum_{n=1}^{\infty} nc_n x^{n-1} - \sum_{n=0}^{\infty} c_n x^n$$

$$= \sum_{n=2}^{\infty} n(n-1)c_n x^n + \sum_{n=2}^{\infty} n(n-1)c_n x^{n-2} + \sum_{n=1}^{\infty} nc_n x^n - \sum_{n=0}^{\infty} c_n x^n$$

$$= 2c_2 x^0 - c_0 x^0 + 6c_3 x + c_1 x - c_1 x + \underbrace{\sum_{n=2}^{\infty} n(n-1)c_n x^n}_{k=n}$$

$$+ \underbrace{\sum_{n=4}^{\infty} n(n-1)c_n x^{n-2}}_{k=n-2} + \underbrace{\sum_{n=2}^{\infty} nc_n x^n}_{k=n} - \underbrace{\sum_{n=2}^{\infty} c_n x^n}_{k=n}$$

$$= 2c_2 - c_0 + 6c_3 x + \sum_{k=2}^{\infty} [k(k-1)c_k + (k+2)(k+1)c_{k+2} + kc_k - c_k]x^k$$

$$= 2c_2 - c_0 + 6c_3 x + \sum_{k=2}^{\infty} [(k+1)(k-1)c_k + (k+2)(k+1)c_{k+2}]x^k = 0.$$

From this identity we conclude that $2c_2 - c_0 = 0, 6c_3 = 0$, and

$$(k + 1)(k - 1)c_k + (k + 2)(k + 1)c_{k+2} = 0.$$

Thus

$$c_2 = \frac{1}{2}c_0$$

$$c_3 = 0$$

$$c_{k+2} = \frac{1 - k}{k + 2}c_k, \quad k = 2, 3, 4, \ldots.$$

Substituting $k = 2, 3, 4, \ldots$ into the last formula gives

$$c_4 = -\frac{1}{4}c_2 = -\frac{1}{2 \cdot 4}c_0 = -\frac{1}{2^2 2!}c_0$$

$$c_5 = -\frac{2}{5}c_3 = 0 \qquad \leftarrow c_3 \text{ is zero}$$

$$c_6 = -\frac{3}{6}c_4 = \frac{3}{2 \cdot 4 \cdot 6}c_0 = \frac{1 \cdot 3}{2^3 3!}c_0$$

$$c_7 = -\frac{4}{7}c_5 = 0 \qquad \leftarrow c_5 \text{ is zero}$$

$$c_8 = -\frac{5}{8}c_6 = -\frac{3 \cdot 5}{2 \cdot 4 \cdot 6 \cdot 8}c_0 = -\frac{1 \cdot 3 \cdot 5}{2^4 4!}c_0$$

$$c_9 = -\frac{6}{9}c_7 = 0, \qquad \leftarrow c_7 \text{ is zero}$$

$$c_{10} = -\frac{7}{10}c_8 = \frac{3 \cdot 5 \cdot 7}{2 \cdot 4 \cdot 6 \cdot 8 \cdot 10}c_0 = \frac{1 \cdot 3 \cdot 5 \cdot 7}{2^5 5!}c_0$$

and so on. Therefore

$$y = c_0 + c_1 x + c_2 x^2 + c_3 x^3 + c_4 x^4 + c_5 x^5 + c_6 x^6 + c_7 x^7 + c_8 x^8 + c_9 x^9 + c_{10} x^{10} + \cdots$$

$$= c_0 \left[1 + \frac{1}{2}x^2 - \frac{1}{2^2 2!}x^4 + \frac{1 \cdot 3}{2^3 3!}x^6 - \frac{1 \cdot 3 \cdot 5}{2^4 4!}x^8 + \frac{1 \cdot 3 \cdot 5 \cdot 7}{2^5 5!}x^{10} - \cdots \right] + c_1 x$$

$$= c_0 y_1(x) + c_1 y_2(x).$$

The solutions are the polynomial $y_2(x) = x$ and the power series

$$y_1(x) = 1 + \frac{1}{2}x^2 + \sum_{n=2}^{\infty}(-1)^{n-1}\frac{1 \cdot 3 \cdot 5 \cdots (2n - 3)}{2^n n!}x^{2n}, \quad |x| < 1. \quad \blacksquare$$

EXAMPLE 5 Three-Term Recurrence Relation

If we seek a power series solution $y = \sum_{n=0}^{\infty} c_n x^n$ for the differential equation

$$y'' - (1 + x)y = 0,$$

we obtain $c_2 = \frac{1}{2}c_0$ and the three-term recurrence relation

$$c_{k+2} = \frac{c_k + c_{k-1}}{(k + 1)(k + 2)}, \quad k = 1, 2, 3, \ldots.$$

It follows from these two results that all coefficients c_n, for $n \geq 3$, are expressed in terms of *both* c_0 and c_1. To simplify life, we can first choose $c_0 \neq 0, c_1 = 0$; this yields coefficients for one solution expressed entirely in terms of c_0. Next, if we choose $c_0 = 0$, $c_1 \neq 0$, then coefficients for the other solution are expressed in terms of c_1. Using $c_2 = \frac{1}{2}c_0$ in both cases, the recurrence relation for $k = 1, 2, 3, \ldots$ gives

$c_0 \neq 0,\ c_1 = 0$	$c_0 = 0,\ c_1 \neq 0$
$c_2 = \dfrac{1}{2}c_0$	$c_2 = \dfrac{1}{2}c_0 = 0$
$c_3 = \dfrac{c_1 + c_0}{2\cdot 3} = \dfrac{c_0}{2\cdot 3} = \dfrac{c_0}{6}$	$c_3 = \dfrac{c_1 + c_0}{2\cdot 3} = \dfrac{c_1}{2\cdot 3} = \dfrac{c_1}{6}$
$c_4 = \dfrac{c_2 + c_1}{3\cdot 4} = \dfrac{c_0}{2\cdot 3\cdot 4} = \dfrac{c_0}{24}$	$c_4 = \dfrac{c_2 + c_1}{3\cdot 4} = \dfrac{c_1}{3\cdot 4} = \dfrac{c_1}{12}$
$c_5 = \dfrac{c_3 + c_2}{4\cdot 5} = \dfrac{c_0}{4\cdot 5}\left[\dfrac{1}{6} + \dfrac{1}{2}\right] = \dfrac{c_0}{30}$	$c_5 = \dfrac{c_3 + c_2}{4\cdot 5} = \dfrac{c_1}{4\cdot 5\cdot 6} = \dfrac{c_1}{120}$

and so on. Finally, we see that the general solution of the equation is $y = c_0 y_1(x) + c_1 y_2(x)$, where

$$y_1(x) = 1 + \frac{1}{2}x^2 + \frac{1}{6}x^3 + \frac{1}{24}x^4 + \frac{1}{30}x^5 + \cdots$$

and

$$y_2(x) = x + \frac{1}{6}x^3 + \frac{1}{12}x^4 + \frac{1}{120}x^5 + \cdots.$$

Each series converges for all finite values of x.

NONPOLYNOMIAL COEFFICIENTS   The next example illustrates how to find a power series solution about the ordinary point $x_0 = 0$ of a differential equation when its coefficients are not polynomials. In this example we see an application of the multiplication of two power series.

EXAMPLE 6   DE with Nonpolynomial Coefficients

Solve $y'' + (\cos x)y = 0$.

SOLUTION   We see that $x = 0$ is an ordinary point of the equation because, as we have already seen, $\cos x$ is analytic at that point. Using the Maclaurin series for $\cos x$ given in (2), along with the usual assumption $y = \sum_{n=0}^{\infty} c_n x^n$ and the results in (1), we find

$$y'' + (\cos x)y = \sum_{n=2}^{\infty} n(n-1)c_n x^{n-2} + \left(1 - \frac{x^2}{2!} + \frac{x^4}{4!} - \frac{x^6}{6!} + \cdots\right)\sum_{n=0}^{\infty} c_n x^n$$

$$= 2c_2 + 6c_3 x + 12c_4 x^2 + 20c_5 x^3 + \cdots + \left(1 - \frac{x^2}{2!} + \frac{x^4}{4!} + \cdots\right)(c_0 + c_1 x + c_2 x^2 + c_3 x^3 + \cdots)$$

$$= 2c_2 + c_0 + (6c_3 + c_1)x + \left(12c_4 + c_2 - \frac{1}{2}c_0\right)x^2 + \left(20c_5 + c_3 - \frac{1}{2}c_1\right)x^3 + \cdots = 0.$$

It follows that

$$2c_2 + c_0 = 0, \quad 6c_3 + c_1 = 0, \quad 12c_4 + c_2 - \frac{1}{2}c_0 = 0, \quad 20c_5 + c_3 - \frac{1}{2}c_1 = 0,$$

and so on. This gives $c_2 = -\frac{1}{2}c_0$, $c_3 = -\frac{1}{6}c_1$, $c_4 = \frac{1}{12}c_0$, $c_5 = \frac{1}{30}c_1$, ... By grouping terms we arrive at the general solution $y = c_0 y_1(x) + c_1 y_2(x)$, where

$$y_1(x) = 1 - \frac{1}{2}x^2 + \frac{1}{12}x^4 - \cdots \quad \text{and} \quad y_2(x) = x - \frac{1}{6}x^3 + \frac{1}{30}x^5 - \cdots.$$

Because the differential equation has no finite singular points, both power series converge for $|x| < \infty$.

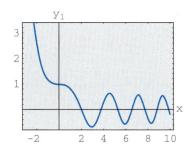

(a) plot of $y_1(x)$ vs. x

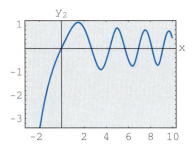

(b) plot of $y_2(x)$ vs. x

FIGURE 6.2 Numerical solution curves for Airy's DE

SOLUTION CURVES The approximate graph of a power series solution $y(x) = \sum_{n=0}^{\infty} c_n x^n$ can be obtained in several ways. We can always resort to graphing the terms in the sequence of partial sums of the series—in other words, the graphs of the polynomials $S_N(x) = \sum_{n=0}^{N} c_n x^n$. For large values of N, $S_N(x)$ should give us an indication of the behavior of $y(x)$ near the ordinary point $x = 0$. We can also obtain an approximate or numerical solution curve by using a solver as we did in Section 4.9. For example, if you carefully scrutinize the series solutions of Airy's equation in Example 3, you should see that $y_1(x)$ and $y_2(x)$ are, in turn, the solutions of the initial-value problems

$$\begin{align} y'' + xy = 0, \quad y(0) = 1, \quad y'(0) = 0, \\ y'' + xy = 0, \quad y(0) = 0, \quad y'(0) = 1. \end{align} \tag{11}$$

The specified initial conditions "pick out" the solutions $y_1(x)$ and $y_2(x)$ from $y = c_0 y_1(x) + c_1 y_2(x)$, since it should be apparent from our basic series assumption $y = \sum_{n=0}^{\infty} c_n x^n$ that $y(0) = c_0$ and $y'(0) = c_1$. Now if your numerical solver requires a system of equations, the substitution $y' = u$ in $y'' + xy = 0$ gives $y'' = u' = -xy$, and so a system of two first-order equations equivalent to Airy's equation is

$$\begin{align} y' &= u \\ u' &= -xy. \end{align} \tag{12}$$

Initial conditions for the system in (12) are the two sets of initial conditions in (11) rewritten as $y(0) = 1$, $u(0) = 0$ and $y(0) = 0$, $u(0) = 1$. The graphs of $y_1(x)$ and $y_2(x)$ shown in Figure 6.2 were obtained with the aid of a numerical solver. The fact that the numerical solution curves appear to be oscillatory is consistent with the fact that Airy's equation appeared in Section 5.1 (page 199) in the form $mx'' + ktx = 0$ as a model of a spring whose "spring constant" $K(t) = kt$ increases with time.

REMARKS

(*i*) In the problems that follow, do not expect to be able to write a solution in terms of summation notation in each case. Even though we can generate as many terms as desired in a series solution $y = \sum_{n=0}^{\infty} c_n x^n$ either through the use of a recurrence relation or, as in Example 6, by multiplication, it might not be possible to deduce any general term for the coefficients c_n. We might have to settle, as we did in Examples 5 and 6, for just writing out the first few terms of the series.

(*ii*) A point x_0 is an ordinary point of a *nonhomogeneous* linear second-order DE $y'' + P(x)y' + Q(x)y = f(x)$ if $P(x)$, $Q(x)$, and $f(x)$ are analytic at x_0. Moreover, Theorem 6.1 extends to such DEs; in other words, we can find power series solutions $y = \sum_{n=0}^{\infty} c_n (x - x_0)^n$ of nonhomogeneous linear DEs in the same manner as in Examples 3–6. See Problem 36 in Exercises 6.1.

EXERCISES 6.1

Answers to selected odd-numbered problems begin on page ANS-7.

6.1.1 REVIEW OF POWER SERIES

In Problems 1–4 find the radius of convergence and interval of convergence for the given power series.

1. $\displaystyle\sum_{n=1}^{\infty} \frac{2^n}{n} x^n$

2. $\displaystyle\sum_{n=0}^{\infty} \frac{(100)^n}{n!}(x+7)^n$

3. $\displaystyle\sum_{k=1}^{\infty} \frac{(-1)^k}{10^k}(x-5)^k$

4. $\displaystyle\sum_{k=0}^{\infty} k!(x-1)^k$

In Problems 5 and 6 the given function is analytic at $x = 0$. Find the first four terms of a power series in x. Perform the multiplication by hand or use a CAS, as instructed.

5. $\sin x \cos x$

6. $e^{-x} \cos x$

In Problems 7 and 8 the given function is analytic at $x = 0$. Find the first four terms of a power series in x. Perform the long division by hand or use a CAS, as instructed. Give the open interval of convergence.

7. $\dfrac{1}{\cos x}$

8. $\dfrac{1-x}{2+x}$

In Problems 9 and 10 rewrite the given power series so that its general term involves x^k.

9. $\displaystyle\sum_{n=1}^{\infty} nc_n x^{n+2}$

10. $\displaystyle\sum_{n=3}^{\infty} (2n-1)c_n x^{n-3}$

In Problems 11 and 12 rewrite the given expression as a single power series whose general term involves x^k.

11. $\displaystyle\sum_{n=1}^{\infty} 2nc_n x^{n-1} + \sum_{n=0}^{\infty} 6c_n x^{n+1}$

12. $\displaystyle\sum_{n=2}^{\infty} n(n-1)c_n x^n + 2\sum_{n=2}^{\infty} n(n-1)c_n x^{n-2} + 3\sum_{n=1}^{\infty} nc_n x^n$

In Problems 13 and 14 verify by direct substitution that the given power series is a particular solution of the indicated differential equation.

13. $y = \displaystyle\sum_{n=1}^{\infty} \frac{(-1)^{n+1}}{n} x^n, \quad (x+1)y'' + y' = 0$

14. $y = \displaystyle\sum_{n=0}^{\infty} \frac{(-1)^n}{2^{2n}(n!)^2} x^{2n}, \quad xy'' + y' + xy = 0$

6.1.2 POWER SERIES SOLUTIONS

In Problems 15 and 16 without actually solving the given differential equation, find a lower bound for the radius of convergence of power series solutions about the ordinary point $x = 0$. About the ordinary point $x = 1$.

15. $(x^2 - 25)y'' + 2xy' + y = 0$

16. $(x^2 - 2x + 10)y'' + xy' - 4y = 0$

In Problems 17–28 find two power series solutions of the given differential equation about the ordinary point $x = 0$.

17. $y'' - xy = 0$

18. $y'' + x^2 y = 0$

19. $y'' - 2xy' + y = 0$

20. $y'' - xy' + 2y = 0$

21. $y'' + x^2 y' + xy = 0$

22. $y'' + 2xy' + 2y = 0$

23. $(x-1)y'' + y' = 0$

24. $(x+2)y'' + xy' - y = 0$

25. $y'' - (x+1)y' - y = 0$

26. $(x^2 + 1)y'' - 6y = 0$

27. $(x^2 + 2)y'' + 3xy' - y = 0$

28. $(x^2 - 1)y'' + xy' - y = 0$

In Problems 29–32 use the power series method to solve the given initial-value problem.

29. $(x-1)y'' - xy' + y = 0, \quad y(0) = -2, y'(0) = 6$

30. $(x+1)y'' - (2-x)y' + y = 0, \quad y(0) = 2, y'(0) = -1$

31. $y'' - 2xy' + 8y = 0, \quad y(0) = 3, y'(0) = 0$

32. $(x^2 + 1)y'' + 2xy' = 0, \quad y(0) = 0, y'(0) = 1$

In Problems 33 and 34 use the procedure in Example 6 to find two power series solutions of the given differential equation about the ordinary point $x = 0$.

33. $y'' + (\sin x)y = 0$

34. $y'' + e^x y' - y = 0$

DISCUSSION/PROJECT PROBLEMS

35. Without actually solving the differential equation $(\cos x)y'' + y' + 5y = 0$, find a lower bound for the radius of convergence of power series solutions about $x = 0$. About $x = 1$.

36. How can the method described in this section be used to find a power series solution of the *nonhomogeneous* equation $y'' - xy = 1$ about the ordinary point $x = 0$? Of $y'' - 4xy' - 4y = e^x$? Carry out your ideas by solving both DEs.

37. Is $x = 0$ an ordinary or a singular point of the differential equation $xy'' + (\sin x)y = 0$? Defend your answer with sound mathematics.

38. For purposes of this problem, ignore the graphs given in Figure 6.2. If Airy's DE is written as $y'' = -xy$, what can we say about the shape of a solution curve if $x > 0$ and $y > 0$? If $x > 0$ and $y < 0$?

COMPUTER LAB ASSIGNMENTS

39. (a) Find two power series solutions for $y'' + xy' + y = 0$ and express the solutions $y_1(x)$ and $y_2(x)$ in terms of summation notation.

    (b) Use a CAS to graph the partial sums $S_N(x)$ for $y_1(x)$. Use $N = 2, 3, 5, 6, 8, 10$. Repeat using the partial sums $S_N(x)$ for $y_2(x)$.

    (c) Compare the graphs obtained in part (b) with the curve obtained using a numerical solver. Use the initial-conditions $y_1(0) = 1$, $y_1'(0) = 0$, and $y_2(0) = 0$, $y_2'(0) = 1$.

    (d) Reexamine the solution $y_1(x)$ in part (a). Express this series as an elementary function. Then use (5) of Section 4.2 to find a second solution of the equation. Verify that this second solution is the same as the power series solution $y_2(x)$.

40. (a) Find one more nonzero term for each of the solutions $y_1(x)$ and $y_2(x)$ in Example 6.

    (b) Find a series solution $y(x)$ of the initial-value problem $y'' + (\cos x)y = 0$, $y(0) = 1$, $y'(0) = 1$.

    (c) Use a CAS to graph the partial sums $S_N(x)$ for the solution $y(x)$ in part (b). Use $N = 2, 3, 4, 5, 6, 7$.

    (d) Compare the graphs obtained in part (c) with the curve obtained using a numerical solver for the initial-value problem in part (b).

6.2 SOLUTIONS ABOUT SINGULAR POINTS

INTRODUCTION: The two differential equations $y'' + xy = 0$ and $xy'' + y = 0$ are similar only in that they are both examples of simple linear second-order DEs with variable coefficients. That is all they have in common. Because $x = 0$ is an *ordinary point* of $y'' + xy = 0$, we saw in the preceding section that there was no problem in finding two distinct power series solutions centered at that point. In contrast, because $x = 0$ is a *singular point* of $xy'' + y = 0$, finding two infinite series—notice that we did not say "power series"—solutions of the equation about that point becomes a more difficult task.

REVIEW MATERIAL: The solution method discussed in this section does not always yield two infinite series solutions. When only one solution is found, we can use the formula given in (5) of Section 4.2 to find a second solution.

A DEFINITION A singular point x_0 of a linear differential equation

$$a_2(x)y'' + a_1(x)y' + a_0(x)y = 0 \tag{1}$$

is further classified as either regular or irregular. The classification again depends on the functions P and Q in the standard form

$$y'' + P(x)y' + Q(x)y = 0. \tag{2}$$

> **DEFINITION 6.2**   Regular and Irregular Singular Points
>
> A singular point x_0 is said to be a **regular singular point** of the differential equation (1) if the functions $p(x) = (x - x_0)P(x)$ and $q(x) = (x - x_0)^2 Q(x)$ are both analytic at x_0. A singular point that is not regular is said to be an **irregular singular point** of the equation.

The second sentence in Definition 6.2 indicates that if one or both of the functions $p(x) = (x - x_0) P(x)$ and $q(x) = (x - x_0)^2 Q(x)$ fail to be analytic at x_0, then x_0 is an irregular singular point.

POLYNOMIAL COEFFICIENTS   As in Section 6.1, we are mainly interested in linear equations (1) where the coefficients $a_2(x)$, $a_1(x)$, and $a_0(x)$ are polynomials with no common factors. We have already seen that if $a_2(x_0) = 0$, then $x = x_0$ is a singular point of (1), since at least one of the rational functions $P(x) = a_1(x)/a_2(x)$ and $Q(x) = a_0(x)/a_2(x)$ in the standard form (2) fails to be analytic at that point. But since $a_2(x)$ is a polynomial and x_0 is one of its zeros, it follows from the Factor Theorem of algebra that $x - x_0$ is a factor of $a_2(x)$. This means that after $a_1(x)/a_2(x)$ and $a_0(x)/a_2(x)$ are reduced to lowest terms, the factor $x - x_0$ must remain, to some positive integer power, in one or both denominators. Now suppose that $x = x_0$ is a singular point of (1) but both the functions defined by the products $p(x) = (x - x_0) P(x)$ and $q(x) = (x - x_0)^2 Q(x)$ are analytic at x_0. We are led to the conclusion that multiplying $P(x)$ by $x - x_0$ and $Q(x)$ by $(x - x_0)^2$ has the effect (through cancellation) that $x - x_0$ no longer appears in either denominator. We can now determine whether x_0 is regular by a quick visual check of denominators: If $x - x_0$ appears *at most* to the first power in the denominator of $P(x)$ and *at most* to the second power in the denominator of $Q(x)$, then $x = x_0$ is a regular singular point. Moreover, observe that if $x = x_0$ is a regular singular point and we multiply (2) by $(x - x_0)^2$, then the original DE can be put into the form

$$(x - x_0)^2 y'' + (x - x_0)p(x)y' + q(x)y = 0, \tag{3}$$

where p and q are analytic at $x = x_0$.

EXAMPLE 1   **Classification of Singular Points**

It should be clear that $x = 2$ and $x = -2$ are singular points of

$$(x^2 - 4)^2 y'' + 3(x - 2)y' + 5y = 0.$$

After dividing the equation by $(x^2 - 4)^2 = (x - 2)^2(x + 2)^2$ and reducing the coefficients to lowest terms, we find that

$$P(x) = \frac{3}{(x - 2)(x + 2)^2} \quad \text{and} \quad Q(x) = \frac{5}{(x - 2)^2(x + 2)^2}.$$

We now test $P(x)$ and $Q(x)$ at each singular point.

For $x = 2$ to be a regular singular point, the factor $x - 2$ can appear at most to the first power in the denominator of $P(x)$ and at most to the second power in the denominator of $Q(x)$. A check of the denominators of $P(x)$ and $Q(x)$ shows that both these conditions are satisfied, so $x = 2$ is a regular singular point. Alternatively, we are led to the same conclusion by noting that both rational functions

$$p(x) = (x - 2)P(x) = \frac{3}{(x + 2)^2} \quad \text{and} \quad q(x) = (x - 2)^2 Q(x) = \frac{5}{(x + 2)^2}$$

are analytic at $x = 2$.

Now since the factor $x - (-2) = x + 2$ appears to the second power in the denominator of $P(x)$, we can conclude immediately that $x = -2$ is an irregular singular point of the equation. This also follows from the fact that

$$p(x) = (x + 2)P(x) = \frac{3}{(x - 2)(x + 2)}$$

is not analytic at $x = -2$.

In Example 1, notice that since $x = 2$ is a regular singular point, the original equation can be written as

$$(x - 2)^2 y'' + (x - 2) \underset{\substack{\uparrow \text{ at } x = 2 \\ p(x) \text{ analytic}}}{\frac{3}{(x + 2)^2}} y' + \underset{\substack{\uparrow \text{ at } x = 2 \\ q(x) \text{ analytic}}}{\frac{5}{(x + 2)^2}} y = 0.$$

As another example, we can see that $x = 0$ is an irregular singular point of $x^3 y'' - 2xy' + 8y = 0$ by inspection of the denominators of $P(x) = -2/x^2$ and $Q(x) = 8/x^3$. On the other hand, $x = 0$ is a regular singular point of $xy'' - 2xy' + 8y = 0$, since $x - 0$ and $(x - 0)^2$ do not even appear in the respective denominators of $P(x) = -2$ and $Q(x) = 8/x$. For a singular point $x = x_0$, any non-negative power of $x - x_0$ less than one (namely, zero) and any nonnegative power less than two (namely, zero and one) in the denominators of $P(x)$ and $Q(x)$, respectively, imply that x_0 is a regular singular point. A singular point can be a complex number. You should verify that $x = 3i$ and $x = -3i$ are two regular singular points of $(x^2 + 9)y'' - 3xy' + (1 - x)y = 0$.

Any second-order Cauchy-Euler equation $ax^2 y'' + bxy' + cy = 0$, a, b, and c real constants, has a regular singular point at $x = 0$. You should verify that two solutions of the Cauchy-Euler equation $x^2 y'' - 3xy' + 4y = 0$ on the interval $(0, \infty)$ are $y_1 = x^2$ and $y_2 = x^2 \ln x$. If we attempted to find a power series solution about the regular singular point $x = 0$ (namely, $y = \sum_{n=0}^{\infty} c_n x^n$), we would succeed in obtaining only the polynomial solution $y_1 = x^2$. The fact that we would not obtain the second solution is not surprising because $\ln x$ (and consequently $y_2 = x^2 \ln x$) is not analytic at $x = 0$—that is, y_2 does not possess a Taylor series expansion centered at $x = 0$.

METHOD OF FROBENIUS To solve a differential equation (1) about a regular singular point, we employ the following theorem due to Frobenius.

THEOREM 6.2	**Frobenius' Theorem**

If $x = x_0$ is a regular singular point of the differential equation (1), then there exists at least one solution of the form

$$y = (x - x_0)^r \sum_{n=0}^{\infty} c_n (x - x_0)^n = \sum_{n=0}^{\infty} c_n (x - x_0)^{n+r}, \tag{4}$$

where the number r is a constant to be determined. The series will converge at least on some interval $0 < x - x_0 < R$.

Notice the words *at least* in the first sentence of Theorem 6.2. This means that, in contrast to Theorem 6.1, Theorem 6.2 gives us no assurance that *two* series solutions of the type indicated in (4) can be found. The **method of Frobenius,** finding series solutions about a regular singular point x_0, is similar to the "method of undetermined series coefficients" of the preceding section in that we substitute $y = \sum_{n=0}^{\infty} c_n (x - x_0)^{n+r}$ into the given differential equation and determine the unknown coefficients c_n by a recurrence relation. However, we have an additional task in this procedure: Before determining the coefficients, we must find the unknown exponent r. If r is found to be a number that is not a non-negative integer, then the corresponding solution $y = \sum_{n=0}^{\infty} c_n (x - x_0)^{n+r}$ is not a power series.

As we did in the discussion of solutions about ordinary points, we shall always assume, for the sake of simplicity in solving differential equations, that the regular singular point is $x = 0$.

EXAMPLE 2   Two Series Solutions

Because $x = 0$ is a regular singular point of the differential equation

$$3xy'' + y' - y = 0, \tag{5}$$

we try to find a solution of the form $y = \sum_{n=0}^{\infty} c_n x^{n+r}$. Now

$$y' = \sum_{n=0}^{\infty} (n + r)c_n x^{n+r-1} \quad \text{and} \quad y'' = \sum_{n=0}^{\infty} (n + r)(n + r - 1)c_n x^{n+r-2},$$

so

$$3xy'' + y' - y = 3\sum_{n=0}^{\infty} (n + r)(n + r - 1)c_n x^{n+r-1} + \sum_{n=0}^{\infty} (n + r)c_n x^{n+r-1} - \sum_{n=0}^{\infty} c_n x^{n+r}$$

$$= \sum_{n=0}^{\infty} (n + r)(3n + 3r - 2)c_n x^{n+r-1} - \sum_{n=0}^{\infty} c_n x^{n+r}$$

$$= x^r \left[r(3r - 2)c_0 x^{-1} + \underbrace{\sum_{n=1}^{\infty} (n + r)(3n + 3r - 2)c_n x^{n-1}}_{k = n-1} - \underbrace{\sum_{n=0}^{\infty} c_n x^n}_{k = n} \right]$$

$$= x^r \left[r(3r - 2)c_0 x^{-1} + \sum_{k=0}^{\infty} [(k + r + 1)(3k + 3r + 1)c_{k+1} - c_k]x^k \right] = 0,$$

which implies that

$$r(3r - 2)c_0 = 0$$

and

$$(k + r + 1)(3k + 3r + 1)c_{k+1} - c_k = 0, \quad k = 0, 1, 2, \ldots.$$

Because nothing is gained by taking $c_0 = 0$, we must then have

$$r(3r - 2) = 0 \tag{6}$$

and

$$c_{k+1} = \frac{c_k}{(k + r + 1)(3k + 3r + 1)}, \quad k = 0, 1, 2, \ldots. \tag{7}$$

When substituted in (7), the two values of r that satisfy the quadratic equation (6), $r_1 = \frac{2}{3}$ and $r_2 = 0$, give two different recurrence relations:

$$r_1 = \frac{2}{3}, \quad c_{k+1} = \frac{c_k}{(3k + 5)(k + 1)}, \quad k = 0, 1, 2, \ldots \tag{8}$$

$$r_2 = 0, \quad c_{k+1} = \frac{c_k}{(k + 1)(3k + 1)}, \quad k = 0, 1, 2, \ldots. \tag{9}$$

From (8) we find

$$c_1 = \frac{c_0}{5 \cdot 1}$$

$$c_2 = \frac{c_1}{8 \cdot 2} = \frac{c_0}{2!5 \cdot 8}$$

$$c_3 = \frac{c_2}{11 \cdot 3} = \frac{c_0}{3!5 \cdot 8 \cdot 11}$$

$$c_4 = \frac{c_3}{14 \cdot 4} = \frac{c_0}{4!5 \cdot 8 \cdot 11 \cdot 14}$$

$$\vdots$$

$$c_n = \frac{c_0}{n!5 \cdot 8 \cdot 11 \cdots (3n + 2)}.$$

From (9) we find

$$c_1 = \frac{c_0}{1 \cdot 1}$$

$$c_2 = \frac{c_1}{2 \cdot 4} = \frac{c_0}{2!1 \cdot 4}$$

$$c_3 = \frac{c_2}{3 \cdot 7} = \frac{c_0}{3!1 \cdot 4 \cdot 7}$$

$$c_4 = \frac{c_3}{4 \cdot 10} = \frac{c_0}{4!1 \cdot 4 \cdot 7 \cdot 10}$$

$$\vdots$$

$$c_n = \frac{c_0}{n!1 \cdot 4 \cdot 7 \cdots (3n - 2)}.$$

Here we encounter something that did not happen when we obtained solutions about an ordinary point; we have what looks to be two different sets of coefficients, but each set contains the *same* multiple c_0. If we omit this term, the series solutions are

$$y_1(x) = x^{2/3}\left[1 + \sum_{n=1}^{\infty} \frac{1}{n!5 \cdot 8 \cdot 11 \cdots (3n + 2)} x^n\right] \tag{10}$$

$$y_2(x) = x^0\left[1 + \sum_{n=1}^{\infty} \frac{1}{n!1 \cdot 4 \cdot 7 \cdots (3n - 2)} x^n\right]. \tag{11}$$

By the ratio test it can be demonstrated that both (10) and (11) converge for all finite values of x—that is, $|x| < \infty$. Also, it should be apparent from the form of these solutions that neither series is a constant multiple of the other, and therefore $y_1(x)$ and $y_2(x)$ are linearly independent on the entire x-axis. Hence by the superposition principle, $y = C_1 y_1(x) + C_2 y_2(x)$ is another solution of (5). On any interval not containing the origin, such as $(0, \infty)$, this linear combination represents the general solution of the differential equation.

INDICIAL EQUATION   Equation (6) is called the **indicial equation** of the problem, and the values $r_1 = \frac{2}{3}$ and $r_2 = 0$ are called the **indicial roots**, or **exponents**, of the singularity $x = 0$. In general, after substituting $y = \sum_{n=0}^{\infty} c_n x^{n+r}$ into the given differential equation and simplifying, the indicial equation is a quadratic equation in r that results from equating the *total coefficient of the lowest power of x to zero*. We solve for the two values of r and substitute these values into a recurrence relation such as (7). Theorem 6.2 guarantees that at least one solution of the assumed series form can be found.

It is possible to obtain the indicial equation in advance of substituting $y = \sum_{n=0}^{\infty} c_n x^{n+r}$ into the differential equation. If $x = 0$ is a regular singular point of (1), then by Definition 6.2 both functions $p(x) = xP(x)$ and $q(x) = x^2 Q(x)$, where P and Q are defined by the standard form (2), are analytic at $x = 0$; that is, the power series expansions

$$p(x) = xP(x) = a_0 + a_1 x + a_2 x^2 + \cdots \quad \text{and} \quad q(x) = x^2 Q(x) = b_0 + b_1 x + b_2 x^2 + \cdots \tag{12}$$

are valid on intervals that have a positive radius of convergence. By multiplying (2) by x^2, we get the form given in (3):

$$x^2 y'' + x[xP(x)]y' + [x^2 Q(x)]y = 0. \tag{13}$$

After substituting $y = \sum_{n=0}^{\infty} c_n x^{n+r}$ and the two series in (12) into (13) and carrying out the multiplication of series, we find the general indicial equation to be

$$r(r - 1) + a_0 r + b_0 = 0, \tag{14}$$

where a_0 and b_0 are as defined in (12). See Problems 13 and 14 in Exercises 6.2.

EXAMPLE 3   **Two Series Solutions**

Solve $2xy'' + (1 + x)y' + y = 0$.

SOLUTION   Substituting $y = \sum_{n=0}^{\infty} c_n x^{n+r}$ gives

$$2xy'' + (1+x)y' + y = 2\sum_{n=0}^{\infty}(n+r)(n+r-1)c_n x^{n+r-1} + \sum_{n=0}^{\infty}(n+r)c_n x^{n+r-1}$$

$$+ \sum_{n=0}^{\infty}(n+r)c_n x^{n+r} + \sum_{n=0}^{\infty}c_n x^{n+r}$$

$$= \sum_{n=0}^{\infty}(n+r)(2n+2r-1)c_n x^{n+r-1} + \sum_{n=0}^{\infty}(n+r+1)c_n x^{n+r}$$

$$= x^r\left[r(2r-1)c_0 x^{-1} + \underbrace{\sum_{n=1}^{\infty}(n+r)(2n+2r-1)c_n x^{n-1}}_{k=n-1} + \underbrace{\sum_{n=0}^{\infty}(n+r+1)c_n x^n}_{k=n}\right]$$

$$= x^r\left[r(2r-1)c_0 x^{-1} + \sum_{k=0}^{\infty}[(k+r+1)(2k+2r+1)c_{k+1} + (k+r+1)c_k]x^k\right],$$

which implies that

$$r(2r-1) = 0 \tag{15}$$

and

$$(k+r+1)(2k+2r+1)c_{k+1} + (k+r+1)c_k = 0, \tag{16}$$

$k = 0, 1, 2, \ldots$. From (15) we see that the indicial roots are $r_1 = \frac{1}{2}$ and $r_2 = 0$.

For $r_1 = \frac{1}{2}$, we can divide by $k + \frac{3}{2}$ in (16) to obtain

$$c_{k+1} = \frac{-c_k}{2(k+1)}, \quad k = 0, 1, 2, \ldots, \tag{17}$$

whereas for $r_2 = 0$, (16) becomes

$$c_{k+1} = \frac{-c_k}{2k+1}, \quad k = 0, 1, 2, \ldots. \tag{18}$$

From (17) we find

$$c_1 = \frac{-c_0}{2 \cdot 1}$$

$$c_2 = \frac{-c_1}{2 \cdot 2} = \frac{c_0}{2^2 \cdot 2!}$$

$$c_3 = \frac{-c_2}{2 \cdot 3} = \frac{-c_0}{2^3 \cdot 3!}$$

$$c_4 = \frac{-c_3}{2 \cdot 4} = \frac{c_0}{2^4 \cdot 4!}$$

$$\vdots$$

$$c_n = \frac{(-1)^n c_0}{2^n n!}.$$

From (18) we find

$$c_1 = \frac{-c_0}{1}$$

$$c_2 = \frac{-c_1}{3} = \frac{c_0}{1 \cdot 3}$$

$$c_3 = \frac{-c_2}{5} = \frac{-c_0}{1 \cdot 3 \cdot 5}$$

$$c_4 = \frac{-c_3}{7} = \frac{c_0}{1 \cdot 3 \cdot 5 \cdot 7}$$

$$\vdots$$

$$c_n = \frac{(-1)^n c_0}{1 \cdot 3 \cdot 5 \cdot 7 \cdots (2n-1)}.$$

Thus for the indicial root $r_1 = \frac{1}{2}$ we obtain the solution

$$y_1(x) = x^{1/2}\left[1 + \sum_{n=1}^{\infty}\frac{(-1)^n}{2^n n!}x^n\right] = \sum_{n=0}^{\infty}\frac{(-1)^n}{2^n n!}x^{n+1/2},$$

where we have again omitted c_0. The series converges for $x \geq 0$; as given, the series is not defined for negative values of x because of the presence of $x^{1/2}$. For $r_2 = 0$, a second solution is

$$y_2(x) = 1 + \sum_{n=1}^{\infty}\frac{(-1)^n}{1 \cdot 3 \cdot 5 \cdot 7 \cdots (2n-1)}x^n, \quad |x| < \infty.$$

On the interval $(0, \infty)$ the general solution is $y = C_1 y_1(x) + C_2 y_2(x)$.

| **EXAMPLE 4**   **Only One Series Solution** |

Solve $xy'' + y = 0$.

SOLUTION   From $xP(x) = 0$, $x^2 Q(x) = x$, and the fact that 0 and x are their own power series centered at 0 we conclude that $a_0 = 0$ and $b_0 = 0$, and so from (14) the indicial equation is $r(r - 1) = 0$. You should verify that the two recurrence relations corresponding to the indicial roots $r_1 = 1$ and $r_2 = 0$ yield exactly the same set of coefficients. In other words, in this case the method of Frobenius produces only a single series solution

$$y_1(x) = \sum_{n=0}^{\infty} \frac{(-1)^n}{n!(n + 1)!} x^{n+1} = x - \frac{1}{2}x^2 + \frac{1}{12}x^3 - \frac{1}{144}x^4 + \cdots .$$

THREE CASES   For the sake of discussion, let us again suppose that $x = 0$ is a regular singular point of equation (1) and that the indicial roots r_1 and r_2 of the singularity are real. When using the method of Frobenius, we distinguish three cases corresponding to the nature of the indicial roots r_1 and r_2. In the first two cases the symbol r_1 denotes the largest of two distinct roots, that is, $r_1 > r_2$. In the last case, $r_1 = r_2$.

CASE I: If r_1 and r_2 are distinct and the difference $r_1 - r_2$ is not a positive integer, then there exist two linearly independent solutions of equation (1) of the form

$$y_1(x) = \sum_{n=0}^{\infty} c_n x^{n+r_1}, \quad c_0 \neq 0, \quad y_2(x) = \sum_{n=0}^{\infty} b_n x^{n+r_2}, \quad b_0 \neq 0.$$

This is the case illustrated in Examples 2 and 3.

Next we assume that the difference of the roots is N, where N is a positive integer. In this case the second solution *may* contain a logarithm.

CASE II: If r_1 and r_2 are distinct and the difference $r_1 - r_2$ is a positive integer, then there exist two linearly independent solutions of equation (1) of the form

$$y_1(x) = \sum_{n=0}^{\infty} c_n x^{n+r_1}, \quad c_0 \neq 0, \tag{19}$$

$$y_2(x) = Cy_1(x) \ln x + \sum_{n=0}^{\infty} b_n x^{n+r_2}, \quad b_0 \neq 0, \tag{20}$$

where C is a constant that could be zero.

Finally, in the last case, the case when $r_1 = r_2$, a second solution will *always* contain a logarithm. The situation is analogous to the solution of a Cauchy-Euler equation when the roots of the auxiliary equation are equal.

CASE III: If r_1 and r_2 are equal, then there always exist two linearly independent solutions of equation (1) of the form

$$y_1(x) = \sum_{n=0}^{\infty} c_n x^{n+r_1}, \quad c_0 \neq 0, \tag{21}$$

$$y_2(x) = y_1(x) \ln x + \sum_{n=1}^{\infty} b_n x^{n+r_1}. \tag{22}$$

FINDING A SECOND SOLUTION   When the difference $r_1 - r_2$ is a positive integer (Case II), we *may* or *may not* be able to find two solutions having the form $y = \sum_{n=0}^{\infty} c_n x^{n+r}$. This is something that we do not know in advance but is determined after we have found the indicial roots and have carefully examined the recurrence relation that defines the coefficients c_n. We just may be lucky enough to find two solutions that involve only powers of x, that is, $y_1(x) = \sum_{n=0}^{\infty} c_n x^{n+r_1}$ (equation (19)) and $y_2(x) = \sum_{n=0}^{\infty} b_n x^{n+r_2}$ (equation (20) with $C = 0$). See Problem 31 in Exercises 6.2. On the other hand, in Example 4 we see that the difference of the indicial roots is a positive integer ($r_1 - r_2 = 1$) and the method of Frobenius failed to give a second series solution. In this situation, equation (20), with $C \neq 0$, indicates what the second solution looks like. Finally, when the difference $r_1 - r_2$ is a zero (Case III), the method of Frobenius fails to give a second series solution; the second solution (22) always contains a logarithm and can be shown to be equivalent to (20) with $C = 1$. One way to obtain the second solution with the logarithmic term is to use the fact that

$$y_2(x) = y_1(x) \int \frac{e^{-\int P(x)\,dx}}{y_1^2(x)}\,dx \tag{23}$$

is also a solution of $y'' + P(x)y' + Q(x)y = 0$ whenever $y_1(x)$ is a known solution. We illustrate how to use (23) in the next example.

EXAMPLE 5   Example 4 Revisited Using a CAS

Find the general solution of $xy'' + y = 0$.

SOLUTION   From the known solution given in Example 4,

$$y_1(x) = x - \frac{1}{2}x^2 + \frac{1}{12}x^3 - \frac{1}{144}x^4 + \cdots,$$

we can construct a second solution $y_2(x)$ using formula (23). Those with the time, energy, and patience can carry out the drudgery of squaring a series, long division, and integration of the quotient by hand. But all these operations can be done with relative ease with the help of a CAS. We give the results:

$$y_2(x) = y_1(x) \int \frac{e^{-\int 0\,dx}}{[y_1(x)]^2}\,dx = y_1(x) \int \frac{dx}{\left[x - \frac{1}{2}x^2 + \frac{1}{12}x^3 - \frac{1}{144}x^4 + \cdots \right]^2}$$

$$= y_1(x) \int \frac{dx}{\left[x^2 - x^3 + \frac{5}{12}x^4 - \frac{7}{72}x^5 + \cdots \right]} \qquad \leftarrow \text{after squaring}$$

$$= y_1(x) \int \left[\frac{1}{x^2} + \frac{1}{x} + \frac{7}{12} + \frac{19}{72}x + \cdots \right] dx \qquad \leftarrow \text{after long division}$$

$$= y_1(x) \left[-\frac{1}{x} + \ln x + \frac{7}{12}x + \frac{19}{144}x^2 + \cdots \right] \qquad \leftarrow \text{after integrating}$$

$$= y_1(x) \ln x + y_1(x) \left[-\frac{1}{x} + \frac{7}{12}x + \frac{19}{144}x^2 + \cdots \right],$$

$$\text{or } y_2(x) = y_1(x) \ln x + \left[-1 - \frac{1}{2}x + \frac{1}{2}x^2 + \cdots \right]. \qquad \leftarrow \text{after multiplying out}$$

On the interval $(0, \infty)$ the general solution is $y = C_1 y_1(x) + C_2 y_2(x)$.

Note that the final form of y_2 in Example 5 matches (20) with $C = 1$; the series in the brackets corresponds to the summation in (20) with $r_2 = 0$.

REMARKS

(*i*) The three different forms of a linear second-order differential equation in (1), (2), and (3) were used to discuss various theoretical concepts. But on a practical level, when it comes to actually solving a differential equation using the method of Frobenius, it is advisable to work with the form of the DE given in (1).

(*ii*) When the difference of indicial roots $r_1 - r_2$ is a positive integer ($r_1 > r_2$), it sometimes pays to iterate the recurrence relation using the smaller root r_2 first. See Problems 31 and 32 in Exercises 6.2.

(*iii*) Because an indicial root r is a solution of a quadratic equation, it could be complex. We shall not, however, investigate this case.

(*iv*) If $x = 0$ is an irregular singular point, then we might not be able to find *any* solution of the DE of form $y = \sum_{n=0}^{\infty} c_n x^{n+r}$.

EXERCISES 6.2

Answers to selected odd-numbered problems begin on page ANS-8.

In Problems 1–10 determine the singular points of the given differential equation. Classify each singular point as regular or irregular.

1. $x^3 y'' + 4x^2 y' + 3y = 0$

2. $x(x + 3)^2 y'' - y = 0$

3. $(x^2 - 9)^2 y'' + (x + 3)y' + 2y = 0$

4. $y'' - \dfrac{1}{x}y' + \dfrac{1}{(x - 1)^3}y = 0$

5. $(x^3 + 4x)y'' - 2xy' + 6y = 0$

6. $x^2(x - 5)^2 y'' + 4xy' + (x^2 - 25)y = 0$

7. $(x^2 + x - 6)y'' + (x + 3)y' + (x - 2)y = 0$

8. $x(x^2 + 1)^2 y'' + y = 0$

9. $x^3(x^2 - 25)(x - 2)^2 y'' + 3x(x - 2)y' + 7(x + 5)y = 0$

10. $(x^3 - 2x^2 + 3x)^2 y'' + x(x - 3)^2 y' - (x + 1)y = 0$

In Problems 11 and 12 put the given differential equation into form (3) for each regular singular point of the equation. Identify the functions $p(x)$ and $q(x)$.

11. $(x^2 - 1)y'' + 5(x + 1)y' + (x^2 - x)y = 0$

12. $xy'' + (x + 3)y' + 7x^2 y = 0$

In Problems 13 and 14, $x = 0$ is a regular singular point of the given differential equation. Use the general form of the indicial equation in (14) to find the indicial roots of the singularity. Without solving, discuss the number of series solutions you would expect to find using the method of Frobenius.

13. $x^2 y'' + \left(\tfrac{5}{3}x + x^2\right)y' - \tfrac{1}{3}y = 0$

14. $xy'' + y' + 10y = 0$

In Problems 15–24, $x = 0$ is a regular singular point of the given differential equation. Show that the indicial roots of the singularity do not differ by an integer. Use the method of Frobenius to obtain two linearly independent series solutions about $x = 0$. Form the general solution on $(0, \infty)$.

15. $2xy'' - y' + 2y = 0$

16. $2xy'' + 5y' + xy = 0$

17. $4xy'' + \tfrac{1}{2}y' + y = 0$

18. $2x^2 y'' - xy' + (x^2 + 1)y = 0$

19. $3xy'' + (2 - x)y' - y = 0$

20. $x^2 y'' - \left(x - \tfrac{2}{9}\right)y = 0$

21. $2xy'' - (3 + 2x)y' + y = 0$

22. $x^2 y'' + xy' + \left(x^2 - \tfrac{4}{9}\right)y = 0$

23. $9x^2 y'' + 9x^2 y' + 2y = 0$

24. $2x^2 y'' + 3xy' + (2x - 1)y = 0$

In Problems 25–30, $x = 0$ is a regular singular point of the given differential equation. Show that the indicial roots of the singularity differ by an integer. Use the method of Frobenius to obtain at least one series solution about $x = 0$.

Use (23) where necessary and a CAS, if instructed, to find a second solution. Form the general solution on $(0, \infty)$.

25. $xy'' + 2y' - xy = 0$

26. $x^2y'' + xy' + \left(x^2 - \frac{1}{4}\right)y = 0$

27. $xy'' - xy' + y = 0$

28. $y'' + \frac{3}{x}y' - 2y = 0$

29. $xy'' + (1 - x)y' - y = 0$

30. $xy'' + y' + y = 0$

In Problems 31 and 32, $x = 0$ is a regular singular point of the given differential equation. Show that the indicial roots of the singularity differ by an integer. Use the recurrence relation found by the method of Frobenius first with the larger root r_1. How many solutions did you find? Next use the recurrence relation with the smaller root r_2. How many solutions did you find?

31. $xy'' + (x - 6)y' - 3y = 0$

32. $x(x - 1)y'' + 3y' - 2y = 0$

33. (a) The differential equation $x^4y'' + \lambda y = 0$ has an irregular singular point at $x = 0$. Show that the substitution $t = 1/x$ yields the DE

$$\frac{d^2y}{dt^2} + \frac{2}{t}\frac{dy}{dt} + \lambda y = 0,$$

which now has a regular singular point at $t = 0$.

(b) Use the method of this section to find two series solutions of the second equation in part (a) about the regular singular point $t = 0$.

(c) Express each series solution of the original equation in terms of elementary functions.

MATHEMATICAL MODEL

34. Buckling of a Tapered Column In Example 3 of Section 5.2, we saw that when a constant vertical compressive force or load P was applied to a thin column of uniform cross section, the deflection $y(x)$ was a solution of the boundary-value problem

$$EI\frac{d^2y}{dx^2} + Py = 0, \quad y(0) = 0, \quad y(L) = 0. \quad (24)$$

The assumption here is that the column is hinged at both ends. The column will buckle or deflect only when the compressive force is a critical load P_n.

(a) In this problem let us assume that the column is of length L, is hinged at both ends, has circular cross-sections, and is tapered as shown in Figure 6.3(a). If the column, a truncated cone, has a linear taper

$y = cx$ as shown in cross section in Figure 6.3(b), the moment of inertia of a cross section with respect to an axis perpendicular to the xy-plane is $I = \frac{1}{4}\pi r^4$, where $r = y$ and $y = cx$. Hence we can write $I(x) = I_0(x/b)^4$, where $I_0 = I(b) = \frac{1}{4}\pi(cb)^4$. Substituting $I(x)$ into the differential equation in (24), we see that the deflection in this case is determined from the BVP

$$x^4\frac{d^2y}{dx^2} + \lambda y = 0, \quad y(a) = 0, \quad y(b) = 0,$$

where $\lambda = Pb^4/EI_0$. Use the results of Problem 33 to find the critical loads P_n for the tapered column. Use an appropriate identity to express the buckling modes $y_n(x)$ as a single function.

(b) Use a CAS to plot the graph of the first buckling mode $y_1(x)$ corresponding to the Euler load P_1 when $b = 11$ and $a = 1$.

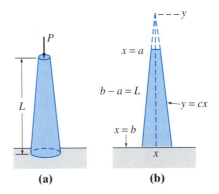

FIGURE 6.3   Tapered column in Problem 34

DISCUSSION/PROJECT PROBLEMS

35. Discuss how you would define a regular singular point for the linear third-order differential equation

$$a_3(x)y''' + a_2(x)y'' + a_1(x)y' + a_0(x)y = 0.$$

36. Each of the differential equations

$$x^3y'' + y = 0 \quad \text{and} \quad x^2y'' + (3x - 1)y' + y = 0$$

has an irregular singular point at $x = 0$. Determine whether the method of Frobenius yields a series solution of each differential equation about $x = 0$. Discuss and explain your findings.

37. We have seen that $x = 0$ is a regular singular point of any Cauchy-Euler equation $ax^2y'' + bxy' + cy = 0$. Are the indicial equation (14) for a Cauchy-Euler equation and its auxiliary equation related? Discuss.

6.3 SPECIAL FUNCTIONS

INTRODUCTION: The two differential equations

$$x^2 y'' + xy' + (x^2 - \nu^2)y = 0 \tag{1}$$

$$(1 - x^2)y'' - 2xy' + n(n + 1)y = 0 \tag{2}$$

occur in advanced studies in applied mathematics, physics, and engineering. They are called **Bessel's equation of order ν** and **Legendre's equation of order n**, respectively. When we solve (1) we shall assume that $\nu \geq 0$, whereas in (2) we shall consider only the case when n is a nonnegative integer.

REVIEW MATERIAL: In this section we will obtain series solutions of each equation about $x = 0$. Observe that the origin is a regular singular point of Bessel's DE (Section 6.2) and an ordinary point of Legendre's DE (Section 6.1).

6.3.1 BESSEL'S EQUATION

THE SOLUTION Because $x = 0$ is a regular singular point of Bessel's equation, we know that there exists at least one solution of the form $y = \sum_{n=0}^{\infty} c_n x^{n+r}$. Substituting the last expression into (1) gives

$$x^2 y'' + xy' + (x^2 - \nu^2)y = \sum_{n=0}^{\infty} c_n(n + r)(n + r - 1)x^{n+r} + \sum_{n=0}^{\infty} c_n(n + r)x^{n+r} + \sum_{n=0}^{\infty} c_n x^{n+r+2} - \nu^2 \sum_{n=0}^{\infty} c_n x^{n+r}$$

$$= c_0(r^2 - r + r - \nu^2)x^r + x^r \sum_{n=1}^{\infty} c_n[(n + r)(n + r - 1) + (n + r) - \nu^2]x^n + x^r \sum_{n=0}^{\infty} c_n x^{n+2}$$

$$= c_0(r^2 - \nu^2)x^r + x^r \sum_{n=1}^{\infty} c_n[(n + r)^2 - \nu^2]x^n + x^r \sum_{n=0}^{\infty} c_n x^{n+2}. \tag{3}$$

From (3) we see that the indicial equation is $r^2 - \nu^2 = 0$, so the indicial roots are $r_1 = \nu$ and $r_2 = -\nu$. When $r_1 = \nu$, (3) becomes

$$x^\nu \sum_{n=1}^{\infty} c_n n(n + 2\nu)x^n + x^\nu \sum_{n=0}^{\infty} c_n x^{n+2}$$

$$= x^\nu \left[(1 + 2\nu)c_1 x + \underbrace{\sum_{n=2}^{\infty} c_n n(n + 2\nu)x^n}_{k = n - 2} + \underbrace{\sum_{n=0}^{\infty} c_n x^{n+2}}_{k = n} \right]$$

$$= x^\nu \left[(1 + 2\nu)c_1 x + \sum_{k=0}^{\infty} [(k + 2)(k + 2 + 2\nu)c_{k+2} + c_k]x^{k+2} \right] = 0.$$

Therefore by the usual argument we can write $(1 + 2\nu)c_1 = 0$ and

$$(k + 2)(k + 2 + 2\nu)c_{k+2} + c_k = 0$$

or

$$c_{k+2} = \frac{-c_k}{(k + 2)(k + 2 + 2\nu)}, \quad k = 0, 1, 2, \ldots. \tag{4}$$

The choice $c_1 = 0$ in (4) implies $c_3 = c_5 = c_7 = \cdots = 0$, so for $k = 0, 2, 4, \ldots$ we find, after letting $k + 2 = 2n$, $n = 1, 2, 3, \ldots$, that

$$c_{2n} = -\frac{c_{2n-2}}{2^2 n(n + \nu)}. \tag{5}$$

Thus $\quad c_2 = -\dfrac{c_0}{2^2 \cdot 1 \cdot (1 + \nu)}$

$$c_4 = -\frac{c_2}{2^2 \cdot 2(2 + \nu)} = \frac{c_0}{2^4 \cdot 1 \cdot 2(1 + \nu)(2 + \nu)}$$

$$c_6 = -\frac{c_4}{2^2 \cdot 3(3 + \nu)} = -\frac{c_0}{2^6 \cdot 1 \cdot 2 \cdot 3(1 + \nu)(2 + \nu)(3 + \nu)}$$

$$\vdots$$

$$c_{2n} = \frac{(-1)^n c_0}{2^{2n} n!(1 + \nu)(2 + \nu) \cdots (n + \nu)}, \quad n = 1, 2, 3, \ldots. \tag{6}$$

It is standard practice to choose c_0 to be a specific value—namely,

$$c_0 = \frac{1}{2^\nu \Gamma(1 + \nu)},$$

where $\Gamma(1 + \nu)$ is the gamma function. See Appendix I. Since this latter function possesses the convenient property $\Gamma(1 + \alpha) = \alpha\Gamma(\alpha)$ we can reduce the indicated product in the denominator of (6) to one term. For example,

$$\Gamma(1 + \nu + 1) = (1 + \nu)\Gamma(1 + \nu)$$

$$\Gamma(1 + \nu + 2) = (2 + \nu)\Gamma(2 + \nu) = (2 + \nu)(1 + \nu)\Gamma(1 + \nu).$$

Hence we can write (6) as

$$c_{2n} = \frac{(-1)^n}{2^{2n+\nu} n!(1 + \nu)(2 + \nu) \cdots (n + \nu)\Gamma(1 + \nu)} = \frac{(-1)^n}{2^{2n+\nu} n!\Gamma(1 + \nu + n)}$$

for $n = 0, 1, 2, \ldots$.

BESSEL FUNCTIONS OF THE FIRST KIND Using the coefficients c_{2n} just obtained and $r = \nu$, a series solution of (1) is $y = \sum_{n=0}^{\infty} c_{2n} x^{2n+\nu}$. This solution is usually denoted by $J_\nu(x)$:

$$J_\nu(x) = \sum_{n=0}^{\infty} \frac{(-1)^n}{n!\Gamma(1 + \nu + n)} \left(\frac{x}{2}\right)^{2n+\nu}. \tag{7}$$

If $\nu \geq 0$, the series converges at least on the interval $[0, \infty)$. Also, for the second exponent $r_2 = -\nu$ we obtain, in exactly the same manner,

$$J_{-\nu}(x) = \sum_{n=0}^{\infty} \frac{(-1)^n}{n!\Gamma(1 - \nu + n)} \left(\frac{x}{2}\right)^{2n-\nu}. \tag{8}$$

The functions $J_\nu(x)$ and $J_{-\nu}(x)$ are called **Bessel functions of the first kind** of order ν and $-\nu$, respectively. Depending on the value of ν, (8) may contain negative powers of x and hence converge on $(0, \infty)$.*

Now some care must be taken in writing the general solution of (1). When $\nu = 0$, it is apparent that (7) and (8) are the same. If $\nu > 0$ and $r_1 - r_2 = \nu - (-\nu) = 2\nu$ is not a positive integer, it follows from Case I of Section 6.2 that $J_\nu(x)$ and $J_{-\nu}(x)$ are linearly independent solutions of (1) on $(0, \infty)$, and so the general solution on the interval is $y = c_1 J_\nu(x) + c_2 J_{-\nu}(x)$. But we also know from Case II of Section 6.2 that when $r_1 - r_2 = 2\nu$ is a positive integer, a second series solution of (1) *may* exist. In this second case we distinguish two possibilities. When $\nu = m = $ positive integer, $J_{-m}(x)$ defined by (8) and $J_m(x)$ are not linearly independent solutions. It can be shown that J_{-m} is a constant multiple of J_m (see Property (i) on page 263). In addition, $r_1 - r_2 = 2\nu$ can be a positive integer when ν is half an odd positive integer. It can

---

*When we replace x by $|x|$, the series given in (7) and (8) converge for $0 < |x| < \infty$.

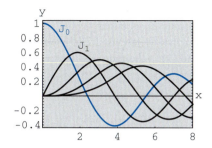

FIGURE 6.4 Bessel functions of the first kind for $n = 0, 1, 2, 3, 4$

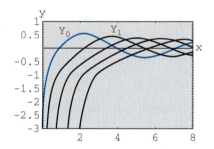

FIGURE 6.5 Bessel functions of the second kind for $n = 0, 1, 2, 3, 4$

be shown in this latter event that $J_\nu(x)$ and $J_{-\nu}(x)$ are linearly independent. In other words, the general solution of (1) on $(0, \infty)$ is

$$y = c_1 J_\nu(x) + c_2 J_{-\nu}(x), \quad \nu \neq \text{integer.} \tag{9}$$

The graphs of $y = J_0(x)$ and $y = J_1(x)$ are given in Figure 6.4.

EXAMPLE 1 **Bessel's Equation of Order $\frac{1}{2}$**

By identifying $\nu^2 = \frac{1}{4}$ and $\nu = \frac{1}{2}$, we can see from (9) that the general solution of the equation $x^2 y'' + xy' + \left(x^2 - \frac{1}{4}\right)y = 0$ on $(0, \infty)$ is $y = c_1 J_{1/2}(x) + c_2 J_{-1/2}(x)$. ∎

BESSEL FUNCTIONS OF THE SECOND KIND If $\nu \neq$ integer, the function defined by the linear combination

$$Y_\nu(x) = \frac{\cos \nu\pi J_\nu(x) - J_{-\nu}(x)}{\sin \nu\pi} \tag{10}$$

and the function $J_\nu(x)$ are linearly independent solutions of (1). Thus another form of the general solution of (1) is $y = c_1 J_\nu(x) + c_2 Y_\nu(x)$, provided $\nu \neq$ integer. As $\nu \to m$, m an integer, (10) has the indeterminate form 0/0. However, it can be shown by L'Hôpital's Rule that $\lim_{\nu \to m} Y_\nu(x)$ exists. Moreover, the function

$$Y_m(x) = \lim_{\nu \to m} Y_\nu(x)$$

and $J_m(x)$ are linearly independent solutions of $x^2 y'' + xy' + (x^2 - m^2)y = 0$. Hence for *any* value of ν the general solution of (1) on $(0, \infty)$ can be written as

$$y = c_1 J_\nu(x) + c_2 Y_\nu(x). \tag{11}$$

$Y_\nu(x)$ is called the **Bessel function of the second kind** of order ν. Figure 6.5 shows the graphs of $Y_0(x)$ and $Y_1(x)$.

EXAMPLE 2 **Bessel's Equation of Order 3**

By identifying $\nu^2 = 9$ and $\nu = 3$ we see from (11) that the general solution of the equation $x^2 y'' + xy' + (x^2 - 9)y = 0$ on $(0, \infty)$ is $y = c_1 J_3(x) + c_2 Y_3(x)$. ∎

DEs SOLVABLE IN TERMS OF BESSEL FUNCTIONS Sometimes it is possible to transform a differential equation into equation (1) by means of a change of variable. We can then express the solution of the original equation in terms of Bessel functions. For example, if we let $t = \alpha x$, $\alpha > 0$, in

$$x^2 y'' + xy' + (\alpha^2 x^2 - \nu^2)y = 0, \tag{12}$$

then by the Chain Rule,

$$\frac{dy}{dx} = \frac{dy}{dt}\frac{dt}{dx} = \alpha\frac{dy}{dt} \quad \text{and} \quad \frac{d^2 y}{dx^2} = \frac{d}{dt}\left(\frac{dy}{dx}\right)\frac{dt}{dx} = \alpha^2\frac{d^2 y}{dt^2}.$$

Accordingly, (12) becomes

$$\left(\frac{t}{\alpha}\right)^2 \alpha^2 \frac{d^2 y}{dt^2} + \left(\frac{t}{\alpha}\right)\alpha\frac{dy}{dt} + (t^2 - \nu^2)y = 0 \quad \text{or} \quad t^2\frac{d^2 y}{dt^2} + t\frac{dy}{dt} + (t^2 - \nu^2)y = 0.$$

The last equation is Bessel's equation of order ν with solution $y = c_1 J_\nu(t) + c_2 Y_\nu(t)$. By resubstituting $t = \alpha x$ in the last expression, we find that the general solution of (12) is

$$y = c_1 J_\nu(\alpha x) + c_2 Y_\nu(\alpha x). \tag{13}$$

Equation (12), called the **parametric Bessel equation of order ν,** and its general solution (13) are very important in the study of certain boundary-value problems involving partial differential equations that are expressed in cylindrical coordinates.

Another equation that bears a resemblance to (1) is the **modified Bessel equation of order ν,**

$$x^2 y'' + xy' - (x^2 + \nu^2)y = 0. \tag{14}$$

This DE can be solved in the manner just illustrated for (12). This time if we let $t = ix$, where $i^2 = -1$, then (14) becomes

$$t^2 \frac{d^2 y}{dt^2} + t \frac{dy}{dt} + (t^2 - \nu^2)y = 0.$$

Because solutions of the last DE are $J_\nu(t)$ and $Y_\nu(t)$, *complex-valued* solutions of equation (14) are $J_\nu(ix)$ and $Y_\nu(ix)$. A real-valued solution, called the **modified Bessel function of the first kind** of order ν is defined in terms of $J_\nu(ix)$:

$$I_\nu(x) = i^{-\nu} J_\nu(ix). \tag{15}$$

See Problem 21 in Exercises 6.3. Analogous to (10), the **modified Bessel function of the second kind** of order $\nu \neq$ integer is defined to be

$$K_\nu(x) = \frac{\pi}{2} \frac{I_{-\nu}(x) - I_\nu(x)}{\sin \nu \pi}, \tag{16}$$

and for integer $\nu = n$,

$$K_n(x) = \lim_{\nu \to n} K_\nu(x).$$

Because I_ν and K_ν are linearly independent on the interval $(0, \infty)$ for any value of v, the general solution of (14) is

$$y = c_1 I_\nu(x) + c_2 K_\nu(x). \tag{17}$$

Yet another equation, important because many DEs fit into its form by appropriate choices of the parameters, is

$$y'' + \frac{1 - 2a}{x} y' + \left(b^2 c^2 x^{2c-2} + \frac{a^2 - p^2 c^2}{x^2} \right) y = 0, \quad p \geq 0. \tag{18}$$

Although we shall not supply the details, the general solution of (18),

$$y = x^a \left[c_1 J_p(bx^c) + c_2 Y_p(bx^c) \right], \tag{19}$$

can be found by means of a change in both the independent and the dependent variables: $z = bx^c$, $y(x) = \left(\dfrac{z}{b} \right)^{a/c} w(z)$. If p is not an integer, then Y_p in (19) can be replaced by J_{-p}.

EXAMPLE 3 Using (18)

Find the general solution of $xy'' + 3y' + 9y = 0$ on $(0, \infty)$.

SOLUTION By writing the given DE as

$$y'' + \frac{3}{x} y' + \frac{9}{x} y = 0,$$

we can make the following identifications with (18):

$$1 - 2a = 3, \quad b^2 c^2 = 9, \quad 2c - 2 = -1, \quad \text{and} \quad a^2 - p^2 c^2 = 0.$$

The first and third equations imply that $a = -1$ and $c = \frac{1}{2}$. With these values the second and fourth equations are satisfied by taking $b = 6$ and $p = 2$. From (19)

we find that the general solution of the given DE on the interval $(0, \infty)$ is
$$y = x^{-1}[c_1 J_2(6x^{1/2}) + c_2 Y_2(6x^{1/2})].$$

EXAMPLE 4   The Aging Spring Revisited

Recall that in Section 5.1 we saw that one mathematical model for the free undamped motion of a mass on an aging spring is given by $mx'' + ke^{-\alpha t}x = 0$, $\alpha > 0$. We are now in a position to find the general solution of the equation. It is left as a problem to show that the change of variables $s = \dfrac{2}{\alpha}\sqrt{\dfrac{k}{m}}\,e^{-\alpha t/2}$ transforms the differential equation of the aging spring into

$$s^2 \frac{d^2 x}{ds^2} + s \frac{dx}{ds} + s^2 x = 0.$$

The last equation is recognized as (1) with $\nu = 0$ and where the symbols x and s play the roles of y and x, respectively. The general solution of the new equation is $x = c_1 J_0(s) + c_2 Y_0(s)$. If we resubstitute s, then the general solution of $mx'' + ke^{-\alpha t}x = 0$ is seen to be

$$x(t) = c_1 J_0\left(\frac{2}{\alpha}\sqrt{\frac{k}{m}}\,e^{-\alpha t/2}\right) + c_2 Y_0\left(\frac{2}{\alpha}\sqrt{\frac{k}{m}}\,e^{-\alpha t/2}\right).$$

See Problems 33 and 39 in Exercises 6.3.

The other model discussed in Section 5.1 of a spring whose characteristics change with time was $mx'' + ktx = 0$. By dividing through by m, we see that the equation $x'' + \dfrac{k}{m}tx = 0$ is Airy's equation $y'' + \alpha^2 xy = 0$. See Example 3 in Section 6.1. The general solution of Airy's differential equation can also be written in terms of Bessel functions. See Problems 34, 35, and 40 in Exercises 6.3.

PROPERTIES   We list below a few of the more useful properties of Bessel functions of order m, $m = 0, 1, 2, \ldots$:

(i) $J_{-m}(x) = (-1)^m J_m(x)$, (ii) $J_m(-x) = (-1)^m J_m(x)$,

(iii) $J_m(0) = \begin{cases} 0, & m > 0 \\ 1, & m = 0, \end{cases}$ (iv) $\displaystyle\lim_{x \to 0^+} Y_m(x) = -\infty.$

Note that Property (ii) indicates that $J_m(x)$ is an even function if m is an even integer and an odd function if m is an odd integer. The graphs of $Y_0(x)$ and $Y_1(x)$ in Figure 6.5 illustrate Property (iv), namely, $Y_m(x)$ is unbounded at the origin. This last fact is not obvious from (10). The solutions of the Bessel equation of order 0 can be obtained by using the solutions $y_1(x)$ in (21) and $y_2(x)$ in (22) of Section 6.2. It can be shown that (21) of Section 6.2 is $y_1(x) = J_0(x)$, whereas (22) of that section is

$$y_2(x) = J_0(x)\ln x - \sum_{k=1}^{\infty} \frac{(-1)^k}{(k!)^2}\left(1 + \frac{1}{2} + \cdots + \frac{1}{k}\right)\left(\frac{x}{2}\right)^{2k}.$$

The Bessel function of the second kind of order 0, $Y_0(x)$, is then defined to be the linear combination $Y_0(x) = \dfrac{2}{\pi}(\gamma - \ln 2)y_1(x) + \dfrac{2}{\pi}y_2(x)$ for $x > 0$. That is,

$$Y_0(x) = \frac{2}{\pi}J_0(x)\left[\gamma + \ln\frac{x}{2}\right] - \frac{2}{\pi}\sum_{k=1}^{\infty} \frac{(-1)^k}{(k!)^2}\left(1 + \frac{1}{2} + \cdots + \frac{1}{k}\right)\left(\frac{x}{2}\right)^{2k},$$

where $\gamma = 0.57721566\ldots$ is **Euler's constant.** Because of the presence of the logarithmic term, it is apparent that $Y_0(x)$ is discontinuous at $x = 0$.

NUMERICAL VALUES The first five nonnegative zeros of $J_0(x)$, $J_1(x)$, $Y_0(x)$, and $Y_1(x)$ are given in Table 6.1. Some additional functional values of these four functions are given in Table 6.2.

TABLE 6.1 Zeros of J_0, J_1, Y_0, and Y_1

$J_0(x)$	$J_1(x)$	$Y_0(x)$	$Y_1(x)$
2.4048	0.0000	0.8936	2.1971
5.5201	3.8317	3.9577	5.4297
8.6537	7.0156	7.0861	8.5960
11.7915	10.1735	10.2223	11.7492
14.9309	13.3237	13.3611	14.8974

TABLE 6.2 Numerical Values of J_0, J_1, Y_0, and Y_1

x	$J_0(x)$	$J_1(x)$	$Y_0(x)$	$Y_1(x)$
0	1.0000	0.0000	—	—
1	0.7652	0.4401	0.0883	−0.7812
2	0.2239	0.5767	0.5104	−0.1070
3	−0.2601	0.3391	0.3769	0.3247
4	−0.3971	−0.0660	−0.0169	0.3979
5	−0.1776	−0.3276	−0.3085	0.1479
6	0.1506	−0.2767	−0.2882	−0.1750
7	0.3001	−0.0047	−0.0259	−0.3027
8	0.1717	0.2346	0.2235	−0.1581
9	−0.0903	0.2453	0.2499	0.1043
10	−0.2459	0.0435	0.0557	0.2490
11	−0.1712	−0.1768	−0.1688	0.1637
12	0.0477	−0.2234	−0.2252	−0.0571
13	0.2069	−0.0703	−0.0782	−0.2101
14	0.1711	0.1334	0.1272	−0.1666
15	−0.0142	0.2051	0.2055	0.0211

DIFFERENTIAL RECURRENCE RELATION Recurrence formulas that relate Bessel functions of different orders are important in theory and in applications. In the next example we derive a **differential recurrence relation.**

EXAMPLE 5   **Derivation Using the Series Definition**

Derive the formula $xJ_\nu'(x) = \nu J_\nu(x) - xJ_{\nu+1}(x)$.

SOLUTION It follows from (7) that

$$xJ_\nu'(x) = \sum_{n=0}^{\infty} \frac{(-1)^n(2n + \nu)}{n!\,\Gamma(1 + \nu + n)} \left(\frac{x}{2}\right)^{2n+\nu}$$

$$= \nu \sum_{n=0}^{\infty} \frac{(-1)^n}{n!\,\Gamma(1 + \nu + n)} \left(\frac{x}{2}\right)^{2n+\nu} + 2 \sum_{n=0}^{\infty} \frac{(-1)^n n}{n!\,\Gamma(1 + \nu + n)} \left(\frac{x}{2}\right)^{2n+\nu}$$

$$= \nu J_\nu(x) + x \underbrace{\sum_{n=1}^{\infty} \frac{(-1)^n}{(n-1)!\,\Gamma(1 + \nu + n)} \left(\frac{x}{2}\right)^{2n+\nu-1}}_{k \,=\, n-1}$$

$$= \nu J_\nu(x) - x \sum_{k=0}^{\infty} \frac{(-1)^k}{k!\,\Gamma(2 + \nu + k)} \left(\frac{x}{2}\right)^{2k+\nu+1} = \nu J_\nu(x) - xJ_{\nu+1}(x).$$

The result in Example 5 can be written in an alternative form. Dividing $xJ_\nu'(x) - \nu J_\nu(x) = -xJ_{\nu+1}(x)$ by x gives

$$J_\nu'(x) - \frac{\nu}{x} J_\nu(x) = -J_{\nu+1}(x).$$

This last expression is recognized as a linear first-order differential equation in $J_\nu(x)$. Multiplying both sides of the equality by the integrating factor $x^{-\nu}$ then yields

$$\frac{d}{dx}[x^{-\nu}J_\nu(x)] = -x^{-\nu}J_{\nu+1}(x). \tag{20}$$

It can be shown in a similar manner that

$$\frac{d}{dx}[x^\nu J_\nu(x)] = x^\nu J_{\nu-1}(x). \tag{21}$$

See Problem 27 in Exercises 6.3. The differential recurrence relations (20) and (21) are also valid for the Bessel function of the second kind $Y_\nu(x)$. Observe that when $\nu = 0$ it follows from (20) that

$$J_0'(x) = -J_1(x) \quad \text{and} \quad Y_0'(x) = -Y_1(x). \tag{22}$$

An application of these results is given in Problem 39 of Exercises 6.3.

SPHERICAL BESSEL FUNCTIONS When the order ν is half an odd integer, that is, $\pm\frac{1}{2}, \pm\frac{3}{2}, \pm\frac{5}{2}, \ldots$, the Bessel functions of the first kind $J_\nu(x)$ can be expressed in terms of the elementary functions $\sin x$, $\cos x$, and powers of x. Such Bessel functions are called **spherical Bessel functions.** Let's consider the case when $\nu = \frac{1}{2}$. From (7),

$$J_{1/2}(x) = \sum_{n=0}^\infty \frac{(-1)^n}{n!\Gamma(1 + \frac{1}{2} + n)} \left(\frac{x}{2}\right)^{2n+1/2}.$$

In view of the property $\Gamma(1 + \alpha) = \alpha\Gamma(\alpha)$ and the fact that $\Gamma(\frac{1}{2}) = \sqrt{\pi}$ the values of $\Gamma(1 + \frac{1}{2} + n)$ for $n = 0$, $n = 1$, $n = 2$, and $n = 3$ are, respectively,

$$\Gamma\left(\frac{3}{2}\right) = \Gamma\left(1 + \frac{1}{2}\right) = \frac{1}{2}\Gamma\left(\frac{1}{2}\right) = \frac{1}{2}\sqrt{\pi}$$

$$\Gamma\left(\frac{5}{2}\right) = \Gamma\left(1 + \frac{3}{2}\right) = \frac{3}{2}\Gamma\left(\frac{3}{2}\right) = \frac{3}{2^2}\sqrt{\pi}$$

$$\Gamma\left(\frac{7}{2}\right) = \Gamma\left(1 + \frac{5}{2}\right) = \frac{5}{2}\Gamma\left(\frac{5}{2}\right) = \frac{5 \cdot 3}{2^3}\sqrt{\pi} = \frac{5 \cdot 4 \cdot 3 \cdot 2 \cdot 1}{2^3 4 \cdot 2}\sqrt{\pi} = \frac{5!}{2^5 2!}\sqrt{\pi}$$

$$\Gamma\left(\frac{9}{2}\right) = \Gamma\left(1 + \frac{7}{2}\right) = \frac{7}{2}\Gamma\left(\frac{7}{2}\right) = \frac{7 \cdot 5}{2^6 \cdot 2!}\sqrt{\pi} = \frac{7 \cdot 6 \cdot 5!}{2^6 \cdot 6 \cdot 2!}\sqrt{\pi} = \frac{7!}{2^7 3!}\sqrt{\pi}.$$

In general, $$\Gamma\left(1 + \frac{1}{2} + n\right) = \frac{(2n + 1)!}{2^{2n+1}n!}\sqrt{\pi}.$$

Hence $$J_{1/2}(x) = \sum_{n=0}^\infty \frac{(-1)^n}{n!\dfrac{(2n+1)!}{2^{2n+1}n!}\sqrt{\pi}} \left(\frac{x}{2}\right)^{2n+1/2} = \sqrt{\frac{2}{\pi x}} \sum_{n=0}^\infty \frac{(-1)^n}{(2n+1)!}x^{2n+1}.$$

Since the infinite series in the last line is the Maclaurin series for $\sin x$, we have shown that

$$J_{1/2}(x) = \sqrt{\frac{2}{\pi x}}\sin x. \tag{23}$$

It is left as an exercise to show that

$$J_{-1/2}(x) = \sqrt{\frac{2}{\pi x}}\cos x. \tag{24}$$

See Problems 31 and 32 in Exercises 6.3.

6.3.2 LEGENDRE'S EQUATION

THE SOLUTION Since $x = 0$ is an ordinary point of Legendre's equation (2), we substitute the series $y = \sum_{k=0}^{\infty} c_k x^k$, shift summation indices, and combine series to get

$$(1 - x^2)y'' - 2xy' + n(n + 1)y = [n(n + 1)c_0 + 2c_2] + [(n - 1)(n + 2)c_1 + 6c_3]x$$

$$+ \sum_{j=2}^{\infty} [(j + 2)(j + 1)c_{j+2} + (n - j)(n + j + 1)c_j]x^j = 0$$

which implies that

$$n(n + 1)c_0 + 2c_2 = 0$$

$$(n - 1)(n + 2)c_1 + 6c_3 = 0$$

$$(j + 2)(j + 1)c_{j+2} + (n - j)(n + j + 1)c_j = 0$$

or

$$c_2 = -\frac{n(n + 1)}{2!}c_0$$

$$c_3 = -\frac{(n - 1)(n + 2)}{3!}c_1$$

$$c_{j+2} = -\frac{(n - j)(n + j + 1)}{(j + 2)(j + 1)}c_j, \quad j = 2, 3, 4, \ldots. \tag{25}$$

If we let j take on the values $2, 3, 4, \ldots$, the recurrence relation (25) yields

$$c_4 = -\frac{(n - 2)(n + 3)}{4 \cdot 3}c_2 = \frac{(n - 2)n(n + 1)(n + 3)}{4!}c_0$$

$$c_5 = -\frac{(n - 3)(n + 4)}{5 \cdot 4}c_3 = \frac{(n - 3)(n - 1)(n + 2)(n + 4)}{5!}c_1$$

$$c_6 = -\frac{(n - 4)(n + 5)}{6 \cdot 5}c_4 = -\frac{(n - 4)(n - 2)n(n + 1)(n + 3)(n + 5)}{6!}c_0$$

$$c_7 = -\frac{(n - 5)(n + 6)}{7 \cdot 6}c_5 = -\frac{(n - 5)(n - 3)(n - 1)(n + 2)(n + 4)(n + 6)}{7!}c_1$$

and so on. Thus for at least $|x| < 1$ we obtain two linearly independent power series solutions:

$$y_1(x) = c_0\left[1 - \frac{n(n + 1)}{2!}x^2 + \frac{(n - 2)n(n + 1)(n + 3)}{4!}x^4\right.$$

$$\left. - \frac{(n - 4)(n - 2)n(n + 1)(n + 3)(n + 5)}{6!}x^6 + \cdots\right],$$

$$y_2(x) = c_1\left[x - \frac{(n - 1)(n + 2)}{3!}x^3 + \frac{(n - 3)(n - 1)(n + 2)(n + 4)}{5!}x^5\right.$$

$$\left. - \frac{(n - 5)(n - 3)(n - 1)(n + 2)(n + 4)(n + 6)}{7!}x^7 + \cdots\right]. \tag{26}$$

Notice that if n is an even integer, the first series terminates, whereas $y_2(x)$ is an infinite series. For example, if $n = 4$, then

$$y_1(x) = c_0\left[1 - \frac{4 \cdot 5}{2!}x^2 + \frac{2 \cdot 4 \cdot 5 \cdot 7}{4!}x^4\right] = c_0\left[1 - 10x^2 + \frac{35}{3}x^4\right].$$

Similarly, when n is an odd integer, the series for $y_2(x)$ terminates with x^n; that is, *when n is a nonnegative integer, we obtain an nth-degree polynomial solution* of Legendre's equation.

Because we know that a constant multiple of a solution of Legendre's equation is also a solution, it is traditional to choose specific values for c_0 or c_1, depending on whether n is an even or odd positive integer, respectively. For $n = 0$ we choose $c_0 = 1$, and for $n = 2, 4, 6, \ldots$,

$$c_0 = (-1)^{n/2} \frac{1 \cdot 3 \cdots (n-1)}{2 \cdot 4 \cdots n},$$

whereas for $n = 1$ we choose $c_1 = 1$, and for $n = 3, 5, 7, \ldots$

$$c_1 = (-1)^{(n-1)/2} \frac{1 \cdot 3 \cdots n}{2 \cdot 4 \cdots (n-1)}.$$

For example, when $n = 4$, we have

$$y_1(x) = (-1)^{4/2} \frac{1 \cdot 3}{2 \cdot 4} \left[1 - 10x^2 + \frac{35}{3} x^4 \right] = \frac{1}{8}(35x^4 - 30x^2 + 3).$$

LEGENDRE POLYNOMIALS These specific nth-degree polynomial solutions are called **Legendre polynomials** and are denoted by $P_n(x)$. From the series for $y_1(x)$ and $y_2(x)$ and from the above choices of c_0 and c_1 we find that the first several Legendre polynomials are

$$P_0(x) = 1, \qquad\qquad P_1(x) = x,$$

$$P_2(x) = \frac{1}{2}(3x^2 - 1), \qquad P_3(x) = \frac{1}{2}(5x^3 - 3x), \qquad (27)$$

$$P_4(x) = \frac{1}{8}(35x^4 - 30x^2 + 3), \quad P_5(x) = \frac{1}{8}(63x^5 - 70x^3 + 15x).$$

Remember, $P_0(x)$, $P_1(x)$, $P_2(x)$, $P_3(x)$, $\ldots$ are, in turn, particular solutions of the differential equations

$$
\begin{aligned}
n = 0: & \quad (1 - x^2)y'' - 2xy' = 0, \\
n = 1: & \quad (1 - x^2)y'' - 2xy' + 2y = 0, \\
n = 2: & \quad (1 - x^2)y'' - 2xy' + 6y = 0, \\
n = 3: & \quad (1 - x^2)y'' - 2xy' + 12y = 0, \\
& \qquad \vdots \qquad\qquad \vdots
\end{aligned}
\qquad (28)
$$

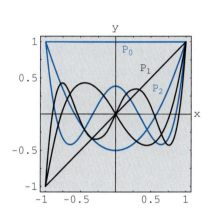

FIGURE 6.6 Legendre polynomials for $n = 0, 1, 2, 3, 4, 5, 6$

The graphs, on the interval $-1 \le x \le 1$, of the six Legendre polynomials in (27) are given in Figure 6.6.

PROPERTIES You are encouraged to verify the following properties using the Legendre polynomials in (27).

$$(i) \ \ P_n(-x) = (-1)^n P_n(x)$$

$$(ii) \ \ P_n(1) = 1 \qquad\qquad (iii) \ \ P_n(-1) = (-1)^n$$

$$(iv) \ \ P_n(0) = 0, \quad n \text{ odd} \qquad (v) \ \ P_n'(0) = 0, \quad n \text{ even}$$

Property (i) indicates, as is apparent in Figure 6.6, that $P_n(x)$ is an even or odd function according to whether n is even or odd.

RECURRENCE RELATION Recurrence relations that relate Legendre polynomials of different degrees are also important in some aspects of their applications. We state, without proof, the three-term recurrence relation

$$(k + 1)P_{k+1}(x) - (2k + 1)xP_k(x) + kP_{k-1}(x) = 0, \qquad (29)$$

which is valid for $k = 1, 2, 3, \ldots$. In (27) we listed the first six Legendre polynomials. If, say, we wish to find $P_6(x)$, we can use (29) with $k = 5$. This relation expresses $P_6(x)$ in terms of the known $P_4(x)$ and $P_5(x)$. See Problem 45 in Exercises 6.3.

Another formula, although not a recurrence relation, can generate the Legendre polynomials by differentiation. **Rodrigues' formula** for these polynomials is

$$P_n(x) = \frac{1}{2^n n!} \frac{d^n}{dx^n} (x^2 - 1)^n, \quad n = 0, 1, 2, \ldots. \tag{30}$$

See Problem 48 in Exercises 6.3.

REMARKS

(*i*) Although we have assumed that the parameter n in Legendre's differential equation $(1 - x^2)y'' - 2xy' + n(n + 1)y = 0$, represented a nonnegative integer, in a more general setting n can represent any real number. Any solution of Legendre's equation is called a **Legendre function.** If n is *not* a nonnegative integer, then both Legendre functions $y_1(x)$ and $y_2(x)$ given in (26) are infinite series convergent on the open interval $-1 < x < 1$ and divergent (unbounded) at $x = \pm 1$. If n is a nonnegative integer, then as we have just seen one of the Legendre functions in (26) is a polynomial and the other is an infinite series convergent for $-1 < x < 1$. You should be aware of the fact that Legendre's equation possesses solutions that are bounded on the *closed* interval $-1 \leq x \leq 1$ only in the case when $n = 0, 1, 2, \ldots$. More to the point, the only Legendre functions that are bounded on the closed interval $-1 \leq x \leq 1$ are the Legendre polynomials $P_n(x)$ or constant multiples of these polynomials. See Problem 47 in Exercises 6.3 and Problem 24 in Chapter 6 in Review.

(*ii*) In the *Remarks* at the end of Section 2.3 we mentioned the branch of mathematics called **special functions.** Perhaps a better appellation for this field of applied mathematics might be *named functions,* since many of the functions studied bear proper names: Bessel functions, Legendre functions, Airy functions, Chebyshev polynomials, Gauss's hypergeometric function, Hermite polynomials, Jacobi polynomials, Laguerre polynomials, Mathieu functions, Weber functions, and so on. Historically, special functions were the byproduct of necessity; someone needed a solution of a very specialized differential equation that arose from an attempt to solve a physical problem.

EXERCISES 6.3

Answers to selected odd-numbered problems begin on page ANS-8.

6.3.1   BESSEL'S EQUATION

In Problems 1–6 use (1) to find the general solution of the given differential equation on $(0, \infty)$.

1. $x^2y'' + xy' + \left(x^2 - \frac{1}{9}\right)y = 0$

2. $x^2y'' + xy' + (x^2 - 1)y = 0$

3. $4x^2y'' + 4xy' + (4x^2 - 25)y = 0$

4. $16x^2y'' + 16xy' + (16x^2 - 1)y = 0$

5. $xy'' + y' + xy = 0$

6. $\frac{d}{dx}[xy'] + \left(x - \frac{4}{x}\right)y = 0$

In Problems 7–10 use (12) to find the general solution of the given differential equation on $(0, \infty)$.

7. $x^2y'' + xy' + (9x^2 - 4)y = 0$

8. $x^2y'' + xy' + \left(36x^2 - \frac{1}{4}\right)y = 0$

9. $x^2y'' + xy' + \left(25x^2 - \frac{4}{9}\right)y = 0$

10. $x^2y'' + xy' + (2x^2 - 64)y = 0$

In Problems 11 and 12 use the indicated change of variable to find the general solution of the given differential equation on $(0, \infty)$.

11. $x^2 y'' + 2xy' + \alpha^2 x^2 y = 0; \quad y = x^{-1/2} v(x)$

12. $x^2 y'' + \left(\alpha^2 x^2 - \nu^2 + \frac{1}{4}\right) y = 0; \quad y = \sqrt{x}\, v(x)$

In Problems 13–20 use (18) to find the general solution of the given differential equation on $(0, \infty)$.

13. $xy'' + 2y' + 4y = 0$ **14.** $xy'' + 3y' + xy = 0$

15. $xy'' - y' + xy = 0$ **16.** $xy'' - 5y' + xy = 0$

17. $x^2 y'' + (x^2 - 2)y = 0$ **18.** $4x^2 y'' + (16x^2 + 1)y = 0$

19. $xy'' + 3y' + x^3 y = 0$

20. $9x^2 y'' + 9xy' + (x^6 - 36)y = 0$

21. Use the series in (7) to verify that $I_\nu(x) = i^{-\nu} J_\nu(ix)$ is a real function.

22. Assume that b in equation (18) can be pure imaginary, that is, $b = \beta i$, $\beta > 0$, $i^2 = -1$. Use this assumption to express the general solution of the given differential equation in terms the modified Bessel functions I_n and K_n.
   (a) $y'' - x^2 y = 0$ **(b)** $xy'' + y' - 7x^3 y = 0$

In Problems 23–26 first use (18) to express the general solution of the given differential equation in terms of Bessel functions. Then use (23) and (24) to express the general solution in terms of elementary functions.

23. $y'' + y = 0$

24. $x^2 y'' + 4xy' + (x^2 + 2)y = 0$

25. $16x^2 y'' + 32xy' + (x^4 - 12)y = 0$

26. $4x^2 y'' - 4xy' + (16x^2 + 3)y = 0$

27. (a) Proceed as in Example 5 to show that
$$x J_\nu'(x) = -\nu J_\nu(x) + x J_{\nu-1}(x).$$
   (*Hint:* Write $2n + \nu = 2(n + \nu) - \nu$.)
   (b) Use the result in part (a) to derive (21).

28. Use the formula obtained in Example 5 along with part (a) of Problem 27 to derive the recurrence relation
$$2\nu J_\nu(x) = x J_{\nu+1}(x) + x J_{\nu-1}(x).$$

In Problems 29 and 30 use (20) or (21) to obtain the given result.

29. $\displaystyle\int_0^x r J_0(r)\,dr = x J_1(x)$

30. $J_0'(x) = J_{-1}(x) = -J_1(x)$

31. Proceed as on page 265 to derive the elementary form of $J_{-1/2}(x)$ given in (24).

32. (a) Use the recurrence relation in Problem 28 along with (23) and (24) to express $J_{3/2}(x)$, $J_{-3/2}(x)$, and $J_{5/2}(x)$ in terms of $\sin x$, $\cos x$, and powers of x.

(b) Use a graphing utility to graph $J_{1/2}(x)$, $J_{-1/2}(x)$, $J_{3/2}(x)$, $J_{-3/2}(x)$, and $J_{5/2}(x)$.

33. Use the change of variables $s = \dfrac{2}{\alpha}\sqrt{\dfrac{k}{m}}\, e^{-\alpha t/2}$ to show that the differential equation of the aging spring $mx'' + ke^{-\alpha t} x = 0$, $\alpha > 0$, becomes
$$s^2 \frac{d^2 x}{ds^2} + s \frac{dx}{ds} + s^2 x = 0.$$

34. Show that $y = x^{1/2} w\left(\frac{2}{3}\alpha x^{3/2}\right)$ is a solution of Airy's differential equation $y'' + \alpha^2 xy = 0$, $x > 0$, whenever w is a solution of Bessel's equation of order $\frac{1}{3}$, that is, $t^2 w'' + tw' + \left(t^2 - \frac{1}{9}\right)w = 0$, $t > 0$. (*Hint:* After differentiating, substituting, and simplifying, then let $t = \frac{2}{3}\alpha x^{3/2}$.)

35. (a) Use the result of Problem 34 to express the general solution of Airy's differential equation for $x > 0$ in terms of Bessel functions.
   (b) Verify the results in part (a) using (18).

36. Use the Table 6.1 to find the first three positive eigenvalues and corresponding eigenfunctions of the boundary-value problem
$$xy'' + y' + \lambda xy = 0,$$
$$y(x),\ y'(x) \text{ bounded as } x \to 0^+, \quad y(2) = 0.$$
(*Hint:* By identifying $\lambda = \alpha^2$, the DE is the parametric Bessel equation of order zero.)

37. (a) Use (18) to show that the general solution of the differential equation $xy'' + \lambda y = 0$ on the interval $(0, \infty)$ is
$$y = c_1 \sqrt{x}\, J_1\left(2\sqrt{\lambda x}\right) + c_2 \sqrt{x}\, Y_1\left(2\sqrt{\lambda x}\right).$$
   (b) Verify by direct substitution that $y = \sqrt{x}\, J_1\left(2\sqrt{x}\right)$ is a particular solution of the DE in the case $\lambda = 1$.

Computer Lab Assignments

38. Use a CAS to graph the modified Bessel functions $I_0(x)$, $I_1(x)$, $I_2(x)$ and $K_0(x)$, $K_1(x)$, $K_2(x)$. Compare these graphs with those shown in Figures 6.4 and 6.5. What major difference is apparent between Bessel functions and the modified Bessel functions?

39. (a) Use the general solution given in Example 4 to solve the IVP
$$4x'' + e^{-0.1t} x = 0, \quad x(0) = 1, \quad x'(0) = -\tfrac{1}{2}.$$
   Also use $J_0'(x) = -J_1(x)$ and $Y_0'(x) = -Y_1(x)$ along with Table 6.1 or a CAS to evaluate coefficients.
   (b) Use a CAS to graph the solution obtained in part (a) over the interval $0 \le t \le \infty$.

40. (a) Use the general solution obtained in Problem 35 to solve the IVP

$$4x'' + tx = 0, \quad x(0.1) = 1, \quad x'(0.1) = -\tfrac{1}{2}.$$

Use a CAS to evaluate coefficients.

(b) Use a CAS to graph the solution obtained in part (a) over the interval $0 \le t \le 200$.

41. Column Bending Under Its Own Weight A uniform thin column of length L, positioned vertically with one end embedded in the ground, will deflect, or bend away, from the vertical under the influence of its own weight when its length or height exceeds a certain critical value. It can be shown that the angular deflection $\theta(x)$ of the column from the vertical at a point $P(x)$ is a solution of the boundary-value problem:

$$EI \frac{d^2\theta}{dx^2} + \delta g(L - x)\theta = 0, \quad \theta(0) = 0, \quad \theta'(L) = 0,$$

where E is Young's modulus, I is the cross-sectional moment of inertia, δ is the constant linear density, and x is the distance along the column measured from its base. See Figure 6.7. The column will bend only for those values of L for which the boundary-value problem has a nontrivial solution.

(a) Restate the boundary-value problem by making the change of variables $t = L - x$. Then use the results of a problem earlier in this exercise set to express the general solution of the differential equation in terms of Bessel functions.

(b) Use the general solution found in part (a) to find a solution of the BVP and an equation which defines the critical length L, that is, the smallest value of L for which the column will start to bend.

(c) With the aid of a CAS, find the critical length L of a solid steel rod of radius $r = 0.05$ in., $\delta g = 0.28\, A$ lb/in., $E = 2.6 \times 10^7$ lb/in.², $A = \pi r^2$, and $I = \tfrac{1}{4}\pi r^4$.

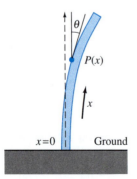

$P(x)$

x

$x = 0$ Ground

FIGURE 6.7 Beam in Problem 41

42. Buckling of a Thin Vertical Column In Example 3 of Section 5.2 we saw that when a constant vertical

compressive force, or load, P was applied to a thin column of uniform cross section and hinged at both ends, the deflection $y(x)$ is a solution of the BVP:

$$EI \frac{d^2y}{dx^2} + Py = 0, \quad y(0) = 0, \quad y(L) = 0.$$

(a) If the bending stiffness factor EI is proportional to x, then $EI(x) = kx$, where k is a constant of proportionality. If $EI(L) = kL = M$ is the maximum stiffness factor, then $k = M/L$ and so $EI(x) = Mx/L$. Use the information in Problem 37 to find a solution of

$$M\frac{x}{L} \frac{d^2y}{dx^2} + Py = 0, \quad y(0) = 0, \quad y(L) = 0$$

if it is known that $\sqrt{x}\, Y_1\!\left(2\sqrt{\lambda x}\right)$ is *not* zero at $x = 0$.

(b) Use Table 6.1 to find the Euler load P_1 for the column.

(c) Use a CAS to graph the first buckling mode $y_1(x)$ corresponding to the Euler load P_1. For simplicity assume that $c_1 = 1$ and $L = 1$.

43. Pendulum of Varying Length For the simple pendulum described on page 222 of Section 5.3, suppose that the rod holding the mass m at one end is replaced by a flexible wire or string and that the wire is strung over a pulley at the point of support O in Figure 5.35. In this manner, while it is in motion in a vertical plane, the mass m can be raised or lowered. In other words, the length $l(t)$ of the pendulum varies with time. Under the same assumptions leading to equation (6) in Section 5.3, it can be shown* that the differential equation for the displacement angle θ is now

$$l\theta'' + 2l'\theta' + g\sin\theta = 0.$$

(a) If l increases at constant rate v and if $l(0) = l_0$, show that a linearization of the foregoing DE is

$$(l_0 + vt)\theta'' + 2v\theta' + g\theta = 0. \tag{31}$$

(b) Make the change of variables $x = (l_0 + vt)/v$ and show that (31) becomes

$$\frac{d^2\theta}{dx^2} + \frac{2}{x}\frac{d\theta}{dx} + \frac{g}{vx}\theta = 0.$$

(c) Use part (b) and (18) to express the general solution of equation (31) in terms of Bessel functions.

(d) Use the general solution obtained in part (c) to solve the initial-value problem consisting of equation (31) and the initial conditions $\theta(0) = \theta_0$, $\theta'(0) = 0$. (*Hints:* To simplify calculations, use a further change of variable $u = \dfrac{2}{v}\sqrt{g(l_0 + vt)} = 2\sqrt{\dfrac{g}{v}}\, x^{1/2}$.

*See *Mathematical Methods in Physical Sciences,* Mary Boas, John Wiley & Sons, Inc., 1966. Also see the article by Borelli, Coleman, and Hobson in *Mathematics Magazine,* vol. 58, no. 2, March 1985.

Also, recall that (20) holds for both $J_1(u)$ and $Y_1(u)$. Finally, the identity

$$J_1(u)Y_2(u) - J_2(u)Y_1(u) = -\frac{2}{\pi u}$$ will be helpful.)

(e) Use a CAS to graph the solution $\theta(t)$ of the IVP in part (d) when $l_0 = 1$ ft, $\theta_0 = \frac{1}{10}$ radian, and $v = \frac{1}{60}$ ft/s. Experiment with the graph using different time intervals such as $[0, 10]$, $[0, 30]$, and so on.

(f) What do the graphs indicate about the displacement angle $\theta(t)$ as the length l of the wire increases with time?

6.3.2 LEGENDRE'S EQUATION

44. (a) Use the explicit solutions $y_1(x)$ and $y_2(x)$ of Legendre's equation given in (26) and the appropriate choice of c_0 and c_1 to find the Legendre polynomials $P_6(x)$ and $P_7(x)$.

(b) Write the differential equations for which $P_6(x)$ and $P_7(x)$ are particular solutions.

45. Use the recurrence relation (29) and $P_0(x) = 1$, $P_1(x) = x$, to generate the next six Legendre polynomials.

46. Show that the differential equation

$$\sin\theta \frac{d^2y}{d\theta^2} + \cos\theta \frac{dy}{d\theta} + n(n+1)(\sin\theta)y = 0$$

can be transformed into Legendre's equation by means of the substitution $x = \cos\theta$.

47. Find the first three positive values of λ for which the problem

$$(1 - x^2)y'' - 2xy' + \lambda y = 0,$$

$$y(0) = 0, \quad y(x), y'(x) \text{ bounded on } [-1,1]$$

has nontrivial solutions.

COMPUTER LAB ASSIGNMENTS

48. For purposes of this problem ignore the list of Legendre polynomials given on page 267 and the graphs given in Figure 6.6. Use Rodrigues' formula (30) to generate the Legendre polynomials $P_1(x)$, $P_2(x), \ldots, P_7(x)$. Use a CAS to carry out the differentiations and simplifications.

49. Use a CAS to graph $P_1(x)$, $P_2(x), \ldots, P_7(x)$ on the interval $[-1, 1]$.

50. Use a root-finding application to find the zeros of $P_1(x)$, $P_2(x), \ldots, P_7(x)$. If the Legendre polynomials are built-in functions of your CAS, find zeros of Legendre polynomials of higher degree. Form a conjecture about the location of the zeros of any Legendre polynomial $P_n(x)$, and then investigate to see whether it is true.

CHAPTER 6 IN REVIEW

Answers to selected odd-numbered problems begin on page ANS-8.

In Problems 1 and 2 answer true or false without referring back to the text.

1. The general solution of $x^2y'' + xy' + (x^2 - 1)y = 0$ is $y = c_1 J_1(x) + c_2 J_{-1}(x)$. _____

2. Because $x = 0$ is an irregular singular point of $x^3y'' - xy' + y = 0$, the DE possesses no solution that is analytic at $x = 0$. _____

3. Both power series solutions of $y'' + \ln(x+1)y' + y = 0$ centered at the ordinary point $x = 0$ are guaranteed to converge for all x in which *one* of the following intervals?

(a) $-\infty < x < \infty$ **(b)** $-1 < x < \infty$
(c) $-\frac{1}{2} \le x \le \frac{1}{2}$ **(d)** $-1 \le x \le 1$

4. $x = 0$ is an ordinary point of a certain linear differential equation. After the assumed solution $y = \sum_{n=0}^{\infty} c_n x^n$ is substituted into the DE, the following algebraic system

is obtained by equating the coefficients of x^0, x^1, x^2, and x^3 to zero:

$$2c_2 + 2c_1 + c_0 = 0$$
$$6c_3 + 4c_2 + c_1 = 0$$
$$12c_4 + 6c_3 + c_2 - \tfrac{1}{3}c_1 = 0$$
$$20c_5 + 8c_4 + c_3 - \tfrac{2}{3}c_2 = 0.$$

Bearing in mind that c_0 and c_1 are arbitrary, write down the first five terms of two power series solutions of the differential equation.

5. Suppose the power series $\sum_{k=0}^{\infty} c_k(x-4)^k$ is known to converge at -2 and diverge at 13. Discuss whether the series converges at -7, 0, 7, 10, and 11. Possible answers are *does, does not, might.*

6. Use the Maclaurin series for $\sin x$ and $\cos x$ along with long division to find the first three nonzero terms of a power series in x for the function $f(x) = \dfrac{\sin x}{\cos x}$.

In Problems 7 and 8 construct a linear second-order differential equation that has the given properties.

7. A regular singular point at $x = 1$ and an irregular singular point at $x = 0$

8. Regular singular points at $x = 1$ and at $x = -3$

In Problems 9–14 use an appropriate infinite series method about $x = 0$ to find two solutions of the given differential equation.

9. $2xy'' + y' + y = 0$

10. $y'' - xy' - y = 0$

11. $(x - 1)y'' + 3y = 0$

12. $y'' - x^2y' + xy = 0$

13. $xy'' - (x + 2)y' + 2y = 0$

14. $(\cos x)y'' + y = 0$

In Problems 15 and 16 solve the given initial-value problem.

15. $y'' + xy' + 2y = 0$, $y(0) = 3$, $y'(0) = -2$

16. $(x + 2)y'' + 3y = 0$, $y(0) = 0$, $y'(0) = 1$

17. Without actually solving the differential equation $(1 - 2 \sin x)y'' + xy = 0$, find a lower bound for the radius of convergence of power series solutions about the ordinary point $x = 0$.

18. Even though $x = 0$ is an ordinary point of the differential equation, explain why it is not a good idea to try to find a solution of the IVP

$$y'' + xy' + y = 0, \quad y(1) = -6, \quad y'(1) = 3$$

of the form $y = \sum_{n=0}^{\infty} c_n x^n$. Using power series, find a better way to solve the problem.

In Problems 19 and 20 investigate whether $x = 0$ is an ordinary point, singular point, or irregular singular point of the given differential equation. (*Hint:* Recall the Maclaurin series for $\cos x$ and e^x.)

19. $xy'' + (1 - \cos x)y' + x^2y = 0$

20. $(e^x - 1 - x)y'' + xy = 0$

21. Note that $x = 0$ is an ordinary point of the differential equation $y'' + x^2y' + 2xy = 5 - 2x + 10x^3$. Use the assumption $y = \sum_{n=0}^{\infty} c_n x^n$ to find the general solution $y = y_c + y_p$ that consists of three power series centered at $x = 0$.

22. The first-order differential equation $dy/dx = x^2 + y^2$ cannot be solved in terms of elementary functions. However, a solution can be expressed in terms of Bessel functions.

(a) Show that the substitution $y = -\dfrac{1}{u}\dfrac{du}{dx}$ leads to the equation $u'' + x^2u = 0$.

(b) Use (18) in Section 6.3 to find the general solution of $u'' + x^2u = 0$.

(c) Use (20) and (21) in Section 6.3 in the forms

$$J_\nu'(x) = \frac{\nu}{x}J_\nu(x) - J_{\nu+1}(x)$$

and

$$J_\nu'(x) = -\frac{\nu}{x}J_\nu(x) + J_{\nu-1}(x)$$

as an aid to show that a one-parameter family of solutions of $dy/dx = x^2 + y^2$ is given by

$$y = x\frac{J_{3/4}\left(\frac{1}{2}x^2\right) - cJ_{-3/4}\left(\frac{1}{2}x^2\right)}{cJ_{1/4}\left(\frac{1}{2}x^2\right) + J_{-1/4}\left(\frac{1}{2}x^2\right)}.$$

23. (a) Use (23) and (24) of Section 6.3 to show that

$$Y_{1/2}(x) = -\sqrt{\frac{2}{\pi x}}\cos x.$$

(b) Use (15) of Section 6.3 to show that

$$I_{1/2}(x) = \sqrt{\frac{2}{\pi x}}\sinh x \quad \text{and} \quad I_{-1/2}(x) = \sqrt{\frac{2}{\pi x}}\cosh x.$$

(c) Use part (b) to show that

$$K_{1/2}(x) = \sqrt{\frac{\pi}{2x}}e^{-x}.$$

24. (a) From (27) and (28) of Section 6.3 we know that when $n = 0$, Legendre's differential equation $(1 - x^2)y'' - 2xy' = 0$ has the polynomial solution $y = P_0(x) = 1$. Use (5) of Section 4.2 to show that a second Legendre function satisfying the DE on the interval $-1 < x < 1$ is

$$y = \frac{1}{2}\ln\left(\frac{1 + x}{1 - x}\right).$$

(b) We also know from (27) and (28) of Section 6.3 that when $n = 1$, Legendre's differential equation $(1 - x^2)y'' - 2xy' + 2y = 0$ possesses the polynomial solution $y = P_1(x) = x$. Use (5) of Section 4.2 to show that a second Legendre function satisfying the DE on the interval $-1 < x < 1$ is

$$y = \frac{x}{2}\ln\left(\frac{1 + x}{1 - x}\right) - 1.$$

(c) Use a graphing utility to graph the logarithmic Legendre functions given in parts (a) and (b).

25. (a) Use binomial series to formally show that

$$(1 - 2xt + t^2)^{-1/2} = \sum_{n=0}^{\infty} P_n(x)t^n.$$

(b) Use the result obtained in part (a) to show that $P_n(1) = 1$ and $P_n(-1) = (-1)^n$. See Properties (*ii*) and (*iii*) on page 267.

DEFEATING TAMARISK

You feel a drop of sweat stinging in your eye as you straighten to stretch your sore back. Your hands hurt, you are terribly thirsty, and the smell of herbicides is strong every time you remove the hot mask you wear. You are fighting a battle against a foreign invader, an interloper from southwest Asia: tamarisk (*Tamarix ramosissima*).

Tamarisk, also called salt cedar, is a reedy plant that was brought to the Americas years ago for its pretty purple flowers and soft leathery foliage. It is now reviled for its thick impenetrable growth and its tendency to lower water tables and crowd out the native cottonwoods and willows of the North American desert Southwest. The battle against it is waged either with fire or with a combination of manual cutting and herbicides. You are responsible for a demonstration project to eliminate tamarisk from a small side canyon of the Escalante River.* You suppose that if your crew can make one pass and cut most of the plants, killing the rest with herbicides, then you should be able to prevent a new invasion with only moderate effort each year. You are at the end of the sweat-and-poison part of the project, and now you must get serious about the maintenance phase.

The question to which you still do not have an answer is this: How much effort do you need to devote to various parts of the canyon each year to keep the tamarisk from returning? You know that the tamarisk will still be going strong down to the Escalante River that flows past the mouth of the canyon, so without constant vigilance the tamarisk will simply spread back up the canyon. You also know that all the cutting in the world will not suffice to eliminate the weed from the canyon entirely. There will always be a few seeds and shoots you missed. Unfortunately, your resources are not limitless. How are you going to manage this battle over time?

FIGURE 1 Tamarisk trees along the Escalante River

Back in your tiny office, you pull out your notes. You have searched the mathematical literature and have found that the spread of tamarisk is commonly modeled by using a nonlinear partial differential equation. On the other hand, that model requires more input information than you have and provides more output information than you want. Instead, you have made some simplifying assumptions to arrive at the linear ordinary differential equation

$$\frac{d^2u}{dx^2} + K(x)u = 0. \tag{1}$$

Here $u = u(x)$ represents the steady population density of tamarisk plants at a given point a distance x from the mouth of the canyon. The variable x is normalized for the length of the canyon. In other words, $x = 0$ at the canyon mouth, and $x = 1$ at the pour-off that ends the canyon.

The function K in equation (1) depends on a number of factors but most notably on the availability of water. You have not done measurements of the water table in your canyon, but you know that there is a lot of water near the surface at

---

*Most of the Escalante River lies within the Glenn Canyon National Recreation Area and the Grand Staircase Escalante National Monument in southern Utah.

the mouth of the canyon, where it meets the Escalante River, and also at the head of the canyon, where there is a pool at the base of a pour-off. To model that, you simply take $K(x) = 4(x - \frac{1}{2})^2$.

All ecological problems are plagued by difficulty in estimating parameters, and the initial conditions for this problem provide no exception to that rule. At the mouth of the canyon, that is, at $x = 0$, there is a fence line to keep cattle out of the canyon. You patrol that periodically and carry a small bottle of herbicide when you do. You can keep the tamarisk level along the fence at zero, but you cannot prevent the seeds from blowing across the fence into the canyon. You can't be sure about the rate at which tamarisk moves into the canyon, but $u'(0) = 0.25$ seems like a reasonable ballpark estimate. Thus, the basic initial-value problem for the steady distribution of tamarisk in the canyon is

$$u'' + 4(x - \tfrac{1}{2})^2 u = 0, \quad u(0) = 0, \quad u'(0) = 0.25. \tag{2}$$

Now Section 6.1 of this text tells how to find a power series solution of this problem. In this case you feel that all you really need is a working approximation to the solution. In other words, you plan to find a series solution of the initial-value problem but compute only the first few terms of that series, arriving at a polynomial approximation to the solution that is good close to $x = 0$, that is near the canyon mouth.

PROBLEM 1. Because $x = 0$ is an ordinary point of the DE in (2), Theorem 6.1 of this text guarantees that a power series solution centered at the point can be found. Substitute $u = \sum_{n=0}^{\infty} a_n x^n$ into the DE and find the coefficients a_0, a_1, a_2, a_3, and a_4. Then form the fourth-degree polynomial $u_4(x) = \sum_{n=0}^{4} a_n x^n$.

PROBLEM 2 (CD). Using your result from Problem 1, compute $d(x) = u_4'' + 4(x - \frac{1}{2})^2 u_4$. Use the **Tamarisk Tool** on the *DE tools* CD to plot $u_4(x)$ and $d(x)$ on the same coordinate axes. What do the values of $d(x)$ tell us? You should notice that $d(x)$ is very small near the origin and grows in magnitude farther from the origin. What does this mean?

You have to apply a treatment to some part of the canyon each year. In other words, you and a small crew of underpaid part-timers will go out for a few days and attack any tamarisk that returns to the canyon. You don't have much time or money for it, so you cannot do the thorough job that you did this year. You model your treatment by reducing the steady population by an amount $c(x)$, where $c(x)$ is a control function that you must determine. All you know is that resources are limited, so

$$\int_0^1 c(x)\, dx = 0.3. \tag{3}$$

Thus, you could spend all your time on the control at the mouth of the canyon, in which case $c(x)$ might look like

$$c(x) = \begin{cases} 0.3n, & x < 1/n \\ 0, & x \geq 1/n, \end{cases}$$

for some $n > 0$. On the other hand, you might try to use the time allotted to sweep uniformly all the way up the canyon, in which case $c(x) = 0.3$. The result is that the differential equation becomes

$$u'' + 4(x - \tfrac{1}{2})^2 u + c(x) = 0. \tag{4}$$

At first it is worrisome to you that your approximate solution u_4 in Problem 1 seems to be accurate near $x = 0$, that is, at the mouth of the canyon, but not elsewhere. Then it occurs to you that your best strategy is probably going to be one involving a lot of control near the mouth of the canyon, to keep down new tamarisk growth, and less up-canyon. You hope that by attacking the tamarisk vigorously near the Escalante River, you can prevent tamarisk from spreading from there up the canyon. Therefore you propose a control function

$$c(x) = he^{-rx}. \tag{5}$$

This will focus the largest part of your effort near the mouth of the canyon while allowing a little work on keeping the upper reaches of the canyon clear.

PROBLEM 3.   Use (5) in (3) and then solve for h in terms of r. What do you do if $r = 0$?

At this point, you must choose r so as to minimize the amount of tamarisk in the canyon.

PROBLEM 4.   Assume a series solution $v = \sum_{n=0}^{\infty} b_n x^n$ for the DE in (4). Use the same initial conditions as in (2). Then proceed as in Problem 1 and find coefficients b_0, b_1, b_2, b_3, and b_4 so that $v_4(x) = \sum_{n=0}^{4} b_n x^n$ is an approximate polynomial solution of degree four. Use the results of Problem 3 to express $v_4(x)$ in terms of r.

PROBLEM 5.   What sort of choice of r seems to lead to less tamarisk in the canyon: small r or larger r? What does this mean in terms of where your crew does its work?

You cannot help but be curious about the effect your treatment near the mouth of the canyon will have on the beautiful pool at the head of the canyon.

PROBLEM 6 (CD).   Use the **Tamarisk Tool** on the *DE Tools* CD to estimate the number of terms of the series you must compute to get reasonably accurate answers near $x = 1$. Make sure you click the "Problems 6 & 8" checkbox at the lower right of the tool. What do you think "reasonably accurate" means? Use $r = 2$.

There is one solution that nobody is considering. If you could change the initial conditions to $u(0) = 0$, $u'(0) = 0$, then once tamarisk was eliminated from the canyon, it would never return.

PROBLEM 7.   What would be required on the ground to get $u'(0) = 0$?

Turning exotic species loose in the wild is like opening Pandora's box. You personally will expend a good deal of sweat trying to keep your little canyon free of tamarisk, but you know that will be as nothing compared to your efforts to maintain funding to support the effort. The more numbers you bring to back up your claims, the better your chances will be.

STILL CURIOUS?

PROBLEM 8 (CD).   How does the accuracy of your solution depend on the value of r you choose? Use the **Tamarisk Tool** on the *DE Tools* CD to estimate the number of terms of the series you must keep to get accurate answers at $x = 1$. For example, try to keep the function $d(x)$ as computed in Problem 2 less than 0.02 everywhere, and find the number of terms you must keep in order for your solution to remain within this tolerance for $r = 1, 2, \ldots, 6$. Plot the number of terms against r. Explain the reason for the shape of the graph.

Driving function 1:

Response 1:

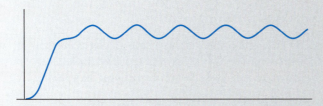

Driving function 2:

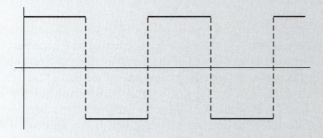

Response 2:

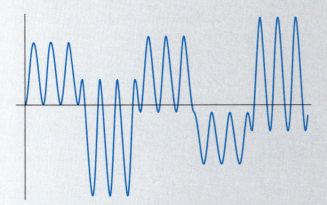

The IVP $x'' + \omega^2 x = F(t)$, $x(0) = 0$, $x'(0) = 0$ describes the displacement of a mass in an undamped spring/mass system. Top to bottom: A driving function $F(t)$ and the corresponding solution $x(t)$, or response of the system. See Section 7.4.

7

THE LAPLACE TRANSFORM

In the linear mathematical models for a physical system such as a spring/mass system or a series electrical circuit, the input or driving function, represents either an external force $f(t)$ or an impressed voltage $E(t)$. In Section 5.1 we considered problems in which the function f and E were continuous. However, discontinuous driving functions are not uncommon. For example, the driving function on a spring/mass system could be piecewise defined and continuous as in the first figure to the left, or piecewise defined, discontinuous, and periodic such as the square wave or "meander function" shown in the third figure to the left. Solving the differential equation of the system could be difficult using the techniques of Chapter 4. The Laplace transform studied in this chapter is an invaluable tool that simplifies the solution of problems such as these.

7.1   DEFINITION OF THE LAPLACE TRANSFORM

INTRODUCTION: In elementary calculus you learned that differentiation and integration are *transforms;* this means, roughly speaking, that these operations transform a function into another function. For example, the function $f(x) = x^2$ is transformed, in turn, into a linear function and a family of cubic polynomial functions by the operations of differentiation and integration: $\frac{d}{dx}x^2 = 2x$ and $\int x^2\, dx = \frac{1}{3}x^3 + c$. Moreover, these two transforms possess the **linearity property** that the transform of a linear combination of functions is a linear combination of the transforms. For α and β constants,

$$\frac{d}{dx}[\alpha f(x) + \beta g(x)] = \alpha f'(x) + \beta g'(x)$$

and

$$\int [\alpha f(x) + \beta g(x)]\, dx = \alpha \int f(x)\, dx + \beta \int g(x)\, dx$$

provided that each derivative and integral exists. In this section we examine a special type of integral transform called the **Laplace transform.** In addition to possessing the linearity property, the Laplace transform has many other interesting properties that make it very useful in solving linear initial-value problems.

REVIEW MATERIAL: Because piecewise continuous functions and improper integrals of the type $\int_a^\infty f(x)\, dx$ play a central role in this chapter, it is imperative that you carefully review these concepts.

INTEGRAL TRANSFORM If $f(x, y)$ is a function of two variables, then a definite integral of f with respect to one of the variables leads to a function of the other variable. For example, by holding y constant, we see that $\int_1^2 2xy^2\, dx = 3y^2$. Similarly, a definite integral such as $\int_a^b K(s, t)f(t)\, dt$ transforms a function f of the variable t into a function F of the variable s. We are particularly interested in an **integral transform,** where the interval of integration is the unbounded interval $[0, \infty)$. If $f(t)$ is defined for $t \ge 0$, then the improper integral $\int_0^\infty K(s, t)f(t)\, dt$ is defined as a limit:

$$\int_0^\infty K(s, t)f(t)\, dt = \lim_{b\to\infty} \int_0^b K(s, t)f(t)\, dt. \tag{1}$$

If the limit in (1) exists, then we say that the integral exists or is **convergent;** if the limit does not exist, the integral does not exist and is **divergent.** The limit in (1) will, in general, exist for only certain values of the variable s.

A DEFINITION The function $K(s, t)$ in (1) is called the **kernel** of the transform. The choice $K(s, t) = e^{-st}$ as the kernel gives us an especially important integral transform.

DEFINITION 7.1   Laplace Transform

Let f be a function defined for $t \ge 0$. Then the integral

$$\mathcal{L}\{f(t)\} = \int_0^\infty e^{-st} f(t)\, dt \tag{2}$$

is said to be the **Laplace transform** of f, provided that the integral converges.

When the defining integral (2) converges, the result is a function of s. In general discussion we shall use a lowercase letter to denote the function being transformed and the corresponding capital letter to denote its Laplace transform—for example,

$$\mathcal{L}\{f(t)\} = F(s), \quad \mathcal{L}\{g(t)\} = G(s), \quad \mathcal{L}\{y(t)\} = Y(s).$$

EXAMPLE 1   Applying Definition 7.1

Evaluate $\mathcal{L}\{1\}$.

SOLUTION   From (2),

$$\mathcal{L}\{1\} = \int_0^\infty e^{-st}(1)\, dt = \lim_{b\to\infty} \int_0^b e^{-st}\, dt$$

$$= \lim_{b\to\infty} \frac{-e^{-st}}{s}\Big|_0^b = \lim_{b\to\infty} \frac{-e^{-sb}+1}{s} = \frac{1}{s}$$

provided $s > 0$. In other words, when $s > 0$, the exponent $-sb$ is negative and $e^{-sb} \to 0$ as $b \to \infty$. The integral diverges for $s < 0$.

The use of the limit sign becomes somewhat tedious, so we shall adopt the notation $\big|_0^\infty$ as a shorthand for writing $\lim_{b\to\infty} (\ )\big|_0^b$. For example,

$$\mathcal{L}\{1\} = \int_0^\infty e^{-st}(1)\, dt = \frac{-e^{-st}}{s}\Big|_0^\infty = \frac{1}{s}, \quad s > 0.$$

At the upper limit, it is understood that we mean $e^{-st} \to 0$ as $t \to \infty$ for $s > 0$.

EXAMPLE 2   Applying Definition 7.1

Evaluate $\mathcal{L}\{t\}$.

SOLUTION   From Definition 7.1 we have $\mathcal{L}\{t\} = \int_0^\infty e^{-st}\, t\, dt$. Integrating by parts and using $\lim_{t\to\infty} te^{-st} = 0$, $s > 0$, along with the result from Example 1, we obtain

$$\mathcal{L}\{t\} = \frac{-te^{-st}}{s}\Big|_0^\infty + \frac{1}{s}\int_0^\infty e^{-st}\, dt = \frac{1}{s}\mathcal{L}\{1\} = \frac{1}{s}\left(\frac{1}{s}\right) = \frac{1}{s^2}.$$

EXAMPLE 3   Applying Definition 7.1

Evaluate $\mathcal{L}\{e^{-3t}\}$.

SOLUTION   From Definition 7.1 we have

$$\mathcal{L}\{e^{-3t}\} = \int_0^\infty e^{-st} e^{-3t}\, dt = \int_0^\infty e^{-(s+3)t}\, dt$$

$$= \frac{-e^{-(s+3)t}}{s+3}\Big|_0^\infty$$

$$= \frac{1}{s+3}, \quad s > -3.$$

The result follows from the fact that $\lim_{t\to\infty} e^{-(s+3)t} = 0$ for $s + 3 > 0$ or $s > -3$.

EXAMPLE 4 Applying Definition 7.1

Evaluate $\mathscr{L}\{\sin 2t\}$.

SOLUTION From Definition 7.1 and integration by parts we have

$$\mathscr{L}\{\sin 2t\} = \int_0^\infty e^{-st} \sin 2t \, dt = \frac{-e^{-st} \sin 2t}{s} \bigg|_0^\infty + \frac{2}{s} \int_0^\infty e^{-st} \cos 2t \, dt$$

$$= \frac{2}{s} \int_0^\infty e^{-st} \cos 2t \, dt, \quad s > 0$$

$$\underset{t\to\infty}{\lim} e^{-st} \cos 2t = 0, \, s > 0 \qquad\qquad \text{Laplace transform of } \sin 2t$$
$$\downarrow \qquad\qquad\qquad\qquad\qquad\qquad\qquad\qquad \downarrow$$

$$= \frac{2}{s}\left[\frac{-e^{-st}\cos 2t}{s}\bigg|_0^\infty - \frac{2}{s}\int_0^\infty e^{-st}\sin 2t \, dt\right]$$

$$= \frac{2}{s^2} - \frac{4}{s^2}\mathscr{L}\{\sin 2t\}.$$

At this point we have an equation with $\mathscr{L}\{\sin 2t\}$ on both sides of the equality. Solving for that quantity yields the result

$$\mathscr{L}\{\sin 2t\} = \frac{2}{s^2 + 4}, \quad s > 0. \qquad\qquad \blacksquare$$

$\mathscr{L}$ IS A LINEAR TRANSFORM For a linear combination of functions we can write

$$\int_0^\infty e^{-st}[\alpha f(t) + \beta g(t)] \, dt = \alpha \int_0^\infty e^{-st} f(t) \, dt + \beta \int_0^\infty e^{-st} g(t) \, dt$$

whenever both integrals converge for $s > c$. Hence it follows that

$$\mathscr{L}\{\alpha f(t) + \beta g(t)\} = \alpha\mathscr{L}\{f(t)\} + \beta\mathscr{L}\{g(t)\} = \alpha F(s) + \beta G(s). \qquad (3)$$

Because of the property given in (3), $\mathscr{L}$ is said to be a **linear transform.** For example, from Examples 1 and 2

$$\mathscr{L}\{1 + 5t\} = \mathscr{L}\{1\} + 5\mathscr{L}\{t\} = \frac{1}{s} + \frac{5}{s^2},$$

and from Examples 3 and 4

$$\mathscr{L}\{4e^{-3t} - 10\sin 2t\} = 4\mathscr{L}\{e^{-3t}\} - 10\mathscr{L}\{\sin 2t\} = \frac{4}{s + 3} - \frac{20}{s^2 + 4}.$$

We state the generalization of some of the preceding examples by means of the next theorem. From this point on we shall also refrain from stating any restrictions on s; it is understood that s is sufficiently restricted to guarantee the convergence of the appropriate Laplace transform.

THEOREM 7.1   **Transforms of Some Basic Functions**

$$\textbf{(a)} \; \mathscr{L}\{1\} = \frac{1}{s}$$

$$\textbf{(b)} \; \mathscr{L}\{t^n\} = \frac{n!}{s^{n+1}}, \quad n = 1, 2, 3, \ldots \qquad \textbf{(c)} \; \mathscr{L}\{e^{at}\} = \frac{1}{s - a}$$

$$\textbf{(d)} \; \mathscr{L}\{\sin kt\} = \frac{k}{s^2 + k^2} \qquad\qquad \textbf{(e)} \; \mathscr{L}\{\cos kt\} = \frac{s}{s^2 + k^2}$$

$$\textbf{(f)} \; \mathscr{L}\{\sinh kt\} = \frac{k}{s^2 - k^2} \qquad\qquad \textbf{(g)} \; \mathscr{L}\{\cosh kt\} = \frac{s}{s^2 - k^2}$$

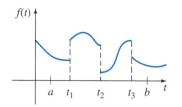

FIGURE 7.1 Piecewise continuous function

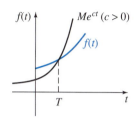

FIGURE 7.2 f is of exponential order c.

SUFFICIENT CONDITIONS FOR EXISTENCE OF $\mathscr{L}\{f(t)\}$ The integral that defines the Laplace transform does not have to converge. For example, neither $\mathscr{L}\{1/t\}$ nor $\mathscr{L}\{e^{t^2}\}$ exists. Sufficient conditions guaranteeing the existence of $\mathscr{L}\{f(t)\}$ are that f be piecewise continuous on $[0, \infty)$ and that f be of exponential order for $t > T$. Recall that a function f is **piecewise continuous** on $[0, \infty)$ if, in any interval $0 \leq a \leq t \leq b$, there are at most a finite number of points t_k, $k = 1, 2, \ldots, n$ $(t_{k-1} < t_k)$ at which f has finite discontinuities and is continuous on each open interval $t_{k-1} < t < t_k$. See Figure 7.1. The concept of **exponential order** is defined in the following manner.

DEFINITION 7.2   **Exponential Order**

A function f is said to be of **exponential order** c if there exist constants c, $M > 0$, and $T > 0$ such that $|f(t)| \leq Me^{ct}$ for all $t > T$.

If f is an *increasing* function, then the condition $|f(t)| \leq Me^{ct}$, $t > T$, simply states that the graph of f on the interval (T, ∞) does not grow faster than the graph of the exponential function Me^{ct}, where c is a positive constant. See Figure 7.2. The functions $f(t) = t$, $f(t) = e^{-t}$, and $f(t) = 2 \cos t$ are all of exponential order $c = 1$ for $t > 0$ since we have, respectively,

$$|t| \leq e^t, \quad |e^{-t}| \leq e^t, \quad \text{and} \quad |2 \cos t| \leq 2e^t.$$

A comparison of the graphs on the interval $(0, \infty)$ is given in Figure 7.3.

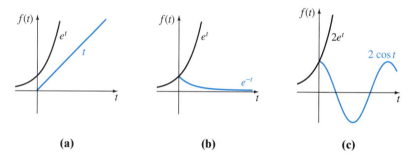

(a)   **(b)**   **(c)**

FIGURE 7.3 Three functions of exponential order $c = 1$

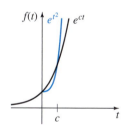

FIGURE 7.4 e^{t^2} is not of exponential order

A function such as $f(t) = e^{t^2}$ is not of exponential order since as shown in Figure 7.4, its graph grows faster than any positive linear power of e for $t > c > 0$.

A positive integral power of t is always of exponential order since, for $c > 0$,

$$|t^n| \leq Me^{ct} \quad \text{or} \quad \left| \frac{t^n}{e^{ct}} \right| \leq M \quad \text{for } t > T$$

is equivalent to showing that $\lim_{t \to \infty} t^n/e^{ct}$ is finite for $n = 1, 2, 3, \ldots$. The result follows by n applications of L'Hôpital's Rule.

THEOREM 7.2   **Sufficient Conditions for Existence**

If f is piecewise continuous on $[0, \infty)$ and of exponential order c, then $\mathscr{L}\{f(t)\}$ exists for $s > c$.

PROOF By the additive property of definite integrals we can write

$$\mathscr{L}\{f(t)\} = \int_0^T e^{-st} f(t) \, dt + \int_T^\infty e^{-st} f(t) \, dt = I_1 + I_2.$$

The integral I_1 exists because it can be written as a sum of integrals over intervals on which $e^{-st} f(t)$ is continuous. Now since f is of exponential order, there exist constants c, $M > 0$, $T > 0$ so that $|f(t)| \leq Me^{ct}$ for $t > T$. We can then write

$$|I_2| \leq \int_T^\infty |e^{-st}f(t)|\, dt \leq M \int_T^\infty e^{-st}e^{ct}dt = M \int_T^\infty e^{-(s-c)t}dt = M\frac{e^{-(s-c)T}}{s-c}$$

for $s > c$. Since $\int_T^\infty Me^{-(s-c)t}\, dt$ converges, the integral $\int_T^\infty |e^{-st}f(t)|\, dt$ converges by the comparison test for improper integrals. This, in turn, implies that I_2 exists for $s > c$. The existence of I_1 and I_2 implies that $\mathscr{L}\{f(t)\} = \int_0^\infty e^{-st}f(t)\, dt$ exists for $s > c$. ∎

EXAMPLE 5   **Transform of a Piecewise Continuous Function**

Evaluate $\mathscr{L}\{f(t)\}$ where $f(t) = \begin{cases} 0, & 0 \leq t < 3 \\ 2, & t \geq 3. \end{cases}$

SOLUTION   The function f, shown in Figure 7.5, is piecewise continuous and of exponential order for $t > 0$. Since f is defined in two pieces, $\mathscr{L}\{f(t)\}$ is expressed as the sum of two integrals:

$$\mathscr{L}\{f(t)\} = \int_0^\infty e^{-st}f(t)\, dt = \int_0^3 e^{-st}(0)\, dt + \int_3^\infty e^{-st}(2)\, dt$$

$$= 0 + \frac{2e^{-st}}{-s}\Big|_3^\infty$$

$$= \frac{2e^{-3s}}{s}, \quad s > 0. \qquad ∎$$

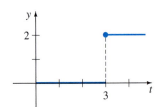

FIGURE 7.5   Piecewise continuous function

We conclude this section with an additional bit of theory related to the types of functions of s that we will, generally, be working with. The next theorem indicates that not every arbitrary function of s is a Laplace transform of a piecewise continuous function of exponential order.

THEOREM 7.3   **Behavior of $F(s)$ as $s \to \infty$**

If f is piecewise continuous on $(0, \infty)$ and of exponential order and $F(s) = \mathscr{L}\{f(t)\}$, then $\lim_{s \to \infty} F(s) = 0$.

PROOF   Since f is of exponential order, there exist constants γ, $M_1 > 0$, and $T > 0$ so that $|f(t)| \leq M_1 e^{\gamma t}$ for $t > T$. Also, since f is piecewise continuous on the interval $0 \leq t \leq T$, it is necessarily bounded on the interval; that is, $|f(t)| \leq M_2 = M_2 e^{0t}$. If M denotes the maximum of the set $\{M_1, M_2\}$ and c denotes the maximum of $\{0, \gamma\}$, then

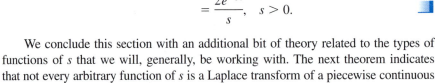

for $s > c$. As $s \to \infty$, we have $|F(s)| \to 0$, and so $F(s) = \mathscr{L}\{f(t)\} \to 0$. ∎

REMARKS

(*i*) Throughout this chapter we shall be concerned primarily with functions that are both piecewise continuous and of exponential order. We note, however, that these two conditions are sufficient but not necessary for the existence of a Laplace transform. The function $f(t) = t^{-1/2}$ is not piecewise continuous on the interval $[0, \infty)$, but its Laplace transform exists. See Problem 42 in Exercises 7.1.

(*ii*) As a consequence of Theorem 7.3 we can say that functions of s such as $F_1(s) = 1$ and $F_2(s) = s/(s+1)$ are not the Laplace transforms of piecewise continuous functions of exponential order, since $F_1(s) \nrightarrow 0$ and $F_2(s) \nrightarrow 0$ as $s \to \infty$. But you should not conclude from this that $F_1(s)$ and $F_2(s)$ are *not* Laplace transforms. There are other kinds of functions.

EXERCISES 7.1

Answers to selected odd-numbered problems begin on page ANS-9.

In Problems 1–18 use Definition 7.1 to find $\mathscr{L}\{f(t)\}$.

1. $f(t) = \begin{cases} -1, & 0 \le t < 1 \\ 1, & t \ge 1 \end{cases}$

2. $f(t) = \begin{cases} 4, & 0 \le t < 2 \\ 0, & t \ge 2 \end{cases}$

3. $f(t) = \begin{cases} t, & 0 \le t < 1 \\ 1, & t \ge 1 \end{cases}$

4. $f(t) = \begin{cases} 2t + 1, & 0 \le t < 1 \\ 0, & t \ge 1 \end{cases}$

5. $f(t) = \begin{cases} \sin t, & 0 \le t < \pi \\ 0, & t \ge \pi \end{cases}$

6. $f(t) = \begin{cases} 0, & 0 \le t < \pi/2 \\ \cos t, & t \ge \pi/2 \end{cases}$

7.

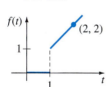

FIGURE 7.6 Graph for Problem 7

8.

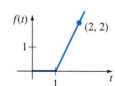

FIGURE 7.7 Graph for Problem 8

9.

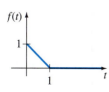

FIGURE 7.8 Graph for Problem 9

10.

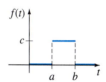

FIGURE 7.9 Graph for Problem 10

11. $f(t) = e^{t+7}$　　　　**12.** $f(t) = e^{-2t-5}$

13. $f(t) = te^{4t}$　　　　**14.** $f(t) = t^2 e^{-2t}$

15. $f(t) = e^{-t}\sin t$　　　**16.** $f(t) = e^t \cos t$

17. $f(t) = t\cos t$　　　　**18.** $f(t) = t\sin t$

In Problems 19–36 use Theorem 7.1 to find $\mathscr{L}\{f(t)\}$.

19. $f(t) = 2t^4$　　　　**20.** $f(t) = t^5$

21. $f(t) = 4t - 10$　　　**22.** $f(t) = 7t + 3$

23. $f(t) = t^2 + 6t - 3$　　**24.** $f(t) = -4t^2 + 16t + 9$

25. $f(t) = (t+1)^3$　　　**26.** $f(t) = (2t-1)^3$

27. $f(t) = 1 + e^{4t}$　　　**28.** $f(t) = t^2 - e^{-9t} + 5$

29. $f(t) = (1 + e^{2t})^2$　　**30.** $f(t) = (e^t - e^{-t})^2$

31. $f(t) = 4t^2 - 5\sin 3t$　**32.** $f(t) = \cos 5t + \sin 2t$

33. $f(t) = \sinh kt$　　　**34.** $f(t) = \cosh kt$

35. $f(t) = e^t \sinh t$　　　**36.** $f(t) = e^{-t}\cosh t$

In Problems 37–40 find $\mathscr{L}\{f(t)\}$ by first using a trigonometric identity.

37. $f(t) = \sin 2t \cos 2t$　**38.** $f(t) = \cos^2 t$

39. $f(t) = \sin(4t + 5)$　　**40.** $f(t) = 10\cos\left(t - \dfrac{\pi}{6}\right)$

41. One definition of the **gamma function** is given by the improper integral $\Gamma(\alpha) = \int_0^\infty t^{\alpha-1}e^{-t}\,dt$, $\alpha > 0$.
 (a) Show that $\Gamma(\alpha + 1) = \alpha\Gamma(\alpha)$.
 (b) Show that $\mathscr{L}\{t^\alpha\} = \dfrac{\Gamma(\alpha+1)}{s^{\alpha+1}}$, $\alpha > -1$.

42. Use the fact that $\Gamma\left(\frac{1}{2}\right) = \sqrt{\pi}$ and Problem 41 to find the Laplace transform of

    (a) $f(t) = t^{-1/2}$ **(b)** $f(t) = t^{1/2}$ **(c)** $f(t) = t^{3/2}$.

DISCUSSION/PROJECT PROBLEMS

43. Make up a function $F(t)$ that is of exponential order but where $f(t) = F'(t)$ is not of exponential order. Make up a function f that is not of exponential order but whose Laplace transform exists.

44. Suppose that $\mathscr{L}\{f_1(t)\} = F_1(s)$ for $s > c_1$ and that $\mathscr{L}\{f_2(t)\} = F_2(s)$ for $s > c_2$. When does $\mathscr{L}\{f_1(t) + f_2(t)\} = F_1(s) + F_2(s)$?

45. Figure 7.4 suggests, but does not prove, that the function $f(t) = e^{t^2}$ is not of exponential order. How does the observation that $t^2 > \ln M + ct$, for $M > 0$

and t sufficiently large, show that $e^{t^2} > Me^{ct}$ for any c?

46. Use part (c) of Theorem 7.1 to show that $\mathscr{L}\{e^{(a+ib)t}\} = \dfrac{s - a + ib}{(s - a)^2 + b^2}$, where a and b are real and $i^2 = -1$. Show how Euler's formula (page 144) can then be used to deduce the results

$$\mathscr{L}\{e^{at}\cos bt\} = \frac{s - a}{(s - a)^2 + b^2}$$

and $$\mathscr{L}\{e^{at}\sin bt\} = \frac{b}{(s - a)^2 + b^2}.$$

47. Under what conditions is a linear function $f(x) = mx + b$, $m \neq 0$, a linear transform?

48. The proof of part (b) of Theorem 4.1 requires the use of mathematical induction. Show that if $\mathscr{L}\{t^{n-1}\} = (n - 1)!/s^n$ is *assumed* to be true, then $\mathscr{L}\{t^n\} = n!/s^{n+1}$ follows.

7.2 INVERSE TRANSFORMS AND TRANSFORMS OF DERIVATIVES

INTRODUCTION: In this section we take a few small steps into an investigation of how the Laplace transform can be used to solve certain types of equations for an unknown function. We begin the discussion with the concept of the inverse Laplace transform or, more precisely, the inverse of a Laplace transform $F(s)$. After some important preliminary background on the Laplace transform of derivatives dy/dt, d^2y/dt^2, ..., we then illustrate how both the Laplace transform and the inverse Laplace transform come into play in solving some simple ordinary differential equations.

REVIEW MATERIAL: Solving equations by the Laplace transform necessitates the evaluation of an inverse Laplace transform; this, in turn, frequently requires subtle algebraic manipulations and the decomposition of a rational expression into partial fractions. Although we discuss here some of the algebra of partial fractions and some shortcuts, it still might be a good idea for you to consult other sources (such as a calculus text or a current precalculus text) for a more comprehensive review of this theory.

7.2.1 INVERSE TRANSFORMS

THE INVERSE PROBLEM If $F(s)$ represents the Laplace transform of a function $f(t)$, that is, $\mathscr{L}\{f(t)\} = F(s)$, we then say $f(t)$ is the **inverse Laplace transform** of $F(s)$ and write $f(t) = \mathscr{L}^{-1}\{F(s)\}$. For example, from Examples 1, 2, and 3 of Section 7.1 we have, respectively:

Transform   **Inverse Transform**

$$\mathscr{L}\{1\} = \frac{1}{s} \qquad\qquad 1 = \mathscr{L}^{-1}\left\{\frac{1}{s}\right\}$$

$$\mathscr{L}\{t\} = \frac{1}{s^2} \qquad\qquad t = \mathscr{L}^{-1}\left\{\frac{1}{s^2}\right\}$$

$$\mathscr{L}\{e^{-3t}\} = \frac{1}{s+3} \qquad\qquad e^{-3t} = \mathscr{L}^{-1}\left\{\frac{1}{s+3}\right\}$$

We shall see shortly that in the application of the Laplace transform to equations we are not able to determine an unknown function $f(t)$ directly; rather, we are able to solve for the Laplace transform $F(s)$ of $f(t)$; but from that knowledge we ascertain f by computing $f(t) = \mathscr{L}^{-1}\{F(s)\}$. The idea is simply this: Suppose $F(s) = \dfrac{-2s+6}{s^2+4}$ is a Laplace transform; find a function $f(t)$ such that $\mathscr{L}\{f(t)\} = F(s)$. We shall show how to solve this last problem in Example 2.

For future reference the analogue of Theorem 7.1 for the inverse transform is presented as our next theorem.

THEOREM 7.4   **Some Inverse Transforms**

$$\textbf{(a)} \;\; 1 = \mathscr{L}^{-1}\left\{\frac{1}{s}\right\}$$

$$\textbf{(b)} \;\; t^n = \mathscr{L}^{-1}\left\{\frac{n!}{s^{n+1}}\right\}, \quad n = 1, 2, 3, \ldots \qquad \textbf{(c)} \;\; e^{at} = \mathscr{L}^{-1}\left\{\frac{1}{s-a}\right\}$$

$$\textbf{(d)} \;\; \sin kt = \mathscr{L}^{-1}\left\{\frac{k}{s^2+k^2}\right\} \qquad\qquad \textbf{(e)} \;\; \cos kt = \mathscr{L}^{-1}\left\{\frac{s}{s^2+k^2}\right\}$$

$$\textbf{(f)} \;\; \sinh kt = \mathscr{L}^{-1}\left\{\frac{k}{s^2-k^2}\right\} \qquad\qquad \textbf{(g)} \;\; \cosh kt = \mathscr{L}^{-1}\left\{\frac{s}{s^2-k^2}\right\}$$

When evaluating inverse transforms, it often happens that a function of s under consideration does not match *exactly* the form of a Laplace transform $F(s)$ given in a table. It may be necessary to "fix up" the function of s by multiplying and dividing by an appropriate constant.

EXAMPLE 1   **Applying Theorem 7.4**

Evaluate   **(a)** $\mathscr{L}^{-1}\left\{\dfrac{1}{s^5}\right\}$ **(b)** $\mathscr{L}^{-1}\left\{\dfrac{1}{s^2+7}\right\}$.

SOLUTION   **(a)** To match the form given in part (b) of Theorem 7.4, we identify $n + 1 = 5$ or $n = 4$ and then multiply and divide by $4!$:

$$\mathscr{L}^{-1}\left\{\frac{1}{s^5}\right\} = \frac{1}{4!}\,\mathscr{L}^{-1}\left\{\frac{4!}{s^5}\right\} = \frac{1}{24}t^4.$$

(b) To match the form given in part (d) of Theorem 7.4, we identify $k^2 = 7$, and so $k = \sqrt{7}$. We fix up the expression by multiplying and dividing by $\sqrt{7}$:

$$\mathscr{L}^{-1}\left\{\frac{1}{s^2+7}\right\} = \frac{1}{\sqrt{7}}\,\mathscr{L}^{-1}\left\{\frac{\sqrt{7}}{s^2+7}\right\} = \frac{1}{\sqrt{7}}\sin\sqrt{7}t.$$

$\mathscr{L}^{-1}$ **IS A LINEAR TRANSFORM**   The inverse Laplace transform is also a linear transform; that is, for constants α and β

$$\mathscr{L}^{-1}\{\alpha F(s) + \beta G(s)\} = \alpha\mathscr{L}^{-1}\{F(s)\} + \beta\mathscr{L}^{-1}\{G(s)\}, \tag{1}$$

where F and G are the transforms of some functions f and g. Like (2) of Section 7.1, (1) extends to any finite linear combination of Laplace transforms.

EXAMPLE 2   Termwise Division and Linearity

Evaluate $\mathscr{L}^{-1}\left\{\dfrac{-2s + 6}{s^2 + 4}\right\}$.

SOLUTION   We first rewrite the given function of s as two expressions by means of termwise division and then use (1):

$$\mathscr{L}^{-1}\left\{\frac{-2s + 6}{s^2 + 4}\right\} = \mathscr{L}^{-1}\left\{\overset{\underset{\text{termwise division}}{\downarrow}}{\frac{-2s}{s^2 + 4}} + \frac{6}{s^2 + 4}\right\} = -2\mathscr{L}^{-1}\left\{\frac{s}{s^2 + 4}\right\} + \frac{6}{2}\,\overset{\underset{\text{linearity and fixing up constants}}{\downarrow}}{\mathscr{L}^{-1}}\left\{\frac{2}{s^2 + 4}\right\} \tag{2}$$

$$= -2\cos 2t + 3\sin 2t. \quad \leftarrow \text{ parts (e) and (d)}$$
$$\text{of Theorem 7.4 with } k = 2$$

See the **SRSM** for a review of partial fractions.

PARTIAL FRACTIONS   Partial fractions play an important role in finding inverse Laplace transforms. The decomposition of a rational expression into component fractions can be done quickly by means of a single command on most computer algebra systems. Indeed, some CASs have packages that implement Laplace transform and inverse Laplace transform commands. But for those of you without access to such software, we will review in this and subsequent sections some of the basic algebra in the important cases where the denominator of a Laplace transform $F(s)$ contains distinct linear factors, repeated linear factors, and quadratic polynomials with no real factors. Although we shall examine each of these cases as this chapter develops, it still might be a good idea for you to consult either a calculus text or a current precalculus text for a more comprehensive review of this theory.

The following example illustrates partial fraction decomposition in the case when the denominator of $F(s)$ is factorable into *distinct linear factors*.

EXAMPLE 3   Partial Fractions: Distinct Linear Factors

Evaluate $\mathscr{L}^{-1}\left\{\dfrac{s^2 + 6s + 9}{(s - 1)(s - 2)(s + 4)}\right\}$.

SOLUTION   There exist unique real constants A, B, and C so that

$$\frac{s^2 + 6s + 9}{(s - 1)(s - 2)(s + 4)} = \frac{A}{s - 1} + \frac{B}{s - 2} + \frac{C}{s + 4}$$

$$= \frac{A(s - 2)(s + 4) + B(s - 1)(s + 4) + C(s - 1)(s - 2)}{(s - 1)(s - 2)(s + 4)}.$$

Since the denominators are identical, the numerators are identical:

$$s^2 + 6s + 9 = A(s - 2)(s + 4) + B(s - 1)(s + 4) + C(s - 1)(s - 2). \tag{3}$$

By comparing coefficients of powers of s on both sides of the equality, we know that (3) is equivalent to a system of three equations in the three unknowns A, B, and C.

However, there is a shortcut for determining these unknowns. If we set $s = 1$, $s = 2$, and $s = -4$ in (3), we obtain, respectively,

$$16 = A(-1)(5), \quad 25 = B(1)(6), \quad \text{and} \quad 1 = C(-5)(-6),$$

and so $A = -\frac{16}{5}$, $B = \frac{25}{6}$, and $C = \frac{1}{30}$. Hence the partial fraction decomposition is

$$\frac{s^2 + 6s + 9}{(s-1)(s-2)(s+4)} = -\frac{16/5}{s-1} + \frac{25/6}{s-2} + \frac{1/30}{s+4}, \tag{4}$$

and thus, from the linearity of $\mathscr{L}^{-1}$ and part (c) of Theorem 7.4,

$$\mathscr{L}^{-1}\left\{\frac{s^2 + 6s + 9}{(s-1)(s-2)(s+4)}\right\} = -\frac{16}{5}\mathscr{L}^{-1}\left\{\frac{1}{s-1}\right\} + \frac{25}{6}\mathscr{L}^{-1}\left\{\frac{1}{s-2}\right\} + \frac{1}{30}\mathscr{L}^{-1}\left\{\frac{1}{s+4}\right\}$$

$$= -\frac{16}{5}e^t + \frac{25}{6}e^{2t} + \frac{1}{30}e^{-4t}. \tag{5} \quad\blacksquare$$

7.2.2 TRANSFORMS OF DERIVATIVES

TRANSFORM A DERIVATIVE As was pointed out in the introduction to this chapter, our immediate goal is to use the Laplace transform to solve differential equations. To that end we need to evaluate quantities such as $\mathscr{L}\{dy/dt\}$ and $\mathscr{L}\{d^2y/dt^2\}$. For example, if f' is continuous for $t \geq 0$, then integration by parts gives

$$\mathscr{L}\{f'(t)\} = \int_0^\infty e^{-st}f'(t)\,dt = e^{-st}f(t)\Big|_0^\infty + s\int_0^\infty e^{-st}f(t)\,dt$$

$$= -f(0) + s\mathscr{L}\{f(t)\}$$

or $\qquad \mathscr{L}\{f'(t)\} = sF(s) - f(0). \tag{6}$

Here we have assumed that $e^{-st}f(t) \to 0$ as $t \to \infty$. Similarly, with the aid of (6),

$$\mathscr{L}\{f''(t)\} = \int_0^\infty e^{-st}f''(t)\,dt = e^{-st}f'(t)\Big|_0^\infty + s\int_0^\infty e^{-st}f'(t)\,dt$$

$$= -f'(0) + s\mathscr{L}\{f'(t)\}$$

$$= s[sF(s) - f(0)] - f'(0) \quad \leftarrow \text{from (6)}$$

or $\qquad \mathscr{L}\{f''(t)\} = s^2F(s) - sf(0) - f'(0). \tag{7}$

In like manner it can be shown that

$$\mathscr{L}\{f'''(t)\} = s^3F(s) - s^2f(0) - sf'(0) - f''(0). \tag{8}$$

The recursive nature of the Laplace transform of the derivatives of a function f should be apparent from the results in (6), (7), and (8). The next theorem gives the Laplace transform of the nth derivative of f. The proof is omitted.

THEOREM 7.5 **Transform of a Derivative**

If $f, f', \dots, f^{(n-1)}$ are continuous on $[0, \infty)$ and are of exponential order and if $f^{(n)}(t)$ is piecewise continuous on $[0, \infty)$, then

$$\mathscr{L}\{f^{(n)}(t)\} = s^nF(s) - s^{n-1}f(0) - s^{n-2}f'(0) - \cdots - f^{(n-1)}(0),$$

where $F(s) = \mathscr{L}\{f(t)\}$.

SOLVING LINEAR ODES It is apparent from the general result given in Theorem 7.5 that $\mathscr{L}\{d^ny/dt^n\}$ depends on $Y(s) = \mathscr{L}\{y(t)\}$ and the $n-1$ derivatives of $y(t)$ evaluated at $t = 0$. This property makes the Laplace transform ideally suited for solving linear initial-value problems in which the differential equation has *constant coefficients*. Such a differential equation is simply a linear combination of terms $y, y', y'', \ldots, y^{(n)}$:

$$a_n \frac{d^n y}{dt^n} + a_{n-1} \frac{d^{n-1} y}{dt^{n-1}} + \cdots + a_0 y = g(t),$$

$$y(0) = y_0,\ y'(0) = y_1,\ \ldots,\ y^{(n-1)}(0) = y_{n-1},$$

where the a_i, $i = 0, 1, \ldots, n$ and $y_0, y_1, \ldots, y_{n-1}$ are constants. By the linearity property the Laplace transform of this linear combination is a linear combination of Laplace transforms:

$$a_n \mathscr{L}\left\{\frac{d^n y}{dt^n}\right\} + a_{n-1} \mathscr{L}\left\{\frac{d^{n-1} y}{dt^{n-1}}\right\} + \cdots + a_0 \mathscr{L}\{y\} = \mathscr{L}\{g(t)\}. \qquad (9)$$

From Theorem 7.5, (9) becomes

$$\begin{aligned} a_n[s^n Y(s) - s^{n-1} y(0) - \cdots - y^{(n-1)}(0)] \\ + a_{n-1}[s^{n-1} Y(s) - s^{n-2} y(0) - \cdots - y^{(n-2)}(0)] + \cdots + a_0 Y(s) = G(s), \end{aligned} \qquad (10)$$

where $\mathscr{L}\{y(t)\} = Y(s)$ and $\mathscr{L}\{g(t)\} = G(s)$. In other words, *the Laplace transform of a linear differential equation with constant coefficients becomes an algebraic equation in $Y(s)$.* If we solve the general transformed equation (10) for the symbol $Y(s)$, we first obtain $P(s)Y(s) = Q(s) + G(s)$ and then write

$$Y(s) = \frac{Q(s)}{P(s)} + \frac{G(s)}{P(s)}, \qquad (11)$$

where $P(s) = a_n s^n + a_{n-1} s^{n-1} + \cdots + a_0$, $Q(s)$ is a polynomial in s of degree less than or equal to $n - 1$ consisting of the various products of the coefficients a_i, $i = 1, \ldots, n$ and the prescribed initial conditions $y_0, y_1, \ldots, y_{n-1}$, and $G(s)$ is the Laplace transform of $g(t)$.* Typically we put the two terms in (11) over the least common denominator and then decompose the expression into two or more partial fractions. Finally, the solution $y(t)$ of the original initial-value problem is $y(t) = \mathscr{L}^{-1}\{Y(s)\}$, where the inverse transform is done term by term.

The procedure is summarized in the following diagram.

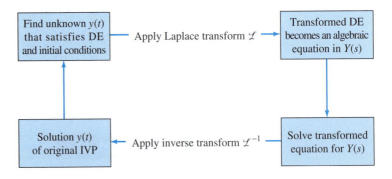

The next example illustrates the foregoing method of solving DEs, as well as partial fraction decomposition in the case when the denominator of $Y(s)$ contains a *quadratic polynomial with no real factors*.

---

*The polynomial $P(s)$ is the same as the nth-degree auxiliary polynomial in (12) in Section 4.3 with the usual symbol m replaced by s.

EXAMPLE 4   Solving a First-Order IVP

Use the Laplace transform to solve the initial-value problem

$$\frac{dy}{dt} + 3y = 13 \sin 2t, \quad y(0) = 6.$$

SOLUTION   We first take the transform of each member of the differential equation:

$$\mathscr{L}\left\{\frac{dy}{dt}\right\} + 3\mathscr{L}\{y\} = 13\mathscr{L}\{\sin 2t\}. \tag{12}$$

From (6), $\mathscr{L}\{dy/dt\} = sY(s) - y(0) = sY(s) - 6$, and from part (d) of Theorem 7.1, $\mathscr{L}\{\sin 2t\} = 2/(s^2 + 4)$, so (12) is the same as

$$sY(s) - 6 + 3Y(s) = \frac{26}{s^2 + 4} \quad \text{or} \quad (s + 3)Y(s) = 6 + \frac{26}{s^2 + 4}.$$

Solving the last equation for $Y(s)$, we get

$$Y(s) = \frac{6}{s + 3} + \frac{26}{(s + 3)(s^2 + 4)} = \frac{6s^2 + 50}{(s + 3)(s^2 + 4)}. \tag{13}$$

Since the quadratic polynomial $s^2 + 4$ does not factor using real numbers, its assumed numerator in the partial fraction decomposition is a linear polynomial in s:

$$\frac{6s^2 + 50}{(s + 3)(s^2 + 4)} = \frac{A}{s + 3} + \frac{Bs + C}{s^2 + 4}.$$

Putting the right-hand side of the equality over a common denominator and equating numerators gives $6s^2 + 50 = A(s^2 + 4) + (Bs + C)(s + 3)$. Setting $s = -3$ then immediately yields $A = 8$. Since the denominator has no more real zeros, we equate the coefficients of s^2 and s: $6 = A + B$ and $0 = 3B + C$. Using the value of A in the first equation gives $B = -2$, and then using this last value in the second equation gives $C = 6$. Thus

$$Y(s) = \frac{6s^2 + 50}{(s + 3)(s^2 + 4)} = \frac{8}{s + 3} + \frac{-2s + 6}{s^2 + 4}.$$

We are not quite finished because the last rational expression still has to be written as two fractions. This was done by termwise division in Example 2. From (2) of that example,

$$y(t) = 8\mathscr{L}^{-1}\left\{\frac{1}{s + 3}\right\} - 2\mathscr{L}^{-1}\left\{\frac{s}{s^2 + 4}\right\} + 3\mathscr{L}^{-1}\left\{\frac{2}{s^2 + 4}\right\}.$$

It follows from parts (c), (d), and (e) of Theorem 7.4 that the solution of the initial-value problem is $y(t) = 8e^{-3t} - 2 \cos 2t + 3 \sin 2t$. ∎

EXAMPLE 5   Solving a Second-Order IVP

Solve $y'' - 3y' + 2y = e^{-4t}$, $y(0) = 1$, $y'(0) = 5$.

SOLUTION   Proceeding as in Example 4, we transform the DE. We take the sum of the transforms of each term, use (6) and (7), use the given initial conditions, use (c) of Theorem 7.4, and then solve for $Y(s)$:

$$\mathscr{L}\left\{\frac{d^2y}{dt^2}\right\} - 3\mathscr{L}\left\{\frac{dy}{dt}\right\} + 2\mathscr{L}\{y\} = \mathscr{L}\{e^{-4t}\}$$

$$s^2Y(s) - sy(0) - y'(0) - 3[sY(s) - y(0)] + 2Y(s) = \frac{1}{s+4}$$

$$(s^2 - 3s + 2)Y(s) = s + 2 + \frac{1}{s+4}$$

$$Y(s) = \frac{s+2}{s^2 - 3s + 2} + \frac{1}{(s^2 - 3s + 2)(s+4)} = \frac{s^2 + 6s + 9}{(s-1)(s-2)(s+4)}. \quad (14)$$

The details of the partial fraction decomposition of $Y(s)$ have already been carried out in Example 3. In view of the results in (4) and (5), we have the solution of the initial-value problem

$$y(t) = \mathscr{L}^{-1}\{Y(s)\} = -\frac{16}{5}e^t + \frac{25}{6}e^{2t} + \frac{1}{30}e^{-4t}.$$

Examples 4 and 5 illustrate the basic procedure for using the Laplace transform to solve a linear initial-value problem, but these examples may appear to demonstrate a method that is not much better than the approach to such problems outlined in Sections 2.3 and 4.3–4.6. Don't draw any negative conclusions from only two examples. Yes, there is a lot of algebra inherent in the use of the Laplace transform, *but* observe that we do not have to use variation of parameters or worry about the cases and algebra in the method of undetermined coefficients. Moreover, since the method incorporates the prescribed initial conditions directly into the solution, there is no need for the separate operation of applying the initial conditions to the general solution $y = c_1y_1 + c_2y_2 + \cdots + c_ny_n + y_p$ of the DE to find specific constants in a particular solution of the IVP.

The Laplace transform has many operational properties. In the sections that follow we will examine some of these properties and see how they enable us to solve problems of greater complexity.

REMARKS

(*i*) The inverse Laplace transform of a function $F(s)$ may not be unique; in other words, it is possible that $\mathscr{L}\{f_1(t)\} = \mathscr{L}\{f_2(t)\}$ and yet $f_1 \neq f_2$. For our purposes this is not anything to be concerned about. If f_1 and f_2 are piecewise continuous on $[0, \infty)$ and of exponential order, then f_1 and f_2 are *essentially* the same. See Problem 44 in Exercises 7.2. However, if f_1 and f_2 are continuous on $[0, \infty)$ and $\mathscr{L}\{f_1(t)\} = \mathscr{L}\{f_2(t)\}$, then $f_1 = f_2$ on the interval.

(*ii*) This remark is for those of you who will be required to do partial fraction decompositions by hand. There is another way of determining the coefficients in a partial fraction decomposition in the special case when $\mathscr{L}\{f(t)\} = F(s)$ is a rational function of s and the denominator of F is a product of *distinct* linear factors. Let us illustrate by reexamining Example 3. Suppose we multiply both sides of the assumed decomposition

$$\frac{s^2 + 6s + 9}{(s-1)(s-2)(s+4)} = \frac{A}{s-1} + \frac{B}{s-2} + \frac{C}{s+4} \quad (15)$$

by, say, $s - 1$, simplify, and then set $s = 1$. Since the coefficients of B and C on the right-hand side of the equality are zero, we get

$$\frac{s^2 + 6s + 9}{(s - 2)(s + 4)}\bigg|_{s=1} = A \quad \text{or} \quad A = -\frac{16}{5}.$$

Written another way,

$$\frac{s^2 + 6s + 9}{\boxed{(s - 1)}(s - 2)(s + 4)}\bigg|_{s=1} = -\frac{16}{5} = A,$$

where we have shaded, or *covered up,* the factor that canceled when the left-hand side was multiplied by $s - 1$. Now to obtain B and C, we simply evaluate the left-hand side of (15) while covering up, in turn, $s - 2$ and $s + 4$:

$$\frac{s^2 + 6s + 9}{(s - 1)\boxed{(s - 2)}(s + 4)}\bigg|_{s=2} = \frac{25}{6} = B \quad \text{and} \quad \frac{s^2 + 6s + 9}{(s - 1)(s - 2)\boxed{(s + 4)}}\bigg|_{s=-4} = \frac{1}{30} = C.$$

The desired decomposition (15) is given in (4). This special technique for determining coefficients is naturally known as the **cover-up method.**

(*iii*) In this remark we continue our introduction to the terminology of dynamical systems. Because of (9) and (10) the Laplace transform is well adapted to *linear* dynamical systems. The polynomial $P(s) = a_n s^n + a_{n-1} s^{n-1} + \cdots + a_0$ in (11) is the total coefficient of $Y(s)$ in (10) and is simply the left-hand side of the DE with the derivatives $d^k y/dt^k$ replaced by powers s^k, $k = 0, 1, \ldots, n$. It is usual practice to call the reciprocal of $P(s)$—namely, $W(s) = 1/P(s)$—the **transfer function** of the system and write (11) as

$$Y(s) = W(s)Q(s) + W(s)G(s). \tag{16}$$

In this manner we have separated, in an additive sense, the effects on the response that are due to the initial conditions (that is, $W(s)Q(s)$) from those due to the input function g (that is, $W(s)G(s)$). See (13) and (14). Hence the response $y(t)$ of the system is a superposition of two responses:

$$y(t) = \mathscr{L}^{-1}\{W(s)Q(s)\} + \mathscr{L}^{-1}\{W(s)G(s)\} = y_0(t) + y_1(t).$$

If the input is $g(t) = 0$, then the solution of the problem is $y_0(t) = \mathscr{L}^{-1}\{W(s)Q(s)\}$. This solution is called the **zero-input response** of the system. On the other hand, the function $y_1(t) = \mathscr{L}^{-1}\{W(s)G(s)\}$ is the output due to the input $g(t)$. Now if the initial state of the system is the zero state (all the initial conditions are zero), then $Q(s) = 0$, and so the only solution of the initial-value problem is $y_1(t)$. The latter solution is called the **zero-state response** of the system. Both $y_0(t)$ and $y_1(t)$ are particular solutions: $y_0(t)$ is a solution of the IVP consisting of the associated homogeneous equation with the given initial conditions, and $y_1(t)$ is a solution of the IVP consisting of the nonhomogeneous equation with zero initial conditions. In Example 5 we see from (14) that the transfer function is $W(s) = 1/(s^2 - 3s + 2)$, the zero-input response is

$$y_0(t) = \mathscr{L}^{-1}\left\{\frac{s + 2}{(s - 1)(s - 2)}\right\} = -3e^t + 4e^{2t},$$

and the zero-state response is

$$y_1(t) = \mathscr{L}^{-1}\left\{\frac{1}{(s - 1)(s - 2)(s + 4)}\right\} = -\frac{1}{5}e^t + \frac{1}{6}e^{2t} + \frac{1}{30}e^{-4t}.$$

Verify that the sum of $y_0(t)$ and $y_1(t)$ is the solution $y(t)$ in Example 5 and that $y_0(0) = 1$, $y_0'(0) = 5$, whereas $y_1(0) = 0$, $y_1'(0) = 0$.

EXERCISES 7.2

Answers to selected odd-numbered problems begin on page ANS-9.

7.2.1 INVERSE TRANSFORMS

In Problems 1–30 use appropriate algebra and Theorem 7.4 to find the given inverse Laplace transform.

1. $\mathscr{L}^{-1}\left\{\dfrac{1}{s^3}\right\}$

2. $\mathscr{L}^{-1}\left\{\dfrac{1}{s^4}\right\}$

3. $\mathscr{L}^{-1}\left\{\dfrac{1}{s^2} - \dfrac{48}{s^5}\right\}$

4. $\mathscr{L}^{-1}\left\{\left(\dfrac{2}{s} - \dfrac{1}{s^3}\right)^2\right\}$

5. $\mathscr{L}^{-1}\left\{\dfrac{(s+1)^3}{s^4}\right\}$

6. $\mathscr{L}^{-1}\left\{\dfrac{(s+2)^2}{s^3}\right\}$

7. $\mathscr{L}^{-1}\left\{\dfrac{1}{s^2} - \dfrac{1}{s} + \dfrac{1}{s-2}\right\}$

8. $\mathscr{L}^{-1}\left\{\dfrac{4}{s} + \dfrac{6}{s^5} - \dfrac{1}{s+8}\right\}$

9. $\mathscr{L}^{-1}\left\{\dfrac{1}{4s+1}\right\}$

10. $\mathscr{L}^{-1}\left\{\dfrac{1}{5s-2}\right\}$

11. $\mathscr{L}^{-1}\left\{\dfrac{5}{s^2+49}\right\}$

12. $\mathscr{L}^{-1}\left\{\dfrac{10s}{s^2+16}\right\}$

13. $\mathscr{L}^{-1}\left\{\dfrac{4s}{4s^2+1}\right\}$

14. $\mathscr{L}^{-1}\left\{\dfrac{1}{4s^2+1}\right\}$

15. $\mathscr{L}^{-1}\left\{\dfrac{2s-6}{s^2+9}\right\}$

16. $\mathscr{L}^{-1}\left\{\dfrac{s+1}{s^2+2}\right\}$

17. $\mathscr{L}^{-1}\left\{\dfrac{1}{s^2+3s}\right\}$

18. $\mathscr{L}^{-1}\left\{\dfrac{s+1}{s^2-4s}\right\}$

19. $\mathscr{L}^{-1}\left\{\dfrac{s}{s^2+2s-3}\right\}$

20. $\mathscr{L}^{-1}\left\{\dfrac{1}{s^2+s-20}\right\}$

21. $\mathscr{L}^{-1}\left\{\dfrac{0.9s}{(s-0.1)(s+0.2)}\right\}$

22. $\mathscr{L}^{-1}\left\{\dfrac{s-3}{(s-\sqrt{3})(s+\sqrt{3})}\right\}$

23. $\mathscr{L}^{-1}\left\{\dfrac{s}{(s-2)(s-3)(s-6)}\right\}$

24. $\mathscr{L}^{-1}\left\{\dfrac{s^2+1}{s(s-1)(s+1)(s-2)}\right\}$

25. $\mathscr{L}^{-1}\left\{\dfrac{1}{s^3+5s}\right\}$

26. $\mathscr{L}^{-1}\left\{\dfrac{s}{(s+2)(s^2+4)}\right\}$

27. $\mathscr{L}^{-1}\left\{\dfrac{2s-4}{(s^2+s)(s^2+1)}\right\}$

28. $\mathscr{L}^{-1}\left\{\dfrac{1}{s^4-9}\right\}$

29. $\mathscr{L}^{-1}\left\{\dfrac{1}{(s^2+1)(s^2+4)}\right\}$

30. $\mathscr{L}^{-1}\left\{\dfrac{6s+3}{s^4+5s^2+4}\right\}$

7.2.2 TRANSFORMS OF DERIVATIVES

In Problems 31–40 use the Laplace transform to solve the given initial-value problem.

31. $\dfrac{dy}{dt} - y = 1, \quad y(0) = 0$

32. $2\dfrac{dy}{dt} + y = 0, \quad y(0) = -3$

33. $y' + 6y = e^{4t}, \quad y(0) = 2$

34. $y' - y = 2\cos 5t, \quad y(0) = 0$

35. $y'' + 5y' + 4y = 0, \quad y(0) = 1, \quad y'(0) = 0$

36. $y'' - 4y' = 6e^{3t} - 3e^{-t}, \quad y(0) = 1, \quad y'(0) = -1$

37. $y'' + y = \sqrt{2}\sin\sqrt{2}t, \quad y(0) = 10, \quad y'(0) = 0$

38. $y'' + 9y = e^t, \quad y(0) = 0, \quad y'(0) = 0$

39. $2y''' + 3y'' - 3y' - 2y = e^{-t}, \quad y(0) = 0, \quad y'(0) = 0, \quad y''(0) = 1$

40. $y''' + 2y'' - y' - 2y = \sin 3t, \quad y(0) = 0, \quad y'(0) = 0, \quad y''(0) = 1$

The inverse forms of the results in Problem 46 in Exercises 7.1 are

$$\mathscr{L}^{-1}\left\{\dfrac{s-a}{(s-a)^2+b^2}\right\} = e^{at}\cos bt$$

and

$$\mathscr{L}^{-1}\left\{\dfrac{b}{(s-a)^2+b^2}\right\} = e^{at}\sin bt.$$

In Problems 41 and 42 use the Laplace transform and these inverses to solve the given initial-value problem.

41. $y' + y = e^{-3t}\cos 2t, \quad y(0) = 0$

42. $y'' - 2y' + 5y = 0, \quad y(0) = 1, \quad y'(0) = 3$

DISCUSSION/PROJECT PROBLEMS

43. (a) With a slight change in notation the transform in (6) is the same as

$$\mathscr{L}\{f'(t)\} = s\mathscr{L}\{f(t)\} - f(0).$$

With $f(t) = te^{at}$, discuss how this result in conjunction with (c) of Theorem 7.1 can be used to evaluate $\mathscr{L}\{te^{at}\}$.

(b) Proceed as in part (a), but this time discuss how to use (7) with $f(t) = t\sin kt$ in conjunction with (d) and (e) of Theorem 7.1 to evaluate $\mathscr{L}\{t \sin kt\}$.

44. Make up two functions f_1 and f_2 that have the same Laplace transform. Do not think profound thoughts.

45. Reread *Remark* (*iii*) on page 291. Find the zero-input and the zero-state response for the IVP in Problem 36.

46. Suppose $f(t)$ is a function for which $f'(t)$ is piecewise continuous and of exponential order c. Use results in this section and Section 7.1 to justify

$$f(0) = \lim_{s \to \infty} sF(s),$$

where $F(s) = \mathscr{L}\{f(t)\}$. Verify this result with $f(t) = \cos kt$.

7.3 OPERATIONAL PROPERTIES I

INTRODUCTION: It is not convenient to use Definition 7.1 each time we wish to find the Laplace transform of a function $f(t)$. For example, the integration by parts involved in evaluating, say, $\mathscr{L}\{e^t t^2 \sin 3t\}$ is formidable to say the least. In this section and the next we present several labor-saving operational properties of the Laplace transform that enable us to build up a more extensive list of transforms (see the table in Appendix III) without having to resort to the basic definition and integration.

REVIEW MATERIAL: Keep practicing partial fraction decomposition. The concept of completing the square is also important in the discussion that follows.

7.3.1   TRANSLATION ON THE s-AXIS

A TRANSLATION   Evaluating transforms such as $\mathscr{L}\{e^{5t}t^3\}$ and $\mathscr{L}\{e^{-2t}\cos 4t\}$ is straightforward provided that we know (and we do) $\mathscr{L}\{t^3\}$ and $\mathscr{L}\{\cos 4t\}$. In general, if we know the Laplace transform of a function f, $\mathscr{L}\{f(t)\} = F(s)$, it is possible to compute the Laplace transform of an exponential multiple of f—that is, $\mathscr{L}\{e^{at}f(t)\}$—with no additional effort other than *translating*, or *shifting*, the transform $F(s)$ to $F(s - a)$. This result is known as the **first translation theorem** or **first shifting theorem.**

THEOREM 7.6	**First Translation Theorem**

If $\mathscr{L}\{f(t)\} = F(s)$ and a is any real number, then

$$\mathscr{L}\{e^{at}f(t)\} = F(s - a).$$

PROOF   The proof is immediate, since by Definition 7.1

$$\mathscr{L}\{e^{at}f(t)\} = \int_0^\infty e^{-st}e^{at}f(t)\, dt = \int_0^\infty e^{-(s-a)t}f(t)\, dt = F(s - a). \quad \blacksquare$$

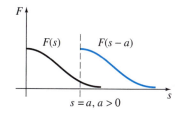

FIGURE 7.10   Shift on s-axis

If we consider s a real variable, then the graph of $F(s - a)$ is the graph of $F(s)$ shifted on the s-axis by the amount $|a|$. If $a > 0$, the graph of $F(s)$ is shifted a units to the right, whereas if $a < 0$, the graph is shifted $|a|$ units to the left. See Figure 7.10. For emphasis it is sometimes useful to use the symbolism

$$\mathscr{L}\{e^{at}f(t)\} = \mathscr{L}\{f(t)\}\big|_{s \to s-a},$$

where $s \to s - a$ means that in the Laplace transform $F(s)$ of $f(t)$ we replace the symbol s wherever it appears by $s - a$.

| EXAMPLE 1 | Using the First Translation Theorem |

Evaluate (a) $\mathscr{L}\{e^{5t}t^3\}$ (b) $\mathscr{L}\{e^{-2t}\cos 4t\}$.

SOLUTION The results follow from Theorems 7.1 and 7.6.

(a) $\mathscr{L}\{e^{5t}t^3\} = \mathscr{L}\{t^3\}\big|_{s \to s-5} = \dfrac{3!}{s^4}\bigg|_{s \to s-5} = \dfrac{6}{(s-5)^4}$

(b) $\mathscr{L}\{e^{-2t}\cos 4t\} = \mathscr{L}\{\cos 4t\}\big|_{s \to s-(-2)} = \dfrac{s}{s^2+16}\bigg|_{s \to s+2} = \dfrac{s+2}{(s+2)^2+16}$ ∎

INVERSE FORM OF THEOREM 7.6 To compute the inverse of $F(s-a)$, we must recognize $F(s)$, find $f(t)$ by taking the inverse Laplace transform of $F(s)$, and then multiply $f(t)$ by the exponential function e^{at}. This procedure can be summarized symbolically in the following manner:

$$\mathscr{L}^{-1}\{F(s-a)\} = \mathscr{L}^{-1}\{F(s)\big|_{s \to s-a}\} = e^{at}f(t), \tag{1}$$

where $f(t) = \mathscr{L}^{-1}\{F(s)\}$.

The first part of the next example illustrates partial fraction decomposition in the case when the denominator of $Y(s)$ contains *repeated linear factors*.

| EXAMPLE 2 | Partial Fractions: Repeated Linear Factors |

Evaluate (a) $\mathscr{L}^{-1}\left\{\dfrac{2s+5}{(s-3)^2}\right\}$ (b) $\mathscr{L}^{-1}\left\{\dfrac{s/2+5/3}{s^2+4s+6}\right\}$.

SOLUTION (a) A repeated linear factor is a term $(s-a)^n$, where a is a real number and n is a positive integer ≥ 2. Recall that if $(s-a)^n$ appears in the denominator of a rational expression, then the assumed decomposition contains n partial fractions with constant numerators and denominators $s-a$, $(s-a)^2$, ..., $(s-a)^n$. Hence with $a = 3$ and $n = 2$ we write

$$\frac{2s+5}{(s-3)^2} = \frac{A}{s-3} + \frac{B}{(s-3)^2}.$$

By putting the two terms on the right-hand side over a common denominator, we obtain the numerator $2s+5 = A(s-3) + B$, and this identity yields $A = 2$ and $B = 11$. Therefore

$$\frac{2s+5}{(s-3)^2} = \frac{2}{s-3} + \frac{11}{(s-3)^2} \tag{2}$$

and

$$\mathscr{L}^{-1}\left\{\frac{2s+5}{(s-3)^2}\right\} = 2\mathscr{L}^{-1}\left\{\frac{1}{s-3}\right\} + 11\mathscr{L}^{-1}\left\{\frac{1}{(s-3)^2}\right\}. \tag{3}$$

Now $1/(s-3)^2$ is $F(s) = 1/s^2$ shifted three units to the right. Since $\mathscr{L}^{-1}\{1/s^2\} = t$, it follows from (1) that

$$\mathscr{L}^{-1}\left\{\frac{1}{(s-3)^2}\right\} = \mathscr{L}^{-1}\left\{\frac{1}{s^2}\bigg|_{s \to s-3}\right\} = e^{3t}t.$$

Finally, (3) is

$$\mathscr{L}^{-1}\left\{\frac{2s+5}{(s-3)^2}\right\} = 2e^{3t} + 11e^{3t}t. \tag{4}$$

(b) To start, observe that the quadratic polynomial $s^2 + 4s + 6$ has no real zeros and so has no real linear factors. In this situation we *complete the square*:

$$\frac{s/2 + 5/3}{s^2 + 4s + 6} = \frac{s/2 + 5/3}{(s + 2)^2 + 2}. \tag{5}$$

Our goal here is to recognize the expression on the right-hand side as some Laplace transform $F(s)$ in which s has been replaced throughout by $s + 2$. What we are trying to do is analogous to working part (b) of Example 1 backwards. The denominator in (5) is already in the correct form—that is, $s^2 + 2$ with s replaced by $s + 2$. However, we must fix up the numerator by manipulating the constants: $\frac{1}{2}s + \frac{5}{3} = \frac{1}{2}(s + 2) + \frac{5}{3} - \frac{2}{2} = \frac{1}{2}(s + 2) + \frac{2}{3}$.

Now by termwise division, the linearity of $\mathscr{L}^{-1}$, parts (e) and (d) of Theorem 7.4, and finally (1),

$$\frac{s/2 + 5/3}{(s + 2)^2 + 2} = \frac{\frac{1}{2}(s + 2) + \frac{2}{3}}{(s + 2)^2 + 2} = \frac{1}{2}\frac{s + 2}{(s + 2)^2 + 2} + \frac{2}{3}\frac{1}{(s + 2)^2 + 2}$$

$$\mathscr{L}^{-1}\left\{\frac{s/2 + 5/3}{s^2 + 4s + 6}\right\} = \frac{1}{2}\mathscr{L}^{-1}\left\{\frac{s + 2}{(s + 2)^2 + 2}\right\} + \frac{2}{3}\mathscr{L}^{-1}\left\{\frac{1}{(s + 2)^2 + 2}\right\}$$

$$= \frac{1}{2}\mathscr{L}^{-1}\left\{\frac{s}{s^2 + 2}\Big|_{s \to s+2}\right\} + \frac{2}{3\sqrt{2}}\mathscr{L}^{-1}\left\{\frac{\sqrt{2}}{s^2 + 2}\Big|_{s \to s+2}\right\} \tag{6}$$

$$= \frac{1}{2}e^{-2t}\cos\sqrt{2}t + \frac{\sqrt{2}}{3}e^{-2t}\sin\sqrt{2}t. \tag{7}$$

EXAMPLE 3   An Initial-Value Problem

Solve $y'' - 6y' + 9y = t^2 e^{3t}$, $y(0) = 2$, $y'(0) = 17$.

SOLUTION   Before transforming the DE, note that its right-hand side is similar to the function in part (a) of Example 1. After using linearity, Theorem 7.6, and the initial conditions, we simplify and then solve for $Y(s) = \mathscr{L}\{f(t)\}$:

$$\mathscr{L}\{y''\} - 6\mathscr{L}\{y'\} + 9\mathscr{L}\{y\} = \mathscr{L}\{t^2 e^{3t}\}$$

$$s^2 Y(s) - sy(0) - y'(0) - 6[sY(s) - y(0)] + 9Y(s) = \frac{2}{(s - 3)^3}$$

$$(s^2 - 6s + 9)Y(s) = 2s + 5 + \frac{2}{(s - 3)^3}$$

$$(s - 3)^2 Y(s) = 2s + 5 + \frac{2}{(s - 3)^3}$$

$$Y(s) = \frac{2s + 5}{(s - 3)^2} + \frac{2}{(s - 3)^5}.$$

The first term on the right-hand side was already decomposed into individual partial fractions in (2) in part (a) of Example 2:

$$Y(s) = \frac{2}{s - 3} + \frac{11}{(s - 3)^2} + \frac{2}{(s - 3)^5}.$$

Thus $$y(t) = 2\mathscr{L}^{-1}\left\{\frac{1}{s - 3}\right\} + 11\mathscr{L}^{-1}\left\{\frac{1}{(s - 3)^2}\right\} + \frac{2}{4!}\mathscr{L}^{-1}\left\{\frac{4!}{(s - 3)^5}\right\}. \tag{8}$$

From the inverse form (1) of Theorem 7.6, the last two terms in (8) are

$$\mathscr{L}^{-1}\left\{\frac{1}{s^2}\Big|_{s \to s-3}\right\} = te^{3t} \quad \text{and} \quad \mathscr{L}^{-1}\left\{\frac{4!}{s^5}\Big|_{s \to s-3}\right\} = t^4 e^{3t}.$$

Thus (8) is $y(t) = 2e^{3t} + 11te^{3t} + \frac{1}{12}t^4 e^{3t}$.

EXAMPLE 4   **An Initial-Value Problem**

Solve $y'' + 4y' + 6y = 1 + e^{-t}$, $y(0) = 0$, $y'(0) = 0$.

SOLUTION

$$\mathcal{L}\{y''\} + 4\mathcal{L}\{y'\} + 6\mathcal{L}\{y\} = \mathcal{L}\{1\} + \mathcal{L}\{e^{-t}\}$$

$$s^2 Y(s) - sy(0) - y'(0) + 4[sY(s) - y(0)] + 6Y(s) = \frac{1}{s} + \frac{1}{s+1}$$

$$(s^2 + 4s + 6)Y(s) = \frac{2s+1}{s(s+1)}$$

$$Y(s) = \frac{2s+1}{s(s+1)(s^2+4s+6)}$$

Since the quadratic term in the denominator does not factor into real linear factors, the partial fraction decomposition for $Y(s)$ is found to be

$$Y(s) = \frac{1/6}{s} + \frac{1/3}{s+1} - \frac{s/2 + 5/3}{s^2 + 4s + 6}.$$

Moreover, in preparation for taking the inverse transform, we already manipulated the last term into the necessary form in part (b) of Example 2. So in view of the results in (6) and (7), we have the solution

$$y(t) = \frac{1}{6}\mathcal{L}^{-1}\left\{\frac{1}{s}\right\} + \frac{1}{3}\mathcal{L}^{-1}\left\{\frac{1}{s+1}\right\} - \frac{1}{2}\mathcal{L}^{-1}\left\{\frac{s+2}{(s+2)^2 + 2}\right\} - \frac{2}{3\sqrt{2}}\mathcal{L}^{-1}\left\{\frac{\sqrt{2}}{(s+2)^2 + 2}\right\}$$

$$= \frac{1}{6} + \frac{1}{3}e^{-t} - \frac{1}{2}e^{-2t}\cos\sqrt{2}t - \frac{\sqrt{2}}{3}e^{-2t}\sin\sqrt{2}t.$$

7.3.2   TRANSLATION ON THE T-AXIS

UNIT STEP FUNCTION   In engineering one frequently encounters functions that are either "off" or "on." For example, an external force acting on a mechanical system or a voltage impressed on a circuit can be turned off after a period of time. It is convenient, then, to define a special function that is the number 0 (off) up to a certain time $t = a$ and then the number 1 (on) after that time. This function is called the **unit step function** or the **Heaviside function.**

DEFINITION 7.3   **Unit Step Function**

The **unit step function** $\mathcal{U}(t - a)$ is defined to be

$$\mathcal{U}(t - a) = \begin{cases} 0, & 0 \le t < a \\ 1, & t \ge a. \end{cases}$$

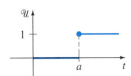

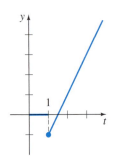

FIGURE 7.11   Graph of unit step function

FIGURE 7.12   Function is $f(t) = (2t - 3)\mathcal{U}(t - 1)$

Notice that we define $\mathcal{U}(t - a)$ only on the nonnegative t-axis, since this is all that we are concerned with in the study of the Laplace transform. In a broader sense $\mathcal{U}(t - a) = 0$ for $t < a$. The graph of $\mathcal{U}(t - a)$ is given in Figure 7.11.

When a function f defined for $t \ge 0$ is multiplied by $\mathcal{U}(t - a)$, the unit step function "turns off" a portion of the graph of that function. For example, consider the function $f(t) = 2t - 3$. To "turn off" the portion of the graph of f on, say, the interval $0 \le t < 1$, we simply form the product $(2t - 3)\mathcal{U}(t - 1)$. See Figure 7.12. In general, the graph of $f(t)\mathcal{U}(t - a)$ is 0 (off) for $0 \le t < a$ and is the portion of the graph of f (on) for $t \ge a$.

The unit step function can also be used to write piecewise-defined functions in a compact form. For example, if we consider the intervals $0 \le t < 2$, $2 \le t < 3$, and $t \ge 3$ and the corresponding values of $\mathcal{U}(t - 2)$ and $\mathcal{U}(t - 3)$, it should be

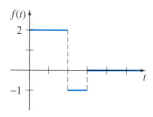

FIGURE 7.13 Function is
$f(t) = 2 - 3\mathcal{U}(t - 2) + \mathcal{U}(t - 3)$

apparent that the piecewise-defined function shown in Figure 7.13 is the same as $f(t) = 2 - 3\mathcal{U}(t - 2) + \mathcal{U}(t - 3)$. Also, a general piecewise-defined function of the type

$$f(t) = \begin{cases} g(t), & 0 \le t < a \\ h(t), & t \ge a \end{cases} \tag{9}$$

is the same as

$$f(t) = g(t) - g(t)\,\mathcal{U}(t - a) + h(t)\,\mathcal{U}(t - a). \tag{10}$$

Similarly, a function of the type

$$f(t) = \begin{cases} 0, & 0 \le t < a \\ g(t), & a \le t < b \\ 0, & t \ge b \end{cases} \tag{11}$$

can be written

$$f(t) = g(t)[\mathcal{U}(t - a) - \mathcal{U}(t - b)]. \tag{12}$$

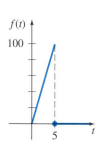

FIGURE 7.14 Function is
$f(t) = 20t - 20t\,\mathcal{U}(t - 5)$

EXAMPLE 5   **A Piecewise-Defined Function**

Express $f(t) = \begin{cases} 20t, & 0 \le t < 5 \\ 0, & t \ge 5 \end{cases}$ in terms of unit step functions. Graph.

SOLUTION   The graph of f is given in Figure 7.14. Now from (9) and (10) with $a = 5$, $g(t) = 20t$, and $h(t) = 0$ we get $f(t) = 20t - 20t\,\mathcal{U}(t - 5)$.

Consider a general function $y = f(t)$ defined for $t \ge 0$. The piecewise-defined function

$$f(t - a)\,\mathcal{U}(t - a) = \begin{cases} 0, & 0 \le t < a \\ f(t - a), & t \ge a \end{cases} \tag{13}$$

plays a significant role in the discussion that follows. As shown in Figure 7.15, for $a > 0$ the graph of the function $y = f(t - a)\,\mathcal{U}(t - a)$ coincides with the graph of $y = f(t - a)$ for $t \ge a$ (which is the *entire* graph of $y = f(t)$, $t \ge 0$ shifted a units to the right on the t-axis), but is identically zero for $0 \le t < a$.

We saw in Theorem 7.6 that an exponential multiple of $f(t)$ results in a translation of the transform $F(s)$ on the s-axis. As a consequence of the next theorem, we see that whenever $F(s)$ is multiplied by an exponential function e^{-as}, $a > 0$, the inverse transform of the product $e^{-as}F(s)$ is the function f shifted along the t-axis in the manner illustrated in Figure 7.15(b). This result, presented next in its direct transform version, is called the **second translation theorem** or **second shifting theorem**.

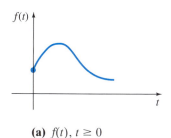

(a) $f(t)$, $t \ge 0$

(b) $f(t - a)\,\mathcal{U}(t - a)$

FIGURE 7.15   Shift on t-axis

THEOREM 7.7   **Second Translation Theorem**

If $F(s) = \mathcal{L}\{f(t)\}$ and $a > 0$, then

$$\mathcal{L}\{f(t - a)\,\mathcal{U}(t - a)\} = e^{-as}F(s).$$

PROOF   By the additive interval property of integrals,

$$\int_0^\infty e^{-st}f(t - a)\,\mathcal{U}(t - a)\,dt$$

can be written as two integrals:

$$\mathcal{L}\{f(t - a)\,\mathcal{U}(t - a)\} = \int_0^a e^{-st}f(t - a)\underbrace{\mathcal{U}(t - a)}_{\substack{\text{zero for} \\ 0 \le t < a}}\,dt + \int_a^\infty e^{-st}f(t - a)\underbrace{\mathcal{U}(t - a)}_{\substack{\text{one for} \\ t \ge a}}\,dt = \int_a^\infty e^{-st}f(t - a)\,dt.$$

Now if we let $v = t - a$, $dv = dt$ in the last integral, then

$$\mathscr{L}\{f(t - a)\,\mathscr{U}(t - a)\} = \int_0^\infty e^{-s(v + a)} f(v)\,dv = e^{-as}\int_0^\infty e^{-sv} f(v)\,dv = e^{-as}\mathscr{L}\{f(t)\}. \quad \blacksquare$$

We often wish to find the Laplace transform of just a unit step function. This can be from either Definition 7.1 or Theorem 7.7. If we identify $f(t) = 1$ in Theorem 7.7, then $f(t - a) = 1$, $F(s) = \mathscr{L}\{1\} = 1/s$, and so

$$\mathscr{L}\{\mathscr{U}(t - a)\} = \frac{e^{-as}}{s}. \tag{14}$$

For example, using (14), the Laplace transform of the function in Figure 7.13 is

$$\mathscr{L}\{f(t)\} = 2\mathscr{L}\{1\} - 3\mathscr{L}\{\mathscr{U}(t - 2)\} + \mathscr{L}\{\mathscr{U}(t - 3)\}$$

$$= 2\frac{1}{s} - 3\frac{e^{-2s}}{s} + \frac{e^{-3s}}{s}.$$

INVERSE FORM OF THEOREM 7.7   If $f(t) = \mathscr{L}^{-1}\{F(s)\}$, the inverse form of Theorem 7.7, $a > 0$, is

$$\mathscr{L}^{-1}\{e^{-as}F(s)\} = f(t - a)\,\mathscr{U}(t - a). \tag{15}$$

EXAMPLE 6   **Using Formula (15)**

Evaluate   **(a)** $\mathscr{L}^{-1}\left\{\dfrac{1}{s - 4}e^{-2s}\right\}$ **(b)** $\mathscr{L}^{-1}\left\{\dfrac{s}{s^2 + 9}e^{-\pi s/2}\right\}$.

SOLUTION   **(a)** With the identifications $a = 2$, $F(s) = 1/(s - 4)$, and $\mathscr{L}^{-1}\{F(s)\} = e^{4t}$, we have from (15)

$$\mathscr{L}^{-1}\left\{\frac{1}{s - 4}e^{-2s}\right\} = e^{4(t-2)}\,\mathscr{U}(t - 2).$$

(b) With $a = \pi/2$, $F(s) = s/(s^2 + 9)$, and $\mathscr{L}^{-1}\{F(s)\} = \cos 3t$, (15) yields

$$\mathscr{L}^{-1}\left\{\frac{s}{s^2 + 9}e^{-\pi s/2}\right\} = \cos 3\left(t - \frac{\pi}{2}\right)\mathscr{U}\left(t - \frac{\pi}{2}\right).$$

The last expression can be simplified somewhat using the addition formula for the cosine. Verify that the result is the same as $-\sin 3t\,\mathscr{U}\left(t - \dfrac{\pi}{2}\right)$. ▪

ALTERNATIVE FORM OF THEOREM 7.7   We are frequently confronted with the problem of finding the Laplace transform of a product of a function g and a unit step function $\mathscr{U}(t - a)$ where the function g lacks the precise shifted form $f(t - a)$ in Theorem 7.7. To find the Laplace transform of $g(t)\mathscr{U}(t - a)$, it is possible to fix up $g(t)$ into the required form $f(t - a)$ by algebraic manipulations. For example, if we wanted to use Theorem 7.7 to find the Laplace transform of $t^2\mathscr{U}(t - 2)$, we would have to force $g(t) = t^2$ into the form $f(t - 2)$. You should work through the details and verify that $t^2 = (t - 2)^2 + 4(t - 2) + 4$ is an identity. Therefore

$$\mathscr{L}\{t^2\mathscr{U}(t - 2)\} = \mathscr{L}\{(t - 2)^2\,\mathscr{U}(t - 2) + 4(t - 2)\,\mathscr{U}(t - 2) + 4\mathscr{U}(t - 2)\},$$

where each term on the right-hand side can now be evaluated by Theorem 7.7. But since these manipulations are time consuming and often not obvious, it is simpler to devise an alternative version of Theorem 7.7. Using Definition 7.1, the definition of $\mathcal{U}(t - a)$, and the substitution $u = t - a$, we obtain

$$\mathcal{L}\{g(t)\,\mathcal{U}(t - a)\} = \int_a^\infty e^{-st} g(t)\,dt = \int_0^\infty e^{-s(u+a)}\, g(u + a)\,du.$$

That is,

$$\mathcal{L}\{g(t)\,\mathcal{U}(t - a)\} = e^{-as}\,\mathcal{L}\{g(t + a)\}. \qquad (16)$$

EXAMPLE 7   Second Translation Theorem—Alternative Form

Evaluate $\mathcal{L}\{\cos t\,\mathcal{U}(t - \pi)\}$.

SOLUTION   With $g(t) = \cos t$ and $a = \pi$, then $g(t + \pi) = \cos(t + \pi) = -\cos t$ by the addition formula for the cosine function. Hence by (16),

$$\mathcal{L}\{\cos t\,\mathcal{U}(t - \pi)\} = -e^{-\pi s}\,\mathcal{L}\{\cos t\} = -\frac{s}{s^2 + 1}e^{-\pi s}. \qquad \blacksquare$$

EXAMPLE 8   An Initial-Value Problem

Solve $y' + y = f(t)$, $y(0) = 5$,   where $f(t) = \begin{cases} 0, & 0 \le t < \pi \\ 3\cos t, & t \ge \pi. \end{cases}$

SOLUTION   The function f can be written as $f(t) = 3\cos t\,\mathcal{U}(t - \pi)$, and so by linearity, the results of Example 7, and the usual partial fractions, we have

$$\mathcal{L}\{y'\} + \mathcal{L}\{y\} = 3\mathcal{L}\{\cos t\,\mathcal{U}(t - \pi)\}$$

$$sY(s) - y(0) + Y(s) = -3\frac{s}{s^2 + 1}e^{-\pi s}$$

$$(s + 1)Y(s) = 5 - \frac{3s}{s^2 + 1}e^{-\pi s}$$

$$Y(s) = \frac{5}{s + 1} - \frac{3}{2}\left[-\frac{1}{s + 1}e^{-\pi s} + \frac{1}{s^2 + 1}e^{-\pi s} + \frac{s}{s^2 + 1}e^{-\pi s}\right]. \qquad (17)$$

Now proceeding as we did in Example 6, it follows from (15) with $a = \pi$ that the inverses of the terms inside the brackets are

$$\mathcal{L}^{-1}\left\{\frac{1}{s + 1}e^{-\pi s}\right\} = e^{-(t - \pi)}\,\mathcal{U}(t - \pi), \quad \mathcal{L}^{-1}\left\{\frac{1}{s^2 + 1}e^{-\pi s}\right\} = \sin(t - \pi)\,\mathcal{U}(t - \pi),$$

and

$$\mathcal{L}^{-1}\left\{\frac{s}{s^2 + 1}e^{-\pi s}\right\} = \cos(t - \pi)\,\mathcal{U}(t - \pi).$$

Thus the inverse of (17) is

$$y(t) = 5e^{-t} + \frac{3}{2}e^{-(t - \pi)}\,\mathcal{U}(t - \pi) - \frac{3}{2}\sin(t - \pi)\,\mathcal{U}(t - \pi) - \frac{3}{2}\cos(t - \pi)\,\mathcal{U}(t - \pi)$$

$$= 5e^{-t} + \frac{3}{2}[e^{-(t - \pi)} + \sin t + \cos t]\,\mathcal{U}(t - \pi) \qquad \leftarrow \text{trigonometric identities}$$

$$= \begin{cases} 5e^{-t}, & 0 \le t < \pi \\ 5e^{-t} + \frac{3}{2}e^{-(t - \pi)} + \frac{3}{2}\sin t + \frac{3}{2}\cos t, & t \ge \pi. \end{cases} \qquad (18)$$

We obtained the graph of (18) shown in Figure 7.16 by using a graphing utility. $\qquad \blacksquare$

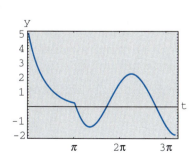

FIGURE 7.16   Graph of function in (18)

BEAMS In Section 5.2 we saw that the static deflection $y(x)$ of a uniform beam of length L carrying load $w(x)$ per unit length is found from the linear fourth-order differential equation

$$EI \frac{d^4y}{dx^4} = w(x), \tag{19}$$

where E is Young's modulus of elasticity and I is a moment of inertia of a cross-section of the beam. The Laplace transform is particularly useful in solving (19) when $w(x)$ is piecewise-defined. However, to use the Laplace transform, we must tacitly assume that $y(x)$ and $w(x)$ are defined on $(0, \infty)$ rather than on $(0, L)$. Note, too, that the next example is a boundary-value problem rather than an initial-value problem.

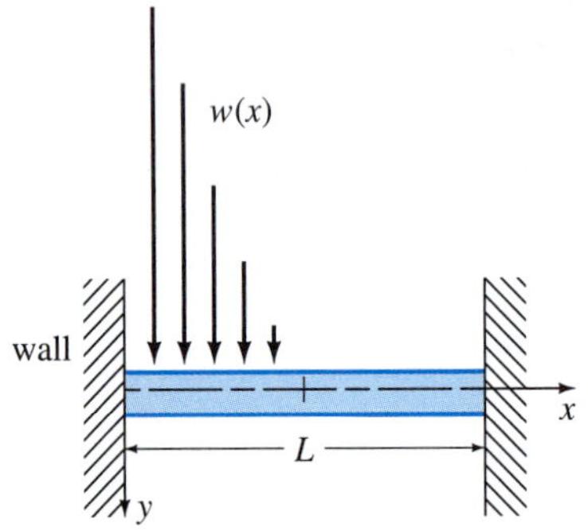

FIGURE 7.17 Embedded beam with variable load

EXAMPLE 9 A Boundary-Value Problem

A beam of length L is embedded at both ends, as shown in Figure 7.17. Find the deflection of the beam when the load is given by

$$w(x) = \begin{cases} w_0\left(1 - \frac{2}{L}x\right), & 0 < x < L/2 \\ 0, & L/2 < x < L. \end{cases}$$

SOLUTION Recall that because the beam is embedded at both ends, the boundary conditions are $y(0) = 0$, $y'(0) = 0$, $y(L) = 0$, $y'(L) = 0$. Now by (10) we can express $w(x)$ in terms of the unit step function:

$$w(x) = w_0\left(1 - \frac{2}{L}x\right) - w_0\left(1 - \frac{2}{L}x\right)\mathcal{U}\left(x - \frac{L}{2}\right)$$

$$= \frac{2w_0}{L}\left[\frac{L}{2} - x + \left(x - \frac{L}{2}\right)\mathcal{U}\left(x - \frac{L}{2}\right)\right].$$

Transforming (19) with respect to the variable x gives

$$EI\left(s^4 Y(s) - s^3 y(0) - s^2 y'(0) - s y''(0) - y'''(0)\right) = \frac{2w_0}{L}\left[\frac{L/2}{s} - \frac{1}{s^2} + \frac{1}{s^2}e^{-Ls/2}\right]$$

or

$$s^4 Y(s) - s y''(0) - y'''(0) = \frac{2w_0}{EIL}\left[\frac{L/2}{s} - \frac{1}{s^2} + \frac{1}{s^2}e^{-Ls/2}\right].$$

If we let $c_1 = y''(0)$ and $c_2 = y'''(0)$, then

$$Y(s) = \frac{c_1}{s^3} + \frac{c_2}{s^4} + \frac{2w_0}{EIL}\left[\frac{L/2}{s^5} - \frac{1}{s^6} + \frac{1}{s^6}e^{-Ls/2}\right],$$

and consequently

$$y(x) = \frac{c_1}{2!}\mathcal{L}^{-1}\left\{\frac{2!}{s^3}\right\} + \frac{c_2}{3!}\mathcal{L}^{-1}\left\{\frac{3!}{s^4}\right\} + \frac{2w_0}{EIL}\left[\frac{L/2}{4!}\mathcal{L}^{-1}\left\{\frac{4!}{s^5}\right\} - \frac{1}{5!}\mathcal{L}^{-1}\left\{\frac{5!}{s^6}\right\} + \frac{1}{5!}\mathcal{L}^{-1}\left\{\frac{5!}{s^6}e^{-Ls/2}\right\}\right]$$

$$= \frac{c_1}{2}x^2 + \frac{c_2}{6}x^3 + \frac{w_0}{60\,EIL}\left[\frac{5L}{2}x^4 - x^5 + \left(x - \frac{L}{2}\right)^5\mathcal{U}\left(x - \frac{L}{2}\right)\right].$$

Applying the conditions $y(L) = 0$ and $y'(L) = 0$ to the last result yields a system of equations for c_1 and c_2:

$$c_1\frac{L^2}{2} + c_2\frac{L^3}{6} + \frac{49w_0L^4}{1920EI} = 0$$

$$c_1 L + c_2\frac{L^2}{2} + \frac{85w_0L^3}{960EI} = 0.$$

Solving, we find $c_1 = 23w_0L^2/960EI$ and $c_2 = -9w_0L/40EI$. Thus the deflection is given by

$$y(x) = \frac{23w_0L^2}{1920EI}x^2 - \frac{3w_0L}{80EI}x^3 + \frac{w_0}{60EIL}\left[\frac{5L}{2}x^4 - x^5 + \left(x - \frac{L}{2}\right)^5 \mathcal{U}\left(x - \frac{L}{2}\right)\right].$$

EXERCISES 7.3

Answers to selected odd-numbered problems begin on page ANS-9.

7.3.1 TRANSLATION ON THE *s*-AXIS

In Problems 1–20 find either $F(s)$ or $f(t)$, as indicated.

1. $\mathcal{L}\{te^{10t}\}$
2. $\mathcal{L}\{te^{-6t}\}$
3. $\mathcal{L}\{t^3e^{-2t}\}$
4. $\mathcal{L}\{t^{10}e^{-7t}\}$
5. $\mathcal{L}\{t(e^t + e^{2t})^2\}$
6. $\mathcal{L}\{e^{2t}(t-1)^2\}$
7. $\mathcal{L}\{e^t \sin 3t\}$
8. $\mathcal{L}\{e^{-2t}\cos 4t\}$
9. $\mathcal{L}\{(1 - e^t + 3e^{-4t})\cos 5t\}$
10. $\mathcal{L}\left\{e^{3t}\left(9 - 4t + 10\sin\frac{t}{2}\right)\right\}$
11. $\mathcal{L}^{-1}\left\{\frac{1}{(s+2)^3}\right\}$
12. $\mathcal{L}^{-1}\left\{\frac{1}{(s-1)^4}\right\}$
13. $\mathcal{L}^{-1}\left\{\frac{1}{s^2 - 6s + 10}\right\}$
14. $\mathcal{L}^{-1}\left\{\frac{1}{s^2 + 2s + 5}\right\}$
15. $\mathcal{L}^{-1}\left\{\frac{s}{s^2 + 4s + 5}\right\}$
16. $\mathcal{L}^{-1}\left\{\frac{2s+5}{s^2 + 6s + 34}\right\}$
17. $\mathcal{L}^{-1}\left\{\frac{s}{(s+1)^2}\right\}$
18. $\mathcal{L}^{-1}\left\{\frac{5s}{(s-2)^2}\right\}$
19. $\mathcal{L}^{-1}\left\{\frac{2s-1}{s^2(s+1)^3}\right\}$
20. $\mathcal{L}^{-1}\left\{\frac{(s+1)^2}{(s+2)^4}\right\}$

In Problems 21–30 use the Laplace transform to solve the given initial-value problem.

21. $y' + 4y = e^{-4t}$, $y(0) = 2$
22. $y' - y = 1 + te^t$, $y(0) = 0$
23. $y'' + 2y' + y = 0$, $y(0) = 1, y'(0) = 1$
24. $y'' - 4y' + 4y = t^3e^{2t}$, $y(0) = 0, y'(0) = 0$
25. $y'' - 6y' + 9y = t$, $y(0) = 0, y'(0) = 1$
26. $y'' - 4y' + 4y = t^3$, $y(0) = 1, y'(0) = 0$

27. $y'' - 6y' + 13y = 0$, $y(0) = 0, y'(0) = -3$
28. $2y'' + 20y' + 51y = 0$, $y(0) = 2, y'(0) = 0$
29. $y'' - y' = e^t \cos t$, $y(0) = 0, y'(0) = 0$
30. $y'' - 2y' + 5y = 1 + t$, $y(0) = 0, y'(0) = 4$

In Problems 31 and 32 use the Laplace transform and the procedure outlined in Example 9 to solve the given boundary-value problem.

31. $y'' + 2y' + y = 0$, $y'(0) = 2, y(1) = 2$
32. $y'' + 8y' + 20y = 0$, $y(0) = 0, y'(\pi) = 0$

33. A 4-pound weight stretches a spring 2 feet. The weight is released from rest 18 inches above the equilibrium position, and the resulting motion takes place in a medium offering a damping force numerically equal to $\frac{7}{8}$ times the instantaneous velocity. Use the Laplace transform to find the equation of motion $x(t)$.

34. Recall that the differential equation for the instantaneous charge $q(t)$ on the capacitor in an *LRC* series circuit is given by

$$L\frac{d^2q}{dt^2} + R\frac{dq}{dt} + \frac{1}{C}q = E(t). \quad (20)$$

See Section 5.1. Use the Laplace transform to find $q(t)$ when $L = 1$ h, $R = 20$ Ω, $C = 0.005$ f, $E(t) = 150$ V, $t > 0$, $q(0) = 0$, and $i(0) = 0$. What is the current $i(t)$?

35. Consider a battery of constant voltage E_0 that charges the capacitor shown in Figure 7.18. Divide equation (20) by L and define $2\lambda = R/L$ and $\omega^2 = 1/LC$. Use the Laplace transform to show that the solution $q(t)$ of $q'' + 2\lambda q' + \omega^2 q = E_0/L$ subject to $q(0) = 0$, $i(0) = 0$ is

$$q(t) = \begin{cases} E_0C\left[1 - e^{-\lambda t}\left(\cosh\sqrt{\lambda^2 - \omega^2}t + \frac{\lambda}{\sqrt{\lambda^2 - \omega^2}}\sinh\sqrt{\lambda^2 - \omega^2}t\right)\right], & \lambda > \omega, \\ E_0C[1 - e^{-\lambda t}(1 + \lambda t)], & \lambda = \omega, \\ E_0C\left[1 - e^{-\lambda t}\left(\cos\sqrt{\omega^2 - \lambda^2}t + \frac{\lambda}{\sqrt{\omega^2 - \lambda^2}}\sin\sqrt{\omega^2 - \lambda^2}t\right)\right], & \lambda < \omega. \end{cases}$$

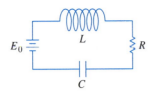

FIGURE 7.18   Series circuit in Problem 35

36. Use the Laplace transform to find the charge $q(t)$ in an RC series circuit when $q(0) = 0$ and $E(t) = E_0 e^{-kt}$, $k > 0$. Consider two cases: $k \neq 1/RC$ and $k = 1/RC$.

7.3.2   TRANSLATION ON THE *T*-AXIS

In Problems 37–48 find either $F(s)$ or $f(t)$, as indicated.

37. $\mathscr{L}\{(t - 1)\,\mathscr{U}(t - 1)\}$ **38.** $\mathscr{L}\{e^{2-t}\,\mathscr{U}(t - 2)\}$

39. $\mathscr{L}\{t\,\mathscr{U}(t - 2)\}$ **40.** $\mathscr{L}\{(3t + 1)\,\mathscr{U}(t - 1)\}$

41. $\mathscr{L}\{\cos 2t\,\mathscr{U}(t - \pi)\}$ **42.** $\mathscr{L}\left\{\sin t\,\mathscr{U}\left(t - \dfrac{\pi}{2}\right)\right\}$

43. $\mathscr{L}^{-1}\left\{\dfrac{e^{-2s}}{s^3}\right\}$ **44.** $\mathscr{L}^{-1}\left\{\dfrac{(1 + e^{-2s})^2}{s + 2}\right\}$

45. $\mathscr{L}^{-1}\left\{\dfrac{e^{-\pi s}}{s^2 + 1}\right\}$ **46.** $\mathscr{L}^{-1}\left\{\dfrac{se^{-\pi s/2}}{s^2 + 4}\right\}$

47. $\mathscr{L}^{-1}\left\{\dfrac{e^{-s}}{s(s + 1)}\right\}$ **48.** $\mathscr{L}^{-1}\left\{\dfrac{e^{-2s}}{s^2(s - 1)}\right\}$

In Problems 49–54 match the given graph with one of the functions in (a)–(f). The graph of $f(t)$ is given in Figure 7.19.

(a) $f(t) - f(t)\,\mathscr{U}(t - a)$
(b) $f(t - b)\,\mathscr{U}(t - b)$
(c) $f(t)\,\mathscr{U}(t - a)$
(d) $f(t) - f(t)\,\mathscr{U}(t - b)$
(e) $f(t)\,\mathscr{U}(t - a) - f(t)\,\mathscr{U}(t - b)$
(f) $f(t - a)\,\mathscr{U}(t - a) - f(t - a)\,\mathscr{U}(t - b)$

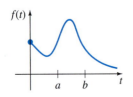

FIGURE 7.19   Graph for Problems 49–54

49.

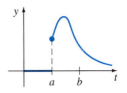

FIGURE 7.20   Graph for Problem 49

50.

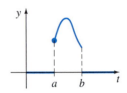

FIGURE 7.21   Graph for Problem 50

51.

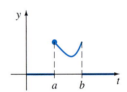

FIGURE 7.22   Graph for Problem 51

52.

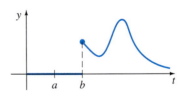

FIGURE 7.23   Graph for Problem 52

53.

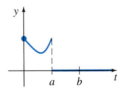

FIGURE 7.24   Graph for Problem 53

54.

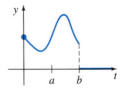

FIGURE 7.25   Graph for Problem 54

In Problems 55–62 write each function in terms of unit step functions. Find the Laplace transform of the given function.

55. $f(t) = \begin{cases} 2, & 0 \le t < 3 \\ -2, & t \ge 3 \end{cases}$

56. $f(t) = \begin{cases} 1, & 0 \le t < 4 \\ 0, & 4 \le t < 5 \\ 1, & t \ge 5 \end{cases}$

57. $f(t) = \begin{cases} 0, & 0 \le t < 1 \\ t^2, & t \ge 1 \end{cases}$

58. $f(t) = \begin{cases} 0, & 0 \le t < 3\pi/2 \\ \sin t, & t \ge 3\pi/2 \end{cases}$

59. $f(t) = \begin{cases} t, & 0 \le t < 2 \\ 0, & t \ge 2 \end{cases}$

60. $f(t) = \begin{cases} \sin t, & 0 \le t < 2\pi \\ 0, & t \ge 2\pi \end{cases}$

61.

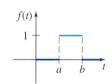

rectangular pulse

FIGURE 7.26 Graph for Problem 61

62.

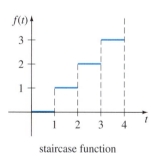

staircase function

FIGURE 7.27 Graph for Problem 62

In Problems 63–70 use the Laplace transform to solve the given initial-value problem.

63. $y' + y = f(t)$, $y(0) = 0$, where $f(t) = \begin{cases} 0, & 0 \le t < 1 \\ 5, & t \ge 1 \end{cases}$

64. $y' + y = f(t)$, $y(0) = 0$, where
$$f(t) = \begin{cases} 1, & 0 \le t < 1 \\ -1, & t \ge 1 \end{cases}$$

65. $y' + 2y = f(t)$, $y(0) = 0$, where
$$f(t) = \begin{cases} t, & 0 \le t < 1 \\ 0, & t \ge 1 \end{cases}$$

66. $y'' + 4y = f(t)$, $y(0) = 0, y'(0) = -1$, where
$$f(t) = \begin{cases} 1, & 0 \le t < 1 \\ 0, & t \ge 1 \end{cases}$$

67. $y'' + 4y = \sin t \; \mathcal{U}(t - 2\pi)$, $y(0) = 1, y'(0) = 0$

68. $y'' - 5y' + 6y = \mathcal{U}(t - 1)$, $y(0) = 0, y'(0) = 1$

69. $y'' + y = f(t)$, $y(0) = 0, y'(0) = 1$, where
$$f(t) = \begin{cases} 0, & 0 \le t < \pi \\ 1, & \pi \le t < 2\pi \\ 0, & t \ge 2\pi \end{cases}$$

70. $y'' + 4y' + 3y = 1 - \mathcal{U}(t - 2) - \mathcal{U}(t - 4) + \mathcal{U}(t - 6)$, $y(0) = 0, y'(0) = 0$

71. Suppose a 32-pound weight stretches a spring 2 feet. If the weight is released from rest at the equilibrium position, find the equation of motion $x(t)$ if an impressed force $f(t) = 20t$ acts on the system for $0 \le t < 5$ and is then removed (see Example 5). Ignore any damping forces. Use a graphing utility to graph $x(t)$ on the interval [0, 10].

72. Solve Problem 71 if the impressed force $f(t) = \sin t$ acts on the system for $0 \le t < 2\pi$ and is then removed.

In Problems 73 and 74 use the Laplace transform to find the charge $q(t)$ on the capacitor in an *RC* series circuit subject to the given conditions.

73. $q(0) = 0$, $R = 2.5 \, \Omega$, $C = 0.08$ f, $E(t)$ given in Figure 7.28

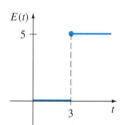

FIGURE 7.28 $E(t)$ in Problem 73

74. $q(0) = q_0$, $R = 10 \, \Omega$, $C = 0.1$ f, $E(t)$ given in Figure 7.29

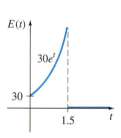

FIGURE 7.29 $E(t)$ in Problem 74

75. (a) Use the Laplace transform to find the current $i(t)$ in a single-loop *LR* series circuit when $i(0) = 0$, $L = 1$ h, $R - 10\ \Omega$, and $E(t)$ is as given in Figure 7.30.

(b) Use a computer graphing program to graph $i(t)$ on the interval $0 \le t \le 6$. Use the graph to estimate i_{max} and i_{min}, the maximum and minimum values of the current.

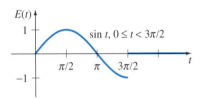

FIGURE 7.30 $E(t)$ in Problem 75

76. (a) Use the Laplace transform to find the charge $q(t)$ on the capacitor in an *RC* series circuit when $q(0) = 0$, $R = 50\ \Omega$, $C = 0.01$ f, and $E(t)$ is as given in Figure 7.31.

(b) Assume that $E_0 = 100$ V. Use a computer graphing program to graph $q(t)$ on the interval $0 \le t \le 6$. Use the graph to estimate q_{max}, the maximum value of the charge.

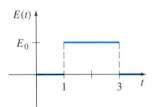

FIGURE 7.31 $E(t)$ in Problem 76

77. A cantilever beam is embedded at its left end and free at its right end. Use the Laplace transform to find the deflection $y(x)$ when the load is given by

$$w(x) = \begin{cases} w_0, & 0 < x < L/2 \\ 0, & L/2 \le x < L. \end{cases}$$

78. Solve Problem 77 when the load is given by

$$w(x) = \begin{cases} 0, & 0 < x < L/3 \\ w_0, & L/3 < x < 2L/3 \\ 0, & 2L/3 < x < L. \end{cases}$$

79. Find the deflection $y(x)$ of a cantilever beam embedded at its left end and free at its right end when the load is as given in Example 9.

80. A beam is embedded at its left end and simply supported at its right end. Find the deflection $y(x)$ when the load is as given in Problem 77.

MATHEMATICAL MODEL

81. Cake Inside an Oven Reread Example 4 in Section 3.1 on the cooling of a cake that is taken out of an oven.

(a) Devise a mathematical model for the temperature of a cake while it is *inside* the oven based on the following assumptions: At $t = 0$ the cake mixture is at the room temperature of $70°$; the oven is not preheated so that at $t = 0$, when the cake mixture is placed into the oven, the temperature inside the oven is also $70°$; the temperature of the oven increases linearly until $t = 4$ minutes, when the desired temperature of $300°$ is attained; the oven temperature is a constant $300°$ for $t \ge 4$.

(b) Use the Laplace transform to solve the initial-value problem in part (a).

DISCUSSION/PROJECT PROBLEMS

82. Discuss how you would fix up each of the following functions so that Theorem 7.7 could be used directly to find the given Laplace transform. Check your answers using (16) of this section.

(a) $\mathcal{L}\{(2t + 1)\mathcal{U}(t - 1)\}$ **(b)** $\mathcal{L}\{e^t \mathcal{U}(t - 5)\}$

(c) $\mathcal{L}\{\cos t\ \mathcal{U}(t - \pi)\}$ **(d)** $\mathcal{L}\{(t^2 - 3t)\mathcal{U}(t - 2)\}$

83. (a) Assume that Theorem 7.6 holds when the symbol a is replaced by ki, where k is a real number and $i^2 = -1$. Show that $\mathcal{L}\{te^{kti}\}$ can be used to deduce

$$\mathcal{L}\{t \cos kt\} = \frac{s^2 - k^2}{(s^2 + k^2)^2}$$

and $$\mathcal{L}\{t \sin kt\} = \frac{2ks}{(s^2 + k^2)^2}.$$

(b) Now use the Laplace transform to solve the initial-value problem $x'' + \omega^2 x = \cos \omega t$, $x(0) = 0$, $x'(0) = 0$.

7.4 OPERATIONAL PROPERTIES II

INTRODUCTION: In this section we develop several more operational properties of the Laplace transform. Specifically, we shall see how to find the transform of a function $f(t)$ that is multiplied by a monomial t^n, the transform of a special type of integral, and the transform of a periodic function. The last two transform properties allow us to solve some equations that we have not encountered up to this point: Volterra integral equations, integrodifferential equations, and ordinary differential equations in which the input function is a periodic piecewise-defined function.

7.4.1 DERIVATIVES OF A TRANSFORM

MULTIPLYING A FUNCTION BY t^n The Laplace transform of the product of a function $f(t)$ with t can be found by differentiating the Laplace transform of $f(t)$. To motivate this result, let us assume that $F(s) = \mathscr{L}\{f(t)\}$ exists and that it is possible to interchange the order of differentiation and integration. Then

$$\frac{d}{ds} F(s) = \frac{d}{ds} \int_0^\infty e^{-st} f(t)\, dt = \int_0^\infty \frac{\partial}{\partial s} [e^{-st} f(t)]\, dt = -\int_0^\infty e^{-st} t f(t)\, dt = -\mathscr{L}\{t f(t)\};$$

that is,
$$\mathscr{L}\{t f(t)\} = -\frac{d}{ds} \mathscr{L}\{f(t)\}.$$

We can use the last result to find the Laplace transform of $t^2 f(t)$:

$$\mathscr{L}\{t^2 f(t)\} = \mathscr{L}\{t \cdot t f(t)\} = -\frac{d}{ds} \mathscr{L}\{t f(t)\} = -\frac{d}{ds}\left(-\frac{d}{ds} \mathscr{L}\{f(t)\}\right) = \frac{d^2}{ds^2} \mathscr{L}\{f(t)\}.$$

The preceding two cases suggest the general result for $\mathscr{L}\{t^n f(t)\}$.

THEOREM 7.8   **Derivatives of Transforms**

If $F(s) = \mathscr{L}\{f(t)\}$ and $n = 1, 2, 3, \ldots$, then

$$\mathscr{L}\{t^n f(t)\} = (-1)^n \frac{d^n}{ds^n} F(s).$$

EXAMPLE 1   **Using Theorem 7.8**

Evaluate $\mathscr{L}\{t \sin kt\}$.

SOLUTION   With $f(t) = \sin kt$, $F(s) = k/(s^2 + k^2)$, and $n = 1$, Theorem 7.8 gives

$$\mathscr{L}\{t \sin kt\} = -\frac{d}{ds} \mathscr{L}\{\sin kt\} = -\frac{d}{ds}\left(\frac{k}{s^2 + k^2}\right) = \frac{2ks}{(s^2 + k^2)^2}.$$

If we want to evaluate $\mathscr{L}\{t^2 \sin kt\}$ and $\mathscr{L}\{t^3 \sin kt\}$, all we need do, in turn, is take the negative of the derivative with respect to s of the result in Example 1 and then take the negative of the derivative with respect to s of $\mathscr{L}\{t^2 \sin kt\}$.

NOTE To find transforms of functions $t^n e^{at}$ we can use either Theorem 7.6 or Theorem 7.8. For example,

Theorem 7.6: $\quad \mathcal{L}\{te^{3t}\} = \mathcal{L}\{t\}_{s \to s-3} = \dfrac{1}{s^2}\bigg|_{s \to s-3} = \dfrac{1}{(s-3)^2}.$

Theorem 7.8: $\quad \mathcal{L}\{te^{3t}\} = -\dfrac{d}{ds}\mathcal{L}\{e^{3t}\} = -\dfrac{d}{ds}\dfrac{1}{s-3} = (s-3)^{-2} = \dfrac{1}{(s-3)^2}.$

EXAMPLE 2   An Initial-Value Problem

Solve $x'' + 16x = \cos 4t$, $x(0) = 0$, $x'(0) = 1$.

SOLUTION The initial-value problem could describe the forced, undamped, and resonant motion of a mass on a spring. The mass starts with an initial velocity of 1 ft/s in the downward direction from the equilibrium position.

Transforming the differential equation gives

$$(s^2 + 16)X(s) = 1 + \frac{s}{s^2 + 16} \quad \text{or} \quad X(s) = \frac{1}{s^2 + 16} + \frac{s}{(s^2 + 16)^2}.$$

Now we just saw in Example 1 that

$$\mathcal{L}^{-1}\left\{\frac{2ks}{(s^2 + k^2)^2}\right\} = t\sin kt, \tag{1}$$

and so with the identification $k = 4$ in (1) and in part (d) of Theorem 7.3, we obtain

$$x(t) = \frac{1}{4}\mathcal{L}^{-1}\left\{\frac{4}{s^2 + 16}\right\} + \frac{1}{8}\mathcal{L}^{-1}\left\{\frac{8s}{(s^2 + 16)^2}\right\}$$

$$= \frac{1}{4}\sin 4t + \frac{1}{8}t\sin 4t. \qquad \blacksquare$$

7.4.2 TRANSFORMS OF INTEGRALS

CONVOLUTION If functions f and g are piecewise continuous on $[0, \infty)$, then a special product, denoted by $f * g$, is defined by the integral

$$f * g = \int_0^t f(\tau)\, g(t - \tau)\, d\tau \tag{2}$$

and is called the **convolution** of f and g. The convolution $f * g$ is a function of t. For example,

$$e^t * \sin t = \int_0^t e^{\tau}\sin(t - \tau)\, d\tau = \frac{1}{2}(-\sin t - \cos t + e^t). \tag{3}$$

It is left as an exercise to show that

$$\int_0^t f(\tau)\, g(t - \tau)\, d\tau = \int_0^t f(t - \tau)\, g(\tau)\, d\tau;$$

that is, $f * g = g * f$. This means that the convolution of two functions is commutative.

It is *not* true that the integral of a product of functions is the product of the integrals. However, it *is* true that the Laplace transform of the special product (2) is the product of the Laplace transform of f and g. This means that it is possible to find the Laplace transform of the convolution of two functions without actually evaluating the integral as we did in (3). The result that follows is known as the **convolution theorem.**

THEOREM 7.9	**Convolution Theorem**

If $f(t)$ and $g(t)$ are piecewise continuous on $[0, \infty)$ and of exponential order, then

$$\mathcal{L}\{f * g\} = \mathcal{L}\{f(t)\}\,\mathcal{L}\{g(t)\} = F(s)G(s).$$

PROOF   Let $\quad F(s) = \mathcal{L}\{f(t)\} = \displaystyle\int_0^\infty e^{-s\tau} f(\tau)\, d\tau$

and $\qquad\qquad G(s) = \mathcal{L}\{g(t)\} = \displaystyle\int_0^\infty e^{-s\beta} g(\beta)\, d\beta.$

Proceeding formally, we have

$$F(s)G(s) = \left(\int_0^\infty e^{-s\tau} f(\tau)\, d\tau\right)\left(\int_0^\infty e^{-s\beta} g(\beta)\, d\beta\right)$$

$$= \int_0^\infty \int_0^\infty e^{-s(\tau+\beta)} f(\tau)g(\beta)\, d\tau\, d\beta$$

$$= \int_0^\infty f(\tau)\, d\tau \int_0^\infty e^{-s(\tau+\beta)} g(\beta)\, d\beta.$$

Holding τ fixed, we let $t = \tau + \beta$, $dt = d\beta$ so that

$$F(s)G(s) = \int_0^\infty f(\tau)\, d\tau \int_\tau^\infty e^{-st} g(t - \tau)\, dt.$$

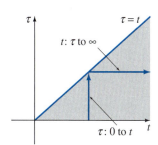

FIGURE 7.32   Changing order of integration from t first to τ first

In the $t\tau$-plane we are integrating over the shaded region in Figure 7.32. Since f and g are piecewise continuous on $[0, \infty)$ and of exponential order, it is possible to interchange the order of integration:

$$F(s)\,G(s) = \int_0^\infty e^{-st}\, dt \int_0^t f(\tau)g(t - \tau)\, d\tau = \int_0^\infty e^{-st}\left\{\int_0^t f(\tau)\, g(t - \tau)\, d\tau\right\}dt = \mathcal{L}\{f * g\}. \quad \blacksquare$$

EXAMPLE 3	**Transform of a Convolution**

Evaluate $\mathcal{L}\left\{\displaystyle\int_0^t e^\tau \sin(t - \tau)\, d\tau\right\}$.

SOLUTION   With $f(t) = e^t$ and $g(t) = \sin t$, the convolution theorem states that the Laplace transform of the convolution of f and g is the product of their Laplace transforms:

$$\mathcal{L}\left\{\int_0^t e^\tau \sin(t - \tau)\, d\tau\right\} = \mathcal{L}\{e^t\} \cdot \mathcal{L}\{\sin t\} = \frac{1}{s - 1} \cdot \frac{1}{s^2 + 1} = \frac{1}{(s - 1)(s^2 + 1)}. \quad \blacksquare$$

INVERSE FORM OF THEOREM 7.9   The convolution theorem is sometimes useful in finding the inverse Laplace transform of the product of two Laplace transforms. From Theorem 7.9 we have

$$\mathcal{L}^{-1}\{F(s)G(s)\} = f * g. \tag{4}$$

Many of the results in the table of Laplace transforms in Appendix III can be derived using (4). For example, in the next example we obtain entry 25 of the table:

$$\mathscr{L}\{\sin kt - kt \cos kt\} = \frac{2k^3}{(s^2 + k^2)^2}. \tag{5}$$

EXAMPLE 4   Inverse Transform as a Convolution

Evaluate $\mathscr{L}^{-1}\left\{\dfrac{1}{(s^2 + k^2)^2}\right\}$.

SOLUTION   Let $F(s) = G(s) = \dfrac{1}{s^2 + k^2}$ so that

$$f(t) = g(t) = \frac{1}{k}\,\mathscr{L}^{-1}\left\{\frac{k}{s^2 + k^2}\right\} = \frac{1}{k}\sin kt.$$

In this case (4) gives

$$\mathscr{L}^{-1}\left\{\frac{1}{(s^2 + k^2)^2}\right\} = \frac{1}{k^2}\int_0^t \sin k\tau \sin k(t - \tau)\,d\tau. \tag{6}$$

With the aid of the trigonometric identity

$$\sin A \cos B = \frac{1}{2}[\cos(A - B) - \cos(A + B)]$$

and the substitutions $A = k\tau$ and $B = k(t - \tau)$ we can carry out the integration in (6):

$$\mathscr{L}^{-1}\left\{\frac{1}{(s^2 + k^2)^2}\right\} = \frac{1}{2k^2}\int_0^t [\cos k(2\tau - t) - \cos kt]\,d\tau$$

$$= \frac{1}{2k^2}\left[\frac{1}{2k}\sin k(2\tau - t) - \tau\cos kt\right]_0^t$$

$$= \frac{\sin kt - kt\cos kt}{2k^3}.$$

Multiplying both sides by $2k^3$ gives the inverse form of (5).   ◼

TRANSFORM OF AN INTEGRAL   When $g(t) = 1$ and $\mathscr{L}\{g(t)\} = G(s) = 1/s$, the convolution theorem implies that the Laplace transform of the integral of f is

$$\mathscr{L}\left\{\int_0^t f(\tau)\,d\tau\right\} = \frac{F(s)}{s}. \tag{7}$$

The inverse form of (7),

$$\int_0^t f(\tau)\,d\tau = \mathscr{L}^{-1}\left\{\frac{F(s)}{s}\right\}, \tag{8}$$

can be used in lieu of partial fractions when s^n is a factor of the denominator and $f(t) = \mathscr{L}^{-1}\{F(s)\}$ is easy to integrate. For example, we know for $f(t) = \sin t$ that $F(s) = 1/(s^2 + 1)$, and so by (8)

$$\mathscr{L}^{-1}\left\{\frac{1}{s(s^2 + 1)}\right\} = \int_0^t \sin \tau\,d\tau = 1 - \cos t$$

$$\mathscr{L}^{-1}\left\{\frac{1}{s^2(s^2 + 1)}\right\} = \int_0^t (1 - \cos \tau)\,d\tau = t - \sin t$$

$$\mathscr{L}^{-1}\left\{\frac{1}{s^3(s^2+1)}\right\} = \int_0^t (\tau - \sin\tau)\,d\tau = \frac{1}{2}t^2 - 1 + \cos t$$

and so on.

VOLTERRA INTEGRAL EQUATION The convolution theorem and the result in (7) are useful in solving other types of equations in which an unknown function appears under an integral sign. In the next example we solve a **Volterra integral equation** for $f(t)$,

$$f(t) = g(t) + \int_0^t f(\tau)\,h(t-\tau)\,d\tau. \tag{9}$$

The functions $g(t)$ and $h(t)$ are known. Notice that the integral in (9) has the convolution form (2) with the symbol h playing the part of g.

EXAMPLE 5 **An Integral Equation**

Solve $f(t) = 3t^2 - e^{-t} - \int_0^t f(\tau)\,e^{t-\tau}d\tau$ for $f(t)$.

SOLUTION In the integral we identify $h(t-\tau) = e^{t-\tau}$ so that $h(t) = e^t$. We take the Laplace transform of each term; in particular, by Theorem 7.9 the transform of the integral is the product of $\mathscr{L}\{f(t)\} = F(s)$ and $\mathscr{L}\{e^t\} = 1/(s-1)$:

$$F(s) = 3\cdot\frac{2}{s^3} - \frac{1}{s+1} - F(s)\cdot\frac{1}{s-1}.$$

After solving the last equation for $F(s)$ and carrying out the partial fraction decomposition, we find

$$F(s) = \frac{6}{s^3} - \frac{6}{s^4} + \frac{1}{s} - \frac{2}{s+1}.$$

The inverse transform then gives

$$f(t) = 3\mathscr{L}^{-1}\left\{\frac{2!}{s^3}\right\} - \mathscr{L}^{-1}\left\{\frac{3!}{s^4}\right\} + \mathscr{L}^{-1}\left\{\frac{1}{s}\right\} - 2\mathscr{L}^{-1}\left\{\frac{1}{s+1}\right\}$$

$$= 3t^2 - t^3 + 1 - 2e^{-t}.$$

SERIES CIRCUITS In a single-loop or series circuit, Kirchhoff's second law states that the sum of the voltage drops across an inductor, resistor, and capacitor is equal to the impressed voltage $E(t)$. Now it is known that the voltage drops across an inductor, resistor, and capacitor are, respectively,

$$L\frac{di}{dt}, \quad Ri(t), \quad \text{and} \quad \frac{1}{C}\int_0^t i(\tau)\,d\tau,$$

where $i(t)$ is the current and $L, R,$ and C are constants. It follows that the current in a circuit, such as that shown in Figure 7.33, is governed by the **integrodifferential equation**

$$L\frac{di}{dt} + Ri(t) + \frac{1}{C}\int_0^t i(\tau)\,d\tau = E(t). \tag{10}$$

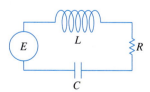

FIGURE 7.33 *LRC* series circuit

| **EXAMPLE 6** | **An Integrodifferential Equation** |

Determine the current $i(t)$ in a single-loop *LRC* circuit when $L = 0.1$ h, $R = 2$ Ω, $C = 0.1$ f, $i(0) = 0$, and the impressed voltage is

$$E(t) = 120t - 120t\, \mathcal{U}(t - 1).$$

SOLUTION With the given data, equation (10) becomes

$$0.1\frac{di}{dt} + 2i + 10\int_0^t i(\tau)\, d\tau = 120t - 120t\, \mathcal{U}(t - 1).$$

Now by (7), $\mathcal{L}\left\{\int_0^t i(\tau)\, d\tau\right\} = I(s)/s$, where $I(s) = \mathcal{L}\{i(t)\}$. Thus the Laplace transform of the integrodifferential equation is

$$0.1sI(s) + 2I(s) + 10\frac{I(s)}{s} = 120\left[\frac{1}{s^2} - \frac{1}{s^2}e^{-s} - \frac{1}{s}e^{-s}\right]. \quad \leftarrow \text{by (16) of Section 7.3}$$

Multiplying this equation by $10s$, using $s^2 + 20s + 100 = (s + 10)^2$, and then solving for $I(s)$ gives

$$I(s) = 1200\left[\frac{1}{s(s + 10)^2} - \frac{1}{s(s + 10)^2}e^{-s} - \frac{1}{(s + 10)^2}e^{-s}\right].$$

By partial fractions,

$$I(s) = 1200\left[\frac{1/100}{s} - \frac{1/100}{s + 10} - \frac{1/10}{(s + 10)^2} - \frac{1/100}{s}e^{-s} \right.$$
$$\left. + \frac{1/100}{s + 10}e^{-s} + \frac{1/10}{(s + 10)^2}e^{-s} - \frac{1}{(s + 10)^2}e^{-s}\right].$$

From the inverse form of the second translation theorem, (15) of Section 7.3, we finally obtain

$$i(t) = 12[1 - \mathcal{U}(t - 1)] - 12[e^{-10t} - e^{-10(t-1)}\mathcal{U}(t - 1)]$$
$$- 120te^{-10t} - 1080(t - 1)e^{-10(t-1)}\mathcal{U}(t - 1).$$

Written as a piecewise-defined function, the current is

$$i(t) = \begin{cases} 12 - 12e^{-10t} - 120te^{-10t}, & 0 \le t < 1 \\ -12e^{-10t} + 12e^{-10(t-1)} - 120te^{-10t} - 1080(t - 1)e^{-10(t-1)}, & t \ge 1. \end{cases}$$

Using this last expression and a CAS, we graph $i(t)$ on each of the two intervals and then combine the graphs. Note in Figure 7.34 that even though the input $E(t)$ is discontinuous, the output or response $i(t)$ is a continuous function.

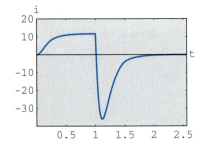

FIGURE 7.34 Graph of current $i(t)$ in Example 6

7.4.3 TRANSFORM OF A PERIODIC FUNCTION

PERIODIC FUNCTION If a periodic function has period T, $T > 0$, then $f(t + T) = f(t)$. The next theorem shows that the Laplace transform of a periodic function can be obtained by integration over one period.

| **THEOREM 7.10** | **Transform of a Periodic Function** |

If $f(t)$ is piecewise continuous on $[0, \infty)$, of exponential order, and periodic with period T, then

$$\mathcal{L}\{f(t)\} = \frac{1}{1 - e^{-sT}}\int_0^T e^{-st} f(t)\, dt.$$

PROOF   Write the Laplace transform of f as two integrals:

$$\mathcal{L}\{f(t)\} = \int_0^T e^{-st} f(t)\, dt + \int_T^\infty e^{-st} f(t)\, dt.$$

When we let $t = u + T$, the last integral becomes

$$\int_T^\infty e^{-st} f(t)\, dt = \int_0^\infty e^{-s(u+T)} f(u + T)\, du = e^{-sT} \int_0^\infty e^{-su} f(u)\, du = e^{-sT}\mathcal{L}\{f(t)\}.$$

Therefore $$\mathcal{L}\{f(t)\} = \int_0^T e^{-st} f(t)\, dt + e^{-sT}\mathcal{L}\{f(t)\}.$$

Solving the equation in the last line for $\mathcal{L}\{f(t)\}$ proves the theorem. ∎

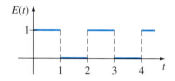

FIGURE 7.35   Square wave

EXAMPLE 7   **Transform of a Periodic Function**

Find the Laplace transform of the periodic function shown in Figure 7.35.

SOLUTION   The function $E(t)$ is called a square wave and has period $T = 2$. On the interval $0 \le t < 2$, $E(t)$ can be defined by

$$E(t) = \begin{cases} 1, & 0 \le t < 1 \\ 0, & 1 \le t < 2 \end{cases}$$

and outside the interval by $f(t + 2) = f(t)$. Now from Theorem 7.10,

$$\mathcal{L}\{E(t)\} = \frac{1}{1 - e^{-2s}} \int_0^2 e^{-st} E(t)\, dt = \frac{1}{1 - e^{-2s}} \left[\int_0^1 e^{-st} \cdot 1\, dt + \int_1^2 e^{-st} \cdot 0\, dt \right]$$

$$= \frac{1}{1 - e^{-2s}} \frac{1 - e^{-s}}{s} \quad \leftarrow 1 - e^{-2s} = (1 + e^{-s})(1 - e^{-s})$$

$$= \frac{1}{s(1 + e^{-s})}. \tag{11}$$

EXAMPLE 8   **A Periodic Impressed Voltage**

The differential equation for the current $i(t)$ in a single-loop LR series circuit is

$$L\frac{di}{dt} + Ri = E(t). \tag{12}$$

Determine the current $i(t)$ when $i(0) = 0$ and $E(t)$ is the square wave function shown in Figure 7.35.

SOLUTION   If we use the result in (11) of the preceding example, the Laplace transform of the DE is

$$LsI(s) + RI(s) = \frac{1}{s(1 + e^{-s})} \quad \text{or} \quad I(s) = \frac{1/L}{s(s + R/L)} \cdot \frac{1}{1 + e^{-s}}. \tag{13}$$

To find the inverse Laplace transform of the last function, we first make use of geometric series. With the identification $x = e^{-s}$, $s > 0$, the geometric series

$$\frac{1}{1 + x} = 1 - x + x^2 - x^3 + \cdots \quad \text{becomes} \quad \frac{1}{1 + e^{-s}} = 1 - e^{-s} + e^{-2s} - e^{-3s} + \cdots.$$

From
$$\frac{1}{s(s+R/L)} = \frac{L/R}{s} - \frac{L/R}{s+R/L}$$

we can then rewrite (13) as

$$I(s) = \frac{1}{R}\left(\frac{1}{s} - \frac{1}{s+R/L}\right)(1 - e^{-s} + e^{-2s} - e^{-3s} + \cdots)$$

$$= \frac{1}{R}\left(\frac{1}{s} - \frac{e^{-s}}{s} + \frac{e^{-2s}}{s} - \frac{e^{-3s}}{s} + \cdots\right) - \frac{1}{R}\left(\frac{1}{s+R/L} - \frac{1}{s+R/L}e^{-s} + \frac{e^{-2s}}{s+R/L} - \frac{e^{-3s}}{s+R/L} + \cdots\right).$$

By applying the form of the second translation theorem to each term of both series, we obtain

$$i(t) = \frac{1}{R}\left(1 - \mathcal{U}(t-1) + \mathcal{U}(t-2) - \mathcal{U}(t-3) + \cdots\right)$$

$$- \frac{1}{R}\left(e^{-Rt/L} - e^{-R(t-1)/L}\,\mathcal{U}(t-1) + e^{-R(t-2)/L}\,\mathcal{U}(t-2) - e^{-R(t-3)/L}\,\mathcal{U}(t-3) + \cdots\right)$$

or, equivalently,

$$i(t) = \frac{1}{R}\left(1 - e^{-Rt/L}\right) + \frac{1}{R}\sum_{n=1}^{\infty}(-1)^n\left(1 - e^{-R(t-n)/L}\right)\mathcal{U}(t-n).$$

To interpret the solution, let us assume for the sake of illustration that $R = 1$, $L = 1$, and $0 \le t < 4$. In this case

$$i(t) = 1 - e^{-t} - (1 - e^{t-1})\,\mathcal{U}(t-1) + (1 - e^{-(t-2)})\,\mathcal{U}(t-2) - (1 - e^{-(t-3)})\,\mathcal{U}(t-3);$$

in other words,

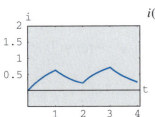

$$i(t) = \begin{cases} 1 - e^{-t}, & 0 \le t < 1 \\ -e^{-t} + e^{-(t-1)}, & 1 \le t < 2 \\ 1 - e^{-t} + e^{-(t-1)} - e^{-(t-2)}, & 2 \le t < 3 \\ -e^{-t} + e^{-(t-1)} - e^{-(t-2)} + e^{-(t-3)}, & 3 \le t < 4. \end{cases}$$

FIGURE 7.36   Graph of current $i(t)$ in Example 8

The graph of $i(t)$ on the interval $0 \le t < 4$, given in Figure 7.36, was obtained with the help of a CAS.

EXERCISES 7.4

Answers to selected odd-numbered problems begin on page ANS-10.

7.4.1   DERIVATIVES OF A TRANSFORM

In Problems 1–8 use Theorem 7.8 to evaluate the given Laplace transform.

1. $\mathcal{L}\{te^{-10t}\}$

2. $\mathcal{L}\{t^3 e^t\}$

3. $\mathcal{L}\{t \cos 2t\}$

4. $\mathcal{L}\{t \sinh 3t\}$

5. $\mathcal{L}\{t^2 \sinh t\}$

6. $\mathcal{L}\{t^2 \cos t\}$

7. $\mathcal{L}\{te^{2t} \sin 6t\}$

8. $\mathcal{L}\{te^{-3t} \cos 3t\}$

In Problems 9–14 use the Laplace transform to solve the given initial-value problem. Use the table of Laplace transforms in Appendix III as needed.

9. $y' + y = t \sin t, \quad y(0) = 0$

10. $y' - y = te^t \sin t, \quad y(0) = 0$

11. $y'' + 9y = \cos 3t, \quad y(0) = 2, \quad y'(0) = 5$

12. $y'' + y = \sin t, \quad y(0) = 1, \quad y'(0) = -1$

13. $y'' + 16y = f(t), \quad y(0) = 0, \quad y'(0) = 1$, where

$$f(t) = \begin{cases} \cos 4t, & 0 \le t < \pi \\ 0, & t \ge \pi \end{cases}$$

14. $y'' + y = f(t)$, $y(0) = 1$, $y'(0) = 0$, where

$$f(t) = \begin{cases} 1, & 0 \le t < \pi/2 \\ \sin t, & t \ge \pi/2 \end{cases}$$

In Problems 15 and 16 use a graphing utility to graph the indicated solution.

15. $y(t)$ of Problem 13 on the interval $0 \le t < 2\pi$

16. $y(t)$ of Problem 14 on the interval $0 \le t < 3\pi$

In some instances the Laplace transform can be used to solve linear differential equations with variable monomial coefficients. In Problems 17 and 18 use Theorem 7.8 to reduce the given differential equation to a linear first-order DE in the transformed function $Y(s) = \mathscr{L}\{y(t)\}$. Solve the first-order DE for $Y(s)$ and then find $y(t) = \mathscr{L}^{-1}\{Y(s)\}$.

See the **SRSM** for an example of this type of problem.

17. $ty'' - y' = 2t^2$, $y(0) = 0$

18. $2y'' + ty' - 2y = 10$,
$y(0) = y'(0) = 0$

7.4.2 TRANSFORMS OF INTEGRALS

In Problems 19–30 use Theorem 7.9 to evaluate the given Laplace transform. Do not evaluate the integral before transforming.

19. $\mathscr{L}\{1 * t^3\}$

20. $\mathscr{L}\{t^2 * te^t\}$

21. $\mathscr{L}\{e^{-t} * e^t \cos t\}$

22. $\mathscr{L}\{e^{2t} * \sin t\}$

23. $\mathscr{L}\left\{\int_0^t e^\tau \, d\tau\right\}$

24. $\mathscr{L}\left\{\int_0^t \cos \tau \, d\tau\right\}$

25. $\mathscr{L}\left\{\int_0^t e^{-\tau} \cos \tau \, d\tau\right\}$

26. $\mathscr{L}\left\{\int_0^t \tau \sin \tau \, d\tau\right\}$

27. $\mathscr{L}\left\{\int_0^t \tau e^{t-\tau} \, d\tau\right\}$

28. $\mathscr{L}\left\{\int_0^t \sin \tau \cos (t - \tau) \, d\tau\right\}$

29. $\mathscr{L}\left\{t \int_0^t \sin \tau \, d\tau\right\}$

30. $\mathscr{L}\left\{t \int_0^t \tau e^{-\tau} \, d\tau\right\}$

In Problems 31–34 use (8) to evaluate the given inverse transform.

31. $\mathscr{L}^{-1}\left\{\dfrac{1}{s(s-1)}\right\}$

32. $\mathscr{L}^{-1}\left\{\dfrac{1}{s^2(s-1)}\right\}$

33. $\mathscr{L}^{-1}\left\{\dfrac{1}{s^3(s-1)}\right\}$

34. $\mathscr{L}^{-1}\left\{\dfrac{1}{s(s-a)^2}\right\}$

35. The table in Appendix III does not contain an entry for

$$\mathscr{L}^{-1}\left\{\frac{8k^3s}{(s^2 + k^2)^3}\right\}.$$

(a) Use (4) along with the results in (5) to evaluate this inverse transform. Use a CAS as an aid in evaluating the convolution integral.

(b) Reexamine your answer to part (a). Could you have obtained the result in a different manner?

36. Use the Laplace transform and the results of Problem 35 to solve the initial-value problem

$$y'' + y = \sin t + t \sin t, \quad y(0) = 0, \quad y'(0) = 0.$$

Use a graphing utility to graph the solution.

In Problems 37–46 use the Laplace transform to solve the given integral equation or integrodifferential equation.

37. $f(t) + \displaystyle\int_0^t (t - \tau) f(\tau) \, d\tau = t$

38. $f(t) = 2t - 4\displaystyle\int_0^t \sin \tau \, f(t - \tau) \, d\tau$

39. $f(t) = te^t + \displaystyle\int_0^t \tau f(t - \tau) \, d\tau$

40. $f(t) + 2\displaystyle\int_0^t f(\tau) \cos (t - \tau) \, d\tau = 4e^{-t} + \sin t$

41. $f(t) + \displaystyle\int_0^t f(\tau) \, d\tau = 1$

42. $f(t) = \cos t + \displaystyle\int_0^t e^{-\tau} f(t - \tau) \, d\tau$

43. $f(t) = 1 + t - \dfrac{8}{3}\displaystyle\int_0^t (\tau - t)^3 f(\tau) \, d\tau$

44. $t - 2f(t) = \displaystyle\int_0^t (e^\tau - e^{-\tau}) f(t - \tau) \, d\tau$

45. $y'(t) = 1 - \sin t - \displaystyle\int_0^t y(\tau) \, d\tau$, $y(0) = 0$

46. $\dfrac{dy}{dt} + 6y(t) + 9\displaystyle\int_0^t y(\tau) \, d\tau = 1$, $y(0) = 0$

In Problems 47 and 48 solve equation (10) subject to $i(0) = 0$ with L, R, C, and $E(t)$ as given. Use a graphing utility to graph the solution on the interval $0 \le t \le 3$.

47. $L = 0.1$ h, $R = 3$ Ω, $C = 0.05$ f,
$E(t) = 100[\mathscr{U}(t - 1) - \mathscr{U}(t - 2)]$

48. $L = 0.005$ h, $R = 1$ Ω, $C = 0.02$ f,
$E(t) = 100[t - (t - 1)\mathscr{U}(t - 1)]$

7.4.3   TRANSFORM OF A PERIODIC FUNCTION

In Problems 49–54 use Theorem 7.10 to find the Laplace transform of the given periodic function.

49.

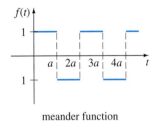

meander function

FIGURE 7.37   Graph for Problem 49

50.

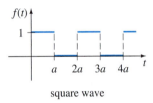

square wave

FIGURE 7.38   Graph for Problem 50

51.

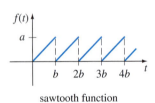

sawtooth function

FIGURE 7.39   Graph for Problem 51

52.

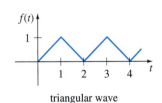

triangular wave

FIGURE 7.40   Graph for Problem 52

53.

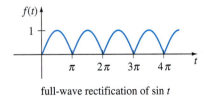

full-wave rectification of sin t

FIGURE 7.41   Graph for Problem 53

54.

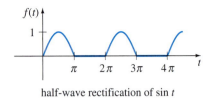

half-wave rectification of sin t

FIGURE 7.42   Graph for Problem 54

In Problems 55 and 56 solve equation (12) subject to $i(0) = 0$ with $E(t)$ as given. Use a graphing utility to graph the solution on the interval $0 \leq t < 4$ in the case when $L = 1$ and $R = 1$.

55. $E(t)$ is the meander function in Problem 49 with amplitude 1 and $a = 1$.

56. $E(t)$ is the sawtooth function in Problem 51 with amplitude 1 and $b = 1$.

In Problems 57 and 58 solve the model for a driven spring/mass system with damping

$$m\frac{d^2x}{dt^2} + \beta\frac{dx}{dt} + kx = f(t), \quad x(0) = 0, \quad x'(0) = 0,$$

where the driving function f is as specified. Use a graphing utility to graph $x(t)$ on the indicated interval.

57. $m = \frac{1}{2}$, $\beta = 1$, $k = 5$, f is the meander function in Problem 49 with amplitude 10, and $a = \pi$, $0 \leq t < 2\pi$.

58. $m = 1$, $\beta = 2$, $k = 1$, f is the square wave in Problem 50 with amplitude 5, and $a = \pi$, $0 \leq t < 4\pi$.

DISCUSSION/PROJECT PROBLEMS

59. Discuss how Theorem 7.8 can be used to find

$$\mathscr{L}^{-1}\left\{\ln\frac{s-3}{s+1}\right\}.$$

60. In Section 6.3 we saw that $ty'' + y' + ty = 0$ is Bessel's equation of order $\nu = 0$. In view of (22) of that section and Table 6.1, a solution of the initial-value problem $ty'' + y' + ty = 0$, $y(0) = 1$, $y'(0) = 0$, is $y = J_0(t)$. Use this result and the procedure outlined in the instructions to Problems 17 and 18 to show that

$$\mathscr{L}\{J_0(t)\} = \frac{1}{\sqrt{s^2+1}}.$$

(*Hint:* You might need to use Problem 46 in Exercises 7.2.)

61. (a) **Laguerre's differential equation**

$$ty'' + (1-t)y' + ny = 0$$

is known to possess polynomial solutions when n is a nonnegative integer. These solutions are naturally called **Laguerre polynomials** and are denoted by $L_n(t)$. Find $y = L_n(t)$, for $n = 0, 1, 2, 3, 4$ if it is known that $L_n(0) = 1$.

(b) Show that

$$\mathscr{L}\left\{\frac{e^t}{n!}\frac{d^n}{dt^n}t^n e^{-t}\right\} = Y(s),$$

where $Y(s) = \mathscr{L}\{y\}$ and $y = L_n(t)$ is a polynomial solution of the DE in part (a). Conclude that

$$L_n(t) = \frac{e^t}{n!}\frac{d^n}{dt^n}t^n e^{-t}, \quad n = 0, 1, 2, \ldots.$$

This last relation for generating the Laguerre polynomials is the analogue of Rodrigues' formula for the Legendre polynomials. See (30) in Section 6.3.

COMPUTER LAB ASSIGNMENTS

62. In this problem you are led through the commands in *Mathematica* that enable you to obtain the symbolic Laplace transform of a differential equation and the solution of the initial-value problem by finding the inverse transform. In *Mathematica* the Laplace transform of a function $y(t)$ is obtained using **LaplaceTransform [y[t], t, s]**. In line two of the syntax we replace **LaplaceTransform [y[t], t, s]** by the symbol **Y**. (*If you do not have* Mathematica, *then adapt the given procedure by finding the corresponding syntax for the CAS you have on hand.*)

Consider the initial-value problem

$$y'' + 6y' + 9y = t\sin t, \quad y(0) = 2, \quad y'(0) = -1.$$

Load the Laplace transform package. Precisely reproduce and then, in turn, execute each line in the following sequence of commands. Either copy the output by hand or print out the results.

diffequat = y″[t] + 6y′[t] + 9y[t] == t Sin[t]
transformdeq = LaplaceTransform [diffequat, t, s] /.
  {y[0] − > 2, y′[0] − > −1,
  LaplaceTransform [y[t], t, s] − > Y}
soln = Solve[transformdeq, Y]//Flatten
Y = Y/.soln
InverseLaplaceTransform[Y, s, t]

63. Appropriately modify the procedure of Problem 62 to find a solution of

$$y''' + 3y' - 4y = 0,$$
$$y(0) = 0, \quad y'(0) = 0, \quad y''(0) = 1.$$

64. The charge $q(t)$ on a capacitor in an *LC* series circuit is given by

$$\frac{d^2 q}{dt^2} + q = 1 - 4\mathscr{U}(t - \pi) + 6\mathscr{U}(t - 3\pi),$$
$$q(0) = 0, \quad q'(0) = 0.$$

Appropriately modify the procedure of Problem 62 to find $q(t)$. Graph your solution.

7.5 THE DIRAC DELTA FUNCTION

INTRODUCTION: In the last paragraph on page 283, we indicated that as an immediate consequence of Theorem 7.3, $F(s) = 1$ cannot be the Laplace transform of a function f that is piecewise continuous on $[0, \infty)$ and of exponential order. In the discussion that follows we are going to introduce a function that is very different from the kinds that you have studied in previous courses. We shall see that there does indeed exist a function, or more precisely a *generalized function*, whose Laplace transform is $F(s) = 1$.

UNIT IMPULSE Mechanical systems are often acted on by an external force (or electromotive force in an electrical circuit) of large magnitude that acts only for a very short period of time. For example, a vibrating airplane wing could be struck by lightning, a mass on a spring could be given a sharp blow by a ball peen hammer, a ball (baseball, golf ball, tennis ball) could be sent soaring when struck violently by some kind of club (baseball bat, golf club, tennis racket). See Figure 7.43. The graph of the piecewise-defined function

FIGURE 7.43 A golf club applies a force of large magnitude on the ball for a very short period of time

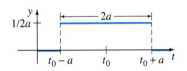

(a) graph of $\delta_a(t - t_0)$

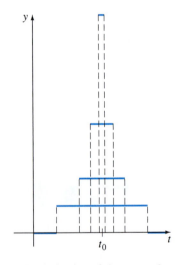

(b) behavior of δ_a as $a \to 0$

FIGURE 7.44 Unit impulse

$$\delta_a(t - t_0) = \begin{cases} 0, & 0 \le t < t_0 - a \\ \dfrac{1}{2a}, & t_0 - a \le t < t_0 + a \\ 0, & t \ge t_0 + a, \end{cases} \qquad (1)$$

$a > 0$, $t_0 > 0$, shown in Figure 7.44(a), could serve as a model for such a force. For a small value of a, $\delta_a(t - t_0)$ is essentially a constant function of large magnitude that is "on" for just a very short period of time, around t_0. The behavior of $\delta_a(t - t_0)$ as $a \to 0$ is illustrated in Figure 7.44(b). The function $\delta_a(t - t_0)$ is called a **unit impulse**, because it possesses the integration property $\displaystyle\int_0^\infty \delta_a(t - t_0)\, dt = 1$.

DIRAC DELTA FUNCTION   In practice it is convenient to work with another type of unit impulse, a "function" that approximates $\delta_a(t - t_0)$ and is defined by the limit

$$\delta(t - t_0) = \lim_{a \to 0} \delta_a(t - t_0). \qquad (2)$$

The latter expression, which is not a function at all, can be characterized by the two properties

$$(i)\ \delta(t - t_0) = \begin{cases} \infty, & t = t_0 \\ 0, & t \neq t_0 \end{cases} \quad \text{and} \quad (ii) \int_0^\infty \delta(t - t_0)\, dt = 1.$$

The unit impulse $\delta(t - t_0)$ is called the **Dirac delta function.**

It is possible to obtain the Laplace transform of the Dirac delta function by the formal assumption that $\mathcal{L}\{\delta(t - t_0)\} = \lim_{a \to 0} \mathcal{L}\{\delta_a(t - t_0)\}$.

THEOREM 7.11   **Transform of the Dirac Delta Function**

For $t_0 > 0$, $\mathcal{L}\{\delta(t - t_0)\} = e^{-st_0}. \qquad (3)$

PROOF   To begin, we can write $\delta_a(t - t_0)$ in terms of the unit step function by virtue of (11) and (12) of Section 7.3:

$$\delta_a(t - t_0) = \frac{1}{2a}[\mathcal{U}(t - (t_0 - a)) - \mathcal{U}(t - (t_0 + a))].$$

By linearity and (14) of Section 7.3 the Laplace transform of this last expression is

$$\mathcal{L}\{\delta_a(t - t_0)\} = \frac{1}{2a}\left[\frac{e^{-s(t_0 - a)}}{s} - \frac{e^{-s(t_0 + a)}}{s}\right] = e^{-st_0}\left(\frac{e^{sa} - e^{-sa}}{2sa}\right). \qquad (4)$$

Since (4) has the indeterminate form $0/0$ as $a \to 0$, we apply L'Hôpital's Rule:

$$\mathcal{L}\{\delta(t - t_0)\} = \lim_{a \to 0} \mathcal{L}\{\delta_a(t - t_0)\} = e^{-st_0} \lim_{a \to 0}\left(\frac{e^{sa} - e^{-sa}}{2sa}\right) = e^{-st_0}. \quad ∎$$

Now when $t_0 = 0$, it seems plausible to conclude from (3) that

$$\mathcal{L}\{\delta(t)\} = 1.$$

The last result emphasizes the fact that $\delta(t)$ is not the usual type of function that we have been considering, since we expect from Theorem 7.3 that $\mathcal{L}\{f(t)\} \to 0$ as $s \to \infty$.

EXAMPLE 1   **Two Initial-Value Problems**

Solve $y'' + y = 4\delta(t - 2\pi)$ subject to

(a) $y(0) = 1$, $y'(0) = 0$ **(b)** $y(0) = 0$, $y'(0) = 0$.

The two initial-value problems could serve as models for describing the motion of a mass on a spring moving in a medium in which damping is negligible. At $t = 2\pi$ the mass is given a sharp blow. In (a) the mass is released from rest 1 unit below the equilibrium position. In (b) the mass is at rest in the equilibrium position.

SOLUTION **(a)** From (3) the Laplace transform of the differential equation is

$$s^2 Y(s) - s + Y(s) = 4e^{-2\pi s} \quad \text{or} \quad Y(s) = \frac{s}{s^2 + 1} + \frac{4e^{-2\pi s}}{s^2 + 1}.$$

Using the inverse form of the second translation theorem, we find

$$y(t) = \cos t + 4 \sin(t - 2\pi)\, \mathcal{U}(t - 2\pi).$$

Since $\sin(t - 2\pi) = \sin t$, the foregoing solution can be written as

$$y(t) = \begin{cases} \cos t, & 0 \le t < 2\pi \\ \cos t + 4 \sin t, & t \ge 2\pi. \end{cases} \tag{5}$$

In Figure 7.45 we see from the graph of (5) that the mass is exhibiting simple harmonic motion until it is struck at $t = 2\pi$. The influence of the unit impulse is to increase the amplitude of vibration to $\sqrt{17}$ for $t > 2\pi$.

(b) In this case the transform of the equation is simply

$$Y(s) = \frac{4e^{-2\pi s}}{s^2 + 1},$$

and so

$$y(t) = 4 \sin(t - 2\pi)\, \mathcal{U}(t - 2\pi)$$

$$= \begin{cases} 0, & 0 \le t < 2\pi \\ 4 \sin t, & t \ge 2\pi. \end{cases} \tag{6}$$

The graph of (6) in Figure 7.46 shows, as we would expect from the initial conditions that the mass exhibits no motion until it is struck at $t = 2\pi$. ∎

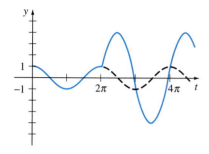

FIGURE 7.45 Mass is struck at $t = 2\pi$

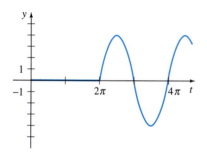

FIGURE 7.46 No motion until mass is struck at $t = 2\pi$

REMARKS

(*i*) If $\delta(t - t_0)$ were a function in the usual sense, then property (*i*) on page 316 would imply $\int_0^\infty \delta(t - t_0)\, dt = 0$ rather than $\int_0^\infty \delta(t - t_0)\, dt = 1$. Because the Dirac delta function did not "behave" like an ordinary function, even though its users produced correct results, it was met initially with great scorn by mathematicians. However, in the 1940s Dirac's controversial function was put on a rigorous footing by the French mathematician Laurent Schwartz in his book *La Théorie de distribution*, and this, in turn, led to an entirely new branch of mathematics known as the **theory of distributions** or **generalized functions**. In this theory, (2) is not an accepted definition of $\delta(t - t_0)$, nor does one speak of a function whose values are either ∞ or 0. Although we shall not pursue this topic any further, suffice it to say that the Dirac delta function is best characterized by its effect on other functions. If f is a continuous function, then

$$\int_0^\infty f(t)\, \delta(t - t_0)\, dt = f(t_0) \tag{7}$$

can be taken as the *definition* of $\delta(t - t_0)$. This result is known as the **sifting property** since $\delta(t - t_0)$ has the effect of sifting the value $f(t_0)$ out of the set of values of f on $[0, \infty)$. Note that property *(ii)* (with $f(t) = 1$) and (3) (with $f(t) = e^{-st}$) are consistent with (7).

(ii) The *Remarks* in Section 7.2 indicated that the transfer function of a general linear *n*th-order differential equation with constant coefficients is $W(s) = 1/P(s)$, where $P(s) = a_n s^n + a_{n-1} s^{n-1} + \cdots + a_0$. The transfer function is the Laplace transform of function $w(t)$, called the **weight function** of a linear system. But $w(t)$ can also be characterized in terms of the discussion at hand. For simplicity let us consider a second-order linear system in which the input is a unit impulse at $t = 0$:

$$a_2 y'' + a_1 y' + a_0 y = \delta(t), \quad y(0) = 0, \quad y'(0) = 0.$$

Applying the Laplace transform and using $\mathcal{L}\{\delta(t)\} = 1$ shows that the transform of the response y in this case is the transfer function

$$Y(s) = \frac{1}{a_2 s^2 + a_1 s + a_0} = \frac{1}{P(s)} = W(s) \quad \text{and} \quad \text{so} \quad y = \mathcal{L}^{-1}\left\{\frac{1}{P(s)}\right\} = w(t).$$

From this we can see, in general, that the weight function $y = w(t)$ of an *n*th-order linear system is the zero-state response of the system to a unit impulse. For this reason $w(t)$ is also called the **impulse response** of the system.

EXERCISES 7.5

Answers to selected odd-numbered problems begin on page ANS-10.

In Problems 1–12 use the Laplace transform to solve the given initial-value problem.

1. $y' - 3y = \delta(t - 2), \quad y(0) = 0$

2. $y' + y = \delta(t - 1), \quad y(0) = 2$

3. $y'' + y = \delta(t - 2\pi), \quad y(0) = 0, y'(0) = 1$

4. $y'' + 16y = \delta(t - 2\pi), \quad y(0) = 0, y'(0) = 0$

5. $y'' + y = \delta\left(t - \frac{1}{2}\pi\right) + \delta\left(t - \frac{3}{2}\pi\right),$
    $y(0) = 0, y'(0) = 0$

6. $y'' + y = \delta(t - 2\pi) + \delta(t - 4\pi), \quad y(0) = 1, y'(0) = 0$

7. $y'' + 2y' = \delta(t - 1), \quad y(0) = 0, y'(0) = 1$

8. $y'' - 2y' = 1 + \delta(t - 2), \quad y(0) = 0, y'(0) = 1$

9. $y'' + 4y' + 5y = \delta(t - 2\pi), \quad y(0) = 0, y'(0) = 0$

10. $y'' + 2y' + y = \delta(t - 1), \quad y(0) = 0, y'(0) = 0$

11. $y'' + 4y' + 13y = \delta(t - \pi) + \delta(t - 3\pi),$
    $y(0) = 1, y'(0) = 0$

12. $y'' - 7y' + 6y = e^t + \delta(t - 2) + \delta(t - 4),$
    $y(0) = 0, y'(0) = 0$

13. A uniform beam of length L carries a concentrated load w_0 at $x = \frac{1}{2}L$. The beam is embedded at its left end and

is free at its right end. Use the Laplace transform to determine the deflection $y(x)$ from

$$EI \frac{d^4 y}{dx^4} = w_0 \delta\left(x - \frac{1}{2}L\right),$$

where $y(0) = 0$, $y'(0) = 0$, $y''(L) = 0$, and $y'''(L) = 0$.

14. Solve the differential equation in Problem 13 subject to $y(0) = 0$, $y'(0) = 0$, $y(L) = 0$, $y'(L) = 0$. In this case the beam is embedded at both ends. See Figure 7.47.

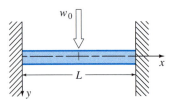

FIGURE 7.47 Beam in Problem 14

DISCUSSION/PROJECT PROBLEMS

15. Someone tells you that the solutions of the two IVPs

$$y'' + 2y' + 10y = 0, \qquad y(0) = 0, y'(0) = 1$$
$$y'' + 2y' + 10y = \delta(t), \quad y(0) = 0, y'(0) = 0$$

are exactly the same. Do you agree or disagree? Defend your answer.

7.6 SYSTEMS OF LINEAR DIFFERENTIAL EQUATIONS

INTRODUCTION: When initial conditions are specified, the Laplace transform of each equation in a system of linear differential equations with constant coefficients reduces the system of DEs to a set of simultaneous algebraic equations in the transformed functions. We solve the system of algebraic equations for each of the transformed functions and then find the inverse Laplace transforms in the usual manner.

REVIEW MATERIAL: In this section you will mainly be expected to solve systems of two equations with two unknowns.

CD: The **Linear Double Pendulum Tool** on the *DE Tools* CD can be used in conjunction with the discussion of that topic on pages 321 and 322.

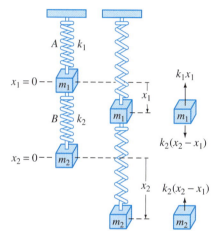

(a) equilibrium   **(b)** motion   **(c)** forces

FIGURE 7.48   Coupled spring/mass system

COUPLED SPRINGS   Two masses m_1 and m_2 are connected to two springs A and B of negligible mass having spring constants k_1 and k_2, respectively. In turn the two springs are attached as shown in Figure 7.48. Let $x_1(t)$ and $x_2(t)$ denote the vertical displacements of the masses from their equilibrium positions. When the system is in motion, spring B is subject to both an elongation and a compression; hence its net elongation is $x_2 - x_1$. Therefore it follows from Hooke's law that springs A and B exert forces $-k_1 x_1$ and $k_2(x_2 - x_1)$, respectively, on m_1. If no external force is impressed on the system and if no damping force is present, then the net force on m_1 is $-k_1 x_1 + k_2(x_2 - x_1)$. By Newton's second law we can write

$$m_1 \frac{d^2 x_1}{dt^2} = -k_1 x_1 + k_2(x_2 - x_1).$$

Similarly, the net force exerted on mass m_2 is due solely to the net elongation of B; that is, $-k_2(x_2 - x_1)$. Hence we have

$$m_2 \frac{d^2 x_2}{dt^2} = -k_2(x_2 - x_1).$$

In other words, the motion of the coupled system is represented by the system of simultaneous second-order differential equations

$$\begin{aligned} m_1 x_1'' &= -k_1 x_1 + k_2(x_2 - x_1) \\ m_2 x_2'' &= -k_2(x_2 - x_1). \end{aligned} \tag{1}$$

In the next example we solve (1) under the assumptions that $k_1 = 6$, $k_2 = 4$, $m_1 = 1$, $m_2 = 1$, and that the masses start from their equilibrium positions with opposite unit velocities.

EXAMPLE 1   Coupled Springs

Solve
$$\begin{aligned} x_1'' + 10x_1 \quad\; - 4x_2 &= 0 \\ -\; 4x_1 + x_2'' + 4x_2 &= 0 \end{aligned} \tag{2}$$

subject to $x_1(0) = 0$, $x_1'(0) = 1$, $x_2(0) = 0$, $x_2'(0) = -1$.

SOLUTION   The Laplace transform of each equation is

$$s^2 X_1(s) - s x_1(0) - x_1'(0) + 10 X_1(s) - 4 X_2(s) = 0$$

$$-4 X_1(s) + s^2 X_2(s) - s x_2(0) - x_2'(0) + 4 X_2(s) = 0,$$

where $X_1(s) = \mathcal{L}\{x_1(t)\}$ and $X_2(s) = \mathcal{L}\{x_2(t)\}$. The preceding system is the same as

$$
\begin{aligned}
(s^2 + 10)X_1(s) - \qquad\quad 4X_2(s) &= 1 \\
-4X_1(s) + (s^2 + 4)X_2(s) &= -1.
\end{aligned}
$$
(3)

Solving (3) for $X_1(s)$ and using partial fractions on the result yields

$$
X_1(s) = \frac{s^2}{(s^2 + 2)(s^2 + 12)} = -\frac{1/5}{s^2 + 2} + \frac{6/5}{s^2 + 12},
$$

and therefore

$$
\begin{aligned}
x_1(t) &= -\frac{1}{5\sqrt{2}}\, \mathcal{L}^{-1}\!\left\{\frac{\sqrt{2}}{s^2 + 2}\right\} + \frac{6}{5\sqrt{12}}\, \mathcal{L}^{-1}\!\left\{\frac{\sqrt{12}}{s^2 + 12}\right\} \\
&= -\frac{\sqrt{2}}{10}\sin\sqrt{2}t + \frac{\sqrt{3}}{5}\sin 2\sqrt{3}t.
\end{aligned}
$$

Substituting the expression for $X_1(s)$ into the first equation of (3) gives

$$
X_2(s) = -\frac{s^2 + 6}{(s^2 + 2)(s^2 + 12)} = -\frac{2/5}{s^2 + 2} - \frac{3/5}{s^2 + 12}
$$

and

$$
\begin{aligned}
x_2(t) &= -\frac{2}{5\sqrt{2}}\, \mathcal{L}^{-1}\!\left\{\frac{\sqrt{2}}{s^2 + 2}\right\} - \frac{3}{5\sqrt{12}}\, \mathcal{L}^{-1}\!\left\{\frac{\sqrt{12}}{s^2 + 12}\right\} \\
&= -\frac{\sqrt{2}}{5}\sin\sqrt{2}t - \frac{\sqrt{3}}{10}\sin 2\sqrt{3}t.
\end{aligned}
$$

Finally, the solution to the given system (2) is

$$
\begin{aligned}
x_1(t) &= -\frac{\sqrt{2}}{10}\sin\sqrt{2}t + \frac{\sqrt{3}}{5}\sin 2\sqrt{3}t \\
x_2(t) &= -\frac{\sqrt{2}}{5}\sin\sqrt{2}t - \frac{\sqrt{3}}{10}\sin 2\sqrt{3}t.
\end{aligned}
$$
(4)

The graphs of x_1 and x_2 in Figure 7.49 reveal the complicated oscillatory motion of each mass. ∎

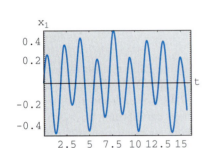

x_1

(a) plot of $x_1(t)$ vs. t

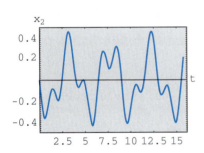

x_2

(b) plot of $x_2(t)$ vs. t

FIGURE 7.49 Displacements of the two masses

NETWORKS In (18) of Section 3.3 we saw the currents $i_1(t)$ and $i_2(t)$ in the network shown in Figure 7.50, containing an inductor, a resistor, and a capacitor, were governed by the system of first-order differential equations

$$
L\frac{di_1}{dt} + Ri_2 = E(t)
$$
(5)
$$
RC\frac{di_2}{dt} + i_2 - i_1 = 0.
$$

We solve this system by the Laplace transform in the next example.

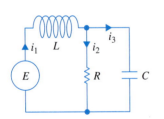

FIGURE 7.50 Electrical network

EXAMPLE 2   An Electrical Network

Solve the system in (5) under the conditions $E(t) = 60$ V, $L = 1$ h, $R = 50\ \Omega$, $C = 10^{-4}$ f, and the currents i_1 and i_2 are initially zero.

SOLUTION We must solve

$$\frac{di_1}{dt} + 50i_2 = 60$$

$$50(10^{-4})\frac{di_2}{dt} + i_2 - i_1 = 0$$

subject to $i_1(0) = 0$, $i_2(0) = 0$.

Applying the Laplace transform to each equation of the system and simplifying gives

$$sI_1(s) + \qquad 50I_2(s) = \frac{60}{s}$$

$$-200I_1(s) + (s + 200)I_2(s) = 0,$$

where $I_1(s) = \mathcal{L}\{i_1(t)\}$ and $I_2(s) = \mathcal{L}\{i_2(t)\}$. Solving the system for I_1 and I_2 and decomposing the results into partial fractions gives

$$I_1(s) = \frac{60s + 12{,}000}{s(s + 100)^2} = \frac{6/5}{s} - \frac{6/5}{s + 100} - \frac{60}{(s + 100)^2}$$

$$I_2(s) = \frac{12{,}000}{s(s + 100)^2} = \frac{6/5}{s} - \frac{6/5}{s + 100} - \frac{120}{(s + 100)^2}.$$

Taking the inverse Laplace transform, we find the currents to be

$$i_1(t) = \frac{6}{5} - \frac{6}{5}e^{-100t} - 60te^{-100t}$$

$$i_2(t) = \frac{6}{5} - \frac{6}{5}e^{-100t} - 120te^{-100t}. \qquad \blacksquare$$

Note that both $i_1(t)$ and $i_2(t)$ in Example 2 tend toward the value $E/R = \frac{6}{5}$ as $t \to \infty$. Furthermore, since the current through the capacitor is $i_3(t) = i_1(t) - i_2(t) = 60te^{-100t}$, we observe that $i_3(t) \to 0$ as $t \to \infty$.

DOUBLE PENDULUM Consider the double-pendulum system consisting of a pendulum attached to a pendulum shown in Figure 7.51. We assume that the system oscillates in a vertical plane under the influence of gravity, that the mass of each rod is negligible, and that no damping forces act on the system. Figure 7.51 also shows that the displacement angle θ_1 is measured (in radians) from a vertical line extending downward from the pivot of the system and that θ_2 is measured from a vertical line extending downward from the center of mass m_1. The positive direction is to the right; the negative direction is to the left. As we might expect from the analysis leading to equation (6) of Section 5.3, the system of differential equations describing the motion is nonlinear:

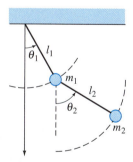

FIGURE 7.51 Double pendulum

$$(m_1 + m_2)l_1^2\theta_1'' + m_2l_1l_2\theta_2''\cos(\theta_1 - \theta_2) + m_2l_1l_2(\theta_2')^2\sin(\theta_1 - \theta_2) + (m_1 + m_2)l_1g\sin\theta_1 = 0$$

$$m_2l_2^2\theta_2'' + m_2l_1l_2\theta_1''\cos(\theta_1 - \theta_2) - m_2l_1l_2(\theta_1')^2\sin(\theta_1 - \theta_2) + m_2l_2g\sin\theta_2 = 0. \qquad (6)$$

But if the displacements $\theta_1(t)$ and $\theta_2(t)$ are assumed to be small, then the approximations $\cos(\theta_1 - \theta_2) \approx 1$, $\sin(\theta_1 - \theta_2) \approx 0$, $\sin\theta_1 \approx \theta_1$, $\sin\theta_2 \approx \theta_2$ enable us to replace system (6) by the linearization

$$(m_1 + m_2)l_1^2\theta_1'' + m_2l_1l_2\theta_2'' + (m_1 + m_2)l_1g\theta_1 = 0$$

$$m_2l_2^2\theta_2'' + m_2l_1l_2\theta_1'' + m_2l_2g\theta_2 = 0. \qquad (7)$$

EXAMPLE 3   Double Pendulum

It is left as an exercise to fill in the details of using the Laplace transform to solve system (7) when $m_1 = 3$, $m_2 = 1$, $l_1 = l_2 = 16$, $\theta_1(0) = 1$, $\theta_2(0) = -1$, $\theta_1'(0) = 0$, and $\theta_2'(0) = 0$. You should find that

$$\theta_1(t) = \frac{1}{4}\cos\frac{2}{\sqrt{3}}t + \frac{3}{4}\cos 2t$$

$$\theta_2(t) = \frac{1}{2}\cos\frac{2}{\sqrt{3}}t - \frac{3}{2}\cos 2t. \tag{8}$$

With the aid of a CAS the positions of the two masses at $t = 0$ and at subsequent times are shown in Figure 7.52. See Problem 21 in Exercises 7.6.

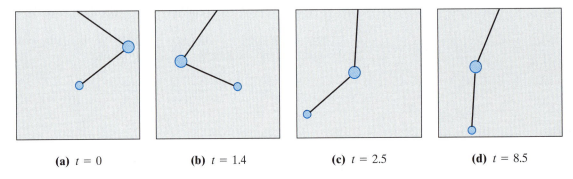

(a) $t = 0$ **(b)** $t = 1.4$ **(c)** $t = 2.5$ **(d)** $t = 8.5$

FIGURE 7.52   Positions of masses on double pendulum at various times

EXERCISES 7.6

Answers to selected odd-numbered problems begin on page ANS-10.

In Problems 1–12 use the Laplace transform to solve the given system of differential equations.

1.
$$\frac{dx}{dt} = -x + y$$
$$\frac{dy}{dt} = 2x$$
$$x(0) = 0, \, y(0) = 1$$

2.
$$\frac{dx}{dt} = 2y + e^t$$
$$\frac{dy}{dt} = 8x - t$$
$$x(0) = 1, \, y(0) = 1$$

3.
$$\frac{dx}{dt} = x - 2y$$
$$\frac{dy}{dt} = 5x - y$$
$$x(0) = -1, \, y(0) = 2$$

4.
$$\frac{dx}{dt} + 3x + \frac{dy}{dt} = 1$$
$$\frac{dx}{dt} - x + \frac{dy}{dt} - y = e^t$$
$$x(0) = 0, \, y(0) = 0$$

5.
$$2\frac{dx}{dt} + \frac{dy}{dt} - 2x = 1$$
$$\frac{dx}{dt} + \frac{dy}{dt} - 3x - 3y = 2$$
$$x(0) = 0, \, y(0) = 0$$

6.
$$\frac{dx}{dt} + x - \frac{dy}{dt} + y = 0$$
$$\frac{dx}{dt} + \frac{dy}{dt} + 2y = 0$$
$$x(0) = 0, \, y(0) = 1$$

7.
$$\frac{d^2x}{dt^2} + x - y = 0$$
$$\frac{d^2y}{dt^2} + y - x = 0$$
$$x(0) = 0, \, x'(0) = -2,$$
$$y(0) = 0, \, y'(0) = 1$$

8.
$$\frac{d^2x}{dt^2} + \frac{dx}{dt} + \frac{dy}{dt} = 0$$
$$\frac{d^2y}{dt^2} + \frac{dy}{dt} - 4\frac{dx}{dt} = 0$$
$$x(0) = 1, \, x'(0) = 0,$$
$$y(0) = -1, \, y'(0) = 5$$

9.
$$\frac{d^2x}{dt^2} + \frac{d^2y}{dt^2} = t^2$$
$$\frac{d^2x}{dt^2} - \frac{d^2y}{dt^2} = 4t$$
$$x(0) = 8, \, x'(0) = 0,$$
$$y(0) = 0, \, y'(0) = 0$$

10.
$$\frac{dx}{dt} - 4x + \frac{d^3y}{dt^3} = 6\sin t$$
$$\frac{dx}{dt} + 2x - 2\frac{d^3y}{dt^3} = 0$$
$$x(0) = 0, \, y(0) = 0,$$
$$y'(0) = 0, \, y''(0) = 0$$

11.
$$\frac{d^2x}{dt^2} + 3\frac{dy}{dt} + 3y = 0$$
$$\frac{d^2x}{dt^2} + 3y = te^{-t}$$
$$x(0) = 0, \, x'(0) = 2, \, y(0) = 0$$

12. $\dfrac{dx}{dt} = 4x - 2y + 2\,\mathscr{U}(t - 1)$

$\dfrac{dy}{dt} = 3x - y + \mathscr{U}(t - 1)$

$x(0) = 0,\, y(0) = \frac{1}{2}$

13. Solve system (1) when $k_1 = 3$, $k_2 = 2$, $m_1 = 1$, $m_2 = 1$ and $x_1(0) = 0$, $x_1'(0) = 1$, $x_2(0) = 1$, $x_2'(0) = 0$.

14. Derive the system of differential equations describing the straight-line vertical motion of the coupled springs shown in Figure 7.53. Use the Laplace transform to solve the system when $k_1 = 1$, $k_2 = 1$, $k_3 = 1$, $m_1 = 1$, $m_2 = 1$ and $x_1(0) = 0$, $x_1'(0) = -1$, $x_2(0) = 0$, $x_2'(0) = 1$.

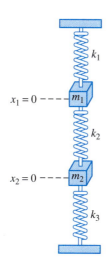

FIGURE 7.53 Coupled springs in Problem 14

15. (a) Show that the system of differential equations for the currents $i_2(t)$ and $i_3(t)$ in the electrical network shown in Figure 7.54 is

$$L_1 \frac{di_2}{dt} + Ri_2 + Ri_3 = E(t)$$

$$L_2 \frac{di_3}{dt} + Ri_2 + Ri_3 = E(t).$$

(b) Solve the system in part (a) if $R = 5\ \Omega$, $L_1 = 0.01$ h, $L_2 = 0.0125$ h, $E = 100$ V, $i_2(0) = 0$, and $i_3(0) = 0$.
(c) Determine the current $i_1(t)$.

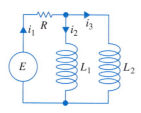

FIGURE 7.54 Network in Problem 15

16. (a) In Problem 12 in Exercises 3.3 you were asked to show that the currents $i_2(t)$ and $i_3(t)$ in the electrical network shown in Figure 7.55 satisfy

$$L\frac{di_2}{dt} + L\frac{di_3}{dt} + R_1 i_2 = E(t)$$

$$-R_1 \frac{di_2}{dt} + R_2 \frac{di_3}{dt} + \frac{1}{C} i_3 = 0.$$

Solve the system if $R_1 = 10\ \Omega$, $R_2 = 5\ \Omega$, $L = 1$ h, $C = 0.2$ f,

$$E(t) = \begin{cases} 120, & 0 \le t < 2 \\ 0, & t \ge 2, \end{cases}$$

$i_2(0) = 0$, and $i_3(0) = 0$.
(b) Determine the current $i_1(t)$.

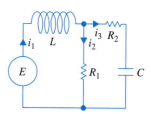

FIGURE 7.55 Network in Problem 16

17. Solve the system given in (17) of Section 3.3 when $R_1 = 6\,\Omega$, $R_2 = 5\,\Omega$, $L_1 = 1$ h, $L_2 = 1$ h, $E(t) = 50 \sin t$ V, $i_2(0) = 0$, and $i_3(0) = 0$.

18. Solve (5) when $E = 60$ V, $L = \frac{1}{2}$ h, $R = 50\ \Omega$, $C = 10^{-4}$ f, $i_1(0) = 0$, and $i_2(0) = 0$.

19. Solve (5) when $E = 60$ V, $L = 2$ h, $R = 50\ \Omega$, $C = 10^{-4}$ f, $i_1(0) = 0$, and $i_2(0) = 0$.

20. (a) Show that the system of differential equations for the charge on the capacitor $q(t)$ and the current $i_3(t)$ in the electrical network shown in Figure 7.56 is

$$R_1 \frac{dq}{dt} + \frac{1}{C} q + R_1 i_3 = E(t)$$

$$L \frac{di_3}{dt} + R_2 i_3 - \frac{1}{C} q = 0.$$

(b) Find the charge on the capacitor when $L = 1$ h, $R_1 = 1\ \Omega$, $R_2 = 1\ \Omega$, $C = 1$ f,

$$E(t) = \begin{cases} 0, & 0 < t < 1 \\ 50e^{-t}, & t \ge 1, \end{cases}$$

$i_3(0) = 0$, and $q(0) = 0$.

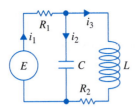

FIGURE 7.56   Network in Problem 20

COMPUTER LAB ASSIGNMENTS

21. (a) Use the Laplace transform and the information given in Example 3 to obtain the solution (8) of the system given in (7).

   (b) Use a graphing utility to graph $\theta_1(t)$ and $\theta_2(t)$ in the $t\theta$-plane. Which mass has extreme displacements of greater magnitude? Use the graphs to estimate the first time that each mass passes through its equilibrium position. Discuss whether the motion of the pendulums is periodic.

(c) Graph $\theta_1(t)$ and $\theta_2(t)$ in the $\theta_1\theta_2$-plane as parametric equations. The curve defined by these parametric equations is called a **Lissajous curve.**

(d) The positions of the masses at $t = 0$ are given in Figure 7.52(a). Note that we have used 1 radian $\approx 57.3°$. Use a calculator or a table application in a CAS to construct a table of values of the angles θ_1 and θ_2 for $t = 1, 2, \ldots, 10$ s. Then plot the positions of the two masses at these times.

(e) Use a CAS to find the first time that $\theta_1(t) = \theta_2(t)$ and compute the corresponding angular value. Plot the positions of the two masses at these times.

(f) Utilize the CAS to draw appropriate lines to simulate the pendulum rods, as in Figure 7.52. Use the animation capability of your CAS to make a "movie" of the motion of the double pendulum from $t = 0$ to $t = 10$ using a time increment of 0.1. (*Hint*: Express the coordinates $(x_1(t), y_1(t))$ and $(x_2(t), y_2(t))$ of the masses m_1 and m_2, respectively, in terms of $\theta_1(t)$ and $\theta_2(t)$.)

CHAPTER 7 IN REVIEW

Answers to selected odd-numbered problems begin on page ANS-10.

In Problems 1 and 2 use the definition of the Laplace transform to find $\mathscr{L}\{f(t)\}$.

1. $f(t) = \begin{cases} t, & 0 \leq t < 1 \\ 2 - t, & t \geq 1 \end{cases}$

2. $f(t) = \begin{cases} 0, & 0 \leq t < 2 \\ 1, & 2 \leq t < 4 \\ 0, & t \geq 4 \end{cases}$

In Problems 3–24 fill in the blanks or answer true or false.

3. If f is not piecewise continuous on $[0, \infty)$, then $\mathscr{L}\{f(t)\}$ will not exist. _____

4. The function $f(t) = (e^t)^{10}$ is not of exponential order. _____

5. $F(s) = s^2/(s^2 + 4)$ is not the Laplace transform of a function that is piecewise continuous and of exponential order. _____

6. If $\mathscr{L}\{f(t)\} = F(s)$ and $\mathscr{L}\{g(t)\} = G(s)$, then $\mathscr{L}^{-1}\{F(s)G(s)\} = f(t)g(t).$ _____

7. $\mathscr{L}\{e^{-7t}\} = $ _____   **8.** $\mathscr{L}\{te^{-7t}\} = $ _____

9. $\mathscr{L}\{\sin 2t\} = $ _____   **10.** $\mathscr{L}\{e^{-3t}\sin 2t\} = $ _____

11. $\mathscr{L}\{t \sin 2t\} = $ _____

12. $\mathscr{L}\{\sin 2t \, \mathcal{U}(t - \pi)\} = $ _____

13. $\mathscr{L}^{-1}\left\{\dfrac{20}{s^6}\right\} = $ _____

14. $\mathscr{L}^{-1}\left\{\dfrac{1}{3s - 1}\right\} = $ _____

15. $\mathscr{L}^{-1}\left\{\dfrac{1}{(s - 5)^3}\right\} = $ _____

16. $\mathscr{L}^{-1}\left\{\dfrac{1}{s^2 - 5}\right\} = $ _____

17. $\mathscr{L}^{-1}\left\{\dfrac{s}{s^2 - 10s + 29}\right\} = $ _____

18. $\mathscr{L}^{-1}\left\{\dfrac{e^{-5s}}{s^2}\right\} = $ _____

19. $\mathscr{L}^{-1}\left\{\dfrac{s + \pi}{s^2 + \pi^2}e^{-s}\right\} = $ _____

20. $\mathscr{L}^{-1}\left\{\dfrac{1}{L^2s^2 + n^2\pi^2}\right\} = $ _____

21. $\mathscr{L}\{e^{-5t}\}$ exists for $s > $ _____.

22. If $\mathscr{L}\{f(t)\} = F(s)$, then $\mathscr{L}\{te^{8t}f(t)\} = $ _____.

23. If $\mathscr{L}\{f(t)\} = F(s)$ and $k > 0$, then $\mathscr{L}\{e^{at}f(t - k)\mathcal{U}(t - k)\} = $ _____.

24. $\mathscr{L}\{\int_0^t e^{a\tau}f(\tau) \, d\tau\} = $ _____ whereas $\mathscr{L}\{e^{at}\int_0^t f(\tau) \, d\tau\} = $ _____.

In Problems 25–28 use the unit step function to find an equation for each graph in terms of the function $y = f(t)$, whose graph is given in Figure 7.57.

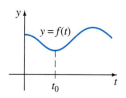

FIGURE 7.57   Graph for Problems 25–28

25.

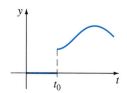

FIGURE 7.58   Graph for Problem 25

26.

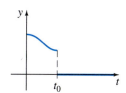

FIGURE 7.59   Graph for Problem 26

27.

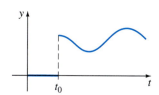

FIGURE 7.60   Graph for Problem 27

28.

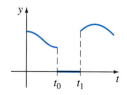

FIGURE 7.61   Graph for Problem 28

In Problems 29–32 express f in terms of unit step functions. Find $\mathscr{L}\{f(t)\}$ and $\mathscr{L}\{e^t f(t)\}$.

29.

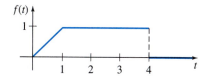

FIGURE 7.62   Graph for Problem 29

30.

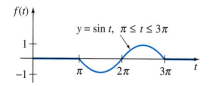

FIGURE 7.63   Graph for Problem 30

31.

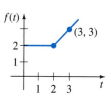

FIGURE 7.64   Graph for Problem 31

32.

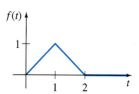

FIGURE 7.65   Graph for Problem 32

In Problems 33–38 use the Laplace transform to solve the given equation.

33. $y'' - 2y' + y = e^t,$ $y(0) = 0, y'(0) = 5$

34. $y'' - 8y' + 20y = te^t,$ $y(0) = 0, y'(0) = 0$

35. $y'' + 6y' + 5y = t - t\,\mathscr{U}(t - 2),$ $y(0) = 1, y'(0) = 0$

36. $y' - 5y = f(t),$ where

$$f(t) = \begin{cases} t^2, & 0 \le t < 1 \\ 0, & t \ge 1 \end{cases}, \quad y(0) = 1$$

37. $y'(t) = \cos t + \displaystyle\int_0^t y(\tau)\cos(t - \tau)\,d\tau,$ $y(0) = 1$

38. $\displaystyle\int_0^t f(\tau)f(t - \tau)\,d\tau = 6t^3$

In Problems 39 and 40 use the Laplace transform to solve each system.

39. $x' + y = t$
$4x + y' = 0$
$x(0) = 1, y(0) = 2$

40. $x'' + y'' = e^{2t}$
$2x' + y'' = -e^{2t}$
$x(0) = 0, \ y(0) = 0,$
$x'(0) = 0, \ y'(0) = 0$

41. The current $i(t)$ in an RC series circuit can be determined from the integral equation

$$Ri + \frac{1}{C}\int_0^t i(\tau)\, d\tau = E(t),$$

where $E(t)$ is the impressed voltage. Determine $i(t)$ when $R = 10\ \Omega$, $C = 0.5$ f, and $E(t) = 2(t^2 + t)$.

42. A series circuit contains an inductor, a resistor, and a capacitor for which $L = \frac{1}{2}$ h, $R = 10\ \Omega$, and $C = 0.01$ f, respectively. The voltage

$$E(t) = \begin{cases} 10, & 0 \le t < 5 \\ 0, & t \ge 5 \end{cases}$$

is applied to the circuit. Determine the instantaneous charge $q(t)$ on the capacitor for $t > 0$ if $q(0) = 0$ and $q'(0) = 0$.

43. A uniform cantilever beam of length L is embedded at its left end ($x = 0$) and free at its right end. Find the deflection $y(x)$ if the load per unit length is given by

$$w(x) = \frac{2w_0}{L}\left[\frac{L}{2} - x + \left(x - \frac{L}{2}\right)\mathscr{U}\left(x - \frac{L}{2}\right)\right].$$

44. When a uniform beam is supported by an elastic foundation, the differential equation for its deflection $y(x)$ is

$$EI\frac{d^4 y}{dx^4} + ky = w(x),$$

where k is the modulus of the foundation and $-ky$ is the restoring force of the foundation that acts in the direction opposite to that of the load $w(x)$. See Figure 7.66. For algebraic convenience suppose the differential equation is written as

$$\frac{d^4 y}{dx^4} + 4a^4 y = \frac{w(x)}{EI},$$

where $a = (k/4EI)^{1/4}$. Assume $L = \pi$ and $a = 1$. Find the deflection $y(x)$ of a beam that is supported on an elastic foundation when

(a) the beam is simply supported at both ends and a constant load w_0 is uniformly distributed along its length,

(b) the beam is embedded at both ends and $w(x)$ is a concentrated load w_0 applied at $x = \pi/2$.

(*Hint:* In both parts of this problem use entries 35 and 36 in the table of Laplace transforms in Appendix III.)

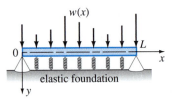

FIGURE 7.66 Beam on elastic foundation in Problem 44

45. (a) Suppose two identical pendulums are coupled by means of a spring with constant k. See Figure 7.67. Under the same assumptions made in the discussion preceding Example 3 in Section 7.6, it can be shown that when the displacement angles $\theta_1(t)$ and $\theta_2(t)$ are small, the system of linear differential equations describing the motion is

$$\theta_1'' + \frac{g}{l}\theta_1 = -\frac{k}{m}(\theta_1 - \theta_2)$$

$$\theta_2'' + \frac{g}{l}\theta_2 = \frac{k}{m}(\theta_1 - \theta_2).$$

Use the Laplace transform to solve the system when $\theta_1(0) = \theta_0$, $\theta_1'(0) = 0$, $\theta_2(0) = \psi_0$, $\theta_2'(0) = 0$, where θ_0 and ψ_0 constants. For convenience, let $\omega^2 = g/l$, $K = k/m$.

(b) Use the solution in part (a) to discuss the motion of the coupled pendulums in the special case when the initial conditions are $\theta_1(0) = \theta_0$, $\theta_1'(0) = 0$, $\theta_2(0) = \theta_0$, $\theta_2'(0) = 0$. When the initial conditions are $\theta_1(0) = \theta_0$, $\theta_1'(0) = 0$, $\theta_2(0) = -\theta_0$, $\theta_2'(0) = 0$.

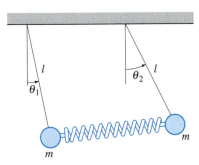

FIGURE 7.67 Coupled pendulums in Problem 45

PROJECT 7

MURDER AT THE MAYFAIR

Dawn at the Mayfair (see Figure 1). The amber glow of streetlights mixed with the violent red flash of police cruisers begins to fade with the rising furnace-orange sun. Detective Diff E. Quasion exits the Mayfair Diner* holding a cup of hot joe in one hand and a summary of the crime scene evidence in the other. Taking a seat on the bumper of his tan LTD, Detective Quasion begins to review the evidence.

At 5:30 A.M. the body of the diner's owner, Joe D. Wood, was found in the walk-in refrigerator in the diner's basement. At 6:00 A.M. the coroner arrived and determined that the core body temperature of the corpse was 85° Fahrenheit. Thirty minutes later the coroner again measured the core body temperature; this time the reading was 84° Fahrenheit. A thermostat inside the refrigerator reads 50° Fahrenheit.

Diff takes a faded yellow legal pad and a ketchup-spattered calculator from the front seat of his cruiser and begins to compute. He knows that Newton's Law of Cooling says that the rate at which an object cools is proportional to the difference between the temperature T of the body at time t and the temperature T_m of the environment or medium surrounding the body. He jots down the differential equation

$$\frac{dT}{dt} = k(T - T_m), \quad t > 0, \tag{1}$$

FIGURE 1   The Mayfair Diner

where k is a constant of proportionality, T and T_m are measured in degrees Fahrenheit, and time t is measured in hours. Because Diff wants to investigate the past using positive values of time, he decides to correspond $t = 0$ with 6:00 A.M., and so, for example, $t = 4$ is 2:00 A.M. After a few scratches on his yellow pad, Diff realizes that with this time convention the constant k in (1) will turn out to be *positive*. Diff jots a reminder to himself that 6:30 A.M. is now $t = -\frac{1}{2}$.

> **PROBLEM 1.**   After much deep thought Diff decides to begin by assuming that Mr. Wood was murdered inside the refrigerator. What is the estimated time of death?

As the cool and quiet dawn gives way to a steamy midsummer morning, Diff begins to sweat and wonders aloud, "But what if he was killed in the diner and then the body was moved into the fridge in a feeble attempt to hide it? How does this change my estimate of the time of death?" He reenters the diner and finds a grease-streaked thermostat above the empty cash register. It reads 70° Fahrenheit.

"But *when* was the body moved?" Diff asks. He decides to leave this question unanswered for now and simply lets h denote the number of hours the body has been in the refrigerator prior to 6:00 A.M. For example, if $h = 6$, then the body was moved into the refrigerator at midnight.

Diff flips a page on his legal pad and begins jotting. As the rapidly cooling coffee begins to do its work, he realizes that a way to model the environmental temperature change caused by the move is with the unit step function $\mathscr{U}(t)$. He writes

$$T_m(t) = 50 + 20\,\mathscr{U}(t - h) \tag{2}$$

---

*The historic Mayfair Diner opened in 1932 in Philadelphia, Pennyslvania. Photo used with permission.

and below it
$$\frac{dT}{dt} = k(T - T_m(t)). \tag{3}$$

PROBLEM 2. Solve the differential equation in (2) using the Laplace transform. Your solution $T(t)$ will depend on h.

Diff's mustard-stained polyester shirt begins to darken with sweat under the blaze of the morning sun. Drained from the heat and the mental exercise, Diff fires up his cruiser and drives to Boodle's Café for another cup of java and a heaping plate of biscuits and gravy. He settles into his favorite faux-leather booth. The intense air-conditioning conspires with his sweat-soaked shirt to raise goose-flesh on his rapidly cooling skin. The chill serves as a gruesome reminder of the tragedy that occurred earlier at the Mayfair.

While Diff waits for his breakfast, he retrieves his legal pad and quickly reviews his calculations. He then carefully constructs a table that relates the refrigeration time h to the time of death.

PROBLEM 3 (CAS). Complete Diff's table. In particular, explain why large values of h give the same approximation to the time of death. The **Newton's Law of Cooling Tool** on the *DE Tools* CD can also be used to estimate these entries graphically.

h	Time Body Moved	Time of Death	h	Time Body Moved	Time of Death
12	6:00 P.M.		6		
11			5		
10			4		
9			3		
8			2		
7					

Shoving away the empty platter, Diff picks up his cell phone to check with his partner Marie. "Any suspects?" Diff asks.

"Yeah," she replies, "we've got three of 'em. The first is the late Mr. Wood's ex-wife, a dancer by the name of Twinkles. She was seen in the Mayfair between 5:00 and 6:00 P.M. engaged in a shouting match with Wood. We haven't found any witnesses that saw Joe after this fight."

"When did she leave?"

"A witness says she left in a hurry a little after 6:00 P.M. The second suspect is an East End bookie who goes by the name of Slim. Slim was in last night asking for Joe around 10:00 P.M. Witnesses say that there was a lot of hand gesturing, like Slim was upset about something. One witness says that Slim stormed into the back room."

"Did anyone see him leave?"

"Yeah. He left quietly around 11:00 P.M. The third suspect is the cook."

"The cook?"

"Yep, the cook. Goes by the name of Shorty. The cashier says she heard Joe and Shorty arguing over the proper way to present a plate of veal scaloppine. She said that Shorty took an unusually long break at 10:30 P.M. He then took off in a huff when the diner closed at 2:00 A.M. Guess that explains why the place was in such a mess."

"Great work, partner. I think I know who to bring in for questioning."

PROBLEM 4. Who does Diff want to question and why?

8

SYSTEMS OF LINEAR FIRST-ORDER DIFFERENTIAL EQUATIONS

Eigenvalues of A are Equal

Eigenvalues of A are Complex

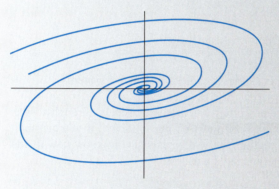

Eigenvalues of A are Pure Imaginary

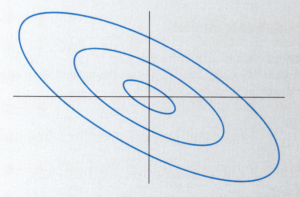

We encountered systems of differential equations in Sections 3.3, 4.8, and 7.6, and were able to solve some of these systems by means of either systematic elimination or the Laplace transform. In this chapter we are going to concentrate on only systems of linear first-order DEs. Although most of the systems considered could be solved using elimination or the Laplace transform, we are going to develop a general theory for these kinds of systems and, in the case of systems with constant coefficients, a method of solution that utilizes some basic concepts from the algebra of matrices. We shall see that this general theory and solution procedure is similar to that of linear higher-order DEs considered in Chapter 4. This material is fundamental to the analysis of systems of nonlinear first-order equations.

A solution $x(t)$, $y(t)$ of a system $\mathbf{X}' = \mathbf{AX}$ of two linear first-order constant-coefficient DEs are parametric equations for a curve in the xy-plane. A collection of such curves is called a phase portrait of the system. The phase portrait depends on the eigenvalues of the matrix $\mathbf{A}$. See pages 340, 344, 347, and 349.

8.1 PRELIMINARY THEORY

INTRODUCTION: Recall that in Section 4.8 we illustrated how to solve systems of n linear differential equations in n unknowns of the form

$$
\begin{aligned}
P_{11}(D)x_1 + P_{12}(D)x_2 + \cdots + P_{1n}(D)x_n &= b_1(t) \\
P_{21}(D)x_1 + P_{22}(D)x_2 + \cdots + P_{2n}(D)x_n &= b_2(t) \\
&\vdots \\
P_{n1}(D)x_1 + P_{n2}(D)x_2 + \cdots + P_{nn}(D)x_n &= b_n(t),
\end{aligned}
\tag{1}
$$

where the P_{ij} were polynomials of various degrees in the differential operator D. In this chapter we confine our study to systems of first-order DEs that are special cases of systems that have the normal form

$$
\begin{aligned}
\frac{dx_1}{dt} &= g_1(t, x_1, x_2, \ldots, x_n) \\
\frac{dx_2}{dt} &= g_2(t, x_1, x_2, \ldots, x_n) \\
&\vdots \\
\frac{dx_n}{dt} &= g_n(t, x_1, x_2, \ldots, x_n).
\end{aligned}
\tag{2}
$$

A system such as (2) of n first-order differential equations is called a **first-order system.**

REVIEW MATERIAL: Matrix notation and properties are used extensively throughout this chapter. It is imperative that you review Appendix II or a text on linear algebra if you are unfamiliar with these concepts. Specifically in this section you will be expected to know how to add matrices, multiply matrices, and evaluate the determinant of a matrix.

LINEAR SYSTEMS When each of the functions $g_1, g_2, \ldots, g_n$ in (2) is linear in the dependent variables $x_1, x_2, \ldots, x_n$, we get the **normal form** of a first-order system of linear equations:

$$
\begin{aligned}
\frac{dx_1}{dt} &= a_{11}(t)x_1 + a_{12}(t)x_2 + \cdots + a_{1n}(t)x_n + f_1(t) \\
\frac{dx_2}{dt} &= a_{21}(t)x_1 + a_{22}(t)x_2 + \cdots + a_{2n}(t)x_n + f_2(t) \\
&\vdots \\
\frac{dx_n}{dt} &= a_{n1}(t)x_1 + a_{n2}(t)x_2 + \cdots + a_{nn}(t)x_n + f_n(t).
\end{aligned}
\tag{3}
$$

We refer to a system of the form given in (3) simply as a **linear system.** We assume that the coefficients a_{ij} as well as the functions f_i are continuous on a common interval I. When $f_i(t) = 0$, $i = 1, 2, \ldots, n$, the linear system (3) is said to be **homogeneous;** otherwise it is **nonhomogeneous.**

MATRIX FORM OF A LINEAR SYSTEM   If $\mathbf{X}$, $\mathbf{A}(t)$, and $\mathbf{F}(t)$ denote the respective matrices

$$\mathbf{X} = \begin{pmatrix} x_1(t) \\ x_2(t) \\ \vdots \\ x_n(t) \end{pmatrix}, \qquad \mathbf{A}(t) = \begin{pmatrix} a_{11}(t) & a_{12}(t) & \cdots & a_{1n}(t) \\ a_{21}(t) & a_{22}(t) & \cdots & a_{2n}(t) \\ \vdots & & & \vdots \\ a_{n1}(t) & a_{n2}(t) & \cdots & a_{nn}(t) \end{pmatrix}, \qquad \mathbf{F}(t) = \begin{pmatrix} f_1(t) \\ f_2(t) \\ \vdots \\ f_n(t) \end{pmatrix},$$

then the system of linear first-order differential equations (3) can be written as

$$\frac{d}{dt} \begin{pmatrix} x_1 \\ x_2 \\ \vdots \\ x_n \end{pmatrix} = \begin{pmatrix} a_{11}(t) & a_{12}(t) & \cdots & a_{1n}(t) \\ a_{21}(t) & a_{22}(t) & \cdots & a_{2n}(t) \\ \vdots & & & \vdots \\ a_{n1}(t) & a_{n2}(t) & \cdots & a_{nn}(t) \end{pmatrix} \begin{pmatrix} x_1 \\ x_2 \\ \vdots \\ x_n \end{pmatrix} + \begin{pmatrix} f_1(t) \\ f_2(t) \\ \vdots \\ f_n(t) \end{pmatrix}$$

or simply
$$\mathbf{X}' = \mathbf{A}\mathbf{X} + \mathbf{F}. \tag{4}$$

If the system is homogeneous, its matrix form is then

$$\mathbf{X}' = \mathbf{A}\mathbf{X}. \tag{5}$$

EXAMPLE 1   Systems Written in Matrix Notation

(a) If $\mathbf{X} = \begin{pmatrix} x \\ y \end{pmatrix}$, then the matrix form of the homogeneous system

$$\begin{aligned} \frac{dx}{dt} &= 3x + 4y \\ \frac{dy}{dt} &= 5x - 7y \end{aligned} \quad \text{is} \quad \mathbf{X}' = \begin{pmatrix} 3 & 4 \\ 5 & -7 \end{pmatrix} \mathbf{X}.$$

(b) If $\mathbf{X} = \begin{pmatrix} x \\ y \\ z \end{pmatrix}$, then the matrix form of the nonhomogeneous system

$$\begin{aligned} \frac{dx}{dt} &= 6x + y + z + t \\ \frac{dy}{dt} &= 8x + 7y - z + 10t \\ \frac{dz}{dt} &= 2x + 9y - z + 6t \end{aligned} \quad \text{is} \quad \mathbf{X}' = \begin{pmatrix} 6 & 1 & 1 \\ 8 & 7 & -1 \\ 2 & 9 & -1 \end{pmatrix} \mathbf{X} + \begin{pmatrix} t \\ 10t \\ 6t \end{pmatrix}.$$

DEFINITION 8.1   Solution Vector

A **solution vector** on an interval I is any column matrix

$$\mathbf{X} = \begin{pmatrix} x_1(t) \\ x_2(t) \\ \vdots \\ x_n(t) \end{pmatrix}$$

whose entries are differentiable functions satisfying the system (4) on the interval.

A solution vector of (4) is, of course, equivalent to n scalar equations $x_1 = \phi_1(t)$, $x_2 = \phi_2(t)$, ..., $x_n = \phi_n(t)$ and can be interpreted geometrically as a set of parametric equations of a space curve. In the important case $n = 2$, the equations $x_1 = \phi_1(t)$, $x_2 = \phi_2(t)$ represent a curve in the x_1x_2-plane. It is common practice to call a curve in the plane a **trajectory** and to call the x_1x_2-plane the **phase plane.** We will come back to these concepts and illustrate them in the next section.

EXAMPLE 2 Verification of Solutions

Verify that on the interval $(-\infty, \infty)$

$$\mathbf{X}_1 = \begin{pmatrix} 1 \\ -1 \end{pmatrix} e^{-2t} = \begin{pmatrix} e^{-2t} \\ -e^{-2t} \end{pmatrix} \quad \text{and} \quad \mathbf{X}_2 = \begin{pmatrix} 3 \\ 5 \end{pmatrix} e^{6t} = \begin{pmatrix} 3e^{6t} \\ 5e^{6t} \end{pmatrix}$$

are solutions of
$$\mathbf{X}' = \begin{pmatrix} 1 & 3 \\ 5 & 3 \end{pmatrix} \mathbf{X}. \tag{6}$$

SOLUTION From $\mathbf{X}_1' = \begin{pmatrix} -2e^{-2t} \\ 2e^{-2t} \end{pmatrix}$ and $\mathbf{X}_2' = \begin{pmatrix} 18e^{6t} \\ 30e^{6t} \end{pmatrix}$ we see that

$$\mathbf{AX}_1 = \begin{pmatrix} 1 & 3 \\ 5 & 3 \end{pmatrix}\begin{pmatrix} e^{-2t} \\ -e^{-2t} \end{pmatrix} = \begin{pmatrix} e^{-2t} - 3e^{-2t} \\ 5e^{-2t} - 3e^{-2t} \end{pmatrix} = \begin{pmatrix} -2e^{-2t} \\ 2e^{-2t} \end{pmatrix} = \mathbf{X}_1',$$

and $\mathbf{AX}_2 = \begin{pmatrix} 1 & 3 \\ 5 & 3 \end{pmatrix}\begin{pmatrix} 3e^{6t} \\ 5e^{6t} \end{pmatrix} = \begin{pmatrix} 3e^{6t} + 15e^{6t} \\ 15e^{6t} + 15e^{6t} \end{pmatrix} = \begin{pmatrix} 18e^{6t} \\ 30e^{6t} \end{pmatrix} = \mathbf{X}_2'.$

Much of the theory of systems of n linear first-order differential equations is similar to that of linear nth-order differential equations.

INITIAL-VALUE PROBLEM Let t_0 denote a point on an interval I and

$$\mathbf{X}(t_0) = \begin{pmatrix} x_1(t_0) \\ x_2(t_0) \\ \vdots \\ x_n(t_0) \end{pmatrix} \quad \text{and} \quad \mathbf{X}_0 = \begin{pmatrix} \gamma_1 \\ \gamma_2 \\ \vdots \\ \gamma_n \end{pmatrix},$$

where the γ_i, $i = 1, 2, \ldots, n$ are given constants. Then the problem

$$\begin{array}{ll} \textit{Solve:} & \mathbf{X}' = \mathbf{A}(t)\mathbf{X} + \mathbf{F}(t) \\ \textit{Subject to:} & \mathbf{X}(t_0) = \mathbf{X}_0 \end{array} \tag{7}$$

is an **initial-value problem** on the interval.

THEOREM 8.1 Existence of a Unique Solution

Let the entries of the matrices $\mathbf{A}(t)$ and $\mathbf{F}(t)$ be functions continuous on a common interval I that contains the point t_0. Then there exists a unique solution of the initial-value problem (7) on the interval.

HOMOGENEOUS SYSTEMS In the next several definitions and theorems we are concerned only with homogeneous systems. Without stating it, we shall always assume that the a_{ij} and the f_i are continuous functions of t on some common interval I.

SUPERPOSITION PRINCIPLE The following result is a **superposition principle** for solutions of linear systems.

THEOREM 8.2	**Superposition Principle**

Let $\mathbf{X}_1, \mathbf{X}_2, \ldots, \mathbf{X}_k$ be a set of solution vectors of the homogeneous system (5) on an interval I. Then the linear combination

$$\mathbf{X} = c_1\mathbf{X}_1 + c_2\mathbf{X}_2 + \cdots + c_k\mathbf{X}_k,$$

where the c_i, $i = 1, 2, \ldots, k$ are arbitrary constants, is also a solution on the interval.

It follows from Theorem 8.2 that a constant multiple of any solution vector of a homogeneous system of linear first-order differential equations is also a solution.

EXAMPLE 3   Using the Superposition Principle

You should practice by verifying that the two vectors

$$\mathbf{X}_1 = \begin{pmatrix} \cos t \\ -\frac{1}{2}\cos t + \frac{1}{2}\sin t \\ -\cos t - \sin t \end{pmatrix} \quad \text{and} \quad \mathbf{X}_2 = \begin{pmatrix} 0 \\ e^t \\ 0 \end{pmatrix}$$

are solutions of the system

$$\mathbf{X}' = \begin{pmatrix} 1 & 0 & 1 \\ 1 & 1 & 0 \\ -2 & 0 & -1 \end{pmatrix} \mathbf{X}. \tag{8}$$

By the superposition principle the linear combination

$$\mathbf{X} = c_1\mathbf{X}_1 + c_2\mathbf{X}_2 = c_1 \begin{pmatrix} \cos t \\ -\frac{1}{2}\cos t + \frac{1}{2}\sin t \\ -\cos t - \sin t \end{pmatrix} + c_2 \begin{pmatrix} 0 \\ e^t \\ 0 \end{pmatrix}$$

is yet another solution of the system.

LINEAR DEPENDENCE AND LINEAR INDEPENDENCE   We are primarily interested in linearly independent solutions of the homogeneous system (5).

DEFINITION 8.2	**Linear Dependence/Independence**

Let $\mathbf{X}_1, \mathbf{X}_2, \ldots, \mathbf{X}_k$ be a set of solution vectors of the homogeneous system (5) on an interval I. We say that the set is **linearly dependent** on the interval if there exist constants $c_1, c_2, \ldots, c_k$, not all zero, such that

$$c_1\mathbf{X}_1 + c_2\mathbf{X}_2 + \cdots + c_k\mathbf{X}_k = \mathbf{0}$$

for every t in the interval. If the set of vectors is not linearly dependent on the interval, it is said to be **linearly independent.**

The case when $k = 2$ should be clear; two solution vectors $\mathbf{X}_1$ and $\mathbf{X}_2$ are linearly dependent if one is a constant multiple of the other, and conversely. For $k > 2$ a set of solution vectors is linearly dependent if we can express at least one solution vector as a linear combination of the remaining vectors.

WRONSKIAN   As in our earlier consideration of the theory of a single ordinary differential equation, we can introduce the concept of the **Wronskian**

determinant as a test for linear independence. We state the following theorem without proof.

THEOREM 8.3 **Criterion for Linearly Independent Solutions**

Let
$$\mathbf{X}_1 = \begin{pmatrix} x_{11} \\ x_{21} \\ \vdots \\ x_{n1} \end{pmatrix}, \quad \mathbf{X}_2 = \begin{pmatrix} x_{12} \\ x_{22} \\ \vdots \\ x_{n2} \end{pmatrix}, \quad \dots, \quad \mathbf{X}_n = \begin{pmatrix} x_{1n} \\ x_{2n} \\ \vdots \\ x_{nn} \end{pmatrix}$$

be n solution vectors of the homogeneous system (5) on an interval I. Then the set of solution vectors is linearly independent on I if and only if the **Wronskian**

$$W(\mathbf{X}_1, \mathbf{X}_2, \dots, \mathbf{X}_n) = \begin{vmatrix} x_{11} & x_{12} & \dots & x_{1n} \\ x_{21} & x_{22} & \dots & x_{2n} \\ \vdots & & & \vdots \\ x_{n1} & x_{n2} & \dots & x_{nn} \end{vmatrix} \neq 0 \tag{9}$$

for every t in the interval.

It can be shown that if $\mathbf{X}_1, \mathbf{X}_2, \dots, \mathbf{X}_n$ are solution vectors of (5), then for every t in I either $W(\mathbf{X}_1, \mathbf{X}_2, \dots, \mathbf{X}_n) \neq 0$ or $W(\mathbf{X}_1, \mathbf{X}_2, \dots, \mathbf{X}_n) = 0$. Thus if we can show that $W \neq 0$ for some t_0 in I, then $W \neq 0$ for every t, and hence the solutions are linearly independent on the interval.

Notice that, unlike our definition of the Wronskian in Section 4.1, here the definition of the determinant (9) does not involve differentiation.

EXAMPLE 4 **Linearly Independent Solutions**

In Example 2 we saw that $\mathbf{X}_1 = \begin{pmatrix} 1 \\ -1 \end{pmatrix} e^{-2t}$ and $\mathbf{X}_2 = \begin{pmatrix} 3 \\ 5 \end{pmatrix} e^{6t}$ are solutions of system (6). Clearly, $\mathbf{X}_1$ and $\mathbf{X}_2$ are linearly independent on the interval $(-\infty, \infty)$ since neither vector is a constant multiple of the other. In addition, we have

$$W(\mathbf{X}_1, \mathbf{X}_2) = \begin{vmatrix} e^{-2t} & 3e^{6t} \\ -e^{-2t} & 5e^{6t} \end{vmatrix} = 8e^{4t} \neq 0$$

for all real values of t.

DEFINITION 8.3 **Fundamental Set of Solutions**

Any set $\mathbf{X}_1, \mathbf{X}_2, \dots, \mathbf{X}_n$ of n linearly independent solution vectors of the homogeneous system (5) on an interval I is said to be a **fundamental set of solutions** on the interval.

THEOREM 8.4 **Existence of a Fundamental Set**

There exists a fundamental set of solutions for the homogeneous system (5) on an interval I.

The next two theorems are the linear system equivalents of Theorems 4.5 and 4.6.

THEOREM 8.5   General Solution—Homogeneous Systems

Let $\mathbf{X}_1, \mathbf{X}_2, \ldots, \mathbf{X}_n$ be a fundamental set of solutions of the homogeneous system (5) on an interval I. Then the **general solution** of the system on the interval is

$$\mathbf{X} = c_1\mathbf{X}_1 + c_2\mathbf{X}_2 + \cdots + c_n\mathbf{X}_n,$$

where the c_i, $i = 1, 2, \ldots, n$ are arbitrary constants.

EXAMPLE 5   General Solution of System (6)

From Example 2 we know that $\mathbf{X}_1 = \begin{pmatrix} 1 \\ -1 \end{pmatrix} e^{-2t}$ and $\mathbf{X}_2 = \begin{pmatrix} 3 \\ 5 \end{pmatrix} e^{6t}$ are linearly independent solutions of (6) on $(-\infty, \infty)$. Hence $\mathbf{X}_1$ and $\mathbf{X}_2$ form a fundamental set of solutions on the interval. The general solution of the system on the interval is then

$$\mathbf{X} = c_1\mathbf{X}_1 + c_2\mathbf{X}_2 = c_1\begin{pmatrix} 1 \\ -1 \end{pmatrix} e^{-2t} + c_2\begin{pmatrix} 3 \\ 5 \end{pmatrix} e^{6t}. \qquad (10)$$

EXAMPLE 6   General Solution of System (8)

The vectors

$$\mathbf{X}_1 = \begin{pmatrix} \cos t \\ -\frac{1}{2}\cos t + \frac{1}{2}\sin t \\ -\cos t - \sin t \end{pmatrix}, \quad \mathbf{X}_2 = \begin{pmatrix} 0 \\ 1 \\ 0 \end{pmatrix} e^t, \quad \mathbf{X}_3 = \begin{pmatrix} \sin t \\ -\frac{1}{2}\sin t - \frac{1}{2}\cos t \\ -\sin t + \cos t \end{pmatrix}$$

are solutions of the system (8) in Example 3 (see Problem 16 in Exercises 8.1). Now

$$W(\mathbf{X}_1, \mathbf{X}_2, \mathbf{X}_3) = \begin{vmatrix} \cos t & 0 & \sin t \\ -\frac{1}{2}\cos t + \frac{1}{2}\sin t & e^t & -\frac{1}{2}\sin t - \frac{1}{2}\cos t \\ -\cos t - \sin t & 0 & -\sin t + \cos t \end{vmatrix} = e^t \neq 0$$

for all real values of t. We conclude that $\mathbf{X}_1$, $\mathbf{X}_2$, and $\mathbf{X}_3$ form a fundamental set of solutions on $(-\infty, \infty)$. Thus the general solution of the system on the interval is the linear combination $\mathbf{X} = c\mathbf{X}_1 + c_2\mathbf{X}_2 + c_3\mathbf{X}_3$; that is,

$$\mathbf{X} = c_1\begin{pmatrix} \cos t \\ -\frac{1}{2}\cos t + \frac{1}{2}\sin t \\ -\cos t - \sin t \end{pmatrix} + c_2\begin{pmatrix} 0 \\ 1 \\ 0 \end{pmatrix} e^t + c_3\begin{pmatrix} \sin t \\ -\frac{1}{2}\sin t - \frac{1}{2}\cos t \\ -\sin t + \cos t \end{pmatrix}.$$

NONHOMOGENEOUS SYSTEMS   For nonhomogeneous systems a **particular solution** $\mathbf{X}_p$ on an interval I is any vector, free of arbitrary parameters, whose entries are functions that satisfy the system (4).

THEOREM 8.6   General Solution — Nonhomogeneous Systems

Let $\mathbf{X}_p$ be a given solution of the nonhomogeneous system (4) on an interval I, and let

$$\mathbf{X}_c = c_1\mathbf{X}_1 + c_2\mathbf{X}_2 + \cdots + c_n\mathbf{X}_n$$

denote the general solution on the same interval of the associated homogeneous system (5). Then the **general solution** of the nonhomogeneous system on the interval is

$$\mathbf{X} = \mathbf{X}_c + \mathbf{X}_p.$$

The general solution $\mathbf{X}_c$ of the associated homogeneous system (5) is called the **complementary function** of the nonhomogeneous system (4).

EXAMPLE 7   General Solution — Nonhomogeneous System

The vector $\mathbf{X}_p = \begin{pmatrix} 3t - 4 \\ -5t + 6 \end{pmatrix}$ is a particular solution of the nonhomogeneous system

$$\mathbf{X}' = \begin{pmatrix} 1 & 3 \\ 5 & 3 \end{pmatrix}\mathbf{X} + \begin{pmatrix} 12t - 11 \\ -3 \end{pmatrix} \tag{11}$$

on the interval $(-\infty, \infty)$. (Verify this.) The complementary function of (11) on the same interval, or the general solution of $\mathbf{X}' = \begin{pmatrix} 1 & 3 \\ 5 & 3 \end{pmatrix}\mathbf{X}$, was seen in (10) of Example 5 to be $\mathbf{X}_c = c_1\begin{pmatrix} 1 \\ -1 \end{pmatrix}e^{-2t} + c_2\begin{pmatrix} 3 \\ 5 \end{pmatrix}e^{6t}$. Hence by Theorem 8.6

$$\mathbf{X} = \mathbf{X}_c + \mathbf{X}_p = c_1\begin{pmatrix} 1 \\ -1 \end{pmatrix}e^{-2t} + c_2\begin{pmatrix} 3 \\ 5 \end{pmatrix}e^{6t} + \begin{pmatrix} 3t - 4 \\ -5t + 6 \end{pmatrix}$$

is the general solution of (11) on $(-\infty, \infty)$. ∎

EXERCISES 8.1

Answers to selected odd-numbered problems begin on page ANS-11.

In Problems 1–6 write the linear system in matrix form.

1. $\dfrac{dx}{dt} = 3x - 5y$

$\dfrac{dy}{dt} = 4x + 8y$

2. $\dfrac{dx}{dt} = 4x - 7y$

$\dfrac{dy}{dt} = 5x$

3. $\dfrac{dx}{dt} = -3x + 4y - 9z$

$\dfrac{dy}{dt} = 6x - y$

$\dfrac{dz}{dt} = 10x + 4y + 3z$

4. $\dfrac{dx}{dt} = x - y$

$\dfrac{dy}{dt} = x + 2z$

$\dfrac{dz}{dt} = -x + z$

5. $\dfrac{dx}{dt} = x - y + z + t - 1$

$\dfrac{dy}{dt} = 2x + y - z - 3t^2$

$\dfrac{dz}{dt} = x + y + z + t^2 - t + 2$

6. $\dfrac{dx}{dt} = -3x + 4y + e^{-t}\sin 2t$

$\dfrac{dy}{dt} = 5x + 9z + 4e^{-t}\cos 2t$

$\dfrac{dz}{dt} = y + 6z - e^{-t}$

In Problems 7–10 write the given system without the use of matrices.

7. $\mathbf{X}' = \begin{pmatrix} 4 & 2 \\ -1 & 3 \end{pmatrix}\mathbf{X} + \begin{pmatrix} 1 \\ -1 \end{pmatrix}e^t$

8. $\mathbf{X}' = \begin{pmatrix} 7 & 5 & -9 \\ 4 & 1 & 1 \\ 0 & -2 & 3 \end{pmatrix}\mathbf{X} + \begin{pmatrix} 0 \\ 2 \\ 1 \end{pmatrix}e^{5t} - \begin{pmatrix} 8 \\ 0 \\ 3 \end{pmatrix}e^{-2t}$

9. $\dfrac{d}{dt}\begin{pmatrix} x \\ y \\ z \end{pmatrix} = \begin{pmatrix} 1 & -1 & 2 \\ 3 & -4 & 1 \\ -2 & 5 & 6 \end{pmatrix}\begin{pmatrix} x \\ y \\ z \end{pmatrix} + \begin{pmatrix} 1 \\ 2 \\ 2 \end{pmatrix}e^{-t} - \begin{pmatrix} 3 \\ -1 \\ 1 \end{pmatrix}t$

10. $\dfrac{d}{dt}\begin{pmatrix} x \\ y \end{pmatrix} = \begin{pmatrix} 3 & -7 \\ 1 & 1 \end{pmatrix}\begin{pmatrix} x \\ y \end{pmatrix} + \begin{pmatrix} 4 \\ 8 \end{pmatrix}\sin t + \begin{pmatrix} t-4 \\ 2t+1 \end{pmatrix}e^{4t}$

In Problems 11–16 verify that the vector $\mathbf{X}$ is a solution of the given system.

11. $\dfrac{dx}{dt} = 3x - 4y$
$\dfrac{dy}{dt} = 4x - 7y;\quad \mathbf{X} = \begin{pmatrix} 1 \\ 2 \end{pmatrix}e^{-5t}$

12. $\dfrac{dx}{dt} = -2x + 5y$
$\dfrac{dy}{dt} = -2x + 4y;\quad \mathbf{X} = \begin{pmatrix} 5\cos t \\ 3\cos t - \sin t \end{pmatrix}e^t$

13. $\mathbf{X}' = \begin{pmatrix} -1 & \frac{1}{4} \\ 1 & -1 \end{pmatrix}\mathbf{X};\quad \mathbf{X} = \begin{pmatrix} -1 \\ 2 \end{pmatrix}e^{-3t/2}$

14. $\mathbf{X}' = \begin{pmatrix} 2 & 1 \\ -1 & 0 \end{pmatrix}\mathbf{X};\quad \mathbf{X} = \begin{pmatrix} 1 \\ 3 \end{pmatrix}e^t + \begin{pmatrix} 4 \\ -4 \end{pmatrix}te^t$

15. $\mathbf{X}' = \begin{pmatrix} 1 & 2 & 1 \\ 6 & -1 & 0 \\ -1 & -2 & -1 \end{pmatrix}\mathbf{X};\quad \mathbf{X} = \begin{pmatrix} 1 \\ 6 \\ -13 \end{pmatrix}$

16. $\mathbf{X}' = \begin{pmatrix} 1 & 0 & 1 \\ 1 & 1 & 0 \\ -2 & 0 & -1 \end{pmatrix}\mathbf{X};\quad \mathbf{X} = \begin{pmatrix} \sin t \\ -\frac{1}{2}\sin t - \frac{1}{2}\cos t \\ -\sin t + \cos t \end{pmatrix}$

In Problems 17–20 the given vectors are solutions of a system $\mathbf{X}' = \mathbf{AX}$. Determine whether the vectors form a fundamental set on $(-\infty, \infty)$.

17. $\mathbf{X}_1 = \begin{pmatrix} 1 \\ 1 \end{pmatrix}e^{-2t},\quad \mathbf{X}_2 = \begin{pmatrix} 1 \\ -1 \end{pmatrix}e^{-6t}$

18. $\mathbf{X}_1 = \begin{pmatrix} 1 \\ -1 \end{pmatrix}e^t,\quad \mathbf{X}_2 = \begin{pmatrix} 2 \\ 6 \end{pmatrix}e^t + \begin{pmatrix} 8 \\ -8 \end{pmatrix}te^t$

19. $\mathbf{X}_1 = \begin{pmatrix} 1 \\ -2 \\ 4 \end{pmatrix} + t\begin{pmatrix} 1 \\ 2 \\ 2 \end{pmatrix},\quad \mathbf{X}_2 = \begin{pmatrix} 1 \\ -2 \\ 4 \end{pmatrix},$

$\mathbf{X}_3 = \begin{pmatrix} 3 \\ -6 \\ 12 \end{pmatrix} + t\begin{pmatrix} 2 \\ 4 \\ 4 \end{pmatrix}$

20. $\mathbf{X}_1 = \begin{pmatrix} 1 \\ 6 \\ -13 \end{pmatrix},\quad \mathbf{X}_2 = \begin{pmatrix} 1 \\ -2 \\ -1 \end{pmatrix}e^{-4t},\quad \mathbf{X}_3 = \begin{pmatrix} 2 \\ 3 \\ -2 \end{pmatrix}e^{3t}$

In Problems 21–24 verify that the vector $\mathbf{X}_p$ is a particular solution of the given system.

21. $\dfrac{dx}{dt} = x + 4y + 2t - 7$
$\dfrac{dy}{dt} = 3x + 2y - 4t - 18;\quad \mathbf{X}_p = \begin{pmatrix} 2 \\ -1 \end{pmatrix}t + \begin{pmatrix} 5 \\ 1 \end{pmatrix}$

22. $\mathbf{X}' = \begin{pmatrix} 2 & 1 \\ 1 & -1 \end{pmatrix}\mathbf{X} + \begin{pmatrix} -5 \\ 2 \end{pmatrix};\quad \mathbf{X}_p = \begin{pmatrix} 1 \\ 3 \end{pmatrix}$

23. $\mathbf{X}' = \begin{pmatrix} 2 & 1 \\ 3 & 4 \end{pmatrix}\mathbf{X} - \begin{pmatrix} 1 \\ 7 \end{pmatrix}e^t;\quad \mathbf{X}_p = \begin{pmatrix} 1 \\ 1 \end{pmatrix}e^t + \begin{pmatrix} 1 \\ -1 \end{pmatrix}te^t$

24. $\mathbf{X}' = \begin{pmatrix} 1 & 2 & 3 \\ -4 & 2 & 0 \\ -6 & 1 & 0 \end{pmatrix}\mathbf{X} + \begin{pmatrix} -1 \\ 4 \\ 3 \end{pmatrix}\sin 3t;\quad \mathbf{X}_p = \begin{pmatrix} \sin 3t \\ 0 \\ \cos 3t \end{pmatrix}$

25. Prove that the general solution of
$$\mathbf{X}' = \begin{pmatrix} 0 & 6 & 0 \\ 1 & 0 & 1 \\ 1 & 1 & 0 \end{pmatrix}\mathbf{X}$$
on the interval $(-\infty, \infty)$ is
$$\mathbf{X} = c_1\begin{pmatrix} 6 \\ -1 \\ -5 \end{pmatrix}e^{-t} + c_2\begin{pmatrix} -3 \\ 1 \\ 1 \end{pmatrix}e^{-2t} + c_3\begin{pmatrix} 2 \\ 1 \\ 1 \end{pmatrix}e^{3t}.$$

26. Prove that the general solution of
$$\mathbf{X}' = \begin{pmatrix} -1 & -1 \\ -1 & 1 \end{pmatrix}\mathbf{X} + \begin{pmatrix} 1 \\ 1 \end{pmatrix}t^2 + \begin{pmatrix} 4 \\ -6 \end{pmatrix}t + \begin{pmatrix} -1 \\ 5 \end{pmatrix}$$
on the interval $(-\infty, \infty)$ is
$$\mathbf{X} = c_1\begin{pmatrix} 1 \\ -1-\sqrt{2} \end{pmatrix}e^{\sqrt{2}t} + c_2\begin{pmatrix} 1 \\ -1+\sqrt{2} \end{pmatrix}e^{-\sqrt{2}t}$$
$$+ \begin{pmatrix} 1 \\ 0 \end{pmatrix}t^2 + \begin{pmatrix} -2 \\ 4 \end{pmatrix}t + \begin{pmatrix} 1 \\ 0 \end{pmatrix}.$$

8.2   HOMOGENEOUS LINEAR SYSTEMS

INTRODUCTION: We saw in Example 5 of Section 8.1 that the general solution of the homogeneous system $\mathbf{X}' = \begin{pmatrix} 1 & 3 \\ 5 & 3 \end{pmatrix}\mathbf{X}$ is $\mathbf{X} = c_1\mathbf{X}_1 + c_2\mathbf{X}_2 = c_1\begin{pmatrix} 1 \\ -1 \end{pmatrix}e^{-2t} + c_2\begin{pmatrix} 3 \\ 5 \end{pmatrix}e^{6t}$. Because both solution vectors have the form $\mathbf{X}_i = \begin{pmatrix} k_1 \\ k_2 \end{pmatrix}e^{\lambda_i t}$, $i = 1, 2$, where k_1, k_2, λ_1, and λ_2 are constants, we are prompted to ask whether we can always find a solution of the form

$$\mathbf{X} = \begin{pmatrix} k_1 \\ k_2 \\ \vdots \\ k_n \end{pmatrix}e^{\lambda t} = \mathbf{K}e^{\lambda t} \tag{1}$$

for the general homogeneous linear first-order system

$$\mathbf{X}' = \mathbf{AX}, \tag{2}$$

where $\mathbf{A}$ is an $n \times n$ matrix of constants,

REVIEW MATERIAL: To find solutions (1) of the homogeneous system (2), we need to find the eigenvalues and eigenvectors of the $n \times n$ coefficient matrix $\mathbf{A}$. Review Section II.3 of Appendix II and the appropriate sections in a linear algebra text on how to compute these quantities by hand. Also see the *Student Resource and Solutions Manual*.

CD: The **Linear Phase Portrait Tool** on the *DE Tools* CD illustrates the relationship between the *eigenvalues* of a 2×2 linear system and its *phase portrait*. See pages 340, 344, 347, and 349.

EIGENVALUES AND EIGENVECTORS   If (1) is to be a solution vector of the homogeneous linear system (2), then $\mathbf{X}' = \mathbf{K}\lambda e^{\lambda t}$, so the system becomes $\mathbf{K}\lambda e^{\lambda t} = \mathbf{AK}e^{\lambda t}$. After dividing out $e^{\lambda t}$ and rearranging, we obtain $\mathbf{AK} = \lambda\mathbf{K}$ or $\mathbf{AK} - \lambda\mathbf{K} = \mathbf{0}$. Since $\mathbf{K} = \mathbf{IK}$, the last equation is the same as

$$(\mathbf{A} - \lambda\mathbf{I})\mathbf{K} = 0. \tag{3}$$

The matrix equation (3) is equivalent to the simultaneous algebraic equations

$$\begin{aligned} (a_{11} - \lambda)k_1 + \quad a_{12}k_2 + \cdots + \quad a_{1n}k_n &= 0 \\ a_{21}k_1 + (a_{22} - \lambda)k_2 + \cdots + \quad a_{2n}k_n &= 0 \\ \vdots \qquad\qquad\qquad\qquad &\;\;\;\vdots \\ a_{n1}k_1 + \quad a_{n2}k_2 + \cdots + (a_{nn} - \lambda)k_n &= 0. \end{aligned}$$

Thus to find a nontrivial solution $\mathbf{X}$ of (2), we must first find a nontrivial solution of the foregoing system; in other words, we must find a nontrivial vector $\mathbf{K}$ that satisfies (3). But for (3) to have solutions other than the obvious solution $k_1 = k_2 = \cdots = k_n = 0$, we must have

$$\det(\mathbf{A} - \lambda\mathbf{I}) = 0.$$

This polynomial equation in λ is called the **characteristic equation** of the matrix $\mathbf{A}$; its solutions are the **eigenvalues** of $\mathbf{A}$. A solution $\mathbf{K} \neq \mathbf{0}$ of (3) corresponding to an eigenvalue λ is called an **eigenvector** of $\mathbf{A}$. A solution of the homogeneous system (2) is then $\mathbf{X} = \mathbf{K}e^{\lambda t}$.

In the discussion that follows we examine three cases: real and distinct eigenvalues (that is, no eigenvalues are equal), repeated eigenvalues, and, finally, complex eigenvalues.

8.2.1 DISTINCT REAL EIGENVALUES

When the $n \times n$ matrix $\mathbf{A}$ possesses n distinct real eigenvalues $\lambda_1, \lambda_2, \ldots, \lambda_n$, then a set of n linearly independent eigenvectors $\mathbf{K}_1, \mathbf{K}_2, \ldots, \mathbf{K}_n$ can always be found and

$$\mathbf{X}_1 = \mathbf{K}_1 e^{\lambda_1 t}, \quad \mathbf{X}_2 = \mathbf{K}_2 e^{\lambda_2 t}, \ldots, \quad \mathbf{X}_n = \mathbf{K}_n e^{\lambda_n t}$$

is a fundamental set of solutions of (2) on $(-\infty, \infty)$.

THEOREM 8.7 General Solution—Homogeneous Systems

Let $\lambda_1, \lambda_2, \ldots, \lambda_n$ be n distinct real eigenvalues of the coefficient matrix $\mathbf{A}$ of the homogeneous system (2), and let $\mathbf{K}_1, \mathbf{K}_2, \ldots, \mathbf{K}_n$ be the corresponding eigenvectors. Then the **general solution** of (2) on the interval $(-\infty, \infty)$ is given by

$$\mathbf{X} = c_1 \mathbf{K}_1 e^{\lambda_1 t} + c_2 \mathbf{K}_2 e^{\lambda_2 t} + \cdots + c_n \mathbf{K}_n e^{\lambda_n t}.$$

EXAMPLE 1 Distinct Eigenvalues

Solve

$$\frac{dx}{dt} = 2x + 3y$$

$$\frac{dy}{dt} = 2x + y. \tag{4}$$

SOLUTION We first find the eigenvalues and eigenvectors of the matrix of coefficients.

From the characteristic equation

$$\det(\mathbf{A} - \lambda \mathbf{I}) = \begin{vmatrix} 2 - \lambda & 3 \\ 2 & 1 - \lambda \end{vmatrix} = \lambda^2 - 3\lambda - 4 = (\lambda + 1)(\lambda - 4) = 0$$

we see that the eigenvalues are $\lambda_1 = -1$ and $\lambda_2 = 4$.

Now for $\lambda_1 = -1$, (3) is equivalent to

$$3k_1 + 3k_2 = 0$$
$$2k_1 + 2k_2 = 0.$$

Thus $k_1 = -k_2$. When $k_2 = -1$, the related eigenvector is

$$\mathbf{K}_1 = \begin{pmatrix} 1 \\ -1 \end{pmatrix}.$$

For $\lambda_2 = 4$ we have

$$-2k_1 + 3k_2 = 0$$
$$2k_1 - 3k_2 = 0$$

so $k_1 = 3k_2/2$, and therefore with $k_2 = 2$ the corresponding eigenvector is

$$\mathbf{K}_2 = \begin{pmatrix} 3 \\ 2 \end{pmatrix}.$$

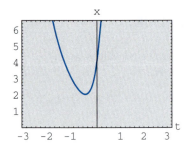

(a) graph of $x = e^{-t} + 3e^{4t}$

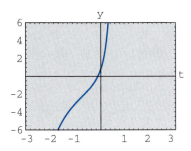

(b) graph of $y = -e^{-t} + 2e^{4t}$

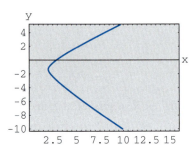

(c) trajectory defined by
$x = e^{-t} + 3e^{4t}$, $y = -e^{-t} + 2e^{4t}$
in the phase plane

FIGURE 8.1   A particular solution from (5) yields three different curves in three different planes

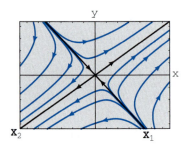

FIGURE 8.2   A phase portrait of system (4)

Since the matrix of coefficients **A** is a 2×2 matrix and since we have found two linearly independent solutions of (4),

$$\mathbf{X}_1 = \begin{pmatrix} 1 \\ -1 \end{pmatrix} e^{-t} \quad \text{and} \quad \mathbf{X}_2 = \begin{pmatrix} 3 \\ 2 \end{pmatrix} e^{4t},$$

we conclude that the general solution of the system is

$$\mathbf{X} = c_1 \mathbf{X}_1 + c_2 \mathbf{X}_2 = c_1 \begin{pmatrix} 1 \\ -1 \end{pmatrix} e^{-t} + c_2 \begin{pmatrix} 3 \\ 2 \end{pmatrix} e^{4t}. \tag{5}$$

PHASE PORTRAIT   You should keep firmly in mind that writing a solution of a system of linear first-order differential equations in terms of matrices is simply an alternative to the method that we employed in Section 4.8, that is, listing the individual functions and the relationship between the constants. If we add the vectors on the right-hand side of (5) and then equate the entries with the corresponding entries in the vector on the left-hand side, we obtain the more familiar statement

$$x = c_1 e^{-t} + 3c_2 e^{4t}, \quad y = -c_1 e^{-t} + 2c_2 e^{4t}.$$

As was pointed out in Section 8.1, we can interpret these equations as parametric equations of curves in the xy-plane or **phase plane.** Each curve, corresponding to specific choices for c_1 and c_2, is called a **trajectory.** For the choice of constants $c_1 = c_2 = 1$ in the solution (5), we see in Figure 8.1 the graph of $x(t)$ in the tx-plane, the graph of $y(t)$ in the ty-plane, and the trajectory consisting of the points $(x(t), y(t))$ in the phase plane. A collection of representative trajectories in the phase plane, as shown in Figure 8.2, is said to be a **phase portrait** of the given linear system. What appears to be *two* black lines in Figure 8.2 are actually *four* black half-lines defined parametrically in the first, second, third, and fourth quadrants by the solutions $\mathbf{X}_2$, $-\mathbf{X}_1$, $-\mathbf{X}_2$, and $\mathbf{X}_1$, respectively. For example, the Cartesian equations $y = \frac{2}{3}x$, $x > 0$ and $y = -x$, $x > 0$, of the half-lines in the first and fourth quadrants were obtained by eliminating the parameter t in the solutions $x = 3e^{4t}$, $y = 2e^{4t}$, and $x = e^{-t}$, $y = -e^{-t}$, respectively. Moreover, each eigenvector can be visualized as a two-dimensional vector lying along one of these half-lines. The eigenvector $\mathbf{K}_2 = \begin{pmatrix} 3 \\ 2 \end{pmatrix}$ lies along $y = \frac{2}{3}x$ in the first quadrant, and $\mathbf{K}_1 = \begin{pmatrix} 1 \\ -1 \end{pmatrix}$ lies along $y = -x$ in the fourth quadrant. Each vector starts at the origin; $\mathbf{K}_2$ terminates at the point $(2, 3)$, and $\mathbf{K}_1$ terminates at $(1, -1)$.

The origin is not only a constant solution $x = 0$, $y = 0$ of every 2×2 homogeneous linear system $\mathbf{X}' = \mathbf{AX}$, but also an important point in the qualitative study of such systems. If we think in physical terms, the arrowheads on each trajectory in Figure 8.2 indicate the direction that a particle with coordinates $(x(t), y(t))$ on that trajectory at time t moves as time increases. Observe that the arrowheads, with the exception of only those on the half-lines in the second and fourth quadrants, indicate that a particle moves away from the origin as time t increases. If we imagine time ranging from $-\infty$ to ∞, then inspection of the solution $x = c_1 e^{-t} + 3c_2 e^{4t}$, $y = -c_1 e^{-t} + 2c_2 e^{4t}$, $c_1 \neq 0$, $c_2 \neq 0$ shows that a trajectory, or moving particle, "starts" asymptotic to one of the half-lines defined by $\mathbf{X}_1$ or $-\mathbf{X}_1$ (since e^{4t} is negligible for $t \to -\infty$) and "finishes" asymptotic to one of the half-lines defined by $\mathbf{X}_2$ and $-\mathbf{X}_2$ (since e^{-t} is negligible for $t \to \infty$).

We note in passing that Figure 8.2 represents a phase portrait that is typical of *all* 2×2 homogeneous linear systems $\mathbf{X}' = \mathbf{AX}$ with real eigenvalues of opposite signs. See Problem 17 in Exercises 8.2. Moreover, phase portraits in the two cases when distinct real eigenvalues have the same algebraic sign are typical of all such 2×2 linear systems; the only difference is that the arrowheads indicate that a particle moves away from the origin on any trajectory as $t \to \infty$ when both λ_1 and λ_2 are positive and moves toward the origin on any trajectory when both λ_1 and λ_2 are negative. Consequently, we call the origin a **repeller** in the case $\lambda_1 > 0$, $\lambda_2 > 0$ and an **attractor** in the case $\lambda_1 < 0$, $\lambda_2 < 0$. See Problem 18 in Exercises 8.2. The origin in Figure 8.2 is neither a repeller nor an attractor. Investigation of the remaining case when $\lambda = 0$ is an eigenvalue of a 2×2 homogeneous linear system is left as an exercise. See Problem 49 in Exercises 8.2.

EXAMPLE 2   Distinct Eigenvalues

Solve

$$\frac{dx}{dt} = -4x + y + z$$

$$\frac{dy}{dt} = x + 5y - z \qquad (6)$$

$$\frac{dz}{dt} = y - 3z.$$

SOLUTION   Using the cofactors of the third row, we find

$$\det(\mathbf{A} - \lambda\mathbf{I}) = \begin{vmatrix} -4 - \lambda & 1 & 1 \\ 1 & 5 - \lambda & -1 \\ 0 & 1 & -3 - \lambda \end{vmatrix} = -(\lambda + 3)(\lambda + 4)(\lambda - 5) = 0,$$

and so the eigenvalues are $\lambda_1 = -3$, $\lambda_2 = -4$, and $\lambda_3 = 5$.

For $\lambda_1 = -3$, Gauss-Jordan elimination gives

$$(\mathbf{A} + 3\mathbf{I}|\mathbf{0}) = \begin{pmatrix} -1 & 1 & 1 & | & 0 \\ 1 & 8 & -1 & | & 0 \\ 0 & 1 & 0 & | & 0 \end{pmatrix} \xrightarrow[\text{operations}]{\text{row}} \begin{pmatrix} 1 & 0 & -1 & | & 0 \\ 0 & 1 & 0 & | & 0 \\ 0 & 0 & 0 & | & 0 \end{pmatrix}.$$

Therefore $k_1 = k_3$ and $k_2 = 0$. The choice $k_3 = 1$ gives an eigenvector and corresponding solution vector

$$\mathbf{K}_1 = \begin{pmatrix} 1 \\ 0 \\ 1 \end{pmatrix}, \quad \mathbf{X}_1 = \begin{pmatrix} 1 \\ 0 \\ 1 \end{pmatrix} e^{-3t}. \qquad (7)$$

Similarly, for $\lambda_2 = -4$,

$$(\mathbf{A} + 4\mathbf{I}|\mathbf{0}) = \begin{pmatrix} 0 & 1 & 1 & | & 0 \\ 1 & 9 & -1 & | & 0 \\ 0 & 1 & 1 & | & 0 \end{pmatrix} \xrightarrow[\text{operations}]{\text{row}} \begin{pmatrix} 1 & 0 & -10 & | & 0 \\ 0 & 1 & 1 & | & 0 \\ 0 & 0 & 0 & | & 0 \end{pmatrix}$$

implies that $k_1 = 10k_3$ and $k_2 = -k_3$. Choosing $k_3 = 1$, we get a second eigenvector and solution vector

$$\mathbf{K}_2 = \begin{pmatrix} 10 \\ -1 \\ 1 \end{pmatrix}, \quad \mathbf{X}_2 = \begin{pmatrix} 10 \\ -1 \\ 1 \end{pmatrix} e^{-4t}. \qquad (8)$$

Finally, when $\lambda_3 = 5$, the augmented matrices

$$(\mathbf{A} + 5\mathbf{I}|\mathbf{0}) = \begin{pmatrix} -9 & 1 & 1 & | & 0 \\ 1 & 0 & -1 & | & 0 \\ 0 & 1 & -8 & | & 0 \end{pmatrix} \xrightarrow[\text{operations}]{\text{row}} \begin{pmatrix} 1 & 0 & -1 & | & 0 \\ 0 & 1 & -8 & | & 0 \\ 0 & 0 & 0 & | & 0 \end{pmatrix}$$

yield
$$\mathbf{K}_3 = \begin{pmatrix} 1 \\ 8 \\ 1 \end{pmatrix}, \quad \mathbf{X}_3 = \begin{pmatrix} 1 \\ 8 \\ 1 \end{pmatrix} e^{5t}. \qquad (9)$$

The general solution of (6) is a linear combination of the solution vectors in (7), (8), and (9):

$$\mathbf{X} = c_1 \begin{pmatrix} 1 \\ 0 \\ 1 \end{pmatrix} e^{-3t} + c_2 \begin{pmatrix} 10 \\ -1 \\ 1 \end{pmatrix} e^{-4t} + c_3 \begin{pmatrix} 1 \\ 8 \\ 1 \end{pmatrix} e^{5t}.$$

USE OF COMPUTERS   Software packages such as MATLAB, *Mathematica, Maple,* and DERIVE can be real time savers in finding eigenvalues and eigenvectors of a matrix **A**.

8.2.2   REPEATED EIGENVALUES

Of course, not all of the n eigenvalues $\lambda_1, \lambda_2, \ldots, \lambda_n$ of an $n \times n$ matrix **A** need be distinct; that is, some of the eigenvalues may be repeated. For example, the characteristic equation of the coefficient matrix in the system

$$\mathbf{X}' = \begin{pmatrix} 3 & -18 \\ 2 & -9 \end{pmatrix} \mathbf{X} \qquad (10)$$

is readily shown to be $(\lambda + 3)^2 = 0$, and therefore $\lambda_1 = \lambda_2 = -3$ is a root of *multiplicity two.* For this value we find the single eigenvector

$$\mathbf{K}_1 = \begin{pmatrix} 3 \\ 1 \end{pmatrix}, \quad \text{so} \quad \mathbf{X}_1 = \begin{pmatrix} 3 \\ 1 \end{pmatrix} e^{-3t} \qquad (11)$$

is one solution of (10). But since we are obviously interested in forming the general solution of the system, we need to pursue the question of finding a second solution.

In general, if m is a positive integer and $(\lambda - \lambda_1)^m$ is a factor of the characteristic equation while $(\lambda - \lambda_1)^{m+1}$ is not a factor, then λ_1 is said to be an **eigenvalue of multiplicity m.** The next three examples illustrate the following cases:

(*i*)   For some $n \times n$ matrices **A** it may be possible to find m linearly independent eigenvectors $\mathbf{K}_1, \mathbf{K}_2, \ldots, \mathbf{K}_m$ corresponding to an eigenvalue λ_1 of multiplicity $m \leq n$. In this case the general solution of the system contains the linear combination

$$c_1 \mathbf{K}_1 e^{\lambda_1 t} + c_2 \mathbf{K}_2 e^{\lambda_1 t} + \cdots + c_m \mathbf{K}_m e^{\lambda_1 t}.$$

(*ii*)   If there is only one eigenvector corresponding to the eigenvalue λ_1 of multiplicity m, then m linearly independent solutions of the form

$$\mathbf{X}_1 = \mathbf{K}_{11} e^{\lambda_1 t}$$
$$\mathbf{X}_2 = \mathbf{K}_{21} t e^{\lambda_1 t} + \mathbf{K}_{22} e^{\lambda_1 t}$$
$$\vdots$$
$$\mathbf{X}_m = \mathbf{K}_{m1} \frac{t^{m-1}}{(m-1)!} e^{\lambda_1 t} + \mathbf{K}_{m2} \frac{t^{m-2}}{(m-2)!} e^{\lambda_1 t} + \cdots + \mathbf{K}_{mm} e^{\lambda_1 t},$$

where $\mathbf{K}_{ij}$ are column vectors, can always be found.

EIGENVALUE OF MULTIPLICITY TWO   We begin by considering eigenvalues of multiplicity two. In the first example we illustrate a matrix for which we can find two distinct eigenvectors corresponding to a double eigenvalue.

EXAMPLE 3   **Repeated Eigenvalues**

Solve $\mathbf{X}' = \begin{pmatrix} 1 & -2 & 2 \\ -2 & 1 & -2 \\ 2 & -2 & 1 \end{pmatrix}\mathbf{X}.$

SOLUTION   Expanding the determinant in the characteristic equation

$$\det(\mathbf{A} - \lambda\mathbf{I}) = \begin{vmatrix} 1 - \lambda & -2 & 2 \\ -2 & 1 - \lambda & -2 \\ 2 & -2 & 1 - \lambda \end{vmatrix} = 0$$

yields $-(\lambda + 1)^2(\lambda - 5) = 0$. We see that $\lambda_1 = \lambda_2 = -1$ and $\lambda_3 = 5$.

For $\lambda_1 = -1$, Gauss-Jordan elimination immediately gives

$$(\mathbf{A} + \mathbf{I}|\mathbf{0}) = \begin{pmatrix} 2 & -2 & 2 & | & 0 \\ -2 & 2 & -2 & | & 0 \\ 2 & -2 & 2 & | & 0 \end{pmatrix} \xrightarrow[\text{operations}]{\text{row}} \begin{pmatrix} 1 & -1 & 0 & | & 0 \\ 0 & 1 & 1 & | & 0 \\ 0 & 0 & 0 & | & 0 \end{pmatrix}.$$

The first row of the last matrix means $k_1 - k_2 + k_3 = 0$ or $k_1 = k_2 - k_3$. The choices $k_2 = 1$, $k_3 = 0$ and $k_2 = 1$, $k_3 = 1$ yield, in turn, $k_1 = 1$ and $k_1 = 0$. Thus two eigenvectors corresponding to $\lambda_1 = -1$ are

$$\mathbf{K}_1 = \begin{pmatrix} 1 \\ 1 \\ 0 \end{pmatrix} \quad \text{and} \quad \mathbf{K}_2 = \begin{pmatrix} 0 \\ 1 \\ 1 \end{pmatrix}.$$

Since neither eigenvector is a constant multiple of the other, we have found two linearly independent solutions,

$$\mathbf{X}_1 = \begin{pmatrix} 1 \\ 1 \\ 0 \end{pmatrix}e^{-t} \quad \text{and} \quad \mathbf{X}_2 = \begin{pmatrix} 0 \\ 1 \\ 1 \end{pmatrix}e^{-t},$$

corresponding to the same eigenvalue. Last, for $\lambda_3 = 5$, the reduction

$$(\mathbf{A} + 5\mathbf{I}|\mathbf{0}) = \begin{pmatrix} -4 & -2 & 2 & | & 0 \\ -2 & -4 & -2 & | & 0 \\ 2 & -2 & -4 & | & 0 \end{pmatrix} \xrightarrow[\text{operations}]{\text{row}} \begin{pmatrix} 1 & 0 & -1 & | & 0 \\ 0 & 1 & 1 & | & 0 \\ 0 & 0 & 0 & | & 0 \end{pmatrix}$$

implies that $k_1 = k_3$ and $k_2 = -k_3$. Picking $k_3 = 1$ gives $k_1 = 1$, $k_2 = -1$; thus a third eigenvector is

$$\mathbf{K}_3 = \begin{pmatrix} 1 \\ -1 \\ 1 \end{pmatrix}.$$

We conclude that the general solution of the system is

$$\mathbf{X} = c_1\begin{pmatrix} 1 \\ 1 \\ 0 \end{pmatrix}e^{-t} + c_2\begin{pmatrix} 0 \\ 1 \\ 1 \end{pmatrix}e^{-t} + c_3\begin{pmatrix} 1 \\ -1 \\ 1 \end{pmatrix}e^{5t}.$$

The matrix of coefficients $\mathbf{A}$ in Example 3 is a special kind of matrix known as a symmetric matrix. An $n \times n$ matrix $\mathbf{A}$ is said to be **symmetric** if its transpose $\mathbf{A}^T$

(where the rows and columns are interchanged) is the same as $\mathbf{A}$—that is, if $\mathbf{A}^T = \mathbf{A}$. It can be proved that if the matrix $\mathbf{A}$ in the system $\mathbf{X}' = \mathbf{AX}$ is symmetric and has real entries, then we can always find n linearly independent eigenvectors $\mathbf{K}_1, \mathbf{K}_2, \ldots, \mathbf{K}_n$, and the general solution of such a system is as given in Theorem 8.7. As illustrated in Example 3, this result holds even when some of the eigenvalues are repeated.

SECOND SOLUTION   Now suppose that λ_1 is an eigenvalue of multiplicity two and that there is only one eigenvector associated with this value. A second solution can be found of the form

$$\mathbf{X}_2 = \mathbf{K}te^{\lambda_1 t} + \mathbf{P}e^{\lambda_1 t}, \tag{12}$$

where
$$\mathbf{K} = \begin{pmatrix} k_1 \\ k_2 \\ \cdot \\ \cdot \\ \cdot \\ k_n \end{pmatrix} \quad \text{and} \quad \mathbf{P} = \begin{pmatrix} p_1 \\ p_2 \\ \cdot \\ \cdot \\ \cdot \\ p_n \end{pmatrix}.$$

To see this, we substitute (12) into the system $\mathbf{X}' = \mathbf{AX}$ and simplify:

$$(\mathbf{AK} - \lambda_1\mathbf{K})te^{\lambda_1 t} + (\mathbf{AP} - \lambda_1\mathbf{P} - \mathbf{K})e^{\lambda_1 t} = \mathbf{0}.$$

Since this last equation is to hold for all values of t, we must have

$$(\mathbf{A} - \lambda_1\mathbf{I})\mathbf{K} = \mathbf{0} \tag{13}$$

and
$$(\mathbf{A} - \lambda_1\mathbf{I})\mathbf{P} = \mathbf{K}. \tag{14}$$

Equation (13) simply states that $\mathbf{K}$ must be an eigenvector of $\mathbf{A}$ associated with λ_1. By solving (13), we find one solution $\mathbf{X}_1 = \mathbf{K}e^{\lambda_1 t}$. To find the second solution $\mathbf{X}_2$, we need only solve the additional system (14) for the vector $\mathbf{P}$.

EXAMPLE 4   **Repeated Eigenvalues**

Find the general solution of the system given in (10).

SOLUTION   From (11) we know that $\lambda_1 = -3$ and that one solution is $\mathbf{X}_1 = \begin{pmatrix} 3 \\ 1 \end{pmatrix} e^{-3t}$. Identifying $\mathbf{K} = \begin{pmatrix} 3 \\ 1 \end{pmatrix}$ and $\mathbf{P} = \begin{pmatrix} p_1 \\ p_2 \end{pmatrix}$, we find from (14) that we must now solve

$$(\mathbf{A} + 3\mathbf{I})\mathbf{P} = \mathbf{K} \quad \text{or} \quad \begin{aligned} 6p_1 - 18p_2 &= 3 \\ 2p_1 - 6p_2 &= 1. \end{aligned}$$

Since this system is obviously equivalent to one equation, we have an infinite number of choices for p_1 and p_2. For example, by choosing $p_1 = 1$ we find $p_2 = \frac{1}{6}$. However, for simplicity, we shall choose $p_1 = \frac{1}{2}$ so that $p_2 = 0$. Hence $\mathbf{P} = \begin{pmatrix} \frac{1}{2} \\ 0 \end{pmatrix}$. Thus from (12) we find $\mathbf{X}_2 = \begin{pmatrix} 3 \\ 1 \end{pmatrix} te^{-3t} + \begin{pmatrix} \frac{1}{2} \\ 0 \end{pmatrix} e^{-3t}$. The general solution of (10) is then $\mathbf{X} = c_1\mathbf{X}_1 + c_2\mathbf{X}_2$ or

$$\mathbf{X} = c_1\begin{pmatrix} 3 \\ 1 \end{pmatrix} e^{-3t} + c_2\left[\begin{pmatrix} 3 \\ 1 \end{pmatrix} te^{-3t} + \begin{pmatrix} \frac{1}{2} \\ 0 \end{pmatrix} e^{-3t}\right].$$

By assigning various values to c_1 and c_2 in the solution in Example 4, we can plot trajectories of the system in (10). A phase portrait of (10) is given in Figure 8.3. The

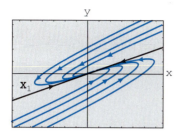

FIGURE 8.3   A phase portrait of system (10)

solutions $\mathbf{X}_1$ and $-\mathbf{X}_1$ determine two half-lines $y = \frac{1}{3}x, x > 0$ and $y = \frac{1}{3}x, x < 0$, respectively, shown in black in the figure. Because the single eigenvalue is negative and $e^{-3t} \to 0$ as $t \to \infty$ on *every* trajectory, we have $(x(t), y(t)) \to (0, 0)$ as $t \to \infty$. This is why the arrowheads in Figure 8.3 indicate that a particle on any trajectory moves toward the origin as time increases and why the origin is an attractor in this case. Moreover, a moving particle or trajectory $x = 3c_1e^{-3t} + c_2\big(3te^{-3t} + \frac{1}{2}e^{-3t}\big)$, $y = c_1e^{-3t} + c_2te^{-3t}, c_2 \neq 0$, approaches $(0, 0)$ tangentially to one of the half-lines as $t \to \infty$. In contrast, when the repeated eigenvalue is positive, the situation is reversed and the origin is a repeller. See Problem 21 in Exercises 8.2. Analogous to Figure 8.2, Figure 8.3 is typical of *all* 2×2 homogeneous linear systems $\mathbf{X}' = \mathbf{A}\mathbf{X}$ that have two repeated negative eigenvalues. See Problem 32 in Exercises 8.2.

EIGENVALUE OF MULTIPLICITY THREE   When the coefficient matrix $\mathbf{A}$ has only one eigenvector associated with an eigenvalue λ_1 of multiplicity three, we can find a second solution of the form (12) and a third solution of the form

$$\mathbf{X}_3 = \mathbf{K}\frac{t^2}{2}e^{\lambda_1 t} + \mathbf{P}te^{\lambda_1 t} + \mathbf{Q}e^{\lambda_1 t}, \tag{15}$$

where

$$\mathbf{K} = \begin{pmatrix} k_1 \\ k_2 \\ \vdots \\ k_n \end{pmatrix}, \quad \mathbf{P} = \begin{pmatrix} p_1 \\ p_2 \\ \vdots \\ p_n \end{pmatrix}, \quad \text{and} \quad \mathbf{Q} = \begin{pmatrix} q_1 \\ q_2 \\ \vdots \\ q_n \end{pmatrix}.$$

By substituting (15) into the system $\mathbf{X}' = \mathbf{A}\mathbf{X}$, we find that the column vectors $\mathbf{K}$, $\mathbf{P}$, and $\mathbf{Q}$ must satisfy

$$(\mathbf{A} - \lambda_1\mathbf{I})\mathbf{K} = \mathbf{0} \tag{16}$$

$$(\mathbf{A} - \lambda_1\mathbf{I})\mathbf{P} = \mathbf{K} \tag{17}$$

and

$$(\mathbf{A} - \lambda_1\mathbf{I})\mathbf{Q} = \mathbf{P}. \tag{18}$$

Of course, the solutions of (16) and (17) can be used in forming the solutions $\mathbf{X}_1$ and $\mathbf{X}_2$.

EXAMPLE 5   **Repeated Eigenvalues**

Solve $\mathbf{X}' = \begin{pmatrix} 2 & 1 & 6 \\ 0 & 2 & 5 \\ 0 & 0 & 2 \end{pmatrix}\mathbf{X}$.

SOLUTION   The characteristic equation $(\lambda - 2)^3 = 0$ shows that $\lambda_1 = 2$ is an eigenvalue of multiplicity three. By solving $(\mathbf{A} - 2\mathbf{I})\mathbf{K} = \mathbf{0}$, we find the single eigenvector

$$\mathbf{K} = \begin{pmatrix} 1 \\ 0 \\ 0 \end{pmatrix}.$$

We next solve the systems $(\mathbf{A} - 2\mathbf{I})\mathbf{P} = \mathbf{K}$ and $(\mathbf{A} - 2\mathbf{I})\mathbf{Q} = \mathbf{P}$ in succession and find that

$$\mathbf{P} = \begin{pmatrix} 0 \\ 1 \\ 0 \end{pmatrix} \quad \text{and} \quad \mathbf{Q} = \begin{pmatrix} 0 \\ -\frac{6}{5} \\ \frac{1}{5} \end{pmatrix}.$$

Using (12) and (15), we see that the general solution of the system is

$$\mathbf{X} = c_1 \begin{pmatrix} 1 \\ 0 \\ 0 \end{pmatrix} e^{2t} + c_2 \left[\begin{pmatrix} 1 \\ 0 \\ 0 \end{pmatrix} te^{2t} + \begin{pmatrix} 0 \\ 1 \\ 0 \end{pmatrix} e^{2t} \right] + c_3 \left[\begin{pmatrix} 1 \\ 0 \\ 0 \end{pmatrix} \frac{t^2}{2} e^{2t} + \begin{pmatrix} 0 \\ 1 \\ 0 \end{pmatrix} te^{2t} + \begin{pmatrix} 0 \\ -\frac{6}{5} \\ \frac{1}{5} \end{pmatrix} e^{2t} \right]. \quad \blacksquare$$

REMARKS

When an eigenvalue λ_1 has multiplicity m, either we can find m linearly independent eigenvectors or the number of corresponding eigenvectors is less than m. Hence the two cases listed on page 342 are not all the possibilities under which a repeated eigenvalue can occur. It can happen, say, that a 5×5 matrix has an eigenvalue of multiplicity five and there exist three corresponding linearly independent eigenvectors. See Problems 31 and 50 in Exercises 8.2.

8.2.3 COMPLEX EIGENVALUES

If $\lambda_1 = \alpha + \beta i$ and $\lambda_2 = \alpha - \beta i$, $\beta > 0$, $i^2 = -1$ are complex eigenvalues of the coefficient matrix $\mathbf{A}$, we can then certainly expect their corresponding eigenvectors to also have complex entries.*

For example, the characteristic equation of the system

$$\frac{dx}{dt} = 6x - y$$

$$\frac{dy}{dt} = 5x + 4y \quad\quad (19)$$

is

$$\det(\mathbf{A} - \lambda \mathbf{I}) = \begin{vmatrix} 6 - \lambda & -1 \\ 5 & 4 - \lambda \end{vmatrix} = \lambda^2 - 10\lambda + 29 = 0.$$

From the quadratic formula we find $\lambda_1 = 5 + 2i$, $\lambda_2 = 5 - 2i$.

Now for $\lambda_1 = 5 + 2i$ we must solve

$$(1 - 2i)k_1 - \quad\quad k_2 = 0$$

$$5k_1 - (1 + 2i)k_2 = 0.$$

Since $k_2 = (1 - 2i)k_1$,[†] the choice $k_1 = 1$ gives the following eigenvector and corresponding solution vector:

$$\mathbf{K}_1 = \begin{pmatrix} 1 \\ 1 - 2i \end{pmatrix}, \quad \mathbf{X}_1 = \begin{pmatrix} 1 \\ 1 - 2i \end{pmatrix} e^{(5 + 2i)t}.$$

In like manner, for $\lambda_2 = 5 - 2i$ we find

$$\mathbf{K}_2 = \begin{pmatrix} 1 \\ 1 + 2i \end{pmatrix}, \quad \mathbf{X}_2 = \begin{pmatrix} 1 \\ 1 + 2i \end{pmatrix} e^{(5 - 2i)t}.$$

---

*When the characteristic equation has real coefficients, complex eigenvalues always appear in conjugate pairs.

[†]Note that the second equation is simply $(1 + 2i)$ times the first.

We can verify by means of the Wronskian that these solution vectors are linearly independent, and so the general solution of (19) is

$$\mathbf{X} = c_1 \begin{pmatrix} 1 \\ 1 - 2i \end{pmatrix} e^{(5+2i)t} + c_2 \begin{pmatrix} 1 \\ 1 + 2i \end{pmatrix} e^{(5-2i)t}. \tag{20}$$

Note that the entries in $\mathbf{K}_2$ corresponding to λ_2 are the conjugates of the entries in $\mathbf{K}_1$ corresponding to λ_1. The conjugate of λ_1 is, of course, λ_2. We write this as $\lambda_2 = \overline{\lambda}_1$ and $\mathbf{K}_2 = \overline{\mathbf{K}}_1$. We have illustrated the following general result.

THEOREM 8.8   **Solutions Corresponding to a Complex Eigenvalue**

Let $\mathbf{A}$ be the coefficient matrix having real entries of the homogeneous system (2), and let $\mathbf{K}_1$ be an eigenvector corresponding to the complex eigenvalue $\lambda_1 = \alpha + i\beta$, α and β real. Then

$$\mathbf{K}_1 e^{\lambda_1 t} \quad \text{and} \quad \overline{\mathbf{K}}_1 e^{\overline{\lambda}_1 t}$$

are solutions of (2).

It is desirable and relatively easy to rewrite a solution such as (20) in terms of real functions. To this end we first use Euler's formula to write

$$e^{(5+2i)t} = e^{5t} e^{2ti} = e^{5t}(\cos 2t + i \sin 2t)$$

$$e^{(5-2i)t} = e^{5t} e^{-2ti} = e^{5t}(\cos 2t - i \sin 2t).$$

Then, after we multiply complex numbers, collect terms, and replace $c_1 + c_2$ by C_1 and $(c_1 - c_2)i$ by C_2, (20) becomes

$$\mathbf{X} = C_1 \mathbf{X}_1 + C_2 \mathbf{X}_2, \tag{21}$$

where

$$\mathbf{X}_1 = \left[\begin{pmatrix} 1 \\ 1 \end{pmatrix} \cos 2t - \begin{pmatrix} 0 \\ -2 \end{pmatrix} \sin 2t \right] e^{5t}$$

and

$$\mathbf{X}_2 = \left[\begin{pmatrix} 0 \\ -2 \end{pmatrix} \cos 2t + \begin{pmatrix} 1 \\ 1 \end{pmatrix} \sin 2t \right] e^{5t}.$$

It is now important to realize that the vectors $\mathbf{X}_1$ and $\mathbf{X}_2$ in (21) constitute a linearly independent set of *real* solutions of the original system. Consequently, we are justified in ignoring the relationship between C_1, C_2 and c_1, c_2, and we can regard C_1 and C_2 as completely arbitrary and real. In other words, the linear combination (21) is an alternative general solution of (19). Moreover, with the real form given in (21) we are able to obtain a phase portrait of the system in (19). From (21) we find $x(t)$ and $y(t)$ to be

$$x = C_1 e^{5t} \cos 2t + C_2 e^{5t} \sin 2t$$

$$y = (C_1 - 2C_2)e^{5t} \cos 2t + (2C_1 + C_2)e^{5t} \sin 2t.$$

By plotting the trajectories $(x(t), y(t))$ for various values of C_1 and C_2, we obtain the phase portrait of (19) shown in Figure 8.4. Because the real part of λ_1 is $5 > 0$, $e^{5t} \to \infty$ as $t \to \infty$. This is why the arrowheads in Figure 8.4 point away from the origin; a particle on any trajectory spirals away from the origin as $t \to \infty$. The origin is a repeller.

The process by which we obtained the real solutions in (21) can be generalized. Let $\mathbf{K}_1$ be an eigenvector of the coefficient matrix $\mathbf{A}$ (with real entries)

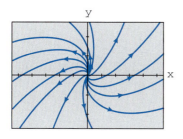

FIGURE 8.4   A phase portrait of system (19)

corresponding to the complex eigenvalue $\lambda_1 = \alpha + i\beta$. Then the solution vectors in Theorem 8.8 can be written as

$$\mathbf{K}_1 e^{\lambda_1 t} = \mathbf{K}_1 e^{\alpha t} e^{i\beta t} = \mathbf{K}_1 e^{\alpha t}(\cos \beta t + i \sin \beta t)$$
$$\overline{\mathbf{K}}_1 e^{\overline{\lambda}_1 t} = \overline{\mathbf{K}}_1 e^{\alpha t} e^{-i\beta t} = \overline{\mathbf{K}}_1 e^{\alpha t}(\cos \beta t - i \sin \beta t).$$

By the superposition principle, Theorem 8.2, the following vectors are also solutions:

$$\mathbf{X}_1 = \frac{1}{2}(\mathbf{K}_1 e^{\lambda_1 t} + \overline{\mathbf{K}}_1 e^{\overline{\lambda}_1 t}) = \frac{1}{2}(\mathbf{K}_1 + \overline{\mathbf{K}}_1) e^{\alpha t} \cos \beta t - \frac{i}{2}(-\mathbf{K}_1 + \overline{\mathbf{K}}_1) e^{\alpha t} \sin \beta t$$

$$\mathbf{X}_2 = \frac{i}{2}(-\mathbf{K}_1 e^{\lambda_1 t} + \overline{\mathbf{K}}_1 e^{\overline{\lambda}_1 t}) = \frac{i}{2}(-\mathbf{K}_1 + \overline{\mathbf{K}}_1) e^{\alpha t} \cos \beta t + \frac{1}{2}(\mathbf{K}_1 + \overline{\mathbf{K}}_1) e^{\alpha t} \sin \beta t.$$

Both $\frac{1}{2}(z + \overline{z}) = a$ and $\frac{i}{2}(-z + \overline{z}) = b$ are *real* numbers for *any* complex number $z = a + ib$. Therefore, the entries in the column vectors $\frac{1}{2}(\mathbf{K}_1 + \overline{\mathbf{K}}_1)$ and $\frac{i}{2}(-\mathbf{K}_1 + \overline{\mathbf{K}}_1)$ are real numbers. By defining

$$\mathbf{B}_1 = \frac{1}{2}(\mathbf{K}_1 + \overline{\mathbf{K}}_1) \quad \text{and} \quad \mathbf{B}_2 = \frac{i}{2}(-\mathbf{K}_1 + \overline{\mathbf{K}}_1), \tag{22}$$

we are led to the following theorem.

THEOREM 8.9   **Real Solutions Corresponding to a Complex Eigenvalue**

Let $\lambda_1 = \alpha + i\beta$ be a complex eigenvalue of the coefficient matrix $\mathbf{A}$ in the homogeneous system (2), and let $\mathbf{B}_1$ and $\mathbf{B}_2$ denote the column vectors defined in (22). Then

$$\mathbf{X}_1 = [\mathbf{B}_1 \cos \beta t - \mathbf{B}_2 \sin \beta t] e^{\alpha t}$$
$$\mathbf{X}_2 = [\mathbf{B}_2 \cos \beta t + \mathbf{B}_1 \sin \beta t] e^{\alpha t} \tag{23}$$

are linearly independent solutions of (2) on $(-\infty, \infty)$.

The matrices $\mathbf{B}_1$ and $\mathbf{B}_2$ in (22) are often denoted by

$$\mathbf{B}_1 = \text{Re}(\mathbf{K}_1) \quad \text{and} \quad \mathbf{B}_2 = \text{Im}(\mathbf{K}_1) \tag{24}$$

since these vectors are, respectively, the *real* and *imaginary* parts of the eigenvector $\mathbf{K}_1$. For example, (21) follows from (23) with

$$\mathbf{K}_1 = \begin{pmatrix} 1 \\ 1 - 2i \end{pmatrix} = \begin{pmatrix} 1 \\ 1 \end{pmatrix} + i \begin{pmatrix} 0 \\ -2 \end{pmatrix},$$

$$\mathbf{B}_1 = \text{Re}(\mathbf{K}_1) = \begin{pmatrix} 1 \\ 1 \end{pmatrix} \quad \text{and} \quad \mathbf{B}_2 = \text{Im}(\mathbf{K}_1) = \begin{pmatrix} 0 \\ -2 \end{pmatrix}.$$

EXAMPLE 6   **Complex Eigenvalues**

Solve the initial-value problem

$$\mathbf{X}' = \begin{pmatrix} 2 & 8 \\ -1 & -2 \end{pmatrix} \mathbf{X}, \quad \mathbf{X}(0) = \begin{pmatrix} 2 \\ -1 \end{pmatrix}. \tag{25}$$

SOLUTION First we obtain the eigenvalues from

$$\det(\mathbf{A} - \lambda\mathbf{I}) = \begin{vmatrix} 2 - \lambda & 8 \\ -1 & -2 - \lambda \end{vmatrix} = \lambda^2 + 4 = 0.$$

The eigenvalues are $\lambda_1 = 2i$ and $\lambda_2 = \overline{\lambda_1} = -2i$. For λ_1, the system

$$(2 - 2i)\, k_1 + 8k_2 = 0$$

$$-k_1 + (-2 - 2i)k_2 = 0$$

gives $k_1 = -(2 + 2i)k_2$. By choosing $k_2 = -1$, we get

$$\mathbf{K}_1 = \begin{pmatrix} 2 + 2i \\ -1 \end{pmatrix} = \begin{pmatrix} 2 \\ -1 \end{pmatrix} + i\begin{pmatrix} 2 \\ 0 \end{pmatrix}.$$

Now from (24) we form

$$\mathbf{B}_1 = \text{Re}(\mathbf{K}_1) = \begin{pmatrix} 2 \\ -1 \end{pmatrix} \quad \text{and} \quad \mathbf{B}_2 = \text{Im}(\mathbf{K}_1) = \begin{pmatrix} 2 \\ 0 \end{pmatrix}.$$

Since $\alpha = 0$, it follows from (23) that the general solution of the system is

$$\mathbf{X} = c_1\left[\begin{pmatrix} 2 \\ -1 \end{pmatrix}\cos 2t - \begin{pmatrix} 2 \\ 0 \end{pmatrix}\sin 2t\right] + c_2\left[\begin{pmatrix} 2 \\ 0 \end{pmatrix}\cos 2t + \begin{pmatrix} 2 \\ -1 \end{pmatrix}\sin 2t\right]$$

$$= c_1\begin{pmatrix} 2\cos 2t - 2\sin 2t \\ -\cos 2t \end{pmatrix} + c_2\begin{pmatrix} 2\cos 2t + 2\sin 2t \\ -\sin 2t \end{pmatrix}. \tag{26}$$

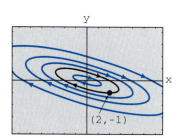

FIGURE 8.5 A phase portrait of system (25)

Some graphs of the curves or trajectories defined by solution (26) of the system are illustrated in the phase portrait in Figure 8.5. Now the initial condition $\mathbf{X}(0) = \begin{pmatrix} 2 \\ -1 \end{pmatrix}$ or, equivalently, $x(0) = 2$ and $y(0) = -1$ yields the algebraic system $2c_1 + 2c_2 = 2$, $-c_1 = -1$, whose solution is $c_1 = 1$, $c_2 = 0$. Thus the solution to the problem is $\mathbf{X} = \begin{pmatrix} 2\cos 2t - 2\sin 2t \\ -\cos 2t \end{pmatrix}$. The specific trajectory defined parametrically by the particular solution $x = 2\cos 2t - 2\sin 2t$, $y = -\cos 2t$ is the black curve in Figure 8.5. Note that this curve passes through $(2, -1)$.

REMARKS

In this section we have examined exclusively homogeneous first-order systems of linear equations in normal form $\mathbf{X}' = \mathbf{A}\mathbf{X}$. But often the mathematical model of a dynamical physical system is a homogeneous second-order system whose normal form is $\mathbf{X}'' = \mathbf{A}\mathbf{X}$. For example, the model for the coupled springs in (1) of Section 7.6,

$$m_1 x_1'' = -k_1 x_1 + k_2(x_2 - x_1)$$

$$m_2 x_2'' = -k_2(x_2 - x_1), \tag{27}$$

can be written as $\quad\quad \mathbf{M}\mathbf{X}'' = \mathbf{K}\mathbf{X},$
where

$$\mathbf{M} = \begin{pmatrix} m_1 & 0 \\ 0 & m_2 \end{pmatrix}, \quad \mathbf{K} = \begin{pmatrix} -k_1 - k_2 & k_2 \\ k_2 & -k_2 \end{pmatrix}, \quad \text{and} \quad \mathbf{X} = \begin{pmatrix} x_1(t) \\ x_2(t) \end{pmatrix}.$$

Since $\mathbf{M}$ is nonsingular, we can solve for $\mathbf{X}''$ as $\mathbf{X}'' = \mathbf{AX}$, where $\mathbf{A} = \mathbf{M}^{-1}\mathbf{K}$. Thus (27) is equivalent to

$$\mathbf{X}'' = \begin{pmatrix} -\dfrac{k_1}{m_1} - \dfrac{k_2}{m_1} & \dfrac{k_2}{m_1} \\[2mm] \dfrac{k_2}{m_2} & -\dfrac{k_2}{m_2} \end{pmatrix} \mathbf{X}. \tag{28}$$

The methods of this section can be used to solve such a system in two ways:

- First, the original system (27) can be transformed into a first-order system by means of substitutions. If we let $x_1' = x_3$ and $x_2' = x_4$, then $x_3' = x_1''$ and $x_4' = x_2''$ and so (27) is equivalent to a system of *four* linear first-order DEs:

$$x_1' = x_3$$
$$x_2' = x_4$$
$$x_3' = -\left(\frac{k_1}{m_1} + \frac{k_2}{m_1}\right)x_1 + \frac{k_2}{m_1}x_2 \quad \text{or} \quad \mathbf{X}' = \begin{pmatrix} 0 & 0 & 1 & 0 \\ 0 & 0 & 0 & 1 \\ -\dfrac{k_1}{m_1} - \dfrac{k_2}{m_1} & \dfrac{k_2}{m_1} & 0 & 0 \\ \dfrac{k_2}{m_2} & -\dfrac{k_2}{m_2} & 0 & 0 \end{pmatrix} \mathbf{X}. \tag{29}$$
$$x_4' = \frac{k_2}{m_2}x_1 - \frac{k_2}{m_2}x_2$$

By finding the eigenvalues and eigenvectors of the coefficient matrix $\mathbf{A}$ in (29), we see that the solution of this first-order system gives the complete state of the physical system—the positions of the masses relative to the equilibrium positions (x_1 and x_2) as well as the velocities of the masses (x_3 and x_4) at time t. See Problem 48(a) in Exercises 8.2.

- Second, because (27) describes free undamped motion, it can be argued that real-valued solutions of the second-order system (28) will have the form

$$\mathbf{X} = \mathbf{V} \cos \omega t \quad \text{and} \quad \mathbf{X} = \mathbf{V} \sin \omega t, \tag{30}$$

where $\mathbf{V}$ is a column matrix of constants. Substituting either of the functions in (30) into $\mathbf{X}'' = \mathbf{AX}$ yields $(\mathbf{A} + \omega^2\mathbf{I})\mathbf{V} = \mathbf{0}$. (Verify.) By identification with (3) of this section we conclude that $\lambda = -\omega^2$ represents an eigenvalue and $\mathbf{V}$ a corresponding eigenvector of $\mathbf{A}$. It can be shown that the eigenvalues $\lambda_i = -\omega_i^2$, $i = 1, 2$ of $\mathbf{A}$ are negative, and so $\omega_i = \sqrt{-\lambda_i}$ is a real number and represents a (circular) frequency of vibration (see (4) of Section 7.6). By superposition of solutions, the general solution of (28) is then

$$\mathbf{X} = c_1\mathbf{V}_1 \cos \omega_1 t + c_2\mathbf{V}_1 \sin \omega_1 t + c_3\mathbf{V}_2 \cos \omega_2 t + c_4\mathbf{V}_2 \sin \omega_2 t$$
$$= (c_1 \cos \omega_1 t + c_2 \sin \omega_1 t)\mathbf{V}_1 + (c_3 \cos \omega_2 t + c_4 \sin \omega_2 t)\mathbf{V}_2, \tag{31}$$

where $\mathbf{V}_1$ and $\mathbf{V}_2$ are, in turn, real eigenvectors of $\mathbf{A}$ corresponding to λ_1 and λ_2.

The result given in (31) generalizes. If $-\omega_1^2, -\omega_2^2, \cdots, -\omega_n^2$ are distinct negative eigenvalues and $\mathbf{V}_1, \mathbf{V}_2, \ldots, \mathbf{V}_n$ are corresponding real eigenvectors of the $n \times n$ coefficient matrix $\mathbf{A}$, then the homogeneous second-order system $\mathbf{X}'' = \mathbf{AX}$ has the general solution

$$\mathbf{X} = \sum_{i=1}^{n} (a_i \cos \omega_i t + b_i \sin \omega_i t)\mathbf{V}_i, \tag{32}$$

where a_i and b_i represent arbitrary constants. See Problem 48(b) in Exercises 8.2.

Answers to selected odd-numbered problems begin on page ANS-11.

8.2.1 DISTINCT REAL EIGENVALUES

In Problems 1–12 find the general solution of the given system.

1. $\dfrac{dx}{dt} = x + 2y$

$\dfrac{dy}{dt} = 4x + 3y$

2. $\dfrac{dx}{dt} = 2x + 2y$

$\dfrac{dy}{dt} = x + 3y$

3. $\dfrac{dx}{dt} = -4x + 2y$

$\dfrac{dy}{dt} = -\dfrac{5}{2}x + 2y$

4. $\dfrac{dx}{dt} = -\dfrac{5}{2}x + 2y$

$\dfrac{dy}{dt} = \dfrac{3}{4}x - 2y$

5. $\mathbf{X}' = \begin{pmatrix} 10 & -5 \\ 8 & -12 \end{pmatrix}\mathbf{X}$

6. $\mathbf{X}' = \begin{pmatrix} -6 & 2 \\ -3 & 1 \end{pmatrix}\mathbf{X}$

7. $\dfrac{dx}{dt} = x + y - z$

$\dfrac{dy}{dt} = 2y$

$\dfrac{dz}{dt} = y - z$

8. $\dfrac{dx}{dt} = 2x - 7y$

$\dfrac{dy}{dt} = 5x + 10y + 4z$

$\dfrac{dz}{dt} = 5y + 2z$

9. $\mathbf{X}' = \begin{pmatrix} -1 & 1 & 0 \\ 1 & 2 & 1 \\ 0 & 3 & -1 \end{pmatrix}\mathbf{X}$

10. $\mathbf{X}' = \begin{pmatrix} 1 & 0 & 1 \\ 0 & 1 & 0 \\ 1 & 0 & 1 \end{pmatrix}\mathbf{X}$

11. $\mathbf{X}' = \begin{pmatrix} -1 & -1 & 0 \\ \frac{3}{4} & -\frac{3}{2} & 3 \\ \frac{1}{8} & \frac{1}{4} & -\frac{1}{2} \end{pmatrix}\mathbf{X}$

12. $\mathbf{X}' = \begin{pmatrix} -1 & 4 & 2 \\ 4 & -1 & -2 \\ 0 & 0 & 6 \end{pmatrix}\mathbf{X}$

In Problems 13 and 14 solve the given initial-value problem.

13. $\mathbf{X}' = \begin{pmatrix} \frac{1}{2} & 0 \\ 1 & -\frac{1}{2} \end{pmatrix}\mathbf{X}, \quad \mathbf{X}(0) = \begin{pmatrix} 3 \\ 5 \end{pmatrix}$

14. $\mathbf{X}' = \begin{pmatrix} 1 & 1 & 4 \\ 0 & 2 & 0 \\ 1 & 1 & 1 \end{pmatrix}\mathbf{X}, \quad \mathbf{X}(0) = \begin{pmatrix} 1 \\ 3 \\ 0 \end{pmatrix}$

COMPUTER LAB ASSIGNMENTS

In Problems 15 and 16 use a CAS or linear algebra software as an aid in finding the general solution of the given system.

15. $\mathbf{X}' = \begin{pmatrix} 0.9 & 2.1 & 3.2 \\ 0.7 & 6.5 & 4.2 \\ 1.1 & 1.7 & 3.4 \end{pmatrix}\mathbf{X}$

16. $\mathbf{X}' = \begin{pmatrix} 1 & 0 & 2 & -1.8 & 0 \\ 0 & 5.1 & 0 & -1 & 3 \\ 1 & 2 & -3 & 0 & 0 \\ 0 & 1 & -3.1 & 4 & 0 \\ -2.8 & 0 & 0 & 1.5 & 1 \end{pmatrix}\mathbf{X}$

17. (a) Use computer software to obtain the phase portrait of the system in Problem 5. If possible, include arrowheads as in Figure 8.2. Also include four half-lines in your phase portrait.

(b) Obtain the Cartesian equations of each of the four half-lines in part (a).

(c) Draw the eigenvectors on your phase portrait of the system.

18. Find phase portraits for the systems in Problems 2 and 4. For each system find any half-line trajectories and include these lines in your phase portrait.

8.2.2 REPEATED EIGENVALUES

In Problems 19–28 find the general solution of the given system.

19. $\dfrac{dx}{dt} = 3x - y$

$\dfrac{dy}{dt} = 9x - 3y$

20. $\dfrac{dx}{dt} = -6x + 5y$

$\dfrac{dy}{dt} = -5x + 4y$

21. $\mathbf{X}' = \begin{pmatrix} -1 & 3 \\ -3 & 5 \end{pmatrix}\mathbf{X}$

22. $\mathbf{X}' = \begin{pmatrix} 12 & -9 \\ 4 & 0 \end{pmatrix}\mathbf{X}$

23. $\dfrac{dx}{dt} = 3x - y - z$

$\dfrac{dy}{dt} = x + y - z$

$\dfrac{dz}{dt} = x - y + z$

24. $\dfrac{dx}{dt} = 3x + 2y + 4z$

$\dfrac{dy}{dt} = 2x + 2z$

$\dfrac{dz}{dt} = 4x + 2y + 3z$

25. $\mathbf{X}' = \begin{pmatrix} 5 & -4 & 0 \\ 1 & 0 & 2 \\ 0 & 2 & 5 \end{pmatrix}\mathbf{X}$

26. $\mathbf{X}' = \begin{pmatrix} 1 & 0 & 0 \\ 0 & 3 & 1 \\ 0 & -1 & 1 \end{pmatrix}\mathbf{X}$

27. $\mathbf{X}' = \begin{pmatrix} 1 & 0 & 0 \\ 2 & 2 & -1 \\ 0 & 1 & 0 \end{pmatrix} \mathbf{X}$ **28.** $\mathbf{X}' = \begin{pmatrix} 4 & 1 & 0 \\ 0 & 4 & 1 \\ 0 & 0 & 4 \end{pmatrix} \mathbf{X}$

In Problems 29 and 30 solve the given initial-value problem.

29. $\mathbf{X}' = \begin{pmatrix} 2 & 4 \\ -1 & 6 \end{pmatrix} \mathbf{X}, \quad \mathbf{X}(0) = \begin{pmatrix} -1 \\ 6 \end{pmatrix}$

30. $\mathbf{X}' = \begin{pmatrix} 0 & 0 & 1 \\ 0 & 1 & 0 \\ 1 & 0 & 0 \end{pmatrix} \mathbf{X}, \quad \mathbf{X}(0) = \begin{pmatrix} 1 \\ 2 \\ 5 \end{pmatrix}$

31. Show that the 5 × 5 matrix

$$\mathbf{A} = \begin{pmatrix} 2 & 1 & 0 & 0 & 0 \\ 0 & 2 & 0 & 0 & 0 \\ 0 & 0 & 2 & 0 & 0 \\ 0 & 0 & 0 & 2 & 1 \\ 0 & 0 & 0 & 0 & 2 \end{pmatrix}$$

has an eigenvalue λ_1 of multiplicity 5. Show that three linearly independent eigenvectors corresponding to λ_1 can be found.

COMPUTER LAB ASSIGNMENTS

32. Find phase portraits for the systems in Problems 20 and 21. For each system find any half-line trajectories and include these lines in your phase portrait.

8.2.3 COMPLEX EIGENVALUES

In Problems 33–44 find the general solution of the given system.

33. $\dfrac{dx}{dt} = 6x - y$

    $\dfrac{dy}{dt} = 5x + 2y$

34. $\dfrac{dx}{dt} = x + y$

    $\dfrac{dy}{dt} = -2x - y$

35. $\dfrac{dx}{dt} = 5x + y$

    $\dfrac{dy}{dt} = -2x + 3y$

36. $\dfrac{dx}{dt} = 4x + 5y$

    $\dfrac{dy}{dt} = -2x + 6y$

37. $\mathbf{X}' = \begin{pmatrix} 4 & -5 \\ 5 & -4 \end{pmatrix} \mathbf{X}$ **38.** $\mathbf{X}' = \begin{pmatrix} 1 & -8 \\ 1 & -3 \end{pmatrix} \mathbf{X}$

39. $\dfrac{dx}{dt} = z$

    $\dfrac{dy}{dt} = -z$

    $\dfrac{dz}{dt} = y$

40. $\dfrac{dx}{dt} = 2x + y + 2z$

    $\dfrac{dy}{dt} = 3x + 6z$

    $\dfrac{dz}{dt} = -4x - 3z$

41. $\mathbf{X}' = \begin{pmatrix} 1 & -1 & 2 \\ -1 & 1 & 0 \\ -1 & 0 & 1 \end{pmatrix} \mathbf{X}$ **42.** $\mathbf{X}' = \begin{pmatrix} 4 & 0 & 1 \\ 0 & 6 & 0 \\ -4 & 0 & 4 \end{pmatrix} \mathbf{X}$

43. $\mathbf{X}' = \begin{pmatrix} 2 & 5 & 1 \\ -5 & -6 & 4 \\ 0 & 0 & 2 \end{pmatrix} \mathbf{X}$ **44.** $\mathbf{X}' = \begin{pmatrix} 2 & 4 & 4 \\ -1 & -2 & 0 \\ -1 & 0 & -2 \end{pmatrix} \mathbf{X}$

In Problems 45 and 46 solve the given initial-value problem.

45. $\mathbf{X}' = \begin{pmatrix} 1 & -12 & -14 \\ 1 & 2 & -3 \\ 1 & 1 & -2 \end{pmatrix} \mathbf{X}, \quad \mathbf{X}(0) = \begin{pmatrix} 4 \\ 6 \\ -7 \end{pmatrix}$

46. $\mathbf{X}' = \begin{pmatrix} 6 & -1 \\ 5 & 4 \end{pmatrix} \mathbf{X}, \quad \mathbf{X}(0) = \begin{pmatrix} -2 \\ 8 \end{pmatrix}$

COMPUTER LAB ASSIGNMENTS

47. Find phase portraits for the systems in Problems 36, 37, and 38.

48. (a) Solve (2) of Section 7.6 using the first method outlined in the *Remarks* (page 349)—that is, express (2) of Section 7.6 as a first-order system of four linear equations. Use a CAS or linear algebra software as an aid in finding eigenvalues and eigenvectors of a 4 × 4 matrix. Then apply the initial conditions to your general solution to obtain (4) of Section 7.6.

    (b) Solve (2) of Section 7.6 using the second method outlined in the *Remarks*—that is, express (2) of Section 7.6 as a second-order system of two linear equations. Assume solutions of the form $\mathbf{X} = \mathbf{V} \sin \omega t$ and $\mathbf{X} = \mathbf{V} \cos \omega t$. Find the eigenvalues and eigenvectors of a 2 × 2 matrix. As in part (a), obtain (4) of Section 7.6.

DISCUSSION/PROJECT PROBLEMS

49. Solve each of the following linear systems.

    (a) $\mathbf{X}' = \begin{pmatrix} 1 & 1 \\ 1 & 1 \end{pmatrix} \mathbf{X}$ **(b)** $\mathbf{X}' = \begin{pmatrix} 1 & 1 \\ -1 & -1 \end{pmatrix} \mathbf{X}$

Find a phase portrait of each system. What is the geometric significance of the line $y = -x$ in each portrait?

50. Consider the 5 × 5 matrix given in Problem 31. Solve the system $\mathbf{X}' = \mathbf{A}\mathbf{X}$ without the aid of matrix methods, but write the general solution using matrix notation. Use the general solution as a basis for a discussion of how the system can be solved using the matrix methods of this section. Carry out your ideas.

51. Obtain a Cartesian equation of the curve defined parametrically by the solution of the linear system in Example 6. Identify the curve passing through $(2, -1)$ in Figure 8.5 (*Hint:* Compute x^2, y^2, and xy.)

52. Examine your phase portraits in Problem 47. Under what conditions will the phase portrait of a 2×2 homogeneous linear system with complex eigenvalues consist of a family of closed curves? consist of a family of spirals? Under what conditions is the origin $(0, 0)$ a repeller? An attractor?

8.3 NONHOMOGENEOUS LINEAR SYSTEMS

INTRODUCTION: The methods of **undetermined coefficients** and **variation of parameters** used in Chapter 4 to find particular solutions of nonhomogeneous linear ODEs can both be adapted to the solution of nonhomogeneous linear systems. Of the two methods, variation of parameters is the more powerful technique. However, there are instances when the method of undetermined coefficients provides a quick means of finding a particular solution.

REVIEW MATERIAL: It might be beneficial to review the basic idea of the version of undetermined coefficients presented in Section 4.4 as well the method of variation of parameters in Section 4.6.

In Section 8.1 we saw that the general solution of a nonhomogeneous linear system $\mathbf{X}' = \mathbf{AX} + \mathbf{F}(t)$ on an interval I is $\mathbf{X} = \mathbf{X}_c + \mathbf{X}_p$ where $\mathbf{X}_c = c_1\mathbf{X}_1 + c_2\mathbf{X}_2 + \cdots + c_n\mathbf{X}_n$ is the complementary function or general solution of the associated homogeneous linear system $\mathbf{X}' = \mathbf{AX}$ and $\mathbf{X}_p$ is any particular solution of the nonhomogeneous system. We just saw how to obtain $\mathbf{X}_c$ in the preceding section when $\mathbf{A}$ was an $n \times n$ matrix of constants; we now consider two methods for obtaining $\mathbf{X}_p$.

8.3.1 UNDETERMINED COEFFICIENTS

THE ASSUMPTIONS As in Section 4.4, the **method of undetermined coefficients** consists of making an educated guess about the form of a particular solution vector $\mathbf{X}_p$; the guess is motivated by the types of functions that make up the entries of the column matrix $\mathbf{F}(t)$. Not surprisingly, the matrix version of undetermined coefficients is applicable to $\mathbf{X}' = \mathbf{AX} + \mathbf{F}(t)$ only when the entries of $\mathbf{A}$ are constants and the entries of $\mathbf{F}(t)$ are constants, polynomials, exponential functions, sines and cosines, or finite sums and products of these functions.

EXAMPLE 1   **Undetermined Coefficients**

Solve the system $\mathbf{X}' = \begin{pmatrix} -1 & 2 \\ -1 & 1 \end{pmatrix}\mathbf{X} + \begin{pmatrix} -8 \\ 3 \end{pmatrix}$ on $(-\infty, \infty)$.

SOLUTION   We first solve the associated homogeneous system

$$\mathbf{X}' = \begin{pmatrix} -1 & 2 \\ -1 & 1 \end{pmatrix}\mathbf{X}.$$

The characteristic equation of the coefficient matrix $\mathbf{A}$,

$$\det(\mathbf{A} - \lambda \mathbf{I}) = \begin{vmatrix} -1 - \lambda & 2 \\ -1 & 1 - \lambda \end{vmatrix} = \lambda^2 + 1 = 0,$$

yields the complex eigenvalues $\lambda_1 = i$ and $\lambda_2 = \overline{\lambda}_1 = -i$. By the procedures of Section 8.2 we find

$$\mathbf{X}_c = c_1 \begin{pmatrix} \cos t + \sin t \\ \cos t \end{pmatrix} + c_2 \begin{pmatrix} \cos t - \sin t \\ -\sin t \end{pmatrix}.$$

Now since $\mathbf{F}(t)$ is a constant vector, we assume a constant particular solution vector $\mathbf{X}_p = \begin{pmatrix} a_1 \\ b_1 \end{pmatrix}$. Substituting this latter assumption into the original system and equating entries leads to

$$0 = -a_1 + 2b_1 - 8$$
$$0 = -a_1 + b_1 + 3.$$

Solving this algebraic system gives $a_1 = 14$ and $b_1 = 11$, and so a particular solution is $\mathbf{X}_p = \begin{pmatrix} 14 \\ 11 \end{pmatrix}$. The general solution of the original system of DEs on the interval $(-\infty, \infty)$ is then $\mathbf{X} = \mathbf{X}_c + \mathbf{X}_p$ or

$$\mathbf{X} = c_1 \begin{pmatrix} \cos t + \sin t \\ \cos t \end{pmatrix} + c_2 \begin{pmatrix} \cos t - \sin t \\ -\sin t \end{pmatrix} + \begin{pmatrix} 14 \\ 11 \end{pmatrix}.$$

EXAMPLE 2   Undetermined Coefficients

Solve the system $\mathbf{X}' = \begin{pmatrix} 6 & 1 \\ 4 & 3 \end{pmatrix} \mathbf{X} + \begin{pmatrix} 6t \\ -10t + 4 \end{pmatrix}$ on $(-\infty, \infty)$.

SOLUTION   The eigenvalues and corresponding eigenvectors of the associated homogeneous system $\mathbf{X}' = \begin{pmatrix} 6 & 1 \\ 4 & 3 \end{pmatrix} \mathbf{X}$ are found to be $\lambda_1 = 2, \lambda_2 = 7, \mathbf{K}_1 = \begin{pmatrix} 1 \\ -4 \end{pmatrix}$, and $\mathbf{K}_2 = \begin{pmatrix} 1 \\ 1 \end{pmatrix}$. Hence the complementary function is

$$\mathbf{X}_c = c_1 \begin{pmatrix} 1 \\ -4 \end{pmatrix} e^{2t} + c_2 \begin{pmatrix} 1 \\ 1 \end{pmatrix} e^{7t}.$$

Now because $\mathbf{F}(t)$ can be written $\mathbf{F}(t) = \begin{pmatrix} 6 \\ -10 \end{pmatrix} t + \begin{pmatrix} 0 \\ 4 \end{pmatrix}$ we shall try to find a particular solution of the system that possesses the *same* form:

$$\mathbf{X}_p = \begin{pmatrix} a_2 \\ b_2 \end{pmatrix} t + \begin{pmatrix} a_1 \\ b_1 \end{pmatrix}.$$

Substituting this last assumption into the given system yields

$$\begin{pmatrix} a_2 \\ b_2 \end{pmatrix} = \begin{pmatrix} 6 & 1 \\ 4 & 3 \end{pmatrix} \left[\begin{pmatrix} a_2 \\ b_2 \end{pmatrix} t + \begin{pmatrix} a_1 \\ b_1 \end{pmatrix} \right] + \begin{pmatrix} 6 \\ -10 \end{pmatrix} t + \begin{pmatrix} 0 \\ 4 \end{pmatrix}$$

or

$$\begin{pmatrix} 0 \\ 0 \end{pmatrix} = \begin{pmatrix} (6a_2 + b_2 + 6)t + 6a_1 + b_1 - a_2 \\ (4a_2 + 3b_2 - 10)t + 4a_1 + 3b_1 - b_2 + 4 \end{pmatrix}.$$

From the last identity we obtain four algebraic equations in four unknowns

$$6a_2 + b_2 + 6 = 0 \qquad 6a_1 + b_1 - a_2 = 0$$
$$\qquad\qquad\qquad\text{and}$$
$$4a_2 + 3b_2 - 10 = 0 \qquad 4a_1 + 3b_1 - b_2 + 4 = 0.$$

Solving the first two equations simultaneously yields $a_2 = -2$ and $b_2 = 6$. We then substitute these values into the last two equations and solve for a_1 and b_1. The results are $a_1 = -\frac{4}{7}$, $b_1 = \frac{10}{7}$. It follows, therefore, that a particular solution vector is

$$\mathbf{X}_p = \begin{pmatrix} -2 \\ 6 \end{pmatrix} t + \begin{pmatrix} -\frac{4}{7} \\ \frac{10}{7} \end{pmatrix}.$$

The general solution of the system on $(-\infty, \infty)$ is $\mathbf{X} = \mathbf{X}_c + \mathbf{X}_p$ or

$$\mathbf{X} = c_1 \begin{pmatrix} 1 \\ -4 \end{pmatrix} e^{2t} + c_2 \begin{pmatrix} 1 \\ 1 \end{pmatrix} e^{7t} + \begin{pmatrix} -2 \\ 6 \end{pmatrix} t + \begin{pmatrix} -\frac{4}{7} \\ \frac{10}{7} \end{pmatrix}.$$

EXAMPLE 3   **Form of $\mathbf{X}_p$**

Determine the form of a particular solution vector $\mathbf{X}_p$ for the system

$$\frac{dx}{dt} = 5x + 3y - 2e^{-t} + 1$$

$$\frac{dy}{dt} = -x + y + e^{-t} - 5t + 7.$$

SOLUTION   Because $\mathbf{F}(t)$ can be written in matrix terms as

$$\mathbf{F}(t) = \begin{pmatrix} -2 \\ 1 \end{pmatrix} e^{-t} + \begin{pmatrix} 0 \\ -5 \end{pmatrix} t + \begin{pmatrix} 1 \\ 7 \end{pmatrix}$$

a natural assumption for a particular solution would be

$$\mathbf{X}_p = \begin{pmatrix} a_3 \\ b_3 \end{pmatrix} e^{-t} + \begin{pmatrix} a_2 \\ b_2 \end{pmatrix} t + \begin{pmatrix} a_1 \\ b_1 \end{pmatrix}.$$

REMARKS

The method of undetermined coefficients for linear systems is not as straightforward as the last three examples would seem to indicate. In Section 4.4 the form of a particular solution y_p was predicated on prior knowledge of the complementary function y_c. The same is true for the formation of $\mathbf{X}_p$. But there are further difficulties: The special rules governing the form of y_p in Section 4.4 do not *quite* carry to the formation of $\mathbf{X}_p$. For example, if $\mathbf{F}(t)$ is a constant vector as in Example 1 and $\lambda = 0$ is an eigenvalue of multiplicity one, then $\mathbf{X}_c$ contains a constant vector. Under the "multiplication rule" on page 155 we would ordinarily try a particular solution of the form $\mathbf{X}_p = \begin{pmatrix} a_1 \\ b_1 \end{pmatrix} t$. This is not the proper assumption for linear systems; it should be $\mathbf{X}_p = \begin{pmatrix} a_2 \\ b_2 \end{pmatrix} t + \begin{pmatrix} a_1 \\ b_1 \end{pmatrix}$. Similarly, in Example 3, if we replace e^{-t} in $\mathbf{F}(t)$ by e^{2t} ($\lambda = 2$ is an eigenvalue), then the correct form of the particular solution vector is

$$\mathbf{X}_p = \begin{pmatrix} a_4 \\ b_4 \end{pmatrix} te^{2t} + \begin{pmatrix} a_3 \\ b_3 \end{pmatrix} e^{2t} + \begin{pmatrix} a_2 \\ b_2 \end{pmatrix} t + \begin{pmatrix} a_1 \\ b_1 \end{pmatrix}.$$

Rather than delving into these difficulties, we turn instead to the method of variation of parameters.

8.3.2 VARIATION OF PARAMETERS

A FUNDAMENTAL MATRIX If $\mathbf{X}_1, \mathbf{X}_2, \ldots, \mathbf{X}_n$ is a fundamental set of solutions of the homogeneous system $\mathbf{X}' = \mathbf{AX}$ on an interval I, then its general solution on the interval is the linear combination $\mathbf{X} = c_1\mathbf{X}_1 + c_2\mathbf{X}_2 + \cdots + c_n\mathbf{X}_n$ or

$$\mathbf{X} = c_1\begin{pmatrix} x_{11} \\ x_{21} \\ \vdots \\ x_{n1} \end{pmatrix} + c_2\begin{pmatrix} x_{12} \\ x_{22} \\ \vdots \\ x_{n2} \end{pmatrix} + \ldots + c_n\begin{pmatrix} x_{1n} \\ x_{2n} \\ \vdots \\ x_{nn} \end{pmatrix} = \begin{pmatrix} c_1x_{11} + c_2x_{12} + \cdots + c_nx_{1n} \\ c_1x_{21} + c_2x_{22} + \cdots + c_nx_{2n} \\ \vdots \\ c_1x_{n1} + c_2x_{n2} + \cdots + c_nx_{nn} \end{pmatrix}. \tag{1}$$

The last matrix in (1) is recognized as the product of an $n \times n$ matrix with an $n \times 1$ matrix. In other words, the general solution (1) can be written as the product

$$\mathbf{X} = \boldsymbol{\Phi}(t)\mathbf{C}, \tag{2}$$

where $\mathbf{C}$ is an $n \times 1$ column vector of arbitrary constants $c_1, c_2, \ldots, c_n$ and the $n \times n$ matrix, whose columns consist of the entries of the solution vectors of the system $\mathbf{X}' = \mathbf{AX}$,

$$\boldsymbol{\Phi}(t) = \begin{pmatrix} x_{11} & x_{12} & \cdots & x_{1n} \\ x_{21} & x_{22} & \cdots & x_{2n} \\ \vdots & & & \vdots \\ x_{n1} & x_{n2} & \cdots & x_{nn} \end{pmatrix},$$

is called a **fundamental matrix** of the system on the interval.

In the discussion that follows we need to use two properties of a fundamental matrix:

- A fundamental matrix $\boldsymbol{\Phi}(t)$ is nonsingular.
- If $\boldsymbol{\Phi}(t)$ is a fundamental matrix of the system $\mathbf{X}' = \mathbf{AX}$, then

$$\boldsymbol{\Phi}'(t) = \mathbf{A}\boldsymbol{\Phi}(t). \tag{3}$$

A reexamination of (9) of Theorem 8.3 shows that det $\boldsymbol{\Phi}(t)$ is the same as the Wronskian $W(\mathbf{X}_1, \mathbf{X}_2, \ldots, \mathbf{X}_n)$. Hence the linear independence of the columns of $\boldsymbol{\Phi}(t)$ on the interval I guarantees that det $\boldsymbol{\Phi}(t) \neq 0$ for every t in the interval. Since $\boldsymbol{\Phi}(t)$ is nonsingular, the multiplicative inverse $\boldsymbol{\Phi}^{-1}(t)$ exists for every t in the interval. The result given in (3) follows immediately from the fact that every column of $\boldsymbol{\Phi}(t)$ is a solution vector of $\mathbf{X}' = \mathbf{AX}$.

VARIATION OF PARAMETERS Analogous to the procedure in Section 4.6 we ask whether it is possible to replace the matrix of constants $\mathbf{C}$ in (2) by a column matrix of functions

$$\mathbf{U}(t) = \begin{pmatrix} u_1(t) \\ u_2(t) \\ \vdots \\ u_n(t) \end{pmatrix} \quad \text{so} \quad \mathbf{X}_p = \boldsymbol{\Phi}(t)\mathbf{U}(t) \tag{4}$$

is a particular solution of the nonhomogeneous system

$$\mathbf{X}' = \mathbf{AX} + \mathbf{F}(t). \tag{5}$$

By the Product Rule the derivative of the last expression in (4) is

$$\mathbf{X}'_p = \boldsymbol{\Phi}(t)\mathbf{U}'(t) + \boldsymbol{\Phi}'(t)\mathbf{U}(t). \tag{6}$$

Note that the order of the products in (6) is very important. Since $\mathbf{U}(t)$ is a column matrix, the products $\mathbf{U}'(t)\mathbf{\Phi}(t)$ and $\mathbf{U}(t)\mathbf{\Phi}'(t)$ are not defined. Substituting (4) and (6) into (5) gives

$$\mathbf{\Phi}(t)\mathbf{U}'(t) + \mathbf{\Phi}'(t)\mathbf{U}(t) = \mathbf{A}\mathbf{\Phi}(t)\mathbf{U}(t) + \mathbf{F}(t). \tag{7}$$

Now if we use (3) to replace $\mathbf{\Phi}'(t)$, (7) becomes

$$\mathbf{\Phi}(t)\mathbf{U}'(t) + \mathbf{A}\mathbf{\Phi}(t)\mathbf{U}(t) = \mathbf{A}\mathbf{\Phi}(t)\mathbf{U}(t) + \mathbf{F}(t)$$

or $$\mathbf{\Phi}(t)\mathbf{U}'(t) = \mathbf{F}(t). \tag{8}$$

Multiplying both sides of equation (8) by $\mathbf{\Phi}^{-1}(t)$ gives

$$\mathbf{U}'(t) = \mathbf{\Phi}^{-1}(t)\mathbf{F}(t) \quad \text{and so} \quad \mathbf{U}(t) = \int \mathbf{\Phi}^{-1}(t)\mathbf{F}(t) \, dt.$$

Since $\mathbf{X}_p = \mathbf{\Phi}(t)\mathbf{U}(t)$, we conclude that a particular solution of (5) is

$$\mathbf{X}_p = \mathbf{\Phi}(t) \int \mathbf{\Phi}^{-1}(t)\mathbf{F}(t) \, dt. \tag{9}$$

To calculate the indefinite integral of the column matrix $\mathbf{\Phi}^{-1}(t)\mathbf{F}(t)$ in (9), we integrate each entry. Thus the general solution of the system (5) is $\mathbf{X} = \mathbf{X}_c + \mathbf{X}_p$ or

$$\mathbf{X} = \mathbf{\Phi}(t)\mathbf{C} + \mathbf{\Phi}(t) \int \mathbf{\Phi}^{-1}(t)\mathbf{F}(t) \, dt. \tag{10}$$

Note that it is not necessary to use a constant of integration in the evaluation of $\int \mathbf{\Phi}^{-1}(t)\mathbf{F}(t) \, dt$ for the same reasons stated in the discussion of variation of parameters in Section 4.6.

EXAMPLE 4   Variation of Parameters

Solve the system

$$\mathbf{X}' = \begin{pmatrix} -3 & 1 \\ 2 & -4 \end{pmatrix}\mathbf{X} + \begin{pmatrix} 3t \\ e^{-t} \end{pmatrix} \tag{11}$$

on $(-\infty, \infty)$.

SOLUTION   We first solve the associated homogeneous system

$$\mathbf{X}' = \begin{pmatrix} -3 & 1 \\ 2 & -4 \end{pmatrix}\mathbf{X}. \tag{12}$$

The characteristic equation of the coefficient matrix is

$$\det(\mathbf{A} - \lambda\mathbf{I}) = \begin{vmatrix} -3 - \lambda & 1 \\ 2 & -4 - \lambda \end{vmatrix} = (\lambda + 2)(\lambda + 5) = 0,$$

so the eigenvalues are $\lambda_1 = -2$ and $\lambda_2 = -5$. By the usual method we find that the eigenvectors corresponding to λ_1 and λ_2 are, respectively, $\mathbf{K}_1 = \begin{pmatrix} 1 \\ 1 \end{pmatrix}$ and $\mathbf{K}_2 = \begin{pmatrix} 1 \\ -2 \end{pmatrix}$. The solution vectors of the system (11) are then

$$\mathbf{X}_1 = \begin{pmatrix} 1 \\ 1 \end{pmatrix}e^{-2t} = \begin{pmatrix} e^{-2t} \\ e^{-2t} \end{pmatrix} \quad \text{and} \quad \mathbf{X}_2 = \begin{pmatrix} 1 \\ -2 \end{pmatrix}e^{-5t} = \begin{pmatrix} e^{-5t} \\ -2e^{-5t} \end{pmatrix}.$$

The entries in $\mathbf{X}_1$ form the first column of $\mathbf{\Phi}(t)$, and the entries in $\mathbf{X}_2$ form the second column of $\mathbf{\Phi}(t)$. Hence

$$\mathbf{\Phi}(t) = \begin{pmatrix} e^{-2t} & e^{-5t} \\ e^{-2t} & -2e^{-5t} \end{pmatrix} \quad \text{and} \quad \mathbf{\Phi}^{-1}(t) = \begin{pmatrix} \frac{2}{3}e^{2t} & \frac{1}{3}e^{2t} \\ \frac{1}{3}e^{5t} & -\frac{1}{3}e^{5t} \end{pmatrix}.$$

From (9) we obtain

$$\mathbf{X}_p = \mathbf{\Phi}(t) \int \mathbf{\Phi}^{-1}(t)\mathbf{F}(t)\, dt = \begin{pmatrix} e^{-2t} & e^{-5t} \\ e^{-2t} & -2e^{-5t} \end{pmatrix} \int \begin{pmatrix} \frac{2}{3}e^{2t} & \frac{1}{3}e^{2t} \\ \frac{1}{3}e^{5t} & -\frac{1}{3}e^{5t} \end{pmatrix} \begin{pmatrix} 3t \\ e^{-t} \end{pmatrix} dt$$

$$= \begin{pmatrix} e^{-2t} & e^{-5t} \\ e^{-2t} & -2e^{-5t} \end{pmatrix} \int \begin{pmatrix} 2te^{2t} + \frac{1}{3}e^{t} \\ te^{5t} - \frac{1}{3}e^{4t} \end{pmatrix} dt$$

$$= \begin{pmatrix} e^{-2t} & e^{-5t} \\ e^{-2t} & -2e^{-5t} \end{pmatrix} \begin{pmatrix} te^{2t} - \frac{1}{2}e^{2t} + \frac{1}{3}e^{t} \\ \frac{1}{5}te^{5t} - \frac{1}{25}e^{5t} - \frac{1}{12}e^{4t} \end{pmatrix}$$

$$= \begin{pmatrix} \frac{6}{5}t - \frac{27}{50} + \frac{1}{4}e^{-t} \\ \frac{3}{5}t - \frac{21}{50} + \frac{1}{2}e^{-t} \end{pmatrix}.$$

Hence from (10) the general solution of (11) on the interval is

$$\mathbf{X} = \begin{pmatrix} e^{-2t} & e^{-5t} \\ e^{-2t} & -2e^{-5t} \end{pmatrix} \begin{pmatrix} c_1 \\ c_2 \end{pmatrix} + \begin{pmatrix} \frac{6}{5}t - \frac{27}{50} + \frac{1}{4}e^{-t} \\ \frac{3}{5}t - \frac{21}{50} + \frac{1}{2}e^{-t} \end{pmatrix}$$

$$= c_1 \begin{pmatrix} 1 \\ 1 \end{pmatrix} e^{-2t} + c_2 \begin{pmatrix} 1 \\ -2 \end{pmatrix} e^{-5t} + \begin{pmatrix} \frac{6}{5} \\ \frac{3}{5} \end{pmatrix} t - \begin{pmatrix} \frac{27}{50} \\ \frac{21}{50} \end{pmatrix} + \begin{pmatrix} \frac{1}{4} \\ \frac{1}{2} \end{pmatrix} e^{-t}. \quad \blacksquare$$

INITIAL-VALUE PROBLEM The general solution of (5) on an interval can be written in the alternative manner

$$\mathbf{X} = \mathbf{\Phi}(t)\mathbf{C} + \mathbf{\Phi}(t) \int_{t_0}^{t} \mathbf{\Phi}^{-1}(s)\mathbf{F}(s)\, ds, \quad (13)$$

where t and t_0 are points in the interval. This last form is useful in solving (5) subject to an initial condition $\mathbf{X}(t_0) = \mathbf{X}_0$, because the limits of integration are chosen so that the particular solution vanishes at $t = t_0$. Substituting $t = t_0$ into (13) yields $\mathbf{X}_0 = \mathbf{\Phi}(t_0)\mathbf{C}$ from which we get $\mathbf{C} = \mathbf{\Phi}^{-1}(t_0)\mathbf{X}_0$. Substituting this last result into (13) gives the following solution of the initial-value problem:

$$\mathbf{X} = \mathbf{\Phi}(t)\mathbf{\Phi}^{-1}(t_0)\mathbf{X}_0 + \mathbf{\Phi}(t) \int_{t_0}^{t} \mathbf{\Phi}^{-1}(s)\mathbf{F}(s)\, ds. \quad (14)$$

EXERCISES 8.3

Answers to selected odd-numbered problems begin on page ANS-12.

8.3.1 UNDETERMINED COEFFICIENTS

In Problems 1–8 use the method of undetermined coefficients to solve the given system.

1. $\dfrac{dx}{dt} = 2x + 3y - 7$

$\dfrac{dy}{dt} = -x - 2y + 5$

2. $\dfrac{dx}{dt} = 5x + 9y + 2$

$\dfrac{dy}{dt} = -x + 11y + 6$

3. $\mathbf{X}' = \begin{pmatrix} 1 & 3 \\ 3 & 1 \end{pmatrix}\mathbf{X} + \begin{pmatrix} -2t^2 \\ t+5 \end{pmatrix}$

4. $\mathbf{X}' = \begin{pmatrix} 1 & -4 \\ 4 & 1 \end{pmatrix}\mathbf{X} + \begin{pmatrix} 4t + 9e^{6t} \\ -t + e^{6t} \end{pmatrix}$

5. $\mathbf{X}' = \begin{pmatrix} 4 & \frac{1}{3} \\ 9 & 6 \end{pmatrix}\mathbf{X} + \begin{pmatrix} -3 \\ 10 \end{pmatrix}e^{t}$

6. $\mathbf{X}' = \begin{pmatrix} -1 & 5 \\ -1 & 1 \end{pmatrix}\mathbf{X} + \begin{pmatrix} \sin t \\ -2\cos t \end{pmatrix}$

7. $\mathbf{X}' = \begin{pmatrix} 1 & 1 & 1 \\ 0 & 2 & 3 \\ 0 & 0 & 5 \end{pmatrix}\mathbf{X} + \begin{pmatrix} 1 \\ -1 \\ 2 \end{pmatrix}e^{4t}$

8. $\mathbf{X}' = \begin{pmatrix} 0 & 0 & 5 \\ 0 & 5 & 0 \\ 5 & 0 & 0 \end{pmatrix}\mathbf{X} + \begin{pmatrix} 5 \\ -10 \\ 40 \end{pmatrix}$

9. Solve $\mathbf{X}' = \begin{pmatrix} -1 & -2 \\ 3 & 4 \end{pmatrix}\mathbf{X} + \begin{pmatrix} 3 \\ 3 \end{pmatrix}$ subject to

$$\mathbf{X}(0) = \begin{pmatrix} -4 \\ 5 \end{pmatrix}.$$

10. (a) The system of differential equations for the currents $i_2(t)$ and $i_3(t)$ in the electrical network shown in Figure 8.6 is

$$\frac{d}{dt}\begin{pmatrix} i_2 \\ i_3 \end{pmatrix} = \begin{pmatrix} -R_1/L_1 & -R_1/L_1 \\ -R_1/L_2 & -(R_1 + R_2)/L_2 \end{pmatrix}\begin{pmatrix} i_2 \\ i_3 \end{pmatrix} + \begin{pmatrix} E/L_1 \\ E/L_2 \end{pmatrix}.$$

Use the method of undetermined coefficients to solve the system if $R_1 = 2\ \Omega$, $R_2 = 3\ \Omega$, $L_1 = 1$ h, $L_2 = 1$ h, $E = 60$ V, $i_2(0) = 0$, and $i_3(0) = 0$.
(b) Determine the current $i_1(t)$.

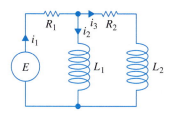

FIGURE 8.6   Network in Problem 10

8.3.2   VARIATION OF PARAMETERS

In Problems 11–30 use variation of parameters to solve the given system.

11. $\dfrac{dx}{dt} = 3x - 3y + 4$

$\dfrac{dy}{dt} = 2x - 2y - 1$

12. $\dfrac{dx}{dt} = 2x - y$

$\dfrac{dy}{dt} = 3x - 2y + 4t$

13. $\mathbf{X}' = \begin{pmatrix} 3 & -5 \\ \frac{3}{4} & -1 \end{pmatrix}\mathbf{X} + \begin{pmatrix} 1 \\ -1 \end{pmatrix}e^{t/2}$

14. $\mathbf{X}' = \begin{pmatrix} 2 & -1 \\ 4 & 2 \end{pmatrix}\mathbf{X} + \begin{pmatrix} \sin 2t \\ 2\cos 2t \end{pmatrix}e^{2t}$

15. $\mathbf{X}' = \begin{pmatrix} 0 & 2 \\ -1 & 3 \end{pmatrix}\mathbf{X} + \begin{pmatrix} 1 \\ -1 \end{pmatrix}e^{t}$

16. $\mathbf{X}' = \begin{pmatrix} 0 & 2 \\ -1 & 3 \end{pmatrix}\mathbf{X} + \begin{pmatrix} 2 \\ e^{-3t} \end{pmatrix}$

17. $\mathbf{X}' = \begin{pmatrix} 1 & 8 \\ 1 & -1 \end{pmatrix}\mathbf{X} + \begin{pmatrix} 12 \\ 12 \end{pmatrix}t$

18. $\mathbf{X}' = \begin{pmatrix} 1 & 8 \\ 1 & -1 \end{pmatrix}\mathbf{X} + \begin{pmatrix} e^{-t} \\ te^{t} \end{pmatrix}$

19. $\mathbf{X}' = \begin{pmatrix} 3 & 2 \\ -2 & -1 \end{pmatrix}\mathbf{X} + \begin{pmatrix} 2e^{-t} \\ e^{-t} \end{pmatrix}$

20. $\mathbf{X}' = \begin{pmatrix} 3 & 2 \\ -2 & -1 \end{pmatrix}\mathbf{X} + \begin{pmatrix} 1 \\ 1 \end{pmatrix}$

21. $\mathbf{X}' = \begin{pmatrix} 0 & -1 \\ 1 & 0 \end{pmatrix}\mathbf{X} + \begin{pmatrix} \sec t \\ 0 \end{pmatrix}$

22. $\mathbf{X}' = \begin{pmatrix} 1 & -1 \\ 1 & 1 \end{pmatrix}\mathbf{X} + \begin{pmatrix} 3 \\ 3 \end{pmatrix}e^{t}$

23. $\mathbf{X}' = \begin{pmatrix} 1 & -1 \\ 1 & 1 \end{pmatrix}\mathbf{X} + \begin{pmatrix} \cos t \\ \sin t \end{pmatrix}e^{t}$

24. $\mathbf{X}' = \begin{pmatrix} 2 & -2 \\ 8 & -6 \end{pmatrix}\mathbf{X} + \begin{pmatrix} 1 \\ 3 \end{pmatrix}\dfrac{e^{-2t}}{t}$

25. $\mathbf{X}' = \begin{pmatrix} 0 & 1 \\ -1 & 0 \end{pmatrix}\mathbf{X} + \begin{pmatrix} 0 \\ \sec t \tan t \end{pmatrix}$

26. $\mathbf{X}' = \begin{pmatrix} 0 & 1 \\ -1 & 0 \end{pmatrix}\mathbf{X} + \begin{pmatrix} 1 \\ \cot t \end{pmatrix}$

27. $\mathbf{X}' = \begin{pmatrix} 1 & 2 \\ -\frac{1}{2} & 1 \end{pmatrix}\mathbf{X} + \begin{pmatrix} \csc t \\ \sec t \end{pmatrix}e^{t}$

28. $\mathbf{X}' = \begin{pmatrix} 1 & -2 \\ 1 & -1 \end{pmatrix}\mathbf{X} + \begin{pmatrix} \tan t \\ 1 \end{pmatrix}$

29. $\mathbf{X}' = \begin{pmatrix} 1 & 1 & 0 \\ 1 & 1 & 0 \\ 0 & 0 & 3 \end{pmatrix}\mathbf{X} + \begin{pmatrix} e^{t} \\ e^{2t} \\ te^{3t} \end{pmatrix}$

30. $\mathbf{X}' = \begin{pmatrix} 3 & -1 & -1 \\ 1 & 1 & -1 \\ 1 & -1 & 1 \end{pmatrix}\mathbf{X} + \begin{pmatrix} 0 \\ t \\ 2e^{t} \end{pmatrix}$

In Problems 31 and 32 use (14) to solve the given initial-value problem.

31. $\mathbf{X}' = \begin{pmatrix} 3 & -1 \\ -1 & 3 \end{pmatrix}\mathbf{X} + \begin{pmatrix} 4e^{2t} \\ 4e^{4t} \end{pmatrix}$, $\mathbf{X}(0) = \begin{pmatrix} 1 \\ 1 \end{pmatrix}$

32. $\mathbf{X}' = \begin{pmatrix} 1 & -1 \\ 1 & -1 \end{pmatrix}\mathbf{X} + \begin{pmatrix} 1/t \\ 1/t \end{pmatrix}$, $\mathbf{X}(1) = \begin{pmatrix} 2 \\ -1 \end{pmatrix}$

33. The system of differential equations for the currents $i_1(t)$ and $i_2(t)$ in the electrical network shown in Figure 8.7 is

$$\frac{d}{dt}\begin{pmatrix} i_1 \\ i_2 \end{pmatrix} = \begin{pmatrix} -(R_1 + R_2)/L_2 & R_2/L_2 \\ R_2/L_1 & -R_2/L_1 \end{pmatrix}\begin{pmatrix} i_1 \\ i_2 \end{pmatrix} + \begin{pmatrix} E/L_2 \\ 0 \end{pmatrix}.$$

Use variation of parameters to solve the system if $R_1 = 8\ \Omega$, $R_2 = 3\ \Omega$, $L_1 = 1$ h, $L_2 = 1$ h, $E(t) = 100 \sin t$ V, $i_1(0) = 0$, and $i_2(0) = 0$.

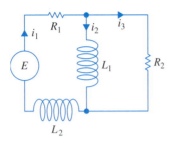

FIGURE 8.7   Network in Problem 33

DISCUSSION/PROJECT PROBLEMS

34. If y_1 and y_2 are linearly independent solutions of the associated homogeneous DE for $y'' + P(x)y' + Q(x)y = f(x)$, show in the case of a nonhomogeneous linear second-order DE that (9) reduces to the form of variation of parameters discussed in Section 4.6

COMPUTER LAB ASSIGNMENTS

35. Solving a nonhomogeneous linear system $\mathbf{X}' = \mathbf{AX} + \mathbf{F}(t)$ by variation of parameters when $\mathbf{A}$ is a 3×3 (or larger) matrix is almost an impossible task to do by hand. Consider the system

$$\mathbf{X}' = \begin{pmatrix} 2 & -2 & 2 & 1 \\ -1 & 3 & 0 & 3 \\ 0 & 0 & 4 & -2 \\ 0 & 0 & 2 & -1 \end{pmatrix} \mathbf{X} + \begin{pmatrix} te^t \\ e^{-t} \\ e^{2t} \\ 1 \end{pmatrix}.$$

(a) Use a CAS or linear algebra software to find the eigenvalues and eigenvectors of the coefficient matrix.

(b) Form a fundamental matrix $\mathbf{\Phi}(t)$ and use the computer to find $\mathbf{\Phi}^{-1}(t)$.

(c) Use the computer to carry out the computations of: $\mathbf{\Phi}^{-1}(t)\mathbf{F}(t)$, $\int \mathbf{\Phi}^{-1}(t)\mathbf{F}(t)\,dt$, $\mathbf{\Phi}(t)\int \mathbf{\Phi}^{-1}(t)\mathbf{F}(t)\,dt$, $\mathbf{\Phi}(t)\mathbf{C}$, and $\mathbf{\Phi}(t)\mathbf{C} + \int \mathbf{\Phi}^{-1}(t)\mathbf{F}(t)\,dt$, where $\mathbf{C}$ is a column matrix of constants c_1, c_2, c_3, and c_4.

(d) Rewrite the computer output for the general solution of the system in the form $\mathbf{X} = \mathbf{X}_c + \mathbf{X}_p$, where $\mathbf{X}_c = c_1\mathbf{X}_1 + c_2\mathbf{X}_2 + c_3\mathbf{X}_3 + c_4\mathbf{X}_4$.

8.4   MATRIX EXPONENTIAL

INTRODUCTION: Matrices can be used in an entirely different manner to solve a system of linear first-order differential equations. Recall that the simple linear first-order differential equation $x' = ax$, where a is constant, has the general solution $x = ce^{at}$. It seems natural then to ask whether we can define a matrix exponential function $e^{\mathbf{A}t}$, where $\mathbf{A}$ is a matrix of constants, so that $e^{\mathbf{A}t}$ is a solution of the system $\mathbf{X}' = \mathbf{AX}$.

HOMOGENEOUS SYSTEMS   We shall now see that it is possible to define a matrix exponential $e^{\mathbf{A}t}$ so that

$$\mathbf{X} = e^{\mathbf{A}t}\mathbf{C} \tag{1}$$

is a solution of the homogeneous system $\mathbf{X}' = \mathbf{AX}$. Here $\mathbf{A}$ is an $n \times n$ matrix of constants, and $\mathbf{C}$ is an $n \times 1$ column matrix of arbitrary constants. Note in (1) that the matrix $\mathbf{C}$ post multiplies $e^{\mathbf{A}t}$ because we want $e^{\mathbf{A}t}$ to be an $n \times n$ matrix. While the complete development of the meaning and theory of the matrix exponential would require a thorough knowledge of matrix algebra, one way of defining $e^{\mathbf{A}t}$ is inspired by the power series representation of the scalar exponential function e^{at}:

$$e^{at} = 1 + at + a^2\frac{t^2}{2!} + \cdots + a^k\frac{t^k}{k!} + \cdots = \sum_{k=0}^{\infty} a^k\frac{t^k}{k!}. \tag{2}$$

The series in (2) converges for all t. Using this series, with 1 replaced by the identity $\mathbf{I}$ and the constant a replaced by an $n \times n$ matrix $\mathbf{A}$ of constants, we arrive at a definition for the $n \times n$ matrix $e^{\mathbf{A}t}$.

> **DEFINITION 8.4**   **Matrix Exponential**
>
> For any $n \times n$ matrix $\mathbf{A}$,
>
> $$e^{\mathbf{A}t} = \mathbf{I} + \mathbf{A}t + \mathbf{A}^2 \frac{t^2}{2!} + \cdots + \mathbf{A}^k \frac{t^k}{k!} + \cdots = \sum_{k=0}^{\infty} \mathbf{A}^k \frac{t^k}{k!}. \qquad (3)$$

It can be shown that the series given in (3) converges to an $n \times n$ matrix for every value of t. Also, $\mathbf{A}^2 = \mathbf{A}\mathbf{A}$, $\mathbf{A}^3 = \mathbf{A}(\mathbf{A}^2)$, and so on.

DERIVATIVE OF $e^{\mathbf{A}t}$ The derivative of the matrix exponential is analogous to the differentiation property of the scalar exponential $\dfrac{d}{dt} e^{at} = ae^{at}$. To justify

$$\frac{d}{dt} e^{\mathbf{A}t} = \mathbf{A}e^{\mathbf{A}t}, \qquad (4)$$

we differentiate (3) term by term:

$$\frac{d}{dt} e^{\mathbf{A}t} = \frac{d}{dt} \left[\mathbf{I} + \mathbf{A}t + \mathbf{A}^2 \frac{t^2}{2!} + \cdots + \mathbf{A}^k \frac{t^k}{k!} + \cdots \right] = \mathbf{A} + \mathbf{A}^2 t + \frac{1}{2!} \mathbf{A}^3 t^2 + \cdots$$

$$= \mathbf{A} \left[\mathbf{I} + \mathbf{A}t + \mathbf{A}^2 \frac{t^2}{2!} + \cdots \right] = \mathbf{A}e^{\mathbf{A}t}.$$

Because of (4), we can now prove that (1) is a solution of $\mathbf{X}' = \mathbf{A}\mathbf{X}$ for every $n \times 1$ vector $\mathbf{C}$ of constants:

$$\mathbf{X}' = \frac{d}{dt} e^{\mathbf{A}t}\mathbf{C} = \mathbf{A}e^{\mathbf{A}t}\mathbf{C} = \mathbf{A}(e^{\mathbf{A}t}\mathbf{C}) = \mathbf{A}\mathbf{X}.$$

$e^{\mathbf{A}t}$ IS A FUNDAMENTAL MATRIX   If we denote the matrix exponential $e^{\mathbf{A}t}$ by the symbol $\boldsymbol{\Psi}(t)$, then (4) is equivalent to the matrix differential equation $\boldsymbol{\Psi}'(t) = \mathbf{A}\boldsymbol{\Psi}(t)$ (see (3) of Section 8.3). In addition, it follows immediately from Definition 8.4 that $\boldsymbol{\Psi}(0) = e^{\mathbf{A}0} = \mathbf{I}$, and so det $\boldsymbol{\Psi}(0) \neq 0$. It turns out that these two properties are sufficient for us to conclude that $\boldsymbol{\Psi}(t)$ is a fundamental matrix of the system $\mathbf{X}' = \mathbf{A}\mathbf{X}$.

NONHOMOGENEOUS SYSTEMS   We saw in (4) of Section 2.4 that the general solution of the single linear first-order differential equation $x' = ax + f(t)$, where a is a constant, can be expressed as

$$x = x_c + x_p = ce^{at} + e^{at} \int_{t_0}^{t} e^{-as} f(s)\, ds.$$

For a nonhomogeneous system of linear first-order differential equations, it can be shown that the general solution of $\mathbf{X}' = \mathbf{A}\mathbf{X} + \mathbf{F}(t)$, where $\mathbf{A}$ is an $n \times n$ matrix of constants, is

$$\mathbf{X} = \mathbf{X}_c + \mathbf{X}_p = e^{\mathbf{A}t}\mathbf{C} + e^{\mathbf{A}t} \int_{t_0}^{t} e^{-\mathbf{A}s} \mathbf{F}(s)\, ds. \qquad (5)$$

Since the matrix exponential $e^{\mathbf{A}t}$ is a fundamental matrix, it is always nonsingular and $e^{-\mathbf{A}s} = (e^{\mathbf{A}s})^{-1}$. In practice, $e^{-\mathbf{A}s}$ can be obtained from $e^{\mathbf{A}t}$ by simply replacing t by $-s$.

COMPUTATION OF $e^{\mathbf{A}t}$ The definition of $e^{\mathbf{A}t}$ given in (3) can, of course, always be used to compute $e^{\mathbf{A}t}$. However, the practical utility of (3) is limited by the fact that the entries in $e^{\mathbf{A}t}$ are power series in t. With a natural desire to work with simple and familiar things, we then try to recognize whether these series define a closed-form function. See Problems 1–4 in Exercises 8.4. Fortunately, there are many alternative ways of computing e^{At}; the following discussion shows how the Laplace transform can be used.

USE OF THE LAPLACE TRANSFORM We saw in (5) that $\mathbf{X} = e^{At}$ is a solution of $\mathbf{X'} = \mathbf{AX}$. Indeed, since $e^{A0} = \mathbf{I}$, $\mathbf{X} = e^{At}$ is a solution of the initial-value problem

$$\mathbf{X'} = \mathbf{AX}, \quad \mathbf{X}(0) = \mathbf{I}. \tag{6}$$

If $\mathbf{x}(s) = \mathcal{L}\{\mathbf{X}(t)\} = \mathcal{L}\{e^{At}\}$, then the Laplace transform of (6) is

$$s\mathbf{x}(s) - \mathbf{X}(0) = \mathbf{Ax}(s) \quad \text{or} \quad (s\mathbf{I} - \mathbf{A})\mathbf{x}(s) = \mathbf{I}.$$

Multiplying the last equation by $(s\mathbf{I} - \mathbf{A})^{-1}$ implies $\mathbf{x}(s) = (s\mathbf{I} - \mathbf{A})^{-1}\mathbf{I} = (s\mathbf{I} - \mathbf{A})^{-1}$. In other words, $\mathcal{L}\{e^{At}\} = (s\mathbf{I} - \mathbf{A})^{-1}$ or

$$e^{At} = \mathcal{L}^{-1}\{(s\mathbf{I} - \mathbf{A})^{-1}\}. \tag{7}$$

EXAMPLE 1 **Matrix Exponential**

Use the Laplace transform to compute e^{At} for $\mathbf{A} = \begin{pmatrix} 1 & -1 \\ 2 & -2 \end{pmatrix}$.

SOLUTION First we compute the matrix $s\mathbf{I} - \mathbf{A}$ and find its inverse:

$$s\mathbf{I} - \mathbf{A} = \begin{pmatrix} s-1 & 1 \\ -2 & s+2 \end{pmatrix},$$

$$(s\mathbf{I} - \mathbf{A})^{-1} = \begin{pmatrix} s-1 & 1 \\ -2 & s+2 \end{pmatrix}^{-1} = \begin{pmatrix} \dfrac{s+2}{s(s+1)} & \dfrac{-1}{s(s+1)} \\[2mm] \dfrac{2}{s(s+1)} & \dfrac{s-1}{s(s+1)} \end{pmatrix}.$$

Then we decompose the entries of the last matrix into partial fractions:

$$(s\mathbf{I} - \mathbf{A})^{-1} = \begin{pmatrix} \dfrac{2}{s} - \dfrac{1}{s+1} & -\dfrac{1}{s} + \dfrac{1}{s+1} \\[2mm] \dfrac{2}{s} - \dfrac{2}{s+1} & -\dfrac{1}{s} + \dfrac{2}{s+1} \end{pmatrix}. \tag{8}$$

It follows from (7) that the inverse Laplace transform of (8) gives the desired result,

$$e^{At} = \begin{pmatrix} 2 - e^{-t} & -1 + e^{-t} \\ 2 - 2e^{-t} & -1 + 2e^{-t} \end{pmatrix}.$$

USE OF COMPUTERS For those willing to momentarily trade understanding for speed of solution, e^{At} can be computed with the aid of computer software. See Problems 27 and 28 in Exercises 8.4.

EXERCISES 8.4

Answers to selected odd-numbered problems begin on page ANS-12.

In Problems 1 and 2 use (3) to compute e^{At} and e^{-At}.

1. $\mathbf{A} = \begin{pmatrix} 1 & 0 \\ 0 & 2 \end{pmatrix}$

2. $\mathbf{A} = \begin{pmatrix} 0 & 1 \\ 1 & 0 \end{pmatrix}$

In Problems 3 and 4 use (3) to compute e^{At}.

3. $\mathbf{A} = \begin{pmatrix} 1 & 1 & 1 \\ 1 & 1 & 1 \\ -2 & -2 & -2 \end{pmatrix}$

4. $\mathbf{A} = \begin{pmatrix} 0 & 0 & 0 \\ 3 & 0 & 0 \\ 5 & 1 & 0 \end{pmatrix}$

In Problems 5–8 use (1) to find the general solution of the given system.

5. $\mathbf{X}' = \begin{pmatrix} 1 & 0 \\ 0 & 2 \end{pmatrix} \mathbf{X}$

6. $\mathbf{X}' = \begin{pmatrix} 0 & 1 \\ 1 & 0 \end{pmatrix} \mathbf{X}$

7. $\mathbf{X}' = \begin{pmatrix} 1 & 1 & 1 \\ 1 & 1 & 1 \\ -2 & -2 & -2 \end{pmatrix} \mathbf{X}$

8. $\mathbf{X}' = \begin{pmatrix} 0 & 0 & 0 \\ 3 & 0 & 0 \\ 5 & 1 & 0 \end{pmatrix} \mathbf{X}$

In Problems 9–12 use (5) to find the general solution of the given system.

9. $\mathbf{X}' = \begin{pmatrix} 1 & 0 \\ 0 & 2 \end{pmatrix} \mathbf{X} + \begin{pmatrix} 3 \\ -1 \end{pmatrix}$

10. $\mathbf{X}' = \begin{pmatrix} 1 & 0 \\ 0 & 2 \end{pmatrix} \mathbf{X} + \begin{pmatrix} t \\ e^{4t} \end{pmatrix}$

11. $\mathbf{X}' = \begin{pmatrix} 0 & 1 \\ 1 & 0 \end{pmatrix} \mathbf{X} + \begin{pmatrix} 1 \\ 1 \end{pmatrix}$

12. $\mathbf{X}' = \begin{pmatrix} 0 & 1 \\ 1 & 0 \end{pmatrix} \mathbf{X} + \begin{pmatrix} \cosh t \\ \sinh t \end{pmatrix}$

13. Solve the system in Problem 7 subject to the initial condition

$$\mathbf{X}(0) = \begin{pmatrix} 1 \\ -4 \\ 6 \end{pmatrix}.$$

14. Solve the system in Problem 9 subject to the initial condition

$$\mathbf{X}(0) = \begin{pmatrix} 4 \\ 3 \end{pmatrix}.$$

In Problems 15–18 use the method of Example 1 to compute $e^{\mathbf{A}t}$ for the coefficient matrix. Use (1) to find the general solution of the given system.

15. $\mathbf{X}' = \begin{pmatrix} 4 & 3 \\ -4 & -4 \end{pmatrix} \mathbf{X}$

16. $\mathbf{X}' = \begin{pmatrix} 4 & -2 \\ 1 & 1 \end{pmatrix} \mathbf{X}$

17. $\mathbf{X}' = \begin{pmatrix} 5 & -9 \\ 1 & -1 \end{pmatrix} \mathbf{X}$

18. $\mathbf{X}' = \begin{pmatrix} 0 & 1 \\ -2 & -2 \end{pmatrix} \mathbf{X}$

Let **P** denote a matrix whose columns are eigenvectors $\mathbf{K}_1, \mathbf{K}_2, \ldots, \mathbf{K}_n$ corresponding to distinct eigenvalues $\lambda_1, \lambda_2, \ldots, \lambda_n$ of an $n \times n$ matrix **A**. Then it can be shown that $\mathbf{A} = \mathbf{PDP}^{-1}$, where **D** is defined by

$$\mathbf{D} = \begin{pmatrix} \lambda_1 & 0 & \cdots & 0 \\ 0 & \lambda_2 & \cdots & 0 \\ \vdots & & & \vdots \\ 0 & 0 & \cdots & \lambda_n \end{pmatrix}. \tag{9}$$

In Problems 19 and 20 verify the foregoing result for the given matrix.

19. $\mathbf{A} = \begin{pmatrix} 2 & 1 \\ -3 & 6 \end{pmatrix}$

20. $\mathbf{A} = \begin{pmatrix} 2 & 1 \\ 1 & 2 \end{pmatrix}$

21. Suppose $\mathbf{A} = \mathbf{PDP}^{-1}$, where **D** is defined as in (9). Use (3) to show that $e^{\mathbf{A}t} = \mathbf{P}e^{\mathbf{D}t}\mathbf{P}^{-1}$.

22. Use (3) to show that

$$e^{\mathbf{D}t} = \begin{pmatrix} e^{\lambda_1 t} & 0 & \cdots & 0 \\ 0 & e^{\lambda_2 t} & \cdots & 0 \\ \vdots & & & \vdots \\ 0 & 0 & \cdots & e^{\lambda_n t} \end{pmatrix},$$

where **D** is defined as in (9).

In Problems 23 and 24 use the results of Problems 19–22 to solve the given system.

23. $\mathbf{X}' = \begin{pmatrix} 2 & 1 \\ -3 & 6 \end{pmatrix} \mathbf{X}$

24. $\mathbf{X}' = \begin{pmatrix} 2 & 1 \\ 1 & 2 \end{pmatrix} \mathbf{X}$

DISCUSSION/PROJECT PROBLEMS

25. Reread the discussion leading to the result given in (7). Does the matrix $s\mathbf{I} - \mathbf{A}$ always have an inverse? Discuss.

26. A matrix **A** is said to be **nilpotent** if there exists some integer m such that $\mathbf{A}^m = \mathbf{0}$. Verify that $\mathbf{A} = \begin{pmatrix} -1 & 1 & 1 \\ -1 & 0 & 1 \\ -1 & 1 & 1 \end{pmatrix}$ is nilpotent. Discuss why it is relatively easy to compute $e^{\mathbf{A}t}$ when **A** is nilpotent. Compute $e^{\mathbf{A}t}$ and then use (1) to solve the system $\mathbf{X}' = \mathbf{AX}$.

COMPUTER LAB ASSIGNMENTS

27. (a) Use (1) to find the general solution of $\mathbf{X}' = \begin{pmatrix} 4 & 2 \\ 3 & 3 \end{pmatrix} \mathbf{X}$. Use a CAS to find $e^{\mathbf{A}t}$. Then use the computer to find eigenvalues and eigenvectors of the coefficient matrix $\mathbf{A} = \begin{pmatrix} 4 & 2 \\ 3 & 3 \end{pmatrix}$ and form the general solution in the manner of Section 8.2. Finally, reconcile the two forms of the general solution of the system.

(b) Use (1) to find the general solution of $\mathbf{X}' = \begin{pmatrix} -3 & -1 \\ 2 & -1 \end{pmatrix}\mathbf{X}$. Use a CAS to find $e^{\mathbf{A}t}$. In the case of complex output, utilize the software to do the simplification; for example, in *Mathematica*, if **m = MatrixExp[A t]** has complex entries, then try the command **Simplify[ComplexExpand[m]]**.

28. Use (1) to find the general solution of
$$\mathbf{X}' = \begin{pmatrix} -4 & 0 & 6 & 0 \\ 0 & -5 & 0 & -4 \\ -1 & 0 & 1 & 0 \\ 0 & 3 & 0 & 2 \end{pmatrix}\mathbf{X}.$$
Use MATLAB or a CAS to find $e^{\mathbf{A}t}$.

CHAPTER 8 IN REVIEW

Answers to selected odd-numbered problems begin on page ANS-13.

In Problems 1 and 2 fill in the blanks.

1. The vector $\mathbf{X} = k\begin{pmatrix} 4 \\ 5 \end{pmatrix}$ is a solution of
$$\mathbf{X}' = \begin{pmatrix} 1 & 4 \\ 2 & -1 \end{pmatrix}\mathbf{X} - \begin{pmatrix} 8 \\ 1 \end{pmatrix}$$
for $k = $ _____.

2. The vector $\mathbf{X} = c_1\begin{pmatrix} -1 \\ 1 \end{pmatrix}e^{-9t} + c_2\begin{pmatrix} 5 \\ 3 \end{pmatrix}e^{7t}$ is solution of the initial-value problem $\mathbf{X}' = \begin{pmatrix} 1 & 10 \\ 6 & -3 \end{pmatrix}\mathbf{X}, \mathbf{X}(0) = \begin{pmatrix} 2 \\ 0 \end{pmatrix}$ for $c_1 = $ _____ and $c_2 = $ _____.

3. Consider the linear system $\mathbf{X}' = \begin{pmatrix} 4 & 6 & 6 \\ 1 & 3 & 2 \\ -1 & -4 & -3 \end{pmatrix}\mathbf{X}$.

Without attempting to solve the system, determine which one of the vectors
$$\mathbf{K}_1 = \begin{pmatrix} 0 \\ 1 \\ 1 \end{pmatrix}, \quad \mathbf{K}_2 = \begin{pmatrix} 1 \\ 1 \\ -1 \end{pmatrix}, \quad \mathbf{K}_3 = \begin{pmatrix} 3 \\ 1 \\ -1 \end{pmatrix}, \quad \mathbf{K}_4 = \begin{pmatrix} 6 \\ 2 \\ -5 \end{pmatrix}$$
is an eigenvector of the coefficient matrix. What is the solution of the system corresponding to this eigenvector?

4. Consider the linear system $\mathbf{X}' = \mathbf{A}\mathbf{X}$ of two differential equations, where $\mathbf{A}$ is a real coefficient matrix. What is the general solution of the system if it is known that $\lambda_1 = 1 + 2i$ is an eigenvalue and $\mathbf{K}_1 = \begin{pmatrix} 1 \\ i \end{pmatrix}$ is a corresponding eigenvector?

In Problems 5–14 solve the given linear system.

5. $\dfrac{dx}{dt} = 2x + y$

$\dfrac{dy}{dt} = -x$

6. $\dfrac{dx}{dt} = -4x + 2y$

$\dfrac{dy}{dt} = 2x - 4y$

7. $\mathbf{X}' = \begin{pmatrix} 1 & 2 \\ -2 & 1 \end{pmatrix}\mathbf{X}$

8. $\mathbf{X}' = \begin{pmatrix} -2 & 5 \\ -2 & 4 \end{pmatrix}\mathbf{X}$

9. $\mathbf{X}' = \begin{pmatrix} 1 & -1 & 1 \\ 0 & 1 & 3 \\ 4 & 3 & 1 \end{pmatrix}\mathbf{X}$

10. $\mathbf{X}' = \begin{pmatrix} 0 & 2 & 1 \\ 1 & 1 & -2 \\ 2 & 2 & -1 \end{pmatrix}\mathbf{X}$

11. $\mathbf{X}' = \begin{pmatrix} 2 & 8 \\ 0 & 4 \end{pmatrix}\mathbf{X} + \begin{pmatrix} 2 \\ 16t \end{pmatrix}$

12. $\mathbf{X}' = \begin{pmatrix} 1 & 2 \\ -\frac{1}{2} & 1 \end{pmatrix}\mathbf{X} + \begin{pmatrix} 0 \\ e^t \tan t \end{pmatrix}$

13. $\mathbf{X}' = \begin{pmatrix} -1 & 1 \\ -2 & 1 \end{pmatrix}\mathbf{X} + \begin{pmatrix} 1 \\ \cot t \end{pmatrix}$

14. $\mathbf{X}' = \begin{pmatrix} 3 & 1 \\ -1 & 1 \end{pmatrix}\mathbf{X} + \begin{pmatrix} -2 \\ 1 \end{pmatrix}e^{2t}$

15. (a) Consider the linear system $\mathbf{X}' = \mathbf{A}\mathbf{X}$ of three first-order differential equations, where the coefficient matrix is
$$\mathbf{A} = \begin{pmatrix} 5 & 3 & 3 \\ 3 & 5 & 3 \\ -5 & -5 & -3 \end{pmatrix}$$
and $\lambda = 2$ is known to be an eigenvalue of multiplicity two. Find two different solutions of the system corresponding to this eigenvalue without using a special formula (such as (12) of Section 8.2).

(b) Use the procedure of part (a) to solve
$$\mathbf{X}' = \begin{pmatrix} 1 & 1 & 1 \\ 1 & 1 & 1 \\ 1 & 1 & 1 \end{pmatrix}\mathbf{X}.$$

16. Verify that $\mathbf{X} = \begin{pmatrix} c_1 \\ c_2 \end{pmatrix}e^t$ is a solution of the linear system
$$\mathbf{X}' = \begin{pmatrix} 1 & 0 \\ 0 & 1 \end{pmatrix}\mathbf{X}$$
for arbitrary constants c_1 and c_2. By hand, draw a phase portrait of the system.

DESIGNING FOR EARTHQUAKES

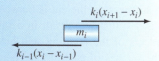

FIGURE 1   Building damage in San Francisco due to the Loma Prieta earthquake of 1989

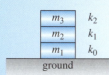

FIGURE 2   Three-story building

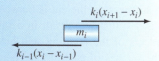

FIGURE 3   Horizontal forces acting on a floor

You are a recently graduated civil engineer. Your friends get to design interesting and exciting things like concrete and earthen dams, skyscrapers, and beautiful suspension bridges. Since you are new, you are stuck doing structural analysis of small buildings and cheap apartments. Still, you take your job seriously because the apartments are going to be built in San Francisco, and if the building is not able to withstand a strong earthquake, it will be on your head—maybe even literally (see Figure 1).

You begin your investigations with some very simplistic assumptions. Each floor has a fixed mass, and the floors are coupled like springs that obey Hooke's Law (see Section 5.1). You also decide to worry only about horizontal displacements of the building's floors. Your first model then is essentially that of a system of coupled springs (see Section 7.6) with each floor being a mass and the connection between two floors playing the role of a spring. You denote the horizontal position and mass of the ith floor by x_i and m_i, respectively. If $x_i = 0$, then floor i is located at its normal position. If an earthquake occurs, then it will drive only the first floor; therefore the forcing term $f(t)$ appears in the force-balance equation for the first floor. In the case of a three story building, this yields a system of three second-order differential equations:

$$m_1 x_1'' = -k_0 x_1 + k_1(x_2 - x_1) + f(t)$$

$$m_2 x_2'' = -k_1(x_2 - x_1) + k_2(x_3 - x_2) \qquad (1)$$

$$m_3 x_3'' = -k_2(x_3 - x_2),$$

where k_i describes the coupling between floor i and floor $i + 1$ and floor 0 is ground level corresponding to k_0. See Figures 2 and 3. We can express system (1) in matrix form as

$$\mathbf{M X}'' = \mathbf{K X} + \mathbf{F}(t) \qquad (2)$$

where

$$\mathbf{X} = \begin{pmatrix} x_1 \\ x_2 \\ x_3 \end{pmatrix}, \quad \mathbf{M} = \begin{pmatrix} m_1 & 0 & 0 \\ 0 & m_2 & 0 \\ 0 & 0 & m_3 \end{pmatrix}, \quad \mathbf{K} = \begin{pmatrix} -(k_0 + k_1) & k_1 & 0 \\ k_1 & -(k_1 + k_2) & k_2 \\ 0 & k_2 & -k_2 \end{pmatrix}, \quad \text{and} \quad \mathbf{F}(t) = \begin{pmatrix} f(t) \\ 0 \\ 0 \end{pmatrix}.$$

The matrix $\mathbf{M}$ is called the mass matrix, and $\mathbf{K}$ is called the stiffness matrix. Since m_i is not zero for all i the matrix $\mathbf{M}$ has an inverse. Thus we can write system (2) in the form

$$\mathbf{X}'' = \mathbf{A X} + \mathbf{M}^{-1}\mathbf{F}(t), \qquad (3)$$

where $\mathbf{A} = \mathbf{M}^{-1}\mathbf{K}$.

In the absence of forcing $\mathbf{F}(t) = \mathbf{0}$ the eigenvalues of $\mathbf{A}$ describe the natural frequencies of the building. In particular, if $\lambda_i < 0$ is an eigenvalue of $\mathbf{A}$, then the corresponding natural frequency $\omega_i = |\sqrt{\lambda_i}|$ and the corresponding natural period is $T_i = 2\pi/\omega_i$. See the *Remarks* at the end of Section 8.1 for a detailed account of this concept.

Your research has shown that earthquakes typically shake the ground with primary periods between 0.5 and 3 seconds. Thus you want to make sure that the natural periods of your building are outside this range. The **Earthquake Tool** on the *DE Tools* CD can be helpful for the next three problems.

PROBLEM 1 (CAS).   Suppose that the mass of each floor in a three-story building is a constant m and that the stiffness coefficients k_i are proportional to the mass of each floor, that is, $k_i = am$. Compute the natural periods of the building and describe how these change with the parameter a. In particular, what is the range of a that ensures that the building will not resonate during a typical earthquake? Should the connections between the floors be very stiff?

PROBLEM 2 (CAS).   Now suppose that you do not have the ability to change the stiffness coefficients except where the building is anchored to the ground. Let $k_i = 20{,}000$, $i = 1, 2$, and $m_i = 7000$, $i = 1, 2, 3$. If $k_0 = 20{,}000$, then is the building in danger of collapse?

PROBLEM 3 (CAS).   Now let's see what we can do by varying the stiffness coefficient k_0 at the ground. This might be accomplished, for example, by adding rebar (reinforcing iron bars) to the foundation. Can the building be made safe in this manner? For this problem reasonable values of k_0 are between 10,000 and 20,000.

STILL CURIOUS?

It is now time for you to design a building that will survive the next earthquake.

PROBLEM 4.   You are told that each floor of a building can have a mass between 5,000 and 10,000 slugs and that each stiffness coefficient is between 10,000 and 20,000 lb/ft. Choose values of m_i and k_i that will result in a safe structure. It is important to note that the residents of the building will have possessions, and these will slightly increase the masses of each floor. Therefore your design should be safe not only at the designed masses, but at masses close to those values as well. Your report should clearly indicate the values of each parameter. In addition, you should indicate, if possible, the range of m_i that continue to be safe. Include graphs or tables, and use the **Earthquake Tool** on the *DE Tools* CD, where appropriate.

Euler Method

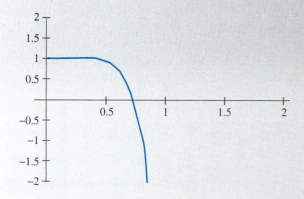

Improved Euler Method

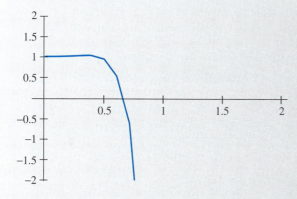

RK4 Method

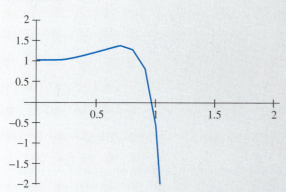

9

Even if it can be shown that a solution of a differential equation exists, we might not be able to exhibit it in explicit or implicit form. In many instances we have to be content with an approximation of the solution. If a solution exists, it represents a set of points in the Cartesian plane. In this chapter we continue to explore the basic idea of Section 2.6—that is, utilizing the differential equation to construct an algorithm to approximate the y-coordinates of points on the actual solution curve. Our concentration in this chapter is primarily on first-order IVPs $dy/dx = f(x, y)$, $y(x_0) = y_0$. We saw in Section 4.9 that numerical procedures developed for first-order DEs extend in a natural way to systems of first-order equations, so we can approximate solutions of a higher-order equation by recasting it as a system of first-order DEs.

Chapter 9 concludes with a method for approximating solutions of linear second-order boundary-value problems.

Which is the more believable numerical solution curve? Top to bottom: Same initial-value problem $y' = 10(y - x^2 + 0.2x - 1)$, $y(0) = 1$, same step size, but different approximation methods. See page 377.

9.1 EULER METHODS AND ERROR ANALYSIS

INTRODUCTION: In Chapter 2 we examined one of the simplest numerical methods for approximating solutions of first-order initial-value problems $y' = f(x, y)$, $y(x_0) = y_0$. Recall that the backbone of Euler's method was the formula

$$y_{n+1} = y_n + hf(x_n, y_n), \tag{1}$$

where f is the function obtained from the differential equation $y' = f(x, y)$. The recursive use of (1) for $n = 0, 1, 2, \ldots$ yields the y-coordinates $y_1, y_2, y_3, \ldots$ of points on successive "tangent lines" to the solution curve at $x_1, x_2, x_3, \ldots$ or $x_n = x_0 + nh$, where h is a constant and is the size of the step between x_n and x_{n+1}. The values $y_1, y_2, y_3, \ldots$ approximate the values of a solution $y(x)$ of the IVP at $x_1, x_2, x_3, \ldots$. But whatever advantage (1) has in its simplicity is lost in the crudeness of its approximations.

REVIEW MATERIAL: We have previously examined Euler's method in Section 2.6. A review of that section is recommended.

A COMPARISON In Problem 4 in Exercises 2.6 you were asked to use Euler's method to obtain the approximate value of $y(1.5)$ for the solution of the initial-value problem $y' = 2xy$, $y(1) = 1$. You should have obtained the analytic solution $y = e^{x^2-1}$ and results similar to those given in Tables 9.1 and 9.2.

TABLE 9.1 Euler's Method with $h = 0.1$

x_n	y_n	Actual value	Abs. error	% Rel. error
1.00	1.0000	1.0000	0.0000	0.00
1.10	1.2000	1.2337	0.0337	2.73
1.20	1.4640	1.5527	0.0887	5.71
1.30	1.8154	1.9937	0.1784	8.95
1.40	2.2874	2.6117	0.3244	12.42
1.50	2.9278	3.4903	0.5625	16.12

TABLE 9.2 Euler's Method with $h = 0.05$

x_n	y_n	Actual value	Abs. error	% Rel. error
1.00	1.0000	1.0000	0.0000	0.00
1.05	1.1000	1.1079	0.0079	0.72
1.10	1.2155	1.2337	0.0182	1.47
1.15	1.3492	1.3806	0.0314	2.27
1.20	1.5044	1.5527	0.0483	3.11
1.25	1.6849	1.7551	0.0702	4.00
1.30	1.8955	1.9937	0.0982	4.93
1.35	2.1419	2.2762	0.1343	5.90
1.40	2.4311	2.6117	0.1806	6.92
1.45	2.7714	3.0117	0.2403	7.98
1.50	3.1733	3.4903	0.3171	9.08

In this case, with a step size $h = 0.1$, a 16% relative error in the calculation of the approximation to $y(1.5)$ is totally unacceptable. At the expense of doubling the number of calculations, some improvement in accuracy is obtained by halving the step size to $h = 0.05$.

ERRORS IN NUMERICAL METHODS In choosing and using a numerical method for the solution of an initial-value problem, we must be aware of the various sources of errors. For some kinds of computation the accumulation of errors might reduce the accuracy of an approximation to the point of making the computation useless. On the other hand, depending on the use to which a numerical solution may be put, extreme accuracy may not be worth the added expense and complication.

One source of error always present in calculations is **round-off error.** This error results from the fact that any calculator or computer can represent numbers using only a finite number of digits. Suppose, for the sake of illustration, that we have a calculator that uses base 10 arithmetic and carries four digits, so that $\frac{1}{3}$ is represented in the calculator as 0.3333 and $\frac{1}{9}$ is represented as 0.1111. If we use this calculator to compute $\left(x^2 - \frac{1}{9}\right)\big/\left(x - \frac{1}{3}\right)$ for $x = 0.3334$, we obtain

$$\frac{(0.3334)^2 - 0.1111}{0.3334 - 0.3333} = \frac{0.1112 - 0.1111}{0.3334 - 0.3333} = 1.$$

With the help of a little algebra, however, we see that

$$\frac{x^2 - \frac{1}{9}}{x - \frac{1}{3}} = \frac{\left(x - \frac{1}{3}\right)\left(x + \frac{1}{3}\right)}{x - \frac{1}{3}} = x + \frac{1}{3},$$

so when $x = 0.3334$, $\left(x^2 - \frac{1}{9}\right)\big/\left(x - \frac{1}{3}\right) \approx 0.3334 + 0.3333 = 0.6667$. This example shows that the effects of round-off error can be quite serious unless some care is taken. One way to reduce the effect of round-off error is to minimize the number of calculations. Another technique on a computer is to use double-precision arithmetic to check the results. In general, round-off error is unpredictable and difficult to analyze, and we will neglect it in the error analysis that follows. We will concentrate on investigating the error introduced by using a formula or algorithm to approximate the values of the solution.

TRUNCATION ERRORS FOR EULER'S METHOD   In the sequence of values $y_1, y_2, y_3, \ldots$ generated from (1), usually the value of y_1 will not agree with the actual solution at x_1—namely, $y(x_1)$—because the algorithm gives only a straight-line approximation to the solution. See Figure 2.33. The error is called the **local truncation error, formula error,** or **discretization error.** It occurs at each step; that is, if we assume that y_n is accurate, then y_{n+1} will contain local truncation error.

To derive a formula for the local truncation error for Euler's method, we use Taylor's formula with remainder. If a function $y(x)$ possesses $k + 1$ derivatives that are continuous on an open interval containing a and x, then

$$y(x) = y(a) + y'(a)\frac{x - a}{1!} + \cdots + y^{(k)}(a)\frac{(x - a)^k}{k!} + y^{(k+1)}(c)\frac{(x - a)^{k+1}}{(k + 1)!},$$

where c is some point between a and x. Setting $k = 1$, $a = x_n$, and $x = x_{n+1} = x_n + h$, we get

$$y(x_{n+1}) = y(x_n) + y'(x_n)\frac{h}{1!} + y''(c)\frac{h^2}{2!}$$

or

$$y(x_{n+1}) = \underbrace{y_n + hf(x_n, y_n)}_{y_{n+1}} + y''(c)\frac{h^2}{2!}.$$

Euler's method (1) is the last formula without the last term; hence the local truncation error in y_{n+1} is

$$y''(c)\frac{h^2}{2!}, \quad \text{where} \quad x_n < c < x_{n+1}.$$

The value of c is usually unknown (it exists theoretically), and so the *exact* error cannot be calculated, but an upper bound on the absolute value of the error is $Mh^2/2!$, where $M = \max_{x_n < x < x_{n+1}} |y''(x)|$.

In discussing errors arising from the use of numerical methods, it is helpful to use the notation $O(h^n)$. To define this concept, we let $e(h)$ denote the error in a numerical calculation depending on h. Then $e(h)$ is said to be of order h^n, denoted by $O(h^n)$, if there exist a constant C and a positive integer n such that $|e(h)| \leq Ch^n$ for h sufficiently small. Thus the local truncation error for Euler's method is $O(h^2)$. We note that, in general, if $e(h)$ in a numerical method is of order h^n and h is halved, the new error is approximately $C(h/2)^n = Ch^n/2^n$; that is, the error is reduced by a factor of $1/2^n$.

EXAMPLE 1   Bound for Local Truncation Errors

Find a bound for the local truncation errors for Euler's method applied to $y' = 2xy$, $y(1) = 1$.

SOLUTION   From the solution $y = e^{x^2-1}$ we get $y'' = (2 + 4x^2)e^{x^2-1}$, so the local truncation error is

$$y''(c)\frac{h^2}{2} = (2 + 4c^2)e^{(c^2-1)}\frac{h^2}{2},$$

where c is between x_n and $x_n + h$. In particular, for $h = 0.1$ we can get an upper bound on the local truncation error for y_1 by replacing c by 1.1:

$$[2 + (4)(1.1)^2]e^{((1.1)^2-1)}\frac{(0.1)^2}{2} = 0.0422.$$

From Table 9.1 we see that the error after the first step is 0.0337, less than the value given by the bound.

Similarly, we can get a bound for the local truncation error for any of the five steps given in Table 9.1 by replacing c by 1.5 (this value of c gives the largest value of $y''(c)$ for any of the steps and may be too generous for the first few steps). Doing this gives

$$[2 + (4)(1.5)^2]e^{((1.5)^2-1)}\frac{(0.1)^2}{2} = 0.1920 \tag{2}$$

as an upper bound for the local truncation error in each step.   ∎

Note that if h is halved to 0.05 in Example 1, then the error bound is 0.0480, about one-fourth as much as shown in (2). This is expected because the local truncation error for Euler's method is $O(h^2)$.

In the above analysis we assumed that the value of y_n was exact in the calculation of y_{n+1}, but it is not because it contains local truncation errors from previous steps. The total error in y_{n+1} is an accumulation of the errors in each of the previous steps. This total error is called the **global truncation error.** A complete analysis of the global truncation error is beyond the scope of this text, but it can be shown that the global truncation error for Euler's method is $O(h)$.

We expect that, for Euler's method, if the step size is halved the error will be approximately halved as well. This is borne out in Tables 9.1 and 9.2 where the absolute error at $x = 1.50$ with $h = 0.1$ is 0.5625 and with $h = 0.05$ is 0.3171, approximately half as large.

In general it can be shown that if a method for the numerical solution of a differential equation has local truncation error $O(h^{\alpha+1})$, then the global truncation error is $O(h^{\alpha})$.

For the remainder of this section and in the subsequent sections we study methods that give significantly greater accuracy than does Euler's method.

IMPROVED EULER'S METHOD   The numerical method defined by the formula

$$y_{n+1} = y_n + h\frac{f(x_n, y_n) + f(x_{n+1}, y_{n+1}^*)}{2}, \tag{3}$$

where

$$y_{n+1}^* = y_n + hf(x_n, y_n), \tag{4}$$

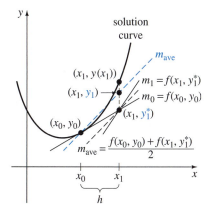

FIGURE 9.1 Slope of colored dashed line is the average of m_0 and m_1

is commonly known as the **improved Euler's method.** To compute y_{n+1} for $n = 0, 1, 2, \ldots$ from (3), we must, at each step, first use Euler's method (4) to obtain an initial estimate y_{n+1}^*. For example, with $n = 0$, (4) gives $y_1^* = y_0 + hf(x_0, y_0)$, and then, knowing this value, we use (3) to get $y_1 = y_0 + h\dfrac{f(x_0, y_0) + f(x_1, y_1^*)}{2}$, where $x_1 = x_0 + h$. These equations can be readily visualized. In Figure 9.1 observe that $m_0 = f(x_0, y_0)$ and $m_1 = f(x_1, y_1^*)$ are slopes of the solid straight lines shown passing through the points (x_0, y_0) and (x_1, y_1^*), respectively. By taking an average of these slopes, that is, $m_{\text{ave}} = \dfrac{f(x_0, y_0) + f(x_1, y_1^*)}{2}$, we obtain the slope of the parallel dashed skew lines. With the first step, rather than advancing along the line through (x_0, y_0) with slope $f(x_0, y_0)$ to the point with y-coordinate y_1^* obtained by Euler's method, we advance instead along the colored dashed line through (x_0, y_0) with slope m_{ave} until we reach x_1. It seems plausible from inspection of the figure that y_1 is an improvement over y_1^*.

In general, the improved Euler's method is an example of a **predictor-corrector method.** The value of y_{n+1}^* given by (4) predicts a value of $y(x_n)$, whereas the value of y_{n+1} defined by formula (3) corrects this estimate.

EXAMPLE 2   **Improved Euler's Method**

Use the improved Euler's method to obtain the approximate value of $y(1.5)$ for the solution of the initial-value problem $y' = 2xy$, $y(1) = 1$. Compare the results for $h = 0.1$ and $h = 0.05$.

SOLUTION   With $x_0 = 1$, $y_0 = 1$, $f(x_n, y_n) = 2x_n y_n$, $n = 0$, and $h = 0.1$, we first compute (4):

$$y_1^* = y_0 + (0.1)(2x_0 y_0) = 1 + (0.1)2(1)(1) = 1.2.$$

We use this last value in (3) along with $x_1 = 1 + h = 1 + 0.1 = 1.1$:

$$y_1 = y_0 + (0.1)\frac{2x_0 y_0 + 2x_1 y_1^*}{2} = 1 + (0.1)\frac{2(1)(1) + 2(1.1)(1.2)}{2} = 1.232.$$

The comparative values of the calculations for $h = 0.1$ and $h = 0.05$ are given in Tables 9.3 and 9.4, respectively.

TABLE 9.3   Improved Euler's Method with $h = 0.1$

x_n	y_n	Actual value	Abs. error	% Rel. error
1.00	1.0000	1.0000	0.0000	0.00
1.10	1.2320	1.2337	0.0017	0.14
1.20	1.5479	1.5527	0.0048	0.31
1.30	1.9832	1.9937	0.0106	0.53
1.40	2.5908	2.6117	0.0209	0.80
1.50	3.4509	3.4904	0.0394	1.13

TABLE 9.4   Improved Euler's Method with $h = 0.05$

x_n	y_n	Actual value	Abs. error	% Rel. error
1.00	1.0000	1.0000	0.0000	0.00
1.05	1.1077	1.1079	0.0002	0.02
1.10	1.2332	1.2337	0.0004	0.04
1.15	1.3798	1.3806	0.0008	0.06
1.20	1.5514	1.5527	0.0013	0.08
1.25	1.7531	1.7551	0.0020	0.11
1.30	1.9909	1.9937	0.0029	0.14
1.35	2.2721	2.2762	0.0041	0.18
1.40	2.6060	2.6117	0.0057	0.22
1.45	3.0038	3.0117	0.0079	0.26
1.50	3.4795	3.4904	0.0108	0.31

A brief word of caution is in order here. We cannot compute all the values of y_n^* first and then substitute these values into formula (3). In other words, we cannot use the data in Table 9.1 to help construct the values in Table 9.3. Why not?

TRUNCATION ERRORS FOR THE IMPROVED EULER'S METHOD The local truncation error for the improved Euler's method is $O(h^3)$. The derivation of this result is similar to the derivation of the local truncation error for Euler's method. Since the local truncation error for the improved Euler's method is $O(h^3)$, the global truncation error is $O(h^2)$. This can be seen in Example 2; when the step size is halved from $h = 0.1$ to $h = 0.05$, the absolute error at $x = 1.50$ is reduced from 0.0394 to 0.0108, a reduction of approximately $\left(\frac{1}{2}\right)^2 = \frac{1}{4}$.

EXERCISES 9.1

Answers to selected odd-numbered problems begin on page ANS-13.

In Problems 1–10 use the improved Euler's method to obtain a four-decimal approximation of the indicated value. First use $h = 0.1$ and then use $h = 0.05$.

1. $y' = 2x - 3y + 1$, $y(1) = 5$; $y(1.5)$

2. $y' = 4x - 2y$, $y(0) = 2$; $y(0.5)$

3. $y' = 1 + y^2$, $y(0) = 0$; $y(0.5)$

4. $y' = x^2 + y^2$, $y(0) = 1$; $y(0.5)$

5. $y' = e^{-y}$, $y(0) = 0$; $y(0.5)$

6. $y' = x + y^2$, $y(0) = 0$; $y(0.5)$

7. $y' = (x - y)^2$, $y(0) = 0.5$; $y(0.5)$

8. $y' = xy + \sqrt{y}$, $y(0) = 1$; $y(0.5)$

9. $y' = xy^2 - \dfrac{y}{x}$, $y(1) = 1$; $y(1.5)$

10. $y' = y - y^2$, $y(0) = 0.5$; $y(0.5)$

11. Consider the initial-value problem $y' = (x + y - 1)^2$, $y(0) = 2$. Use the improved Euler's method with $h = 0.1$ and $h = 0.05$ to obtain approximate values of the solution at $x = 0.5$. At each step compare the approximate value with the actual value of the analytic solution.

12. Although it might not be obvious from the differential equation, its solution could "behave badly" near a point x at which we wish to approximate $y(x)$. Numerical procedures may give widely differing results near this point. Let $y(x)$ be the solution of the initial-value problem $y' = x^2 + y^3$, $y(1) = 1$.
    (a) Use a numerical solver to graph the solution on the interval $[1, 1.4]$.
    (b) Using the step size $h = 0.1$, compare the results obtained from Euler's method with the results from the improved Euler's method in the approximation of $y(1.4)$.

13. Consider the initial-value problem $y' = 2y$, $y(0) = 1$. The analytic solution is $y = e^{2x}$.
    (a) Approximate $y(0.1)$ using one step and Euler's method.
    (b) Find a bound for the local truncation error in y_1.
    (c) Compare the error in y_1 with your error bound.
    (d) Approximate $y(0.1)$ using two steps and Euler's method.
    (e) Verify that the global truncation error for Euler's method is $O(h)$ by comparing the errors in parts (a) and (d).

14. Repeat Problem 13 using the improved Euler's method. Its global truncation error is $O(h^2)$.

15. Repeat Problem 13 using the initial-value problem $y' = x - 2y$, $y(0) = 1$. The analytic solution is
$$y = \frac{1}{2}x - \frac{1}{4} + \frac{5}{4}e^{-2x}.$$

16. Repeat Problem 15 using the improved Euler's method. Its global truncation error is $O(h^2)$.

17. Consider the initial-value problem $y' = 2x - 3y + 1$, $y(1) = 5$. The analytic solution is
$$y(x) = \frac{1}{9} + \frac{2}{3}x + \frac{38}{9}e^{-3(x-1)}.$$
    (a) Find a formula involving c and h for the local truncation error in the nth step if Euler's method is used.
    (b) Find a bound for the local truncation error in each step if $h = 0.1$ is used to approximate $y(1.5)$.
    (c) Approximate $y(1.5)$ using $h = 0.1$ and $h = 0.05$ with Euler's method. See Problem 1 in Exercises 2.6.
    (d) Calculate the errors in part (c) and verify that the global truncation error of Euler's method is $O(h)$.

18. Repeat Problem 17 using the improved Euler's method which has a global truncation error $O(h^2)$.

See Problem 1. You might need to keep more than four decimal places to see the effect of reducing the order of the error.

19. Repeat Problem 17 for the initial-value problem $y' = e^{-y}$, $y(0) = 0$. The analytic solution is $y(x) = \ln(x + 1)$. Approximate $y(0.5)$. See Problem 5 in Exercises 2.6.

20. Repeat Problem 19 using the improved Euler's method, which has global truncation error $O(h^2)$. See

Problem 5. You might need to keep more than four decimal places to see the effect of reducing the order of error.

DISCUSSION/PROJECT PROBLEMS

21. Answer the question "Why not?" that follows the three sentences after Example 2 on page 372.

9.2   RUNGE-KUTTA METHODS

INTRODUCTION: Probably one of the more popular as well as most accurate numerical procedures used in obtaining approximate solutions to a first-order initial-value problem $y' = f(x, y)$, $y(x_0) = y_0$ is the **fourth-order Runge-Kutta method.** As the name suggests, there are Runge-Kutta methods of different orders.

REVIEW MATERIAL: The fourth-order Runge-Kutta method was first mentioned in Section 2.6. See page 83.

RUNGE-KUTTA METHODS   Fundamentally, all Runge-Kutta methods are generalizations of the basic Euler formula (1) of Section 9.1 in that the slope function f is replaced by a weighted average of slopes over the interval $x_n \leq x \leq x_{n+1}$. That is,

$$y_{n+1} = y_n + h \overbrace{(w_1 k_1 + w_2 k_2 + \cdots + w_m k_m)}^{\text{weighted average}}. \tag{1}$$

Here the weights w_i, $i = 1, 2, \ldots, m$, are constants that generally satisfy $w_1 + w_2 + \cdots + w_m = 1$, and each k_i, $i = 1, 2, \ldots, m$, is the function f evaluated at a selected point (x, y) for which $x_n \leq x \leq x_{n+1}$. We shall see that the k_i are defined recursively. The number m is called the **order** of the method. Observe that by taking $m = 1$, $w_1 = 1$, and $k_1 = f(x_n, y_n)$, we get the familiar Euler formula $y_{n+1} = y_n + h f(x_n, y_n)$. Hence Euler's method is said to be a **first-order Runge-Kutta method.**

The average in (1) is not formed willy-nilly, but parameters are chosen so that (1) agrees with a Taylor polynomial of degree m. As we saw in the last section, if a function $y(x)$ possesses $k + 1$ derivatives that are continuous on an open interval containing a and x, then we can write

$$y(x) = y(a) + y'(a)\frac{x - a}{1!} + y''(a)\frac{(x - a)^2}{2!} + \cdots + y^{(k+1)}(c)\frac{(x - a)^{k+1}}{(k + 1)!},$$

where c is some number between a and x. If we replace a by x_n and x by $x_{n+1} = x_n + h$, then the foregoing formula becomes

$$y(x_{n+1}) = y(x_n + h) = y(x_n) + hy'(x_n) + \frac{h^2}{2!}y''(x_n) + \cdots + \frac{h^{k+1}}{(k + 1)!}y^{(k+1)}(c),$$

where c is now some number between x_n and x_{n+1}. When $y(x)$ is a solution of $y' = f(x, y)$ in the case $k = 1$ and the remainder $\frac{1}{2}h^2 y''(c)$ is small, we see that a

Taylor polynomial $y(x_{n+1}) = y(x_n) + hy'(x_n)$ of degree one agrees with the approximation formula of Euler's method

$$y_{n+1} = y_n + hy'_n = y_n + hf(x_n, y_n).$$

A SECOND-ORDER RUNGE-KUTTA METHOD   To further illustrate (1), we consider now a **second-order Runge-Kutta procedure.** This consists of finding constants or parameters w_1, w_2, α, and β so that the formula

$$y_{n+1} = y_n + h(w_1 k_1 + w_2 k_2), \tag{2}$$

where
$$k_1 = f(x_n, y_n)$$

$$k_2 = f(x_n + \alpha h, y_n + \beta h k_1),$$

agrees with a Taylor polynomial of degree two. For our purposes it suffices to say that this can be done whenever the constants satisfy

$$w_1 + w_2 = 1, \quad w_2 \alpha = \frac{1}{2}, \quad \text{and} \quad w_2 \beta = \frac{1}{2}. \tag{3}$$

This is an algebraic system of three equations in four unknowns and has infinitely many solutions:

$$w_1 = 1 - w_2, \quad \alpha = \frac{1}{2w_2}, \quad \text{and} \quad \beta = \frac{1}{2w_2}, \tag{4}$$

where $w_2 \neq 0$. For example, the choice $w_2 = \frac{1}{2}$ yields $w_1 = \frac{1}{2}$, $\alpha = 1$, $\beta = 1$, and so (2) becomes

$$y_{n+1} = y_n + \frac{h}{2}(k_1 + k_2),$$

where
$$k_1 = f(x_n, y_n) \quad \text{and} \quad k_2 = f(x_n + h, y_n + h k_1).$$

Since $x_n + h = x_{n+1}$ and $y_n + hk_1 = y_n + hf(x_n, y_n)$ the foregoing result is recognized to be the improved Euler's method that is summarized in (3) and (4) of Section 9.1.

In view of the fact that $w_2 \neq 0$ can be chosen arbitrarily in (4), there are many possible second-order Runge-Kutta methods. See Problem 2 in Exercises 9.2.

We shall skip any discussion of third-order methods in order to come to the principal point of discussion in this section.

A FOURTH-ORDER RUNGE-KUTTA METHOD   A **fourth-order Runge-Kutta procedure** consists of finding parameters so that the formula

$$y_{n+1} = y_n + h(w_1 k_1 + w_2 k_2 + w_3 k_3 + w_4 k_4), \tag{5}$$

where
$$k_1 = f(x_n, y_n)$$

$$k_2 = f(x_n + \alpha_1 h, y_n + \beta_1 h k_1)$$

$$k_3 = f(x_n + \alpha_2 h, y_n + \beta_2 h k_1 + \beta_3 h k_2)$$

$$k_4 = f(x_n + \alpha_3 h, y_n + \beta_4 h k_1 + \beta_5 h k_2 + \beta_6 h k_3),$$

agrees with a Taylor polynomial of degree four. This results in a system of 11 equations in 13 unknowns. The most commonly used set of values for the parameters yields the following result:

$$y_{n+1} = y_n + \frac{h}{6}\left(k_1 + 2k_2 + 2k_3 + k_4\right),$$

$$k_1 = f(x_n, y_n)$$
$$k_2 = f\left(x_n + \tfrac{1}{2}h, y_n + \tfrac{1}{2}hk_1\right) \qquad (6)$$
$$k_3 = f\left(x_n + \tfrac{1}{2}h, y_n + \tfrac{1}{2}hk_2\right)$$
$$k_4 = f(x_n + h, y_n + hk_3).$$

While other fourth-order formulas are easily derived, the algorithm summarized in (6) is so widely used and recognized as a valuable computational tool it is often referred to as *the* fourth-order Runge-Kutta method or *the classical* Runge-Kutta method. It is (6) that we have in mind, hereafter, when we use the abbreviation *the RK4 method*.

You are advised to look carefully at the formulas in (6); note that k_2 depends on k_1, k_3 depends on k_2, and k_4 depends on k_3. Also, k_2 and k_3 involve approximations to the slope at the midpoint $x_n + \tfrac{1}{2}h$ of the interval $x_n \le x \le x_{n+1}$.

EXAMPLE 1   RK4 Method

Use the RK4 method with $h = 0.1$ to obtain an approximation to $y(1.5)$ for the solution of $y' = 2xy$, $y(1) = 1$.

SOLUTION   For the sake of illustration let us compute the case when $n = 0$. From (6) we find

$$k_1 = f(x_0, y_0) = 2x_0y_0 = 2$$
$$k_2 = f\left(x_0 + \tfrac{1}{2}(0.1), y_0 + \tfrac{1}{2}(0.1)2\right)$$
$$= 2\left(x_0 + \tfrac{1}{2}(0.1)\right)\left(y_0 + \tfrac{1}{2}(0.2)\right) = 2.31$$
$$k_3 = f\left(x_0 + \tfrac{1}{2}(0.1), y_0 + \tfrac{1}{2}(0.1)2.31\right)$$
$$= 2\left(x_0 + \tfrac{1}{2}(0.1)\right)\left(y_0 + \tfrac{1}{2}(0.231)\right) = 2.34255$$
$$k_4 = f(x_0 + (0.1), y_0 + (0.1)2.34255)$$
$$= 2(x_0 + 0.1)(y_0 + 0.234255) = 2.715361$$

and therefore

$$y_1 = y_0 + \frac{0.1}{6}(k_1 + 2k_2 + 2k_3 + k_4)$$

$$= 1 + \frac{0.1}{6}(2 + 2(2.31) + 2(2.34255) + 2.715361) = 1.23367435.$$

The remaining calculations are summarized in Table 9.5, whose entries are rounded to four decimal places.

TABLE 9.5   RK4 Method with $h = 0.1$

x_n	y_n	Actual value	Abs. error	% Rel. error
1.00	1.0000	1.0000	0.0000	0.00
1.10	1.2337	1.2337	0.0000	0.00
1.20	1.5527	1.5527	0.0000	0.00
1.30	1.9937	1.9937	0.0000	0.00
1.40	2.6116	2.6117	0.0001	0.00
1.50	3.4902	3.4904	0.0001	0.00

Inspection of Table 9.5 shows why the fourth-order Runge-Kutta method is so popular. If four-decimal-place accuracy is all that we desire, there is no need to use a smaller step size. Table 9.6 compares the results of applying Euler's, the improved Euler's, and the fourth-order Runge-Kutta methods to the initial-value problem $y' = 2xy$, $y(1) = 1$. (See Tables 9.1 and 9.3.)

TABLE 9.6 $y' = 2xy$, $y(1) = 1$

Comparison of numerical methods with $h = 0.1$				Comparison of numerical methods with $h = 0.05$				
x_n	Euler	Improved Euler	RK4	Actual value	x_n	Euler	Improved Euler	RK4	Actual value
1.00	1.0000	1.0000	1.0000	1.0000	1.00	1.0000	1.0000	1.0000	1.0000
1.10	1.2000	1.2320	1.2337	1.2337	1.05	1.1000	1.1077	1.1079	1.1079
1.20	1.4640	1.5479	1.5527	1.5527	1.10	1.2155	1.2332	1.2337	1.2337
1.30	1.8154	1.9832	1.9937	1.9937	1.15	1.3492	1.3798	1.3806	1.3806
1.40	2.2874	2.5908	2.6116	2.6117	1.20	1.5044	1.5514	1.5527	1.5527
1.50	2.9278	3.4509	3.4902	3.4904	1.25	1.6849	1.7531	1.7551	1.7551
					1.30	1.8955	1.9909	1.9937	1.9937
					1.35	2.1419	2.2721	2.2762	2.2762
					1.40	2.4311	2.6060	2.6117	2.6117
					1.45	2.7714	3.0038	3.0117	3.0117
					1.50	3.1733	3.4795	3.4903	3.4904

TRUNCATION ERRORS FOR THE RK4 METHOD In Section 9.1 we saw that global truncation errors for Euler's method and for the improved Euler's method are, respectively, $O(h)$ and $O(h^2)$. Because the first equation in (6) agrees with a Taylor polynomial of degree four, the local truncation error for this method is $y^{(5)}(c)\, h^5/5!$ or $O(h^5)$, and the global truncation error is thus $O(h^4)$. It is now obvious why Euler's method, the improved Euler's method, and (6) are *first-*, *second-*, and *fourth-order* Runge-Kutta methods, respectively.

EXAMPLE 2 **Bound for Local Truncation Errors**

Find a bound for the local truncation errors for the RK4 method applied to $y' = 2xy$, $y(1) = 1$.

SOLUTION By computing the fifth derivative of the known solution $y(x) = e^{x^2 - 1}$, we get

$$y^{(5)}(c)\frac{h^5}{5!} = (120c + 160c^3 + 32c^5)e^{c^2-1}\frac{h^5}{5!}. \qquad (7)$$

Thus with $c = 1.5$, (7) yields a bound of 0.00028 on the local truncation error for each of the five steps when $h = 0.1$. Note that in Table 9.5 the error in y_1 is much less than this bound.

Table 9.7 gives the approximations to the solution of the initial-value problem at $x = 1.5$ that are obtained from the RK4 method. By computing the value of the analytic solution at $x = 1.5$ we can find the error in these approximations. Because the method is so accurate, many decimal places must be used in the numerical solution to see the effect of halving the step size. Note that when h is halved, from $h = 0.1$ to $h = 0.05$, the error is divided by a factor of about $2^4 = 16$, as expected.

TABLE 9.7 RK4 Method

h	Approx.	Error
0.1	3.49021064	$1.32321089 \times 10^{-4}$
0.05	3.49033382	$9.13776090 \times 10^{-6}$

ADAPTIVE METHODS We have seen that the accuracy of a numerical method for approximating solutions of differential equations can be improved by decreasing the step size h. Of course, this enhanced accuracy is usually obtained at a cost—namely, increased computation time and greater possibility of round-off error. In general, over the interval of approximation there may be subintervals where a relatively large step size suffices and other subintervals where a smaller step is necessary to keep the truncation error within a desired limit. Numerical methods that use a variable step size are called **adaptive methods.** One of the more popular of the adaptive routines is the **Runge-Kutta-Fehlberg method.** Because Fehlberg employed two Runge-Kutta methods of differing orders, a fourth- and a fifth-order method, this algorithm is frequently denoted as the **RKF45 method.***

REMARKS

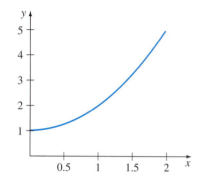

FIGURE 9.2 Solution curve for $y' = 10(y - x^2 + 0.2x - 1)$, $y(0) = 1$.

On the chapter-opening page (page 367) you were asked which of the three given figures represented the most believable numerical solution curve for the IVP $y' = 10(y - x^2 + 0.2x - 1)$, $y(0) = 1$. After seeing the benefits of the RK4 method in this section, you might be inclined to pick the graphic generated by this method. But the correct answer is that all three figures are grossly in error; the correct solution curve is shown in Figure 9.2. Because the IVP involves a linear DE the method of Section 2.3 yields the analytic general solution $y = x^2 + 1 + c_1 e^{10x}$. The initial condition $y(0) = 1$ determines $c_1 = 0$, and so the solution of the problem is simply $y = x^2 + 1$. The basic problem in the Euler, the improved Euler, the RK4, and all similar numerical methods, is that each step along the x-axis after the first uses rounded (that is, incorrect) values. It is equivalent to stipulating an initial condition *different* from $y(0) = 1$ and so determining a value of $c_1 \neq 0$. The e^{10x} term, not present in the actual solution, asserts itself by the buildup of errors in the recursive calculations and quickly overwhelms the values of $x^2 + 1$. The point here is this: Do not blindly accept a numerical solution, even one generated by a "nice" method such as RK4.

---

*The Runge-Kutta method of order four used in RKF45 is *not* the same as that given in (6).

EXERCISES 9.2

Answers to selected odd-numbered problems begin on page ANS-13.

1. Use the RK4 method with $h = 0.1$ to approximate $y(0.5)$, where $y(x)$ is the solution of the initial-value problem $y' = (x + y - 1)^2$, $y(0) = 2$. Compare this approximate value with the actual value obtained in Problem 11 in Exercises 9.1.

2. Assume that $w_2 = \frac{3}{4}$ in (4). Use the resulting second-order Runge-Kutta method to approximate $y(0.5)$, where $y(x)$ is the solution of the initial-value problem in Problem 1. Compare this approximate value with the approximate value obtained in Problem 11 in Exercises 9.1.

In Problems 3–12 use the RK4 method with $h = 0.1$ to obtain a four-decimal approximation of the indicated value.

3. $y' = 2x - 3y + 1$, $y(1) = 5$; $y(1.5)$

4. $y' = 4x - 2y$, $y(0) = 2$; $y(0.5)$

5. $y' = 1 + y^2$, $y(0) = 0$; $y(0.5)$

6. $y' = x^2 + y^2$, $y(0) = 1$; $y(0.5)$

7. $y' = e^{-y}$, $y(0) = 0$; $y(0.5)$

8. $y' = x + y^2$, $y(0) = 0$; $y(0.5)$

9. $y' = (x - y)^2$, $y(0) = 0.5$; $y(0.5)$

10. $y' = xy + \sqrt{y}$, $y(0) = 1$; $y(0.5)$

11. $y' = xy^2 - \dfrac{y}{x}$, $y(1) = 1$; $y(1.5)$

12. $y' = y - y^2$, $y(0) = 0.5$; $y(0.5)$

13. If air resistance is proportional to the square of the instantaneous velocity, then the velocity v of a mass m dropped from a given height is determined from

$$m\frac{dv}{dt} = mg - kv^2, \quad k > 0.$$

Let $v(0) = 0$, $k = 0.125$, $m = 5$ slugs, and $g = 32$ ft/s².
  (a) Use the RK4 method with $h = 1$ to approximate the velocity $v(5)$.
  (b) Use a numerical solver to graph the solution of the IVP on the interval $[0, 6]$.
  (c) Use separation of variables to solve the IVP and then find the actual value $v(5)$.

14. A mathematical model for the area A (in cm²) that a colony of bacteria (*B. dendroides*) occupies is given by

$$\frac{dA}{dt} = A(2.128 - 0.0432A).*$$

Suppose that the initial area is 0.24 cm².
  (a) Use the RK4 method with $h = 0.5$ to complete the following table:

t (days)	1	2	3	4	5
A (observed)	2.78	13.53	36.30	47.50	49.40
A (approximated)					

  (b) Use a numerical solver to graph the solution of the initial-value problem. Estimate the values $A(1)$, $A(2)$, $A(3)$, $A(4)$, and $A(5)$ from the graph.
  (c) Use separation of variables to solve the initial-value problem and compute the actual values $A(1)$, $A(2)$, $A(3)$, $A(4)$, and $A(5)$.

15. Consider the initial-value problem $y' = x^2 + y^3$, $y(1) = 1$. See Problem 12 in Exercises 9.1.
  (a) Compare the results obtained from using the RK4 method over the interval $[1, 1.4]$ with step sizes $h = 0.1$ and $h = 0.05$.
  (b) Use a numerical solver to graph the solution of the initial-value problem on the interval $[1, 1.4]$.

16. Consider the initial-value problem $y' = 2y$, $y(0) = 1$. The analytic solution is $y(x) = e^{2x}$.
  (a) Approximate $y(0.1)$ using one step and the RK4 method.
  (b) Find a bound for the local truncation error in y_1.
  (c) Compare the error in y_1 with your error bound.

  (d) Approximate $y(0.1)$ using two steps and the RK4 method.
  (e) Verify that the global truncation error for the RK4 method is $O(h^4)$ by comparing the errors in parts (a) and (d).

17. Repeat Problem 16 using the initial-value problem $y' = -2y + x$, $y(0) = 1$. The analytic solution is

$$y(x) = \frac{1}{2}x - \frac{1}{4} + \frac{5}{4}e^{-2x}.$$

18. Consider the initial-value problem $y' = 2x - 3y + 1$, $y(1) = 5$. The analytic solution is

$$y(x) = \frac{1}{9} + \frac{2}{3}x + \frac{38}{9}e^{-3(x-1)}.$$

  (a) Find a formula involving c and h for the local truncation error in the nth step if the RK4 method is used.
  (b) Find a bound for the local truncation error in each step if $h = 0.1$ is used to approximate $y(1.5)$.
  (c) Approximate $y(1.5)$ using the RK4 method with $h = 0.1$ and $h = 0.05$. See Problem 3. You will need to carry more than six decimal places to see the effect of reducing the step size.

19. Repeat Problem 18 for the initial-value problem $y' = e^{-y}$, $y(0) = 0$. The analytic solution is $y(x) = \ln(x + 1)$. Approximate $y(0.5)$. See Problem 7.

DISCUSSION/PROJECT PROBLEMS

20. A count of the number of evaluations of the function f used in solving the initial-value problem $y' = f(x, y)$, $y(x_0) = y_0$ is used as a measure of the computational complexity of a numerical method. Determine the number of evaluations of f required for each step of Euler's, the improved Euler's, and the RK4 methods. By considering some specific examples, compare the accuracy of these methods when used with comparable computational complexities.

COMPUTER LAB ASSIGNMENTS

21. The RK4 method for solving an initial-value problem over an interval $[a, b]$ results in a finite set of points that are supposed to approximate points on the graph of the exact solution. In order to expand this set of discrete points to an approximate solution defined at all points on the interval $[a, b]$, we can use an **interpolating function**. This is a function, supported by most computer algebra systems, that agrees with the given data exactly and assumes a smooth transition between data points. These interpolating functions may be polynomials or sets of polynomials joined together smoothly. In *Mathematica* the command

---

*See V. A. Kostitzin, *Mathematical Biology* (London: Harrap, 1939).

y=Interpolation[data] can be used to obtain an interpolating function through the points **data** $= \{\{x_0, y_0\}, \{x_1, y_1\}, \ldots, \{x_n, y_n\}\}$. The interpolating function **y[x]** can now be treated like any other function built into the computer algebra system.

(a) Find the analytic solution of the initial-value problem $y' = -y + 10 \sin 3x$; $y(0) = 0$ on the interval $[0, 2]$. Graph this solution and find its positive roots.

(b) Use the RK4 method with $h = 0.1$ to approximate a solution of the initial-value problem in part (a). Obtain an interpolating function and graph it. Find the positive roots of the interpolating function of the interval $[0, 2]$.

9.3 MULTISTEP METHODS

INTRODUCTION: Euler's, the improved Euler's, and the Runge-Kutta methods are examples of **single-step** or **starting methods.** In these methods each successive value y_{n+1} is computed based only on information about the immediately preceding value y_n. On the other hand, **multistep** or **continuing methods** use the values from several computed steps to obtain the value of y_{n+1}. There are a large number of multistep method formulas for approximating solutions of DEs, but since it is not our intention to survey the vast field of numerical procedures, we will consider only one such method here.

ADAMS-BASHFORTH-MOULTON METHOD   The multistep method discussed in this section is called the fourth-order **Adams-Bashforth-Moulton method.** Like the improved Euler's method it is a predictor-corrector method—that is, one formula is used to predict a value y_{n+1}^*, which in turn is used to obtain a corrected value y_{n+1}. The predictor in this method is the Adams-Bashforth formula

$$y_{n+1}^* = y_n + \frac{h}{24}(55y_n' - 59y_{n-1}' + 37y_{n-2}' - 9y_{n-3}'), \tag{1}$$

$$y_n' = f(x_n, y_n)$$
$$y_{n-1}' = f(x_{n-1}, y_{n-1})$$
$$y_{n-2}' = f(x_{n-2}, y_{n-2})$$
$$y_{n-3}' = f(x_{n-3}, y_{n-3})$$

for $n \geq 3$. The value of y_{n+1}^* is then substituted into the Adams-Moulton corrector

$$y_{n+1} = y_n + \frac{h}{24}(9y_{n+1}' + 19y_n' - 5y_{n-1}' + y_{n-2}')$$

$$y_{n+1}' = f(x_{n+1}, y_{n+1}^*). \tag{2}$$

Notice that formula (1) requires that we know the values of y_0, y_1, y_2, and y_3 to obtain y_4. The value of y_0 is, of course, the given initial condition. The local truncation error of the Adams-Bashforth-Moulton method is $O(h^5)$, the values of y_1, y_2, and y_3 are generally computed by a method with the same error property, such as the fourth-order Runge-Kutta method.

EXAMPLE 1   Adams-Bashforth-Moulton Method

Use the Adams-Bashforth-Moulton method with $h = 0.2$ to obtain an approximation to $y(0.8)$ for the solution of

$$y' = x + y - 1, \quad y(0) = 1.$$

SOLUTION With a step size of $h = 0.2$, $y(0.8)$ will be approximated by y_4. To get started, we use the RK4 method with $x_0 = 0$, $y_0 = 1$, and $h = 0.2$ to obtain

$$y_1 = 1.02140000, \quad y_2 = 1.09181796, \quad y_3 = 1.22210646.$$

Now with the identifications $x_0 = 0$, $x_1 = 0.2$, $x_2 = 0.4$, $x_3 = 0.6$, and $f(x, y) = x + y - 1$, we find

$$y_0' = f(x_0, y_0) = (0) + (1) - 1 = 0$$

$$y_1' = f(x_1, y_1) = (0.2) + (1.02140000) - 1 = 0.22140000$$

$$y_2' = f(x_2, y_2) = (0.4) + (1.09181796) - 1 = 0.49181796$$

$$y_3' = f(x_3, y_3) = (0.6) + (1.22210646) - 1 = 0.82210646.$$

With the foregoing values the predictor (1) then gives

$$y_4^* = y_3 + \frac{0.2}{24}(55y_3' - 59y_2' + 37y_1' - 9y_0') = 1.42535975.$$

To use the corrector (2), we first need

$$y_4' = f(x_4, y_4^*) = 0.8 + 1.42535975 - 1 = 1.22535975.$$

Finally, (2) yields

$$y_4 = y_3 + \frac{0.2}{24}(9y_4' + 19y_3' - 5y_2' + y_1') = 1.42552788. \quad \blacksquare$$

You should verify that the actual value of $y(0.8)$ in Example 1 is $y(0.8) = 1.42554093$. See Problem 1 in Exercises 9.3.

STABILITY OF NUMERICAL METHODS An important consideration in using numerical methods to approximate the solution of an initial-value problem is the stability of the method. Simply stated, a numerical method is **stable** if small changes in the initial condition result in only small changes in the computed solution. A numerical method is said to be **unstable** if it is not stable. The reason that stability considerations are important is that in each step after the first step of a numerical technique we are essentially starting over again with a new initial-value problem, where the initial condition is the approximate solution value computed in the preceding step. Because of the presence of round-off error, this value will almost certainly vary at least slightly from the true value of the solution. Besides round-off error, another common source of error occurs in the initial condition itself; in physical applications the data are often obtained by imprecise measurements.

One possible method for detecting instability in the numerical solution of a specific initial-value problem is to compare the approximate solutions obtained when decreasing step sizes are used. If the numerical method is unstable, the error may actually increase with smaller step sizes. Another way of checking stability is to observe what happens to solutions when the initial condition is slightly perturbed (for example, change $y(0) = 1$ to $y(0) = 0.999$).

For a more detailed and precise discussion of stability, consult a numerical analysis text. In general, all of the methods we have discussed in this chapter have good stability characteristics.

ADVANTAGES/DISADVANTAGES OF MULTISTEP METHODS Many considerations enter into the choice of a method to solve a differential equation numerically. Single-step methods, particularly the RK4 method, are often chosen because of their accuracy and the fact that they are easy to program. However, a major drawback is that the right-hand side of the differential equation must be evaluated many times at each step. For instance, the RK4 method requires four function evaluations for each step. On the other hand, if the function evaluations in the previous step have been calculated and stored, a multistep method requires only one new function evaluation for each step. This can lead to great savings in time and expense.

As an example, solving $y' = f(x, y)$, $y(x_0) = y_0$ numerically using n steps by the fourth-order Runge-Kutta method requires $4n$ function evaluations. The Adams-Bashforth multistep method requires 16 function evaluations for the Runge-Kutta fourth-order starter and $n - 4$ for the n Adams-Bashforth steps, giving a total of $n + 12$ function evaluations for this method. In general the Adams-Bashforth multistep method requires slightly more than a quarter of the number of function evaluations required for the RK4 method. If the evaluation of $f(x, y)$ is complicated, the multistep method will be more efficient.

Another issue involved with multistep methods is how many times the Adams-Moulton corrector formula should be repeated in each step. Each time the corrector is used, another function evaluation is done, and so the accuracy is increased at the expense of losing an advantage of the multistep method. In practice, the corrector is calculated once, and if the value of y_{n+1} is changed by a large amount, the entire problem is restarted using a smaller step size. This is often the basis of the variable step size methods, whose discussion is beyond the scope of this text.

EXERCISES 9.3

Answers to selected odd-numbered problems begin on page ANS-14.

1. Find the analytic solution of the initial-value problem in Example 1. Compare the actual values of $y(0.2)$, $y(0.4)$, $y(0.6)$, and $y(0.8)$ with the approximations y_1, y_2, y_3, and y_4.

2. Write a computer program to implement the Adams-Bashforth-Moulton method.

In Problems 3 and 4 use the Adams-Bashforth-Moulton method to approximate $y(0.8)$, where $y(x)$ is the solution of the given initial-value problem. Use $h = 0.2$ and the RK4 method to compute y_1, y_2, and y_3.

3. $y' = 2x - 3y + 1$, $y(0) = 1$

4. $y' = 4x - 2y$, $y(0) = 2$

In Problems 5–8 use the Adams-Bashforth-Moulton method to approximate $y(1.0)$, where $y(x)$ is the solution of the given initial-value problem. First use $h = 0.2$ and then use $h = 0.1$. Use the RK4 method to compute y_1, y_2, and y_3.

5. $y' = 1 + y^2$, $y(0) = 0$

6. $y' = y + \cos x$, $y(0) = 1$

7. $y' = (x - y)^2$, $y(0) = 0$

8. $y' = xy + \sqrt{y}$, $y(0) = 1$

9.4 HIGHER-ORDER EQUATIONS AND SYSTEMS

INTRODUCTION: So far we have focused on numerical techniques that can be used to approximate the solution of a first-order initial-value problem $y' = f(x, y)$, $y(x_0) = y_0$. To approximate the solution of a second-order initial-value problem we must express a second-order DE as a system of two first-order DEs. To do this, we begin by writing the second-order DE in normal form by solving for y'' in terms of x, y, and y'.

REVIEW MATERIAL: Review Section 1.1 (normal form of second-order DE) and Section 4.9 (second-order DE written as a system of first-order DEs).

CD: The **Numerical Methods Tool** on the *DE Tools* CD illustrates *Euler's method*, the *improved Euler's method*, and the *RK4 method* for various systems of DEs. See pages 384 and 385.

SECOND-ORDER IVPS A second-order initial-value problem

$$y'' = f(x, y, y'), \quad y(x_0) = y_0, \quad y'(x_0) = u_0 \tag{1}$$

can be expressed as an initial-value problem for a system of first order differential equations. If we let $y' = u$, the differential equation in (1) becomes the system

$$\begin{aligned} y' &= u \\ u' &= f(x, y, u). \end{aligned} \tag{2}$$

Since $y'(x_0) = u(x_0)$, the corresponding initial conditions for (2) are then $y(x_0) = y_0$, $u(x_0) = u_0$. The system (2) can now be solved numerically by simply applying a particular numerical method to each first-order differential equation in the system. For example, **Euler's method** applied to the system (2) would be

$$\begin{aligned} y_{n+1} &= y_n + hu_n \\ u_{n+1} &= u_n + hf(x_n, y_n, u_n), \end{aligned} \tag{3}$$

whereas the **fourth-order Runge-Kutta method,** or **RK4 method,** would be

$$y_{n+1} = y_n + \frac{h}{6}(m_1 + 2m_2 + 2m_3 + m_4) \tag{4}$$

$$u_{n+1} = u_n + \frac{h}{6}(k_1 + 2k_2 + 2k_3 + k_4)$$

where

$$\begin{aligned} m_1 &= u_n & k_1 &= f(x_n, y_n, u_n) \\ m_2 &= u_n + \tfrac{1}{2}hk_1 & k_2 &= f\left(x_n + \tfrac{1}{2}h, y_n + \tfrac{1}{2}hm_1, u_n + \tfrac{1}{2}hk_1\right) \\ m_3 &= u_n + \tfrac{1}{2}hk_2 & k_3 &= f\left(x_n + \tfrac{1}{2}h, y_n + \tfrac{1}{2}hm_2, u_n + \tfrac{1}{2}hk_2\right) \\ m_4 &= u_n + hk_3 & k_4 &= f(x_n + h, y_n + hm_3, u_n + hk_3). \end{aligned}$$

In general, we can express every nth-order differential equation $y^{(n)} = f(x, y, y', \ldots, y^{(n-1)})$ as a system of n first-order equations using the substitutions $y = u_1, y' = u_2, y'' = u_3, \ldots, y^{(n-1)} = u_n$.

EXAMPLE 1 Euler's Method

Use Euler's method to obtain the approximate value of $y(0.2)$, where $y(x)$ is the solution of the initial-value problem

$$y'' + xy' + y = 0, \quad y(0) = 1, \quad y'(0) = 2. \tag{5}$$

SOLUTION In terms of the substitution $y' = u$, the equation is equivalent to the system

$$\begin{aligned} y' &= u \\ u' &= -xu - y. \end{aligned}$$

Thus from (3) we obtain

$$\begin{aligned} y_{n+1} &= y_n + hu_n \\ u_{n+1} &= u_n + h[-x_nu_n - y_n]. \end{aligned}$$

Using the step size $h = 0.1$ and $y_0 = 1$, $u_0 = 2$, we find

$$y_1 = y_0 + (0.1)u_0 = 1 + (0.1)2 = 1.2$$

$$u_1 = u_0 + (0.1)[-x_0u_0 - y_0] = 2 + (0.1)[-(0)(2) - 1] = 1.9$$

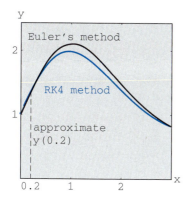

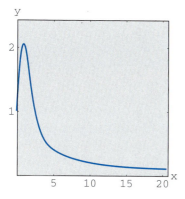

(a) Euler's method (black) and the RK4 method (color)

(b) RK4 method

FIGURE 9.3 Numerical solution curves generated by different methods

$$y_2 = y_1 + (0.1)u_1 = 1.2 + (0.1)(1.9) = 1.39$$

$$u_2 = u_1 + (0.1)[-x_1u_1 - y_1] = 1.9 + (0.1)[-(0.1)(1.9) - 1.2] = 1.761.$$

In other words, $y(0.2) \approx 1.39$ and $y'(0.2) \approx 1.761$.

With the aid of the graphing feature of a numerical solver, in Figure 9.3(a) we compare the solution curve of (5) generated by Euler's method ($h = 0.1$) on the interval [0, 3] with the solution curve generated by the RK4 method ($h = 0.1$). From Figure 9.3(b) it appears that the solution $y(x)$ of (4) has the property that $y(x) \to 0$ and $x \to \infty$.

If desired we can use the method of Section 6.1 to obtain two power series solutions of the differential equation in (5). But unless this method reveals that the DE possesses an elementary solution, we will still only be able to approximate $y(0.2)$ using a partial sum. Reinspection of the infinite series solutions of Airy's differential equation $y'' + xy = 0$, given on page 244, does not reveal the oscillatory behavior of the solutions $y_1(x)$ and $y_2(x)$ exhibited in the graphs in Figure 6.2. Those graphs were obtained from a numerical solver using the RK4 method with a step size of $h = 0.1$.

SYSTEMS REDUCED TO FIRST-ORDER SYSTEMS Using a procedure similar to that just discussed for second-order equations, we can often reduce a system of higher-order differential equations to a system of first-order equations by first solving for the highest-order derivative of each dependent variable and then making appropriate substitutions for the lower-order derivatives.

EXAMPLE 2   **A System Rewritten as a First-Order System**

Write

$$x'' - x' + 5x + 2y'' = e^t$$
$$-2x + y'' + 2y = 3t^2$$

as a system of first-order differential equations.

SOLUTION   Write the system as

$$x'' + 2y'' = e^t - 5x + x'$$
$$y'' = 3t^2 + 2x - 2y$$

and then eliminate y'' by multiplying the second equation by 2 and subtracting. This gives

$$x'' = -9x + 4y + x' + e^t - 6t^2.$$

Since the second equation of the system already expresses the highest-order derivative of y in terms of the remaining functions, we are now in a position to introduce new variables. If we let $x' = u$ and $y' = v$, the expressions for x'' and y'' become, respectively,

$$u' = x'' = -9x + 4y + u + e^t - 6t^2$$
$$v' = y'' = 2x - 2y + 3t^2.$$

The original system can then be written in the form

$$x' = u$$
$$y' = v$$
$$u' = -9x + 4y + u + e^t - 6t^2$$
$$v' = 2x - 2y + 3t^2.$$

It might not always be possible to carry out the reductions illustrated in Example 2.

NUMERICAL SOLUTION OF A SYSTEM The solution of a system of the form

$$\frac{dx_1}{dt} = f_1(t, x_1, x_2, \ldots, x_n)$$

$$\frac{dx_2}{dt} = f_2(t, x_1, x_2, \ldots, x_n)$$

$$\vdots \qquad \vdots$$

$$\frac{dx_n}{dt} = f_n(t, x_1, x_2, \ldots, x_n)$$

can be approximated by a version of Euler's, the Runge-Kutta, or the Adams-Bashforth-Moulton method adapted to the system. For instance, the RK4 method applied to the system

$$x' = f(t, x, y)$$

$$y' = g(t, x, y) \tag{6}$$

$$x(t_0) = x_0, \quad y(t_0) = y_0,$$

looks like this:

$$x_{n+1} = x_n + \frac{h}{6}(m_1 + 2m_2 + 2m_3 + m_4)$$

$$\tag{7}$$

$$y_{n+1} = y_n + \frac{h}{6}(k_1 + 2k_2 + 2k_3 + k_4),$$

where

$$m_1 = f(t_n, x_n, y_n) \qquad\qquad k_1 = g(t_n, x_n, y_n)$$

$$m_2 = f\left(t_n + \tfrac{1}{2}h, x_n + \tfrac{1}{2}hm_1, y_n + \tfrac{1}{2}hk_1\right) \qquad k_2 = g\left(t_n + \tfrac{1}{2}h, x_n + \tfrac{1}{2}hm_1, y_n + \tfrac{1}{2}hk_1\right)$$

$$m_3 = f\left(t_n + \tfrac{1}{2}h, x_n + \tfrac{1}{2}hm_2, y_n + \tfrac{1}{2}hk_2\right) \qquad k_3 = g\left(t_n + \tfrac{1}{2}h, x_n + \tfrac{1}{2}hm_2, y_n + \tfrac{1}{2}hk_2\right) \tag{8}$$

$$m_4 = f(t_n + h, x_n + hm_3, y_n + hk_3) \qquad k_4 = g(t_n + h, x_n + hm_3, y_n + hk_3).$$

EXAMPLE 3   RK4 Method

Consider the initial-value problem

$$x' = 2x + 4y$$

$$y' = -x + 6y$$

$$x(0) = -1, \quad y(0) = 6.$$

Use the RK4 method to approximate $x(0.6)$ and $y(0.6)$. Compare the results for $h = 0.2$ and $h = 0.1$.

SOLUTION We illustrate the computations of x_1 and y_1 with step size $h = 0.2$. With the identifications $f(t, x, y) = 2x + 4y$, $g(t, x, y) = -x + 6y$, $t_0 = 0$, $x_0 = -1$, and $y_0 = 6$, we see from (8) that

$$m_1 = f(t_0, x_0, y_0) = f(0, -1, 6) = 2(-1) + 4(6) = 22$$

$$k_1 = g(t_0, x_0, y_0) = g(0, -1, 6) = -1(-1) + 6(6) = 37$$

TABLE 9.8 $\quad h = 0.2$

t_n	x_n	y_n
0.00	-1.0000	6.0000
0.20	9.2453	19.0683
0.40	46.0327	55.1203
0.60	158.9430	150.8192

TABLE 9.9 $\quad h = 0.1$

t_n	x_n	y_n
0.00	-1.0000	6.0000
0.10	2.3840	10.8883
0.20	9.3379	19.1332
0.30	22.5541	32.8539
0.40	46.5103	55.4420
0.50	88.5729	93.3006
0.60	160.7563	152.0025

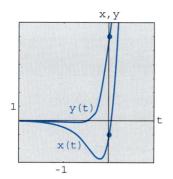

FIGURE 9.4 Numerical solution curves for IVP in Example 3

$$m_2 = f\left(t_0 + \tfrac{1}{2}h, x_0 + \tfrac{1}{2}hm_1, y_0 + \tfrac{1}{2}hk_1\right) = f(0.1, 1.2, 9.7) = 41.2$$

$$k_2 = g\left(t_0 + \tfrac{1}{2}h, x_0 + \tfrac{1}{2}hm_1, y_0 + \tfrac{1}{2}hk_1\right) = g(0.1, 1.2, 9.7) = 57$$

$$m_3 = f\left(t_0 + \tfrac{1}{2}h, x_0 + \tfrac{1}{2}hm_2, y_0 + \tfrac{1}{2}hk_2\right) = f(0.1, 3.12, 11.7) = 53.04$$

$$k_3 = g\left(t_0 + \tfrac{1}{2}h, x_0 + \tfrac{1}{2}hm_2, y_0 + \tfrac{1}{2}hk_2\right) = g(0.1, 3.12, 11.7) = 67.08$$

$$m_4 = f(t_0 + h, x_0 + hm_3, y_0 + hk_3) = f(0.2, 9.608, 19.416) = 96.88$$

$$k_4 = g(t_0 + h, x_0 + hm_3, y_0 + hk_3) = g(0.2, 9.608, 19.416) = 106.888.$$

Therefore from (7) we get

$$x_1 = x_0 + \frac{0.2}{6}\left(m_1 + 2m_2 + 2m_3 + m_4\right)$$

$$= -1 + \frac{0.2}{6}\left(22 + 2(41.2) + 2(53.04) + 96.88\right) = 9.2453$$

$$y_1 = y_0 + \frac{0.2}{6}\left(k_1 + 2k_2 + 2k_3 + k_4\right)$$

$$= 6 + \frac{0.2}{6}\left(37 + 2(57) + 2(67.08) + 106.888\right) = 19.0683,$$

where, as usual, the computed values of x_1 and y_1 are rounded to four decimal places. These numbers give us the approximation $x_1 \approx x(0.2)$ and $y_1 \approx y(0.2)$. The subsequent values, obtained with the aid of a computer, are summarized in Tables 9.8 and 9.9.

You should verify that the solution of the initial-value problem in Example 3 is given by $x(t) = (26t - 1)e^{4t}$, $y(t) = (13t + 6)e^{4t}$. From these equations we see that the actual values $x(0.6) = 160.9384$ and $y(0.6) = 152.1198$ compare favorably with the entries in the last line of Table 9.9. The graph of the solution in a neighborhood of $t = 0$ is shown in Figure 9.4; the graph was obtained from a numerical solver using the RK4 method with $h = 0.1$.

In conclusion, we state Euler's method for the general system (6):

$$x_{n+1} = x_n + hf(t_n, x_n, y_n)$$

$$y_{n+1} = y_n + hg(t_n, x_n, y_n).$$

EXERCISES 9.4

Answers to selected odd-numbered problems begin on page ANS-14.

1. Use Euler's method to approximate $y(0.2)$, where $y(x)$ is the solution of the initial-value problem

$$y'' - 4y' + 4y = 0, \quad y(0) = -2, \quad y'(0) = 1.$$

Use $h = 0.1$. Find the analytic solution of the problem, and compare the actual value of $y(0.2)$ with y_2.

2. Use Euler's method to approximate $y(1.2)$, where $y(x)$ is the solution of the initial-value problem

$$x^2 y'' - 2xy' + 2y = 0, \quad y(1) = 4, \quad y'(1) = 9,$$

where $x > 0$. Use $h = 0.1$. Find the analytic solution of the problem, and compare the actual value of $y(1.2)$ with y_2.

In Problems 3 and 4 repeat the indicated problem using the RK4 method. First use $h = 0.2$ and then use $h = 0.1$.

3. Problem 1

4. Problem 2

5. Use the RK4 method to approximate $y(0.2)$, where $y(x)$ is the solution of the initial-value problem

$$y'' - 2y' + 2y = e^t \cos t, \quad y(0) = 1, \quad y'(0) = 2.$$

First use $h = 0.2$ and then use $h = 0.1$.

6. When $E = 100$ V, $R = 10\ \Omega$, and $L = 1$ h, the system of differential equations for the currents $i_1(t)$ and $i_3(t)$ in the electrical network given in Figure 9.5 is

$$\frac{di_1}{dt} = -20i_1 + 10i_3 + 100$$

$$\frac{di_3}{dt} = 10i_1 - 20i_3,$$

where $i_1(0) = 0$ and $i_3(0) = 0$. Use the RK4 method to approximate $i_1(t)$ and $i_3(t)$ at $t = 0.1, 0.2, 0.3, 0.4$, and 0.5. Use $h = 0.1$. Use a numerical solver to graph the solution on the interval $0 \le t \le 5$. Use the graphs to predict the behavior of $i_1(t)$ and $i_3(t)$ as $t \to \infty$.

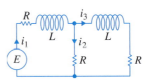

FIGURE 9.5   Network in Problem 6

In Problems 7–12 use the Runge-Kutta method to approximate $x(0.2)$ and $y(0.2)$. First use $h = 0.2$ and then use $h = 0.1$. Use a numerical solver and $h = 0.1$ to graph the solution in a neighborhood of $t = 0$.

7. $x' = 2x - y$
$\quad\ y' = x$
$\quad\ x(0) = 6, \quad y(0) = 2$

8. $x' = x + 2y$
$\quad\ y' = 4x + 3y$
$\quad\ x(0) = 1, \quad y(0) = 1$

9. $x' = -y + t$
$\quad\ y' = x - t$
$\quad\ x(0) = -3, \quad y(0) = 5$

10. $x' = 6x + y + 6t$
$\quad\ y' = 4x + 3y - 10t + 4$
$\quad\ x(0) = 0.5, \quad y(0) = 0.2$

11. $x' + 4x - y' = 7t$
$\quad\ x' + y' - 2y = 3t$
$\quad\ x(0) = 1, \quad y(0) = -2$

12. $\qquad\quad x' + y' = 4t$
$\quad -x' + y' + y = 6t^2 + 10$
$\quad\ x(0) = 3, \quad y(0) = -1$

9.5   SECOND-ORDER BOUNDARY-VALUE PROBLEMS

INTRODUCTION: We just saw in Section 9.4 how to approximate the solution of a *second-order initial-value problem* $y'' = f(x, y, y')$, $y(x_0) = y_0$, $y'(x_0) = u_0$. In this section we examine two methods for approximating a solution of a *second-order boundary-value problem* $y'' = f(x, y, y')$, $y(a) = \alpha$, $y(b) = \beta$. Unlike the procedures used with second-order initial-value problems, the methods of second-order boundary-value problems do not require writing the second-order DE as a system of first-order DEs.

REVIEW MATERIAL: Boundary-value problems were introduced in Section 4.1 (page 128), solved in Exercises 4.3 (Problems 37–40), Exercises 4.4 (Problems 37–40), and discussed in depth in Section 5.2.

FINITE DIFFERENCE APPROXIMATIONS   The Taylor series expansion, centered at a point a, of a function $y(x)$ is

$$y(x) = y(a) + y'(a)\frac{x-a}{1!} + y''(a)\frac{(x-a)^2}{2!} + y'''(a)\frac{(x-a)^3}{3!} + \cdots .$$

If we set $h = x - a$, then the preceding line is the same as

$$y(x) = y(a) + y'(a)\frac{h}{1!} + y''(a)\frac{h^2}{2!} + y'''(a)\frac{h^3}{3!} + \cdots .$$

For the subsequent discussion it is convenient then to rewrite this last expression in two alternative forms:

$$y(x + h) = y(x) + y'(x)h + y''(x)\frac{h^2}{2} + y'''(x)\frac{h^3}{6} + \cdots \tag{1}$$

and $$\quad y(x - h) = y(x) - y'(x)h + y''(x)\frac{h^2}{2} - y'''(x)\frac{h^3}{6} + \cdots . \tag{2}$$

If h is small, we can ignore terms involving h^4, h^5, . . . since these values are negligible. Indeed, if we ignore all terms involving h^2 and higher, then solving (1) and (2), in turn, for $y'(x)$ yields the following approximations for the first derivative:

$$y'(x) \approx \frac{1}{h}[y(x + h) - y(x)] \tag{3}$$

$$y'(x) \approx \frac{1}{h}[y(x) - y(x - h)]. \tag{4}$$

Subtracting (1) and (2) also gives

$$y'(x) \approx \frac{1}{2h}[y(x + h) - y(x - h)]. \tag{5}$$

On the other hand, if we ignore terms involving h^3 and higher, then by adding (1) and (2), we obtain an approximation for the second derivative $y''(x)$:

$$y''(x) \approx \frac{1}{h^2}[y(x + h) - 2y(x) + y(x - h)]. \tag{6}$$

The right-hand sides of (3), (4), (5), and (6) are called **difference quotients.** The expressions

$$y(x + h) - y(x), \quad y(x) - y(x - h), \quad y(x + h) - y(x - h),$$

and $$\qquad\qquad y(x + h) - 2y(x) + y(x - h)$$

are called **finite differences.** Specifically, $y(x + h) - y(x)$ is called a **forward difference,** $y(x) - y(x - h)$ is a **backward difference,** and both $y(x + h) - y(x - h)$ and $y(x + h) - 2y(x) + y(x - h)$ are called **central differences.** The results given in (5) and (6) are referred to as **central difference approximations** for the derivatives y' and y''.

FINITE DIFFERENCE METHOD   Consider now a linear second-order boundary-value problem

$$y'' + P(x)y' + Q(x)y = f(x), \quad y(a) = \alpha, \quad y(b) = \beta. \tag{7}$$

Suppose $a = x_0 < x_1 < x_2 < \cdots < x_{n-1} < x_n = b$ represents a regular partition of the interval $[a, b]$—that is, $x_i = a + ih$, where $i = 0, 1, 2, \ldots, n$ and $h = (b - a)/n$. The points

$$x_1 = a + h, \quad x_2 = a + 2h, \ldots, \quad x_{n-1} = a + (n - 1)h$$

are called **interior mesh points** of the interval $[a, b]$. If we let

$$y_i = y(x_i), \quad P_i = P(x_i), \quad Q_i = Q(x_i), \quad \text{and} \quad f_i = f(x_i)$$

and if y'' and y' in (7) are replaced by the central difference approximations (5) and (6), we get

$$\frac{y_{i+1} - 2y_i + y_{i-1}}{h^2} + P_i\frac{y_{i+1} - y_{i-1}}{2h} + Q_iy_i = f_i$$

or, after simplifying,

$$\left(1 + \frac{h}{2}P_i\right)y_{i+1} + (-2 + h^2Q_i)y_i + \left(1 - \frac{h}{2}P_i\right)y_{i-1} = h^2f_i. \tag{8}$$

The last equation, known as a **finite difference equation,** is an approximation to the differential equation. It enables us to approximate the solution $y(x)$ of (7) at the interior mesh points $x_1, x_2, \ldots, x_{n-1}$ of the interval $[a, b]$. By letting i take on the values $1, 2, \ldots, n - 1$ in (8), we obtain $n - 1$ equations in the $n - 1$ unknowns $y_1, y_2, \ldots, y_{n-1}$. Bear in mind that we know y_0 and y_n since these are the prescribed boundary conditions $y_0 = y(x_0) = y(a) = \alpha$ and $y_n = y(x_n) = y(b) = \beta$.

In Example 1 we consider a boundary-value problem for which we can compare the approximate values found with the actual values of an explicit solution.

EXAMPLE 1   Using the Finite Difference Method

Use the difference equation (8) with $n = 4$ to approximate the solution of the boundary-value problem $y'' - 4y = 0$, $y(0) = 0$, $y(1) = 5$.

SOLUTION   To use (8), we identify $P(x) = 0$, $Q(x) = -4$, $f(x) = 0$, and $h = (1 - 0)/4 = \frac{1}{4}$. Hence the difference equation is

$$y_{i+1} - 2.25y_i + y_{i-1} = 0. \qquad (9)$$

Now the interior points are $x_1 = 0 + \frac{1}{4}, x_2 = 0 + \frac{2}{4}, x_3 = 0 + \frac{3}{4}$, and so for $i = 1, 2$, and 3, (9) yields the following system for the corresponding y_1, y_2, and y_3:

$$y_2 - 2.25y_1 + y_0 = 0$$

$$y_3 - 2.25y_2 + y_1 = 0$$

$$y_4 - 2.25y_3 + y_2 = 0.$$

With the boundary conditions $y_0 = 0$ and $y_4 = 5$, the foregoing system becomes

$$-2.25y_1 + \quad y_2 \qquad\qquad = 0$$

$$y_1 - 2.25y_2 + \quad y_3 = 0$$

$$y_2 - 2.25y_3 = -5.$$

Solving the system gives $y_1 = 0.7256$, $y_2 = 1.6327$, and $y_3 = 2.9479$.

Now the general solution of the given differential equation is $y = c_1 \cosh 2x + c_2 \sinh 2x$. The condition $y(0) = 0$ implies that $c_1 = 0$. The other boundary condition gives c_2. In this way we see that a solution of the boundary-value problem is $y(x) = (5 \sinh 2x)/\sinh 2$. Thus the actual values (rounded to four decimal places) of this solution at the interior points are as follows: $y(0.25) = 0.7184$, $y(0.5) = 1.6201$, and $y(0.75) = 2.9354$.

The accuracy of the approximations in Example 1 can be improved by using a smaller value of h. Of course, the trade-off here is that a smaller value of h necessitates solving a larger system of equations. It is left as an exercise to show that with $h = \frac{1}{8}$, approximations to $y(0.25)$, $y(0.5)$, and $y(0.75)$ are 0.7202, 1.6233, and 2.9386, respectively. See Problem 11 in Exercises 9.5.

EXAMPLE 2   Using the Finite Difference Method

Use the difference equation (8) with $n = 10$ to approximate the solution of

$$y'' + 3y' + 2y = 4x^2, \quad y(1) = 1, \quad y(2) = 6.$$

SOLUTION In this case we identify $P(x) = 3$, $Q(x) = 2$, $f(x) = 4x^2$, and $h = (2 - 1)/10 = 0.1$, and so (8) becomes

$$1.15y_{i+1} - 1.98y_i + 0.85y_{i-1} = 0.04x_i^2. \tag{10}$$

Now the interior points are $x_1 = 1.1$, $x_2 = 1.2$, $x_3 = 1.3$, $x_4 = 1.4$, $x_5 = 1.5$, $x_6 = 1.6$, $x_7 = 1.7$, $x_8 = 1.8$, and $x_9 = 1.9$. For $i = 1, 2, \ldots, 9$ and $y_0 = 1$, $y_{10} = 6$, (10) gives a system of nine equations and nine unknowns:

$$1.15y_2 - 1.98y_1 \qquad\qquad = -0.8016$$
$$1.15y_3 - 1.98y_2 + 0.85y_1 = 0.0576$$
$$1.15y_4 - 1.98y_3 + 0.85y_2 = 0.0676$$
$$1.15y_5 - 1.98y_4 + 0.85y_3 = 0.0784$$
$$1.15y_6 - 1.98y_5 + 0.85y_4 = 0.0900$$
$$1.15y_7 - 1.98y_6 + 0.85y_5 = 0.1024$$
$$1.15y_8 - 1.98y_7 + 0.85y_6 = 0.1156$$
$$1.15y_9 - 1.98y_8 + 0.85y_7 = 0.1296$$
$$\qquad\quad -1.98y_9 + 0.85y_8 = -6.7556.$$

We can solve this large system using Gaussian elimination or, with relative ease, by means of a computer algebra system. The result is found to be $y_1 = 2.4047$, $y_2 = 3.4432$, $y_3 = 4.2010$, $y_4 = 4.7469$, $y_5 = 5.1359$, $y_6 = 5.4124$, $y_7 = 5.6117$, $y_8 = 5.7620$, and $y_9 = 5.8855$. ▌

SHOOTING METHOD Another way of approximating a solution of a boundary-value problem $y'' = f(x, y, y')$, $y(a) = \alpha$, $y(b) = \beta$ is called the **shooting method.** The starting point in this method is the replacement of the boundary-value problem by an initial-value problem

$$y'' = f(x, y, y'), \quad y(a) = \alpha, \quad y'(a) = m_1. \tag{11}$$

The number m_1 in (11) is simply a guess for the unknown slope of the solution curve at the known point $(a, y(a))$. We then apply one of the step-by-step numerical techniques to the second-order equation in (11) to find an approximation β_1 for the value of $y(b)$. If β_1 agrees with the given value $y(b) = \beta$ to some preassigned tolerance, we stop; otherwise the calculations are repeated, starting with a different guess $y'(a) = m_2$ to obtain a second approximation β_2 for $y(b)$. This method can be continued in a trial-and-error manner or the subsequent slopes $m_3, m_4, \ldots$ can be adjusted in some systematic way; linear interpolation is particularly successful when the differential equation in (11) is linear. The procedure is analogous to shooting (the "aim" is the choice of the initial slope) at a target until the bull's-eye $y(b)$ is hit. See Problem 14 in Exercises 9.5.

Of course, underlying the use of these numerical methods is the assumption, which we know is not always warranted, that a solution of the boundary-value problem exists.

REMARKS

The approximation method using finite differences can be extended to boundary-value problems in which the first derivative is specified at a boundary—for example, a problem such as $y'' = f(x, y, y')$, $y'(a) = \alpha$, $y(b) = \beta$. See Problem 13 in Exercises 9.5.

EXERCISES 9.5

Answers to selected odd-numbered problems begin on page ANS-14.

In Problems 1–10 use the finite difference method and the indicated value of n to approximate the solution of the given boundary-value problem.

1. $y'' + 9y = 0$, $y(0) = 4$, $y(2) = 1$; $n = 4$

2. $y'' - y = x^2$, $y(0) = 0$, $y(1) = 0$; $n = 4$

3. $y'' + 2y' + y = 5x$, $y(0) = 0$, $y(1) = 0$; $n = 5$

4. $y'' - 10y' + 25y = 1$, $y(0) = 1$, $y(1) = 0$; $n = 5$

5. $y'' - 4y' + 4y = (x + 1)e^{2x}$,
   $y(0) = 3$, $y(1) = 0$; $n = 6$

6. $y'' + 5y' = 4\sqrt{x}$, $y(1) = 1$, $y(2) = -1$; $n = 6$

7. $x^2 y'' + 3xy' + 3y = 0$, $y(1) = 5$, $y(2) = 0$; $n = 8$

8. $x^2 y'' - xy' + y = \ln x$, $y(1) = 0$, $y(2) = -2$; $n = 8$

9. $y'' + (1 - x)y' + xy = x$, $y(0) = 0$, $y(1) = 2$; $n = 10$

10. $y'' + xy' + y = x$, $y(0) = 1$, $y(1) = 0$; $n = 10$

11. Rework Example 1 using $n = 8$.

12. The electrostatic potential u between two concentric spheres of radius $r = 1$ and $r = 4$ is determined from

$$\frac{d^2 u}{dr^2} + \frac{2}{r}\frac{du}{dr} = 0, \quad u(1) = 50, \quad u(4) = 100.$$

Use the method of this section with $n = 6$ to approximate the solution of this boundary-value problem.

13. Consider the boundary-value problem $y'' + xy = 0$, $y'(0) = 1$, $y(1) = -1$.

    (a) Find the difference equation corresponding to the differential equation. Show that for $i = 0, 1, 2, \ldots, n - 1$ the difference equation yields n equations in $n + 1$ unknowns $y_{-1}, y_0, y_1, y_2, \ldots, y_{n-1}$. Here y_{-1} and y_0 are unknowns since y_{-1} represents an approximation to y at the exterior point $x = -h$ and y_0 is not specified at $x = 0$.

    (b) Use the central difference approximation (5) to show that $y_1 - y_{-1} = 2h$. Use this equation to eliminate y_{-1} from the system in part (a).

    (c) Use $n = 5$ and the system of equations found in parts (a) and (b) to approximate the solution of the original boundary-value problem.

COMPUTER LAB ASSIGNMENTS

14. Consider the boundary-value problem $y'' = y' - \sin(xy)$, $y(0) = 1$, $y(1) = 1.5$. Use the shooting method to approximate the solution of this problem. (The approximation can be obtained using a numerical technique — say, the RK4 method with $h = 0.1$; or, even better, if you have access to a CAS such as *Mathematica* or *Maple*, the **NDSolve** function can be used.)

CHAPTER 9 IN REVIEW

Answers to selected odd-numbered problems begin on page ANS-14.

In Problems 1–4 construct a table comparing the indicated values of $y(x)$ using Euler's method, the improved Euler's method, and the RK4 method. Compute to four rounded decimal places. First use $h = 0.1$ and then use $h = 0.05$.

1. $y' = 2 \ln xy$, $y(1) = 2$;
   $y(1.1), y(1.2), y(1.3), y(1.4), y(1.5)$

2. $y' = \sin x^2 + \cos y^2$, $y(0) = 0$;
   $y(0.1), y(0.2), y(0.3), y(0.4), y(0.5)$

3. $y' = \sqrt{x + y}$, $y(0.5) = 0.5$;
   $y(0.6), y(0.7), y(0.8), y(0.9), y(1.0)$

4. $y' = xy + y^2$, $y(1) = 1$;
   $y(1.1), y(1.2), y(1.3), y(1.4), y(1.5)$

5. Use Euler's method to approximate $y(0.2)$, where $y(x)$ is the solution of the initial-value problem

$y'' - (2x + 1)y = 1$, $y(0) = 3$, $y'(0) = 1$. First use one step with $h = 0.2$, and then repeat the calculations using two steps with $h = 0.1$.

6. Use the Adams-Bashforth-Moulton method to approximate $y(0.4)$, where $y(x)$ is the solution of the initial-value problem $y' = 4x - 2y$, $y(0) = 2$. Use $h = 0.1$ and the RK4 method to compute y_1, y_2, and y_3.

7. Use Euler's method with $h = 0.1$ to approximate $x(0.2)$ and $y(0.2)$, where $x(t)$, $y(t)$ is the solution of the initial-value problem

$$x' = x + y$$

$$y' = x - y$$

$$x(0) = 1, \quad y(0) = 2.$$

8. Use the finite difference method with $n = 10$ to approximate the solution of the boundary-value problem $y'' + 6.55(1 + x)y = 1$, $y(0) = 0$, $y(1) = 0$.

PROJECT 9

THE HAMMER

Your job is to scare people. You design amusement park rides for a living. Typically, you design roller coasters and water slides, and your latest contract is yet another variation on a proven winner. You must design a double "hammer" style ride (see Figure 1(a)).* Passengers are strapped into 16 seats in a gondola located at the end of one of two long arms, the pivot of which is attached to the top of a vertical tower. See Figure 1(b). The arms then begin to swing in opposite directions. The idea is to get the arms to swing in progressively larger oscillations until they just reach the top of their rotation and then disengage the motor and allow the whole apparatus to swing freely for a time. The passenger cabins themselves are mounted on axles so that they spin freely as well. The result is something like a ferris wheel for small oscillations, but when large oscillations are reached, the passengers are subjected to large forces that result from the swinging on the pendulum, combined with smaller spinning motions. Your assignment is to achieve terror and sickness, all in perfect safety.

The ride is basically a pair of driven pendulums that undergo large oscillations, so you will model only one of them. You must first address the driving force required to achieve the large oscillations. You won't worry about the rotation of the cabin for the moment. You have read Section 5.3 so you already know that a nonlinear model for the motion of an undriven pendulum is

$$\frac{d^2\theta}{dt^2} + \frac{g}{l}\sin\theta = 0,$$

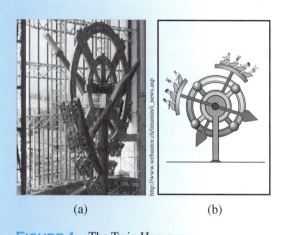

(a)   (b)

FIGURE 1   The Twin Hammer

where g is the constant acceleration due to gravity (32 ft/s^2), l is the length of the pendulum (which you take to be 50 feet), and θ is the angle the pendulum makes with its rest position. In your case there is an engine that drives first one pendulum for part of its cycle and then switches to drive the other while the first swings freely. The engine finally disengages completely after 40 seconds. Thus, you introduce a forcing function of the form

$$F(\theta', t) = \begin{cases} C, & \theta' > 0 \quad \text{and} \quad t < 40 \\ 0, & \text{otherwise.} \end{cases}$$

The differential equation of motion becomes

$$\frac{d^2\theta}{dt^2} + \frac{g}{l}\sin\theta = F(\theta', t). \tag{1}$$

For the engine you plan to use for this apparatus, $C = 0.11$. The 40 seconds that the engine is engaged is a design specification of the company for whom you are working, and has to do with the duration of the ride. You need to find out whether the engine can give a reasonable ride in those circumstances. You could linearize the left-hand side of the DE in (1) (see page 223 of the text) and solve it exactly, but you know that approximation is valid for small oscillations—exactly

---

*Figure 1 shows the 32-passenger Suspended Twin Hammer, one of many innovative thrill rides designed and built by INTAMIN AG, Verenastrasse 37, P.O. Box 95, CH-8832 Wollerau, Switzerland. See www.intaminworldwide.com.

the opposite of the current situation. In this case linearization equals lawsuit. You need instead to solve the problem numerically.

Your first thought is to apply the simple Euler method to the problem, and so the second-order differential equation must be converted to a system of first-order equations.

PROBLEM 1. Show that the DE in (1) is equivalent to the first-order system

$$\theta' = \varphi$$

$$\varphi' = F(\varphi, t) - \frac{g}{l} \sin \theta. \tag{2}$$

You may work this problem using a CAS or calculator, but will probably find it easiest to use the **Hammer Tool** on the *DE Tools* CD.

PROBLEM 2 (CD). Use the **Hammer Tool** on the *DE Tools* CD and the Euler method to solve (2). Use the initial conditions $\theta(0) = 0$, $\varphi(0) = 0$, and a time step of $h = 0.1$ s. Run the method for 40 seconds.

By examining the numerical solution curve in Problem 2, you conclude that the outlook is dark. If your results are to believed, then the riders will be thrown around at an increasing rate. You'll have to use a different engine, or engage it less often as the oscillations increase, or run it for a shorter time. Before you start changing the design radically, it occurs to you to get a better approximation.

PROBLEM 3 (CD). Rework Problem 2, but this time use a step size that is half as large, that is, $h = 0.05$. Plot your numerical solution curve with the one obtained in Problem 2. How do they differ?

You notice that changing the step size seems to improve the results somewhat — that is, people are not jostled around as much. Perhaps if you decrease the step size even more, they will survive. No, that is not right. You try shortening the time step more and more, but the solution always seems eventually to blow up.

What happens after the engine is disengaged? The cabin swings freely then. For the moment you ignored all damping forces, so you have a pretty good idea of how the solutions should behave qualitatively. What does the Euler method show?

PROBLEM 4 (CD). Use the **Hammer Tool** on the *DE Tools* CD and the Euler method to analyze the motion of the pendulum after the engine is disengaged. In particular, use initial conditions after 40 seconds have elapsed; for example, $\theta(40) = 2$, $\varphi(40) = 0$. Use a time step $h = 0.1$ and run the method for only 20 seconds.

Now you are really upset. Even after the engine is turned off, the passengers go around and around with ever increasing speed. If this is true, then the people on this ride will share the distinction of being the first to reach outer space without the aid of a rocket. Clearly, the Euler method is not giving you an accurate picture of the behavior of the ride.

It turns out that the Euler method has more problems than its truncation error that was discussed in Section 9.1. You can shorten the step size as much as you want to improve the local error, but it won't help the *stability,* discussed in Section 9.3, of the numerical method for this problem, and won't prevent it from giving results that do not comply with the laws of physics. Further research indicates that the Runge-Kutta method of order two would have the same problem.

It is more painful, but you can program the RK4 method for this problem.

> **PROBLEM 5 (CD).** Use the **Hammer Tool** on the *DE Tools* CD to compare the results of the Euler method in Problem 2 with the RK4 method using the same step size. Run both methods for 50 seconds. Describe the qualitative difference in the results.

This is more like it. The RK4 method gives a solution of the driven equation that increases initially, reaches a maximal amplitude, and then stays there. This is what you expect and what you want. Perhaps it is an accurate representation of the behavior of the ride, so you can get on with the design. It turns out that for a sufficiently short step size, the RK4 method is numerically stable for this problem. For us this means that it gives results which, while approximate, mimic the behavior of the true solution qualitatively.

> **PROBLEM 6.** Discuss ways that you can verify that the numerical solution you have computed with the RK4 method in Problem 5 is qualitatively accurate. Don't even try to suggest that you can find an exact solution of the equation.

You are now ready to get into the details of the design. If you do your work correctly, the passengers will be stirred, not shaken. The client will be happy, your employer will be happy, and most (but not all) of the riders will be happy. Such is the power of stable numerical methods.

STILL CURIOUS?

> **PROBLEM 7 (CD).** Solve Problem 4 one more time. Now use the Adams-Bashforth predictor method with a step size $h = 0.25$. Do not use the Adams-Moulton corrector. Compare the numerical solution curve obtained from this method with those obtained from the previous two single-step methods. Note that the Adams-Bashforth scheme is of fourth order. How might you explain the results?

10

Plane Autonomous Systems

In Chapter 8 we used matrix techniques to solve systems of linear first-order differential equations. When a system of DEs is not linear, it is usually not possible to find solutions in terms of elementary functions. In this chapter we analyze linear and nonlinear autonomous systems qualitatively. We will show that valuable information about the geometric nature of the solutions of nonlinear systems can be obtained by first analyzing special constant solutions, called critical points, and then searching for periodic solutions. The important concepts of stability will be illustrated with examples from physics and ecology.

Phase Portrait: Nonlinear system

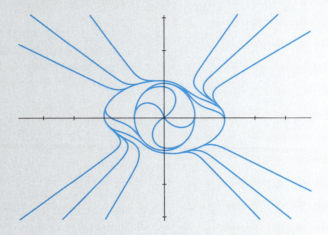

Phase Portrait: Linearization of nonlinear system

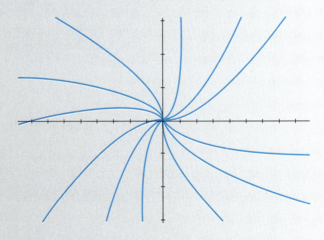

Phase Portrait: Soft spring

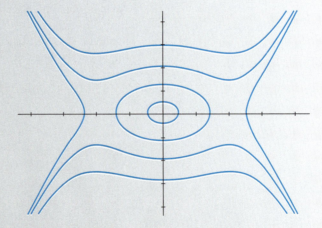

Phase portraits for autonomous systems. The two closed trajectories in the last figure indicate that the DE has periodic solutions. See pages 416 and 417.

10.1 AUTONOMOUS SYSTEMS

INTRODUCTION: We introduced the notions of autonomous first-order DEs, critical points of an autonomous DE, and the stability of a critical point in Section 2.1. This earlier consideration of stability was purposely kept at a fairly intuitive level; it is now time to give the precise definition of this concept. To do this, we need to examine autonomous *systems* of first-order DEs. In this section we define critical points of autonomous systems of two first-order DEs; the autonomous systems can be linear or nonlinear.

REVIEW MATERIAL: A rereading of pages 42–46 of Section 2.1 is highly recommended.

AUTONOMOUS SYSTEMS A system of first-order differential equations is said to be **autonomous** when the system can be written in the form

$$
\begin{aligned}
\frac{dx_1}{dt} &= g_1(x_1, x_2, \ldots, x_n)\\
\frac{dx_2}{dt} &= g_2(x_1, x_2, \ldots, x_n)\\
&\ \ \vdots \qquad\qquad \vdots\\
\frac{dx_n}{dt} &= g_n(x_1, x_2, \ldots, x_n).
\end{aligned}
\tag{1}
$$

Observe that the independent variable t does not appear explicitly on the right-hand side of each differential equation. Compare (1) with the general system given in (2) of Section 8.1.

EXAMPLE 1   A Nonautonomous System

The system of nonlinear first-order differential equations

$$
\frac{dx_1}{dt} = x_1 - 3x_2 + t^2 \quad\overset{t\text{ dependence}}{}
$$

$$
\frac{dx_2}{dt} = t x_1 \sin x_2 \quad\underset{t\text{ dependence}}{}
$$

is *not* autonomous because of the presence of t on the right-hand sides of both DEs. ∎

NOTE When $n = 1$ in (1), a single first-order differential equation takes on the form $dx/dt = g(x)$. This last equation is equivalent to (1) of Section 2.1 with the symbols x and t playing the parts of y and x, respectively. Explicit solutions can be constructed since the differential equation $dx/dt = g(x)$ is separable, and we will make use of this fact to give illustrations of the concepts in this chapter.

SECOND-ORDER DE AS A SYSTEM Any second-order differential equation $x'' = g(x, x')$ can be written as an autonomous system. As we did in Section 4.9, if we let $y = x'$, then $x'' = g(x, x')$ becomes $y' = g(x, y)$. Thus the second-order differential equation becomes the system of two first-order equations

$$
\begin{aligned}
x' &= y\\
y' &= g(x, y).
\end{aligned}
$$

EXAMPLE 2 The Pendulum DE as an Autonomous System

In (6) of Section 5.3 we showed that the displacement angle θ for a pendulum satisfies the nonlinear second-order differential equation

$$\frac{d^2\theta}{dt^2} + \frac{g}{l}\sin\theta = 0.$$

If we let $x = \theta$ and $y = \theta'$, this second-order differential equation can be rewritten as the autonomous system

$$x' = y$$

$$y' = -\frac{g}{l}\sin x.$$

NOTATION If $\mathbf{X}(t)$ and $\mathbf{g}(\mathbf{X})$ denote the respective column vectors

$$\mathbf{X}(t) = \begin{pmatrix} x_1(t) \\ x_2(t) \\ \vdots \\ x_n(t) \end{pmatrix}, \qquad \mathbf{g}(\mathbf{X}) = \begin{pmatrix} g_1(x_1, x_2, \ldots, x_n) \\ g_2(x_1, x_2, \ldots, x_n) \\ \vdots \\ g_n(x_1, x_2, \ldots, x_n) \end{pmatrix},$$

then the autonomous system (1) may be written in the compact **column vector form** $\mathbf{X}' = \mathbf{g}(\mathbf{X})$. The homogeneous linear system $\mathbf{X}' = \mathbf{AX}$ studied in Section 8.2 is an important special case.

In this chapter it is also convenient to write (1) using row vectors. If we let $\mathbf{X}(t) = (x_1(t), x_2(t), \ldots, x_n(t))$ and

$$\mathbf{g}(\mathbf{X}) = (g_1(x_1, x_2, \ldots, x_n), g_2(x_1, x_2, \ldots, x_n), \ldots, g_n(x_1, x_2, \ldots, x_n)),$$

then the autonomous system (1) may also be written in the compact **row vector form** $\mathbf{X}' = \mathbf{g}(\mathbf{X})$. *It should be clear from the context whether we are using column or row vector form; therefore we will not distinguish between $\mathbf{X}$ and $\mathbf{X}^T$, the transpose of $\mathbf{X}$.* In particular, when $n = 2$, it is convenient to use row vector form and write an initial condition as $\mathbf{X}(0) = (x_0, y_0)$.

When the variable t is interpreted as time, we can refer to the system of differential equations in (1) as a **dynamical system** and a solution $\mathbf{X}(t)$ as the **state of the system** or the **response of the system** at time t. With this terminology, a dynamical system is autonomous when the rate $\mathbf{X}'(t)$ at which the system changes depends only on the system's present state $\mathbf{X}(t)$. The linear system $\mathbf{X}' = \mathbf{AX} + \mathbf{F}(t)$ studied in Chapter 8 is then autonomous when $\mathbf{F}(t)$ is constant. In the case $n = 2$ or 3 we can call a solution a **path** or **trajectory** since we may think of $x = x_1(t)$, $y = x_2(t)$, and $z = x_3(t)$ as the parametric equations of a curve.

VECTOR FIELD INTERPRETATION When $n = 2$, the system in (1) is called a **plane autonomous system,** and we write the system as

$$\frac{dx}{dt} = P(x, y)$$

$$\frac{dy}{dt} = Q(x, y). \tag{2}$$

The vector $\mathbf{V}(x, y) = (P(x, y), Q(x, y))$ defines a **vector field** in a region of the plane, and a solution to the system may be interpreted as the resulting path of a particle as it moves through the region. To be more specific, let $\mathbf{V}(x, y) = (P(x, y), Q(x, y))$ denote the velocity of a stream at position (x, y), and suppose that a small particle (such as a cork) is released at a position (x_0, y_0) in the stream. If $\mathbf{X}(t) = (x(t), y(t))$ denotes the position of the particle at time t, then $\mathbf{X}'(t) = (x'(t), y'(t))$ is the velocity vector $\mathbf{V}$. When external forces are not present and frictional forces are neglected, the velocity of the particle at time t is the velocity of the stream at position $\mathbf{X}(t)$:

$$\mathbf{X}'(t) = \mathbf{V}(x(t), y(t)) \quad \text{or} \quad \begin{aligned} \frac{dx}{dt} &= P(x(t), y(t)) \\ \frac{dy}{dt} &= Q(x(t), y(t)). \end{aligned}$$

Thus the path of the particle is a solution to the system that satisfies the initial condition $\mathbf{X}(0) = (x_0, y_0)$. We will frequently call on this simple interpretation of a plane autonomous system to illustrate new concepts.

EXAMPLE 3　　**Plane Autonomous System of a Vector Field**

A vector field for the steady-state flow of a fluid around a cylinder of radius 1 is given by

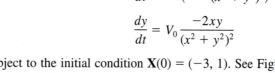

$$\mathbf{V}(x, y) = V_0\left(1 - \frac{x^2 - y^2}{(x^2 + y^2)^2}, \frac{-2xy}{(x^2 + y^2)^2}\right),$$

where V_0 is the speed of the fluid far from the cylinder. If a small cork is released at $(-3, 1)$, the path $\mathbf{X}(t) = (x(t), y(t))$ of the cork satisfies the plane autonomous system

$$\frac{dx}{dt} = V_0\left(1 - \frac{x^2 - y^2}{(x^2 + y^2)^2}\right)$$

$$\frac{dy}{dt} = V_0 \frac{-2xy}{(x^2 + y^2)^2}$$

subject to the initial condition $\mathbf{X}(0) = (-3, 1)$. See Figure 10.1 and Problem 46 in Exercises 2.4. ▪

TYPES OF SOLUTIONS　If $P(x, y)$, $Q(x, y)$, and the first-order partial derivatives $\partial P/\partial x$, $\partial P/\partial y$, $\partial Q/\partial x$, and $\partial Q/\partial y$ are continuous in a region R of the plane, then a solution of the plane autonomous system (2) that satisfies $\mathbf{X}(0) = \mathbf{X}_0$ is unique and of one of three basic types:

(*i*)　A **constant solution** $x(t) = x_0$, $y(t) = y_0$ (or $\mathbf{X}(t) = \mathbf{X}_0$ for all t). A constant solution is called a **critical** or **stationary point.** When the particle is placed at a critical point $\mathbf{X}_0$ (that is, $\mathbf{X}(0) = \mathbf{X}_0$), it remains there indefinitely. For this reason a constant solution is also called an **equilibrium solution.** Note that because $\mathbf{X}'(t) = \mathbf{0}$, a critical point is a solution of the system of algebraic equations

$$P(x, y) = 0$$

$$Q(x, y) = 0.$$

(*ii*)　A solution $x = x(t)$, $y = y(t)$ that defines an **arc**—a plane curve that does *not* cross itself. Thus the curve in Figure 10.2(a) can be a solution to a plane autonomous system, whereas the curve in Figure 10.2(b) cannot be a solution. There would be *two solutions* that start from the point $\mathbf{P}$ of intersection.

(*iii*)　A **periodic solution** $x = x(t)$, $y = y(t)$. A periodic solution is called a **cycle.** If p is the period of the solution, then $\mathbf{X}(t + p) = \mathbf{X}(t)$ and a particle placed on the curve at $\mathbf{X}_0$ will cycle around the curve and return to $\mathbf{X}_0$ in p units of time. See Figure 10.3.

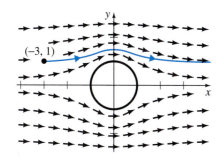

FIGURE 10.1　Vector field of a fluid flow around a circular cylinder

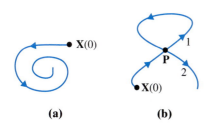

(a)　　　**(b)**

FIGURE 10.2　Curve in (a) is called an arc.

FIGURE 10.3　Periodic solution or cycle

EXAMPLE 4   Finding Critical Points

Find all critical points of each of the following plane autonomous systems:

(a) $x' = -x + y$ (b) $x' = x^2 + y^2 - 6$ (c) $x' = 0.01x(100 - x - y)$

   $y' = x - y$ $y' = x^2 - y$ $y' = 0.05y(60 - y - 0.2x)$

SOLUTION   We find the critical points by setting the right-hand sides of the differential equations equal to zero.

(a) The solution to the system

$$-x + y = 0$$

$$x - y = 0$$

consists of all points on the line $y = x$. Thus there are infinitely many critical points.

(b) To solve the system

$$x^2 + y^2 - 6 = 0$$

$$x^2 - y = 0$$

we substitute the second equation, $x^2 = y$, into the first equation to obtain $y^2 + y - 6 = (y + 3)(y - 2) = 0$. If $y = -3$, then $x^2 = -3$, and so there are no real solutions. If $y = 2$, then $x = \pm\sqrt{2}$, and so the critical points are $\left(\sqrt{2}, 2\right)$ and $\left(-\sqrt{2}, 2\right)$.

(c) Finding the critical points in part (c) requires a careful consideration of cases. The equation $0.01x(100 - x - y) = 0$ implies that $x = 0$ or $x + y = 100$.

   If $x = 0$, then by substituting in $0.05y(60 - y - 0.2x) = 0$, we have $y(60 - y) = 0$. Thus $y = 0$ or 60, so (0, 0) and (0, 60) are critical points.

   If $x + y = 100$, then $0 = y(60 - y - 0.2(100 - y)) = y(40 - 0.8y)$. It follows that $y = 0$ or 50, so (100, 0) and (50, 50) are critical points.

When a plane autonomous system is linear, we can use the methods in Chapter 8 to investigate solutions.

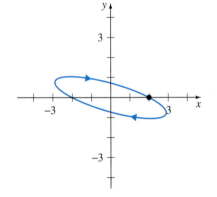

(a) periodic solution

EXAMPLE 5   Discovering Periodic Solutions

Determine whether the given linear system possesses a periodic solution:

(a) $x' = 2x + 8y$ (b) $x' = x + 2y$

   $y' = -x - 2y$ $y' = -\frac{1}{2}x + y$

In each case sketch the graph of the solution that satisfies $\mathbf{X}(0) = (2, 0)$.

SOLUTION   **(a)** In Example 6 of Section 8.2 we used the eigenvalue-eigenvector method to show that

$$x = c_1(2 \cos 2t - 2 \sin 2t) + c_2(2 \cos 2t + 2 \sin 2t)$$

$$y = -c_1 \cos 2t - c_2 \sin 2t.$$

Thus every solution is periodic with period $p = \pi$. The solution satisfying $\mathbf{X}(0) = (2, 0)$ is $x = 2 \cos 2t + 2 \sin 2t$, $y = -\sin 2t$. This solution generates the ellipse shown in Figure 10.4(a).

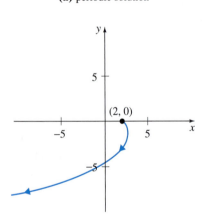

(b) nonperiodic solution

FIGURE 10.4   Solution curves for Example 5

(b) Using the eigenvalue-eigenvector method, we can show that

$$x = 2c_1e^t \cos t + 2c_2e^t \sin t, \quad y = -c_1e^t \sin t + c_2e^t \cos t.$$

Because of the presence of e^t in the general solution, there are no periodic solutions (that is, cycles). The solution satisfying $\mathbf{X}(0) = (2, 0)$ is $x = 2e^t \cos t, y = -e^t \sin t$, and the resulting curve is shown in Figure 10.4(b). ∎

CHANGING TO POLAR COORDINATES Except for the case of constant solutions, it is usually not possible to find explicit expressions for the solutions of a *nonlinear* autonomous system. We can solve some nonlinear systems, however, by changing to polar coordinates. From the formulas $r^2 = x^2 + y^2$ and $\theta = \tan^{-1}(y/x)$ we obtain

$$\frac{dr}{dt} = \frac{1}{r}\left(x\frac{dx}{dt} + y\frac{dy}{dt}\right), \quad \frac{d\theta}{dt} = \frac{1}{r^2}\left(-y\frac{dx}{dt} + x\frac{dy}{dt}\right). \tag{3}$$

We can sometimes use (3) to convert a plane autonomous system in rectangular coordinates to a simpler system in polar coordinates.

EXAMPLE 6 **Changing to Polar Coordinates**

Find the solution of the nonlinear plane autonomous system

$$x' = -y - x\sqrt{x^2 + y^2}$$
$$y' = x - y\sqrt{x^2 + y^2}$$

satisfying the initial condition $\mathbf{X}(0) = (3, 3)$.

SOLUTION Substituting for dx/dt and dy/dt in the expressions for dr/dt and $d\theta/dt$ in (3), we obtain

$$\frac{dr}{dt} = \frac{1}{r}[x(-y - xr) + y(x - yr)] = -r^2$$

$$\frac{d\theta}{dt} = \frac{1}{r^2}[-y(-y - xr) + x(x - yr)] = 1.$$

Since $(3, 3)$ is $\left(3\sqrt{2}, \pi/4\right)$ in polar coordinates, the initial condition $\mathbf{X}(0) = (3, 3)$ becomes $r(0) = 3\sqrt{2}$ and $\theta(0) = \pi/4$. Using separation of variables, we see that the solution of the system is

$$r = \frac{1}{t + c_1}, \quad \theta = t + c_2$$

for $r \neq 0$. (Check this!) Applying the initial condition then gives

$$r = \frac{1}{t + \sqrt{2}/6}, \quad \theta = t + \frac{\pi}{4}.$$

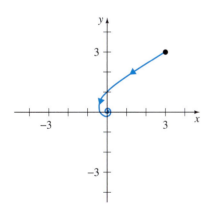

FIGURE 10.5 Solution curve for Example 6

The spiral $r = \dfrac{1}{\theta + \sqrt{2}/6 - \pi/4}$ is sketched in Figure 10.5. ∎

EXAMPLE 7 **Solutions in Polar Coordinates**

When expressed in polar coordinates, a plane autonomous system takes the form

$$\frac{dr}{dt} = 0.5(3 - r)$$

$$\frac{d\theta}{dt} = 1.$$

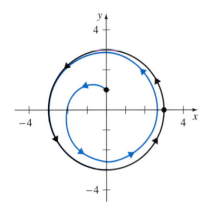

Find and sketch the solutions satisfying $X(0) = (0, 1)$ and $X(0) = (3, 0)$ in rectangular coordinates.

SOLUTION Applying separation of variables to $dr/dt = 0.5(3 - r)$ and integrating $d\theta/dt$ lead to the solution $r = 3 + c_1 e^{-0.5t}$, $\theta = t + c_2$.

If $X(0) = (0, 1)$, then $r(0) = 1$ and $\theta(0) = \pi/2$, and so $c_1 = -2$ and $c_2 = \pi/2$. The solution curve is the spiral $r = 3 - 2e^{-0.5(\theta - \pi/2)}$. Note that as $t \to \infty$, θ increases without bound and r approaches 3.

If $X(0) = (3, 0)$, then $r(0) = 3$ and $\theta(0) = 0$. It follows that $c_1 = c_2 = 0$, so $r = 3$ and $\theta = t$. Hence $x = r \cos \theta = 3 \cos t$ and $y = r \sin \theta = 3 \sin t$, so the solution is periodic. The solution generates a circle of radius 3 about $(0, 0)$. Both solutions are shown in Figure 10.6.

FIGURE 10.6 Curve in black is a periodic solution

EXERCISES 10.1

Answers to selected odd-numbered problems begin on page ANS-14.

In Problems 1–6 write the given nonlinear second-order differential equation as a plane autonomous system. Find all critical points of the resulting system.

1. $x'' + 9 \sin x = 0$

2. $x'' + (x')^2 + 2x = 0$

3. $x'' + x'(1 - x^3) - x^2 = 0$

4. $x'' + 4\dfrac{x}{1 + x^2} + 2x' = 0$

5. $x'' + x = \epsilon x^3$ for $\epsilon > 0$

6. $x'' + x - \epsilon x|x| = 0$ for $\epsilon > 0$

In Problems 7–16 find all critical points of the given plane autonomous system.

7. $x' = x + xy$
$y' = -y - xy$

8. $x' = y^2 - x$
$y' = x^2 - y$

9. $x' = 3x^2 - 4y$
$y' = x - y$

10. $x' = x^3 - y$
$y' = x - y^3$

11. $x' = x\left(10 - x - \frac{1}{2}y\right)$
$y' = y(16 - y - x)$

12. $x' = -2x + y + 10$
$y' = 2x - y - 15\dfrac{y}{y + 5}$

13. $x' = x^2 e^y$
$y' = y(e^x - 1)$

14. $x' = \sin y$
$y' = e^{x-y} - 1$

15. $x' = x(1 - x^2 - 3y^2)$
$y' = y(3 - x^2 - 3y^2)$

16. $x' = -x(4 - y^2)$
$y' = 4y(1 - x^2)$

In Problems 17–22 the given linear system is taken from Exercises 8.2.

(a) Find the general solution and determine whether there are periodic solutions.

(b) Find the solution satisfying the given initial condition.

(c) With the aid of a graphics calculator or computer software graph the solution in part (b) and indicate the direction in which the curve is traversed.

17. $x' = x + 2y$
$y' = 4x + 3y$, $X(0) = (2, -2)$
(Problem 1, Exercises 8.2)

18. $x' = -6x + 2y$
$y' = -3x + y$, $X(0) = (3, 4)$
(Problem 6, Exercises 8.2)

19. $x' = 4x - 5y$
$y' = 5x - 4y$, $X(0) = (4, 5)$
(Problem 37, Exercises 8.2)

20. $x' = x + y$
$y' = -2x - y$, $X(0) = (-2, 2)$
(Problem 34, Exercises 8.2)

21. $x' = 5x + y$
$y' = -2x + 3y$, $X(0) = (-1, 2)$
(Problem 35, Exercises 8.2)

22. $x' = x - 8y$
$y' = x - 3y$, $X(0) = (2, 1)$
(Problem 38, Exercises 8.2)

In Problems 23–26 solve the given nonlinear plane autonomous system by changing to polar coordinates. Describe the geometric behavior of the solution that satisfies the given initial condition(s).

23. $x' = -y - x(x^2 + y^2)^2$
$y' = x - y(x^2 + y^2)^2$, $X(0) = (4, 0)$

24. $x' = y + x(x^2 + y^2)$
$y' = -x + y(x^2 + y^2)$, $\mathbf{X}(0) = (4, 0)$

25. $x' = -y + x(1 - x^2 - y^2)$
$y' = x + y(1 - x^2 - y^2)$, $\mathbf{X}(0) = (1, 0)$, $\mathbf{X}(0) = (2, 0)$
(*Hint:* The resulting differential equation for r is a Bernoulli differential equation. See Section 2.5.)

26. $x' = y - \dfrac{x}{\sqrt{x^2 + y^2}}(4 - x^2 - y^2)$

$y' = -x - \dfrac{y}{\sqrt{x^2 + y^2}}(4 - x^2 - y^2)$,

$\mathbf{X}(0) = (1, 0)$, $\mathbf{X}(0) = (2, 0)$

If a plane autonomous system has a periodic solution, then there must be at least one critical point inside the curve generated by the solution. In Problems 27–30 use this fact together with a numerical solver to investigate the possibility of periodic solutions.

27. $x' = -x + 6y$
$y' = xy + 12$

28. $x' = -x + 6xy$
$y' = -8xy + 2y$

29. $x' = y$
$y' = y(1 - 3x^2 - 2y^2) - x$

30. $x' = xy$
$y' = -1 - x^2 - y^2$

10.2   STABILITY OF LINEAR SYSTEMS

INTRODUCTION: We have seen that a plane autonomous system

$$\frac{dx}{dt} = P(x, y)$$

$$\frac{dy}{dt} = Q(x, y)$$

gives rise to a vector field $\mathbf{V}(x, y) = (P(x, y), Q(x, y))$, and a solution $\mathbf{X} = \mathbf{X}(t)$ of the system can be interpreted as the resulting path of a particle that is initially placed at position $\mathbf{X}(0) = \mathbf{X}_0$. If $\mathbf{X}_0$ is a critical point of the system, then the particle remains stationary. In this section we examine the behavior of solutions when $\mathbf{X}_0$ is chosen *close* to a critical point of the system.

REVIEW MATERIAL: Review Section 10.1, especially Examples 3 and 4.

critical point

(a) locally stable

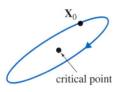

critical point

(b) locally stable

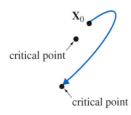

critical point

critical point

(c) unstable

FIGURE 10.7   Critical points

SOME FUNDAMENTAL QUESTIONS   Suppose that $\mathbf{X}_1$ is a critical point of a plane autonomous system and $\mathbf{X} = \mathbf{X}(t)$ is a solution of the system that satisfies $\mathbf{X}(0) = \mathbf{X}_0$. If the solution is interpreted as a path of a moving particle, we are interested in the answers to the following questions when $\mathbf{X}_0$ is placed near $\mathbf{X}_1$:

   (*i*)   Will the particle return to the critical point? More precisely, does $\lim_{t \to \infty} \mathbf{X}(t) = \mathbf{X}_1$?

   (*ii*)   If the particle does *not* return to the critical point, does it remain close to the critical point or move away from the critical point? It is conceivable, for example, that the particle may simply circle the critical point, or it may even return to a different critical point or to no critical point at all. See Figure 10.7.

If in some neighborhood of the critical point case (a) or (b) in Figure 10.7 *always* occurs, we call the critical point **locally stable.** If, however, an initial value $\mathbf{X}_0$ that results in behavior similar to (c) can be found in *any* given neighborhood, we call the critical point **unstable.** These concepts will be made more precise in Section 10.3, where questions (*i*) and (*ii*) will be investigated for nonlinear systems.

STABILITY ANALYSIS   We will first investigate these two stability questions for linear plane autonomous systems and lay the foundation for Section 10.3. The solution methods of Chapter 8 enable us to give a careful geometric analysis of the solutions to

$$x' = ax + by$$
$$y' = cx + dy$$

(1)

in terms of the eigenvalues and eigenvectors of the coefficient matrix

$$\mathbf{A} = \begin{pmatrix} a & b \\ c & d \end{pmatrix}.$$

To ensure that $\mathbf{X}_0 = (0, 0)$ is the only critical point, we will assume that the determinant $\Delta = ad - bc \neq 0$. If $\tau = a + d$ is the trace* of matrix $\mathbf{A}$, then the characteristic equation $\det(\mathbf{A} - \lambda\mathbf{I}) = 0$ may be rewritten as

$$\lambda^2 - \tau\lambda + \Delta = 0.$$

Therefore the eigenvalues of $\mathbf{A}$ are $\lambda = \left(\tau \pm \sqrt{\tau^2 - 4\Delta}\right)/2$, and the usual three cases for these roots occur according to whether $\tau^2 - 4\Delta$ is positive, negative, or zero. In the next example we use a numerical solver to discover the nature of the solutions corresponding to these cases.

EXAMPLE 1   Eigenvalues and the Shape of Solutions

Find the eigenvalues of the linear system

$$x' = -x + y$$
$$y' = cx - y$$

in terms of c, and use a numerical solver to discover the shapes of solutions corresponding to the cases $c = \frac{1}{4}, 4, 0,$ and -9.

SOLUTION The coefficient matrix $\begin{pmatrix} -1 & 1 \\ c & -1 \end{pmatrix}$ has trace $\tau = -2$ and determinant $\Delta = 1 - c$, and so the eigenvalues are

$$\lambda = \frac{\tau \pm \sqrt{\tau^2 - 4\Delta}}{2} = \frac{-2 \pm \sqrt{4 - 4(1 - c)}}{2} = -1 \pm \sqrt{c}.$$

The nature of the eigenvalues is therefore determined by the sign of c.

If $c = \frac{1}{4}$, then the eigenvalues are negative and distinct, $\lambda = -\frac{1}{2}$ and $-\frac{3}{2}$. In Figure 10.8(a) we have used a numerical solver to generate solution curves, or trajectories, that correspond to various initial conditions. Note that, except for the trajectories drawn in black in the figure, the trajectories all appear to approach **0** from a fixed direction. Recall from Chapter 8 that a collection of trajectories in the xy-plane, or **phase plane,** is called a **phase portrait** of the system.

When $c = 4$, the eigenvalues have opposite signs, $\lambda = 1$ and -3, and an interesting phenomenon occurs. All trajectories move away from the origin in a fixed direction except for solutions that start along the single line drawn in black in Figure 10.8(b). We have already seen behavior like this in the phase portrait given in Figure 8.2. Experiment with your numerical solver and verify these observations.

The selection $c = 0$ leads to a single real eigenvalue $\lambda = -1$. This case is very similar to the case $c = \frac{1}{4}$ with one notable exception. All solution curves in Figure 10.8(c) appear to approach **0** from a fixed direction as t increases.

---

*In general, if $\mathbf{A}$ is an $n \times n$ matrix, the **trace** of $\mathbf{A}$ is the sum of the main diagonal entries.

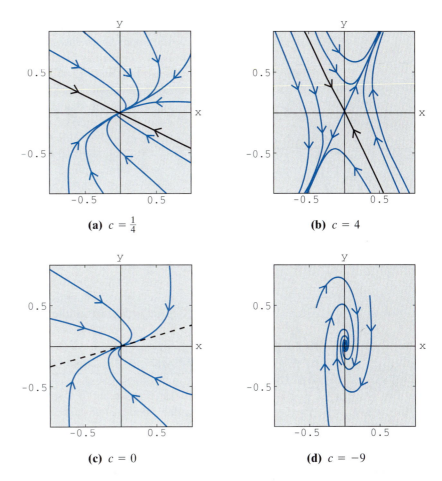

(a) $c = \frac{1}{4}$

(b) $c = 4$

(c) $c = 0$

(d) $c = -9$

FIGURE 10.8   Phase portraits of linear system in Example 1 for various values of c

Finally, when $c = -9$, $\lambda = -1 \pm \sqrt{-9} = -1 \pm 3i$. Thus the eigenvalues are conjugate complex numbers with negative real part -1. Figure 10.8(d) shows that solution curves spiral in toward the origin **0** as t increases.

The behaviors of the trajectories observed in the four phase portraits in Figure 10.8 in Example 1 can be explained using the eigenvalue-eigenvector solution results from Chapter 8.

CASE I: REAL DISTINCT EIGENVALUES ($\tau^2 - 4\Delta > 0$)   According to Theorem 8.7 in Section 8.2, the general solution of (1) is given by

$$\mathbf{X}(t) = c_1 \mathbf{K}_1 e^{\lambda_1 t} + c_2 \mathbf{K}_2 e^{\lambda_2 t}, \tag{2}$$

where λ_1 and λ_2 are the eigenvalues and $\mathbf{K}_1$ and $\mathbf{K}_2$ are the corresponding eigenvectors. Note that $\mathbf{X}(t)$ can also be written as

$$\mathbf{X}(t) = e^{\lambda_1 t}[c_1 \mathbf{K}_1 + c_2 \mathbf{K}_2 e^{(\lambda_2 - \lambda_1)t}]. \tag{3}$$

(a)   **Both eigenvalues negative** ($\tau^2 - 4\Delta > 0$, $\tau < 0$, and $\Delta > 0$)

Stable Node ($\lambda_2 < \lambda_1 < 0$): Since both eigenvalues are negative, it follows from (2) that $\lim_{t \to \infty} \mathbf{X}(t) = \mathbf{0}$. If we assume that $\lambda_2 < \lambda_1$, then $\lambda_2 - \lambda_1 < 0$ and so $e^{(\lambda_2 - \lambda_1)t}$ is an exponential decay function. We may therefore conclude from (3) that $\mathbf{X}(t) \approx c_1 \mathbf{K}_1 e^{\lambda_1 t}$ for large values of t. When $c_1 \neq 0$, $\mathbf{X}(t)$ will approach **0** from one of the two directions determined by the eigenvector $\mathbf{K}_1$ corresponding to λ_1. If $c_1 = 0$, $\mathbf{X}(t) = c_2 \mathbf{K}_2 e^{\lambda_2 t}$ and $\mathbf{X}(t)$ approaches **0** along the line determined by the eigenvector $\mathbf{K}_2$. Figure 10.9 shows a collection of

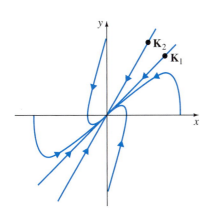

FIGURE 10.9   Stable node

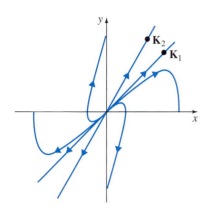

FIGURE 10.10 Unstable node

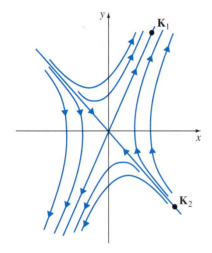

FIGURE 10.11 Saddle point

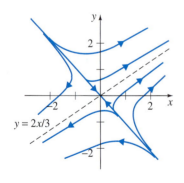

FIGURE 10.12 Saddle point

solution curves around the origin. A critical point is called a **stable node** when both eigenvalues are negative.

(b) **Both eigenvalues positive** ($\tau^2 - 4\Delta > 0$, $\tau > 0$, and $\Delta > 0$)
Unstable Node ($0 < \lambda_2 < \lambda_1$): The analysis for this case is similar to (a). Again from (2), $\mathbf{X}(t)$ becomes unbounded as t increases. Moreover, again assuming that $\lambda_2 < \lambda_1$ and using (3), we see that $\mathbf{X}(t)$ becomes unbounded in one of the directions determined by the eigenvector $\mathbf{K}_1$ (when $c_1 \neq 0$) or along the line determined by the eigenvector $\mathbf{K}_2$ (when $c_1 = 0$). Figure 10.10 shows a typical collection of solution curves. This type of critical point, corresponding to the case when both eigenvalues are positive, is called an **unstable node.**

(c) **Eigenvalues have opposite signs** ($\tau^2 - 4\Delta > 0$ and $\Delta < 0$)
Saddle Point ($\lambda_2 < 0 < \lambda_1$): The analysis of the solutions is identical to (b) with one exception. When $c_1 = 0$, $\mathbf{X}(t) = c_2 \mathbf{K}_2 e^{\lambda_2 t}$, and since $\lambda_2 < 0$, $\mathbf{X}(t)$ will approach $\mathbf{0}$ along the line determined by the eigenvector $\mathbf{K}_2$. If $\mathbf{X}(0)$ does not lie on the line determined by $\mathbf{K}_2$, the line determined by $\mathbf{K}_1$ serves as an asymptote for $\mathbf{X}(t)$. Thus the critical point is unstable even though some solutions approach $\mathbf{0}$ as t increases. This unstable critical point is called a **saddle point.** See Figure 10.11.

EXAMPLE 2   **Real Distinct Eigenvalues**

Classify the critical point $(0, 0)$ of each of the following linear systems $\mathbf{X}' = \mathbf{AX}$ as either a stable node, an unstable node, or a saddle point.

(a) $\mathbf{A} = \begin{pmatrix} 2 & 3 \\ 2 & 1 \end{pmatrix}$ (b) $\mathbf{A} = \begin{pmatrix} -10 & 6 \\ 15 & -19 \end{pmatrix}$

In each case discuss the nature of the solutions in a neighborhood of $(0, 0)$.

SOLUTION   (a) Since the trace $\tau = 3$ and the determinant $\Delta = -4$, the eigenvalues are

$$\lambda = \frac{\tau \pm \sqrt{\tau^2 - 4\Delta}}{2} = \frac{3 \pm \sqrt{3^2 - 4(-4)}}{2} = \frac{3 \pm 5}{2} = 4, -1.$$

The eigenvalues have opposite signs, and so $(0, 0)$ is a saddle point. It is not hard to show (see Example 1, Section 8.2) that eigenvectors corresponding to $\lambda_1 = 4$ and $\lambda_2 = -1$ are

$$\mathbf{K}_1 = \begin{pmatrix} 3 \\ 2 \end{pmatrix} \quad \text{and} \quad \mathbf{K}_2 = \begin{pmatrix} 1 \\ -1 \end{pmatrix},$$

respectively. If $\mathbf{X}(0) = \mathbf{X}_0$ lies on the line $y = -x$, then $\mathbf{X}(t)$ approaches $\mathbf{0}$. For any other initial condition, $\mathbf{X}(t)$ will become unbounded in the directions determined by $\mathbf{K}_1$. In other words, the line $y = \frac{2}{3}x$ serves as an asymptote for all these solution curves. See Figure 10.12.

(b) From $\tau = -29$ and $\Delta = 100$ it follows that the eigenvalues of $\mathbf{A}$ are $\lambda_1 = -4$ and $\lambda_2 = -25$. Both eigenvalues are negative, so $(0, 0)$ is in this case a stable node. Since eigenvectors corresponding to $\lambda_1 = -4$ and $\lambda_2 = -25$ are

$$\mathbf{K}_1 = \begin{pmatrix} 1 \\ 1 \end{pmatrix} \quad \text{and} \quad \mathbf{K}_2 = \begin{pmatrix} 2 \\ -5 \end{pmatrix},$$

respectively, it follows that all solutions approach $\mathbf{0}$ from the direction defined by $\mathbf{K}_1$ except those solutions for which $\mathbf{X}(0) = \mathbf{X}_0$ lies on the line

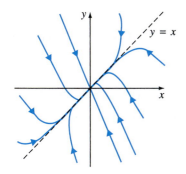

FIGURE 10.13 Stable node

$y = -\frac{5}{2}x$ determined by $\mathbf{K}_2$. These solutions approach $\mathbf{0}$ along $y = -\frac{5}{2}x$. See Figure 10.13. ∎

CASE II: A REPEATED REAL EIGENVALUE ($\tau^2 - 4\Delta = 0$)

DEGENERATE NODES: Recall from Section 8.2 that the general solution takes on one of two different forms depending on whether one or two linearly independent eigenvectors can be found for the repeated eigenvalue λ_1.

(a) **Two linearly independent eigenvectors**

If $\mathbf{K}_1$ and $\mathbf{K}_2$ are two linearly independent eigenvectors corresponding to λ_1, then the general solution is given by

$$\mathbf{X}(t) = c_1\mathbf{K}_1 e^{\lambda_1 t} + c_2\mathbf{K}_2 e^{\lambda_1 t} = (c_1\mathbf{K}_1 + c_2\mathbf{K}_2)e^{\lambda_1 t}.$$

If $\lambda_1 < 0$, then $\mathbf{X}(t)$ approaches $\mathbf{0}$ along the line determined by the vector $c_1\mathbf{K}_1 + c_2\mathbf{K}_2$ and the critical point is called a **degenerate stable node** (see Figure 10.14(a)). The arrows in Figure 10.14(a) are reversed when $\lambda_1 > 0$, and we have a **degenerate unstable node**.

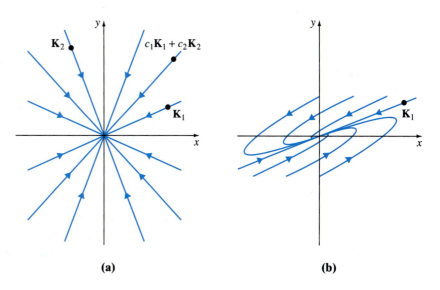

(a) (b)

FIGURE 10.14 Degenerate stable nodes

(b) **A single linearly independent eigenvector**

When only a single linearly independent eigenvector $\mathbf{K}_1$ exists, the general solution is given by

$$\mathbf{X}(t) = c_1\mathbf{K}_1 e^{\lambda_1 t} + c_2(\mathbf{K}_1 t e^{\lambda_1 t} + \mathbf{P}e^{\lambda_1 t}),$$

where $(\mathbf{A} - \lambda_1\mathbf{I})\mathbf{P} = \mathbf{K}_1$ (see Section 8.2, (12)–(14)), and the solution may be rewritten as

$$\mathbf{X}(t) = t e^{\lambda_1 t}\left[c_2\mathbf{K}_1 + \frac{c_1}{t}\mathbf{K}_1 + \frac{c_2}{t}\mathbf{P}\right].$$

If $\lambda_1 < 0$, then $\lim_{t\to\infty} t e^{\lambda_1 t} = 0$ and it follows that $\mathbf{X}(t)$ approaches $\mathbf{0}$ in one of the directions determined by the vector $\mathbf{K}_1$ (see Figure 10.14(b)). The critical point is again called a **degenerate stable node**. When $\lambda_1 > 0$, the solutions look like those in Figure 10.14(b) with the arrows reversed. The line determined by $\mathbf{K}_1$ is an asymptote for *all* solutions. The critical point is again called a **degenerate unstable node**.

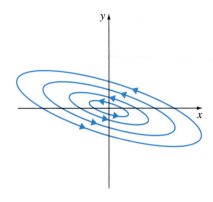

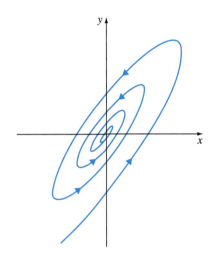

FIGURE 10.15 Center

CASE III: COMPLEX EIGENVALUES $(\tau^2 - 4\Delta < 0)$ If $\lambda_1 = \alpha + i\beta$ and $\lambda_1 = \alpha - i\beta$ are the complex eigenvalues and $\mathbf{K}_1 = \mathbf{B}_1 + i\mathbf{B}_2$ is a complex eigenvector corresponding to λ_1, the general solution can be written as $\mathbf{X}(t) = c_1\mathbf{X}_1(t) + c_2\mathbf{X}_2(t)$, where

$$\mathbf{X}_1(t) = (\mathbf{B}_1 \cos \beta t - \mathbf{B}_2 \sin \beta t)e^{\alpha t}, \quad \mathbf{X}_2(t) = (\mathbf{B}_2 \cos \beta t + \mathbf{B}_1 \sin \beta t)e^{\alpha t}.$$

See equations (23) and (24) in Section 8.2. A solution can therefore be written in the form

$$x(t) = e^{\alpha t}(c_{11}\cos \beta t + c_{12}\sin \beta t), \quad y(t) = e^{\alpha t}(c_{21}\cos \beta t + c_{22}\sin \beta t), \quad (4)$$

and when $\alpha = 0$, we have

$$x(t) = c_{11}\cos \beta t + c_{12}\sin \beta t, \quad y(t) = c_{21}\cos \beta t + c_{22}\sin \beta t. \quad (5)$$

(a) **Pure imaginary roots** $(\tau^2 - 4\Delta < 0, \tau = 0)$
   Center: When $\alpha = 0$, the eigenvalues are pure imaginary and, from (5), all solutions are periodic with period $p = 2\pi/\beta$. Notice that if both c_{12} and c_{21} happened to be 0, then (5) would reduce to

$$x(t) = c_{11}\cos \beta t, \quad y(t) = c_{22}\sin \beta t,$$

   which is a standard parametric representation for the ellipse $x^2/c_{11}^2 + y^2/c_{22}^2 = 1$. By solving the system of equations in (4) for $\cos \beta t$ and $\sin \beta t$ and using the identity $\sin^2\beta t + \cos^2\beta t = 1$, it is possible to show that *all solutions are ellipses* with center at the origin. The critical point $(0, 0)$ is called a **center**, and Figure 10.15 shows a typical collection of solution curves. The ellipses are either *all* traversed in the clockwise direction or *all* traversed in the counterclockwise direction.

(b) **Nonzero real part** $(\tau^2 - 4\Delta < 0, \tau \neq 0)$
   Spiral Points: When $\alpha \neq 0$, the effect of the term $e^{\alpha t}$ in (4) is similar to the effect of the exponential term in the analysis of damped motion given in Section 5.1. When $\alpha < 0$, $e^{\alpha t} \to 0$, and the elliptical-like solution spirals closer and closer to the origin. The critical point is called a **stable spiral point.** When $\alpha > 0$, the effect is the opposite. An elliptical-like solution is driven farther and farther from the origin, and the critical point is now called an **unstable spiral point.** See Figure 10.16.

(a) stable spiral point

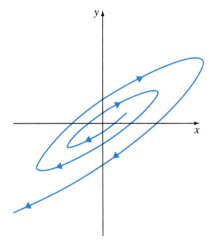

(b) unstable spiral point

FIGURE 10.16 Spiral points

EXAMPLE 3   **Repeated and Complex Eigenvalues**

Classify the critical point $(0, 0)$ of each of the following linear systems $\mathbf{X}' = \mathbf{AX}$:

(a) $\mathbf{A} = \begin{pmatrix} 3 & -18 \\ 2 & -9 \end{pmatrix}$ (b) $\mathbf{A} = \begin{pmatrix} -1 & 2 \\ -1 & 1 \end{pmatrix}$

In each case discuss the nature of the solution that satisfies $\mathbf{X}(0) = (1, 0)$. Determine parametric equations for each solution.

SOLUTION (a) Since $\tau = -6$ and $\Delta = 9$, the characteristic polynomial is $\lambda^2 + 6\lambda + 9 = (\lambda + 3)^2$, and so $(0, 0)$ is a degenerate stable node. For the repeated eigenvalue $\lambda = -3$ we find a single eigenvector $\mathbf{K}_1 = \begin{pmatrix} 3 \\ 1 \end{pmatrix}$, so the solution $\mathbf{X}(t)$ that satisfies $\mathbf{X}(0) = (1, 0)$ approaches $(0, 0)$ from the direction specified by the line $y = x/3$.

(b) Since $\tau = 0$ and $\Delta = 1$, the eigenvalues are $\lambda = \pm i$, and so $(0, 0)$ is a center. The solution $\mathbf{X}(t)$ that satisfies $\mathbf{X}(0) = (1, 0)$ is an ellipse that circles the origin every 2π units of time.

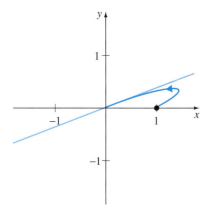

(a) degenerate stable node

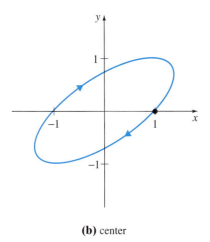

(b) center

FIGURE 10.17 Critical points in Example 3

From Example 4 of Section 8.2 the general solution of the system in (a) is

$$\mathbf{X}(t) = c_1 \begin{pmatrix} 3 \\ 1 \end{pmatrix} e^{-3t} + c_2 \left[\begin{pmatrix} 3 \\ 1 \end{pmatrix} t e^{-3t} + \begin{pmatrix} \frac{1}{2} \\ 0 \end{pmatrix} e^{-3t} \right].$$

The initial condition gives $c_1 = 0$ and $c_2 = 2$, and so $x = (6t + 1)e^{-3t}$, $y = 2te^{-3t}$ are parametric equations for the solution.

The general solution of the system in (b) is

$$\mathbf{X}(t) = c_1 \begin{pmatrix} \cos t + \sin t \\ \cos t \end{pmatrix} + c_2 \begin{pmatrix} \cos t - \sin t \\ -\sin t \end{pmatrix}.$$

The initial condition gives $c_1 = 0$ and $c_2 = 1$, so $x = \cos t - \sin t$, $y = -\sin t$ are parametric equations for the ellipse. Note that $y < 0$ for small positive values of t, and therefore the ellipse is traversed in the clockwise direction.

The solutions of (a) and (b) are shown in Figures 10.17(a) and (b), respectively.

Figure 10.18 conveniently summarizes the results of this section. The general geometric nature of the solutions can be determined by computing the trace and determinant of **A**. In practice, graphs of the solutions are most easily obtained *not* by constructing explicit eigenvalue-eigenvector solutions but rather by generating the solutions using a numerical solver and the Runge-Kutta method for first-order systems.

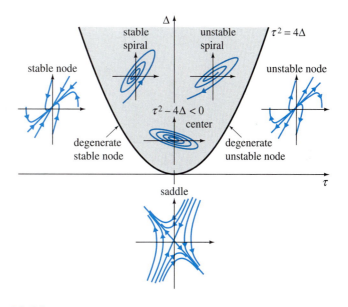

FIGURE 10.18 Geometric summary of Cases I, II, and III

EXAMPLE 4   **Classifying Critical Points**

Classify the critical point $(0, 0)$ of each of the following linear systems $\mathbf{X}' = \mathbf{AX}$:

(a) $\mathbf{A} = \begin{pmatrix} 1.01 & 3.10 \\ -1.10 & -1.02 \end{pmatrix}$ **(b)** $\mathbf{A} = \begin{pmatrix} -a\hat{x} & -ab\hat{x} \\ -cd\hat{y} & -d\hat{y} \end{pmatrix}$

for positive constants a, b, c, d, $\hat{x}$, and $\hat{y}$.

SOLUTION   **(a)** For this matrix $\tau = -0.01$, $\Delta = 2.3798$, so $\tau^2 - 4\Delta < 0$. Using Figure 10.18, we see that $(0, 0)$ is a stable spiral point.

(b) This matrix arises from the Lotka-Volterra competition model, which we will study in Section 10.4. Since $\tau = -(a\hat{x} + d\hat{y})$ and all constants in the matrix are positive, $\tau < 0$. The determinant may be written as $\Delta = ad\hat{x}\hat{y}(1 - bc)$. If $bc > 1$, then $\Delta < 0$ and the critical point is a saddle point. If $bc < 1$, $\Delta > 0$ and the critical point is either a stable node, a degenerate stable node, or a stable spiral point. In all three cases $\lim_{t \to \infty} \mathbf{X}(t) = \mathbf{0}$.

THEOREM 10.1   Stability Criteria for Linear Systems

For a linear plane autonomous system $\mathbf{X}' = \mathbf{AX}$ with det $\mathbf{A} \neq 0$, let $\mathbf{X} = \mathbf{X}(t)$ denote the solution that satisfies the initial condition $\mathbf{X}(0) = \mathbf{X}_0$, where $\mathbf{X}_0 \neq \mathbf{0}$.

   (a) $\lim_{t \to \infty} \mathbf{X}(t) = \mathbf{0}$ if and only if the eigenvalues of $\mathbf{A}$ have negative real parts. This will occur when $\Delta > 0$ and $\tau < 0$.
   (b) $\mathbf{X}(t)$ is periodic if and only if the eigenvalues of $\mathbf{A}$ are pure imaginary. This will occur when $\Delta > 0$ and $\tau = 0$.
   (c) In all other cases, given any neighborhood of the origin, there is at least one $\mathbf{X}_0$ in the neighborhood for which $\mathbf{X}(t)$ becomes unbounded as t increases.

REMARKS

The terminology used to describe the types of critical points varies from text to text. The following table lists many of the alternative terms that you may encounter in your reading.

Term	Alternative Terms
critical point	equilibrium point, singular point, stationary point, rest point
spiral point	focus, focal point, vortex point
stable node or spiral point	attractor, sink
unstable node or spiral point	repeller, source

EXERCISES 10.2

Answers to selected odd-numbered problems begin on page ANS-15.

In Problems 1–8 the general solution of the linear system $\mathbf{X}' = \mathbf{AX}$ is given.
(a) In each case discuss the nature of the solutions in a neighborhood of $(0, 0)$.
(b) With the aid of a graphics calculator or computer software graph the solution that satisfies $\mathbf{X}(0) = (1, 1)$.

1. $\mathbf{A} = \begin{pmatrix} -2 & -2 \\ -2 & -5 \end{pmatrix}$, $\mathbf{X}(t) = c_1 \begin{pmatrix} 2 \\ -1 \end{pmatrix} e^{-t} + c_2 \begin{pmatrix} 1 \\ 2 \end{pmatrix} e^{-6t}$

2. $\mathbf{A} = \begin{pmatrix} -1 & -2 \\ 3 & 4 \end{pmatrix}$, $\mathbf{X}(t) = c_1 \begin{pmatrix} 1 \\ -1 \end{pmatrix} e^{t} + c_2 \begin{pmatrix} -4 \\ 6 \end{pmatrix} e^{2t}$

3. $\mathbf{A} = \begin{pmatrix} 1 & -1 \\ 1 & 1 \end{pmatrix}$, $\mathbf{X}(t) = e^{t} \left[c_1 \begin{pmatrix} -\sin t \\ \cos t \end{pmatrix} + c_2 \begin{pmatrix} \cos t \\ \sin t \end{pmatrix} \right]$

4. $\mathbf{A} = \begin{pmatrix} -1 & -4 \\ 1 & -1 \end{pmatrix}$,

$$\mathbf{X}(t) = e^{-t}\left[c_1\begin{pmatrix} 2\cos 2t \\ \sin 2t \end{pmatrix} + c_2\begin{pmatrix} -2\sin 2t \\ \cos 2t \end{pmatrix}\right]$$

5. $\mathbf{A} = \begin{pmatrix} -6 & 5 \\ -5 & 4 \end{pmatrix}$,

$$\mathbf{X}(t) = c_1\begin{pmatrix} 1 \\ 1 \end{pmatrix}e^{-t} + c_2\left[\begin{pmatrix} 1 \\ 1 \end{pmatrix}te^{-t} + \begin{pmatrix} 0 \\ \frac{1}{5} \end{pmatrix}e^{-t}\right]$$

6. $\mathbf{A} = \begin{pmatrix} 2 & 4 \\ -1 & 6 \end{pmatrix}$,

$$\mathbf{X}(t) = c_1\begin{pmatrix} 2 \\ 1 \end{pmatrix}e^{4t} + c_2\left[\begin{pmatrix} 2 \\ 1 \end{pmatrix}te^{4t} + \begin{pmatrix} 1 \\ 1 \end{pmatrix}e^{4t}\right]$$

7. $\mathbf{A} = \begin{pmatrix} 2 & -1 \\ 3 & -2 \end{pmatrix}$, $\mathbf{X}(t) = c_1\begin{pmatrix} 1 \\ 1 \end{pmatrix}e^{t} + c_2\begin{pmatrix} 1 \\ 3 \end{pmatrix}e^{-t}$

8. $\mathbf{A} = \begin{pmatrix} -1 & 5 \\ -1 & 1 \end{pmatrix}$,

$$\mathbf{X}(t) = c_1\begin{pmatrix} 5\cos 2t \\ \cos 2t - 2\sin 2t \end{pmatrix} + c_2\begin{pmatrix} 5\sin 2t \\ 2\cos 2t + \sin 2t \end{pmatrix}$$

In Problems 9–16 classify the critical point (0, 0) of the given linear system by computing the trace τ and determinant Δ and using Figure 10.18.

9. $x' = -5x + 3y$
  $y' = 2x + 7y$

10. $x' = -5x + 3y$
  $y' = 2x - 7y$

11. $x' = -5x + 3y$
  $y' = -2x + 5y$

12. $x' = -5x + 3y$
  $y' = -7x + 4y$

13. $x' = -\frac{3}{2}x + \frac{1}{4}y$
  $y' = -x - \frac{1}{2}y$

14. $x' = \frac{3}{2}x + \frac{1}{4}y$
  $y' = -x + \frac{1}{2}y$

15. $x' = 0.02x - 0.11y$
  $y' = 0.10x - 0.05y$

16. $x' = 0.03x + 0.01y$
  $y' = -0.01x + 0.05y$

17. Determine conditions on the real constant μ so that (0, 0) is a center for the linear system

$$x' = -\mu x + y$$
$$y' = -x + \mu y.$$

18. Determine a condition on the real constant μ so that (0, 0) is a stable spiral point of the linear system

$$x' = y$$
$$y' = -x + \mu y.$$

19. Show that (0, 0) is always an unstable critical point of the linear system

$$x' = \mu x + y$$
$$y' = -x + y,$$

where μ is a real constant and $\mu \neq -1$. When is (0, 0) an unstable saddle point? When is (0, 0) an unstable spiral point?

20. Let $\mathbf{X} = \mathbf{X}(t)$ be the response of the linear dynamical system

$$x' = \alpha x - \beta y$$
$$y' = \beta x + \alpha y$$

that satisfies the initial condition $\mathbf{X}(0) = \mathbf{X}_0$. Determine conditions on the real constants α and β that will ensure $\lim_{t\to\infty} \mathbf{X}(t) = (0, 0)$. Can (0, 0) be a node or saddle point?

21. Show that the nonhomogeneous linear system $\mathbf{X}' = \mathbf{A}\mathbf{X} + \mathbf{F}$ has a unique critical point $\mathbf{X}_1$ when $\Delta = \det \mathbf{A} \neq 0$. Conclude that if $\mathbf{X} = \mathbf{X}(t)$ is a solution to the nonhomogeneous system, $\tau < 0$ and $\Delta > 0$, then $\lim_{t\to\infty} \mathbf{X}(t) = \mathbf{X}_1$. (*Hint:* $\mathbf{X}(t) = \mathbf{X}_c(t) + \mathbf{X}_1$.)

22. In Example 4(b) show that (0, 0) is a stable node when $bc < 1$.

In Problems 23–26 a nonhomogeneous linear system $\mathbf{X}' = \mathbf{A}\mathbf{X} + \mathbf{F}$ is given.
(a) In each case determine the unique critical point $\mathbf{X}_1$.
(b) Use a numerical solver to determine the nature of the critical point in (a).
(c) Investigate the relationship between $\mathbf{X}_1$ and the critical point (0, 0) of the homogeneous linear system $\mathbf{X}' = \mathbf{A}\mathbf{X}$.

23. $x' = 2x + 3y - 6$
  $y' = -x - 2y + 5$

24. $x' = -5x + 9y + 13$
  $y' = -x - 11y - 23$

25. $x' = 0.1x - 0.2y + 0.35$
  $y' = 0.1x + 0.1y - 0.25$

26. $x' = 3x - 2y - 1$
  $y' = 5x - 3y - 2$

10.3 LINEARIZATION AND LOCAL STABILITY

INTRODUCTION: The key idea in this section is that of linearization. A local linear approximation, or **linearization,** of a differentiable function $f(x)$ at a point $(x_1, f(x_1))$ is the equation of the tangent line to the graph of f at the point $y = f(x_1) + f'(x_1)(x - x_1)$. For x close to x_1 the points on the graph of $f(x)$ are close to the points on the tangent line, so values $y(x)$ obtained from its equation can be used to approximate the corresponding values $f(x)$. In this section we use linearization as a means of analyzing nonlinear DEs and nonlinear systems; the idea is to replace them by linear DEs and linear systems.

REVIEW MATERIAL: The concept of linearization was first introduced in Section 2.6.

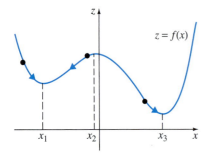

FIGURE 10.19   Bead sliding on graph of $z = f(x)$

SLIDING BEAD   We start this section by refining the stability concepts introduced in Section 10.2 in such a way that they will apply to nonlinear autonomous systems as well. Although the linear system $\mathbf{X}' = \mathbf{AX}$ had only one critical point when det $\mathbf{A} \neq 0$, we saw in Section 10.1 that a nonlinear system may have many critical points. We therefore cannot expect that a particle placed initially at a point $\mathbf{X}_0$ will remain near a given critical point $\mathbf{X}_1$ unless $\mathbf{X}_0$ has been placed sufficiently close to $\mathbf{X}_1$ to begin with. The particle might well be driven to a second critical point. To emphasize this idea, consider the physical system shown in Figure 10.19 in which a bead slides along the curve $z = f(x)$ under the influence of gravity alone. We will show in Section 10.4 that the x-coordinate of the bead satisfies a nonlinear second-order differential equation $x'' = g(x, x')$; therefore letting $y = x'$ satisfies the nonlinear autonomous system

$$x' = y$$

$$y' = g(x, y).$$

If the bead is positioned at $P = (x, f(x))$ and given zero initial velocity, the bead will remain at P provided that $f'(x) = 0$. If the bead is placed near the critical point located at $x = x_1$, it will remain near $x = x_1$ only if its initial velocity does not drive it over the "hump" at $x = x_2$ toward the critical point located at $x = x_3$. Therefore $\mathbf{X}(0) = (x(0), x'(0))$ must be near $(x_1, 0)$.

In the next definition we will denote the distance between two points $\mathbf{X}$ and $\mathbf{Y}$ by $|\mathbf{X} - \mathbf{Y}|$. Recall that if $\mathbf{X} = (x_1, x_2, \ldots, x_n)$ and $\mathbf{Y} = (y_1, y_2, \ldots, y_n)$, then

$$|\mathbf{X} - \mathbf{Y}| = \sqrt{(x_1 - y_1)^2 + (x_2 - y_2)^2 + \cdots + (x_n - y_n)^2}.$$

DEFINITION 10.1   **Stable Critical Points**

Let $\mathbf{X}_1$ be a critical point of an autonomous system, and let $\mathbf{X} = \mathbf{X}(t)$ denote the solution that satisfies the initial condition $\mathbf{X}(0) = \mathbf{X}_0$, where $\mathbf{X}_0 \neq \mathbf{X}_1$. We say that $\mathbf{X}_1$ is a **stable critical point** when, given any radius $\rho > 0$, there is a corresponding radius $r > 0$ such that if the initial position $\mathbf{X}_0$ satisfies $|\mathbf{X}_0 - \mathbf{X}_1| < r$, then the corresponding solution $\mathbf{X}(t)$ satisfies $|\mathbf{X}(t) - \mathbf{X}_1| < \rho$ for all $t > 0$. If, in addition, $\lim_{t \to \infty} \mathbf{X}(t) = \mathbf{X}_1$ whenever $|\mathbf{X}_0 - \mathbf{X}_1| < r$, we call $\mathbf{X}_1$ an **asymptotically stable critical point.**

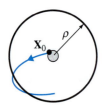

(a) stable

(b) unstable

FIGURE 10.20   Critical points

This definition is illustrated in Figure 10.20(a). Given any disk of radius ρ about the critical point $\mathbf{X}_1$, a solution will remain inside this disk provided that $\mathbf{X}(0) = \mathbf{X}_0$ is selected sufficiently close to $\mathbf{X}_1$. It is *not* necessary that a solution approach the critical point in order for $\mathbf{X}_1$ to be stable. Stable nodes, stable spiral points, and centers are all examples of stable critical points for linear systems. To emphasize that $\mathbf{X}_0$ must be selected close to $\mathbf{X}_1$, the terminology **locally stable critical point** is also used.

By negating Definition 10.1, we obtain the definition of an unstable critical point.

> **DEFINITION 10.2** Unstable Critical Point
>
> Let X_1 be a critical point of an autonomous system, and let $X = X(t)$ denote the solution that satisfies the initial condition $X(0) = X_0$, where $X_0 \neq X_1$. We say that X_1 is an **unstable critical point** if there is a disk of radius $\rho > 0$ with the property that, for any $r > 0$, there is at least one initial position X_0 that satisfies $|X_0 - X_1| < r$ yet the corresponding solution $X(t)$ satisfies $|X(t) - X_1| \geq \rho$ for at least one $t > 0$.

If a critical point X_1 is unstable, no matter how small the neighborhood about X_1, an initial position X_0 can always be found that results in the solution leaving some disk of radius ρ at some future time t. See Figure 10.20(b). Therefore unstable nodes, unstable spiral points, and saddle points are all examples of unstable critical points for linear systems. In Figure 10.19 the critical point $(x_2, 0)$ is unstable. The slightest displacement or initial velocity results in the bead sliding away from the point $(x_2, f(x_2))$.

EXAMPLE 1   A Stable Critical Point

Show that $(0, 0)$ is a stable critical point of the nonlinear plane autonomous system

$$x' = -y - x\sqrt{x^2 + y^2}$$
$$y' = x - y\sqrt{x^2 + y^2}$$

considered in Example 6 of Section 10.1.

SOLUTION   In Example 6 of Section 10.1 we showed that in polar coordinates $r = 1/(t + c_1)$, $\theta = t + c_2$ is the solution of the system. If $X(0) = (r_0, \theta_0)$ is the initial condition in polar coordinates, then

$$r = \frac{r_0}{r_0 t + 1}, \quad \theta = t + \theta_0.$$

Note that $r \leq r_0$ for $t \geq 0$, and r approaches $(0, 0)$ as t increases. Therefore, given $\rho > 0$, a solution that starts less than ρ units from $(0, 0)$ remains within ρ units of the origin for all $t \geq 0$. Hence the critical point $(0, 0)$ is stable and is in fact asymptotically stable. A typical solution is shown in Figure 10.21.

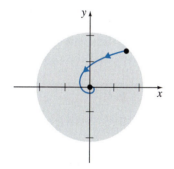

FIGURE 10.21 Asymptotically stable critical point

EXAMPLE 2   An Unstable Critical Point

When expressed in polar coordinates, a plane autonomous system takes the form

$$\frac{dr}{dt} = 0.05r(3 - r)$$

$$\frac{d\theta}{dt} = -1.$$

Show that $(x, y) = (0, 0)$ is an unstable critical point.

SOLUTION   Since $x = r\cos\theta$ and $y = r\sin\theta$, we have

$$\frac{dx}{dt} = -r\sin\theta\frac{d\theta}{dt} + \frac{dr}{dt}\cos\theta$$

$$\frac{dy}{dt} = r\cos\theta\frac{d\theta}{dt} + \frac{dr}{dt}\sin\theta.$$

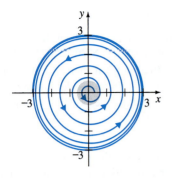

FIGURE 10.22 Unstable critical point

From $dr/dt = 0.05r(3 - r)$ we see that $dr/dt = 0$ when $r = 0$ and can conclude that $(x, y) = (0, 0)$ is a critical point by substituting $r = 0$ into the new system.

The differential equation $dr/dt = 0.05r(3 - r)$ is a logistic equation that can be solved using either separation of variables or equation (5) in Section 3.2. If $r(0) = r_0$ and $r_0 \neq 0$, then

$$r = \frac{3}{1 + c_0 e^{-0.15t}},$$

where $c_0 = (3 - r_0)/r_0$. Since $\lim_{t \to \infty} \dfrac{3}{1 + c_0 e^{-0.15t}} = 3$, it follows that no matter how close to $(0, 0)$ a solution starts, the solution will leave a disk of radius 1 about the origin. Therefore $(0, 0)$ is an unstable critical point. A typical solution that starts near $(0, 0)$ is shown in Figure 10.22.

LINEARIZATION It is rarely possible to determine the stability of a critical point of a nonlinear system by finding explicit solutions, as in Examples 1 and 2. Instead, we replace the term $\mathbf{g}(\mathbf{X})$ in the original autonomous system $\mathbf{X}' = \mathbf{g}(\mathbf{X})$ by a linear term $\mathbf{A}(\mathbf{X} - \mathbf{X}_1)$ that most closely approximates $\mathbf{g}(\mathbf{X})$ in a neighborhood of $\mathbf{X}_1$. This replacement process, called **linearization,** will be illustrated first for the first-order differential equation $x' = g(x)$.

An equation of the tangent line to the curve $y = g(x)$ at $x = x_1$ is $y = g(x_1) + g'(x_1)(x - x_1)$, and if x_1 is a critical point of $x' = g(x)$, we have $x' = g(x) \approx g'(x_1)(x - x_1)$. The general solution to the linear differential equation $x' = g'(x_1)(x - x_1)$ is $x = x_1 + ce^{\lambda_1 t}$, where $\lambda_1 = g'(x_1)$. Thus if $g'(x_1) < 0$, then $x(t)$ approaches x_1. Theorem 10.2 asserts that the same behavior occurs in the original differential equation provided that $x(0) = x_0$ is selected close enough to x_1.

THEOREM 10.2 **Stability Criteria for $x' = g(x)$**

Let x_1 be a critical point of the autonomous differential equation $x' = g(x)$, where g is differentiable at x_1.

   (a) If $g'(x_1) < 0$, then x_1 is an asymptotically stable critical point.
   (b) If $g'(x_1) > 0$, then x_1 is an unstable critical point.

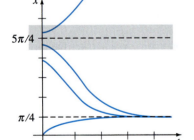

FIGURE 10.23 $\pi/4$ is asymptotically stable and $5\pi/4$ is unstable

EXAMPLE 3 **Stability in a Nonlinear First-Order DE**

Both $x = \pi/4$ and $x = 5\pi/4$ are critical points of the autonomous differential equation $x' = \cos x - \sin x$. This differential equation is difficult to solve explicitly, but we can use Theorem 10.2 to predict the behavior of solutions near these two critical points.

Since $g'(x) = -\sin x - \cos x$, $g'(\pi/4) = -\sqrt{2} < 0$ and $g'(5\pi/4) = \sqrt{2} > 0$. Therefore $x = \pi/4$ is an asymptotically stable critical point, but $x = 5\pi/4$ is unstable. In Figure 10.23 we used a numerical solver to investigate solutions that start near $(0, \pi/4)$ and $(0, 5\pi/4)$. Observe that solution curves that start close to $(0, 5\pi/4)$ quickly move away from the line $x = 5\pi/4$, as predicted.

EXAMPLE 4 **Stability Analysis of the Logistic DE**

Without solving explicitly, analyze the critical points of the logistic differential equation (see Section 3.2) $x' = \dfrac{r}{K}x(K - x)$, where r and K are positive constants.

SOLUTION The two critical points are $x = 0$ and $x = K$, so from $g'(x) = r(K - 2x)/K$ we get $g'(0) = r$ and $g'(K) = -r$. By Theorem 10.2 we conclude that $x = 0$ is an unstable critical point and $x = K$ is an asymptotically stable critical point.

JACOBIAN MATRIX A similar analysis may be carried out for a plane autonomous system. An equation of the tangent plane to the surface $z = g(x, y)$ at $\mathbf{X}_1 = (x_1, y_1)$ is

$$z = g(x_1, y_1) + \frac{\partial g}{\partial x}\bigg|_{(x_1, y_1)} (x - x_1) + \frac{\partial g}{\partial y}\bigg|_{(x_1, y_1)} (y - y_1),$$

and $g(x, y)$ can be approximated by its tangent plane in a neighborhood of $\mathbf{X}_1$.

When $\mathbf{X}_1$ is a critical point of a plane autonomous system, $P(x_1, y_1) = Q(x_1, y_1) = 0$, and we have

$$x' = P(x, y) \approx \frac{\partial P}{\partial x}\bigg|_{(x_1, y_1)} (x - x_1) + \frac{\partial P}{\partial y}\bigg|_{(x_1, y_1)} (y - y_1)$$

$$y' = Q(x, y) \approx \frac{\partial Q}{\partial x}\bigg|_{(x_1, y_1)} (x - x_1) + \frac{\partial Q}{\partial y}\bigg|_{(x_1, y_1)} (y - y_1).$$

The original system $\mathbf{X}' = \mathbf{g}(\mathbf{X})$ may be approximated in a neighborhood of the critical point $\mathbf{X}_1$ by the linear system $\mathbf{X}' = \mathbf{A}(\mathbf{X} - \mathbf{X}_1)$, where

$$\mathbf{A} = \begin{pmatrix} \dfrac{\partial P}{\partial x}\bigg|_{(x_1, y_1)} & \dfrac{\partial P}{\partial y}\bigg|_{(x_1, y_1)} \\ \dfrac{\partial Q}{\partial x}\bigg|_{(x_1, y_1)} & \dfrac{\partial Q}{\partial y}\bigg|_{(x_1, y_1)} \end{pmatrix}.$$

This matrix is called the **Jacobian matrix** at $\mathbf{X}_1$ and is denoted by $\mathbf{g}'(\mathbf{X}_1)$. If we let $\mathbf{H} = \mathbf{X} - \mathbf{X}_1$, then the linear system $\mathbf{X}' = \mathbf{A}(\mathbf{X} - \mathbf{X}_1)$ becomes $\mathbf{H}' = \mathbf{A}\mathbf{H}$, which is the form of the linear system analyzed in Section 10.2. The critical point $\mathbf{X} = \mathbf{X}_1$ for $\mathbf{X}' = \mathbf{A}(\mathbf{X} - \mathbf{X}_1)$ now corresponds to the critical point $\mathbf{H} = \mathbf{0}$ for $\mathbf{H}' = \mathbf{A}\mathbf{H}$. If the eigenvalues of $\mathbf{A}$ have negative real parts, then, by Theorem 10.1, $\mathbf{0}$ is an asymptotically stable critical point for $\mathbf{H}' = \mathbf{A}\mathbf{H}$. If there is an eigenvalue with positive real part, $\mathbf{H} = \mathbf{0}$ is an unstable critical point. Theorem 10.3 asserts that the same conclusions can be made for the critical point $\mathbf{X}_1$ of the original system.

THEOREM 10.3   **Stability Criteria for Plane Autonomous Systems**

Let $\mathbf{X}_1$ be a critical point of the plane autonomous system $\mathbf{X}' = \mathbf{g}(\mathbf{X})$, where $P(x, y)$ and $Q(x, y)$ have continuous first partials in a neighborhood of $\mathbf{X}_1$.

   (a) If the eigenvalues of $\mathbf{A} = \mathbf{g}'(\mathbf{X}_1)$ have negative real part, then $\mathbf{X}_1$ is an asymptotically stable critical point.

   (b) If $\mathbf{A} = \mathbf{g}'(\mathbf{X}_1)$ has an eigenvalue with positive real part, then $\mathbf{X}_1$ is an unstable critical point.

EXAMPLE 5   **Stability Analysis of Nonlinear Systems**

Classify (if possible) the critical points of each of the following plane autonomous systems as stable or unstable:

(a) $x' = x^2 + y^2 - 6$
      $y' = x^2 - y$

(b) $x' = 0.01x(100 - x - y)$
      $y' = 0.05y(60 - y - 0.2x)$

SOLUTION   The critical points of each system were determined in Example 4 of Section 10.1.

(a) The critical points are $\left(\sqrt{2}, 2\right)$ and $\left(-\sqrt{2}, 2\right)$, the Jacobian matrix is

$$\mathbf{g}'(\mathbf{X}) = \begin{pmatrix} 2x & 2y \\ 2x & -1 \end{pmatrix},$$

and so

$$\mathbf{A}_1 = \mathbf{g}'\left(\left(\sqrt{2}, 2\right)\right) = \begin{pmatrix} 2\sqrt{2} & 4 \\ 2\sqrt{2} & -1 \end{pmatrix} \quad \text{and} \quad \mathbf{A}_2 = \mathbf{g}'\left(\left(-\sqrt{2}, 2\right)\right) = \begin{pmatrix} -2\sqrt{2} & 4 \\ -2\sqrt{2} & -1 \end{pmatrix}.$$

Since the determinant of $\mathbf{A}_1$ is negative, $\mathbf{A}_1$ has a positive real eigenvalue. Therefore $\left(\sqrt{2}, 2\right)$ is an unstable critical point. Matrix $\mathbf{A}_2$ has a positive determinant and a negative trace, and so both eigenvalues have negative real parts. It follows that $\left(-\sqrt{2}, 2\right)$ is a stable critical point.

(b) The critical points are $(0, 0)$, $(0, 60)$, $(100, 0)$, and $(50, 50)$, the Jacobian matrix is

$$\mathbf{g}'(\mathbf{X}) = \begin{pmatrix} 0.01(100 - 2x - y) & -0.01x \\ -0.01y & 0.05(60 - 2y - 0.2y) \end{pmatrix},$$

and so

$$\mathbf{A}_1 = \mathbf{g}'((0, 0)) = \begin{pmatrix} 1 & 0 \\ 0 & 3 \end{pmatrix} \qquad \mathbf{A}_2 = \mathbf{g}'((0, 60)) = \begin{pmatrix} 0.4 & 0 \\ -0.6 & -3 \end{pmatrix}$$

$$\mathbf{A}_3 = \mathbf{g}'((100, 0)) = \begin{pmatrix} -1 & -1 \\ 0 & 2 \end{pmatrix} \qquad \mathbf{A}_4 = \mathbf{g}'((50, 50)) = \begin{pmatrix} -0.5 & -0.5 \\ -0.5 & -2.5 \end{pmatrix}.$$

Since the matrix $\mathbf{A}_1$ has a positive determinant and a positive trace, both eigenvalues have positive real parts. Therefore $(0, 0)$ is an unstable critical point. The determinants of matrices $\mathbf{A}_2$ and $\mathbf{A}_3$ are negative, so in each case, one of the eigenvalues is positive. Therefore both $(0, 60)$ and $(100, 0)$ are unstable critical points. Since the matrix $\mathbf{A}_4$ has a positive determinant and a negative trace, $(50, 50)$ is a stable critical point. ∎

In Example 5 we did not compute $\tau^2 - 4\Delta$ (as in Section 10.2) and attempt to further classify the critical points as stable nodes, stable spiral points, saddle points, and so on. For example, for $\mathbf{X}_1 = \left(-\sqrt{2}, 2\right)$ in Example 5(a), $\tau^2 - 4\Delta < 0$, and if the system were linear, we would be able to conclude that $\mathbf{X}_1$ was a stable spiral point. Figure 10.24 shows several solution curves near $\mathbf{X}_1$ that were obtained with a numerical solver, and each solution does *appear* to spiral in toward the critical point.

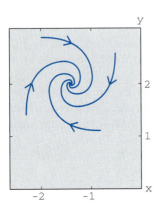

FIGURE 10.24 $\left(-\sqrt{2}, 2\right)$ appears to be a stable spiral point

CLASSIFYING CRITICAL POINTS   It is natural to ask whether we can infer more geometric information about the solutions near a critical point $\mathbf{X}_1$ of a nonlinear autonomous system from an analysis of the critical point of the corresponding linear system. The answer is summarized in Figure 10.25, but you should note the following comments.

(*i*)   In five separate cases (stable node, stable spiral point, unstable spiral point, unstable node, and saddle) the critical point may be categorized like the critical point in the corresponding linear system. The solutions have the same general geometric features as the solutions to the linear system, and the smaller the neighborhood about $\mathbf{X}_1$, the closer the resemblance.

(*ii*)   If $\tau^2 = 4\Delta$ and $\tau > 0$, the critical point $\mathbf{X}_1$ is unstable, but in this borderline case *we are not yet able to decide whether $\mathbf{X}_1$ is an unstable spiral*,

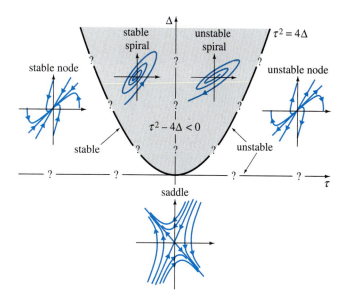

FIGURE 10.25   Geometric summary of some conclusions (see (*i*)), and some unanswered questions (see (*ii*) and (*iii*)) about nonlinear autonomous systems

    unstable node, or degenerate unstable node. Likewise, if $\tau^2 = 4\Delta$ and $\tau < 0$, the critical point $\mathbf{X}_1$ is stable but may be either a stable spiral, a stable node, or a degenerate stable node.

(*iii*) If $\tau = 0$ and $\Delta > 0$, the eigenvalues of $\mathbf{A} = \mathbf{g}'(\mathbf{X})$ are pure imaginary, and in this borderline case $\mathbf{X}_1$ may be either a stable spiral, an unstable spiral, or a center. *It is therefore not yet possible to determine whether* $\mathbf{X}_1$ *is stable or unstable.*

EXAMPLE 6   Classifying Critical Points of a Nonlinear System

Classify each critical point of the plane autonomous system in Example 5(b) as a stable node, a stable spiral point, an unstable spiral point, an unstable node, or a saddle point.

SOLUTION   For the matrix $\mathbf{A}_1$ corresponding to (0, 0), $\Delta = 3$, $\tau = 4$, and so $\tau^2 - 4\Delta = 4$. Therefore (0, 0) is an unstable node. The critical points (0, 60) and (100, 0) are saddles since $\Delta < 0$ in both cases. For matrix $\mathbf{A}_4$, $\Delta > 0$, $\tau < 0$, and $\tau^2 - 4\Delta > 0$. It follows that (50, 50) is a stable node. Experiment with a numerical solver to verify these conclusions. ∎

EXAMPLE 7   Stability Analysis for a Soft Spring

Recall from Section 5.3 that the second-order differential equation $mx'' + kx + k_1x^3 = 0$, for $k > 0$, represents a general model for the free, undamped oscillations of a mass m attached to a nonlinear spring. If $k = 1$ and $k_1 = -1$, the spring is called *soft*, and the plane autonomous system corresponding to the nonlinear second-order differential equation $x'' + x - x^3 = 0$ is

$$x' = y$$
$$y' = x^3 - x.$$

Find and classify (if possible) the critical points.

SOLUTION Since $x^3 - x = x(x^2 - 1)$, the critical points are $(0, 0)$, $(1, 0)$, and $(-1, 0)$. The corresponding Jacobian matrices are

$$\mathbf{A}_1 = \mathbf{g}'((0, 0)) = \begin{pmatrix} 0 & 1 \\ -1 & 0 \end{pmatrix}, \quad \mathbf{A}_2 = \mathbf{g}'((1, 0)) = \mathbf{g}'((-1, 0)) - \begin{pmatrix} 0 & 1 \\ 2 & 0 \end{pmatrix}.$$

Since $\det \mathbf{A}_2 < 0$, critical points $(1, 0)$ and $(-1, 0)$ are both saddle points. The eigenvalues of matrix $\mathbf{A}_1$ are $\pm i$, and according to comment (*iii*), the status of the critical point at $(0, 0)$ remains in doubt. It may be either a stable spiral, an unstable spiral, or a center.

THE PHASE-PLANE METHOD The linearization method, when successful, can provide useful information on the local behavior of solutions near critical points. It is of little help if we are interested in solutions whose initial position $\mathbf{X}(0) = \mathbf{X}_0$ is not close to a critical point or if we wish to obtain a global view of the family of solution curves. The **phase-plane method** is based on the fact that

$$\frac{dy}{dx} = \frac{dy/dt}{dx/dt} = \frac{Q(x, y)}{P(x, y)}$$

and attempts to find y as a function of x using one of the methods available for solving first-order differential equations (Chapter 2). As we show in Examples 8 and 9, the method can sometimes be used to decide whether a critical point such as $(0, 0)$ in Example 7 is a stable spiral, an unstable spiral, or a center.

EXAMPLE 8   Phase-Plane Method

Use the phase-plane method to classify the sole critical point $(0, 0)$ of the plane autonomous system

$$x' = y^2$$
$$y' = x^2.$$

SOLUTION The determinant of the Jacobian matrix

$$\mathbf{g}'(\mathbf{X}) = \begin{pmatrix} 0 & 2y \\ 2x & 0 \end{pmatrix}$$

is 0 at $(0, 0)$, so the nature of the critical point $(0, 0)$ remains in doubt. Using the phase-plane method, we obtain the first-order differential equation

$$\frac{dy}{dx} = \frac{dy/dt}{dx/dt} = \frac{x^2}{y^2},$$

which can be easily solved by separation of variables:

$$\int y^2 \, dy = \int x^2 \, dx \quad \text{or} \quad y^3 = x^3 + c.$$

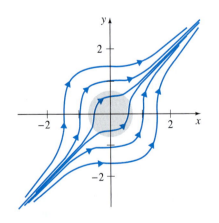

FIGURE 10.26 Phase portrait of nonlinear system in Example 8

If $\mathbf{X}(0) = (0, y_0)$, it follows that $y^3 = x^3 + y_0^3$ or $y = \sqrt[3]{x^3 + y_0^3}$. Figure 10.26 shows a collection of solution curves corresponding to various choices for y_0. The nature of the critical point is clear from this phase portrait: No matter how close to $(0, 0)$ the solution starts, $\mathbf{X}(t)$ moves away from the origin as t increases. The critical point at $(0, 0)$ is therefore unstable.

EXAMPLE 9   Phase-Plane Analysis of a Soft Spring

Use the phase-plane method to determine the nature of the solutions to $x'' + x - x^3 = 0$ in a neighborhood of $(0, 0)$.

SOLUTION If we let $dx/dt = y$, then $dy/dt = x^3 - x$. From this we obtain the first-order differential equation

$$\frac{dy}{dx} = \frac{dy/dt}{dx/dt} = \frac{x^3 - x}{y},$$

which can be solved by separation of variables. Integrating

$$\int y \, dy = \int (x^3 - x) \, dx \quad \text{gives} \quad \frac{y^2}{2} = \frac{x^4}{4} - \frac{x^2}{2} + c.$$

After completing the square, we can write the solution as $y^2 = (x^2 - 1)^2/2 + c_0$. If $\mathbf{X}(0) = (x_0, 0)$, where $0 < x_0 < 1$, then $c_0 = -(x_0^2 - 1)^2/2$, and so

$$y^2 = \frac{(x^2 - 1)^2}{2} - \frac{(x_0^2 - 1)^2}{2} = \frac{(2 - x^2 - x_0^2)(x_0^2 - x^2)}{2}.$$

Note that $y = 0$ when $x = -x_0$. In addition, the right-hand side is positive when $-x_0 < x < x_0$, and so each x has *two* corresponding values of y. The solution $\mathbf{X} = \mathbf{X}(t)$ that satisfies $\mathbf{X}(0) = (x_0, 0)$ is therefore periodic, and so $(0, 0)$ is a center.

Figure 10.27 shows a family of solution curves, or phase portrait, of the original system. We used the original plane autonomous system to determine the directions indicated on each trajectory.

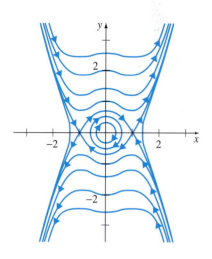

FIGURE 10.27 Phase portrait of nonlinear system in Example 9

EXERCISES 10.3

Answers to selected odd-numbered problems begin on page ANS-15.

1. Show that $(0, 0)$ is an asymptotically stable critical point of the nonlinear autonomous system

$$x' = \alpha x - \beta y + y^2$$
$$y' = \beta x + \alpha y - xy$$

when $\alpha < 0$ and an unstable critical point when $\alpha > 0$. (*Hint:* Switch to polar coordinates.)

2. When expressed in polar coordinates, a plane autonomous system takes the form

$$\frac{dr}{dt} = \alpha r(5 - r)$$

$$\frac{d\theta}{dt} = -1.$$

Show that $(0, 0)$ is an asymptotically stable critical point if and only if $\alpha < 0$.

In Problems 3–10, without solving explicitly, classify the critical points of the given first-order autonomous differential equation as either asymptotically stable or unstable. All constants are assumed to be positive.

3. $\dfrac{dx}{dt} = kx \, (n + 1 - x)$

4. $\dfrac{dx}{dt} = -kx \ln \dfrac{x}{K}, \quad x > 0$

5. $\dfrac{dT}{dt} = k(T - T_0)$

6. $m\dfrac{dv}{dt} = mg - kv$

7. $\dfrac{dx}{dt} = k(\alpha - x)(\beta - x), \quad \alpha > \beta$

8. $\dfrac{dx}{dt} = k(\alpha - x)(\beta - x)(\gamma - x), \quad \alpha > \beta > \gamma$

9. $\dfrac{dP}{dt} = P(a - bP)(1 - cP^{-1}), \quad P > 0, a < bc$

10. $\dfrac{dA}{dt} = k\sqrt{A}\left(K - \sqrt{A}\right), \quad A > 0$

In Problems 11–20 classify (if possible) each critical point of the given plane autonomous system as a stable node, a stable spiral point, an unstable spiral point, an unstable node, or a saddle point.

11. $x' = 1 - 2xy$
$\quad\;\; y' = 2xy - y$

12. $x' = x^2 - y^2 - 1$
$\quad\;\; y' = 2y$

13. $x' = y - x^2 + 2$
$\quad\;\; y' = x^2 - xy$

14. $x' = 2x - y^2$
$\quad\;\; y' = -y + xy$

15. $x' = -3x + y^2 + 2$
$\quad\;\; y' = x^2 - y^2$

16. $x' = xy - 3y - 4$
$\quad\;\; y' = y^2 - x^2$

17. $x' = -2xy$
$\quad\;\; y' = y - x + xy - y^3$

18. $x' = x(1 - x^2 - 3y^2)$
$\quad\;\; y' = y(3 - x^2 - 3y^2)$

19. $x' = x\left(10 - x - \frac{1}{2}y\right)$
$y' = y(16 - y - x)$

20. $x' = -2x + y + 10$
$y' = 2x - y - 15\dfrac{y}{y + 5}$

In Problems 21–26 classify (if possible) each critical point of the given second-order differential equation as a stable node, a stable spiral point, an unstable spiral point, an unstable node, or a saddle point.

21. $\theta'' = (\cos\theta - 0.5)\sin\theta, \quad |\theta| < \pi$

22. $x'' + x = \left(\frac{1}{2} - 3(x')^2\right)x' - x^2$

23. $x'' + x'(1 - x^3) - x^2 = 0$

24. $x'' + 4\dfrac{x}{1 + x^2} + 2x' = 0$

25. $x'' + x = \epsilon x^3$ for $\epsilon > 0$

26. $x'' + x - \epsilon x|x| = 0$ for $\epsilon > 0$ $\left(Hint: \dfrac{d}{dx}x|x| = 2|x|.\right)$

27. Show that the nonlinear second-order differential equation

$$(1 + \alpha^2 x^2)x'' + (\beta + \alpha^2(x')^2)x = 0$$

has a saddle point at $(0, 0)$ when $\beta < 0$.

28. Show that the dynamical system

$$x' = -\alpha x + xy$$
$$y' = 1 - \beta y - x^2$$

has a unique critical point when $\alpha\beta > 1$ and that this critical point is stable when $\beta > 0$.

29. (a) Show that the plane autonomous system

$$x' = -x + y - x^3$$
$$y' = -x - y + y^2$$

has two critical points by sketching the graphs of $-x + y - x^3 = 0$ and $-x - y + y^2 = 0$. Classify the critical point at $(0, 0)$.

(b) Show that the second critical point $\mathbf{X}_1 = (0.88054, 1.56327)$ is a saddle point.

30. (a) Show that $(0, 0)$ is the only critical point of the **Raleigh differential equation**

$$x'' + \epsilon\left(\frac{1}{3}(x')^3 - x'\right) + x = 0.$$

(b) Show that $(0, 0)$ is unstable when $\epsilon > 0$. When is $(0, 0)$ an unstable spiral point?

(c) Show that $(0, 0)$ is stable when $\epsilon < 0$. When is $(0, 0)$ a stable spiral point?

(d) Show that $(0, 0)$ is a center when $\epsilon = 0$.

31. Use the phase-plane method to show that $(0, 0)$ is a center of the nonlinear second-order differential equation $x'' + 2x^3 = 0$.

32. Use the phase-plane method to show that the solution to the nonlinear second-order differential equation $x'' + 2x - x^2 = 0$ that satisfies $x(0) = 1$ and $x'(0) = 0$ is periodic.

33. (a) Find the critical points of the plane autonomous system

$$x' = 2xy$$
$$y' = 1 - x^2 + y^2,$$

and show that linearization gives no information about the nature of these critical points.

(b) Use the phase-plane method to show that the critical points in (a) are both centers.
(*Hint:* Let $u = y^2/x$, and show that $(x - c)^2 + y^2 = c^2 - 1$.)

34. The origin is the only critical point of the nonlinear second-order differential equation $x'' + (x')^2 + x = 0$.

(a) Show that the phase-plane method leads to the Bernoulli differential equation $dy/dx = -y - xy^{-1}$.

(b) Show that the solution satisfying $x(0) = \frac{1}{2}$ and $x'(0) = 0$ is not periodic.

35. A solution of the nonlinear second-order differential equation $x'' + x - x^3 = 0$ satisfies $x(0) = 0$ and $x'(0) = v_0$. Use the phase-plane method to determine when the resulting solution is periodic. (*Hint:* See Example 9.)

36. The nonlinear differential equation $x'' + x = 1 + \epsilon x^2$ arises in the analysis of planetary motion using relativity theory. Classify (if possible) all critical points of the corresponding plane autonomous system.

37. When a nonlinear capacitor is present in an *LRC* circuit, the voltage drop is no longer given by q/C but is more accurately described by $\alpha q + \beta q^3$, where α and β are constants and $\alpha > 0$. Differential equation (34) of Section 5.1 for the free circuit is then replaced by

$$L\frac{d^2q}{dt^2} + R\frac{dq}{dt} + \alpha q + \beta q^3 = 0.$$

Find and classify all critical points of this nonlinear differential equation. (*Hint:* Divide into the two cases $\beta > 0$ and $\beta < 0$.)

38. The nonlinear equation $mx'' + kx + k_1x^3 = 0$, for $k > 0$, represents a general model for the free, undamped oscillations of a mass m attached to a spring. If $k_1 > 0$, the spring is called *hard* (see Example 1 in Section 5.3). Determine the nature of the solutions to $x'' + x + x^3 = 0$ in a neighborhood of $(0, 0)$.

39. The nonlinear equation $\theta'' + \sin\theta = \frac{1}{2}$ can be interpreted as a model for a certain pendulum with a constant driving function.

(a) Show that $(\pi/6, 0)$ and $(5\pi/6, 0)$ are critical points of the corresponding plane autonomous system.

(b) Classify the critical point $(5\pi/6, 0)$ using linearization.

(c) Use the phase-plane method to classify the critical point $(\pi/6, 0)$.

DISCUSSION/PROJECT PROBLEMS

40. (a) Show that $(0, 0)$ is an isolated critical point of the plane autonomous system

$$x' = x^4 - 2xy^3$$
$$y' = 2x^3y - y^4$$

but that linearization gives no useful information about the nature of this critical point.

(b) Use the phase-plane method to show that $x^3 + y^3 = 3cxy$. This classic curve is called a *folium of Descartes*. Parametric equations for a folium are

$$x = \frac{3ct}{1 + t^3}, \quad y = \frac{3ct^2}{1 + t^3}.$$

(*Hint:* The differential equation in x and y is homogeneous.)

(c) Use graphing software or a numerical solver to graph solution curves. Based on your graphs, would you classify the critical point as stable or unstable? Would you classify the critical point as a node, saddle point, center, or spiral point? Explain.

10.4 AUTONOMOUS SYSTEMS AS MATHEMATICAL MODELS

INTRODUCTION: Many applications from physics give rise to nonlinear autonomous second-order differential equations—that is, DEs of the form $x'' = g(x, x')$. For example, in the analysis of free, damped motion of Section 5.1 we assumed that the damping force was proportional to the velocity x' and the resulting model $mx'' = -\beta x' - kx$ is a linear differential equation. But if the magnitude of the damping force is proportional to the square of the velocity, the new differential equation $mx'' = -\beta x' |x'| - kx$ is nonlinear. The corresponding plane autonomous system is nonlinear:

$$x' = y$$

$$y' = -\frac{\beta}{m}y|y| - \frac{k}{m}x.$$

In this section we also analyze the nonlinear pendulum, motion of a bead on a curve, the Lotka-Volterra predator-prey models, and the Lotka-Volterra competition model. Additional models are presented in the exercises.

REVIEW MATERIAL: The results of Section 10.3 will be used to analyze all the mathematical models in this section. A thorough review of that material is recommended.

NONLINEAR PENDULUM In (6) of Section 5.3 we showed that the displacement angle θ for a simple pendulum satisfies the nonlinear second-order differential equation

$$\frac{d^2\theta}{dt^2} + \frac{g}{l}\sin\theta = 0.$$

When we let $x = \theta$ and $y = \theta'$, this second-order differential equation may be rewritten as the dynamical system

$$x' = y$$

$$y' = -\frac{g}{l}\sin x.$$

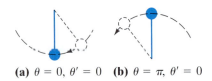

(a) $\theta = 0$, $\theta' = 0$ **(b)** $\theta = \pi$, $\theta' = 0$

FIGURE 10.28 $(0, 0)$ is stable and $(\pi, 0)$ is unstable

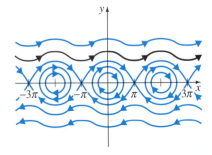

FIGURE 10.29   Phase portrait of pendulum; wavy curves indicate that the pendulum is whirling about its pivot

The critical points are $(\pm k\pi, 0)$, and the Jacobian matrix is easily shown to be

$$g'((\pm k\pi, 0)) = \begin{pmatrix} 0 & 1 \\ (-1)^{k+1}\dfrac{g}{l} & 0 \end{pmatrix}.$$

If $k = 2n + 1$, $\Delta < 0$, and so all critical points $(\pm(2n + 1)\pi, 0)$ are saddle points. In particular, the critical point at $(\pi, 0)$ is unstable as expected. See Figure 10.28. When $k = 2n$, the eigenvalues are pure imaginary, and so the nature of these critical points remains in doubt. Since we have assumed that there are no damping forces acting on the pendulum, we expect that all of the critical points $(\pm 2n\pi, 0)$ are centers. This can be verified using the phase-plane method. From

$$\frac{dy}{dx} = \frac{dy/dt}{dx/dt} = -\frac{g}{l}\frac{\sin x}{y}$$

it follows that $y^2 = (2g/l)\cos x + c$. If $\mathbf{X}(0) = (x_0, 0)$, then $y^2 = (2g/l)(\cos x - \cos x_0)$. Note that $y = 0$ when $x = -x_0$, and that $(2g/l)(\cos x - \cos x_0) > 0$ for $|x| < |x_0| < \pi$. Thus each such x has two corresponding values of y, so the solution $\mathbf{X} = \mathbf{X}(t)$ that satisfies $\mathbf{X}(0) = (x_0, 0)$ is periodic. We may conclude that $(0, 0)$ is a center. Observe that $x = \theta$ increases for solutions that correspond to large initial velocities, such as the one drawn in black in Figure 10.29. In this case the pendulum spins or whirls in complete circles about its pivot.

EXAMPLE 1   **Periodic Solutions of the Pendulum DE**

A pendulum in an equilibrium position with $\theta = 0$ is given an initial angular velocity of ω_0 rad/s. Determine under what conditions the resulting motion is periodic.

SOLUTION   We are asked to examine the solution of the plane autonomous system that satisfies $\mathbf{X}(0) = (0, \omega_0)$. From $y^2 = (2g/l)\cos x + c$ it follows that

$$y^2 = \frac{2g}{l}\left(\cos x - 1 + \frac{l}{2g}\omega_0^2\right).$$

To establish that the solution $\mathbf{X}(t)$ is periodic, it is sufficient to show that there are two x-intercepts $x = \pm x_0$ between $-\pi$ and π and that the right-hand side is positive for $|x| < |x_0|$. Each such x then has two corresponding values of y.

If $y = 0$, $\cos x = 1 - (l/2g)\omega_0^2$, and this equation has two solutions $x = \pm x_0$ between $-\pi$ and π, provided that $1 - (l/2g)\omega_0^2 > -1$. Note that $(2g/l)(\cos x - \cos x_0)$ is then positive for $|x| < |x_0|$. This restriction on the initial angular velocity may be written as $|\omega_0| < 2\sqrt{g/l}$. ∎

NONLINEAR OSCILLATIONS: THE SLIDING BEAD   Suppose, as shown in Figure 10.30, that a bead with mass m slides along a thin wire whose shape is described by the function $z = f(x)$. A wide variety of nonlinear oscillations can be obtained by changing the shape of the wire and by making different assumptions about the forces acting on the bead.

The tangential force $\mathbf{F}$ due to the weight $W = mg$ has magnitude $mg \sin\theta$, and therefore the x-component of $\mathbf{F}$ is $F_x = -mg \sin\theta \cos\theta$. Since $\tan\theta = f'(x)$, we may use the identities $1 + \tan^2\theta = \sec^2\theta$ and $\sin^2\theta = 1 - \cos^2\theta$ to conclude that

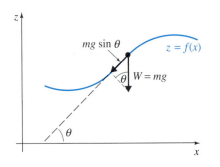

FIGURE 10.30   Some forces acting on sliding bead

$$F_x = -mg \sin\theta \cos\theta = -mg\frac{f'(x)}{1 + [f'(x)]^2}.$$

We assume (as in Section 5.1) that a damping force **D**, acting in the direction opposite to the motion, is a constant multiple of the velocity of the bead. The x-component of **D** is therefore $D_x = -\beta x'$. If we ignore the frictional force between the wire and the bead and assume that no other external forces are impressed on the system, it follows from Newton's second law that

$$mx'' = -mg\frac{f'(x)}{1 + [f'(x)]^2} - \beta x',$$

and the corresponding plane autonomous system is

$$x' = y$$

$$y' = -g\frac{f'(x)}{1 + [f'(x)]^2} - \frac{\beta}{m}y.$$

If $\mathbf{X}_1 = (x_1, y_1)$ is a critical point of the system, $y_1 = 0$ and therefore $f'(x_1) = 0$. The bead must therefore be at rest at a point on the wire where the tangent line is horizontal. When f is twice differentiable, the Jacobian matrix at $\mathbf{X}_1$ is

$$\mathbf{g}'(\mathbf{X}_1) = \begin{pmatrix} 0 & 1 \\ -gf''(x_1) & -\beta/m \end{pmatrix},$$

so $\tau = -\beta/m$, $\Delta = gf''(x_1)$, and $\tau^2 - 4\Delta = \beta^2/m^2 - 4gf''(x_1)$. Using the results of Section 10.3, we can make the following conclusions:

(*i*) $f''(x_1) < 0$:
  A relative maximum therefore occurs at $x = x_1$, and since $\Delta < 0$, an *unstable saddle point* occurs at $\mathbf{X}_1 = (x_1, 0)$.

(*ii*) $f''(x_1) > 0$ and $\beta > 0$:
  A relative minimum therefore occurs at $x = x_1$, and since $\tau < 0$ and $\Delta > 0$, $\mathbf{X}_1 = (x_1, 0)$ is a *stable critical point*. If $\beta^2 > 4gm^2f''(x_1)$, the system is **overdamped** and the critical point is a *stable node*. If $\beta^2 < 4gm^2f''(x_1)$, the system is **underdamped** and the critical point is a *stable spiral point*. The exact nature of the stable critical point is still in doubt if $\beta^2 = 4gm^2f''(x_1)$.

(*iii*) $f''(x_1) > 0$ and the system is undamped ($\beta = 0$):
  In this case the eigenvalues are pure imaginary, but the phase-plane method can be used to show that the critical point is a *center*. Therefore solutions with $\mathbf{X}(0) = (x(0), x'(0))$ near $\mathbf{X}_1 = (x_1, 0)$ are periodic.

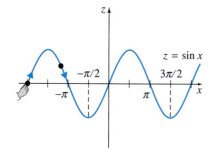

FIGURE 10.31 $-\pi/2$ and $3\pi/2$ are stable.

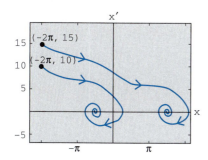

FIGURE 10.32 $\beta = 0.01$

EXAMPLE 2   Bead Sliding Along a Sine Wave

A 10-gram bead slides along the graph of $z = \sin x$. According to conclusion (*ii*), the relative minima at $x_1 = -\pi/2$ and $3\pi/2$ give rise to stable critical points (see Figure 10.31). Since $f''(-\pi/2) = f''(3\pi/2) = 1$, the system will be underdamped provided $\beta^2 < 4gm^2$. If we use SI units, $m = 0.01$ kg and $g = 9.8$ m/s^2, and so the condition for an underdamped system becomes $\beta^2 < 3.92 \times 10^{-3}$.

If $\beta = 0.01$ is the damping constant, both of these critical points are stable spiral points. The two solutions corresponding to initial conditions $\mathbf{X}(0) = (x(0), x'(0)) = (-2\pi, 10)$ and $\mathbf{X}(0) = (-2\pi, 15)$, respectively, were obtained by using a numerical solver and are shown in Figure 10.32. When $x'(0) = 10$, the bead has enough momentum to make it over the hill at $x = -3\pi/2$ but not over the hill at $x = \pi/2$. The bead then approaches the relative minimum based at $x = -\pi/2$. If $x'(0) = 15$, the bead has the momentum to make it over both hills, but then it rocks back and forth in the valley based at $x = 3\pi/2$ and approaches the point

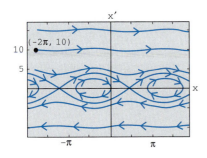

FIGURE 10.33 $\beta = 0$

$(3\pi/2, -1)$ on the wire. Experiment with other initial conditions using your numerical solver.

Figure 10.33 shows a collection of solution curves obtained from a numerical solver for the undamped case. Since $\beta = 0$, the critical points corresponding to $x_1 = -\pi/2$ and $3\pi/2$ are now centers. When $\mathbf{X}(0) = (-2\pi, 10)$, the bead has sufficient momentum to move over *all* hills. The figure also indicates that when the bead is released from rest at a position on the wire between $x = -3\pi/2$ and $x = \pi/2$, the resulting motion is periodic.

LOTKA-VOLTERRA PREDATOR-PREY MODEL   A **predator-prey interaction** between two species occurs when one species (the predator) feeds on a second species (the prey). For example, the snowy owl feeds almost exclusively on a common arctic rodent called a lemming, while a lemming uses arctic tundra plants as its food supply. Interest in using mathematics to help explain predator-prey interactions has been stimulated by the observation of population cycles in many arctic mammals. In the MacKenzie River district of Canada, for example, the principal prey of the lynx is the snowshoe hare, and both populations cycle with a period of about 10 years.

There are many predator-prey models that lead to plane autonomous systems with at least one periodic solution. The first such model was constructed independently by pioneer biomathematicians A. Lotka (1925) and V. Volterra (1926). If x denotes the number of predators and y denotes the number of prey, then the Lotka-Volterra model takes the form

$$x' = -ax + bxy = x(-a + by)$$
$$y' = -cxy + dy = y(-cx + d),$$

where a, b, c, and d are positive constants.

Note that in the absence of predators ($x = 0$), $y' = dy$, and so the number of prey grows exponentially. In the absence of prey, $x' = -ax$, and so the predator population becomes extinct. The term $-cxy$ represents the death rate due to predation. The model therefore assumes that this death rate is directly proportional to the number of possible encounters xy between predator and prey at a particular time t, and the term bxy represents the resulting positive contribution to the predator population.

The critical points of this plane autonomous system are $(0, 0)$ and $(d/c, a/b)$, and the corresponding Jacobian matrices are

$$\mathbf{A}_1 = \mathbf{g}'((0, 0)) = \begin{pmatrix} -a & 0 \\ 0 & d \end{pmatrix} \quad \text{and} \quad \mathbf{A}_2 = \mathbf{g}'((d/c, a/b)) = \begin{pmatrix} 0 & bd/c \\ -ac/b & 0 \end{pmatrix}.$$

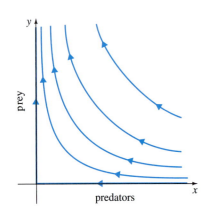

FIGURE 10.34   Solutions near $(0, 0)$

The critical point at $(0, 0)$ is a saddle point, and Figure 10.34 shows a typical profile of solutions that are in the first quadrant and near $(0, 0)$.

Because the matrix $\mathbf{A}_2$ has pure imaginary eigenvalues $\lambda = \pm\sqrt{ad}\,i$, the critical point $(d/c, a/b)$ *may* be a center. This possibility can be investigated using the phase-plane method. Since

$$\frac{dy}{dx} = \frac{y(-cx + d)}{x(-a + by)},$$

we separate variables and obtain

$$\int \frac{-a + by}{y}\, dy = \int \frac{-cx + d}{x}\, dx$$

$$-a \ln y + by = -cx + d \ln x + c_1 \quad \text{or} \quad (x^d e^{-cx})(y^a e^{-by}) = c_0.$$

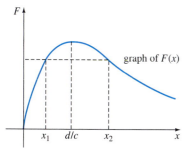

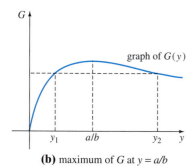

(a) maximum of F at $x = d/c$

(b) maximum of G at $y = a/b$

FIGURE 10.35   Graphs of F and G help to establish properties (1)–(3)

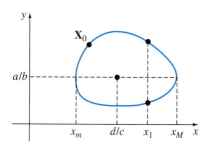

FIGURE 10.36   Periodic solution of Lotka-Volterra model

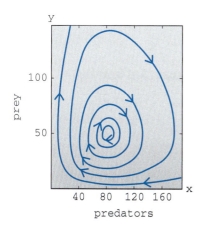

FIGURE 10.37   Phase portrait of Lotka-Volterra model near critical point (80, 50)

The following argument establishes that all solution curves that originate in the first quadrant are periodic.

Typical graphs of the nonnegative functions $F(x) = x^d e^{-cx}$ and $G(y) = y^a e^{-by}$ are shown in Figure 10.35. It is not hard to show that $F(x)$ has an absolute maximum at $x = d/c$, whereas $G(y)$ has an absolute maximum at $y = a/b$. Note that with the exception of 0 and the absolute maximum, F and G each take on all values in their range precisely twice.

These graphs can be used to establish the following properties of a solution curve that originates at a noncritical point (x_0, y_0) in the first quadrant:

1. If $y = a/b$, the equation $F(x)G(y) = c_0$ has exactly two solutions x_m and x_M that satisfy $x_m < d/c < x_M$.

2. If $x_m < x_1 < x_M$ and $x = x_1$, then $F(x)G(y) = c_0$ has exactly two solutions y_1 and y_2 that satisfy $y_1 < a/b < y_2$.

3. If x is outside the interval $[x_m, x_M]$, then $F(x)G(y) = c_0$ has no solutions.

We will give the demonstration of (1) and outline parts (2) and (3) in the exercises. Since $(x_0, y_0) \neq (d/c, a/b)$, $F(x_0)G(y_0) < F(d/c)G(a/b)$. If $y = a/b$, then

$$0 < \frac{c_0}{G(a/b)} = \frac{F(x_0)G(y_0)}{G(a/b)} < \frac{F(d/c)G(a/b)}{G(a/b)} = F(d/c).$$

Therefore $F(x) = c_0/G(a/b)$ has precisely two solutions x_m and x_M that satisfy $x_m < d/c < x_M$. The graph of a typical periodic solution is shown in Figure 10.36.

EXAMPLE 3   **Predator-Prey Population Cycles**

If we let $a = 0.1$, $b = 0.002$, $c = 0.0025$, and $d = 0.2$ in the Lotka-Volterra predator-prey model, the critical point in the first quadrant is $(d/c, a/b) = (80, 50)$, and we know that this critical point is a center. See Figure 10.37, in which we have used a numerical solver to generate these cycles. The closer the initial condition $\mathbf{X}_0$ is to (80, 50), the more the periodic solutions resemble the elliptical solutions to the corresponding linear system. The eigenvalues of $\mathbf{g}'((80, 50))$ are $\lambda = \pm\sqrt{ad}\, i = \pm\sqrt{2}/10\, i$, and so the solutions near the critical point have period $p \approx 10\sqrt{2}\,\pi$, or about 44.4. ∎

LOTKA-VOLTERRA COMPETITION MODEL   A **competitive interaction** occurs when two or more species compete for the food, water, light, and space resources of an ecosystem. The use of one of these resources by one population therefore inhibits the ability of another population to survive and grow. Under what conditions can two competing species coexist? A number of mathematical models have been constructed that offer insights into conditions that permit coexistence. If x denotes the number in species I and y denotes the number in species II, then the Lotka-Volterra model takes the form

$$x' = \frac{r_1}{K_1} x (K_1 - x - \alpha_{12}y)$$

$$y' = \frac{r_2}{K_2} y (K_2 - y - \alpha_{21}x). \tag{1}$$

Note that in the absence of species II ($y = 0$), $x' = (r_1/K_1)x(K_1 - x)$, and so the first population grows logistically and approaches the steady-state population K_1 (see Section 3.3 and Example 4 in Section 10.3). A similar statement holds for species II growing in the absence of species I. The term $-\alpha_{21}xy$ in the second

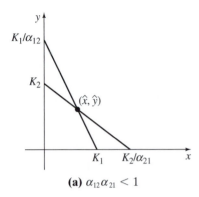

(a) $\alpha_{12}\alpha_{21} < 1$

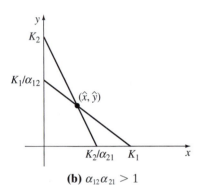

(b) $\alpha_{12}\alpha_{21} > 1$

FIGURE 10.38 Two conditions when critical point $(\hat{x}, \hat{y})$ is in first quadrant

equation stems from the competitive effect of species I on species II. The model therefore assumes that this rate of inhibition is directly proportional to the number of possible competitive pairs xy at a particular time t.

This plane autonomous system has critical points at $(0, 0)$, $(K_1, 0)$, and $(0, K_2)$. When $\alpha_{12}\alpha_{21} \neq 0$, the lines $K_1 - x - \alpha_{12}y = 0$ and $K_2 - y - \alpha_{21}x = 0$ intersect to produce a fourth critical point $\hat{\mathbf{X}} = (\hat{x}, \hat{y})$. Figure 10.38 shows the two conditions under which $(\hat{x}, \hat{y})$ is in the first quadrant. The trace and determinant of the Jacobian matrix at $(\hat{x}, \hat{y})$ are, respectively,

$$\tau = -\hat{x}\frac{r_1}{K_1} - \hat{y}\frac{r_2}{K_2} \quad \text{and} \quad \Delta = (1 - \alpha_{12}\alpha_{21})\hat{x}\hat{y}\frac{r_1 r_2}{K_1 K_2}.$$

In case (a) of Figure 10.38, $K_1/\alpha_{12} > K_2$ and $K_2/\alpha_{21} > K_1$. It follows that $\alpha_{12}\alpha_{21} < 1$, $\tau < 0$, and $\Delta > 0$. Since

$$\tau^2 - 4\Delta = \left(\hat{x}\frac{r_1}{K_1} + \hat{y}\frac{r_2}{K_2}\right)^2 + 4(\alpha_{12}\alpha_{21} - 1)\hat{x}\hat{y}\frac{r_1 r_2}{K_1 K_2}$$

$$= \left(\hat{x}\frac{r_1}{K_1} - \hat{y}\frac{r_2}{K_2}\right)^2 + 4\alpha_{12}\alpha_{21}\hat{x}\hat{y}\frac{r_1 r_2}{K_1 K_2},$$

$\tau^2 - 4\Delta > 0$, and so $(\hat{x}, \hat{y})$ is a stable node. Therefore if $\mathbf{X}(0) = \mathbf{X}_0$ is sufficiently close to $\hat{\mathbf{X}} = (\hat{x}, \hat{y})$, $\lim_{t \to \infty} \mathbf{X}(t) = \hat{\mathbf{X}}$, and we may conclude that coexistence is possible. The demonstration that case (b) leads to a saddle point and the investigation of the nature of critical points at $(0, 0)$, $(K_1, 0)$, and $(0, K_2)$ are left to the exercises.

When the competitive interactions between two species are weak, both of the coefficients α_{12} and α_{21} will be small, so the conditions $K_1/\alpha_{12} > K_2$ and $K_2/\alpha_{21} > K_1$ may be satisfied. This might occur when there is a small overlap in the ranges of two predator species that hunt for a common prey.

EXAMPLE 4   **A Lotka-Volterra Competition Model**

A competitive interaction is described by the Lotka-Volterra competition model

$$x' = 0.004x(50 - x - 0.75y)$$

$$y' = 0.001y(100 - y - 3.0x).$$

Classify all critical points of the system.

SOLUTION   You should verify that critical points occur at $(0, 0)$, $(50, 0)$, $(0, 100)$ and at $(20, 40)$. Since $\alpha_{12}\alpha_{21} = 2.25 > 1$, we have case (b) in Figure 10.38, so the critical point at $(20, 40)$ is a saddle point. The Jacobian matrix is

$$\mathbf{g}'(\mathbf{X}) = \begin{pmatrix} 0.2 - 0.008x - 0.003y & -0.003x \\ -0.003y & 0.1 - 0.002y - 0.003x \end{pmatrix},$$

and we obtain

$$\mathbf{g}'((0, 0)) = \begin{pmatrix} 0.2 & 0 \\ 0 & 0.1 \end{pmatrix}, \quad \mathbf{g}'((50, 0)) = \begin{pmatrix} -0.2 & -0.15 \\ 0 & -0.05 \end{pmatrix}, \quad \mathbf{g}'((0, 100)) = \begin{pmatrix} -0.1 & 0 \\ -0.3 & -0.1 \end{pmatrix}.$$

Therefore $(0, 0)$ is an unstable node, whereas both $(50, 0)$ and $(0, 100)$ are stable nodes. (Check this!)

Coexistence can also occur in the Lotka-Volterra competition model if there is at least one periodic solution lying entirely in the first quadrant. It is possible to show, however, that this model has no periodic solutions.

EXERCISES 10.4

Answers to selected odd-numbered problems begin on page ANS-15.

NONLINEAR PENDULUM

1. A pendulum is released at $\theta = \pi/3$ and is given an initial angular velocity of ω_0 rad/s. Determine under what conditions the resulting motion is periodic.

2. (a) If a pendulum is released from rest at $\theta = \theta_0$, show that the angular velocity is again 0 when $\theta = -\theta_0$.
   (b) The period T of the pendulum is the amount of time needed for θ to change from θ_0 to $-\theta_0$ and back to θ_0. Show that

$$T = \sqrt{\frac{2L}{g}} \int_{-\theta_0}^{\theta_0} \frac{1}{\sqrt{\cos \theta - \cos \theta_0}} \, d\theta.$$

SLIDING BEAD

3. A bead with mass m slides along a thin wire whose shape is described by the function $z = f(x)$. If $\mathbf{X}_1 = (x_1, y_1)$ is a critical point of the plane autonomous system associated with the sliding bead, verify that the Jacobian matrix at $\mathbf{X}_1$ is

$$\mathbf{g}'(\mathbf{X}_1) = \begin{pmatrix} 0 & 1 \\ -gf''(x_1) & -\beta/m \end{pmatrix}.$$

4. A bead with mass m slides along a thin wire whose shape is described by the function $z = f(x)$. When $f'(x_1) = 0$, $f''(x_1) > 0$, and the system is undamped, the critical point $\mathbf{X}_1 = (x_1, 0)$ is a center. Estimate the period of the bead when $x(0)$ is near x_1 and $x'(0) = 0$.

5. A bead is released from the position $x(0) = x_0$ on the curve $z = x^2/2$ with initial velocity $x'(0) = v_0$ cm/s.
   (a) Use the phase-plane method to show that the resulting solution is periodic when the system is undamped.
   (b) Show that the maximum height z_{max} to which the bead rises is given by $z_{max} = \frac{1}{2}[e^{v_0^2/g}(1 + x_0^2) - 1]$.

6. Rework Problem 5 with $z = \cosh x$.

PREDATOR-PREY MODELS

7. (Refer to Figure 10.36.) If $x_m < x_1 < x_M$ and $x = x_1$, show that $F(x)G(y) = c_0$ has exactly two solutions y_1 and y_2 that satisfy $y_1 < a/b < y_2$. (*Hint:* First show that $G(y) = c_0/F(x_1) < G(a/b)$.)

8. From (1) and (3) on page 423, conclude that the maximum number of predators occurs when $y = a/b$.

9. In many fishery science models, the rate at which a species is caught is assumed to be directly proportional to its abundance. If both predator and prey are being exploited in this manner, the Lotka-Volterra differential equations take the form

$$x' = -ax + bxy - \epsilon_1 x$$
$$y' = -cxy + dy - \epsilon_2 y,$$

where ϵ_1 and ϵ_2 are positive constants.
   (a) When $\epsilon_2 < d$, show that there is a new critical point in the first quadrant that is a center.
   (b) **Volterra's principle** states that a moderate amount of exploitation increases the average number of prey and decreases the average number of predators. Is this fisheries model consistent with Volterra's principle?

10. A predator-prey interaction is described by the Lotka-Volterra model

$$x' = -0.1x + 0.02xy$$
$$y' = 0.2y - 0.025xy.$$

   (a) Find the critical point in the first quadrant, and use a numerical solver to sketch some population cycles.
   (b) Estimate the period of the periodic solutions that are close to the critical point in part (a).

COMPETITION MODELS

11. A competitive interaction is described by the Lotka-Volterra competition model

$$x' = 0.08x(20 - 0.4x - 0.3y)$$
$$y' = 0.06y(10 - 0.1y - 0.3x).$$

Find and classify all critical points of the system.

12. In (1) show that $(0, 0)$ is always an unstable node.

13. In (1) show that $(K_1, 0)$ is a stable node when $K_1 > K_2/\alpha_{21}$ and a saddle point when $K_1 < K_2/\alpha_{21}$.

14. Use Problems 12 and 13 to establish that $(0, 0)$, $(K_1, 0)$, and $(0, K_2)$ are unstable when $\hat{\mathbf{X}} = (\hat{x}, \hat{y})$ is a stable node.

15. In (1) show that $\hat{\mathbf{X}} = (\hat{x}, \hat{y})$ is a saddle point when $K_1/\alpha_{12} < K_2$ and $K_2/\alpha_{21} < K_1$.

MISCELLANEOUS MATHEMATICAL MODELS

16. **Damped Pendulum** If we assume that a damping force acts in the direction opposite to the motion of a pendulum and with a magnitude directly proportional to the angular velocity $d\theta/dt$, the displacement angle θ for the pendulum satisfies the nonlinear second-order differential equation

$$ml\frac{d^2\theta}{dt^2} = -mg \sin\theta - \beta\frac{d\theta}{dt}.$$

(a) Write the second-order differential equation as a plane autonomous system. Find all critical points of the system.

(b) Find a condition on m, l, and β that will make $(0, 0)$ a stable spiral point.

17. **Nonlinear Damping** In the analysis of free, damped motion in Section 5.1 we assumed that the damping force was proportional to the velocity x'. Frequently, the magnitude of this damping force is proportional to the square of the velocity, and the new differential equation becomes

$$x'' = -\frac{\beta}{m}x'|x'| - \frac{k}{m}x.$$

(a) Write the second-order differential equation as a plane autonomous system, and find all critical points.

(b) The system is called *overdamped* when $(0, 0)$ is a stable node and is called *underdamped* when $(0, 0)$ is a stable spiral point. Physical considerations suggest that $(0, 0)$ must be an asymptotically stable critical point. Show that the system is necessarily underdamped. $\left(\textit{Hint: } \dfrac{d}{dy}(y|y|) = 2|y|.\right)$

DISCUSSION/PROJECT PROBLEMS

18. A bead with mass m slides along a thin wire whose shape may be described by the function $z = f(x)$. Small stretches of the wire act like an inclined plane, and in mechanics it is assumed that the magnitude of the frictional force between the bead and wire is directly proportional to $mg \cos\theta$ (see Figure 10.30).

(a) Explain why the new differential equation for the x-coordinate of the bead is

$$x'' = g\frac{\mu - f'(x)}{1 + [f'(x)]^2} - \frac{\beta}{m}x'$$

for some positive constant μ.

(b) Investigate the critical points of the corresponding plane autonomous system. Under what conditions is a critical point a saddle point? A stable spiral point?

19. An undamped oscillation satisfies a nonlinear second-order differential equation of the form $x'' + f(x) = 0$, where $f(0) = 0$ and $xf(x) > 0$ for $x \neq 0$ and $-d < x < d$. Use the phase-plane method to investigate whether it is possible for the critical point $(0, 0)$ to be a stable spiral point. $\left(\textit{Hint: } \text{Let } F(x) = \int_0^x f(u)\, du, \text{ and show that } y^2 + 2F(x) = c.\right)$

20. The Lotka-Volterra predator-prey model assumes that in the absence of predators the number of prey grows exponentially. If we make the alternative assumption that the prey population grows logistically, the new system is

$$x' = -ax + bxy$$

$$y' = -cxy + \frac{r}{K}y\,(K - y),$$

where a, b, c, r, and K are positive and $K > a/b$.

(a) Show that the system has critical points at $(0, 0)$, $(0, K)$, and $(\hat{x}, \hat{y})$, where $\hat{y} = a/b$ and

$$c\hat{x} = \frac{r}{K}(K - \hat{y}).$$

(b) Show that the critical points at $(0, 0)$ and $(0, K)$ are saddle points, whereas the critical point at $(\hat{x}, \hat{y})$ is either a stable node or a stable spiral point.

(c) Show that $(\hat{x}, \hat{y})$ is a stable spiral point if

$$\hat{y} < \frac{4bK^2}{r + 4bK}.$$

Explain why this case will occur when the carrying capacity K of the prey is large.

21. The dynamical system

$$x' = \alpha\frac{y}{1 + y}x - x$$

$$y' = -\frac{y}{1 + y}x - y + \beta$$

arises in a model for the growth of microorganisms in a chemostat, a simple laboratory device in which a nutrient from a supply source flows into a growth chamber. In the system, x denotes the concentration of the microorganisms in the growth chamber, y denotes the concentration of nutrients, and $\alpha > 1$ and $\beta > 0$ are constants that can be adjusted by the experimenter. Find conditions on α and β that ensure that the system has a single critical point $(\hat{x}, \hat{y})$ in the first quadrant, and investigate the stability of this critical point.

22. Use the methods of this chapter together with a numerical solver to investigate stability in the nonlinear spring/mass system modeled by

$$x'' + 8x - 6x^3 + x^5 = 0.$$

See Problem 8 in Exercises 5.3.

CHAPTER 10 IN REVIEW

Answers to selected odd-numbered problems begin on page ANS-15.

Answer Problems 1–10 without referring back to the text. Fill in the blank, or answer true or false.

1. The second-order differential equation $x'' + f(x') + g(x) = 0$ can be written as a plane autonomous system. _____

2. If $X = X(t)$ is a solution to a plane autonomous system and $X(t_1) = X(t_2)$ for $t_1 \neq t_2$, then $X(t)$ is a periodic solution. _____

3. If the trace of the matrix A is 0 and $\det A \neq 0$, then the critical point (0, 0) of the linear system $X' = AX$ may be classified as _____.

4. If the critical point (0, 0) of the linear system $X' = AX$ is a stable spiral point, then the eigenvalues of A are _____.

5. If the critical point (0, 0) of the linear system $X' = AX$ is a saddle point and $X = X(t)$ is a solution, then $\lim_{t \to \infty} X(t)$ does not exist. _____

6. If the Jacobian matrix $A = g'(X_1)$ at a critical point of a plane autonomous system has positive trace and determinant, then the critical point X_1 is unstable. _____

7. It is possible to show, using linearization, that a nonlinear plane autonomous system has periodic solutions. _____

8. All solutions to the pendulum equation $\dfrac{d^2\theta}{dt^2} + \dfrac{g}{l}\sin\theta = 0$ are periodic. _____

9. For what value(s) of α does the plane autonomous system
$$x' = \alpha x - 2y$$
$$y' = -\alpha x + y$$
possess periodic solutions? _____

10. For what values of n is $x = n\pi$ an asymptotically stable critical point of the autonomous first-order differential equation $x' = \sin x$? _____

11. Solve the nonlinear plane autonomous system
$$x' = -y - x\left(\sqrt{x^2 + y^2}\right)^3$$
$$y' = x - y\left(\sqrt{x^2 + y^2}\right)^3.$$
by switching to polar coordinates. Describe the geometric behavior of the solution that satisfies the initial condition $X(0) = (1, 0)$.

12. Discuss the geometric nature of the solutions to the linear system $X' = AX$ given that the general solution is

(a) $X(t) = c_1 \begin{pmatrix} 1 \\ 1 \end{pmatrix} e^{-t} + c_2 \begin{pmatrix} 1 \\ -2 \end{pmatrix} e^{-2t}$

(b) $X(t) = c_1 \begin{pmatrix} 1 \\ -1 \end{pmatrix} e^{-t} + c_2 \begin{pmatrix} 1 \\ 2 \end{pmatrix} e^{2t}$

13. Classify the critical point (0, 0) of the given linear system by computing the trace τ and determinant Δ.

(a) $x' = -3x + 4y$ (b) $x' = -3x + 2y$
$y' = -5x + 3y$ $y' = -2x + y$

14. Find and classify (if possible) the critical points of the plane autonomous system
$$x' = x + xy - 3x^2$$
$$y' = 4y - 2xy - y^2.$$

15. Determine the value(s) of α for which (0, 0) is a stable critical point for the plane autonomous system (in polar coordinates)
$$r' = \alpha r$$
$$\theta' = 1.$$

16. Classify the critical point (0, 0) of the plane autonomous system corresponding to the nonlinear second-order differential equation
$$x'' + \mu(x^2 - 1)x' + x = 0,$$
where μ is a real constant.

17. Without solving explicitly, classify (if possible) the critical points of the autonomous first-order differential equation $x' = (x^2 - 1)e^{-x/2}$ as asymptotically stable or unstable.

18. Use the phase-plane method to show that the solutions to the nonlinear second-order differential equation $x'' = -2x\sqrt{(x')^2 + 1}$ that satisfy $x(0) = x_0$ and $x'(0) = 0$ are periodic.

19. In Section 5.1 we assumed that the restoring force F of the spring satisfied Hooke's law $F = ks$, where s is the elongation of the spring and k is a positive constant of proportionality. If we replace this assumption with the nonlinear law $F = ks^3$, the new differential equation for damped motion of the hard spring becomes
$$mx'' = -\beta x' - k(s + x)^3 + mg,$$
where $ks^3 = mg$. The system is called overdamped when (0, 0) is a stable node and is called underdamped when (0, 0) is a stable spiral point. Find new conditions on m, k, and β that will lead to overdamping and underdamping.

20. The rod of a pendulum is attached to a movable joint at a point P and rotates at an angular speed of ω (rad/s) in the plane perpendicular to the rod. See Figure 10.39. As a result the bob of the rotating pendulum experiences an additional centripetal force, and the new differential equation for θ becomes

$$ml\frac{d^2\theta}{dt^2} = \omega^2 ml \sin\theta \cos\theta - mg \sin\theta - \beta\frac{d\theta}{dt}.$$

(a) If $\omega^2 < g/l$, show that $(0, 0)$ is a stable critical point and is the only critical point in the domain $-\pi < \theta < \pi$. Describe what occurs physically when $\theta(0) = \theta_0$, $\theta'(0) = 0$, and θ_0 is small.

(b) If $\omega^2 > g/l$, show that $(0, 0)$ is unstable and there are two additional stable critical points $(\pm\hat{\theta}, 0)$ in the domain $-\pi < \theta < \pi$. Describe what occurs physically when $\theta(0) = \theta_0$, $\theta'(0) = 0$, and θ_0 is small.

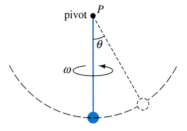

FIGURE 10.39   Rotating pendulum in Problem 20

Partial Sums: $N = 3$

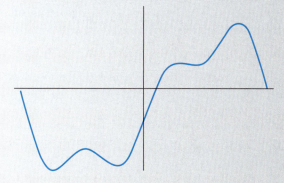

Partial Sums: $N = 8$

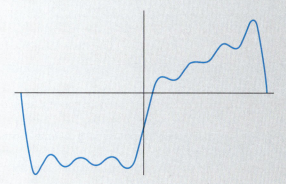

Partial Sums: $N = 22$

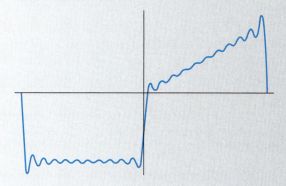

11

ORTHOGONAL FUNCTIONS AND FOURIER SERIES

In calculus you saw that two nonzero vectors are orthogonal when their inner (dot) product is zero. Beyond calculus the notions of vectors, orthogonality, and inner product often lose their geometric interpretation. These concepts have been generalized; it is perfectly common to think of a function as a vector. We can then say that two different functions are orthogonal when their inner product is zero. We shall see in this chapter that the inner product of these vectors (functions) is actually a definite integral.

The concepts of orthogonal functions and the expansion of a given function f in terms of an infinite set of orthogonal functions is fundamental to the material covered in Chapters 12 and 13.

Graphs of partial sums $S_N(x)$ of the Fourier series of the piecewise-defined function

$$f(x) = \begin{cases} x, & 0 < x < 1 \\ -1, & -1 < x < 0. \end{cases}$$

See page 439.

11.1   ORTHOGONAL FUNCTIONS

INTRODUCTION: The concepts of geometric vectors in two and three dimensions, orthogonal or perpendicular vectors, and the inner product of two vectors have been generalized. It is perfectly routine in mathematics to think of a function as a vector. In this section we examine an inner product that is different from the one you studied in calculus. Using this new inner product, we define orthogonal functions and sets of orthogonal functions.

Another topic in a standard calculus course is the expansion of a function f in a power series. In this section we also see how to expand a suitable function f in terms of an infinite set of orthogonal functions.

REVIEW MATERIAL: The notions of generalized vectors and vector spaces can be found in any text on linear algebra.

INNER PRODUCT   Recall that if $\mathbf{u}$ and $\mathbf{v}$ are two vectors in 3-space, then the inner product $(\mathbf{u}, \mathbf{v})$ (in calculus this is written as $\mathbf{u} \cdot \mathbf{v}$) possesses the following properties:

(*i*) $(\mathbf{u}, \mathbf{v}) = (\mathbf{v}, \mathbf{u})$,

(*ii*) $(k\mathbf{u}, \mathbf{v}) = k(\mathbf{u}, \mathbf{v})$, k a scalar,

(*iii*) $(\mathbf{u}, \mathbf{u}) = 0$ if $\mathbf{u} = \mathbf{0}$ and $(\mathbf{u}, \mathbf{u}) > 0$ if $\mathbf{u} \neq \mathbf{0}$,

(*iv*) $(\mathbf{u} + \mathbf{v}, \mathbf{w}) = (\mathbf{u}, \mathbf{w}) + (\mathbf{v}, \mathbf{w})$.

We expect that any generalization of the inner product concept should have these same properties.

Suppose that f_1 and f_2 are functions defined on an interval $[a, b]$.* Since a *definite integral* on $[a, b]$ of the product $f_1(x)f_2(x)$ possesses the foregoing properties $(i)-(iv)$ whenever the integral exists, we are prompted to make the following definition.

DEFINITION 11.1   **Inner Product of Functions**

The **inner product** of two functions f_1 and f_2 on an interval $[a, b]$ is the number

$$(f_1, f_2) = \int_a^b f_1(x) f_2(x)\, dx.$$

ORTHOGONAL FUNCTIONS   Motivated by the fact that two geometric vectors $\mathbf{u}$ and $\mathbf{v}$ are orthogonal whenever their inner product is zero, we define **orthogonal functions** in a similar manner.

DEFINITION 11.2   **Orthogonal Functions**

Two functions f_1 and f_2 are **orthogonal** on an interval $[a, b]$ if

$$(f_1, f_2) = \int_a^b f_1(x) f_2(x)\, dx = 0. \tag{1}$$

---

*The interval could also be $(-\infty, \infty)$, $[0, \infty)$, and so on.

For example, the functions $f_1(x) = x^2$ and $f_2(x) = x^3$ are orthogonal on the interval $[-1, 1]$ since

$$(f_1, f_2) = \int_{-1}^{1} x^2 \cdot x^3 \, dx = \left. \tfrac{1}{6} x^6 \right|_{-1}^{1} = 0.$$

Unlike in vector analysis, where the word *orthogonal* is a synonym for *perpendicular,* in this present context the term *orthogonal* and condition (1) have no geometric significance.

ORTHOGONAL SETS   We are primarily interested in infinite sets of orthogonal functions.

DEFINITION 11.3   **Orthogonal Set**

A set of real-valued functions $\{\phi_0(x), \phi_1(x), \phi_2(x), \ldots\}$ is said to be **orthogonal** on an interval $[a, b]$ if

$$(\phi_m, \phi_n) = \int_{a}^{b} \phi_m(x) \phi_n(x) \, dx = 0, \quad m \neq n. \tag{2}$$

ORTHONORMAL SETS   The norm, or length $\|\mathbf{u}\|$, of a vector $\mathbf{u}$ can be expressed in terms of the inner product. The expression $(\mathbf{u}, \mathbf{u}) = \|\mathbf{u}\|^2$ is called the square norm, and so the norm is $\|\mathbf{u}\| = \sqrt{(\mathbf{u}, \mathbf{u})}$. Similarly, the **square norm** of a function ϕ_n is $\|\phi_n(x)\|^2 = (\phi_n, \phi_n)$, and so the **norm,** or its generalized length, is $\|\phi_n(x)\| = \sqrt{(\phi_n, \phi_n)}$. In other words, the square norm and norm of a function ϕ_n in an orthogonal set $\{\phi_n(x)\}$ are, respectively,

$$\|\phi_n(x)\|^2 = \int_{a}^{b} \phi_n^2(x) \, dx \quad \text{and} \quad \|\phi_n(x)\| = \sqrt{\int_{a}^{b} \phi_n^2(x) \, dx}. \tag{3}$$

If $\{\phi_n(x)\}$ is an orthogonal set of functions on the interval $[a, b]$ with the property that $\|\phi_n(x)\| = 1$ for $n = 0, 1, 2, \ldots$, then $\{\phi_n(x)\}$ is said to be an **orthonormal set** on the interval.

EXAMPLE 1   **Orthogonal Set of Functions**

Show that the set $\{1, \cos x, \cos 2x, \ldots\}$ is orthogonal on the interval $[-\pi, \pi]$.

SOLUTION   If we make the identification $\phi_0(x) = 1$ and $\phi_n(x) = \cos nx$, we must then show that $\int_{-\pi}^{\pi} \phi_0(x) \phi_n(x) \, dx = 0, n \neq 0$, and $\int_{-\pi}^{\pi} \phi_m(x) \phi_n(x) \, dx = 0, m \neq n$. We have, in the first case,

$$(\phi_0, \phi_n) = \int_{-\pi}^{\pi} \phi_0(x) \phi_n(x) \, dx = \int_{-\pi}^{\pi} \cos nx \, dx$$

$$= \left. \frac{1}{n} \sin nx \right|_{-\pi}^{\pi} = \frac{1}{n} [\sin n\pi - \sin(-n\pi)] = 0, \quad n \neq 0,$$

and in the second,

$$(\phi_m, \phi_n) = \int_{-\pi}^{\pi} \phi_m(x)\,\phi_n(x)\,dx$$

$$= \int_{-\pi}^{\pi} \cos mx \cos nx \, dx$$

$$= \frac{1}{2}\int_{-\pi}^{\pi} [\cos(m + n)x + \cos(m - n)x]\,dx \qquad \leftarrow \text{trig identity}$$

$$= \frac{1}{2}\left[\frac{\sin(m + n)x}{m + n} + \frac{\sin(m - n)x}{m - n}\right]_{-\pi}^{\pi} = 0, \quad m \neq n.$$

EXAMPLE 2   Norms

Find the norm of each function in the orthogonal set given in Example 1.

SOLUTION   For $\phi_0(x) = 1$ we have from (3),

$$\|\phi_0(x)\|^2 = \int_{-\pi}^{\pi} dx = 2\pi,$$

so $\|\phi_0(x)\| = \sqrt{2\pi}$. For $\phi_n(x) = \cos nx$, $n > 0$, it follows that

$$\|\phi_n(x)\|^2 = \int_{-\pi}^{\pi} \cos^2 nx \, dx = \frac{1}{2}\int_{-\pi}^{\pi} [1 + \cos 2nx]\,dx = \pi.$$

Thus for $n > 0$, $\|\phi_n(x)\| = \sqrt{\pi}$.

Any orthogonal set of nonzero functions $\{\phi_n(x)\}$, $n = 0, 1, 2, \dots$ can be *normalized*—that is, made into an orthonormal set—by dividing each function by its norm. It follows from Examples 1 and 2 that the set

$$\left\{\frac{1}{\sqrt{2\pi}}, \frac{\cos x}{\sqrt{\pi}}, \frac{\cos 2x}{\sqrt{\pi}}, \dots\right\}$$

is orthonormal on $[-\pi, \pi]$.

We shall make one more analogy between vectors and functions. Suppose $\mathbf{v}_1$, $\mathbf{v}_2$, and $\mathbf{v}_3$ are three mutually orthogonal nonzero vectors in 3-space. Such an orthogonal set can be used as a basis for 3-space; that is, any three-dimensional vector can be written as a linear combination

$$\mathbf{u} = c_1\mathbf{v}_1 + c_2\mathbf{v}_2 + c_3\mathbf{v}_3, \tag{4}$$

where the c_i, $i = 1, 2, 3$, are scalars called the components of the vector. Each component c_i can be expressed in terms of $\mathbf{u}$ and the corresponding vector $\mathbf{v}_i$. To see this, we take the inner product of (4) with $\mathbf{v}_1$:

$$(\mathbf{u}, \mathbf{v}_1) = c_1(\mathbf{v}_1, \mathbf{v}_1) + c_2(\mathbf{v}_2, \mathbf{v}_1) + c_3(\mathbf{v}_3, \mathbf{v}_1) = c_1\|\mathbf{v}_1\|^2 + c_2 \cdot 0 + c_3 \cdot 0.$$

Hence $$c_1 = \frac{(\mathbf{u}, \mathbf{v}_1)}{\|\mathbf{v}_1\|^2}.$$

In like manner we find that the components c_2 and c_3 are given by

$$c_2 = \frac{(\mathbf{u}, \mathbf{v}_2)}{\|\mathbf{v}_2\|^2} \quad \text{and} \quad c_3 = \frac{(\mathbf{u}, \mathbf{v}_3)}{\|\mathbf{v}_3\|^2}.$$

Hence (4) can be expressed as

$$\mathbf{u} = \frac{(\mathbf{u}, \mathbf{v}_1)}{\|\mathbf{v}_1\|^2}\mathbf{v}_1 + \frac{(\mathbf{u}, \mathbf{v}_2)}{\|\mathbf{v}_2\|^2}\mathbf{v}_2 + \frac{(\mathbf{u}, \mathbf{v}_3)}{\|\mathbf{v}_3\|^2}\mathbf{v}_3 = \sum_{n=1}^{3} \frac{(\mathbf{u}, \mathbf{v}_n)}{\|\mathbf{v}_n\|^2}\mathbf{v}_n. \tag{5}$$

ORTHOGONAL SERIES EXPANSION   Suppose $\{\phi_n(x)\}$ is an infinite orthogonal set of functions on an interval $[a, b]$. We ask: If $y = f(x)$ is a function defined on the interval $[a, b]$, is it possible to determine a set of coefficients $c_n, n = 0, 1, 2, \ldots$, for which

$$f(x) = c_0\phi_0(x) + c_1\phi_1(x) + \cdots + c_n\phi_n(x) + \cdots ? \tag{6}$$

As in the foregoing discussion on finding components of a vector, we can find the coefficients c_n by utilizing the inner product. Multiplying (6) by $\phi_m(x)$ and integrating over the interval $[a, b]$ gives

$$\int_a^b f(x)\phi_m(x)\,dx = c_0\int_a^b \phi_0(x)\phi_m(x)\,dx + c_1\int_a^b \phi_1(x)\phi_m(x)\,dx + \cdots + c_n\int_a^b \phi_n(x)\phi_m(x)\,dx + \cdots$$

$$= c_0(\phi_0, \phi_m) + c_1(\phi_1, \phi_m) + \cdots + c_n(\phi_n, \phi_m) + \cdots.$$

By orthogonality each term on the right-hand side of the last equation is zero *except* when $m = n$. In this case we have

$$\int_a^b f(x)\phi_n(x)\,dx = c_n\int_a^b \phi_n^2(x)\,dx.$$

It follows that the required coefficients are

$$c_n = \frac{\int_a^b f(x)\phi_n(x)\,dx}{\int_a^b \phi_n^2(x)\,dx}, \quad n = 0, 1, 2, \ldots.$$

In other words,
$$f(x) = \sum_{n=0}^{\infty} c_n\phi_n(x), \tag{7}$$

where
$$c_n = \frac{\int_a^b f(x)\phi_n(x)\,dx}{\|\phi_n(x)\|^2}. \tag{8}$$

With inner product notation, (7) becomes

$$f(x) = \sum_{n=0}^{\infty} \frac{(f, \phi_n)}{\|\phi_n(x)\|^2}\phi_n(x). \tag{9}$$

Thus (9) is seen to be the functional analogue of the vector result given in (5).

DEFINITION 11.4   **Orthogonal Set/Weight Function**

A set of real-valued functions $\{\phi_0(x), \phi_1(x), \phi_2(x), \ldots\}$ is said to be **orthogonal with respect to a weight function** $w(x)$ on an interval $[a, b]$ if

$$\int_a^b w(x)\phi_m(x)\phi_n(x)\,dx = 0, \quad m \neq n.$$

The usual assumption is that $w(x) > 0$ on the interval of orthogonality $[a, b]$. The set $\{1, \cos x, \cos 2x, \ldots\}$ in Example 1 is orthogonal with respect to the weight function $w(x) = 1$ on the interval $[-\pi, \pi]$.

If $\{\phi_n(x)\}$ is orthogonal with respect to a weight function $w(x)$ on the interval $[a, b]$, then multiplying (6) by $w(x)\phi_n(x)$ and integrating yields

$$c_n = \frac{\int_a^b f(x)w(x)\phi_n(x)\,dx}{\|\phi_n(x)\|^2}, \tag{10}$$

where

$$\|\phi_n(x)\|^2 = \int_a^b w(x)\,\phi_n^2(x)\,dx. \qquad (11)$$

The series (7) with coefficients given by either (8) or (10) is said to be an **orthogonal series expansion** of f or a **generalized Fourier series.**

COMPLETE SETS The procedure outlined for determining the coefficients c_n was *formal;* that is, basic questions about whether or not an orthogonal series expansion such as (7) is actually possible were ignored. Also, to expand f in a series of orthogonal functions, it is certainly necessary that f not be orthogonal to each ϕ_n of the orthogonal set $\{\phi_n(x)\}$. (If f were orthogonal to every ϕ_n, then $c_n = 0$, $n = 0, 1, 2, \ldots$.) To avoid the latter problem, we shall assume, for the remainder of the discussion, that an orthogonal set is **complete.** This means that the only function orthogonal to each member of the set is the zero function.

EXERCISES 11.1

Answers to selected odd-numbered problems begin on page ANS-15.

In Problems 1–6 show that the given functions are orthogonal on the indicated interval.

1. $f_1(x) = x$, $f_2(x) = x^2$; $[-2, 2]$

2. $f_1(x) = x^3$, $f_2(x) = x^2 + 1$; $[-1, 1]$

3. $f_1(x) = e^x$, $f_2(x) = xe^{-x} - e^{-x}$; $[0, 2]$

4. $f_1(x) = \cos x$, $f_2(x) = \sin^2 x$; $[0, \pi]$

5. $f_1(x) = x$, $f_2(x) = \cos 2x$; $[-\pi/2, \pi/2]$

6. $f_1(x) = e^x$, $f_2(x) = \sin x$; $[\pi/4, 5\pi/4]$

In Problems 7–12 show that the given set of functions is orthogonal on the indicated interval. Find the norm of each function in the set.

7. $\{\sin x, \sin 3x, \sin 5x, \ldots\}$; $[0, \pi/2]$

8. $\{\cos x, \cos 3x, \cos 5x, \ldots\}$; $[0, \pi/2]$

9. $\{\sin nx\}$, $n = 1, 2, 3, \ldots$; $[0, \pi]$

10. $\left\{\sin \dfrac{n\pi}{p}x\right\}$, $n = 1, 2, 3, \ldots$; $[0, p]$

11. $\left\{1, \cos \dfrac{n\pi}{p}x\right\}$, $n = 1, 2, 3, \ldots$; $[0, p]$

12. $\left\{1, \cos \dfrac{n\pi}{p}x, \sin \dfrac{m\pi}{p}x\right\}$, $n = 1, 2, 3, \ldots$, $m = 1, 2, 3, \ldots$; $[-p, p]$

In Problems 13 and 14 verify by direct integration that the functions are orthogonal with respect to the indicated weight function on the given interval.

13. $H_0(x) = 1$, $H_1(x) = 2x$, $H_2(x) = 4x^2 - 2$; $w(x) = e^{-x^2}$, $(-\infty, \infty)$

14. $L_0(x) = 1$, $L_1(x) = -x + 1$, $L_2(x) = \dfrac{1}{2}x^2 - 2x + 1$; $w(x) = e^{-x}$, $[0, \infty)$

15. Let $\{\phi_n(x)\}$ be an orthogonal set of functions on $[a, b]$ such that $\phi_0(x) = 1$. Show that $\int_a^b \phi_n(x)\,dx = 0$ for $n = 1, 2, \ldots$.

16. Let $\{\phi_n(x)\}$ be an orthogonal set of functions on $[a, b]$ such that $\phi_0(x) = 1$ and $\phi_1(x) = x$. Show that $\int_a^b (\alpha x + \beta)\phi_n(x)\,dx = 0$ for $n = 2, 3, \ldots$ and any constants α and β.

17. Let $\{\phi_n(x)\}$ be an orthogonal set of functions on $[a, b]$. Show that $\|\phi_m(x) + \phi_n(x)\|^2 = \|\phi_m(x)\|^2 + \|\phi_n(x)\|^2$, $m \neq n$.

18. From Problem 1 we know that $f_1(x) = x$ and $f_2(x) = x^2$ are orthogonal on $[-2, 2]$. Find constants c_1 and c_2 such that $f_3(x) = x + c_1 x^2 + c_2 x^3$ is orthogonal to both f_1 and f_2 on the same interval.

19. The set of functions $\{\sin nx\}$, $n = 1, 2, 3, \ldots$ is orthogonal on the interval $[-\pi, \pi]$. Show that the set is not complete.

20. Suppose f_1, f_2, and f_3 are functions continuous on the interval $[a, b]$. Show that $(f_1 + f_2, f_3) = (f_1, f_3) + (f_2, f_3)$.

DISCUSSION/PROJECT PROBLEMS

21. A real-valued function f is said to be **periodic** with period T if $f(x + T) = f(x)$. For example, 4π is a period of $\sin x$ since $\sin(x + 4\pi) = \sin x$. The smallest value of T for which $f(x + T) = f(x)$ holds is called the **fundamental period** of f. For example, the fundamental period of $f(x) = \sin x$ is $T = 2\pi$. What is the fundamental period of each of the following functions?

(a) $f(x) = \cos 2\pi x$ **(b)** $f(x) = \sin \dfrac{4}{L} x$

(c) $f(x) = \sin x + \sin 2x$ **(d)** $f(x) = \sin 2x + \cos 4x$

(e) $f(x) = \sin 3x + \cos 2x$

(f) $f(x) = A_0 + \sum\limits_{n=1}^{\infty}\left(A_n \cos \dfrac{n\pi}{p} x + B_n \sin \dfrac{n\pi}{p} x\right)$,

A_n and B_n depend only on n

11.2 FOURIER SERIES

INTRODUCTION: We have just seen that if $\{\phi_0(x), \phi_1(x), \phi_2(x), \ldots\}$ is an orthogonal set on an interval $[a, b]$ and if f is a function defined on the same interval, then we can formally expand f in an orthogonal series $c_0\phi_0(x) + c_1\phi_1(x) + c_2\phi_2(x) + \cdots$, where the coefficients c_n are determined by using the inner product concept. The orthogonal set of trigonometric functions

$$\left\{1, \cos\frac{\pi}{p}x, \cos\frac{2\pi}{p}x, \cos\frac{3\pi}{p}x, \ldots, \sin\frac{\pi}{p}x, \sin\frac{2\pi}{p}x, \sin\frac{3\pi}{p}x, \ldots\right\} \tag{1}$$

will be of particular importance later on in the solution of certain kinds of boundary-value problems involving linear partial differential equations. The set (1) is orthogonal on the interval $[-p, p]$.

REVIEW MATERIAL: Reread—or, better, rework—Problem 12 in Exercises 11.1.

A TRIGONOMETRIC SERIES Suppose that f is a function defined on the interval $[-p, p]$ and can be expanded in an orthogonal series consisting of the trigonometric functions in the orthogonal set (1); that is,

$$f(x) = \frac{a_0}{2} + \sum_{n=1}^{\infty}\left(a_n \cos\frac{n\pi}{p}x + b_n \sin\frac{n\pi}{p}x\right). \tag{2}$$

The coefficients $a_0, a_1, a_2, \ldots, b_1, b_2, \ldots$ can be determined in exactly the same manner as in the general discussion of orthogonal series expansions on page 433. Before proceeding, note that we have chosen to write the coefficient of 1 in the set (1) as $\frac{1}{2}a_0$ rather than a_0. This is for convenience only; the formula of a_n will then reduce to a_0 for $n = 0$.

Now integrating both sides of (2) from $-p$ to p gives

$$\int_{-p}^{p} f(x)\, dx = \frac{a_0}{2}\int_{-p}^{p} dx + \sum_{n=1}^{\infty}\left(a_n\int_{-p}^{p}\cos\frac{n\pi}{p}x\, dx + b_n\int_{-p}^{p}\sin\frac{n\pi}{p}x\, dx\right). \tag{3}$$

Since $\cos(n\pi x/p)$ and $\sin(n\pi x/p)$, $n \geq 1$ are orthogonal to 1 on the interval, the right side of (3) reduces to a single term:

$$\int_{-p}^{p} f(x)\, dx = \frac{a_0}{2}\int_{-p}^{p} dx = \frac{a_0}{2}x\,\Big|_{-p}^{p} = pa_0.$$

Solving for a_0 yields

$$a_0 = \frac{1}{p}\int_{-p}^{p} f(x)\, dx. \tag{4}$$

Now we multiply (2) by $\cos(m\pi x/p)$ and integrate:

$$\int_{-p}^{p} f(x) \cos\frac{m\pi}{p}x\,dx = \frac{a_0}{2}\int_{-p}^{p} \cos\frac{m\pi}{p}x\,dx$$

$$+ \sum_{n=1}^{\infty}\left(a_n\int_{-p}^{p} \cos\frac{m\pi}{p}x \cos\frac{n\pi}{p}x\,dx + b_n\int_{-p}^{p}\cos\frac{m\pi}{p}x \sin\frac{n\pi}{p}x\,dx\right). \quad (5)$$

By orthogonality we have

$$\int_{-p}^{p}\cos\frac{m\pi}{p}x\,dx = 0, \quad m > 0, \quad \int_{-p}^{p}\cos\frac{m\pi}{p}x \sin\frac{n\pi}{p}x\,dx = 0,$$

and

$$\int_{-p}^{p}\cos\frac{m\pi}{p}x \cos\frac{n\pi}{p}x\,dx = \begin{cases} 0, & m \neq n \\ p, & m = n. \end{cases}$$

Thus (5) reduces to

$$\int_{-p}^{p} f(x) \cos\frac{n\pi}{p}x\,dx = a_n p,$$

and so

$$a_n = \frac{1}{p}\int_{-p}^{p} f(x) \cos\frac{n\pi}{p}x\,dx. \quad (6)$$

Finally, if we multiply (2) by $\sin(m\pi x/p)$, integrate, and make use of the results

$$\int_{-p}^{p}\sin\frac{m\pi}{p}x\,dx = 0, \quad m > 0, \quad \int_{-p}^{p}\sin\frac{m\pi}{p}x \cos\frac{n\pi}{p}x\,dx = 0,$$

and

$$\int_{-p}^{p}\sin\frac{m\pi}{p}x \sin\frac{n\pi}{p}x\,dx = \begin{cases} 0, & m \neq n \\ p, & m = n, \end{cases}$$

we find that

$$b_n = \frac{1}{p}\int_{-p}^{p} f(x) \sin\frac{n\pi}{p}x\,dx. \quad (7)$$

The trigonometric series (2) with coefficients a_0, a_n, and b_n defined by (4), (6), and (7), respectively, is said to be the **Fourier series** of the function f. The coefficients obtained from (4), (6), and (7) are referred to as **Fourier coefficients** of f.

In finding the coefficients a_0, a_n, and b_n, we assumed that f was integrable on the interval and that (2), as well as the series obtained by multiplying (2) by $\cos(m\pi x/p)$, converged in such a manner as to permit term-by-term integration. Until (2) is shown to be convergent for a given function f, the equality sign is not to be taken in a strict or literal sense. Some texts use the symbol $\sim$ in place of $=$. In view of the fact that most functions in applications are of a type that guarantees convergence of the series, we shall use the equality symbol. We summarize the results:

DEFINITION 11.5   **Fourier Series**

The **Fourier series** of a function f defined on the interval $(-p, p)$ is given by

$$f(x) = \frac{a_0}{2} + \sum_{n=1}^{\infty}\left(a_n \cos\frac{n\pi}{p}x + b_n \sin\frac{n\pi}{p}x\right), \quad (8)$$

where

$$a_0 = \frac{1}{p}\int_{-p}^{p} f(x)\,dx \quad (9)$$

$$a_n = \frac{1}{p}\int_{-p}^{p} f(x) \cos\frac{n\pi}{p}x\,dx \quad (10)$$

$$b_n = \frac{1}{p}\int_{-p}^{p} f(x) \sin\frac{n\pi}{p}x\,dx. \quad (11)$$

EXAMPLE 1   **Expansion in a Fourier Series**

Expand
$$f(x) = \begin{cases} 0, & -\pi < x < 0 \\ \pi - x, & 0 \le x < \pi \end{cases} \qquad (12)$$
in a Fourier series.

SOLUTION   The graph of f is given in Figure 11.1. With $p = \pi$ we have from (9) and (10) that

$$a_0 = \frac{1}{\pi}\int_{-\pi}^{\pi} f(x)\,dx = \frac{1}{\pi}\left[\int_{-\pi}^{0} 0\,dx + \int_{0}^{\pi} (\pi - x)\,dx\right] = \frac{1}{\pi}\left[\pi x - \frac{x^2}{2}\right]_0^{\pi} = \frac{\pi}{2}$$

$$a_n = \frac{1}{\pi}\int_{-\pi}^{\pi} f(x)\cos nx\,dx = \frac{1}{\pi}\left[\int_{-\pi}^{0} 0\,dx + \int_{0}^{\pi} (\pi - x)\cos nx\,dx\right]$$

$$= \frac{1}{\pi}\left[(\pi - x)\frac{\sin nx}{n}\Big|_0^{\pi} + \frac{1}{n}\int_0^{\pi}\sin nx\,dx\right]$$

$$= -\frac{1}{n\pi}\frac{\cos nx}{n}\Big|_0^{\pi} = \frac{1 - (-1)^n}{n^2\pi},$$

where we have used $\cos n\pi = (-1)^n$. In like manner we find from (11) that

$$b_n = \frac{1}{\pi}\int_0^{\pi}(\pi - x)\sin nx\,dx = \frac{1}{n}.$$

Therefore
$$f(x) = \frac{\pi}{4} + \sum_{n=1}^{\infty}\left\{\frac{1 - (-1)^n}{n^2\pi}\cos nx + \frac{1}{n}\sin nx\right\}. \qquad (13) \quad \blacksquare$$

Note that a_n defined by (10) reduces to a_0 given by (9) when we set $n = 0$. But as Example 1 shows, this might not be the case *after* the integral for a_n is evaluated.

CONVERGENCE OF A FOURIER SERIES   The following theorem gives sufficient conditions for convergence of a Fourier series at a point.

***

THEOREM 11.1   **Conditions for Convergence**

Let f and f' be piecewise continuous on the interval $(-p, p)$; that is, let f and f' be continuous except at a finite number of points in the interval and have only finite discontinuities at these points. Then the Fourier series of f on the interval converges to $f(x)$ at a point of continuity. At a point of discontinuity the Fourier series converges to the average

$$\frac{f(x+) + f(x-)}{2},$$

where $f(x+)$ and $f(x-)$ denote the limit of f at x from the right and from the left, respectively.*

***

For a proof of this theorem you are referred to the classic text by Churchill and Brown.[†]

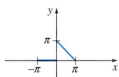

y, π, −π, π, x

FIGURE 11.1   Piecewise-continuous function in Example 1

---

*In other words, for x a point in the interval and $h > 0$,

$$f(x+) = \lim_{h\to 0} f(x + h), \quad f(x-) = \lim_{h\to 0} f(x - h).$$

[†]Ruel V. Churchill and James Ward Brown, *Fourier Series and Boundary Value Problems* (New York: McGraw-Hill).

EXAMPLE 2 Convergence of a Point of Discontinuity

The function (12) in Example 1 satisfies the conditions of Theorem 11.1. Thus for every x in the interval $(-\pi, \pi)$, except at $x = 0$, the series (13) will converge to $f(x)$. At $x = 0$ the function is discontinuous, so the series (13) will converge to

$$\frac{f(0+) + f(0-)}{2} = \frac{\pi + 0}{2} = \frac{\pi}{2}.$$

PERIODIC EXTENSION Observe that each of the functions in the basic set (1) has a different fundamental period*—namely, $2p/n$, $n \geq 1$—but since a positive integer multiple of a period is also a period, we see that all of the functions have in common the period $2p$. (Verify.) Hence the right-hand side of (2) is $2p$-periodic; indeed, $2p$ is the **fundamental period** of the sum. We conclude that a Fourier series not only represents the function on the interval $(-p, p)$, but also gives the **periodic extension** of f outside this interval. We can now apply Theorem 11.1 to the periodic extension of f, or we may assume from the outset that the given function is periodic with period $2p$; that is, $f(x + 2p) = f(x)$. When f is piecewise continuous and the right- and left-hand derivatives exist at $x = -p$ and $x = p$, respectively, then the series (8) converges to the average

$$\frac{f(p-) + f(-p+)}{2}$$

at these endpoints and to this value extended periodically to $\pm 3p$, $\pm 5p$, $\pm 7p$, and so on.

The Fourier series in (13) converges to the periodic extension of (12) on the entire x-axis. At 0, $\pm 2\pi$, $\pm 4\pi$, . . . and at $\pm \pi$, $\pm 3\pi$, $\pm 5\pi$, . . . the series converges to the values

$$\frac{f(0+) + f(0-)}{2} = \frac{\pi}{2} \quad \text{and} \quad \frac{f(\pi-) + f(-\pi+)}{2} = 0,$$

respectively. The solid dots in Figure 11.2 represent the value $\pi/2$.

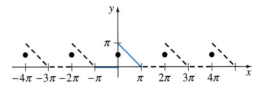

FIGURE 11.2 Periodic extension of function shown in Figure 11.1

SEQUENCE OF PARTIAL SUMS It is interesting to see how the sequence of partial sums $\{S_N(x)\}$ of a Fourier series approximates a function. For example, the first three partial sums of (13) are

$$S_1(x) = \frac{\pi}{4}, \quad S_2(x) = \frac{\pi}{4} + \frac{2}{\pi}\cos x + \sin x, \quad \text{and} \quad S_3(x) = \frac{\pi}{4} + \frac{2}{\pi}\cos x + \sin x + \frac{1}{2}\sin 2x.$$

In Figure 11.3 we have used a CAS to graph the partial sums $S_3(x)$, $S_8(x)$, and $S_{15}(x)$ of (13) on the interval $(-\pi, \pi)$. Figure 11.3(d) shows the periodic extension using $S_{15}(x)$ on $(-4\pi, 4\pi)$.

---

*See Problem 21 in Exercises 11.1.

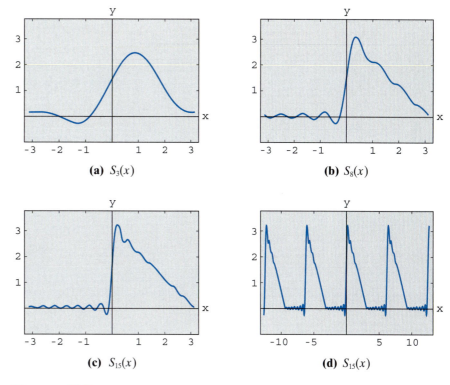

(a) $S_3(x)$ **(b)** $S_8(x)$

(c) $S_{15}(x)$ **(d)** $S_{15}(x)$

FIGURE 11.3   Partial sums of a Fourier series

EXERCISES 11.2

Answers to selected odd-numbered problems begin on page ANS-16.

In Problems 1–16 find the Fourier series of f on the given interval.

1. $f(x) = \begin{cases} 0, & -\pi < x < 0 \\ 1, & 0 \le x < \pi \end{cases}$

2. $f(x) = \begin{cases} -1, & -\pi < x < 0 \\ 2, & 0 \le x < \pi \end{cases}$

3. $f(x) = \begin{cases} 1, & -1 < x < 0 \\ x, & 0 \le x < 1 \end{cases}$

4. $f(x) = \begin{cases} 0, & -1 < x < 0 \\ x, & 0 \le x < 1 \end{cases}$

5. $f(x) = \begin{cases} 0, & -\pi < x < 0 \\ x^2, & 0 \le x < \pi \end{cases}$

6. $f(x) = \begin{cases} \pi^2, & -\pi < x < 0 \\ \pi^2 - x^2, & 0 \le x < \pi \end{cases}$

7. $f(x) = x + \pi, \quad -\pi < x < \pi$

8. $f(x) = 3 - 2x, \quad -\pi < x < \pi$

9. $f(x) = \begin{cases} 0, & -\pi < x < 0 \\ \sin x, & 0 \le x < \pi \end{cases}$

10. $f(x) = \begin{cases} 0, & -\pi/2 < x < 0 \\ \cos x, & 0 \le x < \pi/2 \end{cases}$

11. $f(x) = \begin{cases} 0, & -2 < x < -1 \\ -2, & -1 \le x < 0 \\ 1, & 0 \le x < 1 \\ 0, & 1 \le x < 2 \end{cases}$

12. $f(x) = \begin{cases} 0, & -2 < x < 0 \\ x, & 0 \le x < 1 \\ 1, & 1 \le x < 2 \end{cases}$

13. $f(x) = \begin{cases} 1, & -5 < x < 0 \\ 1 + x, & 0 \le x < 5 \end{cases}$

14. $f(x) = \begin{cases} 2 + x, & -2 < x < 0 \\ 2, & 0 \le x < 2 \end{cases}$

15. $f(x) = e^x, \quad -\pi < x < \pi$

16. $f(x) = \begin{cases} 0, & -\pi < x < 0 \\ e^x - 1, & 0 \le x < \pi \end{cases}$

17. Use the result of Problem 5 to show

$$\frac{\pi^2}{6} = 1 + \frac{1}{2^2} + \frac{1}{3^2} + \frac{1}{4^2} + \cdots$$

and

$$\frac{\pi^2}{12} = 1 - \frac{1}{2^2} + \frac{1}{3^2} - \frac{1}{4^2} + \cdots.$$

18. Use Problem 17 to find a series that gives the numerical value of $\pi^2/8$.

19. Use the result of Problem 7 to show that

$$\frac{\pi}{4} = 1 - \frac{1}{3} + \frac{1}{5} - \frac{1}{7} + \cdots.$$

20. Use the result of Problem 9 to show that

$$\frac{\pi}{4} = \frac{1}{2} + \frac{1}{1 \cdot 3} - \frac{1}{3 \cdot 5} + \frac{1}{5 \cdot 7} - \frac{1}{7 \cdot 9} + \cdots.$$

21. (a) Use the complex exponential form of the cosine and sine,

$$\cos \frac{n\pi}{p} x = \frac{e^{in\pi x/p} + e^{-in\pi x/p}}{2}$$

$$\sin \frac{n\pi}{p} x = \frac{e^{in\pi x/p} - e^{-in\pi x/p}}{2i}$$

to show that (8) can be written in the **complex form**

$$f(x) = \sum_{n=-\infty}^{\infty} c_n e^{in\pi x/p},$$

where $c_0 = a_0/2,$ $c_n = (a_n - ib_n)/2,$ and $c_{-n} = (a_n + ib_n)/2,$ where $n = 1, 2, 3, \ldots.$

(b) Show that c_0, c_n, and c_{-n} of part (a) can be written as one integral

$$c_n = \frac{1}{2p} \int_{-p}^{p} f(x) e^{-in\pi x/p} dx, \quad n = 0, \pm 1, \pm 2, \ldots.$$

22. Use the results of Problem 21 to find the complex form of the Fourier series of $f(x) = e^{-x}$ on the interval $-\pi < x < \pi$.

11.3   FOURIER COSINE AND SINE SERIES

INTRODUCTION: The effort expended in the evaluation of coefficients a_0, a_n, and b_n in expanding a function f in a Fourier series is reduced significantly when f is either an even or an odd function. Recall, a function f is said to be:

even if $f(-x) = f(x)$ and   **odd** if $f(-x) = -f(x).$

On a symmetric interval such as $(-p, p)$, the graph of an even function possesses symmetry with respect to the y-axis whereas the graph of an odd function possesses symmetry with respect to the origin.

EVEN AND ODD FUNCTIONS   It is likely the origin of the words *even* and *odd* derives from the fact that the graphs of polynomial functions that consist of all even powers of x are symmetric with respect to the y-axis and whereas graphs of polynomials that consist of all odd powers of x are symmetric with respect to origin. For example,

$$\underset{\downarrow \text{even integer}}{f(x) = x^2 \text{ is even}} \qquad \text{since } f(-x) = (-x)^2 = x^2 = f(x),$$

$$\underset{\downarrow \text{odd integer}}{f(x) = x^3 \text{ is odd}} \qquad \text{since } f(-x) = (-x)^3 = -x^3 = -f(x).$$

See Figures 11.4 and 11.5. The trigonometric cosine and sine functions are even and odd functions, respectively, since $\cos(-x) = \cos x$ and $\sin(-x) = -\sin x$. The exponential functions $f(x) = e^x$ and $f(x) = e^{-x}$ are neither odd nor even.

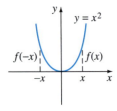

FIGURE 11.4   Even function; graph symmetric with respect to y-axis

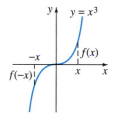

FIGURE 11.5 Odd function; graph symmetric with respect to origin

PROPERTIES   The following theorem lists some properties of even and odd functions.

THEOREM 11.2	**Properties of Even/Odd Functions**

(a)   The product of two even functions is even.
(b)   The product of two odd functions is even.
(c)   The product of an even function and an odd function is odd.
(d)   The sum (difference) of two even functions is even.
(e)   The sum (difference) of two odd functions is odd.
(f)   If f is even, then $\int_{-a}^{a} f(x)\,dx = 2\int_{0}^{a} f(x)\,dx$.
(g)   If f is odd, then $\int_{-a}^{a} f(x)\,dx = 0$.

PROOF OF (B)   Let us suppose that f and g are odd functions. Then we have $f(-x) = -f(x)$ and $g(-x) = -g(x)$. If we define the product of f and g as $F(x) = f(x)g(x)$, then

$$F(-x) = f(-x)\,g(-x) = (-f(x))(-g(x)) = f(x)\,g(x) = F(x).$$

This shows that the product F of two odd functions is an even function. The proofs of the remaining properties are left as exercises. See Problem 48 in Exercises 11.3.   ∎

COSINE AND SINE SERIES   If f is an even function on $(-p, p)$, then in view of the foregoing properties the coefficients (9), (10), and (11) of Section 11.2 become

$$a_0 = \frac{1}{p}\int_{-p}^{p} f(x)\,dx = \frac{2}{p}\int_{0}^{p} f(x)\,dx$$

$$a_n = \frac{1}{p}\int_{-p}^{p} \underbrace{f(x)\cos\frac{n\pi}{p}x}_{\text{even}}\,dx = \frac{2}{p}\int_{0}^{p} f(x)\cos\frac{n\pi}{p}x\,dx$$

$$b_n = \frac{1}{p}\int_{-p}^{p} \underbrace{f(x)\sin\frac{n\pi}{p}x}_{\text{odd}}\,dx = 0.$$

Similarly, when f is odd on the interval $(-p, p)$,

$$a_n = 0, \quad n = 0, 1, 2, \ldots, \quad b_n = \frac{2}{p}\int_{0}^{p} f(x)\sin\frac{n\pi}{p}x\,dx.$$

We summarize the results in the following definition.

DEFINITION 11.6	**Fourier Cosine and Sine Series**

(i)   The Fourier series of an even function on the interval $(-p, p)$ is the **cosine series**

$$f(x) = \frac{a_0}{2} + \sum_{n=1}^{\infty} a_n \cos\frac{n\pi}{p}x, \tag{1}$$

where

$$a_0 = \frac{2}{p}\int_{0}^{p} f(x)\,dx \tag{2}$$

$$a_n = \frac{2}{p}\int_{0}^{p} f(x)\cos\frac{n\pi}{p}x\,dx. \tag{3}$$

EXAMPLE 1　Expansion in a Sine Series

Expand $f(x) = x$, $-2 < x < 2$ in a Fourier series.

SOLUTION　Inspection of Figure 11.6 shows that the given function is odd on the interval $(-2, 2)$, and so we expand f in a sine series. With the identification $2p = 4$ we have $p = 2$. Thus (5), after integration by parts, is

$$b_n = \int_0^2 x \sin \frac{n\pi}{2} x \, dx = \frac{4(-1)^{n+1}}{n\pi}.$$

Therefore
$$f(x) = \frac{4}{\pi} \sum_{n=1}^{\infty} \frac{(-1)^{n+1}}{n} \sin \frac{n\pi}{2} x. \qquad (6)$$

The function in Example 1 satisfies the conditions of Theorem 11.1. Hence the series (6) converges to the function on $(-2, 2)$ and the periodic extension (of period 4) given in Figure 11.7.

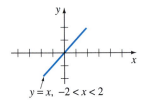

FIGURE 11.6　Odd function in Example 1

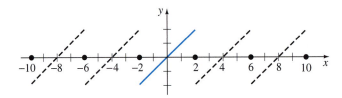

FIGURE 11.7　Periodic extension of function shown in Figure 11.6

EXAMPLE 2　Expansion in a Sine Series

The function $f(x) = \begin{cases} -1, & -\pi < x < 0 \\ 1, & 0 \le x < \pi, \end{cases}$ shown in Figure 11.8 is odd on the interval $(-\pi, \pi)$. With $p = \pi$ we have, from (5),

$$b_n = \frac{2}{\pi} \int_0^{\pi} (1) \sin nx \, dx = \frac{2}{\pi} \frac{1 - (-1)^n}{n},$$

and so
$$f(x) = \frac{2}{\pi} \sum_{n=1}^{\infty} \frac{1 - (-1)^n}{n} \sin nx. \qquad (7)$$

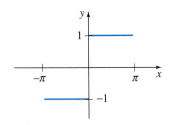

FIGURE 11.8　Odd function in Example 2

GIBBS PHENOMENON　With the aid of a CAS we have plotted the graphs $S_1(x)$, $S_2(x)$, $S_3(x)$, and $S_{15}(x)$ of the partial sums of nonzero terms of (7) in Figure 11.9. As seen in Figure 11.9(d), the graph of $S_{15}(x)$ has pronounced spikes near the discontinuities at $x = 0$, $x = \pi$, $x = -\pi$, and so on. This "overshooting" by the partial sums S_N from the functional values near a point of discontinuity does not smooth out but remains fairly constant, even when the value N is taken to be large. This behavior of a Fourier series near a point at which f is discontinuous is known as the **Gibbs phenomenon.**

The periodic extension of f in Example 2 onto the entire x-axis is a meander function (see page 314).

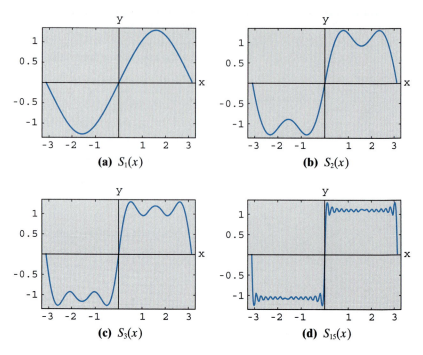

(a) $S_1(x)$ **(b)** $S_2(x)$

(c) $S_3(x)$ **(d)** $S_{15}(x)$

FIGURE 11.9   Partial sums of sine series (7)

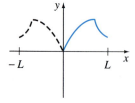

FIGURE 11.10   Even reflection

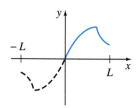

FIGURE 11.11   Odd reflection

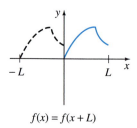

$f(x) = f(x + L)$

FIGURE 11.12   Identity reflection

HALF-RANGE EXPANSIONS   Throughout the preceding discussion it was understood that a function f was defined on an interval with the origin as midpoint—that is, $-p < x < p$. However, in many instances we are interested in representing a function that is defined only for $0 < x < L$ by a trigonometric series. This can be done in many different ways by supplying an arbitrary *definition* of the function on the interval $-L < x < 0$. For brevity we consider the three most important cases. If $y = f(x)$ is defined on the interval $0 < x < L$,

(*i*)   reflect the graph of the function about the y-axis onto $-L < x < 0$; the function is now even on $-L < x < L$ (see Figure 11.10); or

(*ii*)   reflect the graph of the function through the origin onto $-L < x < 0$; the function is now odd on $-L < x < L$ (see Figure 11.11); or

(*iii*)   define f on $-L < x < 0$ by $f(x) = f(x + L)$ (see Figure 11.12).

Note that the coefficients of the series (1) and (4) utilize only the definition of the function on $0 < x < p$ (that is, half of the interval $-p < x < p$). Hence in practice there is no actual need to make the reflections described in (*i*) and (*ii*). If f is defined on $0 < x < L$, we simply identify the half-period as the length of the interval $p = L$. The coefficient formulas (2), (3), and (5) and the corresponding series yield either an even or an odd periodic extension of period $2L$ of the original function. The cosine and sine series obtained in this manner are known as **half-range expansions.** Finally, in case (*iii*) we are defining the functional values on the interval $-L < x < 0$ to be the same as the values on $0 < x < L$. As in the previous two cases, there is no real need to do this. It can be shown that the set of functions in (1) of Section 11.2 is orthogonal on $a \leq x \leq a + 2p$ for any real number a. Choosing $a = -p$, we obtain the limits of integration in (9), (10), and (11) of that section. But for $a = 0$ the limits of integration are from $x = 0$ to $x = 2p$. Thus if f is defined over the interval $0 < x < L$, we identify $2p = L$ or $p = L/2$. The resulting Fourier series will give the periodic extension of f with period L. In this manner the values to which the series converges will be the same on $-L < x < 0$ as on $0 < x < L$.

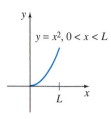

$y = x^2, 0 < x < L$

FIGURE 11.13 Function is neither odd nor even.

EXAMPLE 3 **Expansion in Three Series**

Expand $f(x) = x^2, 0 < x < L$,
(a) in a cosine series **(b)** in a sine series **(c)** in a Fourier series.

SOLUTION The graph of the function is given in Figure 11.13.

(a) We have

$$a_0 = \frac{2}{L}\int_0^L x^2\,dx = \frac{2}{3}L^2, \quad a_n = \frac{2}{L}\int_0^L x^2\cos\frac{n\pi}{L}x\,dx = \frac{4L^2(-1)^n}{n^2\pi^2},$$

where integration by parts was used twice in the evaluation of a_n.

Thus
$$f(x) = \frac{L^2}{3} + \frac{4L^2}{\pi^2}\sum_{n=1}^{\infty}\frac{(-1)^n}{n^2}\cos\frac{n\pi}{L}x. \tag{8}$$

(b) In this case we must again integrate by parts twice:

$$b_n = \frac{2}{L}\int_0^L x^2\sin\frac{n\pi}{L}x\,dx = \frac{2L^2(-1)^{n+1}}{n\pi} + \frac{4L^2}{n^3\pi^3}[(-1)^n - 1].$$

Hence
$$f(x) = \frac{2L^2}{\pi}\sum_{n=1}^{\infty}\left\{\frac{(-1)^{n+1}}{n} + \frac{2}{n^3\pi^2}[(-1)^n - 1]\right\}\sin\frac{n\pi}{L}x. \tag{9}$$

(c) With $p = L/2$, $1/p = 2/L$, and $n\pi/p = 2n\pi/L$, we have

$$a_0 = \frac{2}{L}\int_0^L x^2\,dx = \frac{2}{3}L^2, \quad a_n = \frac{2}{L}\int_0^L x^2\cos\frac{2n\pi}{L}x\,dx = \frac{L^2}{n^2\pi^2},$$

and
$$b_n = \frac{2}{L}\int_0^L x^2\sin\frac{2n\pi}{L}x\,dx = -\frac{L^2}{n\pi}.$$

Therefore
$$f(x) = \frac{L^2}{3} + \frac{L^2}{\pi}\sum_{n=1}^{\infty}\left\{\frac{1}{n^2\pi}\cos\frac{2n\pi}{L}x - \frac{1}{n}\sin\frac{2n\pi}{L}x\right\}. \tag{10}$$

The series (8), (9), and (10) converge to the $2L$-periodic even extension of f, the $2L$-periodic odd extension of f, and the L-periodic extension of f, respectively. The graphs of these periodic extensions are shown in Figure 11.14.

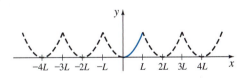

(a) cosine series

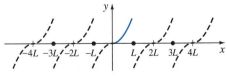

(b) sine series

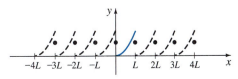

(c) Fourier series

FIGURE 11.14 Same function on $(0, L)$ but different periodic extensions

PERIODIC DRIVING FORCE   Fourier series are sometimes useful in determining a particular solution of a differential equation describing a physical system in which the input or driving force $f(t)$ is periodic. In the next example we find a particular solution of the differential equation

$$m\frac{d^2x}{dt^2} + kx = f(t) \tag{11}$$

by first representing f by a half-range sine expansion and then assuming a particular solution of the form

$$x_p(t) = \sum_{n=1}^{\infty} B_n \sin\frac{n\pi}{p}t. \tag{12}$$

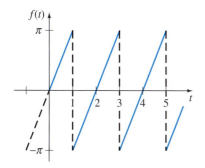

FIGURE 11.15 Periodic forcing function for spring/mass system

EXAMPLE 4   Particular Solution of a DE

An undamped spring/mass system, in which the mass $m = \frac{1}{16}$ slug and the spring constant $k = 4$ lb/ft, is driven by the 2-periodic external force $f(t)$ shown in Figure 11.15. Although the force $f(t)$ acts on the system for $t > 0$, note that if we extend the graph of the function in a 2-periodic manner to the negative t-axis, we obtain an odd function. In practical terms this means that we need only find the half-range sine expansion of $f(t) = \pi t$, $0 < t < 1$. With $p = 1$ it follows from (5) and integration by parts that

$$b_n = 2\int_0^1 \pi t \sin n\pi t\, dt = \frac{2(-1)^{n+1}}{n}.$$

From (11) the differential equation of motion is seen to be

$$\frac{1}{16}\frac{d^2x}{dt^2} + 4x = \sum_{n=1}^{\infty}\frac{2(-1)^{n+1}}{n}\sin n\pi t. \tag{13}$$

To find a particular solution $x_p(t)$ of (13), we substitute (12) into the equation and equate coefficients of $\sin n\pi t$. This yields

$$\left(-\frac{1}{16}n^2\pi^2 + 4\right)B_n = \frac{2(-1)^{n+1}}{n} \quad \text{or} \quad B_n = \frac{32(-1)^{n+1}}{n(64 - n^2\pi^2)}.$$

Thus
$$x_p(t) = \sum_{n=1}^{\infty}\frac{32(-1)^{n+1}}{n(64 - n^2\pi^2)}\sin n\pi t. \tag{14}$$

Observe in the solution (14) that there is no integer $n \geq 1$ for which the denominator $64 - n^2\pi^2$ of B_n is zero. In general, if there *is* a value of n, say N, for which $N\pi/p = \omega$, where $\omega = \sqrt{k/m}$, then the system described by (11) is in a state of pure resonance. In other words, we have pure resonance if the Fourier series expansion of the driving force $f(t)$ contains a term $\sin(N\pi/L)t$ (or $\cos(N\pi/L)t$) that has the same frequency as the free vibrations.

Of course, if the $2p$-periodic extension of the driving force f onto the negative t-axis yields an even function, then we expand f in a cosine series.

EXERCISES 11.3

Answers to selected odd-numbered problems begin on page ANS-16.

In Problems 1–10 determine whether the function is even, odd, or neither.

1. $f(x) = \sin 3x$

2. $f(x) = x \cos x$

3. $f(x) = x^2 + x$

4. $f(x) = x^3 - 4x$

5. $f(x) = e^{|x|}$

6. $f(x) = e^x - e^{-x}$

7. $f(x) = \begin{cases} x^2, & -1 < x < 0 \\ -x^2, & 0 \le x < 1 \end{cases}$

8. $f(x) = \begin{cases} x + 5, & -2 < x < 0 \\ -x + 5, & 0 \le x < 2 \end{cases}$

9. $f(x) = x^3, \quad 0 \le x \le 2$

10. $f(x) = |x^5|$

In Problems 11–24 expand the given function in an appropriate cosine or sine series.

11. $f(x) = \begin{cases} -1, & -\pi < x < 0 \\ 1, & 0 \le x < \pi \end{cases}$

12. $f(x) = \begin{cases} 1, & -2 < x < -1 \\ 0, & -1 < x < 1 \\ 1, & 1 < x < 2 \end{cases}$

13. $f(x) = |x|, \quad -\pi < x < \pi$

14. $f(x) = x, \quad -\pi < x < \pi$

15. $f(x) = x^2, \quad -1 < x < 1$

16. $f(x) = x|x|, \quad -1 < x < 1$

17. $f(x) = \pi^2 - x^2, \quad -\pi < x < \pi$

18. $f(x) = x^3, \quad -\pi < x < \pi$

19. $f(x) = \begin{cases} x - 1, & -\pi < x < 0 \\ x + 1, & 0 \le x < \pi \end{cases}$

20. $f(x) = \begin{cases} x + 1, & -1 < x < 0 \\ x - 1, & 0 \le x < 1 \end{cases}$

21. $f(x) = \begin{cases} 1, & -2 < x < -1 \\ -x, & -1 \le x < 0 \\ x, & 0 \le x < 1 \\ 1, & 1 \le x < 2 \end{cases}$

22. $f(x) = \begin{cases} -\pi, & -2\pi < x < -\pi \\ x, & -\pi \le x < \pi \\ \pi, & \pi \le x < 2\pi \end{cases}$

23. $f(x) = |\sin x|, \quad -\pi < x < \pi$

24. $f(x) = \cos x, \quad -\pi/2 < x < \pi/2$

In Problems 25–34 find the half-range cosine and sine expansions of the given function.

25. $f(x) = \begin{cases} 1, & 0 < x < \frac{1}{2} \\ 0, & \frac{1}{2} \le x < 1 \end{cases}$

26. $f(x) = \begin{cases} 0, & 0 < x < \frac{1}{2} \\ 1, & \frac{1}{2} \le x < 1 \end{cases}$

27. $f(x) = \cos x, \quad 0 < x < \pi/2$

28. $f(x) = \sin x, \quad 0 < x < \pi$

29. $f(x) = \begin{cases} x, & 0 < x < \pi/2 \\ \pi - x, & \pi/2 \le x < \pi \end{cases}$

30. $f(x) = \begin{cases} 0, & 0 < x < \pi \\ x - \pi, & \pi \le x < 2\pi \end{cases}$

31. $f(x) = \begin{cases} x, & 0 < x < 1 \\ 1, & 1 \le x < 2 \end{cases}$

32. $f(x) = \begin{cases} 1, & 0 < x < 1 \\ 2 - x, & 1 \le x < 2 \end{cases}$

33. $f(x) = x^2 + x, \quad 0 < x < 1$

34. $f(x) = x(2 - x), \quad 0 < x < 2$

In Problems 35–38 expand the given function in a Fourier series.

35. $f(x) = x^2, \quad 0 < x < 2\pi$

36. $f(x) = x, \quad 0 < x < \pi$

37. $f(x) = x + 1, \quad 0 < x < 1$

38. $f(x) = 2 - x, \quad 0 < x < 2$

In Problems 39 and 40 proceed as in Example 4 to find a particular solution $x_p(t)$ of equation (11) when $m = 1$, $k = 10$, and the driving force $f(t)$ is as given. Assume that when $f(t)$ is extended to the negative t-axis in a periodic manner, the resulting function is odd.

39. $f(t) = \begin{cases} 5, & 0 < t < \pi \\ -5, & \pi < t < 2\pi \end{cases}; \quad f(t + 2\pi) = f(t)$

40. $f(t) = 1 - t, \quad 0 < t < 2; \quad f(t + 2) = f(t)$

In Problems 41 and 42 proceed as in Example 4 to find a particular solution $x_p(t)$ of equation (11) when $m = \frac{1}{4}$, $k = 12$, and the driving force $f(t)$ is as given. Assume that when $f(t)$ is extended to the negative t-axis in a periodic manner, the resulting function is even.

41. $f(t) = 2\pi t - t^2$, $0 < t < 2\pi$; $f(t + 2\pi) = f(t)$

42. $f(t) = \begin{cases} t, & 0 < t < \frac{1}{2} \\ 1 - t, & \frac{1}{2} < t < 1 \end{cases}$; $f(t + 1) = f(t)$

43. (a) Solve the differential equation in Problem 39, $x'' + 10x = f(t)$, subject to the initial conditions $x(0) = 0$, $x'(0) = 0$.
   (b) Use a CAS to plot the graph of the solution $x(t)$ in part (a).

44. (a) Solve the differential equation in Problem 41, $\frac{1}{4}x'' + 12x = f(t)$, subject to the initial conditions $x(0) = 1$, $x'(0) = 0$.
   (b) Use a CAS to plot the graph of the solution $x(t)$ in part (a).

45. Suppose a uniform beam of length L is simply supported at $x = 0$ and at $x = L$. If the load per unit length is given by $w(x) = w_0 x/L$, $0 < x < L$, then the differential equation for the deflection $y(x)$ is

$$EI \frac{d^4 y}{dx^4} = \frac{w_0 x}{L},$$

where E, I, and w_0 are constants. (See (4) in Section 5.2.)
   (a) Expand $w(x)$ in a half-range sine series.
   (b) Use the method of Example 4 to find a particular solution $y_p(x)$ of the differential equation.

46. Proceed as in Problem 45 to find a particular solution $y_p(x)$ when the load per unit length is as given in Figure 11.16.

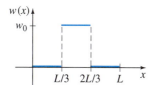

FIGURE 11.16   Graph for Problem 46

47. When a uniform beam is supported by an elastic foundation and subject to a load per unit length $w(x)$, the differential equation for its deflection $y(x)$ is

$$EI \frac{d^4 y}{dx^4} + ky = w(x),$$

where k is the modulus of the foundation. Suppose that the beam and elastic foundation are infinite in length (that is, $-\infty < x < \infty$) and that the load per unit length is the periodic function

$$w(x) = \begin{cases} 0, & -\pi < x < -\pi/2 \\ w_0, & -\pi/2 \le x \le \pi/2, \\ 0 & \pi/2 < x < \pi \end{cases} \quad w(x + 2\pi) = w(x).$$

Use the method of Example 4 to find a particular solution $y_p(x)$ of the differential equation.

DISCUSSION/PROJECT PROBLEMS

48. Prove properties (a), (c), (d), (f), and (g) in Theorem 11.2.

49. There is only one function that is both even and odd. What is it?

50. As we know from Chapter 4, the general solution of the differential equation in Problem 47 is $y = y_c + y_p$. Discuss why we can argue on physical grounds that the solution of Problem 47 is simply y_p. (*Hint:* Consider $y = y_c + y_p$ as $x \to \pm\infty$.)

COMPUTER LAB ASSIGNMENTS

In Problems 51 and 52 use a CAS to plot graphs of partial sums $\{S_N(x)\}$ of the given trigonometric series. Experiment with different values of N and graphs on different intervals of the x-axis. Use your graphs to conjecture a closed-form expression for a function f defined for $0 < x < L$ that is represented by the series.

51. $f(x) = -\dfrac{\pi}{4} + \sum_{n=1}^{\infty} \left[\dfrac{(-1)^n - 1}{n^2 \pi} \cos nx \right.$
$\left. + \dfrac{1 - 2(-1)^n}{n} \sin nx \right]$

52. $f(x) = \dfrac{1}{4} + \dfrac{4}{\pi^2} \sum_{n=1}^{\infty} \dfrac{1}{n^2} \left(1 - \cos \dfrac{n\pi}{2} \right) \cos \dfrac{n\pi}{2} x$

53. Is your answer in Problem 51 or in Problem 52 unique? Give a function f defined on a symmetric interval about the origin $-a < x < a$ that has the same trigonometric series
   (a) as in Problem 51   **(b)** as in Problem 52.

11.4 STURM-LIOUVILLE PROBLEM

INTRODUCTION: In this section we study some special types of boundary-value problems in which the ordinary differential equation in the problem contains a parameter λ. The values of λ for which the BVP possesses nontrivial solutions are called **eigenvalues,** and the corresponding solutions are **eigenfunctions.** Boundary-value problems of this type are especially important throughout Chapters 12 and 13. In this section we also see that there is a connection between orthogonal sets and eigenfunctions of a boundary-value problem.

REVIEW MATERIAL: The concept of eigenvalues and eigenfunctions was first introduced in Section 5.2. A review of that section (especially Example 2 on page 214) and of Section 11.1 is recommended.

REVIEW OF DEs For convenience we present here a brief review of some of the linear ODEs that will occur frequently in the sections and chapters that follow. The symbol α represents a constant.

Constant-coefficient equations	**General solutions**
$y' + \alpha y = 0$	$y = c_1 e^{-\alpha x}$
$y'' + \alpha^2 y = 0, \quad \alpha > 0$	$y = c_1 \cos \alpha x + c_2 \sin \alpha x$
$y'' - \alpha^2 y = 0, \quad \alpha > 0$	$\begin{cases} y = c_1 e^{-\alpha x} + c_2 e^{\alpha x}, \quad \text{or} \\ y = c_1 \cosh \alpha x + c_2 \sinh \alpha x \end{cases}$
Cauchy-Euler equation	**General solutions, $x > 0$**
$x^2 y'' + xy' - \alpha^2 y = 0, \quad \alpha \geq 0$	$\begin{cases} y = c_1 x^{-\alpha} + c_2 x^{\alpha}, \quad \alpha > 0 \\ y = c_1 + c_2 \ln x, \quad\quad \alpha = 0 \end{cases}$
Parametric Bessel equation ($v = 0$)	**General solution, $x > 0$**
$xy'' + y' + \alpha^2 xy = 0,$	$y = c_1 J_0(\alpha x) + c_2 Y_0(\alpha x)$
Legendre's equation **($n = 0, 1, 2, \ldots$)**	**Particular solutions are** **polynomials**
$(1 - x^2)y'' - 2xy' + n(n+1)y = 0,$	$y = P_0(x) = 1,$ $y = P_1(x) = x,$ $y = P_2(x) = \frac{1}{2}(3x^2 - 1), \ldots$

Regarding the two forms of the general solution of $y'' - \alpha^2 y = 0$, we will make use of the following informal rule immediately in Example 1 as well as in future discussions:

This rule will be useful in Chapters 12–14

Use the exponential form $y = c_1 e^{-\alpha x} + c_2 e^{\alpha x}$ when the domain of x is an infinite or semi-infinite interval; use the hyperbolic form $y = c_1 \cosh \alpha x + c_2 \sinh \alpha x$ when the domain of x is a finite interval.

EIGENVALUES AND EIGENFUNCTIONS Orthogonal functions arise in the solution of differential equations. More to the point, an orthogonal set of functions can be generated by solving a certain kind of two-point boundary-value problem involving a linear second-order differential equation containing a parameter λ. In Example 2 of Section 5.2 we saw that the boundary-value problem

$$y'' + \lambda y = 0, \quad y(0) = 0, \quad y(L) = 0, \tag{1}$$

possessed nontrivial solutions only when the parameter λ took on the values $\lambda_n = n^2 \pi^2 / L^2$, $n = 1, 2, 3, \ldots$, called **eigenvalues.** The corresponding nontrivial solutions $y_n = c_2 \sin(n\pi x/L)$, or simply $y_n = \sin(n\pi x/L)$, are called the **eigenfunctions** of the problem. For example, for (1),

BVP: $y'' - 2y = 0, \quad y(0) = 0, \quad y(L) = 0$

$\downarrow$ not an eigenvalue

Trivial solution: $y = 0 \leftarrow$ never an eigenfunction

$\downarrow$ is an eigenvalue ($n = 3$)

BVP: $y'' + \dfrac{9\pi^2}{L^2} y = 0, \quad y(0) = 0, \quad y(L) = 0$

Nontrivial solution: $y_3 = \sin(3\pi x/L) \leftarrow$ eigenfunction

For our purposes in this chapter it is important to recognize that the set of trigonometric functions generated by this BVP, that is, $\{\sin(n\pi x/L)\}$, $n = 1, 2, 3, \ldots$, is an orthogonal set of functions on the interval $[0, L]$ and is used as the basis for the Fourier sine series. See Problem 10 in Exercises 11.1.

EXAMPLE 1   Eigenvalues and Eigenfunctions

Consider the boundary-value problem

$$y'' + \lambda y = 0, \quad y'(0) = 0, \quad y'(L) = 0. \tag{2}$$

As in Example 2 of Section 5.2 there are three possible cases for the parameter λ: zero, negative, or positive; that is, $\lambda = 0$, $\lambda = -\alpha^2 < 0$, and $\lambda = \alpha^2 > 0$, where $\alpha > 0$. The solution of the DEs

$$y'' = 0, \quad \lambda = 0, \tag{3}$$

$$y'' - \alpha^2 y = 0, \quad \lambda = -\alpha^2, \tag{4}$$

$$y'' + \alpha^2 y = 0, \quad \lambda = \alpha^2, \tag{5}$$

are, in turn,

$$y = c_1 + c_2 x, \tag{6}$$

$$y = c_1 \cosh \alpha x + c_2 \sinh \alpha x, \tag{7}$$

$$y = c_1 \cos \alpha x + c_2 \sin \alpha x. \tag{8}$$

When the boundary conditions $y'(0) = 0$ and $y'(L) = 0$ are applied to each of these solutions, (6) yields $y = c_1$, (7) yields only $y = 0$, and (8) yields $y = c_1 \cos \alpha x$ *provided* that $\alpha = n\pi/L$, $n = 1, 2, 3, \ldots$. Since $y = c_1$ satisfies the DE in (3) and the boundary conditions for any *nonzero* choice of c_1, we conclude that $\lambda = 0$ is an eigenvalue. Thus the eigenvalues and corresponding eigenfunctions of the problem are $\lambda_0 = 0$, $y_0 = c_1$, $c_1 \neq 0$, and $\lambda_n = \alpha_n^2 = n^2\pi^2/L^2$, $n = 1, 2, \ldots$, $y_n = c_1 \cos(n\pi x/L)$, $c_1 \neq 0$. We can, if desired, take $c_1 = 1$ in each case. Note also that the eigenfunction $y_0 = 1$ corresponding to the eigenvalue $\lambda_0 = 0$ can be incorporated in the family $y_n = \cos(n\pi x/L)$ by permitting $n = 0$. The set $\{\cos(n\pi x/L)\}$, $n = 0, 1, 2, 3, \ldots$, is orthogonal on the interval $[0, L]$. You are asked to fill in the details of this example in Problem 3 in Exercises 11.4. ∎

REGULAR STURM-LIOUVILLE PROBLEM   The problems in (1) and (2) are special cases of an important general two-point boundary value problem. Let p, q, r, and r' be real-valued functions continuous on an interval $[a, b]$, and let $r(x) > 0$ and $p(x) > 0$ for every x in the interval. Then

Solve: $$\dfrac{d}{dx}[r(x)y'] + (q(x) + \lambda p(x))y = 0 \tag{9}$$

Subject to: $A_1 y(a) + B_1 y'(a) = 0 \tag{10}$

$A_2 y(b) + B_2 y'(b) = 0 \tag{11}$

is said to be a **regular Sturm-Liouville problem.** The coefficients in the boundary conditions (10) and (11) are assumed to be real and independent of λ. In addition, A_1 and B_1 are not both zero, and A_2 and B_2 are not both zero. The boundary-value problems in (1) and (2) are regular Sturm-Liouville problems. From (1) we can identify $r(x) = 1$, $q(x) = 0$, and $p(x) = 1$ in the differential equation (9); in boundary condition (10) we identify $a = 0$, $A_1 = 1$, $B_1 = 0$, and in (11), $b = L$, $A_2 = 1$, $B_2 = 0$. From (2) the identifications would be $a = 0$, $A_1 = 0$, $B_1 = 1$ in (10), $b = L$, $A_2 = 0$, $B_2 = 1$ in (11).

The differential equation (9) is linear and homogeneous. The boundary conditions in (10) and (11), both a linear combination of y and y' *equal to zero at a point*, are also **homogeneous.** A boundary condition such as $A_2 y(b) + B_2 y'(b) = C_2$, where C_2 is a nonzero constant, is **nonhomogeneous.** A boundary-value problem that consists of a homogeneous linear differential equation and homogeneous boundary conditions is, of course, said to be a homogeneous BVP; otherwise, it is nonhomogeneous. The boundary conditions (10) and (11) are referred to as **separated** since each condition involves only a single boundary point.

Because a regular Sturm-Liouville problem is a homogeneous BVP, it always possesses the trivial solution $y = 0$. However, this solution is of no interest to us. As in Example 1, in solving such a problem, we seek numbers λ (eigenvalues) and nontrivial solutions y that depend on λ (eigenfunctions).

PROPERTIES Theorem 11.3 is a list of the more important of the many properties of the regular Sturm-Liouville problem. We shall prove only the last property.

THEOREM 11.3   **Properties of the Regular Sturm-Liouville Problem**

(a) There exist an infinite number of real eigenvalues that can be arranged in increasing order $\lambda_1 < \lambda_2 < \lambda_3 < \cdots < \lambda_n < \cdots$ such that $\lambda_n \to \infty$ as $n \to \infty$.

(b) For each eigenvalue there is only one eigenfunction (except for nonzero constant multiples).

(c) Eigenfunctions corresponding to different eigenvalues are linearly independent.

(d) The set of eigenfunctions corresponding to the set of eigenvalues is orthogonal with respect to the weight function $p(x)$ on the interval $[a, b]$.

PROOF OF (D)   Let y_m and y_n be eigenfunctions corresponding to eigenvalues λ_m and λ_n, respectively. Then

$$\frac{d}{dx}[r(x)y'_m] + (q(x) + \lambda_m p(x))y_m = 0 \qquad (12)$$

$$\frac{d}{dx}[r(x)y'_n] + (q(x) + \lambda_n p(x))y_n = 0. \qquad (13)$$

Multiplying (12) by y_n and (13) by y_m and subtracting the two equations gives

$$(\lambda_m - \lambda_n)p(x)y_m y_n = y_m \frac{d}{dx}[r(x)y'_n] - y_n \frac{d}{dx}[r(x)y'_m].$$

Integrating this last result by parts from $x = a$ to $x = b$ then yields

$$(\lambda_m - \lambda_n)\int_a^b p(x)y_m y_n \, dx = r(b)[y_m(b)y'_n(b) - y_n(b)y'_m(b)] - r(a)[y_m(a)y'_n(a) - y_n(a)y'_m(a)]. \qquad (14)$$

Now the eigenfunctions y_m and y_n must both satisfy the boundary conditions (10) and (11). In particular, from (10) we have

$$A_1 y_m(a) + B_1 y_m'(a) = 0$$
$$A_1 y_n(a) + B_1 y_n'(a) = 0.$$

For this system to be satisfied by A_1 and B_1, not both zero, the determinant of the coefficients must be zero:

$$y_m(a)y_n'(a) - y_n(a)y_m'(a) = 0.$$

A similar argument applied to (11) also gives

$$y_m(b)\,y_n'(b) - y_n(b)\,y_m'(b) = 0.$$

Since both members of the right-hand side of (14) are zero, we have established the orothogonality relation

$$\int_a^b p(x)y_m(x)y_n(x)\,dx = 0, \quad \lambda_m \neq \lambda_n. \tag{15}$$

EXAMPLE 2   A Regular Sturm-Liouville Problem

Solve the boundary-value problem

$$y'' + \lambda y = 0, \quad y(0) = 0, \quad y(1) + y'(1) = 0. \tag{16}$$

SOLUTION We proceed exactly as in Example 1 by considering three cases in which the parameter λ could be zero, negative, or positive: $\lambda = 0$, $\lambda = -\alpha^2 < 0$, and $\lambda = \alpha^2 > 0$, where $\alpha > 0$. The solutions of the DE for these values are listed in (3)–(5). For the cases $\lambda = 0$ and $\lambda = -\alpha^2 < 0$ we find that the BVP in (16) possesses only the trivial solution $y = 0$. For $\lambda = \alpha^2 > 0$ the general solution of the differential equation is $y = c_1 \cos \alpha x + c_2 \sin \alpha x$. Now the condition $y(0) = 0$ implies that $c_1 = 0$ in this solution, so we are left with $y = c_2 \sin \alpha x$. The second boundary condition $y(1) + y'(1) = 0$ is satisfied if

$$c_2 \sin \alpha + c_2 \alpha \cos \alpha = 0.$$

In view of the demand that $c_2 \neq 0$, the last equation can be written

$$\tan \alpha = -\alpha. \tag{17}$$

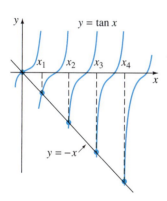

FIGURE 11.17   Positive roots $x_1, x_2, x_3, \ldots$ of $\tan x = -x$

If for a moment we think of (17) as $\tan x = -x$, then Figure 11.17 shows the plausibility that this equation has an infinite number of roots, namely, the x-coordinates of the points where the graph of $y = -x$ intersects the infinite number of branches of the graph of $y = \tan x$. The eigenvalues of the BVP (16) are then $\lambda_n = \alpha_n^2$, where α_n, $n = 1, 2, 3, \ldots$ are the consecutive *positive* roots $\alpha_1, \alpha_2, \alpha_3, \ldots$ of (17). With the aid of a CAS it is easily shown that, to four rounded decimal places, $\alpha_1 = 2.0288$, $\alpha_2 = 4.9132$, $\alpha_3 = 7.9787$, and $\alpha_4 = 11.0855$, and the corresponding solutions are $y_1 = \sin 2.0288x$, $y_2 = \sin 4.9132x$, $y_3 = \sin 7.9787x$, and $y_4 = \sin 11.0855x$. In general, the eigenfunctions of the problem are $\{\sin \alpha_n x\}$, $n = 1, 2, 3, \ldots$.

With the identification $r(x) = 1$, $q(x) = 0$, $p(x) = 1$, $A_1 = 1$, $B_1 = 0$, $A_2 = 1$, $B_2 = 1$ we see that (16) is a regular Sturm-Liouville problem. We conclude that $\{\sin \alpha_n x\}$, $n = 1, 2, 3, \ldots$, is an orthogonal set with respect to the weight function $p(x) = 1$ on the interval $[0, 1]$.

In some circumstances we can prove the orthogonality of solutions of (9) without the necessity of specifying a boundary condition at $x = a$ and at $x = b$.

SINGULAR STURM-LIOUVILLE PROBLEM   There are several other important conditions under which we seek nontrivial solutions of the differential equation (9):

- $r(a) = 0$, and a boundary condition of the type given in (11) is specified at $x = b$;   (18)
- $r(b) = 0$, and a boundary condition of the type given in (10) is specified at $x = a$;   (19)
- $r(a) = r(b) = 0$, and no boundary condition is specified at either $x = a$ or at $x = b$;   (20)
- $r(a) = r(b)$, and boundary conditions $y(a) = y(b)$, $y'(a) = y'(b)$.   (21)

The differential equation (9) along with one of conditions (18)–(20), is said to be a **singular** boundary-value problem. Equation (9) with the conditions specified in (21) is said to be a **periodic** boundary-value problem (the boundary conditions are also said to be periodic). Observe that if, say, $r(a) = 0$, then $x = a$ *may* be a singular point of the differential equation, and consequently, a solution of (9) may become unbounded as $x \rightarrow a$. However, we see from (14) that if $r(a) = 0$, then no boundary condition is required at $x = a$ to prove orthogonality of the eigenfunctions provided that these solutions are bounded at that point. This latter requirement guarantees the existence of the integrals involved. By assuming that the solutions of (9) are bounded on the closed interval $[a, b]$, we can see from inspection of (14) that

- if $r(a) = 0$, then the orthogonality relation (15) holds with no boundary condition specified at $x = a$;   (22)
- if $r(b) = 0$, then the orthogonality relation (15) holds with no boundary condition specified at $x = b$;*   (23)
- if $r(a) = r(b) = 0$, then the orthogonality relation (15) holds with no boundary conditions specified at either $x = a$ or $x = b$;   (24)
- if $r(a) = r(b)$, then the orthogonality relation (15) holds with the periodic boundary conditions $y(a) = y(b)$, $y'(a) = y'(b)$.   (25)

We note that a Sturm-Liouville problem is also singular when the interval under consideration is infinite. See Problems 9 and 10 in Exercises 11.4.

SELF-ADJOINT FORM   By carrying out the indicated differentiation in (9) we see that the differential equation is the same as

$$r(x)y'' + r'(x)y' + (q(x) + \lambda p(x))y = 0. \quad (26)$$

Examination of (26) might lead one to believe, given the coefficient of y' is the derivative of the coefficient of y'', that few differential equations have form (9). On the contrary, if the coefficients are continuous and $a(x) \neq 0$ for all x in some interval, then *any* second-order differential equation

$$a(x)y'' + b(x)y' + (c(x) + \lambda d(x))y = 0 \quad (27)$$

can be recast into the so-called **self-adjoint form** (9). To this end we basically proceed as in Section 2.3, where we rewrote a homogeneous linear first-order equation $a_1(x)y' + a_0(x)y = 0$ in the form $\dfrac{d}{dx}[\mu y] = 0$ by dividing the equation by $a_1(x)$ and then multiplying by the integrating factor $\mu = e^{\int P(x)dx}$, where, assuming no common factors, $P(x) = a_0(x)/a_1(x)$. So first, we divide (27) by $a(x)$. The first two terms are $Y' + \dfrac{b(x)}{a(x)}Y + \cdots$, where for emphasis we have written $Y = y'$. Second, we multiply this equation by the integrating factor $e^{\int(b(x)/a(x))dx}$, where $a(x)$ and $b(x)$ are assumed to have no common factors:

---

*Conditions (22) and (23) are equivalent to choosing $A_1 = 0$, $B_1 = 0$, and $A_2 = 0$, $B_2 = 0$, respectively.

$$e^{\int (b(x)/a(x))dx}Y' + \frac{b(x)}{a(x)}e^{\int (b(x)/a(x))dx}Y + \cdots = \frac{d}{dx}\left[e^{\int (b(x)/a(x))dx}Y\right] + \cdots = \frac{d}{dx}\left[e^{\int (b(x)/a(x))dx}y'\right] + \cdots .$$

$\underbrace{\qquad\qquad\qquad\qquad\qquad\qquad}_{\text{derivative of a product}}$

In summary, by dividing (27) by $a(x)$ and then multiplying by $e^{\int (b(x)/a(x))dx}$, we get

$$e^{\int (b/a)dx}y'' + \frac{b(x)}{a(x)}e^{\int (b/a)dx}y' + \left(\frac{c(x)}{a(x)}e^{\int (b/a)dx} + \lambda\frac{d(x)}{a(x)}e^{\int (b/a)dx}\right)y = 0. \quad (28)$$

Equation (28) is the desired form given in (26) and is the same as (9):

$$\frac{d}{dx}\left[\underbrace{e^{\int (b/a)dx}}_{r(x)}y'\right] + \left(\underbrace{\frac{c(x)}{a(x)}e^{\int (b/a)dx}}_{q(x)} + \lambda\underbrace{\frac{d(x)}{a(x)}e^{\int (b/a)dx}}_{p(x)}\right)y = 0.$$

For example, to express $2y'' + 6y' + \lambda y = 0$ in self-adjoint form, we write $y'' + 3y' + \lambda\frac{1}{2}y = 0$ and then multiply by $e^{\int 3dx} = e^{3x}$. The resulting equation is

$$\underset{\substack{\downarrow\\ r(x)}}{e^{3x}}y'' + \underset{\substack{\downarrow\\ r'(x)}}{3e^{3x}}y' + \lambda\frac{1}{2}\underset{\substack{\downarrow\\ p(x)}}{e^{3x}}y = 0 \quad\text{or}\quad \frac{d}{dx}\left[e^{3x}y'\right] + \lambda\frac{1}{2}e^{3x}y = 0.$$

It is certainly not necessary to put a second-order differential equation (27) into the self-adjoint form (9) to *solve* the DE. For our purposes we use the form given in (9) to determine the weight function $p(x)$ needed in the orthogonality relation (15). The next two examples illustrate orthogonality relations for Bessel functions and for Legendre polynomials.

EXAMPLE 3   Parametric Bessel Equation

In Section 6.3 we saw that the parametric Bessel differential equation of order n is $x^2y'' + xy' + (\alpha^2x^2 - n^2)y = 0$, where n is a fixed nonnegative integer and α is a positive parameter. The general solution of this equation is $y = c_1J_n(\alpha x) + c_2Y_n(\alpha x)$. After dividing the parametric Bessel equation by the lead coefficient x^2 and multiplying the resulting equation by the integrating factor $e^{\int (1/x)dx} = e^{\ln x} = x, x > 0$, we obtain

$$xy'' + y' + \left(\alpha^2x - \frac{n^2}{x}\right)y = 0 \quad\text{or}\quad \frac{d}{dx}[xy'] + \left(\alpha^2x - \frac{n^2}{x}\right)y = 0.$$

By comparing the last result with the self-adjoint form (9), we make the identifications $r(x) = x$, $q(x) = -n^2/x$, $\lambda = \alpha^2$, and $p(x) = x$. Now $r(0) = 0$, and of the two solutions $J_n(\alpha x)$ and $Y_n(\alpha x)$, only $J_n(\alpha x)$ is bounded at $x = 0$. Thus in view of (22) above, the set $\{J_n(\alpha_i x)\}$, $i = 1, 2, 3, \ldots$, is orthogonal with respect to the weight function $p(x) = x$ on the interval $[0, b]$. The orthogonality relation is

$$\int_0^b xJ_n(\alpha_i x)J_n(\alpha_j x)\,dx = 0, \quad \lambda_i \neq \lambda_j, \quad (29)$$

provided that the α_i, and hence the eigenvalues $\lambda_i = \alpha_i^2$, $i = 1, 2, 3, \ldots$ are defined by means of a boundary condition at $x = b$ of the type given in (11):

$$A_2J_n(\alpha b) + B_2\alpha J_n'(\alpha b) = 0.^* \quad (30)$$

---

*The extra factor of α comes from the Chain Rule: $\dfrac{d}{dx}J_n(\alpha x) = J_n'(\alpha x)\dfrac{d}{dx}\alpha x = \alpha J_n'(\alpha x)$.

For any choice of A_2 and B_2, not both zero, it is known that (30) has an infinite number of roots $x_i = \alpha_i b$. The eigenvalues are then $\lambda_i = \alpha_i^2 = (x_i/b)^2$. More will be said about eigenvalues in the next chapter.

EXAMPLE 4   Legendre's Equation

Legendre's differential equation $(1 - x^2)y'' - 2xy' + n(n + 1)y = 0$ is exactly of the form given in (26) with $r(x) = 1 - x^2$ and $r'(x) = -2x$. Hence the self-adjoint form (9) of the differential equation is immediate,

$$\frac{d}{dx}\left[(1 - x^2)y'\right] + n(n + 1)y = 0. \qquad (31)$$

From (31) we can further identify $q(x) = 0$, $\lambda = n(n + 1)$, and $p(x) = 1$. Recall from Section 6.3 that when $n = 0, 1, 2, \ldots$, Legendre's DE possesses polynomial solutions $P_n(x)$. Now we can put the observation that $r(-1) = r(1) = 0$ together with the fact that the Legendre polynomials $P_n(x)$ are the only solutions of (31) that are bounded on the closed interval $[-1, 1]$ to conclude from (24) that the set $\{P_n(x)\}$, $n = 0, 1, 2, \ldots$, is orthogonal with respect to the weight function $p(x) = 1$ on $[-1, 1]$. The orthogonality relation is

$$\int_{-1}^{1} P_m(x)P_n(x)\,dx = 0, \quad m \neq n.$$

EXERCISES 11.4

Answers to selected odd-numbered problems begin on page ANS-17.

In Problems 1 and 2 find the eigenfunctions and the equation that defines the eigenvalues for the given boundary-value problem. Use a CAS to approximate the first four eigenvalues λ_1, λ_2, λ_3, and λ_4. Give the eigenfunctions corresponding to these approximations.

1. $y'' + \lambda y = 0$, $y'(0) = 0, y(1) + y'(1) = 0$

2. $y'' + \lambda y = 0$, $y(0) + y'(0) = 0, y(1) = 0$

3. Consider $y'' + \lambda y = 0$ subject to $y'(0) = 0$, $y'(L) = 0$. Show that the eigenfunctions are
$$\left\{1, \cos\frac{\pi}{L}x, \cos\frac{2\pi}{L}x, \ldots\right\}.$$ This set, which is orthogonal on $[0, L]$, is the basis for the Fourier cosine series.

4. Consider $y'' + \lambda y = 0$ subject to the periodic boundary conditions $y(-L) = y(L)$, $y'(-L) = y'(L)$. Show that the eigenfunctions are
$$\left\{1, \cos\frac{\pi}{L}x, \cos\frac{2\pi}{L}x, \ldots, \sin\frac{\pi}{L}x, \sin\frac{2\pi}{L}x, \sin\frac{3\pi}{L}x, \ldots\right\}.$$
This set, which is orthogonal on $[-L, L]$, is the basis for the Fourier series.

5. Find the square norm of each eigenfunction in Problem 1.

6. Show that for the eigenfunctions in Example 2,
$$\|\sin \alpha_n x\|^2 = \tfrac{1}{2}[1 + \cos^2\alpha_n].$$

7. (a) Find the eigenvalues and eigenfunctions of the boundary-value problem
$$x^2 y'' + xy' + \lambda y = 0, \quad y(1) = 0, \quad y(5) = 0.$$
   (b) Put the differential equation in self-adjoint form.
   (c) Give an orthogonality relation.

8. (a) Find the eigenvalues and eigenfunctions of the boundary-value problem
$$y'' + y' + \lambda y = 0, \quad y(0) = 0, \quad y(2) = 0.$$
   (b) Put the differential equation in self-adjoint form.
   (c) Give an orthogonality relation.

9. **Laguerre's differential equation**
$$xy'' + (1 - x)y' + ny = 0, \quad n = 0, 1, 2, \ldots$$
has polynomial solutions $L_n(x)$. Put the equation in self-adjoint form and give an orthogonality relation.

10. **Hermite's differential equation**
$$y'' - 2xy' + 2ny = 0, \quad n = 0, 1, 2, \ldots$$
has polynomial solutions $H_n(x)$. Put the equation in self-adjoint form and give an orthogonality relation.

11. Consider the regular Sturm-Liouville problem:

$$\frac{d}{dx}\left[(1+x^2)y'\right] + \frac{\lambda}{1+x^2}y = 0,$$
$$y(0) = 0, \quad y(1) = 0.$$

(a) Find the eigenvalues and eigenfunctions of the boundary-value problem. (*Hint:* Let $x = \tan\theta$ and then use the Chain Rule.)

(b) Give an orthogonality relation.

12. (a) Find the eigenfunctions and the equation that defines the eigenvalues for the boundary-value problem

$$x^2y'' + xy' + (\lambda x^2 - 1)y = 0, \quad x > 0,$$
$$y \text{ is bounded at } x = 0, \quad y(3) = 0.$$

Let $\lambda = \alpha^2, \alpha > 0$.

(b) Use Table 6.1 of Section 6.3 to find the approximate values of the first four eigenvalues $\lambda_1, \lambda_2, \lambda_3$, and λ_4.

DISCUSSION/PROJECT PROBLEMS

13. Consider the special case of the regular Sturm-Liouville problem on the interval $[a, b]$:

$$\frac{d}{dx}[r(x)y'] + \lambda p(x)y = 0,$$
$$y'(a) = 0, \quad y'(b) = 0.$$

Is $\lambda = 0$ an eigenvalue of the problem? Defend your answer.

COMPUTER LAB ASSIGNMENTS

14. (a) Give an orthogonality relation for the Sturm-Liouville problem in Problem 1.

(b) Use a CAS as an aid in verifying the orthogonality relation for the eigenfunctions y_1 and y_2 that correspond to the first two eigenvalues λ_1 and λ_2, respectively.

15. (a) Give an orthogonality relation for the Sturm-Liouville problem in Problem 2.

(b) Use a CAS as an aid in verifying the orthogonality relation for the eigenfunctions y_1 and y_2 that correspond to the first two eigenvalues λ_1 and λ_2, respectively.

11.5 BESSEL AND LEGENDRE SERIES

INTRODUCTION: Fourier series, Fourier cosine series, and Fourier sine series are three ways of expanding a function in terms of an orthogonal set of functions. But such expansions are by no means limited to orthogonal sets of *trigonometric* functions. We saw in Section 11.1 that a function f defined on an interval (a, b) could be expanded, at least in a formal manner, in terms of any set of functions $\{\phi_n(x)\}$ that is orthogonal with respect to a weight function on $[a, b]$. Many of these orthogonal series expansions or generalized Fourier series stem from Sturm-Liouville problems which, in turn, arise from attempts to solve linear partial differential equations that serve as models for physical systems. Fourier series and orthogonal series expansions, as well as the two series considered in this section, will appear in the subsequent consideration of these applications in Chapters 12 and 13.

REVIEW MATERIAL: Because the results in Examples 3 and 4 of Section 11.4 will play a major role in the discussion that follows, you are strongly urged to reread those examples in conjunction with (6)–(11) of Section 11.1.

11.5.1 FOURIER-BESSEL SERIES

We saw in Example 3 of Section 11.4 that for a fixed value of n the set of Bessel functions $\{J_n(\alpha_i x)\}$, $i = 1, 2, 3, \ldots$, is orthogonal with respect to the weight function $p(x) = x$ on an interval $[0, b]$ whenever the α_i are defined by means of a boundary condition of the form

$$A_2 J_n(\alpha b) + B_2 \alpha J_n'(\alpha b) = 0. \tag{1}$$

The eigenvalues of the corresponding Sturm-Liouville problem are $\lambda_i = \alpha_i^2$. From (7) and (8) of Section 11.1 the orthogonal series, or generalized Fourier series, expansion of a function f defined on the interval $(0, b)$ in terms of this orthogonal set is

$$f(x) = \sum_{i=1}^{\infty} c_i J_n(\alpha_i x), \tag{2}$$

where
$$c_i = \frac{\int_0^b x J_n(\alpha_i x) f(x)\, dx}{\|J_n(\alpha_i x)\|^2}. \tag{3}$$

The square norm of the function $J_n(\alpha_i x)$ is defined by (11) of Section 11.1.

$$\|J_n(\alpha_i x)\|^2 = \int_0^b x J_n^2(\alpha_i x)\, dx. \tag{4}$$

The series (2) with coefficients (3) is called a **Fourier-Bessel series,** or simply, **a Bessel series.**

DIFFERENTIAL RECURRENCE RELATIONS The differential recurrence relations that were given in (21) and (20) of Section 6.3 are often useful in the evaluation of the coefficients (3). For convenience we reproduce those relations here:

$$\frac{d}{dx}[x^n J_n(x)] = x^n J_{n-1}(x) \tag{5}$$

$$\frac{d}{dx}[x^{-n} J_n(x)] = -x^{-n} J_{n+1}(x). \tag{6}$$

SQUARE NORM The value of the square norm (4) depends on how the eigenvalues $\lambda_i = \alpha_i^2$ are defined. If $y = J_n(\alpha x)$, then we know from Example 3 of Section 11.4 that

$$\frac{d}{dx}[xy'] + \left(\alpha^2 x - \frac{n^2}{x}\right) y = 0.$$

After we multiply by $2xy'$, this equation can be written as

$$\frac{d}{dx}[xy']^2 + (\alpha^2 x^2 - n^2)\frac{d}{dx}[y]^2 = 0.$$

Integrating the last result by parts on $[0, b]$ then gives

$$2\alpha^2 \int_0^b xy^2\, dx = \left([xy']^2 + (\alpha^2 x^2 - n^2)y^2\right)\Big|_0^b.$$

Since $y = J_n(\alpha x)$, the lower limit is zero because $J_n(0) = 0$ for $n > 0$. Furthermore, for $n = 0$ the quantity $[xy']^2 + \alpha^2 x^2 y^2$ is zero at $x = 0$. Thus

$$2\alpha^2 \int_0^b x J_n^2(\alpha x)\, dx = \alpha^2 b^2 [J_n'(\alpha b)]^2 + (\alpha^2 b^2 - n^2)[J_n(\alpha b)]^2, \tag{7}$$

where we have used the Chain Rule to write $y' = \alpha J_n'(\alpha x)$.
   We now consider three cases of (1).

CASE I: If we choose $A_2 = 1$ and $B_2 = 0$, then (1) is

$$J_n(\alpha b) = 0. \tag{8}$$

There are an infinite number of positive roots $x_i = \alpha_i b$ of (8) (see Figure 6.4), which define the α_i as $\alpha_i = x_i/b$. The eigenvalues are positive and are then $\lambda_i = \alpha_i^2 = x_i^2/b^2$. No new eigenvalues result from the negative roots of (8) since $J_n(-x) = (-1)^n J_n(x)$. (See page 263.) The number 0 is not an eigenvalue for any n because $J_n(0) = 0$ for $n = 1, 2, 3, \ldots$ and $J_0(0) = 1$. In other words, if $\lambda = 0$, we get the trivial function (which is never an eigenfunction) for $n = 1, 2, 3, \ldots$, and for $n = 0$, $\lambda = 0$ (or equivalently, $\alpha = 0$) does not satisfy the equation in (8). When (6) is written in the form $xJ_n'(x) = nJ_n(x) - xJ_{n+1}(x)$, it follows from (7) and (8) that the square norm of $J_n(\alpha_i x)$ is

$$\|J_n(\alpha_i x)\|^2 = \frac{b^2}{2} J_{n+1}^2(\alpha_i b). \tag{9}$$

CASE II: If we choose $A_2 = h \geq 0$, and $B_2 = b$, then (1) is

$$hJ_n(\alpha b) + \alpha bJ_n'(\alpha b) = 0. \tag{10}$$

Equation (10) has an infinite number of positive roots $x_i = \alpha_i b$ for each positive integer $n = 1$, 2, 3, ... As before, the eigenvalues are obtained from $\lambda_i = \alpha_i^2 = x_i^2/b^2$. $\lambda = 0$ is not an eigenvalue for $n = 1$, 2, 3, ... Substituting $\alpha_i bJ_n'(\alpha_i b) = -hJ_n(\alpha_i b)$ into (7), we find that the square norm of $J_n(\alpha_i x)$ is now

$$\|J_n(\alpha_i x)\|^2 = \frac{\alpha_i^2 b^2 - n^2 + h^2}{2\alpha_i^2} J_n^2(\alpha_i b). \tag{11}$$

CASE III: If $h = 0$ and $n = 0$ in (10), the α_i are defined from the roots of

$$J_0'(\alpha b) = 0. \tag{12}$$

Even though (12) is just a special case of (10), it is the only situation for which $\lambda = 0$ is an eigenvalue. To see this, observe that for $n = 0$ the result in (6) implies that $J_0'(\alpha b) = 0$ is equivalent to $J_1(\alpha b) = 0$. Since $x_1 = \alpha_1 b = 0$ is root of the last equation, $\alpha_1 = 0$, and because $J_0(0) = 1$ is nontrivial, we conclude from $\lambda_1 = \alpha_1^2 = x_1^2/b^2$ that $\lambda_1 = 0$ is an eigenvalue. But obviously, we cannot use (11) when $\alpha_1 = 0$, $h = 0$, and $n = 0$. However, from the square norm (4),

$$\|1\|^2 = \int_0^b x\, dx = \frac{b^2}{2}. \tag{13}$$

For $\alpha_i > 0$ we can use (11) with $h = 0$ and $n = 0$:

$$\|J_0(\alpha_i x)\|^2 = \frac{b^2}{2} J_0^2(\alpha_i b). \tag{14}$$

The following definition summarizes three forms of the series (2) corresponding to the square norms in the three cases.

DEFINITION 11.7   **Fourier-Bessel Series**

The **Fourier-Bessel series** of a function f defined on the interval $(0, b)$ is given by

(*i*)
$$f(x) = \sum_{i=1}^{\infty} c_i J_n(\alpha_i x) \tag{15}$$

$$c_i = \frac{2}{b^2 J_{n+1}^2(\alpha_i b)} \int_0^b x J_n(\alpha_i x) f(x)\, dx, \tag{16}$$

where the α_i are defined by $J_n(\alpha b) = 0$.

(*ii*)
$$f(x) = \sum_{i=1}^{\infty} c_i J_n(\alpha_i x) \tag{17}$$

$$c_i = \frac{2\alpha_i^2}{(\alpha_i^2 b^2 - n^2 + h^2) J_n^2(\alpha_i b)} \int_0^b x J_n(\alpha_i x) f(x)\, dx, \tag{18}$$

where the α_i are defined by $hJ_n(\alpha b) + \alpha bJ_n'(\alpha b) = 0$.

(iii)
$$f(x) = c_1 + \sum_{i=2}^{\infty} c_i J_0(\alpha_i x) \tag{19}$$

$$c_1 = \frac{2}{b^2} \int_0^b x f(x)\,dx, \quad c_i = \frac{2}{b^2 J_0^2(\alpha_i b)} \int_0^b x J_0(\alpha_i x) f(x)\,dx, \tag{20}$$

where the α_i are defined by $J_0'(\alpha b) = 0$.

CONVERGENCE OF A FOURIER-BESSEL SERIES Sufficient conditions for the convergence of a Fourier-Bessel series are not particularly restrictive.

THEOREM 11.4   **Conditions for Convergence**

If f and f' are piecewise continuous on the open interval $(0, b)$, then a Fourier-Bessel expansion of f converges to $f(x)$ at any point where f is continuous and to the average $[f(x+) + f(x-)]/2$ at a point where f is discontinuous.

EXAMPLE 1   **Expansion in a Fourier-Bessel Series**

Expand $f(x) = x$, $0 < x < 3$, in a Fourier-Bessel series, using Bessel functions of order one that satisfy the boundary condition $J_1(3\alpha) = 0$.

SOLUTION   We use (15) where the coefficients c_i are given by (16) with $b = 3$:

$$c_i = \frac{2}{3^2 J_2^2(3\alpha_i)} \int_0^3 x^2 J_1(\alpha_i x)\,dx.$$

To evaluate this integral, we let $t = \alpha_i x$, $dx = dt/\alpha_i$, $x^2 = t^2/\alpha_i^2$, and use (5) in the form $\dfrac{d}{dt}[t^2 J_2(t)] = t^2 J_1(t)$:

$$c_i = \frac{2}{9\alpha_i^3 J_2^2(3\alpha_i)} \int_0^{3\alpha_i} \frac{d}{dt}[t^2 J_2(t)]\,dt = \frac{2}{\alpha_i J_2(3\alpha_i)}.$$

Therefore the desired expansion is

$$f(x) = 2 \sum_{i=1}^{\infty} \frac{1}{\alpha_i J_2(3\alpha_i)} J_1(\alpha_i x).$$

You are asked to find the first four values of the α_i for the foregoing Fourier-Bessel series in Problem 1 in Exercises 11.5.

EXAMPLE 2   **Expansion in a Fourier-Bessel Series**

If the α_i in Example 1 are defined by $J_1(3\alpha) + \alpha J_1'(3\alpha) = 0$, then the only thing that changes in the expansion is the value of the square norm. Multiplying the boundary condition by 3 gives $3J_1(3\alpha) + 3\alpha J_1'(3\alpha) = 0$, which now matches (10) when $h = 3$, $b = 3$, and $n = 1$. Thus (18) and (17) yield, in turn,

$$c_i = \frac{18\alpha_i J_2(3\alpha_i)}{(9\alpha_i^2 + 8) J_1^2(3\alpha_i)}$$

and

$$f(x) = 18 \sum_{i=1}^{\infty} \frac{\alpha_i J_2(3\alpha_i)}{(9\alpha_i^2 + 8) J_1^2(3\alpha_i)} J_1(\alpha_i x).$$

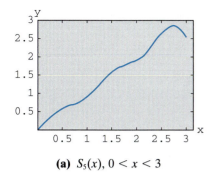

(a) $S_5(x)$, $0 < x < 3$

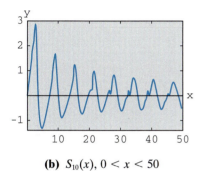

(b) $S_{10}(x)$, $0 < x < 50$

FIGURE 11.18 Graphs of two partial sums of a Fourier-Bessel series

USE OF COMPUTERS Since Bessel functions are "built-in functions" in a CAS, it is a straightforward task to find the approximate values of the α_i and the coefficients c_i in a Fourier-Bessel series. For example, in (10) we can think of $x_i = \alpha_i b$ as a positive root of the equation $h J_n(x) + x J'_n(x) = 0$. Thus in Example 2 we have used a CAS to find the first five positive roots x_i of $3 J_1(x) + x J'_1(x) = 0$, and from these roots we obtain first five values of the α_i: $\alpha_1 = x_1/3 = 0.98320$, $\alpha_2 = x_2/3 = 1.94704$, $\alpha_3 = x_3/3 = 2.95758$, $\alpha_4 = x_4/3 = 3.98538$, and $\alpha_5 = x_5/3 = 5.02078$. Knowing the roots $x_i = 3\alpha_i$ and the α_i, we again use a CAS to calculate the numerical values of $J_2(3\alpha_i)$, $J_1^2(3\alpha_i)$, and finally the coefficients c_i. In this manner we find that the fifth partial sum $S_5(x)$ for the Fourier-Bessel series representation of $f(x) = x$, $0 < x < 3$ in Example 2 is

$$S_5(x) = 4.01844 J_1(0.98320x) - 1.86937 J_1(1.94704x)$$
$$+ 1.07106 J_1(2.95758x) - 0.70306 J_1(3.98538x) + 0.50343 J_1(5.02078x).$$

The graph of $S_5(x)$ on the interval $0 < x < 3$ is shown in Figure 11.18(a). In Figure 11.18(b) we have graphed $S_{10}(x)$ on the interval $0 < x < 50$. Notice that outside the interval of definition $0 < x < 3$ the series does not converge to a periodic extension of f because Bessel functions are not periodic functions. See Problems 11 and 12 in Exercises 11.5.

11.5.2 FOURIER-LEGENDRE SERIES

From Example 4 of Section 11.4 we know that the set of Legendre polynomials $\{P_n(x)\}$, $n = 0, 1, 2, \ldots$, is orthogonal with respect to the weight function $p(x) = 1$ on $[-1, 1]$. Furthermore, it can be proved that the square norm of a polynomial $P_n(x)$ depends on n in the following manner:

$$\|P_n(x)\|^2 = \int_{-1}^{1} P_n^2(x)\, dx = \frac{2}{2n + 1}.$$

The orthogonal series expansion of a function in terms of the Legendre polynomials is summarized in the next definition.

DEFINITION 11.8 **Fourier-Legendre Series**

The **Fourier-Legendre series** of a function f on an interval $(-1, 1)$ is given by

$$f(x) = \sum_{n=0}^{\infty} c_n P_n(x), \qquad (21)$$

where

$$c_n = \frac{2n + 1}{2} \int_{-1}^{1} f(x) P_n(x)\, dx. \qquad (22)$$

CONVERGENCE OF A FOURIER-LEGENDRE SERIES Sufficient conditions for convergence of a Fourier-Legendre series are given in the next theorem.

THEOREM 11.5 **Conditions for Convergence**

If f and f' are piecewise continuous on the open interval $(-1, 1)$, then a Fourier-Legendre expansion of f converges to $f(x)$ at any point where f is continuous and to the average $[f(x+) + f(x-)]/2$ at a point where f is discontinuous.

> **EXAMPLE 3**   **Expansion in a Fourier-Legendre Series**

Write out the first four nonzero terms in the Fourier-Legendre expansion of

$$f(x) = \begin{cases} 0, & -1 < x < 0 \\ 1, & 0 \le x < 1. \end{cases}$$

SOLUTION   The first several Legendre polynomials are listed on page 267. From these and (22) we find

$$c_0 = \frac{1}{2}\int_{-1}^{1} f(x)P_0(x)\,dx = \frac{1}{2}\int_{0}^{1} 1 \cdot 1 \, dx = \frac{1}{2}$$

$$c_1 = \frac{3}{2}\int_{-1}^{1} f(x)P_1(x)\,dx = \frac{3}{2}\int_{0}^{1} 1 \cdot x \, dx = \frac{3}{4}$$

$$c_2 = \frac{5}{2}\int_{-1}^{1} f(x)P_2(x)\,dx = \frac{5}{2}\int_{0}^{1} 1 \cdot \frac{1}{2}(3x^2 - 1)\,dx = 0$$

$$c_3 = \frac{7}{2}\int_{-1}^{1} f(x)P_3(x)\,dx = \frac{7}{2}\int_{0}^{1} 1 \cdot \frac{1}{2}(5x^3 - 3x)\,dx = -\frac{7}{16}$$

$$c_4 = \frac{9}{2}\int_{-1}^{1} f(x)P_4(x)\,dx = \frac{9}{2}\int_{0}^{1} 1 \cdot \frac{1}{8}(35x^4 - 30x^2 + 3)\,dx = 0$$

$$c_5 = \frac{11}{2}\int_{-1}^{1} f(x)P_5(x)\,dx = \frac{11}{2}\int_{0}^{1} 1 \cdot \frac{1}{8}(63x^5 - 70x^3 + 15x)\,dx = \frac{11}{32}.$$

Hence $$f(x) = \frac{1}{2}P_0(x) + \frac{3}{4}P_1(x) - \frac{7}{16}P_3(x) + \frac{11}{32}P_5(x) + \cdots.$$

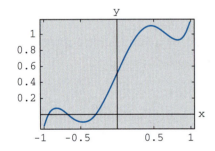

FIGURE 11.19   Partial sum $S_5(x)$ of Fourier-Legendre series

Like the Bessel functions, Legendre polynomials are built-in functions in computer algebra systems such as *Maple* and *Mathematica*, so each of the coefficients just listed can be found using the integration application of such a program. Indeed, using a CAS, we further find that $c_6 = 0$ and $c_7 = -\frac{65}{256}$. The fifth partial sum of the Fourier-Legendre series representation of the function f defined in Example 3 is then

$$S_5(x) = \frac{1}{2}P_0(x) + \frac{3}{4}P_1(x) - \frac{7}{16}P_3(x) + \frac{11}{32}P_5(x) - \frac{65}{256}P_7(x).$$

The graph of $S_5(x)$ on the interval $-1 < x < 1$ is given in Figure 11.19.

ALTERNATIVE FORM OF SERIES   In applications the Fourier-Legendre series appears in an alternative form. If we let $x = \cos\theta$, then $x = 1$ implies $\theta = 0$ whereas $x = -1$ implies $\theta = \pi$. Since $dx = -\sin\theta \, d\theta$, (21) and (22) become, respectively,

$$F(\theta) = \sum_{n=0}^{\infty} c_n P_n(\cos\theta) \qquad\qquad (23)$$

$$c_n = \frac{2n + 1}{2}\int_{0}^{\pi} F(\theta)\, P_n(\cos\theta)\sin\theta \, d\theta, \qquad\qquad (24)$$

where $f(\cos\theta)$ has been replaced by $F(\theta)$.

EXERCISES 11.5

Answers to selected odd-numbered problems begin on page ANS-17.

11.5.1 FOURIER-BESSEL SERIES

In Problems 1 and 2 use Table 6.1 in Section 6.3.

1. Find the first four $\alpha_i > 0$ defined by $J_1(3\alpha) = 0$.

2. Find the first four $\alpha_i \geq 0$ defined by $J_0'(2\alpha) = 0$.

In Problems 3–6 expand $f(x) = 1$, $0 < x < 2$, in a Fourier-Bessel series, using Bessel functions of order zero that satisfy the given boundary condition.

3. $J_0(2\alpha) = 0$ 4. $J_0'(2\alpha) = 0$

5. $J_0(2\alpha) + 2\alpha J_0'(2\alpha) = 0$ 6. $J_0(2\alpha) + \alpha J_0'(2\alpha) = 0$

In Problems 7–10 expand the given function in a Fourier-Bessel series, using Bessel functions of the same order as in the indicated boundary condition.

7. $f(x) = 5x$, $0 < x < 4$,
$3J_1(4\alpha) + 4\alpha J_1'(4\alpha) = 0$

8. $f(x) = x^2$, $0 < x < 1$, $J_2(\alpha) = 0$

9. $f(x) = x^2$, $0 < x < 3$, $J_0'(3\alpha) = 0$ (*Hint:* $t^3 = t^2 \cdot t$.)

10. $f(x) = 1 - x^2$, $0 < x < 1$, $J_0(\alpha) = 0$

COMPUTER LAB ASSIGNMENTS

11. (a) Use a CAS to plot the graph of $y = 3J_1(x) + xJ_1'(x)$ on an interval so that the first five positive x-intercepts of the graph are shown.
    (b) Use the root-finding capability of your CAS to approximate the first five roots x_i of the equation $3J_1(x) + xJ_1'(x) = 0$.
    (c) Use the data obtained in part (b) to find the first five positive values of α_i that satisfy $3J_1(4\alpha) + 4\alpha J_1'(4\alpha) = 0$. (See Problem 7.)
    (d) If instructed, find the first ten positive values of α_i.

12. (a) Use the values of α_i in part (c) of Problem 11 and a CAS to approximate the values of the first five coefficients c_i of the Fourier-Bessel series obtained in Problem 7.
    (b) Use a CAS to plot the graphs of the partial sums $S_N(x)$, $N = 1, 2, 3, 4, 5$ of the Fourier-Bessel series in Problem 7.
    (c) If instructed, plot the graph of the partial sum $S_{10}(x)$ on $0 < x < 4$ and on $0 < x < 50$.

DISCUSSION/PROJECT PROBLEMS

13. If the partial sums in Problem 12 are plotted on a symmetric interval such as $-30 < x < 30$ would the graphs possess any symmetry? Explain.

14. (a) Sketch, by hand, a graph of what you think the Fourier-Bessel series in Problem 3 converges to on the interval $-2 < x < 2$.
    (b) Sketch, by hand, a graph of what you think the Fourier-Bessel series would converge to on the interval $-4 < x < 4$ if the values α_i in Problem 7 were defined by $3J_2(4\alpha) + 4\alpha J_2'(4\alpha) = 0$.

11.5.2 FOURIER-LEGENDRE SERIES

In Problems 15 and 16 write out the first five nonzero terms in the Fourier-Legendre expansion of the given function. If instructed, use a CAS as an aid in evaluating the coefficients. Use a CAS to plot the graph of the partial sum $S_5(x)$.

15. $f(x) = \begin{cases} 0, & -1 < x < 0 \\ x, & 0 < x < 1 \end{cases}$

16. $f(x) = e^x$, $-1 < x < 1$

17. The first three Legendre polynomials are $P_0(x) = 1$, $P_1(x) = x$, and $P_2(x) = \frac{1}{2}(3x^2 - 1)$. If $x = \cos\theta$, then $P_0(\cos\theta) = 1$ and $P_1(\cos\theta) = \cos\theta$. Show that $P_2(\cos\theta) = \frac{1}{4}(3\cos 2\theta + 1)$.

18. Use the results of Problem 17 to find a Fourier-Legendre expansion (23) of $F(\theta) = 1 - \cos 2\theta$.

19. A Legendre polynomial $P_n(x)$ is an even or odd function, depending on whether n is even or odd. Show that if f is an even function on $(-1, 1)$, then (21) and (22) become, respectively,

$$f(x) = \sum_{n=0}^{\infty} c_{2n} P_{2n}(x) \tag{25}$$

$$c_{2n} = (4n + 1) \int_0^1 f(x) P_{2n}(x)\, dx. \tag{26}$$

The series (25) can also be used when f is defined only on the interval $(0, 1)$. The series then represents f on $(0, 1)$ and an even extension of f on the interval $(-1, 0)$.

20. Show that if f is an odd function on $(-1, 1)$, then (21) and (22) become, respectively,

$$f(x) = \sum_{n=0}^{\infty} c_{2n+1} P_{2n+1}(x) \tag{27}$$

$$c_{2n+1} = (4n + 3) \int_0^1 f(x) P_{2n+1}(x)\, dx. \tag{28}$$

The series (27) can also be used when f is defined only on the interval $(0, 1)$. The series then represents f on $(0, 1)$ and an odd extension of f on the interval $(-1, 0)$.

In Problems 21 and 22 write out the first four nonzero terms in the indicated expansion of the given function. What function does the series represent on the interval $(-1, 1)$? Use a CAS to plot the graph of the partial sum $S_4(x)$.

21. $f(x) = x, \quad 0 < x < 1; \quad$ use (25)

22. $f(x) = 1, \quad 0 < x < 1; \quad$ use (27)

DISCUSSION/PROJECT PROBLEMS

23. Discuss: Why is a Fourier-Legendre expansion of a polynomial function that is defined on the interval $(-1, 1)$ necessarily a finite series?

24. Using only your conclusions from Problem 23—that is, do not use (22)—find the finite Fourier-Legendre series of $f(x) = x^2$. The series of $f(x) = x^3$.

CHAPTER 11 IN REVIEW

Answers to selected odd-numbered problems begin on page ANS-17.

In Problems 1–6 fill in the blank or answer true or false without referring back to the text.

1. The functions $f(x) = x^2 - 1$ and $g(x) = x^5$ are orthogonal on $[-\pi, \pi]$. _____

2. The product of an odd function f with an odd function g is _____.

3. To expand $f(x) = |x| + 1, -\pi < x < \pi$, in an appropriate trigonometric series, we would use a _____ series.

4. $y = 0$ is never an eigenfunction of a Sturm-Liouville problem. _____

5. $\lambda = 0$ is never an eigenvalue of a Sturm-Liouville problem. _____

6. If the function $f(x) = \begin{cases} x + 1, & -1 < x < 0 \\ -x, & 0 < x < 1 \end{cases}$ is expanded in a Fourier series, the series will converge to _____ at $x = -1$, to _____ at $x = 0$, and to _____ at $x = 1$.

7. Suppose the function $f(x) = x^2 + 1, \ 0 < x < 3$ is expanded in a Fourier series, a cosine series, and a sine series. Give the value to which each series will converge at $x = 0$.

8. What is the corresponding eigenfunction for the boundary-value problem $y'' + \lambda y = 0, y'(0) = 0, y(\pi/2) = 0$ for $\lambda = 25$?

9. Chebyshev's differential equation

$$(1 - x^2)y'' - xy' + n^2 y = 0$$

has a polynomial solution $y = T_n(x)$ for $n = 0, 1, 2, \ldots$. Specify the weight function $w(x)$ and the interval over which the set of Chebyshev polynomials $\{T_n(x)\}$ is orthogonal. Give an orthogonality relation.

10. The set of Legendre polynomials $\{P_n(x)\}$, where $P_0(x) = 1, P_1(x) = x, \ldots$, is orthogonal with respect to the weight function $w(x) = 1$ on the interval $[-1, 1]$. Explain why $\int_{-1}^{1} P_n(x) \, dx = 0$ for $n > 0$.

11. Without doing any work, explain why the cosine series of $f(x) = \cos^2 x, 0 < x < \pi$ is the finite series $f(x) = \frac{1}{2} + \frac{1}{2} \cos 2x$.

12. (a) Show that the set

$$\left\{ \sin \frac{\pi}{2L} x, \sin \frac{3\pi}{2L} x, \sin \frac{5\pi}{2L} x, \ldots \right\}$$

is orthogonal on the interval $0 \le x \le L$.

(b) Find the norm of each function in part (a). Construct an orthonormal set.

13. Expand $f(x) = |x| - x, -1 < x < 1$ in a Fourier series.

14. Expand $f(x) = 2x^2 - 1, -1 < x < 1$ in a Fourier series.

15. Expand $f(x) = e^x, 0 < x < 1$
(a) in a cosine series   **(b)** in a Fourier series.

16. In Problems 13, 14, and 15, sketch the periodic extension of f to which each series converges.

17. Discuss: Which of the two Fourier series of f in Problem 15 converges to

$$F(x) = \begin{cases} f(x), & 0 < x < 1 \\ f(-x), & -1 < x < 0 \end{cases}$$

on the interval $-1 < x < 1$?

18. Consider the portion of the periodic function f shown in Figure 11.20. Expand f in an appropriate Fourier series.

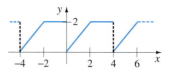

FIGURE 11.20   Graph for Problem 18.

19. Find the eigenvalues and eigenfunctions of the boundary-value problem

$$x^2y'' + xy' + 9\lambda y = 0, \quad y'(1) = 0, y(e) = 0.$$

20. Give an orthogonality relation for the eigenfunctions in Problem 19.

21. Expand $f(x) = \begin{cases} 1, & 0 < x < 2 \\ 0, & 2 < x < 4 \end{cases}$ in a Fourier-Bessel series, using Bessel functions of order zero that satisfy the boundary-condition $J_0(4\alpha) = 0$.

22. Expand $f(x) = x^4$, $-1 < x < 1$ in a Fourier-Legendre series.

12

BOUNDARY-VALUE PROBLEMS IN RECTANGULAR COORDINATES

In this and the next two chapters the emphasis will be on two procedures used in solving partial differential equations that occur frequently in problems involving temperature distributions, vibrations, and potentials. These problems, called boundary-value problems, are described by relatively simple linear second-order PDEs. The thrust of these procedures is to find solutions of a PDE by reducing it to two or more ODEs.

We begin with a method called separation of variables. The application of this method leads us back to the important concepts of Chapter 11 – namely, eigenvalues, eigenfunctions, and the expansion of a function in an infinite series of orthogonal functions.

View: Above the surface from the rear

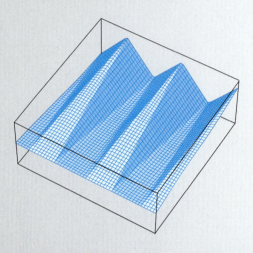

View: Slightly above from right rear

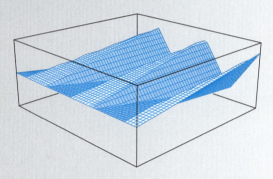

View: Right front

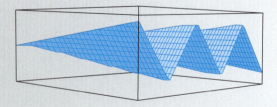

Several 3D viewpoints of the graph of a 50-term partial sum of a series solution representing the twist angle $\theta(x, t)$ of an elastic rod. See page 494–496 and Figure 12.20

12.1 SEPARABLE PARTIAL DIFFERENTIAL EQUATIONS

INTRODUCTION: Partial differential equations (PDEs), like ordinary differential equations (ODEs), are classified as either linear or nonlinear. Analogous to a linear ODE, the dependent variable and its partial derivatives in a linear PDE are only to the first power. For the remaining chapters of this text we shall be interested, for the most part, in *linear second-order* PDEs.

REVIEW MATERIAL: As was mentioned in the chapter introduction, our goal in Chapters 12, 13, and 14 is to find solutions of boundary-value problems. This we do by first finding particular solutions of a linear PDE by reducing it to two or more ODEs. Because the ODEs are linear, logic dictates that a review of Sections 2.3, 4.3, and 4.4 is in order. Also, reread the discussion on "Two Equations Worth Knowing" on page 145.

LINEAR PARTIAL DIFFERENTIAL EQUATION If we let u denote the dependent variable and let x and y denote the independent variables, then the general form of a **linear second-order partial differential equation** is given by

$$A\frac{\partial^2 u}{\partial x^2} + B\frac{\partial^2 u}{\partial x \, \partial y} + C\frac{\partial^2 u}{\partial y^2} + D\frac{\partial u}{\partial x} + E\frac{\partial u}{\partial y} + Fu = G, \tag{1}$$

where the coefficients $A, B, C, \ldots, G$ are functions of x and y. When $G(x, y) = 0$, equation (1) is said to be **homogeneous;** otherwise, it is **nonhomogeneous.** For example, the linear equations

$$\frac{\partial^2 u}{\partial x^2} + \frac{\partial^2 u}{\partial y^2} = 0 \quad \text{and} \quad \frac{\partial^2 u}{\partial x^2} - \frac{\partial u}{\partial y} = xy$$

are homogeneous and nonhomogeneous, respectively.

SOLUTION OF A PDE A **solution** of a linear partial differential equation (1) is a function $u(x, y)$ of two independent variables that possesses all partial derivatives occurring in the equation and that satisfies the equation in some region of the xy-plane.

It is not our intention to examine procedures for finding *general solutions* of linear partial differential equations. Not only is it often difficult to obtain a general solution of a linear second-order PDE, but a general solution is usually not all that useful in applications. Thus our focus throughout will be on finding *particular solutions* of some of the more important linear PDEs—that is, equations that appear in many applications.

SEPARATION OF VARIABLES Although there are several methods that can be tried to find particular solutions of a linear PDE, the one we are interested in at the moment is called the **method of separation of variables.** In this method we seek a particular solution of the form of a *product* of a function of x and a function of y:

$$u(x, y) = X(x)Y(y).$$

With this assumption it is *sometimes* possible to reduce a linear PDE in two variables to two ODEs. To this end we note that

$$\frac{\partial u}{\partial x} = X'Y, \quad \frac{\partial u}{\partial y} = XY', \quad \frac{\partial^2 u}{\partial x^2} = X''Y, \quad \frac{\partial^2 u}{\partial y^2} = XY'',$$

where the primes denote ordinary differentiation.

EXAMPLE 1   **Separation of Variables**

Find product solutions of $\dfrac{\partial^2 u}{\partial x^2} - 4\dfrac{\partial u}{\partial y}$.

SOLUTION   Substituting $u(x, y) = X(x)Y(y)$ into the partial differential equation yields

$$X''Y = 4XY'.$$

After dividing both sides by $4XY$, we have separated the variables:

$$\frac{X''}{4X} = \frac{Y'}{Y}.$$

Since the left-hand side of the last equation is independent of y and is equal to the right-hand side, which is independent of x, we conclude that both sides of the equation are independent of x *and* y. In other words, each side of the equation must be a constant. In practice it is *convenient* to write this real **separation constant** as $-\lambda$ (using λ would lead to the same solutions).

From the two equalities

$$\frac{X''}{4X} = \frac{Y'}{Y} = -\lambda$$

we obtain the two linear ordinary differential equations

$$X'' + 4\lambda X = 0 \quad \text{and} \quad Y' + \lambda Y = 0. \tag{2}$$

Now, as in Example 1 of Section 11.4 we consider three cases for λ: zero, negative, or positive, that is, $\lambda = 0$, $\lambda = -\alpha^2 < 0$, and $\lambda = \alpha^2 > 0$, where $\alpha > 0$.

CASE I   If $\lambda = 0$, then the two ODEs in (2) are

$$X'' = 0 \quad \text{and} \quad Y' = 0.$$

Solving each equation (by, say, integration), we find $X = c_1 + c_2 x$ and $Y = c_3$. Thus a particular product solution of the given PDE is

$$u = XY = (c_1 + c_2 x)c_3 = A_1 + B_1 x, \tag{3}$$

where we have replaced $c_1 c_3$ and $c_2 c_3$ by A_1 and B_1, respectively.

CASE II   If $\lambda = -\alpha^2$, then the DEs in (2) are

$$X'' - 4\alpha^2 X = 0 \quad \text{and} \quad Y' - \alpha^2 Y = 0.$$

From their general solutions

$$X = c_4 \cosh 2\alpha x + c_5 \sinh 2\alpha x \quad \text{and} \quad Y = c_6 e^{\alpha^2 y}$$

we obtain another particular product solution of the PDE,

$$u = XY = (c_4 \cosh 2\alpha x + c_5 \sinh 2\alpha x)c_6 e^{\alpha^2 y}$$

or

$$u = A_2 e^{\alpha^2 y} \cosh 2\alpha x + B_2 e^{\alpha^2 y} \sinh 2\alpha x, \tag{4}$$

where $A_2 = c_4 c_6$ and $B_2 = c_5 c_6$.

CASE III   If $\lambda = \alpha^2$, then the DEs

$$X'' + 4\alpha^2 X = 0 \quad \text{and} \quad Y' + \alpha^2 Y = 0$$

and their general solutions

$$X = c_7 \cos 2\alpha x + c_8 \sin 2\alpha x \quad \text{and} \quad Y = c_9 e^{-\alpha^2 y}$$

give yet another particular solution

$$u = A_3 e^{-\alpha^2 y} \cos 2\alpha x + B_3 e^{-\alpha^2 y} \sin 2\alpha x, \tag{5}$$

where $A_3 = c_7 c_9$ and $B_2 = c_8 c_9$.

It is left as an exercise to verify that (3), (4), and (5) satisfy the given PDE. See Problem 29 in Exercises 12.1.

SUPERPOSITION PRINCIPLE   The following theorem is analogous to Theorem 4.2 and is known as the **superposition principle.**

THEOREM 12.1   **Superposition Principle**

If $u_1, u_2, \ldots, u_k$ are solutions of a homogeneous linear partial differential equation, then the linear combination

$$u = c_1 u_1 + c_2 u_2 + \cdots + c_k u_k,$$

where the c_i, $i = 1, 2, \ldots, k$, are constants, is also a solution.

Throughout the remainder of the chapter we shall assume that whenever we have an infinite set $u_1, u_2, u_3, \ldots$ of solutions of a homogeneous linear equation, we can construct yet another solution u by forming the infinite series

$$u = \sum_{k=1}^{\infty} c_k u_k,$$

where the c_i, $i = 1, 2, \ldots$ are constants.

CLASSIFICATION OF EQUATIONS   A linear second-order partial differential equation in two independent variables with constant coefficients can be classified as one of three types. This classification depends only on the coefficients of the second-order derivatives. Of course, we assume that at least one of the coefficients A, B, and C is not zero.

DEFINITION 12.1   **Classification of Equations**

The linear second-order partial differential equation

$$A \frac{\partial^2 u}{\partial x^2} + B \frac{\partial^2 u}{\partial x \, \partial y} + C \frac{\partial^2 u}{\partial y^2} + D \frac{\partial u}{\partial x} + E \frac{\partial u}{\partial y} + Fu = 0,$$

where A, B, C, D, E, and F are real constants, is said to be

$$\textbf{hyperbolic if} \quad B^2 - 4AC > 0,$$
$$\textbf{parabolic if} \quad B^2 - 4AC = 0,$$
$$\textbf{elliptic if} \quad B^2 - 4AC < 0.$$

EXAMPLE 2   **Classifying Linear Second-Order PDEs**

Classify the following equations:

(a) $3 \dfrac{\partial^2 u}{\partial x^2} = \dfrac{\partial u}{\partial y}$ (b) $\dfrac{\partial^2 u}{\partial x^2} = \dfrac{\partial^2 u}{\partial y^2}$ (c) $\dfrac{\partial^2 u}{\partial x^2} + \dfrac{\partial^2 u}{\partial y^2} = 0$

SOLUTION   (a) By rewriting the given equation as

$$3 \frac{\partial^2 u}{\partial x^2} - \frac{\partial u}{\partial y} = 0$$

we can make the identifications $A = 3$, $B = 0$, and $C = 0$. Since $B^2 - 4AC = 0$, the equation is parabolic.

(b) By rewriting the equation as

$$\frac{\partial^2 u}{\partial x^2} - \frac{\partial^2 u}{\partial y^2} = 0$$

we see that $A = 1$, $B = 0$, $C = -1$, and $B^2 - 4AC = -4(1)(-1) > 0$. The equation is hyperbolic.

(c) With $A = 1$, $B = 0$, $C = 1$, and $B^2 - 4AC = -4(1)(1) < 0$, the equation is elliptic. ■

REMARKS

(*i*) In case you are wondering, separation of variables is not a general method for finding particular solutions; some linear partial differential equations are simply *not* separable. You are encouraged to verify that the assumption $u = XY$ does not lead to a solution for the linear PDE $\partial^2 u / \partial x^2 - \partial u / \partial y = x$.

(*ii*) A detailed explanation of why we would want to classify a linear second-order PDE as hyperbolic, parabolic, or elliptic is beyond the scope of this text, but you should at least be aware that this classification is of practical importance. We are going to solve some PDEs subject to only boundary conditions and others subject to both boundary and initial conditions; the kinds of side conditions that are appropriate for a given equation depend on whether the equation is hyperbolic, parabolic, or elliptic. On a related matter, we shall see in Chapter 15 that numerical-solution methods for linear second-order PDEs differ in conformity with the classification of the equation.

EXERCISES 12.1

Answers to selected odd-numbered problems begin on page ANS-17.

In Problems 1–16 use separation of variables to find, if possible, product solutions for the given partial differential equation.

1. $\dfrac{\partial u}{\partial x} = \dfrac{\partial u}{\partial y}$

2. $\dfrac{\partial u}{\partial x} + 3\dfrac{\partial u}{\partial y} = 0$

3. $u_x + u_y = u$

4. $u_x = u_y + u$

5. $x\dfrac{\partial u}{\partial x} = y\dfrac{\partial u}{\partial y}$

6. $y\dfrac{\partial u}{\partial x} + x\dfrac{\partial u}{\partial y} = 0$

7. $\dfrac{\partial^2 u}{\partial x^2} + \dfrac{\partial^2 u}{\partial x\,\partial y} + \dfrac{\partial^2 u}{\partial y^2} = 0$

8. $y\dfrac{\partial^2 u}{\partial x\,\partial y} + u = 0$

9. $k\dfrac{\partial^2 u}{\partial x^2} - u = \dfrac{\partial u}{\partial t}, \quad k > 0$

10. $k\dfrac{\partial^2 u}{\partial x^2} = \dfrac{\partial u}{\partial t}, \quad k > 0$

11. $a^2\dfrac{\partial^2 u}{\partial x^2} = \dfrac{\partial^2 u}{\partial t^2}$

12. $a^2\dfrac{\partial^2 u}{\partial x^2} = \dfrac{\partial^2 u}{\partial t^2} + 2k\dfrac{\partial u}{\partial t}, \quad k > 0$

13. $\dfrac{\partial^2 u}{\partial x^2} + \dfrac{\partial^2 u}{\partial y^2} = 0$

14. $x^2\dfrac{\partial^2 u}{\partial x^2} + \dfrac{\partial^2 u}{\partial y^2} = 0$

15. $u_{xx} + u_{yy} = u$

16. $a^2 u_{xx} - g = u_{tt}, \quad g$ a constant

In Problems 17–26 classify the given partial differential equation as hyperbolic, parabolic, or elliptic.

17. $\dfrac{\partial^2 u}{\partial x^2} + \dfrac{\partial^2 u}{\partial x\,\partial y} + \dfrac{\partial^2 u}{\partial y^2} = 0$

18. $3\dfrac{\partial^2 u}{\partial x^2} + 5\dfrac{\partial^2 u}{\partial x\,\partial y} + \dfrac{\partial^2 u}{\partial y^2} = 0$

19. $\dfrac{\partial^2 u}{\partial x^2} + 6\dfrac{\partial^2 u}{\partial x\,\partial y} + 9\dfrac{\partial^2 u}{\partial y^2} = 0$

20. $\dfrac{\partial^2 u}{\partial x^2} - \dfrac{\partial^2 u}{\partial x\, \partial y} - 3\dfrac{\partial^2 u}{\partial y^2} = 0$

21. $\dfrac{\partial^2 u}{\partial x^2} = 9\dfrac{\partial^2 u}{\partial x\, \partial y}$

22. $\dfrac{\partial^2 u}{\partial x\, \partial y} - \dfrac{\partial^2 u}{\partial y^2} + 2\dfrac{\partial u}{\partial x} = 0$

23. $\dfrac{\partial^2 u}{\partial x^2} + 2\dfrac{\partial^2 u}{\partial x\, \partial y} + \dfrac{\partial^2 u}{\partial y^2} + \dfrac{\partial u}{\partial x} - 6\dfrac{\partial u}{\partial y} = 0$

24. $\dfrac{\partial^2 u}{\partial x^2} + \dfrac{\partial^2 u}{\partial y^2} = u$

25. $a^2\dfrac{\partial^2 u}{\partial x^2} = \dfrac{\partial^2 u}{\partial t^2}$

26. $k\dfrac{\partial^2 u}{\partial x^2} = \dfrac{\partial u}{\partial t}, \quad k > 0$

In Problems 27 and 28 show that the given partial differential equation possesses the indicated product solution.

27. $k\left(\dfrac{\partial^2 u}{\partial r^2} + \dfrac{1}{r}\dfrac{\partial u}{\partial r}\right) = \dfrac{\partial u}{\partial t};$

$u = e^{-k\alpha^2 t}\big(c_1 J_0(\alpha r) + c_2 Y_0(\alpha r)\big)$

28. $\dfrac{\partial^2 u}{\partial r^2} + \dfrac{1}{r}\dfrac{\partial u}{\partial r} + \dfrac{1}{r^2}\dfrac{\partial^2 u}{\partial \theta^2} = 0;$

$u = (c_1 \cos \alpha\theta + c_2 \sin \alpha\theta)(c_3 r^\alpha + c_4 r^{-\alpha})$

29. Verify that each of the products $u = XY$ in (3), (4), and (5) satisfies the second-order PDE in Example 1.

30. Definition 12.1 generalizes to linear PDEs with coefficients that are functions of x and y. Determine the regions in the xy-plane for which the equation

$$(xy + 1)\dfrac{\partial^2 u}{\partial x^2} + (x + 2y)\dfrac{\partial^2 u}{\partial x\, \partial y} + \dfrac{\partial^2 u}{\partial y^2} + xy^2 u = 0$$

is hyperbolic, parabolic, or elliptic.

12.2 CLASSICAL PDEs AND BOUNDARY-VALUE PROBLEMS

INTRODUCTION: We are not going to solve anything in this section. We are simply going to discuss the types of partial differential equations and boundary-value problems that we will be working with for the remainder of this chapter as well as in Chapters 13–15. The words *boundary-value problem* have a slightly different connotation than they did in Sections 4.1, 4.3, and 5.2. If, say, $u(x, t)$ is a solution of a PDE, where x represents a spatial dimension and t represents time, then we may be able to prescribe the value of u, or $\partial u / \partial x$, or a linear combination of u and $\partial u / \partial x$, at a specified x as well as prescribe u and $\partial u / \partial t$ at a given t (usually, $t = 0$). In other words, a "boundary-value problem" may consist of a PDE along with boundary conditions *and* initial conditions.

CLASSICAL EQUATIONS We shall be concerned principally with applying the method of separation of variables to find product solutions of the following classical equations of mathematical physics

$$k\dfrac{\partial^2 u}{\partial x^2} = \dfrac{\partial u}{\partial t}, \quad k > 0 \tag{1}$$

$$a^2\dfrac{\partial^2 u}{\partial x^2} = \dfrac{\partial^2 u}{\partial t^2} \tag{2}$$

$$\dfrac{\partial^2 u}{\partial x^2} + \dfrac{\partial^2 u}{\partial y^2} = 0 \tag{3}$$

or slight variations of these equations. The PDEs (1), (2), and (3) are known, respectively, as the **one-dimensional heat equation**, the **one-dimensional wave equation**, and the **two-dimensional form of Laplace's equation**. "One-dimensional" in the case of equations (1) and (2) refers to the fact that x denotes a spatial variable, whereas t represents time; "two-dimensional" in (3) means that x and y are both spatial variables. If you compare (1)–(3) with the linear form in Theorem 12.1 (with t playing the part of the symbol y), observe that the heat equation (1) is parabolic, the wave equation (2) is hyperbolic, and Laplace's equation is elliptic. This observation will be important in Chapter 15.

HEAT EQUATION   Equation (1) occurs in the theory of heat flow—that is, heat transferred by conduction in a rod or in a thin wire. The function $u(x, t)$ represents temperature at a point x along the rod at some time t. Problems in mechanical vibrations often lead to the wave equation (2). For purposes of discussion, a solution $u(x, t)$ of (2) will represent the displacement of an idealized string. Finally, a solution $u(x, y)$ of Laplace's equation (3) can be interpreted as the steady-state (that is, time-independent) temperature distribution throughout a thin, two-dimensional plate.

Even though we have to make many simplifying assumptions, it is worthwhile to see how equations such as (1) and (2) arise.

Suppose a thin circular rod of length L has a cross-sectional area A and coincides with the x-axis on the interval $[0, L]$. See Figure 12.1. Let us suppose the following:

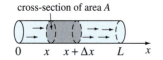

cross-section of area A

FIGURE 12.1   One-dimensional flow of heat

- The flow of heat within the rod takes place only in the x-direction.
- The lateral, or curved, surface of the rod is insulated; that is, no heat escapes from this surface.
- No heat is being generated within the rod.
- The rod is homogeneous; that is, its mass per unit volume ρ is a constant.
- The specific heat γ and thermal conductivity K of the material of the rod are constants.

To derive the partial differential equation satisfied by the temperature $u(x, t)$, we need two empirical laws of heat conduction:

(i)   *The quantity of heat Q in an element of mass m is*

$$Q = \gamma m u, \qquad (4)$$

where u is the temperature of the element.

(ii)   *The rate of heat flow Q_t through the cross-section indicated in Figure 12.1 is proportional to the area A of the cross section and the partial derivative with respect to x of the temperature:*

$$Q_t = -KAu_x. \qquad (5)$$

Since heat flows in the direction of decreasing temperature, the minus sign in (5) is used to ensure that Q_t is positive for $u_x < 0$ (heat flow to the right) and negative for $u_x > 0$ (heat flow to the left). If the circular slice of the rod shown in Figure 12.1 between x and $x + \Delta x$ is very thin, then $u(x, t)$ can be taken as the approximate temperature at each point in the interval. Now the mass of the slice is $m = \rho(A \, \Delta x)$, and so it follows from (4) that the quantity of heat in it is

$$Q = \gamma \rho A \, \Delta x \, u. \qquad (6)$$

Furthermore, when heat flows in the positive x-direction, we see from (5) that heat builds up in the slice at the net rate

$$-KAu_x(x, t) - [-KAu_x(x + \Delta x, t)] = KA \, [u_x(x + \Delta x, t) - u_x(x, t)]. \qquad (7)$$

By differentiating (6) with respect to t, we see that this net rate is also given by

$$Q_t = \gamma \rho A \, \Delta x \, u_t. \qquad (8)$$

Equating (7) and (8) gives

$$\frac{K}{\gamma \rho} \frac{u_x(x + \Delta x, t) - u_x(x, t)}{\Delta x} = u_t. \qquad (9)$$

Finally, by taking the limit of (9) as $\Delta x \to 0$, we obtain (1) in the form*
$(K/\gamma\rho)u_{xx} = u_t$. It is customary to let $k = K/\gamma\rho$ and call this positive constant the
thermal diffusivity.

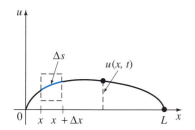

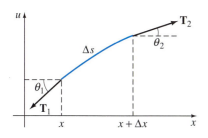

(a) segment of string

(b) enlargement of segment

FIGURE 12.2 Flexible string
anchored at $x = 0$ and $x = L$

WAVE EQUATION Consider a string of length L, such as a guitar string,
stretched taut between two points on the x-axis—say, $x = 0$ and $x = L$. When the
string starts to vibrate, assume that the motion takes place in the xu-plane in such
a manner that each point on the string moves in a direction perpendicular to the
x-axis (transverse vibrations). As is shown in Figure 12.2(a), let $u(x, t)$ denote the
vertical displacement of any point on the string measured from the x-axis for $t > 0$.
We further assume the following:

- The string is perfectly flexible.
- The string is homogeneous; that is, its mass per unit length ρ is a
  constant.
- The displacements u are small in comparison to the length of the string.
- The slope of the curve is small at all points.
- The tension $\mathbf{T}$ acts tangent to the string, and its magnitude T is the same at
  all points.
- The tension is large compared with the force of gravity.
- No other external forces act on the string.

Now in Figure 12.2(b) the tensions $\mathbf{T}_1$ and $\mathbf{T}_2$ are tangent to the ends of the
curve on the interval $[x, x + \Delta x]$. For small θ_1 and θ_2 the net vertical force acting
on the corresponding element Δs of the string is then

$$T \sin \theta_2 - T \sin \theta_1 \approx T \tan \theta_2 - T \tan \theta_1$$
$$= T[u_x(x + \Delta x, t) - u_x(x, t)],^{\dagger}$$

where $T = |\mathbf{T}_1| = |\mathbf{T}_2|$. Now $\rho \Delta s \approx \rho \Delta x$ is the mass of the string on $[x, x + \Delta x]$,
and so Newton's second law gives

$$T[u_x(x + \Delta x, t) - u_x(x, t)] = \rho \Delta x \, u_{tt}$$

or
$$\frac{u_x(x + \Delta x, t) - u_x(x, t)}{\Delta x} = \frac{\rho}{T} u_{tt}.$$

If the limit is taken as $\Delta x \to 0$, the last equation becomes $u_{xx} = (\rho/T)u_{tt}$. This of
course is (2) with $a^2 = T/\rho$.

LAPLACE'S EQUATION Although we shall not present its derivation,
Laplace's equation in two and three dimensions occurs in time-independent problems involving potentials such as electrostatic, gravitational, and velocity in fluid
mechanics. Moreover, a solution of Laplace's equation can also be interpreted as a
steady-state temperature distribution. As illustrated in Figure 12.3, a solution $u(x, y)$
of (3) could represent the temperature that varies from point to point—but not with
time—of a rectangular plate. Laplace's equation in two dimensions and in three
dimensions is abbreviated as $\nabla^2 u = 0$, where

$$\nabla^2 u = \frac{\partial^2 u}{\partial x^2} + \frac{\partial^2 u}{\partial y^2} \quad \text{and} \quad \nabla^2 u = \frac{\partial^2 u}{\partial x^2} + \frac{\partial^2 u}{\partial y^2} + \frac{\partial^2 u}{\partial z^2}$$

are called the two-dimensional **Laplacian** and the three-dimensional Laplacian,
respectively, of a function u.

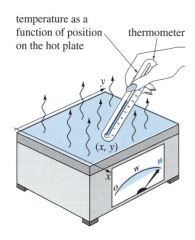

temperature as a
function of position
on the hot plate

thermometer

FIGURE 12.3 Steady-state
temperatures in a rectangular plate

---

*The definition of the second partial derivative is $u_{xx} = \lim\limits_{\Delta x \to 0} \dfrac{u_x(x + \Delta x, t) - u_x(x, t)}{\Delta x}$.

$^{\dagger}\tan \theta_2 = u_x(x + \Delta x, t)$ and $\tan \theta_1 = u_x(x, t)$ are equivalent expressions for slope.

We often wish to find solutions of equations (1), (2), and (3) that satisfy certain side conditions.

INITIAL CONDITIONS   Since solutions of (1) and (2) depend on time t, we can prescribe what happens at $t = 0$; that is, we can give **initial conditions (IC).** If $f(x)$ denotes the initial temperature distribution throughout the rod in Figure 12.1, then a solution $u(x, t)$ of (1) must satisfy the single initial condition $u(x, 0) = f(x)$, $0 < x < L$. On the other hand, for a vibrating string we can specify its initial displacement (or shape) $f(x)$ as well as its initial velocity $g(x)$. In mathematical terms we seek a function $u(x, t)$ satisfying (2) and the two initial conditions:

$$u(x, 0) = f(x), \quad \left.\frac{\partial u}{\partial t}\right|_{t=0} = g(x), \quad 0 < x < L. \tag{10}$$

For example, the string could be plucked, as shown in Figure 12.4, and released from rest ($g(x) = 0$).

BOUNDARY CONDITIONS   The string in Figure 12.4 is secured to the x-axis at $x = 0$ and $x = L$ for all time. We interpret this by the two **boundary conditions (BC):**

$$u(0, t) = 0, \quad u(L, t) = 0, \quad t > 0.$$

Note that in this context the function f in (10) is continuous, and consequently, $f(0) = 0$ and $f(L) = 0$. In general, there are three types of boundary conditions associated with equations (1), (2), and (3). On a boundary we can specify the values of *one* of the following:

$$(i) \quad u, \quad (ii) \quad \frac{\partial u}{\partial n}, \quad \text{or} \quad (iii) \quad \frac{\partial u}{\partial n} + hu, \quad h \text{ a constant.}$$

Here $\partial u/\partial n$ denotes the normal derivative of u (the directional derivative of u in the direction perpendicular to the boundary). A boundary condition of the first type (i) is called a **Dirichlet condition;** a boundary condition of the second type (ii) is called a **Neumann condition;** and a boundary condition of the third type (iii) is known as a **Robin condition.** For example, for $t > 0$ a typical condition at the right-hand end of the rod in Figure 12.1 can be

$$(i)' \quad u(L, t) = u_0, \quad u_0 \text{ a constant,}$$

$$(ii)' \quad \left.\frac{\partial u}{\partial x}\right|_{x=L} = 0, \quad \text{or}$$

$$(iii)' \quad \left.\frac{\partial u}{\partial x}\right|_{x=L} = -h(u(L, t) - u_m), \quad h > 0 \text{ and } u_m \text{ constants.}$$

Condition $(i)'$ simply states that the boundary $x = L$ is held by some means at a constant *temperature* u_0 for all time $t > 0$. Condition $(ii)'$ indicates that the boundary $x = L$ is *insulated.* From the empirical law of heat transfer, the flux of heat across a boundary (that is, the amount of heat per unit area per unit time conducted across the boundary) is proportional to the value of the normal derivative $\partial u/\partial n$ of the temperature u. Thus when the boundary $x = L$ is thermally insulated, no heat flows into or out of the rod, and so

$$\left.\frac{\partial u}{\partial x}\right|_{x=L} = 0.$$

We can interpret $(iii)'$ to mean that *heat is lost* from the right-hand end of the rod by being in contact with a medium, such as air or water, that is held at a constant temperature. From Newton's law of cooling, the outward flux of heat from the rod is proportional to the difference between the temperature $u(L, t)$ at the boundary

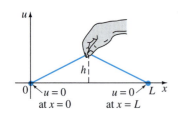

$u = 0$ at $x = 0$ $u = 0$ at $x = L$

FIGURE 12.4   Plucked string

and the temperature u_m of the surrounding medium. We note that if heat is lost from the left-hand end of the rod, the boundary condition is

$$\frac{\partial u}{\partial x}\bigg|_{x=0} = h(u(0, t) - u_m).$$

The change in algebraic sign is consistent with the assumption that the rod is at a higher temperature than the medium surrounding the ends so that $u(0, t) > u_m$ and $u(L, t) > u_m$. At $x = 0$ and $x = L$, the slopes $u_x(0, t)$ and $u_x(L, t)$ must be positive and negative, respectively.

Of course, at the ends of the rod we can specify different conditions at the same time. For example, we could have

$$\frac{\partial u}{\partial x}\bigg|_{x=0} = 0 \quad \text{and} \quad u(L, t) = u_0, \quad t > 0.$$

We note that the boundary condition in $(i)'$ is homogeneous if $u_0 = 0$; if $u_0 \neq 0$, the boundary condition is nonhomogeneous. The boundary condition $(ii)'$ is homogeneous; $(iii)'$ is homogeneous if $u_m = 0$ and nonhomogeneous if $u_m \neq 0$.

BOUNDARY-VALUE PROBLEMS   Problems such as

Solve: $\qquad a^2 \dfrac{\partial^2 u}{\partial x^2} = \dfrac{\partial^2 u}{\partial t^2}, \quad 0 < x < L, \quad t > 0$

Subject to: (BC) $\quad u(0, t) = 0, \quad u(L, t) = 0, \quad t > 0$ $\qquad\qquad$ (11)

$\qquad\qquad\qquad$ (IC) $\quad u(x, 0) = f(x), \quad \dfrac{\partial u}{\partial t}\bigg|_{t=0} = g(x), \quad 0 < x < L$

and

Solve: $\qquad \dfrac{\partial^2 u}{\partial x^2} + \dfrac{\partial^2 u}{\partial y^2} = 0, \quad 0 < x < a, \quad 0 < y < b$

$$ \qquad\qquad\qquad\qquad\qquad\qquad\qquad\qquad\qquad (12) $$

Subject to: (BC) $\begin{cases} \dfrac{\partial u}{\partial x}\bigg|_{x=0} = 0, \quad \dfrac{\partial u}{\partial x}\bigg|_{x=a} = 0, \qquad 0 < y < b \\[2mm] u(x, 0) = 0, \quad u(x, b) = f(x), \quad 0 < x < a \end{cases}$

are called **boundary-value problems.**

MODIFICATIONS   The partial differential equations (1), (2), and (3) must be modified to take into consideration internal or external influences acting on the physical system. More general forms of the one-dimensional heat and wave equations are, respectively,

$$k \frac{\partial^2 u}{\partial x^2} + G(x, t, u, u_x) = \frac{\partial u}{\partial t} \qquad\qquad (13)$$

and

$$a^2 \frac{\partial^2 u}{\partial x^2} + F(x, t, u, u_t) = \frac{\partial^2 u}{\partial t^2}. \qquad\qquad (14)$$

For example, if there is heat transfer from the lateral surface of a rod into a surrounding medium that is held at a constant temperature u_m, then the heat equation (13) is

$$k \frac{\partial^2 u}{\partial x^2} - h(u - u_m) = \frac{\partial u}{\partial t}.$$

In (14) the function F could represent the various forces acting on the string. For example, when external, damping, and elastic restoring forces are taken into account, (14) assumes the form

$$a^2 \frac{\partial^2 u}{\partial x^2} + \underset{\substack{\uparrow \\ \text{external} \\ \text{force}}}{f(x, t)} = \frac{\partial^2 u}{\partial t^2} + \underset{\substack{\uparrow \\ \text{damping} \\ \text{force}}}{c \frac{\partial u}{\partial t}} + \underset{\substack{\uparrow \\ \text{restoring} \\ \text{force}}}{k u}. \tag{15}$$

REMARKS

The analysis of a wide variety of diverse phenomena yields mathematical models (1), (2), or (3) or their generalizations involving a greater number of spatial variables. For example, (1) is sometimes called the **diffusion equation** since the diffusion of dissolved substances in solution is analogous to the flow of heat in a solid. The function $u(x, t)$ satisfying the partial differential equation in this case represents the concentration of the dissolved substance. Similarly, equation (2) arises in the study of the flow of electricity in a long cable or transmission line. In this setting (2) is known as the **telegraph equation.** It can be shown that under certain assumptions the current and the voltage in the line are functions satisfying two equations identical with (2). The wave equation (2) also appears in the theory of high-frequency transmission lines, fluid mechanics, acoustics, and elasticity. Laplace's equation (3) is encountered in the static displacement of membranes.

EXERCISES 12.2

Answers to selected odd-numbered problems begin on page ANS-17.

In Problems 1–4 a rod of length L coincides with the interval $[0, L]$ on the x-axis. Set up the boundary-value problem for the temperature $u(x, t)$.

1. The left end is held at temperature zero, and the right end is insulated. The initial temperature is $f(x)$ throughout.

2. The left end is held at temperature u_0, and the right end is held at temperature u_1. The initial temperature is zero throughout.

3. The left end is held at temperature 100, and there is heat transfer from the right end into the surrounding medium at temperature zero. The initial temperature is $f(x)$ throughout.

4. The ends are insulated, and there is heat transfer from the lateral surface into the surrounding medium at temperature 50. The initial temperature is 100 throughout.

In Problems 5–8 a string of length L coincides with the interval $[0, L]$ on the x-axis. Set up the boundary-value problem for the displacement $u(x, t)$.

5. The ends are secured to the x-axis. The string is released from rest from the initial displacement $x(L - x)$.

6. The ends are secured to the x-axis. Initially, the string is undisplaced but has the initial velocity $\sin(\pi x/L)$.

7. The left end is secured to the x-axis, but the right end moves in a transverse manner according to $\sin \pi t$. The string is released from rest from the initial displacement $f(x)$. For $t > 0$ the transverse vibrations are damped with a force proportional to the instantaneous velocity.

8. The ends are secured to the x-axis, and the string is initially at rest on that axis. An external vertical force proportional to the horizontal distance from the left end acts on the string for $t > 0$.

In Problems 9 and 10 set up the boundary-value problem for the steady-state temperature $u(x, y)$.

9. A thin rectangular plate coincides with the region defined by $0 \le x \le 4$, $0 \le y \le 2$. The left end and the bottom of the plate are insulated. The top of the plate is held at temperature zero, and the right end of the plate is held at temperature $f(y)$.

10. A semi-infinite plate coincides with the region defined by $0 \le x \le \pi$, $y \ge 0$. The left end is held at temperature e^{-y}, and the right end is held at temperature 100 for $0 < y \le 1$ and temperature zero for $y > 1$. The bottom of the plate is held at temperature $f(x)$.

12.3 HEAT EQUATION

INTRODUCTION: Consider a thin rod of length L with an initial temperature $f(x)$ throughout and whose ends are held at temperature zero for all time $t > 0$. If the rod shown in Figure 12.5 satisfies the assumptions given on page 470, then the temperature $u(x, t)$ in the rod is determined from the boundary-value problem

$$k \frac{\partial^2 u}{\partial x^2} = \frac{\partial u}{\partial t}, \quad 0 < x < L, \quad t > 0 \tag{1}$$

$$u(0, t) = 0, \quad u(L, t) = 0, \quad t > 0 \tag{2}$$

$$u(x, 0) = f(x), \quad 0 < x < L. \tag{3}$$

In this section we solve this BVP.

REVIEW MATERIAL: The solution method for the foregoing problem, as well as for all problems in the remainder of this chapter and in Chapter 13, is the method of separation of variables introduced in Section 12.1. In addition, a rereading of Example 2 in Section 5.2 and Example 1 of Section 11.4 is recommended.

FIGURE 12.5 Temperatures in a rod of length L

SOLUTION OF THE BVP To start, we use the product $u(x, t) = X(x)T(t)$ to separate variables in (1). Then, if $-\lambda$ is the separation constant, the two equalities

$$\frac{X''}{X} = \frac{T'}{kT} = -\lambda \tag{4}$$

lead to the two ordinary differential equations

$$X'' + \lambda X = 0 \tag{5}$$

$$T' + k\lambda T = 0. \tag{6}$$

Before solving (5), note that the boundary conditions (2) applied to $u(x, t) = X(x)T(t)$ are

$$u(0, t) = X(0)T(t) = 0 \quad \text{and} \quad u(L, t) = X(L)T(t) = 0.$$

Since it makes sense to expect that $T(t) \neq 0$ for all t, the foregoing equalities hold only if $X(0) = 0$ and $X(L) = 0$. These homogeneous boundary conditions together with the homogeneous DE (5) constitute a regular Sturm-Liouville problem:

$$X'' + \lambda X = 0, \quad X(0) = 0, \quad X(L) = 0. \tag{7}$$

The solution of this BVP was thoroughly discussed in Example 2 of Section 5.2. In that example we considered three possible cases for the parameter λ: zero, negative, or positive. The corresponding solutions of the DEs are, in turn, given by

$$X(x) = c_1 + c_2 x, \qquad \lambda = 0 \tag{8}$$

$$X(x) = c_1 \cosh \alpha x + c_2 \sinh \alpha x, \quad \lambda = -\alpha^2 < 0 \tag{9}$$

$$X(x) = c_1 \cos \alpha x + c_2 \sin \alpha x, \qquad \lambda = \alpha^2 > 0. \tag{10}$$

When the boundary conditions $X(0) = 0$ and $X(L) = 0$ are applied to (8) and (9), these solutions yield only $X(x) = 0$, and so we would have to conclude that $u = 0$. But when $X(0) = 0$ is applied to (10), we find that $c_1 = 0$ and $X(x) = c_2 \sin \alpha x$. The second boundary condition then implies that $X(L) = c_2 \sin \alpha L = 0$. To obtain a nontrivial solution, we must have $c_2 \neq 0$ and $\sin \alpha L = 0$. The last equation is satisfied when $\alpha L = n\pi$ or $\alpha = n\pi/L$. Hence (7) possesses nontrivial solutions when

$\lambda_n = \alpha_n^2 = n^2\pi^2/L^2$, $n = 1, 2, 3, \ldots$. These values of λ are the **eigenvalues** of the problem; the **eigenfunctions** are

$$X(x) = c_2 \sin \frac{n\pi}{L}x, \quad n = 1, 2, 3, \ldots . \tag{11}$$

From (6) we have $T(t) = c_3 e^{-k(n^2\pi^2/L^2)t}$, so

$$u_n = X(x)T(t) = A_n e^{-k(n^2\pi^2/L^2)t} \sin \frac{n\pi}{L}x, \tag{12}$$

where we have replaced the constant $c_2 c_3$ by A_n. Each of the product functions $u_n(x, t)$ given in (12) is a particular solution of the partial differential equation (1), and each $u_n(x, t)$ satisfies both boundary conditions (2) as well. However, for (12) to satisfy the initial condition (3), we would have to choose the coefficient A_n in such a manner that

$$u_n(x, 0) = f(x) = A_n \sin \frac{n\pi}{L}x. \tag{13}$$

In general, we would not expect condition (13) to be satisfied for an arbitrary, but reasonable, choice of f. Therefore we are forced to admit that $u_n(x, t)$ *is not a solution of the given problem*. Now by the superposition principle (Theorem 12.1) the function $u(x, t) = \sum_{n=1}^{\infty} u_n$ or

$$u(x, t) = \sum_{n=1}^{\infty} A_n e^{-k(n^2\pi^2/L^2)t} \sin \frac{n\pi}{L}x \tag{14}$$

must also, although formally, satisfy equation (1) and the conditions in (2). Substituting $t = 0$ into (14) implies that

$$u(x, 0) = f(x) = \sum_{n=1}^{\infty} A_n \sin \frac{n\pi}{L}x.$$

This last expression is recognized as a half-range expansion of f in a sine series. If we make the identification $A_n = b_n$, $n = 1, 2, 3, \ldots$, it follows from (5) of Section 11.3 that

$$A_n = \frac{2}{L} \int_0^L f(x) \sin \frac{n\pi}{L}x \, dx. \tag{15}$$

We conclude that a solution of the boundary-value problem described in (1), (2), and (3) is given by the infinite series

$$u(x, t) = \frac{2}{L} \sum_{n=1}^{\infty} \left(\int_0^L f(x) \sin \frac{n\pi}{L}x \, dx \right) e^{-k(n^2\pi^2/L^2)t} \sin \frac{n\pi}{L}x. \tag{16}$$

In the special case when the initial temperature is $u(x, 0) = 100$, $L = \pi$, and $k = 1$, you should verify that the coefficients (15) are given by

$$A_n = \frac{200}{\pi}\left[\frac{1 - (-1)^n}{n} \right],$$

and that (16) is

$$u(x, t) = \frac{200}{\pi} \sum_{n=1}^{\infty} \left[\frac{1 - (-1)^n}{n} \right] e^{-n^2 t} \sin nx. \tag{17}$$

USE OF COMPUTERS Since u is a function of two variables, the graph of the solution (17) is a surface in 3-space. We could use the 3D-plot application of a computer algebra system to approximate this surface by graphing partial sums $S_n(x, t)$ over a rectangular region defined by $0 \le x \le \pi, 0 \le t \le T$. Alternatively, with the aid of the 2D-plot application of a CAS we can plot the solution $u(x, t)$ on the x-interval $[0, \pi]$ for increasing values of time t. See Figure 12.6(a). In Figure 12.6(b)

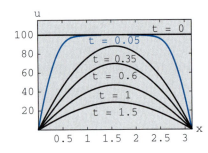

(a) $u(x, t)$ graphed as a function of x for various fixed times

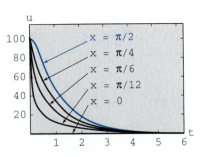

(b) $u(x, t)$ graphed as a function of t for various fixed positions

FIGURE 12.6 Graphs of (17) when one variable is held fixed the solution $u(x, t)$ is graphed on the t-interval $[0, 6]$ for increasing values of x ($x = 0$ is the left end and $x = \pi/2$ is the midpoint of the rod of length $L = \pi$.) Both sets of graphs verify what is apparent in (17)—namely, $u(x, t) \to 0$ as $t \to \infty$.

EXERCISES 12.3

Answers to selected odd-numbered problems begin on page ANS-18.

In Problems 1 and 2 solve the heat equation (1) subject to the given conditions. Assume a rod of length L.

1. $u(0, t) = 0, \quad u(L, t) = 0$

$$u(x, 0) = \begin{cases} 1, & 0 < x < L/2 \\ 0, & L/2 < x < L \end{cases}$$

2. $u(0, t) = 0, \quad u(L, t) = 0$
$u(x, 0) = x(L - x)$

3. Find the temperature $u(x, t)$ in a rod of length L if the initial temperature is $f(x)$ throughout and if the ends $x = 0$ and $x = L$ are insulated.

4. Solve Problem 3 if $L = 2$ and

$$f(x) = \begin{cases} x, & 0 < x < 1 \\ 0, & 1 < x < 2. \end{cases}$$

5. Suppose heat is lost from the lateral surface of a thin rod of length L into a surrounding medium at temperature zero. If the linear law of heat transfer applies, then the heat equation takes on the form $k\partial^2 u/\partial x^2 - hu = \partial u/\partial t$, $0 < x < L$, $t > 0$, h a constant. Find the temperature $u(x, t)$ if the initial temperature is $f(x)$ throughout and the ends $x = 0$ and $x = L$ are insulated. See Figure 12.7.

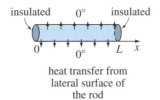

FIGURE 12.7 Rod losing heat in Problem 5

6. Solve Problem 5 if the ends $x = 0$ and $x = L$ are held at temperature zero.

DISCUSSION/PROJECT PROBLEMS

7. Figure 12.6(b) shows the graphs of $u(x, t)$ on the interval $0 \le t \le 6$ for $x = 0$, $x = \pi/12$, $x = \pi/6$, $x = \pi/4$, and $x = \pi/2$. Describe or sketch the graphs of $u(x, t)$ on the same time interval but for the fixed values $x = 3\pi/4$, $x = 5\pi/6$, $x = 11\pi/12$, and $x = \pi$.

COMPUTER LAB ASSIGNMENTS

8. (a) Solve the heat equation (1) subject to

$$u(0, t) = 0, \quad u(100, t) = 0, \quad t > 0$$

$$u(x, 0) = \begin{cases} 0.8x, & 0 \le x \le 50 \\ 0.8(100 - x), & 50 < x \le 100. \end{cases}$$

(b) Use the 3D-plot application of your CAS to graph the partial sum $S_5(x, t)$ consisting of the first five nonzero terms of the solution in part (a) for $0 \le x \le 100$, $0 \le t \le 200$. Assume that $k = 1.6352$. Experiment with various three-dimensional viewing perspectives of the surface (called the **ViewPoint** option in *Mathematica*).

12.4   WAVE EQUATION

INTRODUCTION: We are now in a position to solve the boundary-value problem (11) discussed in Section 12.2. The vertical displacement $u(x, t)$ of the vibrating string of length L shown in Figure 12.2(a) is determined from

$$a^2 \frac{\partial^2 u}{\partial x^2} = \frac{\partial^2 u}{\partial t^2}, \quad 0 < x < L, \quad t > 0 \tag{1}$$

$$u(0, t) = 0, \quad u(L, t) = 0, \quad t > 0 \tag{2}$$

$$u(x, 0) = f(x), \quad \frac{\partial u}{\partial t}\bigg|_{t=0} = g(x), \quad 0 < x < L. \tag{3}$$

REVIEW MATERIAL: Reread pages 471–473 of Section 12.2.

SOLUTION OF THE BVP With the usual assumption that $u(x, t) = X(x)T(t)$, separating variables in (1) gives

$$\frac{X''}{X} = \frac{T''}{a^2 T} = -\lambda$$

so that

$$X'' + \lambda X = 0 \tag{4}$$

$$T'' + a^2 \lambda T = 0. \tag{5}$$

As in the preceding section, the boundary conditions (2) translate into $X(0) = 0$ and $X(L) = 0$. Equation (4) along with these boundary conditions is the regular Sturm-Liouville problem

$$X'' + \lambda X = 0, \quad X(0) = 0, \quad X(L) = 0. \tag{6}$$

Of the usual three possibilities for the parameter, $\lambda = 0$, $\lambda = -\alpha^2 < 0$, and $\lambda = \alpha^2 > 0$, only the last choice leads to nontrivial solutions. Corresponding to $\lambda = \alpha^2$, $\alpha > 0$, the general solution of (4) is

$$X = c_1 \cos \alpha x + c_2 \sin \alpha x.$$

$X(0) = 0$ and $X(L) = 0$ indicate that $c_1 = 0$ and $c_2 \sin \alpha L = 0$. The last equation again implies that $\alpha L = n\pi$ or $\alpha = n\pi/L$. The eigenvalues and corresponding eigenfunctions of (6) are $\lambda_n = n^2\pi^2/L^2$ and $X(x) = c_2 \sin \frac{n\pi}{L} x$, $n = 1, 2, 3, \ldots$. The general solution of the second-order equation (5) is then

$$T(t) = c_3 \cos \frac{n\pi a}{L} t + c_4 \sin \frac{n\pi a}{L} t.$$

By rewriting $c_2 c_3$ as A_n and $c_2 c_4$ as B_n, solutions that satisfy both the wave equation (1) and boundary conditions (2) are

$$u_n = \left(A_n \cos \frac{n\pi a}{L} t + B_n \sin \frac{n\pi a}{L} t \right) \sin \frac{n\pi}{L} x \tag{7}$$

and

$$u(x, t) = \sum_{n=1}^{\infty} \left(A_n \cos \frac{n\pi a}{L} t + B_n \sin \frac{n\pi a}{L} t \right) \sin \frac{n\pi}{L} x. \tag{8}$$

Setting $t = 0$ in (8) and using the initial condition $u(x, 0) = f(x)$ gives

$$u(x, 0) = f(x) = \sum_{n=1}^{\infty} A_n \sin \frac{n\pi}{L} x.$$

Since the last series is a half-range expansion for f in a sine series, we can write $A_n = b_n$:

$$A_n = \frac{2}{L} \int_0^L f(x) \sin \frac{n\pi}{L} x \, dx. \tag{9}$$

To determine B_n, we differentiate (8) with respect to t and then set $t = 0$:

$$\frac{\partial u}{\partial t} = \sum_{n=1}^{\infty} \left(-A_n \frac{n\pi a}{L} \sin \frac{n\pi a}{L} t + B_n \frac{n\pi a}{L} \cos \frac{n\pi a}{L} t \right) \sin \frac{n\pi}{L} x$$

$$\left. \frac{\partial u}{\partial t} \right|_{t=0} = g(x) = \sum_{n=1}^{\infty} \left(B_n \frac{n\pi a}{L} \right) \sin \frac{n\pi}{L} x.$$

For this last series to be the half-range sine expansion of the initial velocity g on the interval, the *total* coefficient $B_n n\pi a/L$ must be given by the form b_n in (5) of Section 11.3, that is,

$$B_n \frac{n\pi a}{L} = \frac{2}{L} \int_0^L g(x) \sin \frac{n\pi}{L} x \, dx$$

from which we obtain

$$B_n = \frac{2}{n\pi a} \int_0^L g(x) \sin \frac{n\pi}{L} x \, dx. \tag{10}$$

The solution of the boundary-value problem (1)–(3) consists of the series (8) with coefficients A_n and B_n defined by (9) and (10), respectively.

We note that when the string is released from *rest*, then $g(x) = 0$ for every x in the interval $0 \le x \le L$, and consequently, $B_n = 0$.

PLUCKED STRING A special case of the boundary-value problem in (1)–(3) is the model of the **plucked string.** We can see the motion of the string by plotting the solution or displacement $u(x, t)$ for increasing values of time t and using the animation feature of a CAS. Some frames of a "movie" generated in this manner are given in Figure 12.8; the initial shape of the string is given in Figure 12.8(a). You are asked to emulate the results given in the figure plotting a sequence of partial sums of (8). See Problems 7 and 22 in Exercises 12.4.

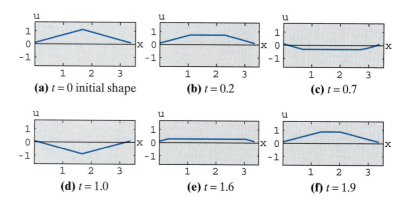

(a) $t = 0$ initial shape   **(b)** $t = 0.2$ **(c)** $t = 0.7$

(d) $t = 1.0$ **(e)** $t = 1.6$ **(f)** $t = 1.9$

FIGURE 12.8 Frames of a CAS "movie"

STANDING WAVES Recall from the derivation of the one-dimensional wave equation in Section 12.2 that the constant a appearing in the solution of the boundary-value problem in (1), (2), and (3) is given by $\sqrt{T/\rho}$, where ρ is mass per unit length and T is the magnitude of the tension in the string. When T is large enough, the vibrating string produces a musical sound. This sound is the result of standing waves. The solution (8) is a superposition of product solutions called **standing waves** or **normal modes:**

$$u(x, t) = u_1(x, t) + u_2(x, t) + u_3(x, t) + \cdots.$$

In view of (6) and (7) of Section 5.1, the product solutions (7) can be written as

$$u_n(x, t) = C_n \sin \left(\frac{n\pi a}{L} t + \phi_n \right) \sin \frac{n\pi}{L} x, \tag{11}$$

(a) first standing wave

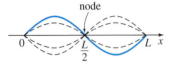

(b) second standing wave

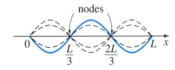

(c) third standing wave

FIGURE 12.9 First three standing waves

where $C_n = \sqrt{A_n^2 + B_n^2}$ and ϕ_n is defined by $\sin \phi_n = A_n/C_n$ and $\cos \phi_n = B_n/C_n$. For $n = 1, 2, 3, \ldots$ the standing waves are essentially the graphs of $\sin(n\pi x/L)$, with a time-varying amplitude given by

$$C_n \sin\left(\frac{n\pi a}{L}t + \phi_n\right).$$

Alternatively, we see from (11) that at a fixed value of x each product function $u_n(x, t)$ represents simple harmonic motion with amplitude $C_n|\sin(n\pi x/L)|$ and frequency $f_n = na/2L$. In other words, each point on a standing wave vibrates with a different amplitude but with the same frequency. When $n = 1$,

$$u_1(x, t) = C_1 \sin\left(\frac{\pi a}{L}t + \phi_1\right) \sin\frac{\pi}{L}x$$

is called the **first standing wave**, the **first normal mode**, or the **fundamental mode of vibration.** The first three standing waves, or normal modes, are shown in Figure 12.9. The dashed graphs represent the standing waves at various values of time. The points in the interval $(0, L)$, for which $\sin(n\pi/L)x = 0$, correspond to points on a standing wave where there is no motion. These points are called **nodes.** For example, in Figures 12.9(b) and 12.9(c) we see that the second standing wave has one node at $L/2$ and the third standing wave has two nodes at $L/3$ and $2L/3$. In general, the nth normal mode of vibration has $n - 1$ nodes.

The frequency

$$f_1 = \frac{a}{2L} = \frac{1}{2L}\sqrt{\frac{T}{\rho}}$$

of the first normal mode is called the **fundamental frequency** or **first harmonic** and is directly related to the pitch produced by a stringed instrument. It is apparent that the greater the tension on the string, the higher the pitch of the sound. The frequencies f_n of the other normal modes, which are integer multiples of the fundamental frequency, are called **overtones.** The second harmonic is the first overtone, and so on.

EXERCISES 12.4

Answers to selected odd-numbered problems begin on page ANS-18.

In Problems 1–8 solve the wave equation (1) subject to the given conditions.

1. $u(0, t) = 0, \quad u(L, t) = 0$

$u(x, 0) = \dfrac{1}{4}x(L - x), \quad \dfrac{\partial u}{\partial t}\Big|_{t=0} = 0$

2. $u(0, t) = 0, \quad u(L, t) = 0$

$u(x, 0) = 0, \quad \dfrac{\partial u}{\partial t}\Big|_{t=0} = x(L - x)$

3. $u(0, t) = 0, \quad u(L, t) = 0$

$u(x, 0)$, given in Figure 12.10, $\dfrac{\partial u}{\partial t}\Big|_{t=0} = 0$

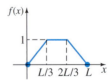

FIGURE 12.10 Initial displacement in Problem 3

4. $u(0, t) = 0, \quad u(\pi, t) = 0$

$u(x, 0) = \frac{1}{6}x(\pi^2 - x^2), \quad \dfrac{\partial u}{\partial t}\Big|_{t=0} = 0$

5. $u(0, t) = 0, \quad u(\pi, t) = 0$

$u(x, 0) = 0, \quad \dfrac{\partial u}{\partial t}\Big|_{t=0} = \sin x$

6. $u(0, t) = 0$, $u(1, t) = 0$

$$u(x, 0) = 0.01 \sin 3\pi x, \quad \left.\frac{\partial u}{\partial t}\right|_{t=0} = 0$$

7. $u(0, t) = 0$, $u(L, t) = 0$

$$u(x, 0) = \begin{cases} \dfrac{2hx}{L}, & 0 < x < \dfrac{L}{2} \\ 2h\left(1 - \dfrac{x}{L}\right), & \dfrac{L}{2} \le x < L \end{cases}, \quad \left.\frac{\partial u}{\partial t}\right|_{t=0} = 0$$

8. $\left.\dfrac{\partial u}{\partial x}\right|_{x=0} = 0$, $\left.\dfrac{\partial u}{\partial x}\right|_{x=L} = 0$

$$u(x, 0) = x, \quad \left.\frac{\partial u}{\partial t}\right|_{t=0} = 0$$

This problem could describe the longitudinal displacement $u(x, t)$ of a vibrating elastic bar. The boundary conditions at $x = 0$ and $x = L$ are called **free-end conditions.** See Figure 12.11.

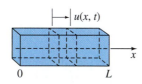

FIGURE 12.11 Vibrating elastic bar in Problem 8

9. A string is stretched and secured on the x-axis at $x = 0$ and $x = \pi$ for $t > 0$. If the transverse vibrations take place in a medium that imparts a resistance proportional to the instantaneous velocity, then the wave equation takes on the form

$$\frac{\partial^2 u}{\partial x^2} = \frac{\partial^2 u}{\partial t^2} + 2\beta \frac{\partial u}{\partial t}, \quad 0 < \beta < 1, \quad t > 0.$$

Find the displacement $u(x, t)$ if the string starts from rest from the initial displacement $f(x)$.

10. Show that a solution of the boundary-value problem

$$\frac{\partial^2 u}{\partial x^2} = \frac{\partial^2 u}{\partial t^2} + u, \quad 0 < x < \pi, \quad t > 0$$

$$u(0, t) = 0, \quad u(\pi, t) = 0, \quad t > 0$$

$$u(x, 0) = \begin{cases} x, & 0 < x < \pi/2 \\ \pi - x, & \pi/2 \le x < \pi \end{cases}$$

$$\left.\frac{\partial u}{\partial t}\right|_{t=0} = 0, \quad 0 < x < \pi$$

is

$$u(x, t) = \frac{4}{\pi} \sum_{k=1}^{\infty} \frac{(-1)^{k+1}}{(2k-1)^2} \sin(2k-1)x \cos \sqrt{(2k-1)^2 + 1}\, t.$$

11. The transverse displacement $u(x, t)$ of a vibrating beam of length L is determined from a fourth-order partial differential equation

$$a^2 \frac{\partial^4 u}{\partial x^4} + \frac{\partial^2 u}{\partial t^2} = 0, \quad 0 < x < L, \quad t > 0.$$

If the beam is **simply supported,** as shown in Figure 12.12, the boundary and initial conditions are

$$u(0, t) = 0, \quad u(L, t) = 0, \quad t > 0$$

$$\left.\frac{\partial^2 u}{\partial x^2}\right|_{x=0} = 0, \quad \left.\frac{\partial^2 u}{\partial x^2}\right|_{x=L} = 0, \quad t > 0$$

$$u(x, 0) = f(x), \quad \left.\frac{\partial u}{\partial t}\right|_{t=0} = g(x), \quad 0 < x < L.$$

Solve for $u(x, t)$. (*Hint:* For convenience use $\lambda = \alpha^4$ when separating variables.)

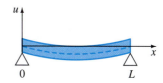

FIGURE 12.12 Simply supported beam in Problem 11

12. If the ends of the beam in Problem 11 are **embedded** at $x = 0$ and $x = L$, the boundary conditions become, for $t > 0$,

$$u(0, t) = 0, \quad u(L, t) = 0$$

$$\left.\frac{\partial u}{\partial x}\right|_{x=0} = 0, \quad \left.\frac{\partial u}{\partial x}\right|_{x=L} = 0.$$

(a) Show that the eigenvalues of the problem are $\lambda_n = x_n^2/L^2$, where x_n, $n = 1, 2, 3, \ldots$, are the positive roots of the equation

$$\cosh x \cos x = 1.$$

(b) Show graphically that the equation in part (a) has an infinite number of roots.

(c) Use a calculator or a computer to find approximations to the first four eigenvalues. Use four decimal places.

13. Consider the boundary-value problem given in (1), (2), and (3) of this section. If $g(x) = 0$ on $0 < x < L$, show that the solution of the problem can be written as

$$u(x, t) = \frac{1}{2}[f(x + at) + f(x - at)].$$

(*Hint:* Use the identity
$$2 \sin \theta_1 \cos \theta_2 = \sin(\theta_1 + \theta_2) + \sin(\theta_1 - \theta_2).)$$

14. The vertical displacement $u(x, t)$ of an infinitely long string is determined from the initial-value problem

$$a^2 \frac{\partial^2 u}{\partial x^2} = \frac{\partial^2 u}{\partial t^2}, \quad -\infty < x < \infty, \quad t > 0$$

$$u(x, 0) = f(x), \quad \frac{\partial u}{\partial t}\bigg|_{t=0} = g(x). \tag{12}$$

This problem can be solved without separating variables.
(a) Show that the wave equation can be put into the form $\partial^2 u / \partial \eta \partial \xi = 0$ by means of the substitutions $\xi = x + at$ and $\eta = x - at$.
(b) Integrate the partial differential equation in part (a), first with respect to η and then with respect to ξ, to show that $u(x, t) = F(x + at) + G(x - at)$, where F and G are arbitrary twice differentiable functions, is a solution of the wave equation. Use this solution and the given initial conditions to show that

$$F(x) = \frac{1}{2}f(x) + \frac{1}{2a}\int_{x_0}^{x} g(s)\,ds + c$$

and $$G(x) = \frac{1}{2}f(x) - \frac{1}{2a}\int_{x_0}^{x} g(s)\,ds - c,$$

where x_0 is arbitrary and c is a constant of integration.
(c) Use the results in part (b) to show that

$$u(x, t) = \frac{1}{2}[f(x + at) + f(x - at)] + \frac{1}{2a}\int_{x-at}^{x+at} g(s)\,ds. \tag{13}$$

Note that when the initial velocity $g(x) = 0$, we obtain

$$u(x, t) = \frac{1}{2}[f(x + at) + f(x - at)], \quad -\infty < x < \infty.$$

This last solution can be interpreted as a super-position of two **traveling waves,** one moving to the right (that is, $\frac{1}{2}f(x - at)$) and one moving to the left $\left(\frac{1}{2}f(x + at)\right)$. Both waves travel with speed a and have the same basic shape as the initial displacement $f(x)$. The form of $u(x, t)$ given in (13) is called **d'Alembert's solution.**

In Problems 15–18 use d'Alembert's solution (13) to solve the initial-value problem in Problem 14 subject to the given initial conditions.

15. $f(x) = \sin x, \quad g(x) = 1$

16. $f(x) = \sin x, \quad g(x) = \cos x$

17. $f(x) = 0, \quad g(x) = \sin 2x$

18. $f(x) = e^{-x^2}, \quad g(x) = 0$

COMPUTER LAB ASSIGNMENTS

19. (a) Use a CAS to plot d'Alembert's solution in Problem 18 on the interval $[-5, 5]$ at the times $t = 0$, $t = 1$, $t = 2$, $t = 3$, and $t = 4$. Superimpose the graphs on one coordinate system. Assume that $a = 1$.
   (b) Use the 3D-plot application of your CAS to plot d'Alembert's solution $u(x, t)$ in Problem 18 for $-5 \le x \le 5$, $0 \le t \le 4$. Experiment with various three-dimensional viewing perspectives of this surface. Choose the perspective of the surface for which you feel the graphs in part (a) are most apparent.

20. A model for an infinitely long string that is initially held at the three points $(-1, 0)$, $(1, 0)$, and $(0, 1)$ and then simultaneously released at all three points at time $t = 0$ is given by (12) with

$$f(x) = \begin{cases} 1 - |x|, & |x| \le 1 \\ 0, & |x| > 1 \end{cases} \quad \text{and} \quad g(x) = 0.$$

   (a) Plot the initial position of the string on the interval $[-6, 6]$.
   (b) Use a CAS to plot d'Alembert's solution (13) on $[-6, 6]$ for $t = 0.2k$, $k = 0, 1, 2, \ldots, 25$. Assume that $a = 1$.
   (c) Use the animation feature of your computer algebra system to make a movie of the solution. Describe the motion of the string over time.

21. An infinitely long string coinciding with the x-axis is struck at the origin with a hammer whose head is 0.2 inch in diameter. A model for the motion of the string is given by (12) with

$$f(x) = 0 \quad \text{and} \quad g(x) = \begin{cases} 1, & |x| \le 0.1 \\ 0, & |x| \ge 0.1. \end{cases}$$

   (a) Use a CAS to plot d'Alembert's solution (13) on $[-6, 6]$ for $t = 0.2k$, $k = 0, 1, 2, \ldots, 25$. Assume that $a = 1$.
   (b) Use the animation feature of your computer algebra system to make a movie of the solution. Describe the motion of the string over time.

22. The model of the vibrating string in Problem 7 is called the **plucked string.** The string is tied to the x-axis at $x = 0$ and $x = L$ and is held at $x = L/2$ at h units above the x-axis. See Figure 12.4. Starting at $t = 0$ the string is released from rest.
   (a) Use a CAS to plot the partial sum $S_6(x, t)$—that is, the first six nonzero terms of your solution—for $t = 0.1k$, $k = 0, 1, 2, \ldots, 20$. Assume that $a = 1$, $h = 1$, and $L = \pi$.
   (b) Use the animation feature of your computer algebra system to make a movie of the solution to Problem 7.

12.5 LAPLACE'S EQUATION

INTRODUCTION: Suppose we wish to find the steady-state temperature $u(x, y)$ in a rectangular plate whose vertical edges $x = 0$ and $x = a$ are insulated, as shown in Figure 12.13. When no heat escapes from the lateral faces of the plate, we solve the following boundary-value problem:

$$\frac{\partial^2 u}{\partial x^2} + \frac{\partial^2 u}{\partial y^2} = 0, \quad 0 < x < a, \quad 0 < y < b \tag{1}$$

$$\left.\frac{\partial u}{\partial x}\right|_{x=0} = 0, \quad \left.\frac{\partial u}{\partial x}\right|_{x=a} = 0, \quad 0 < y < b \tag{2}$$

$$u(x, 0) = 0, \quad u(x, b) = f(x), \quad 0 < x < a. \tag{3}$$

REVIEW MATERIAL: Reread page 471 of Section 12.2 and Example 1 in Section 11.4.

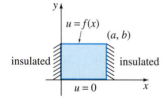

FIGURE 12.13 Steady-state temperatures in a rectangular plate

SOLUTION OF THE BVP With $u(x, y) = X(x)Y(y)$ separation of variables in (1) leads to

$$\frac{X''}{X} = -\frac{Y''}{Y} = -\lambda$$

$$X'' + \lambda X = 0 \tag{4}$$

$$Y'' - \lambda Y = 0. \tag{5}$$

The three homogeneous boundary conditions in (2) and (3) translate into $X'(0) = 0$, $X'(a) = 0$, and $Y(0) = 0$. The Sturm-Liouville problem associated with the equation in (4) is then

$$X'' + \lambda X = 0, \quad X'(0) = 0, \quad X'(a) = 0. \tag{6}$$

Examination of the cases corresponding to $\lambda = 0$, $\lambda = -\alpha^2 < 0$, and $\lambda = \alpha^2 > 0$, where $\alpha > 0$, has already been carried out in Example 1 in Section 11.4.* Here is a brief summary of that analysis.

For $\lambda = 0$, (6) becomes

$$X'' = 0, \quad X'(0) = 0, \quad X'(a) = 0.$$

The solution of the DE is $X = c_1 + c_2 x$. The boundary conditions imply $X = c_1$. By imposing $c_1 \neq 0$, this problem possesses a nontrivial solution. For $\lambda = -\alpha^2 < 0$, (6) possesses only the trivial solution. For $\lambda = \alpha^2 > 0$, (6) becomes

$$X'' + \alpha^2 X = 0, \quad X'(0) = 0, \quad X'(a) = 0.$$

The solution of the DE in this problem is $X = c_1 \cos \alpha x + c_2 \sin \alpha x$. The boundary condition $X'(0) = 0$ implies that $c_2 = 0$, so $X = c_1 \cos \alpha x$. Differentiating this last expression and then setting $x = a$ gives $-c_1 \sin \alpha a = 0$. Since we have assumed that $\alpha > 0$, this last condition is satisfied when $\alpha a = n\pi$ or $\alpha = n\pi/a$, $n = 1, 2, \ldots$. The eigenvalues of (6) are then $\lambda_0 = 0$ and $\lambda_n = \alpha_n^2 = n^2\pi^2/a^2$, $n = 1, 2, \ldots$. If we correspond $\lambda_0 = 0$ with $n = 0$, the eigenfunctions of (6) are

$$X = c_1, \quad n = 0, \quad \text{and} \quad X = c_1 \cos \frac{n\pi}{a} x, \quad n = 1, 2, \ldots.$$

*In that example the symbols y and L play the part of X and a in the current discussion.

We now solve equation (5) subject to the single homogeneous boundary condition $Y(0) = 0$. There are two cases. For $\lambda_0 = 0$, equation (5) is simply $Y'' = 0$; therefore its solution is $Y = c_3 + c_4 y$. But $Y(0) = 0$ implies that $c_3 = 0$, so $Y = c_4 y$. For $\lambda_n = n^2 \pi^2 / a^2$, (5) is $Y'' - \dfrac{n^2 \pi^2}{a^2} Y = 0$. Because $0 < y < b$ is a finite interval, we use (according to the informal rule indicated on page 448) the hyperbolic form of the general solution:

$$Y = c_3 \cosh (n\pi y/a) + c_4 \sinh (n\pi y/a).$$

$Y(0) = 0$ again implies that $c_3 = 0$, so we are left with $Y = c_4 \sinh (n\pi y/a)$.

Thus product solutions $u_n = X(x)Y(y)$ that satisfy the Laplace's equation (1) and the three homogeneous boundary conditions in (2) and (3) are

$$A_0 y, \quad n = 0, \quad \text{and} \quad A_n \sinh \frac{n\pi}{a} y \cos \frac{n\pi}{a} x, \quad n = 1, 2, \ldots,$$

where we have rewritten $c_1 c_4$ as A_0 for $n = 0$ and as A_n for $n = 1, 2, \ldots$.

The superposition principle yields another solution:

$$u(x, y) = A_0 y + \sum_{n=1}^{\infty} A_n \sinh \frac{n\pi}{a} y \cos \frac{n\pi}{a} x. \tag{7}$$

We are now in a position to use the last boundary condition in (3). Substituting $x = b$ in (7) gives

$$u(x, b) = f(x) = A_0 b + \sum_{n=1}^{\infty} \left(A_n \sinh \frac{n\pi}{a} b \right) \cos \frac{n\pi}{a} x,$$

which is a half-range expansion of f in a cosine series. If we make the identifications $A_0 b = a_0/2$ and $A_n \sinh (n\pi b/a) = a_n$, $n = 1, 2, 3, \ldots$, it follows from (2) and (3) of Section 11.3 that

$$2A_0 b = \frac{2}{a} \int_0^a f(x)\, dx$$

$$A_0 = \frac{1}{ab} \int_0^a f(x)\, dx \tag{8}$$

and

$$A_n \sinh \frac{n\pi}{a} b = \frac{2}{a} \int_0^a f(x) \cos \frac{n\pi}{a} x\, dx$$

$$A_n = \frac{2}{a \sinh \dfrac{n\pi}{a} b} \int_0^a f(x) \cos \frac{n\pi}{a} x\, dx. \tag{9}$$

The solution of the boundary-value problem (1)–(3) consists of the series in (7), with coefficients A_0 and A_n defined in (8) and (9), respectively.

DIRICHLET PROBLEM A boundary-value problem in which we seek a solution of an elliptic partial differential equation such as Laplace's equation $\nabla^2 u = 0$, within a bounded region R (in the plane or 3-space) such that u takes on prescribed values on the entire boundary of the region is called a **Dirichlet problem.** In Problem 1 in Exercises 12.5 you are asked to show that the solution of the Dirichlet problem for a rectangular region

$$\frac{\partial^2 u}{\partial x^2} + \frac{\partial^2 u}{\partial y^2} = 0, \quad 0 < x < a, \quad 0 < y < b$$

$$u(0, y) = 0, \quad u(a, y) = 0, \quad 0 < y < b$$

$$u(x, 0) = 0, \quad u(x, b) = f(x), \quad 0 < x < a$$

is

$$u(x, y) = \sum_{n=1}^{\infty} A_n \sinh \frac{n\pi}{a} y \sin \frac{n\pi}{a} x, \quad \text{where} \quad A_n = \frac{2}{a \sinh \frac{n\pi}{a} b} \int_0^a f(x) \sin \frac{n\pi}{a} x \, dx. \quad (10)$$

In the special case when $f(x) = 100$, $a = 1$, $b = 1$, the coefficients A_n in (10) are given by $A_n = 200 \dfrac{1 - (-1)^n}{n\pi \sinh n\pi}$. With the help of a CAS we plotted the surface defined by $u(x, y)$ over the region R: $0 \leq x \leq 1$, $0 \leq y \leq 1$, in Figure 12.14(a). You can see in the figure that the boundary conditions are satisfied; especially note that along $y = 1$, $u = 100$ for $0 \leq x \leq 1$. The **isotherms,** or curves in the rectangular region along which the temperature $u(x, y)$ is constant, can be obtained by using the contour plotting capabilities of a CAS and are illustrated in Figure 12.14(b). The isotherms can also be visualized as the curves of intersection (projected into the xy-plane) of horizontal planes $u = 80$, $u = 60$, and so on, with the surface in Figure 12.14(a). Notice that throughout the region the maximum temperature is $u = 100$ and occurs on the portion of the boundary corresponding to $y = 1$. This is no coincidence. There is a **maximum principle** that states a solution u of Laplace's equation within a bounded region R with boundary B (such as a rectangle, circle, sphere, and so on) takes on its maximum and minimum values on B. In addition, it can be proved that u can have no relative extrema (maxima or minima) in the interior of R. This last statement is clearly borne out by the surface shown in Figure 12.14(a).

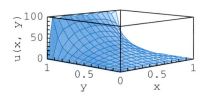

(a) surface

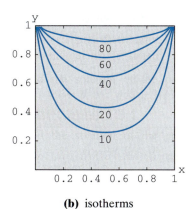

(b) isotherms

FIGURE 12.14 Surface is graph of partial sums when $f(x) = 100$ and $a = b = 1$ in (10)

SUPERPOSITION PRINCIPLE A Dirichlet problem for a rectangle can be readily solved by separation of variables when homogeneous boundary conditions are specified on two *parallel* boundaries. However, the method of separation of variables is not applicable to a Dirichlet problem when the boundary conditions on all four sides of the rectangle are nonhomogeneous. To get around this difficulty, we break the problem

$$\frac{\partial^2 u}{\partial x^2} + \frac{\partial^2 u}{\partial y^2} = 0, \quad 0 < x < a, \quad 0 < y < b$$

$$u(0, y) = F(y), \quad u(a, y) = G(y), \quad 0 < y < b \quad (11)$$

$$u(x, 0) = f(x), \quad u(x, b) = g(x), \quad 0 < x < a$$

into two problems, each of which has homogeneous boundary conditions on parallel boundaries, as shown:

Problem 1	Problem 2
$\dfrac{\partial^2 u_1}{\partial x^2} + \dfrac{\partial^2 u_1}{\partial y^2} = 0$, $0 < x < a$, $\quad 0 < y < b$	$\dfrac{\partial^2 u_2}{\partial x^2} + \dfrac{\partial^2 u_2}{\partial y^2} = 0$, $0 < x < a$, $\quad 0 < y < b$
$u_1(0, y) = 0$, $\quad u_1(a, y) = 0$, $\quad 0 < y < b$	$u_2(0, y) = F(y)$, $\quad u_2(a, y) = G(y)$, $0 < y < b$
$u_1(x, 0) = f(x)$, $\quad u_1(x, b) = g(x)$, $0 < x < a$	$u_2(x, 0) = 0$, $\quad u_2(x, b) = 0$, $\quad 0 < x < a$

Suppose u_1 and u_2 are the solutions of Problems 1 and 2, respectively. If we define $u(x, y) = u_1(x, y) + u_2(x, y)$, it is seen that u satisfies all boundary conditions in the original problem (11). For example,

$$u(0, y) = u_1(0, y) + u_2(0, y) = 0 + F(y) = F(y)$$

$$u(x, b) = u_1(x, b) + u_2(x, b) = g(x) + 0 = g(x)$$

and so on. Furthermore, u is a solution of Laplace's equation by Theorem 12.1. In other words, by solving Problems 1 and 2 and adding their solutions, we have solved the original problem. This additive property of solutions is known as the superposition principle. See Figure 12.15.

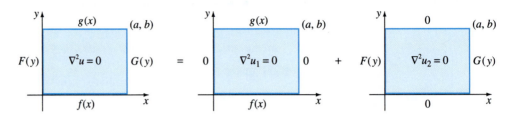

FIGURE 12.15   Solution $u =$ Solution u_1 of Problem 1 + Solution u_2 of Problem 2

We leave as exercises (see Problems 13 and 14 in Exercises 12.5) to show that a solution of Problem 1 is

$$u_1(x, y) = \sum_{n=1}^{\infty} \left\{ A_n \cosh \frac{n\pi}{a} y + B_n \sinh \frac{n\pi}{a} y \right\} \sin \frac{n\pi}{a} x,$$

where

$$A_n = \frac{2}{a} \int_0^a f(x) \sin \frac{n\pi}{a} x \, dx$$

$$B_n = \frac{1}{\sinh \frac{n\pi}{a} b} \left(\frac{2}{a} \int_0^a g(x) \sin \frac{n\pi}{a} x \, dx - A_n \cosh \frac{n\pi}{a} b \right),$$

and that a solution of Problem 2 is

$$u_2(x, y) = \sum_{n=1}^{\infty} \left\{ A_n \cosh \frac{n\pi}{b} x + B_n \sinh \frac{n\pi}{b} x \right\} \sin \frac{n\pi}{b} y,$$

where

$$A_n = \frac{2}{b} \int_0^b F(y) \sin \frac{n\pi}{b} y \, dy$$

$$B_n = \frac{1}{\sinh \frac{n\pi}{b} a} \left(\frac{2}{b} \int_0^b G(y) \sin \frac{n\pi}{b} y \, dy - A_n \cosh \frac{n\pi}{b} a \right).$$

EXERCISES 12.5

Answers to selected odd-numbered problems begin on page ANS-18.

In Problems 1–10 solve Laplace's equation (1) for a rectangular plate subject to the given boundary conditions.

1. $u(0, y) = 0$, $u(a, y) = 0$
    $u(x, 0) = 0$, $u(x, b) = f(x)$

2. $u(0, y) = 0$, $u(a, y) = 0$
    $\left. \dfrac{\partial u}{\partial y} \right|_{y=0} = 0$, $u(x, b) = f(x)$

3. $u(0, y) = 0$, $u(a, y) = 0$
    $u(x, 0) = f(x)$, $u(x, b) = 0$

4. $\left. \dfrac{\partial u}{\partial x} \right|_{x=0} = 0$, $\left. \dfrac{\partial u}{\partial x} \right|_{x=a} = 0$
    $u(x, 0) = x$, $u(x, b) = 0$

5. $u(0, y) = 0$, $u(1, y) = 1 - y$
    $\left. \dfrac{\partial u}{\partial y} \right|_{y=0} = 0$, $\left. \dfrac{\partial u}{\partial y} \right|_{y=1} = 0$

6. $u(0, y) = g(y)$, $\left. \dfrac{\partial u}{\partial x} \right|_{x=1} = 0$
    $\left. \dfrac{\partial u}{\partial y} \right|_{y=0} = 0$, $\left. \dfrac{\partial u}{\partial y} \right|_{y=\pi} = 0$

7. $\left. \dfrac{\partial u}{\partial x} \right|_{x=0} = u(0, y)$, $u(\pi, y) = 1$
    $u(x, 0) = 0$, $u(x, \pi) = 0$

8. $u(0, y) = 0,$ $u(1, y) = 0$

$\left.\dfrac{\partial u}{\partial y}\right|_{y=0} = u(x, 0),$ $u(x, 1) = f(x)$

9. $u(0, y) = 0,$ $u(1, y) = 0$
$u(x, 0) = 100,$ $u(x, 1) = 200$

10. $u(0, y) = 10y,$ $\left.\dfrac{\partial u}{\partial x}\right|_{x=1} = -1$

$u(x, 0) = 0,$ $u(x, 1) = 0$

In Problems 11 and 12 solve Laplace's equation (1) for the given semi-infinite plate extending in the positive y-direction. In each case assume that $u(x, y)$ is bounded as $y \to \infty$.

11.

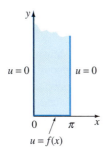

FIGURE 12.16   Plate in Problem 11

12.

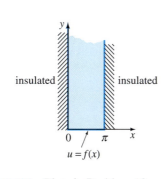

FIGURE 12.17   Plate in Problem 12

In Problems 13 and 14 solve Laplace's equation (1) for a rectangular plate subject to the given boundary conditions.

13. $u(0, y) = 0,$ $u(a, y) = 0$
$u(x, 0) = f(x),$ $u(x, b) = g(x)$

14. $u(0, y) = F(y),$ $u(a, y) = G(y)$
$u(x, 0) = 0,$ $u(x, b) = 0$

In Problems 15 and 16 use the superposition principle to solve Laplace's equation (1) for a square plate subject to the given boundary conditions.

15. $u(0, y) = 1,$ $u(\pi, y) = 1$
$u(x, 0) = 0,$ $u(x, \pi) = 1$

16. $u(0, y) = 0,$ $u(2, y) = y(2 - y)$

$u(x, 0) = 0,$ $u(x, 2) = \begin{cases} x, & 0 < x < 1 \\ 2 - x, & 1 \leq x < 2 \end{cases}$

DISCUSSION/PROJECT PROBLEMS

17. (a) In Problem 1 suppose that $a = b = \pi$ and $f(x) = 100x(\pi - x)$. Without using the solution $u(x, y)$, sketch, by hand, what the surface would look like over the rectangular region defined by $0 \leq x \leq \pi, 0 \leq y \leq \pi$.
(b) What is the maximum value of the temperature u for $0 \leq x \leq \pi, 0 \leq y \leq \pi$?
(c) Use the information in part (a) to compute the coefficients for your answer in Problem 1. Then use the 3D-plot application of your CAS to graph the partial sum $S_5(x, y)$ consisting of the first five nonzero terms of the solution in part (a) for $0 \leq x \leq \pi, 0 \leq y \leq \pi$. Use different perspectives and then compare with your sketch from part (a).

18. In Problem 16 what is the maximum value of the temperature u for $0 \leq x \leq 2, 0 \leq y \leq 2$?

COMPUTER LAB ASSIGNMENTS

19. (a) Use the contour-plot application of your CAS to graph the isotherms $u = 170, 140, 110, 80, 60, 30$ for the solution of Problem 9. Use the partial sum $S_5(x, y)$ consisting of the first five nonzero terms of the solution.
(b) Use the 3D-plot application of your CAS to graph the partial sum $S_5(x, y)$.

20. Use the contour-plot application of your CAS to graph the isotherms $u = 2, 1, 0.5, 0.2, 0.1, 0.05, 0, -0.05$ for the solution of Problem 10. Use the partial sum $S_5(x, y)$ consisting of the first five nonzero terms of the solution.

12.6 NONHOMOGENEOUS BOUNDARY-VALUE PROBLEMS

INTRODUCTION: A boundary-value problem is said to be **nonhomogeneous** if either the partial differential equation or the boundary conditions are nonhomogeneous. The method of separation of variables employed in the preceding three sections may not be applicable to a nonhomogeneous boundary-value problem *directly*. However, in the first of the two techniques examined in this section we employ a change of variable that transforms a nonhomogeneous boundary-value problem into two problems: one a relatively simple BVP for an ODE and the other a homogeneous BVP for a PDE. The latter problem is solvable by separation of variables. The second technique is basically a frontal attack on the BVP using orthogonal series expansions.

REVIEW MATERIAL: Review Sections 12.3–12.5.

NONHOMOGENEOUS BVPs When heat is generated at a rate r within a rod of finite length, the heat equation takes on the form

$$k\frac{\partial^2 u}{\partial x^2} + r = \frac{\partial u}{\partial t}, \quad 0 < x < L, \quad t > 0. \tag{1}$$

Equation (1) is nonhomogeneous and is readily shown not to be separable. On the other hand, suppose we wish to solve the homogeneous heat equation $ku_{xx} = u_t$ when the boundary conditions at $x = 0$ and $x = L$ are nonhomogeneous—say, the boundaries are held at nonzero temperatures: $u(0, t) = u_0$ and $u(L, t) = u_1$. Even though the substitution $u(x, t) = X(t)T(t)$ separates $ku_{xx} = u_t$, we quickly find ourselves at an impasse in determining eigenvalues and eigenfunctions since no conclusion can be drawn about $X(0)$ and $X(L)$ from $u(0, t) = X(0)T(t) = u_0$ and $u(L, t) = X(L)T(t) = u_1$.

What follows are two solution methods that are distinguished by different types of nonhomogeneous BVPs.

METHOD 1 Consider a BVP involving a *time-independent nonhomogeneous equation* and *time-independent boundary conditions* such as

$$k\frac{\partial^2 u}{\partial x^2} + F(x) = \frac{\partial u}{\partial t}, \quad 0 < x < L, \quad t > 0$$

$$u(0, t) = u_0, \quad u(L, t) = u_1, \quad t > 0 \tag{2}$$

$$u(x, 0) = f(x), \quad 0 < x < L,$$

where u_0 and u_1 are constants. By changing the dependent variable u to a new dependent variable v by the substitution $u(x, t) = v(x, t) + \psi(x)$, the problem in (2) can be reduced to two problems:

Problem A: $\{k\psi'' + F(x) = 0, \quad \psi(0) = u_0, \quad \psi(L) = u_1$

Problem B: $\begin{cases} k\dfrac{\partial^2 v}{\partial x^2} = \dfrac{\partial v}{\partial t}, \\ v(0, t) = 0, \quad v(L, t) = 0 \\ v(x, 0) = f(x) - \psi(x) \end{cases}$

Notice that Problem A involves an ODE that can be solved by integration, whereas Problem B is a homogeneous BVP that is solvable by the usual separation of variables. A solution of the original problem (2) is the sum of the solutions of Problems A and B.

The following example illustrates this first method.

EXAMPLE 1 **Using Method 1**

Suppose r is a positive constant. Solve equation (1) subject to

$$u(0, t) = 0, \quad u(1, t) = u_0, \quad t > 0$$

$$u(x, 0) = f(x), \quad 0 < x < 1.$$

SOLUTION Both the partial differential equation and the boundary condition at $x = 1$ are nonhomogeneous. If we let $u(x, t) = v(x, t) + \psi(x)$, then

$$\frac{\partial^2 u}{\partial x^2} = \frac{\partial^2 v}{\partial x^2} + \psi'' \quad \text{and} \quad \frac{\partial u}{\partial t} = \frac{\partial v}{\partial t}.$$

Substituting these results into (1) gives

$$k\frac{\partial^2 v}{\partial x^2} + k\psi'' + r = \frac{\partial v}{\partial t}. \tag{3}$$

Equation (3) reduces to a homogeneous equation if we demand that ψ satisfy

$$k\psi'' + r = 0 \quad \text{or} \quad \psi'' = -\frac{r}{k}.$$

Integrating the last equation twice reveals that

$$\psi(x) = -\frac{r}{2k}x^2 + c_1 x + c_2. \tag{4}$$

Furthermore,
$$u(0, t) = v(0, t) + \psi(0) = 0$$
$$u(1, t) = v(1, t) + \psi(1) = u_0.$$

We have $v(0, t) = 0$ and $v(1, t) = 0$, provided that

$$\psi(0) = 0 \quad \text{and} \quad \psi(1) = u_0.$$

Applying the latter two conditions to (4) gives, in turn, $c_2 = 0$ and $c_1 = r/2k + u_0$. Consequently,

$$\psi(x) = -\frac{r}{2k}x^2 + \left(\frac{r}{2k} + u_0\right)x.$$

Finally, the initial condition $u(x, 0) = v(x, 0) + \psi(x)$ implies that $v(x, 0) = u(x, 0) - \psi(x) = f(x) - \psi(x)$. Thus to determine $v(x, t)$, we solve the new boundary-value problem

$$k\frac{\partial^2 v}{\partial x^2} = \frac{\partial v}{\partial t}, \quad 0 < x < 1, \quad t > 0$$

$$v(0, t) = 0, \quad v(1, t) = 0, \quad t > 0$$

$$v(x, 0) = f(x) + \frac{r}{2k}x^2 - \left(\frac{r}{2k} + u_0\right)x, \quad 0 < x < 1$$

by separation of variables. In the usual manner we find

$$v(x, t) = \sum_{n=1}^{\infty} A_n e^{-kn^2\pi^2 t} \sin n\pi x,$$

where

$$A_n = 2\int_0^1 \left[f(x) + \frac{r}{2k}x^2 - \left(\frac{r}{2k} + u_0\right)x \right] \sin n\pi x \, dx. \tag{5}$$

A solution of the original problem is obtained by adding $\psi(x)$ and $v(x, t)$:

$$u(x, t) = -\frac{r}{2k}x^2 + \left(\frac{r}{2k} + u_0\right)x + \sum_{n=1}^{\infty} A_n e^{-kn^2\pi^2 t} \sin n\pi x, \tag{6}$$

where the coefficients A_n are defined in (5).

Observe in (6) that $u(x, t) \rightarrow \psi(x)$ as $t \rightarrow \infty$. In the context of solving forms of the heat equation, ψ is called a **steady-state solution.** Since $v(x, t) \rightarrow 0$ as $t \rightarrow \infty$, it is called a **transient solution.**

METHOD 2 Another type of problem involves a *time-dependent nonhomogeneous equation* and *homogeneous boundary conditions.* Unlike Method 1, where $u(x, t)$ is found by solving two separate problems, it is possible to find the entire solution of a problem such as

$$k \frac{\partial^2 u}{\partial x^2} + F(x, t) = \frac{\partial u}{\partial t}, \quad 0 < x < L, \quad t > 0$$

$$u(0, t) = 0, \quad u(L, t) = 0, \quad t > 0 \tag{7}$$

$$u(x, 0) = f(x), \quad 0 < x < L,$$

by making the assumption that time-dependent coefficients $u_n(t)$ and $F_n(t)$ can be found such that both $u(x, t)$ and $F(x, t)$ in (7) can be expanded in the series

$$u(x, t) = \sum_{n=1}^{\infty} u_n(t) \sin \frac{n\pi}{L} x \quad \text{and} \quad F(x, t) = \sum_{n=1}^{\infty} F_n(t) \sin \frac{n\pi}{L} x, \tag{8}$$

where $\sin(n\pi x/L)$, $n = 1, 2, 3, \ldots$, are the eigenfunctions of $X'' + \lambda X = 0$, $X(0) = 0$, $X(L) = 0$ corresponding to the eigenvalues $\lambda_n = \alpha_n^2 = n^2\pi^2/L^2$. The latter problem would have been obtained had separation of variables been applied to the associated homogeneous PDE in (7). In (8), observe that the assumed form for $u(x, t)$ already satisfies the boundary conditions in (7). The basic idea here is to substitute the first series in (8) into the nonhomogeneous PDE in (7), collect terms, and equate the resulting series with the actual series expansion found for $F(x, t)$.

The next example illustrates this method.

EXAMPLE 2   **Using Method 2**

Solve $\dfrac{\partial^2 u}{\partial x^2} + (1 - x)\sin t = \dfrac{\partial u}{\partial t}, \quad 0 < x < 1, \quad t > 0$

$$u(0, t) = 0, \quad u(1, t) = 0, \quad t > 0,$$

$$u(x, 0) = 0, \quad 0 < x < 1.$$

SOLUTION   With $k = 1$, $L = 1$, the eigenvalues and eigenfunctions of $X'' + \lambda X = 0$, $X(0) = 0$, $X(1) = 0$ are found to be $\lambda_n = \alpha_n^2 = n^2\pi^2$ and $\sin n\pi x$, $n = 1, 2, 3, \ldots$. If we assume that

$$u(x, t) = \sum_{n=1}^{\infty} u_n(t) \sin n\pi x, \tag{9}$$

then the formal partial derivatives of u are

$$\frac{\partial^2 u}{\partial x^2} = \sum_{n=1}^{\infty} u_n(t)(-n^2\pi^2) \sin n\pi x \quad \text{and} \quad \frac{\partial u}{\partial t} = \sum_{n=1}^{\infty} u_n'(t) \sin n\pi x. \tag{10}$$

Now the assumption that we can write $F(x, t) = (1 - x)\sin t$ as

$$(1 - x)\sin t = \sum_{n=1}^{\infty} F_n(t) \sin n\pi x$$

implies that

$$F_n(t) = \frac{2}{1}\int_0^1 (1-x)\sin t \sin n\pi x\, dx = 2\sin t \int_0^1 (1-x)\sin n\pi x\, dx = \frac{2}{n\pi}\sin t.$$

Hence,
$$(1-x)\sin t = \sum_{n=1}^{\infty} \frac{2}{n\pi}\sin t \sin n\pi x. \qquad (11)$$

Substituting the series in (10) and (11) into $u_t - u_{xx} = (1-x)\sin t$, we get

$$\sum_{n=1}^{\infty}\left[u_n'(t) + n^2\pi^2 u_n(t)\right]\sin n\pi x = \sum_{n=1}^{\infty}\frac{2\sin t}{n\pi}\sin n\pi x.$$

To determine $u_n(t)$, we now equate the coefficients of $\sin n\pi x$ on each side of the preceding equality:

$$u_n'(t) + n^2\pi^2 u_n(t) = \frac{2\sin t}{n\pi}.$$

This last equation is a linear first-order ODE whose solution is

$$u_n(t) = \frac{2}{n\pi}\left[\frac{n^2\pi^2\sin t - \cos t}{n^4\pi^4 + 1}\right] + C_n e^{-n^2\pi^2 t},$$

where C_n denotes the arbitrary constant. Therefore the assumed form of $u(x, t)$ in (9) can be written as the sum of two series:

$$u(x, t) = \sum_{n=1}^{\infty}\frac{2}{n\pi}\left[\frac{n^2\pi^2\sin t - \cos t}{n^4\pi^4 + 1}\right]\sin n\pi x + \sum_{n=1}^{\infty}C_n e^{-n^2\pi^2 t}\sin n\pi x. \qquad (12)$$

Finally, we apply the initial condition $u(x, 0) = 0$ to (12). By rewriting the resulting expression as one series,

$$0 = \sum_{n=1}^{\infty}\left[\frac{-2}{n\pi(n^4\pi^4 + 1)} + C_n\right]\sin n\pi x,$$

we conclude from this identity that the total coefficient of $\sin n\pi x$ must be zero, and so

$$C_n = \frac{2}{n\pi(n^4\pi^4 + 1)}.$$

Hence from (12) we see that a solution of the given problem is

$$u(x, t) = \frac{2}{\pi}\sum_{n=1}^{\infty}\frac{n^2\pi^2\sin t - \cos t}{n(n^4\pi^4 + 1)}\sin n\pi x + \frac{2}{\pi}\sum_{n=1}^{\infty}\frac{1}{n(n^4\pi^4 + 1)}e^{-n^2\pi^2 t}\sin n\pi x. \qquad \blacksquare$$

EXERCISES 12.6

Answers to selected odd-numbered problems begin on page ANS-19.

In Problems 1–12 use Method 1 of this section to solve the given boundary-value problem.

In Problems 1 and 2 solve the heat equation $ku_{xx} = u_t$, $0 < x < 1$, $t > 0$, subject to the given conditions.

1. $u(0, t) = 100, \quad u(1, t) = 100$
$u(x, 0) = 0$

2. $u(0, t) = u_0, \quad u(1, t) = 0$
$u(x, 0) = f(x)$

In Problems 3 and 4 solve the partial differential equation (1) subject to the given conditions.

3. $u(0, t) = u_0, \quad u(1, t) = u_0$
$u(x, 0) = 0$

4. $u(0, t) = u_0, \quad u(1, t) = u_1$
$u(x, 0) = f(x)$

5. Solve the boundary-value problem

$$k\frac{\partial^2 u}{\partial x^2} + Ae^{-\beta x} = \frac{\partial u}{\partial t}, \quad \beta > 0, \quad 0 < x < 1, \quad t > 0$$

$$u(0, t) = 0, \quad u(1, t) = 0, \quad t > 0$$

$$u(x, 0) = f(x), \quad 0 < x < 1.$$

The partial differential equation is a form of the heat equation when heat is generated within a thin rod from radioactive decay of the material.

6. Solve the boundary-value problem

$$k\frac{\partial^2 u}{\partial x^2} - hu = \frac{\partial u}{\partial t}, \quad 0 < x < \pi, \quad t > 0$$

$$u(0, t) = 0, \quad u(\pi, t) = u_0, \quad t > 0$$

$$u(x, 0) = 0, \quad 0 < x < \pi.$$

The partial differential equation is a form of the heat equation when heat is lost by radiation from the lateral surface of a thin rod into a medium at temperature zero.

7. Find a steady-state solution $\psi(x)$ of the boundary-value problem

$$k\frac{\partial^2 u}{\partial x^2} - h(u - u_0) = \frac{\partial u}{\partial t}, \quad 0 < x < 1, \quad t > 0$$

$$u(0, t) = u_0, \quad u(1, t) = 0, \quad t > 0$$

$$u(x, 0) = f(x), \quad 0 < x < 1.$$

8. Find a steady-state solution $\psi(x)$ if the rod in Problem 7 is semi-infinite extending in the positive x-direction, radiates from its lateral surface into a medium of temperature zero, and

$$u(0, t) = u_0, \quad \lim_{x \to \infty} u(x, t) = 0, \quad t > 0$$

$$u(x, 0) = f(x), \quad x > 0.$$

9. When a vibrating string is subjected to an external vertical force that varies with the horizontal distance from the left end, the wave equation takes on the form

$$a^2\frac{\partial^2 u}{\partial x^2} + Ax = \frac{\partial^2 u}{\partial t^2},$$

where A is a constant. Solve this partial differential equation subject to

$$u(0, t) = 0, \quad u(1, t) = 0, \quad t > 0$$

$$u(x, 0) = 0, \quad \frac{\partial u}{\partial t}\bigg|_{t=0} = 0, \quad 0 < x < 1.$$

10. A string initially at rest on the x-axis is secured on the x-axis at $x = 0$ and $x = 1$. If the string is allowed to fall under its own weight for $t > 0$, the displacement $u(x, t)$ satisfies

$$a^2\frac{\partial^2 u}{\partial x^2} - g = \frac{\partial^2 u}{\partial t^2}, \quad 0 < x < 1, \quad t > 0,$$

where g is the acceleration of gravity. Solve for $u(x, t)$.

11. Find the steady-state temperature $u(x, y)$ in the semi-infinite plate shown in Figure 12.18. Assume that the temperature is bounded as $x \to \infty$. (*Hint:* Try $u(x, y) = v(x, y) + \psi(y)$.)

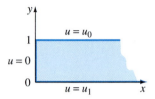

FIGURE 12.18   Plate in Problem 11

12. The partial differential equation

$$\frac{\partial^2 u}{\partial x^2} + \frac{\partial^2 u}{\partial y^2} = -h,$$

where $h > 0$ is a constant, is known as **Poisson's equation** and occurs in many problems involving electrical potential. Solve the equation subject to the conditions

$$u(0, y) = 0, \quad u(\pi, y) = 1, \quad y > 0$$

$$u(x, 0) = 0, \quad 0 < x < \pi.$$

In Problems 13–16 use Method 2 of this section to solve the given boundary-value problem.

13. $\dfrac{\partial^2 u}{\partial x^2} + xe^{-3t} = \dfrac{\partial u}{\partial t}, \quad 0 < x < \pi, \quad t > 0$

$u(0, t) = 0, \quad u(\pi, t) = 0, \quad t > 0$

$u(x, 0) = 0, \quad 0 < x < \pi$

14. $\dfrac{\partial^2 u}{\partial x^2} + xe^{-3t} = \dfrac{\partial u}{\partial t}, \quad 0 < x < \pi, \quad t > 0$

$\dfrac{\partial u}{\partial x}\bigg|_{x=0} = 0, \quad \dfrac{\partial u}{\partial x}\bigg|_{x=\pi} = 0, \quad t > 0$

$u(x, 0) = 0, \quad 0 < x < \pi$

15. $\dfrac{\partial^2 u}{\partial x^2} - 1 + x - x\cos t = \dfrac{\partial u}{\partial t}, \quad 0 < x < 1, \quad t > 0$

$u(0, t) = 0, \quad u(1, t) = 0, \quad t > 0$

$u(x, 0) = x(1 - x), \quad 0 < x < 1$

16. $\dfrac{\partial^2 u}{\partial x^2} + \cos t \sin x = \dfrac{\partial^2 u}{\partial t^2}, \quad 0 < x < \pi, \quad t > 0$

$u(0, t) = 0, \quad u(\pi, t) = 0, \quad t > 0,$

$u(x, 0) = 0, \quad \dfrac{\partial u}{\partial t}\bigg|_{t=0} = 0, \quad 0 < x < \pi$

12.7   ORTHOGONAL SERIES EXPANSIONS

INTRODUCTION: For certain types of boundary conditions the method of separation of variables and the superposition principle lead to an expansion of a function in a trigonometric series that is *not* a Fourier series. To solve the problems in this section, we will utilize the concept of orthogonal series expansions or generalized Fourier series.

REVIEW MATERIAL: The results in (7)–(11) of Section 11.1 form the backbone of the discussion that follows. We suggest you review that material.

EXAMPLE 1   **Using Orthogonal Series Expansions**

The temperature in a rod of unit length in which there is heat transfer from its right boundary into a surrounding medium kept at a constant temperature zero is determined from

$$k\frac{\partial^2 u}{\partial x^2} = \frac{\partial u}{\partial t}, \quad 0 < x < 1, \quad t > 0$$

$$u(0, t) = 0, \quad \left.\frac{\partial u}{\partial x}\right|_{x=1} = -hu(1, t), \quad h > 0, \quad t > 0$$

$$u(x, 0) = 1, \quad 0 < x < 1.$$

Solve for $u(x, t)$.

SOLUTION   Proceeding as in Section 12.3 with $u(x, t) = X(x)T(t)$ and using $-\lambda$ as the separation constant, we find the separated equations and boundary conditions to be, respectively,

$$X'' + \lambda X = 0 \tag{1}$$

$$T' + k\lambda T = 0 \tag{2}$$

$$X(0) = 0 \quad \text{and} \quad X'(1) = -hX(1). \tag{3}$$

Equation (1) and the homogeneous boundary conditions (3) make up a regular Sturm-Liouville problem:

$$X'' + \lambda X = 0, \quad X(0) = 0, \quad X'(1) + hX(1) = 0. \tag{4}$$

By analyzing the usual three cases in which λ is zero, negative, or positive, we find that only the last case will yield nontrivial solutions. Thus with $\lambda = \alpha^2 > 0$, $\alpha > 0$, the general solution of the DE in (4) is

$$X(x) = c_1 \cos \alpha x + c_2 \sin \alpha x. \tag{5}$$

The first boundary condition in (4) immediately gives $c_1 = 0$. Applying the second condition in (4) to $X(x) = c_2 \sin \alpha x$ yields

$$\alpha \cos \alpha + h \sin \alpha = 0 \quad \text{or} \quad \tan \alpha = -\frac{\alpha}{h}. \tag{6}$$

From the analysis in Example 2 of Section 11.4 we know that the last equation in (6) has an infinite number of roots. If the consecutive positive roots are denoted α_n, $n = 1, 2, 3, \ldots$, then the eigenvalues of the problem are $\lambda_n = \alpha_n^2$, and the corresponding eigenfunctions are $X(x) = c_2 \sin \alpha_n x$, $n = 1, 2, 3, \ldots$. The solution of the first-order DE (2) is $T(t) = c_3 e^{-k\alpha_n^2 t}$, and so

$$u_n = XT = A_n e^{-k\alpha_n^2 t} \sin \alpha_n x \quad \text{and} \quad u(x, t) = \sum_{n=1}^{\infty} A_n e^{-k\alpha_n^2 t} \sin \alpha_n x.$$

Now at $t = 0$, $u(x, 0) = 1$, $0 < x < 1$, so that

$$1 = \sum_{n=1}^{\infty} A_n \sin \alpha_n x. \tag{7}$$

The series in (7) is not a Fourier sine series; rather, it is an expansion of $u(x, 0) = 1$ in terms of the orthogonal functions arising from the regular Sturm-Liouville problem (4). It follows that the set of eigenfunctions $\{\sin \alpha_n x\}$, $n = 1, 2, 3, \ldots$, where the α's are defined by $\tan \alpha = -\alpha/h$, is orthogonal with respect to the weight function $p(x) = 1$ on the interval $[0, 1]$. Matching (7) with (7) of Section 11.1, it follows from (8) of that section, with $f(x) = 1$ and $\phi_n(x) = \sin \alpha_n x$, that the coefficients A_n are given by

$$A_n = \frac{\int_0^1 \sin \alpha_n x \, dx}{\int_0^1 \sin^2 \alpha_n x \, dx}. \tag{8}$$

To evaluate the square norm of each of the eigenfunctions, we use a trigonometric identity:

$$\int_0^1 \sin^2 \alpha_n x \, dx = \frac{1}{2} \int_0^1 (1 - \cos 2\alpha x) \, dx = \frac{1}{2}\left(1 - \frac{1}{2\alpha_n} \sin 2\alpha_n\right). \tag{9}$$

Using the double-angle formula $\sin 2\alpha_n = 2 \sin \alpha_n \cos \alpha_n$ and the first equation in (6) in the form $\alpha_n \cos \alpha_n = -h \sin \alpha_n$, we simplify (9) to

$$\int_0^1 \sin^2 \alpha_n x \, dx = \frac{1}{2h}\left(h + \cos^2 \alpha_n\right).$$

Also

$$\int_0^1 \sin \alpha_n x \, dx = -\frac{1}{\alpha_n} \cos \alpha_n x \Big|_0^1 = \frac{1}{\alpha_n}(1 - \cos \alpha_n).$$

Consequently, (8) becomes

$$A_n = \frac{2h(1 - \cos \alpha_n)}{\alpha_n(h + \cos^2 \alpha_n)}.$$

Finally, a solution of the boundary-value problem is

$$u(x, t) = 2h \sum_{n=1}^{\infty} \frac{1 - \cos \alpha_n}{\alpha_n(h + \cos^2 \alpha_n)} e^{-k\alpha_n^2 t} \sin \alpha_n x.$$

EXAMPLE 2   Using Orthogonal Series Expansions

The twist angle $\theta(x, t)$ of a torsionally vibrating shaft of unit length is determined from

$$a^2 \frac{\partial^2 \theta}{\partial x^2} = \frac{\partial^2 \theta}{\partial t^2}, \quad 0 < x < 1, \quad t > 0$$

$$\theta(0, t) = 0, \quad \frac{\partial \theta}{\partial x}\Big|_{x=1} = 0, \quad t > 0$$

$$\theta(x, 0) = x, \quad \frac{\partial \theta}{\partial t}\Big|_{t=0} = 0, \quad 0 < x < 1.$$

See Figure 12.19. The boundary condition at $x = 1$ is called a free-end condition. Solve for $\theta(x, t)$.

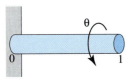

FIGURE 12.19 Twisted shaft

SOLUTION   Proceeding as in Section 12.4 with $\theta(x, t) = X(x)T(t)$ and using $-\lambda$ once again as the separation constant, the separated equations and boundary conditions are

$$X'' + \lambda X = 0 \tag{10}$$

$$T'' + a^2 \lambda T = 0 \tag{11}$$

$$X(0) = 0 \quad \text{and} \quad X'(1) = 0. \tag{12}$$

A regular Sturm-Liouville problem in this case consists of equation (10) and the homogeneous boundary conditions in (12):

$$X'' + \lambda X = 0, \quad X(0) = 0, \quad X'(1) = 0. \tag{13}$$

As in Example 1, (13) possesses nontrivial solutions only for $\lambda = \alpha^2 > 0$, $\alpha > 0$. The boundary conditions $X(0) = 0$ and $X'(1) = 0$ applied to the general solution

$$X(x) = c_1 \cos \alpha x + c_2 \sin \alpha x \tag{14}$$

give, in turn, $c_1 = 0$ and $c_2 \cos \alpha = 0$. Since the cosine function is zero at odd multiples of $\pi/2$, $\alpha = (2n - 1)\pi/2$, and the eigenvalues of (13) are $\lambda_n = \alpha_n^2 = (2n - 1)^2 \pi^2/4, n = 1, 2, 3, \ldots$. The solution of the second-order DE (11) is $T(t) = c_3 \cos a\alpha_n t + c_4 \sin a\alpha_n t$. The initial condition $T'(0) = 0$ gives $c_4 = 0$, so

$$\theta_n = XT = A_n \cos a\left(\frac{2n - 1}{2}\right)\pi t \sin\left(\frac{2n - 1}{2}\right)\pi x.$$

To satisfy the remaining initial condition, we form

$$\theta(x, t) = \sum_{n=1}^{\infty} A_n \cos a\left(\frac{2n - 1}{2}\right)\pi t \sin\left(\frac{2n - 1}{2}\right)\pi x. \tag{15}$$

When $t = 0$, we must have, for $0 < x < 1$,

$$\theta(x, 0) = x = \sum_{n=1}^{\infty} A_n \sin\left(\frac{2n - 1}{2}\right)\pi x. \tag{16}$$

As in Example 1, the set of eigenfunctions $\left\{ \sin\left(\dfrac{2n - 1}{2}\right)\pi x \right\}$, $n = 1, 2, 3, \ldots$, is orthogonal with respect to the weight function $p(x) = 1$ on the interval $[0, 1]$. Although the series in (16) looks like a Fourier sine series, it is not, because the argument of the sine function is not an integer multiple of $\pi x/L$ (here $L = 1$). The series again is an orthogonal series expansion or generalized Fourier series. Hence from (8) of Section 11.1 the coefficients in (16) are

$$A_n = \frac{\displaystyle\int_0^1 x \sin\left(\frac{2n - 1}{2}\right)\pi x \, dx}{\displaystyle\int_0^1 \sin^2\left(\frac{2n - 1}{2}\right)\pi x \, dx}.$$

Carrying out the two integrations, we arrive at

$$A_n = \frac{8(-1)^{n+1}}{(2n - 1)^2 \pi^2}.$$

The twist angle is then

$$\theta(x, t) = \frac{8}{\pi^2} \sum_{n=1}^{\infty} \frac{(-1)^{n+1}}{(2n - 1)^2} \cos a\left(\frac{2n - 1}{2}\right)\pi t \sin\left(\frac{2n - 1}{2}\right)\pi x. \tag{17}$$

We can use a CAS to plot $\theta(x, t)$ defined in (17) either as a three-dimensional surface or as two-dimensional curves by holding one of the variables constant. In

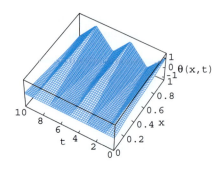

FIGURE 12.20 Surface is the graph of a partial sum of (17) with $a = 1$

Figure 12.20 we have plotted the surface defined by $\theta(x, t)$ over the rectangular region $0 \leq x \leq 1$, $0 \leq t \leq 10$. The cross sections of this surface are interesting. In Figure 12.21 we have plotted θ as a function of time t on the interval $0 \leq t \leq 10$ using four specified values of x and a partial sum of (17) (with $a = 1$). As can be seen in the four parts of Figure 12.21, the twist angle of each cross section of the rod oscillates back and forth (positive and negative values of θ) as time t increases. Figure 12.21(d) portrays what we would intuitively expect in the absence of any damping, the end of the rod $x = 1$ is displaced initially 1 radian ($\theta(1, 0) = 1$); when in motion this end oscillates indefinitely between its maximum displacement of 1 radian and minimum displacement of -1 radian. The graphs in Figure 12.21(a)–(c) show what appears to be a "pausing" behavior of θ at its maximum (minimum) displacement of each of the specified cross sections before changing direction and heading toward its minimum (maximum). This behavior diminishes as $x \to 1$.

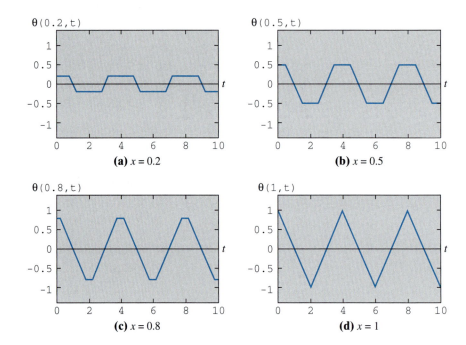

FIGURE 12.21 Angular displacements θ as a function of time at various cross sections of the rod

EXERCISES 12.7

Answers to selected odd-numbered problems begin on page ANS-19.

1. In Example 1 find the temperature $u(x, t)$ when the left end of the rod is insulated.

2. Solve the boundary-value problem

$$k\frac{\partial^2 u}{\partial x^2} = \frac{\partial u}{\partial t}, \quad 0 < x < 1, t > 0$$

$$u(0, t) = 0, \quad \left.\frac{\partial u}{\partial x}\right|_{x=1} = -h(u(1, t) - u_0), \ h > 0, \ t > 0$$

$$u(x, 0) = f(x), \quad 0 < x < 1.$$

3. Find the steady-state temperature for a rectangular plate for which the boundary conditions are

$$u(0, y) = 0, \quad \left.\frac{\partial u}{\partial x}\right|_{x=a} = -hu(a, y), \quad 0 < y < b$$

$$u(x, 0) = 0, \quad u(x, b) = f(x), \quad 0 < x < a.$$

4. Solve the boundary-value problem

$$\frac{\partial^2 u}{\partial x^2} + \frac{\partial^2 u}{\partial y^2} = 0, \quad 0 < y < 1, \quad x > 0$$

$$u(0, y) = u_0, \quad \lim_{x \to \infty} u(x, y) = 0, \quad 0 < y < 1$$

$$\left.\frac{\partial u}{\partial y}\right|_{y=0} = 0, \quad \left.\frac{\partial u}{\partial y}\right|_{y=1} = -hu(x, 1), \quad h > 0, \quad x > 0.$$

5. Find the temperature $u(x, t)$ in a rod of length L if the initial temperature is $f(x)$ throughout and if the end $x = 0$ is kept at temperature zero and the end $x = L$ is insulated.

6. Solve the boundary-value problem

$$a^2 \frac{\partial^2 u}{\partial x^2} = \frac{\partial^2 u}{\partial t^2}, \quad 0 < x < L, \quad t > 0$$

$$u(0, t) = 0, \quad E \frac{\partial u}{\partial x}\Big|_{x=L} = F_0, \quad t > 0$$

$$u(x, 0) = 0, \quad \frac{\partial u}{\partial t}\Big|_{t=0} = 0, \quad 0 < x < L.$$

The solution $u(x, t)$ represents the longitudinal displacement of a vibrating elastic bar that is anchored at its left end and is subjected to a constant force of magnitude F_0 at its right end. See Figure 12.11 in Exercises 12.4. E is a constant called the modulus of elasticity.

7. Solve the boundary-value problem

$$\frac{\partial^2 u}{\partial x^2} + \frac{\partial^2 u}{\partial y^2} = 0, \quad 0 < x < 1, \quad 0 < y < 1$$

$$\frac{\partial u}{\partial x}\Big|_{x=0} = 0, \quad u(1, y) = u_0, \quad 0 < y < 1$$

$$u(x, 0) = 0, \quad \frac{\partial u}{\partial y}\Big|_{y=1} = 0, \quad 0 < x < 1.$$

8. The initial temperature in a rod of unit length is $f(x)$ throughout. There is heat transfer from both ends, $x = 0$ and $x = 1$, into a surrounding medium kept at a constant temperature zero. Show that

$$u(x, t) = \sum_{n=1}^{\infty} A_n e^{-k\alpha_n^2 t}(\alpha_n \cos \alpha_n x + h \sin \alpha_n x),$$

where

$$A_n = \frac{2}{(\alpha_n^2 + 2h + h^2)} \int_0^1 f(x)(\alpha_n \cos \alpha_n x + h \sin \alpha_n x)\, dx.$$

The eigenvalues are $\lambda_n = \alpha_n^2$, $n = 1, 2, 3, \ldots$, where the α_n are the consecutive positive roots of $\tan \alpha = 2\alpha h/(\alpha^2 - h^2)$.

9. Use Method 2 of Section 12.6 to solve the boundary-value problem

$$k \frac{\partial^2 u}{\partial x^2} + x e^{-2t} = \frac{\partial u}{\partial t}, \quad 0 < x < 1, \quad t > 0$$

$$u(0, t) = 0, \quad \frac{\partial u}{\partial x}\Big|_{x=1} = -u(1, t), \quad t > 0$$

$$u(x, 0) = 0, \quad 0 < x < 1.$$

COMPUTER LAB ASSIGNMENTS

10. A vibrating cantilever beam is embedded at its left end ($x = 0$) and free at its right end ($x = 1$). See Figure 12.22. The transverse displacement $u(x, t)$ of the beam is determined from the boundary-value problem

$$\frac{\partial^4 u}{\partial x^4} + \frac{\partial^2 u}{\partial t^2} = 0, \quad 0 < x < 1, \quad t > 0$$

$$u(0, t) = 0, \quad \frac{\partial u}{\partial x}\Big|_{x=0} = 0, \quad t > 0$$

$$\frac{\partial^2 u}{\partial x^2}\Big|_{x=1} = 0, \quad \frac{\partial^3 u}{\partial x^3}\Big|_{x=1} = 0, \quad t > 0$$

$$u(x, 0) = f(x), \quad \frac{\partial u}{\partial t}\Big|_{t=0} = g(x), \quad 0 < x < 1.$$

Use a CAS to find approximations to the first two positive eigenvalues of the problem. (*Hint:* See Problems 11 and 12 in Exercises 12.4.)

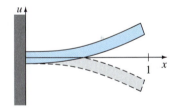

FIGURE 12.22   Vibrating cantilever beam in Problem 10

11. (a) Find an equation that defines the eigenvalues when the ends of the beam in Problem 10 are embedded at $x = 0$ and $x = 1$.
    (b) Use a CAS to find approximations to the first two positive eigenvalues.

12.8 HIGHER-DIMENSIONAL PROBLEMS

INTRODUCTION: Up to now we have solved problems involving the one-dimensional heat and wave equations. In this section we show how to extend the method of separation of variables to problems involving the two-dimensional versions of these partial differential equations.

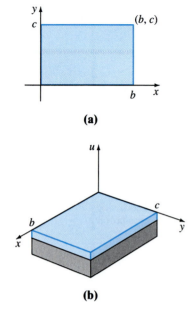

FIGURE 12.23 (a) Rectangular plate and (b) rectangular membrane

HEAT AND WAVE EQUATIONS IN TWO DIMENSIONS Suppose the rectangular region in Figure 12.23(a) is a thin plate in which the temperature u is a function of time t and position (x, y). Then, under suitable conditions, $u(x, y, t)$ can be shown to satisfy the **two-dimensional heat equation**

$$k\left(\frac{\partial^2 u}{\partial x^2} + \frac{\partial^2 u}{\partial y^2}\right) = \frac{\partial u}{\partial t}. \tag{1}$$

On the other hand, suppose Figure 12.23(b) represents a rectangular frame over which a thin flexible membrane has been stretched (a rectangular drum). If the membrane is set in motion, then its displacement u, measured from the xy-plane (transverse vibrations), is also a function of t and position (x, y). When the vibrations are small, free, and undamped, $u(x, y, t)$ satisfies the **two-dimensional wave equation**

$$a^2\left(\frac{\partial^2 u}{\partial x^2} + \frac{\partial^2 u}{\partial y^2}\right) = \frac{\partial^2 u}{\partial t^2}. \tag{2}$$

To separate variables in (1) and (2), we assume a product solution of the form $u(x, y, t) = X(x)Y(y)T(t)$. We note that

$$\frac{\partial^2 u}{\partial x^2} = X''YT, \quad \frac{\partial^2 u}{\partial y^2} = XY''T, \quad \text{and} \quad \frac{\partial u}{\partial t} = XYT'.$$

As we see next, with appropriate boundary conditions, boundary-value problems involving (1) and (2) lead to the concept of Fourier series in two variables.

EXAMPLE 1 **Temperatures in a Plate**

Find the temperature $u(x, y, t)$ in the plate shown in Figure 12.23(a) if the initial temperature is $f(x, y)$ throughout and if the boundaries are held at temperature zero for time $t > 0$.

SOLUTION We must solve

$$k\left(\frac{\partial^2 u}{\partial x^2} + \frac{\partial^2 u}{\partial y^2}\right) = \frac{\partial u}{\partial t}, \quad 0 < x < b, \quad 0 < y < c, \quad t > 0$$

subject to

$$u(0, y, t) = 0, \quad u(b, y, t) = 0, \quad 0 < y < c, t > 0$$

$$u(x, 0, t) = 0, \quad u(x, c, t) = 0, \quad 0 < x < b, t > 0$$

$$u(x, y, 0) = f(x, y), \quad 0 < x < b, \quad 0 < y < c.$$

Substituting $u(x, y, t) = X(x)Y(y)T(t)$, we get

$$k(X''YT + XY''T) = XYT' \quad \text{or} \quad \frac{X''}{X} = -\frac{Y''}{Y} + \frac{T'}{kT}. \tag{3}$$

Since the left-hand side of the last equation in (3) depends only on x and the right side depends only on y and t, we must have both sides equal to a constant $-\lambda$:

$$\frac{X''}{X} = -\frac{Y''}{Y} + \frac{T'}{kT} = -\lambda$$

and so

$$X'' + \lambda X = 0 \tag{4}$$

$$\frac{Y''}{Y} = \frac{T'}{kT} + \lambda. \tag{5}$$

By the same reasoning, if we introduce another separation constant $-\mu$ in (5), then

$$\frac{Y''}{Y} = -\mu \quad \text{and} \quad \frac{T'}{kT} + \lambda = -\mu$$

yield
$$Y'' + \mu Y = 0 \quad \text{and} \quad T' + k(\lambda + \mu)T = 0. \tag{6}$$

Now the homogeneous boundary conditions

$$\left.\begin{array}{ll} u(0, y, t) = 0, & u(b, y, t) = 0 \\ u(x, 0, t) = 0, & u(x, c, t) = 0 \end{array}\right\} \quad \text{imply that} \quad \begin{cases} X(0) = 0, & X(b) = 0 \\ Y(0) = 0, & Y(c) = 0. \end{cases}$$

Thus we have two Sturm-Liouville problems

$$X'' + \lambda X = 0, \quad X(0) = 0, \quad X(b) = 0 \tag{7}$$

and
$$Y'' + \mu Y = 0, \quad Y(0) = 0, \quad Y(c) = 0. \tag{8}$$

The usual consideration of cases ($\lambda = 0$, $\lambda = \alpha^2 > 0$, $\lambda = -\alpha^2 < 0$, $\mu = 0$, and so on) leads to two independent sets of eigenvalues

$$\lambda_m = \frac{m^2 \pi^2}{b^2} \quad \text{and} \quad \mu_n = \frac{n^2 \pi^2}{c^2}.$$

The corresponding eigenfunctions are

$$X(x) = c_2 \sin\frac{m\pi}{b} x, \quad m = 1, 2, 3 \ldots, \quad \text{and} \quad Y(y) = c_4 \sin\frac{n\pi}{c} y, \quad n = 1, 2, 3, \ldots. \tag{9}$$

After substituting the known values of λ_n and μ_n in the first-order DE in (6), its general solution is found to be $T(t) = c_5 e^{-k[(m\pi/b)^2 + (n\pi/c)^2]t}$. A product solution of the two-dimensional heat equation that satisfies the four homogeneous boundary conditions is then

$$u_{mn}(x, y, t) = A_{mn} e^{-k[(m\pi/b)^2 + (n\pi/c)^2]t} \sin\frac{m\pi}{b} x \sin\frac{n\pi}{c} y,$$

where A_{mn} is an arbitrary constant. Because we have two sets of eigenvalues, we are prompted to try the superposition principle in the form of a double sum

$$u(x, y, t) = \sum_{m=1}^{\infty} \sum_{n=1}^{\infty} A_{mn} e^{-k[(m\pi/b)^2 + (n\pi/c)^2]t} \sin\frac{m\pi}{b} x \sin\frac{n\pi}{c} y. \tag{10}$$

At $t = 0$ we must have

$$u(x, y, 0) = f(x, y) = \sum_{m=1}^{\infty} \sum_{n=1}^{\infty} A_{mn} \sin\frac{m\pi}{b} x \sin\frac{n\pi}{c} y. \tag{11}$$

We can find the coefficients A_{mn} by multiplying the double sum (11) by the product $\sin(m\pi x/b) \sin(n\pi y/c)$ and integrating over the rectangle $0 \le x \le b$, $0 \le y \le c$. It follows that

$$A_{mn} = \frac{4}{bc} \int_0^c \int_0^b f(x, y) \sin\frac{m\pi}{b} x \sin\frac{n\pi}{c} y \, dx \, dy. \tag{12}$$

Thus the solution of the BVP consists of (10) with the A_{mn} defined in (12). ∎

The series (11) with coefficients (12) is called a **sine series in two variables** or a **double sine series**. We summarize next the **cosine series in two variables**.

The **double cosine series** of a function $f(x, y)$ defined over a rectangular region $0 \le x \le b, 0 \le y \le c$ is given by

$$f(x, y) = A_{00} + \sum_{m=1}^{\infty} A_{m0} \cos \frac{m\pi}{b} x + \sum_{n=1}^{\infty} A_{0n} \cos \frac{n\pi}{c} y$$

$$+ \sum_{m=1}^{\infty} \sum_{n=1}^{\infty} A_{mn} \cos \frac{m\pi}{b} x \cos \frac{n\pi}{c} y,$$

where

$$A_{00} = \frac{1}{bc} \int_0^c \int_0^b f(x, y) \, dxdy$$

$$A_{m0} = \frac{2}{bc} \int_0^c \int_0^b f(x, y) \cos \frac{m\pi}{b} x \, dxdy$$

$$A_{0n} = \frac{2}{bc} \int_0^c \int_0^b f(x, y) \cos \frac{n\pi}{c} y \, dxdy$$

$$A_{mn} = \frac{4}{bc} \int_0^c \int_0^b f(x, y) \cos \frac{m\pi}{b} x \cos \frac{n\pi}{c} y \, dxdy.$$

For a problem leading to a double-cosine series see Problem 2 in Exercises 12.8.

EXERCISES 12.8

Answers to selected odd-numbered problems begin on page ANS-19.

In Problems 1 and 2 solve the heat equation (1) subject to the given conditions.

1. $u(0, y, t) = 0$, $\quad u(\pi, y, t) = 0$
$u(x, 0, t) = 0$, $\quad u(x, \pi, t) = 0$
$u(x, y, 0) = u_0$

2. $\left. \dfrac{\partial u}{\partial x} \right|_{x=0} = 0$, $\quad \left. \dfrac{\partial u}{\partial x} \right|_{x=1} = 0$

$\left. \dfrac{\partial u}{\partial y} \right|_{y=0} = 0$, $\quad \left. \dfrac{\partial u}{\partial y} \right|_{y=1} = 0$

$u(x, y, 0) = xy$

In Problems 3 and 4 solve the wave equation (2) subject to the given conditions.

3. $u(0, y, t) = 0$, $\quad u(\pi, y, t) = 0$
$u(x, 0, t) = 0$, $\quad u(x, \pi, t) = 0$
$u(x, y, 0) = xy(x - \pi)(y - \pi)$
$\left. \dfrac{\partial u}{\partial t} \right|_{t=0} = 0$

4. $u(0, y, t) = 0$, $\quad u(b, y, t) = 0$
$u(x, 0, t) = 0$, $\quad u(x, c, t) = 0$
$u(x, y, 0) = f(x, y)$
$\left. \dfrac{\partial u}{\partial t} \right|_{t=0} = g(x, y)$

The steady-state temperature $u(x, y, z)$ in the rectangular parallelepiped shown in Figure 12.24 satisfies Laplace's equation in three dimensions:

$$\frac{\partial^2 u}{\partial x^2} + \frac{\partial^2 u}{\partial y^2} + \frac{\partial^2 u}{\partial z^2} = 0. \tag{13}$$

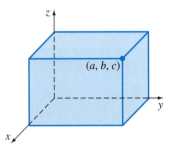

FIGURE 12.24 Rectangular parallelepiped in Problems 5 and 6

5. Solve Laplace's equation (13) if the top ($z = c$) of the parallelepiped is kept at temperature $f(x, y)$ and the remaining sides are kept at temperature zero.

6. Solve Laplace's equation (13) if the bottom ($z = 0$) of the parallelepiped is kept at temperature $f(x, y)$ and the remaining sides are kept at temperature zero.

Answers to selected odd-numbered problems begin on page ANS-19.

1. Use separation of variables to find product solutions of

$$\frac{\partial^2 u}{\partial x\, \partial y} = u.$$

2. Use separation of variables to find product solutions of

$$\frac{\partial^2 u}{\partial x^2} + \frac{\partial^2 u}{\partial y^2} + 2\frac{\partial u}{\partial x} + 2\frac{\partial u}{\partial y} = 0.$$

Is it possible to choose a separation constant so that both X and Y are oscillatory functions?

3. Find a steady-state solution $\psi(x)$ of the boundary-value problem

$$k\frac{\partial^2 u}{\partial x^2} = \frac{\partial u}{\partial t}, \quad 0 < x < \pi, \quad t > 0,$$

$$u(0, t) = u_0, \quad -\frac{\partial u}{\partial x}\Big|_{x=\pi} = u(\pi, t) - u_1, \quad t > 0$$

$$u(x, 0) = 0, \quad 0 < x < \pi.$$

4. Give a physical interpretation for the boundary conditions in Problem 3.

5. At $t = 0$ a string of unit length is stretched on the positive x-axis. The ends of the string $x = 0$ and $x = 1$ are secured on the x-axis for $t > 0$. Find the displacement $u(x, t)$ if the initial velocity $g(x)$ is as given in Figure 12.25.

FIGURE 12.25 Initial velocity $g(x)$ in Problem 5

6. The partial differential equation

$$\frac{\partial^2 u}{\partial x^2} + x^2 = \frac{\partial^2 u}{\partial t^2}$$

is a form of the wave equation when an external vertical force proportional to the square of the horizontal distance from the left end is applied to the string. The string is secured at $x = 0$ one unit above the x-axis and on the x-axis at $x = 1$ for $t > 0$. Find the displacement $u(x, t)$ if the string starts from rest from the initial displacement $f(x)$.

7. Find the steady-state temperature $u(x, y)$ in the square plate shown in Figure 12.26.

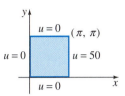

FIGURE 12.26 Square plate in Problem 7

8. Find the steady-state temperature $u(x, y)$ in the semi-infinite plate shown in Figure 12.27.

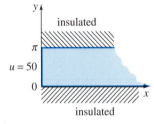

FIGURE 12.27 Semi-infinite plate in Problem 8

9. Solve Problem 8 if the boundaries $y = 0$ and $y = \pi$ are held at temperature zero for all time.

10. Find the temperature $u(x, t)$ in the infinite plate of width $2L$ shown in Figure 12.28 if the initial temperature is u_0 throughout. (*Hint:* $u(x, 0) = u_0$, $-L < x < L$ is an even function of x.)

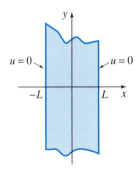

FIGURE 12.28 Infinite plate in Problem 10

11. Solve the boundary-value problem

$$\frac{\partial^2 u}{\partial x^2} = \frac{\partial u}{\partial t}, \quad 0 < x < \pi, \quad t > 0$$

$$u(0, t) = 0, \quad u(\pi, t) = 0, \quad t > 0$$

$$u(x, 0) = \sin x, \quad 0 < x < \pi.$$

12. Solve the boundary-value problem

$$\frac{\partial^2 u}{\partial x^2} + \sin x = \frac{\partial u}{\partial t}, \quad 0 < x < \pi, \quad t > 0$$

$$u(0, t) = 400, \quad u(\pi, t) = 200, \quad t > 0$$

$$u(x, 0) = 400 + \sin x, \quad 0 < x < \pi.$$

13. Find a formal series solution of the problem

$$\frac{\partial^2 u}{\partial x^2} + 2\frac{\partial u}{\partial x} = \frac{\partial^2 u}{\partial t^2} + 2\frac{\partial u}{\partial t} + u, \quad 0 < x < \pi, \quad t > 0$$

$$u(0, t) = 0, \quad u(\pi, t) = 0, \quad t > 0$$

$$\left.\frac{\partial u}{\partial t}\right|_{t=0} = 0, \quad 0 < x < \pi.$$

14. The concentration $c(x, t)$ of a substance that both diffuses in a medium and is convected by the currents in the medium satisfies the partial differential equation

$$k\frac{\partial^2 c}{\partial x^2} - h\frac{\partial c}{\partial x} = \frac{\partial c}{\partial t}, \quad k \text{ and } h \text{ constants.}$$

Solve the PDE subject to

$$c(0, t) = 0, \quad c(1, t) = 0, \quad t > 0$$

$$c(x, 0) = c_0, \quad 0 < x < 1,$$

where c_0 is a constant.

13

BOUNDARY-VALUE PROBLEMS IN OTHER COORDINATE SYSTEMS

13.1 Polar Coordinates
13.2 Polar and Cylindrical Coordinates
13.3 Spherical Coordinates
CHAPTER 13 IN REVIEW

All the boundary-value problems that we have considered up to this point were expressed only in terms of a rectangular coordinate system. But if we wished to find, say, temperatures in a circular plate, in a circular cylinder, or in a sphere, we would naturally try to describe the problem in terms of polar coordinates, cylindrical coordinates, or spherical coordinates, respectively. In this chapter we shall see that by trying to solve BVPs in these latter three coordinate systems by the method of separation of variables, the theory of Fourier-Bessel series and Fourier-Legendre series is put to practical use.

Time: $t = 0.3$ s

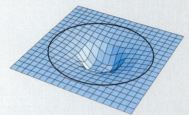

Time: $t = 0.6$ s

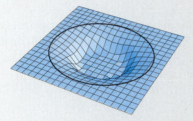

Time: $t = 1.2$ s

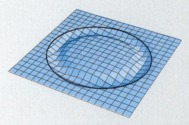

Time: $t = 5.0$ s

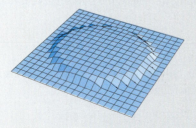

Graphs, for various times, of a 5-term partial sum of a series solution representing the displacement $u(r, t)$ of a circular membrane. See pages 510 and 511.

13.1 POLAR COORDINATES

INTRODUCTION: Because only steady-state temperature problems in polar coordinates are considered in this section, the first thing we must do is convert the familiar Laplace's equation in rectangular coordinates to polar coordinates.

REVIEW MATERIAL: For boundary-value problems in this section, the method of separation of variables leads to a Cauchy-Euler differential equation. We recommend that you review the forms of the general solutions of Cauchy-Euler DEs in Section 4.7. Also see "Review of DEs" in Section 11.4.

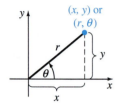

FIGURE 13.1 Polar coordinates of a point (x, y) are (r, θ)

LAPLACIAN IN POLAR COORDINATES The relationships between polar coordinates in the plane and rectangular coordinates are given by

$$x = r \cos \theta, \quad y = r \sin \theta, \quad \text{and} \quad r^2 = x^2 + y^2, \quad \tan \theta = \frac{y}{x}.$$

See Figure 13.1. The first pair of equations transforms polar coordinates (r, θ) into rectangular coordinates (x, y); the second pair of equations enables us to transform rectangular coordinates into polar coordinates. These equations also make it possible to convert the two-dimensional Laplacian $\nabla^2 u = \partial^2 u/\partial x^2 + \partial^2 u/\partial y^2$ into polar coordinates. You are encouraged to work through the details of the Chain Rule and show that

$$\frac{\partial u}{\partial x} = \frac{\partial u}{\partial r}\frac{\partial r}{\partial x} + \frac{\partial u}{\partial \theta}\frac{\partial \theta}{\partial x} = \cos \theta \frac{\partial u}{\partial r} - \frac{\sin \theta}{r}\frac{\partial u}{\partial \theta}$$

$$\frac{\partial u}{\partial y} = \frac{\partial u}{\partial r}\frac{\partial r}{\partial y} + \frac{\partial u}{\partial \theta}\frac{\partial \theta}{\partial y} = \sin \theta \frac{\partial u}{\partial r} + \frac{\cos \theta}{r}\frac{\partial u}{\partial \theta}$$

$$\frac{\partial^2 u}{\partial x^2} = \cos^2 \theta \frac{\partial^2 u}{\partial r^2} - \frac{2 \sin \theta \cos \theta}{r}\frac{\partial^2 u}{\partial r \partial \theta} + \frac{\sin^2 \theta}{r^2}\frac{\partial^2 u}{\partial \theta^2} + \frac{\sin^2 \theta}{r}\frac{\partial u}{\partial r} + \frac{2 \sin \theta \cos \theta}{r^2}\frac{\partial u}{\partial \theta} \quad (1)$$

$$\frac{\partial^2 u}{\partial y^2} = \sin^2 \theta \frac{\partial^2 u}{\partial r^2} + \frac{2 \sin \theta \cos \theta}{r}\frac{\partial^2 u}{\partial r \partial \theta} + \frac{\cos^2 \theta}{r^2}\frac{\partial^2 u}{\partial \theta^2} + \frac{\cos^2 \theta}{r}\frac{\partial u}{\partial r} - \frac{2 \sin \theta \cos \theta}{r^2}\frac{\partial u}{\partial \theta}. \quad (2)$$

Adding (1) and (2) and simplifying yields the Laplacian of u in polar coordinates

$$\nabla^2 u = \frac{\partial^2 u}{\partial r^2} + \frac{1}{r}\frac{\partial u}{\partial r} + \frac{1}{r^2}\frac{\partial^2 u}{\partial \theta^2}.$$

In this section we focus only on boundary-value problems involving Laplace's equation $\nabla^2 u = 0$ in polar coordinates:

$$\frac{\partial^2 u}{\partial r^2} + \frac{1}{r}\frac{\partial u}{\partial r} + \frac{1}{r^2}\frac{\partial^2 u}{\partial \theta^2} = 0. \quad (3)$$

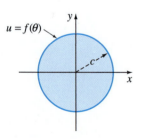

FIGURE 13.2 Dirichlet problem for a circle

Our first example is a Dirichlet problem for a circular disk. We wish to solve Laplace's equation (3) for the steady-state temperature $u(r, \theta)$ in a circular disk or plate of radius c when the temperature on the circumference is $u(c, \theta) = f(\theta)$, $0 < \theta < 2\pi$. See Figure 13.2. It is assumed that the two faces of the plate are insulated. This seemingly simple problem is unlike any we encountered in the previous chapter.

> **EXAMPLE 1**   **Steady Temperatures in a Circular Plate**

Solve Laplace's equation (3) subject to $u(c, \theta) = f(\theta), 0 < \theta < 2\pi$.

SOLUTION   Before attempting separation of variables, we note that the single boundary condition is nonhomogeneous. In other words, there are no explicit conditions in the statement of the problem that enable us to determine either the coefficients in the solutions of the separated ODEs or the required eigenvalues. However, there are some *implicit* conditions.

First, our physical intuition leads us to expect that the temperature $u(r, \theta)$ should be continuous and therefore bounded inside the circle $r = c$. In addition, the temperature $u(r, \theta)$ should be single-valued; this means that the value of u should be the same at a specified point in the circle regardless of the polar description of that point. Because $(r, \theta + 2\pi)$ is an equivalent description of the point (r, θ), we must have $u(r, \theta) = u(r, \theta + 2\pi)$. That is, $u(r, \theta)$ must be periodic in θ with period 2π. If we seek a product solution $u = R(r)\Theta(\theta)$, then $\Theta(\theta)$ needs to be 2π-periodic.

With all this in mind we choose to write the separation constant in the separation of variables as λ:

$$\frac{r^2R'' + rR'}{R} = -\frac{\Theta''}{\Theta} = \lambda.$$

The separated equations are then

$$r^2R'' + rR' - \lambda R = 0 \tag{4}$$

$$\Theta'' + \lambda\Theta = 0. \tag{5}$$

We are seeking a solution of the problem

$$\Theta'' + \lambda\Theta = 0, \quad \Theta(\theta) = \Theta(\theta + 2\pi). \tag{6}$$

Although (6) is not a regular Sturm-Liouville problem, nonetheless the problem generates eigenvalues and eigenfunctions. The latter form an orthogonal set on the interval $[0, 2\pi]$.

Of the three possible general solutions of (5),

$$\Theta(\theta) = c_1 + c_2\theta, \qquad\qquad \lambda = 0 \tag{7}$$

$$\Theta(\theta) = c_1\cosh\alpha\theta + c_2\sinh\alpha\theta, \quad \lambda = -\alpha^2 < 0 \tag{8}$$

$$\Theta(\theta) = c_1\cos\alpha\theta + c_2\sin\alpha\theta, \quad \lambda = \alpha^2 > 0 \tag{9}$$

we can dismiss (8) as inherently nonperiodic unless $c_1 = c_2 = 0$. Similarly, solution (7) is nonperiodic unless we define $c_2 = 0$. The remaining constant solution $\Theta(\theta) = c_1, c_1 \neq 0$, can be assigned any period, and so $\lambda = 0$ is an eigenvalue. Finally, solution (9) will be 2π-periodic if we take $\alpha = n$, where $n = 1, 2, \ldots$.* The eigenvalues of (6) are then $\lambda_0 = 0$ and $\lambda_n = n^2$, $n = 1, 2, \ldots$. If we correspond $\lambda_0 = 0$ with $n = 0$, the eigenfunctions of (6) are

$$\Theta(\theta) = c_1, \quad n = 0, \quad \text{and} \quad \Theta(\theta) = c_1\cos n\theta + c_2\sin n\theta, \quad n = 1, 2, \ldots.$$

When $\lambda_n = n^2, n = 0, 1, 2, \ldots$, the solutions of the Cauchy-Euler DE (4) are

$$R(r) = c_3 + c_4\ln r, \quad n = 0, \tag{10}$$

$$R(r) = c_3r^n + c_4r^{-n}, \quad n = 1, 2, \ldots. \tag{11}$$

Now observe in (11) that $r^{-n} = 1/r^n$. In either of the solutions (10) or (11) we must define $c_4 = 0$ to guarantee that the solution u is bounded at the center of the plate

---

*For example, note that $\cos n(\theta + 2\pi) = \cos(n\theta + 2n\pi) = \cos n\theta$.

(which is $r = 0$). Thus product solutions $u_n = R(r)\Theta(\theta)$ for Laplace's equation in polar coordinates are

$$u_0 = A_0, \quad n = 0, \quad \text{and} \quad u_n = r^n(A_n \cos n\theta + B_n \sin n\theta), \quad n = 1, 2, \ldots,$$

where we have replaced $c_3 c_1$ by A_0 for $n = 0$ and by A_n for $n = 1, 2, \ldots$; the combination $c_3 c_2$ has been replaced by B_n. The superposition principle then gives

$$u(r, \theta) = A_0 + \sum_{n=1}^{\infty} r^n(A_n \cos n\theta + B_n \sin n\theta). \tag{12}$$

By applying the boundary condition at $r = c$ to (12), we recognize

$$f(\theta) = A_0 + \sum_{n=1}^{\infty} c^n(A_n \cos n\theta + B_n \sin n\theta)$$

as an expansion of f in a full Fourier series. Consequently, we can make the identifications

$$A_0 = \frac{a_0}{2}, \quad c^n A_n = a_n, \quad \text{and} \quad c^n B_n = b_n.$$

That is,

$$A_0 = \frac{1}{2\pi} \int_0^{2\pi} f(\theta) \, d\theta \tag{13}$$

$$A_n = \frac{1}{c^n \pi} \int_0^{2\pi} f(\theta) \cos n\theta \, d\theta \tag{14}$$

$$B_n = \frac{1}{c^n \pi} \int_0^{2\pi} f(\theta) \sin n\theta \, d\theta. \tag{15}$$

The solution of the problem consists of the series given in (12), where the coefficients A_0, A_n, and B_n are defined in (13), (14), and (15).

Observe in Example 1 that corresponding to each *positive* eigenvalue $\lambda_n = n^2$, $n = 1, 2, \ldots$, there are two different eigenfunctions—namely, $\cos n\theta$ and $\sin n\theta$. In this situation the eigenvalues are sometimes called **double eigenvalues.**

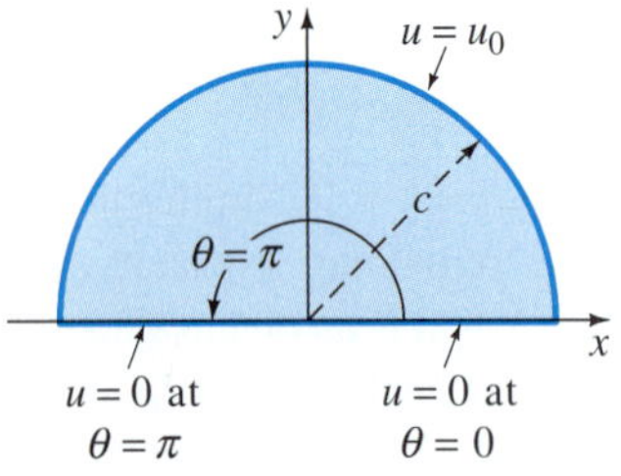

FIGURE 13.3 Semicircular plate in Example 2

EXAMPLE 2   **Steady Temperatures in a Semicircular Plate**

Find the steady-state temperature $u(r, \theta)$ in the semicircular plate shown in Figure 13.3.

SOLUTION   The boundary-value problem is

$$\frac{\partial^2 u}{\partial r^2} + \frac{1}{r}\frac{\partial u}{\partial r} + \frac{1}{r^2}\frac{\partial^2 u}{\partial \theta^2} = 0, \quad 0 < \theta < \pi, \quad 0 < r < c$$

$$u(c, \theta) = u_0, \quad 0 < \theta < \pi,$$

$$u(r, 0) = 0, \quad u(r, \pi) = 0, \quad 0 < r < c.$$

Defining $u = R(r)\Theta(\theta)$ and separating variables gives

$$\frac{r^2 R'' + r R'}{R} = -\frac{\Theta''}{\Theta} = \lambda$$

and

$$r^2 R'' + r R' - \lambda R = 0 \tag{16}$$

$$\Theta'' + \lambda \Theta = 0. \tag{17}$$

The homogeneous conditions stipulated at the boundaries $\theta = 0$ and $\theta = \pi$ translate into $\Theta(0) = 0$ and $\Theta(\pi) = 0$. These conditions together with equation (17) constitute a regular Sturm-Liouville problem:

$$\Theta'' + \lambda\Theta = 0, \quad \Theta(0) = 0, \quad \Theta(\pi) = 0. \tag{18}$$

This familiar problem* possesses eigenvalues $\lambda_n = n^2$ and eigenfunctions $\Theta(\theta) = c_2 \sin n\theta$, $n = 1, 2, \ldots$. Also, by replacing λ by n^2, the solution of (16) is $R(r) = c_3 r^n + c_4 r^{-n}$. The reasoning used in Example 1, namely, that we expect a solution u of the problem to be bounded at $r = 0$, prompts us to define $c_4 = 0$. Therefore $u_n = R(r)\Theta(\theta) = A_n r^n \sin n\theta$, and

$$u(r, \theta) = \sum_{n=1}^{\infty} A_n r^n \sin n\theta.$$

The remaining boundary condition at $r = c$ gives the sine series

$$u_0 = \sum_{n=1}^{\infty} A_n c^n \sin n\theta.$$

Consequently,

$$A_n c^n = \frac{2}{\pi} \int_0^{\pi} u_0 \sin n\theta \, d\theta,$$

and so

$$A_n = \frac{2u_0}{\pi c^n} \frac{1 - (-1)^n}{n}.$$

Hence the solution of the problem is given by

$$u(r, \theta) = \frac{2u_0}{\pi} \sum_{n=1}^{\infty} \frac{1 - (-1)^n}{n} \left(\frac{r}{c}\right)^n \sin n\theta.$$

---

*The problem in (18) is Example 2 of Section 5.2 with $L = \pi$.

EXERCISES 13.1

Answers to selected odd-numbered problems begin on page ANS-20.

In Problems 1–4 find the steady-state temperature $u(r, \theta)$ in a circular plate of radius $r = 1$ if the temperature on the circumference is as given.

1. $u(1, \theta) = \begin{cases} u_0, & 0 < \theta < \pi \\ 0, & \pi < \theta < 2\pi \end{cases}$

2. $u(1, \theta) = \begin{cases} \theta, & 0 < \theta < \pi \\ \pi - \theta, & \pi < \theta < 2\pi \end{cases}$

3. $u(1, \theta) = 2\pi\theta - \theta^2$, $\quad 0 < \theta < 2\pi$

4. $u(1, \theta) = \theta$, $\quad 0 < \theta < 2\pi$

5. Solve the exterior Dirichlet problem for a circular disk of radius c if $u(c, \theta) = f(\theta)$, $0 < \theta < 2\pi$. In other words, find the steady-state temperature $u(r, \theta)$ in a plate that coincides with the entire xy-plane in which a circular hole of radius c has been cut out around the origin and the temperature on the circumference of the hole is $f(\theta)$. (*Hint:* Assume that the temperature is bounded as $r \to \infty$.)

6. Find the steady-state temperature in the quarter-circular plate shown in Figure 13.4.

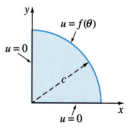

FIGURE 13.4 Quarter-circular plate in Problem 6

7. If the boundaries $\theta = 0$ and $\theta = \pi/2$ in Figure 13.4 are insulated, we then have, respectively,

$$\left.\frac{\partial u}{\partial \theta}\right|_{\theta=0} = 0, \quad \left.\frac{\partial u}{\partial \theta}\right|_{\theta=\pi/2} = 0.$$

Find the steady-state temperature if

$$u(c, \theta) = \begin{cases} 1, & 0 < \theta < \pi/4 \\ 0, & \pi/4 < \theta < \pi/2. \end{cases}$$

8. Find the steady-state temperature in the infinite wedge-shaped plate shown in Figure 13.5. (*Hint:* Assume that the temperature is bounded as $r \to 0$ and as $r \to \infty$.)

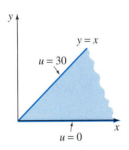

FIGURE 13.5   Wedge-shaped plate in Problem 8

9. Find the steady-state temperature $u(r, \theta)$ in the circular ring shown in Figure 13.6. (*Hint:* Proceed as in Example 1.)

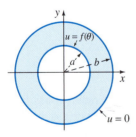

FIGURE 13.6   Ring-shaped plate in Problem 9

10. If the boundary conditions for the circular ring in Figure 13.6 are $u(a, \theta) = u_0$, $u(b, \theta) = u_1$, $0 < \theta < 2\pi$, u_0 and u_1 constants, show that the steady-state temperature is given by

$$u(r, \theta) = \frac{u_0 \ln(r/b) - u_1 \ln(r/a)}{\ln(a/b)}.$$

(*Hint:* Try a solution of the form $u(r, \theta) = v(r, \theta) + \psi(r)$.)

11. Find the steady-state temperature $u(r, \theta)$ in a semicircular ring if

$$u(a, \theta) = \theta(\pi - \theta), \quad u(b, \theta) = 0, \quad 0 < \theta < \pi$$

$$u(r, 0) = 0, \qquad\qquad u(r, \pi) = 0, \quad a < r < b.$$

12. Find the steady-state temperature $u(r, \theta)$ in a semicircular plate of radius $r = 1$ if

$$u(1, \theta) = u_0, \quad 0 < \theta < \pi$$

$$u(r, 0) = 0, \quad u(r, \pi) = u_0, \quad 0 < r < 1,$$

u_0 a constant.

13. Find the steady-state temperature $u(r, \theta)$ in a semicircular plate of radius $r = 2$ if

$$u(2, \theta) = \begin{cases} u_0, & 0 < \theta < \pi/2 \\ 0, & \pi/2 < \theta < \pi, \end{cases}$$

u_0 a constant, and the edges $\theta = 0$ and $\theta = \pi$ are insulated.

DISCUSSION/PROJECT PROBLEMS

14. Consider the circular ring shown in Figure 13.6. Discuss how the steady-state temperature $u(r, \theta)$ can be found when the boundary conditions are $u(a, \theta) = f(\theta)$, $u(b, \theta) = g(\theta)$, $0 \le \theta \le 2\pi$.

15. Carry out your ideas from Problem 14 to find the steady-state temperature $u(r, \theta)$ in the circular ring shown in Figure 13.6. when the boundary conditions are $u\left(\frac{1}{2}, \theta\right) = 100(1 + 0.5 \cos \theta)$, $u(1, \theta) = 200$, $0 \le \theta \le 2\pi$.

COMPUTER LAB ASSIGNMENTS

16. (a) Find the series solution for $u(r, \theta)$ in Example 1 when

$$u(1, \theta) = \begin{cases} 100, & 0 < \theta < \pi \\ 0, & \pi < \theta < 2\pi. \end{cases}$$

(See Problem 1.)
   (b) Use a CAS or a graphing utility to plot the partial sum $S_5(r, \theta)$ consisting of the first five nonzero terms of the solution in part (a) for $r = 0.9$, $r = 0.7$, $r = 0.5$, $r = 0.3$, and $r = 0.1$. Superimpose the graphs on the same coordinate axes.
   (c) Approximate the temperatures $u(0.9, 1.3)$, $u(0.7, 2)$, $u(0.5, 3.5)$, $u(0.3, 4)$, $u(0.1, 5.5)$. Then approximate $u(0.9, 2\pi - 1.3)$, $u(0.7, 2\pi - 2)$, $u(0.5, 2\pi - 3.5)$, $u(0.3, 2\pi - 4)$, $u(0.1, 2\pi - 5.5)$.
   (d) What is the temperature at the center of the circular plate? Why is it appropriate to call this value the average temperature in the plate? (*Hint:* Look at the graphs in part (b) and look at the numbers in part (c).)

13.2 POLAR AND CYLINDRICAL COORDINATES

INTRODUCTION: In this section we consider boundary-value problems involving forms of the heat and wave equations in polar coordinates and a form of Laplace's equation in cylindrical coordinates. There is a commonality throughout the examples and exercises: Each boundary-value problem in this section possesses radial symmetry.

REVIEW MATERIAL: The types of boundary-value problems that are considered in this section give rise to the parametric Bessel differential equation, Bessel functions, and Fourier-Bessel series. Because there are three forms of a Fourier-Bessel series we recommend that you review the summary of those series in Definition 11.7 of Section 11.5. Also review Section 6.3.

RADIAL SYMMETRY The two-dimensional heat and wave equations

$$k\left(\frac{\partial^2 u}{\partial x^2} + \frac{\partial^2 u}{\partial y^2}\right) = \frac{\partial u}{\partial t} \quad \text{and} \quad a^2\left(\frac{\partial^2 u}{\partial x^2} + \frac{\partial^2 u}{\partial y^2}\right) = \frac{\partial^2 u}{\partial t^2}$$

expressed in polar coordinates are, in turn,

$$k\left(\frac{\partial^2 u}{\partial r^2} + \frac{1}{r}\frac{\partial u}{\partial r} + \frac{1}{r^2}\frac{\partial^2 u}{\partial \theta^2}\right) = \frac{\partial u}{\partial t} \quad \text{and} \quad a^2\left(\frac{\partial^2 u}{\partial r^2} + \frac{1}{r}\frac{\partial u}{\partial r} + \frac{1}{r^2}\frac{\partial^2 u}{\partial \theta^2}\right) = \frac{\partial^2 u}{\partial t^2}, \quad (1)$$

where $u = u(r, \theta, t)$. To solve a boundary-value problem involving either of these equations by separation of variables, we must define $u = R(r)\Theta(\theta)T(t)$. As in Section 12.8, this assumption leads to multiple infinite series. See Problem 12 in Exercises 13.2. In the discussion that follows we shall consider the simpler, but still important, problems that possess **radial symmetry**—that is, problems in which the unknown function u is independent of the angular coordinate θ. In this case the heat and wave equations in (1) take, respectively, the forms

$$k\left(\frac{\partial^2 u}{\partial r^2} + \frac{1}{r}\frac{\partial u}{\partial r}\right) = \frac{\partial u}{\partial t} \quad \text{and} \quad a^2\left(\frac{\partial^2 u}{\partial r^2} + \frac{1}{r}\frac{\partial u}{\partial r}\right) = \frac{\partial^2 u}{\partial t^2}, \quad (2)$$

where $u = u(r, t)$. Vibrations described by the second equation in (2) are said to be **radial vibrations.**

The first example deals with the free undamped radial vibrations of a thin circular membrane. We assume that the displacements are small and that the motion is such that each point on the membrane moves in a direction perpendicular to the xy-plane (transverse vibrations)—that is, the u-axis is perpendicular to the xy-plane. A physical model to keep in mind while working through this example is a vibrating drumhead.

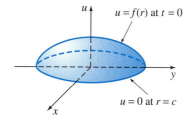

FIGURE 13.7 Initial displacement of a circular membrane in Example 1

| **EXAMPLE 1** | **Radial Vibrations of a Circular Membrane** |

Find the displacement $u(r, t)$ of a circular membrane of radius c clamped along its circumference if its initial displacement is $f(r)$ and its initial velocity is $g(r)$. See Figure 13.7.

SOLUTION The boundary-value problem to be solved is

$$a^2\left(\frac{\partial^2 u}{\partial r^2} + \frac{1}{r}\frac{\partial u}{\partial r}\right) = \frac{\partial^2 u}{\partial t^2}, \quad 0 < r < c, \quad t > 0$$

$$u(c, t) = 0, \quad t > 0,$$

$$u(r, 0) = f(r), \quad \left.\frac{\partial u}{\partial t}\right|_{t=0} = g(r), \quad 0 < r < c.$$

Substituting $u = R(r)T(t)$ into the partial differential equation and separating variables gives

$$\frac{R'' + \frac{1}{r}R'}{R} = \frac{T''}{a^2 T} = -\lambda. \tag{3}$$

Note that in (3) we have returned to our usual separation constant $-\lambda$. The two equations obtained from (3) are

$$rR'' + R' + \lambda rR = 0 \tag{4}$$

and
$$T'' + a^2\lambda T = 0. \tag{5}$$

Because of the vibrational nature of the problem, equation (5) suggests that we use only $\lambda = \alpha^2 > 0$, $\alpha > 0$, since this choice leads to periodic functions. Also, take a second look at equation (4); it is *not* a Cauchy-Euler equation but is the parametric Bessel equation of order $\nu = 0$, that is, $rR'' + R' + \alpha^2 rR = 0$. From (13) of Section 6.3 the general solution of the last equation is

$$R = c_1 J_0(\alpha r) + c_2 Y_0(\alpha r). \tag{6}$$

The general solution of the familiar equation (5) is

$$T = c_3 \cos a\alpha t + c_4 \sin a\alpha t.$$

Now recall that $Y_0(\alpha r) \to -\infty$ as $r \to 0^+$, so the implicit assumption that the displacement $u(r, t)$ should be bounded at $r = 0$ forces us to define $c_2 = 0$ in (6). Thus $R = c_1 J_0(\alpha r)$.

Since the boundary condition $u(c, t) = 0$ is equivalent to $R(c) = 0$, we must have $c_1 J_0(\alpha c) = 0$. We rule out $c_1 = 0$ (this would lead to a trivial solution of the PDE), so consequently,

$$J_0(\alpha c) = 0. \tag{7}$$

If $x_n = \alpha_n c$ are the positive roots of (7), then $\alpha_n = x_n/c$, and so the eigenvalues of the problem are $\lambda_n = \alpha_n^2 = x_n^2/c^2$, and the eigenfunctions are $c_1 J_0(\alpha_n r)$. Product solutions that satisfy the partial differential equation and the boundary conditions are

$$u_n = R(r)T(t) = (A_n \cos a\alpha_n t + B_n \sin a\alpha_n t) J_0(\alpha_n r), \tag{8}$$

where we have done the usual relabeling of constants. The superposition principle gives

$$u(r, t) = \sum_{n=1}^{\infty} (A_n \cos a\alpha_n t + B_n \sin a\alpha_n t) J_0(\alpha_n r). \tag{9}$$

The given initial conditions determine the coefficients A_n and B_n.

Setting $t = 0$ in (9) and using $u(r, 0) = f(r)$ gives

$$f(r) = \sum_{n=1}^{\infty} A_n J_0(\alpha_n r). \tag{10}$$

The last result is recognized as the Fourier-Bessel expansion of the function f on the interval $(0, c)$. Hence by a direct comparison of (7) and (10) with (8) and (15) of Section 11.5 we can identify the coefficients A_n with those given in (16) of Section 11.5:

$$A_n = \frac{2}{c^2 J_1^2(\alpha_n c)} \int_0^c r J_0(\alpha_n r) f(r)\, dr. \tag{11}$$

Next, we differentiate (9) with respect to t, set $t = 0$, and use $u_t(r, 0) = g(r)$:

$$g(r) = \sum_{n=1}^{\infty} a\alpha_n B_n J_0(\alpha_n r).$$

This is now a Fourier-Bessel expansion of the function g. By identifying the total coefficient $a\alpha_n B_n$ with (16) of Section 11.5, we can write

$$B_n = \frac{2}{a\alpha_n c^2 J_1^2(\alpha_n c)} \int_0^c r J_0(\alpha_n r) g(r) \, dr. \qquad (12)$$

Finally, the solution of the original boundary-value problem is the series in (9) with coefficients A_n and B_n defined in (11) and (12). ◾

STANDING WAVES Analogous to (11) of Section 12.4, the product solutions (8) are called **standing waves.** For $n = 1, 2, 3, \ldots$ the standing waves are basically the graph of $J_0(\alpha_n r)$ with the time varying amplitude

$$A_n \cos a\alpha_n t + B_n \sin a\alpha_n t.$$

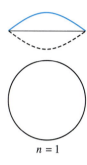

$n = 1$

(a)

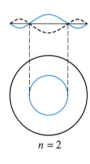

$n = 2$

(b)

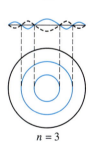

$n = 3$

(c)

FIGURE 13.8 Standing waves

The standing waves at different values of time are represented by the dashed graphs in Figure 13.8. The zeros of each standing wave in the interval $(0, c)$ are the roots of $J_0(\alpha_n r) = 0$ and correspond to the set of points on a standing wave where there is no motion. The set of points is called a **nodal line.** If (as in Example 1) the positive roots of $J_0(\alpha_n c) = 0$ are denoted by x_n, then $x_n = \alpha_n c$ implies that $\alpha_n = x_n/c$ and consequently the zeros of the standing wave are determined from

$$J_0(\alpha_n r) = J_0\left(\frac{x_n}{c} r\right) = 0.$$

Now from Table 6.1 the first three positive zeros of J_0 are (approximately) $x_1 = 2.4$, $x_2 = 5.5$, and $x_3 = 8.7$. Thus for $n = 1$ the first positive root of

$$J_0\left(\frac{x_1}{c} r\right) = 0 \quad \text{is} \quad \frac{2.4}{c} r = 2.4 \quad \text{or} \quad r = c.$$

Since we are seeking zeros of the standing waves in the open interval $(0, c)$, the last result means that the first standing wave has no nodal line. For $n = 2$ the first two positive roots of

$$J_0\left(\frac{x_2}{c} r\right) = 0 \quad \text{are determined from} \quad \frac{5.5}{c} r = 2.4 \quad \text{and} \quad \frac{5.5}{c} r = 5.5.$$

Thus the second standing wave has one nodal line defined by $r = x_1 c/x_2 = 2.4c/5.5$. Note that $r \approx 0.44c < c$. For $n = 3$ a similar analysis shows that there are two nodal lines defined by $r = x_1 c/x_3 = 2.4c/8.7$ and $r = x_2 c/x_3 = 5.5c/8.7$. In general, the nth standing wave has $n - 1$ nodal lines $r = x_1 c/x_n, r = x_2 c/x_n, \ldots, r = x_{n-1} c/x_n$. Since $r = constant$ is an equation of a circle in polar coordinates, we see in Figure 13.8 that the nodal lines of a standing wave are concentric circles.

USE OF COMPUTERS It is possible to see the effect of a single drumbeat for the model solved in Example 1 by means of the animation capabilities of a computer algebra system. In Problem 13 in Exercises 13.2 you are asked to find the solution given in (6) when

$$c = 1, \quad f(r) = 0, \quad \text{and} \quad g(r) = \begin{cases} -v_0, & 0 \le r < b \\ 0, & b \le r < 1. \end{cases}$$

Some frames of a "movie" of the vibrating drumhead are given in Figure 13.9.

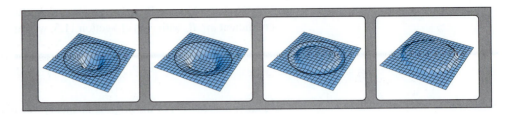

FIGURE 13.9   Frames of a CAS "movie"

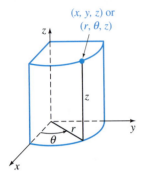

FIGURE 13.10   Cylindrical coordinates of a point (x, y, z) are (r, θ, z).

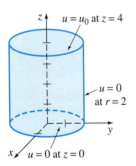

FIGURE 13.11   Circular cylinder in Example 2

LAPLACIAN IN CYLINDRICAL COORDINATES   In Figure 13.10 we can see that the relationship between the cylindrical coordinates of a point in space and its rectangular coordinates is given by

$$x = r \cos \theta, \quad y = r \sin \theta, \quad z = z.$$

It follows immediately from the derivation of the Laplacian in polar coordinates (see Section 13.1) that the Laplacian of a function u in cylindrical coordinates is

$$\nabla^2 u = \frac{\partial^2 u}{\partial r^2} + \frac{1}{r} \frac{\partial u}{\partial r} + \frac{1}{r^2} \frac{\partial^2 u}{\partial \theta^2} + \frac{\partial^2 u}{\partial z^2}.$$

EXAMPLE 2   **Steady Temperatures in a Circular Cylinder**

Find the steady-state temperature u in the circular cylinder shown in Figure 13.11.

SOLUTION   The boundary conditions suggest that the temperature u has radial symmetry. Accordingly, $u(r, z)$ is determined from

$$\frac{\partial^2 u}{\partial r^2} + \frac{1}{r} \frac{\partial u}{\partial r} + \frac{\partial^2 u}{\partial z^2} = 0, \quad 0 < r < 2, \quad 0 < z < 4$$

$$u(2, z) = 0, \quad 0 < z < 4$$

$$u(r, 0) = 0, \quad u(2, z) = u_0, \quad 0 < r < 2.$$

Using $u = R(r)Z(z)$ and separating variables gives

$$\frac{R'' + \frac{1}{r} R'}{R} = -\frac{Z''}{Z} = -\lambda \tag{13}$$

and

$$rR'' + R' + \lambda r R = 0 \tag{14}$$

$$Z'' - \lambda Z = 0. \tag{15}$$

We choose the separation constant to be $\lambda = \alpha^2 > 0$ (the choice $\lambda = -\alpha^2 < 0$ would, in view of equation (15), result in a condition that there is no reason to expect, namely, a solution $u(r, z)$ that is periodic in z). The solution of (14) is

$$R(r) = c_1 J_0(\alpha r) + c_2 Y_0(\alpha r),$$

and since the solution of (15) is defined on the finite interval $[0, 4]$, we write its general solution as

$$Z(z) = c_3 \cosh \alpha z + c_4 \sinh \alpha z.$$

As in Example 1, the assumption that the temperature u is bounded at $r = 0$ demands that $c_2 = 0$. The condition $u(2, z) = 0$ implies that $R(2) = 0$. This equation,

$$J_0(2\alpha) = 0, \tag{16}$$

defines the positive eigenvalues $\lambda_n = \alpha_n^2$ of the problem. Finally, $Z(0) = 0$ implies that $c_3 = 0$. Hence we have $R(r) = c_1 J_0(\alpha_n r)$, $Z(z) = c_4 \sinh \alpha_n z$, and

$$u_n = R(r)Z(z) = A_n \sinh \alpha_n z J_0(\alpha_n r)$$

$$u(r, z) = \sum_{n=1}^{\infty} A_n \sinh \alpha_n z J_0(\alpha_n r).$$

The remaining boundary condition at $z = 4$ then yields the Fourier-Bessel series

$$u_0 = \sum_{n=1}^{\infty} A_n \sinh 4\alpha_n J_0(\alpha_n r),$$

so that in view of the defining equation (16) the coefficients are given by (16) of Section 11.5,

$$A_n \sinh 4\alpha_n = \frac{2u_0}{2^2 J_1^2(2\alpha_n)} \int_0^2 r J_0(\alpha_n r)\, dr.$$

To evaluate the last integral, we first use the substitution $t = \alpha_n r$, followed by $\frac{d}{dt}[tJ_1(t)] = tJ_0(t)$. From

$$A_n \sinh 4\alpha_n = \frac{u_0}{2\alpha_n^2 J_1^2(2\alpha_n)} \int_0^{2\alpha_n} \frac{d}{dt}[tJ_1(t)]\, dt = \frac{u_0}{\alpha_n J_1(2\alpha_n)}$$

we get

$$A_n = \frac{u_0}{\alpha_n \sinh 4\alpha_n J_1(2\alpha_n)}.$$

Thus the temperature in the cylinder is

$$u(r, z) = u_0 \sum_{n=1}^{\infty} \frac{1}{\alpha_n \sinh 4\alpha_n J_1(2\alpha_n)} \sinh \alpha_n z J_0(\alpha_n r).$$

EXERCISES 13.2

Answers to selected odd-numbered problems begin on page ANS-20.

1. Find the displacement $u(r, t)$ in Example 1 if $f(r) = 0$ and the circular membrane is given an initial unit velocity in the upward direction.

2. A circular membrane of unit radius 1 is clamped along its circumference. Find the displacement $u(r, t)$ if the membrane starts from rest from the initial displacement $f(r) = 1 - r^2$, $0 < r < 1$. (*Hint:* See Problem 10 in Exercises 11.5.)

3. Find the steady-state temperature $u(r, z)$ in the cylinder in Example 2 if the boundary conditions are $u(2, z) = 0$, $0 < z < 4$, $u(r, 0) = u_0$, $u(r, 4) = 0$, $0 < r < 2$.

4. If the lateral side of the cylinder in Example 2 is insulated, then

$$\left.\frac{\partial u}{\partial r}\right|_{r=2} = 0, \quad 0 < z < 4.$$

(a) Find the steady-state temperature $u(r, z)$ when $u(r, 4) = f(r)$, $0 < r < 2$.

(b) Show that the steady-state temperature in part (a) reduces to $u(r, z) = u_0 z/4$ when $f(r) = u_0$. (*Hint:* Use (12) of Section 11.5.)

5. The temperature in a circular plate of radius c is determined from the boundary-value problem

$$k\left(\frac{\partial^2 u}{\partial r^2} + \frac{1}{r}\frac{\partial u}{\partial r}\right) = \frac{\partial u}{\partial t}, \quad 0 < r < c, \quad t > 0$$

$$u(c, t) = 0, \quad t > 0$$

$$u(r, 0) = f(r), \quad 0 < r < c.$$

Solve for $u(r, t)$.

6. Solve Problem 5 if the edge $r = c$ of the plate is insulated.

7. When there is heat transfer from the lateral side of an infinite circular cylinder of unit radius (see Figure 13.12) into a surrounding medium at temperature zero, the temperature inside the cylinder is determined from

$$k\left(\frac{\partial^2 u}{\partial r^2} + \frac{1}{r}\frac{\partial u}{\partial r}\right) = \frac{\partial u}{\partial t}, \quad 0 < r < 1, \quad t > 0$$

$$\left.\frac{\partial u}{\partial r}\right|_{r=1} = -hu(1, t), \quad h > 0, \quad t > 0$$

$$u(r, 0) = f(r), \quad 0 < r < 1.$$

Solve for $u(r, t)$.

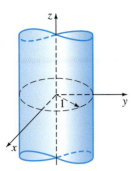

FIGURE 13.12   Infinite cylinder in Problem 7

8. Find the steady-state temperature $u(r, z)$ in a semi-infinite cylinder of unit radius $(z \geq 0)$ if there is heat transfer from its lateral side into a surrounding medium at temperature zero and if the temperature of the base $z = 0$ is held at a constant temperature u_0.

9. A circular plate is a composite of two different materials in the form of concentric circles. See Figure 13.13. The temperature in the plate is determined from the boundary-value problem

$$\frac{\partial^2 u}{\partial r^2} + \frac{1}{r}\frac{\partial u}{\partial r} = \frac{\partial u}{\partial t}, \quad 0 < r < 2, \quad t > 0$$

$$u(2, t) = 100, \quad t > 0$$

$$u(r, 0) = \begin{cases} 200, & 0 < r < 1 \\ 100, & 1 < r < 2. \end{cases}$$

Solve for $u(r, t)$. (*Hint:* Let $u(r, t) = v(r, t) + \psi(r)$.)

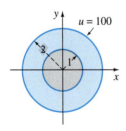

FIGURE 13.13   Composite circular plate in Problem 9

10. Solve the boundary-value problem

$$\frac{\partial^2 u}{\partial r^2} + \frac{1}{r}\frac{\partial u}{\partial r} + \beta = \frac{\partial u}{\partial t}, \quad 0 < r < 1, \quad t > 0$$

$$u(1, t) = 0, \quad t > 0$$

$$u(r, 0) = 0, \quad 0 < r < 1.$$

Assume β is a constant.

11. The horizontal displacement $u(x, t)$ of a heavy uniform chain of length L oscillating in a vertical plane satisfies the partial differential equation

$$g\frac{\partial}{\partial x}\left(x\frac{\partial u}{\partial x}\right) = \frac{\partial^2 u}{\partial t^2}, \quad 0 < x < L, \quad t > 0.$$

See Figure 13.14.

(a) Using $-\lambda$ as a separation constant, show that the ordinary differential equation in the spatial variable x is $xX'' + X' + \lambda X = 0$. Solve this equation by means of the substitution $x = \tau^2/4$.

(b) Use the result of part (a) to solve the given partial differential equation subject to

$$u(L, t) = 0, \quad t > 0$$

$$u(x, 0) = f(x), \quad \left.\frac{\partial u}{\partial t}\right|_{t=0} = 0, \quad 0 < x < L.$$

(*Hint:* Assume the oscillations at the free end $x = 0$ are finite.)

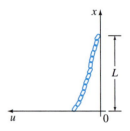

FIGURE 13.14   Oscillating chain in Problem 11

12. In this problem we consider the general case—that is, with θ dependence—of the vibrating circular membrane of radius c:

$$a^2\left(\frac{\partial^2 u}{\partial r^2} + \frac{1}{r}\frac{\partial u}{\partial r} + \frac{1}{r^2}\frac{\partial^2 u}{\partial \theta^2}\right) = \frac{\partial^2 u}{\partial t^2}, \quad 0 < r < c, \quad t > 0$$

$$u(c, \theta, t) = 0, \quad 0 < \theta < 2\pi, \quad t > 0$$

$$u(r, \theta, 0) = f(r, \theta), \quad 0 < r < c, \quad 0 < \theta < 2\pi$$

$$\left.\frac{\partial u}{\partial t}\right|_{t=0} = g(r, \theta), \quad 0 < r < c, \quad 0 < \theta < 2\pi.$$

(a) Assume that $u = R(r)\Theta(\theta)T(t)$ and that the separation constants are $-\lambda$ and $-\nu$. Show that the separated differential equations are

$$T'' + a^2\lambda T = 0, \quad \Theta'' + \nu\Theta = 0$$

$$r^2 R'' + rR' + (\lambda r^2 - \nu)R = 0.$$

(b) Let $\lambda = \alpha^2$ and $\nu = \beta^2$ and solve the separated equations.

(c) Determine the eigenvalues and eigenfunctions of the problem.

(d) Use the superposition principle to determine a multiple series solution. Do not attempt to evaluate the coefficients.

COMPUTER LAB ASSIGNMENTS

13. Consider an idealized drum consisting of a thin membrane stretched over a circular frame of unit radius. When such a drum is struck at its center, one hears a sound that is frequently described as a dull thud rather than a melodic tone. We can model a single drumbeat using the boundary-value problem solved in Example 1.

(a) Find the solution $u(r, t)$ given in (6) when $c = 1$, $f(r) = 0$, and

$$g(r) = \begin{cases} -v_0, & 0 \le r < b \\ 0, & b \le r < 1. \end{cases}$$

(b) Show that the frequency of the standing wave $u_n(r, t)$ is $f_n = a\alpha_n / 2\pi$, where α_n is the nth positive zero of $J_0(x)$. Unlike the solution of the one-dimensional wave equation in Section 12.4, the frequencies are not integer multiples of the fundamental frequency f_1. Show that $f_2 \approx 2.295 f_1$ and $f_3 \approx 3.598 f_1$. We say that the drumbeat produces **anharmonic overtones.** As a result, the displacement function $u(r, t)$ is not periodic, and so our ideal drum cannot produce a sustained tone.

(c) Let $a = 1$, $b = \frac{1}{4}$, and $v_0 = 1$ in your solution in part (a). Use a CAS to graph the fifth partial sum $S_5(r, t)$ for the times $t = 0, 0.1, 0.2, 0.3, \ldots, 5.9$, 6.0 on the interval $-1 \le r \le 1$. Use the animation capabilities of your CAS to produce a movie of these vibrations.

(d) For a greater challenge, use the 3D-plot application of your CAS to make a movie of the motion of the circular drum head that is shown in cross section in part (c). (*Hint:* There are several ways of proceeding. For a fixed time, either graph u as a function of x and y using $r = \sqrt{x^2 + y^2}$ or use the equivalent of *Mathematica's* **CylindricalPlot3D.**)

14. (a) Consider Example 1 with $a = 1$, $c = 10$, $g(r) = 0$, and $f(r) = 1 - r/10$, $0 < r < 10$. Use a CAS as an aid in finding the numerical values of the first three eigenvalues $\lambda_1, \lambda_2, \lambda_3$ of the boundary-value problem and the first three coefficients A_1, A_2, A_3 of the solution $u(r, t)$ given in (6). Write the third partial sum $S_3(r, t)$ of the series solution.

(b) Use a CAS to plot the graph of $S_3(r, t)$ for $t = 0, 4$, 10, 12, 20.

15. Solve Problem 5 with boundary conditions $u(c, t) = 200$, $u(r, 0) = 0$. With these imposed conditions, one would expect intuitively that at any interior point of the plate, $u(r, t) \to 200$ as $t \to \infty$. Assume that $c = 10$ and that the plate is cast iron so that $k = 0.1$ (approximately). Use a CAS as an aid in finding the numerical values of the first five eigenvalues $\lambda_1, \lambda_2, \lambda_3$, λ_4, λ_5 of the boundary-value problem and the five coefficients A_1, A_2, A_3, A_4, A_5 in the solution $u(r, t)$. Let the corresponding approximate solution be denoted by $S_5(r, t)$. Plot $S_5(5, t)$ and $S_5(0, t)$ on a sufficiently large time interval $0 \le t \le T$. Use the plots of $S_5(5, t)$ and $S_5(0, t)$ to estimate the times (in seconds) for which $u(5, t) \approx 100$ and $u(0, t) \approx 100$. Repeat for $u(5, t) \approx 200$ and $u(0, t) \approx 200$.

13.3   SPHERICAL COORDINATES

INTRODUCTION: We conclude our examination of boundary-value problems in different coordinate systems by next considering problems involving the heat, wave, and Laplace's equation in spherical coordinates.

REVIEW MATERIAL: The types of boundary-value problems considered in this section give rise to Legendre's differential equation and Fourier-Legendre series. Review Definition 11.8 for Fourier-Legendre series, *but* the form of the series that is important to us in this section is that given in (23) and (24) of Section 11.5.

LAPLACIAN IN SPHERICAL COORDINATES As shown in Figure 13.15, a point in 3-space is described in terms of rectangular coordinates and in spherical coordinates. The rectangular coordinates x, y, and z of the point are related to its spherical coordinates r, θ, and ϕ through the equations

$$x = r \sin \theta \cos \phi, \quad y = r \sin \theta \sin \phi, \quad z = r \cos \theta. \tag{1}$$

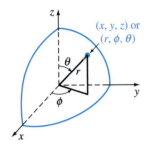

FIGURE 13.15 Spherical coordinates of a point (x, y, z) are (r, θ, ϕ).

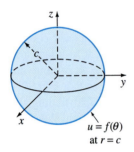

FIGURE 13.16 Dirichlet problem for a sphere

By using the equations in (1), it can be shown that the Laplacian $\nabla^2 u$ in the spherical coordinate system is

$$\nabla^2 u = \frac{\partial^2 u}{\partial r^2} + \frac{2}{r}\frac{\partial u}{\partial r} + \frac{1}{r^2 \sin^2\theta}\frac{\partial^2 u}{\partial \phi^2} + \frac{1}{r^2}\frac{\partial^2 u}{\partial \theta^2} + \frac{\cot\theta}{r^2}\frac{\partial u}{\partial \theta}. \tag{2}$$

As you might imagine, problems involving (2) can be quite formidable. Consequently, we shall consider only a few of the simpler problems that are independent of the azimuthal angle ϕ.

The next example is a Dirichlet problem for a sphere.

EXAMPLE 1   Steady Temperatures in a Sphere

Find the steady-state temperature $u(r, \theta)$ within the sphere shown in Figure 13.16.

SOLUTION   The temperature is determined from

$$\frac{\partial^2 u}{\partial r^2} + \frac{2}{r}\frac{\partial u}{\partial r} + \frac{1}{r^2}\frac{\partial^2 u}{\partial \theta^2} + \frac{\cot\theta}{r^2}\frac{\partial u}{\partial \theta} = 0, \quad 0 < r < c, \quad 0 < \theta < \pi$$

$$u(c, \theta) = f(\theta), \quad 0 < \theta < \pi.$$

If $u = R(r)\Theta(\theta)$, the partial differential equation separates as

$$\frac{r^2 R'' + 2rR'}{R} = -\frac{\Theta'' + \cot\theta\,\Theta'}{\Theta} = \lambda,$$

and so

$$r^2 R'' + 2rR' - \lambda R = 0 \tag{3}$$

$$\sin\theta\,\Theta'' + \cos\theta\,\Theta' + \lambda\sin\theta\,\Theta = 0. \tag{4}$$

After we substitute $x = \cos\theta, 0 \le \theta \le \pi$, (4) becomes

$$(1 - x^2)\frac{d^2\Theta}{dx^2} - 2x\frac{d\Theta}{dx} + \lambda\Theta = 0, \quad -1 \le x \le 1. \tag{5}$$

The latter equation is a form of Legendre's equation (see Problem 46 in Exercises 6.3). Now the only solutions of (5) that are continuous and have continuous derivatives on the closed interval $[-1, 1]$ are the Legendre polynomials $P_n(x)$ corresponding to $\lambda = n(n + 1), n = 0, 1, 2, \ldots$. Thus we take the solutions of (4) to be

$$\Theta = P_n(\cos\theta).$$

Furthermore, when $\lambda = n(n + 1)$, the general solution of the Cauchy-Euler equation (3) is

$$R = c_1 r^n + c_2 r^{-(n+1)}.$$

Since we again expect $u(r, \theta)$ to be bounded at $r = 0$, we define $c_2 = 0$. Hence $u_n = A_n r^n P_n(\cos\theta)$, and

$$u(r, \theta) = \sum_{n=0}^{\infty} A_n r^n P_n(\cos\theta).$$

At $r = c$,

$$f(\theta) = \sum_{n=0}^{\infty} A_n c^n P_n(\cos\theta).$$

Therefore $A_n c^n$ are the coefficients of the Fourier-Legendre series (23) of Section 11.5:

$$A_n = \frac{2n + 1}{2c^n}\int_0^{\pi} f(\theta)P_n(\cos\theta)\sin\theta\,d\theta.$$

It follows that the solution is

$$u(r, \theta) = \sum_{n=0}^{\infty}\left(\frac{2n + 1}{2}\int_0^{\pi} f(\theta)\,P_n(\cos\theta)\sin\theta\,d\theta\right)\left(\frac{r}{c}\right)^n P_n(\cos\theta). \qquad \blacksquare$$

Answers to selected odd-numbered problems begin on page ANS-20.

1. Solve the BVP in Example 1 if

$$f(\theta) = \begin{cases} 50, & 0 < \theta < \pi/2 \\ 0, & \pi/2 < \theta < \pi. \end{cases}$$

Write out the first four nonzero terms of the series solution. (*Hint:* See Example 3 in Section 11.5.)

2. The solution $u(r, \theta)$ in Example 1 of this section could also be interpreted as the potential inside the sphere due to a charge distribution $f(\theta)$ on its surface. Find the potential outside the sphere.

3. Find the solution of the problem in Example 1 if $f(\theta) = \cos\theta, 0 < \theta < \pi$. (*Hint:* $P_1(\cos\theta) = \cos\theta$. Use orthogonality.)

4. Find the solution of the problem in Example 1 if $f(\theta) = 1 - \cos 2\theta, 0 < \theta < \pi$. (*Hint:* See Problem 18 in Exercises 11.5.)

5. Find the steady-state temperature $u(r, \theta)$ within a hollow sphere $a < r < b$ if its inner surface $r = a$ is kept at temperature $f(\theta)$ and its outer surface $r = b$ is kept at temperature zero. The sphere in the first octant is shown in Figure 13.17.

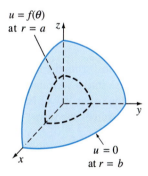

FIGURE 13.17   Hollow sphere in Problem 5

6. The steady-state temperature in a hemisphere of radius $r = c$ is determined from

$$\frac{\partial^2 u}{\partial r^2} + \frac{2}{r}\frac{\partial u}{\partial r} + \frac{1}{r^2}\frac{\partial^2 u}{\partial \theta^2} + \frac{\cot\theta}{r^2}\frac{\partial u}{\partial \theta} = 0, \quad 0 < r < c, \quad 0 < \theta < \frac{\pi}{2}$$

$$u\left(r, \frac{\pi}{2}\right) = 0, \quad 0 < r < c$$

$$u(r, \theta) = f(\theta), \quad 0 < \theta < \frac{\pi}{2}$$

Solve for $u(r, \theta)$. (*Hint:* $P_n(0) = 0$ only if n is odd. Also see Problem 20 in Exercises 11.5.)

7. Solve Problem 6 when the base of the hemisphere is insulated; that is,

$$\left.\frac{\partial u}{\partial \theta}\right|_{\theta = \pi/2} = 0, \quad 0 < r < c.$$

8. Solve Problem 6 for $r > c$.

9. The time-dependent temperature within a sphere of unit radius is determined from

$$\frac{\partial^2 u}{\partial r^2} + \frac{2}{r}\frac{\partial u}{\partial r} = \frac{\partial u}{\partial t}, \quad 0 < r < 1, \quad t > 0$$

$$u(1, t) = 100, \quad t > 0$$

$$u(r, 0) = 0, \quad 0 < r < 1.$$

Solve for $u(r, t)$. (*Hint:* Verify that the left-hand side of the partial differential equation can be written as $\frac{1}{r}\frac{\partial^2}{\partial r^2}(ru)$. Let $ru(r, t) = v(r, t) + \psi(r)$. Use only functions that are bounded as $r \to 0$.)

10. A uniform solid sphere of radius 1 at an initial constant temperature u_0 throughout is dropped into a large container of fluid that is kept at a constant temperature u_1 ($u_1 > u_0$) for all time. See Figure 13.18. Since there is heat transfer across the boundary $r = 1$, the temperature $u(r, t)$ in the sphere is determined from the boundary-value problem

$$\frac{\partial^2 u}{\partial r^2} + \frac{2}{r}\frac{\partial u}{\partial r} = \frac{\partial u}{\partial t}, \quad 0 < r < 1, \quad t > 0$$

$$\left.\frac{\partial u}{\partial r}\right|_{r=1} = -h(u(1, t) - u_1), \quad 0 < h < 1$$

$$u(r, 0) = u_0, \quad 0 < r < 1.$$

Solve for $u(r, t)$. (*Hint:* Proceed as in Problem 9.)

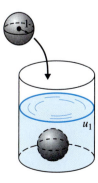

FIGURE 13.18   Container of fluid in Problem 10

11. Solve the boundary-value problem involving spherical vibrations:

$$a^2\left(\frac{\partial^2 u}{\partial r^2} + \frac{2}{r}\frac{\partial u}{\partial r}\right) = \frac{\partial^2 u}{\partial t^2}, \quad 0 < r < c, \quad t > 0$$

$$u(c, t) = 0, \quad t > 0$$

$$u(r, 0) = f(r), \quad \frac{\partial u}{\partial t}\Big|_{t=0} = g(r), \quad 0 < r < c.$$

(*Hint:* Verify that the left side of the partial differential equation is $a^2 \dfrac{1}{r}\dfrac{\partial^2}{\partial r^2}(ru)$. Let $v(r, t) = ru(r, t)$.)

12. A conducting sphere of radius $r = c$ is grounded and placed in a uniform electric field that has intensity E in the z-direction. The potential $u(r, \theta)$ outside the sphere is determined from the boundary-value problem

$$\frac{\partial^2 u}{\partial r^2} + \frac{2}{r}\frac{\partial u}{\partial r} + \frac{1}{r^2}\frac{\partial^2 u}{\partial \theta^2} + \frac{\cot \theta}{r^2}\frac{\partial u}{\partial \theta} = 0, \quad r > c, \ 0 < \theta < \pi$$

$$u(c, \theta) = 0, \quad 0 < \theta < \pi$$

$$\lim_{r \to \infty} u(r, \theta) = -Ez = -Er\cos\theta.$$

Show that

$$u(r, \theta) = -Er\cos\theta + E\frac{c^3}{r^2}\cos\theta.$$

(*Hint:* Explain why $\int_0^\pi \cos\theta \, P_n(\cos\theta) \sin\theta \, d\theta = 0$ for all nonnegative integers except $n = 1$. See (24) of Section 11.5.)

CHAPTER 13 IN REVIEW

Answers to selected odd-numbered problems begin on page ANS-20.

1. Find the steady-state temperature $u(r, \theta)$ in a circular plate of radius c if the temperature on the circumference is given by

$$u(c, \theta) = \begin{cases} u_0, & 0 < \theta < \pi \\ -u_0, & \pi < \theta < 2\pi. \end{cases}$$

2. Find the steady-state temperature in the circular plate in Problem 1 if

$$u(c, \theta) = \begin{cases} 1, & 0 < \theta < \pi/2 \\ 0, & \pi/2 < \theta < 3\pi/2 \\ 1, & 3\pi/2 < \theta < 2\pi. \end{cases}$$

3. Find the steady-state temperature $u(r, \theta)$ in a semicircular plate of radius 1 if

$$u(1, \theta) = u_0(\pi\theta - \theta^2), \quad 0 < \theta < \pi$$

$$u(r, 0) = 0, \quad u(r, \pi) = 0, \quad 0 < r < 1.$$

4. Find the steady-state temperature $u(r, \theta)$ in the semicircular plate in Problem 3 if $u(1, \theta) = \sin\theta, 0 < \theta < \pi$.

5. Find the steady-state temperature $u(r, \theta)$ in the plate shown in Figure 13.19.

6. Find the steady-state temperature $u(r, \theta)$ in the infinite plate shown in Figure 13.20.

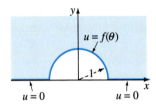

FIGURE 13.20 Infinite plate in Problem 6

7. Suppose heat is lost from the flat surfaces of a very thin circular unit disk into a surrounding medium at temperature zero. If the linear law of heat transfer applies, the heat equation assumes the form

$$\frac{\partial^2 u}{\partial r^2} + \frac{1}{r}\frac{\partial u}{\partial r} - hu = \frac{\partial u}{\partial t}, \quad h > 0, \quad 0 < r < 1, \quad t > 0$$

See Figure 13.21. Find the temperature $u(r, t)$ if the edge $r = 1$ is kept at temperature zero and if initially the temperature of the plate is unity throughout.

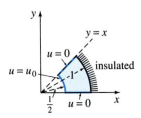

FIGURE 13.19 Wedge-shaped plate in Problem 5

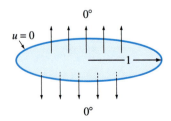

FIGURE 13.21 Circular plate in Problem 7

8. Suppose x_k is a positive zero of J_0. Show that a solution of the boundary-value problem

$$a^2\left(\frac{\partial^2 u}{\partial r^2} + \frac{1}{r}\frac{\partial u}{\partial r}\right) = \frac{\partial^2 u}{\partial t^2}, \quad 0 < r < 1, \quad t > 0$$

$$u(1, t) = 0, \quad t > 0$$

$$u(r, 0) = u_0 J_0(x_k r), \quad \frac{\partial u}{\partial t}\bigg|_{t=0} = 0, \quad 0 < r < 1$$

is $u(r, t) = u_0 J_0(x_k r) \cos a x_k t$.

9. Find the steady-state temperature $u(r, z)$ in the cylinder in Figure 13.11 if the lateral side is kept at temperature zero, the top $z = 4$ is kept at temperature 50, and the base $z = 0$ is insulated.

10. Solve the boundary-value problem

$$\frac{\partial^2 u}{\partial r^2} + \frac{1}{r}\frac{\partial u}{\partial r} + \frac{\partial^2 u}{\partial z^2} = 0, \quad 0 < r < 1, \quad 0 < z < 1$$

$$\frac{\partial u}{\partial r}\bigg|_{r=1} = 0, \quad 0 < z < 1$$

$$u(r, 0) = f(r), \quad u(r, 1) = g(r), \quad 0 < r < 1.$$

11. Find the steady-state temperature $u(r, \theta)$ in a sphere of unit radius if the surface is kept at

$$u(1, \theta) = \begin{cases} 100 & 0 < \theta < \pi/2 \\ -100 & \pi/2 < \theta < \pi. \end{cases}$$

(*Hint:* See Problem 22 in Exercises 11.5.)

12. Solve the boundary-value problem

$$\frac{\partial^2 u}{\partial r^2} + \frac{2}{r}\frac{\partial u}{\partial r} = \frac{\partial^2 u}{\partial t^2}, \quad 0 < r < 1, \quad t > 0$$

$$\frac{\partial u}{\partial r}\bigg|_{r=1} = 0, \quad t > 0$$

$$u(r, 0) = f(r), \quad \frac{\partial u}{\partial t}\bigg|_{t=0} = g(r), \quad 0 < r < 1.$$

(*Hint:* Proceed as in Problems 9 and 10 in Exercises 13.3, but let $v(r, t) = ru(r, t)$. See Section 12.7.)

13. The function $u(x) = Y_0(\alpha a)J_0(\alpha x) - J_0(\alpha a)Y_0(\alpha x)$, $a > 0$ is a solution of the parametric Bessel equation

$$x^2 \frac{d^2 u}{dx^2} + x \frac{du}{dx} + \alpha^2 x^2 u = 0$$

on the interval $[a, b]$. If the eigenvalues $\lambda_n = \alpha_n^2$ are defined by the positive roots of the equation

$$Y_0(\alpha a)J_0(\alpha b) - J_0(\alpha a)Y_0(\alpha b) = 0,$$

show that the functions

$$u_m(x) = Y_0(\alpha_m a)J_0(\alpha_m x) - J_0(\alpha_m a)Y_0(\alpha_m x)$$

$$u_n(x) = Y_0(\alpha_n a)J_0(\alpha_n x) - J_0(\alpha_n a)Y_0(\alpha_n x)$$

are orthogonal with respect to the weight function $p(x) = x$ on the interval $[a, b]$; that is,

$$\int_a^b x u_m(x) u_n(x)\, dx = 0, \quad m \neq n.$$

(*Hint:* Follow the procedure on page 450.)

14. Use the results of Problem 13 to solve the following boundary-value problem for the temperature $u(r, t)$ in a circular ring:

$$\frac{\partial^2 u}{\partial r^2} + \frac{1}{r}\frac{\partial u}{\partial r} = \frac{\partial u}{\partial t}, \quad a < r < b, \quad t > 0$$

$$u(a, t) = 0, \quad u(b, t) = 0, \quad t > 0$$

$$u(r, 0) = f(r), \quad a < r < b.$$

15. Discuss how to solve

$$\frac{\partial^2 u}{\partial r^2} + \frac{1}{r}\frac{\partial u}{\partial r} + \frac{\partial^2 u}{\partial z^2} = 0, \quad 0 < r < c, \quad 0 < z < L$$

with the boundary conditions given in Figure 13.22. Carry out your ideas and find $u(r, z)$. (*Hint:* Review (11) of Section 12.5.)

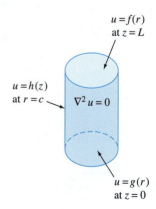

$u = f(r)$ at $z = L$

$u = h(z)$ at $r = c$

$\nabla^2 u = 0$

$u = g(r)$ at $z = 0$

FIGURE 13.22 Cylinder in Problem 15

14

INTEGRAL TRANSFORM METHOD

The method of separation of variables is a powerful, but not universally applicable, method for solving boundary-value problems. If the PDE is nonhomogeneous, or if the boundary conditions are time dependent, or if the domain of the spatial variable is infinite $(-\infty, \infty)$ or semi-infinite (a, ∞), we may be able to use an integral transform to solve the problem. In Section 14.2 we shall solve problems that involve the heat and wave equations by means of the familiar Laplace transform. In Section 14.4 three new integral transforms—Fourier transforms—will be introduced and used.

Partial Integral: $b = 4$

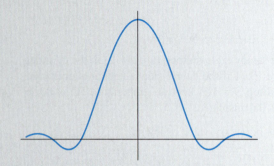

Partial Integral: $b = 6$

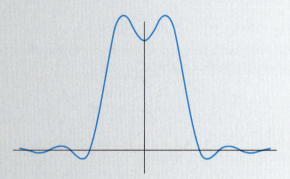

Partial Integral: $b = 15$

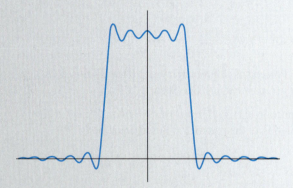

Graphs of partial integrals $F_b(x)$ of a Fourier integral representation of the piecewise-defined function.

$$f(x) = \begin{cases} 1, & |x| < 1 \\ 0, & |x| > 1. \end{cases}$$

See pages 533 and 534.

14.1 ERROR FUNCTION

INTRODUCTION: There are many functions in mathematics that are defined in terms of an integral. For example, in many traditional calculus texts the natural logarithm is defined in the following manner $\ln x = \int_1^x dt/t, x > 0$. In earlier chapters we saw, albeit briefly, the error function $\mathrm{erf}(x)$, the complementary error function $\mathrm{erfc}(x)$, the sine integral function $\mathrm{Si}(x)$, the Fresnel sine integral $S(x)$, and the gamma function $\Gamma(\alpha)$; all of these functions are defined by means of an integral. Before applying the Laplace transform to boundary-value problems, we need to know a little more about the error function and the complementary error function. In this section we examine the graphs and a few of the more obvious properties of $\mathrm{erf}(x)$ and $\mathrm{erfc}(x)$.

REVIEW MATERIAL: The error function was first introduced in (14) of Section 2.3.

PROPERTIES AND GRAPHS The definitions of the **error function** $\mathrm{erf}(x)$ and **complementary error function** $\mathrm{erfc}(x)$ are, respectively,

$$\mathrm{erf}(x) = \frac{2}{\sqrt{\pi}} \int_0^x e^{-u^2}\, du \quad \text{and} \quad \mathrm{erfc}(x) = \frac{2}{\sqrt{\pi}} \int_x^\infty e^{-u^2}\, du. \tag{1}$$

With the aid of polar coordinates it can be demonstrated that

$$\int_0^\infty e^{-u^2}\, du = \frac{\sqrt{\pi}}{2} \quad \text{or} \quad \frac{2}{\sqrt{\pi}} \int_0^\infty e^{-u^2}\, du = 1.$$

Thus from the additive interval property of definite integrals, $\int_0^\infty = \int_0^x + \int_x^\infty$, the last result can be written as

$$\frac{2}{\sqrt{\pi}} \left[\int_0^x e^{-u^2}\, du + \int_x^\infty e^{-u^2}\, du \right] = 1.$$

This shows that $\mathrm{erf}(x)$ and $\mathrm{erfc}(x)$ are related by the identity

$$\mathrm{erf}(x) + \mathrm{erfc}(x) = 1. \tag{2}$$

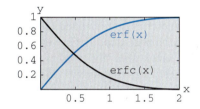

FIGURE 14.1 Graphs of $\mathrm{erf}(x)$ and $\mathrm{erfc}(x)$ for $x \geq 0$

The graphs of $\mathrm{erf}(x)$ and $\mathrm{erfc}(x)$ for $x \geq 0$ are given in Figure 14.1. Note that $\mathrm{erf}(0) = 0$, $\mathrm{erfc}(0) = 1$ and that $\mathrm{erf}(x) \to 1$, $\mathrm{erfc}(x) \to 0$ as $x \to \infty$. Other numerical values of $\mathrm{erf}(x)$ and $\mathrm{erfc}(x)$ can be obtained from a CAS or tables. In tables the error function is often referred to as the **probability integral.** The domain of $\mathrm{erf}(x)$ and of $\mathrm{erfc}(x)$ is $(-\infty, \infty)$. In Problem 11 in Exercises 14.1 you are asked to obtain the graph of each function on this interval and to deduce a few additional properties.

Table 14.1, of Laplace transforms, will be useful in the exercises in the next section. The proofs of these results are complicated and will not be given.

TABLE 14.1

$f(t), a > 0$	$\mathcal{L}\{f(t)\} = F(s)$	$f(t), a > 0$	$\mathcal{L}\{f(t)\} = F(s)$
1. $\dfrac{1}{\sqrt{\pi t}}\, e^{-a^2/4t}$	$\dfrac{e^{-a\sqrt{s}}}{\sqrt{s}}$	**4.** $2\sqrt{\dfrac{t}{\pi}}\, e^{-a^2/4t} - a\,\mathrm{erfc}\!\left(\dfrac{a}{2\sqrt{t}}\right)$	$\dfrac{e^{-a\sqrt{s}}}{s\sqrt{s}}$
2. $\dfrac{a}{2\sqrt{\pi t^3}}\, e^{-a^2/4t}$	$e^{-a\sqrt{s}}$	**5.** $e^{ab}e^{b^2 t}\,\mathrm{erfc}\!\left(b\sqrt{t} + \dfrac{a}{2\sqrt{t}}\right)$	$\dfrac{e^{-a\sqrt{s}}}{\sqrt{s}\,(\sqrt{s} + b)}$
3. $\mathrm{erfc}\!\left(\dfrac{a}{2\sqrt{t}}\right)$	$\dfrac{e^{-a\sqrt{s}}}{s}$	**6.** $-e^{ab}e^{b^2 t}\,\mathrm{erfc}\!\left(b\sqrt{t} + \dfrac{a}{2\sqrt{t}}\right) + \mathrm{erfc}\!\left(\dfrac{a}{2\sqrt{t}}\right)$	$\dfrac{b\, e^{-a\sqrt{s}}}{s\,(\sqrt{s} + b)}$

EXERCISES 14.1

Answers to selected odd-numbered problems begin on page ANS-21.

1. (a) Show that $\text{erf}(\sqrt{t}) = \dfrac{1}{\sqrt{\pi}}\displaystyle\int_0^t \dfrac{e^{-\tau}}{\sqrt{\tau}}\, d\tau$.

(b) Use the convolution theorem and the results of Problems 41 and 42 in Exercises 7.1 to show that

$$\mathscr{L}\{\text{erf}(\sqrt{t})\} = \dfrac{1}{s\sqrt{s+1}}.$$

2. Use the result of Problem 1 to show that

$$\mathscr{L}\{\text{erfc}(\sqrt{t})\} = \dfrac{1}{s}\left[1 - \dfrac{1}{\sqrt{s+1}}\right].$$

3. Use the result of Problem 1 to show that

$$\mathscr{L}\{e^t\,\text{erf}(\sqrt{t})\} = \dfrac{1}{\sqrt{s}\,(s-1)}.$$

4. Use the result of Problem 2 to show that

$$\mathscr{L}\{e^t\,\text{erfc}(\sqrt{t})\} = \dfrac{1}{\sqrt{s}\,(\sqrt{s}+1)}.$$

5. Let C, G, R, and x be constants. Use Table 14.1 to show that

$$\mathscr{L}^{-1}\left\{\dfrac{C}{Cs+G}\left(1 - e^{-x\sqrt{RCs+RG}}\right)\right\} = e^{-Gt/C}\,\text{erf}\left(\dfrac{x}{2}\sqrt{\dfrac{RC}{t}}\right).$$

6. Let a be a constant. Show that

$$\mathscr{L}^{-1}\left\{\dfrac{\sinh a\sqrt{s}}{s\sinh\sqrt{s}}\right\} = \sum_{n=0}^{\infty}\left[\text{erf}\left(\dfrac{2n+1+a}{2\sqrt{t}}\right) - \text{erf}\left(\dfrac{2n+1-a}{2\sqrt{t}}\right)\right].$$

(*Hint:* Use the exponential definition of the hyperbolic sine. Expand $1/(1-e^{-2\sqrt{s}})$ in a geometric series.)

7. Use the Laplace transform and Table 14.1 to solve the integral equation

$$y(t) = 1 - \int_0^t \dfrac{y(\tau)}{\sqrt{t-\tau}}\, d\tau.$$

8. Use the third and fifth entries in Table 14.1 to derive the sixth entry.

9. Show that $\displaystyle\int_a^b e^{-u^2}\, du = \dfrac{\sqrt{\pi}}{2}[\text{erf}(b) - \text{erf}(a)]$.

10. Show that $\displaystyle\int_{-a}^a e^{-u^2}\, du = \sqrt{\pi}\,\text{erf}(a)$.

COMPUTER LAB ASSIGNMENTS

11. The functions $\text{erf}(x)$ and $\text{erfc}(x)$ are defined for $x < 0$. Use a CAS to superimpose the graphs of $\text{erf}(x)$ and $\text{erfc}(x)$ on the same axes for $-10 \le x \le 10$. Do the graphs possess any symmetry? What are $\lim_{x\to-\infty}\text{erf}(x)$ and $\lim_{x\to-\infty}\text{erfc}(x)$?

14.2   APPLICATIONS OF THE LAPLACE TRANSFORM

INTRODUCTION: The Laplace transform of a function $f(t)$, $t \ge 0$, is defined to be $\mathscr{L}\{f(t)\} = \int_0^\infty e^{-st} f(t)\, dt$ whenever the improper integral converges. This integral transforms the function $f(t)$ into a function F of the transform parameter s, that is, $\mathscr{L}\{f(t)\} = F(s)$. Similar to Chapter 7, where the Laplace transform was used mainly to solve linear ordinary differential equations, in this section we use the Laplace transform to solve linear partial differential equations. But in contrast to Chapter 7, where the Laplace transform reduced a linear ODE with constant coefficients to an algebraic equation, in this section we see that a linear PDE with constant coefficients is transformed into an ODE.

REVIEW MATERIAL: In this section we assume a working knowledge of the operational properties of the Laplace transform (Sections 7.2–7.4) as well as the ability to solve linear second-order initial-value problems (Sections 4.3 and 4.4).

TRANSFORM OF A FUNCTION OF TWO VARIABLES The boundary-value problems considered in this section will involve either the one-dimensional wave and heat equations or slight variations of these equations. These PDEs involve an unknown function of two independent variables $u(x, t)$, where the variable t represents time $t \geq 0$. The Laplace transform of the function $u(x, t)$ with respect to t is defined by

$$\mathscr{L}\{u(x, t)\} = \int_0^\infty e^{-st} u(x, t) \, dt,$$

where x is treated as a parameter. We continue the convention of using capital letters to denote the Laplace transform of a function by writing $\mathscr{L}\{u(x, t)\} = U(x, s)$.

TRANSFORM OF PARTIAL DERIVATIVES The transforms of the partial derivatives $\partial u/\partial t$ and $\partial^2 u/\partial t^2$ follow analogously from (6) and (7) of Section 7.2:

$$\mathscr{L}\left\{\frac{\partial u}{\partial t}\right\} = sU(x, s) - u(x, 0), \tag{1}$$

$$\mathscr{L}\left\{\frac{\partial^2 u}{\partial t^2}\right\} = s^2 U(x, s) - su(x, 0) - u_t(x, 0). \tag{2}$$

Because we are transforming with respect to t, we further suppose that it is legitimate to interchange integration and differentiation in the transform of $\partial^2 u/\partial x^2$:

$$\mathscr{L}\left\{\frac{\partial^2 u}{\partial x^2}\right\} = \int_0^\infty e^{-st} \frac{\partial^2 u}{\partial x^2} \, dt = \int_0^\infty \frac{\partial^2}{\partial x^2} [e^{-st} u(x, t)] \, dt = \frac{d^2}{dx^2} \int_0^\infty e^{-st} u(x, t) \, dt = \frac{d^2}{dx^2} \mathscr{L}\{u(x, t)\};$$

that is,

$$\mathscr{L}\left\{\frac{\partial^2 u}{\partial x^2}\right\} = \frac{d^2 U}{dx^2}. \tag{3}$$

In view of (1) and (2) we see that the Laplace transform is suited to problems with initial conditions—namely, those problems associated with the heat equation or the wave equation.

EXAMPLE 1   Laplace Transform of a PDE

Find the Laplace transform of the wave equation $a^2 \dfrac{\partial^2 u}{\partial x^2} = \dfrac{\partial^2 u}{\partial t^2}, \, t > 0.$

SOLUTION   From (2) and (3),

$$\mathscr{L}\left\{a^2 \frac{\partial^2 u}{\partial x^2}\right\} = \mathscr{L}\left\{\frac{\partial^2 u}{\partial t^2}\right\}$$

becomes $a^2 \dfrac{d^2}{dx^2} \mathscr{L}\{u(x, t)\} = s^2 \mathscr{L}\{u(x, t)\} - su(x, 0) - u_t(x, 0)$

or $a^2 \dfrac{d^2 U}{dx^2} - s^2 U = -su(x, 0) - u_t(x, 0).$ \tag{4}

The Laplace transform with respect to t of either the wave equation or the heat equation eliminates that variable, and for the one-dimensional equations the transformed equations are then *ordinary differential equations* in the spatial variable x. In solving a transformed equation, we treat s as a parameter.

EXAMPLE 2   Using the Laplace Transform to Solve a BVP

Solve
$$\frac{\partial^2 u}{\partial x^2} = \frac{\partial^2 u}{\partial t^2}, \quad 0 < x < 1, \quad t > 0$$

subject to
$$u(0, t) = 0, \quad u(1, t) = 0, \quad t > 0$$

$$u(x, 0) = 0, \quad \left.\frac{\partial u}{\partial t}\right|_{t=0} = \sin \pi x, \quad 0 < x < 1.$$

SOLUTION The partial differential equation is recognized as the wave equation with $a = 1$. From (4) and the given initial conditions the transformed equation is

$$\frac{d^2 U}{dx^2} - s^2 U = -\sin \pi x, \tag{5}$$

where $U(x, s) = \mathscr{L}\{u(x, t)\}$. Since the boundary conditions are functions of t, we must also find their Laplace transforms:

$$\mathscr{L}\{u(0, t)\} = U(0, s) = 0 \quad \text{and} \quad \mathscr{L}\{u(1, t)\} = U(1, s) = 0. \tag{6}$$

The results in (6) are boundary conditions for the ordinary differential equation (5). Since (5) is defined over a finite interval, its complementary function is

$$U_c(x, s) = c_1 \cosh sx + c_2 \sinh sx.$$

The method of undetermined coefficients yields a particular solution

$$U_p(x, s) = \frac{1}{s^2 + \pi^2} \sin \pi x.$$

Hence
$$U(x, s) = c_1 \cosh sx + c_2 \sinh sx + \frac{1}{s^2 + \pi^2} \sin \pi x.$$

But the conditions $U(0, s) = 0$ and $U(1, s) = 0$ yield, in turn, $c_1 = 0$ and $c_2 = 0$. We conclude that

$$U(x, s) = \frac{1}{s^2 + \pi^2} \sin \pi x$$

$$u(x, t) = \mathscr{L}^{-1}\left\{\frac{1}{s^2 + \pi^2} \sin \pi x\right\} = \frac{1}{\pi} \sin \pi x \, \mathscr{L}^{-1}\left\{\frac{\pi}{s^2 + \pi^2}\right\}.$$

Therefore
$$u(x, t) = \frac{1}{\pi} \sin \pi x \sin \pi t.$$

EXAMPLE 3   Using the Laplace Transform to Solve a BVP

A very long string is initially at rest on the nonnegative x-axis. The string is secured at $x = 0$, and its distant right end slides down a frictionless vertical support. The string is set in motion by letting it fall under its own weight. Find the displacement $u(x, t)$.

SOLUTION Since the force of gravity is taken into consideration, it can be shown that the wave equation has the form

$$a^2 \frac{\partial^2 u}{\partial x^2} - g = \frac{\partial^2 u}{\partial t^2}, \quad x > 0, \quad t > 0.$$

Here g represents the constant acceleration due to gravity. The boundary and initial conditions are, respectively,

$$u(0, t) = 0, \quad \lim_{x \to \infty} \frac{\partial u}{\partial x} = 0, \quad t > 0$$

$$u(x, 0) = 0, \quad \frac{\partial u}{\partial t}\bigg|_{t=0} = 0, \quad x > 0.$$

The second boundary condition, $\lim_{x \to \infty} \partial u/\partial x = 0$, indicates that the string is horizontal at a great distance from the left end. Now from (2) and (3),

$$\mathcal{L}\left\{a^2 \frac{\partial^2 u}{\partial x^2}\right\} - \mathcal{L}\{g\} = \mathcal{L}\left\{\frac{\partial^2 u}{\partial t^2}\right\}$$

becomes
$$a^2 \frac{d^2 U}{dx^2} - \frac{g}{s} = s^2 U - su(x, 0) - u_t(x, 0)$$

or, in view of the initial conditions,

$$\frac{d^2 U}{dx^2} - \frac{s^2}{a^2} U = \frac{g}{a^2 s}.$$

The transforms of the boundary conditions are

$$\mathcal{L}\{u(0, t)\} = U(0, s) = 0 \quad \text{and} \quad \mathcal{L}\left\{\lim_{x \to \infty} \frac{\partial u}{\partial x}\right\} = \lim_{x \to \infty} \frac{dU}{dx} = 0.$$

With the aid of undetermined coefficients, the general solution of the transformed equation is found to be

$$U(x, s) = c_1 e^{-(x/a)s} + c_2 e^{(x/a)s} - \frac{g}{s^3}.$$

The boundary condition $\lim_{x \to \infty} dU/dx = 0$ implies that $c_2 = 0$, and $U(0, s) = 0$ gives $c_1 = g/s^3$. Therefore

$$U(x, s) = \frac{g}{s^3} e^{-(x/a)s} - \frac{g}{s^3}.$$

Now by the second translation theorem we have

$$u(x, t) = \mathcal{L}^{-1}\left\{\frac{g}{s^3} e^{-(x/a)s} - \frac{g}{s^3}\right\} = \frac{1}{2} g \left(t - \frac{x}{a}\right)^2 \mathcal{U}\left(t - \frac{x}{a}\right) - \frac{1}{2} gt^2$$

or
$$u(x, t) = \begin{cases} -\dfrac{1}{2} gt^2, & 0 \le t < \dfrac{x}{a} \\ -\dfrac{g}{2a^2}(2axt - x^2), & t \ge \dfrac{x}{a}. \end{cases}$$

To interpret the solution, let us suppose that $t > 0$ is fixed. For $0 \le x \le at$, the string is the shape of a parabola passing through $(0, 0)$ and $\left(at, -\frac{1}{2} gt^2\right)$. For $x > at$, the string is described by the horizontal line $u = -\frac{1}{2} gt^2$. See Figure 14.2.

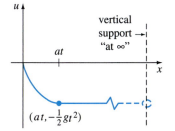

FIGURE 14.2 "Infinitely long" string falling under its own weight

Observe that the problem in the next example could be solved by the procedure in Section 12.6. The Laplace transform provides an alternative solution.

EXAMPLE 4 **A Solution in Terms of erf(x)**

Solve the heat equation

$$\frac{\partial^2 u}{\partial x^2} = \frac{\partial u}{\partial t}, \quad 0 < x < 1, \quad t > 0$$

subject to

$$u(0, t) = 0, \quad u(1, t) = u_0, \quad t > 0$$

$$u(x, 0) = 0, \quad 0 < x < 1.$$

SOLUTION From (1) and (3) and the given initial condition,

$$\mathscr{L}\left\{\frac{\partial^2 u}{\partial x^2}\right\} = \mathscr{L}\left\{\frac{\partial u}{\partial t}\right\}$$

becomes

$$\frac{d^2 U}{dx^2} - sU = 0. \tag{7}$$

The transforms of the boundary conditions are

$$U(0, s) = 0 \quad \text{and} \quad U(1, s) = \frac{u_0}{s}. \tag{8}$$

Since we are concerned with a finite interval on the x-axis, we choose to write the general solution of (7) as

$$U(x, s) = c_1 \cosh\left(\sqrt{s}x\right) + c_2 \sinh\left(\sqrt{s}x\right).$$

Applying the two boundary conditions in (8) yields $c_1 = 0$ and $c_2 = u_0/\left(s \sinh \sqrt{s}\right)$, respectively. Thus

$$U(x, s) = u_0 \frac{\sinh\left(\sqrt{s}x\right)}{s \sinh \sqrt{s}}.$$

Now the inverse transform of the latter function cannot be found in most tables. However, by writing

$$\frac{\sinh\left(\sqrt{s}x\right)}{s \sinh \sqrt{s}} = \frac{e^{\sqrt{s}x} - e^{-\sqrt{s}x}}{s\left(e^{\sqrt{s}} - e^{-\sqrt{s}}\right)} = \frac{e^{(x-1)\sqrt{s}} - e^{-(x+1)\sqrt{s}}}{s\left(1 - e^{-2\sqrt{s}}\right)}$$

and using the geometric series

$$\frac{1}{1 - e^{-2\sqrt{s}}} = \sum_{n=0}^{\infty} e^{-2n\sqrt{s}}$$

we find

$$\frac{\sinh\left(\sqrt{s}x\right)}{s \sinh \sqrt{s}} = \sum_{n=0}^{\infty}\left[\frac{e^{-(2n+1-x)\sqrt{s}}}{s} - \frac{e^{-(2n+1+x)\sqrt{s}}}{s}\right].$$

If we assume that the inverse Laplace transform can be done term by term, it follows from entry 3 of Table 14.1 that

$$u(x, t) = u_0 \mathscr{L}^{-1}\left\{\frac{\sinh\left(\sqrt{s}x\right)}{s \sinh \sqrt{s}}\right\}$$

$$= u_0 \sum_{n=0}^{\infty}\left[\mathscr{L}^{-1}\left\{\frac{e^{-(2n+1-x)\sqrt{s}}}{s}\right\} - \mathscr{L}^{-1}\left\{\frac{e^{-(2n+1+x)\sqrt{s}}}{s}\right\}\right]$$

$$= u_0 \sum_{n=0}^{\infty}\left[\operatorname{erfc}\left(\frac{2n + 1 - x}{2\sqrt{t}}\right) - \operatorname{erfc}\left(\frac{2n + 1 + x}{2\sqrt{t}}\right)\right]. \tag{9}$$

The solution (9) can be rewritten in terms of the error function using $\operatorname{erfc}(x) = 1 - \operatorname{erf}(x)$:

$$u(x, t) = u_0 \sum_{n=0}^{\infty}\left[\operatorname{erf}\left(\frac{2n + 1 + x}{2\sqrt{t}}\right) - \operatorname{erf}\left(\frac{2n + 1 - x}{2\sqrt{t}}\right)\right]. \tag{10} \quad \blacksquare$$

Figure 14.3(a), obtained with the aid of the 3D-plot application in a CAS, shows the surface over the rectangular region $0 \le x \le 1$, $0 \le t \le 6$ defined by the partial sum $S_{10}(x, t)$ of the solution (10) with $u_0 = 100$. It is apparent from the surface and the accompanying two-dimensional graphs that at a fixed value of x (the curve of intersection of a plane slicing the surface perpendicular to the x-axis on the interval $0 \leq x \leq 1$) the temperature $u(x, t)$ increases rapidly to a constant value as time increases. See Figures 14.3(b) and 14.3(c). For a fixed time (the curve of intersection of a plane slicing the surface perpendicular to the t-axis) the temperature $u(x, t)$ naturally increases from 0 to 100. See Figures 14.3(d) and 14.3(e).

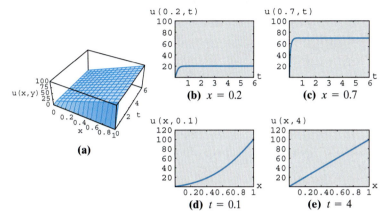

(a)

(b) $x = 0.2$

(c) $x = 0.7$

(d) $t = 0.1$

(e) $t = 4$

FIGURE 14.3 Graph of solution given in (10). In (b) and (c) x is held constant. In (d) and (e) t is held constant

EXERCISES 14.2

Answers to selected odd-numbered problems begin on page ANS-21.

In the following problems use tables as necessary.

1. A string is stretched along the x-axis between $(0, 0)$ and $(L, 0)$. Find the displacement $u(x, t)$ if the string starts from rest in the initial position $A \sin(\pi x/L)$.

2. Solve the boundary-value problem

$$\frac{\partial^2 u}{\partial x^2} = \frac{\partial^2 u}{\partial t^2}, \quad 0 < x < 1, \quad t > 0$$

$$u(0, t) = 0, \quad u(1, t) = 0$$

$$u(x, 0) = 0, \quad \left.\frac{\partial u}{\partial t}\right|_{t=0} = 2 \sin \pi x + 4 \sin 3\pi x.$$

3. The displacement of a semi-infinite elastic string is determined from

$$a^2 \frac{\partial^2 u}{\partial x^2} = \frac{\partial^2 u}{\partial t^2}, \quad x > 0, \quad t > 0$$

$$u(0, t) = f(t), \quad \lim_{x \to \infty} u(x, t) = 0, \quad t > 0$$

$$u(x, 0) = 0, \quad \left.\frac{\partial u}{\partial t}\right|_{t=0} = 0, \quad x > 0.$$

Solve for $u(x, t)$.

4. Solve the boundary-value problem in Problem 3 when

$$f(t) = \begin{cases} \sin \pi t, & 0 \leq t \leq 1 \\ 0, & t > 1. \end{cases}$$

Sketch the displacement $u(x, t)$ for $t > 1$.

5. In Example 3 find the displacement $u(x, t)$ when the left end of the string at $x = 0$ is given an oscillatory motion described by $f(t) = A \sin \omega t$.

6. The displacement $u(x, t)$ of a string that is driven by an external force is determined from

$$\frac{\partial^2 u}{\partial x^2} + \sin \pi x \sin \omega t = \frac{\partial^2 u}{\partial t^2}, \quad 0 < x < 1, \quad t > 0$$

$$u(0, t) = 0, \quad u(1, t) = 0, \quad t > 0$$

$$u(x, 0) = 0, \quad \left.\frac{\partial u}{\partial t}\right|_{t=0} = 0, \quad 0 < x < 1.$$

Solve for $u(x, t)$.

7. A uniform bar is clamped at $x = 0$ and is initially at rest. If a constant force F_0 is applied to the free end at $x = L$, the longitudinal displacement $u(x, t)$ of a cross-section of the bar is determined from

$$a^2 \frac{\partial^2 u}{\partial x^2} = \frac{\partial^2 u}{\partial t^2}, \quad 0 < x < L, \quad t > 0$$

$$u(0, t) = 0, \quad \left. E \frac{\partial u}{\partial x}\right|_{x=L} = F_0, \quad E \text{ a constant}, \quad t > 0$$

$$u(x, 0) = 0, \quad \left.\frac{\partial u}{\partial t}\right|_{t=0} = 0, \quad 0 < x < L.$$

Solve for $u(x, t)$. (*Hint:* Expand $1/(1 + e^{-2sL/a})$ in a geometric series.)

8. A uniform semi-infinite elastic beam moving along the x-axis with a constant velocity $-v_0$ is brought to a stop by hitting a wall at time $t = 0$. See Figure 14.4. The longitudinal displacement $u(x, t)$ is determined from

$$a^2 \frac{\partial^2 u}{\partial x^2} = \frac{\partial^2 u}{\partial t^2}, \quad x > 0, \quad t > 0$$

$$u(0, t) = 0, \quad \lim_{x \to \infty} \frac{\partial u}{\partial x} = 0, \quad t > 0$$

$$u(x, 0) = 0, \quad \frac{\partial u}{\partial t}\bigg|_{t=0} = -v_0, \quad x > 0.$$

Solve for $u(x, t)$.

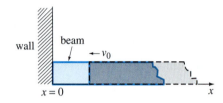

FIGURE 14.4   Moving elastic beam in Problem 8

9. Solve the boundary-value problem

$$\frac{\partial^2 u}{\partial x^2} = \frac{\partial^2 u}{\partial t^2}, \quad x > 0, \quad t > 0$$

$$u(0, t) = 0, \quad \lim_{x \to \infty} u(x, t) = 0, \quad t > 0$$

$$u(x, 0) = xe^{-x}, \quad \frac{\partial u}{\partial t}\bigg|_{t=0} = 0, \quad x > 0.$$

10. Solve the boundary-value problem

$$\frac{\partial^2 u}{\partial x^2} = \frac{\partial^2 u}{\partial t^2}, \quad x > 0, \quad t > 0$$

$$u(0, t) = 1, \quad \lim_{x \to \infty} u(x, t) = 0, \quad t > 0$$

$$u(x, 0) = e^{-x}, \quad \frac{\partial u}{\partial t}\bigg|_{t=0} = 0, \quad x > 0.$$

In Problems 11–18 use the Laplace transform to solve the heat equation $u_{xx} = u_t$, $x > 0$, $t > 0$, subject to the given conditions.

11. $u(0, t) = u_0$, $\lim_{x \to \infty} u(x, t) = u_1$, $u(x, 0) = u_1$

12. $u(0, t) = u_0$, $\lim_{x \to \infty} \frac{u(x, t)}{x} = u_1$, $u(x, 0) = u_1 x$

13. $\frac{\partial u}{\partial x}\bigg|_{x=0} = u(0, t)$, $\lim_{x \to \infty} u(x, t) = u_0$, $u(x, 0) = u_0$

14. $\frac{\partial u}{\partial x}\bigg|_{x=0} = u(0, t) - 50$, $\lim_{x \to \infty} u(x, t) = 0$, $u(x, 0) = 0$

15. $u(0, t) = f(t)$, $\lim_{x \to \infty} u(x, t) = 0$, $u(x, 0) = 0$

(*Hint:* Use the convolution theorem.)

16. $\frac{\partial u}{\partial x}\bigg|_{x=0} = -f(t)$, $\lim_{x \to \infty} u(x, t) = 0$, $u(x, 0) = 0$

17. $u(0, t) = 60 + 40 \,\mathcal{U}(t - 2)$, $\lim_{x \to \infty} u(x, t) = 60$, $u(x, 0) = 60$

18. $u(0, t) = \begin{cases} 20, & 0 < t < 1 \\ 0, & t \geq 1 \end{cases}$, $\lim_{x \to \infty} u(x, t) = 100$, $u(x, 0) = 100$

19. Solve the boundary-value problem

$$\frac{\partial^2 u}{\partial x^2} = \frac{\partial u}{\partial t}, \quad -\infty < x < 1, \quad t > 0$$

$$\frac{\partial u}{\partial x}\bigg|_{x=1} = 100 - u(1, t), \quad \lim_{x \to -\infty} u(x, t) = 0, \quad t > 0$$

$$u(x, 0) = 0, \quad -\infty < x < 1.$$

20. Show that a solution of the boundary-value problem

$$k \frac{\partial^2 u}{\partial x^2} + r = \frac{\partial u}{\partial t}, \quad x > 0, \quad t > 0$$

$$u(0, t) = 0, \quad \lim_{x \to \infty} \frac{\partial u}{\partial x} = 0, \quad t > 0$$

$$u(x, 0) = 0, \quad x > 0,$$

where r is a constant, is given by

$$u(x, t) = rt - r \int_0^t \text{erfc}\left(\frac{x}{2\sqrt{k\tau}}\right) d\tau.$$

21. A rod of length L is held at a constant temperature u_0 at its ends $x = 0$ and $x = L$. If the rod's initial temperature is $u_0 + u_0 \sin(x\pi/L)$, solve the heat equation $u_{xx} = u_t$, $0 < x < L$, $t > 0$ for the temperature $u(x, t)$.

22. If there is a heat transfer from the lateral surface of a thin wire of length L into a medium at constant temperature u_m, then the heat equation takes on the form

$$k \frac{\partial^2 u}{\partial x^2} - h(u - u_m) = \frac{\partial u}{\partial t}, \quad 0 < x < L, \quad t > 0,$$

where h is a constant. Find the temperature $u(x, t)$ if the initial temperature is a constant u_0 throughout and the ends $x = 0$ and $x = L$ are insulated.

23. A rod of unit length is insulated at $x = 0$ and is kept at temperature zero at $x = 1$. If the initial temperature of the rod is a constant u_0, solve $ku_{xx} = u_t$, $0 < x < 1$, $t > 0$ for the temperature $u(x, t)$. (*Hint:* Expand $1/\left(1 + e^{-2\sqrt{s/k}}\right)$ in a geometric series.)

24. An infinite porous slab of unit width is immersed in a solution of constant concentration c_0. A dissolved substance in the solution diffuses into the slab. The concentration $c(x, t)$ in the slab is determined from

$$D\frac{\partial^2 c}{\partial x^2} = \frac{\partial c}{\partial t}, \quad 0 < x < 1, \quad t > 0$$

$$c(0, t) = c_0, \quad c(1, t) = c_0, \quad t > 0$$

$$c(x, 0) = 0, \quad 0 < x < 1,$$

where D is a constant. Solve for $c(x, t)$.

25. A very long telephone transmission line is initially at a constant potential u_0. If the line is grounded at $x = 0$ and insulated at the distant right end, then the potential $u(x, t)$ at a point x along the line at time t is determined from

$$\frac{\partial^2 u}{\partial x^2} - RC\frac{\partial u}{\partial t} - RGu = 0, \quad x > 0, \quad t > 0$$

$$u(0, t) = 0, \quad \lim_{x \to \infty}\frac{\partial u}{\partial x} = 0, \quad t > 0$$

$$u(x, 0) = u_0, \quad x > 0,$$

where R, C, and G are constants known as resistance, capacitance, and conductance, respectively. Solve for $u(x, t)$. (*Hint:* See Problem 5 in Exercises 14.1.)

26. Show that a solution of the boundary-value problem

$$\frac{\partial^2 u}{\partial x^2} - hu = \frac{\partial u}{\partial t}, \quad x > 0, \quad t > 0, \quad h \text{ constant}$$

$$u(0, t) = u_0, \quad \lim_{x \to \infty} u(x, t) = 0, \quad t > 0$$

$$u(x, 0) = 0, \quad x > 0$$

is $\quad u(x, t) = \dfrac{u_0 x}{2\sqrt{\pi}} \displaystyle\int_0^t \dfrac{e^{-h\tau - x^2/4\tau}}{\tau^{3/2}} \, d\tau.$

27. Starting at $t = 0$, a concentrated load of magnitude F_0 moves with a constant velocity v_0 along a semi-infinite string. In this case the wave equation becomes

$$a^2\frac{\partial^2 u}{\partial x^2} = \frac{\partial^2 u}{\partial t^2} + F_0\delta\left(t - \frac{x}{v_0}\right),$$

where $\delta(t - x/v_0)$ is the Dirac delta function. Solve the above PDE subject to

$$u(0, t) = 0, \quad \lim_{x \to \infty} u(x, t) = 0, \quad t > 0$$

$$u(x, 0) = 0, \quad \frac{\partial u}{\partial t}\Big|_{t=0} = 0, \quad x > 0$$

(a) when $v_0 \neq a$ (b) when $v_0 = a$.

COMPUTER LAB ASSIGNMENTS

28. (a) The temperature in a semi-infinite solid is modeled by the boundary-value problem

$$k\frac{\partial^2 u}{\partial x^2} = \frac{\partial u}{\partial t}, \quad x > 0, \quad t > 0$$

$$u(0, t) = u_0, \quad \lim_{x \to \infty} u(x, t) = 0, \quad t > 0$$

$$u(x, 0) = 0, \quad x > 0.$$

Solve for $u(x, t)$. Use the solution to determine analytically the value of $\lim_{t \to \infty} u(x, t), x > 0$.

(b) Use a CAS to graph $u(x, t)$ over the rectangular region $0 \leq x \leq 10, 0 \leq t \leq 15$. Assume that $u_0 = 100$ and $k = 1$. Indicate the two boundary conditions and initial condition on your graph. Use 2D and 3D plots of $u(x, t)$ to verify your answer to part (a).

29. (a) In Problem 28 if there is a constant flux of heat into the solid at its left-hand boundary, then the boundary condition is $\dfrac{\partial u}{\partial x}\Big|_{x=0} = -A, A > 0, t > 0.$
Solve for $u(x, t)$. Use the solution to determine analytically the value of $\lim_{t \to \infty} u(x, t), x > 0$.

(b) Use a CAS to graph $u(x, t)$ over the rectangular region $0 \leq x \leq 10, 0 \leq t \leq 15$. Assume that $u_0 = 100$ and $k = 1$. Use 2D and 3D plots of $u(x, t)$ to verify your answer to part (a).

30. Humans gather most of their information on the outside world through sight and sound. But many creatures use chemical signals as their primary means of communication; for example, honeybees, when alarmed, emit a substance and fan their wings feverishly to relay the warning signal to the bees that attend to the queen. These molecular messages between members of the same species are called pheromones. The signals may be carried by moving air or water or by a *diffusion process* in which the random movement of gas molecules transports the chemical away from its source. Figure 14.5 shows an ant emitting an alarm chemical into the still air of a tunnel. If $c(x, t)$ denotes the concentration of the chemical x centimeters from the source at time t, then $c(x, t)$ satisfies

$$k\frac{\partial^2 c}{\partial x^2} = \frac{\partial c}{\partial t}, \quad x > 0, \quad t > 0$$

and k is a positive constant. The emission of pheromones as a discrete pulse gives rise to a boundary condition of the form

$$\frac{\partial c}{\partial x}\Big|_{x=0} = -A\delta(t),$$

where $\delta(t)$ is the Dirac delta function.
(a) Solve the boundary-value problem if it is further known that

$$c(x, 0) = 0, \quad x > 0 \quad \text{and} \quad \lim_{x \to \infty} c(x, t) = 0, \quad t > 0.$$

(b) Use a CAS to graph the solution in part (a) for $x \geq 0$ at the fixed times $t = 0.1$, $t = 0.5$, $t = 1$, $t = 2$, and $t = 5$.

(c) For any fixed time t, show that $\int_0^\infty c(x, t) \, dx = Ak$. Thus Ak represents the total amount of chemical discharged.

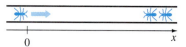

FIGURE 14.5   Ant responding to chemical signal in Problem 30

14.3   FOURIER INTEGRAL

INTRODUCTION: We used Fourier series in preceding chapters to represent a function f defined on a finite interval such as $(-p, p)$ or $(0, L)$. When f and f' are piecewise continuous on such an interval, a Fourier series represents the function on the interval and converges to the periodic extension of f outside the interval. In this way we are justified in saying that Fourier series are associated only with *periodic* functions. We shall now derive, in a nonrigorous fashion, a means of representing certain kinds of *nonperiodic* functions that are defined on either an infinite interval $(-\infty, \infty)$ or a semi-infinite interval $(0, \infty)$.

REVIEW MATERIAL: The Fourier integral has different forms that are analogous to the four forms of Fourier series given in Definitions 11.5 and 11.6 and Problem 21 in Exercises 11.2. A review of these various forms is recommended.

FOURIER SERIES TO FOURIER INTEGRAL   Suppose a function f is defined on $(-p, p)$. If we use the integral definitions of the coefficients (9), (10), and (11) of Section 11.2 in (8) of that section, then the Fourier series of f on the interval is

$$f(x) = \frac{1}{2p} \int_{-p}^{p} f(t) \, dt + \frac{1}{p} \sum_{n=1}^{\infty} \left[\left(\int_{-p}^{p} f(t) \cos \frac{n\pi}{p} t \, dt \right) \cos \frac{n\pi}{p} x + \left(\int_{-p}^{p} f(t) \sin \frac{n\pi}{p} t \, dt \right) \sin \frac{n\pi}{p} x \right]. \quad (1)$$

If we let $\alpha_n = n\pi/p$, $\Delta\alpha = \alpha_{n+1} - \alpha_n = \pi/p$, then (1) becomes

$$f(x) = \frac{1}{2\pi} \left(\int_{-p}^{p} f(t) \, dt \right) \Delta\alpha + \frac{1}{\pi} \sum_{n=1}^{\infty} \left[\left(\int_{-p}^{p} f(t) \cos \alpha_n t \, dt \right) \cos \alpha_n x + \left(\int_{-p}^{p} f(t) \sin \alpha_n t \, dt \right) \sin \alpha_n x \right] \Delta\alpha. \quad (2)$$

We now expand the interval $(-p, p)$ by letting $p \to \infty$. Since $p \to \infty$ implies that $\Delta\alpha \to 0$, the limit of (2) has the form $\lim_{\Delta\alpha \to 0} \sum_{n=1}^{\infty} F(\alpha_n) \, \Delta\alpha$, which is suggestive of the definition of the integral $\int_0^\infty F(\alpha) \, d\alpha$. Thus if $\int_{-\infty}^{\infty} f(t) \, dt$ exists, the limit of the first term in (2) is zero, and the limit of the sum becomes

$$f(x) = \frac{1}{\pi} \int_0^\infty \left[\left(\int_{-\infty}^{\infty} f(t) \cos \alpha t \, dt \right) \cos \alpha x + \left(\int_{-\infty}^{\infty} f(t) \sin \alpha t \, dt \right) \sin \alpha x \right] d\alpha. \quad (3)$$

The result given in (3) is called the **Fourier integral** of f on $(-\infty, \infty)$. As the following summary shows, the basic structure of the Fourier integral is reminiscent of that of a Fourier series.

> **DEFINITION 14.1   Fourier Integral**
>
> The **Fourier integral** of a function f defined on the interval $(-\infty, \infty)$ is given by
>
> $$f(x) = \frac{1}{\pi} \int_0^\infty [A(\alpha) \cos \alpha x + B(\alpha) \sin \alpha x] \, d\alpha, \tag{4}$$
>
> where
> $$A(\alpha) = \int_{-\infty}^\infty f(x) \cos \alpha x \, dx \tag{5}$$
>
> $$B(\alpha) = \int_{-\infty}^\infty f(x) \sin \alpha x \, dx. \tag{6}$$

CONVERGENCE OF A FOURIER INTEGRAL   Sufficient conditions under which a Fourier integral converges to $f(x)$ are similar to, but slightly more restrictive than, the conditions for a Fourier series.

> **THEOREM 14.1   Conditions for Convergence**
>
> Let f and f' be piecewise continuous on every finite interval, and let f be absolutely integrable on $(-\infty, \infty)$.* Then the Fourier integral of f on the interval converges to $f(x)$ at a point of continuity. At a point of discontinuity the Fourier integral will converge to the average
>
> $$\frac{f(x+) + f(x-)}{2},$$
>
> where $f(x+)$ and $f(x-)$ denote the limit of f at x from the right and from the left, respectively.

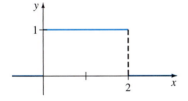

FIGURE 14.6 Piecewise-continuous function defined on $(-\infty, \infty)$

> **EXAMPLE 1   Fourier Integral Representation**

Find the Fourier integral representation of the function

$$f(x) = \begin{cases} 0, & x < 0 \\ 1, & 0 < x < 2 \\ 0, & x > 2. \end{cases}$$

SOLUTION   The function, whose graph is shown in Figure 14.6, satisfies the hypotheses of Theorem 14.1. Hence from (5) and (6) we have at once

$$A(\alpha) = \int_{-\infty}^\infty f(x) \cos \alpha x \, dx$$

$$= \int_{-\infty}^0 f(x) \cos \alpha x \, dx + \int_0^2 f(x) \cos \alpha x \, dx + \int_2^\infty f(x) \cos \alpha x \, dx$$

$$= \int_0^2 \cos \alpha x \, dx = \frac{\sin 2\alpha}{\alpha}$$

$$B(\alpha) = \int_{-\infty}^\infty f(x) \sin \alpha x \, dx = \int_0^2 \sin \alpha x \, dx = \frac{1 - \cos 2\alpha}{\alpha}.$$

---

*This means that the integral $\int_{-\infty}^\infty |f(x)| \, dx$ converges.

Substituting these coefficients into (4) then gives

$$f(x) = \frac{1}{\pi} \int_0^\infty \left[\left(\frac{\sin 2\alpha}{\alpha} \right) \cos \alpha x + \left(\frac{1 - \cos 2\alpha}{\alpha} \right) \sin \alpha x \right] d\alpha.$$

When we use trigonometric identities, the last integral simplifies to

$$f(x) = \frac{2}{\pi} \int_0^\infty \frac{\sin \alpha \cos \alpha (x - 1)}{\alpha} d\alpha. \tag{7}$$

The Fourier integral can be used to evaluate integrals. For example, it follows from Theorem 14.1 that (7) converges to $f(1) = 1$; that is,

$$\frac{2}{\pi} \int_0^\infty \frac{\sin \alpha}{\alpha} d\alpha = 1 \quad \text{and so} \quad \int_0^\infty \frac{\sin \alpha}{\alpha} d\alpha = \frac{\pi}{2}.$$

The latter result is worthy of special note since it cannot be obtained in the "usual" manner; the integrand $(\sin x)/x$ does not possess an antiderivative that is an elementary function.

COSINE AND SINE INTEGRALS   When f is an even function on the interval $(-\infty, \infty)$, then the product $f(x) \cos \alpha x$ is also an even function whereas $f(x) \sin \alpha x$ is an odd function. As a consequence of property (g) of Theorem 11.2, $B(\alpha) = 0$, and so (4) becomes

$$f(x) = \frac{2}{\pi} \int_0^\infty \left(\int_0^\infty f(t) \cos \alpha t \, dt \right) \cos \alpha x \, d\alpha.$$

Here we have also used property (f) of Theorem 11.2 to write

$$\int_{-\infty}^\infty f(t) \cos \alpha t \, dt = 2 \int_0^\infty f(t) \cos \alpha t \, dt.$$

Similarly, when f is an odd function on $(-\infty, \infty)$, products $f(x) \cos \alpha x$ and $f(x) \sin \alpha x$ are odd and even functions, respectively. Therefore $A(\alpha) = 0$, and

$$f(x) = \frac{2}{\pi} \int_0^\infty \left(\int_0^\infty f(t) \sin \alpha t \, dt \right) \sin \alpha x \, d\alpha.$$

We summarize in the following definition.

DEFINITION 14.2   **Fourier Cosine and Sine Integrals**

(i)   The Fourier integral of an even function on the interval $(-\infty, \infty)$ is the **cosine integral**

$$f(x) = \frac{2}{\pi} \int_0^\infty A(\alpha) \cos \alpha x \, d\alpha, \tag{8}$$

where

$$A(\alpha) = \int_0^\infty f(x) \cos \alpha x \, dx. \tag{9}$$

(ii)   The Fourier integral of an odd function on the interval $(-\infty, \infty)$ is the **sine integral**

$$f(x) = \frac{2}{\pi} \int_0^\infty B(\alpha) \sin \alpha x \, d\alpha, \tag{10}$$

where

$$B(\alpha) = \int_0^\infty f(x) \sin \alpha x \, dx. \tag{11}$$

EXAMPLE 2   **Cosine Integral Representation**

Find the Fourier integral representation of the function

$$f(x) = \begin{cases} 1, & |x| < a \\ 0, & |x| > a. \end{cases}$$

SOLUTION   It is apparent from Figure 14.7 that f is an even function. Hence we represent f by the Fourier cosine integral (8). From (9) we obtain

$$A(\alpha) = \int_0^\infty f(x) \cos \alpha x \, dx = \int_0^a f(x) \cos \alpha x \, dx + \int_a^\infty f(x) \cos \alpha x \, dx = \int_0^a \cos \alpha x \, dx = \frac{\sin a\alpha}{\alpha},$$

and so

$$f(x) = \frac{2}{\pi} \int_0^\infty \frac{\sin a\alpha \cos \alpha x}{\alpha} \, d\alpha. \qquad (12)$$

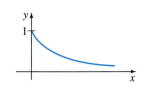

The integrals (8) and (10) can be used when f is neither odd nor even and defined only on the half-line $(0, \infty)$. In this case (8) represents f on the interval $(0, \infty)$ and its even (but not periodic) extension to $(-\infty, 0)$, whereas (10) represents f on $(0, \infty)$ and its odd extension to the interval $(-\infty, 0)$. The next example illustrates this concept.

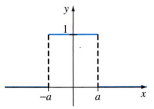

FIGURE 14.7 Piecewise-continuous even function defined on $(-\infty, \infty)$

EXAMPLE 3   **Cosine and Sine Integral Representations**

Represent $f(x) = e^{-x}, x > 0$
(a) by a cosine integral   **(b)** by a sine integral.

SOLUTION   The graph of the function is given in Figure 14.8.

(a) Using integration by parts, we find

$$A(\alpha) = \int_0^\infty e^{-x} \cos \alpha x \, dx = \frac{1}{1 + \alpha^2}.$$

Therefore the cosine integral of f is

$$f(x) = \frac{2}{\pi} \int_0^\infty \frac{\cos \alpha x}{1 + \alpha^2} \, d\alpha. \qquad (13)$$

(b) Similarly, we have

$$B(\alpha) = \int_0^\infty e^{-x} \sin \alpha x \, dx = \frac{\alpha}{1 + \alpha^2}.$$

The sine integral of f is then

$$f(x) = \frac{2}{\pi} \int_0^\infty \frac{\alpha \sin \alpha x}{1 + \alpha^2} \, d\alpha. \qquad (14)$$

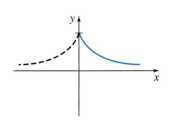

FIGURE 14.8 Function defined on $(0, \infty)$

Figure 14.9 shows the graphs of the functions and their extensions represented by the two integrals.

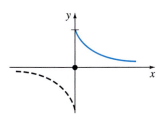

(a) cosine integral

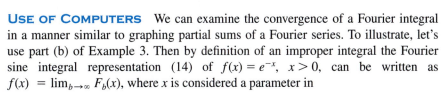

(b) sine integral

USE OF COMPUTERS   We can examine the convergence of a Fourier integral in a manner similar to graphing partial sums of a Fourier series. To illustrate, let's use part (b) of Example 3. Then by definition of an improper integral the Fourier sine integral representation (14) of $f(x) = e^{-x}$, $x > 0$, can be written as $f(x) = \lim_{b \to \infty} F_b(x)$, where x is considered a parameter in

$$F_b(x) = \frac{2}{\pi} \int_0^b \frac{\alpha \sin \alpha x}{1 + \alpha^2} \, d\alpha. \qquad (15)$$

FIGURE 14.9 (a) is the even extension of f; (b) is the odd extension of f

Now the idea is this: Since the Fourier sine integral (14) converges, for a specified value of $b > 0$ the graph of the **partial integral** $F_b(x)$ in (15) will be an approximation to the graph of f in Figure 14.9(b). The graphs of $F_b(x)$ for $b = 5$ and $b = 20$ given in Figure 14.10 were obtained by using *Mathematica* and its **NIntegrate** application. See Problem 21 in Exercises 14.3.

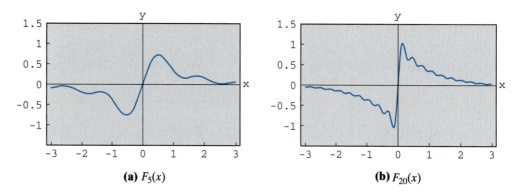

(a) $F_5(x)$ **(b)** $F_{20}(x)$

FIGURE 14.10   Convergence of $F_b(x)$ to $f(x)$ in Example 3(b) as $b \to \infty$

COMPLEX FORM   The Fourier integral (4) also possesses an equivalent **complex form,** or **exponential form,** that is analogous to the complex form of a Fourier series (see Problem 21 in Exercises 11.2). If (5) and (6) are substituted into (4), then

$$f(x) = \frac{1}{\pi} \int_0^\infty \int_{-\infty}^\infty f(t) \left[\cos \alpha t \cos \alpha x + \sin \alpha t \sin \alpha x\right] dt \, d\alpha$$

$$= \frac{1}{\pi} \int_0^\infty \int_{-\infty}^\infty f(t) \cos \alpha(t - x) \, dt \, d\alpha$$

$$= \frac{1}{2\pi} \int_{-\infty}^\infty \int_{-\infty}^\infty f(t) \cos \alpha(t - x) \, dt \, d\alpha \tag{16}$$

$$= \frac{1}{2\pi} \int_{-\infty}^\infty \int_{-\infty}^\infty f(t)[\cos \alpha(t - x) + i \sin \alpha(t - x)] \, dt \, d\alpha \tag{17}$$

$$= \frac{1}{2\pi} \int_{-\infty}^\infty \int_{-\infty}^\infty f(t) e^{i\alpha(t-x)} \, dt \, d\alpha$$

$$= \frac{1}{2\pi} \int_{-\infty}^\infty \left(\int_{-\infty}^\infty f(t) e^{i\alpha t} \, dt \right) e^{-i\alpha x} \, d\alpha. \tag{18}$$

We note that (16) follows from the fact that the integrand is an even function of α. In (17) we have simply added zero to the integrand;

$$i \int_{-\infty}^\infty \int_{-\infty}^\infty f(t) \sin \alpha(t - x) \, dt \, d\alpha = 0$$

because the integrand is an odd function of α. The integral in (18) can be expressed as

$$f(x) = \frac{1}{2\pi} \int_{-\infty}^\infty C(\alpha) e^{-i\alpha x} \, d\alpha, \tag{19}$$

where

$$C(\alpha) = \int_{-\infty}^\infty f(x) e^{i\alpha x} \, dx. \tag{20}$$

This latter form of the Fourier integral will be put to use in the next section when we return to the solution of boundary-value problems.

EXERCISES 14.3

Answers to selected odd-numbered problems begin on page ANS-21.

In Problems 1–6 find the Fourier integral representation of the given function.

1. $f(x) = \begin{cases} 0, & x < -1 \\ -1, & -1 < x < 0 \\ 2, & 0 < x < 1 \\ 0, & x > 1 \end{cases}$

2. $f(x) = \begin{cases} 0, & x < \pi \\ 4, & \pi < x < 2\pi \\ 0, & x > 2\pi \end{cases}$

3. $f(x) = \begin{cases} 0, & x < 0 \\ x, & 0 < x < 3 \\ 0, & x > 3 \end{cases}$

4. $f(x) = \begin{cases} 0, & x < 0 \\ \sin x, & 0 \le x \le \pi \\ 0, & x > \pi \end{cases}$

5. $f(x) = \begin{cases} 0, & x < 0 \\ e^{-x}, & x > 0 \end{cases}$

6. $f(x) = \begin{cases} e^x, & |x| < 1 \\ 0, & |x| > 1 \end{cases}$

In Problems 7–12 represent the given function by an appropriate cosine or sine integral.

7. $f(x) = \begin{cases} 0, & x < -1 \\ -5, & -1 < x < 0 \\ 5, & 0 < x < 1 \\ 0, & x > 1 \end{cases}$

8. $f(x) = \begin{cases} 0, & |x| < 1 \\ \pi, & 1 < |x| < 2 \\ 0, & |x| > 2 \end{cases}$

9. $f(x) = \begin{cases} |x|, & |x| < \pi \\ 0, & |x| > \pi \end{cases}$ 10. $f(x) = \begin{cases} x, & |x| < \pi \\ 0, & |x| > \pi \end{cases}$

11. $f(x) = e^{-|x|} \sin x$ 12. $f(x) = xe^{-|x|}$

In Problems 13–16 find the cosine and sine integral representations of the given function.

13. $f(x) = e^{-kx}, \quad k > 0, \quad x > 0$

14. $f(x) = e^{-x} - e^{-3x}, \quad x > 0$

15. $f(x) = xe^{-2x}, \quad x > 0$

16. $f(x) = e^{-x} \cos x, \quad x > 0$

In Problems 17 and 18 solve the given integral equation for the function f.

17. $\displaystyle\int_0^\infty f(x) \cos \alpha x \, dx = e^{-\alpha}$

18. $\displaystyle\int_0^\infty f(x) \sin \alpha x \, dx = \begin{cases} 1, & 0 < \alpha < 1 \\ 0, & \alpha > 1 \end{cases}$

19. **(a)** Use (7) to show that

$$\int_0^\infty \frac{\sin 2x}{x} \, dx = \frac{\pi}{2}.$$

   (*Hint:* α is a dummy variable of integration.)

   (b) Show in general that for $k > 0$,

$$\int_0^\infty \frac{\sin kx}{x} \, dx = \frac{\pi}{2}.$$

20. Use the complex form (19) to find the Fourier integral representation of $f(x) = e^{-|x|}$. Show that the result is the same as that obtained from (8).

COMPUTER LAB ASSIGNMENTS

21. While the integral (12) can be graphed in the same manner discussed on page 534 to obtain Figure 14.10, it can also be expressed in terms of a special function that is built into a CAS.

   (a) Use a trigonometric identity to show that an alternative form of the Fourier integral representation (12) of the function f in Example 2 (with $a = 1$) is

$$f(x) = \frac{1}{\pi} \int_0^\infty \frac{\sin \alpha(x+1) - \sin \alpha(x-1)}{\alpha} \, d\alpha.$$

   (b) As a consequence of part (a), $f(x) = \displaystyle\lim_{b \to \infty} F_b(x)$, where

$$F_b(x) = \frac{1}{\pi} \int_0^b \frac{\sin \alpha(x+1) - \sin \alpha(x-1)}{\alpha} \, d\alpha.$$

   Show that the last integral can be written as

$$F_b(x) = \frac{1}{\pi} [\text{Si}(b(x+1)) - \text{Si}(b(x-1))],$$

   where $\text{Si}(x)$ is the **sine integral function.** See Problem 49 in Exercises 2.3.

   (c) Use a CAS and the sine integral form of $F_b(x)$ obtained in part (b) to verify the graphs in the chapter opener (page 520) on the interval $-3 \le x \le 3$ for $b = 4$, 6, and 15. Then graph $F_b(x)$ for larger values of $b > 0$.

14.4 FOURIER TRANSFORMS

INTRODUCTION: So far in this text we have studied and used only one integral transform: the Laplace transform. But in Section 14.3 we saw that the Fourier integral had three alternative forms: the cosine integral, the sine integral, and the complex or exponential form. In this section we take these three forms of the Fourier integral and develop them into three new integral transforms, not surprisingly, called **Fourier transforms.** In addition, we expand on the concept of a transform pair, that is, an integral transform and its inverse. We also see that the inverse of an integral transform is itself another integral transform.

REVIEW MATERIAL: See Definition 14.2 and (19) and (20) in Section 14.3.

TRANSFORM PAIRS The Laplace transform $F(s)$ of a function $f(t)$ is defined by an integral, but up to now we have been using the symbolic representation $f(t) = \mathcal{L}^{-1}\{F(s)\}$ to denote the inverse Laplace transform of $F(s)$. Actually, the inverse Laplace transform is also an integral transform.

If $\mathcal{L}\{f(t)\} = \int_0^\infty e^{-st} f(t)\, dt = F(s)$, then the inverse Laplace transform is

$$\mathcal{L}^{-1}\{F(s)\} = \frac{1}{2\pi i} \int_{\gamma - i\infty}^{\gamma + i\infty} e^{st} F(s)\, ds = f(t).$$

The last integral is called a **contour integral;** it evaluation requires the use of complex variables and is beyond the scope of this text. The point here is this: integral transforms appear in **transform pairs.** If $f(x)$ is transformed into $F(\alpha)$ by an **integral transform**

$$F(\alpha) = \int_a^b f(x) K(\alpha, x)\, dx,$$

then the function f can be recovered by another integral transform

$$f(x) = \int_c^d F(\alpha) H(\alpha, x)\, d\alpha,$$

called the **inverse transform.** The functions K and H in the integrands are called the **kernels** of their respective transforms. We identify $K(s, t) = e^{-st}$ as the kernel of the Laplace transform and $H(s, t) = e^{st}/2\pi i$ as the kernel of the inverse Laplace transform.

FOURIER TRANSFORM PAIRS The Fourier integral is the source of three new integral transforms. From (20)–(19), (11)–(10), and (9)–(8) of Section 14.3 we are prompted to define the following **Fourier transform pairs.**

DEFINITION 14.3	Fourier Transform Pairs

(i)	Fourier transform:	$\mathcal{F}\{f(x)\} = \displaystyle\int_{-\infty}^{\infty} f(x) e^{i\alpha x}\, dx = F(\alpha)$	(1)
	Inverse Fourier transform:	$\mathcal{F}^{-1}\{F(\alpha)\} = \dfrac{1}{2\pi} \displaystyle\int_{-\infty}^{\infty} F(\alpha) e^{-i\alpha x}\, d\alpha = f(x)$	(2)

(ii) Fourier sine transform:	$\mathscr{F}_s\{f(x)\} = \int_0^\infty f(x) \sin \alpha x \, dx = F(\alpha)$	(3)
Inverse Fourier sine transform:	$\mathscr{F}_s^{-1}\{F(\alpha)\} = \dfrac{2}{\pi} \int_0^\infty F(\alpha) \sin \alpha x \, d\alpha = f(x)$	(4)
(iii) Fourier cosine transform:	$\mathscr{F}_c\{f(x)\} = \int_0^\infty f(x) \cos \alpha x \, dx = F(\alpha)$	(5)
Inverse Fourier cosine transform:	$\mathscr{F}_c^{-1}\{F(\alpha)\} = \dfrac{2}{\pi} \int_0^\infty F(\alpha) \cos \alpha x \, d\alpha = f(x)$	(6)

EXISTENCE   The conditions under which (1), (3), and (5) exist are more stringent than those for the Laplace transform. For example, you should verify that $\mathscr{F}\{1\}$, $\mathscr{F}_s\{1\}$, and $\mathscr{F}_c\{1\}$ do not exist. Sufficient conditions for existence are that f be absolutely integrable on the appropriate interval and that f and f' be piecewise continuous on every finite interval.

OPERATIONAL PROPERTIES   Since our immediate goal is to apply these new transforms to boundary-value problems, we need to examine the transforms of derivatives.

FOURIER TRANSFORM   Suppose that f is continuous and absolutely integrable on the interval $(-\infty, \infty)$ and f' is piecewise continuous on every finite interval. If $f(x) \to 0$ as $x \to \pm\infty$, then integration by parts gives

$$\mathscr{F}\{f'(x)\} = \int_{-\infty}^\infty f'(x) e^{i\alpha x} \, dx$$

$$= f(x) \, e^{i\alpha x} \Big|_{-\infty}^\infty - i\alpha \int_{-\infty}^\infty f(x) e^{i\alpha x} \, dx$$

$$= -i\alpha \int_{-\infty}^\infty f(x) e^{i\alpha x} \, dx,$$

that is, $\mathscr{F}\{f'(x)\} = -i\alpha \, F(\alpha).$ (7)

Similarly, under the added assumptions that f' is continuous on $(-\infty, \infty)$, $f''(x)$ is piecewise continuous on every finite interval, and $f'(x) \to 0$ as $x \to \pm\infty$, we have

$$\mathscr{F}\{f''(x)\} = (-i\alpha)^2 \, \mathscr{F}\{f(x)\} = -\alpha^2 F(\alpha).$$ (8)

It is important to be aware that the sine and cosine transforms are not suitable for transforming the first derivative (or, for that matter, any derivative of *odd* order). It is readily shown that

$$\mathscr{F}_s\{f'(x)\} = -\alpha\mathscr{F}_c\{f(x)\} \quad \text{and} \quad \mathscr{F}_c\{f'(x)\} = \alpha\mathscr{F}_s\{f(x)\} - f(0).$$

The difficulty is apparent; the transform of $f'(x)$ is not expressed in terms of the original integral transform.

FOURIER SINE TRANSFORM   Suppose that f and f' are continuous, f is absolutely integrable on the interval $[0, \infty)$, and f'' is piecewise continuous on every finite interval. If $f \to 0$ and $f' \to 0$ as $x \to \infty$, then

$$\mathscr{F}_s\{f''(x)\} = \int_0^\infty f''(x) \sin \alpha x \, dx$$

$$= f'(x) \sin \alpha x \Big|_0^\infty - \alpha \int_0^\infty f'(x) \cos \alpha x \, dx$$

$$= -\alpha \left[f(x) \cos \alpha x \Big|_0^\infty + \alpha \int_0^\infty f(x) \sin \alpha x \, dx \right]$$

$$= \alpha f(0) - \alpha^2 \mathscr{F}_s\{f(x)\},$$

that is, $$\mathscr{F}_s\{f''(x)\} = -\alpha^2 F(\alpha) + \alpha f(0). \qquad (9)$$

FOURIER COSINE TRANSFORM   Under the same assumptions that lead to (9) we find the Fourier cosine transform of $f''(x)$ to be

$$\mathscr{F}_c\{f''(x)\} = -\alpha^2 F(\alpha) - f'(0). \qquad (10)$$

A natural question is "How do we know which transform to use on a given boundary-value problem?" Clearly, to use a Fourier transform, the domain of the variable to be eliminated must be $(-\infty, \infty)$. To utilize a sine or cosine transform, the domain of at least one of the variables in the problem must be $[0, \infty)$. But the determining factor in choosing between the sine transform and the cosine transform is the type of boundary condition specified at zero.

In the examples that follow we shall assume without further mention that both u and $\partial u/\partial x$ (or $\partial u/\partial y$) approach zero as $x \to \pm\infty$. This is not a major restriction since these conditions hold in most applications.

Remember this when working Exercises 14.4.

EXAMPLE 1   Using the Fourier Transform

Solve the heat equation $k \dfrac{\partial^2 u}{\partial x^2} = \dfrac{\partial u}{\partial t}$, $-\infty < x < \infty, t > 0$ subject to

$$u(x, 0) = f(x), \quad \text{where} \quad f(x) = \begin{cases} u_0, & |x| < 1 \\ 0, & |x| > 1. \end{cases}$$

SOLUTION   The problem can be interpreted as finding the temperature $u(x, t)$ in an infinite rod. Because the domain of x is the infinite interval $(-\infty, \infty)$, we use the Fourier transform (1) and define

$$\mathscr{F}\{u(x, t)\} = \int_{-\infty}^\infty u(x, t) e^{i\alpha x} \, dx = U(\alpha, t).$$

If we transform the partial differential equation and use (8),

$$\mathscr{F}\left\{ k \frac{\partial^2 u}{\partial x^2} \right\} = \mathscr{F}\left\{ \frac{\partial u}{\partial t} \right\}$$

yields $$-k\alpha^2 U(\alpha, t) = \frac{dU}{dt} \quad \text{or} \quad \frac{dU}{dt} + k\alpha^2 U(\alpha, t) = 0.$$

Solving the last equation gives $U(\alpha, t) = c e^{-k\alpha^2 t}$. Now the transform of the initial condition is

$$\mathscr{F}\{u(x, 0)\} = \int_{-\infty}^\infty f(x) e^{i\alpha x} \, dx = \int_{-1}^1 u_0 e^{i\alpha x} \, dx = u_0 \frac{e^{i\alpha} - e^{-i\alpha}}{i\alpha}.$$

This result is the same as $U(\alpha, 0) = 2u_0\dfrac{\sin \alpha}{\alpha}$. Applying this condition to the solution $U(\alpha, t)$ gives $U(\alpha, 0) = c = (2u_0 \sin \alpha)/\alpha$, and so

$$U(\alpha, t) = 2u_0\frac{\sin \alpha}{\alpha} e^{-k\alpha^2 t}.$$

It then follows from the inversion integral (2) that

$$u(x, t) = \frac{u_0}{\pi} \int_{-\infty}^{\infty} \frac{\sin \alpha}{\alpha} e^{-k\alpha^2 t} e^{-i\alpha x}\, d\alpha.$$

The last expression can be simplified somewhat by using Euler's formula $e^{-i\alpha x} = \cos \alpha x - i \sin \alpha x$ and noting that $\displaystyle\int_{-\infty}^{\infty} \frac{\sin \alpha}{\alpha} e^{-k\alpha^2 t} \sin \alpha x\, d\alpha = 0$ since the integrand is an odd function of α. Hence we finally have

$$u(x, t) = \frac{u_0}{\pi} \int_{-\infty}^{\infty} \frac{\sin \alpha \cos \alpha x}{\alpha} e^{-k\alpha^2 t}\, d\alpha. \tag{11}$$

It is left to the reader to show that the solution (11) can be expressed in terms of the error function. See Problem 23 in Exercises 14.4.

EXAMPLE 2 Using the Cosine Transform

The steady-state temperature in a semi-infinite plate is determined from

$$\frac{\partial^2 u}{\partial x^2} + \frac{\partial^2 u}{\partial y^2} = 0, \quad 0 < x < \pi, \quad y > 0$$

$$u(0, y) = 0, \quad u(\pi, y) = e^{-y}, \quad y > 0$$

$$\left.\frac{\partial u}{\partial y}\right|_{y=0} = 0, \quad 0 < x < \pi.$$

Solve for $u(x, y)$.

SOLUTION The domain of the variable y and the prescribed condition at $y = 0$ indicate that the Fourier cosine transform is suitable for the problem. We define

$$\mathscr{F}_c\{u(x, y)\} = \int_0^{\infty} u(x, y) \cos \alpha y\, dy = U(x, \alpha).$$

In view of (10), $\qquad \mathscr{F}_c\left\{\dfrac{\partial^2 u}{\partial x^2}\right\} + \mathscr{F}_c\left\{\dfrac{\partial^2 u}{\partial y^2}\right\} = \mathscr{F}_c\{0\}$

becomes $\qquad \dfrac{d^2 U}{dx^2} - \alpha^2 U(x, \alpha) - u_y(x, 0) = 0 \quad$ or $\quad \dfrac{d^2 U}{dx^2} - \alpha^2 U = 0.$

Since the domain of x is a finite interval, we choose to write the solution of the ordinary differential equation as

$$U(x, \alpha) = c_1 \cosh \alpha x + c_2 \sinh \alpha x. \tag{12}$$

Now $\mathscr{F}_c\{u(0, y)\} = \mathscr{F}_c\{0\}$ and $\mathscr{F}_c\{u(\pi, y)\} = \mathscr{F}_c\{e^{-y}\}$ are in turn equivalent to

$$U(0, \alpha) = 0 \quad \text{and} \quad U(\pi, \alpha) = \frac{1}{1 + \alpha^2}.$$

When we apply these latter conditions, the solution (12) gives $c_1 = 0$ and $c_2 = 1/[(1 + \alpha^2) \sinh \alpha\pi]$. Therefore

$$U(x, \alpha) = \frac{\sinh \alpha x}{(1 + \alpha^2) \sinh \alpha\pi},$$

and so from (6) we arrive at

$$u(x, y) = \frac{2}{\pi} \int_0^\infty \frac{\sinh \alpha x}{(1 + \alpha^2) \sinh \alpha\pi} \cos \alpha y \, d\alpha . \qquad (13) \quad \blacksquare$$

Had $u(x, 0)$ been given in Example 2 rather than $u_y(x, 0)$, then the sine transform would have been appropriate.

EXERCISES 14.4

Answers to selected odd-numbered problems begin on page ANS-22.

In Problems 1–21 use the Fourier integral transforms of this section to solve the given boundary-value problem. Make assumptions about boundedness where necessary.

1. $k \dfrac{\partial^2 u}{\partial x^2} = \dfrac{\partial u}{\partial t}, \quad -\infty < x < \infty, \quad t > 0$

$u(x, 0) = e^{-|x|}, \quad -\infty < x < \infty$

2. $k \dfrac{\partial^2 u}{\partial x^2} = \dfrac{\partial u}{\partial t}, \quad -\infty < x < \infty \quad t > 0$

$u(x, 0) = \begin{cases} 0, & x < -1 \\ -100, & -1 < x < 0 \\ 100, & 0 < x < 1 \\ 0, & x > 1 \end{cases}$

3. Find the temperature $u(x, t)$ in a semi-infinite rod if $u(0, t) = u_0, t > 0$ and $u(x, 0) = 0, x > 0$.

4. Use the result $\displaystyle\int_0^\infty \frac{\sin \alpha x}{\alpha} \, d\alpha = \frac{\pi}{2}, x > 0$ to show that the solution of Problem 3 can be written as

$$u(x, t) = u_0 - \frac{2u_0}{\pi} \int_0^\infty \frac{\sin \alpha x}{\alpha} e^{-k\alpha^2 t} \, d\alpha.$$

5. Find the temperature $u(x, t)$ in a semi-infinite rod if $u(0, t) = 0, t > 0$, and

$$u(x, 0) = \begin{cases} 1, & 0 < x < 1 \\ 0, & x > 1. \end{cases}$$

6. Solve Problem 3 if the condition at the left boundary is

$$\left.\frac{\partial u}{\partial x}\right|_{x=0} = -A, \quad t > 0,$$

where A is a constant.

7. Solve Problem 5 if the end $x = 0$ is insulated.

8. Find the temperature $u(x, t)$ in a semi-infinite rod if $u(0, t) = 1, t > 0$, and $u(x, 0) = e^{-x}, x > 0$.

9. **(a)** $a^2 \dfrac{\partial^2 u}{\partial x^2} = \dfrac{\partial^2 u}{\partial t^2}, \quad -\infty < x < \infty, \quad t > 0$

$u(x, 0) = f(x), \quad \left.\dfrac{\partial u}{\partial t}\right|_{t=0} = g(x), \quad -\infty < x < \infty$

(b) If $g(x) = 0$, show that the solution of part (a) can be written as $u(x, t) = \frac{1}{2}[f(x + at) + f(x - at)]$.

10. Find the displacement $u(x, t)$ of a semi-infinite string if

$$u(0, t) = 0, \quad t > 0$$

$$u(x, 0) = xe^{-x}, \quad \left.\frac{\partial u}{\partial t}\right|_{t=0} = 0, \quad x > 0$$

11. Solve the problem in Example 2 if the boundary conditions at $x = 0$ and $x = \pi$ are reversed: $u(0, y) = e^{-y}$, $u(\pi, y) = 0, y > 0$.

12. Solve the problem in Example 2 if the boundary condition at $y = 0$ is $u(x, 0) = 1, 0 < x < \pi$.

13. Find the steady-state temperature $u(x, y)$ in a plate defined by $x \geq 0, y \geq 0$ if the boundary $x = 0$ is insulated and, at $y = 0$,

$$u(x, 0) = \begin{cases} 50, & 0 < x < 1 \\ 0, & x > 1. \end{cases}$$

14. Solve Problem 13 if the boundary condition at $x = 0$ is $u(0, y) = 0, y > 0$.

15. $\dfrac{\partial^2 u}{\partial x^2} + \dfrac{\partial^2 u}{\partial y^2} = 0, \quad x > 0, \quad 0 < y < 2$

$u(0, y) = 0, \quad 0 < y < 2$

$u(x, 0) = f(x), \quad u(x, 2) = 0, \quad x > 0$

16. $\dfrac{\partial^2 u}{\partial x^2} + \dfrac{\partial^2 u}{\partial y^2} = 0,\quad 0 < x < \pi,\quad y > 0$

$u(0, y) = f(y),\quad \left.\dfrac{\partial u}{\partial x}\right|_{x=\pi} = 0,\quad y > 0$

$\left.\dfrac{\partial u}{\partial y}\right|_{y=0} = 0,\quad 0 < x < \pi$

In Problems 17 and 18 find the steady-state temperature in the plate given in the figure. (*Hint:* One way of proceeding is to express Problems 17 and 18 as two- and three-boundary-value problems, respectively. Use the superposition principle (see Section 12.5).)

17.

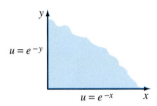

FIGURE 14.11   Plate in Problem 17

18.

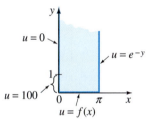

FIGURE 14.12   Plate in Problem 18

19. Use the result $\mathscr{F}\{e^{-x^2/4p^2}\} = 2\sqrt{\pi}\,p\,e^{-p^2\alpha^2}$ to solve the boundary-value problem

$$k\dfrac{\partial^2 u}{\partial x^2} = \dfrac{\partial u}{\partial t},\quad -\infty < x < \infty,\quad t > 0$$

$$u(x, 0) = e^{-x^2},\quad -\infty < x < \infty.$$

20. If $\mathscr{F}\{f(x)\} = F(\alpha)$ and $\mathscr{F}\{g(x)\} = G(\alpha)$, then the **convolution theorem** for the Fourier transform is given by

$$\int_{-\infty}^{\infty} f(\tau)g(x - \tau)\,d\tau = \mathscr{F}^{-1}\{F(\alpha)G(\alpha)\}.$$

Use this result and $\mathscr{F}\{e^{-x^2/4p^2}\} = 2\sqrt{\pi}\,p\,e^{-p^2\alpha^2}$ to show that a solution of the boundary-value problem

$$k\dfrac{\partial^2 u}{\partial x^2} = \dfrac{\partial u}{\partial t},\quad -\infty < x < \infty,\quad t > 0$$

$$u(x, 0) = f(x),\quad -\infty < x < \infty$$

is

$$u(x, t) = \dfrac{1}{2\sqrt{k\pi t}}\int_{-\infty}^{\infty} f(\tau)e^{-(x-\tau)^2/4kt}\,d\tau.$$

21. Use the transform $\mathscr{F}\{e^{-x^2/4p^2}\}$ given in Problem 19 to find the steady-state temperature in the infinite strip shown in Figure 14.13.

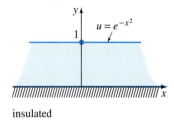

insulated

FIGURE 14.13   Infinite strip in Problem 21

22. The solution of Problem 14 can be integrated. Use entries 42 and 43 of the table in Appendix III to show that

$$u(x, y) = \dfrac{100}{\pi}\left[\arctan\dfrac{x}{y} - \dfrac{1}{2}\arctan\dfrac{x+1}{y} - \dfrac{1}{2}\arctan\dfrac{x-1}{y}\right].$$

23. Use the solution given in Problem 20 to rewrite the solution of Example 1 in an alternative integral form. Then use the change of variables $v = (x - \tau)/2\sqrt{kt}$ and the results of Problem 9 in Exercises 14.1 to show that the solution of Example 1 can be expressed as

$$u(x, t) = \dfrac{u_0}{2}\left[\operatorname{erf}\left(\dfrac{x+1}{2\sqrt{kt}}\right) - \operatorname{erf}\left(\dfrac{x-1}{2\sqrt{kt}}\right)\right].$$

COMPUTER LAB ASSIGNMENTS

24. Assume that $u_0 = 100$ and $k = 1$ in the solution in Problem 23. Use a CAS to graph $u(x, t)$ over the rectangular region $-4 \leq x \leq 4,\ 0 \leq t \leq 6$. Use a 2D plot to superimpose the graphs of $u(x, t)$ for $t = 0.05, 0.125,$ $0.5, 1, 2, 4, 6,$ and 15 on the interval $-4 \leq x \leq 4$. Use the graphs to conjecture the values of $\lim_{t \to \infty} u(x, t)$ and $\lim_{x \to \infty} u(x, t)$. Then prove these results analytically using the properties of $\operatorname{erf}(x)$.

CHAPTER 14 IN REVIEW

Answers to selected odd-numbered problems begin on page ANS-22.

In Problems 1–15 solve the given boundary-value problem by an appropriate integral transform. Make assumptions about boundedness where necessary.

1. $\dfrac{\partial^2 u}{\partial x^2} + \dfrac{\partial^2 u}{\partial y^2} = 0, \quad x > 0, \quad 0 < y < \pi$

$\left. \dfrac{\partial u}{\partial x} \right|_{x=0} = 0, \quad 0 < y < \pi$

$u(x, 0) = 0, \quad \left. \dfrac{\partial u}{\partial y} \right|_{y=\pi} = e^{-x}, \quad x > 0$

2. $\dfrac{\partial^2 u}{\partial x^2} = \dfrac{\partial u}{\partial t}, \quad 0 < x < 1, \quad t > 0$

$u(0, t) = 0, \quad u(1, t) = 0, \quad t > 0$

$u(x, 0) = 50 \sin 2\pi x, \quad 0 < x < 1$

3. $\dfrac{\partial^2 u}{\partial x^2} - hu = \dfrac{\partial u}{\partial t}, \quad h > 0, \quad x > 0, \quad t > 0$

$u(0, t) = 0, \quad \lim_{x \to \infty} \dfrac{\partial u}{\partial x} = 0, \quad t > 0$

$u(x, 0) = u_0, \quad x > 0$

4. $\dfrac{\partial u}{\partial t} - \dfrac{\partial^2 u}{\partial x^2} = e^{-|x|}, \quad -\infty < x < \infty, \quad t > 0$

$u(x, 0) = 0, \quad -\infty < x < \infty$

5. $\dfrac{\partial^2 u}{\partial x^2} = \dfrac{\partial u}{\partial t}, \quad x > 0, \quad t > 0$

$u(0, t) = t, \quad \lim_{x \to \infty} u(x, t) = 0$

$u(x, 0) = 0, \quad x > 0$ (*Hint:* Use Theorem 7.9.)

6. $\dfrac{\partial^2 u}{\partial x^2} = \dfrac{\partial^2 u}{\partial t^2}, \quad 0 < x < 1, \quad t > 0$

$u(0, t) = 0, \quad u(1, t) = 0, \quad t > 0$

$u(x, 0) = \sin \pi x, \quad \left. \dfrac{\partial u}{\partial t} \right|_{t=0} = -\sin \pi x, \quad 0 < x < 1$

7. $k\dfrac{\partial^2 u}{\partial x^2} = \dfrac{\partial u}{\partial t}, \quad -\infty < x < \infty, \quad t > 0$

$u(x, 0) = \begin{cases} 0, & x < 0 \\ u_0, & 0 < x < \pi \\ 0, & x > \pi \end{cases}$

8. $\dfrac{\partial^2 u}{\partial x^2} + \dfrac{\partial^2 u}{\partial y^2} = 0, \quad 0 < x < \pi, \quad y > 0$

$u(0, y) = 0, \quad u(\pi, y) = \begin{cases} 0, & 0 < y < 1 \\ 1, & 1 < y < 2 \\ 0, & y > 2 \end{cases}$

$\left. \dfrac{\partial u}{\partial y} \right|_{y=0} = 0, \quad 0 < x < \pi$

9. $\dfrac{\partial^2 u}{\partial x^2} + \dfrac{\partial^2 u}{\partial y^2} = 0, \quad x > 0, \quad y > 0$

$u(0, y) = \begin{cases} 50, & 0 < y < 1 \\ 0, & y > 1 \end{cases}$

$u(x, 0) = \begin{cases} 100, & 0 < x < 1 \\ 0, & x > 1 \end{cases}$

10. $\dfrac{\partial^2 u}{\partial x^2} + r = \dfrac{\partial u}{\partial t}, \quad 0 < x < 1, \quad t > 0$

$\left. \dfrac{\partial u}{\partial x} \right|_{x=0} = 0, \quad u(1, t) = 0, \quad t > 0$

$u(x, 0) = 0, \quad 0 < x < 1$

11. $\dfrac{\partial^2 u}{\partial x^2} + \dfrac{\partial^2 u}{\partial y^2} = 0, \quad x > 0, \quad 0 < y < \pi$

$u(0, y) = A, \quad 0 < y < \pi$

$\left. \dfrac{\partial u}{\partial y} \right|_{y=0} = 0, \quad \left. \dfrac{\partial u}{\partial y} \right|_{y=\pi} = Be^{-x}, \quad x > 0$

12. $\dfrac{\partial^2 u}{\partial x^2} = \dfrac{\partial u}{\partial t}, \quad 0 < x < 1, \quad t > 0$

$u(0, t) = u_0, \quad u(1, t) = u_0, \quad t > 0$

$u(x, 0) = 0, \quad 0 < x < 1$

(*Hint:* Use the identity

$$\sinh (x - y) = \sinh x \cosh y - \cosh x \sinh y,$$

and then use Problem 6 in Exercises 14.1.)

13. $k\dfrac{\partial^2 u}{\partial x^2} = \dfrac{\partial u}{\partial t}, \quad -\infty < x < \infty, \quad t > 0$

$u(x, 0) = \begin{cases} 0, & x < 0 \\ e^{-x}, & x > 0 \end{cases}$

14. $\dfrac{\partial^2 u}{\partial x^2} = \dfrac{\partial u}{\partial t}, \quad x > 0, \quad t > 0$

$\left. \dfrac{\partial u}{\partial x} \right|_{x=0} = -50, \quad \lim_{x \to \infty} u(x, t) = 100, \quad t > 0$

$u(x, 0) = 100, \quad x > 0$

15. Show that a solution of the BVP

$$\dfrac{\partial^2 u}{\partial x^2} + \dfrac{\partial^2 u}{\partial y^2} = 0, \quad -\infty < x < \infty, \quad 0 < y < 1$$

$$\left. \dfrac{\partial u}{\partial y} \right|_{y=0} = 0, \quad u(x, 1) = f(x), \quad -\infty < x < \infty$$

is $u(x, y) = \dfrac{1}{\pi} \displaystyle\int_0^\infty \int_{-\infty}^\infty f(t) \dfrac{\cosh \alpha y \cos \alpha(t - x)}{\cosh \alpha} \, dt \, d\alpha.$

15

NUMERICAL SOLUTIONS OF PARTIAL DIFFERENTIAL EQUATIONS

15.1 Laplace's Equation
15.2 Heat Equation
15.3 Wave Equation

In Section 9.5 we saw that one way of approximating a solution of a second-order boundary-value problem was to work with a finite difference equation replacement of the ordinary differential equation. The difference equation was constructed by replacing the ordinary derivatives d^2y/dx^2 and dy/dx by difference quotients. The same idea carries over to BVPs involving partial differential equations. In the succeeding sections of this chapter we shall form a difference equation replacement for Laplace's equation, the heat equation, and the wave equation by replacing the partial derivatives $\partial^2 u/\partial x^2$, $\partial^2 u/\partial y^2$, $\partial^2 u/\partial t^2$, and $\partial u/\partial t$ by difference quotients.

Surface: Partial Sum $S_N(x, y),\ N = 5$

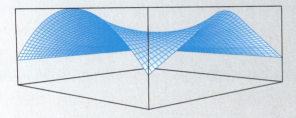

Surface: $u = e^{-\pi^2 t}\sin(\pi x)$

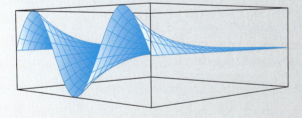

Surface: $u = \sin(\pi x)\cos(2\pi t)$

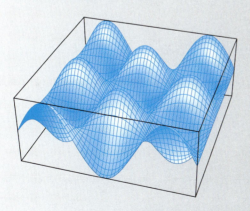

Three solution surfaces of BVPs. These solutions will be approximated by the methods of this chapter. See pages 545, 552, and 557.

15.1 LAPLACE'S EQUATION

INTRODUCTION: Recall from Section 12.1 that linear second-order PDEs in two independent variables are classified as *elliptic, parabolic,* and *hyperbolic.* Roughly, elliptic PDEs involve partial derivatives with respect to spatial variables only and as a consequence, solutions of such equations are determined by boundary conditions alone. Parabolic and hyperbolic equations involve partial derivatives with respect to both spatial and time variables, and so solutions of such equations generally are determined from boundary and initial conditions. A solution of an elliptic PDE (such as Laplace's equation) can describe a physical system whose state is in equilibrium (steady-state); a solution of a parabolic PDE (such as the heat equation) can describe a diffusional state, whereas a hyperbolic PDE (such as the wave equation) can describe a vibrational state.

In this section we begin our discussion with approximation methods appropriate for elliptic equations. Our focus will be on the simplest but probably the most important PDE of the elliptic type, Laplace's equation.

REVIEW MATERIAL: Review Sections 9.5, 12.1, 12.2, and 12.5.

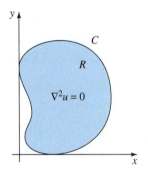

FIGURE 15.1 Planar region R with boundary C

DIFFERENCE EQUATION REPLACEMENT Suppose that we are seeking a solution $u(x, y)$ of Laplace's equation

$$\frac{\partial^2 u}{\partial x^2} + \frac{\partial^2 u}{\partial y^2} = 0 \tag{1}$$

in a planar region R that is bounded by some curve C. See Figure 15.1. Analogous to (6) of Section 9.5, by using the central differences

$$u(x + h, y) - 2u(x, y) + u(x - h, y) \quad \text{and} \quad u(x, y + h) - 2u(x, y) + u(x, y - h),$$

approximations for the second partial derivatives u_{xx} and u_{yy} can be obtained using the difference quotients

$$\frac{\partial^2 u}{\partial x^2} \approx \frac{1}{h^2}[u(x + h, y) - 2u(x, y) + u(x - h, y)] \tag{2}$$

$$\frac{\partial^2 u}{\partial y^2} \approx \frac{1}{h^2}[u(x, y + h) - 2u(x, y) + u(x, y - h)]. \tag{3}$$

By adding (2) and (3), we obtain a **five-point approximation** to the Laplacian:

$$\frac{\partial^2 u}{\partial x^2} + \frac{\partial^2 u}{\partial y^2} \approx \frac{1}{h^2}[u(x + h, y) + u(x, y + h) + u(x - h, y) + u(x, y - h) - 4u(x, y)].$$

Hence we can replace Laplace's equation (1) by the difference equation

$$u(x + h, y) + u(x, y + h) + u(x - h, y) + u(x, y - h) - 4u(x, y) = 0. \tag{4}$$

If we adopt the notation $u(x, y) = u_{ij}$ and

$$u(x + h, y) = u_{i+1,j}, \quad u(x, y + h) = u_{i,j+1}$$

$$u(x - h, y) = u_{i-1,j}, \quad u(x, y - h) = u_{i,j-1},$$

then (4) becomes

$$u_{i+1,j} + u_{i,j+1} + u_{i-1,j} + u_{i,j-1} - 4u_{ij} = 0. \tag{5}$$

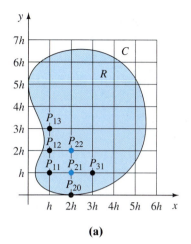

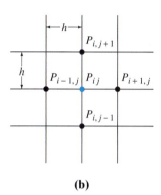

FIGURE 15.2 Region R overlaid with rectangular grid

To understand (5) a little better, suppose a rectangular grid consisting of horizontal lines spaced h units apart and vertical lines spaced h units apart is placed over the region R. The number h is called the **mesh size**. See Figure 15.2(a). The points $P_{ij} = P(ih, jh)$, where i and j are integers, of intersection of the horizontal and vertical lines, are called **mesh points** or **lattice points**. A mesh point is an **interior point** if its four nearest neighboring mesh points are points of R. Points in R or on C that are not interior points are called **boundary points**. For example, in Figure 15.2(a) we have

$$P_{20} = P(2h, 0), \quad P_{11} = P(h, h), \quad P_{21} = P(2h, h), \quad P_{22} = P(2h, 2h),$$

and so on. Of the points just listed, P_{21} and P_{22} are interior points, whereas P_{20} and P_{11} are boundary points. In Figure 15.2(a) interior points are the dots shown in color, and the boundary points are shown in black. Now from (5) we see that

$$u_{ij} = \frac{1}{4}\left[u_{i+1,j} + u_{i,j+1} + u_{i-1,j} + u_{i,j-1} \right], \tag{6}$$

and so, as can be seen in Figure 15.2(b), the value u_{ij} at an interior mesh point of R is the average of the values of u at four neighboring mesh points. The neighboring points $P_{i+1,j}$, $P_{i,j+1}$, $P_{i-1,j}$, and $P_{i,j-1}$ correspond to the four points on the compass E, N, W, and S, respectively.

DIRICHLET PROBLEM Recall that in the **Dirichet problem** for Laplace's equation $\nabla^2 u = 0$ the values of $u(x, y)$ are prescribed on the boundary of a region R. The basic idea is to find an approximate solution to Laplace's equation at interior mesh points by replacing the partial differential equation at these points by the difference equation (5). Hence the approximate values of u at the mesh points—namely, the u_{ij}—are related to each other and, possibly, to known values of u if a mesh point lies on the boundary. In this manner we obtain a system of linear algebraic equations that we solve for the unknown u_{ij}. The following example illustrates the method for a square region.

EXAMPLE 1   **A BVP Revisited**

In Problem 16 of Exercises 12.5 you were asked to solve the boundary-value problem

$$\frac{\partial^2 u}{\partial x^2} + \frac{\partial^2 u}{\partial y^2} = 0, \quad 0 < x < 2, \quad 0 < y < 2$$

$$u(0, y) = 0, \quad u(2, y) = y(2 - y), \quad 0 < y < 2$$

$$u(x, 0) = 0, \quad u(x, 2) = \begin{cases} x, & 0 < x < 1 \\ 2 - x, & 1 \le x < 2. \end{cases}$$

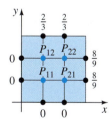

FIGURE 15.3 Square region R for Example 1

utilizing the superposition principle. To apply the present numerical method, let us start with a mesh size of $h = \frac{2}{3}$. As we see in Figure 15.3, that choice yields four interior points and eight boundary points. The numbers listed next to the boundary points are the exact values of u obtained from the specified condition along that boundary. For example, at $P_{31} = P(3h, h) = P\left(2, \frac{2}{3}\right)$ we have $x = 2$ and $y = \frac{2}{3}$, and so the condition $u(2, y)$ gives $u\left(2, \frac{2}{3}\right) = \frac{2}{3}\left(2 - \frac{2}{3}\right) = \frac{8}{9}$. Similarly, at $P_{13} = P\left(\frac{2}{3}, 2\right)$ the condition $u(x, 2)$ gives $u\left(\frac{2}{3}, 2\right) = \frac{2}{3}$. We now apply (5) at each interior point. For example, at P_{11} we have $i = 1$ and $j = 1$, so (5) becomes

$$u_{21} + u_{12} + u_{01} + u_{10} - 4u_{11} = 0.$$

Since $u_{01} = u\left(0, \frac{2}{3}\right) = 0$ and $u_{10} = u\left(\frac{2}{3}, 0\right) = 0$, the foregoing equation becomes $-4u_{11} + u_{21} + u_{12} = 0$. Repeating this, in turn, at P_{21}, P_{12}, and P_{22} we get three additional equations:

$$-4u_{11} + u_{21} + u_{12} = 0$$

$$u_{11} - 4u_{21} + u_{22} = -\frac{8}{9}$$

$$u_{11} - 4u_{12} + u_{22} = -\frac{2}{3} \qquad (7)$$

$$u_{21} + u_{12} - 4u_{22} = -\frac{14}{9}.$$

Using a computer algebra system to solve the system, we find the approximate values at the four interior points to be

$$u_{11} = \frac{7}{36} = 0.1944, \quad u_{21} = \frac{5}{12} = 0.4167, \quad u_{12} = \frac{13}{36} = 0.3611, \quad u_{22} = \frac{7}{12} = 0.5833. \quad \blacksquare$$

As in the discussion of ordinary differential equations, we expect that a smaller value of h will improve the accuracy of the approximation. However, using a smaller mesh size means, of course, that there are more interior mesh points, and correspondingly there is a much larger system of equations to be solved. For a *square* region whose length of side is L, a mesh size of $h = L/n$ will yield a total of $(n-1)^2$ interior mesh points. In Example 1, for $n = 8$ the mesh size is a reasonable $h = \frac{2}{8} = \frac{1}{4}$, but the number of interior points is $(8-1)^2 = 49$. Thus we have 49 equations in 49 unknowns. In the next example we use a mesh size of $h = \frac{1}{2}$.

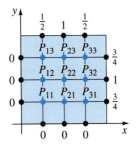

FIGURE 15.4   Region R in Example 1 with additional mesh points

EXAMPLE 2   **Example 1 with More Mesh Points**

As we see in Figure 15.4, with $n = 4$ a mesh size $h = \frac{2}{4} = \frac{1}{2}$ for the square in Example 1 gives $3^2 = 9$ interior mesh points. Applying (5) at these points and using the indicated boundary conditions, we get nine equations in nine unknowns. So that you can verify the results, we give the system in an unsimplified form:

$$u_{21} + u_{12} + 0 + 0 - 4u_{11} = 0$$

$$u_{31} + u_{22} + u_{11} + 0 - 4u_{21} = 0$$

$$\frac{3}{4} + u_{32} + u_{21} + 0 - 4u_{31} = 0$$

$$u_{22} + u_{13} + u_{11} + 0 - 4u_{12} = 0$$

$$u_{32} + u_{23} + u_{12} + u_{21} - 4u_{22} = 0 \qquad (8)$$

$$1 + u_{33} + u_{22} + u_{31} - 4u_{32} = 0$$

$$u_{23} + \frac{1}{2} + 0 + u_{12} - 4u_{13} = 0$$

$$u_{33} + 1 + u_{13} + u_{22} - 4u_{23} = 0$$

$$\frac{3}{4} + \frac{1}{2} + u_{23} + u_{32} - 4u_{33} = 0.$$

In this case a CAS yields

$$u_{11} = \frac{7}{64} = 0.1094, \quad u_{21} = \frac{51}{224} = 0.2277, \quad u_{31} = \frac{177}{448} = 0.3951$$

$$u_{12} = \frac{47}{224} = 0.2098, \quad u_{22} = \frac{13}{32} = 0.4063, \quad u_{32} = \frac{135}{224} = 0.6027$$

$$u_{13} = \frac{145}{448} = 0.3237, \quad u_{23} = \frac{131}{224} = 0.5848, \quad u_{33} = \frac{39}{64} = 0.6094.$$

After we simplify (8), it is interesting to note that the 9×9 matrix of coefficients is

$$\begin{vmatrix} -4 & 1 & 0 & 1 & 0 & 0 & 0 & 0 & 0 \\ 1 & -4 & 1 & 0 & 1 & 0 & 0 & 0 & 0 \\ 0 & 1 & -4 & 0 & 0 & 1 & 0 & 0 & 0 \\ 1 & 0 & 0 & -4 & 1 & 0 & 1 & 0 & 0 \\ 0 & 1 & 0 & 1 & -4 & 1 & 0 & 1 & 0 \\ 0 & 0 & 1 & 0 & 1 & -4 & 0 & 0 & 1 \\ 0 & 0 & 0 & 1 & 0 & 0 & -4 & 1 & 0 \\ 0 & 0 & 0 & 0 & 1 & 0 & 1 & -4 & 1 \\ 0 & 0 & 0 & 0 & 0 & 1 & 0 & 1 & -4 \end{vmatrix}. \tag{9}$$

This is an example of a **sparse matrix** in that a large percentage of the entries are zeros. The matrix (9) is also an example of a **banded matrix.** These kinds of matrices are characterized by the properties that the entries on the main diagonal and on diagonals (or bands) parallel to the main diagonal are all nonzero.

GAUSS-SEIDEL ITERATION Problems requiring approximations to solutions of partial differential equations invariably lead to large systems of linear algebraic equations. It is not uncommon to have to solve systems involving hundreds of equations. Although a direct method of solution such as Gaussian elimination leaves unchanged the zero entries outside the bands in a matrix such as (9), it does fill in the positions between the bands with nonzeros. Since storing very large matrices uses up a large portion of computer memory, it is usual practice to solve a large system in an indirect manner. One popular indirect method is called **Gauss-Seidel iteration.**

We shall illustrate this method for the system in (7). For the sake of simplicity we replace the double-subscripted variables u_{11}, u_{21}, u_{12}, and u_{22} by x_1, x_2, x_3, and x_4, respectively.

EXAMPLE 3 Gauss-Seidel Iteration

Step 1: *Solve each equation for the variables on the main diagonal of the system.* That is, in (7) solve the first equation for x_1, the second equation for x_2, and so on:

$$\begin{aligned} x_1 &= 0.25x_2 + 0.25x_3 \\ x_2 &= 0.25x_1 + 0.25x_4 + 0.2222 \\ x_3 &= 0.25x_1 + 0.25x_4 + 0.1667 \\ x_4 &= 0.25x_2 + 0.25x_3 + 0.3889. \end{aligned} \tag{10}$$

These equations can be obtained directly by using (6) rather than (5) at the interior points.

Step 2: *Iterations.* We start by making an initial guess for the values of x_1, x_2, x_3, and x_4. If this were simply a system of linear equations and we knew nothing about the solution, we could start with $x_1 = 0$, $x_2 = 0$, $x_3 = 0$, $x_4 = 0$. But since the solution of (10) represents approximations to a solution of a boundary-value problem, it would seem reasonable to use as the initial guess for the values of $x_1 = u_{11}$, $x_2 = u_{21}$, $x_3 = u_{12}$, and $x_4 = u_{22}$ the average of all the boundary conditions. In this case the average of the numbers at the eight boundary points shown in Figure 15.3 is approximately 0.4. Thus our initial guess is $x_1 = 0.4$, $x_2 = 0.4$, $x_3 = 0.4$, and $x_4 = 0.4$. Iterations of the Gauss-Seidel method use the x values as soon as they are computed. Note that the first equation in (10) depends only on x_2 and x_3; thus substituting $x_2 = 0.4$ and $x_3 = 0.4$ gives $x_1 = 0.2$. Since the second and third equations depend on x_1 and x_4, we use the newly calculated values $x_1 = 0.2$ and $x_4 = 0.4$ to obtain $x_2 = 0.3722$ and $x_3 = 0.3167$. The fourth equation depends on x_2 and x_3, so we use the new values $x_2 = 0.3722$ and $x_3 = 0.3167$ to get $x_4 = 0.5611$. In summary, the first iteration has given the values

$$x_1 = 0.2, \quad x_2 = 0.3722, \quad x_3 = 0.3167, \quad x_4 = 0.5611.$$

Note how close these numbers are already to the actual values given at the end of Example 1.

The second iteration starts with substituting $x_2 = 0.3722$ and $x_3 = 0.3167$ into the first equation. This gives $x_1 = 0.1722$. From $x_1 = 0.1722$ and the last computed value of x_4 (namely, $x_4 = 0.5611$), the second and third equations give, in turn, $x_2 = 0.4055$ and $x_3 = 0.3500$. Using these two values, we find from the fourth equation that $x_4 = 0.5678$. At the end of the second iteration we have

$$x_1 = 0.1722, \quad x_2 = 0.4055, \quad x_3 = 0.3500, \quad x_4 = 0.5678.$$

The third through seventh iterations are summarized in Table 15.1.

TABLE 15.1

Iteration	3rd	4th	5th	6th	7th
x_1	0.1889	0.1931	0.1941	0.1944	0.1944
x_2	0.4139	0.4160	0.4165	0.4166	0.4166
x_3	0.3584	0.3605	0.3610	0.3611	0.3611
x_4	0.5820	0.5830	0.5833	0.5833	0.5833

NOTE To apply Gauss-Seidel iteration to a general system of n linear equations in n unknowns, the variable x_i must actually appear in the ith equation of the system. Moreover, after each equation is solved for x_i, $i = 1, 2, \ldots, n$, the resulting system has the form $\mathbf{X} = \mathbf{AX} + \mathbf{B}$, where all the entries on the main diagonal of $\mathbf{A}$ are zero.

REMARKS

(*i*) In the examples given in this section the values of u_{ij} were determined by using known values of u at boundary points. But what do we do if the region is such that boundary points do not coincide with the actual boundary C of the region R? In this case the required values can be obtained by interpolation.

(*ii*) It is sometimes possible to cut down the number of equations to solve by using symmetry. Consider the rectangular region $0 \leq x \leq 2$, $0 \leq y \leq 1$, shown in Figure 15.5. The boundary conditions are $u = 0$ along the boundaries $x = 0$, $x = 2$, $y = 1$, and $u = 100$ along $y = 0$. The region is symmetric

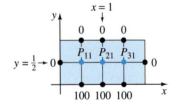

FIGURE 15.5 Rectangular region R

about the lines $x = 1$ and $y = \frac{1}{2}$, and the interior points P_{11} and P_{31} are equidistant from the neighboring boundary points at which the specified values of u are the same. Consequently, we assume that $u_{11} = u_{31}$, and so the system of three equations in three unknowns reduces to two equations in two unknowns. See Problem 2 in Exercises 15.1.

(*iii*) In the context of approximating a solution to Laplace's equation, the iteration technique illustrated in Example 3 is often referred to as **Liebman's method.**

(*iv*) Although it may not be noticeable on a computer, but convergence of Gauss-Seidel iteration, or Liebman's method, might not be particularly fast. Also, in a more general setting, Gauss-Seidel iteration might not converge at all. For conditions that are sufficient to guarantee convergence of Gauss-Seidel iteration, you are encouraged to consult texts on numerical analysis.

EXERCISES 15.1

Answers to selected odd-numbered problems begin on page ANS-22.

In Problems 1–8 use a computer as a computation aid.

In Problems 1–4 use (5) to approximate the solution of Laplace's equation at the interior points of the given region. Use symmetry when possible.

1. $u(0, y) = 0$, $u(3, y) = y(2 - y)$, $0 < y < 2$
$u(x, 0) = 0$, $u(x, 2) = x(3 - x)$, $0 < x < 3$
mesh size: $h = 1$

2. $u(0, y) = 0$, $u(2, y) = 0$, $0 < y < 1$
$u(x, 0) = 100$, $u(x, 1) = 0$, $0 < x < 2$
mesh size: $h = \frac{1}{2}$

3. $u(0, y) = 0$, $u(1, y) = 0$, $0 < y < 1$
$u(x, 0) = 0$, $u(x, 1) = \sin \pi x$, $0 < x < 1$
mesh size: $h = \frac{1}{3}$

4. $u(0, y) = 108y^2(1 - y)$, $u(1, y) = 0$, $0 < y < 1$
$u(x, 0) = 0$, $u(x, 1) = 0$, $0 < x < 1$
mesh size: $h = \frac{1}{3}$

In Problems 5 and 6 use (6) and Gauss-Seidel iteration to approximate the solution of Laplace's equation at the interior points of a unit square. Use the mesh size $h = \frac{1}{4}$. In Problem 5 the boundary conditions are given; in Problem 6 the values of u at boundary points are given in Figure 15.6.

5. $u(0, y) = 0$, $u(1, y) = 100y$, $0 < y < 1$
$u(x, 0) = 0$, $u(x, 1) = 100x$, $0 < x < 1$

6.

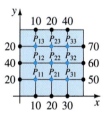

FIGURE 15.6 Region for Problem 6

7. (a) In Problem 12 of Exercises 12.6 you solved a potential problem using a special form of Poisson's equation $\dfrac{\partial^2 u}{\partial x^2} + \dfrac{\partial^2 u}{\partial y^2} = f(x, y)$. Show that the difference equation replacement for Poisson's equation is

$$u_{i+1, j} + u_{i, j+1} + u_{i-1, j} + u_{i, j-1} - 4u_{ij} = h^2 f(x, y).$$

(b) Use the result in part (a) to approximate the solution of the Poisson equation $\dfrac{\partial^2 u}{\partial x^2} + \dfrac{\partial^2 u}{\partial y^2} = -2$ at the interior points of the region in Figure 15.7. The mesh size is $h = \frac{1}{2}$, $u = 1$ at every point along *ABCD,* and $u = 0$ at every point along *DEFGA.* Use symmetry and, if necessary, Gauss-Seidel iteration.

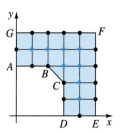

FIGURE 15.7 Region for Problem 7

8. Use the result in part (a) of Problem 7 to approximate the solution of the Poisson equation $\dfrac{\partial^2 u}{\partial x^2} + \dfrac{\partial^2 u}{\partial y^2} = -64$ at the interior points of the region in Figure 15.8. The mesh size is $h = \frac{1}{8}$, and $u = 0$ at every point on the boundary of the region. If necessary, use Gauss-Seidel iteration.

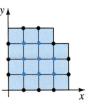

FIGURE 15.8 Region for Problem 8

15.2 HEAT EQUATION

INTRODUCTION: The basic idea in the following discussion is the same as in Section 15.1, we approximate a solution of a PDE—this time a parabolic PDE—by replacing the equation with a finite difference equation. But unlike in the preceding section we shall consider *two* finite-difference approximation methods for parabolic partial differential equations: one called an explicit method and the other called an implicit method.

For the sake of definiteness, we consider only the one-dimensional heat equation.

REVIEW MATERIAL: Review Sections 9.5, 12.1, 12.2, 12.3, and 15.1.

DIFFERENCE EQUATION REPLACEMENT To approximate a solution $u(x, t)$ of the one-dimensional heat equation

$$c \frac{\partial^2 u}{\partial x^2} = \frac{\partial u}{\partial t} \tag{1}$$

we again replace each derivative by a difference quotient. By using the central difference approximation (2) of Section 15.1,

$$\frac{\partial^2 u}{\partial x^2} \approx \frac{1}{h^2} [u(x + h, t) - 2u(x, t) + u(x - h, t)]$$

and the forward difference approximation (3) of Section 9.5,

$$\frac{\partial u}{\partial t} \approx \frac{1}{h} [u(x, t + h) - u(x, t)]$$

equation (1) becomes

$$\frac{c}{h^2} [u(x + h, t) - 2u(x, t) + u(x - h, t)] = \frac{1}{k} [u(x, t + k) - u(x, t)]. \tag{2}$$

If we let $\lambda = ck/h^2$ and

$$u(x, t) = u_{ij}, \quad u(x + h, t) = u_{i+1, j}, \quad u(x - h, t) = u_{i-1, j}, \quad u(x, t + k) = u_{i, j+1},$$

then, after simplifying, (2) is

$$u_{i, j+1} = \lambda u_{i+1, j} + (1 - 2\lambda) u_{ij} + \lambda u_{i-1, j}. \tag{3}$$

In the case of the heat equation (1), typical boundary conditions are $u(0, t) = u_1$, $u(a, t) = u_2, t > 0$, and an initial condition is $u(x, 0) = f(x)$, $0 < x < a$. The function f can be interpreted as the initial temperature distribution in a homogeneous rod extending from $x = 0$ to $x = a$; u_1 and u_2 can be interpreted as constant temperatures at the endpoints of the rod. Although we shall not prove it, the boundary-value problem consisting of (1) and these two boundary conditions and one initial condition has a unique solution when f is continuous on the closed interval $[0, a]$. This latter condition will

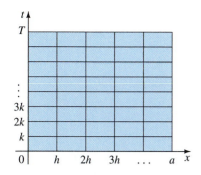

FIGURE 15.9 Rectangular region in *xt*-plane

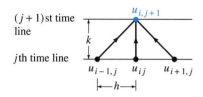

FIGURE 15.10 *u* at $t = j + 1$ is determined from three values of *u* at $t = j$

be assumed, and so we replace the initial condition by $u(x, 0) = f(x)$, $0 \le x \le a$. Moreover, instead of working with the semi-infinite region in the *xt*-plane defined by the inequalities $0 \le x \le a$, $t \ge 0$, we use a rectangular region defined by $0 \le x \le a$, $0 \le t \le T$, where T is some specified value of time. Over this region we place a rectangular grid consisting of vertical lines h units apart and horizontal lines k units apart. See Figure 15.9. If we choose two positive integers n and m and define

$$h = \frac{a}{n} \quad \text{and} \quad k = \frac{T}{m},$$

then the vertical and horizontal grid lines are defined by

$$x_i = ih, \quad i = 0, 1, 2, \ldots, n \quad \text{and} \quad t_j = jk, \quad j = 0, 1, 2, \ldots, m.$$

As illustrated in Figure 15.10, the plan here is to use formula (3) to estimate the values of the solution $u(x, t)$ at the points on the $(j + 1)$st time line using only values from the jth time line. For example, the values on the first time line ($j = 1$) depend on the initial condition $u_{i,0} = u(x_i, 0) = f(x_i)$ given on the zeroth time ($j = 0$). This kind of numerical procedure is called an **explicit finite difference method.**

| **EXAMPLE 1** | **Using the Finite Difference Method** |

Consider the boundary-value problem

$$\frac{\partial^2 u}{\partial x^2} = \frac{\partial u}{\partial t}, \quad 0 < x < 1, \quad 0 < t < 0.5$$

$$u(0, t) = 0, \quad u(1, t) = 0, \quad 0 \le t \le 0.5$$

$$u(x, 0) = \sin \pi x, \quad 0 \le x \le 1.$$

First we identify $c = 1$, $a = 1$, and $T = 0.5$. If we choose, say, $n = 5$ and $m = 50$, then $h = 1/5 = 0.2$, $k = 0.5/50 = 0.01$, $\lambda = 0.25$,

$$x_i = i\frac{1}{5}, \quad i = 0, 1, 2, 3, 4, 5, \quad t_j = j\frac{1}{100}, \quad j = 0, 1, 2, \ldots, 50.$$

Thus (3) becomes

$$u_{i,j+1} = 0.25(u_{i+1,j} + 2u_{ij} + u_{i-1,j}).$$

By setting $j = 0$ in this formula, we get a formula for the approximations to the temperature u on the first time line:

$$u_{i,1} = 0.25(u_{i+1,0} + 2u_{i,0} + u_{i-1,0}).$$

If we then let $i = 1, \ldots, 4$ in the last equation, we obtain, in turn,

$$u_{11} = 0.25(u_{20} + 2u_{10} + u_{00})$$
$$u_{21} = 0.25(u_{30} + 2u_{20} + u_{10})$$
$$u_{31} = 0.25(u_{40} + 2u_{30} + u_{20})$$
$$u_{41} = 0.25(u_{50} + 2u_{40} + u_{30}).$$

The first equation in this list is interpreted as

$$u_{11} = 0.25(u(x_2, 0) + 2u(x_1, 0) + u(0, 0))$$
$$= 0.25(u(0.4, 0) + 2u(0.2, 0) + u(0, 0)).$$

From the initial condition $u(x, 0) = \sin \pi x$ the last line becomes

$$u_{11} = 0.25(0.951056516 + 2(0.587785252) + 0) = 0.531656755.$$

This number represents an approximation to the temperature $u(0.2, 0.01)$.

Since it would require a rather large table of over 200 entries to summarize all the approximations over the rectangular grid determined by h and k, we give only selected values in Table 15.2.

TABLE 15.2 Explicit Difference Equation Approximation with $h = 0.2$, $k = 0.01$, $\lambda = 0.25$

Time	$x = 0.20$	$x = 0.40$	$x = 0.60$	$x = 0.80$
0.00	0.5878	0.9511	0.9511	0.5878
0.10	0.2154	0.3486	0.3486	0.2154
0.20	0.0790	0.1278	0.1278	0.0790
0.30	0.0289	0.0468	0.0468	0.0289
0.40	0.0106	0.0172	0.0172	0.0106
0.50	0.0039	0.0063	0.0063	0.0039

TABLE 15.3

Actual	Approx.
$u(0.4, 0.05) = 0.5806$	$u_{25} = 0.5758$
$u(0.6, 0.06) = 0.5261$	$u_{36} = 0.5208$
$u(0.2, 0.10) = 0.2191$	$u_{1,10} = 0.2154$
$u(0.8, 0.14) = 0.1476$	$u_{4,14} = 0.1442$

You should verify, using the methods of Chapter 12, that an exact solution of the boundary-value problem in Example 1 is given by $u(x, t) = e^{-\pi^2 t} \sin \pi x$. Using this solution, we compare in Table 15.3 a sample of actual values with their corresponding approximations.

STABILITY These approximations are comparable to the exact values and are accurate enough for some purposes. But there is a problem with the foregoing method. Recall that a numerical method is **unstable** if round-off errors or any other errors grow too rapidly as the computations proceed. The numerical procedure illustrated in Example 1 can exhibit this kind of behavior. It can be proved that the procedure is stable if λ is less than or equal to 0.5 but unstable otherwise. To obtain $\lambda = 0.25 \le 0.5$ in Example 1, we had to choose the value $k = 0.01$; the necessity of using very small step sizes in the time direction is the principal fault of this method. You are urged to work Problem 12 in Exercises 15.2 and witness the predictable instability when $\lambda = 1$.

CRANK-NICHOLSON METHOD There are **implicit finite difference methods** for solving parabolic partial differential equations. These methods require that we solve a system of equations to determine the approximate values of u on the $(j + 1)$st time line. However, implicit methods do not suffer from instability problems.

The algorithm introduced by J. Crank and P. Nicholson in 1947 is used mostly for solving the heat equation. The algorithm consists of replacing the second partial derivative in $c\dfrac{\partial^2 u}{\partial x^2} = \dfrac{\partial u}{\partial t}$ by an average of two central difference quotients, one evaluated at t and the other at $t + k$:

$$\frac{c}{2}\left[\frac{u(x+h,t)-2u(x,t)+u(x-h,t)}{h^2} + \frac{u(x+h,t+k)-2u(x,t+k)+u(x-h,t+k)}{h^2}\right]$$
$$= \frac{1}{k}[u(x,t+k)-u(x,t)]. \quad (4)$$

If we again define $\lambda = ck/h^2$, then after rearranging terms, we can write (4) as

$$-u_{i-1,j+1} + \alpha u_{i,j+1} - u_{i+1,j+1} = u_{i+1,j} - \beta u_{ij} + u_{i-1,j}, \quad (5)$$

where $\alpha = 2(1 + 1/\lambda)$ and $\beta = 2(1 - 1/\lambda)$, $j = 0, 1, \ldots, m - 1$, and $i = 1, 2, \ldots, n - 1$.

For each choice of j the difference equation (5) for $i = 1, 2, \ldots, n - 1$ gives $n - 1$ equations in $n - 1$ unknowns $u_{i,j+1}$. Because of the prescribed boundary conditions, the values of $u_{i,j+1}$ are known for $i = 0$ and for $i = n$. For example, in the case $n = 4$ the system of equations for determining the approximate values of u on the $(j + 1)$st time line is

$$-u_{0,j+1} + \alpha u_{1,j+1} - u_{2,j+1} = u_{2,j} - \beta u_{1,j} + u_{0,j}$$

$$-u_{1,j+1} + \alpha u_{2,j+1} - u_{3,j+1} = u_{3,j} - \beta u_{2,j} + u_{1,j}$$

$$-u_{2,j+1} + \alpha u_{3,j+1} - u_{4,j+1} = u_{4,j} - \beta u_{3,j} + u_{2,j}$$

or

$$\alpha u_{1,j+1} - u_{2,j+1} = b_1$$

$$-u_{1,j+1} + \alpha u_{2,j+1} - u_{3,j+1} = b_2 \qquad (6)$$

$$- u_{2,j+1} + \alpha u_{3,j+1} = b_3,$$

where

$$b_1 = u_{2,j} - \beta u_{1,j} + u_{0,j} + u_{0,j+1}$$

$$b_2 = u_{3,j} - \beta u_{2,j} + u_{1,j}$$

$$b_3 = u_{4,j} - \beta u_{3,j} + u_{2,j} + u_{4,j+1}.$$

In general, if we use the difference equation (5) to determine values of u on the $(j + 1)$st time line, we need to solve a linear system $\mathbf{AX} = \mathbf{B}$, where the coefficient matrix $\mathbf{A}$ is a **tridiagonal matrix,**

$$\mathbf{A} = \begin{pmatrix} \alpha & -1 & 0 & 0 & 0 & \cdots & & 0 \\ -1 & \alpha & -1 & 0 & 0 & & & 0 \\ 0 & -1 & \alpha & -1 & 0 & & & 0 \\ 0 & 0 & -1 & \alpha & -1 & & & 0 \\ \vdots & & & & & \ddots & & \vdots \\ 0 & 0 & 0 & 0 & 0 & & \alpha & -1 \\ 0 & 0 & 0 & 0 & 0 & \cdots & -1 & \alpha \end{pmatrix},$$

and the entries of the column matrix $\mathbf{B}$ are

$$b_1 = u_{2,j} - \beta u_{1,j} + u_{0,j} + u_{0,j+1}$$

$$b_2 = u_{3,j} - \beta u_{2,j} + u_{1,j}$$

$$b_3 = u_{4,j} - \beta u_{3,j} + u_{2,j}$$

$$\vdots$$

$$b_{n-1} = u_{n,j} - \beta u_{n-1,j} + u_{n-2,j} + u_{n,j+1}.$$

EXAMPLE 2   **Using the Crank-Nicholson Method**

Use the Crank-Nicholson method to approximate the solution of the boundary-value problem

$$0.25 \frac{\partial^2 u}{\partial x^2} = \frac{\partial u}{\partial t}, \quad 0 < x < 2, \quad 0 < t < 0.3$$

$$u(0, t) = 0, \quad u(2, t) = 0, \quad 0 \le t \le 0.3$$

$$u(x, 0) = \sin \pi x, \quad 0 \le x \le 2,$$

using $n = 8$ and $m = 30$.

SOLUTION From the identifications $a = 2$, $T = 0.3$, $h = \frac{1}{4} = 0.25$, $k = \frac{1}{100} = 0.01$, and $c = 0.25$ we get $\lambda = 0.04$. With the aid of a computer we get the results in Table 15.4. As in Example 1 the entries in this table represent only a selected number from the 210 approximations over the rectangular grid determined by h and k.

TABLE 15.4 Crank-Nicholson Method with $h = 0.25$, $k = 0.01$, $\lambda = 0.25$

Time	$x = 0.25$	$x = 0.50$	$x = 0.75$	$x = 1.00$	$x = 1.25$	$x = 1.50$	$x = 1.75$
0.00	0.7071	1.0000	0.7071	0.0000	−0.7071	−1.0000	−0.7071
0.05	0.6289	0.8894	0.6289	0.0000	−0.6289	−0.8894	−0.6289
0.10	0.5594	0.7911	0.5594	0.0000	−0.5594	−0.7911	−0.5594
0.15	0.4975	0.7036	0.4975	0.0000	−0.4975	−0.7036	−0.4975
0.20	0.4425	0.6258	0.4425	0.0000	−0.4425	−0.6258	−0.4425
0.25	0.3936	0.5567	0.3936	0.0000	−0.3936	−0.5567	−0.3936
0.30	0.3501	0.4951	0.3501	0.0000	−0.3501	−0.4951	−0.3501

TABLE 15.5

Actual	Approx.
$u(0.75, 0.05) = 0.6250$	$u_{35} = 0.6289$
$u(0.50, 0.20) = 0.6105$	$u_{2, 20} = 0.6259$
$u(0.25, 0.10) = 0.5525$	$u_{1, 10} = 0.5594$

Like Example 1, the boundary-value problem in Example 2 possesses an exact solution given by $u(x, t) = e^{-\pi^2 t/4} \sin \pi x$. The sample comparisons listed in Table 15.5 show that the absolute errors are of the order 10^{-2} or 10^{-3}. Smaller errors can be obtained by decreasing either h or k.

EXERCISES 15.2

Answers to selected odd-numbered problems begin on page ANS-22.

In Problems 1–12 use a computer as a computation aid.

1. Use the difference equation (3) to approximate the solution of the boundary-value problem

$$\frac{\partial^2 u}{\partial x^2} = \frac{\partial u}{\partial t}, \quad 0 < x < 2, \quad 0 < t < 1$$

$$u(0, t) = 0, \quad u(2, t) = 0, \quad 0 \leq t \leq 1$$

$$u(x, 0) = \begin{cases} 1, & 0 \leq x \leq 1 \\ 0, & 1 < x \leq 2. \end{cases}$$

Use $n = 8$ and $m = 40$.

2. Using the Fourier series solution obtained in Problem 1 of Exercises 12.3, with $L = 2$, one can sum the first 20 terms to estimate the values for $u(0.25, 0.1)$, $u(1, 0.5)$, and $u(1.5, 0.8)$ for the solution $u(x, t)$ of Problem 1 above. A student wrote a computer program to do this and obtained the results $u(0.25, 0.1) = 0.3794$, $u(1, 0.5) = 0.1854$, and $u(1.5, 0.8) = 0.0623$. Assume that these results are accurate for all digits given. Compare these values with the approximations obtained in Problem 1 above. Find the absolute errors in each case.

3. Solve Problem 1 by the Crank-Nicholson method with $n = 8$ and $m = 40$. Use the values for $u(0.25, 0.1)$, $u(1, 0.5)$, and $u(1.5, 0.8)$ given in Problem 2 to compute the absolute errors.

4. Repeat Problem 1 using $n = 8$ and $m = 20$. Use the values for $u(0.25, 0.1)$, $u(1, 0.5)$, and $u(1.5, 0.8)$ given in Problem 2 to compute the absolute errors. Why are the approximations so inaccurate in this case?

5. Solve Problem 1 by the Crank-Nicholson method with $n = 8$ and $m = 20$. Use the values for $u(0.25, 0.1)$, $u(1, 0.5)$, and $u(1.5, 0.8)$ given in Problem 2 to compute the absolute errors. Compare the absolute errors with those obtained in Problem 4.

6. It was shown in Section 12.2 that if a rod of length L is made of a material with thermal conductivity K, specific heat γ, and density ρ, the temperature $u(x, t)$ satisfies the partial differential equation

$$\frac{K}{\gamma \rho} \frac{\partial^2 u}{\partial x^2} = \frac{\partial u}{\partial t}, \quad 0 < x < L.$$

Consider the boundary-value problem consisting of the foregoing equation and the following conditions:

$$u(0, t) = 0, \quad u(L, t) = 0, \quad 0 \leq t \leq 10$$

$$u(x, 0) = f(x), \quad 0 \leq x \leq L.$$

Use the difference equation (3) in this section with $n = 10$ and $m = 10$ to approximate the solution of the boundary-value problem when

(a) $L = 20$, $K = 0.15$, $\rho = 8.0$, $\gamma = 0.11$, $f(x) = 30$
(b) $L = 50$, $K = 0.15$, $\rho = 8.0$, $\gamma = 0.11$, $f(x) = 30$
(c) $L = 20$, $K = 1.10$, $\rho = 2.7$, $\gamma = 0.22$,
$\quad f(x) = 0.5x(20 - x)$

(d) $L = 100$, $K = 1.04$, $\rho = 10.6$, $\gamma = 0.06$,

$$f(x) = \begin{cases} 0.8x, & 0 \le x \le 50 \\ 0.8(100 - x), & 50 < x \le 100 \end{cases}$$

7. Solve Problem 6 by the Crank-Nicholson method with $n = 10$ and $m = 10$.

8. Repeat Problem 6 if the endpoint temperatures are $u(0, t) = 0$, $u(L, t) = 20$, $0 \le t \le 10$.

9. Solve Problem 8 by the Crank-Nicholson method.

10. Consider the boundary-value problem in Example 2. Assume that $n = 4$.

   (a) Find the new value of λ.

   (b) Use the Crank-Nicholson difference equation (5) to find the system of equations for u_{11}, u_{21}, and u_{31}—that is, the approximate values of u on the first time line. (*Hint:* Set $j = 0$ in (5), and let i take on the values 1, 2, 3.)

   (c) Solve the system of three equations without the aid of a computer program. Compare your results with the corresponding entries in Table 15.4.

11. Consider a rod whose length is $L = 20$ for which $K = 1.05$, $\rho = 10.6$, and $\gamma = 0.056$. Suppose

$$u(0, t) = 20, \quad u(20, t) = 30$$

$$u(x, 0) = 50.$$

   (a) Use the method outlined in Section 12.6 to find the steady-state solution $\psi(x)$.

   (b) Use the Crank-Nicholson method to approximate the temperatures $u(x, t)$ for $0 \le t \le T_{max}$. Select T_{max} large enough to allow the temperatures to approach the steady-state values. Compare the approximations for $t = T_{max}$ with the values of $\psi(x)$ found in part (a).

12. Use the difference equation (3) to approximate the solution of the boundary-value problem

$$\frac{\partial^2 u}{\partial x^2} = \frac{\partial u}{\partial t}, \quad 0 < x < 1, \quad 0 < t < 1$$

$$u(0, t) = 0, \quad u(1, t) = 0, \quad 0 \le t \le 1$$

$$u(x, 0) = \sin \pi x, \quad 0 \le x \le 1.$$

Use $n = 5$ and $m = 25$.

15.3 WAVE EQUATION

INTRODUCTION: In this section we approximate a solution of the one-dimensional wave equation using the finite difference method used in the preceding two sections. The one-dimensional wave equation is the prototype hyperbolic partial differential equation.

REVIEW MATERIAL: Review Sections 9.5, 12.1, 12.2, 12.4, and 15.2.

DIFFERENCE EQUATION REPLACEMENT Suppose $u(x, t)$ represents a solution of the one-dimensional wave equation

$$c^2 \frac{\partial^2 u}{\partial x^2} = \frac{\partial^2 u}{\partial t^2}. \tag{1}$$

Using two central differences,

$$\frac{\partial^2 u}{\partial x^2} \approx \frac{1}{h^2} [u(x + h, t) - 2u(x, t) + u(x - h, t)]$$

$$\frac{\partial^2 u}{\partial t^2} \approx \frac{1}{k^2} [u(x, t + k) - 2u(x, t) + u(x, t - k)]$$

we replace equation (1) by

$$\frac{c^2}{h^2} [u(x + h, t) - 2u(x, t) + u(x - h, t)] = \frac{1}{k^2} [u(x, t + k) - 2u(x, t) + u(x, t - k)]. \tag{2}$$

We solve (2) for $u(x, t + k)$, which is $u_{i, j+1}$. If $\lambda = ck/h$, then (2) yields

$$u_{i, j+1} = \lambda^2 u_{i+1, j} + 2(1 - \lambda^2)u_{ij} + \lambda^2 u_{i-1, j} - u_{i, j-1} \tag{3}$$

for $i = 1, 2, \ldots, n - 1$ and $j = 1, 2, \ldots, m - 1$.

In the case in which the wave equation (1) is a model for the vertical displacements $u(x, t)$ of a vibrating string, typical boundary conditions are $u(0, t) = 0$, $u(a, t) = 0$, $t > 0$, and initial conditions are $u(x, 0) = f(x)$, $\partial u/\partial t|_{t=0} = g(x)$, $0 < x < a$. The functions f and g can be interpreted as the initial position and the initial velocity of the string. The numerical method based on equation (3), like the first method considered in Section 15.2, is an explicit finite difference method. As before, we use the difference equation (3) to approximate the solution $u(x, t)$ of (1), using the boundary and initial conditions, over a rectangular region in the xt-plane defined by the inequalities $0 \le x \le a$, $0 \le t \le T$, where T is some specified value of time. If n and m are positive integers and

$$h = \frac{a}{n} \quad \text{and} \quad k = \frac{T}{m},$$

the vertical and horizontal grid lines on this region are defined by

$$x_i = ih, \quad i = 0, 1, 2, \ldots, n \quad \text{and} \quad t_j = jk, \quad j = 0, 1, 2, \ldots, m.$$

FIGURE 15.11 u at $t = j + 1$ is determined from three values of u at $t = j$ and one value at $t = j - 1$

As shown in Figure 15.11, (3) enables us to obtain the approximation $u_{i,j+1}$ on the $(j + 1)$st time line from the values indicated on the jth and $(j - 1)$st time lines. Moreover, we use

$$u_{0,j} = u(0, jk) = 0, \quad u_{n,j} = u(a, jk) = 0 \quad \leftarrow \text{boundary conditions}$$

and

$$u_{i,0} = u(x_i, 0) = f(x_i). \quad \leftarrow \text{initial condition}$$

There is one minor problem in getting started. You can see from (3) that for $j = 1$ we need to know the values of $u_{i,1}$ (that is, the estimates of u on the first time line) in order to find $u_{i,2}$. But from Figure 15.11, with $j = 0$, we see that the values of $u_{i,1}$ on the first time line depend on the values of $u_{i,0}$ on the zeroth time line and on the values of $u_{i,-1}$. To compute these latter values, we make use of the initial-velocity condition $u_t(x, 0) = g(x)$. At $t = 0$ it follows from (5) of Section 9.5 that

$$g(x_i) = u_t(x_i, 0) \approx \frac{u(x_i, k) - u(x_i, -k)}{2k}. \tag{4}$$

To make sense of the term $u(x_i, -k) = u_{i,-1}$ in (4), we have to imagine $u(x, t)$ extended backward in time. It follows from (4) that

$$u(x_i, -k) \approx u(x_i, k) - 2kg(x_i).$$

This last result suggests that we define

$$u_{i,-1} = u_{i,1} - 2kg(x_i) \tag{5}$$

in the iteration of (3). By substituting (5) into (3) when $j = 0$, we get the special case

$$u_{i,1} = \frac{\lambda^2}{2}(u_{i+1,0} + u_{i-1,0}) + (1 - \lambda^2)u_{i,0} + kg(x_i). \tag{6}$$

EXAMPLE 1   Using the Finite Difference Method

Approximate the solution of the boundary-value problem

$$4\frac{\partial^2 u}{\partial x^2} = \frac{\partial^2 u}{\partial t^2}, \quad 0 < x < 1, \quad 0 < t < 1$$

$$u(0, t) = 0, \quad u(1, t) = 0, \quad 0 \le t \le 1$$

$$u(x, 0) = \sin \pi x, \quad \left.\frac{\partial u}{\partial t}\right|_{t=0} = 0, \quad 0 \le x \le 1,$$

using (3) with $n = 5$ and $m = 20$.

SOLUTION We make the identifications $c = 2$, $a = 1$, and $T = 1$. With $n = 5$ and $m = 20$ we get $h = \frac{1}{5} = 0.2$, $k = \frac{1}{20} = 0.05$, and $\lambda = 0.5$. Thus, with $g(x) = 0$, equations (6) and (3) become, respectively,

$$u_{i,1} = 0.125(u_{i+1,0} + u_{i-1,0}) + 0.75u_{i,0} \tag{7}$$

$$u_{i,j+1} = 0.25u_{i+1,j} + 1.5u_{ij} + 0.25u_{i-1,j} - u_{i,j-1}. \tag{8}$$

For $i = 1, 2, 3, 4$, equation (7) yields the following values for the $u_{i,1}$ on the first time line:

$$
\begin{aligned}
u_{11} &= 0.125(u_{20} + u_{00}) + 0.75u_{10} = 0.55972100 \\
u_{21} &= 0.125(u_{30} + u_{10}) + 0.75u_{20} = 0.90564761 \\
u_{31} &= 0.125(u_{40} + u_{20}) + 0.75u_{30} = 0.90564761 \\
u_{41} &= 0.125(u_{50} + u_{30}) + 0.75u_{40} = 0.55972100.
\end{aligned}
\tag{9}
$$

Note that the results given in (9) were obtained from the initial condition $u(x, 0) = \sin \pi x$. For example, $u_{20} = \sin(0.2\pi)$, and so on. Now $j = 1$ in (8) gives

$$u_{i,2} = 0.25u_{i+1,1} + 1.5u_{i,1} + 0.25u_{i-1,1} - u_{i,0},$$

and so for $i = 1, 2, 3, 4$ we get

$$
\begin{aligned}
u_{12} &= 0.25u_{21} + 1.5u_{11} + 0.25u_{01} - u_{10} \\
u_{22} &= 0.25u_{31} + 1.5u_{21} + 0.25u_{11} - u_{20} \\
u_{32} &= 0.25u_{41} + 1.5u_{31} + 0.25u_{21} - u_{30} \\
u_{42} &= 0.25u_{51} + 1.5u_{41} + 0.25u_{31} - u_{40}.
\end{aligned}
$$

Using the boundary conditions, the initial conditions, and the data obtained in (9), we get from these equations the approximations for u on the second time line. These last results and an abbreviation of the remaining calculations are given in Table 15.6.

TABLE 15.6 Explicit Difference Equation Approximation with $h = 0.2$, $k = 0.05$, $\lambda = 0.5$

Time	$x = 0.20$	$x = 0.40$	$x = 0.60$	$x = 0.80$
0.00	0.5878	0.9511	0.9511	0.5878
0.10	0.4782	0.7738	0.7738	0.4782
0.20	0.1903	0.3080	0.3080	0.1903
0.30	−0.1685	−0.2727	−0.2727	−0.1685
0.40	−0.4645	−0.7516	−0.7516	−0.4645
0.50	−0.5873	−0.9503	−0.9503	−0.5873
0.60	−0.4912	−0.7947	−0.7947	−0.4912
0.70	−0.2119	−0.3428	−0.3428	−0.2119
0.80	0.1464	0.2369	0.2369	0.1464
0.90	0.4501	0.7283	0.7283	0.4501
1.00	0.5860	0.9482	0.9482	0.5860

It is readily verified that the exact solution of the BVP in Example 1 is $u(x, t) = \sin \pi x \cos 2\pi t$. With this function we can compare actual values with approximations. For example, some selected comparisons are given in Table 15.7. As you can see in the table, the approximations are in the same "ballpark" as the actual values, but the accuracy is not particularly impressive. We can, however, obtain more accurate results. The accuracy of the algorithm varies with the choice of λ. Of course, λ is determined by the choice of integers n and m, which in turn determine the values of the step sizes h and k. It can be proved that the best accuracy is always obtainable from this method when the ratio $\lambda = kc/h$ is equal to one—in other words, when the step in the time direction is $k = h/c$. For example, the choice $n = 8$ and $m = 16$ yields $h = \frac{1}{8}$, $k = \frac{1}{16}$, and $\lambda = 1$. The sample values listed in Table 15.8 clearly show the improved accuracy.

TABLE 15.7

Actual	Approx.
$u(0.4, 0.25) = 0$	$u_{25} = 0.0185$
$u(0.6, 0.3) = -0.2939$	$u_{36} = -0.2727$
$u(0.2, 0.5) = -0.5878$	$u_{1,10} = -0.5873$
$u(0.8, 0.7) = -0.1816$	$u_{4,14} = -0.2119$

TABLE 15.8

Actual	Approx.
$u(0.25, 0.3125) = -0.2706$	$u_{25} = -0.2706$
$u(0.375, 0.375) = -0.6533$	$u_{36} = -0.6533$
$u(0.125, 0.625) = -0.2706$	$u_{1,10} = -0.2706$

STABILITY   We note in conclusion that this explicit finite difference method for the wave equation is stable when $\lambda \leq 1$ and unstable when $\lambda > 1$.

EXERCISES 15.3

Answers to selected odd-numbered problems begin on page ANS-25.

In Problems 1, 3, 5, and 6 use a computer as a computation aid.

1. Use the difference equation (3) to approximate the solution of the boundary-value problem

$$c^2 \frac{\partial^2 u}{\partial x^2} = \frac{\partial^2 u}{\partial t^2}, \quad 0 < x < a, \quad 0 < t < T$$

$$u(0, t) = 0, \quad u(a, t) = 0, \quad 0 \leq t \leq T$$

$$u(x, 0) = f(x), \quad \frac{\partial u}{\partial t}\Big|_{t=0} = 0, \quad 0 \leq x \leq a$$

when

(a) $c = 1, a = 1, T = 1, f(x) = x(1 - x); n = 4$ and $m = 10$
(b) $c = 1, \ a = 2, \ T = 1, \ f(x) = e^{-16(x-1)^2}; \ n = 5$ and $m = 10$
(c) $c = \sqrt{2}, a = 1, T = 1,$

$$f(x) = \begin{cases} 0, & 0 \leq x \leq 0.5 \\ 0.5, & 0.5 < x \leq 1 \end{cases};$$

$n = 10$ and $m = 25$.

2. Consider the boundary-value problem

$$\frac{\partial^2 u}{\partial x^2} = \frac{\partial^2 u}{\partial t^2}, \quad 0 < x < 1, \quad 0 < t < 0.5$$

$$u(0, t) = 0, \quad u(1, t) = 0, \quad 0 \leq t \leq 0.5$$

$$u(x, 0) = \sin \pi x, \quad \frac{\partial u}{\partial t}\Big|_{t=0} = 0, \quad 0 \leq x \leq 1.$$

(a) Use the methods of Chapter 12 to verify that the solution of the problem is $u(x, t) = \sin \pi x \cos \pi t$.
(b) Use the method of this section to approximate the solution of the problem without the aid of a computer program. Use $n = 4$ and $m = 5$.
(c) Compute the absolute error at each interior grid point.

3. Approximate the solution of the boundary-value problem in Problem 2 using a computer program with
(a) $n = 5, m = 10$ **(b)** $n = 5, m = 20$.

4. Given the boundary-value problem

$$\frac{\partial^2 u}{\partial x^2} = \frac{\partial^2 u}{\partial t^2}, \quad 0 < x < 1, \quad 0 < t < 1$$

$$u(0, t) = 0, \quad u(1, t) = 0, \quad 0 \leq t \leq 1$$

$$u(x, 0) = x(1 - x), \quad \frac{\partial u}{\partial t}\Big|_{t=0} = 0, \quad 0 \leq x \leq 1,$$

use $h = k = \frac{1}{5}$ in equation (6) to compute the values of $u_{i,1}$ by hand.

5. It was shown in Section 12.2 that the equation of a vibrating string is

$$\frac{T}{\rho} \frac{\partial^2 u}{\partial x^2} = \frac{\partial^2 u}{\partial t^2},$$

where T is the constant magnitude of the tension in the string and ρ is its mass per unit length. Suppose a string of length 60 centimeters is secured to the x-axis at its ends and is released from rest from the initial displacement

$$f(x) = \begin{cases} 0.01x, & 0 \le x \le 30 \\ 0.30 - \dfrac{x - 30}{100}, & 30 < x \le 60. \end{cases}$$

Use the difference equation (3) in this section to approximate the solution of the boundary-value problem

when $h = 10$, $k = 5\sqrt{\rho/T}$ and where $\rho = 0.0225$ g/cm, $T = 1.4 \times 10^7$ dynes. Use $m = 50$.

6. Repeat Problem 5 using

$$f(x) = \begin{cases} 0.2x, & 0 \le x \le 15 \\ 0.30 - \dfrac{x - 15}{150}, & 15 < x \le 60 \end{cases}$$

and $h = 10$, $k = 2.5\sqrt{\rho/T}$. Use $m = 50$.

CHAPTER 15 IN REVIEW

Answers to selected odd-numbered problems begin on page ANS-27.

1. Consider the boundary-value problem

$$\frac{\partial^2 u}{\partial x^2} + \frac{\partial^2 u}{\partial y^2} = 0, \quad 0 < x < 2, \quad 0 < y < 1$$

$$u(0, y) = 0, \quad u(2, y) = 50, \quad 0 < y < 1$$

$$u(x, 0) = 0, \quad u(x, 1) = 0, \quad 0 < x < 2.$$

Approximate the solution of the differential equation at the interior points of the region with mesh size $h = \frac{1}{2}$. Use Gaussian elimination or Gauss-Seidel iteration.

2. Solve Problem 1 using mesh size $h = \frac{1}{4}$. Use Gauss-Seidel iteration.

3. Consider the boundary-value problem

$$\frac{\partial^2 u}{\partial x^2} = \frac{\partial u}{\partial t}, \quad 0 < x < 1, \quad 0 < t < 0.05$$

$$u(0, t) = 0, \quad u(1, t) = 0, \quad t > 0$$

$$u(x, 0) = x, \quad 0 < x < 1.$$

(a) Note that the initial temperature $u(x, 0) = x$ indicates that the temperature at the right boundary $x = 1$ should be $u(1, 0) = 1$, whereas the boundary conditions imply that $u(1, 0) = 0$. Write a computer program for the explicit finite difference

method so that the boundary conditions prevail for all times considered, including $t = 0$. Use the program to complete Table 15.9.

(b) Modify your computer program so that the initial condition prevails at the boundaries at $t = 0$. Use this program to complete Table 15.10.

(c) Are Tables 15.9 and 15.10 related in any way? Use a larger time interval if necessary.

TABLE 15.9

Time	x = 0.00	x = 0.20	x = 0.40	x = 0.60	x = 0.80	x = 1.00
0.00	0.0000	0.2000	0.4000	0.6000	0.8000	0.0000
0.01	0.0000					0.0000
0.02	0.0000					0.0000
0.03	0.0000					0.0000
0.04	0.0000					0.0000
0.05	0.0000					0.0000

TABLE 15.10

Time	x = 0.00	x = 0.20	x = 0.40	x = 0.60	x = 0.80	x = 1.00
0.00	0.0000	0.2000	0.4000	0.6000	0.8000	1.0000
0.01	0.0000					0.0000
0.02	0.0000					0.0000
0.03	0.0000					0.0000
0.04	0.0000					0.0000
0.05	0.0000					0.0000

GAMMA FUNCTION

Euler's integral definition of the **gamma function** is

$$\Gamma(x) = \int_0^\infty t^{x-1} e^{-t} dt. \tag{1}$$

Convergence of the integral requires that $x - 1 > -1$ or $x > 0$. The recurrence relation

$$\Gamma(x + 1) = x\Gamma(x), \tag{2}$$

which we saw in Section 6.3, can be obtained from (1) with integration by parts. Now when $x = 1$, $\Gamma(1) = \int_0^\infty e^{-t} dt = 1$, and thus (2) gives

$$\Gamma(2) = 1\Gamma(1) = 1$$

$$\Gamma(3) = 2\Gamma(2) = 2 \cdot 1$$

$$\Gamma(4) = 3\Gamma(3) = 3 \cdot 2 \cdot 1$$

and so on. In this manner it is seen that when n is a positive integer, $\Gamma(n + 1) = n!$. For this reason the gamma function is often called the **generalized factorial function.**

Although the integral form (1) does not converge for $x < 0$, it can be shown by means of alternative definitions that the gamma function is defined for all real and complex numbers *except* $x = -n$, $n = 0, 1, 2, \ldots$. As a consequence, (2) is actually valid for $x \neq -n$. The graph of $\Gamma(x)$, considered as a function of a real variable x, is as given in Figure I.1. Observe that the nonpositive integers correspond to vertical asymptotes of the graph.

In Problems 31 and 32 of Exercises 6.3 we utilized the fact that $\Gamma\left(\frac{1}{2}\right) = \sqrt{\pi}$. This result can be derived from (1) by setting $x = \frac{1}{2}$:

$$\Gamma\left(\tfrac{1}{2}\right) = \int_0^\infty t^{-1/2} e^{-t} dt. \tag{3}$$

When we let $t = u^2$, (3) can be written as $\Gamma\left(\frac{1}{2}\right) = 2 \int_0^\infty e^{-u^2} du$. But $\int_0^\infty e^{-u^2} du = \int_0^\infty e^{-v^2} dv$, and so

$$\left[\Gamma\left(\tfrac{1}{2}\right)\right]^2 = \left(2 \int_0^\infty e^{-u^2} du\right)\left(2 \int_0^\infty e^{-v^2} dv\right) = 4 \int_0^\infty \int_0^\infty e^{-(u^2+v^2)} du\, dv.$$

Switching to polar coordinates $u = r\cos\theta$, $v = r\sin\theta$ enables us to evaluate the double integral:

$$4 \int_0^\infty \int_0^\infty e^{-(u^2+v^2)} du\, dv = 4 \int_0^{\pi/2} \int_0^\infty e^{-r^2} r\, dr\, d\theta = \pi.$$

Hence

$$\left[\Gamma\left(\tfrac{1}{2}\right)\right]^2 = \pi \quad \text{or} \quad \Gamma\left(\tfrac{1}{2}\right) = \sqrt{\pi}. \tag{4}$$

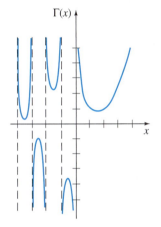

FIGURE I.1 Graph of $\Gamma(x)$ for x neither 0 nor a negative integer

> **EXAMPLE 1**   **Value of $\Gamma\left(-\frac{1}{2}\right)$**

Evaluate $\Gamma\left(-\frac{1}{2}\right)$.

SOLUTION   In view of (2) and (4) it follows that, with $x = -\frac{1}{2}$,

$$\Gamma\left(\tfrac{1}{2}\right) = -\tfrac{1}{2}\Gamma\left(-\tfrac{1}{2}\right).$$

Therefore

$$\Gamma\left(-\tfrac{1}{2}\right) = -2\Gamma\left(\tfrac{1}{2}\right) = -2\sqrt{\pi}.$$

EXERCISES FOR APPENDIX I

Answers to odd-numbered problems begin on page ANS-27.

1. Evaluate.
   (a) $\Gamma(5)$ **(b)** $\Gamma(7)$
   (c) $\Gamma\left(-\frac{3}{2}\right)$ **(d)** $\Gamma\left(-\frac{5}{2}\right)$

2. Use (1) and the fact that $\Gamma\left(\frac{6}{5}\right) = 0.92$ to evaluate $\displaystyle\int_0^\infty x^5 e^{-x^5}\, dx$. (*Hint:* Let $t = x^5$.)

3. Use (1) and the fact that $\Gamma\left(\frac{5}{3}\right) = 0.89$ to evaluate $\displaystyle\int_0^\infty x^4 e^{-x^3}\, dx$.

4. Evaluate $\displaystyle\int_0^1 x^3 \left(\ln\frac{1}{x}\right)^3 dx$. (*Hint:* Let $t = -\ln x$.)

5. Use the fact that $\Gamma(x) > \displaystyle\int_0^1 t^{x-1} e^{-t}\, dt$ to show that $\Gamma(x)$ is unbounded as $x \to 0^+$.

6. Use (1) to derive (2) for $x > 0$.

INTRODUCTION TO MATRICES

II.1   BASIC DEFINITIONS AND THEORY

DEFINITION II.1   **Matrix**

A **matrix A** is any rectangular array of numbers or functions:

$$\mathbf{A} = \begin{pmatrix} a_{11} & a_{12} & \cdots & a_{1n} \\ a_{21} & a_{22} & \cdots & a_{2n} \\ \vdots & & & \vdots \\ a_{m1} & a_{m2} & \cdots & a_{mn} \end{pmatrix}. \tag{1}$$

If a matrix has m rows and n columns, we say that its **size** is m by n (written $m \times n$). An $n \times n$ matrix is called a **square matrix** of order n.

The element, or entry, in the ith row and jth column of an $m \times n$ matrix $\mathbf{A}$ is written a_{ij}. An $m \times n$ matrix $\mathbf{A}$ is then abbreviated as $\mathbf{A} = (a_{ij})_{m \times n}$ or simply $\mathbf{A} = (a_{ij})$. A 1×1 matrix is simply one constant or function.

DEFINITION II.2   **Equality of Matrices**

Two $m \times n$ matrices $\mathbf{A}$ and $\mathbf{B}$ are **equal** if $a_{ij} = b_{ij}$ for each i and j.

DEFINITION II.3   **Column Matrix**

A **column matrix X** is any matrix having n rows and one column:

$$\mathbf{X} = \begin{pmatrix} b_{11} \\ b_{21} \\ \vdots \\ b_{n1} \end{pmatrix} = (b_{i1})_{n \times 1}.$$

A column matrix is also called a **column vector** or simply a **vector.**

DEFINITION II.4   **Multiples of Matrices**

A **multiple** of a matrix $\mathbf{A}$ is defined to be

$$k\mathbf{A} = \begin{pmatrix} ka_{11} & ka_{12} & \cdots & ka_{1n} \\ ka_{21} & ka_{22} & \cdots & ka_{2n} \\ \vdots & & & \vdots \\ ka_{m1} & ka_{m2} & \cdots & ka_{mn} \end{pmatrix} = (ka_{ij})_{m \times n},$$

where k is a constant or a function.

EXAMPLE 1   **Multiples of Matrices**

(a) $5\begin{pmatrix} 2 & -3 \\ 4 & -1 \\ \frac{1}{5} & 6 \end{pmatrix} = \begin{pmatrix} 10 & -15 \\ 20 & -5 \\ 1 & 30 \end{pmatrix}$ (b) $e^t \begin{pmatrix} 1 \\ -2 \\ 4 \end{pmatrix} = \begin{pmatrix} e^t \\ -2e^t \\ 4e^t \end{pmatrix}$

We note in passing that for any matrix $\mathbf{A}$ the product $k\mathbf{A}$ is the same as $\mathbf{A}k$. For example,

$$e^{-3t}\begin{pmatrix} 2 \\ 5 \end{pmatrix} = \begin{pmatrix} 2e^{-3t} \\ 5e^{-3t} \end{pmatrix} = \begin{pmatrix} 2 \\ 5 \end{pmatrix}e^{-3t}.$$

DEFINITION II.5   **Addition of Matrices**

The **sum** of two $m \times n$ matrices $\mathbf{A}$ and $\mathbf{B}$ is defined to be the matrix

$$\mathbf{A} + \mathbf{B} = (a_{ij} + b_{ij})_{m \times n}.$$

In other words, when adding two matrices of the same size, we add the corresponding elements.

EXAMPLE 2   **Matrix Addition**

The sum of $\mathbf{A} = \begin{pmatrix} 2 & -1 & 3 \\ 0 & 4 & 6 \\ -6 & 10 & -5 \end{pmatrix}$ and $\mathbf{B} = \begin{pmatrix} 4 & 7 & -8 \\ 9 & 3 & 5 \\ 1 & -1 & 2 \end{pmatrix}$ is

$$\mathbf{A} + \mathbf{B} = \begin{pmatrix} 2+4 & -1+7 & 3+(-8) \\ 0+9 & 4+3 & 6+5 \\ -6+1 & 10+(-1) & -5+2 \end{pmatrix} = \begin{pmatrix} 6 & 6 & -5 \\ 9 & 7 & 11 \\ -5 & 9 & -3 \end{pmatrix}.$$

EXAMPLE 3   **A Matrix Written as a Sum of Column Matrices**

The single matrix $\begin{pmatrix} 3t^2 - 2e^t \\ t^2 + 7t \\ 5t \end{pmatrix}$ can be written as the sum of three column vectors:

$$\begin{pmatrix} 3t^2 - 2e^t \\ t^2 + 7t \\ 5t \end{pmatrix} = \begin{pmatrix} 3t^2 \\ t^2 \\ 0 \end{pmatrix} + \begin{pmatrix} 0 \\ 7t \\ 5t \end{pmatrix} + \begin{pmatrix} -2e^t \\ 0 \\ 0 \end{pmatrix} = \begin{pmatrix} 3 \\ 1 \\ 0 \end{pmatrix}t^2 + \begin{pmatrix} 0 \\ 7 \\ 5 \end{pmatrix}t + \begin{pmatrix} -2 \\ 0 \\ 0 \end{pmatrix}e^t.$$

The **difference** of two $m \times n$ matrices is defined in the usual manner: $\mathbf{A} - \mathbf{B} = \mathbf{A} + (-\mathbf{B})$, where $-\mathbf{B} = (-1)\mathbf{B}$.

DEFINITION II.6   **Multiplication of Matrices**

Let $\mathbf{A}$ be a matrix having m rows and n columns and $\mathbf{B}$ be a matrix having n rows and p columns. We define the **product AB** to be the $m \times p$ matrix

$$\mathbf{AB} = \begin{pmatrix} a_{11} & a_{12} & \cdots & a_{1n} \\ a_{21} & a_{22} & \cdots & a_{2n} \\ \vdots & & & \vdots \\ a_{m1} & a_{m2} & \cdots & a_{mn} \end{pmatrix} \begin{pmatrix} b_{11} & b_{12} & \cdots & b_{1p} \\ b_{21} & b_{22} & \cdots & b_{2p} \\ \vdots & & & \vdots \\ b_{n1} & b_{n2} & \cdots & b_{np} \end{pmatrix}$$

$$= \begin{pmatrix} a_{11}b_{11} + a_{12}b_{21} + \cdots + a_{1n}b_{n1} & \cdots & a_{11}b_{1p} + a_{12}b_{2p} + \cdots + a_{1n}b_{np} \\ a_{21}b_{11} + a_{22}b_{21} + \cdots + a_{2n}b_{n1} & \cdots & a_{21}b_{1p} + a_{22}b_{2p} + \cdots + a_{2n}b_{np} \\ \vdots & & \vdots \\ a_{m1}b_{11} + a_{m2}b_{21} + \cdots + a_{mn}b_{n1} & \cdots & a_{m1}b_{1p} + a_{m2}b_{2p} + \cdots + a_{mn}b_{np} \end{pmatrix}$$

$$= \left(\sum_{k=1}^{n} a_{ik}b_{kj} \right)_{m \times p}.$$

Note carefully in Definition II.6 that the product $\mathbf{AB} = \mathbf{C}$ is defined only when the number of columns in the matrix $\mathbf{A}$ is the same as the number of rows in $\mathbf{B}$. The size of the product can be determined from

$$\mathbf{A}_{m \times n} \mathbf{B}_{n \times p} = \mathbf{C}_{m \times p}.$$

Also, you might recognize that the entries in, say, the ith row of the final matrix $\mathbf{AB}$ are formed by using the component definition of the inner, or dot, product of the ith row of $\mathbf{A}$ with each of the columns of $\mathbf{B}$.

EXAMPLE 4   **Multiplication of Matrices**

(a) For $\mathbf{A} = \begin{pmatrix} 4 & 7 \\ 3 & 5 \end{pmatrix}$ and $\mathbf{B} = \begin{pmatrix} 9 & -2 \\ 6 & 8 \end{pmatrix}$,

$$\mathbf{AB} = \begin{pmatrix} 4 \cdot 9 + 7 \cdot 6 & 4 \cdot (-2) + 7 \cdot 8 \\ 3 \cdot 9 + 5 \cdot 6 & 3 \cdot (-2) + 5 \cdot 8 \end{pmatrix} = \begin{pmatrix} 78 & 48 \\ 57 & 34 \end{pmatrix}.$$

(b) For $\mathbf{A} = \begin{pmatrix} 5 & 8 \\ 1 & 0 \\ 2 & 7 \end{pmatrix}$ and $\mathbf{B} = \begin{pmatrix} -4 & -3 \\ 2 & 0 \end{pmatrix}$,

$$\mathbf{AB} = \begin{pmatrix} 5 \cdot (-4) + 8 \cdot 2 & 5 \cdot (-3) + 8 \cdot 0 \\ 1 \cdot (-4) + 0 \cdot 2 & 1 \cdot (-3) + 0 \cdot 0 \\ 2 \cdot (-4) + 7 \cdot 2 & 2 \cdot (-3) + 7 \cdot 0 \end{pmatrix} = \begin{pmatrix} -4 & -15 \\ -4 & -3 \\ 6 & -6 \end{pmatrix}.$$

In general, *matrix multiplication is not commutative;* that is, $\mathbf{AB} \neq \mathbf{BA}$. Observe in part (a) of Example 4 that $\mathbf{BA} = \begin{pmatrix} 30 & 53 \\ 48 & 82 \end{pmatrix}$, whereas in part (b) the product $\mathbf{BA}$ is not defined, since Definition II.6 requires that the first matrix (in this case $\mathbf{B}$) have the same number of columns as the second matrix has rows.

We are particularly interested in the product of a square matrix and a column vector.

EXAMPLE 5   **Multiplication of Matrices**

(a) $\begin{pmatrix} 2 & -1 & 3 \\ 0 & 4 & 5 \\ 1 & -7 & 9 \end{pmatrix}\begin{pmatrix} -3 \\ 6 \\ 4 \end{pmatrix} = \begin{pmatrix} 2\cdot(-3)+(-1)\cdot6+3\cdot4 \\ 0\cdot(-3)+4\cdot6+5\cdot4 \\ 1\cdot(-3)+(-7)\cdot6+9\cdot4 \end{pmatrix} = \begin{pmatrix} 0 \\ 44 \\ -9 \end{pmatrix}$

(b) $\begin{pmatrix} -4 & 2 \\ 3 & 8 \end{pmatrix}\begin{pmatrix} x \\ y \end{pmatrix} = \begin{pmatrix} -4x+2y \\ 3x+8y \end{pmatrix}$

MULTIPLICATIVE IDENTITY   For a given positive integer n, the $n \times n$ matrix

$$I = \begin{pmatrix} 1 & 0 & 0 & \cdots & 0 \\ 0 & 1 & 0 & \cdots & 0 \\ \vdots & & & & \vdots \\ 0 & 0 & 0 & \cdots & 1 \end{pmatrix}$$

is called the **multiplicative identity matrix.** It follows from Definition II.6 that for any $n \times n$ matrix A.

$$AI = IA = A.$$

Also, it is readily verified that if X is an $n \times 1$ column matrix, then $IX = X$.

ZERO MATRIX   A matrix consisting of all zero entries is called a **zero matrix** and is denoted by 0. For example,

$$0 = \begin{pmatrix} 0 \\ 0 \end{pmatrix}, \quad 0 = \begin{pmatrix} 0 & 0 \\ 0 & 0 \end{pmatrix}, \quad 0 = \begin{pmatrix} 0 & 0 \\ 0 & 0 \\ 0 & 0 \end{pmatrix},$$

and so on. If A and 0 are $m \times n$ matrices, then

$$A + 0 = 0 + A = A.$$

ASSOCIATIVE LAW   Although we shall not prove it, matrix multiplication is **associative.** If A is an $m \times p$ matrix, B a $p \times r$ matrix, and C an $r \times n$ matrix, then

$$A(BC) = (AB)C$$

is an $m \times n$ matrix.

DISTRIBUTIVE LAW   If all products are defined, multiplication is **distributive** over addition:

$$A(B + C) = AB + AC \quad \text{and} \quad (B + C)A = BA + CA.$$

DETERMINANT OF A MATRIX   Associated with every *square* matrix A of constants is a number called the **determinant of the matrix,** which is denoted by det A.

EXAMPLE 6   **Determinant of a Square Matrix**

For $A = \begin{pmatrix} 3 & 6 & 2 \\ 2 & 5 & 1 \\ -1 & 2 & 4 \end{pmatrix}$ we expand det A by cofactors of the first row:

$$\det A = \begin{vmatrix} 3 & 6 & 2 \\ 2 & 5 & 1 \\ -1 & 2 & 4 \end{vmatrix} = 3\begin{vmatrix} 5 & 1 \\ 2 & 4 \end{vmatrix} - 6\begin{vmatrix} 2 & 1 \\ -1 & 4 \end{vmatrix} + 2\begin{vmatrix} 2 & 5 \\ -1 & 2 \end{vmatrix}$$

$$= 3(20 - 2) - 6(8 + 1) + 2(4 + 5) = 18.$$

It can be proved that a determinant det **A** can be expanded by cofactors using any row or column. If det **A** has a row (or a column) containing many zero entries, then wisdom dictates that we expand the determinant by that row (or column).

DEFINITION II.7   **Transpose of a Matrix**

The **transpose** of the $m \times n$ matrix (1) is the $n \times m$ matrix $\mathbf{A}^T$ given by

$$\mathbf{A}^T = \begin{pmatrix} a_{11} & a_{21} & \cdots & a_{m1} \\ a_{12} & a_{22} & \cdots & a_{m2} \\ \vdots & & & \vdots \\ a_{1n} & a_{2n} & \cdots & a_{mn} \end{pmatrix}.$$

In other words, the rows of a matrix **A** become the columns of its transpose $\mathbf{A}^T$.

EXAMPLE 7   **Transpose of a Matrix**

(a) The transpose of $\mathbf{A} = \begin{pmatrix} 3 & 6 & 2 \\ 2 & 5 & 1 \\ -1 & 2 & 4 \end{pmatrix}$ is $\mathbf{A}^T = \begin{pmatrix} 3 & 2 & -1 \\ 6 & 5 & 2 \\ 2 & 1 & 4 \end{pmatrix}$.

(b) If $\mathbf{X} = \begin{pmatrix} 5 \\ 0 \\ 3 \end{pmatrix}$, then $\mathbf{X}^T = (5 \quad 0 \quad 3)$.

DEFINITION II.8   **Multiplicative Inverse of a Matrix**

Let **A** be an $n \times n$ matrix. If there exists an $n \times n$ matrix **B** such that

$$\mathbf{AB} = \mathbf{BA} = \mathbf{I},$$

where **I** is the multiplicative identity, then **B** is said to be the **multiplicative inverse of A** and is denoted by $\mathbf{B} = \mathbf{A}^{-1}$.

DEFINITION II.9   **Nonsingular/Singular Matrices**

Let **A** be an $n \times n$ matrix. If det $\mathbf{A} \neq 0$, then **A** is said to be **nonsingular.** If det $\mathbf{A} = 0$, then **A** is said to be **singular.**

The following theorem gives a necessary and sufficient condition for a square matrix to have a multiplicative inverse.

THEOREM II.1   **Nonsingularity Implies A Has an Inverse**

An $n \times n$ matrix **A** has a multiplicative inverse $\mathbf{A}^{-1}$ if and only if **A** is nonsingular.

The following theorem gives one way of finding the multiplicative inverse for a nonsingular matrix.

> ### THEOREM II.2   A Formula for the Inverse of a Matrix
>
> Let $\mathbf{A}$ be an $n \times n$ nonsingular matrix and let $C_{ij} = (-1)^{i+j} M_{ij}$, where M_{ij} is the determinant of the $(n-1) \times (n-1)$ matrix obtained by deleting the ith row and jth column from $\mathbf{A}$. Then
>
> $$\mathbf{A}^{-1} = \frac{1}{\det \mathbf{A}} (C_{ij})^T. \tag{2}$$

Each C_{ij} in Theorem II.2 is simply the **cofactor** (signed minor) of the corresponding entry a_{ij} in $\mathbf{A}$. Note that the transpose is utilized in formula (2).

For future reference we observe in the case of a 2×2 nonsingular matrix

$$\mathbf{A} = \begin{pmatrix} a_{11} & a_{12} \\ a_{21} & a_{22} \end{pmatrix}$$

that $C_{11} = a_{22}$, $C_{12} = -a_{21}$, $C_{21} = -a_{12}$, and $C_{22} = a_{11}$. Thus

$$\mathbf{A}^{-1} = \frac{1}{\det \mathbf{A}} \begin{pmatrix} a_{22} & -a_{21} \\ -a_{12} & a_{11} \end{pmatrix}^T = \frac{1}{\det \mathbf{A}} \begin{pmatrix} a_{22} & -a_{12} \\ -a_{21} & a_{11} \end{pmatrix}. \tag{3}$$

For a 3×3 nonsingular matrix

$$\mathbf{A} = \begin{pmatrix} a_{11} & a_{12} & a_{13} \\ a_{21} & a_{22} & a_{23} \\ a_{31} & a_{32} & a_{33} \end{pmatrix},$$

$$C_{11} = \begin{vmatrix} a_{22} & a_{23} \\ a_{32} & a_{33} \end{vmatrix}, \quad C_{12} = -\begin{vmatrix} a_{21} & a_{23} \\ a_{31} & a_{33} \end{vmatrix}, \quad C_{13} = \begin{vmatrix} a_{21} & a_{22} \\ a_{31} & a_{32} \end{vmatrix},$$

and so on. Carrying out the transposition gives

$$\mathbf{A}^{-1} = \frac{1}{\det \mathbf{A}} \begin{pmatrix} C_{11} & C_{21} & C_{31} \\ C_{12} & C_{22} & C_{32} \\ C_{13} & C_{23} & C_{33} \end{pmatrix}. \tag{4}$$

EXAMPLE 8   Inverse of a 2 × 2 Matrix

Find the multiplicative inverse for $\mathbf{A} = \begin{pmatrix} 1 & 4 \\ 2 & 10 \end{pmatrix}$.

SOLUTION   Since $\det \mathbf{A} = 10 - 8 = 2 \neq 0$, $\mathbf{A}$ is nonsingular. It follows from Theorem II.1 that $\mathbf{A}^{-1}$ exists. From (3) we find

$$\mathbf{A}^{-1} = \frac{1}{2} \begin{pmatrix} 10 & -4 \\ -2 & 1 \end{pmatrix} = \begin{pmatrix} 5 & -2 \\ -1 & \frac{1}{2} \end{pmatrix}. \qquad \blacksquare$$

Not every square matrix has a multiplicative inverse. The matrix $\mathbf{A} = \begin{pmatrix} 2 & 2 \\ 3 & 3 \end{pmatrix}$ is singular since $\det \mathbf{A} = 0$. Hence $\mathbf{A}^{-1}$ does not exist.

EXAMPLE 9   Inverse of a 3 × 3 Matrix

Find the multiplicative inverse for $\mathbf{A} = \begin{pmatrix} 2 & 2 & 0 \\ -2 & 1 & 1 \\ 3 & 0 & 1 \end{pmatrix}$.

SOLUTION Since det $\mathbf{A} = 12 \neq 0$, the given matrix is nonsingular. The cofactors corresponding to the entries in each row of det $\mathbf{A}$ are

$$C_{11} = \begin{vmatrix} 1 & 1 \\ 0 & 1 \end{vmatrix} = 1 \qquad C_{12} = -\begin{vmatrix} -2 & 1 \\ 3 & 1 \end{vmatrix} = 5 \qquad C_{13} = \begin{vmatrix} -2 & 1 \\ 3 & 0 \end{vmatrix} = -3$$

$$C_{21} = -\begin{vmatrix} 2 & 0 \\ 0 & 1 \end{vmatrix} = -2 \quad C_{22} = \begin{vmatrix} 2 & 0 \\ 3 & 1 \end{vmatrix} = 2 \qquad C_{23} = -\begin{vmatrix} 2 & 2 \\ 3 & 0 \end{vmatrix} = 6$$

$$C_{31} = \begin{vmatrix} 2 & 0 \\ 1 & 1 \end{vmatrix} = 2 \qquad C_{32} = -\begin{vmatrix} 2 & 0 \\ -2 & 1 \end{vmatrix} = -2 \quad C_{33} = \begin{vmatrix} 2 & 2 \\ -2 & 1 \end{vmatrix} = 6.$$

If follows from (4) that

$$\mathbf{A}^{-1} = \frac{1}{12}\begin{pmatrix} 1 & -2 & 2 \\ 5 & 2 & -2 \\ -3 & 6 & 6 \end{pmatrix} = \begin{pmatrix} \frac{1}{12} & -\frac{1}{6} & \frac{1}{6} \\ \frac{5}{12} & \frac{1}{6} & -\frac{1}{6} \\ -\frac{1}{4} & \frac{1}{2} & \frac{1}{2} \end{pmatrix}.$$

You are urged to verify that $\mathbf{A}^{-1}\mathbf{A} = \mathbf{A}\mathbf{A}^{-1} = \mathbf{I}$.

Formula (2) presents obvious difficulties for nonsingular matrices larger than 3×3. For example, to apply (2) to a 4×4 matrix, we would have to calculate *sixteen* 3×3 determinants.* In the case of a large matrix there are more efficient ways of finding $\mathbf{A}^{-1}$. The curious reader is referred to any text in linear algebra.

Since our goal is to apply the concept of a matrix to systems of linear first-order differential equations, we need the following definitions.

DEFINITION II.10 **Derivative of a Matrix of Functions**

If $\mathbf{A}(t) = (a_{ij}(t))_{m \times n}$ is a matrix whose entries are functions differentiable on a common interval, then

$$\frac{d\mathbf{A}}{dt} = \left(\frac{d}{dt}a_{ij}\right)_{m \times n}.$$

DEFINITION II.11 **Integral of a Matrix of Functions**

If $\mathbf{A}(t) = (a_{ij}(t))_{m \times n}$ is a matrix whose entries are functions continuous on a common interval containing t and t_0, then

$$\int_{t_0}^{t} \mathbf{A}(s)\,ds = \left(\int_{t_0}^{t} a_{ij}(s)\,ds\right)_{m \times n}.$$

To differentiate (integrate) a matrix of functions, we simply differentiate (integrate) each entry. The derivative of a matrix is also denoted by $\mathbf{A}'(t)$.

EXAMPLE 10 **Derivative/Integral of a Matrix**

If $\quad \mathbf{X}(t) = \begin{pmatrix} \sin 2t \\ e^{3t} \\ 8t - 1 \end{pmatrix}, \quad$ then $\quad \mathbf{X}'(t) = \begin{pmatrix} \frac{d}{dt}\sin 2t \\ \frac{d}{dt}e^{3t} \\ \frac{d}{dt}(8t-1) \end{pmatrix} = \begin{pmatrix} 2\cos 2t \\ 3e^{3t} \\ 8 \end{pmatrix}$

*Strictly speaking, a determinant is a number, but it is sometimes convenient to refer to a determinant as if it were an array.

and
$$\int_0^t \mathbf{X}(s)\,ds = \begin{pmatrix} \int_0^t \sin 2s\,ds \\ \int_0^t e^{3s}\,ds \\ \int_0^t (8s \quad 1)\,ds \end{pmatrix} = \begin{pmatrix} -\frac{1}{2}\cos 2t + \frac{1}{2} \\ \frac{1}{3}e^{3t} - \frac{1}{3} \\ 4t^2 - t \end{pmatrix}.$$

II.2 GAUSSIAN AND GAUSS-JORDAN ELIMINATION

Matrices are an invaluable aid in solving algebraic systems of n linear equations in n unknowns,

$$\begin{aligned}
a_{11}x_1 + a_{12}x_2 + \cdots + a_{1n}x_n &= b_1 \\
a_{21}x_1 + a_{22}x_2 + \cdots + a_{2n}x_n &= b_2 \\
&\vdots \\
a_{n1}x_1 + a_{n2}x_2 + \cdots + a_{nn}x_n &= b_n.
\end{aligned} \tag{5}$$

If $\mathbf{A}$ denotes the matrix of coefficients in (5), we know that Cramer's rule could be used to solve the system whenever det $\mathbf{A} \neq 0$. However, that rule requires a herculean effort if $\mathbf{A}$ is larger than 3×3. The procedure that we shall now consider has the distinct advantage of being not only an efficient way of handling large systems, but also a means of solving consistent systems (5) in which det $\mathbf{A} = 0$ and a means of solving m linear equations in n unknowns.

DEFINITION II.12 Augmented Matrix

The **augmented matrix** of the system (5) is the $n \times (n + 1)$ matrix

$$\begin{pmatrix}
a_{11} & a_{12} & \cdots & a_{1n} & b_1 \\
a_{21} & a_{22} & \cdots & a_{2n} & b_2 \\
\vdots & & & & \vdots \\
a_{n1} & a_{n2} & \cdots & a_{nn} & b_n
\end{pmatrix}.$$

If $\mathbf{B}$ is the column matrix of the b_i, $i = 1, 2, \ldots, n$, the augmented matrix of (5) is denoted by $(\mathbf{A}|\mathbf{B})$.

ELEMENTARY ROW OPERATIONS   Recall from algebra that we can transform an algebraic system of equations into an equivalent system (that is, one having the same solution) by multiplying an equation by a nonzero constant, interchanging the positions of any two equations in a system, and adding a nonzero constant multiple of an equation to another equation. These operations on equations in a system are, in turn, equivalent to **elementary row operations** on an augmented matrix:

(*i*)   Multiply a row by a nonzero constant.

(*ii*)   Interchange any two rows.

(*iii*)   Add a nonzero constant multiple of one row to any other row.

ELIMINATION METHODS   To solve a system such as (5) using an augmented matrix, we use either **Gaussian elimination** or the **Gauss-Jordan elimination method.** In the former method we carry out a succession of elementary row operations until we arrive at an augmented matrix in **row-echelon form:**

(*i*)   The first nonzero entry in a nonzero row is 1.

(*ii*)   In consecutive nonzero rows the first entry 1 in the lower row appears to the right of the first 1 in the higher row.

(*iii*)   Rows consisting of all 0's are at the bottom of the matrix.

In the Gauss-Jordan method the row operations are continued until we obtain an augmented matrix that is in **reduced row-echelon form.** A reduced row-echelon matrix has the same three properties listed above in addition to the following one:

(*iv*)   A column containing a first entry 1 has 0's everywhere else.

EXAMPLE 11   Row-Echelon/Reduced Row-Echelon Form

(a) The augmented matrices

$$\begin{pmatrix} 1 & 5 & 0 & | & 2 \\ 0 & 1 & 0 & | & -1 \\ 0 & 0 & 0 & | & 0 \end{pmatrix} \quad \text{and} \quad \begin{pmatrix} 0 & 0 & 1 & -6 & 2 & | & 2 \\ 0 & 0 & 0 & 0 & 1 & | & 4 \end{pmatrix}$$

are in row-echelon form. You should verify that the three criteria are satisfied.

(b) The augmented matrices

$$\begin{pmatrix} 1 & 0 & 0 & | & 7 \\ 0 & 1 & 0 & | & -1 \\ 0 & 0 & 0 & | & 0 \end{pmatrix} \quad \text{and} \quad \begin{pmatrix} 0 & 0 & 1 & -6 & 0 & | & -6 \\ 0 & 0 & 0 & 0 & 1 & | & 4 \end{pmatrix}$$

are in reduced row-echelon form. Note that the remaining entries in the columns containing a leading entry 1 are all 0's.

Note that in Gaussian elimination we stop once we have obtained *an* augmented matrix in row-echelon form. In other words, by using different sequences of row operations we may arrive at different row-echelon forms. This method then requires the use of back-substitution. In Gauss-Jordan elimination we stop when we have obtained *the* augmented matrix in reduced row-echelon form. Any sequence of row operations will lead to the same augmented matrix in reduced row-echelon form. This method does not require back-substitution; the solution of the system will be apparent by inspection of the final matrix. In terms of the equations of the original system, our goal in both methods is simply to make the coefficient of x_1 in the first equation* equal to 1 and then use multiples of that equation to eliminate x_1 from other equations. The process is repeated on the other variables.

To keep track of the row operations on an augmented matrix, we utilize the following notation:

Symbol	Meaning
R_{ij}	Interchange rows i and j
cR_i	Multiply the ith row by the nonzero constant c
$cR_i + R_j$	Multiply the ith row by c and add to the jth row

EXAMPLE 12   Solution by Elimination

Solve
$$2x_1 + 6x_2 + \ x_3 = 7$$
$$x_1 + 2x_2 - \ x_3 = -1$$
$$5x_1 + 7x_2 - 4x_3 = 9$$

using **(a)** Gaussian elimination and **(b)** Gauss-Jordan elimination.

---

*We can always interchange equations so that the first equation contains the variable x_1.

SOLUTION **(a)** Using row operations on the augmented matrix of the system, we obtain

$$\begin{pmatrix} 2 & 6 & 1 & | & 7 \\ 1 & 2 & -1 & | & -1 \\ 5 & 7 & -4 & | & 9 \end{pmatrix} \xrightarrow{R_{12}} \begin{pmatrix} 1 & 2 & -1 & | & -1 \\ 2 & 6 & 1 & | & 7 \\ 5 & 7 & -4 & | & 9 \end{pmatrix} \xrightarrow[-5R_1 + R_3]{-2R_1 + R_2} \begin{pmatrix} 1 & 2 & -1 & | & -1 \\ 0 & 2 & 3 & | & 9 \\ 0 & -3 & 1 & | & 14 \end{pmatrix}$$

$$\xrightarrow{\frac{1}{2}R_2} \begin{pmatrix} 1 & 2 & -1 & | & -1 \\ 0 & 1 & \frac{3}{2} & | & \frac{9}{2} \\ 0 & -3 & 1 & | & 14 \end{pmatrix} \xrightarrow{3R_2 + R_3} \begin{pmatrix} 1 & 2 & -1 & | & -1 \\ 0 & 1 & \frac{3}{2} & | & \frac{9}{2} \\ 0 & 0 & \frac{11}{2} & | & \frac{55}{2} \end{pmatrix} \xrightarrow{\frac{2}{11}R_3} \begin{pmatrix} 1 & 2 & -1 & | & -1 \\ 0 & 1 & \frac{3}{2} & | & \frac{9}{2} \\ 0 & 0 & 1 & | & 5 \end{pmatrix}.$$

The last matrix is in row-echelon form and represents the system

$$x_1 + 2x_2 - x_3 = -1$$

$$x_2 + \frac{3}{2}x_3 = \frac{9}{2}$$

$$x_3 = 5.$$

Substituting $x_3 = 5$ into the second equation then gives $x_2 = -3$. Substituting both these values back into the first equation finally yields $x_1 = 10$.

(b) We start with the last matrix above. Since the first entries in the second and third rows are 1's, we must, in turn, make the remaining entries in the second and third columns 0's:

$$\begin{pmatrix} 1 & 2 & -1 & | & -1 \\ 0 & 1 & \frac{3}{2} & | & \frac{9}{2} \\ 0 & 0 & 1 & | & 5 \end{pmatrix} \xrightarrow{-2R_2 + R_1} \begin{pmatrix} 1 & 0 & -4 & | & -10 \\ 0 & 1 & \frac{3}{2} & | & \frac{9}{2} \\ 0 & 0 & 1 & | & 5 \end{pmatrix} \xrightarrow[-\frac{3}{2}R_3 + R_2]{4R_3 + R_1} \begin{pmatrix} 1 & 0 & 0 & | & 10 \\ 0 & 1 & 0 & | & -3 \\ 0 & 0 & 1 & | & 5 \end{pmatrix}.$$

The last matrix is now in reduced row-echelon form. Because of what the matrix means in terms of equations, it is evident that the solution of the system is $x_1 = 10$, $x_2 = -3$, $x_3 = 5$.

EXAMPLE 13   Gauss-Jordan Elimination

Solve

$$x + 3y - 2z = -7$$

$$4x + y + 3z = 5$$

$$2x - 5y + 7z = 19.$$

SOLUTION   We solve the system using Gauss-Jordan elimination:

$$\begin{pmatrix} 1 & 3 & -2 & | & -7 \\ 4 & 1 & 3 & | & 5 \\ 2 & -5 & 7 & | & 19 \end{pmatrix} \xrightarrow[-2R_1 + R_3]{-4R_1 + R_2} \begin{pmatrix} 1 & 3 & -2 & | & -7 \\ 0 & -11 & 11 & | & 33 \\ 0 & -11 & 11 & | & 33 \end{pmatrix}$$

$$\xrightarrow[-\frac{1}{11}R_3]{-\frac{1}{11}R_2} \begin{pmatrix} 1 & 3 & -2 & | & -7 \\ 0 & 1 & -1 & | & -3 \\ 0 & 1 & -1 & | & -3 \end{pmatrix} \xrightarrow[-R_2 + R_3]{-3R_2 + R_1} \begin{pmatrix} 1 & 0 & 1 & | & 1 \\ 0 & 1 & -1 & | & -3 \\ 0 & 0 & 0 & | & 0 \end{pmatrix}.$$

In this case the last matrix in reduced row-echelon form implies that the original system of three equations in three unknowns is really equivalent to two equations in three unknowns. Since only z is common to both equations (the nonzero rows), we can assign its values arbitrarily. If we let $z = t$, where t represents any real number, then we see that the system has infinitely many solutions: $x = 2 - t$, $y = -3 + t$, $z = t$. Geometrically, these equations are the parametric equations for the line of intersection of the planes $x + 0y + z = 2$ and $0x + y - z = 3$.

USING ROW OPERATIONS TO FIND AN INVERSE   Because of the number of determinants that must be evaluated, formula (2) in Theorem II.2 is seldom used to find the inverse when the matrix $\mathbf{A}$ is large. In the case of 3×3 or larger matrices, the method described in the next theorem is a particularly efficient means for finding $\mathbf{A}^{-1}$.

> ### THEOREM II.3   Finding $\mathbf{A}^{-1}$ Using Elementary Row Operations
>
> If an $n \times n$ matrix $\mathbf{A}$ can be transformed into the $n \times n$ identity $\mathbf{I}$ by a sequence of elementary row operations, then $\mathbf{A}$ is nonsingular. The same sequence of operations that transforms $\mathbf{A}$ into the identity $\mathbf{I}$ will also transform $\mathbf{I}$ into $\mathbf{A}^{-1}$.

It is convenient to carry out these row operations on $\mathbf{A}$ and $\mathbf{I}$ simultaneously by means of an $n \times 2n$ matrix obtained by augmenting $\mathbf{A}$ with the identity $\mathbf{I}$ as shown here:

$$(\mathbf{A} \mid \mathbf{I}) = \begin{pmatrix} a_{11} & a_{12} & \cdots & a_{1n} & 1 & 0 & \cdots & 0 \\ a_{21} & a_{22} & \cdots & a_{2n} & 1 & 0 & \cdots & 0 \\ \vdots & & & \vdots & \vdots & & & \vdots \\ a_{n1} & a_{n2} & \cdots & a_{nn} & 0 & 0 & \cdots & 1 \end{pmatrix}.$$

The procedure for finding $\mathbf{A}^{-1}$ is outlined in the following diagram:

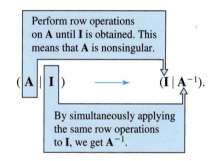

Perform row operations on $\mathbf{A}$ until $\mathbf{I}$ is obtained. This means that $\mathbf{A}$ is nonsingular.

$(\mathbf{A} \mid \mathbf{I}) \longrightarrow (\mathbf{I} \mid \mathbf{A}^{-1})$.

By simultaneously applying the same row operations to $\mathbf{I}$, we get $\mathbf{A}^{-1}$.

EXAMPLE 14   Inverse by Elementary Row Operations

Find the multiplicative inverse for $\mathbf{A} = \begin{pmatrix} 2 & 0 & 1 \\ -2 & 3 & 4 \\ -5 & 5 & 6 \end{pmatrix}$.

SOLUTION   We shall use the same notation as we did when we reduced an augmented matrix to reduced row-echelon form:

$$\left(\begin{array}{ccc|ccc} 2 & 0 & 1 & 1 & 0 & 0 \\ -2 & 3 & 4 & 0 & 1 & 0 \\ -5 & 5 & 6 & 0 & 0 & 1 \end{array} \right) \xrightarrow{\frac{1}{2}R_1} \left(\begin{array}{ccc|ccc} 1 & 0 & \frac{1}{2} & \frac{1}{2} & 0 & 0 \\ -2 & 3 & 4 & 0 & 1 & 0 \\ -5 & 5 & 6 & 0 & 0 & 1 \end{array} \right) \xrightarrow[5R_1 + R_3]{2R_1 + R_2} \left(\begin{array}{ccc|ccc} 1 & 0 & \frac{1}{2} & \frac{1}{2} & 0 & 0 \\ 0 & 3 & 5 & 1 & 1 & 0 \\ 0 & 5 & \frac{17}{2} & \frac{5}{2} & 0 & 1 \end{array} \right)$$

$$\begin{array}{c}\tfrac{1}{3}R_2 \\ \tfrac{1}{5}R_3 \\ \longrightarrow \end{array}\left(\begin{array}{ccc|ccc}1 & 0 & \tfrac{1}{2} & \tfrac{1}{2} & 0 & 0 \\ 0 & 1 & \tfrac{5}{3} & \tfrac{1}{3} & \tfrac{1}{3} & 0 \\ 0 & 1 & \tfrac{17}{10} & \tfrac{1}{2} & 0 & \tfrac{1}{5}\end{array}\right) \xrightarrow{-R_2+R_3} \left(\begin{array}{ccc|ccc}1 & 0 & \tfrac{1}{2} & \tfrac{1}{2} & 0 & 0 \\ 0 & 1 & \tfrac{5}{3} & \tfrac{1}{3} & \tfrac{1}{3} & 0 \\ 0 & 0 & \tfrac{1}{30} & \tfrac{1}{6} & -\tfrac{1}{3} & \tfrac{1}{5}\end{array}\right)$$

$$\xrightarrow{30R_3} \left(\begin{array}{ccc|ccc}1 & 0 & \tfrac{1}{2} & \tfrac{1}{2} & 0 & 0 \\ 0 & 1 & \tfrac{5}{3} & \tfrac{1}{3} & \tfrac{1}{3} & 0 \\ 0 & 0 & 1 & 5 & -10 & 6\end{array}\right) \begin{array}{c}-\tfrac{1}{3}R_3+R_1 \\ -\tfrac{5}{3}R_3+R_2 \\ \longrightarrow \end{array} \left(\begin{array}{ccc|ccc}1 & 0 & 0 & -2 & 5 & -3 \\ 0 & 1 & 0 & -8 & 17 & -10 \\ 0 & 0 & 1 & 5 & -10 & 6\end{array}\right).$$

Because **I** appears to the left of the vertical line, we conclude that the matrix to the right of the line is

$$\mathbf{A}^{-1} = \begin{pmatrix} -2 & 5 & -3 \\ -8 & 17 & -10 \\ 5 & -10 & 6 \end{pmatrix}.$$

If row reduction of $(\mathbf{A}\,|\,\mathbf{I})$ leads to the situation

$$(\mathbf{A}\,|\,\mathbf{I}) \xrightarrow[\text{operations}]{\text{row}} (\mathbf{B}\,|\,\mathbf{C}),$$

where the matrix **B** contains a row of zeros, then necessarily **A** is singular. Since further reduction of **B** always yields another matrix with a row of zeros, we can never transform **A** into **I**.

II.3 THE EIGENVALUE PROBLEM

Gauss-Jordan elimination can be used to find the eigenvectors of a square matrix.

DEFINITION II.13 **Eigenvalues and Eigenvectors**

Let **A** be an $n \times n$ matrix. A number λ is said to be an **eigenvalue** of **A** if there exists a *nonzero* solution vector **K** of the linear system

$$\mathbf{AK} = \lambda\mathbf{K}. \tag{6}$$

The solution vector **K** is said to be an **eigenvector** corresponding to the eigenvalue λ.

The word *eigenvalue* is a combination of German and English terms adapted from the German word *eigenwert,* which, translated literally, is "proper value." Eigenvalues and eigenvectors are also called **characteristic values** and **characteristic vectors,** respectively.

EXAMPLE 15 **Eigenvector of a Matrix**

Verify that $\mathbf{K} = \begin{pmatrix} 1 \\ -1 \\ 1 \end{pmatrix}$ is an eigenvector of the matrix

$$\mathbf{A} = \begin{pmatrix} 0 & -1 & -3 \\ 2 & 3 & 3 \\ -2 & 1 & 1 \end{pmatrix}.$$

SOLUTION   By carrying out the multiplication $\mathbf{AK}$, we see that

$$\mathbf{AK} = \begin{pmatrix} 0 & -1 & -3 \\ 2 & 3 & 3 \\ -2 & 1 & 1 \end{pmatrix} \begin{pmatrix} 1 \\ -1 \\ 1 \end{pmatrix} = \begin{pmatrix} -2 \\ 2 \\ -2 \end{pmatrix} = (-2) \begin{pmatrix} 1 \\ -1 \\ 1 \end{pmatrix} = \overset{\text{eigenvalue}}{\underset{\downarrow}{(-2)}} \mathbf{K}.$$

We see from the preceding line and Definition II.13 that $\lambda = -2$ is an eigenvalue of $\mathbf{A}$.

Using properties of matrix algebra, we can write (6) in the alternative form

$$(\mathbf{A} - \lambda \mathbf{I})\mathbf{K} = \mathbf{0}. \tag{7}$$

where $\mathbf{I}$ is the multiplicative identity. If we let

$$\mathbf{K} = \begin{pmatrix} k_1 \\ k_2 \\ \vdots \\ k_n \end{pmatrix},$$

then (7) is the same as

$$\begin{aligned} (a_{11} - \lambda)k_1 + \quad & a_{12}k_2 + \cdots + \quad & a_{1n}k_n = 0 \\ a_{21}k_1 + (a_{22} - \lambda)k_2 + \cdots + \quad & a_{2n}k_n = 0 \\ & \vdots & \vdots \\ a_{n1}k_1 + \quad & a_{n2}k_2 + \cdots + (a_{nn} - \lambda)k_n = 0 \end{aligned} \tag{8}$$

Although an obvious solution of (8) is $k_1 = 0, k_2 = 0, \ldots, k_n = 0$, we are seeking only nontrivial solutions. It is known that a homogeneous system of n linear equations in n unknowns (that is, $b_i = 0, i = 1, 2, \ldots, n$ in (5)) has a nontrivial solution if and only if the determinant of the coefficient matrix is equal to zero. Thus to find a nonzero solution $\mathbf{K}$ for (7), we must have

$$\det(\mathbf{A} - \lambda \mathbf{I}) = 0. \tag{9}$$

Inspection of (8) shows that the expansion of $\det(\mathbf{A} - \lambda \mathbf{I})$ by cofactors results in an nth-degree polynomial in λ. The equation (9) is called the **characteristic equation** of $\mathbf{A}$. Thus *the eigenvalues of $\mathbf{A}$ are the roots of the characteristic equation.* To find an eigenvector corresponding to an eigenvalue λ, we simply solve the system of equations $(\mathbf{A} - \lambda \mathbf{I})\mathbf{K} = \mathbf{0}$ by applying Gauss-Jordan elimination to the augmented matrix $(\mathbf{A} - \lambda \mathbf{I} | \mathbf{0})$.

EXAMPLE 16   Eigenvalues/Eigenvectors

Find the eigenvalues and eigenvectors of $\mathbf{A} = \begin{pmatrix} 1 & 2 & 1 \\ 6 & -1 & 0 \\ -1 & -2 & -1 \end{pmatrix}$.

SOLUTION   To expand the determinant in the characteristic equation, we use the cofactors of the second row:

$$\det(\mathbf{A} - \lambda \mathbf{I}) = \begin{vmatrix} 1 - \lambda & 2 & 1 \\ 6 & -1 - \lambda & 0 \\ -1 & -2 & -1 - \lambda \end{vmatrix} = -\lambda^3 - \lambda^2 + 12\lambda = 0.$$

From $-\lambda^3 - \lambda^2 + 12\lambda = -\lambda(\lambda + 4)(\lambda - 3) = 0$ we see that the eigenvalues are $\lambda_1 = 0, \lambda_2 = -4,$ and $\lambda_3 = 3$. To find the eigenvectors, we must now reduce $(\mathbf{A} - \lambda \mathbf{I} | \mathbf{0})$ three times corresponding to the three distinct eigenvalues.

For $\lambda_1 = 0$ we have

$$(\mathbf{A} - 0\mathbf{I} \mid \mathbf{0}) = \begin{pmatrix} 1 & 2 & 1 & | & 0 \\ 6 & -1 & 0 & | & 0 \\ -1 & -2 & -1 & | & 0 \end{pmatrix} \xrightarrow[R_1 + R_3]{-6R_1 + R_2} \begin{pmatrix} 1 & 2 & 1 & | & 0 \\ 0 & -13 & -6 & | & 0 \\ 0 & 0 & 0 & | & 0 \end{pmatrix}$$

$$\xrightarrow{-\frac{1}{13}R_2} \begin{pmatrix} 1 & 2 & 1 & | & 0 \\ 0 & 1 & \frac{6}{13} & | & 0 \\ 0 & 0 & 0 & | & 0 \end{pmatrix} \xrightarrow{-2R_2 + R_1} \begin{pmatrix} 1 & 0 & \frac{1}{13} & | & 0 \\ 0 & 1 & \frac{6}{13} & | & 0 \\ 0 & 0 & 0 & | & 0 \end{pmatrix}.$$

Thus we see that $k_1 = -\frac{1}{13}k_3$ and $k_2 = -\frac{6}{13}k_3$, Choosing $k_3 = -13$, we get the eigenvector*

$$\mathbf{K}_1 = \begin{pmatrix} 1 \\ 6 \\ -13 \end{pmatrix}.$$

For $\lambda_2 = -4$,

$$(\mathbf{A} + 4\mathbf{I} \mid \mathbf{0}) = \begin{pmatrix} 5 & 2 & 1 & | & 0 \\ 6 & 3 & 0 & | & 0 \\ -1 & -2 & 3 & | & 0 \end{pmatrix} \xrightarrow[R_{31}]{-R_3} \begin{pmatrix} 1 & 2 & -3 & | & 0 \\ 6 & 3 & 0 & | & 0 \\ 5 & 2 & 1 & | & 0 \end{pmatrix}$$

$$\xrightarrow[-5R_1 + R_3]{-6R_1 + R_2} \begin{pmatrix} 1 & 2 & -3 & | & 0 \\ 0 & -9 & 18 & | & 0 \\ 0 & -8 & 16 & | & 0 \end{pmatrix} \xrightarrow[-\frac{1}{8}R_3]{-\frac{1}{9}R_2} \begin{pmatrix} 1 & 2 & -3 & | & 0 \\ 0 & 1 & -2 & | & 0 \\ 0 & 1 & -2 & | & 0 \end{pmatrix} \xrightarrow[-R_2 + R_3]{-2R_2 + R_1} \begin{pmatrix} 1 & 0 & 1 & | & 0 \\ 0 & 1 & -2 & | & 0 \\ 0 & 0 & 0 & | & 0 \end{pmatrix}$$

implies $k_1 = -k_3$ and $k_2 = 2k_3$. Choosing $k_3 = 1$ then yields the second eigenvector

$$\mathbf{K}_2 = \begin{pmatrix} -1 \\ 2 \\ 1 \end{pmatrix}.$$

Finally, for $\lambda_3 = 3$ Gauss-Jordan elimination gives

$$(\mathbf{A} - 3\mathbf{I} \mid \mathbf{0}) = \begin{pmatrix} -2 & 2 & 1 & | & 0 \\ 6 & -4 & 0 & | & 0 \\ -1 & -2 & -4 & | & 0 \end{pmatrix} \xrightarrow[\text{operations}]{\text{row}} \begin{pmatrix} 1 & 0 & 1 & | & 0 \\ 0 & 1 & \frac{3}{2} & | & 0 \\ 0 & 0 & 0 & | & 0 \end{pmatrix},$$

and so $k_1 = -k_3$ and $k_2 = -\frac{3}{2}k_3$. The choice of $k_3 = -2$ leads to the third eigenvector:

$$\mathbf{K}_3 = \begin{pmatrix} 2 \\ 3 \\ -2 \end{pmatrix}.$$

When an $n \times n$ matrix $\mathbf{A}$ possesses n distinct eigenvalues $\lambda_1, \lambda_2, \dots, \lambda_n$, it can be proved that a set of n linearly independent[†] eigenvectors $\mathbf{K}_1, \mathbf{K}_2, \dots, \mathbf{K}_n$ can be found. However, when the characteristic equation has repeated roots, it may not be possible to find n linearly independent eigenvectors for $\mathbf{A}$.

---

*Of course k_3 could be chosen as any nonzero number. In other words, a nonzero constant multiple of an eigenvector is also an eigenvector.

†Linear independence of column vectors is defined in exactly the same manner as for functions.

EXAMPLE 17	**Eigenvalues/Eigenvectors**

Find the eigenvalues and eigenvectors of $\mathbf{A} = \begin{pmatrix} 3 & 4 \\ -1 & 7 \end{pmatrix}$.

SOLUTION From the characteristic equation

$$\det(\mathbf{A} - \lambda\mathbf{I}) = \begin{vmatrix} 3 - \lambda & 4 \\ -1 & 7 - \lambda \end{vmatrix} = (\lambda - 5)^2 = 0$$

we see that $\lambda_1 = \lambda_2 = 5$ is an eigenvalue of multiplicity two. In the case of a 2×2 matrix there is no need to use Gauss-Jordan elimination. To find the eigenvector(s) corresponding to $\lambda_1 = 5$, we resort to the system $(\mathbf{A} - 5\mathbf{I}|\mathbf{0})$ in its equivalent form

$$-2k_1 + 4k_2 = 0$$

$$-k_1 + 2k_2 = 0.$$

It is apparent from this system that $k_1 = 2k_2$. Thus if we choose $k_2 = 1$, we find the single eigenvector

$$\mathbf{K}_1 = \begin{pmatrix} 2 \\ 1 \end{pmatrix}.$$

EXAMPLE 18	**Eigenvalues/Eigenvectors**

Find the eigenvalues and eigenvectors of $\mathbf{A} = \begin{pmatrix} 9 & 1 & 1 \\ 1 & 9 & 1 \\ 1 & 1 & 9 \end{pmatrix}$.

SOLUTION The characteristic equation

$$\det(\mathbf{A} - \lambda\mathbf{I}) = \begin{vmatrix} 9 - \lambda & 1 & 1 \\ 1 & 9 - \lambda & 1 \\ 1 & 1 & 9 - \lambda \end{vmatrix} = -(\lambda - 11)(\lambda - 8)^2 = 0$$

shows that $\lambda_1 = 11$ and that $\lambda_2 = \lambda_3 = 8$ is an eigenvalue of multiplicity two.
   For $\lambda_1 = 11$ Gauss-Jordan elimination gives

$$(\mathbf{A} - 11\mathbf{I}|\mathbf{0}) = \begin{pmatrix} -2 & 1 & 1 & | & 0 \\ 1 & -2 & 1 & | & 0 \\ 1 & 1 & -2 & | & 0 \end{pmatrix} \xrightarrow[\text{operations}]{\text{row}} \begin{pmatrix} 1 & 0 & -1 & | & 0 \\ 0 & 1 & -1 & | & 0 \\ 0 & 0 & 0 & | & 0 \end{pmatrix}.$$

Hence $k_1 = k_3$ and $k_2 = k_3$. If $k_3 = 1$, then

$$\mathbf{K}_1 = \begin{pmatrix} 1 \\ 1 \\ 1 \end{pmatrix}.$$

   Now for $\lambda_2 = 8$ we have

$$(\mathbf{A} - 8\mathbf{I}|\mathbf{0}) = \begin{pmatrix} 1 & 1 & 1 & | & 0 \\ 1 & 1 & 1 & | & 0 \\ 1 & 1 & 1 & | & 0 \end{pmatrix} \xrightarrow[\text{operations}]{\text{row}} \begin{pmatrix} 1 & 1 & 1 & | & 0 \\ 0 & 0 & 0 & | & 0 \\ 0 & 0 & 0 & | & 0 \end{pmatrix}.$$

In the equation $k_1 + k_2 + k_3 = 0$ we are free to select two of the variables arbitrarily. Choosing, on the one hand, $k_2 = 1$, $k_3 = 0$ and, on the other, $k_2 = 0$, $k_3 = 1$, we obtain two linearly independent eigenvectors

$$\mathbf{K}_2 = \begin{pmatrix} -1 \\ 1 \\ 0 \end{pmatrix} \quad \text{and} \quad \mathbf{K}_3 = \begin{pmatrix} -1 \\ 0 \\ 1 \end{pmatrix}.$$

Exercises for Appendix II

Answers to odd-numbered problems begin on page ANS-27.

II.1   Basic Definitions and Theory

1. If $\mathbf{A} = \begin{pmatrix} 4 & 5 \\ -6 & 9 \end{pmatrix}$ and $\mathbf{B} = \begin{pmatrix} -2 & 6 \\ 8 & -10 \end{pmatrix}$, find

(a) $\mathbf{A} + \mathbf{B}$ (b) $\mathbf{B} - \mathbf{A}$ (c) $2\mathbf{A} + 3\mathbf{B}$

2. If $\mathbf{A} = \begin{pmatrix} -2 & 0 \\ 4 & 1 \\ 7 & 3 \end{pmatrix}$ and $\mathbf{B} = \begin{pmatrix} 3 & -1 \\ 0 & 2 \\ -4 & -2 \end{pmatrix}$, find

(a) $\mathbf{A} - \mathbf{B}$ (b) $\mathbf{B} - \mathbf{A}$ (c) $2(\mathbf{A} + \mathbf{B})$

3. If $\mathbf{A} = \begin{pmatrix} 2 & -3 \\ -5 & 4 \end{pmatrix}$ and $\mathbf{B} = \begin{pmatrix} -1 & 6 \\ 3 & 2 \end{pmatrix}$, find

(a) $\mathbf{AB}$ (b) $\mathbf{BA}$ (c) $\mathbf{A}^2 = \mathbf{AA}$ (d) $\mathbf{B}^2 = \mathbf{BB}$

4. If $\mathbf{A} = \begin{pmatrix} 1 & 4 \\ 5 & 10 \\ 8 & 12 \end{pmatrix}$ and $\mathbf{B} = \begin{pmatrix} -4 & 6 & -3 \\ 1 & -3 & 2 \end{pmatrix}$, find

(a) $\mathbf{AB}$ (b) $\mathbf{BA}$

5. If $\mathbf{A} = \begin{pmatrix} 1 & -2 \\ -2 & 4 \end{pmatrix}$, $\mathbf{B} = \begin{pmatrix} 6 & 3 \\ 2 & 1 \end{pmatrix}$, and $\mathbf{C} = \begin{pmatrix} 0 & 2 \\ 3 & 4 \end{pmatrix}$, find

(a) $\mathbf{BC}$ (b) $\mathbf{A}(\mathbf{BC})$ (c) $\mathbf{C}(\mathbf{BA})$ (d) $\mathbf{A}(\mathbf{B} + \mathbf{C})$

6. If $\mathbf{A} = (5 \quad -6 \quad 7)$, $\mathbf{B} = \begin{pmatrix} 3 \\ 4 \\ -1 \end{pmatrix}$, and

$\mathbf{C} = \begin{pmatrix} 1 & 2 & 4 \\ 0 & 1 & -1 \\ 3 & 2 & 1 \end{pmatrix}$, find

(a) $\mathbf{AB}$ (b) $\mathbf{BA}$ (c) $(\mathbf{BA})\mathbf{C}$ (d) $(\mathbf{AB})\mathbf{C}$

7. If $\mathbf{A} = \begin{pmatrix} 4 \\ 8 \\ -10 \end{pmatrix}$ and $\mathbf{B} = (2 \quad 4 \quad 5)$, find

(a) $\mathbf{A}^T\mathbf{A}$ (b) $\mathbf{B}^T\mathbf{B}$ (c) $\mathbf{A} + \mathbf{B}^T$

8. If $\mathbf{A} = \begin{pmatrix} 1 & 2 \\ 2 & 4 \end{pmatrix}$ and $\mathbf{B} = \begin{pmatrix} -2 & 3 \\ 5 & 7 \end{pmatrix}$, find

(a) $\mathbf{A} + \mathbf{B}^T$ (b) $2\mathbf{A}^T - \mathbf{B}^T$ (c) $\mathbf{A}^T(\mathbf{A} - \mathbf{B})$

9. If $\mathbf{A} = \begin{pmatrix} 3 & 4 \\ 8 & 1 \end{pmatrix}$ and $\mathbf{B} = \begin{pmatrix} 5 & 10 \\ -2 & -5 \end{pmatrix}$, find

(a) $(\mathbf{AB})^T$ (b) $\mathbf{B}^T\mathbf{A}^T$

10. If $\mathbf{A} = \begin{pmatrix} 5 & 9 \\ -4 & 6 \end{pmatrix}$ and $\mathbf{B} = \begin{pmatrix} -3 & 11 \\ -7 & 2 \end{pmatrix}$, find

(a) $\mathbf{A}^T + \mathbf{B}^T$ (b) $(\mathbf{A} + \mathbf{B})^T$

In Problems 11–14 write the given sum as a single column matrix.

11. $4\begin{pmatrix} -1 \\ 2 \end{pmatrix} - 2\begin{pmatrix} 2 \\ 8 \end{pmatrix} + 3\begin{pmatrix} -2 \\ 3 \end{pmatrix}$

12. $3t\begin{pmatrix} 2 \\ t \\ -1 \end{pmatrix} + (t-1)\begin{pmatrix} -1 \\ -t \\ 3 \end{pmatrix} - 2\begin{pmatrix} 3t \\ 4 \\ -5t \end{pmatrix}$

13. $\begin{pmatrix} 2 & -3 \\ 1 & 4 \end{pmatrix}\begin{pmatrix} -2 \\ 5 \end{pmatrix} - \begin{pmatrix} -1 & 6 \\ -2 & 3 \end{pmatrix}\begin{pmatrix} -7 \\ 2 \end{pmatrix}$

14. $\begin{pmatrix} 1 & -3 & 4 \\ 2 & 5 & -1 \\ 0 & -4 & -2 \end{pmatrix}\begin{pmatrix} t \\ 2t-1 \\ -t \end{pmatrix} + \begin{pmatrix} -t \\ 1 \\ 4 \end{pmatrix} - \begin{pmatrix} 2 \\ 8 \\ -6 \end{pmatrix}$

In Problems 15–22 determine whether the given matrix is singular or nonsingular. If it is nonsingular, find $\mathbf{A}^{-1}$ using Theorem II.2.

15. $\mathbf{A} = \begin{pmatrix} -3 & 6 \\ -2 & 4 \end{pmatrix}$ **16.** $\mathbf{A} = \begin{pmatrix} 2 & 5 \\ 1 & 4 \end{pmatrix}$

17. $\mathbf{A} = \begin{pmatrix} 4 & 8 \\ -3 & -5 \end{pmatrix}$ **18.** $\mathbf{A} = \begin{pmatrix} 7 & 10 \\ 2 & 2 \end{pmatrix}$

19. $\mathbf{A} = \begin{pmatrix} 2 & 1 & 0 \\ -1 & 2 & 1 \\ 1 & 2 & 1 \end{pmatrix}$ **20.** $\mathbf{A} = \begin{pmatrix} 3 & 2 & 1 \\ 4 & 1 & 0 \\ -2 & 5 & -1 \end{pmatrix}$

21. $\mathbf{A} = \begin{pmatrix} 2 & 1 & 1 \\ 1 & -2 & -3 \\ 3 & 2 & 4 \end{pmatrix}$ **22.** $\mathbf{A} = \begin{pmatrix} 4 & 1 & -1 \\ 6 & 2 & -3 \\ -2 & -1 & 2 \end{pmatrix}$

In Problems 23 and 24 show that the given matrix is nonsingular for every real value of t. Find $\mathbf{A}^{-1}(t)$ using Theorem II.2.

23. $\mathbf{A}(t) = \begin{pmatrix} 2e^{-t} & e^{4t} \\ 4e^{-t} & 3e^{4t} \end{pmatrix}$

24. $\mathbf{A}(t) = \begin{pmatrix} 2e^t \sin t & -2e^t \cos t \\ e^t \cos t & e^t \sin t \end{pmatrix}$

In Problems 25–28 find $d\mathbf{X}/dt$.

25. $\mathbf{X} = \begin{pmatrix} 5e^{-t} \\ 2e^{-t} \\ -7e^{-t} \end{pmatrix}$ **26.** $\mathbf{X} = \begin{pmatrix} \frac{1}{2} \sin 2t - 4 \cos 2t \\ -3 \sin 2t + 5 \cos 2t \end{pmatrix}$

27. $\mathbf{X} = 2 \begin{pmatrix} 1 \\ -1 \end{pmatrix} e^{2t} + 4 \begin{pmatrix} 2 \\ 1 \end{pmatrix} e^{-3t}$ **28.** $\mathbf{X} = \begin{pmatrix} 5te^{2t} \\ t \sin 3t \end{pmatrix}$

29. Let $\mathbf{A}(t) = \begin{pmatrix} e^{4t} & \cos \pi t \\ 2t & 3t^2 - 1 \end{pmatrix}$. Find

(a) $\dfrac{d\mathbf{A}}{dt}$ (b) $\displaystyle\int_0^2 \mathbf{A}(t)\,dt$ (c) $\displaystyle\int_0^t \mathbf{A}(s)\,ds$

30. Let $\mathbf{A}(t) = \begin{pmatrix} \dfrac{1}{t^2 + 1} & 3t \\ t^2 & t \end{pmatrix}$ and $\mathbf{B}(t) = \begin{pmatrix} 6t & 2 \\ 1/t & 4t \end{pmatrix}$. Find

(a) $\dfrac{d\mathbf{A}}{dt}$ (b) $\dfrac{d\mathbf{B}}{dt}$

(c) $\displaystyle\int_0^1 \mathbf{A}(t)\,dt$ (d) $\displaystyle\int_1^2 \mathbf{B}(t)\,dt$

(e) $\mathbf{A}(t)\mathbf{B}(t)$ (f) $\dfrac{d}{dt}\mathbf{A}(t)\mathbf{B}(t)$

(g) $\displaystyle\int_1^t \mathbf{A}(s)\mathbf{B}(s)\,ds$

II.2 GAUSSIAN AND GAUSS-JORDAN ELIMINATION

In Problems 31–38 solve the given system of equations by either Gaussian elimination or Gauss-Jordan elimination.

31. $\begin{aligned} x + y - 2z &= 14 \\ 2x - y + z &= 0 \\ 6x + 3y + 4z &= 1 \end{aligned}$ **32.** $\begin{aligned} 5x - 2y + 4z &= 10 \\ x + y + z &= 9 \\ 4x - 3y + 3z &= 1 \end{aligned}$

33. $\begin{aligned} y + z &= -5 \\ 5x + 4y - 16z &= -10 \\ x - y - 5z &= 7 \end{aligned}$ **34.** $\begin{aligned} 3x + y + z &= 4 \\ 4x + 2y - z &= 7 \\ x + y - 3z &= 6 \end{aligned}$

35. $\begin{aligned} 2x + y + z &= 4 \\ 10x - 2y + 2z &= -1 \\ 6x - 2y + 4z &= 8 \end{aligned}$ **36.** $\begin{aligned} x + 2z &= 8 \\ x + 2y - 2z &= 4 \\ 2x + 5y - 6z &= 6 \end{aligned}$

37. $\begin{aligned} x_1 + x_2 - x_3 - x_4 &= -1 \\ x_1 + x_2 + x_3 + x_4 &= 3 \\ x_1 - x_2 + x_3 - x_4 &= 3 \\ 4x_1 + x_2 - 2x_3 + x_4 &= 0 \end{aligned}$ **38.** $\begin{aligned} 2x_1 + x_2 + x_3 &= 0 \\ x_1 + 3x_2 + x_3 &= 0 \\ 7x_1 + x_2 + 3x_3 &= 0 \end{aligned}$

In Problems 39 and 40 use Gauss-Jordan elimination to demonstrate that the given system of equations has no solution.

39. $\begin{aligned} x + 2y + 4z &= 2 \\ 2x + 4y + 3z &= 1 \\ x + 2y - z &= 7 \end{aligned}$ **40.** $\begin{aligned} x_1 + x_2 - x_3 + 3x_4 &= 1 \\ x_2 - x_3 - 4x_4 &= 0 \\ x_1 + 2x_2 - 2x_3 - x_4 &= 6 \\ 4x_1 + 7x_2 - 7x_3 &= 9 \end{aligned}$

In Problems 41–46 use Theorem II.3 to find $\mathbf{A}^{-1}$ for the given matrix or show that no inverse exists.

41. $\mathbf{A} = \begin{pmatrix} 4 & 2 & 3 \\ 2 & 1 & 0 \\ -1 & -2 & 0 \end{pmatrix}$ **42.** $\mathbf{A} = \begin{pmatrix} 2 & 4 & -2 \\ 4 & 2 & -2 \\ 8 & 10 & -6 \end{pmatrix}$

43. $\mathbf{A} = \begin{pmatrix} -1 & 3 & 0 \\ 1 & -2 & 1 \\ 0 & 1 & 2 \end{pmatrix}$ **44.** $\mathbf{A} = \begin{pmatrix} 1 & 2 & 3 \\ 0 & 1 & 4 \\ 0 & 0 & 8 \end{pmatrix}$

45. $\mathbf{A} = \begin{pmatrix} 1 & 2 & 3 & 1 \\ -1 & 0 & 2 & 1 \\ 2 & 1 & -3 & 0 \\ 1 & 1 & 2 & 1 \end{pmatrix}$ **46.** $\mathbf{A} = \begin{pmatrix} 1 & 0 & 0 & 0 \\ 0 & 0 & 1 & 0 \\ 0 & 0 & 0 & 1 \\ 0 & 1 & 0 & 0 \end{pmatrix}$

11.3 THE EIGENVALUE PROBLEM

In Problems 47–54 find the eigenvalues and eigenvectors of the given matrix.

47. $\begin{pmatrix} -1 & 2 \\ -7 & 8 \end{pmatrix}$ **48.** $\begin{pmatrix} 2 & 1 \\ 2 & 1 \end{pmatrix}$

49. $\begin{pmatrix} -8 & -1 \\ 16 & 0 \end{pmatrix}$ **50.** $\begin{pmatrix} 1 & 1 \\ \frac{1}{4} & 1 \end{pmatrix}$

51. $\begin{pmatrix} 5 & -1 & 0 \\ 0 & -5 & 9 \\ 5 & -1 & 0 \end{pmatrix}$ **52.** $\begin{pmatrix} 3 & 0 & 0 \\ 0 & 2 & 0 \\ 4 & 0 & 1 \end{pmatrix}$

53. $\begin{pmatrix} 0 & 4 & 0 \\ -1 & -4 & 0 \\ 0 & 0 & -2 \end{pmatrix}$ **54.** $\begin{pmatrix} 1 & 6 & 0 \\ 0 & 2 & 1 \\ 0 & 1 & 2 \end{pmatrix}$

In Problems 55 and 56 show that the given matrix has complex eigenvalues. Find the eigenvectors of the matrix.

55. $\begin{pmatrix} -1 & 2 \\ -5 & 1 \end{pmatrix}$

56. $\begin{pmatrix} 2 & -1 & 0 \\ 5 & 2 & 4 \\ 0 & 1 & 2 \end{pmatrix}$

MISCELLANEOUS PROBLEMS

57. If $A(t)$ is a 2×2 matrix of differentiable functions and $X(t)$ is a 2×1 column matrix of differentiable functions, prove the product rule

$$\frac{d}{dt}[A(t)X(t)] = A(t)X'(t) + A'(t)X(t).$$

58. Derive formula (3). (*Hint:* Find a matrix

$$B = \begin{pmatrix} b_{11} & b_{12} \\ b_{21} & b_{22} \end{pmatrix}$$

for which $AB = I$. Solve for $b_{11}, b_{12}, b_{21},$ and b_{22}. Then show that $BA = I$.)

59. If A is nonsingular and $AB = AC$, show that $B = C$.

60. If A and B are nonsingular, show that $(AB)^{-1} = B^{-1}A^{-1}$.

61. Let A and B be $n \times n$ matrices. In general, is

$$(A + B)^2 = A^2 + 2AB + B^2?$$

62. A square matrix A is said to be a **diagonal matrix** if all its entries off the main diagonal are zero—that is, $a_{ij} = 0$, $i \neq j$. The entries a_{ii} on the main diagonal may or may not be zero. The multiplicative identity matrix I is an example of a diagonal matrix.
(a) Find the inverse of the 2×2 diagonal matrix

$$A = \begin{pmatrix} a_{11} & 0 \\ 0 & a_{22} \end{pmatrix}$$

when $a_{11} \neq 0$, $a_{22} \neq 0$.
(b) Find the inverse of a 3×3 diagonal matrix A whose main diagonal entries a_{ii} are all nonzero.
(c) In general, what is the inverse of an $n \times n$ diagonal matrix A whose main diagonal entries a_{ii} are all nonzero?

APPENDIX III

LAPLACE TRANSFORMS

$f(t)$	$\mathscr{L}\{f(t)\} = F(s)$
1. 1	$\dfrac{1}{s}$
2. t	$\dfrac{1}{s^2}$
3. t^n	$\dfrac{n!}{s^{n+1}}$, $\quad n$ a positive integer
4. $t^{-1/2}$	$\sqrt{\dfrac{\pi}{s}}$
5. $t^{1/2}$	$\dfrac{\sqrt{\pi}}{2s^{3/2}}$
6. t^α	$\dfrac{\Gamma(\alpha+1)}{s^{\alpha+1}}$, $\quad \alpha > -1$
7. $\sin kt$	$\dfrac{k}{s^2 + k^2}$
8. $\cos kt$	$\dfrac{s}{s^2 + k^2}$
9. $\sin^2 kt$	$\dfrac{2k^2}{s(s^2 + 4k^2)}$
10. $\cos^2 kt$	$\dfrac{s^2 + 2k^2}{s(s^2 + 4k^2)}$
11. e^{at}	$\dfrac{1}{s - a}$
12. $\sinh kt$	$\dfrac{k}{s^2 - k^2}$
13. $\cosh kt$	$\dfrac{s}{s^2 - k^2}$
14. $\sinh^2 kt$	$\dfrac{2k^2}{s(s^2 - 4k^2)}$
15. $\cosh^2 kt$	$\dfrac{s^2 - 2k^2}{s(s^2 - 4k^2)}$
16. te^{at}	$\dfrac{1}{(s - a)^2}$
17. $t^n e^{at}$	$\dfrac{n!}{(s - a)^{n+1}}$, $\quad n$ a positive integer

$f(t)$	$\mathcal{L}\{f(t)\} = F(s)$
18. $e^{at} \sin kt$	$\dfrac{k}{(s-a)^2 + k^2}$
19. $e^{at} \cos kt$	$\dfrac{s-a}{(s-a)^2 + k^2}$
20. $e^{at} \sinh kt$	$\dfrac{k}{(s-a)^2 - k^2}$
21. $e^{at} \cosh kt$	$\dfrac{s-a}{(s-a)^2 - k^2}$
22. $t \sin kt$	$\dfrac{2ks}{(s^2 + k^2)^2}$
23. $t \cos kt$	$\dfrac{s^2 - k^2}{(s^2 + k^2)^2}$
24. $\sin kt + kt \cos kt$	$\dfrac{2ks^2}{(s^2 + k^2)^2}$
25. $\sin kt - kt \cos kt$	$\dfrac{2k^3}{(s^2 + k^2)^2}$
26. $t \sinh kt$	$\dfrac{2ks}{(s^2 - k^2)^2}$
27. $t \cosh kt$	$\dfrac{s^2 + k^2}{(s^2 - k^2)^2}$
28. $\dfrac{e^{at} - e^{bt}}{a - b}$	$\dfrac{1}{(s-a)(s-b)}$
29. $\dfrac{ae^{at} - be^{bt}}{a - b}$	$\dfrac{s}{(s-a)(s-b)}$
30. $1 - \cos kt$	$\dfrac{k^2}{s(s^2 + k^2)}$
31. $kt - \sin kt$	$\dfrac{k^3}{s^2(s^2 + k^2)}$
32. $\dfrac{a \sin bt - b \sin at}{ab\,(a^2 - b^2)}$	$\dfrac{1}{(s^2 + a^2)(s^2 + b^2)}$
33. $\dfrac{\cos bt - \cos at}{a^2 - b^2}$	$\dfrac{s}{(s^2 + a^2)(s^2 + b^2)}$
34. $\sin kt \sinh kt$	$\dfrac{2k^2 s}{s^4 + 4k^4}$
35. $\sin kt \cosh kt$	$\dfrac{k(s^2 + 2k^2)}{s^4 + 4k^4}$
36. $\cos kt \sinh kt$	$\dfrac{k(s^2 - 2k^2)}{s^4 + 4k^4}$
37. $\cos kt \cosh kt$	$\dfrac{s^3}{s^4 + 4k^4}$

$f(t)$	$\mathcal{L}\{f(t)\} = F(s)$
38. $J_0(kt)$	$\dfrac{1}{\sqrt{s^2 + k^2}}$
39. $\dfrac{e^{bt} - e^{at}}{t}$	$\ln \dfrac{s - a}{s - b}$
40. $\dfrac{2(1 - \cos kt)}{t}$	$\ln \dfrac{s^2 + k^2}{s^2}$
41. $\dfrac{2(1 - \cosh kt)}{t}$	$\ln \dfrac{s^2 - k^2}{s^2}$
42. $\dfrac{\sin at}{t}$	$\arctan\left(\dfrac{a}{s}\right)$
43. $\dfrac{\sin at \cos bt}{t}$	$\dfrac{1}{2}\arctan \dfrac{a + b}{s} + \dfrac{1}{2}\arctan \dfrac{a - b}{s}$
44. $\dfrac{1}{\sqrt{\pi t}}\, e^{-a^2/4t}$	$\dfrac{e^{-a\sqrt{s}}}{\sqrt{s}}$
45. $\dfrac{a}{2\sqrt{\pi t^3}}\, e^{-a^2/4t}$	$e^{-a\sqrt{s}}$
46. $\operatorname{erfc}\left(\dfrac{a}{2\sqrt{t}}\right)$	$\dfrac{e^{-a\sqrt{s}}}{s}$
47. $2\sqrt{\dfrac{t}{\pi}}\, e^{-a^2/4t} - a\operatorname{erfc}\left(\dfrac{a}{2\sqrt{t}}\right)$	$\dfrac{e^{-a\sqrt{s}}}{s\sqrt{s}}$
48. $e^{ab}e^{b^2 t}\operatorname{erfc}\left(b\sqrt{t} + \dfrac{a}{2\sqrt{t}}\right)$	$\dfrac{e^{-a\sqrt{s}}}{\sqrt{s}(\sqrt{s} + b)}$
49. $-e^{ab}e^{b^2 t}\operatorname{erfc}\left(b\sqrt{t} + \dfrac{a}{2\sqrt{t}}\right)$ $+ \operatorname{erfc}\left(\dfrac{a}{2\sqrt{t}}\right)$	$\dfrac{be^{-a\sqrt{s}}}{s(\sqrt{s} + b)}$
50. $\delta(t)$	1
51. $\delta(t - t_0)$	e^{-st_0}
52. $e^{at}f(t)$	$F(s - a)$
53. $f(t - a)\,\mathcal{U}(t - a)$	$e^{-as}F(s)$
54. $\mathcal{U}(t - a)$	$\dfrac{e^{-as}}{s}$
55. $f^{(n)}(t)$	$s^n F(s) - s^{(n-1)}f(0) - \cdots - f^{(n-1)}(0)$
56. $t^n f(t)$	$(-1)^n \dfrac{d^n}{ds^n}F(s)$
57. $\displaystyle\int_0^t f(\tau)g(t - \tau)\,d\tau$	$F(s)G(s)$

ANSWERS FOR SELECTED ODD-NUMBERED PROBLEMS

EXERCISES 1.1 (PAGE 10)

1. linear, second order   3. linear, fourth order
5. nonlinear, second order   7. linear, third order
9. linear in x but nonlinear in y
15. domain of function is $-2 \leq x < \infty$; largest interval of definition for solution is $-2 < x < \infty$.
17. domain of function is the set of real numbers except $x = 2$ and $x = -2$; largest intervals of definition for solution are $-\infty < x < -2$, $-2 < x < 2$, or $2 < x < \infty$.
19. $X = \dfrac{e^t - 1}{e^t - 2}$ defined on $(-\infty, \ln 2)$ or on $(\ln 2, \infty)$
27. (a) $m = -2$ (b) $m = 2, m = 3$
29. $y = 2$
31. no constant solutions

EXERCISES 1.2 (PAGE 16)

1. $y = 1/(1 - 4e^{-x})$
3. $y = 1/(x^2 - 1); 1 < x < \infty$
5. $y = 1/(x^2 + 1); -\infty < x < \infty$
7. $x = -\cos t + 8 \sin t$
9. $x = \frac{\sqrt{3}}{4}\cos t + \frac{1}{4}\sin t$ 11. $y = \frac{3}{2}e^x - \frac{1}{2}e^{-x}$
13. $y = 5e^{-x-1}$ 15. $y = 0, y = x^3$
17. half-planes defined by either $y > 0$ or $y < 0$
19. half-planes defined by either $x > 0$ or $x < 0$
21. the regions defined by $y > 2$, $y < -2$, or $-2 < y < 2$
23. any region not containing $(0, 0)$
25. yes
27. no
29. (a) $y = cx$
   (b) any rectangular region not touching the y-axis
   (c) No, the function is not differentiable at $x = 0$.
31. (b) $y = 1/(1 - x)$ on $(-\infty, 1)$;
   $y = -1/(x + 1)$ on $(-1, \infty)$

EXERCISES 1.3 (PAGE 28)

1. $\dfrac{dP}{dt} = kP + r; \dfrac{dP}{dt} = kP - r$

3. $\dfrac{dP}{dt} = k_1 P - k_2 P^2$

7. $\dfrac{dx}{dt} = kx(1000 - x)$

9. $\dfrac{dA}{dt} + \dfrac{1}{100}A = 0; A(0) = 50$

11. $\dfrac{dA}{dt} + \dfrac{7}{600 - t}A = 6$ 13. $\dfrac{dh}{dt} = -\dfrac{c\pi}{450}\sqrt{h}$

15. $L\dfrac{di}{dt} + Ri = E(t)$ 17. $m\dfrac{dv}{dt} = mg - kv^2$

19. $m\dfrac{d^2x}{dt^2} = -kx$ 21. $x\dfrac{d^2x}{dt^2} + \left(\dfrac{dx}{dt}\right)^2 + 32x = 160$

23. $\dfrac{d^2r}{dt^2} + \dfrac{gR^2}{r^2} = 0$ 25. $\dfrac{dA}{dt} = k(M - A), k > 0$

27. $\dfrac{dx}{dt} + kx = r, k > 0$

29. $\dfrac{dy}{dx} = \dfrac{-x + \sqrt{x^2 + y^2}}{y}$

CHAPTER 1 IN REVIEW (PAGE 34)

1. $\dfrac{dy}{dx} = ky$ 3. $y'' + k^2 y = 0$
5. $y'' - 2y' + y = 0$ 7. (a), (d)
9. (b)   11. (b)
13. $y = c_1$ and $y = c_2 e^x$, c_1 and c_2 constants
15. $y' = x^2 + y^2$
17. (a) The domain is the set of all real numbers.
   (b) either $(-\infty, 0)$ or $(0, \infty)$
19. For $x_0 = -1$ the interval is $(-\infty, 0)$, and for $x_0 = 2$ the interval is $(0, \infty)$.
21. (c) $y = \begin{cases} -x^2, & x < 0 \\ x^2, & x \geq 0 \end{cases}$ 23. $(-\infty, \infty)$
25. $(0, \infty)$ 27. $y_0 = -3, y_1 = 0$
29. $\dfrac{dP}{dt} = k(P - 200 + 10t)$

EXERCISES 2.1 (PAGE 46)

21. 0 is asymptotically stable (attractor); 3 is unstable (repeller).
23. 2 is semi-stable.
25. -2 is unstable (repeller), 0 is semi-stable, 2 is asymptotically stable (attractor).
27. -1 is asymptotically stable (attractor); 0 is unstable (repeller).
39. (a) mg/k (b) $\sqrt{mg/k}$

EXERCISES 2.2 (PAGE 54)

1. $y = -\frac{1}{5}\cos 5x + c$ 3. $y = \frac{1}{3}e^{-3x} + c$
5. $y = cx^4$ 7. $-3e^{-2y} = 2e^{3x} + c$
9. $\frac{1}{3}x^3 \ln x - \frac{1}{9}x^3 = \frac{1}{2}y^2 + 2y + \ln|y| + c$
11. $4 \cos y = 2x + \sin 2x + c$
13. $(e^x + 1)^{-2} + 2(e^y + 1)^{-1} = c$
15. $S = ce^{kr}$ 17. $P = \dfrac{ce^t}{1 + ce^t}$
19. $(y + 3)^5 e^x = c(x + 4)^5 e^y$ 21. $y = \sin\left(\frac{1}{2}x^2 + c\right)$

23. $x = \tan\left(4t - \frac{3}{4}\pi\right)$

25. $y = \dfrac{e^{-(1+1/x)}}{x}$

27. $y = \frac{1}{2}x + \frac{1}{2}\sqrt{3}\sqrt{1 - x^2}$

29. (a) $y = 2,\ y = -2,\ y = 2\dfrac{3 - e^{4x-1}}{3 + e^{4x-1}}$

33. $y = 1$

35. $y = 1 + \frac{1}{10}\tan\left(\frac{1}{10}x\right)$

39. (a) $y = -\sqrt{x^2 + x - 1}$ **(c)** $\left(-\infty, -\frac{1}{2} - \frac{1}{2}\sqrt{5}\right)$

47. $y(x) = (4h/L^2)x^2 + a$

EXERCISES 2.3 (PAGE 65)

1. $y = ce^{5x},\ -\infty < x < \infty$

3. $y = \frac{1}{4}e^{3x} + ce^{-x},\ -\infty < x < \infty$

5. $y = \frac{1}{3} + ce^{-x^3},\ -\infty < x < \infty$

7. $y = x^{-1}\ln x + cx^{-1},\ x > 0$

9. $y = cx - x\cos x,\ x > 0$

11. $y = \frac{1}{7}x^3 - \frac{1}{5}x + cx^{-4},\ x > 0$

13. $y = \frac{1}{2}x^{-2}e^x + cx^{-2}e^{-x},\ x > 0$

15. $x = 2y^6 + cy^4,\ y > 0$

17. $y = \sin x + c\cos x,\ -\pi/2 < x < \pi/2$

19. $(x + 1)e^x y = x^2 + c,\ x > -1$

21. $(\sec\theta + \tan\theta)r = \theta - \cos\theta + c,\ -\pi/2 < x < \pi/2$

23. $y = e^{-3x} + cx^{-1}e^{-3x},\ 0 < x < \infty$

25. $y = x^{-1}e^x + (2 - e)x^{-1},\ x > 0$

27. $i = \dfrac{E}{R} + \left(i_0 - \dfrac{E}{R}\right)e^{-Rt/L},\ -\infty < t < \infty$

29. $(x + 1)y = x\ln x - x + 21,\ x > 0$

31. $y = \begin{cases} \frac{1}{2}(1 - e^{-2x}), & 0 \le x \le 3 \\ \frac{1}{2}(e^6 - 1)e^{-2x}, & x > 3 \end{cases}$

33. $y = \begin{cases} \frac{1}{2} + \frac{3}{2}e^{-x^2}, & 0 \le x < 1 \\ \left(\frac{1}{2}e + \frac{3}{2}\right)e^{-x^2}, & x \ge 1 \end{cases}$

35. $y = \begin{cases} 2x - 1 + 4e^{-2x}, & 0 \le x \le 1 \\ 4x^2\ln x + (1 + 4e^{-2})x^2, & x > 1 \end{cases}$

37. $y = e^{x^2-1} + \frac{1}{2}\sqrt{\pi}\,e^{x^2}(\mathrm{erf}(x) - \mathrm{erf}(1))$

47. $E(t) = E_0 e^{-(t-4)/RC}$

EXERCISES 2.4 (PAGE 73)

1. $x^2 - x + \frac{3}{2}y^2 + 7y = c$ **3.** $\frac{5}{2}x^2 + 4xy - 2y^4 = c$

5. $x^2 y^2 - 3x + 4y = c$ **7.** not exact

9. $xy^3 + y^2\cos x - \frac{1}{2}x^2 = c$

11. not exact

13. $xy - 2xe^x + 2e^x - 2x^3 = c$

15. $x^3 y^3 - \tan^{-1} 3x = c$

17. $-\ln|\cos x| + \cos x\sin y = c$

19. $t^4 y - 5t^3 - ty + y^3 = c$

21. $\frac{1}{3}x^3 + x^2 y + xy^2 - y = \frac{4}{3}$

23. $4ty + t^2 - 5t + 3y^2 - y = 8$

25. $y^2\sin x - x^3 y - x^2 + y\ln y - y = 0$

27. $k = 10$ **29.** $x^2 y^2\cos x = c$

31. $x^2 y^2 + x^3 = c$ **33.** $3x^2 y^3 + y^4 = c$

35. $-2ye^{3x} + \frac{10}{3}e^{3x} + x = c$

37. $e^{y^2}(x^2 + 4) = 20$

39. (c) $y_1(x) = -x^2 - \sqrt{x^4 - x^3 + 4}$

$\qquad y_2(x) = -x^2 + \sqrt{x^4 - x^3 + 4}$

45. (a) $v(x) = 8\sqrt{\dfrac{x}{3} - \dfrac{9}{x^2}}$ **(b)** 12.7 ft/s

EXERCISES 2.5 (PAGE 78)

1. $y + x\ln|x| = cx$

3. $(x - y)\ln|x - y| = y + c(x - y)$

5. $x + y\ln|x| = cy$

7. $\ln(x^2 + y^2) + 2\tan^{-1}(y/x) = c$

9. $4x = y(\ln|y| - c)^2$ **11.** $y^3 + 3x^3\ln|x| = 8x^3$

13. $\ln|x| = e^{y/x} - 1$ **15.** $y^3 = 1 + cx^{-3}$

17. $y^{-3} = x + \frac{1}{3} + ce^{3x}$ **19.** $e^{t/y} = ct$

21. $y^{-3} = -\frac{9}{5}x^{-1} + \frac{49}{5}x^{-6}$

23. $y = -x - 1 + \tan(x + c)$

25. $2y - 2x + \sin 2(x + y) = c$

27. $4(y - 2x + 3) = (x + c)^2$

29. $-\cot(x + y) + \csc(x + y) = x + \sqrt{2} - 1$

35. (b) $y = \dfrac{2}{x} + \left(-\frac{1}{4}x + cx^{-3}\right)^{-1}$

EXERCISES 2.6 (PAGE 84)

1. $y_2 = 2.9800,\quad y_4 = 3.1151$

3. $y_{10} = 2.5937,\quad y_{20} = 2.6533;\ y = e^x$

5. $y_5 = 0.4198,\quad y_{10} = 0.4124$

7. $y_5 = 0.5639,\quad y_{10} = 0.5565$

9. $y_5 = 1.2194,\quad y_{10} = 1.2696$

13. Euler: $y_{10} = 3.8191,\quad y_{20} = 5.9363$

$\qquad$ RK4: $y_{10} = 42.9931,\quad y_{20} = 84.0132$

CHAPTER 2 IN REVIEW (PAGE 84)

1. $-A/k$, a repeller for $k > 0$, an attractor for $k < 0$

3. $\dfrac{dy}{dx} = (y - 1)^2(y - 3)^3$

5. semi-stable for n even and unstable for n odd; semi-stable for n even and asymptotically stable for n odd.

9. $2x + \sin 2x = 2\ln(y^2 + 1) + c$

11. $(6x + 1)y^3 = -3x^3 + c$

13. $Q = ct^{-1} + \frac{1}{25}t^4(-1 + 5\ln t)$

15. $y = \frac{1}{4} + c(x^2 + 4)^{-4}$

17. $y = \csc x,\ \pi < x < 2\pi$

19. (b) $y = \frac{1}{4}\left(x + 2\sqrt{y_0} - x_0\right)^2,\ x \ge x_0 - 2\sqrt{y_0}$

EXERCISES 3.1 (PAGE 98)

1. 7.9 yr; 10 yr   **3.** 760; approximately 11 persons/yr

5. 11 h   **7.** 136.5 h

9. $I(15) = 0.00098 I_0$ or approximately 0.1% of I_0

11. 15,600 years

13. $T(1) = 36.76°$ F; approximately 3.06 min

15. approximately 82.1 s; approximately 145.7 s

17. $390°$

19. $A(t) = 200 - 170e^{-t/50}$

21. $A(t) = 1000 - 1000e^{-t/100}$

23. $A(t) = 1000 - 10t - \frac{1}{10}(100 - t)^2$;   100 min

25. 64.38 lb

27. $i(t) = \frac{3}{5} - \frac{3}{5}e^{-500t}$; $i \to \frac{3}{5}$ as $t \to \infty$

29. $q(t) = \frac{1}{100} - \frac{1}{100}e^{-50t}$; $i(t) = \frac{1}{2}e^{-50t}$

31. $i(t) = \begin{cases} 60 - 60e^{-t/10}, & 0 \le t \le 20 \\ 60(e^2 - 1)e^{-t/10}, & t > 20 \end{cases}$

33. (a) $v(t) = \dfrac{mg}{k} + \left(v_0 - \dfrac{mg}{k}\right)e^{-kt/m}$

  (b) $v \to \dfrac{mg}{k}$ as $t \to \infty$

  (c) $s(t) = \dfrac{mg}{k}t - \dfrac{m}{k}\left(v_0 - \dfrac{mg}{k}\right)e^{-kt/m}$
  $+ \dfrac{m}{k}\left(v_0 - \dfrac{mg}{k}\right) + s_0$

37. (a) $v(t) = \dfrac{\rho g}{4k}\left(\dfrac{k}{\rho}t + r_0\right) - \dfrac{\rho g r_0}{4k}\left(\dfrac{r_0}{\dfrac{k}{\rho}t + r_0}\right)^3$

  (c) $33\frac{1}{3}$ min

39. (a) $P(t) = P_0 e^{(k_1 - k_2)t}$

41. (a) As $t \to \infty$, $x(t) \to r/k$.

  (b) $x(t) = r/k - (r/k)e^{-kt}$; $(\ln 2)/k$

EXERCISES 3.2 (PAGE 108)

1. (a) $N = 2000$

  (b) $N(t) = \dfrac{2000\,e^t}{1999 + e^t}$; $N(10) = 1834$

3. 1,000,000;   52.9 mo

5. (b) $P(t) = \dfrac{4(P_0 - 1) - (P_0 - 4)e^{-3t}}{(P_0 - 1) - (P_0 - 4)e^{-3t}}$

  (c) For $0 < P_0 < 1$, time of extinction is
  $t = -\dfrac{1}{3}\ln\dfrac{4(P_0 - 1)}{P_0 - 4}$.

7. $P(t) = \dfrac{5}{2} + \dfrac{\sqrt{3}}{2}\tan\left[-\dfrac{\sqrt{3}}{2}t + \tan^{-1}\left(\dfrac{2P_0 - 5}{\sqrt{3}}\right)\right]$;

  time of extinction is
  $t = \dfrac{2}{\sqrt{3}}\left[\tan^{-1}\dfrac{5}{\sqrt{3}} + \tan^{-1}\left(\dfrac{2P_0 - 5}{\sqrt{3}}\right)\right]$

9. 29.3 g; $X \to 60$ as $t \to \infty$; 0 g of A and 30 g of B

11. (a) $h(t) = \left(\sqrt{H} - \dfrac{4A_h}{A_w}t\right)^2$; I is $0 \le t \le \sqrt{H}A_w/4A_h$

  (b) $576\sqrt{10}$ s or 30.36 min

13. (a) approximately 858.65 s or 14.31 min

  (b) 243 s or 4.05 min

15. (a) $v(t) = \sqrt{\dfrac{mg}{k}}\tanh\left(\sqrt{\dfrac{kg}{m}}t + c_1\right)$

  where $c_1 = \tanh^{-1}\left(\sqrt{\dfrac{k}{mg}}v_0\right)$

(b) $\sqrt{\dfrac{mg}{k}}$

  (c) $s(t) = \dfrac{m}{k}\ln\cosh\left(\sqrt{\dfrac{kg}{m}}t + c_1\right) + c_2$,
  where $c_2 = -(m/k)\ln\cosh c_1$

17. (a) $m\dfrac{dv}{dt} = mg - kv^2 - \rho V$,

  where ρ is the weight density of water

  (b) $v(t) = \sqrt{\dfrac{mg - \rho V}{k}}\tanh\left(\dfrac{\sqrt{kmg - k\rho V}}{m}t + c_1\right)$

  (c) $\sqrt{\dfrac{mg - \rho V}{k}}$

19. (a) $W = 0$ and $W = 2$

  (b) $W(x) = 2\,\text{sech}^2(x - c_1)$

  (c) $W(x) = 2\,\text{sech}^2 x$

EXERCISES 3.3 (PAGE 117)

1. $x(t) = x_0\,e^{-\lambda_1 t}$

  $y(t) = \dfrac{x_0\lambda_1}{\lambda_2 - \lambda_1}(e^{-\lambda_1 t} - e^{-\lambda_2 t})$

  $z(t) = x_0\left(1 - \dfrac{\lambda_2}{\lambda_2 - \lambda_1}e^{-\lambda_1 t} + \dfrac{\lambda_1}{\lambda_2 - \lambda_1}e^{-\lambda_2 t}\right)$

3. 5, 20, 147 days. The time when $y(t)$ and $z(t)$ are the same makes sense because most of A and half of B are gone, so half of C should have been formed.

5. $\dfrac{dx_1}{dt} = 6 - \dfrac{2}{25}x_1 + \dfrac{1}{50}x_2$

  $\dfrac{dx_2}{dt} = \dfrac{2}{25}x_1 - \dfrac{2}{25}x_2$

7. (a) $\dfrac{dx_1}{dt} = 3\dfrac{x_2}{100 - t} - 2\dfrac{x_1}{100 + t}$

  $\dfrac{dx_2}{dt} = 2\dfrac{x_1}{100 + t} - 3\dfrac{x_2}{100 - t}$

  (b) $x_1(t) + x_2(t) = 150$; $x_2(30) \approx 47.4$ lb

13. $L_1\dfrac{di_2}{dt} + (R_1 + R_2)i_2 + R_1 i_3 = E(t)$

  $L_2\dfrac{di_3}{dt} + R_1 i_2 + (R_1 + R_3)i_3 = E(t)$

15. $i(0) = i_0$, $s(0) = n - i_0$, $r(0) = 0$

CHAPTER 3 IN REVIEW (PAGE 120)

1. $dP/dt = 0.15P$

3. $P(45) = 8.99$ billion

5. $x = 10\ln\left(\dfrac{10 + \sqrt{100 - y^2}}{y}\right) - \sqrt{100 - y^2}$

7. (a) $\dfrac{BT_1 + T_2}{1 + B}$, $\dfrac{BT_1 + T_2}{1 + B}$

  (b) $T(t) = \dfrac{BT_1 + T_2}{1 + B} + \dfrac{T_1 - T_2}{1 + B}e^{k(1+B)t}$

9. $i(t) = \begin{cases} 4t - \frac{1}{5}t^2, & 0 \le t < 10 \\ 20, & t \ge 10 \end{cases}$

11. $h(t) = \left(\sqrt{2} - 0.00000163t\right)^2$

13. No

15. $x(t) = \dfrac{\alpha c_1 e^{\alpha k_1 t}}{1 + c_1 e^{\alpha k_1 t}}, \quad y(t) = c_2(1 + c_1 e^{\alpha k_1 t})^{k_2/k_1}$

17. $x = -y + 1 + c_2 e^{-y}$

EXERCISES 4.1 (PAGE 137)

1. $y = \frac{1}{2} e^x - \frac{1}{2} e^{-x}$

3. $y = 3x - 4x \ln x$

9. $(-\infty, 2)$

11. (a) $y = \dfrac{e}{e^2 - 1}(e^x - e^{-x})$ **(b)** $y = \dfrac{\sinh x}{\sinh 1}$

13. (a) $y = e^x \cos x - e^x \sin x$
 (b) no solution
 (c) $y = e^x \cos x + e^{-\pi/2} e^x \sin x$
 (d) $y = c_2 e^x \sin x$, where c_2 is arbitrary

15. dependent   **17.** dependent

19. dependent   **21.** independent

23. The functions satisfy the DE and are linearly independent on the interval since $W(e^{-3x}, e^{4x}) = 7e^x \neq 0$; $y = c_1 e^{-3x} + c_2 e^{4x}$.

25. The functions satisfy the DE and are linearly independent on the interval since $W(e^x \cos 2x, e^x \sin 2x) = 2e^{2x} \neq 0$; $y = c_1 e^x \cos 2x + c_2 e^x \sin 2x$.

27. The functions satisfy the DE and are linearly independent on the interval since $W(x^3, x^4) = x^6 \neq 0$; $y = c_1 x^3 + c_2 x^4$.

29. The functions satisfy the DE and are linearly independent on the interval since $W(x, x^{-2}, x^{-2} \ln x) = 9x^{-6} \neq 0$; $y = c_1 x + c_2 x^{-2} + c_3 x^{-2} \ln x$.

35. (b) $y_p = x^2 + 3x + 3e^{2x}$; $\quad y_p = -2x^2 - 6x - \frac{1}{3} e^{2x}$

EXERCISES 4.2 (PAGE 141)

1. $y_2 = xe^{2x}$ **3.** $y_2 = \sin 4x$

5. $y_2 = \sinh x$ **7.** $y_2 = xe^{2x/3}$

9. $y_2 = x^4 \ln|x|$ **11.** $y_2 = 1$

13. $y_2 = x \cos(\ln x)$ **15.** $y_2 = x^2 + x + 2$

17. $y_2 = e^{2x}, y_p = -\frac{1}{2}$ **19.** $y_2 = e^{2x}, y_p = \frac{5}{2} e^{3x}$

EXERCISES 4.3 (PAGE 147)

1. $y = c_1 + c_2 e^{-x/4}$ **3.** $y = c_1 e^{3x} + c_2 e^{-2x}$

5. $y = c_1 e^{-4x} + c_2 xe^{-4x}$ **7.** $y = c_1 e^{2x/3} + c_2 e^{-x/4}$

9. $y = c_1 \cos 3x + c_2 \sin 3x$

11. $y = e^{2x}(c_1 \cos x + c_2 \sin x)$

13. $y = e^{-x/3}\left(c_1 \cos \frac{1}{3}\sqrt{2}\,x + c_2 \sin \frac{1}{3}\sqrt{2}\,x\right)$

15. $y = c_1 + c_2 e^{-x} + c_3 e^{5x}$

17. $y = c_1 e^{-x} + c_2 e^{3x} + c_3 xe^{3x}$

19. $u = c_1 e^t + e^{-t}(c_2 \cos t + c_3 \sin t)$

21. $y = c_1 e^{-x} + c_2 xe^{-x} + c_3 x^2 e^{-x}$

23. $y = c_1 + c_2 x + e^{-x/2}\left(c_3 \cos \frac{1}{2}\sqrt{3}\,x + c_4 \sin \frac{1}{2}\sqrt{3}\,x\right)$

25. $y = c_1 \cos \frac{1}{2}\sqrt{3}\,x + c_2 \sin \frac{1}{2}\sqrt{3}\,x$
 $\quad + c_3 x \cos \frac{1}{2}\sqrt{3}\,x + c_4 x \sin \frac{1}{2}\sqrt{3}\,x$

27. $u = c_1 e^r + c_2 re^r + c_3 e^{-r} + c_4 re^{-r} + c_5 e^{-5r}$

29. $y = 2 \cos 4x - \frac{1}{2} \sin 4x$

31. $y = -\frac{1}{3} e^{-(t-1)} + \frac{1}{3} e^{5(t-1)}$

33. $y = 0$ **35.** $y = \frac{5}{36} - \frac{5}{36} e^{-6x} + \frac{1}{6} xe^{-6x}$

37. $y = e^{5x} - xe^{5x}$ **39.** $y = 0$

41. $y = \dfrac{1}{2}\left(1 - \dfrac{5}{\sqrt{3}}\right)e^{-\sqrt{3}x} + \dfrac{1}{2}\left(1 + \dfrac{5}{\sqrt{3}}\right)e^{\sqrt{3}x};$

$\qquad y = \cosh \sqrt{3}x + \dfrac{5}{\sqrt{3}} \sinh \sqrt{3}x$

EXERCISES 4.4 (PAGE 158)

1. $y = c_1 e^{-x} + c_2 e^{-2x} + 3$

3. $y = c_1 e^{5x} + c_2 xe^{5x} + \frac{6}{5} x + \frac{3}{5}$

5. $y = c_1 e^{-2x} + c_2 xe^{-2x} + x^2 - 4x + \frac{7}{2}$

7. $y = c_1 \cos \sqrt{3}x + c_2 \sin \sqrt{3}x + \left(-4x^2 + 4x - \frac{4}{3}\right)e^{3x}$

9. $y = c_1 + c_2 e^x + 3x$

11. $y = c_1 e^{x/2} + c_2 xe^{x/2} + 12 + \frac{1}{2}x^2 e^{x/2}$

13. $y = c_1 \cos 2x + c_2 \sin 2x - \frac{3}{4}x \cos 2x$

15. $y = c_1 \cos x + c_2 \sin x - \frac{1}{2}x^2 \cos x + \frac{1}{2}x \sin x$

17. $y = c_1 e^x \cos 2x + c_2 e^x \sin 2x + \frac{1}{4} xe^x \sin 2x$

19. $y = c_1 e^{-x} + c_2 xe^{-x} - \frac{1}{2} \cos x$
 $\qquad + \frac{12}{25} \sin 2x - \frac{9}{25} \cos 2x$

21. $y = c_1 + c_2 x + c_3 e^{6x} - \frac{1}{4}x^2 - \frac{6}{37} \cos x + \frac{1}{37} \sin x$

23. $y = c_1 e^x + c_2 xe^x + c_3 x^2 e^x - x - 3 - \frac{2}{3}x^3 e^x$

25. $y = c_1 \cos x + c_2 \sin x + c_3 x \cos x + c_4 x \sin x$
 $\qquad + x^2 - 2x - 3$

27. $y = \sqrt{2} \sin 2x - \frac{1}{2}$

29. $y = -200 + 200 e^{-x/5} - 3x^2 + 30x$

31. $y = -10 e^{-2x} \cos x + 9 e^{-2x} \sin x + 7 e^{-4x}$

33. $x = \dfrac{F_0}{2\omega^2} \sin \omega t - \dfrac{F_0}{2\omega} t \cos \omega t$

35. $y = 11 - 11 e^x + 9 xe^x + 2x - 12 x^2 e^x + \frac{1}{2} e^{5x}$

37. $y = 6 \cos x - 6(\cot 1) \sin x + x^2 - 1$

39. $y = \dfrac{-4 \sin \sqrt{3}x}{\sin \sqrt{3} + \sqrt{3} \cos \sqrt{3}} + 2x$

41. $y = \begin{cases} \cos 2x + \frac{5}{6} \sin 2x + \frac{1}{3} \sin x, & 0 \le x \le \pi/2 \\ \frac{2}{3} \cos 2x + \frac{5}{6} \sin 2x, & x > \pi/2 \end{cases}$

EXERCISES 4.5 (PAGE 166)

1. $(3D - 2)(3D + 2)y = \sin x$

3. $(D - 6)(D + 2)y = x - 6$

5. $D(D + 5)^2 y = e^x$

7. $(D - 1)(D - 2)(D + 5)y = xe^{-x}$

9. $D(D + 2)(D^2 - 2D + 4)y = 4$

15. D^4 **17.** $D(D - 2)$

19. $D^2 + 4$ **21.** $D^3(D^2 + 16)$

23. $(D + 1)(D - 1)^3$ **25.** $D(D^2 - 2D + 5)$

27. $1, x, x^2, x^3, x^4$ **29.** $e^{6x}, e^{-3x/2}$

31. $\cos \sqrt{5}x, \sin \sqrt{5}x$ **33.** $1, e^{5x}, xe^{5x}$

35. $y = c_1 e^{-3x} + c_2 e^{3x} - 6$ **37.** $y = c_1 + c_2 e^{-x} + 3x$

39. $y = c_1 e^{-2x} + c_2 xe^{-2x} + \frac{1}{2}x + 1$

41. $y = c_1 + c_2 x + c_3 e^{-x} + \frac{2}{3}x^4 - \frac{8}{3}x^3 + 8x^2$

43. $y = c_1 e^{-3x} + c_2 e^{4x} + \frac{1}{7} x e^{4x}$

45. $y = c_1 e^{-x} + c_2 e^{3x} - e^x + 3$

47. $y = c_1 \cos 5x + c_2 \sin 5x + \frac{1}{4} \sin x$

49. $y = c_1 e^{-3x} + c_2 x e^{-3x} - \frac{1}{49} x e^{4x} + \frac{2}{343} e^{4x}$

51. $y = c_1 e^{-x} + c_2 e^x + \frac{1}{6} x^3 e^x - \frac{1}{4} x^2 e^x + \frac{1}{4} x e^x - 5$

53. $y = e^x (c_1 \cos 2x + c_2 \sin 2x) + \frac{1}{3} e^x \sin x$

55. $y = c_1 \cos 5x + c_2 \sin 5x - 2x \cos 5x$

57. $y = e^{-x/2} \left(c_1 \cos \dfrac{\sqrt{3}}{2} x + c_2 \sin \dfrac{\sqrt{3}}{2} x \right)$
$\qquad + \sin x + 2 \cos x - x \cos x$

59. $y = c_1 + c_2 x + c_3 e^{-8x} + \frac{11}{256} x^2 + \frac{7}{32} x^3 - \frac{1}{16} x^4$

61. $y = c_1 e^x + c_2 x e^x + c_3 x^2 e^x + \frac{1}{6} x^3 e^x + x - 13$

63. $y = c_1 + c_2 x + c_3 e^x + c_4 x e^x + \frac{1}{2} x^2 e^x + \frac{1}{2} x^2$

65. $y = \frac{5}{8} e^{-8x} + \frac{5}{8} e^{8x} - \frac{1}{4}$

67. $y = -\frac{41}{125} + \frac{41}{125} e^{5x} - \frac{1}{10} x^2 + \frac{9}{25} x$

69. $y = -\pi \cos x - \frac{11}{3} \sin x - \frac{8}{3} \cos 2x + 2x \cos x$

71. $y = 2 e^{2x} \cos 2x - \frac{3}{64} e^{2x} \sin 2x + \frac{1}{8} x^3 + \frac{3}{16} x^2 + \frac{3}{32} x$

EXERCISES 4.6 (PAGE 172)

1. $y = c_1 \cos x + c_2 \sin x + x \sin x + \cos x \ln|\cos x|$

3. $y = c_1 \cos x + c_2 \sin x - \frac{1}{2} x \cos x$

5. $y = c_1 \cos x + c_2 \sin x + \frac{1}{2} - \frac{1}{6} \cos 2x$

7. $y = c_1 e^x + c_2 e^{-x} + \frac{1}{2} x \sinh x$

9. $y = c_1 e^{2x} + c_2 e^{-2x} + \frac{1}{4} \left(e^{2x} \ln|x| - e^{-2x} \displaystyle\int_{x_0}^{x} \frac{e^{4t}}{t}\, dt \right),$
$\qquad x_0 > 0$

11. $y = c_1 e^{-x} + c_2 e^{-2x} + (e^{-x} + e^{-2x}) \ln(1 + e^x)$

13. $y = c_1 e^{-2x} + c_2 e^{-x} - e^{-2x} \sin e^x$

15. $y = c_1 e^{-t} + c_2 t e^{-t} + \frac{1}{2} t^2 e^{-t} \ln t - \frac{3}{4} t^2 e^{-t}$

17. $y = c_1 e^x \sin x + c_2 e^x \cos x + \frac{1}{3} x e^x \sin x$
$\qquad + \frac{1}{3} e^x \cos x \ln|\cos x|$

19. $y = \frac{1}{4} e^{-x/2} + \frac{3}{4} e^{x/2} + \frac{1}{8} x^2 e^{x/2} - \frac{1}{4} x e^{x/2}$

21. $y = \frac{4}{9} e^{-4x} + \frac{25}{36} e^{2x} - \frac{1}{4} e^{-2x} + \frac{1}{9} e^{-x}$

23. $y = c_1 x^{-1/2} \cos x + c_2 x^{-1/2} \sin x + x^{-1/2}$

25. $y = c_1 + c_2 \cos x + c_3 \sin x - \ln|\cos x|$
$\qquad - \sin x \ln|\sec x + \tan x|$

EXERCISES 4.7 (PAGE 178)

1. $y = c_1 x^{-1} + c_2 x^2$

3. $y = c_1 + c_2 \ln x$

5. $y = c_1 \cos(2 \ln x) + c_2 \sin(2 \ln x)$

7. $y = c_1 x^{(2-\sqrt{6})} + c_2 x^{(2+\sqrt{6})}$

9. $y = c_1 \cos\left(\frac{1}{5} \ln x\right) + c_2 \sin\left(\frac{1}{5} \ln x\right)$

11. $y = c_1 x^{-2} + c_2 x^{-2} \ln x$

13. $y = x^{-1/2}\left[c_1 \cos\left(\frac{1}{6} \sqrt{3} \ln x\right) + c_2 \sin\left(\frac{1}{6} \sqrt{3} \ln x\right) \right]$

15. $y = c_1 x^3 + c_2 \cos(\sqrt{2} \ln x) + c_3 \sin(\sqrt{2} \ln x)$

17. $y = c_1 + c_2 x + c_3 x^2 + c_4 x^{-3}$

19. $y = c_1 + c_2 x^5 + \frac{1}{5} x^5 \ln x$

21. $y = c_1 x + c_2 x \ln x + x (\ln x)^2$

23. $y = c_1 x^{-1} + c_2 x - \ln x$

25. $y = 2 - 2 x^{-2}$ **27.** $y = \cos(\ln x) + 2 \sin(\ln x)$

29. $y = \frac{3}{4} - \ln x + \frac{1}{4} x^2$ **31.** $y = c_1 x^{-10} + c_2 x^2$

33. $y = c_1 x^{-1} + c_2 x^{-8} + \frac{1}{30} x^2$

35. $y = x^2 [c_1 \cos(3 \ln x) + c_2 \sin(3 \ln x)] + \frac{4}{13} + \frac{3}{10} x$

37. $y = 2(-x)^{1/2} - 5(-x)^{1/2} \ln(-x),\ x < 0$

EXERCISES 4.8 (PAGE 182)

1. $x = c_1 e^t + c_2 t e^t$
$\quad y = (c_1 - c_2) e^t + c_2 t e^t$

3. $x = c_1 \cos t + c_2 \sin t + t + 1$
$\quad y = c_1 \sin t - c_2 \cos t + t - 1$

5. $x = \frac{1}{2} c_1 \sin t + \frac{1}{2} c_2 \cos t - 2 c_3 \sin \sqrt{6} t - 2 c_4 \cos \sqrt{6} t$
$\quad y = c_1 \sin t + c_2 \cos t + c_3 \sin \sqrt{6} t + c_4 \cos \sqrt{6} t$

7. $x = c_1 e^{2t} + c_2 e^{-2t} + c_3 \sin 2t + c_4 \cos 2t + \frac{1}{5} e^t$
$\quad y = c_1 e^{2t} + c_2 e^{-2t} - c_3 \sin 2t - c_4 \cos 2t - \frac{1}{5} e^t$

9. $x = c_1 - c_2 \cos t + c_3 \sin t + \frac{17}{15} e^{3t}$
$\quad y = c_1 + c_2 \sin t + c_3 \cos t - \frac{4}{15} e^{3t}$

11. $x = c_1 e^t + c_2 e^{-t/2} \cos \frac{1}{2} \sqrt{3} t + c_3 e^{-t/2} \sin \frac{1}{2} \sqrt{3} t$
$\quad y = \left(-\frac{3}{2} c_2 - \frac{1}{2} \sqrt{3} c_3 \right) e^{-t/2} \cos \frac{1}{2} \sqrt{3} t$
$\qquad + \left(\frac{1}{2} \sqrt{3} c_2 - \frac{3}{2} c_3 \right) e^{-t/2} \sin \frac{1}{2} \sqrt{3} t$

13. $x = c_1 e^{4t} + \frac{4}{3} e^t$
$\quad y = -\frac{3}{4} c_1 e^{4t} + c_2 + 5 e^t$

15. $x = c_1 + c_2 t + c_3 e^t + c_4 e^{-t} - \frac{1}{2} t^2$
$\quad y = (c_1 - c_2 + 2) + (c_2 + 1) t + c_4 e^{-t} - \frac{1}{2} t^2$

17. $x = c_1 e^t + c_2 e^{-t/2} \sin \frac{1}{2} \sqrt{3} t + c_3 e^{-t/2} \cos \frac{1}{2} \sqrt{3} t$
$\quad y = c_1 e^t + \left(-\frac{1}{2} c_2 - \frac{1}{2} \sqrt{3} c_3 \right) e^{-t/2} \sin \frac{1}{2} \sqrt{3} t$
$\qquad + \left(\frac{1}{2} \sqrt{3} c_2 - \frac{1}{2} c_3 \right) e^{-t/2} \cos \frac{1}{2} \sqrt{3} t$
$\quad z = c_1 e^t + \left(-\frac{1}{2} c_2 + \frac{1}{2} \sqrt{3} c_3 \right) e^{-t/2} \sin \frac{1}{2} \sqrt{3} t$
$\qquad + \left(-\frac{1}{2} \sqrt{3} c_2 - \frac{1}{2} c_3 \right) e^{-t/2} \cos \frac{1}{2} \sqrt{3} t$

19. $x = -6 c_1 e^{-t} - 3 c_2 e^{-2t} + 2 c_3 e^{3t}$
$\quad y = c_1 e^{-t} + c_2 e^{-2t} + c_3 e^{3t}$
$\quad z = 5 c_1 e^{-t} + c_2 e^{-2t} + c_3 e^{3t}$

21. $x = e^{-3t+3} - t e^{-3t+3}$
$\quad y = -e^{-3t+3} + 2 t e^{-3t+3}$

23. $m x'' = 0$
$\quad m y'' = -mg;$
$\quad x = c_1 t + c_2$
$\quad y = -\frac{1}{2} g t^2 + c_3 t + c_4$

EXERCISES 4.9 (PAGE 188)

3. $y = \ln|\cos(c_1 - x)| + c_2$

5. $y = \dfrac{1}{c_1^2} \ln|c_1 x + 1| - \dfrac{1}{c_1} x + c_2$

7. $\frac{1}{3} y^3 - c_1 y = x + c_2$

9. $y = \tan\left(\frac{1}{4} \pi - \frac{1}{2} x\right),\ -\frac{1}{2} \pi < x < \frac{3}{2} \pi$

11. $y = -\dfrac{1}{c_1} \sqrt{1 - c_1^2 x^2} + c_2$

13. $y = 1 + x + \frac{1}{2} x^2 + \frac{1}{2} x^3 + \frac{1}{6} x^4 + \frac{1}{10} x^5 + \cdots$

15. $y = 1 + x - \frac{1}{2} x^2 + \frac{2}{3} x^3 - \frac{1}{4} x^4 + \frac{7}{60} x^5 + \cdots$

17. $y = -\sqrt{1 - x^2}$

CHAPTER 4 IN REVIEW (PAGE 189)

1. $y = 0$

3. false

5. $(-\infty, 0); (0, \infty)$

7. $y = c_1 e^{3x} + c_2 e^{-5x} + c_3 x e^{-5x} + c_4 e^x + c_5 x e^x + c_6 x^2 e^x$;
$y = c_1 x^3 + c_2 x^{-5} + c_3 x^{-5} \ln x + c_4 x + c_5 x \ln x$
$\quad + c_6 x (\ln x)^2$

9. $y = c_1 e^{(1+\sqrt{3})x} + c_2 e^{(1-\sqrt{3})x}$

11. $y = c_1 + c_2 e^{-5x} + c_3 x e^{-5x}$

13. $y = c_1 e^{-x/3} + e^{-3x/2}\left(c_2 \cos \frac{1}{2}\sqrt{7}x + c_3 \sin \frac{1}{2}\sqrt{7}x\right)$

15. $y = e^{3x/2}\left(c_2 \cos \frac{1}{2}\sqrt{11}x + c_3 \sin \frac{1}{2}\sqrt{11}x\right) + \frac{4}{5}x^3 + \frac{36}{25}x^2$
$\quad + \frac{46}{125}x - \frac{222}{625}$

17. $y = c_1 + c_2 e^{2x} + c_3 e^{3x} + \frac{1}{5}\sin x - \frac{1}{5}\cos x + \frac{4}{3}x$

19. $y = e^x(c_1 \cos x + c_2 \sin x)$
$\quad - e^x \cos x \ln|\sec x + \tan x|$

21. $y = c_1 x^{-1/3} + c_2 x^{1/2}$

23. $y = c_1 x^2 + c_2 x^3 + x^4 - x^2 \ln x$

25. (a) $y = c_1 \cos \omega x + c_2 \sin \omega x + A \cos \alpha x$
$\quad + B \sin \alpha x, \quad \omega \neq \alpha;$
$y = c_1 \cos \omega x + c_2 \sin \omega x + A x \cos \omega x$
$\quad + B x \sin \omega x, \quad \omega = \alpha$

(b) $y = c_1 e^{-\omega x} + c_2 e^{\omega x} + A e^{\alpha x}, \omega \neq \alpha;$
$y = c_1 e^{-\omega x} + c_2 e^{\omega x} + A x e^{\omega x}, \omega = \alpha$

27. (a) $y = c_1 \cosh x + c_2 \sinh x + c_3 x \cosh x$
$\quad + c_4 x \sinh x$

(b) $y_p = A x^2 \cosh x + B x^2 \sinh x$

29. $y = e^{x-\pi} \cos x$

31. $y = \frac{13}{4} e^x - \frac{5}{4} e^{-x} - x - \frac{1}{2} \sin x$

33. $y = x^2 + 4$

37. $x = -c_1 e^t - \frac{3}{2} c_2 e^{2t} + \frac{5}{2}$
$y = c_1 e^t + c_2 e^{2t} - 3$

39. $x = c_1 e^t + c_2 e^{5t} + t e^t$
$y = -c_1 e^t + 3 c_2 e^{5t} - t e^t + 2 e^t$

EXERCISES 5.1 (PAGE 207)

1. $\dfrac{\sqrt{2}\,\pi}{8}$

3. $x(t) = -\frac{1}{4} \cos 4\sqrt{6}\, t$

5. (a) $x\left(\frac{\pi}{12}\right) = -\frac{1}{4}; x\left(\frac{\pi}{8}\right) = -\frac{1}{2}; x\left(\frac{\pi}{6}\right) = -\frac{1}{4};$
$x\left(\frac{\pi}{4}\right) = \frac{1}{2}; x\left(\frac{9\pi}{32}\right) = \frac{\sqrt{2}}{4}$

(b) 4 ft/s; downward

(c) $t = \dfrac{(2n+1)\pi}{16}, n = 0, 1, 2, \ldots$

7. (a) the 20-kg mass

(b) the 20-kg mass; the 50-kg mass

(c) $t = n\pi, n = 0, 1, 2, \ldots$; at the equilibrium position; the 50-kg mass is moving upward whereas the 20-kg mass is moving upward when n is even and downward when n is odd.

9. $x(t) = \frac{1}{2} \cos 2t + \frac{3}{4} \sin 2t = \dfrac{\sqrt{13}}{4} \sin(2t + 0.5880)$

11. (a) $x(t) = -\frac{2}{3} \cos 10t + \frac{1}{2} \sin 10t$
$= \frac{5}{6} \sin(10t - 0.927)$

(b) $\dfrac{5}{6}$ ft; $\dfrac{\pi}{5}$

(c) 15 cycles

(d) 0.721 s

(e) $\dfrac{(2n+1)\pi}{20} + 0.0927, n = 0, 1, 2, \ldots$

(f) $x(3) = -0.597$ ft   **(g)** $x'(3) = -5.814$ ft/s

(h) $x''(3) = 59.702$ ft/s^2 **(i)** $\pm 8\frac{1}{3}$ ft/s

(j) $0.1451 + \dfrac{n\pi}{5}; 0.3545 + \dfrac{n\pi}{5}, n = 0, 1, 2, \ldots$

(k) $0.3545 + \dfrac{n\pi}{5}, n = 0, 1, 2, \ldots$

13. 120 lb/ft; $x(t) = \dfrac{\sqrt{3}}{12} \sin 8\sqrt{3}\, t$

17. (a) above   **(b)** heading upward

19. (a) below   **(b)** heading upward

21. $\frac{1}{4}$ s; $\frac{1}{2}$ s, $x\left(\frac{1}{2}\right) = e^{-2}$; that is, the weight is approximately 0.14 ft below the equilibrium position.

23. (a) $x(t) = \frac{4}{3} e^{-2t} - \frac{1}{3} e^{-8t}$

(b) $x(t) = -\frac{2}{3} e^{-2t} + \frac{5}{3} e^{-8t}$

25. (a) $x(t) = e^{-2t}\left(-\cos 4t - \frac{1}{2}\sin 4t\right)$

(b) $x(t) = \dfrac{\sqrt{5}}{2} e^{-2t} \sin(4t + 4.249)$

(c) $t = 1.294$ s

27. (a) $\beta > \frac{5}{2}$ **(b)** $\beta = \frac{5}{2}$ **(c)** $0 < \beta < \frac{5}{2}$

29. $x(t) = e^{-t/2}\left(-\dfrac{4}{3}\cos\dfrac{\sqrt{47}}{2}t - \dfrac{64}{3\sqrt{47}}\sin\dfrac{\sqrt{47}}{2}t\right)$
$\quad + \dfrac{10}{3}(\cos 3t + \sin 3t)$

31. $x(t) = \frac{1}{4} e^{-4t} + t e^{-4t} - \frac{1}{4}\cos 4t$

33. $x(t) = -\frac{1}{2}\cos 4t + \frac{9}{4}\sin 4t + \frac{1}{2} e^{-2t}\cos 4t$
$\quad - 2 e^{-2t}\sin 4t$

35. (a) $m\dfrac{d^2x}{dt^2} = -k(x - h) - \beta\dfrac{dx}{dt}$ or
$\dfrac{d^2x}{dt^2} + 2\lambda\dfrac{dx}{dt} + \omega^2 x = \omega^2 h(t)$,
where $2\lambda = \beta/m$ and $\omega^2 = k/m$

(b) $x(t) = e^{-2t}\left(-\frac{56}{13}\cos 2t - \frac{72}{13}\sin 2t\right) + \frac{56}{13}\cos t$
$\quad + \frac{32}{13}\sin t$

37. $x(t) = -\cos 2t - \frac{1}{8}\sin 2t + \frac{3}{4}t\sin 2t + \frac{5}{4}t\cos 2t$

39. (b) $\dfrac{F_0}{2\omega} t \sin \omega t$

45. 4.568 C; 0.0509 s

47. $q(t) = 10 - 10 e^{-3t}(\cos 3t + \sin 3t)$
$i(t) = 60 e^{-3t} \sin 3t$; 10.432 C

49. $q_p = \frac{100}{13}\sin t + \frac{150}{13}\cos t$
$i_p = \frac{100}{13}\cos t - \frac{150}{13}\sin t$

53. $q(t) = -\frac{1}{2} e^{-10t}(\cos 10t + \sin 10t) + \frac{3}{2}; \frac{3}{2}$ C

57. $q(t) = \left(q_0 - \dfrac{E_0 C}{1 - \gamma^2 LC}\right)\cos\dfrac{t}{\sqrt{LC}}$
$\quad + \sqrt{LC}\, i_0 \sin\dfrac{t}{\sqrt{LC}} + \dfrac{E_0 C}{1 - \gamma^2 LC}\cos \gamma t$

$$i(t) = i_0 \cos \frac{t}{\sqrt{LC}} - \frac{1}{\sqrt{LC}} \left(q_0 - \frac{E_0 C}{1 - \gamma^2 LC} \right) \sin \frac{t}{\sqrt{LC}}$$
$$- \frac{E_0 C \gamma}{1 - \gamma^2 LC} \sin \gamma t$$

When $r = 1$,
$$y(x) = \frac{1}{2} \left[\frac{1}{2a} (x^2 - a^2) + \frac{1}{a} \ln \frac{a}{x} \right]$$
(c) The paths intersect when $r < 1$.

EXERCISES 5.2 (PAGE 217)

1. (a) $y(x) = \dfrac{w_0}{24EI} (6L^2x^2 - 4Lx^3 + x^4)$

3. (a) $y(x) = \dfrac{w_0}{48EI} (3L^2x^2 - 5Lx^3 + 2x^4)$

5. (a) $y(x) = \dfrac{w_0}{360EI} (7L^4x - 10L^2x^3 + 3x^5)$
   (c) $x \approx 0.51933$, $y_{max} \approx 0.234799$

7. $y(x) = -\dfrac{w_0 EI}{P^2} \cosh \sqrt{\dfrac{P}{EI}} x$

$$+ \left(\frac{w_0 EI}{P^2} \sinh \sqrt{\frac{P}{EI}} L - \frac{w_0 L\sqrt{EI}}{P\sqrt{P}} \right) \frac{\sinh \sqrt{\frac{P}{EI}} x}{\cosh \sqrt{\frac{P}{EI}} L}$$

$$+ \frac{w_0}{2P} x^2 + \frac{w_0 EI}{P^2}$$

9. $\lambda_n = n^2$, $n = 1, 2, 3, \ldots$; $y = \sin nx$

11. $\lambda_n = \dfrac{(2n-1)^2 \pi^2}{4L^2}$, $n = 1, 2, 3, \ldots$;

   $y = \cos \dfrac{(2n-1)\pi x}{2L}$

13. $\lambda_n = n^2$, $n = 0, 1, 2, \ldots$; $y = \cos nx$

15. $\lambda_n = \dfrac{n^2 \pi^2}{25}$, $n = 1, 2, 3, \ldots$; $y = e^{-x} \sin \dfrac{n\pi x}{5}$

17. $\lambda_n = n^2$, $n = 1, 2, 3, \ldots$; $y = \sin(n \ln x)$
19. $\lambda_n = n^4 \pi^4$, $n = 1, 2, 3, \ldots$; $y = \sin n\pi x$
21. $x = L/4$, $x = L/2$, $x = 3L/4$

25. $\omega_n = \dfrac{n\pi \sqrt{T}}{L\sqrt{\rho}}$, $n = 1, 2, 3, \ldots$; $y = \sin \dfrac{n\pi x}{L}$

27. $u(r) = \left(\dfrac{u_0 - u_1}{b - a} \right) \dfrac{ab}{r} + \dfrac{u_1 b - u_0 a}{b - a}$

EXERCISES 5.3 (PAGE 226)

7. $\dfrac{d^2 x}{dt^2} + x = 0$

15. (a) 5 ft   .(b) $4\sqrt{10}$ ft/s   (c) $0 \le t \le \frac{3}{8}\sqrt{10}$; 7.5 ft

17. (a) $xv \dfrac{dv}{dx} + v^2 = 32x$ (b) $\frac{1}{2}v^2 x^2 - \frac{32}{3} x^3 = -288$
   (c) approximately 0.66 s

19. (a) $xy'' = r\sqrt{1 + (y')^2}$.
   When $t = 0$, $x = a$, $y = 0$, $dy/dx = 0$.
   (b) When $r \ne 1$,
$$y(x) = \frac{a}{2} \left[\frac{1}{1 + r} \left(\frac{x}{a} \right)^{1+r} - \frac{1}{1 - r} \left(\frac{x}{a} \right)^{1-r} \right]$$
$$+ \frac{ar}{1 - r^2}$$

CHAPTER 5 IN REVIEW (PAGE 230)

1. 8 ft
3. $\frac{5}{4}$ m
5. False; there could be an impressed force driving the system.
7. overdamped
9. $y = 0$ since $\lambda = 8$ is not an eigenvalue
11. 14.4 lb   13. $x(t) = -\frac{2}{3} e^{-2t} + \frac{1}{3} e^{-4t}$
15. $0 < m \le 2$ 17. $\gamma = \frac{8}{3}\sqrt{3}$
19. $x(t) = e^{-4t} \left(\frac{26}{17} \cos 2\sqrt{2}\, t + \frac{28}{17} \sqrt{2} \sin 2\sqrt{2}\, t \right) + \frac{8}{17} e^{-t}$
21. (a) $q(t) = -\frac{1}{150} \sin 100t + \frac{1}{75} \sin 50t$
   (b) $i(t) = -\frac{2}{3} \cos 100t + \frac{2}{3} \cos 50t$
   (c) $t = \dfrac{n\pi}{50}$, $n = 0, 1, 2, \ldots$

25. $m \dfrac{d^2 x}{dt^2} + kx = 0$

EXERCISES 6.1 (PAGE 248)

1. $R = \frac{1}{2}$, $\left[-\frac{1}{2}, \frac{1}{2} \right)$
3. $R = 10$, $(-5, 15)$
5. $x - \frac{2}{3} x^3 + \frac{2}{15} x^5 - \frac{4}{315} x^7 + \cdots$
7. $1 + \frac{1}{2} x^2 + \frac{5}{24} x^4 + \frac{61}{720} x^6 + \cdots$, $(-\pi/2, \pi/2)$
9. $\displaystyle\sum_{k=3}^{\infty} (k - 2) c_{k-2} x^k$

11. $2c_1 + \displaystyle\sum_{k=1}^{\infty} [2(k + 1) c_{k+1} + 6c_{k-1}] x^k$

15. 5; 4

17. $y_1(x) = c_0 \left[1 + \dfrac{1}{3 \cdot 2} x^3 + \dfrac{1}{6 \cdot 5 \cdot 3 \cdot 2} x^6 \right.$

$$\left. + \dfrac{1}{9 \cdot 8 \cdot 6 \cdot 5 \cdot 3 \cdot 2} x^9 + \cdots \right]$$

$y_2(x) = c_1 \left[x + \dfrac{1}{4 \cdot 3} x^4 + \dfrac{1}{7 \cdot 6 \cdot 4 \cdot 3} x^7 \right.$

$$\left. + \dfrac{1}{10 \cdot 9 \cdot 7 \cdot 6 \cdot 4 \cdot 3} x^{10} + \cdots \right]$$

19. $y_1(x) = c_0 \left[1 - \dfrac{1}{2!} x^2 - \dfrac{3}{4!} x^4 - \dfrac{21}{6!} x^6 - \cdots \right]$

$y_2(x) = c_1 \left[x + \dfrac{1}{3!} x^3 + \dfrac{5}{5!} x^5 + \dfrac{45}{7!} x^7 + \cdots \right]$

21. $y_1(x) = c_0 \left[1 - \dfrac{1}{3!} x^3 + \dfrac{4^2}{6!} x^6 - \dfrac{7^2 \cdot 4^2}{9!} x^9 + \cdots \right]$

$y_2(x) = c_1 \left[x - \dfrac{2^2}{4!} x^4 + \dfrac{5^2 \cdot 2^2}{7!} x^7 \right.$

$$\left. - \dfrac{8^2 \cdot 5^2 \cdot 2^2}{10!} x^{10} + \cdots \right]$$

23. $y_1(x) = c_0; y_2(x) = c_1 \sum\limits_{n=1}^{\infty} \dfrac{1}{n} x^n$

25. $y_1(x) = c_0\left[1 + \frac{1}{2}x^2 + \frac{1}{6}x^3 + \frac{1}{6}x^4 + \cdots\right]$
$y_2(x) = c_1\left[x + \frac{1}{2}x^2 + \frac{1}{2}x^3 + \frac{1}{4}x^4 + \cdots\right]$

27. $y_1(x) = c_0\left[1 + \frac{1}{4}x^2 - \frac{7}{4\cdot4!}x^4 + \frac{23\cdot7}{8\cdot6!}x^6 - \cdots\right]$
$y_2(x) = c_1\left[x - \frac{1}{6}x^3 + \frac{14}{2\cdot5!}x^5 - \frac{34\cdot14}{4\cdot7!}x^7 - \cdots\right]$

29. $y(x) = -2\left[1 + \frac{1}{2!}x^2 + \frac{1}{3!}x^3 + \frac{1}{4!}x^4 + \cdots\right] + 6x$
$= 8x - 2e^x$

31. $y(x) = 3 - 12x^2 + 4x^4$

33. $y_1(x) = c_0\left[1 - \frac{1}{6}x^3 + \frac{1}{120}x^5 + \cdots\right]$
$y_2(x) = c_1\left[x - \frac{1}{12}x^4 + \frac{1}{180}x^6 + \cdots\right]$

EXERCISES 6.2 (PAGE 257)

1. $x = 0$, irregular singular point

3. $x = -3$, regular singular point;
$x = 3$, irregular singular point

5. $x = 0, 2i, -2i$, regular singular points

7. $x = -3, 2$, regular singular points

9. $x = 0$, irregular singular point;
$x = -5, 5, 2$, regular singular points

11. for $x = 1$: $p(x) = 5$, $q(x) = \dfrac{x(x-1)^2}{x+1}$
for $x = -1$: $p(x) = \dfrac{5(x+1)}{x-1}$, $q(x) = x^2 + x$

13. $r_1 = \frac{1}{3}, r_2 = -1$

15. $r_1 = \frac{3}{2}, r_2 = 0$
$y(x) = C_1x^{3/2}\left[1 - \frac{2}{5}x + \frac{2^2}{7\cdot5\cdot2}x^2\right.$
$\left. - \frac{2^3}{9\cdot7\cdot5\cdot3!}x^3 + \cdots\right]$
$+ C_2\left[1 + 2x - 2x^2 + \frac{2^3}{3\cdot3!}x^3 - \cdots\right]$

17. $r_1 = \frac{7}{8}, r_2 = 0$
$y(x) = C_1x^{7/8}\left[1 - \frac{2}{15}x + \frac{2^2}{23\cdot15\cdot2}x^2\right.$
$\left. - \frac{2^3}{31\cdot23\cdot15\cdot3!}x^3 + \cdots\right]$
$+ C_2\left[1 - 2x + \frac{2^2}{9\cdot2}x^2\right.$
$\left. - \frac{2^3}{17\cdot9\cdot3!}x^3 + \cdots\right]$

19. $r_1 = \frac{1}{3}, r_2 = 0$
$y(x) = C_1x^{1/3}\left[1 + \frac{1}{3}x + \frac{1}{3^2\cdot2}x^2\right.$
$\left. + \frac{1}{3^3\cdot3!}x^3 + \cdots\right]$
$+ C_2\left[1 + \frac{1}{2}x + \frac{1}{5\cdot2}x^2 + \frac{1}{8\cdot5\cdot2}x^3 + \cdots\right]$

21. $r_1 = \frac{5}{2}, r_2 = 0$
$y(x) = C_1x^{5/2}\left[1 + \frac{2\cdot2}{7}x + \frac{2^2\cdot3}{9\cdot7}x^2\right.$
$\left. + \frac{2^3\cdot4}{11\cdot9\cdot7}x^3 + \cdots\right]$
$+ C_2\left[1 + \frac{1}{3}x - \frac{1}{6}x^2 - \frac{1}{6}x^3 - \cdots\right]$

23. $r_1 = \frac{2}{3}, r_2 = \frac{1}{3}$
$y(x) = C_1x^{2/3}\left[1 - \frac{1}{2}x + \frac{5}{28}x^2 - \frac{1}{21}x^3 + \cdots\right]$
$+ C_2x^{1/3}\left[1 - \frac{1}{2}x + \frac{1}{5}x^2 - \frac{7}{120}x^3 + \cdots\right]$

25. $r_1 = 0, r_2 = -1$
$y(x) = C_1\sum\limits_{n=0}^{\infty} \dfrac{1}{(2n+1)!}x^{2n} + C_2x^{-1}\sum\limits_{n=0}^{\infty} \dfrac{1}{(2n)!}x^{2n}$
$= C_1x^{-1}\sum\limits_{n=0}^{\infty} \dfrac{1}{(2n+1)!}x^{2n+1} + C_2x^{-1}\sum\limits_{n=0}^{\infty} \dfrac{1}{(2n)!}x^{2n}$
$= \dfrac{1}{x}[C_1\sinh x + C_2\cosh x]$

27. $r_1 = 1, r_2 = 0$
$y(x) = C_1x + C_2\left[x\ln x - 1 + \frac{1}{2}x^2\right.$
$\left. + \frac{1}{12}x^3 + \frac{1}{72}x^4 + \cdots\right]$

29. $r_1 = r_2 = 0$
$y(x) = C_1y(x) + C_2\left[y_1(x)\ln x + y_1(x)\left(-x + \frac{1}{4}x^2\right.\right.$
$\left.\left. - \frac{1}{3\cdot3!}x^3 + \frac{1}{4\cdot4!}x^4 - \cdots\right)\right]$
where $y_1(x) = \sum\limits_{n=0}^{\infty} \dfrac{1}{n!}x^n = e^x$

33. (b) $y_1(t) = \sum\limits_{n=0}^{\infty} \dfrac{(-1)^n}{(2n+1)!}(\sqrt{\lambda}\,t)^{2n} = \dfrac{\sin(\sqrt{\lambda}\,t)}{\sqrt{\lambda}\,t}$
$y_2(t) = t^{-1}\sum\limits_{n=0}^{\infty} \dfrac{(-1)^n}{(2n)!}(\sqrt{\lambda}\,t)^{2n} = \dfrac{\cos(\sqrt{\lambda}\,t)}{t}$

(c) $y = c_1x\sin\left(\dfrac{\sqrt{\lambda}}{x}\right) + c_2x\cos\left(\dfrac{\sqrt{\lambda}}{x}\right)$

EXERCISES 6.3 (PAGE 268)

1. $y = c_1J_{1/3}(x) + c_2J_{-1/3}(x)$

3. $y = c_1J_{5/2}(x) + c_2J_{-5/2}(x)$

5. $y = c_1J_0(x) + c_2Y_0(x)$

7. $y = c_1J_2(3x) + c_2Y_2(3x)$

9. $y = c_1J_{2/3}(5x) + c_2J_{-2/3}(5x)$

11. $y = c_1x^{-1/2}J_{1/2}(\alpha x) + c_2x^{-1/2}J_{-1/2}(\alpha x)$

13. $y = x^{-1/2}\left[c_1J_1(4x^{1/2}) + c_2Y_1(4x^{1/2})\right]$

15. $y = x\left[c_1J_1(x) + c_2Y_1(x)\right]$

17. $y = x^{1/2}\left[c_1J_{3/2}(x) + c_2Y_{3/2}(x)\right]$

19. $y = x^{-1}\left[c_1J_{1/2}(\frac{1}{2}x^2) + c_2J_{-1/2}(\frac{1}{2}x^2)\right]$

23. $y = x^{1/2}\left[c_1J_{1/2}(x) + c_2J_{-1/2}(x)\right]$
$= C_1\sin x + C_2\cos x$

25. $y = x^{-1/2}\left[c_1J_{1/2}(\frac{1}{8}x^2) + c_2J_{-1/2}(\frac{1}{8}x^2)\right]$
$= C_1x^{-3/2}\sin(\frac{1}{8}x^2) + C_2x^{-3/2}\cos(\frac{1}{8}x^2)$

35. $y = c_1x^{1/2}J_{1/3}(\frac{2}{3}\alpha x^{3/2}) + c_2x^{1/2}J_{-1/3}(\frac{2}{3}\alpha x^{3/2})$

45. $P_2(x)$, $P_3(x)$, $P_4(x)$, and $P_5(x)$ are given in the text,
$P_6(x) = \frac{1}{16}(231x^6 - 315x^4 + 105x^2 - 5)$,
$P_7(x) = \frac{1}{16}(429x^7 - 693x^5 + 315x^3 - 35x)$

47. $\lambda_1 = 2$, $\lambda_2 = 12$, $\lambda_3 = 30$

CHAPTER 6 IN REVIEW (PAGE 271)

1. False

3. $-\frac{1}{2} \le x \le \frac{1}{2}$

7. $x^2(x-1)y'' + y' + y = 0$

9. $r_1 = \frac{1}{2}$, $r_2 = 0$
$y_1(x) = C_1 x^{1/2}\left[1 - \frac{1}{3}x + \frac{1}{30}x^2 - \frac{1}{630}x^3 + \cdots\right]$
$y_2(x) = C_2\left[1 - x + \frac{1}{6}x^2 - \frac{1}{90}x^3 + \cdots\right]$

11. $y_1(x) = c_0\left[1 + \frac{3}{2}x^2 + \frac{1}{2}x^3 + \frac{5}{8}x^4 + \cdots\right]$
$y_2(x) = c_1\left[x + \frac{1}{2}x^3 + \frac{1}{4}x^4 + \cdots\right]$

13. $r_1 = 3$, $r_2 = 0$
$y_1(x) = C_1 x^3\left[1 + \frac{1}{4}x + \frac{1}{20}x^2 + \frac{1}{120}x^3 + \cdots\right]$
$y_2(x) = C_2\left[1 + x + \frac{1}{2}x^2\right]$

15. $y(x) = 3\left[1 - x^2 + \frac{1}{3}x^4 - \frac{1}{15}x^6 + \cdots\right]$
$- 2\left[x - \frac{1}{2}x^3 + \frac{1}{8}x^5 - \frac{1}{48}x^7 + \cdots\right]$

17. $\frac{1}{6}\pi$

19. $x = 0$ is an ordinary point

21. $y(x) = c_0\left[1 - \frac{1}{3}x^3 + \frac{1}{3^2 \cdot 2!}x^6 - \frac{1}{3^3 \cdot 3!}x^9 + \cdots\right]$
$+ c_1\left[x - \frac{1}{4}x^4 + \frac{1}{4 \cdot 7}x^7\right.$
$- \frac{1}{4 \cdot 7 \cdot 10}x^{10} + \cdots\left] + \left[\frac{5}{2}x^2 - \frac{1}{3}x^3\right.\right.$
$\left.+ \frac{1}{3^2 \cdot 2!}x^6 - \frac{1}{3^3 \cdot 3!}x^9 + \cdots\right]$

EXERCISES 7.1 (PAGE 283)

1. $\frac{2}{s}e^{-s} - \frac{1}{s}$

3. $\frac{1}{s^2} - \frac{1}{s^2}e^{-s}$

5. $\frac{1 + e^{-\pi s}}{s^2 + 1}$

7. $\frac{1}{s}e^{-s} + \frac{1}{s^2}e^{-s}$

9. $\frac{1}{s} - \frac{1}{s^2} + \frac{1}{s^2}e^{-s}$

11. $\frac{e^7}{s - 1}$

13. $\frac{1}{(s - 4)^2}$

15. $\frac{1}{s^2 + 2s + 2}$

17. $\frac{s^2 - 1}{(s^2 + 1)^2}$

19. $\frac{48}{s^5}$

21. $\frac{4}{s^2} - \frac{10}{s}$

23. $\frac{2}{s^3} + \frac{6}{s^2} - \frac{3}{s}$

25. $\frac{6}{s^4} + \frac{6}{s^3} + \frac{3}{s^2} + \frac{1}{s}$

27. $\frac{1}{s} + \frac{1}{s - 4}$

29. $\frac{1}{s} + \frac{2}{s - 2} + \frac{1}{s - 4}$

31. $\frac{8}{s^3} - \frac{15}{s^2 + 9}$

33. Use $\sinh kt = \dfrac{e^{kt} - e^{-kt}}{2}$ to show that
$$\mathcal{L}\{\sinh kt\} = \frac{k}{s^2 - k^2}.$$

35. $\dfrac{1}{2(s - 2)} - \dfrac{1}{2s}$

37. $\dfrac{2}{s^2 + 16}$

39. $\dfrac{4\cos 5 + (\sin 5)s}{s^2 + 16}$

EXERCISES 7.2 (PAGE 292)

1. $\frac{1}{2}t^2$

3. $t - 2t^4$

5. $1 + 3t + \frac{3}{2}t^2 + \frac{1}{6}t^3$

7. $t - 1 + e^{2t}$

9. $\frac{1}{4}e^{-t/4}$

11. $\frac{5}{7}\sin 7t$

13. $\cos\dfrac{t}{2}$

15. $2\cos 3t - 2\sin 3t$

17. $\frac{1}{3} - \frac{1}{3}e^{-3t}$

19. $\frac{3}{4}e^{-3t} + \frac{1}{4}e^t$

21. $0.3e^{0.1t} + 0.6e^{-0.2t}$

23. $\frac{1}{2}e^{2t} - e^{3t} + \frac{1}{2}e^{6t}$

25. $\frac{1}{5} - \frac{1}{5}\cos\sqrt{5}t$

27. $-4 + 3e^{-t} + \cos t + 3\sin t$

29. $\frac{1}{3}\sin t - \frac{1}{6}\sin 2t$

31. $y = -1 + e^t$

33. $y = \frac{1}{10}e^{4t} + \frac{19}{10}e^{-6t}$

35. $y = \frac{4}{3}e^{-t} - \frac{1}{3}e^{-4t}$

37. $y = 10\cos t + 2\sin t - \sqrt{2}\sin\sqrt{2}t$

39. $y = -\frac{8}{9}e^{-t/2} + \frac{1}{9}e^{-2t} + \frac{5}{18}e^t + \frac{1}{2}e^{-t}$

41. $y = \frac{1}{4}e^{-t} - \frac{1}{4}e^{-3t}\cos 2t + \frac{1}{4}e^{-3t}\sin 2t$

EXERCISES 7.3 (PAGE 301)

1. $\dfrac{1}{(s - 10)^2}$

3. $\dfrac{6}{(s + 2)^4}$

5. $\dfrac{1}{(s - 2)^2} + \dfrac{2}{(s - 3)^2} + \dfrac{1}{(s - 4)^2}$

7. $\dfrac{3}{(s - 1)^2 + 9}$

9. $\dfrac{s}{s^2 + 25} - \dfrac{s - 1}{(s - 1)^2 + 25} + 3\dfrac{s + 4}{(s + 4)^2 + 25}$

11. $\frac{1}{2}t^2 e^{-2t}$

13. $e^{3t}\sin t$

15. $e^{-2t}\cos t - 2e^{-2t}\sin t$

17. $e^{-t} - te^{-t}$

19. $5 - t - 5e^{-t} - 4te^{-t} - \frac{3}{2}t^2 e^{-t}$

21. $y = te^{-4t} + 2e^{-4t}$

23. $y = e^{-t} + 2te^{-t}$

25. $y = \frac{1}{9}t + \frac{2}{27} - \frac{2}{27}e^{3t} + \frac{10}{9}te^{3t}$

27. $y = -\frac{3}{2}e^{3t}\sin 2t$

29. $y = \frac{1}{2} - \frac{1}{2}e^t\cos t + \frac{1}{2}e^t\sin t$

31. $y = (e + 1)te^{-t} + (e - 1)e^{-t}$

33. $x(t) = -\dfrac{3}{2}e^{-7t/2}\cos\dfrac{\sqrt{15}}{2}t - \dfrac{7\sqrt{15}}{10}e^{-7t/2}\sin\dfrac{\sqrt{15}}{2}t$

37. $\dfrac{e^{-s}}{s^2}$

39. $\dfrac{e^{-2s}}{s^2} + 2\dfrac{e^{-2s}}{s}$

41. $\dfrac{s}{s^2 + 4}e^{-\pi s}$

43. $\frac{1}{2}(t - 2)^2\,\mathcal{U}(t - 2)$

45. $-\sin t\,\mathcal{U}(t - \pi)$

47. $\mathcal{U}(t - 1) - e^{-(t-1)}\mathcal{U}(t - 1)$

49. (c)

51. (f)

53. (a)

55. $f(t) = 2 - 4\mathcal{U}(t - 3)$; $\mathcal{L}\{f(t)\} = \dfrac{2}{s} - \dfrac{4}{s}e^{-3s}$

57. $f(t) = t^2\,\mathcal{U}(t - 1)$; $\mathcal{L}\{f(t)\} = 2\dfrac{e^{-s}}{s^3} + 2\dfrac{e^{-s}}{s^2} + \dfrac{e^{-s}}{s}$

59. $f(t) = t - t\,\mathcal{U}(t - 2)$; $\mathcal{L}\{f(t)\} = \dfrac{1}{s^2} - \dfrac{e^{-2s}}{s^2} - 2\dfrac{e^{-2s}}{s}$

61. $f(t) = \mathscr{U}(t - a) - \mathscr{U}(t - b)$; $\mathscr{L}\{f(t)\} = \dfrac{e^{-as}}{s} - \dfrac{e^{-bs}}{s}$

63. $y = [5 - 5e^{-(t-1)}]\,\mathscr{U}(t - 1)$

65. $y = -\frac{1}{4} + \frac{1}{2}t + \frac{1}{4}e^{-2t} - \frac{1}{4}\mathscr{U}(t - 1)$
$\quad - \frac{1}{2}(t - 1)\,\mathscr{U}(t - 1) + \frac{1}{4}e^{-2(t-1)}\,\mathscr{U}(t - 1)$

67. $y = \cos 2t - \frac{1}{6}\sin 2(t - 2\pi)\,\mathscr{U}(t - 2\pi)$
$\quad + \frac{1}{3}\sin(t - 2\pi)\,\mathscr{U}(t - 2\pi)$

69. $y = \sin t + [1 - \cos(t - \pi)]\mathscr{U}(t - \pi)$
$\quad - [1 - \cos(t - 2\pi)]\,\mathscr{U}(t - 2\pi)$

71. $x(t) = \frac{5}{4}t - \frac{5}{16}\sin 4t - \frac{5}{4}(t - 5)\,\mathscr{U}(t - 5)$
$\quad + \frac{5}{16}\sin 4(t - 5)\,\mathscr{U}(t - 5) - \frac{25}{4}\mathscr{U}(t - 5)$
$\quad + \frac{25}{4}\cos 4(t - 5)\,\mathscr{U}(t - 5)$

73. $q(t) = \frac{2}{5}\mathscr{U}(t - 3) - \frac{2}{5}e^{-5(t-3)}\,\mathscr{U}(t - 3)$

75. (a) $i(t) = \dfrac{1}{101}e^{-10t} - \dfrac{1}{101}\cos t + \dfrac{10}{101}\sin t$
$\quad - \dfrac{10}{101}e^{-10(t-3\pi/2)}\,\mathscr{U}\!\left(t - \dfrac{3\pi}{2}\right)$
$\quad + \dfrac{10}{101}\cos\!\left(t - \dfrac{3\pi}{2}\right)\mathscr{U}\!\left(t - \dfrac{3\pi}{2}\right)$
$\quad + \dfrac{1}{101}\sin\!\left(t - \dfrac{3\pi}{2}\right)$

(b) $i_{max} \approx 0.1$ at $t \approx 1.6$, $i_{min} \approx -0.1$ at $t \approx 4.7$

77. $y(x) = \dfrac{w_0 L^2}{16EI}x^2 - \dfrac{w_0 L}{12EI}x^3 + \dfrac{w_0}{24EI}x^4$
$\quad - \dfrac{w_0}{24EI}\left(x - \dfrac{L}{2}\right)^4 \mathscr{U}\!\left(x - \dfrac{L}{2}\right)$

79. $y(x) = \dfrac{w_0 L^2}{48EI}x^2 - \dfrac{w_0 L}{24EI}x^3$
$\quad + \dfrac{w_0}{60EI}\left[\dfrac{5L}{2}x^4 - x^5 + \left(x - \dfrac{L}{2}\right)^5 \mathscr{U}\!\left(x - \dfrac{L}{2}\right)\right]$

81. (a) $\dfrac{dT}{dt} = k\big(T - 70 - 57.5t - (230 - 57.5t)\mathscr{U}(t - 4)\big)$

Exercises 7.4 (Page 312)

1. $\dfrac{1}{(s + 10)^2}$

3. $\dfrac{s^2 - 4}{(s^2 + 4)^2}$

5. $\dfrac{6s^2 + 2}{(s^2 - 1)^3}$

7. $\dfrac{12s - 24}{[(s - 2)^2 + 36]^2}$

9. $y = -\frac{1}{2}e^{-t} + \frac{1}{2}\cos t - \frac{1}{2}t\cos t + \frac{1}{2}t\sin t$

11. $y = 2\cos 3t + \frac{5}{3}\sin 3t + \frac{1}{6}t\sin 3t$

13. $y = \frac{1}{4}\sin 4t + \frac{1}{8}t\sin 4t$
$\quad - \frac{1}{8}(t - \pi)\sin 4(t - \pi)\mathscr{U}(t - \pi)$

17. $y = \frac{2}{3}t^3 + c_1 t^2$

19. $\dfrac{6}{s^5}$

21. $\dfrac{s - 1}{(s + 1)[(s - 1)^2 + 1]}$

23. $\dfrac{1}{s(s - 1)}$

25. $\dfrac{s + 1}{s[(s + 1)^2 + 1]}$

27. $\dfrac{1}{s^2(s - 1)}$

29. $\dfrac{3s^2 + 1}{s^2(s^2 + 1)^2}$

31. $e^t - 1$

33. $e^t - \frac{1}{2}t^2 - t - 1$

37. $f(t) = \sin t$

39. $f(t) = -\frac{1}{8}e^{-t} + \frac{1}{8}e^t + \frac{3}{4}te^t + \frac{1}{4}t^2 e^t$

41. $f(t) = e^{-t}$

43. $f(t) = \frac{3}{8}e^{2t} + \frac{1}{8}e^{-2t} + \frac{1}{2}\cos 2t + \frac{1}{4}\sin 2t$

45. $y(t) = \sin t - \frac{1}{2}t\sin t$

47. $i(t) = 100[e^{-10(t-1)} - e^{-20(t-1)}]\mathscr{U}(t - 1)$
$\quad - 100[e^{-10(t-2)} - e^{-20(t-2)}]\mathscr{U}(t - 2)$

49. $\dfrac{1 - e^{-as}}{s(1 + e^{-as})}$

51. $\dfrac{a}{s}\left(\dfrac{1}{bs} - \dfrac{1}{e^{bs} - 1}\right)$

53. $\dfrac{\coth(\pi s/2)}{s^2 + 1}$

55. $i(t) = \dfrac{1}{R}\left(1 - e^{-Rt/L}\right)$
$\quad + \dfrac{2}{R}\sum_{n=1}^{\infty}(-1)^n\left(1 - e^{-R(t-n)/L}\right)\mathscr{U}(t - n)$

57. $x(t) = 2(1 - e^{-t}\cos 3t - \frac{1}{3}e^{-t}\sin 3t)$
$\quad + 4\sum_{n=1}^{\infty}(-1)^n\big[1 - e^{-(t-n\pi)}\cos 3(t - n\pi)$
$\quad\quad - \frac{1}{3}e^{-(t-n\pi)}\sin 3(t - n\pi)\big]\mathscr{U}(t - n\pi)$

Exercises 7.5 (Page 318)

1. $y = e^{3(t-2)}\,\mathscr{U}(t - 2)$

3. $y = \sin t + \sin t\,\mathscr{U}(t - 2\pi)$

5. $y = -\cos t\,\mathscr{U}\!\left(t - \dfrac{\pi}{2}\right) + \cos t\,\mathscr{U}\!\left(t - \dfrac{3\pi}{2}\right)$

7. $y = \frac{1}{2} - \frac{1}{2}e^{-2t} + \left[\frac{1}{2} - \frac{1}{2}e^{-2(t-1)}\right]\mathscr{U}(t - 1)$

9. $y = e^{-2(t-2\pi)}\sin t\,\mathscr{U}(t - 2\pi)$

11. $y = e^{-2t}\cos 3t + \frac{2}{3}e^{-2t}\sin 3t$
$\quad + \frac{1}{3}e^{-2(t-\pi)}\sin 3(t - \pi)\,\mathscr{U}(t - \pi)$
$\quad + \frac{1}{3}e^{-2(t-3\pi)}\sin 3(t - 3\pi)\,\mathscr{U}(t - 3\pi)$

13. $y(x) = \begin{cases} \dfrac{P_0}{EI}\left(\dfrac{L}{4}x^2 - \dfrac{1}{6}x^3\right), & 0 \le x < \dfrac{L}{2} \\[2mm] \dfrac{P_0 L^2}{4EI}\left(\dfrac{1}{2}x - \dfrac{L}{12}\right), & \dfrac{L}{2} \le x \le L \end{cases}$

Exercises 7.6 (Page 322)

1. $x = -\frac{1}{3}e^{-2t} + \frac{1}{3}e^t$
$\quad y = \frac{1}{3}e^{-2t} + \frac{2}{3}e^t$

3. $x = -\cos 3t - \frac{5}{3}\sin 3t$
$\quad y = 2\cos 3t - \frac{7}{3}\sin 3t$

5. $x = -2e^{3t} + \frac{5}{2}e^{2t} - \frac{1}{2}$
$\quad y = \frac{8}{3}e^{3t} - \frac{5}{2}e^{2t} - \frac{1}{6}$

7. $x = -\frac{1}{2}t - \frac{3}{4}\sqrt{2}\sin\sqrt{2}\,t$
$\quad y = -\frac{1}{2}t + \frac{3}{4}\sqrt{2}\sin\sqrt{2}\,t$

9. $x = 8 + \dfrac{2}{3!}t^3 + \dfrac{1}{4!}t^4$
$\quad y = -\dfrac{2}{3!}t^3 + \dfrac{1}{4!}t^4$

11. $x = \frac{1}{2}t^2 + t + 1 - e^{-t}$
$\quad y = -\frac{1}{3} + \frac{1}{3}e^{-t} + \frac{1}{3}te^{-t}$

13. $x_1 = \dfrac{1}{5}\sin t + \dfrac{2\sqrt{6}}{15}\sin\sqrt{6}\,t + \dfrac{2}{5}\cos t - \dfrac{2}{5}\cos\sqrt{6}\,t$
$\quad x_2 = \dfrac{2}{5}\sin t - \dfrac{\sqrt{6}}{15}\sin\sqrt{6}\,t + \dfrac{4}{5}\cos t + \dfrac{1}{5}\cos\sqrt{6}\,t$

15. (b) $i_2 = \frac{100}{9} - \frac{100}{9} e^{-900t}$

$i_3 = \frac{80}{9} - \frac{80}{9} e^{-900t}$

(c) $i_1 = 20 - 20e^{-900t}$

17. $i_2 = -\frac{20}{13} e^{-2t} + \frac{375}{1469} e^{-15t} + \frac{145}{113} \cos t + \frac{85}{113} \sin t$

$i_3 = \frac{30}{13} e^{-2t} + \frac{250}{1469} e^{-15t} - \frac{280}{113} \cos t + \frac{810}{113} \sin t$

19. $i_1 = \frac{6}{5} - \frac{6}{5} e^{-100t} \cosh 50\sqrt{2}\, t - \frac{9\sqrt{2}}{5} e^{-100t} \sinh 50\sqrt{2}\, t$

$i_2 = \frac{6}{5} - \frac{6}{5} e^{-100t} \cosh 50\sqrt{2}\, t - \frac{6\sqrt{2}}{5} e^{-100t} \sinh 50\sqrt{2}\, t$

CHAPTER 7 IN REVIEW (PAGE 324)

1. $\dfrac{1}{s^2} - \dfrac{2}{s^2} e^{-s}$ **3.** false

5. true **7.** $\dfrac{1}{s + 7}$

9. $\dfrac{2}{s^2 + 4}$ **11.** $\dfrac{4s}{(s^2 + 4)^2}$

13. $\frac{1}{6} t^5$ **15.** $\frac{1}{2} t^2 e^{5t}$

17. $e^{5t} \cos 2t + \frac{5}{2} e^{5t} \sin 2t$

19. $\cos \pi(t - 1)\,\mathcal{U}(t - 1) + \sin \pi(t - 1)\,\mathcal{U}(t - 1)$

21. -5 **23.** $e^{-k(s-a)} F(s - a)$

25. $f(t)\,\mathcal{U}(t - t_0)$ **27.** $f(t - t_0)\,\mathcal{U}(t - t_0)$

29. $f(t) = t - (t - 1)\,\mathcal{U}(t - 1) - \mathcal{U}(t - 4)$;

$\mathcal{L}\{f(t)\} = \dfrac{1}{s^2} - \dfrac{1}{s^2} e^{-s} - \dfrac{1}{s} e^{-4s}$;

$\mathcal{L}\{e^t f(t)\} = \dfrac{1}{(s - 1)^2} - \dfrac{1}{(s - 1)^2} e^{-(s-1)}$

$\qquad\qquad - \dfrac{1}{s - 1} e^{-4(s-1)}$

31. $f(t) = 2 + (t - 2)\,\mathcal{U}(t - 2)$;

$\mathcal{L}\{f(t)\} = \dfrac{2}{s} + \dfrac{1}{s^2} e^{-2s}$;

$\mathcal{L}\{e^t f(t)\} = \dfrac{2}{s - 1} + \dfrac{1}{(s - 1)^2} e^{-2(s-1)}$

33. $y = 5te^t + \frac{1}{2} t^2 e^t$

35. $y = -\frac{6}{25} + \frac{1}{5} t + \frac{3}{2} e^{-t} - \frac{13}{50} e^{-5t} - \frac{4}{25} \mathcal{U}(t - 2)$

$\qquad - \frac{1}{5}(t - 2)\,\mathcal{U}(t - 2) + \frac{1}{4} e^{-(t-2)}\,\mathcal{U}(t - 2)$

$\qquad - \frac{9}{100} e^{-5(t-2)}\,\mathcal{U}(t - 2)$

37. $y = 1 + t + \frac{1}{2} t^2$

39. $x = -\frac{1}{4} + \frac{9}{8} e^{-2t} + \frac{1}{8} e^{2t}$

$y = t + \frac{9}{4} e^{-2t} - \frac{1}{4} e^{2t}$

41. $i(t) = -9 + 2t + 9e^{-t/5}$

43. $y(x) = \dfrac{w_0}{12 EIL}\left[-\dfrac{1}{5} x^5 + \dfrac{L}{2} x^4 - \dfrac{L^2}{2} x^3 + \dfrac{L^3}{4} x^2 \right.$

$\qquad\qquad \left. + \dfrac{1}{5}\left(x - \dfrac{L}{2} \right)^5 \mathcal{U}\left(x - \dfrac{L}{2} \right) \right]$

45. (a) $\theta_1(t) = \dfrac{\theta_0 + \psi_0}{2} \cos \omega t + \dfrac{\theta_0 - \psi_0}{2} \cos \sqrt{\omega^2 + 2Kt}$

$\theta_2(t) = \dfrac{\theta_0 + \psi_0}{2} \cos \omega t - \dfrac{\theta_0 - \psi_0}{2} \cos \sqrt{\omega^2 + 2Kt}$

EXERCISES 8.1 (PAGE 336)

1. $\mathbf{X}' = \begin{pmatrix} 3 & -5 \\ 4 & 8 \end{pmatrix} \mathbf{X}$, where $\mathbf{X} = \begin{pmatrix} x \\ y \end{pmatrix}$

3. $\mathbf{X}' = \begin{pmatrix} -3 & 4 & -9 \\ 6 & -1 & 0 \\ 10 & 4 & 3 \end{pmatrix} \mathbf{X}$, where $\mathbf{X} = \begin{pmatrix} x \\ y \\ z \end{pmatrix}$

5. $\mathbf{X}' = \begin{pmatrix} 1 & -1 & 1 \\ 2 & 1 & -1 \\ 1 & 1 & 1 \end{pmatrix} \mathbf{X} + \begin{pmatrix} 0 \\ -3t^2 \\ t^2 \end{pmatrix} + \begin{pmatrix} t \\ 0 \\ -t \end{pmatrix} + \begin{pmatrix} -1 \\ 0 \\ 2 \end{pmatrix}$,

where $\mathbf{X} = \begin{pmatrix} x \\ y \\ z \end{pmatrix}$

7. $\dfrac{dx}{dt} = 4x + 2y + e^t$

$\dfrac{dy}{dt} = -x + 3y - e^t$

9. $\dfrac{dx}{dt} = x - y + 2z + e^{-t} - 3t$

$\dfrac{dy}{dt} = 3x - 4y + z + 2e^{-t} + t$

$\dfrac{dz}{dt} = -2x + 5y + 6z + 2e^{-t} - t$

17. Yes; $W(\mathbf{X}_1, \mathbf{X}_2) = -2e^{-8t} \neq 0$ implies that $\mathbf{X}_1$ and $\mathbf{X}_2$ are linearly independent on $(-\infty, \infty)$.

19. No; $W(\mathbf{X}_1, \mathbf{X}_2, \mathbf{X}_3) = 0$ for every t. The solution vectors are linearly dependent on $(-\infty, \infty)$. Note that $\mathbf{X}_3 = 2\mathbf{X}_1 + \mathbf{X}_2$.

EXERCISES 8.2 (PAGE 351)

1. $\mathbf{X} = c_1 \begin{pmatrix} 1 \\ 2 \end{pmatrix} e^{5t} + c_2 \begin{pmatrix} 1 \\ -1 \end{pmatrix} e^{-t}$

3. $\mathbf{X} = c_1 \begin{pmatrix} 2 \\ 1 \end{pmatrix} e^{-3t} + c_2 \begin{pmatrix} 2 \\ 5 \end{pmatrix} e^t$

5. $\mathbf{X} = c_1 \begin{pmatrix} 5 \\ 2 \end{pmatrix} e^{8t} + c_2 \begin{pmatrix} 1 \\ 4 \end{pmatrix} e^{-10t}$

7. $\mathbf{X} = c_1 \begin{pmatrix} 1 \\ 0 \\ 0 \end{pmatrix} e^t + c_2 \begin{pmatrix} 2 \\ 3 \\ 1 \end{pmatrix} e^{2t} + c_3 \begin{pmatrix} 1 \\ 0 \\ 2 \end{pmatrix} e^{-t}$

9. $\mathbf{X} = c_1 \begin{pmatrix} -1 \\ 0 \\ 1 \end{pmatrix} e^{-t} + c_2 \begin{pmatrix} 1 \\ 4 \\ 3 \end{pmatrix} e^{3t} + c_3 \begin{pmatrix} 1 \\ -1 \\ 3 \end{pmatrix} e^{-2t}$

11. $\mathbf{X} = c_1 \begin{pmatrix} 4 \\ 0 \\ -1 \end{pmatrix} e^{-t} + c_2 \begin{pmatrix} -12 \\ 6 \\ 5 \end{pmatrix} e^{-t/2} + c_3 \begin{pmatrix} 4 \\ 2 \\ -1 \end{pmatrix} e^{-3t/2}$

13. $\mathbf{X} = 3 \begin{pmatrix} 1 \\ 1 \end{pmatrix} e^{t/2} + 2 \begin{pmatrix} 0 \\ 1 \end{pmatrix} e^{-t/2}$

19. $\mathbf{X} = c_1 \begin{pmatrix} 1 \\ 3 \end{pmatrix} + c_2 \left[\begin{pmatrix} 1 \\ 3 \end{pmatrix} t + \begin{pmatrix} \frac{1}{4} \\ -\frac{1}{4} \end{pmatrix} \right]$

21. $\mathbf{X} = c_1 \begin{pmatrix} 1 \\ 1 \end{pmatrix} e^{2t} + c_2 \left[\begin{pmatrix} 1 \\ 1 \end{pmatrix} te^{2t} + \begin{pmatrix} -\frac{1}{3} \\ 0 \end{pmatrix} e^{2t} \right]$

23. $\mathbf{X} = c_1 \begin{pmatrix} 1 \\ 1 \\ 1 \end{pmatrix} e^t + c_2 \begin{pmatrix} 1 \\ 1 \\ 0 \end{pmatrix} e^{2t} + c_3 \begin{pmatrix} 1 \\ 0 \\ 1 \end{pmatrix} e^{2t}$

25. $\mathbf{X} = c_1 \begin{pmatrix} -4 \\ -5 \\ 2 \end{pmatrix} + c_2 \begin{pmatrix} 2 \\ 0 \\ -1 \end{pmatrix} e^{5t}$

$+ c_3 \left[\begin{pmatrix} 2 \\ 0 \\ -1 \end{pmatrix} te^{5t} + \begin{pmatrix} -\frac{1}{2} \\ -\frac{1}{2} \\ -1 \end{pmatrix} e^{5t} \right]$

27. $\mathbf{X} = c_1 \begin{pmatrix} 0 \\ 1 \\ 1 \end{pmatrix} e^t + c_2 \left[\begin{pmatrix} 0 \\ 1 \\ 1 \end{pmatrix} te^t + \begin{pmatrix} 0 \\ 1 \\ 0 \end{pmatrix} e^t \right]$

$+ c_3 \left[\begin{pmatrix} 0 \\ 1 \\ 1 \end{pmatrix} \frac{t^2}{2} e^t + \begin{pmatrix} 0 \\ 1 \\ 0 \end{pmatrix} te^t + \begin{pmatrix} \frac{1}{2} \\ 0 \\ 0 \end{pmatrix} e^t \right]$

29. $\mathbf{X} = -7 \begin{pmatrix} 2 \\ 1 \end{pmatrix} e^{4t} + 13 \begin{pmatrix} 2t+1 \\ t+1 \end{pmatrix} e^{4t}$

31. Corresponding to the eigenvalue $\lambda_1 = 2$ of multiplicity five, the eigenvectors are

$$\mathbf{K}_1 = \begin{pmatrix} 1 \\ 0 \\ 0 \\ 0 \\ 0 \end{pmatrix}, \qquad \mathbf{K}_2 = \begin{pmatrix} 0 \\ 0 \\ 1 \\ 0 \\ 0 \end{pmatrix}, \qquad \mathbf{K}_3 = \begin{pmatrix} 0 \\ 0 \\ 0 \\ 1 \\ 0 \end{pmatrix}.$$

33. $\mathbf{X} = c_1 \begin{pmatrix} \cos t \\ 2\cos t + \sin t \end{pmatrix} e^{4t} + c_2 \begin{pmatrix} \sin t \\ 2\sin t - \cos t \end{pmatrix} e^{4t}$

35. $\mathbf{X} = c_1 \begin{pmatrix} \cos t \\ -\cos t - \sin t \end{pmatrix} e^{4t} + c_2 \begin{pmatrix} \sin t \\ -\sin t + \cos t \end{pmatrix} e^{4t}$

37. $\mathbf{X} = c_1 \begin{pmatrix} 5\cos 3t \\ 4\cos 3t + 3\sin 3t \end{pmatrix} + c_2 \begin{pmatrix} 5\sin 3t \\ 4\sin 3t - 3\cos 3t \end{pmatrix}$

39. $\mathbf{X} = c_1 \begin{pmatrix} 1 \\ 0 \\ 0 \end{pmatrix} + c_2 \begin{pmatrix} -\cos t \\ \cos t \\ \sin t \end{pmatrix} + c_3 \begin{pmatrix} \sin t \\ -\sin t \\ \cos t \end{pmatrix}$

41. $\mathbf{X} = c_1 \begin{pmatrix} 0 \\ 2 \\ 1 \end{pmatrix} e^t + c_2 \begin{pmatrix} \sin t \\ \cos t \\ \cos t \end{pmatrix} e^t + c_3 \begin{pmatrix} \cos t \\ -\sin t \\ -\sin t \end{pmatrix} e^t$

43. $\mathbf{X} = c_1 \begin{pmatrix} 28 \\ -5 \\ 25 \end{pmatrix} e^{2t} + c_2 \begin{pmatrix} 4\cos 3t - 3\sin 3t \\ -5\cos 3t \\ 0 \end{pmatrix} e^{-2t}$

$+ c_3 \begin{pmatrix} 3\cos 3t + 4\sin 3t \\ -5\sin 3t \\ 0 \end{pmatrix} e^{-2t}$

45. $\mathbf{X} = -\begin{pmatrix} 25 \\ -7 \\ 6 \end{pmatrix} e^t - \begin{pmatrix} \cos 5t - 5\sin 5t \\ \cos 5t \\ \cos 5t \end{pmatrix}$

$+ 6 \begin{pmatrix} 5\cos 5t + \sin 5t \\ \sin 5t \\ \sin 5t \end{pmatrix}$

1. $\mathbf{X} = c_1 \begin{pmatrix} -1 \\ 1 \end{pmatrix} e^{-t} + c_2 \begin{pmatrix} -3 \\ 1 \end{pmatrix} e^t + \begin{pmatrix} -1 \\ 3 \end{pmatrix}$

3. $\mathbf{X} = c_1 \begin{pmatrix} 1 \\ -1 \end{pmatrix} e^{-2t} + c_2 \begin{pmatrix} 1 \\ 1 \end{pmatrix} e^{4t} + \begin{pmatrix} -\frac{1}{4} \\ \frac{3}{4} \end{pmatrix} t^2$

$+ \begin{pmatrix} \frac{1}{4} \\ -\frac{1}{4} \end{pmatrix} t + \begin{pmatrix} -2 \\ \frac{3}{4} \end{pmatrix}$

5. $\mathbf{X} = c_1 \begin{pmatrix} 1 \\ -3 \end{pmatrix} e^{3t} + c_2 \begin{pmatrix} 1 \\ 9 \end{pmatrix} e^{7t} + \begin{pmatrix} \frac{55}{36} \\ -\frac{19}{4} \end{pmatrix} e^t$

7. $\mathbf{X} = c_1 \begin{pmatrix} 1 \\ 0 \\ 0 \end{pmatrix} e^t + c_2 \begin{pmatrix} 1 \\ 1 \\ 0 \end{pmatrix} e^{2t} + c_3 \begin{pmatrix} 1 \\ 2 \\ 2 \end{pmatrix} e^{5t} - \begin{pmatrix} \frac{3}{2} \\ \frac{7}{2} \\ 2 \end{pmatrix} e^{4t}$

9. $\mathbf{X} = 13 \begin{pmatrix} 1 \\ -1 \end{pmatrix} e^t + 2 \begin{pmatrix} -4 \\ 6 \end{pmatrix} e^{2t} + \begin{pmatrix} -9 \\ 6 \end{pmatrix}$

11. $\mathbf{X} = c_1 \begin{pmatrix} 1 \\ 1 \end{pmatrix} + c_2 \begin{pmatrix} 3 \\ 2 \end{pmatrix} e^t - \begin{pmatrix} 11 \\ 11 \end{pmatrix} t - \begin{pmatrix} 15 \\ 10 \end{pmatrix}$

13. $\mathbf{X} = c_1 \begin{pmatrix} 2 \\ 1 \end{pmatrix} e^{t/2} + c_2 \begin{pmatrix} 10 \\ 3 \end{pmatrix} e^{3t/2} - \begin{pmatrix} \frac{13}{2} \\ \frac{13}{4} \end{pmatrix} te^{t/2} - \begin{pmatrix} \frac{15}{2} \\ \frac{9}{4} \end{pmatrix} e^{t/2}$

15. $\mathbf{X} = c_1 \begin{pmatrix} 2 \\ 1 \end{pmatrix} e^t + c_2 \begin{pmatrix} 1 \\ 1 \end{pmatrix} e^{2t} + \begin{pmatrix} 3 \\ 3 \end{pmatrix} e^t + \begin{pmatrix} 4 \\ 2 \end{pmatrix} te^t$

17. $\mathbf{X} = c_1 \begin{pmatrix} 4 \\ 1 \end{pmatrix} e^{3t} + c_2 \begin{pmatrix} -2 \\ 1 \end{pmatrix} e^{-3t} + \begin{pmatrix} -12 \\ 0 \end{pmatrix} t - \begin{pmatrix} \frac{4}{3} \\ \frac{4}{3} \end{pmatrix}$

19. $\mathbf{X} = c_1 \begin{pmatrix} 1 \\ -1 \end{pmatrix} e^t + c_2 \begin{pmatrix} -t \\ \frac{1}{2} - t \end{pmatrix} e^t + \begin{pmatrix} \frac{1}{2} \\ -2 \end{pmatrix} e^{-t}$

21. $\mathbf{X} = c_1 \begin{pmatrix} \cos t \\ \sin t \end{pmatrix} + c_2 \begin{pmatrix} \sin t \\ -\cos t \end{pmatrix} + \begin{pmatrix} \cos t \\ \sin t \end{pmatrix} t$

$+ \begin{pmatrix} -\sin t \\ \cos t \end{pmatrix} \ln|\cos t|$

23. $\mathbf{X} = c_1 \begin{pmatrix} \cos t \\ \sin t \end{pmatrix} e^t + c_2 \begin{pmatrix} \sin t \\ -\cos t \end{pmatrix} e^t + \begin{pmatrix} \cos t \\ \sin t \end{pmatrix} te^t$

25. $\mathbf{X} = c_1 \begin{pmatrix} \cos t \\ -\sin t \end{pmatrix} + c_2 \begin{pmatrix} \sin t \\ \cos t \end{pmatrix} + \begin{pmatrix} \cos t \\ -\sin t \end{pmatrix} t$

$+ \begin{pmatrix} -\sin t \\ \sin t \tan t \end{pmatrix} - \begin{pmatrix} \sin t \\ \cos t \end{pmatrix} \ln|\cos t|$

27. $\mathbf{X} = c_1 \begin{pmatrix} 2\sin t \\ \cos t \end{pmatrix} e^t + c_2 \begin{pmatrix} 2\cos t \\ -\sin t \end{pmatrix} e^t + \begin{pmatrix} 3\sin t \\ \frac{3}{2}\cos t \end{pmatrix} te^t$

$+ \begin{pmatrix} \cos t \\ -\frac{1}{2}\sin t \end{pmatrix} e^t \ln|\sin t| + \begin{pmatrix} 2\cos t \\ -\sin t \end{pmatrix} e^t \ln|\cos t|$

29. $\mathbf{X} = c_1 \begin{pmatrix} 1 \\ -1 \\ 0 \end{pmatrix} + c_2 \begin{pmatrix} 1 \\ 1 \\ 0 \end{pmatrix} e^{2t} + c_3 \begin{pmatrix} 0 \\ 0 \\ 1 \end{pmatrix} e^{3t}$

$+ \begin{pmatrix} -\frac{1}{4}e^{2t} + \frac{1}{2}te^{2t} \\ -e^t + \frac{1}{4}e^{2t} + \frac{1}{2}te^{2t} \\ \frac{1}{2}t^2 e^{3t} \end{pmatrix}$

31. $\mathbf{X} = \begin{pmatrix} 2 \\ 2 \end{pmatrix} te^{2t} + \begin{pmatrix} -1 \\ 1 \end{pmatrix} e^{2t} + \begin{pmatrix} -2 \\ 2 \end{pmatrix} te^{4t} + \begin{pmatrix} 2 \\ 0 \end{pmatrix} e^{4t}$

33. $\begin{pmatrix} i_1 \\ i_2 \end{pmatrix} = 2\begin{pmatrix} 1 \\ 3 \end{pmatrix}e^{-2t} + \dfrac{6}{29}\begin{pmatrix} 3 \\ -1 \end{pmatrix}e^{-12t} - \dfrac{4}{29}\begin{pmatrix} 19 \\ 42 \end{pmatrix}\cos t$

$\qquad + \dfrac{4}{29}\begin{pmatrix} 83 \\ 69 \end{pmatrix}\sin t$

EXERCISES 8.4 (PAGE 362)

1. $e^{\mathbf{A}t} = \begin{pmatrix} e^t & 0 \\ 0 & e^{2t} \end{pmatrix}$; $\quad e^{-\mathbf{A}t} = \begin{pmatrix} e^{-t} & 0 \\ 0 & e^{-2t} \end{pmatrix}$

3. $e^{\mathbf{A}t} = \begin{pmatrix} t+1 & t & t \\ t & t+1 & t \\ -2t & -2t & -2t+1 \end{pmatrix}$

5. $\mathbf{X} = c_1\begin{pmatrix} 1 \\ 0 \end{pmatrix}e^t + c_2\begin{pmatrix} 0 \\ 1 \end{pmatrix}e^{2t}$

7. $\mathbf{X} = c_1\begin{pmatrix} t+1 \\ t \\ -2t \end{pmatrix} + c_2\begin{pmatrix} t \\ t+1 \\ -2t \end{pmatrix} + c_3\begin{pmatrix} t \\ t \\ -2t+1 \end{pmatrix}$

9. $\mathbf{X} = c_3\begin{pmatrix} 1 \\ 0 \end{pmatrix}e^t + c_4\begin{pmatrix} 0 \\ 1 \end{pmatrix}e^{2t} + \begin{pmatrix} -3 \\ \frac{1}{2} \end{pmatrix}$

11. $\mathbf{X} = c_1\begin{pmatrix} \cosh t \\ \sinh t \end{pmatrix} + c_2\begin{pmatrix} \sinh t \\ \cosh t \end{pmatrix} - \begin{pmatrix} 1 \\ 1 \end{pmatrix}$

13. $\mathbf{X} = \begin{pmatrix} t+1 \\ t \\ -2t \end{pmatrix} - 4\begin{pmatrix} t \\ t+1 \\ -2t \end{pmatrix} + 6\begin{pmatrix} t \\ t \\ -2t+1 \end{pmatrix}$

15. $e^{\mathbf{A}t} = \begin{pmatrix} \frac{3}{2}e^{2t} - \frac{1}{2}e^{-2t} & \frac{3}{4}e^{2t} - \frac{3}{4}e^{-2t} \\ -e^{2t} + e^{-2t} & -\frac{1}{2}e^{2t} + \frac{3}{2}e^{-2t} \end{pmatrix}$;

$\mathbf{X} = c_1\begin{pmatrix} \frac{3}{2}e^{2t} - \frac{1}{2}e^{-2t} \\ -e^{2t} + e^{-2t} \end{pmatrix} + c_2\begin{pmatrix} \frac{3}{4}e^{2t} - \frac{3}{4}e^{-2t} \\ -\frac{1}{2}e^{2t} + \frac{3}{2}e^{-2t} \end{pmatrix}$ or

$\mathbf{X} = c_3\begin{pmatrix} 3 \\ -2 \end{pmatrix}e^{2t} + c_4\begin{pmatrix} 1 \\ -2 \end{pmatrix}e^{-2t}$

17. $e^{\mathbf{A}t} = \begin{pmatrix} e^{2t} + 3te^{2t} & -9te^{2t} \\ te^{2t} & e^{2t} - 3te^{2t} \end{pmatrix}$;

$\mathbf{X} = c_1\begin{pmatrix} 1+3t \\ t \end{pmatrix}e^{2t} + c_2\begin{pmatrix} -9t \\ 1-3t \end{pmatrix}e^{2t}$

23. $\mathbf{X} = c_1\begin{pmatrix} \frac{3}{2}e^{3t} - \frac{1}{2}e^{5t} \\ \frac{3}{2}e^{3t} - \frac{3}{2}e^{5t} \end{pmatrix} + c_2\begin{pmatrix} -\frac{1}{2}e^{3t} + \frac{1}{2}e^{5t} \\ -\frac{1}{2}e^{3t} + \frac{3}{2}e^{5t} \end{pmatrix}$ or

$\mathbf{X} = c_3\begin{pmatrix} 1 \\ 1 \end{pmatrix}e^{3t} + c_4\begin{pmatrix} 1 \\ 3 \end{pmatrix}e^{5t}$

CHAPTER 8 IN REVIEW (PAGE 364)

1. $k = \frac{1}{3}$

5. $\mathbf{X} = c_1\begin{pmatrix} 1 \\ -1 \end{pmatrix}e^t + c_2\left[\begin{pmatrix} 1 \\ -1 \end{pmatrix}te^t + \begin{pmatrix} 0 \\ 1 \end{pmatrix}e^t\right]$

7. $\mathbf{X} = c_1\begin{pmatrix} \cos 2t \\ -\sin 2t \end{pmatrix}e^t + c_2\begin{pmatrix} \sin 2t \\ \cos 2t \end{pmatrix}e^t$

9. $\mathbf{X} = c_1\begin{pmatrix} -2 \\ 3 \\ 1 \end{pmatrix}e^{2t} + c_2\begin{pmatrix} 0 \\ 1 \\ 1 \end{pmatrix}e^{4t} + c_1\begin{pmatrix} 7 \\ 12 \\ -16 \end{pmatrix}e^{-3t}$

11. $\mathbf{X} = c_1\begin{pmatrix} 1 \\ 0 \end{pmatrix}e^{2t} + c_2\begin{pmatrix} 4 \\ 1 \end{pmatrix}e^{4t} + \begin{pmatrix} 16 \\ -4 \end{pmatrix}t + \begin{pmatrix} 11 \\ -1 \end{pmatrix}$

13. $\mathbf{X} = c_1\begin{pmatrix} \cos t \\ \cos t - \sin t \end{pmatrix} + c_2\begin{pmatrix} \sin t \\ \sin t + \cos t \end{pmatrix} - \begin{pmatrix} 1 \\ 1 \end{pmatrix}$

$\qquad + \begin{pmatrix} \sin t \\ \sin t + \cos t \end{pmatrix}\ln|\csc t - \cot t|$

15. (b) $\mathbf{X} = c_1\begin{pmatrix} -1 \\ 1 \\ 0 \end{pmatrix} + c_2\begin{pmatrix} -1 \\ 0 \\ 1 \end{pmatrix} + c_1\begin{pmatrix} 1 \\ 1 \\ 1 \end{pmatrix}e^{3t}$

EXERCISES 9.1 (PAGE 372)

1. for $h = 0.1$, $y_5 = 2.0801$; for $h = 0.05$, $y_{10} = 2.0592$
3. for $h = 0.1$, $y_5 = 0.5470$; for $h = 0.05$, $y_{10} = 0.5465$
5. for $h = 0.1$, $y_5 = 0.4053$; for $h = 0.05$, $y_{10} = 0.4054$
7. for $h = 0.1$, $y_5 = 0.5503$; for $h = 0.05$, $y_{10} = 0.5495$
9. for $h = 0.1$, $y_5 = 1.3260$; for $h = 0.05$, $y_{10} = 1.3315$
11. for $h = 0.1$, $y_5 = 3.8254$; for $h = 0.05$, $y_{10} = 3.8840$;
at $x = 0.5$ the actual value is $y(0.5) = 3.9082$
13. (a) $y_1 = 1.2$
(b) $y''(c)\dfrac{h^2}{2} = 4e^{2c}\dfrac{(0.1)^2}{2} = 0.02e^{2c} \le 0.02e^{0.2}$
$\qquad = 0.0244$
(c) Actual value is $y(0.1) = 1.2214$. Error is 0.0214.
(d) If $h = 0.05$, $y_2 = 1.21$.
(e) Error with $h = 0.1$ is 0.0214. Error with $h = 0.05$ is 0.0114.
15. (a) $y_1 = 0.8$
(b) $y''(c)\dfrac{h^2}{2} = 5e^{-2c}\dfrac{(0.1)^2}{2} = 0.025e^{-2c} \le 0.025$
$\qquad$ for $0 \le c \le 0.1$.
(c) Actual value is $y(0.1) = 0.8234$. Error is 0.0234.
(d) If $h = 0.05$, $y_2 = 0.8125$.
(e) Error with $h = 0.1$ is 0.0234. Error with $h = 0.05$ is 0.0109.
17. (a) Error is $19h^2e^{-3(c-1)}$.
(b) $y''(c)\dfrac{h^2}{2} \le 19(0.1)^2(1) = 0.19$
(c) If $h = 0.1$, $y_5 = 1.8207$.
If $h = 0.05$, $y_{10} = 1.9424$.
(d) Error with $h = 0.1$ is 0.2325. Error with $h = 0.05$ is 0.1109.
19. (a) Error is $\dfrac{1}{(c+1)^2}\dfrac{h^2}{2}$.
(b) $\left|y''(c)\dfrac{h^2}{2}\right| \le (1)\dfrac{(0.1)^2}{2} = 0.005$
(c) If $h = 0.1$, $y_5 = 0.4198$. If $h = 0.05$, $y_{10} = 0.4124$.
(d) Error with $h = 0.1$ is 0.0143. Error with $h = 0.05$ is 0.0069.

EXERCISES 9.2 (PAGE 377)

1. $y_5 = 3.9078$; actual value is $y(0.5) = 3.9082$

3. $y_5 = 2.0533$ **5.** $y_5 = 0.5463$

7. $y_5 = 0.4055$ **9.** $y_5 = 0.5493$

11. $y_5 = 1.3333$

13. (a) 35.7678

     (c) $v(t) = \sqrt{\dfrac{mg}{k}} \tanh \sqrt{\dfrac{kg}{m}}\, t; \quad v(5) = 35.7678$

15. (a) for $h = 0.1$, $y_4 = 903.0282$;

     for $h = 0.05$, $y_8 = 1.1 \times 10^{15}$

17. (a) $y_1 = 0.82341667$

     (b) $y^{(5)}(c)\dfrac{h^5}{5!} = 40e^{-2c}\dfrac{h^5}{5!} \leq 40e^{2(0)}\dfrac{(0.1)^5}{5!}$

             $= 3.333 \times 10^{-6}$

     (c) Actual value is $y(0.1) = 0.8234134413$. Error is $3.225 \times 10^{-6} \leq 3.333 \times 10^{-6}$.

     (d) If $h = 0.05$, $y_2 = 0.82341363$.

     (e) Error with $h = 0.1$ is 3.225×10^{-6}. Error with $h = 0.05$ is 1.854×10^{-7}.

19. (a) $y^{(5)}(c)\dfrac{h^5}{5!} = \dfrac{24}{(c+1)^5}\dfrac{h^5}{5!}$

     (b) $\dfrac{24}{(c+1)^5}\dfrac{h^5}{5!} \leq 24\dfrac{(0.1)^5}{5!} = 2.0000 \times 10^{-6}$

     (c) From calculation with $h = 0.1$, $y_5 = 0.40546517$. From calculation with $h = 0.05$, $y_{10} = 0.40546511$.

EXERCISES 9.3 (PAGE 381)

1. $y(x) = -x + e^x$; actual values are $y(0.2) = 1.0214$, $y(0.4) = 1.0918$, $y(0.6) = 1.2221$, $y(0.8) = 1.4255$; approximations are given in Example 1

3. $y_4 = 0.7232$

5. for $h = 0.2$, $y_5 = 1.5569$;   for $h = 0.1$, $y_{10} = 1.5576$

7. for $h = 0.2$, $y_5 = 0.2385$;   for $h = 0.1$, $y_{10} = 0.2384$

EXERCISES 9.4 (PAGE 385)

1. $y(x) = -2e^{2x} + 5xe^{2x}$; $y(0.2) = -1.4918$, $y_2 = -1.6800$

3. $y_1 = -1.4928$, $y_2 = -1.4919$

5. $y_1 = 1.4640$, $y_2 = 1.4640$

7. $x_1 = 8.3055$, $y_1 = 3.4199$; $x_2 = 8.3055$, $y_2 = 3.4199$

9. $x_1 = -3.9123$, $y_1 = 4.2857$; $x_2 = -3.9123$, $y_2 = 4.2857$

11. $x_1 = 0.4179$, $y_1 = -2.1824$; $x_2 = 0.4173$, $y_2 = -2.1821$

EXERCISES 9.5 (PAGE 390)

1. $y_1 = -5.6774$, $y_2 = -2.5807$, $y_3 = 6.3226$

3. $y_1 = -0.2259$, $y_2 = -0.3356$, $y_3 = -0.3308$, $y_4 = -0.2167$

5. $y_1 = 3.3751$, $y_2 = 3.6306$, $y_3 = 3.6448$, $y_4 = 3.2355$, $y_5 = 2.1411$

7. $y_1 = 3.8842$, $y_2 = 2.9640$, $y_3 = 2.2064$, $y_4 = 1.5826$, $y_5 = 1.0681$, $y_6 = 0.6430$, $y_7 = 0.2913$

9. $y_1 = 0.2660$, $y_2 = 0.5097$, $y_3 = 0.7357$, $y_4 = 0.9471$, $y_5 = 1.1465$, $y_6 = 1.3353$, $y_7 = 1.5149$, $y_8 = 1.6855$, $y_9 = 1.8474$

11. $y_1 = 0.3492$, $y_2 = 0.7202$, $y_3 = 1.1363$, $y_4 = 1.6233$, $y_5 = 2.2118$, $y_6 = 2.9386$, $y_7 = 3.8490$

13. (c) $y_0 = -2.2755$, $y_1 = -2.0755$, $y_2 = -1.8589$, $y_3 = -1.6126$, $y_4 = -1.3275$

CHAPTER 9 IN REVIEW (PAGE 390)

1. Comparison of Numerical Methods with $h = 0.1$

x_n	Euler	Improved Euler	RK4
1.10	2.1386	2.1549	2.1556
1.20	2.3097	2.3439	2.3454
1.30	2.5136	2.5672	2.5695
1.40	2.7504	2.8246	2.8278
1.50	3.0201	3.1157	3.1197

Comparison of Numerical Methods with $h = 0.05$

x_n	Euler	Improved Euler	RK4
1.10	2.1469	2.1554	2.1556
1.20	2.3272	2.3450	2.3454
1.30	2.5410	2.5689	2.5695
1.40	2.7883	2.8269	2.8278
1.50	3.0690	3.1187	3.1197

3. Comparison of Numerical Methods with $h = 0.1$

x_n	Euler	Improved Euler	RK4
0.60	0.6000	0.6048	0.6049
0.70	0.7095	0.7191	0.7194
0.80	0.8283	0.8427	0.8431
0.90	0.9559	0.9752	0.9757
1.00	1.0921	1.1163	1.1169

Comparison of Numerical Methods with $h = 0.05$

x_n	Euler	Improved Euler	RK4
0.60	0.6024	0.6049	0.6049
0.70	0.7144	0.7194	0.7194
0.80	0.8356	0.8431	0.8431
0.90	0.9657	0.9757	0.9757
1.00	1.1044	1.1170	1.1169

5. $h = 0.2$: $y(0.2) \approx 3.2$; $h = 0.1$: $y(0.2) \approx 3.23$

7. $x(0.2) \approx 1.62$, $y(0.2) \approx 1.84$

EXERCISES 10.1 (PAGE 400)

1. $x' = y$
   $y' = -9 \sin x$; critical points at $(\pm n\pi, 0)$
3. $x' = y$
   $y' = x^2 + y(x^3 - 1)$; critical point at $(0, 0)$
5. $x' = y$
   $y' = \epsilon x^3 - x$;

   critical points at $(0, 0)$, $\left(\frac{1}{\sqrt{\epsilon}}, 0\right), \left(-\frac{1}{\sqrt{\epsilon}}, 0\right)$

7. $(0, 0)$ and $(-1, -1)$
9. $(0, 0)$ and $\left(\frac{4}{3}, \frac{4}{3}\right)$
11. $(0, 0)$, $(10, 0)$, $(0, 16)$, and $(4, 12)$
13. $(0, y)$, y arbitrary
15. $(0, 0)$, $(0, 1)$, $(0, -1)$, $(1, 0)$, $(-1, 0)$
17. **(a)** $x = c_1 e^{5t} - c_2 e^{-t}$ **(b)** $x = -2e^{-t}$
    $y = 2c_1 e^{5t} + c_2 e^{-t}$ $y = 2e^{-t}$
19. **(a)** $x = c_1(4 \cos 3t - 3 \sin 3t) + c_2(4 \sin 3t + 3 \cos 3t)$
    $y = c_1(5 \cos 3t) + c_2(5 \sin 3t)$
    (b) $x = 4 \cos 3t - 3 \sin 3t$
    $y = 5 \cos 3t$
21. **(a)** $x = c_1(\sin t - \cos t)e^{4t} + c_2(-\sin t - \cos t)e^{4t}$
    $y = 2c_1(\cos t)e^{4t} + 2c_2(\sin t)e^{4t}$
    (b) $x = (\sin t - \cos t)e^{4t}$
    $y = 2(\cos t)e^{4t}$
23. $r = \dfrac{1}{\sqrt[4]{4t + c_1}}, \theta = t + c_2$; $r = 4\dfrac{1}{\sqrt[4]{1024t + 1}}, \theta = t$;
    the solution spirals toward the origin as t increases.
25. $r = \dfrac{1}{\sqrt{1 + c_1 e^{-2t}}}, \theta = t + c_2$; $r = 1$, $\theta = t$ (or $x = \cos t$
    and $y = \sin t$) is the solution that satisfies $\mathbf{X}(0) = (1, 0)$;
    $r = \dfrac{1}{\sqrt{1 - \frac{3}{4}e^{-2t}}}, \quad \theta = t$ is the solution that satisfies
    $\mathbf{X}(0) = (2, 0)$. This solution spirals toward the circle
    $r = 1$ as t increases.
27. There are no critical points and therefore no periodic solutions.
29. There appears to be a periodic solution enclosing the critical point $(0, 0)$.

EXERCISES 10.2 (PAGE 408)

1. **(a)** If $\mathbf{X}(0) = \mathbf{X}_0$ lies on the line $y = 2x$, then $\mathbf{X}(t)$ approaches $(0, 0)$ along this line. For all other initial conditions, $\mathbf{X}(t)$ approaches $(0, 0)$ from the direction determined by the line $y = -x/2$.
3. **(a)** All solutions are unstable spirals that become unbounded as t increases.
5. **(a)** All solutions approach $(0, 0)$ from the direction specified by the line $y = x$.
7. **(a)** If $\mathbf{X}(0) = \mathbf{X}_0$ lies on the line $y = 3x$, then $\mathbf{X}(t)$ approaches $(0, 0)$ along this line. For all other initial conditions, $\mathbf{X}(t)$ becomes unbounded and $y = x$ serves as the asymptote.
9. saddle point   11. saddle point
13. degenerate stable node   15. stable spiral

17. $|\mu| < 1$
19. $\mu < -1$ for a saddle point; $-1 < \mu < 3$ for an unstable spiral point
23. **(a)** $(-3, 4)$
    (b) unstable node or saddle point
    (c) $(0, 0)$ is a saddle point.
25. **(a)** $\left(\frac{1}{2}, 2\right)$
    (b) unstable spiral point
    (c) $(0, 0)$ is an unstable spiral point.

EXERCISES 10.3 (PAGE 417)

1. $r = r_0 e^{\alpha t}$
3. $x = 0$ is unstable; $x = n + 1$ is asymptotically stable.
5. $T = T_0$ is unstable.
7. $x = \alpha$ is unstable; $x = \beta$ is asymptotically stable.
9. $P = a/b$ is asymptotically stable; $P = c$ is unstable.
11. $\left(\frac{1}{2}, 1\right)$ is a stable spiral point.
13. $(\sqrt{2}, 0)$ and $(-\sqrt{2}, 0)$ are saddle points; $\left(\frac{1}{2}, -\frac{7}{4}\right)$ is a stable spiral point.
15. $(1, 1)$ is a stable node; $(1, -1)$ is a saddle point; $(2, 2)$ is a saddle point; $(2, -2)$ is an unstable spiral point.
17. $(0, -1)$ is a saddle point; $(0, 0)$ is unclassified; $(0, 1)$ is stable but we are unable to classify further.
19. $(0, 0)$ is an unstable node; $(10, 0)$ is a saddle point; $(0, 16)$ is a saddle point; $(4, 12)$ is a stable node.
21. $\theta = 0$ is a saddle point. It is not possible to classify either $\theta = \pi/3$ or $\theta = -\pi/3$.
23. It is not possible to classify $x = 0$.
25. It is not possible to classify $x = 0$, but $x = 1/\sqrt{\epsilon}$ and $x = -1/\sqrt{\epsilon}$ and are each saddle points.
29. **(a)** $(0, 0)$ is a stable spiral point.
33. **(a)** $(1, 0), (-1, 0)$
35. $|v_0| < \frac{1}{2}\sqrt{2}$
37. If $\beta > 0$, $(0, 0)$ is the only critical point and is stable. If $\beta < 0$, $(0, 0)$, $(\hat{x}, 0)$, and $(-\hat{x}, 0)$, where $\hat{x}^2 = -\alpha/\beta$, are critical points. $(0, 0)$ is stable, while $(\hat{x}, 0)$, and $(-\hat{x}, 0)$ are each saddle points.
39. **(b)** $(5\pi/6, 0)$ is a saddle point.
    (c) $(\pi/6, 0)$ is a center.

EXERCISES 10.4 (PAGE 425)

1. $|\omega_0| < \sqrt{3g/L}$
5. **(a)** First show that $y^2 = v_0^2 + g \ln\left(\dfrac{1 + x^2}{1 + x_0^2}\right)$.
9. **(a)** The new critical point is $(d/c - \epsilon_2/c, a/b + \epsilon_1/b)$.
   (b) yes
11. $(0, 0)$ is an unstable node, $(0, 100)$ is a stable node, $(50, 0)$ is a stable node, and $(20, 40)$ is a saddle point.
17. **(a)** $(0, 0)$ is the only critical point.

CHAPTER 10 IN REVIEW (PAGE 427)

1. true   3. a center or a saddle point
5. false   7. false

9. $\alpha = -1$

11. $r = 1/\sqrt[3]{3t + 1}$, $\theta = t$. The solution curve spirals toward the origin.

13. (a) center

(b) degenerate stable node

15. $(0, 0)$ is a stable critical point for $\alpha \leq 0$.

17. $x = 1$ is unstable; $x = -1$ is asymptotically stable.

19. The system is overdamped when $\beta^2 > 12\,kms^2$ and underdamped when $\beta^2 < 12\,kms^2$.

EXERCISES 11.1 (PAGE 434)

7. $\frac{1}{2}\sqrt{\pi}$

9. $\sqrt{\pi/2}$

11. $\|1\| = \sqrt{p}$; $\|\cos(n\pi x/p)\| = \sqrt{p/2}$

21. (a) $T = 1$ (b) $T = \pi L/2$
(c) $T = 2\pi$ (d) $T = \pi$
(e) $T = 2\pi$ (f) $T = 2p$

EXERCISES 11.2 (PAGE 439)

1. $f(x) = \dfrac{1}{2} + \dfrac{1}{\pi} \sum_{n=1}^{\infty} \dfrac{1 - (-1)^n}{n} \sin nx$

3. $f(x) = \dfrac{3}{4} + \sum_{n=1}^{\infty}\left\{\dfrac{(-1)^n - 1}{n^2\pi^2}\cos n\pi x - \dfrac{1}{n\pi}\sin n\pi x\right\}$

5. $f(x) = \dfrac{\pi^2}{6} + \sum_{n=1}^{\infty}\left\{\dfrac{2(-1)^n}{n^2}\cos nx\right.$

$\left. + \left(\dfrac{(-1)^{n+1}\pi}{n} + \dfrac{2}{\pi n^3}[(-1)^n - 1]\right)\sin nx\right\}$

7. $f(x) = \pi + 2\sum_{n=1}^{\infty}\dfrac{(-1)^{n+1}}{n}\sin nx$

9. $f(x) = \dfrac{1}{\pi} + \dfrac{1}{2}\sin x + \dfrac{1}{\pi}\sum_{n=2}^{\infty}\dfrac{(-1)^n + 1}{1 - n^2}\cos nx$

11. $f(x) = -\dfrac{1}{4} + \dfrac{1}{\pi}\sum_{n=1}^{\infty}\left\{-\dfrac{1}{n}\sin\dfrac{n\pi}{2}\cos\dfrac{n\pi}{2}x\right.$

$\left. + \dfrac{3}{n}\left(1 - \cos\dfrac{n\pi}{2}\right)\sin\dfrac{n\pi}{2}x\right\}$

13. $f(x) = \dfrac{9}{4} + 5\sum_{n=1}^{\infty}\left\{\dfrac{(-1)^n - 1}{n^2\pi^2}\cos\dfrac{n\pi}{5}x\right.$

$\left. + \dfrac{(-1)^{n+1}}{n\pi}\sin\dfrac{n\pi}{5}x\right\}$

15. $f(x) = \dfrac{2\sinh\pi}{\pi}\left[\dfrac{1}{2} + \sum_{n=1}^{\infty}\dfrac{(-1)^n}{1 + n^2}(\cos nx - n\sin nx)\right]$

19. Set $x = \pi/2$.

EXERCISES 11.3 (PAGE 446)

1. odd   **3.** neither even nor odd
5. even   **7.** odd
9. neither even nor odd

11. $f(x) = \dfrac{2}{\pi}\sum_{n=1}^{\infty}\dfrac{1 - (-1)^n}{n}\sin nx$

13. $f(x) = \dfrac{\pi}{2} + \dfrac{2}{\pi}\sum_{n=1}^{\infty}\dfrac{(-1)^n - 1}{n^2}\cos nx$

15. $f(x) = \dfrac{1}{3} + \dfrac{4}{\pi^2}\sum_{n=1}^{\infty}\dfrac{(-1)^n}{n^2}\cos n\pi x$

17. $f(x) = \dfrac{2\pi^2}{3} + 4\sum_{n=1}^{\infty}\dfrac{(-1)^{n+1}}{n^2}\cos nx$

19. $f(x) = \dfrac{2}{\pi}\sum_{n=1}^{\infty}\dfrac{1 - (-1)^n(1 + \pi)}{n}\sin nx$

21. $f(x) = \dfrac{3}{4} + \dfrac{4}{\pi^2}\sum_{n=1}^{\infty}\dfrac{\cos\dfrac{n\pi}{2} - 1}{n^2}\cos\dfrac{n\pi}{2}x$

23. $f(x) = \dfrac{2}{\pi} + \dfrac{2}{\pi}\sum_{n=2}^{\infty}\dfrac{1 + (-1)^n}{1 - n^2}\cos nx$

25. $f(x) = \dfrac{1}{2} + \dfrac{2}{\pi}\sum_{n=1}^{\infty}\dfrac{\sin\dfrac{n\pi}{2}}{n}\cos n\pi x$

$f(x) = \dfrac{2}{\pi}\sum_{n=1}^{\infty}\dfrac{1 - \cos\dfrac{n\pi}{2}}{n}\sin n\pi x$

27. $f(x) = \dfrac{2}{\pi} + \dfrac{4}{\pi}\sum_{n=1}^{\infty}\dfrac{(-1)^n}{1 - 4n^2}\cos 2nx$

$f(x) = \dfrac{8}{\pi}\sum_{n=1}^{\infty}\dfrac{n}{4n^2 - 1}\sin 2nx$

29. $f(x) = \dfrac{\pi}{4} + \dfrac{2}{\pi}\sum_{n=1}^{\infty}\dfrac{2\cos\dfrac{n\pi}{2} - (-1)^n - 1}{n^2}\cos nx$

$f(x) = \dfrac{4}{\pi}\sum_{n=1}^{\infty}\dfrac{\sin\dfrac{n\pi}{2}}{n^2}\sin nx$

31. $f(x) = \dfrac{3}{4} + \dfrac{4}{\pi^2}\sum_{n=1}^{\infty}\dfrac{\cos\dfrac{n\pi}{2} - 1}{n^2}\cos\dfrac{n\pi}{2}x$

$f(x) = \sum_{n=1}^{\infty}\left\{\dfrac{4}{n^2\pi^2}\sin\dfrac{n\pi}{2} - \dfrac{2}{n\pi}(-1)^n\right\}\sin\dfrac{n\pi}{2}x$

33. $f(x) = \dfrac{5}{6} + \dfrac{2}{\pi^2}\sum_{n=1}^{\infty}\dfrac{3(-1)^n - 1}{n^2}\cos n\pi x$

$f(x) = 4\sum_{n=1}^{\infty}\left\{\dfrac{(-1)^{n+1}}{n\pi} + \dfrac{(-1)^n - 1}{n^3\pi^3}\right\}\sin n\pi x$

35. $f(x) = \dfrac{4\pi^2}{3} + 4\sum_{n=1}^{\infty}\left\{\dfrac{1}{n^2}\cos nx - \dfrac{\pi}{n}\sin nx\right\}$

37. $f(x) = \dfrac{3}{2} - \dfrac{1}{\pi}\sum_{n=1}^{\infty}\dfrac{1}{n}\sin 2n\pi x$

39. $x_p(t) = \dfrac{10}{\pi}\sum_{n=1}^{\infty}\dfrac{1 - (-1)^n}{n(10 - n^2)}\sin nt$

41. $x_p(t) = \dfrac{\pi^2}{18} + 16 \displaystyle\sum_{n=1}^{\infty} \dfrac{1}{n^2(n^2 - 48)} \cos nt$

43. $x(t) = \dfrac{10}{\pi} \displaystyle\sum_{n=1}^{\infty} \dfrac{1 - (-1)^n}{10 - n^2}\left[\dfrac{1}{n} \sin nt - \dfrac{1}{\sqrt{10}} \sin \sqrt{10}\,t\right]$

45. (b) $y_p(x) = \dfrac{2w_0 L^4}{EI\pi^5} \displaystyle\sum_{n=1}^{\infty} \dfrac{(-1)^{n+1}}{n^5} \sin \dfrac{n\pi}{L} x$

47. $y_p(x) = \dfrac{w_0}{2k} + \dfrac{2w_0}{\pi} \displaystyle\sum_{n=1}^{\infty} \dfrac{\sin(n\pi/2)}{n(EIn^4 + k)} \cos nx$

EXERCISES 11.4 (PAGE 454)

1. $y = \cos \alpha_n x$; α defined by $\cot \alpha = \alpha$;
$\lambda_1 = 0.7402$, $\lambda_2 = 11.7349$,
$\lambda_3 = 41.4388$, $\lambda_4 = 90.8082$
$y_1 = \cos 0.8603x$, $y_2 = \cos 3.4256x$,
$y_3 = \cos 6.4373x$, $y_4 = \cos 9.5293x$

5. $\frac{1}{2}[1 + \sin^2 \alpha_n]$

7. (a) $\lambda_n = \left(\dfrac{n\pi}{\ln 5}\right)^2$, $y_n = \sin\left(\dfrac{n\pi}{\ln 5} \ln x\right)$, $n = 1, 2, 3, \ldots$

(b) $\dfrac{d}{dx}[xy'] + \dfrac{\lambda}{x}y = 0$

(c) $\displaystyle\int_1^5 \dfrac{1}{x} \sin\left(\dfrac{m\pi}{\ln 5} \ln x\right) \sin\left(\dfrac{n\pi}{\ln 5} \ln x\right) dx = 0, m \neq n$

9. $\dfrac{d}{dx}[xe^{-x}y'] + ne^{-x}y = 0$;

$\displaystyle\int_0^{\infty} e^{-x} L_m(x) L_n(x)\, dx = 0, \; m \neq n$

11. (a) $\lambda_n = 16n^2$, $y_n = \sin(4n \tan^{-1}x)$, $n = 1, 2, 3, \ldots$

(b) $\displaystyle\int_0^1 \dfrac{1}{1 + x^2} \sin(4m \tan^{-1} x) \sin(4n \tan^{-1} x)\, dx = 0, \; m \neq n$

EXERCISES 11.5 (PAGE 461)

1. $\alpha_1 = 1.277$, $\alpha_2 = 2.339$, $\alpha_3 = 3.391$, $\alpha_4 = 4.441$

3. $f(x) = \displaystyle\sum_{i=1}^{\infty} \dfrac{1}{\alpha_i J_1(2\alpha_i)} J_0(\alpha_i x)$

5. $f(x) = 4 \displaystyle\sum_{i=1}^{\infty} \dfrac{\alpha_i J_1(2\alpha_i)}{(4\alpha_i^2 + 1)J_0^2(2\alpha_i)} J_0(\alpha_i x)$

7. $f(x) = 20 \displaystyle\sum_{i=1}^{\infty} \dfrac{\alpha_i J_2(4\alpha_i)}{(2\alpha_i^2 + 1)J_1^2(4\alpha_i)} J_1(\alpha_i x)$

9. $f(x) = \dfrac{9}{2} - 4 \displaystyle\sum_{i=1}^{\infty} \dfrac{J_2(3\alpha_i)}{\alpha_i^2 J_0^2(3\alpha_i)} J_0(\alpha_i x)$

15. $f(x) = \frac{1}{4}P_0(x) + \frac{1}{2}P_1(x) + \frac{5}{16}P_2(x) - \frac{3}{32}P_4(x) + \cdots$

21. $f(x) = \frac{1}{2}P_0(x) + \frac{5}{8}P_2(x) - \frac{3}{16}P_4(x) + \cdots$,
$f(x) = |x|$ on $(-1, 1)$

CHAPTER 11 IN REVIEW (PAGE 462)

1. true   **3.** cosine
5. false   **7.** 5.5, 1, 0

9. $\dfrac{1}{\sqrt{1 - x^2}}$, $-1 \leq x \leq 1$,

$\displaystyle\int_{-1}^{1} \dfrac{1}{\sqrt{1 - x^2}} T_m(x) T_n(x)\, dx = 0, m \neq n$

13. $f(x) = \dfrac{1}{2} + \dfrac{2}{\pi} \displaystyle\sum_{n=1}^{\infty}\left\{\dfrac{1}{n^2\pi}[(-1)^n - 1] \cos n\pi x\right.$
$\left. + \dfrac{2}{n}(-1)^n \sin n\pi x\right\}$

15. $f(x) = 1 - e^{-1} + 2 \displaystyle\sum_{n=1}^{\infty} \dfrac{1 - (-1)^n e^{-1}}{1 + n^2\pi^2} \cos n\pi x$,

$f(x) = \displaystyle\sum_{n=1}^{\infty} \dfrac{2n\pi[1 - (-1)^n e^{-1}]}{1 + n^2\pi^2} \sin n\pi x$

19. $\lambda_n = \dfrac{(2n - 1)^2 \pi^2}{36}$, $n = 1, 2, 3, \ldots,$

$y_n = \cos\left(\dfrac{2n - 1}{2} \pi \ln x\right)$

21. $f(x) = \dfrac{1}{4} \displaystyle\sum_{i=1}^{\infty} \dfrac{J_1(2\alpha_i)}{\alpha_i J_1^2(4\alpha_i)} J_0(\alpha_i x)$

EXERCISES 12.1 (PAGE 468)

1. The possible cases can be summarized in one form
$u = c_1 e^{c_2(x+y)}$, where c_1 and c_2 are constants.

3. $u = c_1 e^{y + c_2(x - y)}$

5. $u = c_1 (xy)^{c_2}$

7. not separable

9. $u = e^{-t}(A_1 e^{k\alpha^2 t} \cosh \alpha x + B_1 e^{k\alpha^2 t} \sinh \alpha x)$
$u = e^{-t}(A_2 e^{-k\alpha^2 t} \cos \alpha x + B_2 e^{-k\alpha^2 t} \sin \alpha x)$
$u = e^{-t}(A_3 x + B_3)$

11. $u = (c_1 \cosh \alpha x + c_2 \sinh \alpha x)(c_3 \cosh \alpha at + c_4 \sinh \alpha at)$
$u = (c_5 \cos \alpha x + c_6 \sin \alpha x)(c_7 \cos \alpha at + c_8 \sin \alpha at)$
$u = (c_9 x + c_{10})(c_{11} t + c_{12})$

13. $u = (c_1 \cosh \alpha x + c_2 \sinh \alpha x)(c_3 \cos \alpha y + c_4 \sin \alpha y)$
$u = (c_5 \cos \alpha x + c_6 \sin \alpha x)(c_7 \cosh \alpha y + c_8 \sinh \alpha y)$
$u = (c_9 x + c_{10})(c_{11} y + c_{12})$

15. For $\lambda = \alpha^2 > 0$ there are three possibilities:
(i) For $0 < \alpha^2 < 1$,

$u = (c_1 \cosh \alpha x + c_2 \sinh \alpha x)\left(c_3 \cosh \sqrt{1 - \alpha^2}\,y + c_4 \sinh \sqrt{1 - \alpha^2}\,y\right)$

(ii) For $\alpha^2 > 1$,

$u = (c_1 \cosh \alpha x + c_2 \sinh \alpha x)\left(c_3 \cos \sqrt{\alpha^2 - 1}\,y + c_4 \sin \sqrt{\alpha^2 - 1}\,y\right)$

(iii) For $\alpha^2 = 1$,

$u = (c_1 \cosh x + c_2 \sinh x)(c_3 y + c_4)$

The results for the case $\lambda = -\alpha^2$ are similar. For $\lambda = 0$,

$u = (c_1 x + c_2)(c_3 \cosh y + c_4 \sinh y)$

17. elliptic   **19.** parabolic
21. hyperbolic   **23.** parabolic
25. hyperbolic

EXERCISES 12.2 (PAGE 474)

1. $k\dfrac{\partial^2 u}{\partial x^2} = \dfrac{\partial u}{\partial t}, \quad 0 < x < L, t > 0$

$u(0, t) = 0, \quad \dfrac{\partial u}{\partial x}\Big|_{x=L} = 0, t > 0$

$u(x, 0) = f(x), \quad 0 < x < L$

3. $k\dfrac{\partial^2 u}{\partial x^2} = \dfrac{\partial u}{\partial t}, 0 < x < L, t > 0$

$u(0, t) = 100, \dfrac{\partial u}{\partial x}\Big|_{x=L} = -hu(L, t), t > 0$

$u(x, 0) = f(x), 0 < x < L$

5. $a^2\dfrac{\partial^2 u}{\partial x^2} = \dfrac{\partial^2 u}{\partial t^2}, 0 < x < L, t > 0$

$u(0, t) = 0, u(L, t) = 0, t > 0$

$u(x, 0) = x(L - x), \dfrac{\partial u}{\partial t}\Big|_{t=0} = 0, 0 < x < L$

7. $a^2\dfrac{\partial^2 u}{\partial x^2} - 2\beta\dfrac{\partial u}{\partial t} = \dfrac{\partial^2 u}{\partial t^2}, 0 < x < L, t > 0$

$u(0, t) = 0, u(L, t) = \sin \pi t, t > 0$

$u(x, 0) = f(x), \dfrac{\partial u}{\partial t}\Big|_{t=0} = 0, 0 < x < L$

9. $\dfrac{\partial^2 u}{\partial x^2} + \dfrac{\partial^2 u}{\partial y^2} = 0, 0 < x < 4, 0 < y < 2$

$\dfrac{\partial u}{\partial x}\Big|_{x=0} = 0, u(4, y) = f(y), 0 < y < 2$

$\dfrac{\partial u}{\partial y}\Big|_{y=0} = 0, u(x, 2) = 0, 0 < x < 4$

EXERCISES 12.3 (PAGE 477)

1. $u(x, t) = \dfrac{2}{\pi}\sum_{n=1}^{\infty}\left(\dfrac{-\cos\dfrac{n\pi}{2} + 1}{n}\right)e^{-k(n^2\pi^2/L^2)t}\sin\dfrac{n\pi}{L}x$

3. $u(x, t) = \dfrac{1}{L}\int_0^L f(x)\,dx$

$+ \dfrac{2}{L}\sum_{n=1}^{\infty}\left(\int_0^L f(x)\cos\dfrac{n\pi}{L}x\,dx\right)e^{-k(n^2\pi^2/L^2)t}\cos\dfrac{n\pi}{L}x$

5. $u(x, t) = e^{-ht}\left[\dfrac{1}{L}\int_0^L f(x)\,dx\right.$

$\left. + \dfrac{2}{L}\sum_{n=1}^{\infty}\left(\int_0^L f(x)\cos\dfrac{n\pi}{L}x\,dx\right)e^{-k(n^2\pi^2/L^2)t}\cos\dfrac{n\pi}{L}x\right]$

EXERCISES 12.4 (PAGE 480)

1. $u(x, t) = \dfrac{L^2}{\pi^3}\sum_{n=1}^{\infty}\dfrac{1 - (-1)^n}{n^3}\cos\dfrac{n\pi a}{L}t\sin\dfrac{n\pi}{L}x$

3. $u(x, t) = \dfrac{6\sqrt{3}}{\pi^2}\left(\cos\dfrac{\pi a}{L}t\sin\dfrac{\pi}{L}x\right.$

$- \dfrac{1}{5^2}\cos\dfrac{5\pi a}{L}t\sin\dfrac{5\pi}{L}x$

$\left. + \dfrac{1}{7^2}\cos\dfrac{7\pi a}{L}t\sin\dfrac{7\pi}{L}x - \cdots\right)$

5. $u(x, t) = \dfrac{1}{a}\sin at \sin x$

7. $u(x, t) = \dfrac{8h}{\pi^2}\sum_{n=1}^{\infty}\dfrac{\sin\dfrac{n\pi}{2}}{n^2}\cos\dfrac{n\pi a}{L}t\sin\dfrac{n\pi}{L}x$

9. $u(x, t) = e^{-\beta t}\sum_{n=1}^{\infty}A_n\left\{\cos q_n t + \dfrac{\beta}{q_n}\sin q_n t\right\}\sin nx,$

where $A_n = \dfrac{2}{\pi}\int_0^\pi f(x)\sin nx\,dx$ and $q_n = \sqrt{n^2 - \beta^2}$

11. $u(x, t) = \sum_{n=1}^{\infty}\left(A_n\cos\dfrac{n^2\pi^2}{L^2}at + B_n\sin\dfrac{n^2\pi^2}{L^2}at\right)$

$\times \sin\dfrac{n\pi}{L}x,$

where $A_n = \dfrac{2}{L}\int_0^L f(x)\sin\dfrac{n\pi}{L}x\,dx$

$B_n = \dfrac{2L}{n^2\pi^2 a}\int_0^L g(x)\sin\dfrac{n\pi}{L}x\,dx$

15. $u(x, t) = \sin x \cos 2at + t$

17. $u(x, t) = \dfrac{1}{2a}\sin 2x \sin 2at$

EXERCISES 12.5 (PAGE 481)

1. $u(x, y) = \dfrac{2}{a}\sum_{n=1}^{\infty}\left(\dfrac{1}{\sinh\dfrac{n\pi}{a}b}\int_0^a f(x)\sin\dfrac{n\pi}{a}x\,dx\right)$

$\times \sinh\dfrac{n\pi}{a}y\sin\dfrac{n\pi}{a}x$

3. $u(x, y) = \dfrac{2}{a}\sum_{n=1}^{\infty}\left(\dfrac{1}{\sinh\dfrac{n\pi}{a}b}\int_0^a f(x)\sin\dfrac{n\pi}{a}x\,dx\right)$

$\times \sinh\dfrac{n\pi}{a}(b - y)\sin\dfrac{n\pi}{a}x$

5. $u(x, y) = \dfrac{1}{2}x + \dfrac{2}{\pi^2}\sum_{n=1}^{\infty}\dfrac{1 - (-1)^n}{n^2\sinh n\pi}\sinh n\pi x\cos n\pi y$

7. $u(x, y) = \dfrac{2}{\pi}\sum_{n=1}^{\infty}\dfrac{[1 - (-1)^n]}{n}$

$\times \dfrac{n\cosh nx + \sinh nx}{n\cosh n\pi + \sinh n\pi}\sin ny$

9. $u(x, y) = \sum_{n=1}^{\infty}(A_n\cosh n\pi y + B_n\sinh n\pi y)\sin n\pi x,$

where $A_n = 200\dfrac{[1 - (-1)^n]}{n\pi}$

$B_n = 200\dfrac{[1 - (-1)^n]}{n\pi}\dfrac{[2 - \cosh n\pi]}{\sinh n\pi}$

11. $u(x, y) = \dfrac{2}{\pi}\sum_{n=1}^{\infty}\left(\int_0^\pi f(x)\sin nx\,dx\right)e^{-ny}\sin nx$

13. $u(x, y) = \sum_{n=1}^{\infty}\left(A_n\cosh\dfrac{n\pi}{a}y + B_n\sinh\dfrac{n\pi}{a}y\right)\sin\dfrac{n\pi}{a}x,$

where $A_n = \dfrac{2}{a}\displaystyle\int_0^a f(x)\sin\dfrac{n\pi}{a}x\,dx$

$B_n = \dfrac{1}{\sinh\dfrac{n\pi}{a}b}\left(\dfrac{2}{a}\displaystyle\int_0^a g(x)\sin\dfrac{n\pi}{a}x\,dx - A_n\cosh\dfrac{n\pi}{a}b\right)$

15. $u = u_1 + u_2$, where

$u_1(x,y) = \dfrac{2}{\pi}\displaystyle\sum_{n=1}^{\infty}\dfrac{1-(-1)^n}{n\sinh n\pi}\sinh ny\sin nx$

$u_2(x,y) = \dfrac{2}{\pi}\displaystyle\sum_{n=1}^{\infty}\dfrac{[1-(-1)^n]}{n}$

$\times\dfrac{\sinh nx + \sinh n(\pi - x)}{\sinh n\pi}\sin ny$

EXERCISES 12.6 (PAGE 491)

1. $u(x,t) = 100 + \dfrac{200}{\pi}\displaystyle\sum_{n=1}^{\infty}\dfrac{(-1)^n - 1}{n}e^{-kn^2\pi^2 t}\sin n\pi x$

3. $u(x,t) = u_0 - \dfrac{r}{2k}x(x-1) + 2\displaystyle\sum_{n=1}^{\infty}\left[\dfrac{u_0}{n\pi} + \dfrac{r}{kn^3\pi^3}\right]$

$\times[(-1)^n - 1]e^{-kn^2\pi^2 t}\sin n\pi x$

5. $u(x,t) = \psi(x) + \displaystyle\sum_{n=1}^{\infty}A_n e^{-kn^2\pi^2 t}\sin n\pi x$,

where $\psi(x) = \dfrac{A}{k\beta^2}[-e^{-\beta x} + (e^{-\beta} - 1)x + 1]$

and $A_n = 2\displaystyle\int_0^1 [f(x) - \psi(x)]\sin n\pi x\,dx$

7. $\psi(x) = u_0\left(1 - \dfrac{\sinh\sqrt{h/k}\,x}{\sinh\sqrt{h/k}}\right)$

9. $u(x,t) = \dfrac{A}{6a^2}(x - x^3)$

$+ \dfrac{2A}{a^2\pi^3}\displaystyle\sum_{n=1}^{\infty}\dfrac{(-1)^n}{n^3}\cos n\pi at\sin n\pi x$

11. $u(x,y) = (u_0 - u_1)y + u_1$

$+ \dfrac{2}{\pi}\displaystyle\sum_{n=1}^{\infty}\dfrac{u_0(-1)^n - u_1}{n}e^{-n\pi x}\sin n\pi y$

13. $u(x,t) = 2\displaystyle\sum_{n=1}^{\infty}\dfrac{(-1)^{n+1}}{n(n^2 - 3)}e^{-3t}\sin nx$

$+ 2\displaystyle\sum_{n=1}^{\infty}\dfrac{(-1)^n}{n(n^2 - 3)}e^{-n^2 t}\sin nx$

15. $u(x,t) = \displaystyle\sum_{n=1}^{\infty}\dfrac{2}{n\pi}\left[-\dfrac{1}{n^2\pi^2} + (-1)^n\dfrac{n^2\pi^2\cos t - \sin t}{n^4\pi^4 + 1}\right]\sin n\pi x$

$+ \displaystyle\sum_{n=1}^{\infty}\left[\dfrac{4 - 2(-1)^n}{n^3\pi^3}\right.$

$\left. - (-1)^n\dfrac{2n\pi}{n^4\pi^4 + 1}\right]e^{-n^2\pi^2 t}\sin n\pi x$

EXERCISES 12.7 (PAGE 496)

1. $u(x,t) = 2h\displaystyle\sum_{n=1}^{\infty}\dfrac{\sin\alpha_n}{\alpha_n(h + \sin^2\alpha_n)}e^{-k\alpha_n^2 t}\cos\alpha_n x$, where

the α_n are the consecutive positive roots of $\cot\alpha = \alpha/h$

3. $u(x,y) = \displaystyle\sum_{n=1}^{\infty}A_n\sinh\alpha_n y\sin\alpha_n x$, where

$A_n = \dfrac{2h}{\sinh\alpha_n b(ah + \cos^2\alpha_n a)}\displaystyle\int_0^a f(x)\sin\alpha_n x\,dx$

and the α_n are the consecutive positive roots of $\tan\alpha a = -\alpha/h$

5. $u(x,t) = \displaystyle\sum_{n=1}^{\infty}A_n e^{-k(2n-1)^2\pi^2 t/4L^2}\sin\left(\dfrac{2n-1}{2L}\right)\pi x$, where

$A_n = \dfrac{2}{L}\displaystyle\int_0^L f(x)\sin\left(\dfrac{2n-1}{2L}\right)\pi x\,dx$

7. $u(x,y) = \dfrac{4u_0}{\pi}\displaystyle\sum_{n=1}^{\infty}\dfrac{1}{(2n-1)\cosh\left(\dfrac{2n-1}{2}\right)\pi}$

$\times\cosh\left(\dfrac{2n-1}{2}\right)\pi x\sin\left(\dfrac{2n-1}{2}\right)\pi y$

9. $u(x,t) = \displaystyle\sum_{n=1}^{\infty}\dfrac{4\sin\alpha_n}{\alpha_n^2(k\alpha_n^2 - 2)(1 + \cos^2\alpha_n)}$

$\times(e^{-2t} - e^{-k\alpha_n^2 t})\sin\alpha_n x$

EXERCISES 12.8 (PAGE 500)

1. $u(x,y,t) = \displaystyle\sum_{m=1}^{\infty}\sum_{n=1}^{\infty}A_{mn}e^{-k(m^2+n^2)t}\sin mx\sin ny$,

where $A_{mn} = \dfrac{4u_0}{mn\pi^2}[1 - (-1)^m][1 - (-1)^n]$

3. $u(x,y,t) = \displaystyle\sum_{m=1}^{\infty}\sum_{n=1}^{\infty}A_{mn}\sin mx\sin ny\cos a\sqrt{m^2 + n^2}\,t$,

where $A_{mn} = \dfrac{16}{m^3 n^3\pi^2}[(-1)^m - 1][(-1)^n - 1]$

5. $u(x,y,z) = \displaystyle\sum_{m=1}^{\infty}\sum_{n=1}^{\infty}A_{mn}\sinh\omega_{mn}z\sin\dfrac{m\pi}{a}x\sin\dfrac{n\pi}{b}y$,

where $\omega_{mn} = \sqrt{(m\pi/a)^2 + (n\pi/b)^2}$

$A_{mn} = \dfrac{4}{ab\sinh(c\omega_{mn})}\displaystyle\int_0^b\int_0^a f(x,y)$

$\times\sin\dfrac{m\pi}{a}x\sin\dfrac{n\pi}{b}y\,dx\,dy$

CHAPTER 12 IN REVIEW (PAGE 501)

1. $u = c_1 e^{(c_2 x + y/c_2)}$

3. $\psi(x) = u_0 + \dfrac{(u_1 - u_0)}{1 + \pi}x$

5. $u(x,t) = \dfrac{2h}{\pi^2 a}\displaystyle\sum_{n=1}^{\infty}\dfrac{\cos\dfrac{n\pi}{4} - \cos\dfrac{3n\pi}{4}}{n^2}\sin n\pi at\sin n\pi x$

7. $u(x,y) = \dfrac{100}{\pi}\displaystyle\sum_{n=1}^{\infty}\dfrac{1 - (-1)^n}{n\sinh n\pi}\sinh nx\sin ny$

9. $u(x,y) = \dfrac{100}{\pi}\displaystyle\sum_{n=1}^{\infty}\dfrac{1 - (-1)^n}{n}e^{-nx}\sin ny$

11. $u(x, t) = e^{-t} \sin x$

13. $u(x, t) = e^{-(x+t)} \sum_{n=1}^{\infty} A_n \left[\sqrt{n^2 + 1} \cos \sqrt{n^2 + 1}\, t \right.$
$$\left. + \sin \sqrt{n^2 + 1}\, t \right] \sin nx$$

EXERCISES 13.1 (PAGE 507)

1. $u(r, \theta) = \dfrac{u_0}{2} + \dfrac{u_0}{\pi} \sum_{n=1}^{\infty} \dfrac{1 - (-1)^n}{n} r^n \sin n\theta$

3. $u(r, \theta) = \dfrac{2\pi^2}{3} - 4\sum_{n=1}^{\infty} \dfrac{r^n}{n^2} \cos n\theta$

5. $u(r, \theta) = A_0 + \sum_{n=1}^{\infty} r^{-n}(A_n \cos n\theta + B_n \sin n\theta)$,

where $A_0 = \dfrac{1}{2\pi} \int_0^{2\pi} f(\theta)\, d\theta$

$A_n = \dfrac{c^n}{\pi} \int_0^{2\pi} f(\theta) \cos n\theta\, d\theta$

$B_n = \dfrac{c^n}{\pi} \int_0^{2\pi} f(\theta) \sin n\theta\, d\theta$

7. $u(r, \theta) = \dfrac{1}{2} + \dfrac{2}{\pi} \sum_{n=1}^{\infty} \dfrac{\sin \dfrac{n\pi}{2}}{n} \left(\dfrac{r}{c}\right)^{2n} \cos 2n\theta$

9. $u(r, \theta) = A_0 \ln\left(\dfrac{r}{b}\right) + \sum_{n=1}^{\infty} \left[\left(\dfrac{b}{r}\right)^n - \left(\dfrac{r}{b}\right)^n \right]$
$$\times\, [A_n \cos n\theta + B_n \sin n\theta],$$

where $A_0 \ln\left(\dfrac{a}{b}\right) = \dfrac{1}{2\pi} \int_0^{2\pi} f(\theta)\, d\theta$

$\left[\left(\dfrac{b}{a}\right)^n - \left(\dfrac{a}{b}\right)^n \right] A_n = \dfrac{1}{\pi} \int_0^{2\pi} f(\theta) \cos n\theta\, d\theta$

$\left[\left(\dfrac{b}{a}\right)^n - \left(\dfrac{a}{b}\right)^n \right] B_n = \dfrac{1}{\pi} \int_0^{2\pi} f(\theta) \sin n\theta\, d\theta$

11. $u(r, \theta) = \dfrac{4}{\pi} \sum_{n=1}^{\infty} \dfrac{1 - (-1)^n}{n^3} \dfrac{r^{2n} - b^{2n}}{a^{2n} - b^{2n}} \left(\dfrac{a}{r}\right)^n \sin n\theta$

13. $u(r, \theta) = \dfrac{u_0}{2} + \dfrac{2u_0}{\pi} \sum_{n=1}^{\infty} \dfrac{\sin \dfrac{n\pi}{2}}{n} \left(\dfrac{r}{2}\right)^n \cos n\theta$

EXERCISES 13.2 (PAGE 513)

1. $u(r, t) = \dfrac{2}{ac} \sum_{n=1}^{\infty} \dfrac{\sin \alpha_n at}{\alpha_n^2 J_1(\alpha_n c)} J_0(\alpha_n r)$

3. $u(r, z) = u_0 \sum_{n=1}^{\infty} \dfrac{\sinh \alpha_n (4 - z)}{\alpha_n \sinh 4\alpha_n J_1(2\alpha_n)} J_0(\alpha_n r)$

5. $u(r, t) = \sum_{n=1}^{\infty} A_n J_0(\alpha_n r) e^{-ka_n^2 t}$,

where $A_n = \dfrac{2}{c^2 J_1^2(\alpha_n c)} \int_0^c r J_0(\alpha_n r) f(r)\, dr$

7. $u(r, t) = \sum_{n=1}^{\infty} A_n J_0(\alpha_n r) e^{-ka_n^2 t}$,

where $A_n = \dfrac{2\alpha_n^2}{(\alpha_n^2 + h^2) J_0^2(\alpha_n)} \int_0^1 r J_0(\alpha_n r) f(r)\, dr$

9. $u(r, t) = 100 + 50 \sum_{n=1}^{\infty} \dfrac{J_1(\alpha_n) J_0(\alpha_n r)}{\alpha_n J_1^2(2\alpha_n)} e^{-\alpha_n^2 t}$

11. (b) $u(x, t) = \sum_{n=1}^{\infty} A_n \cos\left(\alpha_n \sqrt{g}\, t\right) J_0\!\left(2\alpha_n \sqrt{x}\right)$,

where $A_n = \dfrac{2}{L J_1^2\!\left(2\alpha_n \sqrt{L}\right)} \int_0^{\sqrt{L}} v J_0(2\alpha_n v) f(v^2)\, dv$

EXERCISES 13.3 (PAGE 517)

1. $u(r, \theta) = 50 \left[\dfrac{1}{2} P_0(\cos \theta) + \dfrac{3}{4}\left(\dfrac{r}{c}\right) P_1(\cos \theta) \right.$
$$\left. - \dfrac{7}{16}\left(\dfrac{r}{c}\right)^3 P_3(\cos \theta) + \dfrac{11}{32}\left(\dfrac{r}{c}\right)^5 P_5(\cos \theta) + \cdots \right]$$

3. $u(r, \theta) = \dfrac{r}{c} \cos \theta$

5. $u(r, \theta) = \sum_{n=0}^{\infty} A_n \dfrac{b^{2n+1} - r^{2n+1}}{b^{2n+1} r^{n+1}} P_n(\cos \theta)$, where

$\dfrac{b^{2n+1} - a^{2n+1}}{b^{2n+1} a^{n+1}} A_n = \dfrac{2n + 1}{2} \int_0^\pi f(\theta) P_n(\cos \theta) \sin \theta\, d\theta$

7. $u(r, \theta) = \sum_{n=0}^{\infty} A_{2n} r^{2n} P_{2n}(\cos \theta)$, where

$A_{2n} = \dfrac{4n + 1}{c^{2n}} \int_0^{\pi/2} f(\theta) P_{2n}(\cos \theta) \sin \theta\, d\theta$

9. $u(r, t) = 100 + \dfrac{200}{\pi r} \sum_{n=1}^{\infty} \dfrac{(-1)^n}{n} e^{-n^2 \pi^2 t} \sin n\pi r$

11. $u(r, t) = \dfrac{1}{r} \sum_{n=1}^{\infty} \left(A_n \cos \dfrac{n\pi a}{c} t + B_n \sin \dfrac{n\pi a}{c} t \right) \sin \dfrac{n\pi}{c} r$,

where $A_n = \dfrac{2}{c} \int_0^c r f(r) \sin \dfrac{n\pi}{c} r\, dr$,

$B_n = \dfrac{2}{n\pi a} \int_0^c r g(r) \sin \dfrac{n\pi}{c} r\, dr$

CHAPTER 13 IN REVIEW (PAGE 518)

1. $u(r, \theta) = \dfrac{2u_0}{\pi} \sum_{n=1}^{\infty} \dfrac{1 - (-1)^n}{n} \left(\dfrac{r}{c}\right)^n \sin n\theta$

3. $u(r, \theta) = \dfrac{4u_0}{\pi} \sum_{n=1}^{\infty} \dfrac{1 - (-1)^n}{n^3} r^n \sin n\theta$

5. $u(r, \theta) = \dfrac{2u_0}{\pi} \sum\limits_{n=1}^{\infty} \dfrac{r^{4n} + r^{-4n}}{2^{4n} + 2^{-4n}} \dfrac{1 - (-1)^n}{n} \sin 4n\theta$

7. $u(r, t) = 2e^{-ht} \sum\limits_{n=1}^{\infty} \dfrac{1}{\alpha_n J_1(\alpha_n)} J_0(\alpha_n r)\, e^{-\alpha_n^2 t}$

9. $u(r, z) = 50 \sum\limits_{n=1}^{\infty} \dfrac{\cosh \alpha_n z}{\alpha_n \cosh 4\alpha_n J_1(2\alpha_n)} J_0(\alpha_n r)$

11. $u(r, \theta) = 100\left[\dfrac{3}{2} rP_1(\cos\theta) - \dfrac{7}{8} r^3 P_3(\cos\theta)\right.$
$\left. + \dfrac{11}{16} r^5 P_5(\cos\theta) + \cdots\right]$

EXERCISES 14.1 (PAGE 521)

1. (a) Let $\tau = u^2$ in the integral $\mathrm{erf}(\sqrt{t})$.
7. $y(t) = e^{\pi t}\,\mathrm{erfc}(\sqrt{\pi t})$
9. Use the property $\displaystyle\int_0^b - \int_0^a = \int_0^b + \int_a^0$.

EXERCISES 14.2 (PAGE 527)

1. $u(x, t) = A \cos\dfrac{a\pi t}{L} \sin\dfrac{\pi x}{L}$

3. $u(x, t) = f\left(t - \dfrac{x}{a}\right)\mathcal{U}\left(t - \dfrac{x}{a}\right)$

5. $u(x, t) = \left[\dfrac{1}{2} g\left(t - \dfrac{x}{a}\right)^2 + A \sin\omega\left(t - \dfrac{x}{a}\right)\right]$
$\times \mathcal{U}\left(t - \dfrac{x}{a}\right) - \dfrac{1}{2} gt^2$

7. $u(x, t) = a\dfrac{F_0}{E}\sum\limits_{n=0}^{\infty}(-1)^n\left\{\left(t - \dfrac{2nL + L - x}{a}\right)\right.$
$\times \mathcal{U}\left(t - \dfrac{2nL + L - x}{a}\right)$
$- \left(t - \dfrac{2nL + L + x}{a}\right)$
$\left.\times \mathcal{U}\left(t - \dfrac{2nL + L + x}{a}\right)\right\}$

9. $u(x, t) = (t - x)\sinh(t - x)\mathcal{U}(t - x)$
$+ xe^{-x}\cosh t - e^{-x}t \sinh t$

11. $u(x, t) = u_1 + (u_0 - u_1)\,\mathrm{erfc}\left(\dfrac{x}{2\sqrt{t}}\right)$

13. $u(x, t) = u_0\left[1 - \left\{\mathrm{erfc}\left(\dfrac{x}{2\sqrt{t}}\right)\right.\right.$
$\left.\left. - e^{x+t}\,\mathrm{erfc}\left(\sqrt{t} + \dfrac{x}{2\sqrt{t}}\right)\right\}\right]$

15. $u(x, t) = \dfrac{x}{2\sqrt{\pi}}\int_0^t \dfrac{f(t - \tau)}{\tau^{3/2}} e^{-x^2/4\tau}\, d\tau$

17. $u(x, t) = 60 + 40\,\mathrm{erfc}\left(\dfrac{x}{2\sqrt{t - 2}}\right)\mathcal{U}(t - 2)$

19. $u(x, t) = 100\left[-e^{1-x+t}\,\mathrm{erfc}\left(\sqrt{t} + \dfrac{1 - x}{2\sqrt{t}}\right)\right.$
$\left. + \mathrm{erfc}\left(\dfrac{1 - x}{2\sqrt{t}}\right)\right]$

21. $u(x, t) = u_0 + u_0 e^{-(\pi^2/L^2)t}\sin\left(\dfrac{\pi}{L}x\right)$

23. $u(x, t) = u_0 - u_0\sum\limits_{n=0}^{\infty}(-1)^n\left[\mathrm{erfc}\left(\dfrac{2n + 1 - x}{2\sqrt{kt}}\right)\right.$
$\left. + \mathrm{erfc}\left(\dfrac{2n + 1 + x}{2\sqrt{kt}}\right)\right]$

25. $u(x, t) = u_0 e^{-Gt/C}\,\mathrm{erf}\left(\dfrac{x}{2}\sqrt{\dfrac{RC}{t}}\right)$

27. (a) $u(x, t) = \dfrac{v_0^2 F_0}{a^2 - v_0^2}\left[\left(t - \dfrac{x}{v_0}\right)\mathcal{U}\left(t - \dfrac{x}{v_0}\right)\right.$
$\left. - \left(t - \dfrac{x}{a}\right)\mathcal{U}\left(t - \dfrac{x}{a}\right)\right]$

(b) $u(x, t) = -\dfrac{xF_0}{2a}\mathcal{U}\left(t - \dfrac{x}{a}\right)$

EXERCISES 14.3 (PAGE 534)

1. $f(x) = \dfrac{1}{\pi}\int_0^{\infty} \dfrac{\sin\alpha\cos\alpha x + 3(1 - \cos\alpha)\sin\alpha x}{\alpha}\, d\alpha$

3. $f(x) = \dfrac{1}{\pi}\int_0^{\infty}[A(\alpha)\cos\alpha x + B(\alpha)\sin\alpha x]\, d\alpha,$
where $A(\alpha) = \dfrac{3\alpha\sin 3\alpha + \cos 3\alpha - 1}{\alpha^2}$
$B(\alpha) = \dfrac{\sin 3\alpha - 3\alpha\cos 3\alpha}{\alpha^2}$

5. $f(x) = \dfrac{1}{\pi}\int_0^{\infty} \dfrac{\cos\alpha x + \alpha\sin\alpha x}{1 + \alpha^2}\, d\alpha$

7. $f(x) = \dfrac{10}{\pi}\int_0^{\infty} \dfrac{(1 - \cos\alpha)\sin\alpha x}{\alpha}\, d\alpha$

9. $f(x) = \dfrac{2}{\pi}\int_0^{\infty} \dfrac{(\pi\alpha\sin\pi\alpha + \cos\pi\alpha - 1)\cos\alpha x}{\alpha^2}\, d\alpha$

11. $f(x) = \dfrac{4}{\pi}\int_0^{\infty} \dfrac{\alpha\sin\alpha x}{4 + \alpha^4}\, d\alpha$

13. $f(x) = \dfrac{2k}{\pi}\int_0^{\infty} \dfrac{\cos\alpha x}{k^2 + \alpha^2}\, d\alpha$
$f(x) = \dfrac{2}{\pi}\int_0^{\infty} \dfrac{\alpha\sin\alpha x}{k^2 + \alpha^2}\, d\alpha$

15. $f(x) = \dfrac{2}{\pi}\int_0^{\infty} \dfrac{(4 - \alpha^2)\cos\alpha x}{(4 + \alpha^2)^2}\, d\alpha$
$f(x) = \dfrac{8}{\pi}\int_0^{\infty} \dfrac{\alpha\sin\alpha x}{(4 + \alpha^2)^2}\, d\alpha$

17. $f(x) = \dfrac{2}{\pi}\dfrac{1}{1 + x^2},\quad x > 0$

19. Let $x = 2$ in (7). Use a trigonometric identity and replace α by x. In part (b) make the change of variable $2x = kt$.

21. $u(x, y) = \dfrac{1}{2\sqrt{\pi}} \displaystyle\int_{-\infty}^{\infty} \dfrac{e^{-\alpha^2/4} \cosh \alpha y}{\cosh \alpha} e^{-i\alpha x} \, d\alpha$

$\qquad = \dfrac{1}{2\sqrt{\pi}} \displaystyle\int_{-\infty}^{\infty} \dfrac{e^{-\alpha^2/4} \cosh \alpha y}{\cosh \alpha} \cos \alpha x \, d\alpha$

EXERCISES 14.4 (PAGE 540)

1. $u(x, t) = \dfrac{1}{\pi} \displaystyle\int_{-\infty}^{\infty} \dfrac{e^{-k\alpha^2 t}}{1 + \alpha^2} e^{-i\alpha x} \, d\alpha$

$\qquad = \dfrac{1}{\pi} \displaystyle\int_{-\infty}^{\infty} \dfrac{\cos \alpha x}{1 + \alpha^2} e^{-k\alpha^2 t} \, d\alpha$

3. $u(x, t) = \dfrac{2u_0}{\pi} \displaystyle\int_{-\infty}^{\infty} \dfrac{1 - e^{-k\alpha^2 t}}{\alpha} \sin \alpha x \, d\alpha$

5. $u(x, t) = \dfrac{2}{\pi} \displaystyle\int_{0}^{\infty} \dfrac{1 - \cos \alpha}{\alpha} e^{-k\alpha^2 t} \sin \alpha x \, d\alpha$

7. $u(x, t) = \dfrac{2}{\pi} \displaystyle\int_{0}^{\infty} \dfrac{\sin \alpha}{\alpha} e^{-k\alpha^2 t} \cos \alpha x \, d\alpha$

9. (a) $u(x, t) = \dfrac{1}{2\pi} \displaystyle\int_{-\infty}^{\infty} \left(F(\alpha) \cos \alpha a t \right.$

$\qquad \left. + \, G(\alpha) \dfrac{\sin \alpha a t}{\alpha a} \right) e^{-i\alpha x} \, d\alpha$

11. $u(x, y) = \dfrac{2}{\pi} \displaystyle\int_{0}^{\infty} \dfrac{\sinh \alpha(\pi - x)}{(1 + \alpha^2) \sinh \alpha \pi} \cos \alpha y \, d\alpha$

13. $u(x, y) = \dfrac{100}{\pi} \displaystyle\int_{0}^{\infty} \dfrac{\sin \alpha}{\alpha} e^{-\alpha y} \cos \alpha x \, d\alpha$

15. $u(x, y) = \dfrac{2}{\pi} \displaystyle\int_{0}^{\infty} F(\alpha) \dfrac{\sinh \alpha(2 - y)}{\sinh 2\alpha} \sin \alpha x \, d\alpha$

17. $u(x, y) = \dfrac{2}{\pi} \displaystyle\int_{0}^{\infty} \dfrac{\alpha}{1 + \alpha^2} [e^{-\alpha x} \sin \alpha y + e^{-\alpha y} \sin \alpha x] \, d\alpha$

19. $u(x, t) = \dfrac{1}{\sqrt{1 + 4kt}} e^{-x^2/(1 + 4kt)}$

CHAPTER 14 IN REVIEW (PAGE 541)

1. $u(x, y) = \dfrac{2}{\pi} \displaystyle\int_{0}^{\infty} \dfrac{\sinh \alpha y}{\alpha(1 + \alpha^2) \cosh \alpha \pi} \cos \alpha x \, d\alpha$

3. $u(x, t) = u_0 e^{-ht} \operatorname{erf}\left(\dfrac{x}{2\sqrt{t}} \right)$

5. $u(x, t) = \displaystyle\int_{0}^{t} \operatorname{erfc}\left(\dfrac{x}{2\sqrt{\tau}} \right) d\tau$

7. $u(x, t) = \dfrac{u_0}{2\pi} \displaystyle\int_{-\infty}^{\infty} \dfrac{\sin \alpha(\pi - x) + \sin \alpha x}{\alpha} e^{-k\alpha^2 t} \, d\alpha$

9. $u(x, y) = \dfrac{100}{\pi} \displaystyle\int_{0}^{\infty} \left(\dfrac{1 - \cos \alpha}{\alpha} \right)$

$\qquad \times [e^{-\alpha x} \sin \alpha y + 2e^{-\alpha y} \sin \alpha x] \, d\alpha$

11. $u(x, y) = \dfrac{2}{\pi} \displaystyle\int_{0}^{\infty} \left(\dfrac{B \cosh \alpha y}{(1 + \alpha^2) \sinh \alpha \pi} + \dfrac{A}{\alpha} \right) \sin \alpha x \, d\alpha$

13. $u(x, y) = \dfrac{1}{2\pi} \displaystyle\int_{-\infty}^{\infty} \dfrac{\cos \alpha x + \alpha \sin \alpha x}{1 + \alpha^2} e^{-k\alpha^2 t} \, d\alpha$

EXERCISES 15.1 (PAGE 549)

1. $u_{11} = \frac{11}{15}, u_{21} = \frac{14}{15}$

3. $u_{11} = u_{21} = \sqrt{3}/16, u_{22} = u_{12} = 3\sqrt{3}/16$

5. $u_{21} = u_{12} = 12.50, u_{31} = u_{13} = 18.75, u_{32} = u_{23} = 37.50,$
$\quad u_{11} = 6.25, u_{22} = 25.00, u_{33} = 56.25$

7. (b) $u_{14} = u_{41} = 0.5427, u_{24} = u_{42} = 0.6707,$
$\qquad u_{34} = u_{43} = 0.6402, u_{33} = 0.4451, u_{44} = 0.9451$

EXERCISES 15.2 (PAGE 554)

The tables in this section give a selection of the total number of approximations.

1.

Time	x = 0.25	x = 0.50	x = 0.75	x = 1.00	x = 1.25	x = 1.50	x = 1.75
0.000	1.0000	1.0000	1.0000	1.0000	0.0000	0.0000	0.0000
0.100	0.3728	0.6288	0.6800	0.5904	0.3840	0.2176	0.0768
0.200	0.2248	0.3942	0.4708	0.4562	0.3699	0.2517	0.1239
0.300	0.1530	0.2752	0.3448	0.3545	0.3101	0.2262	0.1183
0.400	0.1115	0.2034	0.2607	0.2757	0.2488	0.1865	0.0996
0.500	0.0841	0.1545	0.2002	0.2144	0.1961	0.1487	0.0800
0.600	0.0645	0.1189	0.1548	0.1668	0.1534	0.1169	0.0631
0.700	0.0499	0.0921	0.1201	0.1297	0.1196	0.0914	0.0494
0.800	0.0387	0.0715	0.0933	0.1009	0.0931	0.0712	0.0385
0.900	0.0301	0.0555	0.0725	0.0785	0.0725	0.0554	0.0300
1.000	0.0234	0.0432	0.0564	0.0610	0.0564	0.0431	0.0233

3.

Time	x = 0.25	x = 0.50	x = 0.75	x = 1.00	x = 1.25	x = 1.50	x = 1.75
0.000	1.0000	1.0000	1.0000	1.0000	0.0000	0.0000	0.0000
0.100	0.4015	0.6577	0.7084	0.5837	0.3753	0.1871	0.0684
0.200	0.2430	0.4198	0.4921	0.4617	0.3622	0.2362	0.1132
0.300	0.1643	0.2924	0.3604	0.3626	0.3097	0.2208	0.1136
0.400	0.1187	0.2150	0.2725	0.2843	0.2528	0.1871	0.0989
0.500	0.0891	0.1630	0.2097	0.2228	0.2020	0.1521	0.0814
0.600	0.0683	0.1256	0.1628	0.1746	0.1598	0.1214	0.0653
0.700	0.0530	0.0976	0.1270	0.1369	0.1259	0.0959	0.0518
0.800	0.0413	0.0762	0.0993	0.1073	0.0989	0.0755	0.0408
0.900	0.0323	0.0596	0.0778	0.0841	0.0776	0.0593	0.0321
1.000	0.0253	0.0466	0.0609	0.0659	0.0608	0.0465	0.0252

Absolute errors are approximately 2.2×10^{-2}, 3.7×10^{-2}, 1.3×10^{-2}.

5.

Time	x = 0.25	x = 0.50	x = 0.75	x = 1.00	x = 1.25	x = 1.50	x = 1.75
0.00	1.0000	1.0000	1.0000	1.0000	0.0000	0.0000	0.0000
0.10	0.3972	0.6551	0.7043	0.5883	0.3723	0.1955	0.0653
0.20	0.2409	0.4171	0.4901	0.4620	0.3636	0.2385	0.1145
0.30	0.1631	0.2908	0.3592	0.3624	0.3105	0.2220	0.1145
0.40	0.1181	0.2141	0.2718	0.2840	0.2530	0.1876	0.0993
0.50	0.0888	0.1625	0.2092	0.2226	0.2020	0.1523	0.0816
0.60	0.0681	0.1253	0.1625	0.1744	0.1597	0.1214	0.0654
0.70	0.0528	0.0974	0.1268	0.1366	0.1257	0.0959	0.0518
0.80	0.0412	0.0760	0.0991	0.1071	0.0987	0.0754	0.0408
0.90	0.0322	0.0594	0.0776	0.0839	0.0774	0.0592	0.0320
1.00	0.0252	0.0465	0.0608	0.0657	0.0607	0.0464	0.0251

Absolute errors are approximately 1.8×10^{-2}, 3.7×10^{-2}, 1.3×10^{-2}.

7. (a)

Time	x = 2.00	x = 4.00	x = 6.00	x = 8.00	x = 10.00	x = 12.00	x = 14.00	x = 16.00	x = 18.00
0.00	30.0000	30.0000	30.0000	30.0000	30.0000	30.0000	30.0000	30.0000	30.0000
2.00	27.6450	29.9037	29.9970	29.9999	30.0000	29.9999	29.9970	29.9037	27.6450
4.00	25.6452	29.6517	29.9805	29.9991	29.9999	29.9991	29.9805	29.6517	25.6452
6.00	23.9347	29.2922	29.9421	29.9963	29.9996	29.9963	29.9421	29.2922	23.9347
8.00	22.4612	28.8606	29.8782	29.9898	29.9986	29.9898	29.8782	28.8606	22.4612
10.00	21.1829	28.3831	29.7878	29.9782	29.9964	29.9782	29.7878	28.3831	21.1829

(b)

Time	x = 5.00	x = 10.00	x = 15.00	x = 20.00	x = 25.00	x = 30.00	x = 35.00	x = 40.00	x = 45.00
0.00	30.0000	30.0000	30.0000	30.0000	30.0000	30.0000	30.0000	30.0000	30.0000
2.00	29.5964	29.9973	30.0000	30.0000	30.0000	30.0000	30.0000	29.9973	29.5964
4.00	29.2036	29.9893	29.9999	30.0000	30.0000	30.0000	29.9999	29.9893	29.2036
6.00	28.8212	29.9762	29.9997	30.0000	30.0000	30.0000	29.9997	29.9762	28.8213
8.00	28.4490	29.9585	29.9992	30.0000	30.0000	30.0000	29.9993	29.9585	28.4490
10.00	28.0864	29.9363	29.9986	30.0000	30.0000	30.0000	29.9986	29.9363	28.0864

(c)

Time	$x = 2.00$	$x = 4.00$	$x = 6.00$	$x = 8.00$	$x = 10.00$	$x = 12.00$	$x = 14.00$	$x = 16.00$	$x = 18.00$
0.00	18.0000	32.0000	42.0000	48.0000	50.0000	48.0000	42.0000	32.0000	18.0000
2.00	15.3312	28.5348	38.3465	44.3067	46.3001	44.3067	38.3465	28.5348	15.3312
4.00	13.6371	25.6867	34.9416	40.6988	42.6453	40.6988	34.9416	25.6867	13.6371
6.00	12.3012	23.2863	31.8624	37.2794	39.1273	37.2794	31.8624	23.2863	12.3012
8.00	11.1659	21.1877	29.0757	34.0984	35.8202	34.0984	29.0757	21.1877	11.1659
10.00	10.1665	19.3143	26.5439	31.1662	32.7549	31.1662	26.5439	19.3143	10.1665

(d)

Time	$x = 10.00$	$x = 20.00$	$x = 30.00$	$x = 40.00$	$x = 50.00$	$x = 60.00$	$x = 70.00$	$x = 80.00$	$x = 90.00$
0.00	8.0000	16.0000	24.0000	32.0000	40.0000	32.0000	24.0000	16.0000	8.0000
2.00	8.0000	16.0000	23.9999	31.9918	39.4932	31.9918	23.9999	16.0000	8.0000
4.00	8.0000	16.0000	23.9993	31.9686	39.0175	31.9686	23.9993	16.0000	8.0000
6.00	8.0000	15.9999	23.9978	31.9323	38.5701	31.9323	23.9978	15.9999	8.0000
8.00	8.0000	15.9998	23.9950	31.8844	38.1483	31.8844	23.9950	15.9998	8.0000
10.00	8.0000	15.9996	23.9908	31.8265	37.7498	31.8265	23.9908	15.9996	8.0000

9. (a)

Time	$x = 2.00$	$x = 4.00$	$x = 6.00$	$x = 8.00$	$x = 10.00$	$x = 12.00$	$x = 14.00$	$x = 16.00$	$x = 18.00$
0.00	30.0000	30.0000	30.0000	30.0000	30.0000	30.0000	30.0000	30.0000	30.0000
2.00	27.6450	29.9037	29.9970	29.9999	30.0000	30.0000	29.9990	29.9679	29.2150
4.00	25.6452	29.6517	29.9805	29.9991	30.0000	29.9997	29.9935	29.8839	28.5484
6.00	23.9347	29.2922	29.9421	29.9963	29.9997	29.9988	29.9807	29.7641	27.9782
8.00	22.4612	28.8606	29.8782	29.9899	29.9991	29.9966	29.9594	29.6202	27.4870
10.00	21.1829	28.3831	29.7878	29.9783	29.9976	29.9927	29.9293	29.4610	27.0610

(b)

Time	$x = 5.00$	$x = 10.00$	$x = 15.00$	$x = 20.00$	$x = 25.00$	$x = 30.00$	$x = 35.00$	$x = 40.00$	$x = 45.00$
0.00	30.0000	30.0000	30.0000	30.0000	30.0000	30.0000	30.0000	30.0000	30.0000
2.00	29.5964	29.9973	30.0000	30.0000	30.0000	30.0000	30.0000	29.9991	29.8655
4.00	29.2036	29.9893	29.9999	30.0000	30.0000	30.0000	30.0000	29.9964	29.7345
6.00	28.8212	29.9762	29.9997	30.0000	30.0000	30.0000	29.9999	29.9921	29.6071
8.00	28.4490	29.9585	29.9992	30.0000	30.0000	30.0000	29.9997	29.9862	29.4830
10.00	28.0864	29.9363	29.9986	30.0000	30.0000	30.0000	29.9995	29.9788	29.3621

(c)

Time	$x = 2.00$	$x = 4.00$	$x = 6.00$	$x = 8.00$	$x = 10.00$	$x = 12.00$	$x = 14.00$	$x = 16.00$	$x = 18.00$
0.00	18.0000	32.0000	42.0000	48.0000	50.0000	48.0000	42.0000	32.0000	18.0000
2.00	15.3312	28.5350	38.3477	44.3130	46.3327	44.4671	39.0872	31.5755	24.6930
4.00	13.6381	25.6913	34.9606	40.7728	42.9127	41.5716	37.4340	31.7086	25.6986
6.00	12.3088	23.3146	31.9546	37.5566	39.8880	39.1565	35.9745	31.2134	25.7128
8.00	11.1946	21.2785	29.3217	34.7092	37.2109	36.9834	34.5032	30.4279	25.4167
10.00	10.2377	19.5150	27.0178	32.1929	34.8117	34.9710	33.0338	29.5224	25.0019

(d)

Time	$x = 10.00$	$x = 20.00$	$x = 30.00$	$x = 40.00$	$x = 50.00$	$x = 60.00$	$x = 70.00$	$x = 80.00$	$x = 90.00$
0.00	8.0000	16.0000	24.0000	32.0000	40.0000	32.0000	24.0000	16.0000	8.0000
2.00	8.0000	16.0000	23.9999	31.9918	39.4932	31.9918	24.0000	16.0102	8.6333
4.00	8.0000	16.0000	23.9993	31.9686	39.0175	31.9687	24.0002	16.0391	9.2272
6.00	8.0000	15.9999	23.9978	31.9323	38.5701	31.9324	24.0005	16.0845	9.7846
8.00	8.0000	15.9998	23.9950	31.8844	38.1483	31.8846	24.0012	16.1441	10.3084
10.00	8.0000	15.9996	23.9908	31.8265	37.7499	31.8269	24.0023	16.2160	10.8012

11. (a) $\psi(x) = \frac{1}{2}x + 20$

(b)

Time	$x = 4.00$	$x = 8.00$	$x = 12.00$	$x = 16.00$
0.00	50.0000	50.0000	50.0000	50.0000
10.00	32.7433	44.2679	45.4228	38.2971
30.00	26.9487	32.1409	34.0874	32.9644
50.00	24.1178	27.4348	29.4296	30.1207
70.00	22.8995	25.4560	27.4554	28.8998
90.00	22.3817	24.6176	26.6175	28.3817
110.00	22.1619	24.2620	26.2620	28.1619
130.00	22.0687	24.1112	26.1112	28.0687
150.00	22.0291	24.0472	26.0472	28.0291
170.00	22.0124	24.0200	26.0200	28.0124
190.00	22.0052	24.0085	26.0085	28.0052
210.00	22.0022	24.0036	26.0036	28.0022
230.00	22.0009	24.0015	26.0015	28.0009
250.00	22.0004	24.0007	26.0007	28.0004
270.00	22.0002	24.0003	26.0003	28.0002
290.00	22.0001	24.0001	26.0001	28.0001
310.00	22.0000	24.0001	26.0001	28.0000
330.00	22.0000	24.0000	26.0000	28.0000
350.00	22.0000	24.0000	26.0000	28.0000

EXERCISES 15.3 (PAGE 557)

The tables in this section give a selection of the total number of approximations.

1. (a)

Time	$x = 0.25$	$x = 0.50$	$x = 0.75$
0.00	0.1875	0.2500	0.1875
0.20	0.1491	0.2100	0.1491
0.40	0.0556	0.0938	0.0556
0.60	−0.0501	−0.0682	−0.0501
0.80	−0.1361	−0.2072	−0.1361
1.00	−0.1802	−0.2591	−0.1802

(b)

Time	$x = 0.4$	$x = 0.8$	$x = 1.2$	$x = 1.6$
0.00	0.0032	0.5273	0.5273	0.0032
0.20	0.0652	0.4638	0.4638	0.0652
0.40	0.2065	0.3035	0.3035	0.2065
0.60	0.3208	0.1190	0.1190	0.3208
0.80	0.3094	−0.0180	−0.0180	0.3094
1.00	0.1450	−0.0768	−0.0768	0.1450

(c)

Time	$x = 0.1$	$x = 0.2$	$x = 0.3$	$x = 0.4$	$x = 0.5$	$x = 0.6$	$x = 0.7$	$x = 0.8$	$x = 0.9$
0.00	0.0000	0.0000	0.0000	0.0000	0.0000	0.5000	0.5000	0.5000	0.5000
0.12	0.0000	0.0000	0.0082	0.1126	0.3411	0.1589	0.3792	0.3710	0.0462
0.24	0.0071	0.0657	0.2447	0.3159	0.1735	0.2463	−0.1266	−0.3056	−0.0625
0.36	0.1623	0.3197	0.2458	0.1657	0.0877	−0.2853	−0.2843	−0.2104	−0.2887
0.48	0.1965	0.1410	0.1149	−0.1216	−0.3593	−0.2381	−0.1977	−0.1715	0.0800
0.60	−0.2194	−0.2069	−0.3875	−0.3411	−0.1901	−0.1662	−0.0666	0.1140	−0.0446
0.72	−0.3003	−0.6865	−0.5097	−0.3230	−0.1585	0.0156	0.0893	−0.0874	0.0384
0.84	−0.2647	−0.1633	−0.3546	−0.3214	−0.1763	−0.0954	−0.1249	0.0665	−0.0386
0.96	0.3012	0.1081	0.1380	−0.0487	−0.2974	−0.3407	−0.1250	−0.1548	0.0092

3. (a)

Time	$x = 0.2$	$x = 0.4$	$x = 0.6$	$x = 0.8$
0.00	0.5878	0.9511	0.9511	0.5878
0.10	0.5599	0.9059	0.9059	0.5599
0.20	0.4788	0.7748	0.7748	0.4788
0.30	0.3524	0.5701	0.5701	0.3524
0.40	0.1924	0.3113	0.3113	0.1924
0.50	0.0142	0.0230	0.0230	0.0142

(b)

Time	$x = 0.2$	$x = 0.4$	$x = 0.6$	$x = 0.8$
0.00	0.5878	0.9511	0.9511	0.5878
0.05	0.5808	0.9397	0.9397	0.5808
0.10	0.5599	0.9060	0.9060	0.5599
0.15	0.5257	0.8507	0.8507	0.5257
0.20	0.4790	0.7750	0.7750	0.4790
0.25	0.4209	0.6810	0.6810	0.4209
0.30	0.3527	0.5706	0.5706	0.3527
0.35	0.2761	0.4467	0.4467	0.2761
0.40	0.1929	0.3122	0.3122	0.1929
0.45	0.1052	0.1701	0.1701	0.1052
0.50	0.0149	0.0241	0.0241	0.0149

5.

Time	$x = 10$	$x = 20$	$x = 30$	$x = 40$	$x = 50$
0.00000	0.1000	0.2000	0.3000	0.2000	0.1000
0.60134	0.0984	0.1688	0.1406	0.1688	0.0984
1.20268	0.0226	−0.0121	0.0085	−0.0121	0.0226
1.80401	−0.1271	−0.1347	−0.1566	−0.1347	−0.1271
2.40535	−0.0920	−0.2292	−0.2571	−0.2292	−0.0920
3.00669	−0.0932	−0.1445	−0.2018	−0.1445	−0.0932
3.60803	−0.0284	−0.0205	0.0336	−0.0205	−0.0284
4.20936	0.1064	0.1555	0.1265	0.1555	0.1064
4.81070	0.1273	0.2060	0.2612	0.2060	0.1273
5.41204	0.0625	0.1689	0.2038	0.1689	0.0625
6.01338	0.0436	0.0086	−0.0080	0.0086	0.0436
6.61472	−0.0931	−0.1364	−0.1578	−0.1364	−0.0931
7.21605	−0.1436	−0.2173	−0.2240	−0.2173	−0.1436
7.81739	−0.0625	−0.1644	−0.2247	−0.1644	−0.0625
8.41873	−0.0287	−0.0192	−0.0085	−0.0192	−0.0287
9.02007	0.0654	0.1332	0.1755	0.1332	0.0654
9.62140	0.1540	0.2189	0.2089	0.2189	0.1540

Note: Time is expressed in milliseconds.

CHAPTER 15 IN REVIEW (PAGE 558)

1. $u_{11} = 0.8929$, $u_{21} = 3.5714$, $u_{31} = 13.3929$

3. (a)

$x = 0.20$	$x = 0.40$	$x = 0.60$	$x = 0.80$
0.2000	0.4000	0.6000	0.8000
0.2000	0.4000	0.6000	0.5500
0.2000	0.4000	0.5375	0.4250
0.2000	0.3844	0.4750	0.3469
0.1961	0.3609	0.4203	0.2922
0.1883	0.3346	0.3734	0.2512

(b)

$x = 0.20$	$x = 0.40$	$x = 0.60$	$x = 0.80$
0.2000	0.4000	0.6000	0.8000
0.2000	0.4000	0.6000	0.8000
0.2000	0.4000	0.6000	0.5500
0.2000	0.4000	0.5375	0.4250
0.2000	0.3844	0.4750	0.3469
0.1961	0.3609	0.4203	0.2922

(c) Yes; the table in part (b) is the table in part (a) shifted downward.

EXERCISES FOR APPENDIX I (PAGE APP-2)

1. (a) 24   **(b)** 720   **(c)** $\dfrac{4\sqrt{\pi}}{3}$ **(d)** $-\dfrac{8\sqrt{\pi}}{15}$

3. 0.297

EXERCISES FOR APPENDIX II (PAGE APP-18)

1. (a) $\begin{pmatrix} 2 & 11 \\ 2 & -1 \end{pmatrix}$ **(b)** $\begin{pmatrix} -6 & 1 \\ 14 & -19 \end{pmatrix}$

(c) $\begin{pmatrix} 2 & 28 \\ 12 & -12 \end{pmatrix}$

3. (a) $\begin{pmatrix} -11 & 6 \\ 17 & -22 \end{pmatrix}$ **(b)** $\begin{pmatrix} -32 & 27 \\ -4 & -1 \end{pmatrix}$

(c) $\begin{pmatrix} 19 & -18 \\ -30 & 31 \end{pmatrix}$ **(d)** $\begin{pmatrix} 19 & 6 \\ 3 & 22 \end{pmatrix}$

5. (a) $\begin{pmatrix} 9 & 24 \\ 3 & 8 \end{pmatrix}$ **(b)** $\begin{pmatrix} 3 & 8 \\ -6 & -16 \end{pmatrix}$

(c) $\begin{pmatrix} 0 & 0 \\ 0 & 0 \end{pmatrix}$ **(d)** $\begin{pmatrix} -4 & -5 \\ 8 & 10 \end{pmatrix}$

7. (a) 180   **(b)** $\begin{pmatrix} 4 & 8 & 10 \\ 8 & 16 & 20 \\ 10 & 20 & 25 \end{pmatrix}$

(c) $\begin{pmatrix} 6 \\ 12 \\ -5 \end{pmatrix}$

9. (a) $\begin{pmatrix} 7 & 38 \\ 10 & 75 \end{pmatrix}$ **(b)** $\begin{pmatrix} 7 & 38 \\ 10 & 75 \end{pmatrix}$

11. $\begin{pmatrix} -14 \\ 1 \end{pmatrix}$

13. $\begin{pmatrix} -38 \\ -2 \end{pmatrix}$

15. singular

17. nonsingular; $\mathbf{A}^{-1} = \dfrac{1}{4}\begin{pmatrix} -5 & -8 \\ 3 & 4 \end{pmatrix}$

19. nonsingular; $\mathbf{A}^{-1} = \dfrac{1}{2}\begin{pmatrix} 0 & -1 & 1 \\ 2 & 2 & -2 \\ -4 & -3 & 5 \end{pmatrix}$

21. nonsingular; $\mathbf{A}^{-1} = -\dfrac{1}{9}\begin{pmatrix} -2 & -2 & -1 \\ -13 & 5 & 7 \\ 8 & -1 & -5 \end{pmatrix}$

23. $\mathbf{A}^{-1}(t) = \dfrac{1}{2e^{3t}}\begin{pmatrix} 3e^{4t} & -e^{4t} \\ -4e^{-t} & 2e^{-t} \end{pmatrix}$

25. $\dfrac{d\mathbf{X}}{dt} = \begin{pmatrix} -5e^{-t} \\ -2e^{-t} \\ 7e^{-t} \end{pmatrix}$

27. $\dfrac{d\mathbf{X}}{dt} = 4\begin{pmatrix} 1 \\ -1 \end{pmatrix}e^{2t} - 12\begin{pmatrix} 2 \\ 1 \end{pmatrix}e^{-3t}$

29. (a) $\begin{pmatrix} 4e^{4t} & -\pi\sin\pi t \\ 2 & 6t \end{pmatrix}$ **(b)** $\begin{pmatrix} \frac{1}{4}e^8 - \frac{1}{4} & 0 \\ 4 & 6 \end{pmatrix}$

(c) $\begin{pmatrix} \frac{1}{4}e^{4t} - \frac{1}{4} & (1/\pi)\sin\pi t \\ t^2 & t^3 - t \end{pmatrix}$

31. $x = 3$, $y = 1$, $z = -5$

33. $x = 2 + 4t$, $y = -5 - t$, $z = t$

35. $x = -\frac{1}{2}$, $y = \frac{3}{2}$, $z = \frac{7}{2}$

37. $x_1 = 1$, $x_2 = 0$, $x_3 = 2$, $x_4 = 0$

41. $\mathbf{A}^{-1} = \begin{pmatrix} 0 & \frac{2}{3} & \frac{1}{3} \\ 0 & -\frac{1}{3} & -\frac{2}{3} \\ \frac{1}{3} & -\frac{2}{3} & 0 \end{pmatrix}$

43. $\mathbf{A}^{-1} = \begin{pmatrix} 5 & 6 & -3 \\ 2 & 2 & -1 \\ -1 & -1 & 1 \end{pmatrix}$

45. $\mathbf{A}^{-1} = \begin{pmatrix} -\frac{1}{2} & -\frac{2}{3} & -\frac{1}{6} & \frac{7}{6} \\ 1 & \frac{1}{3} & \frac{1}{3} & -\frac{4}{3} \\ 0 & -\frac{1}{3} & -\frac{1}{3} & \frac{1}{3} \\ -\frac{1}{2} & 1 & \frac{1}{2} & \frac{1}{2} \end{pmatrix}$

47. $\lambda_1 = 6, \lambda_2 = 1, \mathbf{K}_1 = \begin{pmatrix} 2 \\ 7 \end{pmatrix}, \mathbf{K}_2 = \begin{pmatrix} 1 \\ 1 \end{pmatrix}$

49. $\lambda_1 = \lambda_2 = -4, \mathbf{K}_1 = \begin{pmatrix} 1 \\ -4 \end{pmatrix}$

51. $\lambda_1 = 0, \lambda_2 = 4, \lambda_3 = -4,$

$$\mathbf{K}_1 = \begin{pmatrix} 9 \\ 45 \\ 25 \end{pmatrix}, \mathbf{K}_2 = \begin{pmatrix} 1 \\ 1 \\ 1 \end{pmatrix}, \mathbf{K}_3 = \begin{pmatrix} 1 \\ 9 \\ 1 \end{pmatrix}$$

53. $\lambda_1 = \lambda_2 = \lambda_3 = -2,$

$$\mathbf{K}_1 = \begin{pmatrix} 2 \\ -1 \\ 0 \end{pmatrix}, \mathbf{K}_2 = \begin{pmatrix} 0 \\ 0 \\ 1 \end{pmatrix}$$

55. $\lambda_1 = 3i, \lambda_2 = -3i,$

$$\mathbf{K}_1 = \begin{pmatrix} 1 - 3i \\ 5 \end{pmatrix}, \mathbf{K}_2 = \begin{pmatrix} 1 + 3i \\ 5 \end{pmatrix}$$

INDEX

TABLE OF LAPLACE TRANSFORMS

$f(t)$	$\mathscr{L}\{f(t)\} = F(s)$
1. 1	$\dfrac{1}{s}$
2. t	$\dfrac{1}{s^2}$
3. t^n	$\dfrac{n!}{s^{n+1}}$, $\quad n$ a positive integer
4. $t^{-1/2}$	$\sqrt{\dfrac{\pi}{s}}$
5. $t^{1/2}$	$\dfrac{\sqrt{\pi}}{2s^{3/2}}$
6. t^α	$\dfrac{\Gamma(\alpha+1)}{s^{\alpha+1}}$, $\quad \alpha > -1$
7. $\sin kt$	$\dfrac{k}{s^2+k^2}$
8. $\cos kt$	$\dfrac{s}{s^2+k^2}$
9. $\sin^2 kt$	$\dfrac{2k^2}{s(s^2+4k^2)}$
10. $\cos^2 kt$	$\dfrac{s^2+2k^2}{s(s^2+4k^2)}$
11. e^{at}	$\dfrac{1}{s-a}$
12. $\sinh kt$	$\dfrac{k}{s^2-k^2}$
13. $\cosh kt$	$\dfrac{s}{s^2-k^2}$
14. $\sinh^2 kt$	$\dfrac{2k^2}{s(s^2-4k^2)}$
15. $\cosh^2 kt$	$\dfrac{s^2-2k^2}{s(s^2-4k^2)}$
16. te^{at}	$\dfrac{1}{(s-a)^2}$
17. $t^n e^{at}$	$\dfrac{n!}{(s-a)^{n+1}}$, $\quad n$ a positive integer
18. $e^{at}\sin kt$	$\dfrac{k}{(s-a)^2+k^2}$
19. $e^{at}\cos kt$	$\dfrac{s-a}{(s-a)^2+k^2}$

$f(t)$	$\mathscr{L}\{f(t)\} = F(s)$
20. $e^{at}\sinh kt$	$\dfrac{k}{(s-a)^2-k^2}$
21. $e^{at}\cosh kt$	$\dfrac{s-a}{(s-a)^2-k^2}$
22. $t\sin kt$	$\dfrac{2ks}{(s^2+k^2)^2}$
23. $t\cos kt$	$\dfrac{s^2-k^2}{(s^2+k^2)^2}$
24. $\sin kt + kt\cos kt$	$\dfrac{2ks^2}{(s^2+k^2)^2}$
25. $\sin kt - kt\cos kt$	$\dfrac{2k^3}{(s^2+k^2)^2}$
26. $t\sinh kt$	$\dfrac{2ks}{(s^2-k^2)^2}$
27. $t\cosh kt$	$\dfrac{s^2+k^2}{(s^2-k^2)^2}$
28. $\dfrac{e^{at}-e^{bt}}{a-b}$	$\dfrac{1}{(s-a)(s-b)}$
29. $\dfrac{ae^{at}-be^{bt}}{a-b}$	$\dfrac{s}{(s-a)(s-b)}$
30. $1-\cos kt$	$\dfrac{k^2}{s(s^2+k^2)}$
31. $kt-\sin kt$	$\dfrac{k^3}{s^2(s^2+k^2)}$
32. $\dfrac{a\sin bt - b\sin at}{ab(a^2-b^2)}$	$\dfrac{1}{(s^2+a^2)(s^2+b^2)}$
33. $\dfrac{\cos bt - \cos at}{a^2-b^2}$	$\dfrac{s}{(s^2+a^2)(s^2+b^2)}$
34. $\sin kt \sinh kt$	$\dfrac{2k^2 s}{s^4+4k^4}$
35. $\sin kt \cosh kt$	$\dfrac{k(s^2+2k^2)}{s^4+4k^4}$
36. $\cos kt \sinh kt$	$\dfrac{k(s^2-2k^2)}{s^4+4k^4}$
37. $\cos kt \cosh kt$	$\dfrac{s^3}{s^4+4k^4}$
38. $J_0(kt)$	$\dfrac{1}{\sqrt{s^2+k^2}}$